FIFTH EDITION

UNIVERSE

Stars and Galaxies

FIFTH EDITION

UNIVERSE

Stars and Galaxies

Roger A. Freedman
University of California, Santa Barbara

Robert M. Geller
University of California, Santa Barbara

William J. Kaufmann III
San Diego State University

W.H. Freeman and Company
A Macmillan Higher Education Company

Publisher:	*Jessica Fiorillo*
Acquisitions Editor:	*Alicia Brady*
Development Editor:	*Brittany Murphy*
Senior Media and Supplements Editor:	*Amy Thorne*
Assistant Editor:	*Courtney Lyons*
Associate Director of Marketing:	*Debbie Clare*
Marketing Assistant:	*Samantha Zimbler*
Project Editor:	*Kerry O'Shaughnessy*
Production Manager:	*Julia DeRosa*
Cover and Text Designer:	*Victoria Tomaselli*
Illustration Coordinator:	*Janice Donnola*
Illustrations:	*Dragonfly Media Group, George Kelvin*
Photo Editors:	*Robin Fadool, Bianca Moscatelli*
Photo Researcher:	*Deborah Anderson*
Composition:	*Sheridan Sellers*
Printing and Binding:	*RR Donnelley*

Library of Congress Control Number: 2013950516
ISBN-13: 978-1-4641-3527-9
ISBN-10: 1-4641-3527-4

Printed in the United States of America

First printing

W. H. Freeman and Company, 41 Madison Avenue, New York, NY 10010
Houndmills, Basingstoke RG21 6XS, England
www.whfreeman.com

Credit is given to the following sources for photographs:

Cover: Small Magellanic Cloud near the Milky Way: NASA/CXC/Univ. of Potsdam/L. Oskinova et al., optical: NASA/STScI, infrared: NASA/JPL-Caltech.

Title Page: *(from left)* Cat's Eye Nebula: NASA/UIUC/Y. Chu et al., optical: NASA/HST; solar prominence: SOHO/Extreme Ultraviolet Imaging Telescope, ESA, NASA; The Lunar Far Side: The Side Never Seen from Earth: NASA/Goddard; Hale Bopp: Andrew M. Sorenson; Enceladus ice geysers: Cassini Imaging Team, SSI, JPL, ESA, NASA; Helix Nebula: NASA/JPL-Caltech/Univ. of Arizona; Sombrero Galaxy: NASA/JPL-Caltech/R. Kennicutt (University of Arizona) and the SINGS Team.

Cosmic Connections: p. 16: Sun: NOAO/NASA; galaxies: Hubble Heritage Team (STScI/AURA)/NASA; Earth: NASA Goddard Space Flight Center, image by Reto Stöckli; galaxy clusters: NOAO/NASA; Milky Way: NASA/JPL-Caltech/R. Hurt (SSC); galaxy map: T. H. Jarrett, J. Carpenter, & R. Hurt/2MASS; pp. 192–193: NASA; p. 437: Sun: NOAO; Leonids: Walter Pacholka, Astropics/Science Source; p. 622: NASA; p. 655: spiral galaxy: NASA and ESA; globular cluster: S. Kafka and K. Honeycutt, Indiana University/WIYN/NOAO/NSF; p. 690: 1. Bob and Bill Twardy/Adam Block/NOAO/AURA/NSF; 2. NASA, ESA, and the Hubble Heritage Team (STScI/AURA)-ESA/Hubble Collaboration; 3. NASA/JPL-Caltech/Z. Wang (Harvard-Smithsonian CfA); p. 709: *top:* Accretion disks, Alfred T. Kamajian; *bottom:* Accretion disk simulation, Michael Owen and John Blondin, North Carolina State University.

To Lee Johnson Kaufmann and
Caroline Robillard-Freedman,
strong survivors

and to the memory of
PFC Richard Freedman, AUS
and S/Sgt. Ann Kazmierczak Freedman, WAC

About the Authors

Roger A. Freedman is on the faculty of the Department of Physics at the University of California, Santa Barbara. He grew up in San Diego, California, and was an undergraduate at the University of California campuses in San Diego and Los Angeles. He did his doctoral research in nuclear theory and its astrophysical applications at Stanford University under the direction of Professor J. Dirk Walecka. Dr. Freedman joined the faculty at UCSB in 1981 after three years of teaching and doing research at the University of Washington. Dr. Freedman holds a commercial pilot's license, and when not teaching or writing he can frequently be found flying with his wife, Caroline. He has flown across the United States and Canada. (Photo courtesy of Caroline J. Robillard)

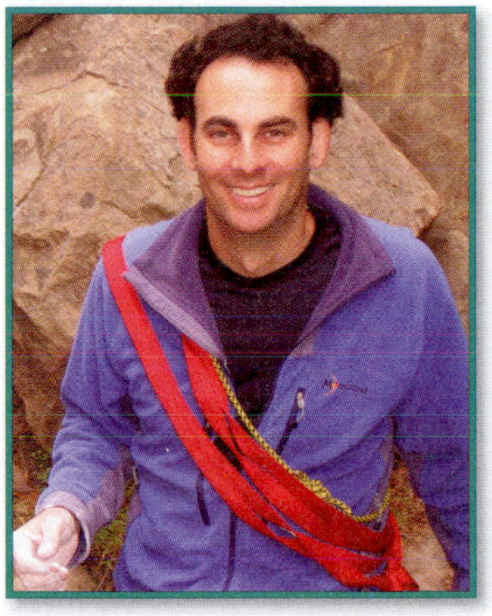

Robert M. Geller teaches and conducts research in astrophysics at the University of California, Santa Barbara, where he also obtained his Ph.D. His doctoral research was in observational cosmology under Professor Robert Antonucci. Using data from the Hubble Space Telescope, he is currently involved in a search for bursts of light that are predicted to occur when a supermassive black hole consumes a star. His other project, in biomedicine, explores the use of magnetotactic bacteria to enhance the effectiveness of radiation therapy in treating cancer. Dr. Geller also has a strong emphasis on education, and he received the Distinguished Teaching Award at UCSB in 2003. His hobbies include rock climbing, and he built an unusual telescope with lenses made of water. (Photo courtesy of Richard Rouse)

William J. Kaufmann III was the author of the first four editions of *Universe*. Dr. Kaufmann earned his bachelor's degree magna cum laude in physics from Adelphi University in 1963, a master's degree in physics from Rutgers in 1965, and a Ph.D. in astrophysics from Indiana University in 1968. At 27, he became the youngest director of any major planetarium in the United States when he took the helm of the Griffith Observatory in Los Angeles. During his career, he also held positions at San Diego State University, UCLA, Caltech, and the University of Illinois. A prolific author, his many books include *Black Holes and Warped Spacetime*, *Relativity and Cosmology*, *Planets and Moons*, *Stars and Nebulas*, and *Galaxies and Quasars*. Dr. Kaufmann died in 1994.

Contents Overview

Chapters 9–15 can be found in *Universe,* 10th edition, of which this book is an abbreviated version.

Contents

IV Galaxies and Cosmology

Preface

Astronomy is one of the most dynamic and exciting areas of modern science. Its recent discoveries often capture widespread interest, and its subjects—the planets, stars, and the universe—inspire awe and inquiry. From observations of previously unseen worlds in the outer depths of our solar system to new evidence of how the most massive stars can collapse into black holes, astronomy provides a wide array of opportunities to show students the continuing process of science.

With so many possible concepts to explore, students can easily get overwhelmed. *Universe* has been developed to help students focus on and learn core concepts. Through clear, highlighted explanations of key topics, and multiple interactions with material, the text fosters an appreciative comprehension of the subject.

COSMIC CONNECTIONS figures summarize key ideas visually

Many students learn more through visual presentations than from reading long passages of text. To help these students, large figures, called **Cosmic Connections**, appear in most chapters of *Universe*. Cosmic Connections give an overview or summary of a particular important topic in a chapter. Subjects range from the variety of modern telescopes to the formation of the solar system; from the scale of distances in the universe to the life cycles of stars; and from the influence of gravitational tidal forces to the evolution of the universe after the Big Bang. Cosmic Connections convey the sense of excitement and adventure that lead students to study astronomy in the first place.

Enhanced, animated versions of the Cosmic Connections are available within the textbook's associated multimedia resources.

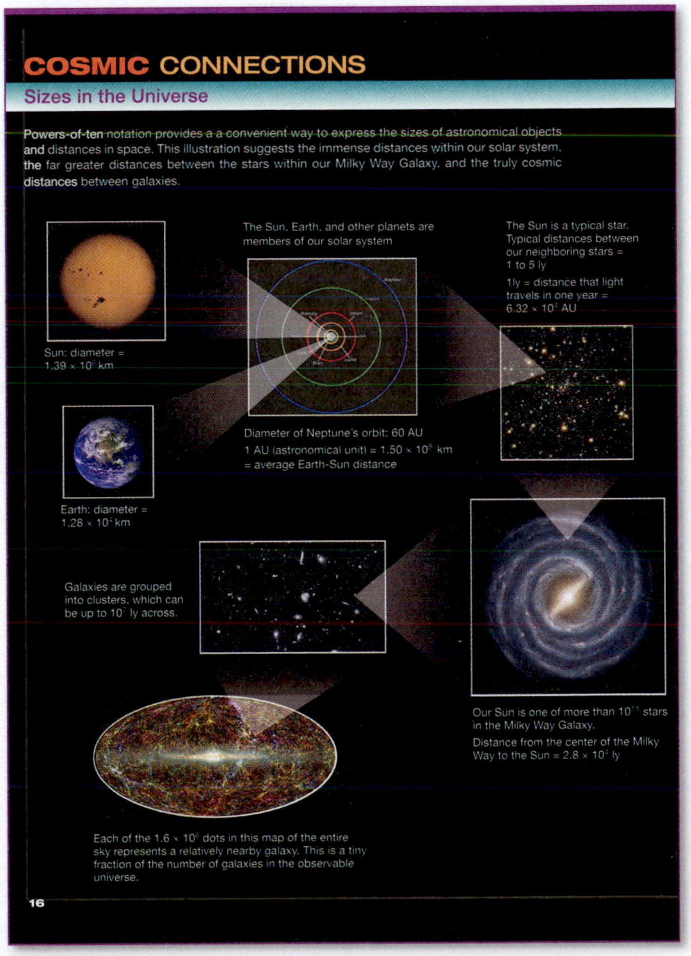

COSMIC CONNECTIONS
Sizes in the Universe
Powers-of-ten notation provides a a convenient way to express the sizes of astronomical objects and distances in space. This illustration suggests the immense distances within our solar system, the far greater distances between the stars within our Milky Way Galaxy, and the truly cosmic distances between galaxies.
Sun: diameter = 1.39 × 10⁶ km
The Sun, Earth, and other planets are members of our solar system
The Sun is a typical star. Typical distances between our neighboring stars = 1 to 5 ly
1ly = distance that light travels in one year = 6.32 × 10⁴ AU
Earth: diameter = 1.28 × 10⁴ km
Diameter of Neptune's orbit: 60 AU
1 AU (astronomical unit) = 1.50 × 10⁸ km = average Earth-Sun distance
Galaxies are grouped into clusters, which can be up to 10⁷ ly across.
Our Sun is one of more than 10¹¹ stars in the Milky Way Galaxy.
Distance from the center of the Milky Way to the Sun = 2.8 × 10⁴ ly
Each of the 1.6 × 10⁶ dots in this map of the entire sky represents a relatively nearby galaxy. This is a tiny fraction of the number of galaxies in the observable universe.
16

NEW! CONCEPTCHECK AND CALCULATIONCHECK QUESTIONS allow students to go beyond passive reading

Included at the end of each section, thought-provoking **ConceptCheck** questions provide immediate assessment by going beyond reading comprehension, often asking students to draw conclusions informed by the text, calling for applied thinking and synthesis of concepts.

> **CONCEPTCHECK 4-1**
>
> If Mars is moving retrograde, will it rise above the eastern horizon or above the western horizon?

> **CONCEPTCHECK 11-4**
>
> Microwaves are emitted from Venus's hot surface. What do 1.35-cm microwaves tell us about the presence of water on the surface of Venus and in its thick atmosphere?
>
> *Answer appears at the end of the chapter.*

Similar to ConceptCheck questions and focusing on mathematics, **CalculationCheck** questions give students the opportunity to test themselves by solving different mathematical problems associated

CALCULATIONCHECK 5-2

Which wavelength of light would our Sun emit most if its temperature were twice its current temperature of 5800 K?

Answers appear at the end of the chapter.

with the chapter conceps. They appear only within sections where relevant mathematical reasoning is present. Answers to all ConceptCheck and CalculationCheck questions are provided at the back of the book.

4-5 Galileo's discoveries with a telescope strongly supported a heliocentric model

Galileo used the cutting-edge technology of the 1600s to radically transform our picture of the universe

When Dutch opticians invented the telescope during the first decade of the seventeenth century, astronomy was changed forever. The scholar who used this new tool to amass convincing evidence that the planets orbit the Sun, not Earth, was the Italian mathematician and physical scientist Galileo Galilei (Figure 4-12).

MOTIVATION STATEMENTS connect detailed discussion to the broader picture

As students read a science text, they often wonder "What is the key idea? How does it relate to the discussion on previous pages? What motivated scientists to make these observations and deductions?" Free-standing **Motivation Statements** address these questions. These brief sentences show students how a section's material fits in with the larger picture of astronomy presented in the chapter. Some statements explore the scientific method. Others compare topics in one section with those in another section or chapter; still other statements give a quick summary of a section's conclusion.

CAUTIONS confront misconceptions

CAUTION! Note that the quantity *c* is the speed of light *in a vacuum*. Light travels more slowly through air, water, glass, or any other transparent substance than it does in a vacuum. In our study of astronomy, however, we will almost always consider light traveling through the vacuum (or near-vacuum) of space.

Many people think that Earth is closer to the Sun in summer than in winter and that the phases of the Moon are caused by Earth's shadow falling on the Moon. However, these "common sense" ideas are incorrect (as explained in Chapters 2 and 3). Throughout *Universe*, paragraphs marked by the **Caution** icon alert the reader to such common conceptual pitfalls.

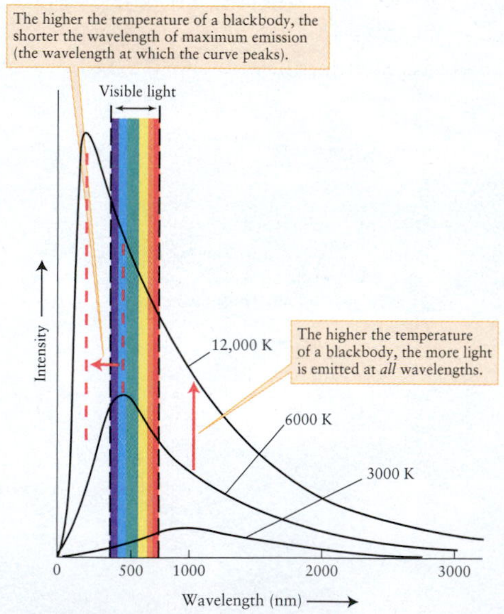

The higher the temperature of a blackbody, the shorter the wavelength of maximum emission (the wavelength at which the curve peaks).
Visible light
12,000 K
The higher the temperature of a blackbody, the more light is emitted at *all* wavelengths.
6000 K
3000 K
Intensity
0
500
1000
2000
3000
Wavelength (nm)

ANIMATION 5-1

FIGURE 5-12

Blackbody Curves Each of these curves shows the intensity of light at every wavelength that is emitted by a blackbody (an idealized case of a dense object) at a particular temperature. The rainbow-colored band shows the range of visible wavelengths. The vertical scale has been compressed so that all three curves can be seen; the peak intensity for the 12,000-K curve is actually about 1000 times greater than the peak intensity for the 3000-K curve.

ANALOGIES make astronomy relatable to students

When learning new astronomical ideas, it can be helpful to relate them to more familiar experiences on Earth. Throughout *Universe*, **Analogy** paragraphs make these connections. For example, the motions of the planets can be related to children on a merry-go-round, and the bending of light through a telescope lens is similar to the path of a car driving from firm ground onto sand.

ANALOGY Figure skaters make use of the conservation of angular momentum. When a spinning skater pulls her arms and legs in close to her body, the rate at which she spins automatically increases (Figure 8-7). Even if you are not a figure skater, you can demonstrate this by sitting on a rotating office chair. Sit with your arms outstretched and hold a weight, like a brick or a full water bottle, in either hand. Now use your feet to start your body and the chair rotating, lift your feet off the ground, and then pull your arms inward. Your rotation will speed up quite noticeably.

EXPLANATORY ART helps students visualize concepts

Many **figures** in the text utilize balloon captions to help students interpret complex figures. Select **photos** also contain brief labels that point out the most important relevant features.

NEW! FOLD-OUT PHOTOS highlight some of astronomy's most majestic sights

Throughout the text are special gatefold **photo spreads**, which take advantage of extra space to highlight awe-inspiring astronomy photography in full detail.

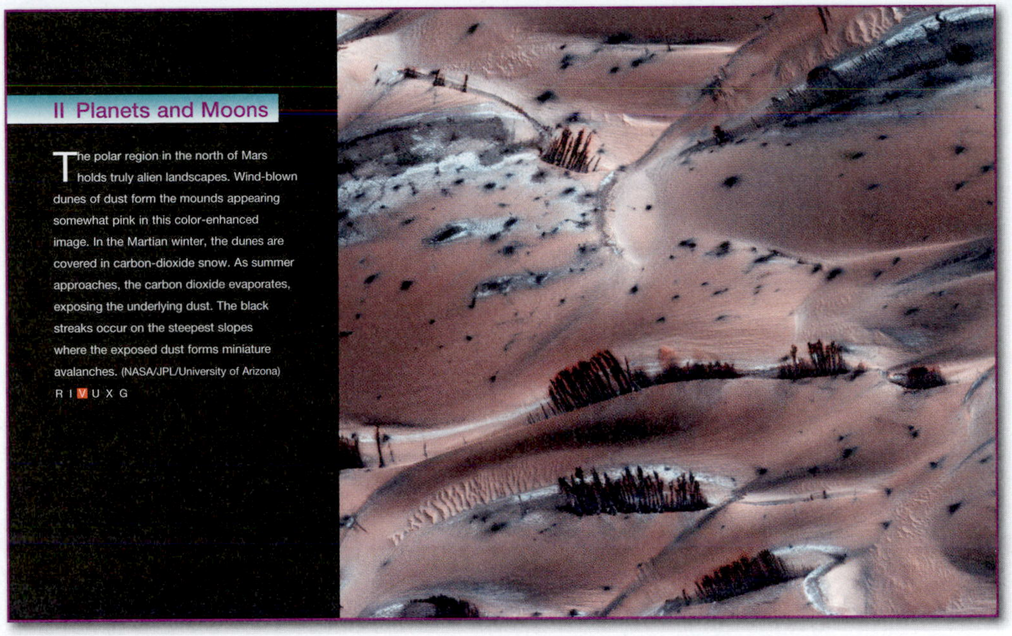

II Planets and Moons

The polar region in the north of Mars holds truly alien landscapes. Wind-blown dunes of dust form the mounds appearing somewhat pink in this color-enhanced image. In the Martian winter, the dunes are covered in carbon-dioxide snow. As summer approaches, the carbon dioxide evaporates, exposing the underlying dust. The black streaks occur on the steepest slopes where the exposed dust forms miniature avalanches. (NASA/JPL/University of Arizona)

R I V U X G

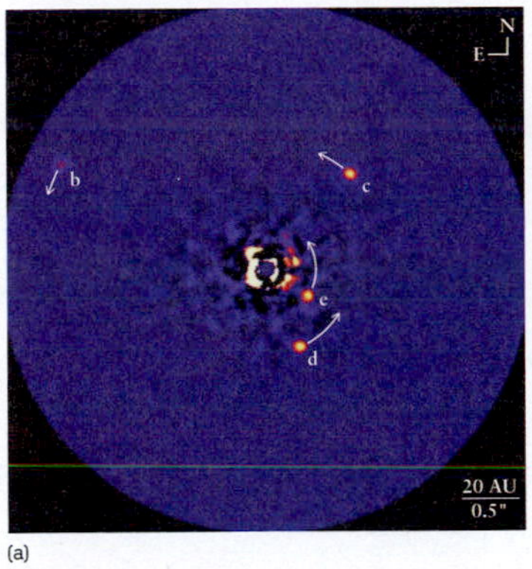

FIGURE 8-15 R I V U X G

Imaging Extrasolar Planets (a) Four planets can be seen orbiting the star HR 8799. The star is about 129 light-years from Earth and can be seen with the naked eye on a clear night. In capturing this image, special techniques make the star less visible to reduce its glare and reveal its planets. The white arrows indicate possible trajectories that each star might take over the next ten years. (b) About 170 light-years away, the star 2M1207 and

WAVELENGTH TABS astronomers observe the sky in many forms of light

Astronomers rely on telescopes that are sensitive to nonvisible forms of light. To help students appreciate these different kinds of observation, all of the photos in *Universe* appear with wavelength tabs. The highlighted letter on each tab indicates whether the image was made with **R**adio waves, **I**nfrared radiation, **V**isible light, **U**ltraviolet light, **X**-rays, or **G**amma rays.

TOOLS OF THE ASTRONOMER'S TRADE AND S.T.A.R. a problem-solving rubric

All the worked examples in *Universe* (found in boxes called **Tools of the Astronomer's Trade**) follow a logical and consistent sequence of steps called **S.T.A.R.**: assess the **S**ituation, select the **T**ools, find the **A**nswer, and **R**eview the answer and explore its significance. Feedback from instructors indicates that when students follow these steps in their own work, they more rapidly develop important problem-solving skills.

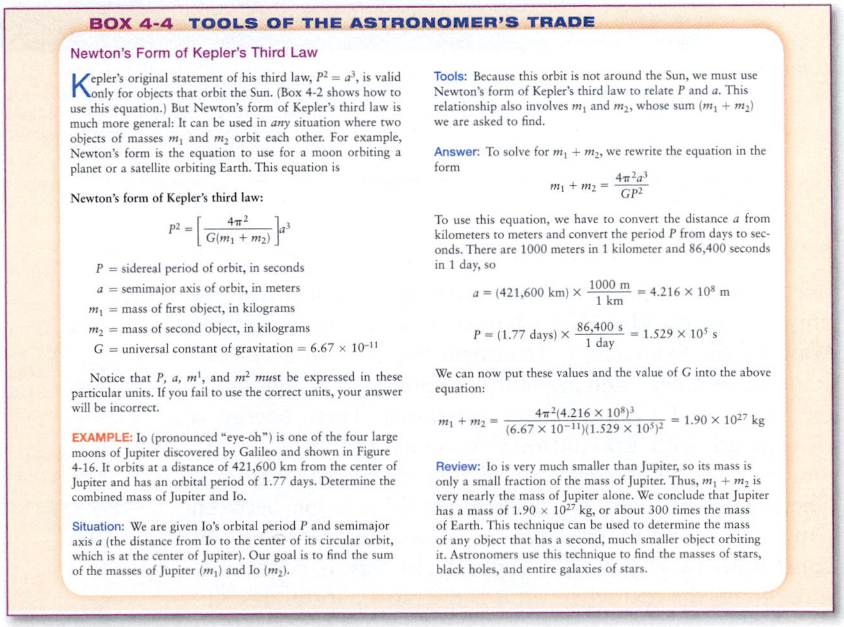

BOX 4-4 TOOLS OF THE ASTRONOMER'S TRADE

Newton's Form of Kepler's Third Law

Kepler's original statement of his third law, $P^2 = a^3$, is valid only for objects that orbit the Sun. (Box 4-2 shows how to use this equation.) But Newton's form of Kepler's third law is much more general: It can be used in *any* situation where two objects of masses m_1 and m_2 orbit each other. For example, Newton's form is the equation to use for a moon orbiting a planet or a satellite orbiting Earth. This equation is

Newton's form of Kepler's third law:

$$P^2 = \left[\frac{4\pi^2}{G(m_1 + m_2)}\right]a^3$$

P = sidereal period of orbit, in seconds
a = semimajor axis of orbit, in meters
m_1 = mass of first object, in kilograms
m_2 = mass of second object, in kilograms
G = universal constant of gravitation = 6.67×10^{-11}

Notice that P, a, m^1, and m^2 *must* be expressed in these particular units. If you fail to use the correct units, your answer will be incorrect.

EXAMPLE: Io (pronounced "eye-oh") is one of the four large moons of Jupiter discovered by Galileo and shown in Figure 4-16. It orbits at a distance of 421,600 km from the center of Jupiter and has an orbital period of 1.77 days. Determine the combined mass of Jupiter and Io.

Situation: We are given Io's orbital period P and semimajor axis a (the distance from Io to the center of its circular orbit, which is at the center of Jupiter). Our goal is to find the sum of the masses of Jupiter (m_1) and Io (m_2).

Tools: Because this orbit is not around the Sun, we must use Newton's form of Kepler's third law to relate P and a. This relationship also involves m_1 and m_2, whose sum ($m_1 + m_2$) we are asked to find.

Answer: To solve for $m_1 + m_2$, we rewrite the equation in the form

$$m_1 + m_2 = \frac{4\pi^2 a^3}{GP^2}$$

To use this equation, we have to convert the distance a from kilometers to meters and convert the period P from days to seconds. There are 1000 meters in 1 kilometer and 86,400 seconds in 1 day, so

$$a = (421{,}600 \text{ km}) \times \frac{1000 \text{ m}}{1 \text{ km}} = 4.216 \times 10^8 \text{ m}$$

$$P = (1.77 \text{ days}) \times \frac{86{,}400 \text{ s}}{1 \text{ day}} = 1.529 \times 10^5 \text{ s}$$

We can now put these values and the value of G into the above equation:

$$m_1 + m_2 = \frac{4\pi^2(4.216 \times 10^8)^3}{(6.67 \times 10^{-11})(1.529 \times 10^5)^2} = 1.90 \times 10^{27} \text{ kg}$$

Review: Io is very much smaller than Jupiter, so its mass is only a small fraction of the mass of Jupiter. Thus, $m_1 + m_2$ is very nearly the mass of Jupiter alone. We conclude that Jupiter has a mass of 1.90×10^{27} kg, or about 300 times the mass of Earth. This technique can be used to determine the mass of any object that has a second, much smaller object orbiting it. Astronomers use this technique to find the masses of stars, black holes, and entire galaxies of stars.

BOX 5-3 ASTRONOMY DOWN TO EARTH

Photons at the Supermarket

A beam of light can be regarded as a stream of tiny packets of energy called photons. The Planck relationships $E = hc/\lambda$ and $E = h\nu$ can be used to relate the energy E carried by a photon to its wavelength λ and frequency ν.

As an example, the laser bar-code scanners used at stores and supermarkets emit orange-red light of wavelength 633 nm. To calculate the energy of a single photon of this light, we must first express the wavelength in meters. A nanometer (nm) is equal to 10^{-9} m, so the wavelength is

$$\lambda = (633 \text{ nm}) \left(\frac{10^{-9} \text{ m}}{1 \text{ nm}} \right) = 633 \times 10^{-9} \text{ m} = 6.33 \times 10^{-7} \text{ m}$$

Then, using the Planck formula $E = hc/\lambda$, we find that the energy of a single photon is

$$E = \frac{hc}{\lambda} = \frac{(6.625 \times 10^{-34} \text{ J s})(3 \times 10^{8} \text{ m/s})}{6.33 \times 10^{-7} \text{ m}} = 3.14 \times 10^{-19} \text{ J}$$

This amount of energy is very small. The laser in a typical bar-code scanner emits 10^{-3} joule of light energy per second, so the number of photons emitted per second is

$$\frac{10^{-3} \text{ joule per second}}{3.14 \times 10^{-19} \text{ joule per photon}} = 3.2 \times 10^{15} \text{ photons per second}$$

This number is so large that the laser beam seems like a continuous flow of energy rather than a stream of little energy packets.

ASTRONOMY DOWN TO EARTH reveals the universal applicability of the laws of nature

Astronomy Down to Earth boxes illustrate how the same principles astronomers use to explain celestial phenomena can also explain everyday behavior here on Earth, from the color of the sky to why diet soft drink cans float in water.

Up-to-date information shows the cutting edge of astronomy

Universe is up to date with the latest astronomical discoveries, ideas, and images. These include:

- New information on the formation of solar system (Chapter 8)
- *Kepler*'s discoveries of extrasolar planets (Chapter 8)
- The best evidence from *Curiosity* for the long-term flow of water on Mars (Chapter 11)
- Water geysers revealing a large subsurface ocean on Enceladus (Chapter 13)
- Cryovolcanism on Titan (Chapter 13)
- The history-making 2013 asteroid impact in Russia (Chapter 15)
- *Dawn*'s exploration of the asteroid Vesta (Chapter 15)
- New models of supernovae explosions (Chapter 20)
- Gamma-ray bubbles in the Milky Way (Chapter 22)
- Discovery of the Higgs particles (Chapter 26)
- Microtunnels found in a Martian meteorite consistent with microbial origins (Chapter 27)

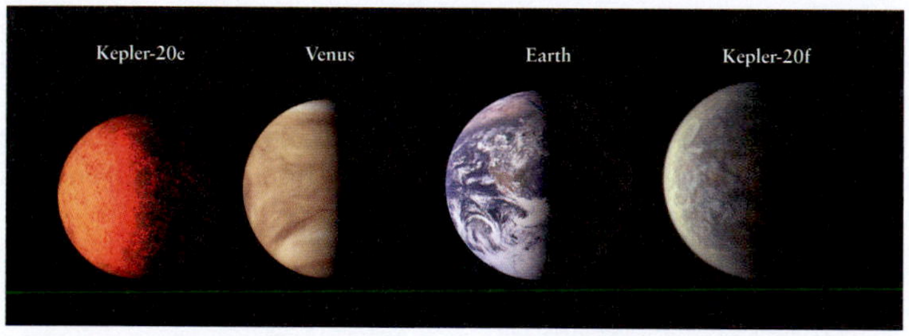

FIGURE 8-20
Earth-Size Exoplanets Kepler-20 is a star with five known exoplanets. The exoplanets Kepler-20e and Kepler-20f are illustrated in comparison to Earth and Venus. Kepler-20e (at 0.05 AU) has a radius just 0.87 times smaller than Earth, while Kepler-20f (at 0.1 AU) is 1.03 times larger than Earth. However, these objects orbit extremely close to their star and are much too hot for liquid water on their surfaces. (Ames/JPL-Caltech/NASA)

Energy, Stellar Remnant Cores, and Active Galactic Nuclei

Energy is developed as an overarching concept in astronomy and is introduced in **Chapter 4, Gravitation and the Waltz of the Planets.** Our treatment begins with an understanding of kinetic and gravitational potential energy, aided by figures and ConceptCheck questions. Then, energy is considered with gravitational systems and provides an intuitive basis for a variety of orbital properties. In **Chapter 5, The Nature of Light,** we develop the connection between temperature and thermal energy, as well as the energy of light. With these powerful concepts, energy is considered throughout the text wherever it is the most natural description of phenomena.

4-7 Describing orbits with energy and gravity

TUTORIAL 4-3

Imagine a cannonball shot into the air; it arcs through the sky before it crashes back to the ground. If you want to shoot the cannonball into orbit around Earth (like a satellite), you might guess that the ball must be
This guess is correct, and next
orbits in terms of gravity and en

5-8 Spectral lines are produced when an electron jumps from one energy level to another within an atom

Niels Bohr explained spectral lines with a radical new model of atom

Niels Bohr began his study of the connection between atomic spectra and atomic structure by trying to understand the structure of hydrogen, the simplest and lightest of the elements. (As we dis-

Discussions of white dwarfs and neutron stars are combined into **Chapter 20, Stellar Evolution: The Deaths of Stars.** As the only two possible stellar remnant cores, these objects complement each other physically and pedagogically. The depth of treatment on neutron stars is reduced as part of this approach, and the full original chapter, **Neutron Stars**, is available online.

Chapter 24, Quasars and Active Galaxies, has been significantly revised. The historical development of this field is followed when that leads to the clearest explanation of the topic, and when the history itself is confusing, we continue the treatment with our modern understanding of active galaxies. The revised chapter avoids many of the complicated historical names originally given to objects that are now simply referred to as active galaxies. With these changes, the reader can more easily understand and enjoy these intriguing objects.

Planetary Themes: Comparative Planetology and Systems Approach

The two chapters on comparative planetology—**Chapter 7, Our Solar System,** and **Chapter 8, The Origin of the Solar System**—have been revised and updated with the latest information. They contain explanations of the official distinctions between planets and dwarf planets, as well as discussions of new discoveries about objects in our own solar system. Figures elucidating comparative planetology appear throughout the planetary coverage. In addition to its comparative planetology theme, *Universe* takes a unique *systems* approach to planetary surfaces and atmospheres: each planet's surface and atmosphere are discussed in tandem, where other texts address them separately. This underscores their interrelationships and builds a deeper understanding of the planets' overall characteristics.

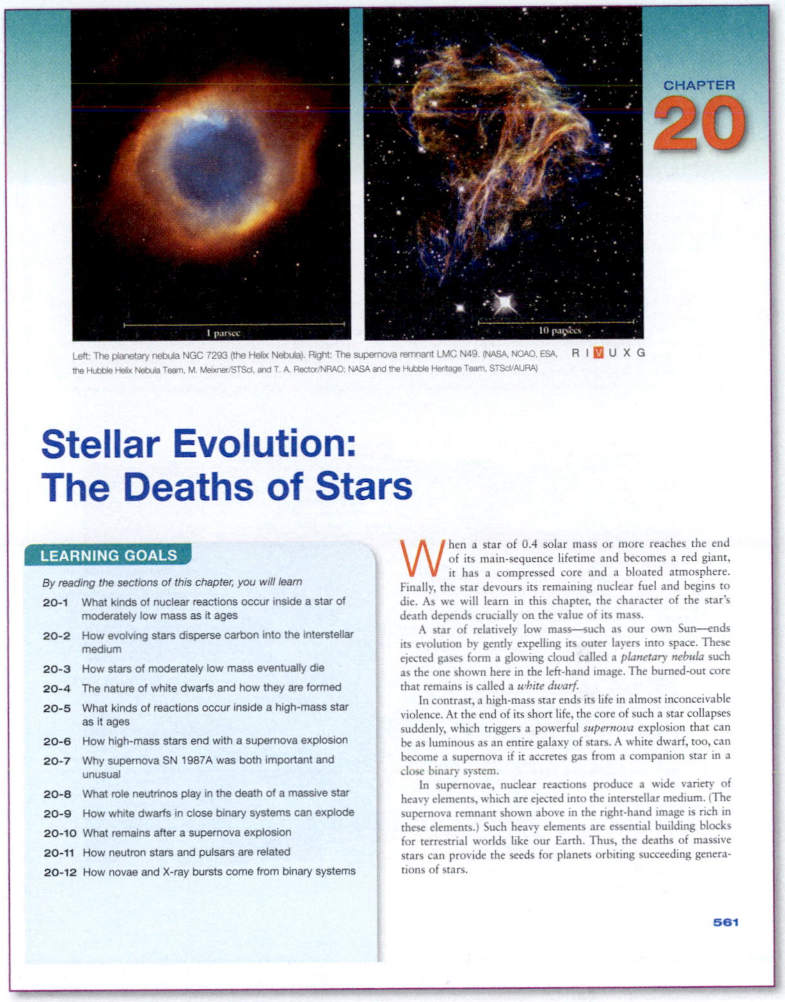
CHAPTER 20

1 parsec

10 parsecs

Left: The planetary nebula NGC 7293 (the Helix Nebula). Right: The supernova remnant LMC N49. (NASA, NOAO, ESA, the Hubble Helix Nebula Team, M. Meixner/STScI, and T. A. Rector/NRAO; NASA and the Hubble Heritage Team, STScI/AURA) R I V U X G

Stellar Evolution: The Deaths of Stars

LEARNING GOALS

By reading the sections of this chapter, you will learn

20-1 What kinds of nuclear reactions occur inside a star of moderately low mass as it ages

20-2 How evolving stars disperse carbon into the interstellar medium

20-3 How stars of moderately low mass eventually die

20-4 The nature of white dwarfs and how they are formed

20-5 What kinds of reactions occur inside a high-mass star as it ages

20-6 How high-mass stars end with a supernova explosion

20-7 Why supernova SN 1987A was both important and unusual

20-8 What role neutrinos play in the death of a massive star

20-9 How white dwarfs in close binary systems can explode

20-10 What remains after a supernova explosion

20-11 How neutron stars and pulsars are related

20-12 How novae and X-ray bursts come from binary systems

When a star of 0.4 solar mass or more reaches the end of its main-sequence lifetime and becomes a red giant, it has a compressed core and a bloated atmosphere. Finally, the star devours its remaining nuclear fuel and begins to die. As we will learn in this chapter, the character of the star's death depends crucially on the value of its mass.

A star of relatively low mass—such as our own Sun—ends its evolution by gently expelling its outer layers into space. These ejected gases form a glowing cloud called a *planetary nebula* such as the one shown here in the left-hand image. The burned-out core that remains is called a *white dwarf*.

In contrast, a high-mass star ends its life in almost inconceivable violence. At the end of its short life, the core of such a star collapses suddenly, which triggers a powerful *supernova* explosion that can be as luminous as an entire galaxy of stars. A white dwarf, too, can become a supernova if it accretes gas from a companion star in a close binary system.

In supernovae, nuclear reactions produce a wide variety of heavy elements, which are ejected into the interstellar medium. (The supernova remnant shown above in the right-hand image is rich in these elements.) Such heavy elements are essential building blocks for terrestrial worlds like our Earth. Thus, the deaths of massive stars can provide the seeds for planets orbiting succeeding generations of stars.

561

End-of-chapter material provides students with opportunities to review and assess what they have learned and opportunities to extend their knowledge

Key Words

A list of **Key Words** appears at the end of each chapter, along with the number of the page where each term is introduced.

Key Ideas

Students can get the most benefit from these brief chapter summaries by using them in conjunction with the notes they take while reading.

Questions

Items from these sections are designed to be assigned as homework or used as jumping-off points for class discussion. Some questions ask students to analyze images in the text or evaluate how the mass media portray concepts in astronomy. Advanced questions are accompanied by a **Problem-Solving Tips and Tools** box to

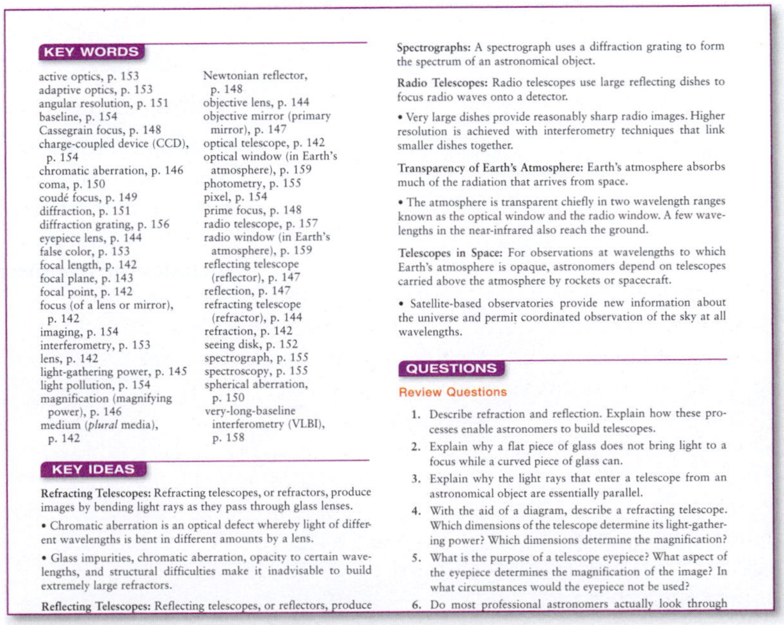
KEY WORDS

active optics, p. 153
adaptive optics, p. 153
angular resolution, p. 151
baseline, p. 154
Cassegrain focus, p. 148
charge-coupled device (CCD), p. 154
chromatic aberration, p. 146
coma, p. 150
coudé focus, p. 149
diffraction, p. 151
diffraction grating, p. 156
eyepiece lens, p. 144
false color, p. 153
focal length, p. 142
focal plane, p. 143
focal point, p. 142
focus (of a lens or mirror), p. 142
imaging, p. 154
interferometry, p. 153
lens, p. 142
light-gathering power, p. 145
light pollution, p. 154
magnification (magnifying power), p. 146
medium (*plural* media), p. 142
Newtonian reflector, p. 148
objective lens, p. 144
objective mirror (primary mirror), p. 147
optical telescope, p. 142
optical window (in Earth's atmosphere), p. 159
photometry, p. 155
pixel, p. 154
prime focus, p. 148
radio telescope, p. 157
radio window (in Earth's atmosphere), p. 159
reflecting telescope (reflector), p. 147
reflection, p. 147
refracting telescope (refractor), p. 144
refraction, p. 142
seeing disk, p. 152
spectrograph, p. 155
spectroscopy, p. 155
spherical aberration, p. 150
very-long-baseline interferometry (VLBI), p. 158

KEY IDEAS

Refracting Telescopes: Refracting telescopes, or refractors, produce images by bending light rays as they pass through glass lenses.

• Chromatic aberration is an optical defect whereby light of different wavelengths is bent in different amounts by a lens.

• Glass impurities, chromatic aberration, opacity to certain wavelengths, and structural difficulties make it inadvisable to build extremely large refractors.

Reflecting Telescopes: Reflecting telescopes, or reflectors, produce

Spectrographs: A spectrograph uses a diffraction grating to form the spectrum of an astronomical object.

Radio Telescopes: Radio telescopes use large reflecting dishes to focus radio waves onto a detector.

• Very large dishes provide reasonably sharp radio images. Higher resolution is achieved with interferometry techniques that link smaller dishes together.

Transparency of Earth's Atmosphere: Earth's atmosphere absorbs much of the radiation that arrives from space.

• The atmosphere is transparent chiefly in two wavelength ranges known as the optical window and the radio window. A few wavelengths in the near-infrared also reach the ground.

Telescopes in Space: For observations at wavelengths to which Earth's atmosphere is opaque, astronomers depend on telescopes carried above the atmosphere by rockets or spacecraft.

• Satellite-based observatories provide new information about the universe and permit coordinated observation of the sky at all wavelengths.

QUESTIONS

Review Questions

1. Describe refraction and reflection. Explain how these processes enable astronomers to build telescopes.
2. Explain why a flat piece of glass does not bring light to a focus while a curved piece of glass can.
3. Explain why the light rays that enter a telescope from an astronomical object are essentially parallel.
4. With the aid of a diagram, describe a refracting telescope. Which dimensions of the telescope determine its light-gathering power? Which dimensions determine the magnification?
5. What is the purpose of a telescope eyepiece? What aspect of the eyepiece determines the magnification of the image? In what circumstances would the eyepiece not be used?
6. Do most professional astronomers actually look through

INSIGHTS

SPACE

Reading the Red Planet

At 10:31 P.M. Pacific time on August 5, 2012, NASA's *Curiosity* rover began the first direct search for habitable environments on Mars

BY JOHN P. GROTZINGER AND ASHWIN VASAVADA
(from John P. Grotzinger and Ashwin Vasavada's "Reading the Red Planet," *Scientific American*, July 2012)

All science begins in a *Star Trek* mode: go where no one has gone before and discover new things without knowing in advance what they might be. As researchers complete their initial surveys and accumulate a long list of questions, they shift to a Sherlock Holmes mode: formulate specific hypotheses and develop ways to test them. The exploration of Mars is now about to make this transition. Orbiters have made global maps of geographic features and composition, and landers have pieced together the broad outlines of the planet's geologic history. It is time to get more sophisticated.

Our team has built the Mars Science Laboratory, also known as the *Curiosity* rover, on the hypothesis that Mars was once a habitable planet. The rover carries an analytic laboratory to test that hypothesis and find out what happened

Weather station will measure environmental variables and issue daily reports, providing the first ever continuous record of Martian meteorology. Apart from its inherent interest, the weather report will guide rover operations.
Color cameras can image landscapes and rock and soil textures in high-definition resolution. Those textures help scientists to reconstruct the processes that formed the rock or soil, perhaps including the action of liquid water. One of the cameras is mounted on the bottom of the rover, looking downward, and created a movie of the descent and landing.
CheMin instrument beams X-rays through fine powders to create a diffraction pattern that definitively identifies minerals of all types. Spectrometers on previous landers were limited in scope to, for example, iron-bearing minerals.
Active neutron spectrometer will search for water in rocks and soil underneath the rover.
Robot arm, reaching out as far as 2 m, holds 30 kg of gadgetry to drill holes and pulverize rocks. A set of sieves sorts powder for the onboard lab instruments.
Laser-induced breakdown spectrometer will burn holes in rocks and soil up to seven meters away and remotely sense their chemical composition.
Sample analysis at Mars (SAM) instrument suite can perform chemical analysis. It bakes powder in small ovens with combustion or chemical solvents to release gases, which the gas chromatograph/mass spectrometer and gas analyzer will examine, looking especially for organic carbon. It also can directly sample the atmosphere.
Radiation sensor will monitor solar and cosmic radiation.
Alpha-particle X-ray spectrometer will perform in situ determination of rock and soil chemistry.

(Don Foley)

provide guidance. Web/e-Book questions challenge students to work with animations and interactive modules on the *Universe* Web site or e-Book (described below) or to research topics on the World Wide Web. Near the end of the book is a section of **Answers to Selected Questions.**

Scientific American Articles

Chosen by the textbook authors to illuminate core concepts in the text, these brief, relevant selections demonstrate the process of science and discovery. Revealing recent developments related to specific topics in the text, the articles also provide touchstones for classroom discussion.

Guest Essays

Several chapters in *Universe* end with essays written by scientists involved with some of the recent discoveries described in the text. These include essays by Scott Sheppard on Pluto and the Kuiper belt and Kevin Plaxco on astrobiology. For a full list of essays, see the Contents Overview on pages ix–x.

A revised selection of Observing Projects

The download code that comes free (bundled upon request) with this text grants 18 months of access to the *Starry Night*™ planetarium software from Simulation Curriculum Corp. Every chapter of *Universe* includes one or more **Observing Projects** that use this software package. Many projects are accompanied by an **Observing Tips and Tools** box to provide help and guidance.

60. Use the *Starry Night*™ program to examine simulations of various features that appear on the surface of the Sun. Select **Favourites > Explorations > Sun** to show a simulated view of the visible surface of the Sun as it might appear from a spacecraft. **Stop** time flow and use the **Location Scroller** to examine this surface. (**a**) Which layer of the Sun's atmosphere is shown in this part of the simulation? (**b**) List the different features that are visible in this view of the Sun's surface.

Three versions of the text meet the needs of different instructors

In addition to the complete 27-chapter version of *Universe*, two shorter versions are also available. *Universe: The Solar System* includes Chapters 1–16 and 27; it omits the chapters on stars, stellar evolution, galaxies, and cosmology. *Universe: Stars and Galaxies* includes Chapters 1–8 and 16–27; it omits the detailed chapters on the solar system but includes the overview of the solar system in Chapters 7 and 8.

Multimedia

Universe was developed to more deeply integrate learning resources, study aids, and multimedia tools into the use of the textbook. A variety of teaching and learning options provides an array of choices for students and instructors in their use of these materials, which were created based on input and contributions from a large number of faculty.

PREMIUM MULTIMEDIA RESOURCES

Electronic Versions

The *Universe* is offered in two electronic versions. One is an **Interactive e-Book**, available as part of LaunchPad, described below. The other is a PDF-based **e-Book from CourseSmart**, available through www.coursesmart.com. These options are provided to offer students and instructors flexibility in their use of course materials.

CourseSmart e-Book

Universe **CourseSmart e-Book** offers the complete text in an easy-to-use, flexible format. Students can choose to view the CourseSmart e-Book online or to download it to their computer or a mobile device, such as iPad, iPhone, or Android device. To help students study and mirror the experience of a printed textbook, CourseSmart e-Books feature notetaking, highlighting, and bookmark features.

ONLINE LEARNING OPTIONS

Universe supports instructors with a variety of online learning preferences. Its rich array of resources and platforms provides solutions according to each instructor's teaching method. Students can also access the resources through the **Companion Web Site.**

LaunchPad: Because Technology Should Never Get in the Way

At W. H. Freeman, we are committed to providing online instructional materials that meet the needs of instructors and students in powerful, yet simple ways—powerful enough to dramatically enhance teaching and learning, yet simple enough to use right away.

We have taken what we've learned from thousands of instructors and hundreds of thousands of students to create a new generation of technology: **LaunchPad**. LaunchPad offers our acclaimed content curated and organized for easy assignment in a breakthrough user interface in which power and simplicity go hand in hand.

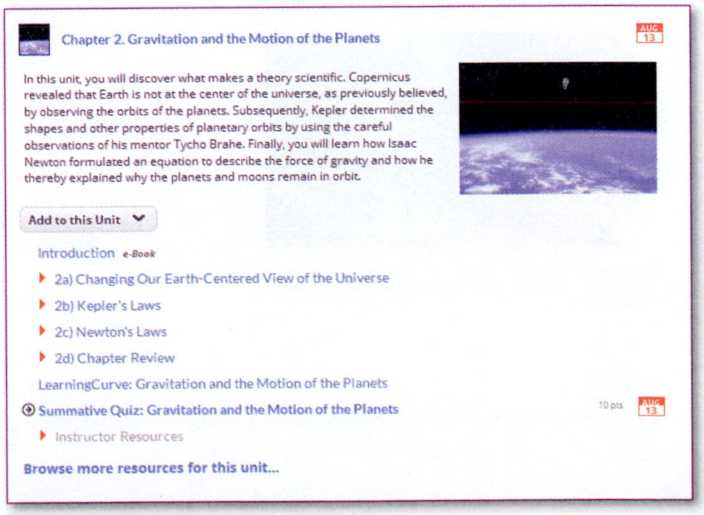

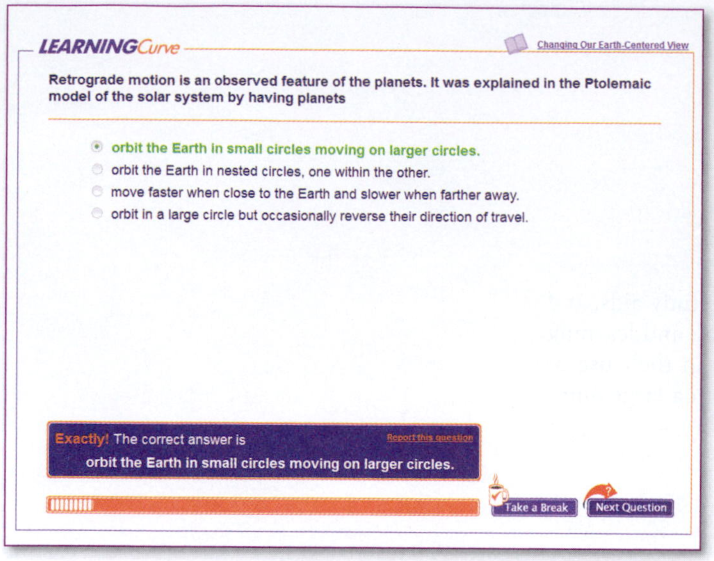

Curated LaunchPad Units make class prep a whole lot easier

Combining a curated collection of video, tutorials, animations, projects, multimedia activities and exercises, and e-Book content, LaunchPad's interactive units give instructors a building block to use as is, or as a starting point for their own learning units. An entire unit's worth of work can be assigned in seconds, drastically saving the amount of time it takes to have the course up and running.

LearningCurve

• Powerful adaptive quizzing, a gamelike format, direct links to the e-Book, instant feedback, and the promise of better grades make using **LearningCurve** a no-brainer.

• Customized quizzing tailored to the text adapts to students' responses and provides material at different difficulty levels and topics based on student performance. Students love the simple yet powerful system and instructors can access class reports to help refine lecture content.

Interactive e-Book

The **Interactive e-Book** is a complete online version of the textbook with easy access to rich multimedia resources that complete student understanding. All text, graphics, tables, boxes, and end-of-chapter resources are included in the e-Book, and the e-Book provides instructors and students with powerful functionality to tailor their course resources to fit their needs.

• Quick, intuitive navigation to any section or subsection

• Full-text search, including the Glossary and Index

• Sticky-note feature allows users to place notes anywhere on the screen, and choose the note color for easy categorization.

• "Top-note" feature allows users to place a prominent note at the top of the page to provide a more significant alert or reminder.

• Text highlighting, in a variety of colors

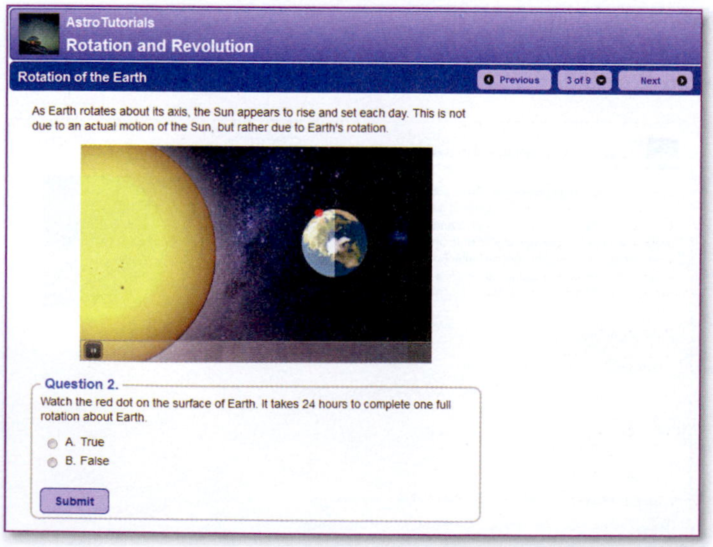

Astronomy Tutorials

• These self-guided, concept-driven experiential walkthroughs engage students in the process of scientific discovery as they make observations, draw conclusions, and apply their knowledge. Astronomy tutorials combine multimedia resources, activities, and quizzes.

Image Map Activities

• These activities use figures and photographs from the text to assess key ideas, helping students to develop their visual literacy. Students must click the appropriate section(s) of the image and answer corresponding questions.

Animations, Videos, Interactive Exercises, Flashcards

• Other LaunchPad resources highlight key concepts in introductory astronomy.

Assignments for Online Quizzing, Homework, and Self-Study

• Instructors can create and assign automatically graded homework and quizzes from the complete test bank, which is preloaded in LaunchPad. All quiz results feed directly into the instructors's gradebook.

Scientific American Newsfeed

• To demonstrate the continued process of science and the exciting new developments in astronomy, the *Scientific American* Newsfeed delivers regularly updated material from the well-known magazine. Articles, podcasts, news briefs, and videos on subjects related to astronomy are selected for inclusion by *Scientific American*'s editors. The newsfeed provides several updates per week, and instructors can archive or assign the content they find most valuable.

Gradebook

• The included **gradebook** quickly and easily allows instructors to look up performance **metrics** for a whole class, for individual students and for individual assignments. Having ready access to this information can help both in lecture prep and in making office hours more productive and efficient.

WebAssign Premium: Trusted Homework Management, Enhanced Learning Tools

www.webassign.net

For instructors interested in online homework management, **WebAssign Premium** features a time-tested, secure, online environment already used by millions of students worldwide. Featuring algorithmic problem-generation and supported by a wealth of astronomy-specific learning tools, WebAssign Premium for *Universe* presents instructors with a powerful assignment manager and study environment. WebAssign Premium provides the following resources:

• **Algorithmically generated problems:** Students receive homework problems containing unique values for computation, encouraging them to work out the problems on their own.

• Innovative **ImageActive** questions, which allow students to interact directly with figures to answer questions. These problems are most often critical thinking questions, which ask students to make connections across concepts and apply their knowledge.

• **Complete access to the interactive e-Book**, from a live table of contents as well as from relevant problem statements.

• Links to Tutorials, e-Book sections, animations, videos, and other interactive tools as hints and feedback to ensure a clearer understanding of the problems and the concepts they reinforce.

Sapling Learning

www.saplinglearning.com

Developed by educators with both online expertise and extensive classroom experience, Sapling Learning provides highly effective interactive homework and instruction that improve student learning outcomes for the problem-solving disciplines. Sapling Learning offers an enjoyable teaching and effective learning experience that is distinctive in three important ways:

• **Ease of use:** Sapling Learning's easy-to-use interface keeps students engaged in problem-solving, not struggling with the software.

• **Targeted instructional content:** Sapling Learning increases student engagement and comprehension by delivering immediate feedback and targeted instructional content.

• **Unsurpassed service and support:** Sapling Learning makes teaching more enjoyable by providing a dedicated Masters- and Ph.D.-level colleague to service instructors' unique needs throughout the course, including content customization.

We offer bundled packages with all versions of our texts that include Sapling Learning Online Homework.

Student Companion Web Site

The *Universe* Book Companion Web site, accessed at www.whfreeman.com/universe10e, provides a range of tools for student self-study and review. They include:

• **Online self-study quizzes** with instant feedback referring to specific sections in the text to help students study, review, and prepare for exams. Instructors can access results through an online database or they can have them e-mailed directly to their accounts.

• **Animations** of key figures

• **NASA videos** of important processes and phenomena

• **Vocabulary and concept-review flashcards**

• **Interactive exercises**, based on text illustrations, help students grasp the vocabulary in context.

Starry Night™ Planetarium Software

***Starry Night*™** is a brilliantly realistic planetarium software package. It is designed for easy use by anyone with an interest in the night sky. See the sky from anywhere on Earth or lift off and visit any solar system body or any location up to 20,000 light years away. View 2,500,000 stars along with more than 170 deep-space objects like galaxies, star clusters, and nebulae. You can travel 15,000 years in time, check out the view from the International Space Station, and see planets up close from any one of their moons. Included are stunning OpenGL graphics. You can also print handy star charts to explore the sky out of doors. This version of ***Starry Night*™** also contains student exercises specific to the Freeman version of the software for use with *Universe*. ***Starry Night*™** is available via online download by an access code packaged with the text at no extra charge upon instructor request.

Observing Projects Using *Starry Night*™

ISBN 1-4641-2502-3
by Marcel Bergman, T. Alan Clark, and William J. F. Wilson, University of Calgary
Available for packaging with the text and compatible with both PC and Mac, this book contains a variety of comprehensive lab activities for ***Starry Night*™** planetarium software.

Test Bank CD-ROM

Windows and Mac versions on one disc, ISBN 1-4641-2496-5
by Thomas Krause, Towson University, T. Alan Clark, and William J. F. Wilson, University of Calgary
More than 3,500 multiple-choice questions are section-referenced. The easy-to-use CD-ROM provides test questions in a format that lets instructors add, edit, resequence, and print questions to suit their needs.

Online Course Materials

Blackboard, Moodle, Sakai, Canvas
As a service for adopters, we will provide content files in the appropriate online course

format, including the instructor and student resources for this text. The files can be used as is or can be customized to fit specific needs. Course outlines, prebuilt quizzes, links, activities, and a whole array of materials are included.

PowerPoint Lecture Presentations

A set of online lecture presentations created in PowerPoint allows instructors to tailor their lectures to suit their own needs using images and notes from the textbook. These presentations are available on the instructor portion of the companion Web site and within the LaunchPad.

Acknowledgments

We would like to thank our colleagues who have carefully scrutinized the manuscript of this edition. *Universe* is a stronger and better textbook because of their conscientious efforts.

Robert Antonucci, *University of California, Santa Barbara*
Nahum Arav, *Virginia Tech University*
Anca Constantin, *James Madison University*
John Cottle, *University of California, Santa Barbara*
Peter Detterline, *Kutztown University*
Tamara Davis, *University of Queensland*
Jacqueline Dunn, *Midwestern State University*
Martin Fisk, *Oregon State University*
Andy Howell, *University of California, Santa Barbara*
William Keel, *University of Alabama*
Charles Kerton, *Iowa State University*
Erik Kubik, *Providence College*
Peter Meinhold, *University of California, Santa Barbara*
Anatoly Miroshnichenko, *University of North Carolina, Greensboro*
Patrick Motl, *Indiana University, Kokomo*
Krishna Mukherjee, *Slippery Rock University*
Fritz Osell, *Northern Oklahoma College*
Stan Peale, *University of California, Santa Barbara*
Nicolas Pereyra, *University of Texas, Pan American*
Ylva Pihlstrom, *University of New Mexico*
Kevin Plaxco, *University of California, Santa Barbara*
Raghavan Rangarajan, *Physical Research Laboratory, Ahmedabad, India*
Sean Raymond, *Laboratoire d'Astrophysique de Bordeaux*
Ian Redmount, *Saint Louis University*
Adam Rengstorf, *Purdue University, Calumet*
Andrew Rivers, *Northwestern University*
Robert Rosner, *University of Chicago*
Alan Rubin, *University of California, Los Angeles*
Paul Schmidtke, *Arizona State University*
Frank Spera, *University of California, Santa Barbara*
Jack Sweeney, *Salem State University*
Nicole Vogt, *New Mexico State University*
Lauren White, NASA/JPL
Mark Whittle, *University of Virginia*
Michael E. Wysession, *Washington University, St. Louis*

We would also like to renew our thanks to those who have reviewed past editions of *Universe:*

Ann Bragg, *Marietta College*
Richard Bowman, *Bridgewater College*
Robert Braunstein, *Northern Virginia Community College, Loudoun*
Thomas Campbell, *Allen Community College*
Brian Carter, *Grossmont College*
Stephen Case, *Olivet Nazarene University*
Demian H. J. Cho, *Kenyon College*
Richard A. Christie, *Okanagan College*
Eric Collins, *California State University, Northridge*
George J. Corso, *DePaul University*
Jamie Day, *Transylvania University*
Steven Desch, *Arizona State University*
James Dickinson, *Clackamas Community College*
Jeanne Digel, *Canada College*
Edward Dingler, *Southwest Virginia Community College*
David Duluk, Los Angeles *City College/Glendale Community College*
Jean W. Dupon, *Menlo College*
Stephanie Fawcett, *McKendree University*
Kent Fisher, *Columbus State Community College*
Juhan Frank, *Louisiana State University*
Tony George, *Columbia Basin College*
Erika Gibb, *University of Missouri, St. Louis*
Daniel Greenberger, *City College of New York*
Wayne K. Guinn, *Lon Morris College*
Joshua Gundersen, *University of Miami*
Hunt Guitar, *New Mexico State University*
Kathleen A. Harper, *Denison University*
Lynn Higgs, *University of Utah*
Melinda Hutson, *Portland Community College*
Douglas R. Ingram, *Texas Christian University*
Adam G. Jensen, *University of Colorado, Boulder*
Darell Johnson, *Missouri Western State University*
Michael Joner, *Brigham Young University*
Lauren Jones, *Denison University*
Steve Kawaler, *Iowa State University*
Charles KeRton, *Iowa State University*
Kevin Kimberlin, *Bradley University*
H. S. Krawczynski, *Washington University, St. Louis*
Lauren Likkel, *University of Wisconsin, Eau Claire*
Dennis Machnik, *Plymouth State University*
Franck Marchis, *University of California, Berkeley*
Michele M. Montgomery, *University of Central Florida*
Edward M. Murphy, *University of Virginia*
Donna Naples, *University of Pittsburgh*
Gerald H. Newsom, *The Ohio State University*
Brian Oetiker, *Sam Houston State University*
Ronald P. Olowin, *Saint Mary's College*
Michael J. O'Shea, *Kansas State University*
J. Douglas Patterson, *Johnson County Community* College
Charles Peterson, *University of Missouri*

Richard Rand, *University of New Mexico*
Mike Reynolds, *Florida Community College*
Frederick Ringwald, *California State University, Fresno*
Todd Rimkus, *Marymount University*
Paul Robinson, *Westchester Community College*
Louis Rubbo, *Coastal Carolina University*
Kyla Scarborough, *Pittsburg State University*
Doug Showell, *University of Nebraska, Omaha*
Caroline Simpson, *Florida International University*
Glenn Spiczak, *University of Wisconsin, River Falls*
James D. Stickler, *Allegany College of Maryland*
Larry K. Smith, *Snow College*
Chris Taylor, *California State University, Sacramento*
Charles M. Telesco, *University of Florida*
Dale Trapp, *Concordia University, St. Paul*

Many others have participated in the preparation of this book and we thank them for their efforts. We are grateful for the guidance we have received from Alicia Brady, our acquisitions editor. We're also deeply grateful to our development editor on *Universe*, Brittany Murphy, and for the contributions of assistant editor, Courtney Lyons. Amy Thorne, the media and supplements editor, supervised the creation of the superb multimedia. Special thanks go to Kerry O'Shaughnessy, who skillfully kept all three versions of this book on track and on target. Vicki Tomaselli deserves credit for the design of the book. We also thank Janice Donnola and Dragonfly Media for coordinating and producing the excellent artwork, and extend our gratitude to photo editor Robin Fadool and photo researcher Deborah Anderson for finding just the right photographs. Louise Ketz deserves a medal for wading through the prose and copyediting it into proper English.

On a personal note, Roger Freedman would like to thank his father, Richard Freedman, for first cultivating his interest in space many years ago, and for his father's personal contributions to the exploration of the universe as an engineer for the Atlas and Centaur launch vehicle programs. Most of all, Roger thanks his charming wife, Caroline, for putting up with his long nights slaving over the computer!

Robert Geller would like to extend special thanks to Robert Antonucci (aka Ski). Ski has helped with numerous topics throughout this textbook, was Robert's Ph.D. thesis advisor, and—very conveniently—his next door neighbor. While some discussions take the usual form of pointing to data plots on the computer, many discussions were literally over the fence. In addition to benefiting from Ski's seemingly unlimited expertise, he has made astronomy FUN from the first day Robert was a Teaching Assistant for his class.

Robert Geller would also like to thank his parents for nurturing a love of science. Although some home appliances that were taken apart in the name of science were never put back together, he always had their support. Most of all, Robert would like to thank his family for their support through years of long hours researching and writing for this textbook. His wife, Susanne, has even contributed scientific results to this book as part of the team that discovered the Higgs particle in 2012; if her next project is successful, she can add dark matter to her list. With the completion of this edition, Robert and his daughter, Zoe, will have some time to use their telescope, fish, and surf together.

Although we have made a concerted effort to make this edition error-free, some mistakes may have crept in unbidden. We would appreciate hearing from anyone who finds an error or wishes to comment on the text.

Roger A. Freedman
Department of Physics
University of California, Santa Barbara
Santa Barbara CA 93106
airboy@physics.ucsb.edu

Robert M. Geller
Department of Physics
University of California, Santa Barbara
Santa Barbara CA 93106
rhmg@physics.ucsb.edu

To the Student

HOW TO GET THE MOST FROM *UNIVERSE*

If you're like most students just opening this textbook, you're enrolled in one of the few science courses you'll take in college. As you study astronomy, you'll probably do relatively little reading compared to a literature or history course—at least in terms of the number of pages. But your readings will be packed with information, much of it new to you and (we hope) exciting. You can't read this textbook like a novel and expect to learn much from it. Don't worry, though. We wrote this book with you in mind. In this section, we'll suggest how Universe can help you succeed in your astronomy course, and take you on a guided tour of the book and media.

Apply these techniques to studying astronomy

• **Read before each lecture** You'll get the most out of your astronomy course if you read each chapter before hearing a lecture about its subject matter. That way, many of the topics will already be clear in your mind, and you'll understand the lecture better. You'll be able to spend more of your listening and note-taking time on the more challenging ideas presented in the lecture.

• **Take notes as you read and make use of office hours** Keep a notebook handy as you read, and write down the key points of each section so that you can review them later. If any parts of the section don't seem clear on first reading, make a note of them, too, including the page numbers. Once you've gone through the chapter, reread it with special emphasis on the ideas that gave you trouble the first time. If you're still unsure after the lecture, consult your instructor, either during office hours or after class. Bring your notes with you so your instructor can see which concepts are giving you trouble. Once your instructor has helped clarify things for you, revise your notes so you'll remember your newfound insights. You'll end up with a chapter summary in your own words. This will be a tremendous help when studying for exams!

• **Make use of your fellow students** Many students find it useful to form study groups for astronomy. You can hash out challenging topics with each other and have a good time while you're doing it. But make sure that you write up your homework by yourself, because the penalties for copying or plagiarizing other students' work can be severe in the extreme. Some students find individual assistance useful, too. If you think a tutor will be helpful, link up with one early. Getting a tutor late in the course, in the belief that you'll be able to catch up with what you missed earlier on, is almost always a lost cause.

• **Take advantage of the Web site and e-Book** Take some time to explore the Universe Web site (www.whfreeman.com/universe10e). There you'll find review materials, animations, videos, interactive exercises, flashcards, and many other features keyed to chapters in *Universe*. All of these features are designed to help you learn and enjoy astronomy, so make sure to take full advantage of them. You can also use the e-Book version of this textbook, which combines the complete content of the book and the Web site in a convenient online format.

- **Try astronomy for yourself with your star charts** At the back of this book you'll find a set of star charts for each month of the year in the northern hemisphere. (For a set of southern hemisphere star charts, see the *Universe* Web site.) Star charts can get you started with your own observations of the universe. Hold the chart overhead in the same orientation as the compass points, with southern horizon toward the south and western horizon toward the west. (To save strain on your arms, you may want to cut these pages out of the book.) Depending on the version of this textbook that your instructor requested, this book may also include access to the easy-to-use *Starry Night*™ planetarium program, which you can use to view the sky on any date and time as seen from anywhere on Earth.

Here's the most important advice of all

We haven't mentioned the most important thing you should do when studying astronomy: Have fun! Of all the different kinds of scientists, astronomers are among the most excited about what they do and what they study. Let some of that excitement about the universe rub off on you, and you'll have a great time with this course and with this textbook.

In preparing this edition of *Universe*, we've tried very hard to make it the kind of textbook that a student like you will find useful. We're very interested in your comments and opinions! Please feel free to e-mail or write to us and we will respond personally.

Best wishes for success in your studies!

Roger A. Freedman
Department of Physics
University of California, Santa Barbara
Santa Barbara CA 93106
airboy@physics.ucsb.edu

Robert M. Geller
Department of Physics
University of California, Santa Barbara
Santa Barbara CA 93106
rhmg@physics.ucsb.edu

I Introducing Astronomy

Many details are revealed by looking at ultraviolet light emitted by the Sun. The bright arches result from magnetic fields poking out of the solar surface. Electrons accelerated by these magnetic fields then emit ultraviolet light. Part of these features can also be seen in visible light, where they are called solar flares. More than just emitting light, solar flares can hurl mountain-sized quantities of matter from the Sun. Sometimes the solar flares are directed at Earth, where they can damage communication satellites and pose risks to astronauts. (NASA)

R I V U X G

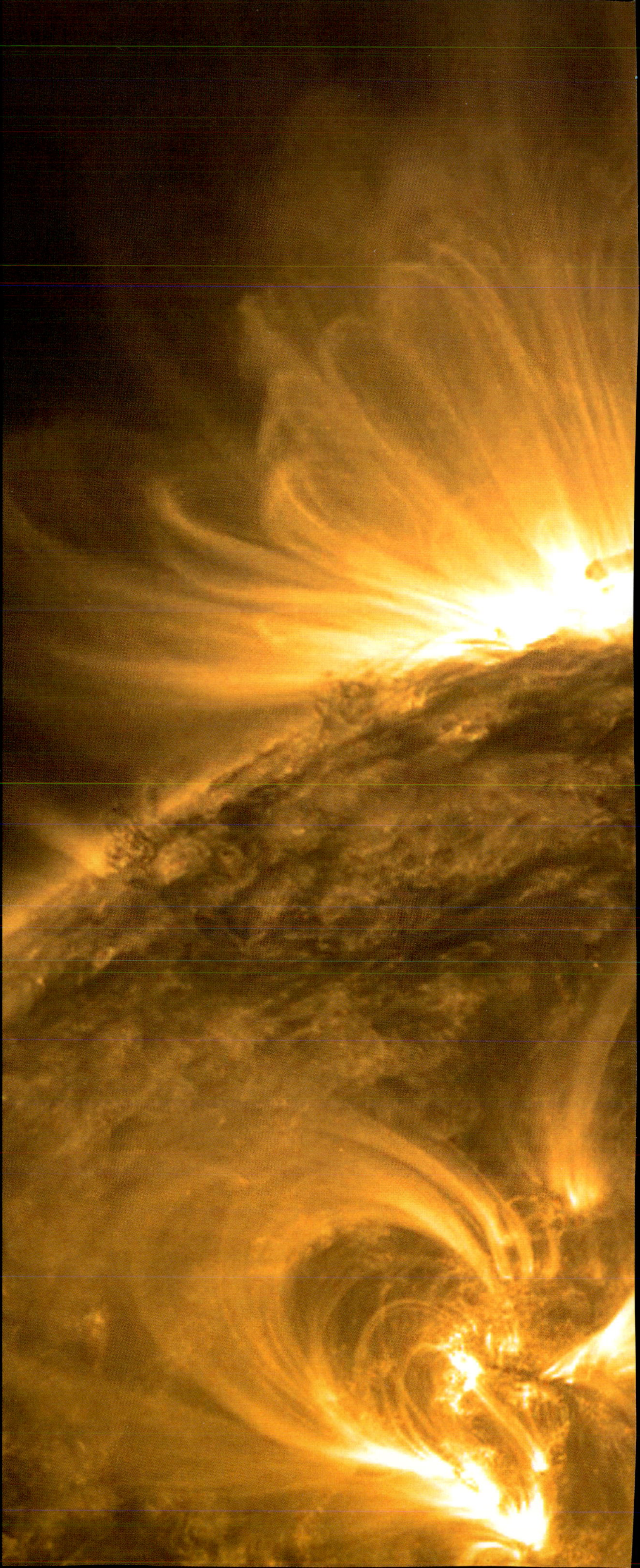

This tall pillar of hydrogen and dust extends a few light-years and lies within the Eagle Nebula about 7000 light-years away. There are several fingerlike structures near the top that contain newly forming solar systems within their tips. This dense column is slowly eroding as ultraviolet light from nearby stars breaks apart its gas and dust. There are several of these majestic star-forming structures next to each other, and they are often called the Pillars of Creation. (NASA)

R I V U X G

CHAPTER

1

The night sky seen from Mauna Kea in Hawaii. The feature extending across the sky is the Milky Way, consisting of hundreds of billions of stars. The Milky Way is the galaxy in which we reside. (John Hook/Flickr/Getty Images) R I **V** U X G

Astronomy and the Universe

LEARNING GOALS

By reading the sections of this chapter, you will learn

1-1 What distinguishes the methods of science from other human activities

1-2 How exploring other planets provides insight into the origins of the solar system and the nature of our Earth

1-3 Stars have a life cycle—they form, evolve over millions or billions of years, and die

1-4 Stars are grouped into galaxies, which are found throughout the universe

1-5 How astronomers measure position and size of a celestial object

1-6 How to express very large or very small numbers in convenient notation

1-7 Why astronomers use different units to measure distances in space

1-8 What astronomy can tell us about our place in the universe

Imagine yourself looking skyward on a clear, dark, moonless night, far from the glare of city lights. As you gaze upward, you see a panorama that no poet's words can truly describe and that no artist's brush could truly capture. Literally thousands of stars are scattered from horizon to horizon, many of them grouped into a luminous band called the Milky Way (which extends up and down across the middle of this photograph). As you watch, the entire spectacle swings slowly overhead from east to west as the night progresses.

For thousands of years people have looked up at the heavens and contemplated the universe. Like our ancestors, we find our thoughts turning to profound questions as we gaze at the stars. How was the universe created? Where did Earth, the Moon, and the Sun come from? What are the planets and stars made of? And how do we fit in? What is our place in the cosmic scope of space and time?

Wondering about the universe is a key part of what makes us human. Our curiosity, our desire to explore and discover, and, most important, our ability to reason about what we have discovered are qualities that distinguish us from other animals. The study of the stars transcends all boundaries of culture, geography, and politics. In a literal sense, astronomy is a universal subject—its subject is the entire universe.

1-1 To understand the universe, astronomers use the laws of physics to construct testable theories and models

Astronomy has a rich heritage that dates back to the myths and legends of antiquity. Centuries ago, the heavens were thought to be populated with demons and heroes, gods and goddesses. Astronomical phenomena were explained as the result of supernatural forces and divine intervention.

The course of civilization was greatly affected by a profound realization: *The universe is comprehensible.* For example, ancient Greek astronomers discovered that by observing the heavens and carefully reasoning about what they saw, they could learn something about how the universe operates. As we shall see in Chapter 3, ancient Greek astronomers measured the size of Earth and were able to understand and predict eclipses without appealing to supernatural forces. Modern science is a direct descendant of astronomy, which had many contributors from the Middle East, Africa, Asia, Central America, and, eventually, Greece.

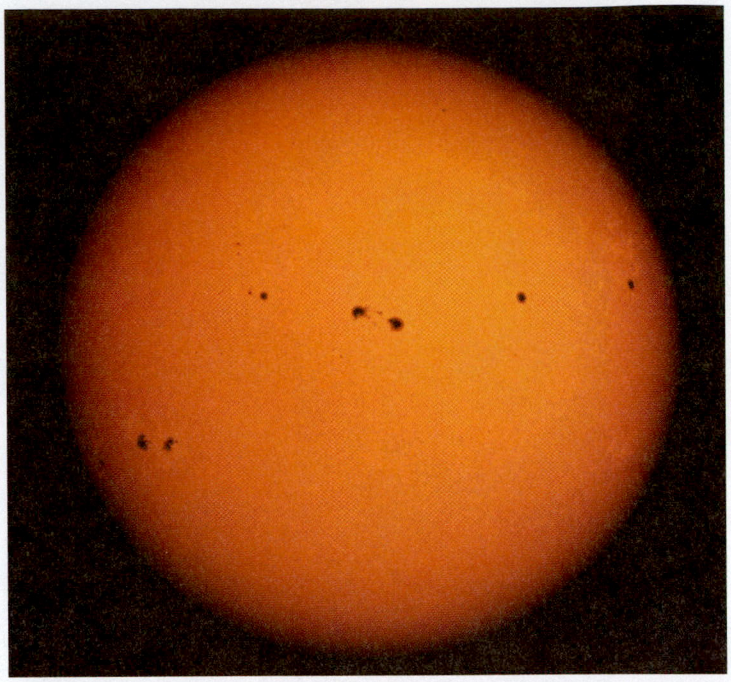

FIGURE 1-1 R I V U X G

Our Star, the Sun The Sun is a typical star. Its diameter is about 1.39 million kilometers (roughly a million miles), and its surface temperature is about 5500°C (10,000°F). A detailed scientific model of the Sun tells us that it draws its energy from nuclear reactions occurring at its center, where the temperature is about 15 million degrees Celsius. (NSO/AURA/NSF)

The Scientific Method

TUTORIAL 1-1 Like art, music, or any other human creative activity, science makes use of intuition and experience. But the approach used by scientists to explore physical reality differs from other forms of intellectual endeavor in that it is based fundamentally on *observation, logic,* and *skepticism.* This approach, called the **scientific method,** requires that our ideas about the world around us be consistent with what we actually observe.

Hypotheses, models, theories, and laws are essential parts of the scientific way of knowing

The scientific method goes something like this: A scientist trying to understand some observed phenomenon proposes a **hypothesis,** which is a collection of ideas that seems to explain what is observed. It is in developing hypotheses that scientists are at their most creative, imaginative, and intuitive. But their hypotheses must always agree with existing observations and experiments, because a discrepancy with what is observed implies that the hypothesis is wrong. (The exception is if the scientist thinks that the existing results are wrong and can give compelling evidence to show that they are wrong.) The scientist then uses logic to work out the implications of the hypothesis and to make predictions that can be tested. A hypothesis is on firm ground only after it has accurately forecast the results of new experiments or observations. (In practice, scientists typically go through these steps in a less linear fashion than we have described.)

Scientists describe reality in terms of **models,** which are hypotheses that have withstood observational or experimental tests. A model tells us about the properties and behavior of some object or phenomenon. A familiar example is a model of the atom, which scientists picture as electrons orbiting a central nucleus. Another example, which we will encounter in Chapter 18, is a model that tells us about physical conditions (for example, temperature, pressure, and density) in the interior of the Sun (Figure 1-1). A well-developed model uses mathematics—one of the most powerful tools for logical thinking—to make detailed predictions. For example, a successful model of the Sun's interior should describe what the values of temperature, pressure, and density are at each depth within the Sun, as well as the relations between these quantities. For this reason, mathematics is one of the most important tools used by scientists.

A body of related hypotheses can be pieced together into a self-consistent description of nature called a **theory.** An example from Chapter 4 is the theory that the planets are held in their orbits around the Sun by the Sun's gravitational force (Figure 1-2). Without models and theories there is no understanding and no science, only collections of facts.

CAUTION! In everyday language the word "theory" is often used to mean an idea that looks good on paper, but has little to do with reality. In science, however, a good theory is one that explains reality very well and that can be applied to explain new observations. An excellent example is the theory of gravitation (Chapter 4), which was devised by the English scientist Isaac Newton in the late 1600s to explain the orbits of the six planets known at that time. When astronomers of later centuries discovered the planets Uranus and Neptune and the dwarf planet Pluto, they found that these planets also moved in accordance with Newton's theory. The same theory describes the motions of satellites around Earth as well as the orbits of planets around other stars.

An important part of a scientific theory is its ability to make predictions that can be tested by other scientists. If the predictions are verified by observation that lends support to the theory and

FIGURE 1-2
Planets Orbiting the Sun An example of a scientific theory is the idea that Earth and planets orbit the Sun due to the Sun's gravitational attraction. This theory is universally accepted because it makes predictions that have been tested and confirmed by observation. (The Sun and planets are actually much smaller than this illustration would suggest.) (Detlev Van Ravenswaay/Science Photo Library)

suggests that it might be correct. If the predictions are *not* verified, the theory needs to be modified or completely replaced. For example, an old theory held that the Sun and planets orbit around a stationary Earth. This theory led to certain predictions that could be checked by observation, as we will see in Chapter 4. In the early 1600s the Italian scientist Galileo Galilei used one of the first telescopes to show that these predictions were incorrect. As a result, the theory of a stationary Earth was rejected, eventually to be replaced by the modern model shown in Figure 1-2 in which Earth and other planets orbit the Sun.

An idea that *cannot* be tested by observation or experiment does not qualify as a scientific theory. An example is the idea that there is a little man living in your refrigerator who turns the inside light on or off when you open and close the door. The little man is invisible, weightless, and makes no sound, so you cannot detect his presence. While this is an amusing idea, it cannot be tested and so cannot be considered science.

Skepticism is an essential part of the scientific method. New hypotheses must be able to withstand the close scrutiny of other scientists. The more radical the hypothesis, the more skepticism and critical evaluation it will receive from the scientific community, because the general rule in science is that **extraordinary claims require extraordinary evidence.** That is why scientists as a rule do not accept claims that people have been abducted by aliens and taken aboard UFOs. The evidence presented for these claims is flimsy, secondhand, and unverifiable.

At the same time, scientists must be open-minded. They must be willing to discard long-held ideas if these ideas fail to agree with new observations and experiments, provided the new data have survived critical review. (If an alien spacecraft really did land on Earth, scientists would be the first to accept that aliens existed—provided they could take a careful look at the spacecraft and its occupants.) That is why the validity of scientific knowledge can be temporary. With new evidence, some ideas only need modification, while occasionally, some ideas may need to be entirely replaced. As you go through this book, you will encounter many instances where new observations have transformed our understanding of Earth, the planets, the Sun and stars, and indeed the very structure of the universe.

Theories that accurately describe the workings of physical reality have a significant effect on civilization. For example, basing his conclusions in part on observations of how the planets orbit the Sun, Isaac Newton deduced a set of fundamental principles that describe how *all* objects move. These theoretical principles, which we will encounter in Chapter 4, work equally well on Earth as they do in the most distant corner of the universe. They represent our first complete, coherent description of how objects move in the physical universe. **Newtonian mechanics** had an immediate practical application in the construction of machines, buildings, and bridges. It is no coincidence that the Industrial Revolution followed hard on the heels of these theoretical and mathematical advances inspired by astronomy.

Newtonian mechanics and other physical theories have stood the test of time and been shown to have great and general validity. Proven theories of this kind are collectively referred to as the **laws of physics.** Thus, the most reliable theories with the broadest applicability can eventually be considered laws of physics. Astronomers use these laws to interpret and understand their observations of the universe. The laws governing light and its relationship to matter are of particular importance, because the only information we can gather about distant stars and galaxies is in the light that we receive from them. Using the physical laws that describe how objects absorb and emit light, astronomers have measured the temperature of the Sun and even learned what the Sun is made of. By analyzing starlight in the same way, they have discovered that our own Sun is a rather ordinary star and that the observable universe may contain 10 billion trillion stars just like the Sun.

Technology in Science

An important part of science is the development of new tools for research and new techniques of observation. As an example, until fairly recently everything we knew about the distant universe was based on visible light. Astronomers would peer through telescopes to observe and analyze visible starlight. By the end of the nineteenth century, however, scientists had begun to discover forms of light invisible to the human eye: X-rays, gamma rays, radio waves, microwaves, and ultraviolet and infrared radiation.

As we will see in Chapter 6, in recent years astronomers have constructed telescopes that can detect such nonvisible forms of light (Figure 1-3). These instruments give us views of the universe vastly different from anything our eyes can see. These new views have allowed us to see through the atmospheres of distant planets, to study the thin but incredibly violent gas that surrounds our Sun, and even to observe new solar systems being formed around distant stars. Aided by high-technology telescopes, today's astronomers carry on the program of careful observation and logical analysis begun thousands of years ago by their ancient Greek predecessors.

FIGURE 1-3 R I **V** U X G

A Telescope in Space Because it orbits outside Earth's atmosphere in the near-vacuum of space, the Hubble Space Telescope (HST) can detect not only visible light but also ultraviolet and near-infrared light coming from distant stars and galaxies. These forms of nonvisible light are absorbed by our atmosphere and hence are difficult or impossible to detect with a telescope on Earth's surface. This photo of HST was taken by the crew of the space shuttle *Columbia* after a servicing mission in 2002. (Courtesy of Scientific American/NASA/AAT)

CONCEPTCHECK 1-1

Which is held in higher regard by professional astronomers: a hypothesis or a theory? Explain your answer.

Answer appears at the end of the chapter.

1-2 By exploring the planets, astronomers uncover clues about the formation of the solar system

> Studying planetary science gives us a better perspective on our own unique Earth

The science of astronomy allows our intellects to voyage across the cosmos. We can think of three stages in this voyage: from Earth to other parts of the solar system, from the solar system to the stars, and from stars to galaxies and the grand scheme of the universe.

The star we call the Sun and all the celestial bodies that orbit the Sun—including Earth, the other planets, all their various moons, and smaller bodies such as asteroids and comets—make up the **solar system.** Since the 1960s a series of unmanned spacecraft has been sent to explore each of the planets (Figure 1-4). Using the remote "eyes" of such spacecraft, we have flown over Mercury's cratered surface, peered beneath Venus's poisonous cloud cover, and discovered enormous canyons and extinct volcanoes on Mars. We have found active volcanoes on a moon of Jupiter, probed the atmosphere of Saturn's moon Titan, seen the rings of Uranus up close, and looked down on the active atmosphere of Neptune.

Along with rocks brought back by the *Apollo* astronauts from the Moon (the only world beyond Earth visited by humans), new information from spacecraft has revolutionized our understanding of the origin and evolution of the solar system. We have come to realize that many of the planets and their satellites were shaped by collisions with other objects. Craters on the Moon and on many other worlds are the relics of innumerable impacts by bits of interplanetary rock. The Moon may itself be the result of a catastrophic collision between Earth and a planet-sized object shortly after the solar system was formed. Such a collision could have torn sufficient material from the primordial Earth to create the Moon.

The oldest objects found on Earth are **meteorites,** chemically distinct bits of interplanetary debris that sometimes fall to our planet's surface. By using radioactive age-dating techniques, scientists have found that the oldest meteorites are 4.56 billion years old—older than any rocks found on Earth or the Moon. The conclusion is that our entire solar system, including the Sun and planets, formed 4.56 billion years ago. The few thousand years of recorded human history is no more than the twinkling of an eye compared to the long history of our solar system.

The discoveries that we have made in our journeys across the solar system are directly relevant to the quality of human life on our own planet. Until recently, our understanding of geology, weather, and climate was based solely on data from Earth. Since

FIGURE 1-4
The Sun and Planets to Scale This montage of images from various spacecraft and ground-based telescopes shows the relative sizes of the planets and the Sun. The Sun is so large compared to the planets that only a portion of it fits into this illustration. The distances from the Sun to each planet are not shown to scale; the actual distance from the Sun to Earth, for instance, is 12,000 times greater than Earth's diameter. (Calvin J. Hamilton and NASA/JPL)

the advent of space exploration, however, we have been able to compare and contrast other worlds with our own. This new knowledge gives us valuable insight into our origins, the nature of our planetary home, and the limits of our natural resources.

1-3 By studying stars and nebulae, astronomers discover how stars are born, grow old, and die

The nearest of all stars to Earth is the Sun. Although humans have used the Sun's warmth since the dawn of our species, it was only in the 1920s and 1930s that physicists figured out how the Sun shines. At the center of the Sun, thermonuclear reactions—so called because they require extremely high temperatures—convert hydrogen (the Sun's primary constituent) into helium. This violent process releases a vast amount of energy, which eventually makes its way to the Sun's surface and escapes as light (see Figure 1-1). Thus, through nuclear reactions, hydrogen acts as a "fuel" for stars. All the stars you can see in the nighttime sky also shine by nuclear reactions (Figure 1-5). By 1950 physicists could reproduce such nuclear reactions here on Earth in the form of a hydrogen bomb (Figure 1-6). In the future, converting hydrogen into helium might someday offer a cleaner method for the production of nuclear energy.

Because nuclear reactions consume the original material of which stars are made, the way an engine consumes its fuel, stars cannot last forever. Rather, they must form, evolve, and eventually die.

FIGURE 1-5 R I **V** U X G
Stars like Grains of Sand This Hubble Space Telescope image shows thousands of stars in the constellation Sagittarius. Each star shines because of thermonuclear reactions that release energy in its interior. Different colors indicate stars with different surface temperatures: stars with the hottest surfaces appear blue, while those with the coolest surfaces appear red. (The Hubble Heritage Team, AURA/STScI/NASA)

FIGURE 1-6 R I **V** U X G
A Thermonuclear Explosion A hydrogen bomb uses the same physical principle as the thermonuclear reactions at the Sun's center: the conversion of matter into energy by nuclear reactions. This thermonuclear detonation on October 31, 1952, had an energy output equivalent to 10.4 million tons of TNT (almost 1000 times greater than the nuclear bomb detonated over Hiroshima in World War II). This is a mere ten-billionth of the amount of energy released by the Sun in one second. (Defense Nuclear Agency)

CAUTION! Astronomers often use biological terms such as "birth" and "death" to describe stages in the evolution of inanimate objects like stars. Keep in mind that such terms are used only as *analogies,* which help us visualize these stages. They are not to be taken literally!

The Life Stories of Stars

The rate at which stars emit energy in the form of light tells us how rapidly they are consuming their nuclear "fuel" (hydrogen), and hence how long they can continue to shine before reaching the end of their life spans. More massive stars have more hydrogen, and thus, more nuclear "fuel," but consume it at such a prodigious rate that they live out their lives in just a few million years. Less massive stars have less material to consume, but their nuclear reactions proceed so slowly that their life spans are measured in billions of years. (Our own star, the Sun, is in early middle age: It is 4.56 billion years old, with a lifetime of 12.5 billion years.)

While no astronomer can watch a single star go through all of its life stages, we have been able to piece together the life stories of stars by observing many different stars at different points in their life cycles. Important pieces of the puzzle have been discovered by studying huge clouds of interstellar gas, called **nebulae** (singular **nebula**), which are found scattered across the sky. Within some nebulae, such as the Orion Nebula shown in Figure 1-7, stars are born from the material of the nebula itself. Other nebulae reveal what happens when nuclear reactions stop and a star dies. Some stars that are far more massive than the Sun end their lives with a spectacular detonation called a **supernova** (plural **supernovae**) that blows the star apart. The Crab Nebula (Figure 1-8) is a striking example of a remnant left behind by a supernova.

Studying the life cycles of stars is crucial for understanding our own origins

FIGURE 1-7 R I V U X G
The Orion Nebula—Birthplace of Stars This beautiful nebula is a stellar "nursery" where stars are formed out of the nebula's gas. Intense ultraviolet light from newborn stars excites the surrounding gas and causes it to glow. Many of the stars embedded in this nebula are less than a million years old, a brief interval in the lifetime of a typical star. The Orion Nebula is some 1500 light-years from Earth and is about 30 light-years across. (NASA, ESA, M. Robberto/STScI/ESA, and the Hubble Space Telescope Orion Treasury Project Team)

FIGURE 1-8 R I V U X G
The Crab Nebula—Wreckage of an Exploded Star When a dying star exploded in a supernova, it left behind this elegant funeral shroud of glowing gases blasted violently into space. A thousand years after the explosion these gases are still moving outward at about 1800 kilometers per second (roughly 4 million miles per hour). The Crab Nebula is 6500 light-years from Earth and about 13 light-years across. (NASA, ESA, J. Hester and A. Loll/Arizona State University)

Dying stars can produce some of the strangest objects in the sky. Some dead stars become **pulsars,** which spin rapidly at rates of tens or hundreds of rotations per second. And some stars end their lives as almost inconceivably dense objects called **black holes,** whose gravity is so powerful that nothing—not even light—can escape. Even though a black hole itself emits essentially no radiation, a number of black holes have been discovered beyond our solar system by Earth-orbiting telescopes. This is done by detecting the X-rays emitted by gas falling toward a black hole.

During their death throes, stars return the gas of which they are made to interstellar space. (Figure 1-8 shows these expelled gases expanding away from the site of a supernova explosion.) This gas contains heavy elements—that is, elements heavier than hydrogen and helium—that were created during the star's lifetime by nuclear reactions in its interior. Interstellar space thus becomes enriched with newly manufactured atoms and molecules. The Sun and its planets were formed from interstellar material that was enriched in this way. This means that the atoms of iron and nickel that make up Earth, as well as the carbon in our bodies and the oxygen we

FIGURE 1-9 R I V U X G

A Galaxy This spectacular galaxy, called M63, contains about a hundred billion stars. M63 has a diameter of about 60,000 light-years and is located about 35 million light-years from Earth. Along this galaxy's spiral arms you can see a number of glowing clumps. Like the Orion Nebula in our own Milky Way Galaxy (see Figure 1-7), these are sites of active star formation. (2004–2013 R. Jay GaBany, Cosmotography.com)

breathe, were created deep inside ancient stars. By studying stars and their evolution, we are really studying our own origins.

CONCEPTCHECK 1-2

If a star were twice as massive as our Sun, would it shine for a longer or shorter span of time? Why?

Answer appears at the end of the chapter.

1-4 By observing galaxies, astronomers learn about the origin and fate of the universe

Stars are not spread uniformly across the universe but are grouped together in huge assemblages called **galaxies.** Galaxies come in a wide range of shapes and sizes. Our Sun is just one star in a galaxy we call the Milky Way (see the figure that opens this chapter). A typical galaxy, like our Milky Way, contains several hundred billion stars. Some galaxies are much smaller, containing only a few million stars. Others are monstrosities that devour neighboring galaxies in a process called "galactic cannibalism."

Our Milky Way Galaxy has arching spiral arms like those of the galaxy shown in Figure 1-9. These arms are particularly active sites of star formation. In recent years, astronomers have discovered a mysterious object at the center of the Milky Way with a mass millions of times greater than that of our Sun. It now seems certain that this curious object is an enormous black hole.

Some of the most intriguing galaxies appear to be in the throes of violent convulsions and are rapidly expelling matter. The centers of these strange galaxies, which may harbor even more massive black holes, are often powerful sources of X-rays and radio waves.

Even more awesome sources of energy are found still deeper in space. Often located at distances so great that their light takes billions of years to reach Earth, we find the mysterious **quasars.** Although in some ways quasars look like nearby stars (Figure 1-10), they are among the most distant and most luminous objects in the sky. A typical quasar shines with the brilliance of a hundred galaxies. Detailed observations of quasars imply that they draw their energy from material falling into enormous black holes.

Galaxies and the Expanding Universe

The motions of distant galaxies reveal that they are moving away from us and from each other. In other words, the universe is *expanding.* Extrapolating into the past, we learn that the universe must have been born from an incredibly dense state some 13.7 billion years ago. A variety of evidence indicates that at that moment—the beginning of time—the universe began with a cosmic explosion, known as the **Big Bang,** which occurred throughout all space.

The motions of distant galaxies motivate the ideas of the expanding universe and the Big Bang

Thanks to the combined efforts of astronomers and physicists, we are making steady advances in understanding the nature and history of the universe. This understanding may reveal the origin of some of the most basic properties of physical reality. Studying the most remote galaxies is also helping to answer questions about the

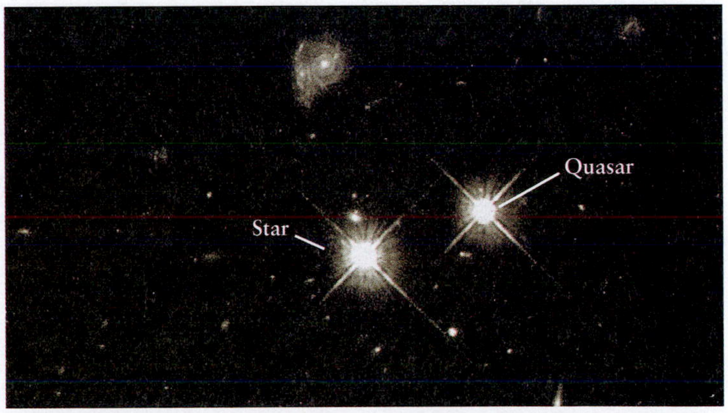

FIGURE 1-10 R I V U X G

A Quasar The two bright starlike objects in this image look almost identical, but they are dramatically different. The object on the left is indeed a star that lies a few hundred light-years from Earth. But the "star" on the right is actually a quasar about 9 billion light-years away. To appear so bright even though they are so distant, quasars like this one must be some of the most luminous objects in the universe. The other objects in this image are galaxies like that in Figure 1-9. (Charles Steidel, California Institute of Technology; and NASA) ultimate fate of the universe. Such studies suggest that the expansion of the universe will continue forever, and it is actually gaining speed.

The work of unraveling the deepest mysteries of the universe requires specialized tools, including telescopes, spacecraft, and computers. But for many purposes, the most useful device for studying the universe is the human brain itself. Our goal in this book is to help you use *your* brain to share in the excitement of scientific discovery.

In the remainder of this chapter we introduce some of the key concepts and mathematics that we will use in subsequent chapters. Study these carefully, for you will use them over and over again throughout your own study of astronomy.

1-5 Astronomers use angles to denote the positions and apparent sizes of objects in the sky

Whether they study planets, stars, galaxies, or the very origins of the universe, astronomers must know where to point their telescopes. For this reason, an important part of astronomy is keeping track of the positions of objects in the sky. A system for measuring angles is an essential part of this aspect of astronomy (Figure 1-11).

An **angle** measures the opening between two lines that meet at a point. A basic unit to express angles is the **degree,** designated by the symbol °. A full circle is divided into 360°, and a right angle measures 90° (Figure 1-11a). As Figure 1-11b shows, if you draw lines from your eye to each of the two "pointer stars" in the Big Dipper, the angle between these lines—that is, the **angular distance** between these two stars—is about 5°. (In Chapter 2 we will see that these two stars "point" to Polaris, the North Star.) The angular distance between the stars that make up the top and bottom of the Southern Cross, which is visible from south of the equator, is about 6° (Figure 1-11c).

Angles are a tool that we will use throughout our study of astronomy

Astronomers also use angles to describe the apparent size of a celestial object—that is, how wide the object appears in the sky. For example, the angle covered by the diameter of the full moon is about ½° (Figure 1-11a). We therefore say that the **angular diameter** (or **angular size**) of the Moon is ½°. Alternatively, astronomers say that the Moon **subtends,** or extends over, an angle of ½°. Ten full moons could fit side by side between the two pointer stars in the Big Dipper.

The average adult human hand held at arm's length provides a means of estimating angles, as Figure 1-12 shows. For example, a fist covers an angle of about 10°, whereas a fingertip is about 1° wide. You can use various segments of your index finger extended to arm's length to estimate angles a few degrees across.

To talk about smaller angles, we subdivide the degree into 60 **arcminutes** (also called minutes of arc), which is commonly abbreviated as 60 arcmin or 60′. An arcminute is further subdivided into 60 **arcseconds** (or seconds of arc), usually written as 60 arcsec or 60″. Thus,

$$1° = 60 \text{ arcmin} = 60'$$

$$1' = 60 \text{ arcsec} = 60''$$

For example, on January 1, 2007, the planet Saturn had an angular diameter of 19.6 arcsec as viewed from Earth. That is a convenient, precise statement of how big the planet appeared in Earth's sky on that date. (Because this angular diameter is so small, to the naked eye Saturn appears simply as a point of light. To see any detail on Saturn, such as the planet's rings, requires a telescope.)

If we know the angular size of an object as well as the distance to that object, we can determine the actual linear size of the object

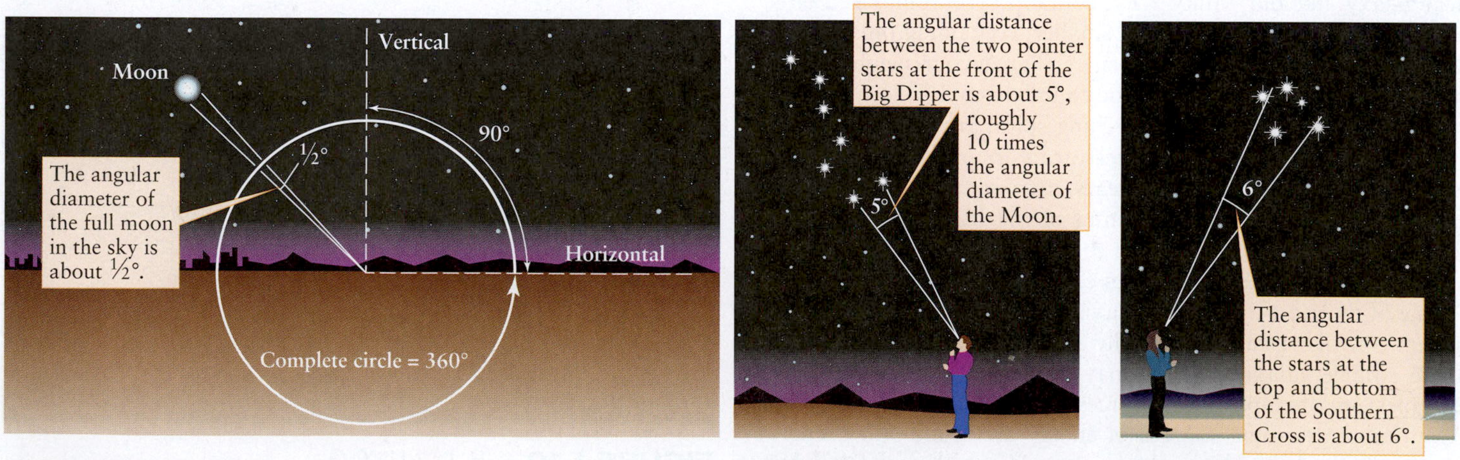

(a) Measuring angles in the sky (b) Angular distances in the northern hemisphere (c) Angular distances in the southern hemisphere

FIGURE 1-11
Measuring Angles **(a)** Angles are measured in degrees (°). There are 360° in a complete circle and 90° in a right angle. For example, the angle between the vertical direction (directly above you) and the horizontal direction (toward the horizon) is 90°. The angular diameter of the full moon in the sky is about ½°. **(b)** The seven bright stars that make up the Big Dipper can be seen from anywhere in the northern hemisphere. The angular distance between the two "pointer stars" at the front of the Big Dipper is about 5°. **(c)** The four bright stars that make up the Southern Cross can be seen from anywhere in the southern hemisphere. The angular distance between the stars at the top and bottom of the cross is about 6°.

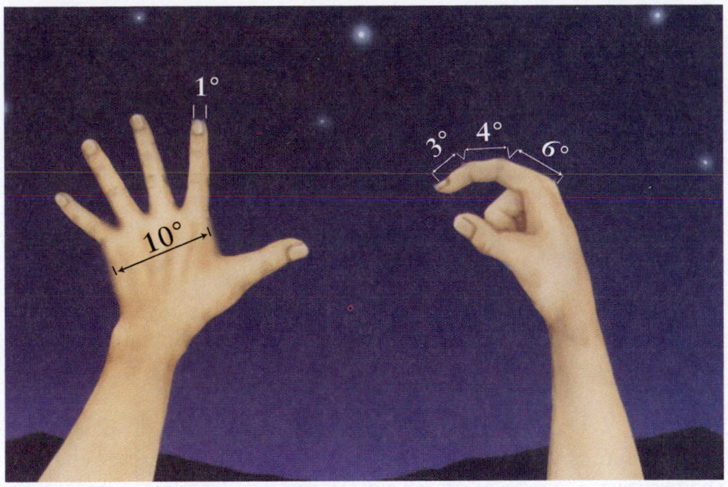
1°
3°
4°
6°
10°

FIGURE 1-12

Estimating Angles with Your Hand The adult human hand extended to arm's length can be used to estimate angular distances and angular sizes in the sky.

(measured in kilometers or miles, for example). Box 1-1 describes how this is done.

CONCEPTCHECK 1-3

Is it possible for a basketball to look bigger than the Moon? Smaller?

Answer appears at the end of the chapter.

1-6 Powers-of-ten notation is a useful shorthand system for writing numbers

Learning powers-of-ten notation will help you deal with very large and very small numbers

Astronomy is a subject of extremes. Astronomers investigate the largest structures in the universe, including galaxies and clusters of galaxies. But they must also study atoms and atomic nuclei, among the smallest objects in the universe, in order to explain how and why stars shine. They also study conditions

BOX 1-1 TOOLS OF THE ASTRONOMER'S TRADE

The Small-Angle Formula

You can estimate the angular sizes of objects in the sky with your hand and fingers (see Figure 1-12). Using rather more sophisticated equipment, astronomers can measure angular sizes to a fraction of an arcsecond. Keep in mind, however, that *angular* size is not the same as *actual* size. As an example, if you extend your arm while looking at a full moon, you can completely cover the Moon with your thumb. That's because from your perspective, your thumb has a larger angular size (that is, it subtends a larger angle) than the Moon. But the actual size of your thumb (about 2 centimeters) is much less than the actual diameter of the Moon (more than 3000 kilometers).

The accompanying figure shows how the angular size of an object is related to its linear size. Part (**a**) of the figure shows that for a given angular size, the more distant the object, the larger its actual size. For example, your fingertip held at arm's length covers the full moon, but the Moon is much farther away and is far larger in linear size. Part (**b**) shows that for a given linear size, the angular size decreases the farther away the object. This is why a car looks smaller and smaller as it drives away from you.

We can put these relationships together into a single mathematical expression called the **small-angle formula.** Suppose that an object subtends an angle α (the Greek letter alpha) and is at a distance d from the observer, as in part (**c**) of the figure. If the angle α is small, as is almost always the case for objects in the sky, the width or linear size (D) of the object is given by the following expression:

The small-angle formula

$$D = \frac{\alpha d}{206{,}265}$$

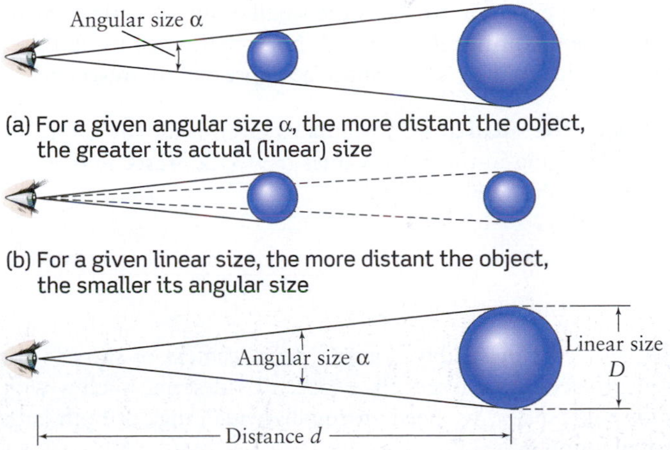
Angular size α
(a) For a given angular size α, the more distant the object, the greater its actual (linear) size
(b) For a given linear size, the more distant the object, the smaller its angular size
Angular size α
Linear size D
Distance d

(**a**) Two objects that have the same angular size may have different linear sizes if they are at different distances from the observer. (**b**) For an object of a given linear size, the angular size is smaller the farther the object is from the observer. (**c**) The small-angle formula relates the linear size D of an object to its angular size α and its distance d from the observer.

D = width or linear size of an object

α = angular size of the object, in arcsec

d = distance to the object

The number 206,265 is required in the formula so that the units of α are in arcseconds. (206,265 is the number of

(continued on the next page)

BOX 1-1 *(continued)*

arcseconds in a complete 360° circle divided by the number 2π.) As long as the same units are used for D and d, any units for linear distance can be used (km, light-years, etc.).

The following examples show two different ways to use the small-angle formula. In both examples we follow a four-step process: Evaluate the ***situation*** given in the example, decide which ***tools*** are needed to solve the problem, use those tools to find the ***answer*** to the problem, and ***review*** the result to see what it tells you. Throughout this book, we'll use these same four steps in *all* examples that require the use of formulas. We encourage you to follow this four-step process when solving problems for homework or exams. You can remember these steps by their acronym: ***S.T.A.R.***

EXAMPLE: On December 11, 2006, Jupiter was 944 million kilometers from Earth and had an angular diameter of 31.2 arcsec. From this information, calculate the actual diameter of Jupiter in kilometers.

Situation: The astronomical object in this example is Jupiter, and we are given its distance d and its angular size π (the same as angular diameter). Our goal is to find Jupiter's diameter D.

Tools: The equation to use is the small-angle formula, which relates the quantities d, α, and D. Note that when using this formula, the angular size α must be expressed in arcseconds.

Answer: The small-angle formula as given is an equation for D. Plugging in the given values $\alpha = 31.2$ arcsec and $d = 944$ million kilometers,

$$D = \frac{31.2 \times 944{,}000{,}000 \text{ km}}{206{,}265} = 143{,}000 \text{ km}$$

Because the distance d to Jupiter is given in kilometers, the diameter D is also in kilometers.

Review: Does our answer make sense? From Appendix 2 at the back of this book, the equatorial diameter of Jupiter measured by spacecraft flybys is 142,984 kilometers, so our calculated answer is very close.

EXAMPLE: Under excellent conditions, a telescope on Earth can see details with an angular size as small as 1 arcsec. What is the greatest distance at which you could see details as small as 1.7 meters (the height of a typical person) under these conditions?

Situation: Now the object in question is a person, whose linear size D we are given. Our goal is to find the distance d at which the person has an angular size α equal to 1 arcsec.

Tools: Again we use the small-angle formula to relate d, α, and D.

Answer: We first rewrite the formula to solve for the distance d, then plug in the given values $D = 1.7$ m and $\alpha = 1$ arcsec:

$$d = \frac{206{,}265D}{\alpha} = \frac{206{,}265 \times 1.7 \text{ m}}{1} = 350{,}000 \text{ m} = 350 \text{ km}$$

Review: This value is much less than the distance to the Moon, which is 384,000 kilometers. Thus, even the best telescope on Earth could not be used to see an astronaut walking on the surface of the Moon.

ranging from the incredibly hot and dense centers of stars to the frigid near-vacuum of interstellar space. To describe such a wide range of phenomena, we need an equally wide range of both large and small numbers.

Powers-of-Ten Notation: Large Numbers

Astronomers avoid such confusing terms as "a million billion billion" by using a standard shorthand system called **powers-of-ten notation.** All the cumbersome zeros that accompany a large number are consolidated into one term consisting of 10 followed by an **exponent,** which is written as a superscript. The exponent indicates how many zeros you would need to write out the long form of the number. Thus,

$10^0 = 1$ (one)

$10^1 = 10$ (ten)

$10^2 = 100$ (one hundred)

$10^3 = 1000$ (one thousand)

$10^4 = 10{,}000$ (ten thousand)

$10^6 = 1{,}000{,}000$ (one million)

$10^9 = 1{,}000{,}000{,}000$ (one billion)

$10^{12} = 1{,}000{,}000{,}000{,}000$ (one trillion)

and so forth. The exponent also tells you how many tens must be multiplied together to give the desired number, which is why the exponent is also called the **power of ten.** For example, ten thousand can be written as 10^4 ("ten to the fourth" or "ten to the fourth power") because $10^4 = 10 \times 10 \times 10 \times 10 = 10{,}000$.

In powers-of-ten notation, numbers are written as a figure between 1 and 10 multiplied by the appropriate power of ten. The approximate distance between Earth and the Sun, for example, can be written as 1.5×10^8 kilometers (or 1.5×10^8 km, for short). Once you get used to it, this is more convenient than writing "150,000,000 kilometers" or "one hundred and fifty million kilometers." (The same number could also be written as 15×10^7 or 0.15×10^9, but the preferred form is *always* to have the first figure be between 1 and 10.)

Calculators and Powers-of-Ten Notation

Most electronic calculators use a shorthand for powers-of-ten notation. To enter the number 1.5×10^8, you first enter 1.5, then press a key labeled "EXP" or "EE," then enter the exponent 8. (The EXP or EE key takes care of the "× 10" part of the expression.) The number will then appear on your calculator's display as "1.5 E 8," "1.5 8," or some variation of this; typically the "× 10" is not displayed as such. There are some variations from one kind of calculator to another, so you should spend a few minutes reading over your calculator's instruction manual to make sure you know the correct procedure for working with numbers in powers-of-ten notation. You will be using this notation continually in your study of astronomy, so this is time well spent.

CAUTION! Confusion can result from the way that calculators display powers-of-ten notation. Since 1.5×10^8 is displayed as "1.5 8" or "1.5 E 8," it is not uncommon to think that 1.5×10^8 is the same as 1.5^8. That is not correct, however; 1.5^8 is equal to 1.5 multiplied by itself 8 times, or 25.63, which is not even close to $150{,}000{,}000 = 1.5 \times 10^8$. Another, not uncommon, mistake is to write 1.5×10^8 as 15^8. If you are inclined to do this, perhaps you are thinking that you can multiply 1.5 by 10, then tack on the exponent later. Another process that does not work: 15^8 is equal to 15 multiplied by itself 8 times, or 2,562,890,625, which again is nowhere near 1.5×10^8. Reading over the manual for your calculator will help you to avoid these common errors.

Powers-of-Ten Notation: Small Numbers

You can use powers-of-ten notation for numbers that are less than one by using a minus sign in front of the exponent. A negative exponent tells you to *divide* by the appropriate number of tens. For example, 10^{-2} ("ten to the minus two") means to divide by 10 twice, so $10^{-2} = 1/10 \times 1/10 = 1/100 = 0.01$. This same idea tells us how to interpret other negative powers of ten:

$10^0 = 1$ (one)

$10^{-1} = 1/10 = 0.1$ (one tenth)

$10^{-2} = 1/10 \times 1/10 = 1/10^2 = 0.01$ (one hundredth)

$10^{-3} = 1/10 \times 1/10 \times 1/10 = 1/10^3 = 0.001$ (one thousandth)

$10^{-4} = 1/10 \times 1/10 \times 1/10 \times 1/10 = 1/10^4 = 0.0001$ (one ten-thousandth)

$10^{-6} = 1/10 \times 1/10 \times 1/10 \times 1/10 \times 1/10 \times 1/10 = 1/10^6 = 0.000001$ (one millionth)

$10^{-12} = 1/10 \times 1/10 \times 1/10 \times 1/10 \times 1/10 \times 1/10 \times 1/10 \times 1/10 \times 1/10 \times 1/10 \times 1/10 \times 1/10 = 1/10^{12} = 0.000000000001$ (one trillionth)

and so forth.

As these examples show, negative exponents tell you how many tenths must be multiplied together to give the desired number. For example, one ten-thousandth, or 0.0001, can be written as 10^{-4} ("ten to the minus four") because $10^{-4} = 1/10 \times 1/10 \times 1/10 \times 1/10 = 0.0001$.

A useful shortcut in converting a decimal to powers-of-ten notation is to notice where the decimal point is. For example, the decimal point in 0.0001 is four places to the left of the "1," so the exponent is –4, that is, $0.0001 = 10^{-4}$.

You can also use powers-of-ten notation to express a number like 0.00245, which is not a multiple of 1/10. For example, $0.00245 = 2.45 \times 0.001 = 2.45 \times 10^{-4}$. (Again, the standard for powers-of-ten notation is that the first figure is a number between 1 and 10.) This notation is particularly useful when dealing with very small numbers. A good example is the diameter of a hydrogen atom, which is much more convenient to state in powers-of-ten notation (1.1×10^{-10} meter, or 1.1×10^{-10} m) than as a decimal (0.00000000011 m) or a fraction (110 trillionths of a meter.)

Because it bypasses all the awkward zeros, powers-of-ten notation is ideal for describing the size of objects as small as atoms or as big as galaxies (Figure 1-13). Box 1-2 explains how powers-of-ten notation also makes it easy to multiply and divide numbers that are very large or very small.

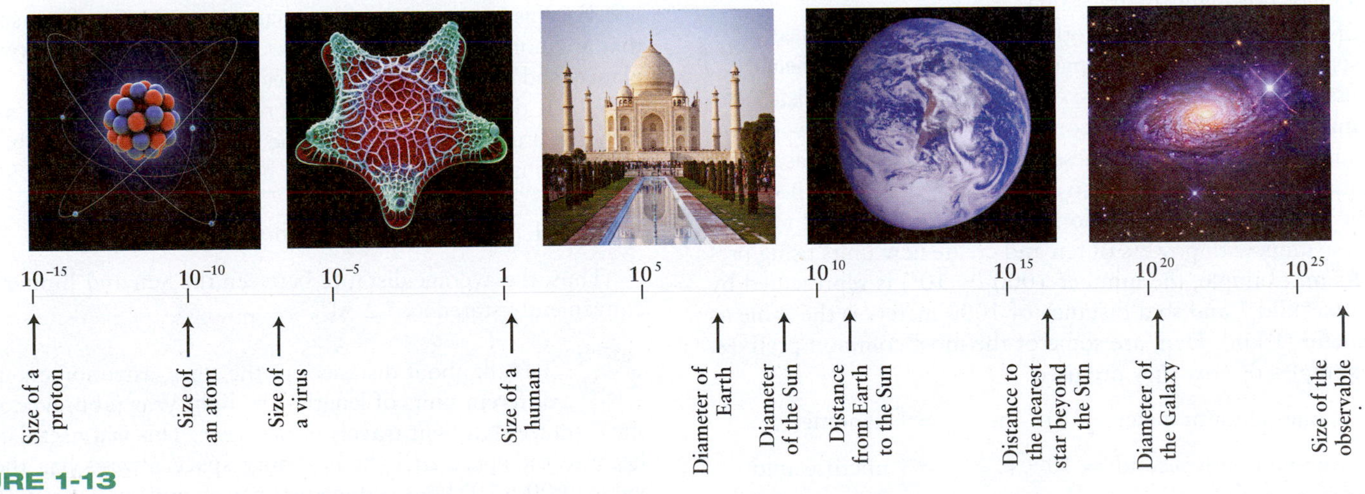

FIGURE 1-13
Examples of Powers-of-Ten Notation This scale gives the sizes of objects in meters, ranging from subatomic particles at the left to the entire observable universe at the right. The protons and neutrons in the illustration on the left form the nucleus of an atom; the electrons orbit around a hundred thousand times farther out. Second from left is the cell wall of a single-celled aquatic organism called a diatom, 10^{-4} meter (0.1 millimeter) in size. At the center is the Taj Mahal, about 60 meters tall and within reach of our unaided senses. On the right is the planet Earth, about 10^7 meters in diameter. At the far right is a galaxy, 10^{21} meters (100,000 light-years) in diameter. (Andrzej Wojcicki/Science Photo Library/Getty Images; Steve Gschmeissner/Science Photo Library/Getty; iStockphoto/Thinkstock; NASA/JPL; 2004–2013 R. Jay GaBany, Cosmotography.com)

BOX 1-2 TOOLS OF THE ASTRONOMER'S TRADE

Arithmetic with Powers-of-Ten Notation

Using powers-of-ten notation makes it easy to multiply numbers. For example, suppose you want to multiply 100 by 1000. If you use ordinary notation, you have to write a lot of zeros:

$$100 \times 1000 = 100{,}000 \text{ (one hundred thousand)}$$

By converting these numbers to powers-of-ten notation, we can write this same multiplication more compactly as

$$10^2 \times 10^3 = 10^5$$

Because $2 + 3 = 5$, we are led to the following general rule for *multiplying* numbers expressed in terms of powers of ten: Simply *add* the exponents.

EXAMPLE: $10^4 \times 10^3 = 10^{4+3} = 10^7$.

To *divide* numbers expressed in terms of powers of ten, remember that $10^{-1} = 1/10$, $10^{-2} = 1/100$, and so on. The general rule for any exponent n is

$$10^{-n} = \frac{1}{10^n}$$

In other words, dividing by 10^n is the same as multiplying by 10^{-n}. To carry out a division, you first transform it into multiplication by changing the sign of the exponent, and then carry out the multiplication by adding the exponents.

EXAMPLE: $\frac{10^4}{10^6} = 10^4 \times 10^{-6} = 10^{4+(-6)} = 10^{4-6} = 10^{-2}$

Usually a computation involves numbers like 3.0×10^{10}, that is, an ordinary number multiplied by a factor of 10 with an exponent. In such cases, to perform multiplication or division, you can treat the numbers separately from the factors of 10^n.

EXAMPLE: We can redo the first numerical example from Box 1-1 in a straightforward manner by using exponents:

$$D = \frac{31.2 \times 944{,}000{,}000 \text{ km}}{206{,}265}$$

$$= \frac{3.12 \times 10 \times 9.44 \times 10^8}{2.06265 \times 10^5} \text{ km}$$

$$= \frac{3.12 \times 9.44 \times 10^{1+8-5}}{2.06265} \text{ km} = 14.3 \times 10^4 \text{ km}$$

$$= 1.43 \times 10 \times 10^4 \text{ km} = 1.43 \times 10^5 \text{ km}$$

1-7 Astronomical distances are often measured in astronomical units, light-years, or parsecs

Specialized units make it easier to comprehend immense cosmic distances

Astronomers use many of the same units of measurement as do other scientists. They often measure lengths in meters (abbreviated m), masses in kilograms (kg), and time in seconds (s). (You can read more about these units of measurement, as well as techniques for converting between different sets of units, in Box 1-3.)

Like other scientists, astronomers often find it useful to combine these units with powers of ten and create new units using prefixes. As an example, the number 1000 ($= 10^3$) is represented by the prefix "kilo," and so a distance of 1000 meters is the same as 1 kilometer (1 km). Here are some of the most common prefixes, with examples of how they are used:

one-billionth meter	$= 10^{-9}$ m	= 1 nanometer
one-millionth second	$= 10^{-6}$ s	= 1 microsecond
one-thousandth arcsecond	$= 10^{-3}$ arcsec	= 1 milliarcsecond
one-hundredth meter	$= 10^{-2}$ m	= 1 centimeter
one thousand meters	$= 10^3$ m	= 1 kilometer
one million tons	$= 10^6$ tons	= 1 megaton

In principle, we could express all sizes and distances in astronomy using units based on the meter. Indeed, we will use kilometers to give the diameters of Earth and the Moon, as well as the Earth-Moon distance. But, while a kilometer (roughly equal to three-fifths of a mile) is an easy distance for humans to visualize, a megameter (10^6 m) is not. For this reason, astronomers have devised units of measurement that are more appropriate for the tremendous distances between the planets and the far greater distances between the stars.

When discussing distances across the solar system, astronomers use a unit of length called the **astronomical unit** (abbreviated **AU**). This is the average distance between Earth and the Sun:

$$1 \text{ AU} = 1.496 \times 10^8 \text{ km} = 92.96 \text{ million miles}$$

Thus, the average distance between the Sun and Jupiter can be conveniently stated as 5.2 AU.

ANIMATION 1-1

To talk about distances to the stars, astronomers use two different units of length. The **light-year** (abbreviated **ly**) is the distance that light travels in one year. This is a useful concept because the speed of light in empty space always has the same value, 3.00×10^5 km/s (kilometers per second) or 1.86×10^5 mi/s (miles per second). In terms of kilometers or astronomical units, one light-year is given by

$$1 \text{ ly} = 9.46 \times 10^{12} \text{ km} = 63{,}240 \text{ AU}$$

This distance is roughly equal to 6 trillion miles.

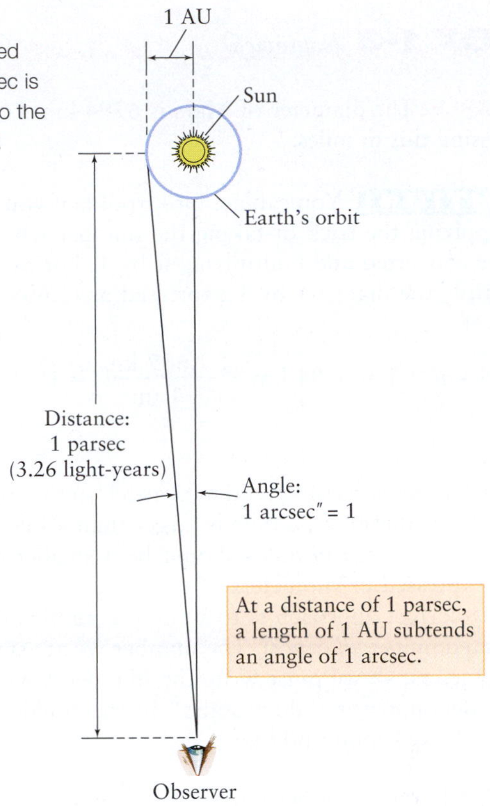

FIGURE 1-14
A Parsec The parsec, a unit of length commonly used by astronomers, is equal to 3.26 light-years. The parsec is defined as the distance at which 1 AU perpendicular to the observer's line of sight subtends an angle of 1 arcsec.

CAUTION! Keep in mind that despite its name, the light-year is a unit of distance and *not* a unit of time. As an example, Proxima Centauri, the nearest star other than the Sun, is a distance of 4.2 light-years from Earth. This means that light takes 4.2 years to travel to us from Proxima Centauri.

Physicists often measure interstellar distances in light-years because the speed of light is one of nature's most important numbers. But many astronomers prefer to use another unit of length, the *parsec,* because its definition is closely related to a method of measuring distances to the stars.

Imagine taking a journey far into space, beyond the orbits of the outer planets. As you look back toward the Sun, Earth's orbit subtends, or extends over, a smaller angle in the sky the farther you are from the Sun. As Figure 1-14 shows, the distance at which 1 AU subtends an angle of 1 arcsec is defined as 1 **parsec** (abbreviated **pc**):

$$1 \text{ pc} = 3.09 \times 10^{13} \text{ km} = 3.26 \text{ ly}$$

BOX 1-3 TOOLS OF THE ASTRONOMER'S TRADE

Units of Length, Time, and Mass

To understand and appreciate the universe, we need to describe phenomena not only on the large scales of galaxies but also on the submicroscopic scale of the atom. Astronomers generally use units that are best suited to the topic at hand. For example, interstellar distances are conveniently expressed in either light-years or parsecs, whereas the diameters of the planets are more comfortably presented in kilometers.

Most scientists prefer to use a version of the metric system called the International System of Units, abbreviated SI (after the French name Système International). In **SI units,** length is measured in meters (m), time is measured in seconds (s), and mass (a measure of the amount of material in an object) is measured in kilograms (kg). How are these basic units related to other measures?

When discussing objects on a human scale, sizes and distances can also be expressed in millimeters (mm), centimeters (cm), and kilometers (km). These units of length are related to the meter as follows:

$$1 \text{ millimeter} = 0.001 \text{ m} = 10^{-3} \text{ m}$$
$$1 \text{ centimeter} = 0.01 \text{ m} = 10^{-2} \text{ m}$$
$$1 \text{ kilometer} = 1000 \text{ m} = 10^{3} \text{ m}$$

A useful set of conversions for the English system is

$$1 \text{ in} = 2.54 \text{ cm}$$
$$1 \text{ ft} = 0.3048 \text{ m}$$
$$1 \text{ mi} = 1.609 \text{ km}$$

Each of these equalities can also be written as a fraction equal to 1. For example, you can write

$$\frac{0.3048 \text{ m}}{1 \text{ ft}} = 1 \quad \text{or} \quad \frac{1 \text{ ft}}{0.3048 \text{ m}} = 1$$

Fractions like this are useful for converting a quantity from one set of units to another. For example, the *Saturn V* rocket used to send astronauts to the Moon stands about 363 ft tall. How can we convert this height to meters? The trick is to remember that a quantity does not change if you multiply it by 1. Expressing the number 1 by the fraction (0.3048 m)/(1 ft), we can write the height of the rocket as

$$363 \text{ ft} \times 1 = 363 \text{ ft} \times \frac{0.3048 \text{ m}}{1 \text{ ft}} = 111 \frac{\text{ft} \times \text{m}}{\text{ft}} = 111 \text{ m}$$

(continued on the next page)

BOX 1-3 *(continued)*

EXAMPLE: The diameter of Mars is 6794 km. Let's try expressing this in miles.

CAUTION! You can get into trouble if you are careless in applying the trick of taking the number whose units are to be converted and multiplying it by 1. For example, if we multiply the diameter by 1 expressed as (1.609 km)/(1 mi), we get

$$6794 \text{ km} \times 1 = 6794 \text{ km} \times \frac{1.609 \text{ km}}{1 \text{ mi}} = 10{,}930 \frac{\text{km}^2}{\text{mi}}$$

The unwanted units of km did not cancel, so this answer cannot be right. Furthermore, a mile is larger than a kilometer, so the diameter expressed in miles should be a smaller number than when expressed in kilometers.

The correct approach is to write the number 1 so that the unwanted units *will* cancel. The number we are starting with is in kilometers, so we must write the number 1 with kilometers in the denominator ("downstairs" in the fraction). Thus, we express 1 as (1 mi)/(1.609 km):

$$6794 \text{ km} \times 1 = 6794 \text{ km} \times \frac{1 \text{ mi}}{1.609 \text{ km}}$$

$$= 4222 \text{ km} \times \frac{\text{mi}}{\text{km}} = 4222 \text{ mi}$$

Now the units of km cancel as they should, and the distance in miles is a smaller number than in kilometers (as it must be).

When discussing very small distances such as the size of an atom, astronomers often use the micrometer (μm) or the nanometer (nm). These are related to the meter as follows:

$$1 \text{ micrometer} = 1 \text{ μm} = 10^{-6} \text{ m}$$

$$1 \text{ nanometer} = 1 \text{ nm} = 10^{-9} \text{ m}$$

Thus, 1 μm = 10^3 nm. (Note that the micrometer is often called the micron.)

The basic unit of time is the second (s). It is related to other units of time as follows:

$$1 \text{ minute (min)} = 60 \text{ s}$$

$$1 \text{ hour (h)} = 3600 \text{ s}$$

$$1 \text{ day (d)} = 86{,}400 \text{ s}$$

$$1 \text{ year (y)} = 3.156 \times 10^7 \text{ s}$$

In the SI system, speed is properly measured in meters per second (m/s). Quite commonly, however, speed is also expressed in km/s and mi/h:

$$1 \text{ km/s} = 10^3 \text{ m/s}$$

$$1 \text{ km/s} = 2237 \text{ mi/h}$$

$$1 \text{ mi/h} = 0.447 \text{ m/s}$$

$$1 \text{ mi/h} = 1.47 \text{ ft/s}$$

In addition to using kilograms, astronomers sometimes express mass in grams (g) and in solar masses ($M_\odot$), where the subscript ⊙ is the symbol denoting the Sun. It is especially convenient to use solar masses when discussing the masses of stars and galaxies. These units are related to each other as follows:

$$1 \text{ kg} = 1000 \text{ g}$$

$$1 M_\odot = 1.99 \times 10^{30} \text{ kg}$$

CAUTION! You may be wondering why we have not given a conversion between kilograms and pounds (lb). The reason is that these units do not refer to the same physical quantity! A kilogram is a unit of *mass,* which is a measure of the amount of material in an object. By contrast, a pound is a unit of *weight,* which tells you how strongly gravity pulls on that object's material. Consider a person who weighs 110 pounds on Earth, corresponding to a mass of 50 kg. Gravity is only about one-sixth as strong on the Moon as it is on Earth, so on the Moon this person would weigh only one-sixth of 110 pounds, or about 18 pounds. But that person's mass of 50 kg is the same on the Moon; wherever you go in the universe, you take all of your material along with you. We will explore the relationship between mass and weight in Chapter 4.

The distance from Earth to Proxima Centauri can be stated as 1.3 pc or as 4.2 ly. Whether you choose to use parsecs or light-years is a matter of personal taste.

For even greater distances, astronomers commonly use **kiloparsecs** and **megaparsecs** (abbreviated **kpc** and **Mpc**). As we saw before, these prefixes simply mean "thousand" and "million," respectively:

$$1 \text{ kiloparsec} = 1 \text{ kpc} - 1000 \text{ pc} = 10^3 \text{ pc}$$

$$1 \text{ megaparsec} = 1 \text{ Mpc} = 1{,}000{,}000 \text{ pc} = 10^6 \text{ pc}$$

For example, the distance from Earth to the center of our Milky Way Galaxy is about 8 kpc, and the galaxy shown in Figure 1-9 is about 11 Mpc away.

Some astronomers prefer to talk about thousands or millions of light-years rather than kiloparsecs and megaparsecs. Once again, the choice is a matter of personal taste. As a general rule, astronomers use whatever measuring sticks seem best suited for the issue at hand and do not restrict themselves to one system of measurement. For example, an astronomer might say that the supergiant star Antares has a diameter of 860 million kilometers and is located at a distance of 185 parsecs from Earth. The *Cosmic Connections* figure on the following page shows where these different systems are useful.

1-8 Astronomy is an adventure of the human mind

An underlying theme of this book is that the behavior of the universe is governed by underlying laws of nature. It is not a hodgepodge of unrelated things behaving in unpredictable ways. Rather, we find strong evidence that fundamental laws of physics govern the nature of the universe and the behavior of everything in it. These unifying concepts enable us to explore realms far removed from our earthly experience. Thus, a scientist can do experiments in a laboratory to determine the properties of light or the behavior of atoms and then use this knowledge to investigate the structure of the universe. Through careful testing, the laws of physics discovered on Earth have been found to apply throughout the universe at the most distant locations and in the distant past. Indeed, Albert Einstein expressed the view that **"the eternal mystery of the world is its comprehensibility."**

Studying the universe benefits our lives on Earth

The discovery of fundamental laws of nature has had a profound influence on humanity. These laws have led to an immense number of practical applications that have fundamentally transformed commerce, medicine, entertainment, transportation, and other aspects of our lives. In particular, space technology has given us instant contact with any point on the globe through communication satellites, precise navigation to any point on Earth using signals from the satellites of the Global Positioning System (GPS), and accurate weather forecasts from meteorological satellites (Figure 1-15).

As important as the applications of science are, the pursuit of scientific knowledge for its own sake is no less important. We are fortunate to live in an age in which this pursuit is in full flower. Just as explorers such as Columbus and Magellan discovered the true size of our planet in the fifteenth and sixteenth centuries, astronomers of the twenty-first century are exploring the universe to an extent that is unparalleled in human history. Indeed, even the voyages into space imagined by such great science fiction writers as Jules Verne and H. G. Wells pale in comparison to today's reality. Over a few short decades, humans have walked on the Moon, sent robot spacecraft to dig into the Martian soil and explore the satellites of Saturn, and used the most powerful telescopes ever built to probe the limits of the observable universe. Never before has so much been revealed in so short a time.

As you proceed through this book, you will learn about the tools that scientists use to explore the natural world, as well as what they observe with these tools. But, most important, you will see how astronomers build from their observations an understanding of the universe in which we live. It is this search for understanding that makes science more than merely a collection of data and elevates it to one of the great adventures of the human mind.

FIGURE 1-15 R I **V** U X G
A Hurricane Seen from Space This image of Hurricane Frances was made on September 2, 2004, by GOES-12 (*Geostationary Operational Environmental Satellite 12*). A geostationary satellite like GOES-12 orbits around Earth's equator every 24 hours, the same length of time it takes the planet to make a complete rotation. Hence, this satellite remains over the same spot on Earth, from which it can monitor the weather continuously. By tracking hurricanes from orbit, GOES-12 makes it much easier for meteorologists to give early warning of these immense storms. The resulting savings in lives and property more than pay for the cost of the satellite. (NASA, NOAA)

It is an adventure that will continue as long as there are mysteries in the universe—an adventure we hope you will come to appreciate and share.

KEY WORDS

Terms preceded by an asterisk (*) are discussed in the Boxes.

angle, p. 8
angular diameter (angular size), p. 8
angular distance, p. 8
arcminute (′, minute of arc), p. 8
arcsecond (″, second of arc), p. 8
astronomical unit (AU), p. 12
Big Bang, p. 7
black hole, p. 6
degree (°), p. 8
exponent, p. 10
galaxy, p. 7
hypothesis, p. 2
kiloparsec (kpc), p. 14
laws of physics, p. 3
light-year (ly), p. 12
megaparsec (Mpc), p. 14
meteorite, p. 4
model, p. 2
nebula (*plural* nebulae), p. 6
Newtonian mechanics, p. 3
parsec (pc), p. 13
power of ten, p. 10
powers-of-ten notation, p. 10
pulsar, p. 6
quasar, p. 7
scientific method, p. 2
*SI units, p. 13
*small-angle formula, p. 9
solar system, p. 4
subtend (an angle), p. 8
supernova (*plural* supernovae), p. 6
theory, p. 2

COSMIC CONNECTIONS

Sizes in the Universe

Powers-of-ten notation provides a a convenient way to express the sizes of astronomical objects and distances in space. This illustration suggests the immense distances within our solar system, the far greater distances between the stars within our Milky Way Galaxy, and the truly cosmic distances between galaxies.

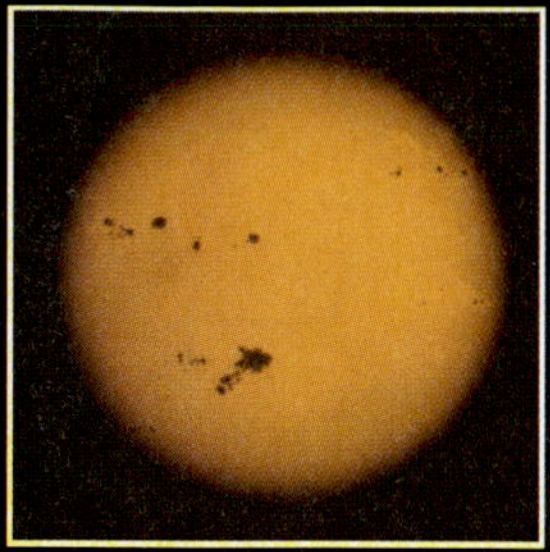

Sun: diameter = 1.39×10^6 km

The Sun, Earth, and other planets are members of our solar system

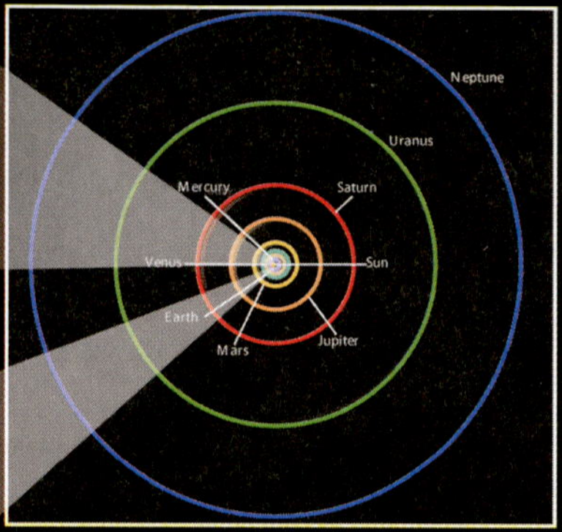

Diameter of Neptune's orbit: 60 AU

1 AU (astronomical unit) = 1.50×10^8 km = average Earth-Sun distance

Earth: diameter = 1.28×10^4 km

The Sun is a typical star. Typical distances between our neighboring stars = 1 to 5 ly

1ly = distance that light travels in one year = 6.32×10^4 AU

Our Sun is one of more than 10^{11} stars in the Milky Way Galaxy.

Distance from the center of the Milky Way to the Sun = 2.8×10^4 ly

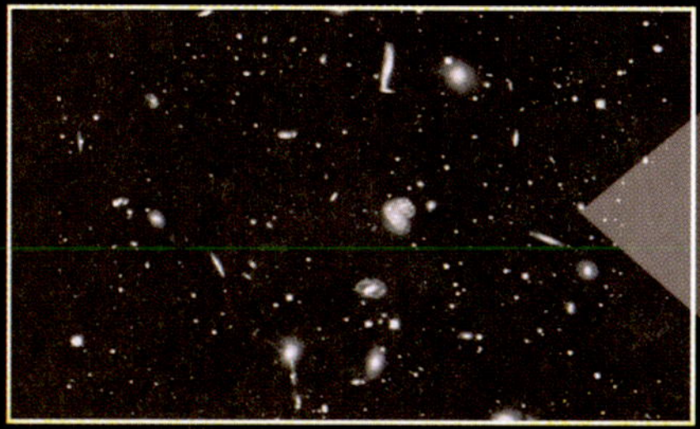

Galaxies are grouped into clusters, which can be up to 10^7 ly across.

Each of the 1.6×10^6 dots in this map of the entire sky represents a relatively nearby galaxy. This is a tiny fraction of the number of galaxies in the observable universe.

KEY IDEAS

Astronomy, Science, and the Nature of the Universe: The universe is comprehensible. The scientific method is a procedure for formulating hypotheses about the universe. Hypotheses are tested by observation or experimentation in order to build consistent models or theories that accurately describe phenomena in nature.

Observations of the heavens have helped scientists discover some of the fundamental laws of physics. The laws of physics are in turn used by astronomers to interpret their observations.

The Solar System: Exploration of the planets provides information about the origin and evolution of the solar system, as well as about the history and resources of Earth.

Stars and Nebulae: Studying the stars and nebulae helps us learn about the origin and history of the Sun and the solar system.

Galaxies: Observations of galaxies tell us about the origin and history of the universe.

Angles: Astronomers use angles to denote the positions and sizes of objects in the sky. The size of an angle is measured in degrees, arcminutes, and arcseconds.

Powers-of-Ten Notation is a convenient shorthand system for writing numbers. It allows very large and very small numbers to be expressed in a compact form.

Units of Distance: Astronomers use a variety of distance units. These include the astronomical unit (the average distance from Earth to the Sun), the light-year (the distance that light travels in one year), and the parsec.

QUESTIONS

Review Questions

1. TUTORIAL 1-1 What is the difference between a hypothesis and a theory?
2. What is the difference between a theory and a law of physics?
3. How are scientific theories tested?
4. Describe the role that skepticism plays in science.
5. Describe one reason why it is useful to have telescopes in space.
6. What caused the craters on the Moon?
7. What are meteorites? Why are they important for understanding the history of the solar system?
8. What makes the Sun and stars shine?
9. What role do nebulae like the Orion Nebula play in the life stories of stars?
10. What is the difference between a solar system and a galaxy?
11. What are degrees, arcminutes, and arcseconds used for? What are the relationships among these units of measure?
12. How many arcseconds equal 1°?
13. With the aid of a diagram, explain what it means to say that the Moon subtends an angle of ½°.
14. What is an exponent? How are exponents used in powers-of-ten notation?
15. What are the advantages of using powers-of-ten notation?
16. Write the following numbers using powers-of-ten notation: (a) ten million, (b) sixty thousand, (c) four one-thousandths, (d) thirty-eight billion, (e) your age in months.
17. How is an astronomical unit (AU) defined? Give an example of a situation in which this unit of measure would be convenient to use.
18. What is the advantage to the astronomer of using the light-year as a unit of distance?
19. What is a parsec? How is it related to a kiloparsec and to a megaparsec?
20. Give the word or phrase that corresponds to the following standard abbreviations: (a) km, (b) cm, (c) s, (d) km/s, (e) mi/h, (f) m, (g) m/s, (h) h, (i) y, (j) g, (k) kg. Which of these are units of speed? (*Hint:* You may have to refer to a dictionary. All of these abbreviations should be part of your working vocabulary.)
21. In the original (1977) *Star Wars* movie, Han Solo praises the speed of his spaceship by saying, "It's the ship that made the Kessel run in less than 12 parsecs!" Explain why this statement is obvious misinformation.
22. A reporter once described a light-year as "the time it takes light to reach us traveling at the speed of light." How would you correct this statement?

Advanced Questions

Questions preceded by an asterisk (*) are discussed in the Boxes.

Problem-solving tips and tools

The small-angle formula, given in Box 1-1, relates the size of an astronomical object to the angle it subtends. Box 1-3 illustrates how to convert from one unit of measure to another. An object traveling at speed v for a time t covers a distance d given by $d = vt$; for example, a car traveling at 90 km/h (v) for 3 h (t) covers a distance $d = (90 \text{ km/h})(3 \text{ h}) = 270$ km. Similarly, the time t required to cover a given distance d at speed v is $t = d/v$; for example, if $d = 270$ km and $v = 90$ km/h, then $t = (270 \text{ km})/(90 \text{ km/h}) = 3$ h.

23. What is the meaning of the letters **R I V U X G** that appear with some of the figures in this chapter? Why in each case is one of the letters highlighted? (*Hint:* See the Preface that precedes Chapter 1.)
24. The diameter of the Sun is 1.4×10^{11} cm, and the distance to the nearest star, Proxima Centauri, is 4.2 ly. Suppose you want to build an exact scale model of the Sun and Proxima Centauri, and you are using a ball 30 cm in diameter to represent the Sun. In your scale model, how far away would Proxima Centauri be from the Sun? Give your answer in kilometers, using powers-of-ten notation.

25. How many Suns would it take, laid side by side, to reach the nearest star? Use powers-of-ten notation. (*Hint:* See the preceding question.)

26. A hydrogen atom has a radius of about 5×10^{-9} cm. The radius of the observable universe is about 14 billion light-years. How many times larger than a hydrogen atom is the observable universe? Use powers-of-ten notation.

27. The Sun's mass is 1.99×10^{30} kg, three-quarters of which is hydrogen. The mass of a hydrogen atom is 1.67×10^{-27} kg. How many hydrogen atoms does the Sun contain? Use powers-of-ten notation.

28. The average distance from Earth to the Sun is 1.496×10^8 km. Express this distance **(a)** in light-years and **(b)** in parsecs. Use powers-of-ten notation. **(c)** Are light-years or parsecs useful units for describing distances of this size? Explain.

29. The speed of light is 3.00×10^8 m/s. How long does it take light to travel from the Sun to Earth? Give your answer in seconds, using powers-of-ten notation. (*Hint:* See the preceding question.)

30. When the *Voyager 2* spacecraft sent back pictures of Neptune during its flyby of that planet in 1989, the spacecraft's radio signals traveled for 4 hours at the speed of light to reach Earth. How far away was the spacecraft? Give your answer in kilometers, using powers-of-ten notation. (*Hint:* See the preceding question.)

31. The star Altair is 5.15 pc from Earth. **(a)** What is the distance to Altair in kilometers? Use powers-of-ten notation. **(b)** How long does it take for light emanating from Altair to reach Earth? Give your answer in years. (*Hint:* You do not need to know the value of the speed of light.)

32. The age of the universe is about 13.7 billion years. What is this age in seconds? Use powers-of-ten notation.

*33. Explain where the number 206,265 in the small-angle formula comes from.

*34. At what distance would a person have to hold a European 2-euro coin (which has a diameter of about 2.6 cm) in order for the coin to subtend an angle of **(a)** 1°? **(b)** 1 arcmin? **(c)** 1 arcsec? Give your answers in meters.

*35. A person with good vision can see details that subtend an angle of as small as 1 arcminute. If two dark lines on an eye chart are 2 millimeters apart, how far can such a person be from the chart and still be able to tell that there are two distinct lines? Give your answer in meters.

*36. The average distance to the Moon is 384,000 km, and the Moon subtends an angle of ½°. Use this information to calculate the diameter of the Moon in kilometers.

*37. Suppose your telescope can give you a clear view of objects and features that subtend angles of at least 2 arcsec. What is the diameter in kilometers of the smallest crater you can see on the Moon? (*Hint:* See the preceding question.)

*38. On April 18, 2006, the planet Venus was a distance of 0.869 AU from Earth. The diameter of Venus is 12,104 km. What was the angular size of Venus as seen from Earth on April 18, 2006? Give your answer in arcminutes.

*39. **(a)** Use the information given in the caption to Figure 1-7 to determine the angular size of the Orion Nebula. Give your answer in degrees. **(b)** How does the angular diameter of the Orion Nebula compare to the angular diameter of the Moon?

Discussion Questions

40. Scientists assume that "reality is rational." Discuss what this means and the thinking behind it.

41. All scientific knowledge is inherently provisional. Discuss whether this is a weakness or a strength of the scientific method.

42. How do astronomical observations differ from those of other sciences?

Web/eBook Questions

43. Use the links given in the *Universe* Web site or eBook, Chapter 1, to learn about the Orion Nebula (Figure 1-7). Can the nebula be seen with the naked eye? Does the nebula stand alone, or is it part of a larger cloud of interstellar material? What has been learned by examining the Orion Nebula with telescopes sensitive to infrared light?

44. Use the links given in the *Universe* Web site or eBook, Chapter 1, to learn more about the Crab Nebula (Figure 1-8). When did observers on Earth see the supernova that created this nebula? Does the nebula emit any radiation other than visible light? What kind of object is at the center of the nebula?

ACTIVITIES

Observing Projects

45. On a dark, clear, moonless night, can you see the Milky Way from where you live? If so, briefly describe its appearance. If not, what seems to be interfering with your ability to see the Milky Way?

46. Look up at the sky on a clear, cloud-free night. Is the Moon in the sky? If so, does it interfere with your ability to see the fainter stars? Why do you suppose astronomers prefer to schedule their observations on nights when the Moon is not in the sky?

47. Look up at the sky on a clear, cloud-free night and note the positions of a few prominent stars relative to such reference markers as rooftops, telephone poles, and treetops. Also note the location from where you make your observations. A few hours later, return to that location and again note the positions of the same bright stars that you observed earlier. How have their positions changed? From these changes, can you deduce the general direction in which the stars appear to be moving?

48. If you have access to the *Starry Night*™ planetarium software, install it on your computer. There are several guides to the use of this software. As an initial introduction, you can run through the step-by-step basics of the program by clicking the **Sky Guide** tab to the left of the main screen and then clicking the **Starry Night basics** hyperlink at the bottom of the **Sky Guide** pane. A more comprehensive guide is available by choosing the **Student Exercises** hyperlink and then the

Tutorial hyperlink. A User's Guide to this software is available under the **Help** menu. As a start, you can use this program to determine when the Moon is visible today from your location. If the viewing location in the *Starry Night*™ control panel is not set to your location, select **Set Home Location ...** in the **File** menu (on a Macintosh, this command is found under the **Starry Night** menu). Click the **List** tab in the **Home Location** dialog box; then select the name of your city or town and click the **Save As Home Location** button. Next, use the hand tool to explore the sky and search for the Moon by moving your viewpoint around the sky. (Click and drag the mouse to achieve this motion.) If the Moon is not easily seen in your sky at this time, click the **Find** tab at the top left of the main view. The **Find** pane that opens should contain a list of solar system objects. Ensure that there is no text in the edit box at the top of the **Find** pane. If the message "Search all Databases" is not displayed below this edit box, then click the magnifying glass icon in the edit box and select **Search All** from the drop-down menu that appears. Click the + symbol to the left of the listing for **Earth** to display **The Moon** and double-click on this entry in the list in order to center the view upon the Moon. (If a message is displayed indicating that "the Moon is not currently visible from your location," click on the **Best Time** button to advance to a more suitable time). You will see that the Moon can be seen in the daytime as well as at night. Note that the **Time Flow Rate** is set to **1x**, indicating that time is running forward at the normal rate. Note also the phase of the Moon. (**a**) Estimate how long it will take before the Moon reaches its full phase. Set the **Time Flow Rate** to **1 minutes.** (**b**) Find the time of moonset at your location. (**c**) Determine which, if any, of the following planets are visible tonight: Mercury, Venus, Mars, Jupiter, and Saturn. (*Hint:* Use the **Find** pane and click on each planet in turn to explore the positions of these objects.) Feel free to experiment with the many features of *Starry Night*™.

49. Use the *Starry Night*™ program to measure angular spacing between stars in the sky. Open **Favourites > Explorations > N Pole** to display the northern sky from Calgary, Canada, at a latitude of 51°. This view shows several asterisms, or groups of stars, outlined and labeled with their common names. The stars in the Big Dipper asterism outline the shape of a "dipper," used for scooping water from a barrel. The two stars in the Big Dipper on the opposite side of the scoop from the handle, Merak and Dubhe, can be seen to point to the brightest star in the Little Dipper, the Pole Star. The Pole Star is close to the North Celestial Pole, the point in the sky directly above the north pole of Earth. It is thus a handy aid to navigation for northern hemisphere observers because it indicates the approximate direction of true north. Measure the spacing between the two "pointer stars" in the Big Dipper and then the spacing between the Pole Star and the closest of the pointer stars, Dubhe. (*Hint:* These measurements are best made by activating the angular separation tool from the cursor selection control on the left side of the toolbar.) (**a**) What is the angular distance between the pointer stars Merak and Dubhe? (**b**) What is the angular spacing, or separation, between Dubhe (the pointer star at the end of the Big Dipper) and the Pole Star? (**c**) Approximately how many pointer-star spacings are there between Dubhe and the Pole Star?

Click the **Play** button in the toolbar. Notice that the Pole Star will appear to remain fixed in the sky as time progresses because it lies very close to the North Celestial Pole. Select **Edit > Undo Time Flow** or **File > Revert** from the menu to return to the initial view. Select **View > Celestial Guides > Celestial Poles** from the menu to indicate the position of the North Celestial Pole on the screen. Right-click on the Pole Star (Ctrl-click on a Macintosh) and select **Centre** from the drop down menu to center the view on the Pole Star. Zoom in and use the angular separation tool to measure the angular spacing between the Pole Star and the North Celestial Pole. (**d**) What is the angular separation between the Pole Star and the North Celestial Pole? (**e**) Select **File > Revert** from the menu and use the angular separation tool to measure the angle between the Pole Star and the horizon at Calgary. What is the relationship between this angle and the latitude of Calgary (51°)?

Collaborative Exercises

50. A scientific theory is fundamentally different than the everyday use of the word "theory." List and describe any three scientific theories of your choice and creatively imagine an additional three hypothetical theories that are not scientific. Briefly describe what is scientific and what is nonscientific about each of these theories.

51. Angles describe how far apart two objects appear to an observer. From where you are currently sitting, estimate the angular distance between the floor and the ceiling at the front of the room you are sitting in, the angular distance between the two people sitting closest to you, and the angular size of a clock or an exit sign on the wall. Draw sketches to illustrate each answer and describe how each of your answers would change if you were standing in the very center of the room.

52. Astronomers use powers of ten to describe the distances to objects. List an object or place that is located at very roughly each of the following distances from you: 10^{-2} m, 10^{0} m, 10^{1} m, 10^{3} m, 10^{7} m, 10^{10} m, and 10^{20} m.

ANSWERS

ConceptChecks

ConceptCheck 1-1: While a hypothesis is a testable idea that seems to explain an observation about nature, a scientific theory represents a set of well-tested and internally consistent hypotheses that are able to successfully and repeatedly predict the outcome of experiments and observations.

ConceptCheck 1-2: Shorter lifetime. The larger star has twice as much mass as the Sun. However, it consumes its nuclear fuel much more than twice as fast, so that its lifetime is shorter than the Sun.

ConceptCheck 1-3: Yes. The apparent size of an object is given by how wide it appears (in other words, the angle that it subtends). By moving a basketball near and far, it can appear very large or very small.

Why Astronomy? by Sandra M. Faber

As you study astronomy, you may ask, "Why am I studying this subject? What good is it for people in general and for me in particular?" Admittedly, astronomy does not offer the same practical benefits as other sciences, so how can it be important to your life?

On the most basic level, I think of astronomy as providing the ultimate background for human history. Recorded history goes back about 3000 years. For knowledge of the time before that, we consult archeologists and anthropologists about early human history and paleontologists, biologists, and geologists about the evolution of life and of our planet—altogether going back some five billion years. Astronomy tells us about the time before that, the ten billion years or so when the Sun, solar system, and Milky Way Galaxy formed, and even about the origin of the universe in the Big Bang. Knowledge of astronomy is part of a well-educated person's view of history.

Astronomy challenges our belief system and impels us to put our "philosophical house" in order. For example, the Bible says the world and everything in it were created in six days by the hand of God. However, according to the ancient Egyptians, Earth arose spontaneously from the infinite waters of the eternal universe, called Nun. Alaskan legends teach that the world was created by the conscious imaginings of a deity named Father Raven.

Modern astronomy, supported by physics and observations, differs from these stories of the creation of Earth. Astronomers believe that the Sun formed about five billion years ago by gravitational collapse from a dense cloud of interstellar gas and dust. At the same time, and over a period of several hundred thousand years, the planets condensed within the swirling solar nebula. Astronomers have actually seen young stars form in this way.

At issue here, really, is the question of how we are to gain information about the nature of the physical world—whether by revelation and intuition or by logic and observation. Where science stops and faith begins is a thorny issue for everyone, but particularly for astronomers—and for astronomy students.

Astronomy cultivates our notions about cosmic time and cosmic evolution. Given the short span of human life, it is all too easy to overlook that the universe is a dynamic place. This idea implies fragility—if something can change, it might even some day disappear. For instance, in another five billion years or so, the Sun will swell up and brighten to 1000 times its present luminosity, incinerating Earth in the process. This is far enough in the future that neither you nor I need to feel any personal responsibility for preparing to meet this challenge. However, other cosmic catastrophes will inevitably occur before then. Earth will be hit by a sizable piece of space debris—craters show that this happens every few million years or so. Enormous volcanic eruptions have occurred in the past and will certainly occur again. Another Ice Age is virtually certain to begin within the next 20,000 years, unless we first cook Earth ourselves by burning too much fossil fuel.

Such common notions as the inevitability of human progress, the desirability of endless economic growth, and Earth's ability to support its human population are all based on limited experience—they will probably not prove viable in the long run. Consequently, we must rethink who we are as a species and what is our proper activity on Earth. These long-term problems involve the whole human race and are vital to our survival and well-being. Astronomy is essential to developing a perspective on human existence and its relation to the cosmos.

Many astronomers believe that the ultimate, proper concept of "home" for the human race is our universe. It seems increasingly likely that a large number of other universes exist, with the vast majority incapable of harboring intelligent life as we know it. The parallel with Earth is striking. Among the solar system planets, only Earth can support human life. Among the great number of planets in our galaxy, only a small fraction may be such that we can call them home. The fraction of hospitable universes is likely to be smaller still. It seems, then, that our universe is the ultimate "home," a sanctuary in a vast sea of inhospitable universes.

I began this essay by talking about history and ended with issues that border on the ethical and religious. Astronomy is like that: It offers a modern-day version of Genesis—and of the Apocalypse, too. I hope that during this course you will be able to take time out to contemplate the broader implications of what you are studying. This is one of the rare opportunities in life to think about who you are and where you and the human race are going. Don't miss it.

Sandra M. Faber is professor of astronomy at the University of California, Santa Cruz, and astronomer at Lick Observatory. She chaired the now legendary group of astronomers called the Seven Samurai, who surveyed the nearest 400 elliptical galaxies and discovered a new mass concentration, called the Great Attractor. Dr. Faber received the Bok Prize from Harvard University in 1978, was elected to the National Academy of Sciences in 1985, and in 1986 won the coveted Dannie Heineman Prize from the American Astronomical Society, in recognition of her sustained body of especially influential astronomical research.

CHAPTER

2

Earth's rotation makes stars appear to trace out circles in the sky. (Gemini Observatory) R I V U X G

Knowing the Heavens

LEARNING GOALS

By reading the sections of this chapter, you will learn

2-1 The importance of astronomy in ancient civilizations around the world

2-2 That regions of the sky are divided around groups of stars called constellations

2-3 How the sky changes from night to night

2-4 How astronomers locate objects in the sky

2-5 What causes the seasons

2-6 The effect of changes in the direction of Earth's axis of rotation

2-7 The role of astronomy in measuring time

2-8 How the modern calendar developed

It is a clear night at the Gemini North Observatory atop Mauna Kea, a dormant volcano on the island of Hawaii. As you gaze toward the north, as in this time-exposure photograph, you find that the stars are not motionless. Rather, they move in counterclockwise circles around a fixed point above the northern horizon. Stars close to this point never dip below the horizon, while stars farther from the fixed point rise in the east and set in the west. These motions fade from view when the Sun rises in the east and illuminates the sky. The Sun, too, arcs across the sky in the same manner as the stars. At day's end, when the Sun sets in the west, the panorama of stars is revealed for yet another night.

These observations are at the heart of *naked-eye astronomy*—the sort that requires no equipment but human vision. Naked-eye astronomy cannot tell us what the Sun is made of or how far away the stars are. For such purposes we need tools such as the Gemini Telescope, housed within the dome shown in the photograph. But by studying naked-eye astronomy, you will learn the answers to equally profound questions such as why there are seasons, why the night sky is different at different times of year, and why the night sky looks different in Australia than in North America. In discovering the answers to these questions, you will learn how Earth moves through space and will begin to understand our true place in the cosmos.

2-1 Naked-eye astronomy had an important place in civilizations of the past

Positional astronomy—the study of the positions of objects in the sky and how these positions change—has roots that extend far back in time. Four to five thousand years ago, the inhabitants of the British Isles erected stone structures, such as Stonehenge, that suggest a preoccupation with the motions of the sky. Alignments of these stones appear to show the changing locations where the Sun rose and set at key times during the year. Stoneworks of a different sort but with a similar astronomical purpose are found in the Americas, from the southwestern United States to the Andes of Peru. The ancestral Puebloans (also called Anasazi) of modern-day New Mexico, Arizona, Utah, and Colorado created stone carvings that were illuminated by the Sun on the first days of summer or winter (Figure 2-1). At the Incan city of Machu Picchu in Peru, a narrow window carved in a rock 2 meters thick looks out on the sunrise only on December 21 each year. It is thought that the need to keep track of seasons for farming—to know when to sow seeds—provided at least one motivation for early astronomical alignments.

> The astronomical knowledge of ancient peoples is the foundation of modern astronomy

Other ancient peoples designed buildings with astronomical orientations. The great Egyptian pyramids, built around 3000 B.C.E., are oriented north-south and east-west with an accuracy much better than 1 degree. Similar alignments are found in the grand tomb of Shih Huang Ti (259 B.C.E.–210 B.C.E.), the first emperor of China.

Evidence of a highly sophisticated understanding of astronomy can be found in the written records of the Mayan civilization of Central America. Mayan astronomers deduced by observation that the apparent motions of the planet Venus follow a cycle that repeats every 584 days. They also developed a technique for calculating the position of Venus on different dates. The Maya believed that Venus was associated with war, so such calculations were important for choosing the most promising dates on which to attack an enemy.

These archaeological discoveries bear witness to an awareness of naked-eye astronomy by the peoples of many cultures. Many of the concepts of modern positional astronomy come to us from these ancients, including the idea of dividing the sky into constellations.

FIGURE 2-1 R I V U X G
The Sun Dagger at Chaco Canyon On the first day of winter, rays of sunlight passing between stone slabs bracket a spiral stone carving, or petroglyph, at Chaco Canyon in New Mexico. A single band of light strikes the center of the spiral on the first day of summer. This astronomically aligned petroglyph and others were carved by the ancestral Puebloan culture between 850 and 1250 C.E. (Courtesy Karl Kernberger)

2-2 Eighty-eight constellations cover the entire sky

Looking at the sky on a clear, dark night, you might think that you can see millions of stars. Actually, the unaided human eye can detect only about 6000 stars. Because half of the sky is below the horizon at any one time, you can see at most about 3000 stars. When ancient peoples looked at these thousands of stars, they sometimes imagined that groupings of stars traced out pictures in the sky. Astronomers still refer to these groupings, called **constellations** (from the Latin for "group of stars").

> The constellations provide a convenient framework for stating the position of an object in the heavens

You may already be familiar with some of these pictures or patterns in the sky, such as the Big Dipper, which is actually part of the large constellation Ursa Major (the Great Bear). Many constellations, such as Orion in Figure 2-2, have names derived from the myths and legends of antiquity. Although some star groupings vaguely resemble the figures they are supposed to represent (see Figure 2-2c), most do not.

The term "constellation" has a broader definition in present-day astronomy. On modern star charts, the entire sky is divided into 88 regions, each of which is called a constellation. For example, the constellation Orion is now defined to be an irregular patch of sky whose borders are shown in Figure 2-2b. When astronomers refer to the "Great Nebula" M42 in Orion, they mean that as seen from Earth this nebula appears to be within Orion's patch of sky. Some constellations cover large areas of the sky (Ursa Major being one of the biggest) and others very small areas (Crux, the Southern Cross, being the smallest). But because

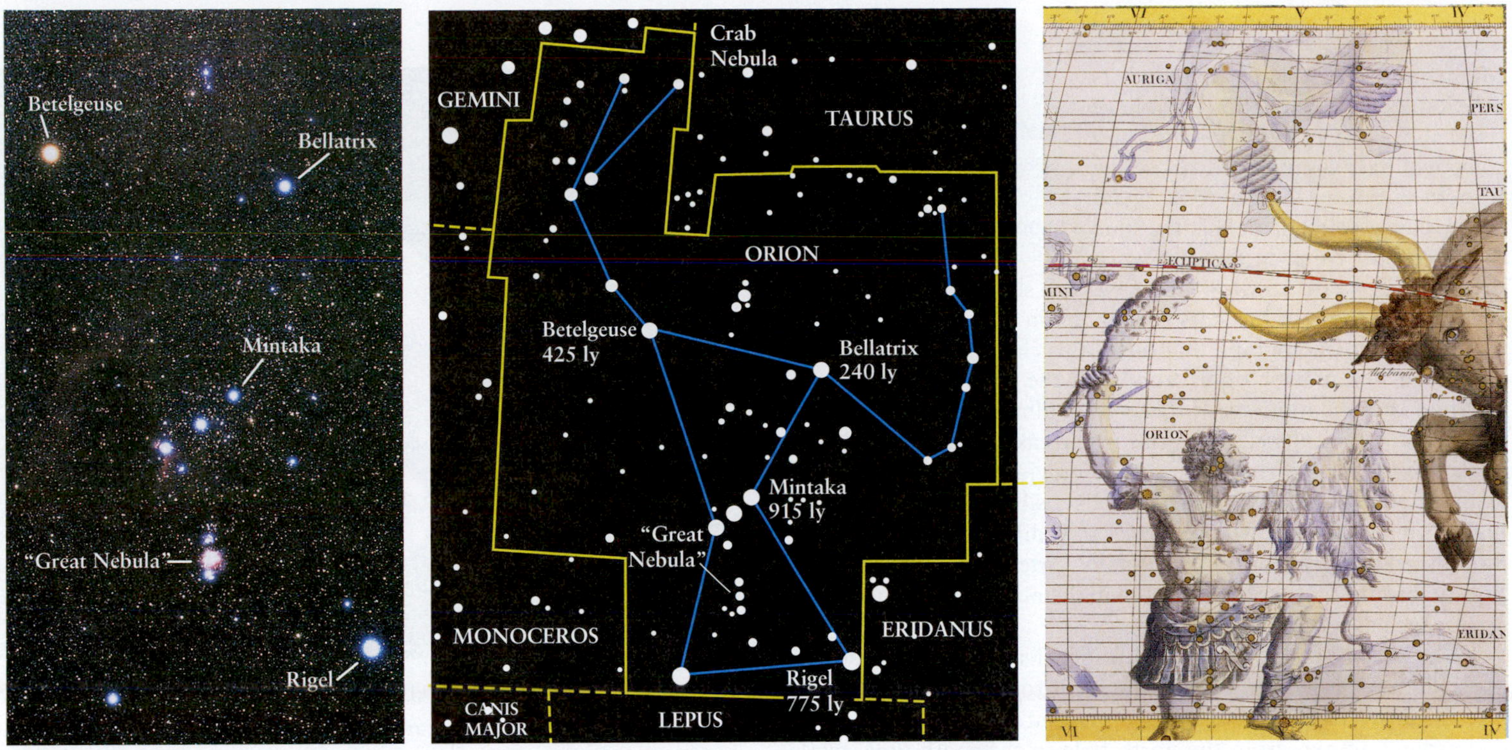

(a) (b) (c)

FIGURE 2-2 R I **V** U X G

Three Views of Orion The constellation Orion is easily seen on nights from December through March. **(a)** This photograph of Orion shows many more stars than can be seen with the naked eye. **(b)** A portion of a modern star atlas shows the distances in light-years (ly) to some of the stars in Orion. The yellow lines show the borders between Orion and its neighboring constellations (labeled in capital letters). **(c)** This fanciful drawing from a star atlas published in 1835 shows Orion the Hunter as well as other celestial creatures. (a: Eckhard Slawik/Science Source; c: Stapleton Collection/Corbis)

the modern constellations cover the entire sky, every star lies in one constellation or another.

CAUTION! When you look at a constellation's star pattern, it is tempting to conclude that you are seeing a group of stars that are all relatively close together. In fact, most of these stars are nowhere near one another. As an example, Figure 2-2b shows the distances in light-years from Earth to four stars in Orion. Although Bellatrix (Arabic for "the Amazon") and Mintaka ("the belt") appear to be close to each other, Mintaka is actually more than 600 light-years farther away from us. The two stars only *appear* to be close because they are in nearly the same direction as seen from Earth. The same illusion often appears when you see an airliner's lights at night. It is very difficult to tell how far away a single bright light is, which is why you can mistake an airliner a few kilometers away for a star trillions of times more distant.

LOOKING DEEPER 2.1 The star names shown in Figure 2-2b are from the Arabic language. For example, Betelgeuse means "armpit," which makes sense when you look at the star atlas drawing in Figure 2-2c. Other types of names are also used for stars. For example, Betelgeuse is also known as α Orionis because it is the brightest star in Orion (α, or alpha, is the first letter in the Greek alphabet).

CAUTION! A number of unscrupulous commercial firms offer to name a star for you for a fee. The money that they charge you for this "service" is real, but the star names are not; none of these names is recognized by astronomers. If you want to use astronomy to commemorate your name or the name of a friend or relative, consider making a donation to your local planetarium or science museum. The money will be put to much better use!

CONCEPTCHECK 2-1

If Jupiter is reported to be in the constellation of Taurus the Bull, does Jupiter need to be within the outline of the bull's body? Why or why not?

Answer appears at the end of the chapter.

2-3 The appearance of the sky changes during the course of the night and from one night to the next

Go outdoors soon after dark, find a spot away from bright lights, and note the patterns of stars in the sky. Do the same a few hours later. You will find that the entire pattern of stars (as well as the

Moon, if it is visible) has shifted its position. New constellations will have risen above the eastern horizon, and some will have disappeared below the western horizon. If you look again before dawn, you will see that the stars that were just rising in the east when the night began are now low in the western sky. This daily motion, or **diurnal motion,** of the stars is apparent in time-exposure photographs (see the photograph that opens this chapter).

By understanding the motions of Earth through space, we can understand why the Sun and stars appear to move in the sky

If you repeat your observations on the following night, you will find that the motions of the sky are almost but not quite the same. The same constellations rise in the east and set in the west, but a few minutes earlier than on the previous night. If you look again after a month, the constellations visible at a given time of night (say, midnight) will be noticeably different, and after six months you will see an almost totally different set of constellations. Only after a year has passed will the night sky have the same appearance as when you began.

Why does the sky go through diurnal motion? Why do the constellations slowly shift from one night to the next? As we will see, the answer to the first question is that Earth *rotates* once a day around an axis from the north pole to the south pole, while the answer to the second question is that Earth also *revolves* once a year around the Sun.

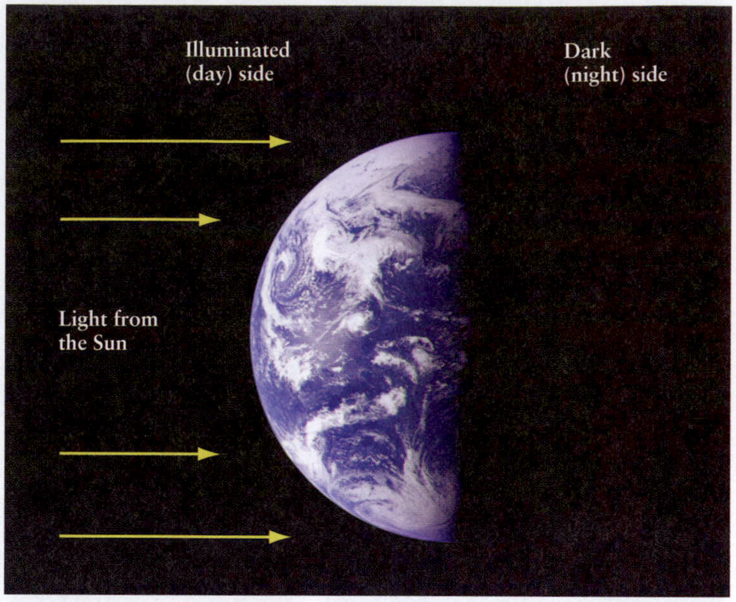

FIGURE 2-3 R I V U X G
Day and Night on Earth At any moment, half of Earth is illuminated by the Sun. As Earth rotates from west to east, your location moves from the dark (night) hemisphere into the illuminated (day) hemisphere and back again. This image was recorded in 1992 by the *Galileo* spacecraft as it was en route to Jupiter. (JPL/NASA)

Diurnal Motion and Earth's Rotation

To understand diurnal motion, note that at any given moment it is daytime on the half of Earth illuminated by the Sun and nighttime on the other half (Figure 2-3). Earth rotates from west to east, making one complete rotation every 24 hours, which is why there is a daily cycle of day and night. Because of this rotation, stars appear to us to rise in the east and set in the west, as do the Sun and Moon.

Figure 2-4 helps to further explain diurnal motion. It shows two views of Earth as seen from a point above the north pole.

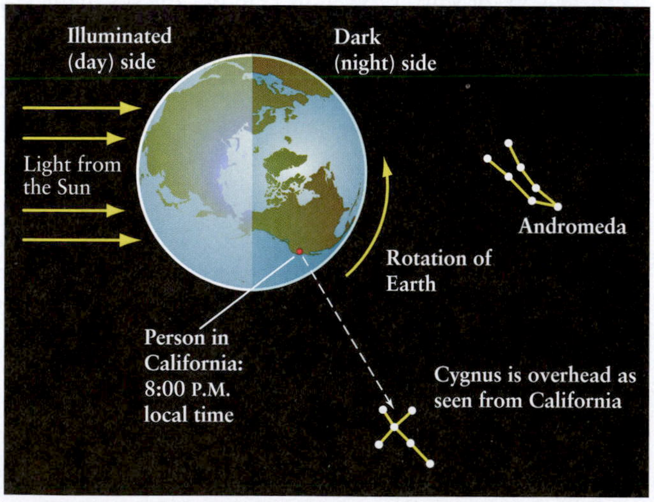

(a) Earth as seen from above the north pole

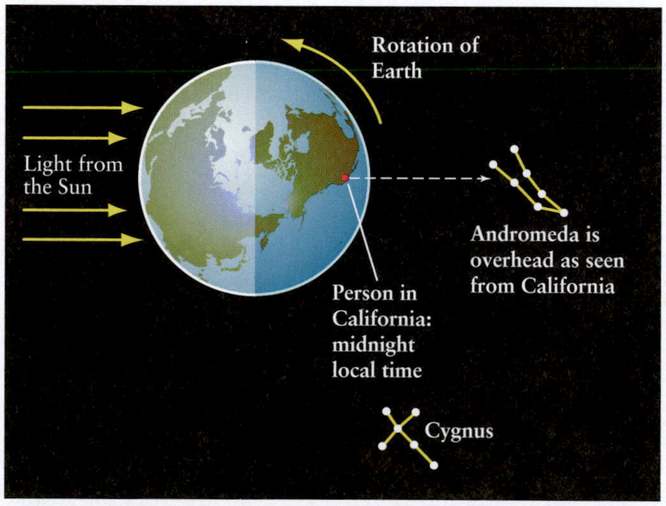

(b) 4 hours (one-sixth of a complete rotation) later

ANIMATION 2-1

FIGURE 2-4
Why Diurnal Motion Happens The diurnal (daily) motion of the stars, the Sun, and the Moon is a consequence of Earth's rotation. **(a)** This drawing shows Earth from a vantage point above the north pole. In this drawing, for a person in California the local time is 8:00 P.M. and the constellation Cygnus is directly overhead. **(b)** Four hours later, Earth has made one-sixth of a complete rotation to the east. As seen from Earth, the entire sky appears to have rotated to the west by one-sixth of a complete rotation. It is now midnight in California, and the constellation directly over California is Andromeda.

At the instant shown in Figure 2-4a, it is day in Asia but night in most of North America and Europe. Figure 2-4b shows Earth four hours later. Four hours is one-sixth of a complete 24-hour day, so Earth has made one-sixth of a rotation between Figures 2-4a and 2-4b. Europe is now in the illuminated half of Earth (the Sun has risen in Europe), while Alaska has moved from the illuminated to the dark half of Earth (the Sun has set in Alaska). For a person in California, in Figure 2-4a the time is 8:00 P.M. and the constellation Cygnus (the Swan) is directly overhead. Four hours later, the constellation over California is Andromeda (named for a mythological princess). Because Earth rotates from west to east, it appears to us on Earth that the entire sky rotates around us in the opposite direction, from east to west.

CONCEPTCHECK 2-2

People in which of the following cities in North America experience sunrise first: New York, San Francisco, Chicago, or Denver? Explain in terms of Earth's rotation.

CALCULATIONCHECK 2-1

If the constellation of Cygnus rises along the eastern horizon at sunset, at what time will it be highest above the southern horizon?

Answers appear at the end of the chapter.

Yearly Motion and Earth's Orbit

We described earlier that in addition to the diurnal motion of the sky, the constellations visible in the night sky also change slowly over the course of a year. This happens because Earth orbits, or revolves around, the Sun (Figure 2-5). Over the course of a year, Earth makes one complete orbit, and the darkened, nighttime side of Earth gradually turns toward different parts of the heavens. For example, as seen from the northern hemisphere, at midnight in late July the constellation Cygnus is close to overhead; at midnight in late September the constellation Andromeda is close to overhead; and at midnight in late November the constellation Perseus (commemorating a mythological hero) is close to overhead. If you follow a particular star on successive evenings, you will find that it rises approximately 4 minutes earlier each night, or 2 hours earlier each month.

CONCEPTCHECK 2-3

If Earth suddenly rotated on its axis 3 times faster than it does now, then how many times would the Sun rise and set each year?

Answer appears at the end of the chapter.

Constellations and the Night Sky

Constellations can help you find your way around the sky. For example, if you live in the northern hemisphere, you can use the Big Dipper in Ursa Major to find the north direction by drawing a straight line through the two stars at the front of the Big Dipper's bowl (Figure 2-6). The first moderately bright star you come to is Polaris, also called the North Star because it is located almost directly over Earth's north pole. If you draw a line from Polaris straight down to the horizon, you will find the north direction.

As Figure 2-6 shows, by following the handle of the Big Dipper you can locate the bright reddish star Arcturus in Boötes (the Shepherd) and the prominent bluish star Spica in Virgo (the Virgin). The saying "Follow the arc to Arcturus and speed to Spica" may help you remember these stars, which are conspicuous in the evening sky during the spring and summer.

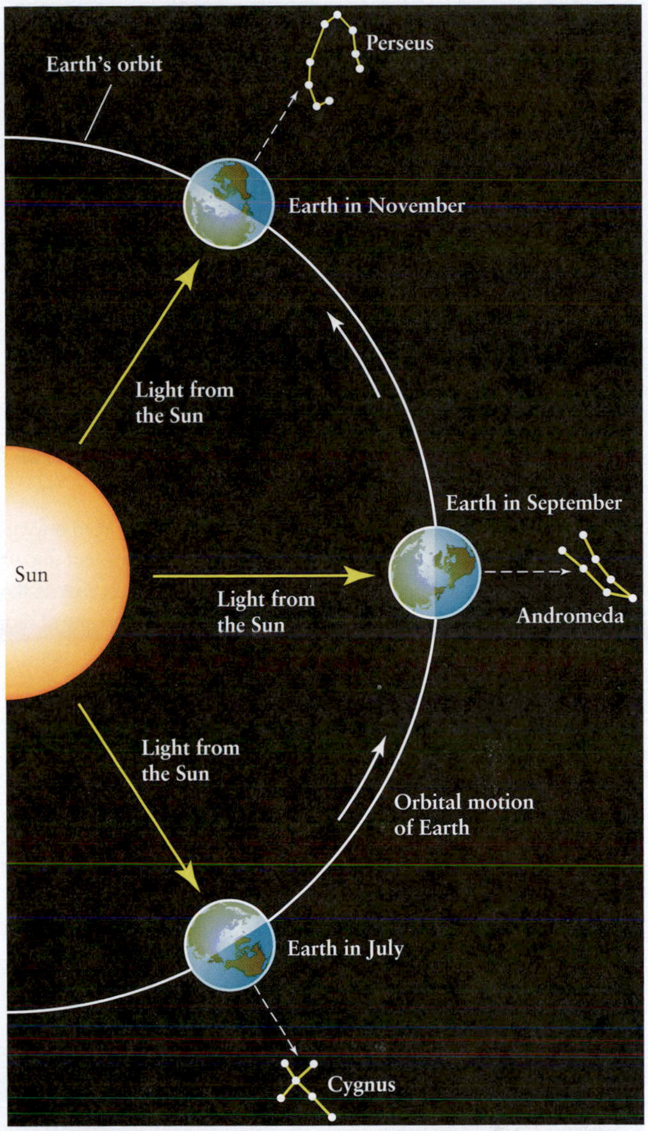

FIGURE 2-5

Why the Night Sky Changes During the Year As Earth orbits around the Sun, the nighttime side of Earth gradually turns toward different parts of the sky. Hence, the particular stars that you see in the night sky are different at different times of the year. This figure shows which constellation is overhead at midnight local time—when the Sun is on the opposite side of Earth from your location—during different months for observers at midnorthern latitudes (including the United States). If you want to view the constellation Andromeda, the best time of the year to do it is in late September, when Andromeda is nearly overhead at midnight.

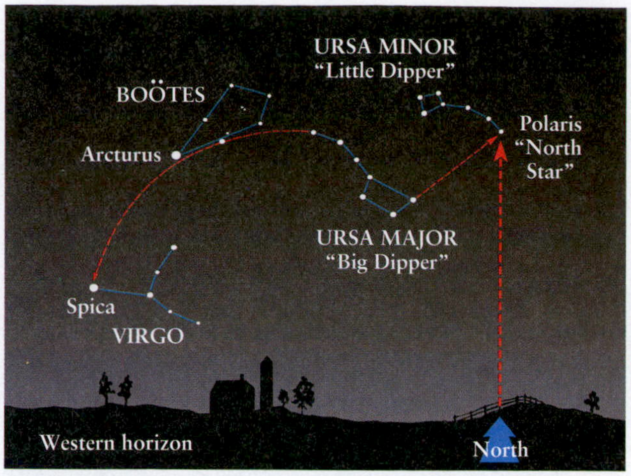

FIGURE 2-6

The Big Dipper as a Guide The North Star can be seen from anywhere in the northern hemisphere on any night of the year. This star chart shows how the Big Dipper can be used to point out the North Star as well as the brightest stars in two other constellations. The chart shows the sky at around 11 P.M. (daylight savings time) on August 1. Due to Earth's orbital motion around the Sun, you will see this same view at 1 A.M. on July 1 and at 9 P.M. on September 1. The angular distance from Polaris to Spica is 102°.

During winter in the northern hemisphere, you can see some of the brightest stars in the sky. Many of them are in the vicinity of the "winter triangle" (Figure 2-7), which connects bright stars in the constellations of Orion (the Hunter), Canis Major (the Large Dog), and Canis Minor (the Small Dog).

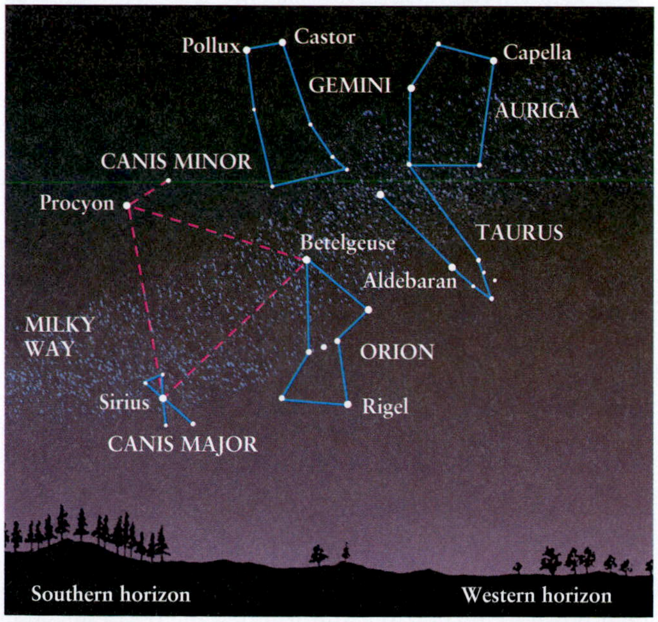

FIGURE 2-7

The "Winter Triangle" This star chart shows the view toward the southwest on a winter evening in the northern hemisphere (around midnight on January 1, 10 P.M. on February 1, or 8 P.M. on March 1). Three of the brightest stars in the sky make up the "winter triangle," which is about 26° on a side. In addition to the constellations involved in the triangle, the chart shows the prominent constellations Gemini (the Twins), Auriga (the Charioteer), and Taurus (the Bull).

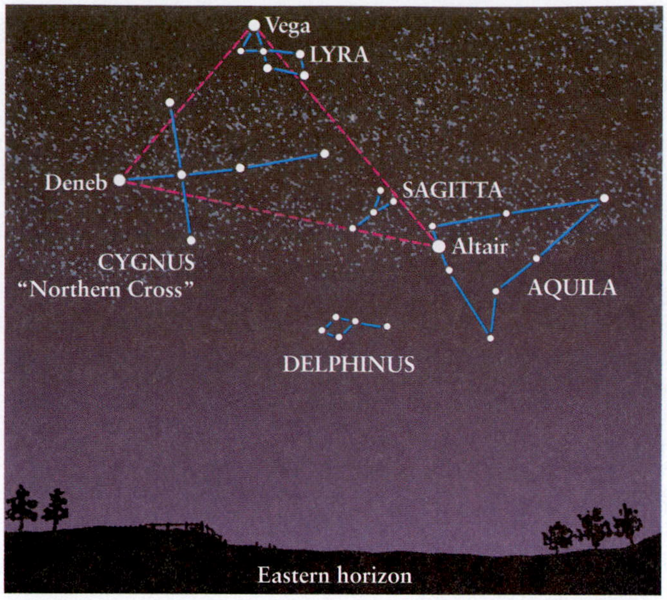

FIGURE 2-8

The "Summer Triangle" This star chart shows the eastern sky as it appears in the evening during spring and summer in the northern hemisphere (around 1 A.M. daylight savings time on June 1, around 11 P.M. on July 1, and around 9 P.M. on August 1). The angular distance from Deneb to Altair is about 38°. The constellations Sagitta (the Arrow) and Delphinus (the Dolphin) are much fainter than the three constellations that make up the triangle.

A similar feature, the "summer triangle," graces the summer sky in the northern hemisphere. This triangle connects the brightest stars in Lyra (the Harp), Cygnus (the Swan), and Aquila (the Eagle) (Figure 2-8). A conspicuous portion of the Milky Way forms a beautiful background for these constellations, which are nearly overhead during the middle of summer at midnight.

A wonderful tool to help you find your way around the night sky is the planetarium program *Starry Night*™, which is on the CD-ROM that accompanies certain print copies of this book. (You can also obtain *Starry Night*™ separately.) In addition, at the end of this book you will find a set of selected star charts for the evening hours of all 12 months of the year. You may find stargazing an enjoyable experience, and *Starry Night*™ and the star charts will help you identify many well-known constellations.

Note that all the star charts in this section and at the end of this book are drawn for an observer in the northern hemisphere. If you live in the southern hemisphere, you can see constellations that are not visible from the northern hemisphere, and vice versa. In the next section we will see why this is so.

2-4 It is convenient to imagine that the stars are located on a celestial sphere

TUTORIAL 2-1

Many ancient societies believed that all the stars are the same distance from Earth. They imagined the stars to be bits of fire imbedded in the inner surface of an immense hollow sphere, called the **celestial sphere,** with Earth at its center. In this picture of the universe, Earth was fixed and did not rotate. Instead, the entire celestial sphere rotated once a day around Earth from east to west, thereby causing the diurnal motion of the sky. The picture of a rotating celestial sphere fit well with naked-eye observations, and for its time was a useful model of how the universe works. (We discussed the role of models in science in Section 1-1).

The concept of the imaginary celestial sphere helps us visualize the motions of stars in the sky

Today's astronomers know that this simple model of the universe is not correct. Diurnal motion is due to the rotation of Earth, not the rest of the universe. Furthermore, as we learned when discussing the constellations in Section 2-2, the stars are not all at the same distance from Earth. Indeed, the stars that you can see with the naked eye range from 4.2 to more than 1000 light-years away, and telescopes allow us to see objects at distances of billions of light-years.

Thus, astronomers now recognize that the celestial sphere is an *imaginary* object that has no basis in physical reality. Nonetheless, the celestial sphere model remains a useful tool of positional astronomy. If we imagine, as did the ancients, that Earth is stationary and that the celestial sphere rotates around us, it is relatively easy to specify the directions to different objects in the sky and to visualize the motions of these objects. Figure 2-9 depicts the celestial sphere, with Earth at its center. (A truly proportional drawing would show the celestial sphere as being millions of times larger than Earth.) We picture the stars as points of light that are fixed on the inner surface of the celestial sphere.

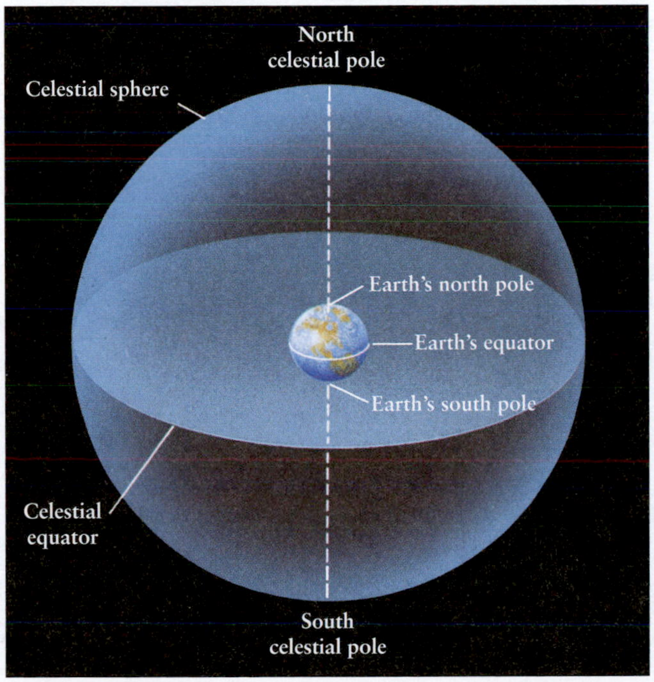

ANIMATION 2-2

FIGURE 2-9

The Celestial Sphere The celestial sphere is the apparent sphere of the sky. The view in this figure is from the outside of this (wholly imaginary) sphere. Earth is at the center of the celestial sphere, so our view is always of the inside of the sphere. The celestial equator and poles are the projections of Earth's equator and axis of rotation out into space. The celestial poles are therefore located directly over Earth's poles.

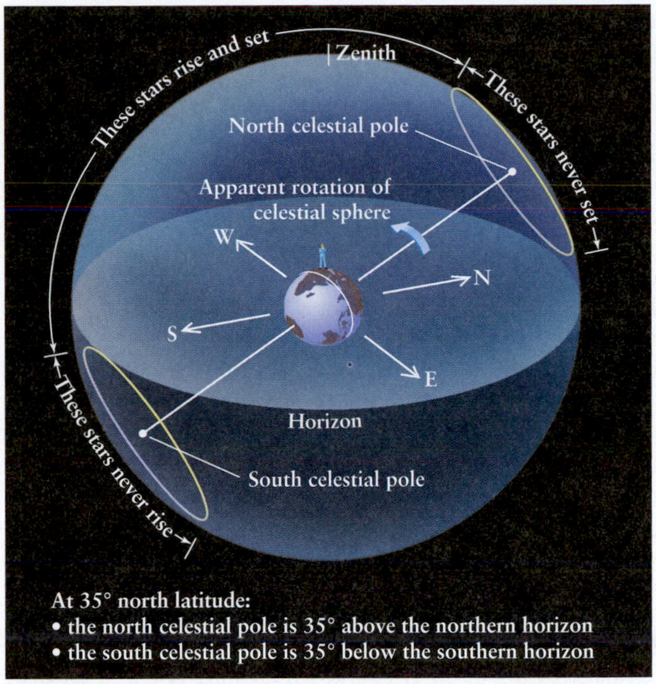

FIGURE 2-10

The View from 35° North Latitude To an observer at 35° north latitude (roughly the latitude of Los Angeles, Atlanta, Tel Aviv, and Tokyo), the north celestial pole is always 35° above the horizon. Stars within 35° of the north celestial pole are circumpolar; they trace out circles around the north celestial pole during the course of the night and are always above the horizon on any night of the year. Stars within 35° of the south celestial pole are always below the horizon and can never be seen from this latitude. Stars that lie between these two extremes rise in the east and set in the west.

Other features on the celestial sphere result from **projections.** The projection of a point on Earth is made by extending an imaginary line perpendicular to the surface of Earth until it intersects the celestial sphere. If we project Earth's north and south poles into space, we obtain the **north celestial pole** and the **south celestial pole.** Thus, the two celestial poles are where Earth's axis of rotation intersects the celestial sphere (see Figure 2-9). The star Polaris is less than 1° away from the north celestial pole, which is why it is called the North Star or the Pole Star. If we project Earth's equator out into space, we obtain the **celestial equator.** The celestial equator divides the sky into northern and southern hemispheres, just as Earth's equator divides Earth into two hemispheres.

The point in the sky directly overhead an observer anywhere on Earth is called that observer's **zenith.** The zenith and celestial sphere are shown in Figure 2-10 for an observer located at 35° north latitude (that is, at a location on Earth's surface 35° north of the equator). The zenith is shown at the top of Figure 2-10, so Earth and the celestial sphere appear "tipped" compared to Figure 2-9. At any time, an observer can see only half of the celestial sphere; the other half is below the horizon, hidden by the body of

Earth. The hidden half of the celestial sphere is darkly shaded in Figure 2-10.

Motions of the Celestial Sphere

For an observer anywhere in the northern hemisphere, including the observer in Figure 2-10, the north celestial pole is always above the horizon. As Earth turns from west to east, it appears to the observer that the celestial sphere turns from east to west. Stars sufficiently near the north celestial pole revolve around the pole, never rising or setting. Such stars are called **circumpolar.** For example, as seen from North America or Europe, Polaris is a circumpolar star and can be seen at any time of night on any night of the year. The photograph that opens this chapter shows the circular trails of stars around the north celestial pole as seen from Hawaii (at 20° north latitude). Stars near the south celestial pole revolve around that pole but always remain below the horizon of an observer in the northern hemisphere. Hence, these stars can never be seen by the observer in Figure 2-10. Stars between those two limits rise in the east and set in the west.

CAUTION! Keep in mind that which stars are circumpolar, which stars never rise, and which stars rise and set depends on the latitude from which you view the heavens. As an example, for an observer at 35° south latitude (roughly the latitude of Sydney, Cape Town, and Buenos Aires), the roles of the north and south celestial poles are the opposite of those shown in Figure 2-10: Objects close to the *south* celestial pole are circumpolar, that is, they revolve around that pole and never rise or set. For an observer in the southern hemisphere, stars close to the *north* celestial pole are always below the horizon and can never be seen. Hence, astronomers in Australia, South Africa, and Argentina never see the North Star but are able to see other stars that are forever hidden from North American or European observers.

For observers at most locations on Earth, stars rise in the east and set in the west at an angle to the horizon. To see why this is so, notice that the rotation of the celestial sphere carries stars across the sky in paths that are parallel to the celestial equator (Figure 2-11).

CONCEPTCHECK 2-4

Where would you need to be standing on Earth for the celestial equator to pass through your zenith?

Answer appears at the end of the chapter.

How do we describe the location of a star? Using the celestial equator and poles, we can define a coordinate system based on angles to specify the position of any star on the celestial sphere. As Box 2-1 describes, the most commonly used coordinate system uses two angles, *right ascension* and *declination,* that are analogous to longitude and latitude on Earth. These coordinates tell us in what direction we should look to see the star. To locate the star's true position in three-dimensional space, we must also know the distance to the star.

2-5 The seasons are caused by the tilt of Earth's axis of rotation

TUTORIAL 2-2 In addition to rotating on its axis every 24 hours, Earth revolves around the Sun—that is, it *orbits* the Sun—in about 365¼ days (see Figure 2-5). As we travel with Earth around its orbit, we experience the annual cycle of seasons. But why *are* there seasons? Furthermore, the seasons are opposite in the northern and southern hemispheres. For

Earth's tilt causes variations in the concentration of sunlight

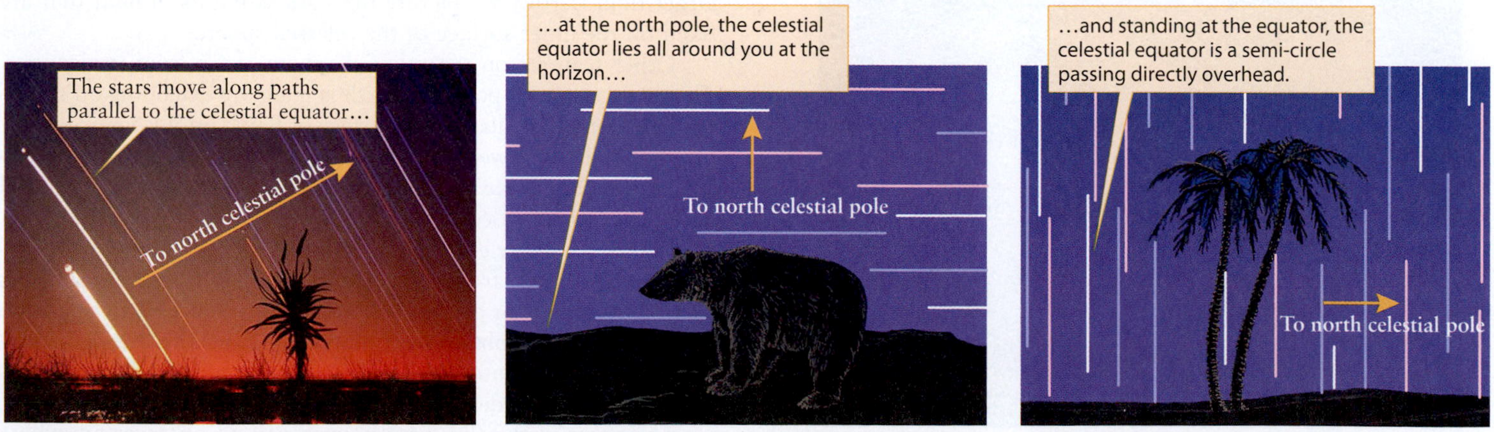

The stars move along paths parallel to the celestial equator…
To north celestial pole
…at the north pole, the celestial equator lies all around you at the horizon…
To north celestial pole
…and standing at the equator, the celestial equator is a semi-circle passing directly overhead.
To north celestial pole

(a) At middle northern latitudes (b) At the north pole (c) At the equator

ANIMATION 2-3 **FIGURE 2-11** R I V U X G

The Apparent Motion of Stars at Different Latitudes

As Earth rotates, stars appear to rotate around us along paths that are parallel to the celestial equator. **(a)** As shown in this long time exposure, at most locations on Earth the rising and setting motions are at an angle to the horizon that depends on the latitude. **(b)** At the north pole (latitude 90° north) the stars appear to move parallel to the horizon. **(c)** At the equator (latitude 0°) the stars rise and set along vertical paths. (a: David Miller/David Malin Images)

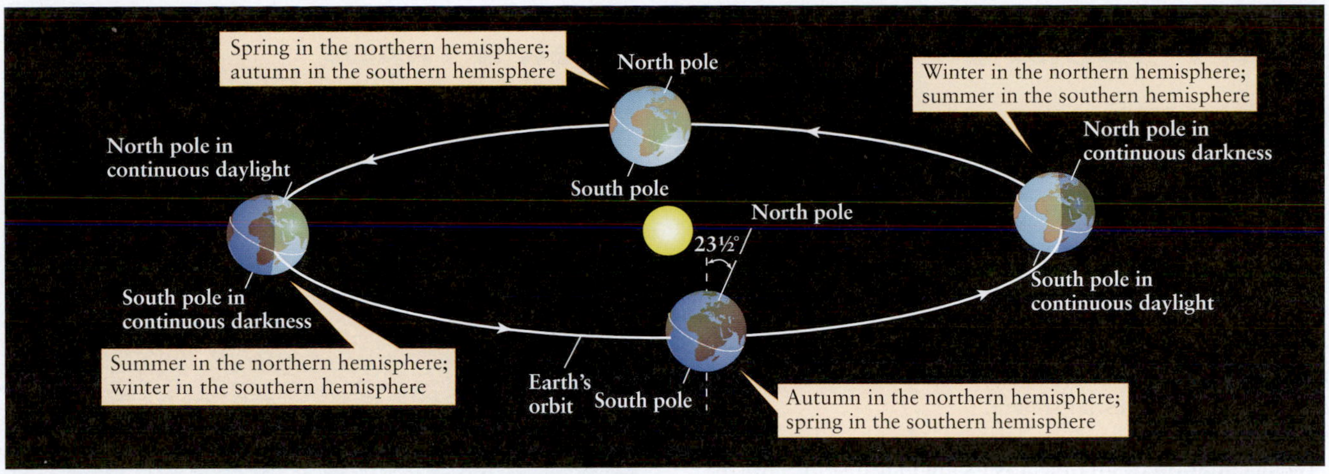

ANIMATION 2-4

FIGURE 2-12
The Seasons Earth's axis of rotation is inclined 23½° away from the perpendicular to the plane of Earth's orbit. The north pole points in the same direction as it orbits the Sun, near the star Polaris. Consequently, the amount of solar illumination and the number of daylight hours at any location on Earth vary in a regular pattern throughout the year. This is the origin of the seasons.

example, February is midwinter in North America but midsummer in Australia. Why should this be?

The Origin of the Seasons

As we know from common experience, there are warmer temperatures in summer. This warmth is due to an increased amount of sunlight striking the ground beneath your feet and the surrounding areas. But what causes the concentration of sunlight to change through the seasons?

The reason why we have seasons, and why they are different in different hemispheres, is that Earth's axis of rotation is not perpendicular to the plane of Earth's orbit. Instead, as Figure 2-12 shows, the axis is tilted about 23½° away from the perpendicular. Earth maintains this tilt as it orbits the Sun, with Earth's north pole always pointing in the same direction throughout the year. As we saw earlier, this direction just happens to have a star (named Polaris) within 1°. (This stability is a hallmark of all rotating objects. A top will not fall over as long as it is spinning, and the rotating wheels of a motorcycle help to keep the rider upright.)

During part of the year, when Earth is in the part of its orbit shown on the left side of Figure 2-12, the northern hemisphere is tilted toward the Sun. As Earth spins on its axis, a point in the northern hemisphere spends more than 12 hours in the sunlight. Thus, the days there are long and the nights are short, and it is summer in the northern hemisphere. The summer is hot not only because of the extended daylight hours but also because the Sun is high in the northern hemisphere's sky. As a result, sunlight strikes the ground at a nearly perpendicular angle that heats the ground efficiently (Figure 2-13a). During this same time of year in the southern hemisphere, the days are short and the nights are long, because a point in this hemisphere spends fewer than 12 hours a day in the sunlight. The Sun is low in the sky, so sunlight strikes the

(a) The Sun in summer

(b) The Sun in winter

FIGURE 2-13
Solar Energy in Summer and Winter
At different times of the year, sunlight strikes the ground at different angles. **(a)** In summer, sunlight is concentrated and the days are also longer, which further increases the heating. **(b)** In winter, the sunlight is less concentrated, the days are short, and little heating of the ground takes place. This accounts for the low temperatures in winter.

BOX 2-1 TOOLS OF THE ASTRONOMER'S TRADE

Celestial Coordinates

Your *latitude* and *longitude* describe where on Earth's surface you are located. The latitude of your location denotes how far north or south of the equator you are, and the longitude of your location denotes how far west or east you are of an imaginary circle that runs from the north pole to the south pole through the Royal Observatory in Greenwich, England. In an analogous way, astronomers use coordinates called *declination* and *right ascension* to describe the position of a planet, star, or galaxy on the celestial sphere.

Declination is analogous to latitude. As the illustration shows, the **declination** of an object is its angular distance north or south of the celestial equator, measured along a circle passing through both celestial poles. Like latitude, it is measured in degrees, arcminutes, and arcseconds (see Section 1-5).

Right ascension is analogous to longitude. It is measured from a line that runs between the north and south celestial poles and passes through a point on the celestial equator called the *vernal equinox* (shown as a red dot in the illustration). This point is one of two locations where the Sun crosses the celestial equator during its apparent annual motion, as we discuss in Section 2-5. In Earth's northern hemisphere, spring officially begins when the Sun reaches the vernal equinox in late March. The **right ascension** of an object is the angular distance from the vernal equinox eastward along the celestial equator to the circle used in measuring its declination (see illustration). Astronomers measure right ascension in *time* units (hours, minutes, and seconds), corresponding to the time required for the celestial sphere to rotate through this angle. For example, suppose there is a star at your zenith right now with right ascension $6^h\ 0^m\ 0^s$. Two hours and 30 minutes from now, there will be a different object at your zenith with right ascension $8^h\ 30^m\ 0^s$.

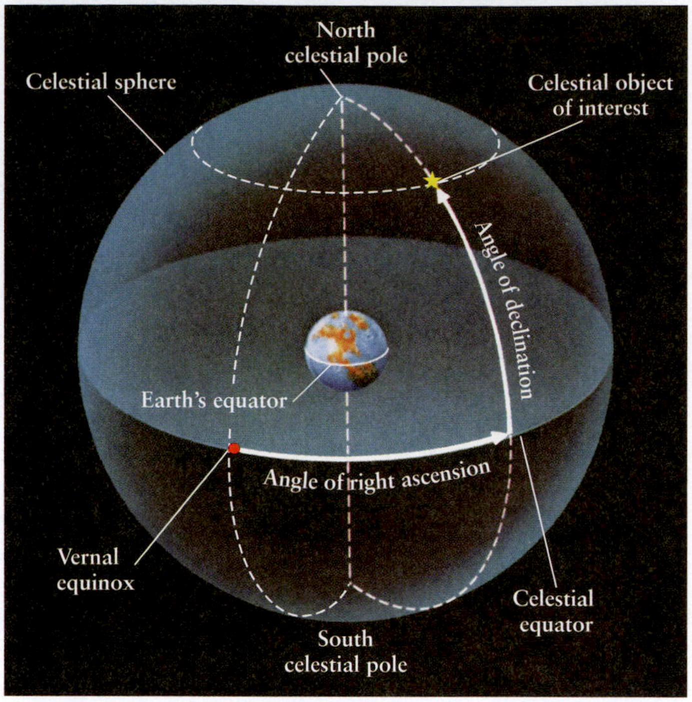

surface at a grazing angle that causes little heating (Figure 2-13b), and it is winter in the southern hemisphere.

Half a year later, Earth is in the part of its orbit shown on the right side of Figure 2-12. Now the situation is reversed, with winter in the northern hemisphere (which is now tilted away from the Sun) and summer in the southern hemisphere. During spring and autumn, the two hemispheres receive roughly equal amounts of illumination from the Sun, and daytime and nighttime are of roughly equal length everywhere on Earth.

CAUTION! A common misconception is that the seasons are caused by variations in the distance from Earth to the Sun. According to this idea, Earth is closer to the Sun in summer and farther away in winter. But in fact, Earth's orbit around the Sun is very nearly circular, and the Earth-Sun distance varies only about 3 percent over the course of a year. (Earth's orbit only *looks* elongated in Figure 2-12 because this illustration shows the orbit from a side view; it would be circular if viewed overhead.) We are slightly closer to the Sun in January than in July, but this small variation has little influence on the cycle of the seasons. Also, if the seasons were really caused by variations in the Earth-Sun distance, the seasons would be the same in both hemispheres!

CONCEPTCHECK 2-5

If Earth's axis were not tilted, but rather was straight up and down compared to the plane created by the path of Earth's orbit, would observers near Earth's north pole still observe periods where the Sun never rises and the Sun never sets?

Answer appears at the end of the chapter.

The coordinates of the bright star Rigel for the year 2000 are R.A. = $5^h\ 14^m\ 32.2^s$, Decl. = $-8°\ 12'\ 06''$. (R.A. and Decl. are abbreviations for right ascension and declination.) A minus sign on the declination indicates that the star is south of the celestial equator; a plus sign (or no sign at all) indicates that an object is north of the celestial equator. As we discuss in Box 2-2, right ascension helps determine the best time to observe a particular object.

It is important to state the year for which a star's right ascension and declination are valid. This is so because of precession, which we discuss in Section 2-6.

EXAMPLE: What are the coordinates of a star that lies exactly halfway between the vernal equinox and the south celestial pole?

Situation: Our goal is to find the right ascension and declination of the star in question.

Tools: We use the definitions depicted in the figure.

Answer: Since the circle used to measure this star's declination passes through the vernal equinox, this star's right ascension is R.A. = $0^h\ 0^m\ 0^s$. The angle between the celestial equator and south celestial pole is $90°\ 0'\ 0''$, so the declination of this star is Decl. = $-45°\ 0'\ 0''$.

Review: The declination in this example is negative because the star is in the southern half of the celestial sphere.

EXAMPLE: At midnight local time you see a star with R.A. = $2^h\ 30^m\ 0^s$ at your zenith. When will you see a star at your zenith with R.A. = $21^h\ 0^m\ 0^s$?

Situation: If you held your finger stationary over a globe of Earth, the longitude of the point directly under your finger would change as you rotated the globe. In the same way, the right ascension of the point directly over your head (the zenith) changes as the celestial sphere rotates. We use this concept to determine the time in question.

Tools: We use the idea that a change in right ascension of 24^h corresponds to an elapsed time of 24 hours and a complete rotation of the celestial sphere.

Answer: The time required for the sky to rotate through the angle between the stars is the difference in their right ascensions: $21^h\ 0^m\ 0^s - 2^h\ 30^m\ 0^s = 18^h\ 30^m\ 0^s$. So the second star will be at your zenith 18½ hours after the first one, or at 6:30 P.M. the following evening.

Review: Our answer was based on the idea that the celestial sphere makes *exactly* one complete rotation in 24 hours. If this were so, from one night to the next each star would be in exactly the same position at a given time. But because of the way that we measure time, the celestial sphere makes slightly more than one complete rotation in 24 hours. (We explore the reasons for this in Box 2-2.) As a result, our answer is in error by about 3 minutes. For our purposes, this is a small enough error that we can ignore it.

How the Sun Moves on the Celestial Sphere

The plane of Earth's orbit around the Sun is called the **ecliptic plane** (Figure 2-14a). The ecliptic plane is easy to picture if you imagine hovering above the solar system, but is described quite differently when viewed from Earth. As a result of Earth's annual motion around the Sun, it appears to us (observing from Earth) that the Sun slowly changes its position on the celestial sphere over the course of a year. (Since we don't see the Sun and stars at the same time, this motion relative to the stars isn't readily familiar.) The circular path that the Sun appears to trace out against the background of stars is called the **ecliptic** (Figure 2-14b). The plane of this path is the same as the ecliptic plane. (The name *ecliptic* suggests that the path traced out by the Sun has something to do with eclipses. We will discuss the connection in Chapter 3.) Because there are 365¼ days in a year and 360° in a circle, the Sun appears to move along the ecliptic at a rate of about 1° per day. This motion is from west to east, that is, in the direction opposite to the apparent motion of the celestial sphere.

ANALOGY Note that at the same time that the Sun is making its yearlong trip around the ecliptic, the entire celestial sphere is rotating around us once per day. You can envision the celestial sphere as a merry-go-round rotating clockwise and the Sun as a restless child who is walking slowly around the merry-go-round's rim in the counterclockwise direction. During the time it takes the child to make a round trip, the merry-go-round rotates 365¼ times.

The ecliptic plane is *not* the same as the plane of Earth's equator, thanks to the 23½° tilt of Earth's rotation axis shown in Figure 2-12. As a result, the ecliptic and the celestial

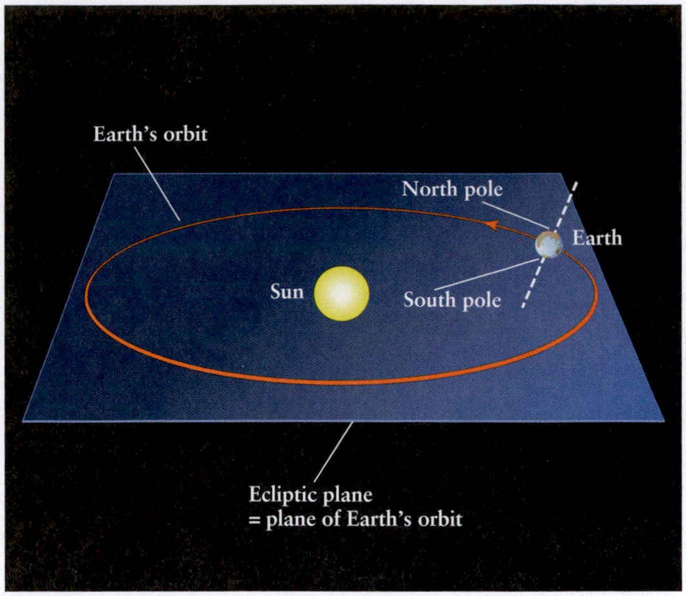

(a) In reality Earth orbits the Sun once a year

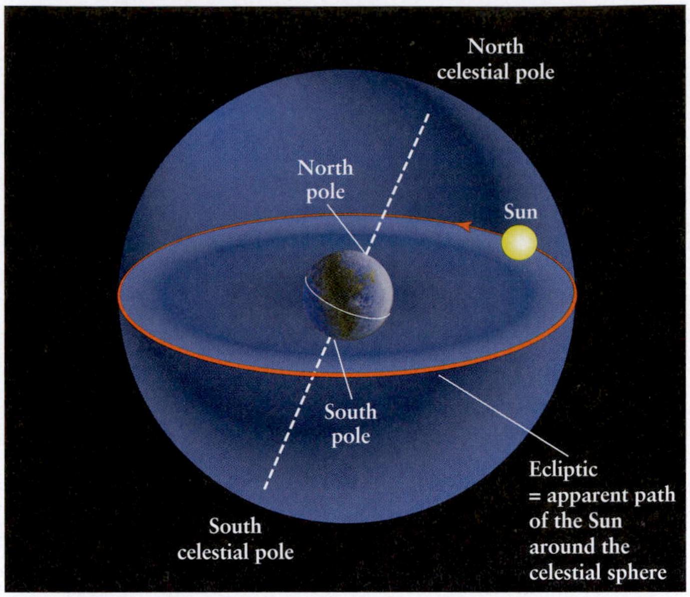

(b) It appears from Earth that the Sun travels around the celestial sphere once a year

FIGURE 2-14

The Ecliptic Plane and the Ecliptic **(a)** The ecliptic plane is the plane in which Earth moves around the Sun. **(b)** As seen from Earth, the Sun appears to move around the celestial sphere along a circular path called the ecliptic. Earth takes a year to complete one orbit around the Sun, so as seen by us the Sun takes a year to make a complete trip around the ecliptic.

equator are inclined to each other by that same 23½° angle (Figure 2-15).

CONCEPTCHECK 2-6

How long does the Sun take to move from being next to a bright star all the way around the celestial sphere and back to that same bright star?

Answer appears at the end of the chapter.

Equinoxes and Solstices

The ecliptic and the celestial equator intersect at only two points, which are exactly opposite each other on the celestial sphere. Each point is called an **equinox** (from the Latin for "equal night"), because when the Sun appears at either of these points, day and night are each about 12 hours long at all locations on Earth. The term "equinox" is also used to refer to the date on which the Sun passes through one of these special points on the ecliptic.

On about March 21 of each year, the Sun passes northward across the celestial equator at the **vernal equinox.** This marks the beginning of spring in the northern hemisphere ("vernal" is from the Latin for "spring"). On about September 22, the Sun moves southward across the celestial equator at the **autumnal equinox,** marking the moment when fall begins in the northern hemisphere (see Figure 2-15). Since the seasons are opposite in the northern and southern hemispheres, for Australians, South Africans, and South Americans the vernal equinox actually marks the beginning of autumn. The names of the equinoxes come from astronomers of the past who lived north of the equator.

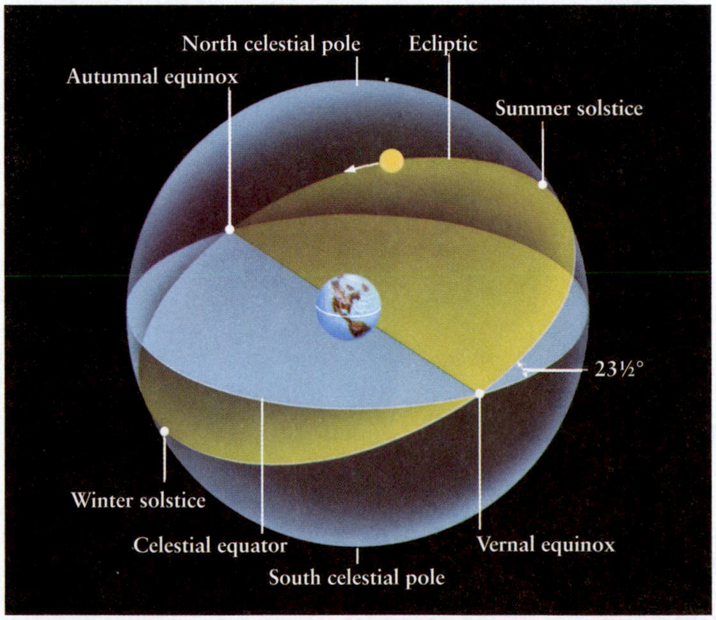

FIGURE 2-15

The Ecliptic, Equinoxes, and Solstices This illustration of the celestial sphere is similar to Figure 2-14b, but is drawn with the north celestial pole at the top and the celestial equator running through the middle. The ecliptic is inclined to the celestial equator by 23½° because of the tilt of Earth's axis of rotation. It intersects the celestial equator at two points, called equinoxes. The northernmost point on the ecliptic is the summer solstice, and the southernmost point is the winter solstice. The Sun is shown in its approximate position for August 1.

Between the vernal and autumnal equinoxes lie two other significant locations along the ecliptic. The point on the ecliptic farthest north of the celestial equator is called the **summer solstice.** "Solstice" is from the Latin for "solar standstill," and it is at the summer solstice that the Sun stops moving northward on the celestial sphere. At this point, the Sun is as far north of the celestial equator as it can get. It marks the location of the Sun at the moment summer begins in the northern hemisphere (about June 21). At the beginning of the northern hemisphere's winter (about December 21), the Sun is farthest south of the celestial equator at a point called the **winter solstice** (see Figure 2-15).

Because the Sun's position on the celestial sphere varies slowly over the course of a year, its daily path across the sky (due to Earth's rotation) also varies with the seasons (Figure 2-16). On the first day of spring or the first day of fall, when the Sun is at one of the equinoxes, the Sun rises directly in the east and sets directly in the west.

When the northern hemisphere is tilted away from the Sun and it is winter in the northern hemisphere, the Sun rises in the southeast. Daylight lasts for fewer than 12 hours as the Sun skims low over the southern horizon and sets in the southwest. Northern hemisphere nights are longest when the Sun is at the winter solstice.

The closer you get to the north pole, the shorter the winter days and the longer the winter nights. In fact, anywhere within 23½° of the north pole (that is, north of latitude 90° – 23½° = 66½° N) the Sun is below the horizon for 24 continuous hours at least one day of the year. The circle around Earth at 66½° north latitude is called the **Arctic Circle** (Figure 2-17). The corresponding region around the south pole is bounded by the **Antarctic Circle** at 66½° south latitude. At the time of the winter solstice, explorers south of the Antarctic Circle enjoy "the midnight sun," or 24 hours of continuous daylight.

During summer in the northern hemisphere, when the northern hemisphere is tilted toward the Sun, the Sun rises in the northeast and sets in the northwest. The Sun is at its northernmost position at the summer solstice, giving the northern hemisphere

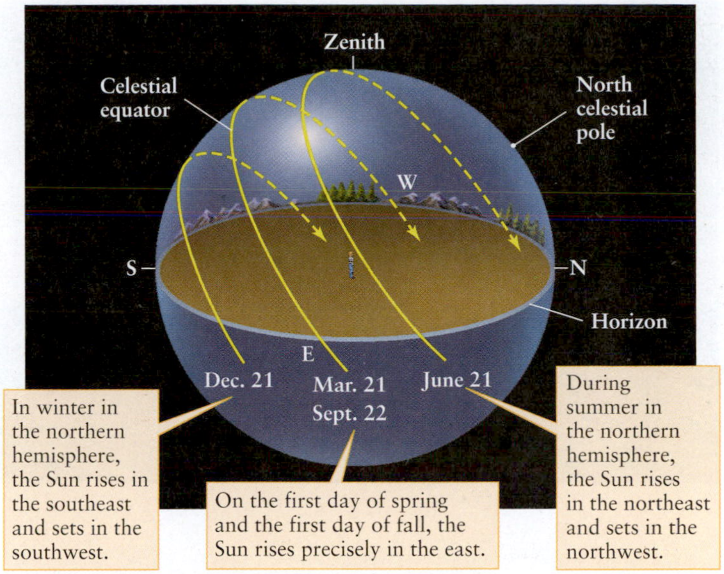

Zenith
Celestial equator
North celestial pole
W
S
N
Horizon
E
Dec. 21
Mar. 21
Sept. 22
June 21
In winter in the northern hemisphere, the Sun rises in the southeast and sets in the southwest.
On the first day of spring and the first day of fall, the Sun rises precisely in the east.
During summer in the northern hemisphere, the Sun rises in the northeast and sets in the northwest.

ANIMATION 2-5

FIGURE 2-16

The Sun's Daily Path Across the Sky This drawing shows the apparent path of the Sun during the course of a day on four different dates. Like Figure 2-10, this drawing is for an observer at 35° north latitude.

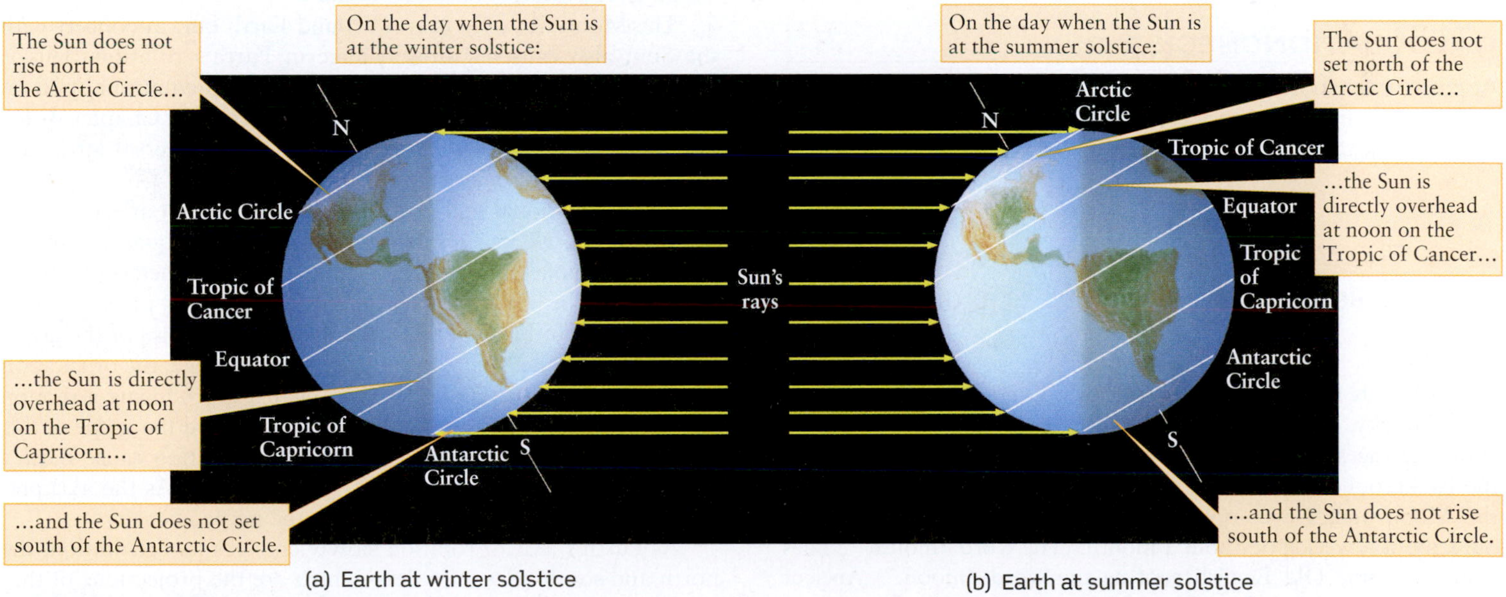

The Sun does not rise north of the Arctic Circle...
On the day when the Sun is at the winter solstice:
N
Arctic Circle
Tropic of Cancer
Equator
...the Sun is directly overhead at noon on the Tropic of Capricorn...
Tropic of Capricorn
Antarctic Circle
S
...and the Sun does not set south of the Antarctic Circle.
Sun's rays
On the day when the Sun is at the summer solstice:
The Sun does not set north of the Arctic Circle...
Arctic Circle
Tropic of Cancer
Equator
...the Sun is directly overhead at noon on the Tropic of Cancer...
Tropic of Capricorn
Antarctic Circle
...and the Sun does not rise south of the Antarctic Circle.
(a) Earth at winter solstice
(b) Earth at summer solstice

FIGURE 2-17

Tropics and Circles Four important latitudes on Earth are the Arctic Circle (66½° north latitude), Tropic of Cancer (23½° north latitude), Tropic of Capricorn (23½° south latitude), and Antarctic Circle (66½° south latitude). These drawings show the significance of these latitudes when the Sun is **(a)** at the winter solstice and **(b)** at the summer solstice.

FIGURE 2-18 R I V U X G
The Midnight Sun This time-lapse photograph was taken on July 19, 1985, at 69° north latitude in northeast Alaska. At this latitude, the Sun is above the horizon continuously (that is, it is circumpolar) from mid-May to the end of July. (Doug Plummer/Science Photo Library)

the greatest number of daylight hours. At the summer solstice the Sun does not set at all north of the Arctic Circle (Figure 2-18) and does not rise at all south of the Antarctic Circle.

The variations of the seasons are much less pronounced close to the equator. Between the **Tropic of Capricorn** at 23½° south latitude and the **Tropic of Cancer** at 23½° north latitude, the Sun is directly overhead—that is, at the zenith—at high noon at least one day a year. Outside of the tropics, the Sun is never directly overhead, but is always either south of the zenith (as seen from locations north of the Tropic of Cancer) or north of the zenith (as seen from south of the Tropic of Capricorn).

CONCEPTCHECK 2-7

How often each year does an observer standing on Earth's equator experience no shadow during the noontime sun?

CALCULATIONCHECK 2-2

Approximately how many days are there between the summer solstice and the March equinox?

Answers appear at the end of the chapter.

2-6 The Moon helps to cause precession, a slow, conical motion of Earth's axis of rotation

The Moon is by far the brightest and most obvious naked-eye object in the nighttime sky. Like the Sun, the Moon slowly changes its position relative to the background stars; unlike the Sun, the Moon makes a complete trip around the celestial sphere in only about 4 weeks, or about a month. (The word "month" comes from the same Old English root as the word "moon.") Ancient astronomers realized that this motion occurs because the Moon orbits Earth in roughly 4 weeks. In 1 hour, the Moon moves on the celestial sphere by about ½°, or roughly its own angular size.

Precession causes the apparent positions of the stars to slowly change over the centuries

The Moon's path on the celestial sphere is never far from the Sun's path (that is, the ecliptic). This is because the plane of the Moon's orbit around Earth is inclined only slightly from the plane of Earth's orbit around the Sun (the ecliptic plane shown in Figure 2-14a). The Moon's path varies somewhat from one month to the next, but always remains within a band called the **zodiac** that extends about 8° on either side of the ecliptic. Twelve famous constellations—Aries, Taurus, Gemini, Cancer, Leo, Virgo, Libra, Scorpius, Sagittarius, Capricornus, Aquarius, and Pisces—lie along the zodiac. The Moon is generally found in one of these 12 constellations. (Thanks to a redrawing of constellation boundaries in the mid-twentieth century, the zodiac actually passes through a thirteenth constellation—Ophiuchus, the Serpent Bearer—between Scorpius and Sagittarius.) As it moves along its orbit, the Moon appears north of the celestial equator for about two weeks and then south of the celestial equator for about the next two weeks. We will learn more about the Moon's motion, as well as why the Moon goes through phases, in Chapter 3.

The Moon not only moves around Earth but, in concert with the Sun, also causes a slow change in Earth's rotation. This is because both the Sun and the Moon exert a gravitational pull on Earth. We will learn much more about gravity in Chapter 4; for now, all we need is the idea that gravity is a universal attraction of matter for other matter.

The gravitational pull of the Sun and the Moon affects Earth's rotation because Earth is slightly fatter across the equator than it is from pole to pole: Its equatorial diameter is 43 kilometers (27 miles) larger than the diameter measured from pole to pole. Earth is therefore said to have an "equatorial bulge." Because of the gravitational pull of the Moon and the Sun on this bulge, the orientation of Earth's axis of rotation gradually changes, producing a motion called **precession.** Although the details are beyond the scope of this book, the combined actions of gravity and rotation cause Earth's axis to trace out a circle in the sky (Figure 2-19). As the axis precesses, it remains tilted about 23½° to the perpendicular.

As Earth's axis of rotation slowly changes its orientation, the north and south celestial poles—which are the projections of that axis onto the celestial sphere—change their positions relative to the stars. At present, the north celestial pole lies within 1° of the star Polaris, which is why Polaris is the North Star. But 5000 years ago, the north celestial pole was closest to the star Thuban in the constellation of Draco (the Dragon). Thus, that star and not

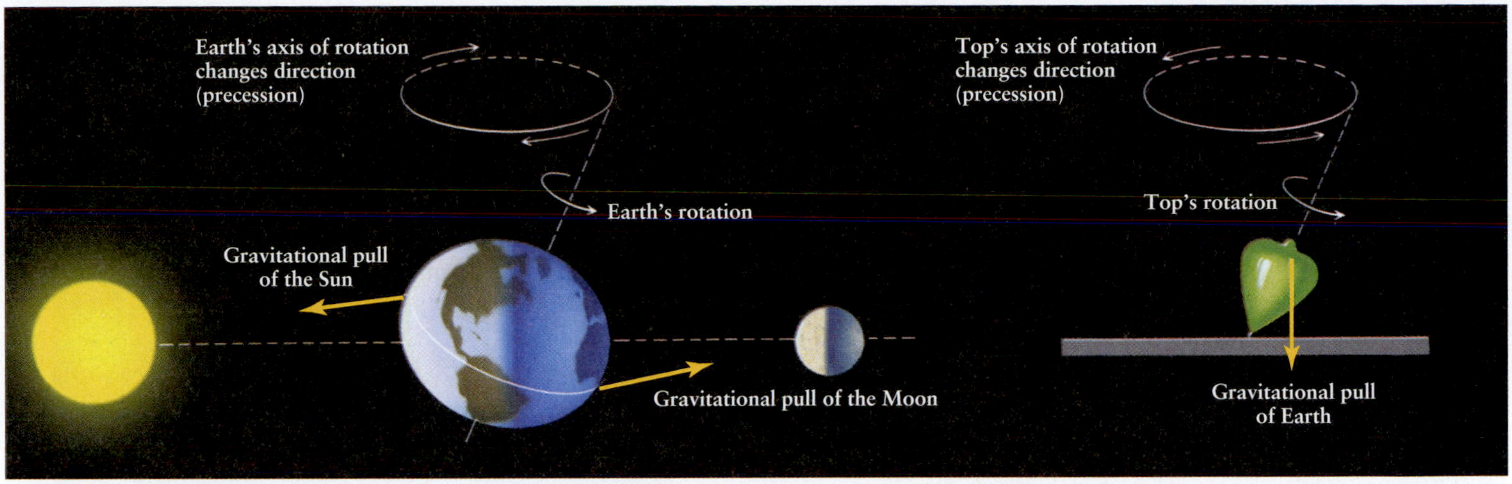

ANIMATION 2-6

FIGURE 2-19

Precession Because Earth's rotation axis is tilted, the gravitational pull of the Moon and the Sun on Earth's equatorial bulge together cause Earth to precess. As Earth precesses, its axis of rotation slowly traces out a circle in the sky, like the shifting axis of a spinning top.

Polaris was the North Star. And 12,000 years from now, the North Star will be the bright star Vega in Lyra (the Harp). It takes 26,000 years for the north celestial pole to complete one full precessional circle around the sky (Figure 2-20). The south celestial pole executes a similar circle in the southern sky.

Precession also causes Earth's equatorial plane to change its orientation. Because this plane defines the location of the celestial equator in the sky, the celestial equator precesses as well. The intersections of the celestial equator and the ecliptic define the equinoxes (see Figure 2-15), so these key locations in the sky also shift slowly from year to year. For this reason, the precession of Earth is also called the **precession of the equinoxes.** The first person to detect the precession of the equinoxes, in the second century B.C.E., was the Greek astronomer Hipparchus, who compared his own observations with those of Babylonian astronomers three centuries earlier. Today, the vernal equinox is located in the constellation Pisces (the Fishes). Two thousand years ago, it was in Aries (the Ram). Around the year 2600 C.E, the vernal equinox will move into Aquarius (the Water Bearer).

CAUTION! Astrological terms like the "Age of Aquarius" involve boundaries in the sky that are not recognized by astronomers and are generally not even related to the positions of the constellations. For example, most astrologers would call a person born on March 21, 1988, an "Aries" because the Sun was supposedly in the direction of that constellation on March 21. But due to precession, the Sun was actually in the constellation Pisces on that date! Indeed, astrology is not a science at all, but merely a collection of superstitions and hokum. Its practitioners use some of the terminology of astronomy but reject the logical thinking that is at the heart of science. James Randi has more to say about astrology and other pseudosciences in his essay "Why Astrology Is Not Science" at the end of this chapter.

The astronomer's system of locating heavenly bodies by their right ascension and declination, discussed in Box 2-1, is tied to the positions of the celestial equator and the vernal equinox. Because of precession, these positions are changing, and thus the coordinates of stars in the sky are also constantly changing. These changes are very small and gradual, but they add up over the years. To cope with this difficulty, astronomers always make

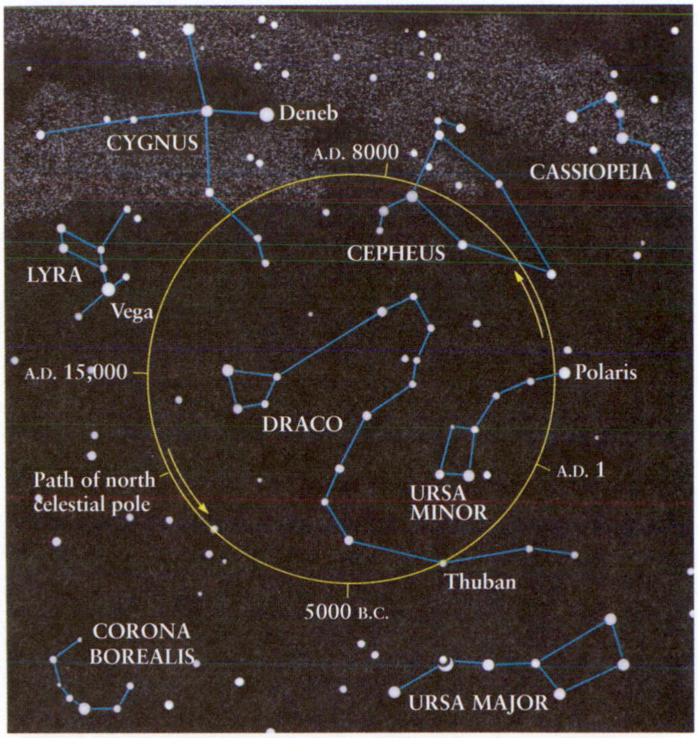

FIGURE 2-20

Precession and the Path of the North Celestial Pole As Earth precesses, the north celestial pole slowly traces out a circle among the northern constellations. At present, the north celestial pole is near the moderately bright star Polaris, which serves as the North Star. Twelve thousand years from now the bright star Vega will be the North Star.

note of the date (called the **epoch**) for which a particular set of coordinates is precisely correct. Consequently, star catalogs and star charts are periodically updated. Most current catalogs and star charts are prepared for the epoch 2000. The coordinates in these reference books, which are precise for January 1, 2000, will require very little correction over the next few decades.

2-7 Positional astronomy plays an important role in keeping track of time

Astronomers have traditionally been responsible for telling time. This is because we want the system of timekeeping used in everyday life to reflect the position of the Sun in the sky. Thousands of years ago, the sundial was invented to keep track of **apparent solar time.** To obtain more accurate measurements, astronomers use the **meridian.** As Figure 2-21 shows, this is a north-south circle on the celestial sphere that passes through the zenith (the point directly overhead) and both celestial poles. *Local noon* is defined as when the Sun crosses the **upper meridian,** which is the half of the meridian above the horizon. At *local midnight,* the Sun crosses the **lower meridian,** the half of the meridian below the horizon; this crossing cannot be observed directly.

Ancient scholars developed a system of timekeeping based on the Sun

The crossing of the meridian by any object in the sky is called a **meridian transit** of that object. If the crossing occurs above the horizon, it is an *upper* meridian transit. When an object makes a meridian transit, it is not necessarily at the zenith directly overhead, but is at the *highest point above the horizon* that you will see it from your location.

An **apparent solar day** is the interval between two successive upper meridian transits of the Sun as observed from any fixed spot on Earth. Stated less formally, an apparent solar day is the time from one local noon to the next local noon, or from when the Sun is highest in the sky to when it is again highest in the sky.

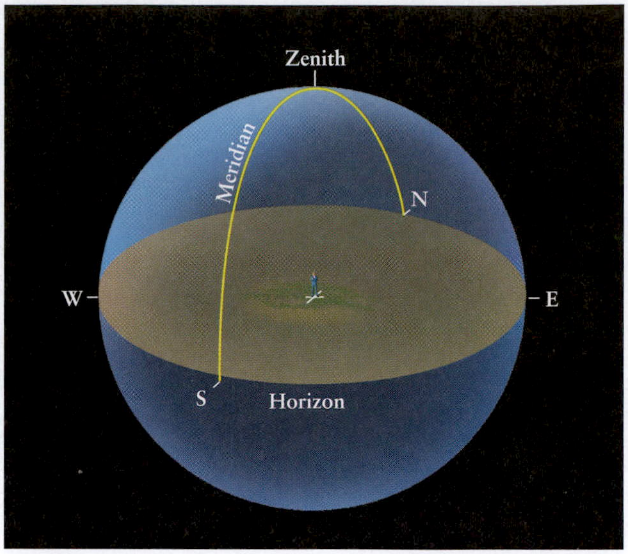

FIGURE 2-21
The Meridian The meridian is a circle on the celestial sphere that passes through the observer's zenith (the point directly overhead) and the north and south points on the observer's horizon. The passing of celestial objects across the meridian can be used to measure time. The upper meridian is the part above the horizon, and the lower meridian (not shown) is the part below the horizon.

The Sun as a Timekeeper

Unfortunately, the Sun is not a good timekeeper (Figure 2-22). The length of an apparent solar day (as measured by a device such as an hourglass) varies from one time of year to another. There are two main reasons why this is so, both having to do with the way in which Earth orbits the Sun.

The first reason is that Earth's orbit is not a perfect circle; rather, it is an ellipse, as Figure 2-22a shows in exaggerated form. As we will learn in Chapter 4, Earth moves more rapidly along its orbit when it is near the Sun than when it is farther away. Hence, the Sun appears to us to move more than 1° per day along the ecliptic in January, when Earth is nearest the Sun, and less than 1° per day in July, when Earth is farthest from the Sun. By itself, this effect would cause the apparent solar day to be longer in January than in July.

The second reason why the Sun is not a good timekeeper is the 23½° angle between the ecliptic and the celestial equator (see Figure 2-15). As Figure 2-22b shows, this causes a significant part of the Sun's apparent motion when near the equinoxes to be in a north-south direction. The net daily eastward progress in the sky is then somewhat foreshortened. At the summer and winter solstices, by contrast, the Sun's motion is parallel to the celestial equator. Thus, there is no comparable foreshortening around the beginning of summer or winter. This effect by itself would make the apparent solar day shorter in March and September than in June or December. Combining these effects with those due to Earth's noncircular orbit, we find that the length of the apparent solar day varies in a complicated fashion over the course of a year.

To avoid these difficulties, astronomers invented an imaginary object called the **mean sun** that moves along the celestial equator at a uniform rate. (In science and mathematics, "mean" is a synonym for "average.") The mean sun is sometimes slightly ahead of the real Sun in the sky, sometimes behind. As a result, mean solar time and apparent solar time can differ by as much as a quarter of an hour at certain times of the year.

Because the mean sun moves at a constant rate, it serves as a fine timekeeper. A **mean solar day** is the interval between successive upper meridian transits of the mean sun. It is exactly 24 hours long, the average length of an apparent solar day. One 24-hour day as measured by your alarm clock or wristwatch is a mean solar day.

Time zones were invented for convenience in commerce, transportation, and communication. In a time zone, all clocks and watches are set to the mean solar time for a meridian of longitude that runs approximately through the center of the zone. Time zones around the world are generally centered on meridians of longitude at 15° intervals. In most cases, going from one time zone to the next requires you to change the time on your wristwatch by exactly 1 hour. The time zones for most of North America are shown in Figure 2-23.

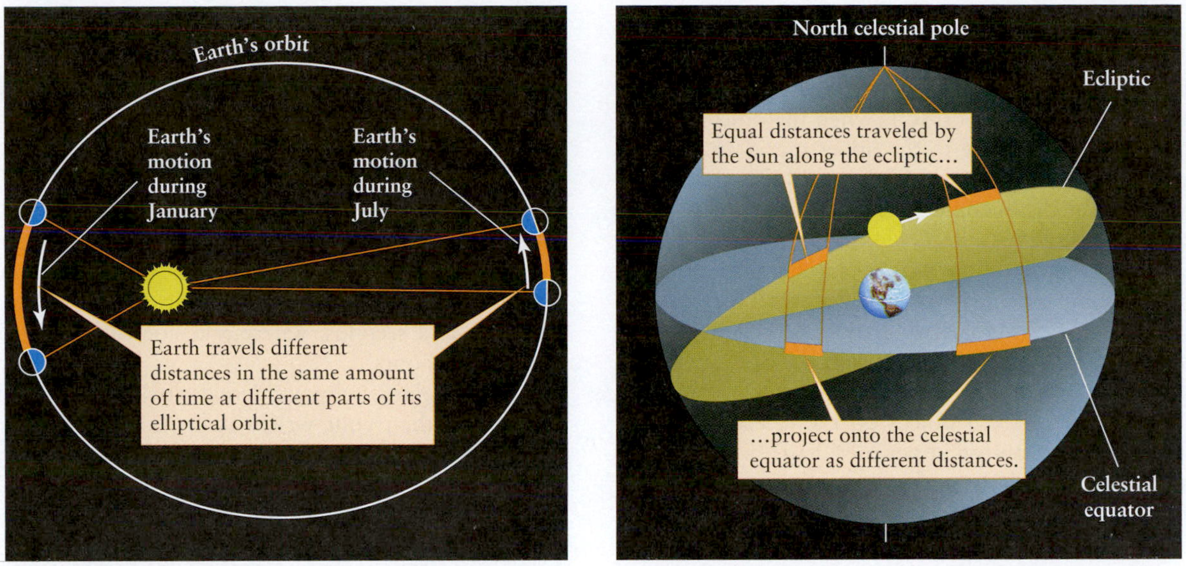

FIGURE 2-22

Why the Sun Is a Poor Timekeeper There are two main reasons that the Sun is a poor timekeeper. **(a)** Earth's speed along its orbit varies during the year. It moves fastest when closest to the Sun in January and slowest when farthest from the Sun in July. Hence, the apparent speed of the Sun along the ecliptic is not constant. **(b)** Because of the tilt of Earth's rotation axis, the ecliptic is inclined with respect to the celestial equator. Therefore, the projection of the Sun's daily progress (shown in orange) along the ecliptic onto the celestial equator (shown in blue) varies during the year. This causes further variations in the length of the apparent solar day.

In order to coordinate their observations with colleagues elsewhere around the globe, astronomers often keep track of time using Coordinated Universal Time, somewhat confusingly abbreviated UTC or UT. This is the time in a zone that includes Greenwich, England, a seaport just outside of London where the first internationally accepted time standard was kept. (UTC was formerly known as Greenwich Mean Time.) UTC is on a 24-hour system, with no A.M. or P.M. In North America, Eastern Standard Time (EST) is 5 hours different from UTC; 9:00 A.M. EST is 14:00 UTC. Coordinated Universal Time is also used by aviators and sailors, who regularly travel from one time zone to another.

Although it is natural to want our clocks and method of timekeeping to be related to the Sun, astronomers often use a system that is based on the apparent motion of the stars. This system, called **sidereal time**, is useful when aiming a telescope. Most observatories are therefore equipped with a clock that measures sidereal time, as discussed in Box 2-2.

2-8 Astronomical observations led to the development of the modern calendar

Our calendar is complex because a year does not contain a whole number of days

Just as the day is a natural unit of time based on Earth's rotation, the year is a natural unit of time based on Earth's revolution about the Sun. However, nature has not arranged things for our convenience. The year does not divide into exactly 365 whole days. Ancient astronomers realized that the length of a year is approximately 365¼ days, motivating the Roman emperor Julius Caesar to establish the system of "leap years" to account for this extra quarter of a day. By adding an extra day to the calendar every four years, he hoped to ensure that seasonal astronomical

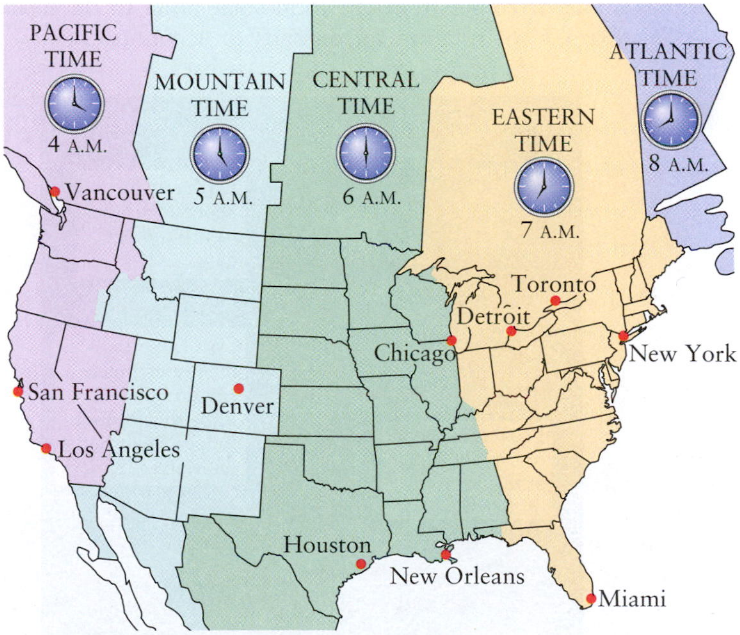

FIGURE 2-23

Time Zones in North America For convenience, Earth is divided into 24 time zones, generally centered on 15° intervals of longitude around the globe. There are four time zones across the continental United States, making for a 3-hour time difference between New York and California.

Sidereal Time

TUTORIAL 2-3

If you want to observe a particular object in the heavens, the ideal time to do so is when the object is high in the sky, on or close to the upper meridian.

Timing observations to occur when an object is high in the sky minimizes the distorting effects of Earth's atmosphere, which increase as you view objects closer to the horizon. For astronomers who study the Sun, this means making observations at local noon, which is not too different from noon as determined using mean solar time. For astronomers who observe planets, stars, or galaxies, however, the optimum time to observe depends on the particular object to be studied. Given the location of a given object on the celestial sphere, when will that object be on the upper meridian?

To answer this question, astronomers use *sidereal time* rather than solar time. It is different from the time on your wristwatch. In fact, a *sidereal clock* and an ordinary clock even tick at different rates, because they are based on different astronomical objects. Ordinary clocks are related to the position of the Sun, while sidereal clocks are based on the position of the vernal equinox, the location from which right ascension is measured. (See Box 2-1 for a discussion of right ascension.)

Regardless of where the Sun is, midnight sidereal time at your location is defined to be when the vernal equinox crosses your upper meridian. (Like solar time, sidereal time depends on where you are on Earth.) A **sidereal day** is the time between two successive upper meridian passages of the vernal equinox. By contrast, an apparent solar day is the time between two successive upper meridian crossings of the Sun. The illustration shows why these two kinds of day are not equal. Because Earth orbits the Sun, Earth must make one complete rotation plus about 1° to get from one local solar noon to the next. This extra 1° of rotation corresponds to 4 minutes of time,

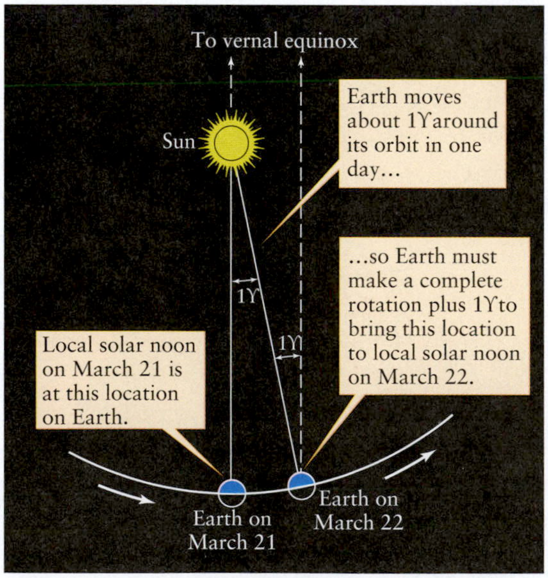

To vernal equinox
Sun
Earth moves about 1ϒ around its orbit in one day...
...so Earth must make a complete rotation plus 1ϒ to bring this location to local solar noon on March 22.
1ϒ
1ϒ
Local solar noon on March 21 is at this location on Earth.
Earth on March 21
Earth on March 22

ANIMATION 2-7

A month's motion of Earth along its orbit

which is the amount by which a solar day exceeds a sidereal day. To be precise:

$$1 \text{ sidereal day} = 23^{h}\,56^{m}\,4.091^{s}$$

where the hours, minutes, and seconds are in mean solar time.

One day according to your wristwatch is one mean solar day, which is exactly 24 hours of solar time long. A **sidereal clock** measures sidereal time in terms of sidereal hours, minutes, and seconds, where one sidereal day is divided into 24 sidereal hours.

This explains why a sidereal clock ticks at a slightly different rate than your wristwatch. As a result, at some times of the year a sidereal clock will show a very different time than an ordinary clock. (At local noon on March 21, when the Sun is at the vernal equinox, a sidereal clock will say that it is midnight. Do you see why?)

We can now answer the question in the opening paragraph. The vernal equinox, whose celestial coordinates are R.A. = $0^{h}\,0^{m}\,0^{s}$, Decl. = $0°\,0'\,0''$, crosses the upper meridian at midnight sidereal time (0:00). The autumnal equinox, which is on the opposite side of the celestial sphere at R.A. = $12^{h}\,0^{m}\,0^{s}$, Decl. = $0°\,0'\,0''$, crosses the upper meridian 12 sidereal hours later at noon sidereal time (12:00). As these examples illustrate, *any* object crosses the upper meridian when the sidereal time is equal to the object's right ascension. That is why astronomers measure right ascension in units of time rather than degrees, and why right ascension is always given in sidereal hours, minutes, and seconds.

EXAMPLE: Suppose you want to observe the bright star Spica (Figure 2-6), which has epoch 2000 coordinates R.A. = $13^{h}\,25^{m}\,11.6^{s}$, Decl. = $-11°\,9'\,41''$. What is the best time to do this?

Situation: Our goal is to find the time when Spica passes through your upper meridian, where it can best be observed.

Tools: We use the idea that a celestial object is on your upper meridian when the sidereal time equals the object's right ascension.

Answer: Based on the given right ascension of Spica, it will be best placed for observation when the sidereal time at your location is about 13:25. (Note that sidereal time is measured using a 24-hour clock.)

Review: By itself, our answer doesn't tell you what time on your wristwatch (which measures mean solar time) is best for observing Spica. That's why most observatories are equipped with a sidereal clock.

While sidereal time is extremely useful in astronomy, mean solar time is still the best method of timekeeping for most earthbound purposes. All time measurements in this book are expressed in mean solar time unless otherwise stated.

events, such as the beginning of spring, would occur on the same date year after year.

Caesar's system would have been perfect if the year were exactly 365¼ days long and if there were no precession. Unfortunately, this is not the case. To be more accurate, astronomers now use several different types of years. For example, the **sidereal year** is defined to be the time required for the Sun to return to the same position with respect to the stars. It is equal to 365.2564 mean solar days, or $365^{d}\ 6^{h}\ 9^{m}\ 10^{s}$.

The sidereal year is the orbital period of Earth around the Sun, but it is *not* the year on which we base our calendar. Like Caesar, most people want annual events—in particular, the first days of the seasons—to fall on the same date each year. For example, we want the first day of spring to occur on March 21. But spring begins when the Sun is at the vernal equinox, and the vernal equinox moves slowly against the background stars because of precession. Therefore, to set up a calendar we use the **tropical year,** which is equal to the time needed for the Sun to return to the vernal equinox. This period is equal to 365.2422 mean solar days, or $365^{d}\ 5^{h}\ 48^{m}\ 46^{s}$. Because of precession, the tropical year is 20 minutes and 24 seconds shorter than the sidereal year.

Caesar's assumption that the tropical year equals 365¼ days was off by 11 minutes and 14 seconds. This tiny error adds up to about three days every four centuries. Although Caesar's astronomical advisers were aware of the discrepancy, they felt that it was too small to matter. However, by the sixteenth century the first day of spring was occurring on March 11.

The Roman Catholic Church became concerned because Easter kept shifting to progressively earlier dates. To straighten things out, Pope Gregory XIII instituted a calendar reform in 1582. He began by dropping 10 days (October 4, 1582, was followed by October 15, 1582), which brought the first day of spring back to March 21. Next, he modified Caesar's system of leap years.

Caesar had added February 29 to every calendar year that is evenly divisible by four. Thus, for example, 2008, 2012, and 2016 are all leap years with 366 days. But we have seen that this system produces an error of about three days every four centuries. To solve the problem, Pope Gregory decreed that only the century years evenly divisible by 400 should be leap years. For example, the years 1700, 1800, and 1900 (which would have been leap years according to Caesar) were not leap years in the improved Gregorian system, but the year 2000, which can be divided evenly by 400, *was* a leap year.

We use the Gregorian system today. It assumes that the year is 365.2425 mean solar days long, which is very close to the true length of the tropical year. In fact, the error is only one day in every 3300 years, which won't cause any problems for a long time.

KEY WORDS

Terms preceded by an asterisk (*) are discussed in the Boxes.

Antarctic Circle, p. 33
apparent solar day, p. 36
apparent solar time, p. 36
Arctic Circle, p. 33
autumnal equinox, p. 32
celestial equator, p. 27
celestial sphere, p. 27
circumpolar, p. 28
constellation, p. 22
*declination, p. 30
diurnal motion, p. 24
ecliptic, p. 31
ecliptic plane, p. 31
epoch, p. 36
equinox, p. 32
lower meridian, p. 36
mean solar day, p. 36
mean sun, p. 36
meridian, p. 36
meridian transit, p. 36
north celestial pole, p. 27
positional astronomy, p. 22
precession, p. 34
precession of the equinoxes, p. 35
projection, p. 27
*right ascension, p. 30
*sidereal clock, p. 38
*sidereal day, p. 38
sidereal time, p. 37
sidereal year, p. 39
south celestial pole, p. 27
summer solstice, p. 33
time zone, p. 36
tropical year, p. 39
Tropic of Cancer, p. 34
Tropic of Capricorn, p. 34
upper meridian, p. 36
vernal equinox, p. 32
winter solstice, p. 33
zenith, p. 27
zodiac, p. 34

KEY IDEAS

Ideas preceded by an asterisk (*) are discussed in the Boxes.

Constellations and the Celestial Sphere: It is convenient to imagine the stars fixed to the celestial sphere with Earth at its center.

• The surface of the celestial sphere is divided into 88 regions called constellations.

Diurnal (Daily) Motion of the Celestial Sphere: The celestial sphere appears to rotate around Earth once in each 24-hour period. In fact, it is actually Earth that is rotating.

• The poles and equator of the celestial sphere are determined by extending the axis of rotation and the equatorial plane of Earth out to the celestial sphere.

• *The positions of objects on the celestial sphere are described by specifying their right ascension (in time units) and declination (in angular measure).

Seasons and the Tilt of Earth's Axis: Earth's axis of rotation is tilted at an angle of about 23½° from the perpendicular to the plane of Earth's orbit.

• The seasons are caused by the tilt of Earth's axis.

• Over the course of a year, the Sun appears to move around the celestial sphere along a path called the ecliptic. The ecliptic is inclined to the celestial equator by about 23½°.

• The ecliptic crosses the celestial equator at two points in the sky, the vernal and autumnal equinoxes. The northernmost point that the Sun reaches on the celestial sphere is the summer solstice, and the southernmost point is the winter solstice.

Precession: The orientation of Earth's axis of rotation changes slowly, a phenomenon called precession.

• Precession is caused by the gravitational pull of the Sun and Moon on Earth's equatorial bulge.

• Precession of Earth's axis causes the positions of the equinoxes and celestial poles to shift slowly.

• *Because the system of right ascension and declination is tied to the position of the vernal equinox, the date (or epoch) of observation must be specified when giving the position of an object in the sky.

Timekeeping: Astronomers use several different means of keeping time.

• Apparent solar time is based on the apparent motion of the Sun across the celestial sphere, which varies over the course of the year.

• Mean solar time is based on the motion of an imaginary mean sun along the celestial equator, which produces a uniform mean solar day of 24 hours. Ordinary watches and clocks measure mean solar time.

• *Sidereal time is based on the apparent motion of the celestial sphere.

The Calendar: The tropical year is the period between two passages of the Sun across the vernal equinox. Leap year corrections are needed because the tropical year is not exactly 365 days. The sidereal year is the actual orbital period of Earth.

QUESTIONS

Review Questions

1. Describe three structures or carvings made by past civilizations that show an understanding of astronomy.
2. How are constellations useful to astronomers? How many stars are not part of any constellation?
3. A fellow student tells you that only those stars in Figure 2-2b that are connected by blue lines are part of the constellation Orion. How would you respond?
4. Why is a particular star overhead at 10:00 P.M. on a given night rather than two hours later at midnight; how does this relate to Earth's orbit? Why are different stars overhead at midnight on June 1 rather than at midnight on December 1?
5. TUTORIAL 2-1 What is the celestial sphere? Why is this ancient concept still useful today?
6. Imagine that someone suggests sending a spacecraft to land on the surface of the celestial sphere. How would you respond to such a suggestion?
7. What is the celestial equator? How is it related to Earth's equator? How are the north and south celestial poles related to Earth's axis of rotation?
8. Where would you have to look to see your zenith? Where on Earth would you have to be for the celestial equator to pass through your zenith? Where on Earth would you have to be for the south celestial pole to be at your zenith?
9. How many degrees is the angle from the horizon to the zenith? Does your answer depend on what point on the horizon you choose?
10. Why can't a person in Antarctica use the Big Dipper to find the north direction?
11. Is there any place on Earth where you could see the north celestial pole on the northern horizon? If so, where? Is there any place on Earth where you could see the north celestial pole on the western horizon? If so, where? Explain your answers.
12. How do the stars appear to move over the course of the night as seen from the north pole? As seen from the equator? Why are these two motions different?
13. TUTORIAL 2-2 Using a diagram, explain why the tilt of Earth's axis relative to Earth's orbit causes the seasons as we orbit the Sun.
14. Give two reasons why it is warmer in summer than in winter.
15. What is the ecliptic plane? What is the ecliptic?
16. Why is the ecliptic tilted with respect to the celestial equator? Does the Sun appear to move along the ecliptic, the celestial equator, or neither? By about how many degrees does the Sun appear to move on the celestial sphere each day?
17. Where on Earth do you have to be in order to see the north celestial pole directly overhead? What is the maximum possible elevation of the Sun above the horizon at that location? On what date can this maximum elevation be observed?
18. What are the vernal and the autumnal equinoxes? What are the summer and winter solstices? How are these four points related to the ecliptic and the celestial equator?
19. At what point on the horizon does the vernal equinox rise? Where on the horizon does it set? (*Hint:* See Figure 2-16.)
20. How does the daily path of the Sun across the sky change with the seasons? Why does it change?
21. Where on Earth do you have to be in order to see the Sun at the zenith? As seen from such a location, will the Sun be at the zenith every day? Explain your reasoning.
22. What is precession of the equinoxes? What causes it? How long does it take for the vernal equinox to move 1° along the ecliptic?
23. What is the (fictitious) mean sun? What path does it follow on the celestial sphere? Why is it a better timekeeper than the actual Sun in the sky?
24. Why is it convenient to divide Earth into time zones?
25. Why is the time given by a sundial not necessarily the same as the time on your wristwatch?
26. What is the difference between the sidereal year and the tropical year? Why are these two kinds of year slightly different in length? Why are calendars based on the tropical year?
27. When is the next leap year? Was 2000 a leap year? Will 2100 be a leap year?

Advanced Questions

Questions preceded by an asterisk (*) are discussed in the Boxes.

Problem-solving tips and tools

To help you visualize the heavens, it is worth taking the time to become familiar with various types of star charts. These include the simple star charts at the end of this book, the monthly star charts published in such magazines as *Sky & Telescope* and *Astronomy*, and the more detailed maps of the heavens found in star atlases.

One of the best ways to understand the sky and its motions is to use the ***Starry Night***™ computer program on the CD-ROM that accompanies certain printed copies of this book. This easy-to-use program allows you to view the sky on any date and at any time, as seen from any point on Earth, and to animate the sky to visualize its diurnal and annual motions.

You may also find it useful to examine a planisphere, a device consisting of two rotatable disks. The bottom disk shows all the stars in the sky (for a particular latitude), and the top one is opaque with a transparent oval window through which only some of the stars can be seen. By rotating the top disk, you can immediately see which constellations are above the horizon at any time of the year. A planisphere is a convenient tool to carry with you when you are out observing the night sky.

28. On November 1 at 8:30 P.M. you look toward the eastern horizon and see the bright star Bellatrix (shown in Figure 2-2b) rising. At approximately what time will Bellatrix rise one week later, on November 8?

29. Figure 2-4 shows the situation on September 21, when Cygnus is highest in the sky at 8:00 P.M. local time and Andromeda is highest in the sky at midnight. But as Figure 2-5 shows, on July 21 Cygnus is highest in the sky at midnight. On July 21, at approximately what local time is Andromeda highest in the sky? Explain your reasoning.

30. Figure 2-5 shows which constellations are high in the sky (for observers in the northern hemisphere) in the months of July, September, and November. From this figure, would you be able to see Perseus at midnight on May 15? Draw a picture to justify your answer.

31. Figure 2-6 shows the appearance of Polaris, the Little Dipper, and the Big Dipper at 11 P.M. (daylight saving time) on August 1. Sketch how these objects would appear on this same date at (**a**) 8 P.M. and (**b**) 2 A.M. Include the horizon in your sketches, and indicate the north direction.

32. Figure 2-6 shows the appearance of the sky near the North Star at 11 P.M. (daylight saving time) on August 1. Explain why the sky has this same appearance at 1 A.M. on July 1 and at 9 P.M. on September 1.

33. The time-exposure photograph that opens this chapter shows the trails made by individual stars as the celestial sphere appears to rotate around Earth. (**a**) For approximately what length of time was the camera shutter left open to take this photograph? (**b**) The stars in this photograph (taken in Hawaii, at roughly 20° north latitude) appear to rotate around one of the celestial poles. Which celestial pole is it? As seen from this location, do the stars move clockwise or counterclockwise around this celestial pole? (**c**) If you were at 20° south latitude, which celestial pole could you see? In which direction would you look to see it? As seen from this location, do the stars move clockwise or counterclockwise around this celestial pole?

34. (**a**) Redraw Figure 2-10 for an observer at the north pole. (*Hint:* The north celestial pole is directly above this observer.) (**b**) Redraw Figure 2-10 for an observer at the equator. (*Hint:* The celestial equator passes through this observer's zenith.) (**c**) Using Figure 2-10 and your drawings from (a) and (b), justify the following rule, long used by navigators: The latitude of an observer in the northern hemisphere is equal to the angle in the sky between that observer's horizon and the north celestial pole. (**d**) State the rule that corresponds to (c) for an observer in the southern hemisphere.

35. The photograph that opens this chapter was taken next to the Gemini North Observatory atop Mauna Kea in Hawaii. The telescope is at longitude 155° 28′ 09″ west and latitude 19° 49′ 26″ north. (**a**) By making measurements on the photograph, find the approximate angular width and angular height of the photo. (**b**) How far (in degrees, arcminutes, and arcseconds) from the south celestial pole can a star be and still be circumpolar as seen from the Gemini North Observatory?

36. The Gemini North Observatory shown in the photograph that opens this chapter is located in Hawaii, roughly 20° north of the equator. Its near-twin, the Gemini South Observatory, is located roughly 30° south of the equator in Chile. Why is it useful to have telescopes in both the northern and southern hemispheres?

37. Is there any place on Earth where all the visible stars are circumpolar? If so, where? Is there any place on Earth where none of the visible stars is circumpolar? If so, where? Explain your answers.

38. This image of Earth was made by the *Galileo* spacecraft while en route to Jupiter. South America is at the center of the image and Antarctica is at the bottom of the image. (**a**) In which month of the year was this image made? Explain your reasoning. (**b**) When this image was made, was Earth relatively close to the Sun or relatively distant from the Sun? Explain your reasoning.

R I V U X G (NASA/JPL)

39. Figure 2-16 shows the daily path of the Sun across the sky on March 21, June 21, September 22, and December 21 for an observer at 35° north latitude. Sketch drawings of this kind for (**a**) an observer at 35° south latitude; (**b**) an observer at the equator; and (**c**) an observer at the north pole.

40. Suppose that you live at a latitude of 40° N. What is the elevation (angle) of the Sun above the southern horizon at noon (**a**) at the time of the vernal equinox? (**b**) at the time of the winter solstice? Explain your reasoning. Include a drawing as part of your explanation.

41. In the northern hemisphere, houses are designed to have "southern exposure," that is, with the largest windows on the southern side of the house. But in the southern hemisphere houses are designed to have "northern exposure." Why are houses designed this way, and why is there a difference between the hemispheres?.

42. The city of Mumbai (formerly Bombay) in India is 19° north of the equator. On how many days of the year, if any, is the Sun at the zenith at midday as seen from Mumbai? Explain your answer.

43. Ancient records show that 2000 years ago, the stars of the constellation Crux (the Southern Cross) were visible in the southern sky from Greece. Today, however, these stars cannot be seen from Greece. What accounts for this change?

44. The Great Pyramid at Giza has a tunnel that points toward the north celestial pole. At the time the pyramid was built, around 2600 B.C.E., toward which star did it point? Toward which star does this same tunnel point today? (See Figure 2-20.)

45. The photo shows a statue of the Greek god Atlas. The globe that Atlas is holding represents the celestial sphere, with depictions of several important constellations and the celestial equator. Although the statue dates from around 150 C.E., it has been proposed that the arrangement of constellations depicts the sky as it was mapped in an early star atlas that dates from 129 B.C.E. Explain the reasoning that could lead to such a proposal.

R I V U X G (Scala/Art Resource, NY)

46. Unlike western Europe, Imperial Russia did not use the revised calendar instituted by Pope Gregory XIII. Explain why the Russian Revolution, which started on November 7, 1917, according to the modern calendar, is called the October Revolution in Russia. What was this date according to the Russian calendar at the time? Explain your answer.

***47.** What is the right ascension of a star that is on the meridian at midnight at the time of the autumnal equinox? Explain your answer.

***48.** The coordinates on the celestial sphere of the summer solstice are R.A. = $6^h\,0^m\,0^s$, Decl. = $+23°\,27'$. What are the right ascension and declination of the winter solstice? Explain your answer.

***49.** Because 24 hours of right ascension takes you all the way around the celestial equator, $24^h = 360°$, what is the angle in the sky (measured in degrees) between a star with R.A. = $8^h\,0^m\,0^s$, Decl. = $0°\,0'\,0''$ and a second star with R.A. = $11^h\,20^m\,0^s$, Decl. = $0°\,0'\,0''$? Explain your answer.

***50.** On a certain night, the first star in Advanced Question 49 passes through the zenith at 12:30 A.M. local time. At what time will the second star pass through the zenith? Explain your answer.

***51.** At local noon on March 21, when the Sun is at the vernal equinox, a sidereal clock will say that it is midnight. Explain why.

***52.** (**a**) What is the sidereal time when the vernal equinox rises? (**b**) On what date is the sidereal time nearly equal to the solar time? Explain your answer.

***53.** How would the sidereal and solar days change (**a**) if Earth's rate of rotation increased, (**b**) if Earth's rate of rotation decreased, and (**c**) if Earth's rotation were retrograde (that is, if Earth rotated about its axis opposite to the direction in which it revolves about the Sun)?

Discussion Questions

54. Examine a list of the 88 constellations. Are there any constellations whose names obviously date from modern times? Where are these constellations located? Why do you suppose they do not have archaic names?

55. Describe how the seasons would be different if Earth's axis of rotation, rather than having its present 23½° tilt, were tilted (**a**) by 0° or (**b**) by 90°.

56. In William Shakespeare's *Julius Caesar* (act 3, scene 1), Caesar says:

> *But I am constant as the northern star,*
> *Of whose true-fix'd and resting quality*
> *There is no fellow in the firmament.*

Translate Caesar's statement about the "northern star" into modern astronomical language. Is the northern star truly "constant"? Was the northern star the same in Shakespeare's time (1564–1616) as it is today?

Web/eBook Questions

57. Search the World Wide Web for information about the national flags of Australia, New Zealand, and Brazil and the state flag of Alaska. Which stars are depicted on these flags? Explain any similarities or differences among these flags.

58. Some people say that on the date that the Sun is at the vernal equinox, and only on this date, you can stand a raw egg on end. Others say that there is nothing special about the vernal equinox, and that with patience you can stand a raw egg on end on any day of the year. Search the World Wide Web for information about this story and for hints about how to stand an egg on end. Use these hints to try the experiment yourself on a day when the Sun is *not* at the vernal equinox. What do you conclude about the connection between eggs and equinoxes?

59. Use the U.S. Naval Observatory Web site to find the times of sunset and sunrise on (**a**) your next birthday and (**b**) the date this assignment is due. (**c**) Are the times the same for the two dates? Explain why or why not.

ACTIVITIES

Observing Projects

Observing tips and tools

Moonlight is so bright that it interferes with seeing the stars. For the best view of the constellations, do your observing when the Moon is below the horizon. You can find the times of moonrise and moonset in your local newspaper or on the World Wide Web. Each monthly issue of the magazines *Sky & Telescope* and *Astronomy* includes much additional observing information.

60. On a clear, cloud-free night, use the star charts at the end of this book to see how many constellations of the zodiac you can identify. Which ones were easy to find? Which were difficult? Are the zodiacal constellations the most prominent ones in the sky?

61. Examine the star charts that are published monthly in such popular astronomy magazines as *Sky & Telescope* and *Astronomy.* How do they differ from the star charts at the end of this book? On a clear, cloud-free night, use one of these star charts to locate the celestial equator and the ecliptic. Note the inclination of the Milky Way to the ecliptic and celestial equator. The Milky Way traces out the plane of our galaxy. What do your observations tell you about the orientation of Earth and its orbit relative to the galaxy's plane?

62. Suppose you wake up before dawn and want to see which constellations are in the sky. Explain how the star charts at the end of this book can be quite useful, even though chart times are given only for the evening hours. Which chart most closely depicts the sky at 4:00 A.M. on the morning that this assignment is due? Set your alarm clock for 4:00 A.M. to see if you are correct.

63. Use *Starry Night*™ to observe the diurnal motion of the sky. First, set *Starry Night*™ to display the sky as seen from where you live, if you have not already done so. To do this, select **File > Set Home Location...** (**Starry Night > Set Home Location** on a Macintosh) and click on the **List** tab to find the name of your city or town. Highlight the name and note the latitude of your location as given in the list and click the **Save As Home Location** button. Select **Options > Other Options > Local Horizon...** from the menu. In the **Local Horizon Options** dialog box, click on the radio button labeled **Flat** in the **Horizon** style section and click **OK.** For viewers in the northern hemisphere, press the "N" key (or click the **N** button in the **Gaze** section of the toolbar) to set the gaze direction to the northern sky. If your location is in the southern hemisphere, press the "S" key (or click the **S** button in the Gaze section of the toolbar) to set the gaze direction to the south. Select **Hide Daylight** under the **View** menu to view the present sky without daylight. Select **View > Constellations > Astronomical** and **View > Constellations > Labels** to display the constellation patterns on the sky. In the toolbar, click on the **Time Flow Rate** control and set the time step to **1 minute.** Then click the **Play** button to run time forward. (The rapid motions of artificial Earth-orbiting satellites can prove irritating in this view. You can remove these satellites by clicking on **View > Solar System** and turning off **Satellites**). (**a**) Do the stars appear to rotate clockwise or counterclockwise? Explain this observation in terms of Earth's rotation. (**b**) Are any of the stars circumpolar, that is, do they stay above your horizon for the full 24 hours of a day? If some stars at your location are circumpolar, adjust time and locate a star that moves very close to the horizon during its diurnal motion. Click the **Stop** button and right-click (Ctrl-click on a Macintosh) on the star and then select **Show Info** from the contextual menu to open the **Info** pane. Expand the **Position in Sky** layer and note the star's declination (its N-S position on the sky with reference to a coordinate system whose zero value is the projection of Earth's equator). (**c**) How is this limiting declination, above which stars are circumpolar, related to your latitude, noted above?

    (i) The limiting declination is equal to the latitude of the observer's location.

    (ii) The limiting declination is equal to (90° – latitude).

    (iii) There is no relationship between this limiting declination and the observer's latitude.

    (**d**) Now center your field of view on the southern horizon (if you live in the northern hemisphere) or the northern horizon (if you live in the southern hemisphere) and click **Play** to resume time flow. Describe what you see. Are any of these stars circumpolar?

64. Use the *Starry Night*™ program to observe the Sun's motion on the celestial sphere. Select **Favourites > Explorations > Sun** from the menu. The view shows the entire celestial sphere as if you were at the center of a transparent Earth

on January 1, 2010. The view is centered upon the Sun and shows the ecliptic, the celestial equator, and the boundary and name of the constellation in which the Sun is located. With the **Time Flow Rate** set to **8 hours,** click the **Play** button. Observe the Sun for a full year of simulated time. The motion of Earth in its orbit causes this apparent motion. (**a**) How does the Sun appear to move against the background stars? (**b**) What path does the Sun follow and does it ever change direction? (**c**) Through which constellations does the Sun appear to move over the course of a full year? In the toolbar, click the **Now** button to go to the current date and time. (**d**) In which constellation is the Sun located today? The Sun (and therefore this constellation) is high in the sky at midday. (**e**) Approximately how long do you think it will take for this constellation to be high in the sky at midnight?

65. Use *Starry Night*™ to demonstrate the reason for seasonal variations on Earth at mid-latitudes. A common misconception is that summertime is warmer because Earth is closer to the Sun in the summer. The real reason is that the tilt of the spin axis of Earth to its orbital plane places the Sun at a higher angle in the sky in the summer than in the winter. Thus, sunlight hits Earth's surface at a less oblique angle in summertime than in winter, thereby depositing greater heat. Open **Favourites > Explorations > Seasonal Variations** to view the southern sky in daylight from Calgary, Canada, at a latitude of 51°N, on December 21, 2013, at 12:38 P.M., local standard time, when the Sun is at its highest angle on that day. Form a table of values of Sun altitude and Sun–Earth distance as a function of date in the year. To find these values, move the cursor over the Sun to reveal the **Info** panel. (Ensure that the relevant information is displayed by opening **File > Preferences > Cursor Tracking (HUD)** and clicking on **Altitude** and **Distance from Observer**.) Note the date, Sun altitude, and distance in your table. Advance the **Date** to March 21, 2014, and then to June 21, 2014, noting the values of these parameters. (**a**) On which of the three dates, in winter, spring, and summer respectively, is the Sun at the highest altitude in the Calgary sky? (**b**) From the values in your table, what is the Sun altitude at midday in December in Calgary? (**c**) How does the Sun altitude at midday on March 21 relate to the latitude of Calgary?

(i) The Sun's altitude is equal to the latitude of Calgary.

(ii) The Sun's altitude is 90° minus the latitude of Calgary.

(iii) The Sun's altitude is not related to the latitude of the location of the observer.

(**d**) On which of these three observing dates is Earth closest to the Sun? (**e**) On which of these three dates is Earth farthest from the Sun?

ANSWERS

ConceptChecks

ConceptCheck 2-1: No, Jupiter does not need to be one of the stars in the asterism that makes the outline of the bull's body. Jupiter would only need to be within the somewhat rectangular boundary of the constellation of Taurus.

ConceptCheck 2-2: Earth rotates from west to east. As locations on Earth rotate from the dark, nighttime side of the planet into the bright, daytime side of the planet, the easternmost cities experience sunrise first. New York is the farthest east of the cities listed, so the sun rises there first.

ConceptCheck 2-3: The Sun would still only rise and set once each day (since that defines what a day is), but each year would have 3 times as many days ($365 \times 3 = 1095$ days each year), and each day would be 3 times shorter.

ConceptCheck 2-4: The celestial equator is a projection, or an extension, of Earth's equator out into the sky. In order for this imaginary line to pass directly overhead, one would need to be standing somewhere on Earth's equator.

ConceptCheck 2-5: No, they would not; in this imaginary scenario, even near the north pole the Sun would rise and set every 12 hours all year long. Earth's tilted axis means that observers near Earth's north pole will experience six months when the Sun never rises (when it is tilted away from the Sun) and then six months when the Sun never sets (when it is tilted toward the Sun).

ConceptCheck 2-6: It takes one year. The Sun slowly moves through the sky a little each day, taking 365¼ days to return to the same place it was one year earlier.

ConceptCheck 2-7: To cast no shadow, the Sun must be directly overhead. As the Sun's position on the celestial sphere slowly moves back and forth between the northern and southern solstice points over the course of a year, the noontime Sun will be directly overhead (and will cast no shadow) for an observer at Earth's equator only twice each year, on the March and September equinoxes.

CalculationChecks

CalculationCheck 2-1: Earth rotates once in 24 hours, so when the stars of Cygnus rise in the east at sunset, they take about 12 hours to go from one side of the sky to the other. As a result, it takes about one half of that time, or 6 hours, for stars of Cygnus to move halfway across the sky. Thus, Cygnus will be highest in the sky around midnight.

CalculationCheck 2-2: The summer solstice occurs on about June 21 and the March equinox occurs about March 21, so there are about 9 months or 270 days between these two events.

Why Astrology Is Not Science by James Randi

I'm involved in the strange business of telling folks what they should already know. I meet audiences who believe in all sorts of impossible things, often despite their education and intelligence. My job is to explain how science differs from the unproven, illogical assumptions of pseudoscience—and why it matters. Perhaps my best example is the difference between astronomy and astrology.

Both astrology and astronomy arose from the wonders of the night sky, from the stars to comets, planets, the Sun, and the Moon. Surely, humans have long reasoned, there must be some meaning in their motions. Surely the Moon's effect on tides hints at hidden "causes" for strange events. *Judiciary* (literally "judging") astrology therefore attempted to foretell the future—our earthly future. To serve it, *horary* (literally "hourly") astrology carefully tracked the heavens.

It is the latter that has become astronomy. Thanks to its process of careful measurement and testing, we now understand more about the true nature of the starry universe than astrologers could ever have imagined. With the birth of a new science, astronomers had a logical framework based on physical causes and systematic observations.

Astrology remains a popular delusion. Far too many believe today that patterns in the sky govern our lives. They accept the vague tendencies and portents of seers who cast horoscopes. They shouldn't. Just a glance at the tenets of astrology provides ample evidence of its absurdity.

An individual is said to be born under a sign. To the astrologer, the Sun was located "in" that sign at the moment of birth. (Stars are not seen in the daytime, but no matter—a calculation tells where the Sun is.) Each sign takes its name from a constellation, a totally imaginary figure invented for our convenience in referring to stars. Different cultures have different mythical figures up there, and so different schools of astrology assign different meanings to the signs they use.

In the spirit of equal-opportunity swindling, astrologers divide up the year fairly, ignoring variations in the size of constellations. Since Libra is tiny, while Virgo is huge, they chop some of the sky off Virgo and add it—along with bits of Scorpio—to bring Libra up to size. The Sun could well be declared "in" Libra when it is actually outside that constellation.

It gets worse. Science constantly challenges itself and changes. The rules of astrology could not, although they were made up thousands of years ago, and since then the "fixed" stars have moved. In particular, precession of the equinoxes has shifted objects in the sky relative to our calendar. The constellations have changed but astrology has not. If you were born August 7, you are said to be a Leo, but the Sun that day was really in the same part of the sky as the constellation Cancer.

With a theory like this to back it up, we should not be surprised at the bottom line: *A pseudoscience does not work*. Test after test has checked its predictions, and the result is always the same. One such investigator is Shawn Carlson of the University of California, San Diego. As he put it in *Nature* magazine, astrology is "a hopeless cause." Johannes Kepler, the pioneering astronomer, himself cast horoscopes, but they are little remembered today. Owen Gingerich, a historian of science at Harvard, puts it well: Kepler was the astrologer who destroyed astrology.

Astronomy works—it works very well indeed—which isn't easy. Because we humans tend to find what we want in any body of data, it takes science's careful process of observation, creative insight, and critical thinking to understand and predict changes in nature. As I write, a transit of Ganymede is due next Thursday at 21:47:20. At exactly that time, the satellite of Jupiter will cross in front of its planet as seen from Earth, and yet most of us will never know it. Still other moons of Jupiter may hold fresh clues to the formation of our entire solar system and the conditions for life elsewhere.

For most people, astronomy has too little fantasy or money in it, and they will never experience the beauty in its predictions. The dedicated labors of generations of scientists have enabled us to perform a genuine wonder.

James ("the Amazing") Randi works tirelessly to expose trickery so that others can relish the greater wonder of science. As a magician, he has had his own television show and an enormous public following. As a lecturer, he addresses teachers, students, and others worldwide. His newsletter and column for *The Skeptic* are key resources for educators. His many books include *Flim-Flam!, The Faith Healers,* and *The Mask of Nostradamus,* about a legendary con man with secrets of his own.

Mr. Randi is the founder of the James Randi Educational Foundation, and his one million dollar prize for "the performance of any paranormal event . . . under proper observing conditions" has gone unclaimed for more than 25 years. An amateur archeologist and astronomer as well, he lives in Florida with several untalented parrots and the occasional visiting magus.

CHAPTER

3

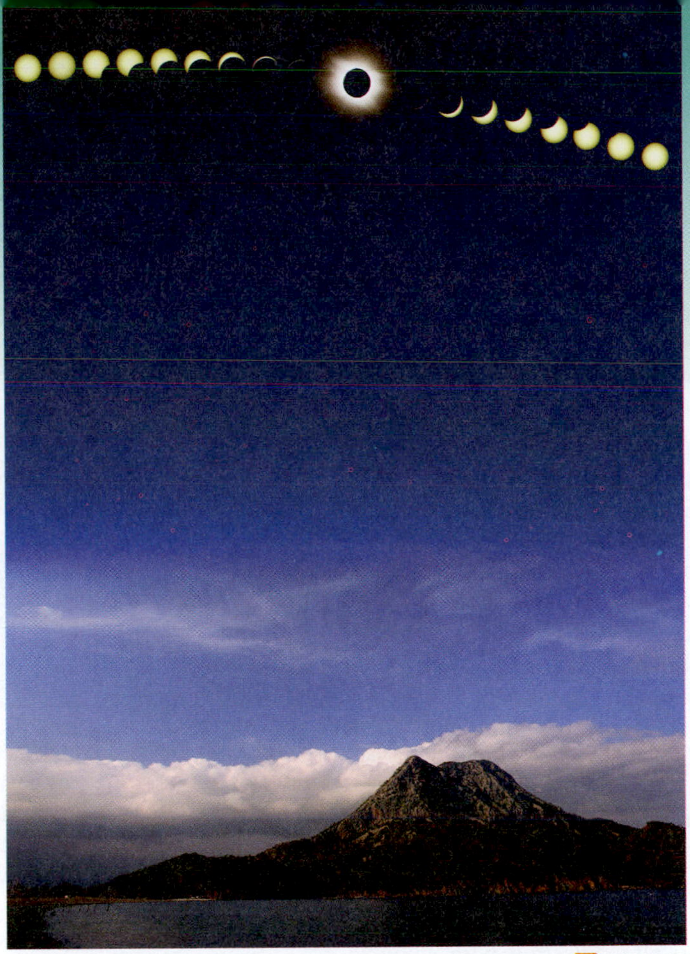

The Sun in total eclipse, March 29, 2006. R I V U X G
(Stefan Seip)

Eclipses and the Motion of the Moon

LEARNING GOALS

By reading the sections of this chapter, you will learn

3-1 Why we see the Moon go through phases

3-2 Why we always see the same side of the Moon

3-3 The differences between lunar and solar eclipses

3-4 Why not all lunar eclipses are total eclipses

3-5 Why solar eclipses are visible only from certain special locations on Earth

3-6 How ancient Greek astronomers deduced the sizes of Earth, the Moon, and the Sun

On March 29, 2006, a rare cosmic spectacle—a total solar eclipse—was visible along a narrow corridor that extended from the coast of Brazil through equatorial Africa and into central Asia. As shown in this digital composite taken in Turkey, the Moon slowly moved over the disk of the Sun. (Time flows from left to right in this image.) For a few brief minutes the Sun was totally covered, darkening the sky and revealing the Sun's thin outer atmosphere, or corona, which glows with an unearthly pearlescent light.

Such eclipses can be seen only on specific dates from special locations on Earth, so not everyone will ever see the Moon cover the Sun in this way. But anyone can find the Moon in the sky and observe how its appearance changes from night to night, from new moon to full moon and back again, and how the times when the Moon rises and sets differ noticeably from one night to the next.

In this chapter our subject is how the Moon moves as seen from Earth. We will explore why the Moon goes through a regular cycle of phases, and how the Moon's orbit around Earth leads to solar eclipses as well as lunar eclipses. We will also see how ancient astronomers used their observations of the Moon to determine the size and shape of Earth, as well as other features of the solar system. Thus, the Moon—which has always loomed large in the minds of poets, lovers, and dreamers—has also played a key role in the development of our modern picture of the universe.

3-1 The phases of the Moon are caused by its orbital motion

You can tell the Moon's position relative to Earth and the Sun by observing its phase

As seen from Earth, both the Sun and the Moon appear to move from west to east on the celestial sphere—that is, relative to the background of stars—but they move at very different rates. The Sun takes one year to make a complete trip around the imaginary celestial sphere along the path we call the *ecliptic* (Section 2-5). By comparison, the Moon takes only about four weeks. In the past, these similar motions led people to believe that both the Sun and the Moon orbit around Earth. We now know that only the Moon orbits Earth, while the Earth-Moon system as a whole (Figure 3-1) orbits the Sun. (In Chapter 4 we will learn how this was discovered.)

One key difference between the Sun and the Moon is the nature of the light that we receive from them. The Sun emits its own light. So do the stars, which are objects like the Sun but much farther away, and so does an ordinary lightbulb. By contrast, the light that we see from the Moon is reflected light. This is sunlight that has struck the Moon's surface, bounced off, and ended up in our eyes here on Earth.

CAUTION! You probably associate *reflection* with shiny objects like a mirror or the surface of a still lake. In science, however, the term refers to light bouncing off any object. You see most objects around you by reflected light. When you look at your hand, for example, you are seeing light from the Sun (or from a light fixture) that has been reflected from the skin of your hand and into your eyes. In the same way, moonlight is really sunlight that has been reflected by the Moon's surface.

Understanding the Moon's Phases

TUTORIAL 3-1 Figure 3-1 shows both the Moon and Earth as seen from a spacecraft. When this image was recorded, the Sun was far off to the right. Hence, only the right-hand hemispheres of both worlds were illuminated by the Sun; the left-hand hemispheres were in darkness and are not visible in the picture. In the same way, when we view the Moon from Earth, we see only the half of the Moon that faces the Sun and is illuminated. However, not all of the illuminated half of the Moon is necessarily facing us. As the Moon moves around Earth, from one night to the next, we see different amounts of the illuminated half of the Moon. These different appearances of the Moon are called **lunar phases.**

Figure 3-2 shows the relationship between the lunar phase visible from Earth and the position of the Moon in its orbit. For example, when the Moon is at position A, we see it in roughly the same direction in the sky as the Sun. Hence, the dark hemisphere of the Moon faces Earth. This phase, in which the Moon is barely visible for a day, is called **new moon.** Since a new moon can only be located near the Sun in the sky, it rises around sunrise, and sets around sunset.

As the Moon continues around its orbit from position A in Figure 3-2, more of its illuminated half becomes exposed to our view. The result, shown at position B, is a phase called **waxing crescent moon** ("waxing" is a synonym for "increasing"). About a week after new moon, the Moon is at position C; for a day we

FIGURE 3-1 R I V U X G
Earth and the Moon This picture of Earth and the Moon was taken in 1992 by the *Galileo* spacecraft on its way toward Jupiter. The Sun, which provides the illumination for both Earth and the Moon, was far to the right and out of the camera's field of view when this photograph was taken. (NASA/JPL)

then see half of the Moon's illuminated hemisphere and half of the dark hemisphere. This phase is called **first quarter moon.**

As seen from Earth, a first quarter moon is one-quarter of the way around the celestial sphere from the Sun. It rises and sets about one-quarter of an Earth rotation, or 6 hours, after the Sun does: Moonrise occurs around noon, moonset occurs around midnight, and the moon is highest in the sky around 6 P.M.

CAUTION! Despite the name, a first quarter moon appears to be *half* illuminated, not one-quarter illuminated! The name means that this phase is one-quarter of the way through the complete cycle of lunar phases.

About four days later, the Moon reaches position D in Figure 3-2. Still more of the illuminated hemisphere can now be seen from Earth, giving us the phase called **waxing gibbous moon** ("gibbous" is another word for "swollen"). When you look at the Moon in this phase, as in the waxing crescent and first quarter phases, the illuminated part of the Moon is toward the west. Two weeks after new moon, when the Moon stands opposite the Sun in the sky (position E), we see the fully illuminated hemisphere. This phase is called **full moon.** Because a full moon is opposite the Sun on the celestial sphere, it rises at sunset and sets at sunrise.

Over the following two weeks, we see less and less of the Moon's illuminated hemisphere as it continues along its orbit, and the Moon is said to be *waning* ("decreasing"). While the Moon is waning, its illuminated side is toward the east. The phases are called **waning gibbous moon** (position F), **third quarter moon** (position G, also called *last quarter moon*), and **waning crescent moon** (position H). A third quarter moon appears one-quarter of the way around the celestial sphere from the Sun, but on the opposite side of the celestial sphere from a first quarter moon. Hence, a third quarter moon rises and sets about one-quarter

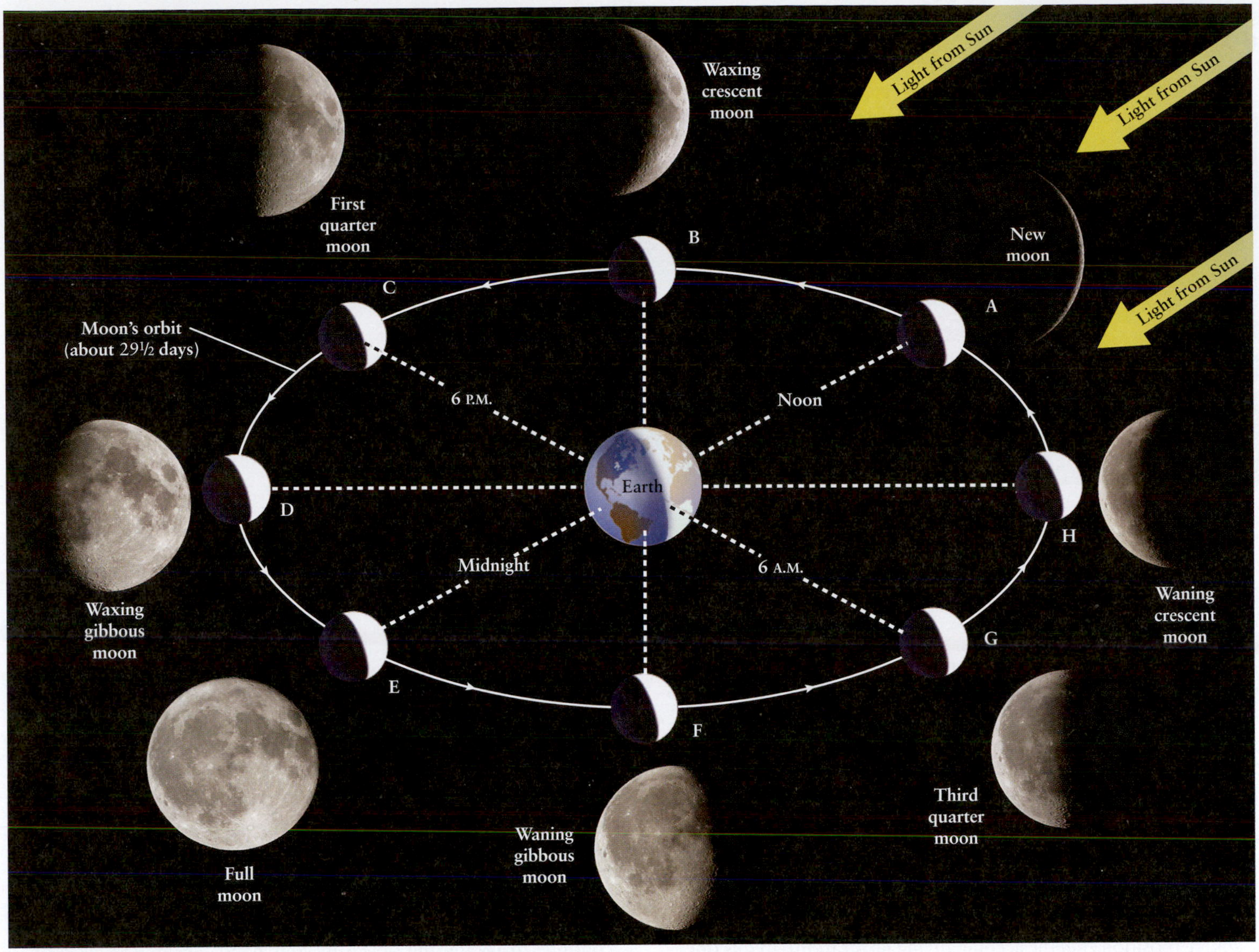

FIGURE 3-2

Why the Moon Goes Through Phases This figure illustrates the Moon at eight positions on its orbit, along with photographs of what the Moon looks like at each position as seen from Earth. The changes in phase occur because light from the Sun illuminates one half of the Moon, and as the Moon orbits Earth we see varying amounts of the Moon's illuminated half. It takes about 29½ days for the Moon to go through a complete cycle of phases, and each of the phases shown is visible for one day during the complete cycle. The times given in the figure for each phase correspond to the time each phase is observed highest in the sky, halfway between its rise and set. (Photographs from Larry Landolfi/Science Source)

Earth rotation, or 6 hours, *before* the Sun: Moonrise is around midnight and moonset is around noon.

The Moon takes about four weeks to complete one orbit around Earth, so it likewise takes about four weeks for a complete cycle of phases from new moon to full moon and back to new moon. Since the Moon's position relative to the Sun on the celestial sphere is constantly changing, and since our system of timekeeping is based on the Sun (see Section 2-7), the times of moonrise and moonset are different on different nights. On average, the Moon rises and sets about an hour later each night.

Figure 3-2 also explains why the Moon is often visible in the daytime, as shown in Figure 3-3. From any location on Earth, about half of the Moon's orbit is visible at any time. For example, if it is midnight at your location, you are in the middle of the dark side of Earth that faces away from the Sun. At that time you can easily see the Moon if it is at position C, D, E, F, or G. If it is midday at your location, you are in the middle of Earth's illuminated side, and the Moon will be easily visible if it is at position A, B, C, G, or H. (The Moon is so bright that it can be seen even against the bright blue sky.) You can see that the Moon is prominent in the midnight sky for about half of its orbit, and prominent in the midday sky for the other half.

CAUTION! A very common misconception about lunar phases is that they are caused by the shadow of *Earth* falling on the Moon. As Figure 3-2 shows, this is not the case at all. Instead, phases are simply the result of our seeing the illuminated half of the Moon at different angles as the Moon moves around its orbit. To help you better visualize how this works, Box 3-1 describes how you can simulate the cycle shown in Figure 3-2 using ordinary objects on Earth. (As we will learn in Section 3-3, Earth's shadow does indeed fall on the Moon on rare occasions. When this happens, we see a lunar eclipse.)

FIGURE 3-3 R I V U X G
The Moon During the Day The Moon can be seen during the daytime as well as at night. The time of day or night when it is visible depends on its phase. (Karl Beath/Gallo Images)

BOX 3-1 ASTRONOMY DOWN TO EARTH

Phases and Shadows

Figure 3-2 shows how the relative positions of Earth, the Moon, and the Sun explain the phases of the Moon. You can visualize lunar phases more clearly by doing a simple experiment here on Earth. All you need are a small round object, such as an orange or a baseball, and a bright source of light, such as a street lamp or the Sun.

In this experiment, you play the role of an observer on Earth looking at the Moon, and the round object plays the role of the Moon. The light source plays the role of the Sun. Hold the object in your right hand with your right arm stretched straight out in front of you, with the object directly between you and the light source (position A in the accompanying illustration). In this orientation the illuminated half of the object faces away from you, like the Moon when it is in its new phase (position A in Figure 3-2).

Now, slowly turn your body to the left so that the object in your hand "orbits" around you (toward positions C, E, and G in the illustration). As you turn, more and more of the illuminated side of the "moon" in your hand becomes visible, and it goes through the same cycle of phases—waxing crescent, first quarter, and waxing gibbous—as does the real Moon. When you have rotated through half a turn so that the light source is directly behind you, you will be looking face on at the illuminated side of the object in your hand. This corresponds to a full moon (position E in Figure 3-2). Make sure your body does not cast a shadow on the "moon" in your hand—that would correspond to a lunar eclipse!

As you continue turning to the left, more of the unilluminated half of the object becomes visible as its phase moves through waning gibbous, third quarter, and waning crescent. When your body has rotated back to the same orientation that you were in originally, the unilluminated half of your handheld "moon" is again facing toward you, and its phase is again new. If you continue to rotate, the object in your hand repeats the cycle of "phases," just as the Moon does as it orbits around Earth.

The experiment works best when there is just one light source around. If there are several light sources, such as in a room with several lamps turned on, the different sources will create multiple shadows, and it will be difficult to see the phases of your handheld "moon." If you do the experiment outdoors using sunlight, you may find that it is best to perform it in the early morning or late afternoon, when shadows are most pronounced and the Sun's rays are nearly horizontal.

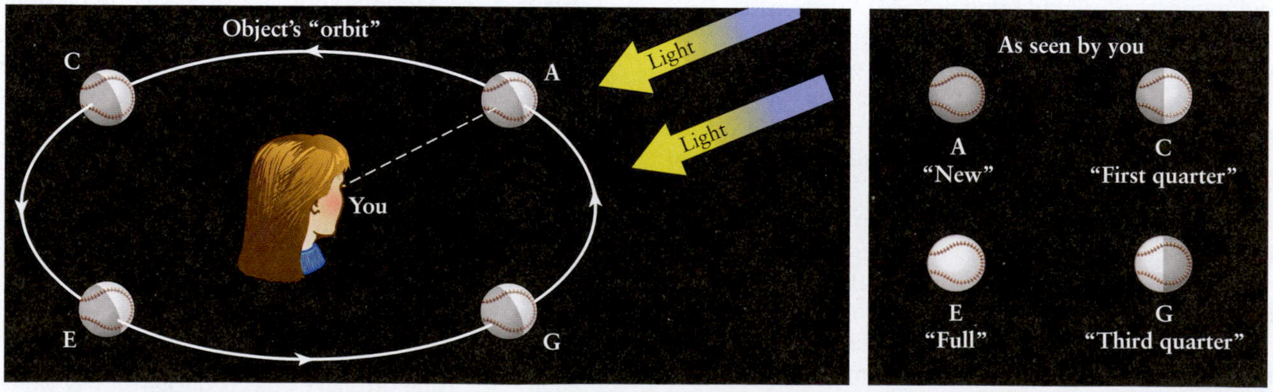

CONCEPTCHECK 3-1

If an observer on Earth sees just a tiny sliver of the crescent moon, how much of the Moon's total surface is being illuminated by the Sun?

CONCEPTCHECK 3-2

If the Moon appears in its waxing crescent phase, how will it appear in two weeks?

Answers appear at the end of the chapter.

3-2 The Moon always keeps the same face toward Earth

The Moon rotates in a special way: It spins exactly once per orbit

Although the phase of the Moon is constantly changing, one aspect of its appearance remains the same: It always keeps essentially the same hemisphere, or face, toward Earth. Thus, you will always see the same craters and mountains on the Moon, no matter when you look at it; the only difference will be the angle at which these surface features are illuminated by the Sun. (You can verify this by carefully examining the photographs of the Moon in Figure 3-2.)

The Moon's Synchronous Rotation

Why is it that we only ever see one face of the Moon? You might think that it is because the Moon does not rotate (unlike Earth, which rotates around an axis that passes from its north pole to its south pole). To see that the Moon must rotate, consider Figure 3-4. This figure shows Earth and the orbiting Moon from a vantage point far above Earth's north pole. In this figure two craters on the lunar surface have been colored, one in red and one in blue. If the Moon did not rotate on its axis, as in Figure 3-4a, sometimes the red crater would be visible from Earth, while at other times the blue crater would be visible. Thus, we would see different parts of the lunar surface over time, which does not happen in reality.

In fact, the Moon always keeps the same face toward us because it *is* rotating, but in a very special way: It takes exactly as long for the Moon to rotate on its axis as it does to make one orbit around Earth. This situation is called **synchronous rotation.** As Figure 3-4b shows, this keeps the crater shown in red always facing Earth, so that we always see the same face of the Moon. In Chapter 4 we will learn why the Moon's rotation and orbital motion are in step with each other.

Is there a permanently "dark side of the Moon?" Not at all. To understand this, consider the red crater in Figure 3-4b. The red crater would spend two weeks (half of a lunar orbit) in darkness, and the next two weeks in sunlight. Thus, no part of the Moon

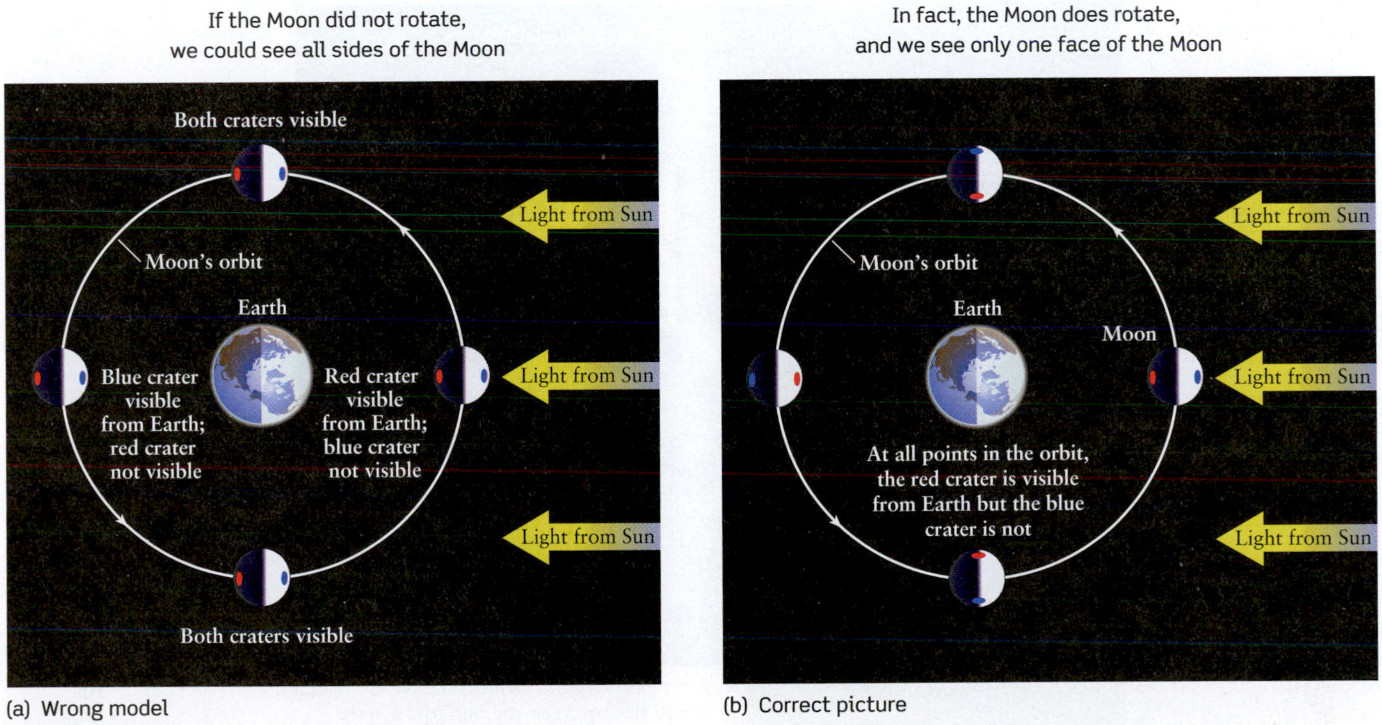

ANIMATION 3-2

FIGURE 3-4

The Moon's Rotation These diagrams show the Moon at four points in its orbit as viewed from high above Earth's north pole. **(a)** The wrong model: If the Moon did not rotate, then at various times the red crater would be visible from Earth while at other times the blue crater would be visible. Over a complete orbit, the entire surface of the Moon would be visible. **(b)** In reality, the Moon rotates on its north-south axis. Because the Moon makes one rotation in exactly the same time that it makes one orbit around Earth, we see only one face of the Moon.

is perpetually in darkness. The side of the Moon that constantly faces away from Earth is properly called the *far* side.

CONCEPTCHECK 3-3

Viewed from space, the far side of the Moon is fully dark at only one point during the lunar orbit. Can you identify this point in Figure 3-4b?

CONCEPTCHECK 3-4

If astronauts landed on the Moon near the center of the visible surface at full moon, how many Earth days would pass before the astronauts experienced darkness on the Moon?

Answers appear at the end of the chapter.

Sidereal and Synodic Months

It takes about four weeks for the Moon to complete one cycle of its phases as seen from Earth. This regular cycle of phases inspired our ancestors to invent the concept of a month. For historical reasons, the calendar we use today has months of differing lengths. Astronomers find it useful to define two other types of months, depending on whether the Moon's motion is measured relative to the stars or to the Sun. Neither corresponds exactly to the familiar months of the calendar.

The **sidereal month** is the time it takes the Moon to complete one full orbit of Earth, as measured *with respect to the stars*. The sidereal period is what you would observe while hovering in space (like the stars), watching the Moon orbit Earth. This true orbital period is equal to about 27.32 days. The **synodic month,** or *lunar month,* is the time it takes the Moon to complete one cycle of phases (that is, from new moon to new moon or from full moon to full moon) and thus is measured *with respect to the Sun* rather than the stars.

How long is a "day" on the Moon? On Earth, a 24-hour day measures the average time between successive sunrises (or sunsets). Therefore, a day on Earth is measured with respect to the Sun and is a synodic day. The lunar "day" is also the time from sunrise to sunrise as seen from the Moon's surface, and this is just what defines the Moon's synodic month; a lunar day is equal to a synodic month.

The synodic month is longer than the sidereal month because Earth is orbiting the Sun while the Moon goes through its phases. As Figure 3-5 shows, the Moon must travel *more* than 360° along its orbit to complete a cycle of phases (for example, from one new moon to the next). Because of this extra distance, the synodic month is equal to about 29.53 days, about two days longer than the sidereal month.

Both the sidereal month and synodic month vary somewhat from one orbit to another, the latter by as much as half a day. The reason is that the Sun's gravity sometimes causes the

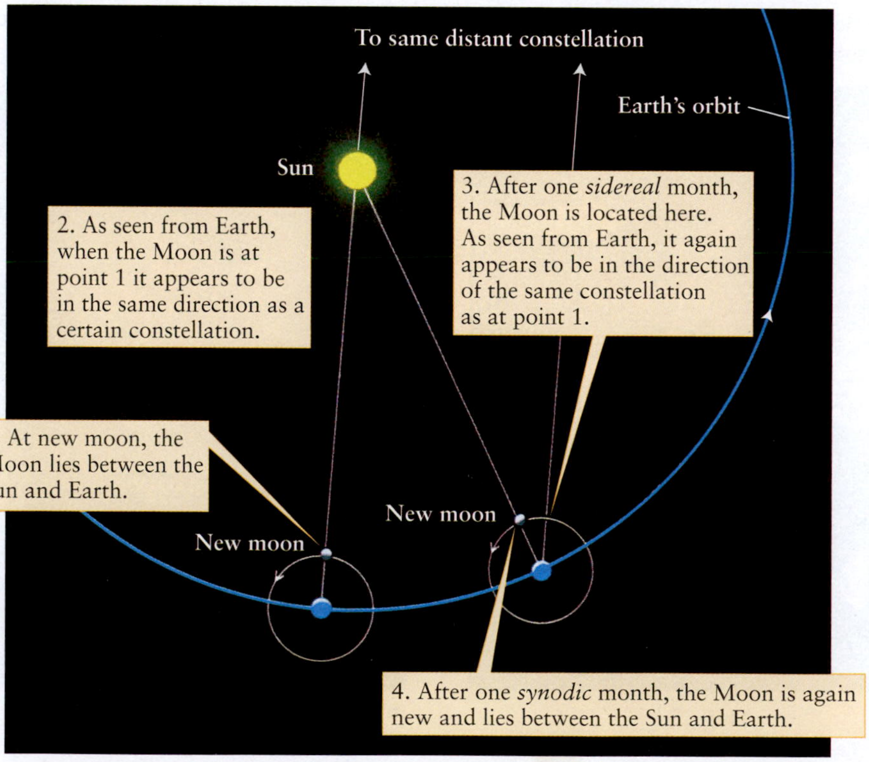
To same distant constellation
Earth's orbit
Sun
2. As seen from Earth, when the Moon is at point 1 it appears to be in the same direction as a certain constellation.
3. After one *sidereal* month, the Moon is located here. As seen from Earth, it again appears to be in the direction of the same constellation as at point 1.
1. At new moon, the Moon lies between the Sun and Earth.
New moon
New moon
4. After one *synodic* month, the Moon is again new and lies between the Sun and Earth.

ANIMATION 3-3

FIGURE 3-5

The Sidereal and Synodic Months The sidereal month is the time the Moon takes to complete one full revolution around Earth with respect to the background stars. However, because Earth is constantly moving along its orbit about the Sun, the Moon must travel through slightly more than 360° of its orbit to get from one new moon to the next. Thus, the synodic month—the time from one new moon to the next—is longer than the sidereal month.

Moon to speed up or slow down slightly in its orbit, depending on the relative positions of the Sun, Moon, and Earth. Furthermore, the Moon's orbit changes slightly from one month to the next.

CONCEPTCHECK 3-5

If Earth was orbiting the Sun much faster than it is now, would the length of time between full moons increase, decrease, or stay the same?

Answer appears at the end of the chapter.

3-3 Eclipses occur only when the Sun and Moon are both on the line of nodes

TUTORIAL 3-2

The tilt of the Moon's orbit makes lunar and solar eclipses rare events

From time to time the Sun, Earth, and Moon all happen to lie along a straight line. When this occurs, the shadow of Earth can fall on the Moon or the shadow of the Moon can fall on Earth. Such phenomena are called **eclipses.** They are perhaps the most dramatic astronomical events that can be seen with the naked eye.

A **lunar eclipse** occurs when the Moon passes through Earth's shadow. This occurs when the Sun, Earth, and Moon are in a straight line, with Earth between the Sun and Moon so that the Moon is at full phase (position E in Figure 3-2). At this point in the Moon's orbit, the face of the Moon seen from Earth would normally be fully illuminated by the Sun. Instead, it appears quite dim because Earth casts a shadow on the Moon.

A **solar eclipse** occurs when Earth passes through the Moon's shadow. As seen from Earth, the Moon moves in front of the Sun. Once again, this can happen only when the Sun, Moon, and Earth are in a straight line. However, for a solar eclipse to occur, the Moon must be between Earth and the Sun. Therefore, a solar eclipse can occur only at new moon (position A in Figure 3-2).

CAUTION! Both new moon and full moon occur at intervals of 29½ days. Hence, you might expect that there would be a solar eclipse every 29½ days, followed by a lunar eclipse about two weeks (half a lunar orbit) later. But in fact, there are only a few solar eclipses and lunar eclipses per year. Solar and lunar eclipses are so infrequent because the plane of the Moon's orbit and the plane of Earth's orbit are not exactly aligned, as Figure 3-6 shows. The angle between the plane of Earth's orbit and the plane of the Moon's orbit is about 5°. Because of this tilt, new moon and full moon usually occur when the Moon is either above or below the plane of Earth's orbit. When the Moon is not in the plane of Earth's orbit, the Sun, Moon, and Earth cannot align perfectly, and an eclipse cannot occur.

In order for the Sun, Earth, and Moon to be lined up for an eclipse, the Moon must lie in the same plane as Earth's orbit around the Sun. As we saw in Section 2-5, this plane is called the *ecliptic plane* because it is the same as the plane of the Sun's apparent path around the sky, or ecliptic (see Figure 2-14). Thus, when an eclipse occurs, the Moon appears from Earth to be on the ecliptic—this is how the ecliptic gets its name.

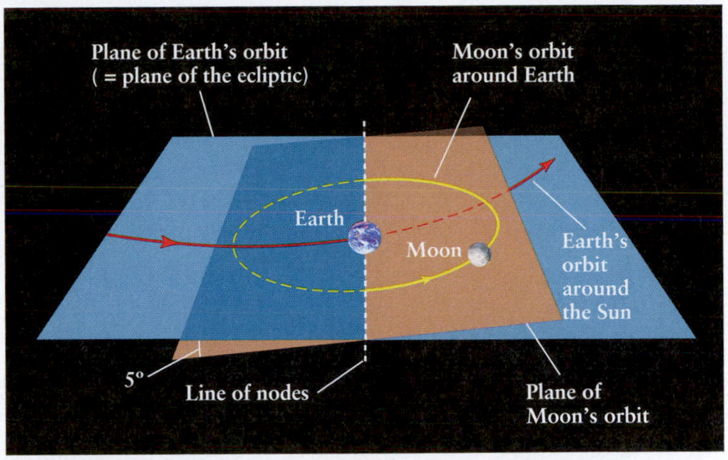

FIGURE 3-6

The Inclination of the Moon's Orbit This drawing shows the Moon's orbit around Earth (in yellow) and part of Earth's orbit around the Sun (in red). The plane of the Moon's orbit (shown in brown) is tilted by about 5° with respect to the plane of Earth's orbit, also called the plane of the ecliptic (shown in blue). These two planes intersect along a line called the line of nodes.

The planes of Earth's orbit and the Moon's orbit intersect along a line called the **line of nodes,** shown in Figure 3-6. The line of nodes passes through Earth and points in a particular direction in space. Eclipses can occur only if the line of nodes points toward the Sun—that is, if the Sun lies on or near the line of nodes—and if, at the same time, the Moon lies on or very near the line of nodes. Only then do the Sun, Earth, and Moon lie in a line straight enough for an eclipse to occur (Figure 3-7). Note that the alignment necessary for eclipses has nothing to do with either the solstices or the equinoxes.

LOOKING DEEPER 3.1

Anyone who wants to predict eclipses must know the orientation of the line of nodes. But the line of nodes is gradually shifting because of the gravitational pull of the Sun on the Moon. As a result, the line of nodes rotates slowly westward. Astronomers calculate such details to fix the dates and times of upcoming eclipses.

There are at least two—but never more than five—solar eclipses each year. The last year in which five solar eclipses occurred was 1935. The fewest eclipses possible (two solar, zero lunar) happened in 1969. Lunar eclipses occur just about as frequently as solar eclipses, but the maximum possible number of eclipses (lunar and solar combined) in a single year is seven.

CONCEPTCHECK 3-6

Why don't lunar eclipses occur each time the Moon reaches full moon phase?

Answer appears at the end of the chapter.

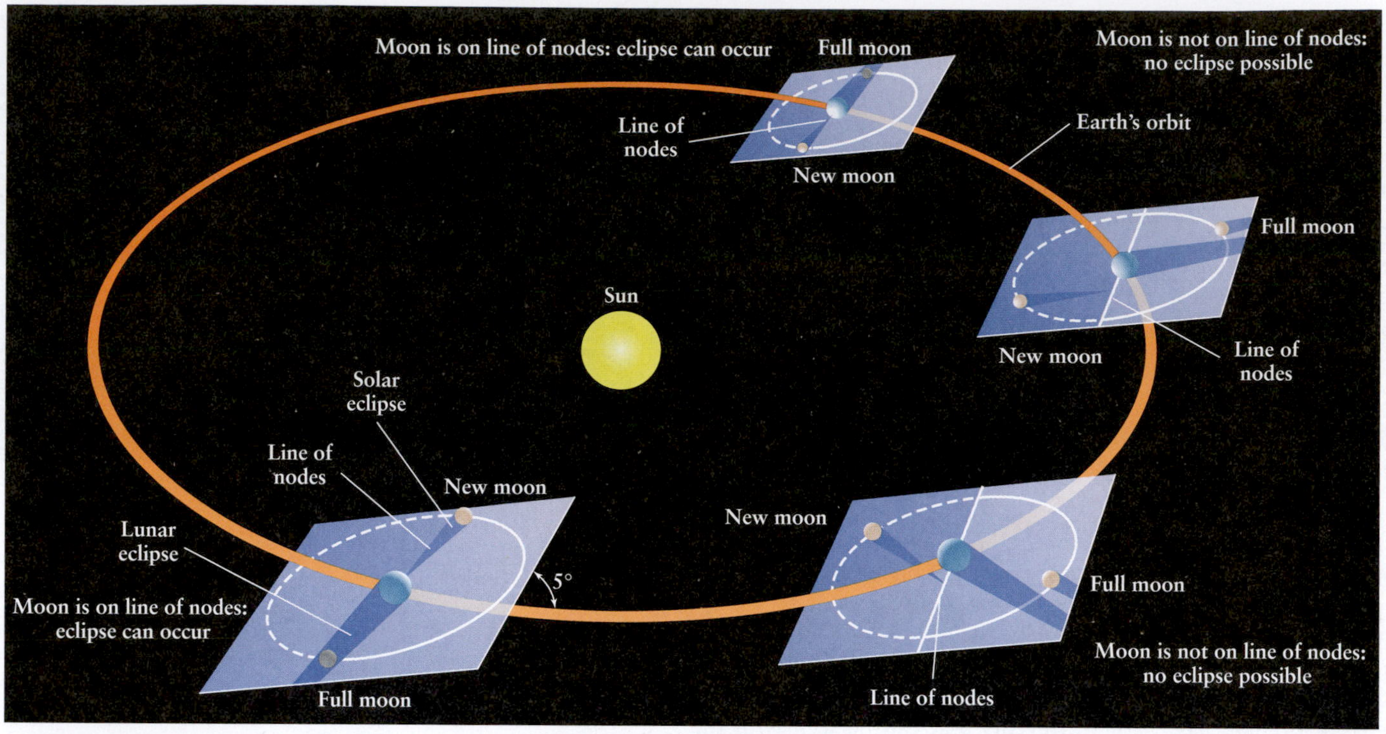

ANIMATION 3-2

FIGURE 3-7

Conditions for Eclipses Eclipses can take place only if the Sun and Moon are both very near to or on the line of nodes. Only then can the Sun, Earth, and Moon all lie along a straight line. A solar eclipse occurs only if the Moon is very near the line of nodes at new moon; a lunar eclipse occurs only if the Moon is very near the line of nodes at full moon. If the Sun and Moon are not near the line of nodes, the Moon's shadow cannot fall on Earth and Earth's shadow cannot fall on the Moon.

3-4 The character of a lunar eclipse depends on the alignment of the Sun, Earth, and Moon

The Sun's tenuous outer atmosphere is revealed during a total solar eclipse

The character of a lunar eclipse depends on exactly how the Moon travels through Earth's shadow. As Figure 3-8 shows, the shadow of Earth has two distinct parts. To picture how the Moon's surface would be illuminated, imagine an observer on the Moon as the Moon passes through the shadow. In the **umbra,** no portion of the Sun's surface can be seen from the Moon: This is the darkest part of the shadow. On the other hand, a portion of the Sun's surface is visible in the **penumbra,** which therefore is not quite as dark. Most people notice a lunar eclipse only if the Moon passes into Earth's

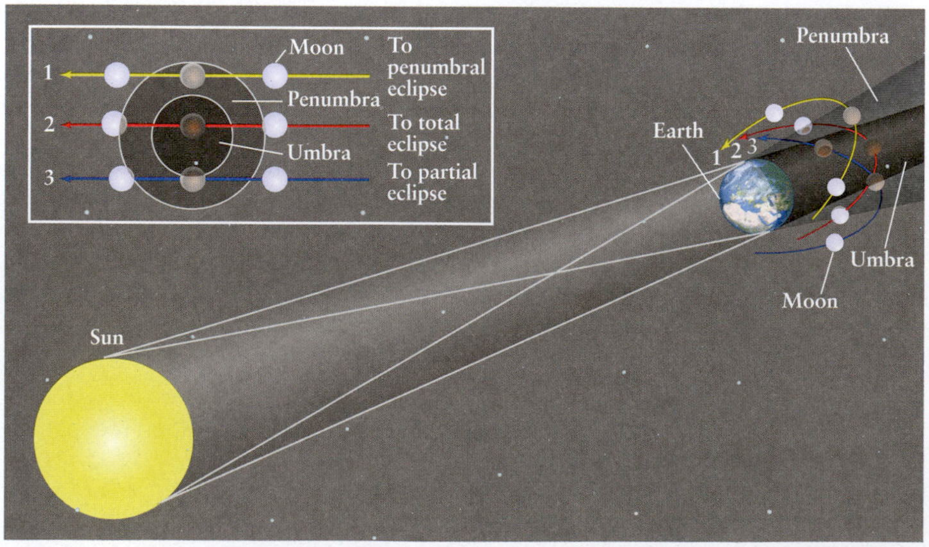

FIGURE 3-8

Three Types of Lunar Eclipse People on the nighttime side of Earth see a lunar eclipse when the Moon moves through Earth's shadow. In the umbra, the darkest part of the shadow, the Sun is completely covered by Earth. The penumbra is less dark because only part of the Sun is covered by Earth. The three paths show the motion of the Moon if the lunar eclipse is penumbral (Path 1, in yellow), total (Path 2, in red), or partial (Path 3, in blue). The inset shows these same paths, along with the umbra and penumbra, as viewed from Earth.

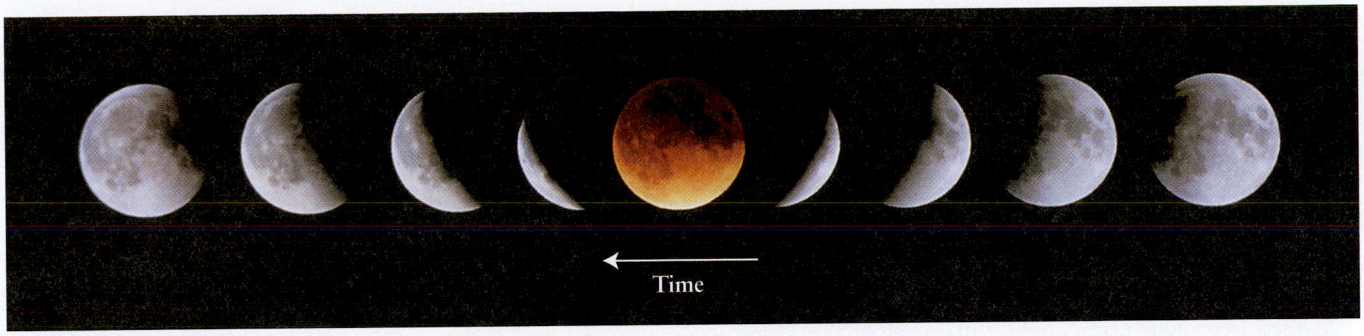

FIGURE 3-9 R I V U X G

A Total Lunar Eclipse This sequence of nine photographs was taken over a 3-hour period during the lunar eclipse of January 20, 2000. The sequence, which runs from right to left, shows the Moon moving through Earth's umbra. During the total phase of the eclipse (shown in the center), the Moon has a distinct reddish color. (Courtesy of Fred Espenak, www.mreclipse.com)

umbra. As this umbral phase of the eclipse begins, a bite appears to have been taken out of the Moon.

The inset in Figure 3-8 shows the different ways in which the Moon can pass into Earth's shadow. When the Moon passes through only Earth's penumbra (Path 1), we see a **penumbral eclipse.** During a penumbral eclipse, Earth blocks only part of the Sun's light and so none of the lunar surface is completely shaded. Because the Moon still looks full but only a little dimmer than usual, penumbral eclipses are easy to miss. If the Moon travels completely into the umbra (Path 2), **a total lunar eclipse** occurs. If only part of the Moon passes through the umbra (Path 3), we see a **partial lunar eclipse.**

If you were on the Moon during a total lunar eclipse, the Sun would be hidden behind Earth. But some sunlight would be visible through the thin ring of atmosphere around Earth, just as you can see sunlight through a person's hair if they stand with their head between your eyes and the Sun. As a result, a small amount of light reaches the Moon during a total lunar eclipse, and so the Moon does not completely disappear from the sky as seen from Earth. Most of the sunlight that passes through Earth's atmosphere is red, and thus the eclipsed Moon glows faintly in reddish hues, as Figure 3-9 shows. (We'll see in Chapter 5 how our atmosphere causes this reddish color.)

Lunar eclipses occur at full moon, when the Moon is directly opposite the Sun in the sky. Hence, a lunar eclipse can be seen at any place on Earth where the Sun is below the horizon (that is, where it is nighttime). A lunar eclipse has the maximum possible duration if the Moon travels directly through the center of the umbra. The Moon's speed through Earth's shadow is roughly 1 kilometer per second (3600 kilometers per hour, or 2280 miles per hour), which means that **totality**—the period when the Moon is completely within Earth's umbra—can last for as long as 1 hour and 42 minutes.

On average, two or three lunar eclipses occur in a year. Table 3-1 lists all nine lunar eclipses from 2012 to 2015. Of all lunar eclipses, roughly one-third are total, one-third are partial, and one-third are penumbral.

TABLE 3-1 Lunar Eclipses, 2012–2015

Date	Type	Where visible	Duration of totality (h = hours, m = minutes)
2012 June 4	Partial	Asia, Australia, Pacific, Americas	—
2012 Nov 28	Penumbral	Europe, eastern Africa, Asia, Australia, Pacific, North America	—
2013 Apr 25	Penumbral	Europe, Africa, Asia, Australia	—
2013 May 25	Penumbral	Americas, Africa	—
2013 Oct 18	Penumbral	Americas, Europe, Africa, Asia	—
2014 Apr 15	Total	Australia, Pacific, Americas	01h 18m
2014 Oct 08	Total	Asia, Australia, Pacific, Americas	59m
2015 Apr 04	Total	Asia, Australia, Pacific, Americas	5m
2015 Sep 28	Total	East Pacific, Americas, Europe, Africa, western Asia	1h 12m

Eclipse predictions by Fred Espenak, NASA/Goddard Space Flight Center.

CONCEPTCHECK 3-7

Is it possible to have a total lunar eclipse without a penumbral eclipse?

Answer appears at the end of the chapter.

3-5 Solar eclipses also depend on the alignment of the Sun, Earth, and Moon

TUTORIAL 3-3

As seen from Earth, the angular diameter of the Moon is almost exactly the same as the angular diameter of the far larger but more distant Sun—about 0.5°. Thanks to this coincidence of nature, the Moon just "fits" over the Sun during a **total solar eclipse.**

Total Solar Eclipses

A lunar eclipse is most impressive when total, but can also be partial or penumbral

A total solar eclipse is a dramatic event. The sky begins to darken, the air temperature falls, and winds increase as the Moon gradually covers more and more of the Sun's disk. All nature responds: Birds go to roost, flowers close their petals, and crickets begin to chirp as if evening had arrived. As the last few rays of sunlight peek out from behind the edge of the Moon and the eclipse becomes total, the landscape around you is bathed in an eerie gray or, less frequently, in shimmering bands of light and dark. Finally, for a few minutes the Moon completely blocks out the dazzling solar disk (Figure 3-10a). The **solar corona**—the Sun's thin, hot outer atmosphere, which is normally too dim to be seen—blazes forth in the darkened daytime sky (Figure 3-10b). It is an awe-inspiring sight.

CAUTION! If you are fortunate enough to see a solar eclipse, keep in mind that the only time when it is safe to look at the Sun is during **totality,** when the solar disk is blocked by the Moon and only the solar corona is visible. Viewing this magnificent spectacle cannot harm you in any way. But you must *never* look directly at the Sun when even a portion of its intensely brilliant disk is exposed. *If you look directly at the Sun at any time without a special filter approved for solar viewing, you will suffer permanent eye damage or blindness.*

To see the remarkable spectacle of a total solar eclipse, you must be inside the darkest part of the Moon's shadow, also called the umbra, where the Moon completely blocks the Sun. Because the Sun and the Moon have nearly the same angular diameter as seen from Earth, only the tip of the Moon's umbra reaches Earth's surface (Figure 3-11). As Earth rotates, the tip of the umbra traces

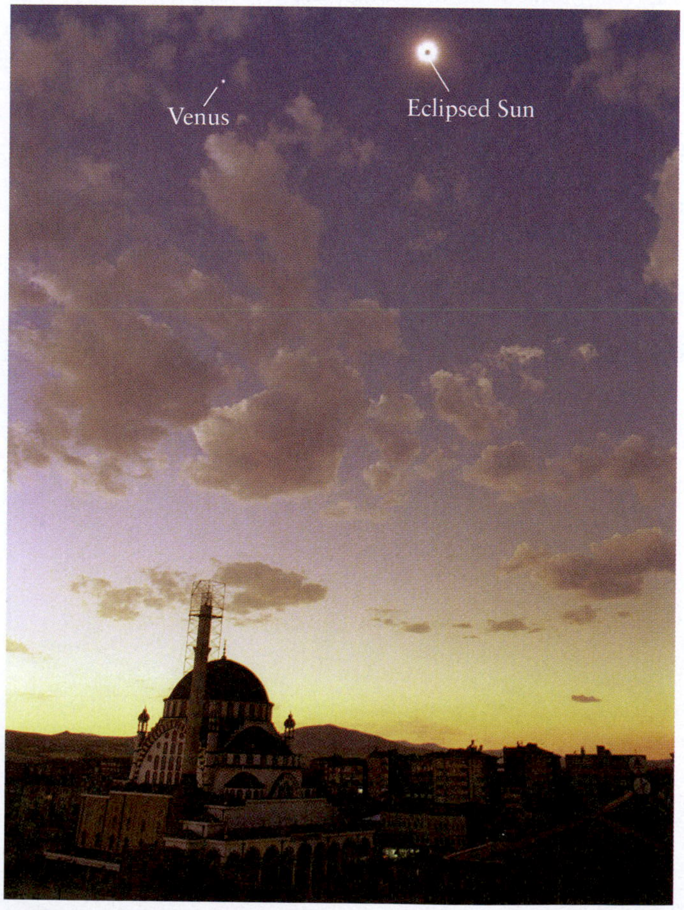

(a)

(b)

FIGURE 3-10 R I V U X G

A Total Solar Eclipse **(a)** This photograph shows the total solar eclipse of August 11, 1999, as seen from Elâziğ, Turkey. The sky is so dark that the planet Venus can be seen to the left of the eclipsed Sun. **(b)** When the Moon completely covers the Sun's disk during a total eclipse, the faint solar corona is revealed. (a: Courtesy of Fred Espenak, www.mreclipse.com; b: © 2008 Miloslav Druckmüller, Martin Dietzel, Peter Aniol, Vojtech Rušin)

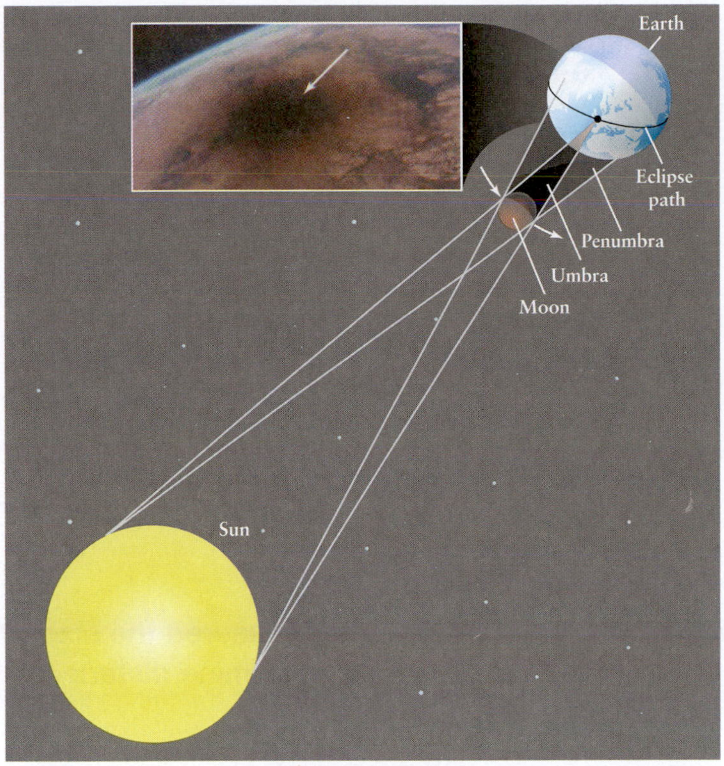

ANIMATION 3-4 **FIGURE 3-11** R I V U X G
The Geometry of a Total Solar Eclipse During a total solar eclipse, the tip of the Moon's umbra reaches Earth's surface. As Earth and the Moon move along their orbits, this tip traces an eclipse path across Earth's surface. People within the eclipse path see a total solar eclipse as the tip moves over them. Anyone within the penumbra sees only a partial eclipse. The inset photograph was taken from the *Mir* space station during the August 11, 1999, total solar eclipse (the same eclipse shown in Figure 3-10). The tip of the umbra appears as a black spot on Earth's surface. At the time the photograph was taken, this spot was 105 km (65 mi) wide and was crossing the cloud-covered English Channel at 3000 km/h (1900 mi/h). (Photograph by Jean-Pierre Haigneré, Centre National d'Etudes Spatiales, France/GSFS/NASA)

an **eclipse path** across Earth's surface. Only those locations within the eclipse path are treated to the spectacle of a total solar eclipse. The inset in Figure 3-11 shows the dark spot on Earth's surface produced by the Moon's umbra.

Partial Solar Eclipses

Immediately surrounding the Moon's umbra is the region of partial shadow called the penumbra. As seen from this area, the Sun's surface appears only partially covered by the Moon. During a solar eclipse, the Moon's penumbra covers a large portion of Earth's surface, and anyone standing inside the penumbra sees a **partial solar eclipse.** Such eclipses are much less interesting events than total solar eclipses, which is why astronomy enthusiasts strive to be inside the eclipse path. If you are within the eclipse path, you will see a partial eclipse before and after the brief period of totality (see the photograph that opens this chapter).

The width of the eclipse path depends primarily on the Earth-Moon distance during totality. The eclipse path is widest if the Moon happens to be at **perigee,** the point in its orbit nearest Earth. In this case the width of the eclipse path can be as great as 270 kilometers (170 miles). In most eclipses, however, the path is much narrower.

CONCEPTCHECK 3-8

Why can a total lunar eclipse be seen by people all over the world whereas total solar eclipses can only be seen from a very limited geographic region?

Answer appears at the end of the chapter.

Annular Solar Eclipses

In some eclipses the Moon's umbra does not reach all the way to Earth's surface. This can happen if the Moon is at or near **apogee,** its farthest position from Earth. In this case, the Moon appears too small to cover the Sun completely. The result is a third type of solar eclipse, called an **annular eclipse.** During an annular eclipse, a thin ring of the Sun is seen around the edge of the Moon (Figure 3-12). The length of the Moon's umbra is

FIGURE 3-12 R I V U X G
An Annular Solar Eclipse This composite of six photographs taken at sunrise in Costa Rica shows the progress of an annular eclipse of the Sun on December 24, 1973. Note that at mideclipse the limb, or outer edge, of the Sun is visible around the Moon. (Photo Researchers/Getty Images)

nearly 5000 kilometers (3100 miles) less than the average distance between the Moon and Earth's surface. Thus, the Moon's shadow often fails to reach Earth even when the Sun, Moon, and Earth are properly aligned for an eclipse. Hence, annular eclipses are slightly more common—as well as far less dramatic—than total eclipses.

Even during a total eclipse, most people along the eclipse path observe totality for only a few moments. Earth's rotation, coupled with the orbital motion of the Moon, causes the umbra to race eastward along the eclipse path at speeds in excess of 1700 kilometers per hour (1060 miles per hour). Because of the umbra's high speed, totality never lasts for more than 7½ minutes. In a typical total solar eclipse, the Sun-Moon-Earth alignment and the Earth-Moon distance are such that totality lasts much less than this maximum.

The details of solar eclipses are calculated well in advance. They are published in such reference books as the *Astronomical Almanac* and are available on the World Wide Web. Figure 3-13 shows the eclipse paths for all total solar eclipses from 1997 to 2020. Table 3-2 lists all the total, annular, and partial eclipses from 2012 to 2015, including the maximum duration of totality for total eclipses.

Ancient astronomers achieved a limited ability to predict eclipses. In those times, religious and political leaders who were able to predict such awe-inspiring events as eclipses must have made a tremendous impression on their followers. One of three priceless manuscripts to survive the devastating Spanish Conquest shows that the Mayan astronomers of Mexico and Guatemala had a fairly reliable method for predicting eclipses. The great Greek astronomer Thales of Miletus is said to have predicted the famous eclipse of 585 B.C.E, which occurred during the middle of a war. The sight was so unnerving that the soldiers put down their arms and declared peace.

In retrospect, it seems that what ancient astronomers actually produced were eclipse "warnings" of various degrees of reliability rather than true predictions. Working with historical records, these astronomers generally sought to discover cycles and regularities from which future eclipses could be anticipated. Box 3-2 describes how you might produce eclipse warnings yourself.

CONCEPTCHECK 3-9

If you had a chance to observe a total solar eclipse and a total lunar eclipse, in general, how much longer would you expect one type to last than the other?

Answer appears at the end of the chapter.

3-6 Ancient astronomers measured the size of Earth and attempted to determine distances to the Sun and Moon

The prediction of eclipses was not the only problem attacked by ancient astronomers. More than 2000 years ago, centuries before

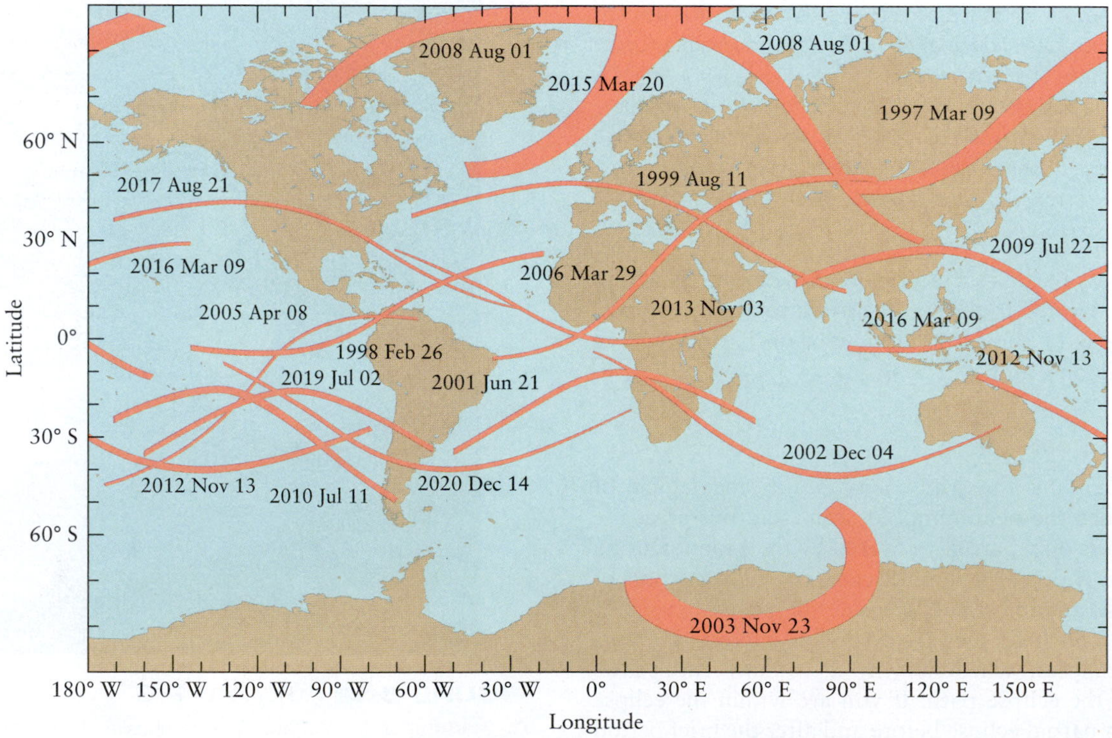

FIGURE 3-13

Eclipse Paths for Total Eclipses, 1997–2020 This map shows the eclipse paths for all 18 total solar eclipses occurring from 1997 through 2020. In each eclipse, the Moon's shadow travels along the eclipse path in a generally eastward direction across Earth's surface. (Courtesy of Fred Espenak, NASA/Goddard Space Flight Center)

TABLE 3-2 Solar Eclipses, 2012–2015

Date	Type	Where visible	Notes
2012 May 20	Annular	Asia, Pacific, North America	94% eclipsed
2012 Nov 13	Total	Australia, New Zealand, southern Pacific, southern South America	Maximum duration of totality 4m 02s
2013 May 10	Annular	Australia, New Zealand, central Pacific	Maximum duration of 06m 03s
2013 Nov 03	Total/Annular	Eastern Americas, southern Europe, Africa	Rare hybrid eclipse with maximum duration of 1m 40s
2014 Apr 29	Annular	Southern Indian Ocean, Australia, Antarctica	98% eclipsed
2015 Mar 20	Total	Iceland, Europe, northern Africa, northern Asia	Maximum duration of 02m 47s
2015 Sep 13	Partial	Southern Africa, southern Indian Ocean, Antarctica	79% eclipsed

Eclipse predictions by Fred Espenak, NASA/Goddard Space Flight Center.

BOX 3-2 TOOLS OF THE ASTRONOMER'S TRADE

Predicting Solar Eclipses

Suppose that you observe a solar eclipse in your hometown and want to figure out when you and your neighbors might see another eclipse. How would you begin?

First, remember that a solar eclipse can occur only if the line of nodes points toward the Sun at the same time that there is a new moon (see Figure 3-7). Second, you must know that it takes 29.53 days (one synodic month) to go from one new moon to the next. Because solar eclipses occur only during new moon, you must wait several whole lunar months for the proper alignment to occur again.

However, there is a complication: The line of nodes gradually shifts its position with respect to the background stars. It takes 346.6 days to move from one alignment of the line of nodes pointing toward the Sun to the next identical alignment. This period is called the **eclipse year.**

Therefore, to predict when you will see another solar eclipse, you need to know how many whole lunar months equal some whole number of eclipse years. This information will tell you how long you will have to wait for the next virtually identical alignment of the Sun, the Moon, and the line of nodes. By trial and error, you find that 223 lunar months is the same length of time as 19 eclipse years, because

$$223 \times 29.53 \text{ days} = 19 \times 346.6 \text{ days} = 6585 \text{ days}$$

This calculation is accurate to within a few hours. A more accurate calculation gives an interval, called the **saros,** that is about one-third of a day longer, or 6585.3 days (18 years, 11.3 days). Eclipses separated by the saros interval are said to form an *eclipse series.*

You might think that you and your neighbors would simply have to wait one full saros interval to go from one solar eclipse to the next. However, because of the extra one-third day, Earth will have rotated by an extra 120° (one-third of a complete rotation) when the next solar eclipse of a particular series occurs. The eclipse path will thus be one-third of the way around the world from you. Therefore, you must wait three full saros intervals (54 years, 34 days) before the eclipse path comes back around to your part of Earth. The illustration shows a series of six solar eclipse paths, each separated from the next by one saros interval.

There is evidence that ancient Babylonian astronomers knew about the saros interval. However, the discovery of the saros is more likely to have come from lunar eclipses than solar eclipses. If you are far from the eclipse path, there is a good chance that you could fail to notice a solar eclipse. Even if half the Sun is covered by the Moon, the remaining solar surface provides enough sunlight for the outdoor illumination not to be greatly diminished. By contrast, anyone on the nighttime side of Earth can see an eclipse of the Moon unless clouds block the view.

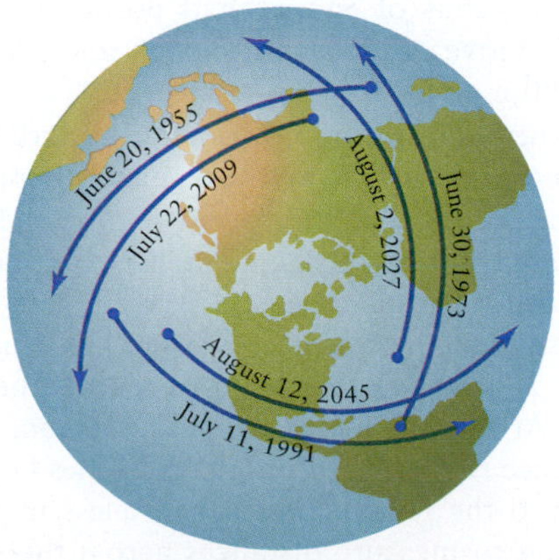

sailors of Columbus's era crossed the oceans, Greek astronomers were fully aware that Earth is not flat. They had come to this conclusion using a combination of observation and logical deduction, much like modern scientists. The Greeks noted that during lunar eclipses, when the Moon passes through Earth's shadow, the edge of the shadow is always circular. Because a sphere is the only shape that always casts a circular shadow from any angle, they concluded that Earth is spherical.

The ideas of geometry made it possible for Greek scholars to estimate cosmic distances

Eratosthenes and the Size of Earth

Around 200 B.C.E., the Greek astronomer Eratosthenes devised a way to measure the circumference of the spherical Earth. It was known that on the date of the summer solstice (the first day of summer; see Section 2-5) in the ancient town of Syene in Egypt, near present-day Aswan, the Sun shone directly down the vertical shafts of water wells. Hence, at local noon on that day, the Sun was at the zenith (see Section 2-4) as seen from Syene. Eratosthenes knew that the Sun never appeared at the zenith at his home in the Egyptian city of Alexandria, which is on the Mediterranean Sea almost due north of Syene. Rather, on the summer solstice in Alexandria, the position of the Sun at local noon was about 7° south of the zenith (Figure 3-14). This angle is about one-fiftieth of a complete circle, so he concluded that the distance from Alexandria to Syene must be about one-fiftieth of Earth's circumference.

In Eratosthenes's day, the distance from Alexandria to Syene was said to be 5000 stades. Therefore, Eratosthenes found Earth's circumference to be

$$50 \times 5000 \text{ stades} = 250{,}000 \text{ stades}$$

Unfortunately, no one today is sure of the exact length of the Greek unit called the stade. One guess is that the stade was about one-sixth of a kilometer, which would mean that Eratosthenes obtained a circumference for Earth of about 42,000 kilometers. This is remarkably close to the modern value of 40,000 kilometers.

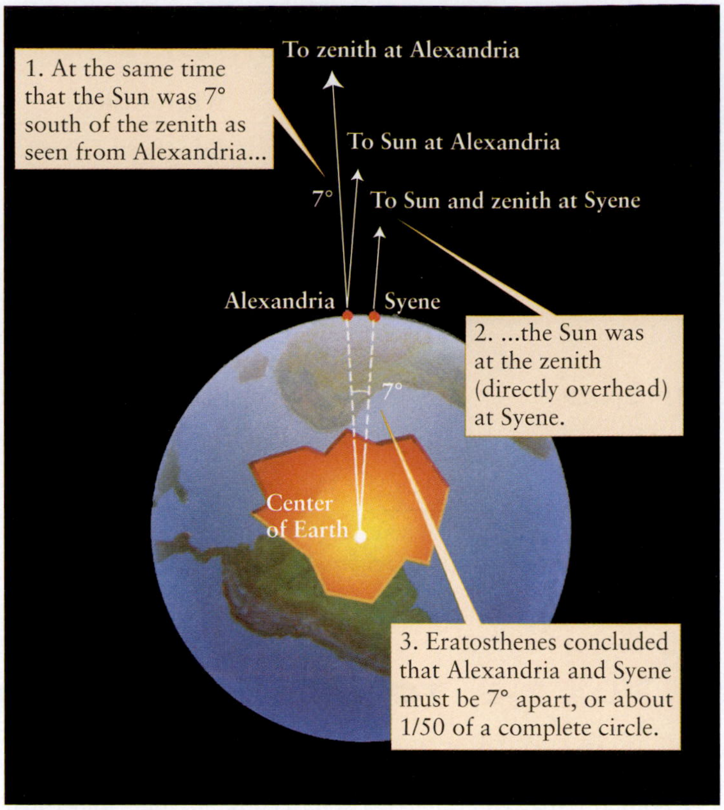

FIGURE 3-14
Eratosthenes's Method of Determining the Diameter of Earth Around 200 B.C.E., Eratosthenes used observations of the Sun's position at noon on the summer solstice to show that Alexandria and Syene were about 7° apart on the surface of Earth. This angle is about one-fiftieth of a circle, so the distance between Alexandria and Syene must be about one-fiftieth of Earth's circumference.

Aristarchus and Distances in the Solar System

Eratosthenes was only one of several brilliant astronomers to emerge from the so-called Alexandrian school, which by his time had a distinguished tradition. One of the first Alexandrian astronomers, Aristarchus of Samos, had proposed a method of determining the relative distances to the Sun and Moon, perhaps as long ago as 280 B.C.E.

Aristarchus knew that the Sun, Moon, and Earth form a right triangle at the moment of first or third quarter moon, with the right angle at the location of the Moon (Figure 3-15). He estimated that, as seen from Earth, the angle between the Moon and the Sun at first and third quarters is 87°, or 3° less than a right angle. Using the rules of geometry, Aristarchus concluded that the Sun is about 20 times farther from us than is the Moon. We now know that Aristarchus erred in measuring angles and that the average distance to the Sun is about 390 times larger than the average distance to the Moon. It is nevertheless impressive that people were trying to measure distances across the solar system more than 2000 years ago.

Aristarchus also made an equally bold attempt to determine the relative sizes of Earth, the Moon, and the Sun. From his observations of how long the Moon takes to move through Earth's shadow during a lunar eclipse, Aristarchus estimated the diameter of Earth to be about 3 times larger than the diameter of the Moon. To determine the diameter of the Sun, Aristarchus simply pointed out that the Sun and the Moon have the same angular size in the sky. Therefore, their diameters must be in the same proportion as their distances (see part a of the figure in Box 1-1). In other words, because Aristarchus thought the Sun to be 20 times farther from Earth than the Moon, he concluded that the Sun must be 20 times larger than the Moon. Once Eratosthenes had measured Earth's circumference, astronomers of the Alexandrian school could estimate the diameters of the Sun and Moon as well as their distances from Earth.

Table 3-3 summarizes some ancient and modern measurements of the sizes of Earth, the Moon, and the Sun and the distances between them. Some of these ancient measurements are far from the modern values. Yet the achievements of our ancestors still stand as impressive applications of observation and reasoning and important steps toward the development of the scientific method.

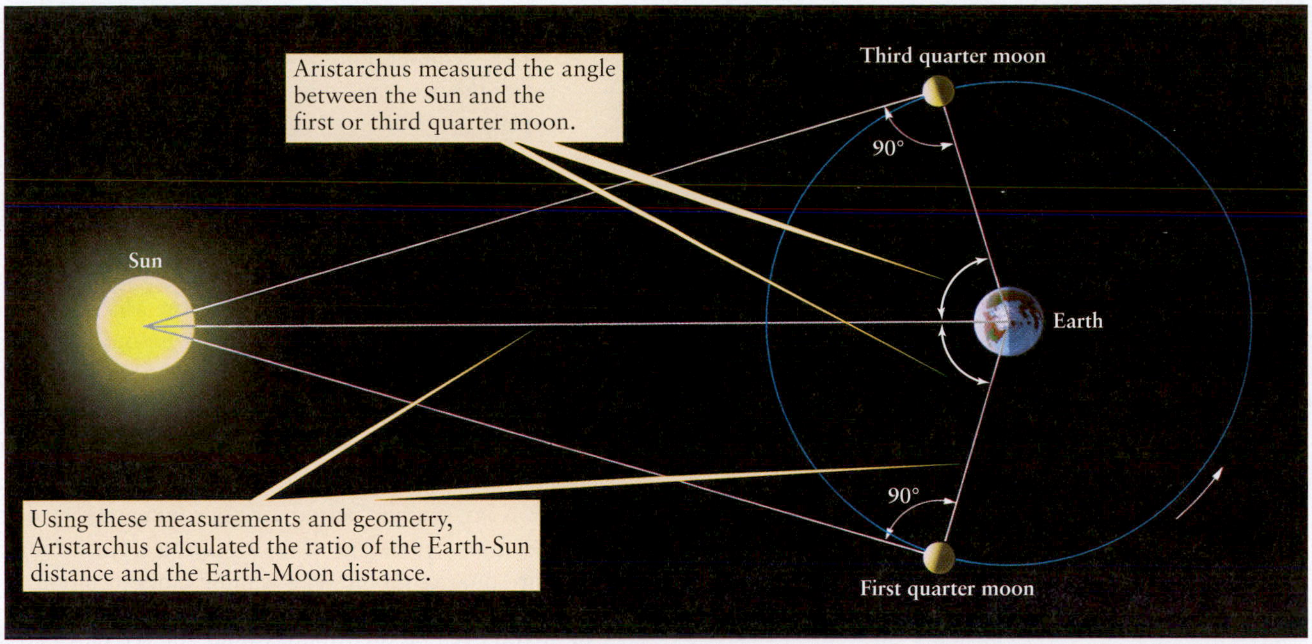
Aristarchus measured the angle between the Sun and the first or third quarter moon.
Third quarter moon
90°
Sun
Earth
90°
Using these measurements and geometry, Aristarchus calculated the ratio of the Earth-Sun distance and the Earth-Moon distance.
First quarter moon

FIGURE 3-15
Aristarchus's Method of Determining Distances to the Sun and Moon Aristarchus knew that the Sun, Moon, and Earth form a right triangle at first and third quarter phases. Using geometrical arguments, he calculated the relative lengths of the sides of these triangles, thereby obtaining the distances to the Sun and Moon.

TABLE 3-3 Comparison of Ancient and Modern Astronomical Measurements

	Ancient (km)	Modern (km)
Earth's diameter	13,000	12,756
Moon's diameter	4,300	3,476
Sun's diameter	9×10^4	1.39×10^6
Earth-Moon distance	4×10^5	3.84×10^5
Earth-Sun distance	10^7	1.50×10^8

CONCEPTCHECK 3-10

If Eratosthenes had found that the Sun's noontime summer solstice altitude at Alexandria was much closer to directly overhead, would he then assume that Earth was larger, smaller, or about the same size?

CALCULATIONCHECK 3-1

If Aristarchus had estimated the Sun to be 100 times farther from Earth than the Moon, how large would he have estimated the Sun to be?

Answers appear at the end of the chapter.

KEY WORDS

Terms preceded by an asterisk (*) are discussed in the Boxes.

annular eclipse, p. 57
apogee, p. 57
eclipse, p. 53
eclipse path, p. 57
*eclipse year, p. 59
first quarter moon, p. 48
full moon, p. 48
line of nodes, p. 53
lunar eclipse, p. 53
lunar phases, p. 48
new moon, p. 48
partial lunar eclipse, p. 55
partial solar eclipse, p. 57
penumbra (*plural* penumbrae), p. 54
penumbral eclipse, p. 55
perigee, p. 57
*saros, p. 59
sidereal month, p. 52
solar corona, p. 56
solar eclipse, p. 53
synchronous rotation, p. 51
synodic month, p. 52
third quarter moon, p. 48
totality (lunar eclipse), p. 56
totality (solar eclipse), p. 56
total lunar eclipse, p. 55
total solar eclipse, p. 56
umbra (*plural* umbrae), p. 54
waning crescent moon, p. 48
waning gibbous moon, p. 48
waxing crescent moon, p. 48
waxing gibbous moon, p. 48

KEY IDEAS

Ideas preceded by an asterisk (*) are discussed in the Boxes.

Lunar Phases: The phases of the Moon occur because light from the Moon is actually reflected sunlight. As the relative positions of Earth, the Moon, and the Sun change, we see more or less of the illuminated half of the Moon.

Length of the Month: Two types of months are used in describing the motion of the Moon.

• With respect to the stars, the Moon completes one orbit around Earth in a sidereal month, averaging 27.32 days.

• The Moon completes one cycle of phases (one orbit around Earth with respect to the Sun) in a synodic month, averaging 29.53 days.

The Moon's Orbit: The plane of the Moon's orbit is tilted by about 5° from the plane of Earth's orbit, or ecliptic.

• The line of nodes is the line where the planes of the Moon's orbit and Earth's orbit intersect. The gravitational pull of the Sun gradually shifts the orientation of the line of nodes with respect to the stars.

Conditions for Eclipses: During a lunar eclipse, the Moon passes through Earth's shadow. During a solar eclipse, Earth passes through the Moon's shadow.

• Lunar eclipses occur at full moon, while solar eclipses occur at new moon.

• Either type of eclipse can occur only when the Sun and Moon are both on or very near the line of nodes. If this condition is not met, Earth's shadow cannot fall on the Moon and the Moon's shadow cannot fall on Earth.

Umbra and Penumbra: The shadow of an object has two parts: the umbra, within which the light source is completely blocked, and the penumbra, where the light source is only partially blocked.

Lunar Eclipses: Depending on the relative positions of the Sun, Moon, and Earth, lunar eclipses may be total (the Moon passes completely into Earth's umbra), partial (only part of the Moon passes into Earth's umbra), or penumbral (the Moon passes only into Earth's penumbra).

Solar Eclipses: Solar eclipses may be total, partial, or annular.

• During a total solar eclipse, the Moon's umbra traces out an eclipse path over Earth's surface as Earth rotates. Observers outside the eclipse path but within the penumbra see only a partial solar eclipse.

• During an annular eclipse, the umbra falls short of Earth, and the outer edge of the Sun's disk is visible around the Moon at mideclipse.

The Moon and Ancient Astronomers: Ancient astronomers such as Aristarchus and Eratosthenes made great progress in determining the sizes and relative distances of Earth, the Moon, and the Sun.

QUESTIONS

Review Questions

Questions preceded by an asterisk (*) are discussed in the Boxes.

1. Explain the difference between sunlight and moonlight.
2. TUTORIAL 3-1 (**a**) Explain why the Moon exhibits phases. (**b**) A common misconception about the Moon's phases is that they are caused by Earth's shadow. Use Figure 3-2 to explain why this is not correct.
3. How would the sequence and timing of lunar phases be affected if the Moon moved around its orbit (**a**) in the same direction, but at twice the speed; (**b**) at the same speed, but in the opposite direction? Explain your answers.
4. At approximately what time does the Moon rise when it is (**a**) a new moon; (**b**) a first quarter moon; (**c**) a full moon; (**d**) a third quarter moon?
5. Astronomers sometimes refer to lunar phases in terms of the *age* of the Moon. This is the time that has elapsed since new moon phase. Thus, the age of a full moon is half of a 29½-day synodic period, or approximately 15 days. Find the approximate age of (**a**) a waxing crescent moon; (**b**) a third quarter moon; (**c**) a waning gibbous moon.
6. If you lived on the Moon, would you see Earth go through phases? If so, would the sequence of phases be the same as those of the Moon as seen from Earth, or would the sequence be reversed? Explain using Figure 3-2.
7. Is the far side of the Moon (the side that can never be seen from Earth) the same as the dark side of the Moon? Explain.
8. (**a**) If you lived on the Moon, would you see the Sun rise and set, or would it always be in the same place in the sky? Explain. (**b**) Would you see Earth rise and set, or would it always be in the same place in the sky? Explain using Figure 3-4.
9. What is the difference between a sidereal month and a synodic month? Which is longer? Why?
10. On a certain date the Moon is in the direction of the constellation Gemini as seen from Earth. When will the Moon next be in the direction of Gemini: one sidereal month later, or one synodic month later? Explain your answer.
11. TUTORIAL 3-2 What is the difference between the umbra and the penumbra of a shadow?
12. Why doesn't a lunar eclipse occur at every full moon and a solar eclipse at every new moon?
13. What is the line of nodes? Why is it important to the subject of eclipses?
14. What is a penumbral eclipse of the Moon? Why do you suppose that it is easy to overlook such an eclipse?
15. Why is the duration of totality different for different total lunar eclipses, as shown in Table 3-1?
16. Can one ever observe an annular eclipse of the Moon? Why or why not?

17. TUTORIAL 3-3 If you were looking at Earth from the side of the Moon that faces Earth, what would you see during (a) a total lunar eclipse; (b) a total solar eclipse? Explain your answers.

18. If there is a total eclipse of the Sun in April, can there be a lunar eclipse three months later in July? Why or why not?

19. Which type of eclipse—lunar or solar—do you think most people on Earth have seen? Why?

20. How is an annular eclipse of the Sun different from a total eclipse of the Sun? What causes this difference?

*21. What is the saros? How did ancient astronomers use it to predict eclipses?

22. How did Eratosthenes measure the size of Earth?

23. How did Aristarchus try to estimate the distance from Earth to the Sun and Moon?

24. How did Aristarchus try to estimate the diameters of the Sun and Moon?

Advanced Questions

Questions preceded by an asterisk (*) are discussed in the Boxes.

Problem-solving tips and tools

To estimate the average angular speed of the Moon along its orbit (that is, how many degrees around its orbit the Moon travels per day), divide 360° by the length of a sidereal month. It is helpful to know that the saros interval of 6585.3 days equals 18 years and 11⅓ days if the interval includes 4 leap years, but is 18 years and 10⅓ days if it includes 5 leap years.

25. The dividing line between the illuminated and unilluminated halves of the Moon is called the *terminator*. The terminator appears curved when there is a crescent or gibbous moon, but appears straight when there is a first quarter or third quarter moon (see Figure 3-2). Describe how you could use these facts to explain to a friend why lunar phases cannot be caused by Earth's shadow falling on the Moon.

26. What is the phase of the Moon if it rises at (a) midnight; (b) sunrise; (c) halfway between sunset and midnight; (d) halfway between noon and sunset? Explain your answers.

27. The Moon is highest in the sky when it crosses the meridian (see Figure 2-21), halfway between the time of moonrise and the time of moonset. At approximately what time does the Moon cross the meridian if it is (a) a new moon; (b) a first quarter moon; (c) a full moon; (d) a third quarter moon? Explain your answers.

28. The Moon is highest in the sky when it crosses the meridian (see Figure 2-21), halfway between the time of moonrise and the time of moonset. What is the phase of the Moon if it is highest in the sky at (a) midnight; (b) sunrise; (c) noon; (d) sunset? Explain your answers.

29. Suppose it is the first day of autumn in the northern hemisphere. What is the phase of the Moon if the Moon is located at (a) the vernal equinox; (b) the summer solstice; (c) the autumnal equinox; (d) the winter solstice? Explain your answers. (*Hint:* Make a drawing showing the relative positions of the Sun, Earth, and Moon. Compare with Figure 3-2.)

30. This photograph of Earth was taken by the crew of the *Apollo 8* spacecraft as they orbited the Moon. A portion of the lunar surface is visible at the right-hand side of the photo. In this photo, Earth is oriented with its north pole approximately at the top. When this photo was taken, was the Moon waxing or waning as seen from Earth? Explain your answer with a diagram.

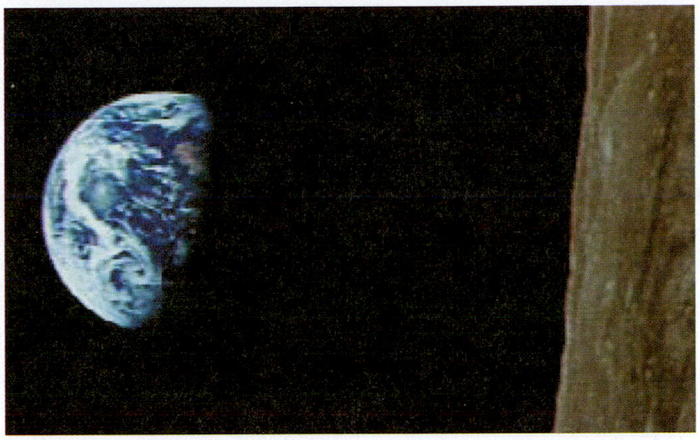

R I **V** U X G (NASA/JSC)

31. (a) The Moon moves noticeably on the celestial sphere over the space of a single night. To show this, calculate how long it takes the Moon to move through an angle equal to its own angular diameter (½°) against the background of stars. Give your answer in hours. (b) Through what angle (in degrees) does the Moon move during a 12-hour night? Can you notice an angle of this size? (*Hint:* See Figure 1-11.).

32. During an occultation, or "covering up," of Jupiter by the Moon, an astronomer notices that it takes the Moon's edge 90 seconds to cover Jupiter's disk completely. If the Moon's motion is assumed to be uniform and the occultation was "central" (that is, center over center), find the angular diameter of Jupiter. (*Hint:* Assume that Jupiter does not appear to move against the background of stars during this brief 90-second interval. You will need to convert the Moon's angular speed from degrees per day to arcseconds per second.)

33. The plane of the Moon's orbit is inclined at a 5° angle from the ecliptic, and the ecliptic is inclined at a 23½° angle from the celestial equator. Could the Moon ever appear at your zenith if you lived at (a) the equator; (b) the south pole? Explain your answers.

34. How many more sidereal months than synodic months are there in a year? Explain your answer.

35. Suppose Earth moved a little faster around the Sun so that it took a bit less than one year to make a complete orbit. If the speed of the Moon's orbit around Earth were unchanged, would the length of the sidereal month be the same, longer, or shorter than it is now? What about the synodic month? Explain your answers.

36. If the Moon revolved about Earth in the same orbit but in the opposite direction, would the synodic month be longer or shorter than the sidereal month? Explain your reasoning.

37. One definition of a "blue moon" is the second full moon within the same calendar month. There is usually only one full moon within a calendar month, so the phrase "once in a blue moon" means "hardly ever." Why are blue moons so rare? Are there any months of the year in which it would be impossible to have two full moons? Explain your answer.

38. You are watching a lunar eclipse from some place on Earth's night side. Will you see the Moon enter Earth's shadow from the east or from the west? Explain your reasoning.

39. The total lunar eclipse of October 28, 2004, was visible from South America. The duration of totality was 1 hour, 21 minutes. Was this total eclipse also visible from Australia, on the opposite side of Earth? Explain your reasoning.

40. During a total solar eclipse, the Moon's umbra moves in a generally eastward direction across Earth's surface. Use a drawing like Figure 3-11 to explain why the motion is eastward, not westward.

41. A total solar eclipse was visible from Africa on March 29, 2006 (see Figure 3-10b). Draw what the eclipse would have looked like as seen from France, to the north of the path of totality. Explain the reasoning behind your drawing.

42. Figures 3-11 and 3-13 show that the path of a total eclipse is quite narrow. Use this to explain why a glow is visible all around the horizon when you are viewing a solar eclipse during totality (see Figure 3-10a).

43. (**a**) Suppose the diameter of the Moon were doubled, but the orbit of the Moon remained the same. Would total solar eclipses be more common, less common, or just as common as they are now? Explain. (**b**) Suppose the diameter of the Moon were halved, but the orbit of the Moon remained the same. Explain why there would be *no* total solar eclipses.

44. Just as the distance from Earth to the Moon varies somewhat as the Moon orbits Earth, the distance from the Sun to Earth changes as Earth orbits the Sun. Earth is closest to the Sun at its *perihelion;* it is farthest from the Sun at its *aphelion.* In order for a total solar eclipse to have the maximum duration of totality, should Earth be at perihelion or aphelion? Assume that the Earth-Moon distance is the same in both situations. As part of your explanation, draw two pictures like Figure 3-11, one with Earth relatively close to the Sun and one with Earth relatively far from the Sun.

*45. On March 29, 2006, residents of northern Africa were treated to a total solar eclipse. (**a**) On what date and over what part of the world will the next total eclipse of that series occur? Explain. (**b**) On what date might you next expect a total eclipse of that series to be visible from northern Africa? Explain.

Discussion Questions

46. Describe the cycle of lunar phases that would be observed if the Moon moved around Earth in an orbit perpendicular to the plane of Earth's orbit. Would it be possible for both solar and lunar eclipses to occur under these circumstances? Explain your reasoning.

47. How would a lunar eclipse look if Earth had no atmosphere? Explain your reasoning.

48. In his 1885 novel *King Solomon's Mines,* H. Rider Haggard described a total solar eclipse that was seen in both South Africa and in the British Isles. Is such an eclipse possible? Why or why not?

49. Why do you suppose that total solar eclipse paths fall more frequently on oceans than on land? (You may find it useful to look at Figure 3-13.)

50. Examine Figure 3-13, which shows all of the total solar eclipses from 1997 to 2020. What are the chances that you might be able to travel to one of the eclipse paths? Do you think you might go through your entire life without ever seeing a total eclipse of the Sun?

Web/eBook Questions

51. ANIMATION 3-1 Access the animation "The Moon's Phases" in Chapter 3 of the *Universe* Web site or eBook. This shows the Earth-Moon system as seen from a vantage point looking down onto the north pole. (**a**) Describe where you would be on the diagram if you are on the equator and the time is 6:00 P.M. (**b**) If it is 6:00 P.M. and you are standing on Earth's equator, would a third quarter moon be visible? Why or why not? If it would be visible, describe its appearance.

52. Search the World Wide Web for information about the next total lunar eclipse. Will the total phase of the eclipse be visible from your location? If not, will the penumbral phase be visible? Draw a picture showing the Sun, Earth, and Moon when the totality is at its maximum duration, and indicate your location on the drawing of Earth.

53. Search the World Wide Web for information about the next total solar eclipse. Through which major cities, if any, does the path of totality pass? What is the maximum duration of totality? At what location is this maximum duration observed? Will this eclipse be visible (even as a partial eclipse) from your location? Draw a picture showing the Sun, Earth, and Moon when the totality is at its maximum duration, and indicate your location on the drawing of Earth.

54. ANIMATION 3-4 Access the animation "A Solar Eclipse Viewed from the Moon" in Chapter 3 of the *Universe* Web site or eBook. This shows the solar eclipse of August 11, 1999, as viewed from the Moon. Using a diagram, explain why the stars and the Moon's shadow move in the directions shown in this animation.

ACTIVITIES

Observing Projects

55. Observe the Moon on each clear night over the course of a month. On each night, note the Moon's location among the constellations and record that location on a star chart that also shows the ecliptic. After a few weeks, your observations will begin to trace the Moon's orbit. Identify the orientation of the line of nodes by marking the points where the Moon's orbit and the ecliptic intersect. On what dates is the Sun near the nodes marked on your star chart? Compare these dates with the dates of the next solar and lunar eclipses.

56. It is quite possible that a lunar eclipse will occur while you are taking this course. Look up the date of the next lunar eclipse in Table 3-1, on the World Wide Web, or in the current issue of a reference such as the *Astronomical Almanac* or *Astronomical Phenomena.* Then make arrangements to observe this lunar eclipse. You can observe the eclipse with the naked eye, but binoculars or a small telescope will enhance your viewing experience. If the eclipse is partial or total, note the times at which the Moon enters and exits Earth's umbra. If the eclipse is penumbral, can you see any changes in the Moon's brightness as the eclipse progresses?

57. Use *Starry Night*™ to demonstrate the phases of the Moon. From the menu, select **Favourites > Explorations > Moon Phases.** In this view, you are looking down upon the southern hemisphere of Earth from a location 69,300 kilometers above Earth's surface with Earth centered in the view. The size of the Moon is greatly exaggerated in this view. The face of the Moon in this diagrammatic representation is configured to show its phase as seen from Earth. The inner green circle represents the Moon's orbit around Earth. The position of the Sun with respect to Earth is shown on the outer green circle, the ecliptic, near to the top of the view. In reality, the Sun would be in this direction from Earth but at a far larger distance. Click the **Play** button and observe the relative positions of the Sun, Earth, and Moon and the associated phases of the Moon. (**a**) Describe the geometry of the Sun, Moon, and Earth when the Moon is new. (**b**) Describe the geometry of the Sun, Moon, and Earth when the phase of the Moon is full. (**c**) When the phase of the Moon is at first or third quarter, what is the approximate angle at the Moon between the Moon-Sun and Moon-Earth lines? (**d**) What general term describes the phase of the Moon when the angle created by the Moon-Sun and Moon-Earth lines at the Moon is less than 90°? (**e**) What general term describes the phase of the Moon when the angle between the Moon-Sun and Moon-Earth lines at the Moon is greater than 90°?

58. Use *Starry Night*™ to study the Moon's path in the sky and the phenomenon of eclipses, both lunar and solar. Select **Favourites > Explorations > Moon Motion** to display the sky as seen from the center of a transparent Earth on August 30, 2010, with daylight removed and the Moon centered in the view. As a guide to the positions and motions in the sky, the celestial equator and the associated equatorial grid are shown. The ecliptic, representing the plane of Earth's orbit extended onto the sky, is also shown. This line also represents the apparent path of the Sun across our sky in the course of a year. The **Time Flow Rate** is set to 1 hour but this can be adjusted if necessary to follow the Moon across the sky. (**a**) Is the Moon on the ecliptic at this time? (**b**) With the view locked onto the Moon, click the **Play** button. In which direction does the Moon move against the background stars? Keep in mind that the celestial equator runs in the east-west direction with east on the left. Note also that the ecliptic intersects this equator at an angle of 23½°. Does the Moon ever change its direction of motion relative to the stars? Why or why not? Describe its path relative to the ecliptic. (**c**) Determine how many days elapse between successive times that the Moon is on the ecliptic. (**d**) Set the **Time** and **Date** to 11 P.M. on July 21, 2009, and set the **Time Flow Rate** to 1 minute. Note that the Moon is almost on the ecliptic and that it is close to the position of the Sun in the sky. **Run time forward** at this 1-minute rate and stop time when the Moon is closest to the Sun. **Zoom** in to a field of view of about 3° and adjust the time in single steps to move the Moon to the position of maximum eclipse. What is the time of maximum eclipse? (**e**) Sketch the Sun and Moon on a piece of paper. Even though the Moon may not be directly over the Sun on the screen, an eclipse is occurring. What type of eclipse is it? Why is the Moon not directly over the Sun at this time? (**f**) Click on the **Events** tab, expand the **Event Filters** layer, and deselect all but the **Lunar and Solar Eclipse Events.** Expand the **Events Browser** layer, set the **Start** and **End** dates to cover the period from January 21, 2009, to January 21, 2010, and click **Find Events.** Right-click on each solar eclipse in turn, choose **View Event, Zoom** out to about 5°, and **Run time forward** and **backward** at a rate of 1 minute until the Moon is again closest to the Sun. Draw the Sun and Moon. Are they separated by the same amount as in part (d)? If not, why not? If there is a difference, what effect does it have on the eclipse?

59. Use *Starry Night*™ to observe a lunar eclipse from a position in space near the Sun. From the main menu select **Favourites > Explorations > Lunar Eclipse.** This view from a position near to the Sun shows Earth and two tan-colored circles. The inner circle represents the umbra of Earth's shadow and the outer circle represents the boundary of the penumbra of Earth's shadow. (If these circles do not appear, click on **Options > Solar System > Planets-Moons...** and click on **Shadow Colour** to adjust the brightness of the umbral and penumbral outlines.) (**a**) What is the Moon's phase? Click the **Play** button and watch as time flows forward and the Moon passes through Earth's shadow. Select **File > Revert** from the menu, move the cursor over the Moon, right-click the mouse and click on **Centre** to center the view on the Moon. **Zoom** in to almost fill the field of view with the Moon and click the **Play** button again to watch the lunar eclipse in detail. Use the

Time controls to determine the duration of totality to the nearest minute. (b) What is the duration of totality for this eclipse? (c) Is it possible for a lunar eclipse to have a longer duration of totality than the one depicted in this simulation? Explain your answer.

Collaborative Exercise

60. Using a bright light source at the center of a darkened room, or a flashlight, use your fist held at arm's length to demonstrate the difference between a full moon and a lunar eclipse. (Use yourself or a classmate as Earth.) How must your fist "orbit" Earth so that lunar eclipses do not happen at every full moon? Create a simple sketch to illustrate your answers.

ANSWERS

ConceptChecks

ConceptCheck 3-1: Because the Sun is always shining on the Moon, the Moon is always half illuminated and half dark. Except in the rare events of eclipses, discussed later, the Moon's apparent changing phases are due to observers on Earth seeing differing amounts of the Moon's half illuminated surface.

ConceptCheck 3-2: The Moon goes through an entire cycle of phases in about four weeks, so if it is currently in the waxing gibbous phase, in one week it will be in the waxing gibbous phase, and after two weeks it will be in the waning gibbous phase.

ConceptCheck 3-3: The Moon is shown at four points of its orbit in Figure 3-4b. In the left position, the far side is fully dark.

ConceptCheck 3-4: If the Moon is full, then after one week, it would reach the third quarter phase, and the point that used to be in the center of the Moon's visible surface at full moon would now fall into darkness that would last for two weeks. This scenario can be seen in two figures: In Figure 3-4b, this scenario is shown by the red dot on the leftmost depiction of the Moon, and the Moon's position a week later at the bottom of the figure. In Figure 3-2, this scenario occurs between positions E and G.

ConceptCheck 3-5: Increase. If Earth was moving around the Sun faster than it is now, Earth would move farther around the Sun during the Moon's orbit and it would take longer for the Moon to reach the position where it was in line with the Sun and Earth, increasing its synodic period. This is illustrated (for the new moon) in Figure 3-5.

ConceptCheck 3-6: Lunar eclipses only occur when the Sun, Earth, and Moon all lie exactly on a line; this line is called the line of nodes. As shown in Figure 3-7, during a full moon the Sun, Earth, and Moon are not necessarily lined up because the Moon orbits in a different plane than the ecliptic. Usually, the full moon is above or below the ecliptic plane of Earth's orbit around the Sun and misses being covered by Earth's shadow.

ConceptCheck 3-7: No. Before the Moon passes into the full shadow (the umbra), where the Sun is completely blocked by Earth, the Moon must pass through the penumbra where only part of the Sun is blocked. This is illustrated by Path 2 in Figure 3-8.

ConceptCheck 3-8: Total solar eclipses are only observed in the small region where the Moon's tiny umbral shadow lands on Earth (see Figure 3-11), whereas when the Moon enters Earth's much larger umbral shadow, anyone on Earth who can see the Moon can observe it in Earth's shadow.

ConceptCheck 3-9: A total lunar eclipse can last for more than an hour whereas a total solar eclipse lasts only a few minutes.

ConceptCheck 3-10: If the noontime position of the Sun on the summer solstice was much closer to being directly overhead, it means that on the curved surface of Earth, the overhead directions at both Syene and Alexandria are pointing in similar directions. This scenario does not occur on Earth and would only do so if Earth were much larger. (Keep in mind that the distance between these two cities is assumed to be constant for this question, no matter what the size of Earth.)

CalculationCheck

CalculationCheck 3-1: Because their diameters must be in the same proportion as their distances, if Aristarchus assumed the Sun to be 100 times farther away when the Sun and Moon appeared to be the same diameter in the sky, he would have proposed that the Sun is 100 times wider than the Moon. Today we know that the Sun is about 400 times farther away from Earth and 400 times wider than the Moon.

Archaeoastronomy and Ethnoastronomy by Mark Hollabaugh

Many years ago I read astronomer John Eddy's *National Geographic* article about the Wyoming Medicine Wheel. A few years later I visited this archaeological site in the Bighorn Mountains and started on the major preoccupation of my career.

Archaeoastronomy combines astronomy and archaeology. You may be familiar with sites such as Stonehenge or the Mayan ruins of the Yucatán. You may not know that there are many archaeological sites in the United States that demonstrate the remarkable understanding a people, usually known as the Anasazi, had about celestial motions (see Figure 2-1). Chaco Canyon, Hovenweep, and Chimney Rock in the Four Corners area of the Southwest preserve ruins from this ancient Pueblo culture.

Moonrise at Chimney Rock in southern Colorado provides a good example of the Anasazi's knowledge of the lunar cycles. If you watched the rising of the Moon for many, many years, you would discover that the Moon's northernmost rising point undergoes an 18.6-year cycle. Dr. McKim Malville of the University of Colorado discovered that the Anasazi who lived there knew of the lunar standstill cycle and watched the northernmost rising of the Moon between the twin rock pillars of Chimney Rock.

My own specialty is the ethnoastronomy of the Lakota, or Teton Sioux, who flourished on the Great Plains of what are now Nebraska, the Dakotas, and Wyoming. Ethnoastronomy combines ethnography with astronomy. As an ethnoastronomer, I am less concerned with physical evidence in the form of ruins and more interested in myths, legends, religious belief, and current practices. In my quest to understand the astronomical thinking and customs of the nineteenth-century Lakota, I have traveled to museums, archives, and libraries in Nebraska, Colorado, Wyoming, South Dakota, and North Dakota. I frequently visit the Pine Ridge and Rosebud Reservations in South Dakota. The Lakota, and other Plains Indians, had a rich tradition of understanding celestial motions and developed an even richer explanation of why things appear the way they do in the sky.

My first professional contribution to ethnoastronomy was in 1996 at the Fifth Oxford International Conference on Astronomy in Culture held in Santa Fe, New Mexico. I had noticed that images of eclipses often appeared in Lakota winter counts—their method of making a historic record of events. The great Leonid meteor shower of 1833 appears in almost every Plains Indian winter count. As I looked at the hides or in ledger books recording these winter counts, I wondered why the Sun, the Moon, and the stars are so common among the Lakota of 150 years ago. My curiosity led me to look deeper: What did the Lakota think about eclipses? Why does their central ritual, the Sun Dance, focus on the Sun? Why do so many legends involve the stars?

The Lakota observed lunar and solar eclipses. Perhaps they felt they had the power to restore the eclipsed Sun or Moon. In August 1869, an Indian agency physician in South Dakota told the Lakota there would be an eclipse. When the Sun disappeared from view, the Lakota began firing their guns in the air. In their minds, they were more powerful than the white doctor because the result of their action was to restore the Sun.

The Lakota used a lunar calendar. Their names for the months came from the world around them. October was the moon of falling leaves. In some years, there actually are 13 new moons, and they often called this extra month the "Lost Moon." The lunar calendar often dictated the timing of their sacred rites. Although the Lakota were never dogmatic about it, they preferred to hold their most important ceremony, the Sun Dance, at the time of the full moon in June, which is when the summer solstice occurs.

Why did the Lakota pay attention to the night sky? Lakota elder Ringing Shield's statement about Polaris, recorded in the late nineteenth century, provides a clue: "One star never moves and it is *wakan*. Other stars move in a circle about it. They are dancing in the dance circle." For the Lakota, a driving force in their culture was a quest to understand the nature of the sacred, or *wakan*. Anything hard to understand or different from the ordinary was *wakan*.

Reaching for the stars, as far away as they are, was a means for the Lakota to bring the incomprehensible universe a bit closer to Earth. Their goal was the same as what Dr. Sandra M. Faber says in her essay "Why Astronomy?" at the end of Chapter 1, "a perspective on human existence and its relation to the cosmos."

Mark Hollabaugh teaches physics and astronomy at Normandale Community College, in Bloomington, Minnesota. A graduate of St. Olaf College, he earned his Ph.D. in science education at the University of Minnesota. He has also taught at St. Olaf College, Augsburg College, and the U.S. Air Force Academy. As a boy he watched the dance of the northern lights and the flash of meteors. Meeting *Apollo 13* commander Jim Lovell and going for a ride in Roger Freedman's airplane are as close as he came to his dream of being an astronaut.

CHAPTER 4

An astronaut orbits Earth attached to the International Space Station's manipulator arm. (NASA) R I V U X G

Gravitation and the Waltz of the Planets

LEARNING GOALS

By reading the sections of this chapter, you will learn

4-1 How ancient astronomers attempted to explain the motions of the planets

4-2 What led Copernicus to a Sun-centered model of planetary motion

4-3 How Tycho's naked-eye observations of the sky revolutionized ideas about the heavens

4-4 How Kepler deduced the shapes of the orbits of the planets

4-5 How Galileo's pioneering observations with a telescope supported a Sun-centered model

4-6 The ideas behind Newton's laws of motion and his law of gravity, which apply to all physical objects

4-7 How energy can help us understand planetary orbits

4-8 What causes ocean tides on Earth

Sixty years ago the idea of humans orbiting Earth or sending spacecraft to other worlds was regarded as science fiction. Today, science fiction has become commonplace reality. Literally thousands of artificial satellites orbit our planet to track weather, relay signals for communications and entertainment, and collect scientific data about Earth and the universe. Humans live and work in Earth orbit (as in the accompanying photograph), have ventured as far as the Moon, and have sent dozens of robotic spacecraft to explore all the planets of the solar systemsphere, or corona, which glows with an unearthly pearlescent light.

While we think of spaceflight as an innovation of the twentieth century, we can trace its origins to a series of scientific revolutions that began in the 1500s. The first of these revolutions overthrew the ancient idea that Earth is an immovable object at the center of the universe, around which the Sun, the Moon, and the planets move. In this chapter we will learn how Nicolaus Copernicus, Tycho Brahe, Johannes Kepler, and Galileo Galilei helped us understand that Earth is itself one of several planets orbiting the Sun.

We will learn, too, about Isaac Newton's revolutionary discovery of why the planets move in the way that they do. This was just one aspect of Newton's immense body of work, which included formulating the fundamental laws of physics and developing a precise mathematical description of the force of gravitation—the force that holds the planets in their orbits.

Gravitation proves to be a truly universal force: It guides spacecraft as they journey across the solar system, and keeps satellites and astronauts in their orbits. In this chapter you will learn more about this important force of nature, which plays a central role in all parts of astronomy.

4-1 Ancient astronomers invented geocentric models to explain planetary motions

Since the dawn of civilization, scholars have attempted to explain the nature of the universe. The ancient Greeks were the first to use the principle that still guides scientists today: *The universe can be described and understood logically.* For example, more than 2500 years ago Pythagoras and his followers put forth the idea that nature can be described with mathematics. About 200 years later, Aristotle asserted that the universe is governed by physical laws. One of the most important tasks before the scholars of ancient Greece was to create a model (see Section 1-1) to explain the motions of objects in the heavens.

The Greek Geocentric Model

> The scholars of ancient Greece imagined that the planets follow an ornate combination of circular paths

Most Greek scholars thought that the Sun, the Moon, the stars, and the planets revolve about a stationary Earth. A model of this kind, in which Earth is at the center of the universe, is called a **geocentric model.** Similar ideas were held by the scholars of ancient China.

Today we recognize that the stars are not merely points of light on an immense celestial sphere. But in fact this is how the ancient Greeks regarded the stars in their geocentric model of the universe. To explain the diurnal motions of the stars, they assumed that the celestial sphere was *real,* and that it rotated around the stationary Earth once a day.

The Sun and Moon both participated in this daily rotation of the sky, which explained their rising and setting motions. To explain why the Sun and Moon both move slowly with respect to the stars, the ancient Greeks imagined that both of these objects orbit around Earth.

ANALOGY Imagine a merry-go-round that rotates clockwise as seen from above, as in Figure 4-1a. As it rotates, two children walk slowly counterclockwise at different speeds around the merry-go-round's platform. Thus, the children rotate along with the merry-go-round and also change their positions with respect to the merry-go-round's wooden horses. This scene is analogous to the way the ancient Greeks pictured the motions of the stars, the Sun, and the Moon. In their model, the celestial sphere rotated to the west around a stationary Earth (Figure 4-1b). The stars rotate along with the celestial sphere just as the wooden horses rotate along with the merry-go-round in Figure 4-1a. The Sun and Moon are analogous to the two children; they both turn westward with the celestial sphere, making one complete turn each day, and also move slowly eastward at different speeds with respect to the stars.

The geocentric model of the heavens also had to explain the motions of the planets. The ancient Greeks and other cultures of that time knew of five planets: Mercury, Venus, Mars, Jupiter, and Saturn, each of which is a bright object in the night sky. For example, when Venus is at its maximum brilliancy, it is 16 times brighter than the brightest star. (By contrast, Uranus and Neptune are quite dim and were not discovered until after the invention of the telescope.)

Like the Sun and Moon, all of the planets rise in the east and set in the west once a day. And like the Sun and Moon, from night to night the planets slowly move on the celestial sphere, that is, with respect to the background of stars. However, the character of this motion on the celestial sphere is quite different for the planets. Both the Sun and the Moon always move from west to east on the celestial sphere, that is, opposite the direction in which the celestial sphere appears to rotate. The Sun follows the path called the ecliptic

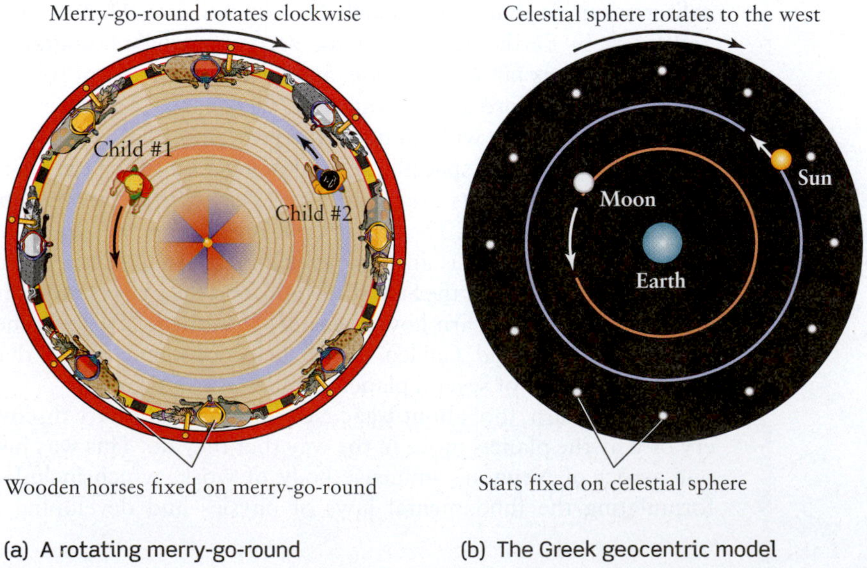

ANIMATION 4-1 **FIGURE 4-1**
A Merry-Go-Round Analogy **(a)** Two children walk at different speeds around a rotating merry-go-round with its wooden horses. **(b)** In an analogous way, the ancient Greeks imagined that the Sun and Moon move around the rotating celestial sphere with its fixed stars. Thus, the Sun and Moon move from east to west across the sky every day and also move slowly eastward from one night to the next relative to the background of stars.

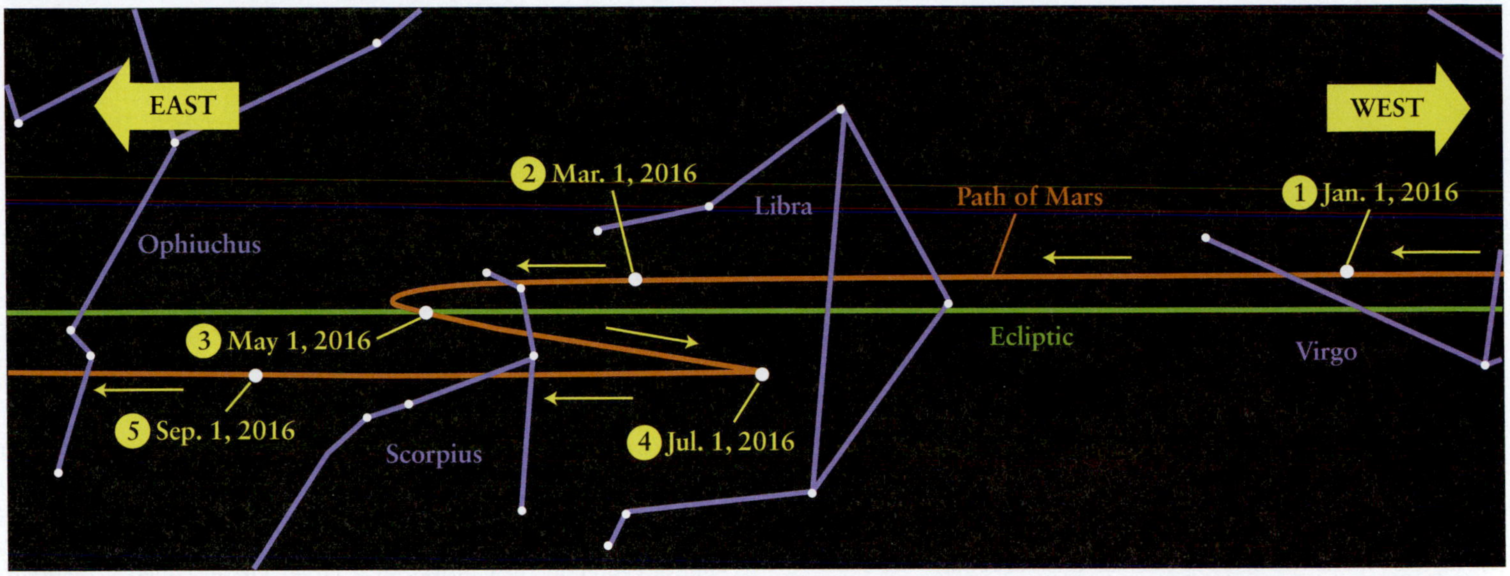

ANIMATION 4-2

FIGURE 4-2

The Path of Mars in 2016 From January 2016 through September 2016, Mars traverses zodiacal constellations from Virgo to Ophiuchus. Mars's motion is direct (from west to east, or from right to left in this figure) most of the time but is retrograde (from east to west, or from left to right in this figure) during May to July 2016. Notice that the speed of Mars relative to the stars is not constant: The planet travels farther across the sky from January 1 to March 1 than it does from March 1 to May 1.

(see Section 2-5), while the Moon follows a path that is slightly inclined to the ecliptic (see Section 3-3). Furthermore, the Sun and the Moon each move at relatively constant speeds around the celestial sphere. (The Moon's speed is faster than that of the Sun: It travels all the way around the celestial sphere in about a month while the Sun takes an entire year.) The planets, too, appear to move along paths that are close to the ecliptic. The difference is that each of the planets appears to wander back and forth on the celestial sphere with varying speed. As an example, Figure 4-2 shows the wandering motion of Mars with respect to the background of stars during 2016. (This figure shows that the name *planet* is well deserved; it comes from a Greek word meaning "wanderer.")

CAUTION! On a map of Earth with north at the top, west is to the left and east is to the right. Why, then, is *east* on the left and *west* on the right in Figure 4-2? The answer is that a map of Earth is a view looking downward at the ground from above, while a star map like Figure 4-2 is a view looking upward at the sky. You can see how east and west are flipped if you first draw a set of north-south and east-west axes on paper, and then hold the paper up against the sky.

Most of the time planets move slowly eastward relative to the stars, just as the Sun and Moon do. This eastward progress is called **direct motion.** For example, Figure 4-2 shows that Mars will be in direct motion from January through April 2016 and from July through September 2016. Occasionally, however, the planet seems to stop and then back up for several weeks or months. This occasional westward movement is called **retrograde motion.** Mars will undergo retrograde motion from May to July 2016 (see Figure 4-2), and will do so again about every 22½ months. All the other planets go through retrograde motion, but at different intervals. In the language of the merry-go-round analogy in Figure 4-1, the Greeks imagined the planets as children walking around the rotating merry-go-round but who keep changing their minds about which direction to walk!

CAUTION! Whether a planet is in direct or retrograde motion, over the course of a single night you will see it rise in the east and set in the west. That is because both direct and retrograde motions are much slower than the apparent daily rotation of the sky. Hence, they are best detected by mapping the position of a planet against the background stars from night to night over a long period. Figure 4-2 is a map of just this sort.

CONCEPTCHECK 4-1

If Mars is moving retrograde, will it rise above the eastern horizon or above the western horizon?

CONCEPTCHECK 4-2

How fast is Earth spinning on its axis in the Greek geocentric model?

Answers appear at the end of the chapter.

The Ptolemaic System

Explaining the nonuniform motions of the five planets was one of the main challenges facing the astronomers of antiquity. The Greeks developed many theories to account for retrograde motion and the loops that the planets trace out against the background stars. One of the most successful and enduring models was originated by Apollonius of Perga and by Hipparchus in the second

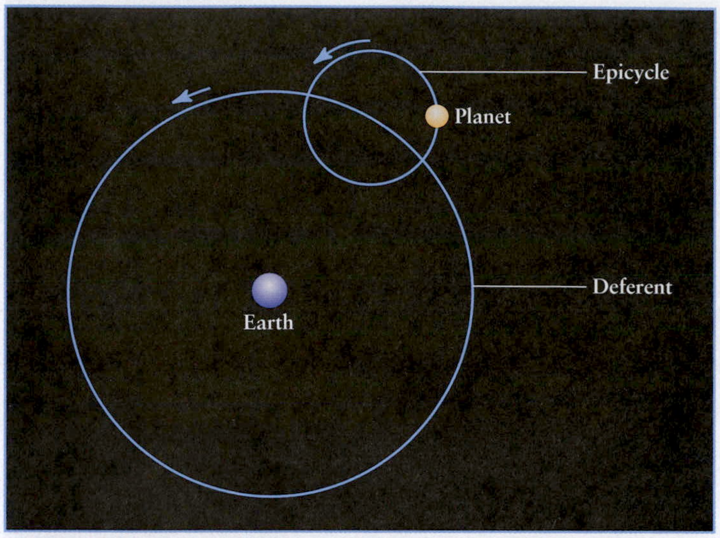

(a) Planetary motion modeled as a combination of circular motions

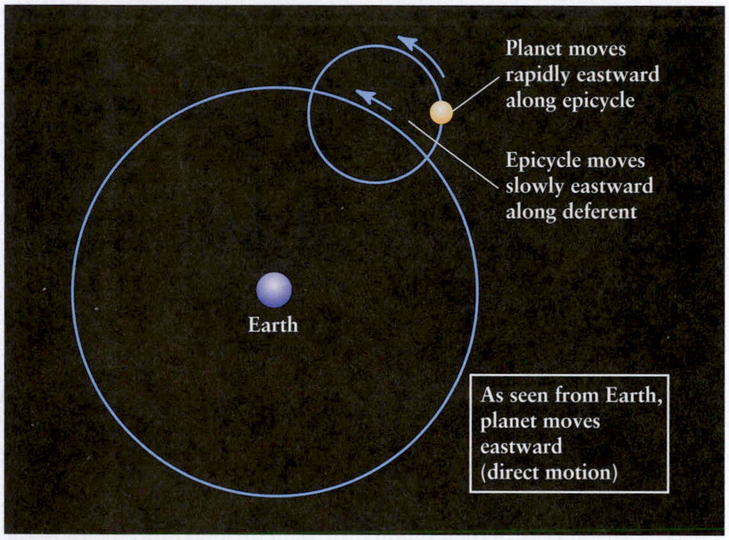

(b) Modeling direct motion

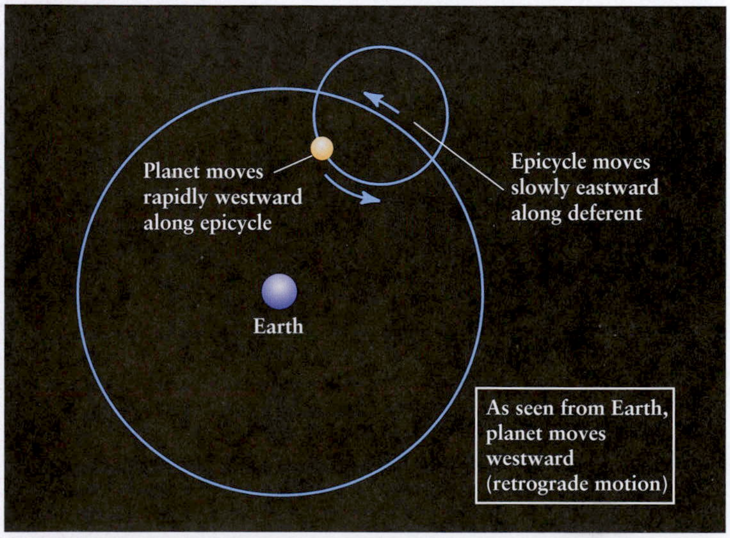

(c) Modeling retrograde motion

ANIMATION 4-3

FIGURE 4-3

A Geocentric Explanation of Retrograde Motion **(a)** The ancient Greeks imagined that each planet moves along an epicycle, which in turn moves along a deferent centered approximately on Earth. The planet moves along the epicycle more rapidly than the epicycle moves along the deferent. **(b)** At most times the eastward motion of the planet on the epicycle adds to the eastward motion of the epicycle on the deferent. Then the planet moves eastward in direct motion as seen from Earth. **(c)** When the planet is on the inside of the deferent, its motion along the epicycle is westward. Because this motion is faster than the eastward motion of the epicycle on the deferent, the planet appears from Earth to be moving westward in retrograde motion.

century B.C.E. and expanded upon by Ptolemy, the last of the great Greek astronomers, during the second century C.E. Figure 4-3a sketches the basic concept, usually called the **Ptolemaic system.** Each planet is assumed to move in a small circle called an **epicycle,** whose center in turn moves in a larger circle, called a **deferent,** which is centered approximately on Earth. Both the epicycle and deferent rotate in the same direction, shown as counterclockwise in Figure 4-3a.

As viewed from Earth, the epicycle moves eastward along the deferent. Most of the time the eastward motion of the planet on its epicycle adds to the eastward motion of the epicycle on the deferent (Figure 4-3b). Then the planet is seen to be in direct (eastward) motion against the background stars. However, when the planet is on the part of its epicycle nearest Earth, the motion of the planet along the epicycle is opposite to the motion of the epicycle along the deferent. The planet therefore appears to slow down and halt its usual eastward movement among the constellations, and actually goes backward in retrograde (westward) motion for a few weeks or months (Figure 4-3c). Thus, the concept of epicycles and deferents enabled Greek astronomers to explain the retrograde loops of the planets.

Using the wealth of astronomical data in the library at Alexandria, including records of planetary positions for hundreds of years, Ptolemy deduced the sizes and rotation rates of the epicycles and deferents needed to reproduce the recorded paths of the planets. After years of tedious work, Ptolemy assembled his calculations into 13 volumes, collectively called the *Almagest.* His work was used to predict the positions and paths of the Sun, Moon, and planets with unprecedented accuracy. In fact, the *Almagest* was so successful that it became the astronomer's bible, and for more than 1000 years, the Ptolemaic system endured as a useful description of the workings of the heavens.

A major problem posed by the Ptolemaic system was a philosophical one: It treated each planet independent of the others. There was no rule in the *Almagest* that related the size and rotation speed of one planet's epicycle and deferent to the corresponding sizes and speeds for other planets. This problem made the Ptolemaic system very unsatisfying to many Islamic and European astronomers of the Middle Ages. They thought that a correct model of the universe should be based on a simple set of underlying principles that applied to all of the planets.

With only limited observational data, the astronomers of the Middle Ages needed a way to judge if a scientific model might be correct. To help in this assessment, a guiding principle called

Occam's razor is often used to judge between competing scientific models. Named after the fourteenth-century English philosopher William of Occam, the principle states: *When two hypotheses can explain the available data, the hypothesis requiring the fewest new assumptions should be favored.* (The "razor" refers to shaving extraneous details or assumptions from an explanation.) In other words, the simplest explanation consistent with available data is most likely to be correct. The geocentric model required many parameters (or new assumptions) to characterize its epicycles and deferents, leaving open the possibility of a more simple theory. In the centuries that followed, Occam's razor would help motivate a simpler yet revolutionary view of the universe.

CONCEPTCHECK 4-3

Do the planets actually stop and change their direction of motion in the Ptolemaic model?

Answer appears at the end of the chapter.

4-2 Nicolaus Copernicus devised the first comprehensive heliocentric model

During the first half of the sixteenth century, a Polish lawyer, physician, canon of the church, and gifted mathematician named Nicolaus Copernicus (Figure 4-4) began to construct a new model of the universe. His model, which placed the Sun at the center, explained the motions of the planets in a more natural way than the Ptolemaic system. As we will see, it also helped lay the foundations of modern physical science. But Copernicus was not the first to conceive a Sun-centered model of planetary motion. In the third century B.C.E., the Greek astronomer Aristarchus suggested such a model as a way to explain retrograde motion.

A Heliocentric Model Explains Retrograde Motion

> The retrograde motion of the planets is a result of our viewing the universe from a moving Earth

Imagine riding on a fast racehorse. As you pass a slowly walking pedestrian, he appears to move backward, even though he is traveling in the same direction as you and your horse. This sort of simple observation inspired Aristarchus to formulate a **heliocentric** (Sun-centered) **model** in which all the planets, including Earth, revolve about the Sun. Different planets take different lengths of time to complete an orbit, so from time to time one planet will overtake another, just as a fast-moving horse overtakes a person on foot. When Earth overtakes Mars, for example, Mars appears to move backward in retrograde motion, as Figure 4-5 shows. Thus, in the heliocentric picture, the occasional retrograde motion of a planet is merely the result of Earth's fast motion.

As we saw in Section 3-6, Aristarchus demonstrated that the Sun is bigger than Earth (see Table 3-3). This made it sensible to imagine Earth orbiting the larger Sun. He also imagined that Earth rotated on its axis once a day, which explained the daily rising and setting of the Sun, Moon, and planets and the diurnal motions of the stars. To explain why the apparent motions of the planets never take them far from the ecliptic, Aristarchus proposed that the orbits of Earth and all the planets must lie in nearly the same plane. (Recall from Section 2-5 that the ecliptic is the projection onto the celestial sphere of the plane of Earth's orbit.)

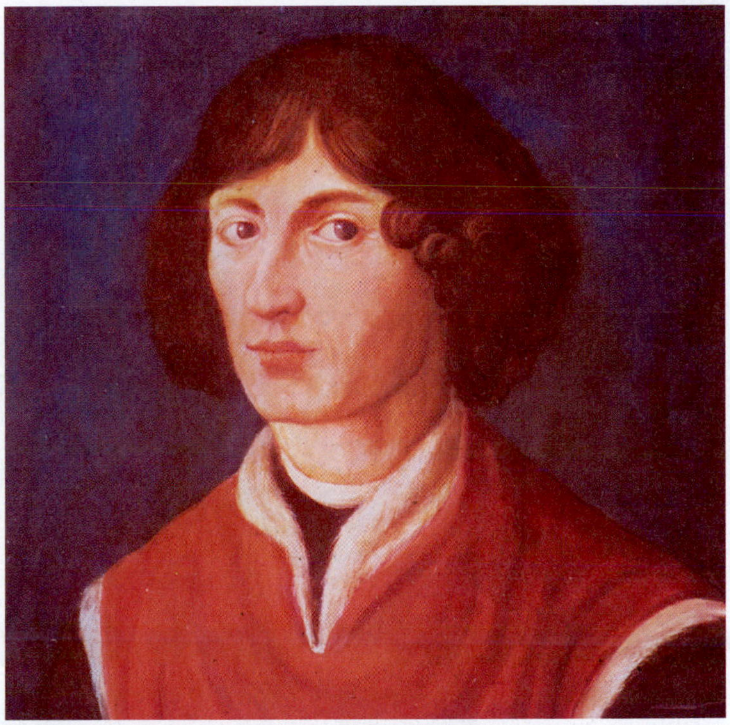

FIGURE 4-4
Nicolaus Copernicus (1473–1543) Copernicus was the first person to work out the details of a heliocentric system in which the planets, including Earth, orbit the Sun. (E. Lessing/Magnum)

This heliocentric model is conceptually much simpler than an Earth-centered system, such as that of Ptolemy, with all its "circles upon circles." In Aristarchus's day, however, the idea of an orbiting, rotating Earth seemed inconceivable, given Earth's apparent stillness and immobility. Nearly 2000 years would pass before a heliocentric model found broad acceptance.

CONCEPTCHECK 4-4

In the heliocentric model, could an imaginary observer on the Sun look out and see planets moving in retrograde motion?

Answer appears at the end of the chapter.

Copernicus and the Arrangement of the Planets

In the years after 1500, Copernicus came to realize that a heliocentric model has several advantages beyond providing a natural explanation of retrograde motion. In the Ptolemaic system, the arrangement of the planets—that is, which are close to Earth and which are far away—was chosen in large part by guesswork. By using a heliocentric model, Copernicus could determine the arrangement of the planets without ambiguity.

Copernicus realized that because Mercury and Venus are always observed fairly near the Sun in the sky, their orbits must be smaller than Earth's. Planets in such orbits are called **inferior planets**

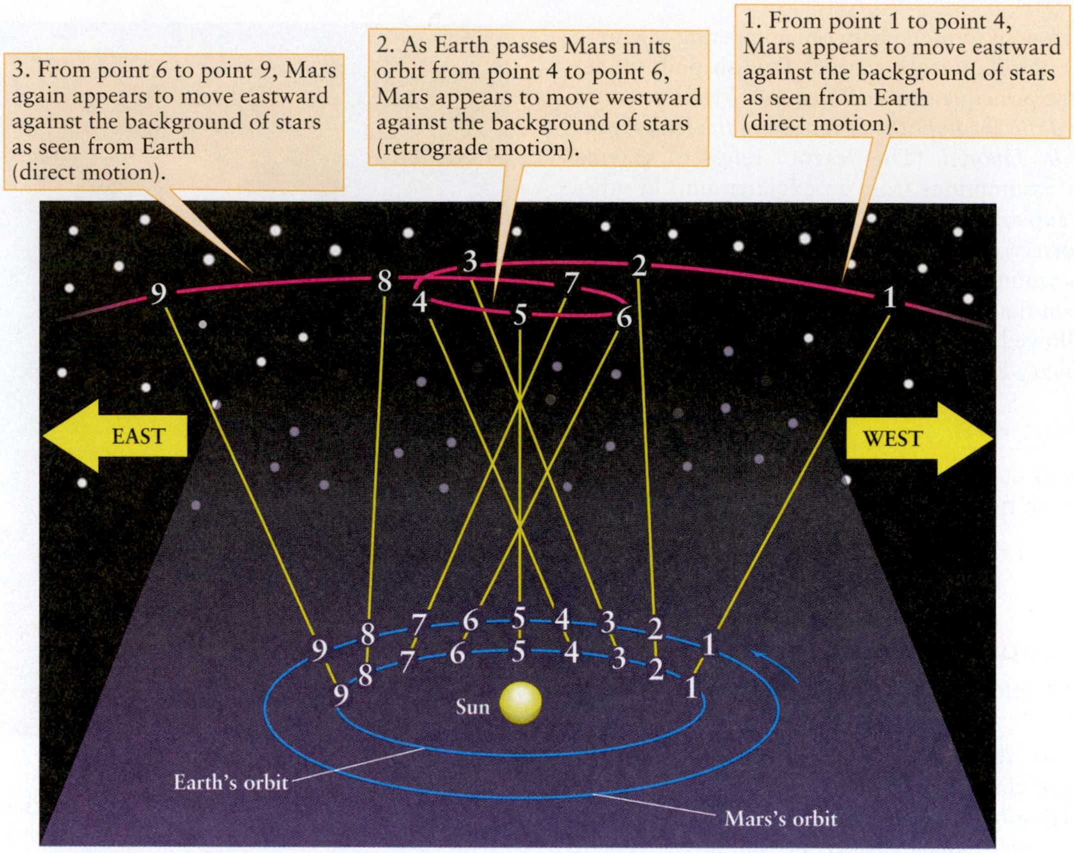
3. From point 6 to point 9, Mars again appears to move eastward against the background of stars as seen from Earth (direct motion).
2. As Earth passes Mars in its orbit from point 4 to point 6, Mars appears to move westward against the background of stars (retrograde motion).
1. From point 1 to point 4, Mars appears to move eastward against the background of stars as seen from Earth (direct motion).
EAST
WEST
Sun
Earth's orbit
Mars's orbit

ANIMATION 4-4

FIGURE 4-5

A Heliocentric Explanation of Retrograde Motion In the heliocentric model of Aristarchus, Earth and the other planets orbit the Sun. Earth travels around the Sun more rapidly than Mars. Consequently, as Earth overtakes and passes this slower-moving planet, Mars appears for a few months (from points 4 through 6) to fall behind and move backward with respect to the background of stars.

(Figure 4-6). The other visible planets—Mars, Jupiter, and Saturn—are sometimes seen on the side of the celestial sphere opposite the Sun, so these planets appear high above the horizon at midnight (when the Sun is far below the horizon). When this happens, Earth must lie between the Sun and these planets. Copernicus therefore concluded that the orbits of Mars, Jupiter, and Saturn must be larger than Earth's orbit. Hence, these planets are called **superior planets.**

Uranus and Neptune, as well as Pluto and a number of other small bodies called asteroids that also orbit the Sun, were discovered after the telescope was invented (and after the death of Copernicus). All of these can be seen at times in the midnight sky, so these also have orbits larger than that of Earth.

The heliocentric model also explains why planets appear in different parts of the sky on different dates. Both inferior planets (Mercury and Venus) go through cycles: The planet is seen in the west after sunset for several weeks or months, then for several weeks or months in the east before sunrise, and then in the west after sunset again.

Figure 4-6 shows the reason for this cycle. When Mercury or Venus is visible after sunset, it is near **greatest eastern elongation.** (The angle between the Sun and a planet as viewed from Earth is called the planet's **elongation.**) The planet's position in the sky is as far east of the Sun as possible, so it appears above the western horizon after sunset (that is, to the east of the Sun) and is often called an "evening star." At **greatest western elongation,** Mercury or Venus is as far west of the Sun as it can possibly be. It then rises before the Sun, gracing the predawn sky as a "morning star" in the east. When Mercury or Venus is at **inferior conjunction,** it is between us and the Sun, and it moves from the evening sky into the morning sky over weeks or months. At **superior conjunction,** when the planet is on the opposite side of the Sun, it moves back into the evening sky.

A superior planet such as Mars, whose orbit is larger than Earth's, is best seen in the night sky when it is at **opposition.** At this point in its orbit the planet is in the part of the sky opposite the Sun and is highest in the sky at midnight. This is also when the planet appears brightest, because it is closest to us. But when a superior planet like Mars is located behind the Sun at **conjunction,** it is above the horizon during the daytime and thus is not well placed for nighttime viewing.

CONCEPTCHECK 4-5

How many times is Mars at inferior conjunction during one orbit around the Sun?

Answer appears at the end of the chapter.

Planetary Periods and Orbit Sizes

The Ptolemaic system has no simple rules relating the motion of one planet to another. But Copernicus showed that there *are* such

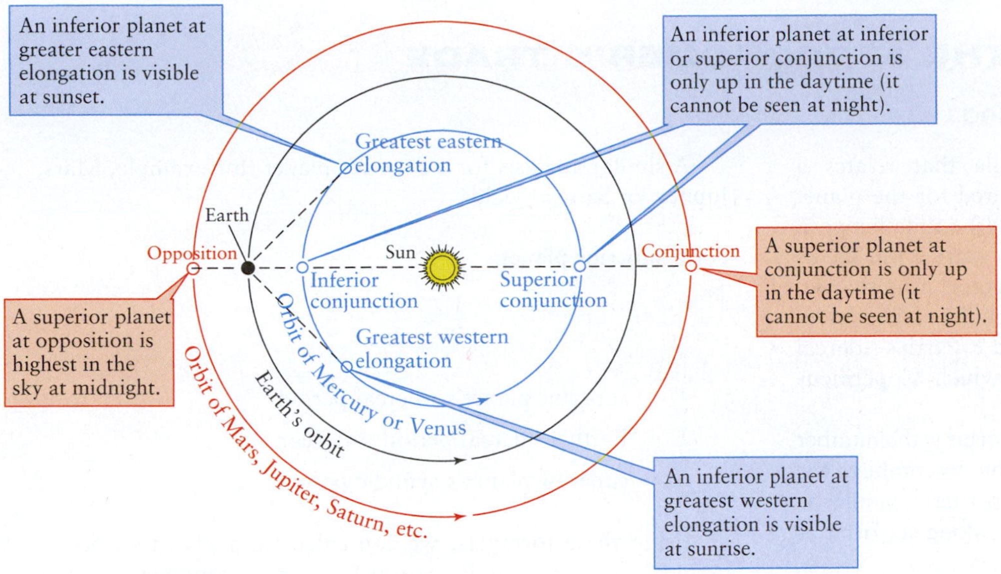
An inferior planet at greater eastern elongation is visible at sunset.
An inferior planet at inferior or superior conjunction is only up in the daytime (it cannot be seen at night).
Greatest eastern elongation
Earth
Opposition
Sun
Inferior conjunction
Superior conjunction
Conjunction
A superior planet at conjunction is only up in the daytime (it cannot be seen at night).
A superior planet at opposition is highest in the sky at midnight.
Greatest western elongation
Orbit of Mercury or Venus
Earth's orbit
Orbit of Mars, Jupiter, Saturn, etc.
An inferior planet at greatest western elongation is visible at sunrise.

FIGURE 4-6
Planetary Orbits and Configurations When and where in the sky a planet can be seen from Earth depends on the size of its orbit and its location on that orbit. Inferior planets have orbits smaller than Earth's, while superior planets have orbits larger than Earth's. (Note that in this figure you are looking down onto the solar system from a point far above Earth's northern hemisphere.)

rules in a heliocentric model. In particular, he found a correspondence between the time a planet takes to complete one orbit—that is, its **period**—and the size of the orbit.

Determining the period of a planet takes some care, because Earth, from which we must make the observations, is also moving. Realizing this, Copernicus was careful to distinguish between two different periods of each planet. The **synodic period** is the time that elapses between two successive identical configurations as seen from Earth—from one opposition to the next, for example, or from one conjunction to the next. The **sidereal period** is the true orbital period of a planet, the time it takes the planet to complete one full orbit of the Sun relative to the stars.

The synodic period of a planet can be determined by observing the sky from Earth, but the sidereal period has to be found by calculation. Copernicus figured out how to do this (Box 4-1). Table 4-1 shows the results for all of the planets.

To find a relationship between the sidereal period of a planet and the size of its orbit, Copernicus still had to determine the relative distances of the planets from the Sun. He devised a straightforward geometric method of determining the relative distances of the planets from the Sun using trigonometry. His answers turned out to be remarkably close to the modern values, as shown in Table 4-2.

The distances in Table 4-2 are given in terms of the astronomical unit (AU), which is the average distance from Earth to the Sun (Section 1-7). The astronomical unit is 1 AU = 1.496×10^8 km (92.96 million miles). However, Copernicus did not know the precise value of this distance, so he could only determine the *relative* sizes of the orbits of the planets. (Using the astronomical unit, the average distance from the Sun to each of the planets in kilometers can be determined from Table 4-2. A table of these distances is given in Appendix 2.)

By comparing Tables 4-1 and 4-2, you can see the unifying relationship between planetary orbits in the Copernican model: The farther a planet is from the Sun, the longer it takes to travel around its orbit (that is, the longer its sidereal period). That is so for *two* reasons: (1) the larger the orbit, the farther a planet must

TABLE 4-2 Average Distances of the Planets from the Sun

Planet	Copernican value (AU*)	Modern value (AU)
Mercury	0.38	0.39
Venus	0.72	0.72
Earth	1.00	1.00
Mars	1.52	1.52
Jupiter	5.22	5.20
Saturn	9.07	9.55
Uranus	—	19.19
Neptune	—	30.07

**1 AU = 1 astronomical unit = average distance from Earth to the Sun.*

TABLE 4-1 Synodic and Sidereal Periods of the Planets

Planet	Synodic period	Sidereal period
Mercury	116 days	88 days
Venus	584 days	225 days
Earth	—	1.0 year
Mars	780 days	1.9 years
Jupiter	399 days	11.9 years
Saturn	378 days	29.5 years
Uranus	370 days	84.1 years
Neptune	368 days	164.9 years

BOX 4-1 TOOLS OF THE ASTRONOMER'S TRADE

Relating Synodic and Sidereal Periods

We can derive a mathematical formula that relates a planet's sidereal period (the time required for the planet to complete one orbit) to its synodic period (the time between two successive identical configurations). To start with, let us consider an inferior planet (Mercury or Venus) orbiting the Sun as shown in the accompanying figure. Let P be the planet's sidereal period, S the planet's synodic period, and E Earth's sidereal period or sidereal year (see Section 2-8), which Copernicus knew to be nearly 365¼ days.

The rate at which Earth moves around its orbit is the number of degrees around the orbit divided by the time to complete the orbit, or $360°/E$ (equal to a little less than 1° per day). Similarly, the rate at which the inferior planet moves along its orbit is $360°/P$.

During a given time interval, the angular distance that Earth moves around its orbit is its rate, $(360°/E)$, multiplied by the length of the time interval. Thus, during a time S, or one synodic period of the inferior planet, Earth covers an angular distance of $(360°/E)S$ around its orbit. In that same time, the inferior planet covers an angular distance of $(360°/P)S$. Note, however, that the inferior planet has gained one full lap on Earth, and hence has covered 360° more than Earth has (see the figure). Thus, $(360°/P)S = (360°/E)S + 360°$. Dividing each term of this equation by $360°S$ gives

For an inferior planet:

$$\frac{1}{P} = \frac{1}{E} + \frac{1}{S}$$

P = inferior planet's sidereal period

E = Earth's sidereal period = 1 year

S = inferior planet's synodic period

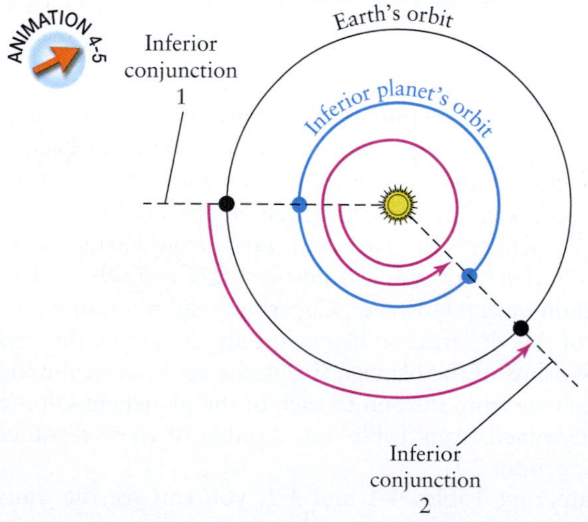

A similar analysis for a superior planet (for example, Mars, Jupiter, or Saturn) yields

For a superior planet:

$$\frac{1}{P} = \frac{1}{E} - \frac{1}{S}$$

P = superior planet's sidereal period

E = Earth's sidereal period = 1 year

S = superior planet's synodic period

Using these formulas, we can calculate a planet's sidereal period P from its synodic period S. Often astronomers express P, E, and S in terms of years, by which they mean Earth years of approximately 365.26 days.

EXAMPLE: Jupiter has an observed synodic period of 398.9 days, or 1.092 years. What is its sidereal period?

Situation: Our goal is to find the sidereal period of Jupiter, a superior planet.

Tools: Since Jupiter is a superior planet, we use the second of the two equations given above to determine the sidereal period P.

Answer: We are given Earth's sidereal period $E = 1$ year and Jupiter's synodic period $S = 1.092$ years. Using the equation $1/P = 1/E - 1/S$,

$$\frac{1}{P} = \frac{1}{1} - \frac{1}{1.092} = 0.08425,$$

so

$$P = \frac{1}{0.08425} = 11.87 \text{ years}$$

Review: Our answer means that it takes 11.87 years for Jupiter to complete one full orbit of the Sun. This sidereal period is greater than Earth's 1-year sidereal period because Jupiter's orbit is larger than Earth's orbit. Jupiter's synodic period of 1.092 years is the time from one opposition to the next, or the time that elapses from when Earth overtakes Jupiter to when it next overtakes Jupiter. This synodic period is so much shorter than the sidereal period because Jupiter moves quite slowly around its orbit. Earth overtakes it a little less often than once per Earth orbit, that is, at intervals of a little bit more than a year.

travel to complete an orbit; and (2) the larger the orbit, the slower a planet moves. For example, Mercury, with its small orbit, moves at an average speed of 47.9 km/s (107,000 mi/h). Saturn travels around its large orbit much more slowly, at an average speed of 9.64 km/s (21,600 mi/h). The older Ptolemaic model offers no such simple relations between the motions of different planets.

CONCEPTCHECK 4-6

What causes the planets to stop and change their direction of motion through the sky in the heliocentric model?

CONCEPTCHECK 4-7

Why is Jupiter's sidereal period longer than its synodic period?

Answers appear at the end of the chapter.

The Shapes of Orbits in the Copernican Model

At first, Copernicus assumed that Earth travels around the Sun along a circular path. He found, however, that perfectly circular orbits could not accurately describe the paths of the other planets, so he had to add an epicycle to each planet. These epicycles were *not* added to explain retrograde motion, which Copernicus realized was because of the differences in orbital speeds of different planets, as shown in Figure 4-5. Rather, the small epicycles helped Copernicus account for slight variations in each planet's speed along its orbit.

Even though he clung to the old notion that orbits must be made up of circles, Copernicus had shown that a heliocentric model could explain the motions of the planets. He compiled his ideas and calculations into a book entitled *De revolutionibus orbium coelestium* (On the Revolutions of the Celestial Spheres), which was published in 1543, the year of his death.

For several decades after Copernicus, most astronomers saw little reason to change their allegiance from the older geocentric model of Ptolemy. The predictions that the Copernican model makes for the apparent positions of the planets are, on average, no better or worse than those of the Ptolemaic model. The test of Occam's razor does not really favor either model, because both use a combination of circles to describe each planet's motion.

More concrete evidence was needed to convince scholars to abandon the old, comfortable idea of a stationary Earth at the center of the universe. The story of how this evidence was accumulated begins nearly 30 years after the death of Copernicus, when a young Danish astronomer pondered the nature of a new star in the heavens.

CONCEPTCHECK 4-8

Why was Copernicus's model more accurate than Ptolemy's model?

Answer appears at the end of the chapter.

4-3 Tycho Brahe's astronomical observations disproved ancient ideas about the heavens

On November 11, 1572, a bright star suddenly appeared in the constellation Cassiopeia. At first, it was even brighter than Venus, but then it began to grow dim. After 18 months, it faded from view. Modern astronomers recognize this event as a supernova explosion, the violent death of a massive star.

A supernova explosion and a comet revealed to Tycho that our universe is more dynamic than had been imagined

In the sixteenth century, however, the vast majority of scholars held with the ancient teachings of Aristotle and Plato, who had argued that the heavens are permanent and unalterable. Consequently, the "new star" of 1572 could not really be a star at all, because the heavens do not change; it must instead be some sort of bright object quite near Earth, perhaps not much farther away than the clouds overhead.

The 25-year-old Danish astronomer Tycho Brahe (1546–1601) realized that straightforward observations might reveal the distance to the new star. It is common experience that when you walk from one place to another, nearby objects appear to change position against the background of more distant objects. This phenomenon, whereby the apparent position of an object changes because of the motion of the observer, is called **parallax.** If the new star was nearby, then its position should shift against the background stars over the course of a single night because Earth's rotation changes our viewpoint. Figure 4-7 shows this predicted shift. (Actually, Tycho believed that the heavens rotate about Earth, as in the Ptolemaic model, but the net effect is the same.)

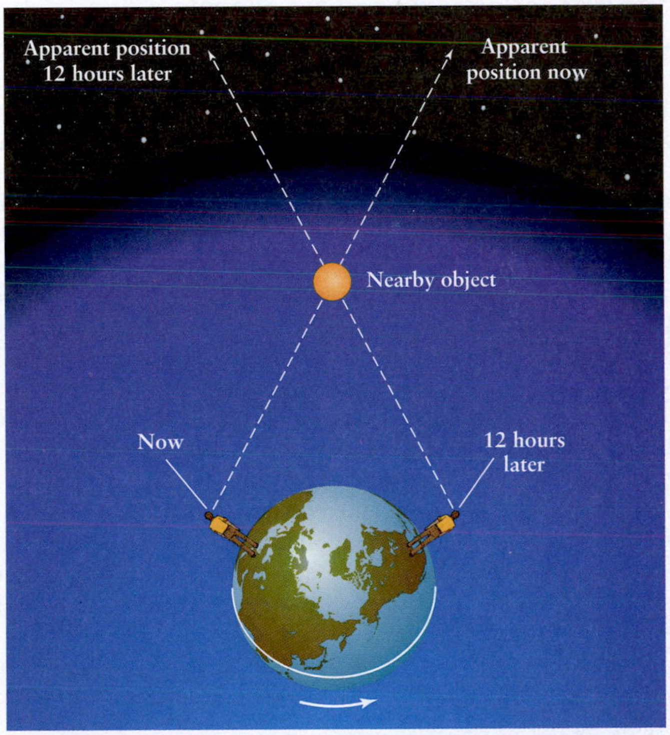

FIGURE 4-7

A Nearby Object Shows a Parallax Shift Tycho Brahe argued that if an object is near Earth, an observer would have to look in different directions to see that object over the course of a night and its position relative to the background stars would change. Tycho failed to measure such changes for a supernova in 1572 and a comet in 1577. He therefore concluded that these objects were far from Earth.

Tycho's careful observations failed to disclose any parallax. The farther away an object is, the less it appears to shift against the background as we change our viewpoint, and the smaller the parallax. Hence, the new star had to be quite far away, farther from Earth than anyone had imagined. This was the first evidence that the "unchanging" stars were in fact changeable.

Tycho also attempted to measure the parallax of a bright comet that appeared in 1577 and, again, found it too small to measure. Thus, the comet also had to be far beyond Earth. Furthermore, because the comet's position relative to the stars changed from night to night, its motion was more like a planet than a star. But most scholars of Tycho's time taught that the motions of the planets had existed in unchanging form since the beginning of the universe. If new objects such as comets could appear and disappear within the realm of the planets, the conventional notions of planetary motions needed to be revised.

Tycho's observations showed that the heavens are by no means pristine and unchanging. This discovery flew in the face of nearly 2000 years of astronomical thought. In support of these revolutionary observations, the king of Denmark financed the construction of two magnificent observatories for Tycho's use on the island of Hven, just off the Danish coast. The two observatories—Uraniborg ("heavenly castle") and Stjerneborg ("star castle")—allowed Tycho to design and have built a set of naked-eye astronomical instruments vastly superior in quality to any earlier instruments (Figure 4-8).

FIGURE 4-8
Tycho Brahe (1546–1601) Observing This contemporary illustration shows Tycho Brahe with some of the state-of-the-art measuring apparatus at Uraniborg, one of the two observatories that he built under the patronage of Frederik II of Denmark. (This magnificent observatory lacked a telescope, which had not yet been invented.) The data that Tycho collected were crucial to the development of astronomy in the years after his death. (Detlev van Ravenswaay/Science Source)

Tycho and the Positions of the Planets

With this state-of-the-art equipment, Tycho proceeded to measure the positions of stars and planets with unprecedented accuracy. In addition, he and his assistants were careful to make several observations of the same star or planet with different instruments, in order to identify any errors that might be caused by the instruments themselves. This painstaking approach to mapping the heavens revolutionized the practice of astronomy and is used by astronomers today.

A key goal of Tycho's observations during this period was to test the ideas Copernicus had proposed decades earlier about Earth going around the Sun. Tycho argued that if Earth was in motion, then nearby stars should appear to shift their positions with respect to background stars as we orbit the Sun. Tycho failed to detect any such parallax, and he concluded that Earth was at rest and the Copernican system was wrong.

On this point Tycho was in error, for nearby stars do in fact shift their positions as he had suggested. But even the nearest stars are so far away that the shifts in their positions are less than an arcsecond, too small to be seen with the naked eye. Tycho would have needed a telescope to detect the parallax that he was looking for, but the telescope was not invented until 1608, years after his death in 1601. Indeed, the first accurate determination of stellar parallax was not made until 1838.

Although he remained convinced that a stationary Earth was at the center of the universe, Tycho nonetheless made a tremendous contribution toward putting the heliocentric model on a solid foundation. From 1576 to 1597, he used his instruments to make comprehensive measurements of the positions of the planets with an accuracy of 1 arcminute. These measurements are as well as can be done with the naked eye and were far superior to any earlier measurements. Within the reams of data that Tycho compiled lay the truth about the motions of the planets. The person who would extract this truth was a German mathematician who became Tycho's assistant in 1600, a year before the great astronomer's death. His name was Johannes Kepler (Figure 4-9).

4-4 Johannes Kepler proposed elliptical paths for the planets about the Sun

TUTORIAL 4-1

Kepler's ideas apply not just to the planets, but to all orbiting celestial objects

The task that Johannes Kepler took on at the beginning of the seventeenth century was to find a model of planetary motion that agreed completely with Tycho's extensive and very accurate observations of planetary positions. To do this, Kepler found that he had to break with an ancient prejudice about planetary motions.

Elliptical Orbits and Kepler's First Law

Astronomers had long assumed that heavenly objects move in circles, which were considered the most perfect and harmonious of all geometric shapes. They believed that if a perfect God resided in heaven along with the stars and planets, then the motions of these objects must be perfect too. Against this context, Kepler dared to try to explain planetary motions with noncircular curves. In

FIGURE 4-9
Johannes Kepler (1571–1630) By analyzing Tycho Brahe's detailed records of planetary positions, Kepler developed three general principles, called Kepler's laws, that describe how the planets move about the Sun. Kepler was the first to realize that the orbits of the planets are ellipses and not circles. (Erich Lessing/Art Resource)

particular, he found that he had the best success with a particular kind of curve called an **ellipse.**

You can draw an ellipse by using a loop of string, two thumbtacks, and a pencil, as shown in Figure 4-10a. Each thumbtack in the figure is at a **focus** (plural **foci**) of the ellipse; an ellipse has two foci. The longest diameter of an ellipse, called the **major axis,** passes through both foci. Half of that distance is called the **semimajor axis** and is usually designated by the letter *a*. A circle is a special case of an ellipse in which the two foci are at the same point (this corresponds to using only a single thumbtack in Figure 4-10b). The semimajor axis of a circle is equal to its radius.

By assuming that planetary orbits were ellipses, Kepler found, to his delight, that he could make his theoretical calculations match precisely to Tycho's observations. This important discovery, first published in 1609, is now called **Kepler's first law:**

The orbit of a planet about the Sun is an ellipse with the Sun at one focus.

The semimajor axis *a* of a planet's orbit is the average distance between the planet and the Sun.

CAUTION! The Sun is at one focus of a planet's elliptical orbit, but there is *nothing* at the other focus. This "empty focus" has geometrical significance, because it helps to define the shape of the ellipse, but plays no other role.

Ellipses come in different shapes, depending on the elongation of the ellipse. The shape of an ellipse is described by its **eccentricity,** designated by the letter *e*. Figure 4-10b shows a few examples of ellipses with different eccentricities. The value of *e* can range from 0 (a circle) to just under 1 (nearly a straight line). The greater the eccentricity, the more elongated the ellipse. Because a circle is a special case of an ellipse, it is possible to have a perfectly circular orbit. But all of the objects that orbit the Sun have orbits that are at least slightly elliptical. The most circular of any planetary orbit is that of Venus, with an eccentricity of just 0.007; Mercury's orbit has an eccentricity of 0.206, and a number of small bodies called comets move in very elongated orbits with eccentricities of less than 1.

CONCEPTCHECK 4-9

Which orbit is closer to being circular—Venus's orbit, with $e = 0.007$, or Mars's orbit, with $e = 0.093$?

Answer appears at the end of the chapter.

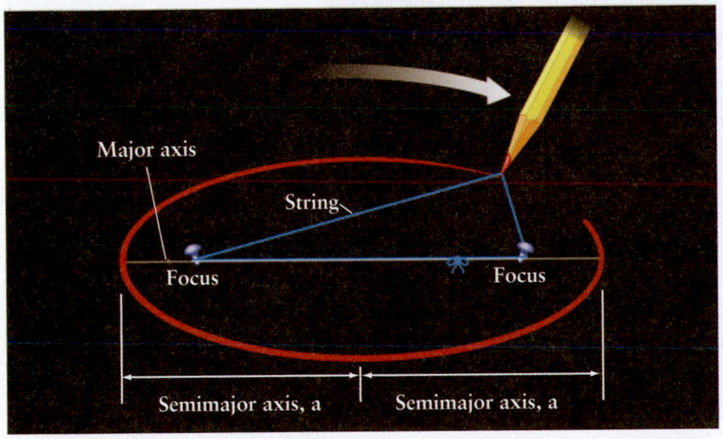

(a) The geometry of an ellipse

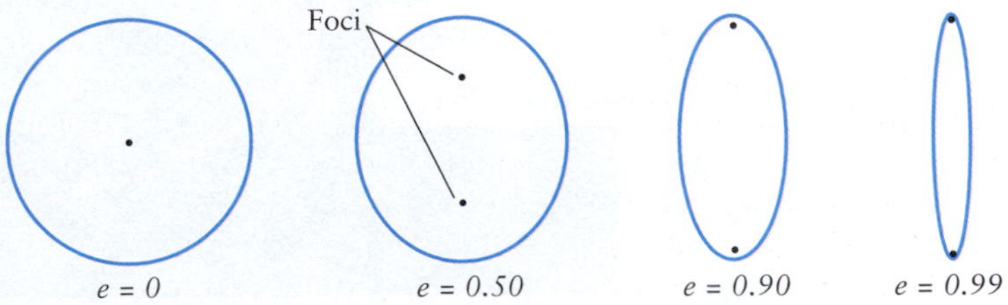

(b) Ellipses with different eccentricities

FIGURE 4-10
Ellipses **(a)** To draw an ellipse, use two thumbtacks to secure the ends of a piece of string, then use a pencil to pull the string taut. If you move the pencil while keeping the string taut, the pencil traces out an ellipse. The thumbtacks are located at the two foci of the ellipse. The major axis is the greatest distance across the ellipse; the semimajor axis is half of this distance. **(b)** A series of ellipses with the same major axis but different eccentricities. An ellipse can have any eccentricity from $e = 0$ (a circle) to just under $e = 1$ (virtually a straight line).

Orbital Speeds and Kepler's Second Law

Once he knew the shape of a planet's orbit, Kepler was ready to describe exactly *how* it moves on that orbit. As a planet travels in an elliptical orbit, its distance from the Sun varies. Kepler realized that the speed of a planet also varies along its orbit. A planet moves most rapidly when it is nearest the Sun, at a point on its orbit called **perihelion.** Conversely, a planet moves most slowly when it is farthest from the Sun, at a point called **aphelion** (Figure 4-11). After much trial and error, Kepler found a way to describe just how a planet's speed varies as it moves along its orbit. Figure 4-11 illustrates this discovery, referred to as **Kepler's second law:**

A line joining a planet and the Sun sweeps out equal areas in equal intervals of time.

This relationship is also called the **law of equal areas.** In the idealized case of a circular orbit, a planet would have to move at a constant speed around the orbit in order to satisfy Kepler's second law. In the case of an elliptical orbit, a planet speeds up when it is closer to the Sun, and slows down when it is farther away.

CONCEPTCHECK 4-10

According to Kepler's second law, at what point in a communications satellite's elliptical orbit around Earth will it move the slowest?

Answer appears at the end of the chapter.

ANALOGY An analogy for Kepler's second law is a twirling ice skater holding weights in each hand. If the skater moves the weights closer to her body by pulling her arms straight in, her rate of spin increases and the weights move faster; if she extends her arms so the weights move away from her body, her rate of spin decreases and the weights slow down. Just like the weights, a planet in an elliptical orbit travels at a higher speed when it moves closer to the Sun (toward perihelion) and travels at a lower speed when it moves away from the Sun (toward aphelion).

Orbital Periods and Kepler's Third Law

Kepler's second law describes how the speed of a given planet changes as it orbits the Sun. Kepler also deduced from Tycho's data a relationship that can be used to compare the motions of *different* planets. Published in 1618 and now called **Kepler's third law,** it states a relationship between the size of a planet's orbit and the time the planet takes to go once around the Sun:

The square of the sidereal period of a planet is directly proportional to the cube of the semimajor axis of the orbit.

Kepler's third law says that the larger a planet's orbit—that is, the larger the semimajor axis, or average distance from the planet to the Sun—the longer the sidereal period, which is the time it takes the planet to complete an orbit. From Kepler's third law one can show that the larger the semimajor axis, the slower the average speed at which the planet moves around its orbit. (By contrast,

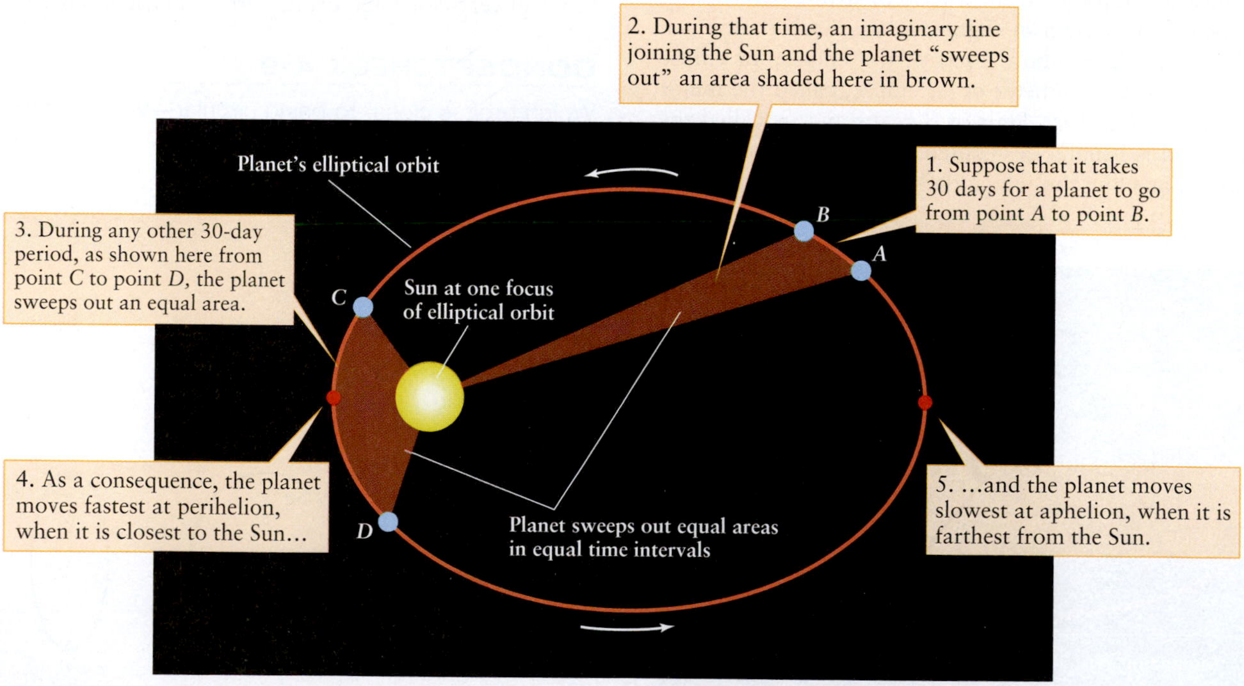

2. During that time, an imaginary line joining the Sun and the planet "sweeps out" an area shaded here in brown.
Planet's elliptical orbit
1. Suppose that it takes 30 days for a planet to go from point A to point B.
3. During any other 30-day period, as shown here from point C to point D, the planet sweeps out an equal area.
B
A
C
Sun at one focus of elliptical orbit
4. As a consequence, the planet moves fastest at perihelion, when it is closest to the Sun…
D
Planet sweeps out equal areas in equal time intervals
5. …and the planet moves slowest at aphelion, when it is farthest from the Sun.

ANIMATION 4-6

FIGURE 4-11
Kepler's First and Second Laws According to Kepler's first law, a planet travels around the Sun along an elliptical orbit with the Sun at one focus. According to his second law, a planet moves fastest when closest to the Sun (at perihelion) and slowest when farthest from the Sun (at aphelion). As the planet moves, an imaginary line joining the planet and the Sun sweeps out equal areas in equal intervals of time (from A to B or from C to D). By using these laws in his calculations, Kepler found an excellent fit to the apparent motions of the planets.

Kepler's *second* law describes how the speed of a given planet is sometimes faster and sometimes slower than its average speed.) This qualitative relationship between orbital size and orbital speed is just what Aristarchus and Copernicus used to explain retrograde motion, as we saw in Section 4-2. Kepler's great contribution was to make this relationship a quantitative one.

It is useful to restate Kepler's third law as an equation. If a planet's sidereal period P is measured in years and the length of its semimajor axis a is measured in astronomical units (AU), where 1 AU is the average distance from Earth to the Sun (see Section 1-7), then Kepler's third law is

Kepler's third law

$$P^2 = a^3$$

P = planet's sidereal period, in years

a = planet's semimajor axis, in AU

If you know either the sidereal period of a planet or the semimajor axis of its orbit, you can find the other quantity using this equation. Box 4-2 gives some examples of how this is done.

We can verify Kepler's third law for all of the planets, including those that were discovered after Kepler's death, using data

BOX 4-2 TOOLS OF THE ASTRONOMER'S TRADE

Using Kepler's Third Law

Kepler's third law relates the sidereal period P of an object orbiting the Sun to the semimajor axis a of its orbit:

$$P^2 = a^3$$

You must keep two essential points in mind when working with this equation:

1. The period P *must* be measured in years, and the semimajor axis a *must* be measured in astronomical units (AU). Otherwise you will get nonsensical results.

2. This equation applies *only* to the special case of an object, like a planet, that orbits the Sun. If you want to analyze the orbit of the Moon around Earth, of a spacecraft around Mars, or of a planet around a distant star, you must use a different, generalized form of Kepler's third law. We discuss this alternative equation in Section 4-7 and Box 4-4.

EXAMPLE: The average distance from Venus to the Sun is 0.72 AU. Use this to determine the sidereal period of Venus.

Situation: The average distance from Venus to the Sun is the semimajor axis a of the planet's orbit. Our goal is to calculate the planet's sidereal period P.

Tools: To relate a and P we use Kepler's third law, $P^2 = a^3$.

Answer: We first cube the semimajor axis (multiply it by itself twice):

$$a^3 = (0.72)^3 = 0.72 \times 0.72 \times 0.72 = 0.373$$

According to Kepler's third law this is also equal to P^2, the square of the sidereal period. So, to find P, we have to "undo" the square, that is, take the square root. Using a calculator, we find

$$P = \sqrt{P^2} = \sqrt{0.373} = 0.61$$

Review: The sidereal period of Venus is 0.61 years, or a bit more than seven Earth months. This result makes sense: A planet with a smaller orbit than Earth's (an inferior planet) must have a shorter sidereal period than Earth.

EXAMPLE: A certain small asteroid (a rocky body a few tens of kilometers across) takes eight years to complete one orbit around the Sun. Find the semimajor axis of the asteroid's orbit.

Situation: We are given the sidereal period $P = 8$ years, and are to determine the semimajor axis a.

Tools: As in the preceding example, we relate a and P using Kepler's third law, $P^2 = a^3$.

Answer: We first square the period:

$$P^2 = 8^2 = 8 \times 8 = 64$$

From Kepler's third law, 64 is also equal to a^3. To determine a, we must take the *cube root* of a^3, that is, find the number whose cube is 64. If your calculator has a cube root function, denoted by the symbol $\sqrt[3]{\ }$, you can use it to find that the cube root of 64 is 4: $\sqrt[3]{64} = 4$. Otherwise, you can determine by trial and error that the cube of 4 is 64:

$$4^3 = 4 \times 4 \times 4 = 64$$

Because the cube of 4 is 64, it follows that the cube root of 64 is 4 (taking the cube root "undoes" the cube).

With either technique you find that the orbit of this asteroid has semimajor axis $a = 4$ AU.

Review: The period is greater than 1 year, so the semimajor axis is greater than 1 AU. Note that $a = 4$ AU is intermediate between the orbits of Mars and Jupiter (see Table 4-3). Many asteroids are known with semimajor axes in this range, forming a region in the solar system called the asteroid belt.

TABLE 4-3 A Demonstration of Kepler's Third Law ($P^2 = a^3$)

Planet	Sidereal period P (years)	Semimajor axis a (AU)	P^2	a^3
Mercury	0.24	0.39	0.06	0.06
Venus	0.61	0.72	0.37	0.37
Earth	1.00	1.00	1.00	1.00
Mars	1.88	1.52	3.53	3.51
Jupiter	11.86	5.20	140.7	140.6
Saturn	29.46	9.55	867.9	871.0
Uranus	84.10	19.19	7,072	7,067
Neptune	164.86	30.07	27,180	27,190

Kepler's third law states that $P^2 = a^3$ for each of the planets. The last two columns of this table demonstrate that this relationship holds true to a very high level of accuracy.

from Tables 4-1 and 4-2. If Kepler's third law is correct, for each planet the numerical values of P^2 and a^3 should be equal. This result is indeed true to very high accuracy, as Table 4-3 shows.

CONCEPTCHECK 4-11

The space shuttle typically orbited Earth at an altitude of 300 km whereas the International Space Station orbits Earth at an altitude of 450 km. Although the space shuttle took less time to orbit Earth, which orbiter moved at a faster speed?

CALCULATIONCHECK 4-1

If Pluto's orbit has a semimajor axis of 39.5 AU, how long does it take Pluto to orbit the Sun once?

Answers appear at the end of the chapter.

The Significance of Kepler's Laws

Kepler's laws are a landmark in the history of astronomy. They made it possible to calculate the motions of the planets with better accuracy than any geocentric model ever had, and they helped to justify the idea of a heliocentric model. Kepler's laws also pass the test of Occam's razor, for they are simpler in every way than the schemes of Ptolemy or Copernicus, both of which used a complicated combination of circles.

But the significance of Kepler's laws goes beyond understanding planetary orbits. These same laws are also obeyed by spacecraft orbiting Earth, by two stars revolving about each other in a binary star system, and even by galaxies in their orbits about each other. Throughout this book, we shall use Kepler's laws in a wide range of situations.

As impressive as Kepler's accomplishments were, he did not prove that the planets orbit the Sun, nor was he able to explain *why* planets move in accordance with his three laws. These advances were made by two other figures who loom large in the history of astronomy: Galileo Galilei and Isaac Newton.

CONCEPTCHECK 4-12

Do Kepler's laws of planetary motion apply only to the planets?

Answer appears at the end of the chapter.

4-5 Galileo's discoveries with a telescope strongly supported a heliocentric model

Galileo used the cutting-edge technology of the 1600s to radically transform our picture of the universe

When Dutch opticians invented the telescope during the first decade of the seventeenth century, astronomy was changed forever. The scholar who used this new tool to amass convincing evidence that the planets orbit the Sun, not Earth, was the Italian mathematician and physical scientist Galileo Galilei (Figure 4-12).

While Galileo did not invent the telescope, he was the first to point one of these new devices toward the sky and to publish his observations. Beginning in 1610, he saw sights of which no one

FIGURE 4-12
Galileo Galilei (1564–1642) Galileo was one of the first people to use a telescope to observe the heavens. He discovered craters on the Moon, sunspots on the Sun, the phases of Venus, and four moons orbiting Jupiter. His observations strongly suggested that Earth orbits the Sun, not vice versa. (Erich Lessing/Art Resource)

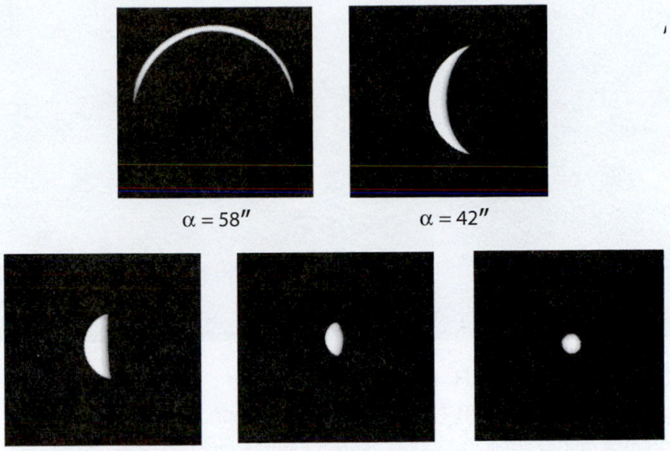

FIGURE 4-13 R I V U X G

The Phases of Venus This series of photographs shows how the appearance of Venus changes as it moves along its orbit. The number below each view is the angular diameter α of the planet in arcseconds. Venus has the largest angular diameter when it is a crescent, and the smallest angular diameter when it is gibbous (nearly full). (New Mexico State University Observatory)

had ever dreamed. He discovered mountains on the Moon, sunspots on the Sun, and the rings of Saturn, and he was the first to see that the Milky Way is not a featureless band of light but rather "a mass of innumerable stars."

The Phases of Venus

One of Galileo's most important discoveries with the telescope was that Venus exhibits phases like those of the Moon (Figure 4-13). Galileo also noticed that the apparent size of Venus as seen through his telescope was related to the planet's phase. Venus appears small at gibbous phase and largest at crescent phase. Galileo also noted a correlation between the phases of Venus and the planet's angular distance from the Sun.

Figure 4-14 shows that these relationships are entirely compatible with a heliocentric model in which Earth and Venus both go around the Sun. They are also completely *incompatible* with the Ptolemaic system, in which the Sun and Venus both orbit Earth. To explain why Venus is never seen very far from the Sun, the Ptolemaic model had to assume that the deferents of Venus and of the Sun move together in lockstep, with the epicycle of Venus centered on a straight line between Earth and the Sun (Figure 4-15). In this model, Venus was never on the opposite side of the Sun from Earth, and so it could never have shown the gibbous phases that Galileo observed.

CONCEPTCHECK 4-13

Which phase will Venus be in when it is at its maximum distance from Earth?

Answer appears at the end of the chapter.

The Moons of Jupiter

Galileo also found more unexpected evidence supporting the ideas of Copernicus. In 1610 Galileo discovered four moons, now

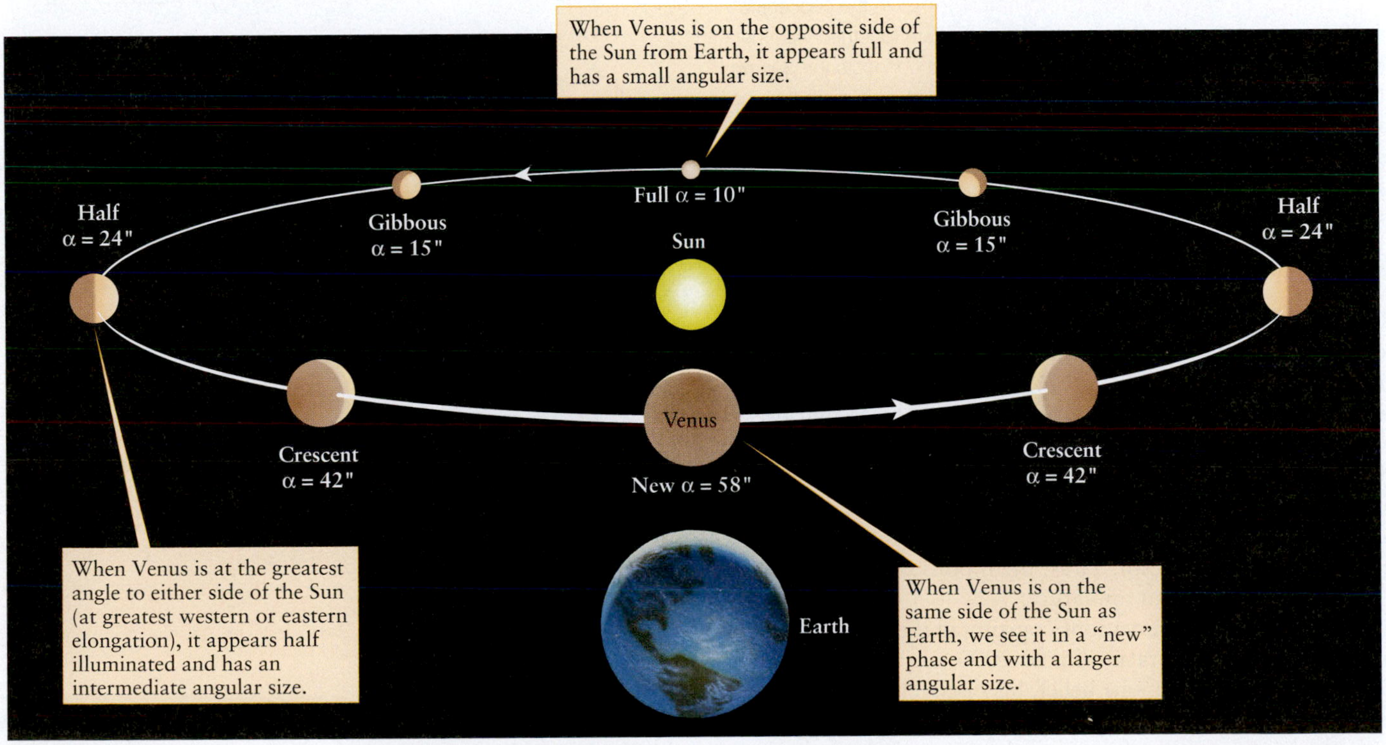

FIGURE 4-14

The Changing Appearance of Venus Explained in a Heliocentric Model A heliocentric model, in which Earth and Venus both orbit the Sun, provides a natural explanation for the changing appearance of Venus shown in Figure 4-13.

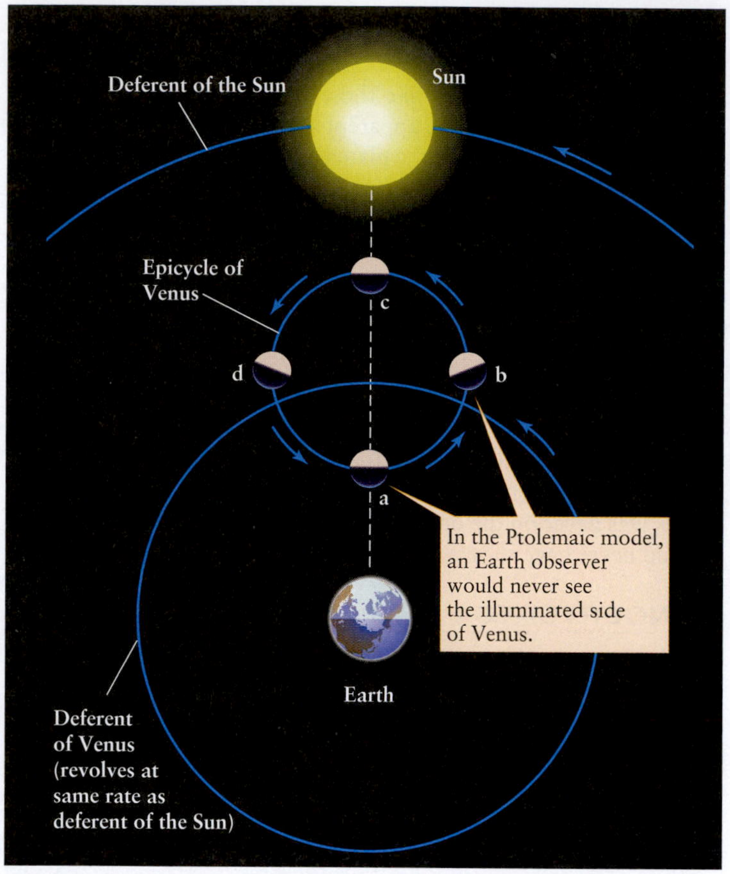

FIGURE 4-15

The Appearance of Venus in the Ptolemaic Model In the geocentric Ptolemaic model the deferents of Venus and the Sun rotate together, with the epicycle of Venus centered on a line (shown dashed) that connects the Sun and Earth. In this model an Earth observer would never see Venus as more than half illuminated. (At positions a and c, Venus appears in a "new" phase; at positions b and d, it appears as a crescent. Compare with Figure 3-2, which shows the phases of the Moon.) Because Galileo saw Venus in nearly fully illuminated phases, he concluded that the Ptolemaic model must be incorrect.

FIGURE 4-16 R I V U X G

Jupiter and Its Largest Moons This photograph, taken by an amateur astronomer with a small telescope, shows the four Galilean satellites alongside an overexposed image of Jupiter. Each satellite is bright enough to be seen with the unaided eye, were it not overwhelmed by the glare of Jupiter. (Rev. Ronald Royer/Science Source)

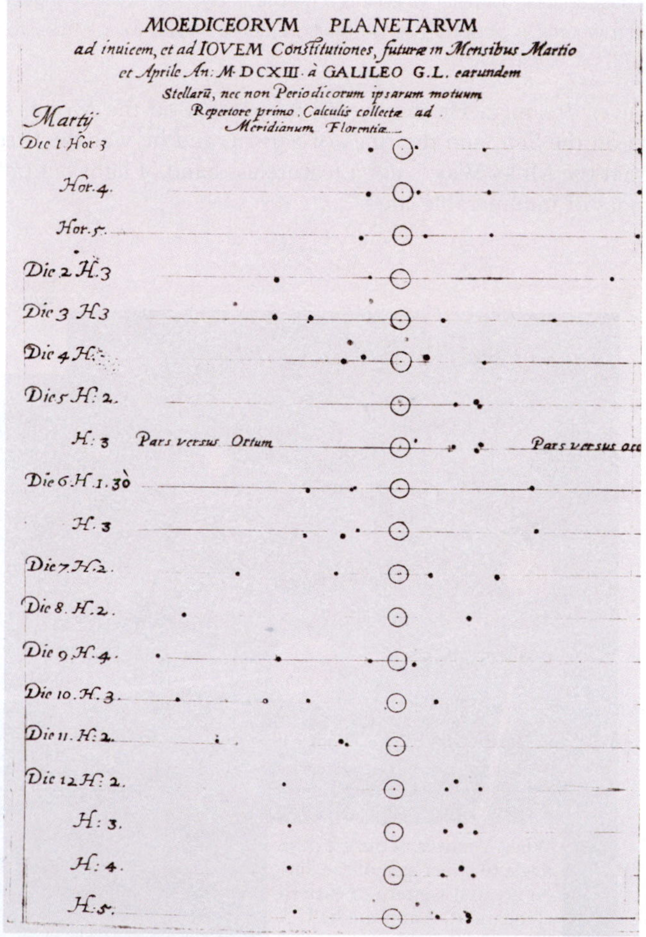

FIGURE 4-17

Early Observations of Jupiter's Moons In 1610 Galileo discovered four "stars" that move back and forth across Jupiter from one night to the next. He concluded that these are four moons that orbit Jupiter, much as our Moon orbits Earth. The circle represents Jupiter and the dots its moons. Compare the drawing numbered 7 with the photograph in Figure 4-16. (Royal Astronomical Society/Science Source)

called the Galilean satellites, orbiting Jupiter (Figure 4-16). He realized that they were orbiting Jupiter because they appeared to move back and forth from one side of the planet to the other. Figure 4-17 shows confirming observations made by Jesuit observers in 1620. Astronomers soon realized that the larger the orbit of one of the moons around Jupiter, the slower that moon moves and the longer it takes that moon to travel around its orbit. These are the same relationships that Copernicus deduced for the motions of the planets around the Sun. Thus, the moons of Jupiter behave like a Copernican system in miniature.

Galileo's telescopic observations constituted the first fundamentally new astronomical data in almost 2000 years. Contradicting prevailing opinion and religious belief, his discoveries strongly suggested a heliocentric structure of the universe. The Roman Catholic Church, which was a powerful political force in Italy and whose doctrine at the time placed Earth at the center of the universe, warned Galileo not to advocate a heliocentric model. He nonetheless persisted and was sentenced to spend the last years of his life under house arrest "for vehement suspicion of heresy." Nevertheless, there was no turning back. (The Roman Catholic Church lifted its ban against Galileo's heliocentric ideas in the 1700s.)

While Galileo's observations showed convincingly that the Ptolemaic model was entirely wrong and that a heliocentric model is the more nearly correct one, he was unable to provide a complete explanation of why Earth should orbit the Sun and not vice versa. The first person who was able to provide such an explanation was the Englishman Isaac Newton, born on Christmas Day of 1642, a dozen years after the death of Kepler and the same year that Galileo died. While Kepler and Galileo revolutionized our understanding of planetary motions, Newton's contribution was far greater: He deduced the basic laws that govern all motions on Earth as well as in the heavens.

CONCEPTCHECK 4-14

Why had Jupiter's moons not been observed prior to Galileo's time?

Answer appears at the end of the chapter.

4-6 Newton formulated laws of motion and gravity that describe fundamental properties of physical reality

TUTORIAL 4-2

Until the mid-seventeenth century, virtually all attempts to describe the motions of the heavens were *empirical*, or based directly on data and observations. From Ptolemy to Kepler, astronomers would adjust their ideas and calculations by trial and error until they ended up with answers that agreed with observation.

> The same laws of motion that hold sway on Earth apply throughout the universe

Isaac Newton (Figure 4-18) introduced a new approach. He began with three quite general statements, now called Newton's laws of motion. These laws, deduced from experimental observation, apply to all forces and all objects. Newton then showed that Kepler's three laws follow logically from these laws of motion and from a formula for the force of gravity that he derived from observation.

In other words, Kepler's laws are not just an empirical description of the motions of the planets, but a direct consequence of the fundamental laws of physical matter. Using this deeper insight into the nature of motions in the heavens, Newton and his successors were able to accurately describe not just the orbits of the planets but also the orbits of the Moon and comets.

Newton's First Law

Newton's laws of motion describe objects on Earth as well as in the heavens. Thus, we can understand each of these laws by considering the motions of objects around us. We begin with **Newton's first law of motion:**

An object remains at rest, or moves in a straight line at a constant speed, unless acted upon by a net outside force.

By **force** we mean any push or pull that acts on the object. An *outside* force is one that is exerted on the object by something other than the object itself. The net, or total, outside force is the combined effect of all of the individual outside forces that act on the object.

FIGURE 4-18
Isaac Newton (1642–1727) Using mathematical techniques that he devised, Isaac Newton formulated the law of universal gravitation and demonstrated that the planets orbit the Sun according to simple mechanical rules. (Corbis Images)

Right now, you are demonstrating the first part of Newton's first law—remaining at rest. As you sit in your chair reading this passage, there are two outside forces acting on you: The force of gravity pulls you downward, and the chair pushes up on you. These two forces are of equal strength but of opposite direction, so their effects cancel one another—there is no *net* outside force. Hence, your body remains at rest as stated in Newton's first law. If you try to lift yourself out of your chair by grabbing your knees and pulling up, you will remain at rest because this force is not an outside force: It comes from your body.

CAUTION! It is easy to confuse the *net* (or total) outside force on an object (central to Newton's first law) with *individual* outside forces on an object. If you want to make a book move across the floor in a straight line at a *constant speed,* you must continually push on it. You might therefore think that your push is a net outside force. But another force also acts on the book—the force of friction as the book rubs across the floor. The force of your push and the force of friction combine to make the *net* outside force. If you push the book at a constant speed, then the force of your push exactly balances the force of friction, so there is no net outside force. If you stop pushing, there will be nothing to balance the effects of friction. Then the friction will be a net outside force and the book will slow to a stop.

Newton's first law tells us that if no net outside force acts on a moving object, it can only move in a straight line and at a constant speed. This means that a net outside force *must* be acting on the planets since they don't move in straight lines but instead move around elliptical paths. Another way to see that planetary orbits require a net outside force is that a planet would fly off into space at a constant speed along a straight line if there were no net outside force acting on it. Because planets don't fly off, Newton concluded that a force *must* act continuously on the planets to keep them in their elliptical orbits.

CONCEPTCHECK 4-15

Imagine a 10-kg rock speeding through empty space at 200 m/s, so far away from other objects that there is no gravitational force (or any other outside forces) exerted on the rock. Describe the rock's motion.

Answer appears at the end of the chapter.

Newton's Second Law

Newton's second law describes how the motion of an object *changes* if there is a net outside force acting on it. To appreciate Newton's second law, we must first understand three quantities that describe motion—speed, velocity, and acceleration.

Speed is a measure of how fast an object is moving. Speed and direction of motion together constitute an object's **velocity.** Compared with a car driving north at 100 km/h (62 mi/h), a car driving east at 100 km/h has the same speed but a different velocity. We can restate Newton's first law to say that an object has a constant velocity (its speed and direction of motion do not change) if no net outside force acts on the object.

Acceleration is the rate at which velocity changes. Because velocity involves both speed and direction, acceleration can result from changes in either. Contrary to popular use of the term, acceleration does not simply mean speeding up. A car is accelerating if it is speeding up, and it is also accelerating if it is slowing down or turning (that is, changing the direction in which it is moving).

You can verify these statements about acceleration if you think about the sensations of riding in a car. If the car is moving with a constant velocity (in a straight line at a constant speed), you feel the same, aside from vibrations, as if the car were not moving at all. But you can feel it when the car accelerates in any way: You feel thrown back in your seat if the car speeds up, thrown forward if the car slows down, and thrown sideways if the car changes direction in a tight turn. In Box 4-3 we discuss the reasons for these sensations, along with other applications of Newton's laws to everyday life.

An apple falling from a tree is a good example of acceleration that involves only an increase in speed. Initially, at the moment the stem breaks, the apple's speed is zero. After 1 second, its downward speed is 9.8 meters per second, or 9.8 m/s (32 feet per second, or 32 ft/s). After 2 seconds, the apple's speed is twice this, or 19.6 m/s. After 3 seconds, the speed is 29.4 m/s. Because the apple's speed increases by 9.8 m/s for each second of free fall, the rate of acceleration is 9.8 meters per second per second, or 9.8 m/s^2 (32 ft/s^2). Thus, Earth's gravity gives the apple a constant acceleration of 9.8 m/s^2 downward, toward the center of Earth.

A planet revolving about the Sun along a perfectly circular orbit is an example of acceleration that involves change of direction only. As the planet moves along its orbit, its speed remains constant. Nevertheless, the planet is continuously being accelerated because its direction of motion is continuously changing.

Newton's second law of motion says that in order to give an object an acceleration (that is, to change its velocity), a net outside force *must* act on the object. To be specific, this law says that the acceleration of an object is proportional to the net outside force acting on the object. That is, the harder you push on an object, the greater the resulting acceleration. This law can be succinctly stated as an equation. If a net outside force F acts on an object of mass m, the object will experience an acceleration a such that

Newton's second law

$$F = ma$$

F = net outside force on an object

m = mass of object

a = acceleration of object

The **mass** of an object is a measure of the total amount of material in the object. It is usually expressed in kilograms (kg) or grams (g). For example, the mass of the Sun is 2×10^{30} kg, the mass of a hydrogen atom is 1.7×10^{227} kg, and the mass of an average adult is 75 kg. The Sun, a hydrogen atom, and a person have these masses regardless of where they happen to be in the universe.

CAUTION! It is important not to confuse the concepts of mass and weight. **Weight** is the force of gravity that acts on an object and, like any force, is usually expressed in pounds or newtons (1 newton = 0.225 pounds). For example, astronauts feel lighter on the Moon because they weigh less in the Moon's weaker gravity, but an astronaut's mass on the Moon is the same as her mass on Earth.

We can use Newton's second law to relate mass and weight. We have seen that the acceleration caused by Earth's gravity is 9.8 m/s^2. When a 50-kg swimmer falls from a diving board, the only outside force acting on her as she falls is her weight. Thus, from Newton's second law ($F = ma$), her weight is equal to her mass multiplied by the acceleration due to gravity:

$$50 \text{ kg} \times 9.8 \text{ m/s}^2 = 490 \text{ newtons} = 110 \text{ pounds}$$

Note that this answer is correct only when the swimmer is on Earth. She would weigh less on the Moon, where the pull of gravity is weaker, and more on Jupiter, where the gravitational pull is stronger. Floating deep in space, she would have no weight at all; she would be "weightless." Nevertheless, in all these circumstances, she would always have exactly the same mass, because mass is an inherent property of matter unaffected by details of the environment. Whenever we describe the properties of planets, stars, or galaxies, we speak of their masses, never of their weights.

We have seen that a planet is continually accelerating as it orbits the Sun. From Newton's second law, this means that there must be a net outside force that acts continually on each of the planets. As we will see in the next section, this force is the gravitational attraction of the Sun.

BOX 4-3 ASTRONOMY DOWN TO EARTH

Newton's Laws in Everyday Life

In our study of astronomy, we use Newton's three laws of motion to help us understand the motions of objects in the heavens. But you can see applications of Newton's laws every day in the world around you. By considering these everyday applications, we can gain insight into how Newton's laws apply to celestial events that are far removed from ordinary human experience.

Newton's *first* law, or principle of inertia, says that an object at rest naturally tends to remain at rest and that an object in motion naturally tends to remain in motion. This law explains the sensations that you feel when riding in an automobile. When you are waiting at a red light, your car and your body are both at rest. When the light turns green and you press on the gas pedal, the car accelerates forward but your body attempts to stay where it was. Hence, the seat of the accelerating car pushes forward into your body, and it feels as though you are being pushed back in your seat.

Once the car is up to cruising speed, your body wants to keep moving in a straight line at this cruising speed. If the car makes a sharp turn to the left, the right side of the car will move toward you. Thus, you will feel as though you are being thrown to the car's right side (the side on the outside of the turn). If you bring the car to a sudden stop by pressing on the brakes, your body will continue moving forward until the seat belt stops you. In this case, it feels as though you are being thrown toward the front of the car.

Newton's *second* law states that the net outside force on an object equals the product of the object's mass and its acceleration. You can accelerate a crumpled-up piece of paper to a pretty good speed by throwing it with a moderate force. But if you try to throw a heavy rock by using the same force, the acceleration will be much less because the rock has much more mass than the crumpled paper. Because of the smaller acceleration, the rock will leave your hand moving at only a slow speed.

Automobile airbags are based on the relationship between force and acceleration. It takes a large force to bring a fast-moving object suddenly to rest because this requires a large acceleration. In a collision, the driver of a car not equipped with airbags is jerked to a sudden stop and the large forces that act can cause major injuries. But if the car has airbags that deploy in an accident, the driver's body will slow down more gradually as it contacts the airbag, and the driver's acceleration will be less. (Remember that *acceleration* can refer to slowing down as well as to speeding up.) Hence, the force on the driver and the chance of injury will both be greatly reduced.

Newton's *third* law, the principle of action and reaction, explains how a car can accelerate at all. It is not correct to say that the engine pushes the car forward, because Newton's second law tells us that it takes a force acting from outside the car to make the car accelerate. Rather, the engine makes the wheels and tires turn, and the tires push backward on the ground. (You can see this backward force in action when a car drives through wet ground and sprays mud backward from the tires.) From Newton's third law, the ground must exert an equally large forward force on the car, and this is the force that pushes the car forward.

You use the same principles when you walk: You push backward on the ground with your foot, and the ground pushes forward on you. Icy pavement or a freshly waxed floor have greatly reduced friction. In these situations, your feet and the surface under you can exert only weak forces on each other, and it is much harder to walk.

Newton's Third Law

The last of Newton's general laws of motion is called **Newton's third law of motion:**

Whenever one object exerts a force on a second object, the second object exerts an oppositely directed force of equal strength on the first object.

Newton's third law is sometimes described in terms of actions and reactions: When two objects interact by exerting forces on each other, these action and reaction forces are equal in magnitude but opposite in direction.

For example, consider your weight; Earth pulls down on you with a gravitational force equal to your weight. But you also pull back up on Earth with a gravitational force of equal strength. In another example, consider a bat hitting a ball. Clearly, the bat exerts a large force on the ball when hitting a home run (this can be called an action force). However, the ball exerts a force of *equal strength* on the bat as well (this can be called a reaction force). It might seem totally counterintuitive that a small, passive ball could exert an equally large force on a bat that has someone's full swing behind it. However, imagine a spring between the ball and bat during the hit; a spring is a very good approximation for the deformed ball. Regardless of which object moves in order to compress the spring, the spring pushes outward with an equal force on both ends (outward on the ball and bat).

CONCEPTCHECK 4-16

Two sumo wrestlers push against each other during a match. One wrestler is much larger than the other. The larger wrestler's feet remain on the floor while the smaller wrestler's feet slip as he is accelerated in a push right out of the ring. Compare the force that the larger wrestler exerts on the smaller wrestler to the force the smaller wrestler exerts on the larger wrestler.

CONCEPTCHECK 4-17

In midair after stepping off a diving board, a diver is pulled down to the water by her weight. However, the diver pulls up on Earth with a force of equal strength, so why doesn't Earth move an equal amount as the diver?

Answers appears at the end of the chapter.

In these examples, you can think of each force in these pairs as a reaction to the other force, which is the origin of the phrase "action and reaction."

Newton realized that because the Sun is exerting a force on each planet to keep it in orbit, each planet must also be exerting an equal and opposite force on the Sun. However, the planets are much less massive than the Sun (for example, Earth has only 1/300,000 of the Sun's mass). Therefore, although the Sun's force on a planet is the same as the planet's force on the Sun, the planet's much smaller mass gives it a much larger acceleration, according to Newton's second law. This is why the planets circle the Sun instead of vice versa. Thus, Newton's laws reveal the reason for our heliocentric solar system.

The Law of Universal Gravitation

Tie a ball to one end of a piece of string, hold the other end of the string in your hand, and whirl the ball around in a circle. As the ball "orbits" your hand, it is continuously accelerating because its velocity is changing. (Even if its speed is constant, its direction of motion is changing.) In accordance with Newton's second law, this can happen only if the ball is continuously acted on by an outside force—the pull of the string. The pull is directed along the string toward your hand. In the same way, Newton saw, the force that keeps a planet in orbit around the Sun is a pull that always acts toward the Sun. That pull is **gravity**, or **gravitational force.**

Newton's law of gravitation is truly universal: It applies to falling apples as well as to planets and galaxies

Newton's discovery about the forces that act on planets led him to suspect that the force of gravity pulling a falling apple straight down to the ground is fundamentally the same as the force on a planet that is always directed straight at the Sun. In other words, gravity is the force that shapes the orbits of the planets. What is more, he was able to determine how the force of gravity depends on the distance between the Sun and the planet. His result was a law of gravitation that could apply to the motion of distant planets as well as to the flight of a football on Earth. Using this law, Newton achieved the remarkable goal of deducing Kepler's laws from fundamental principles of nature.

To see how Newton reasoned, think again about a ball attached to a string. If you use a short string, so that the ball orbits in a small circle, and whirl the ball around your hand at a high speed, you will find that you have to pull fairly hard on the string (Figure 4-19a). But if you use a longer string, so that the ball moves in a larger orbit, and if you make the ball orbit your hand at a slow speed, you only have to exert a light tug on

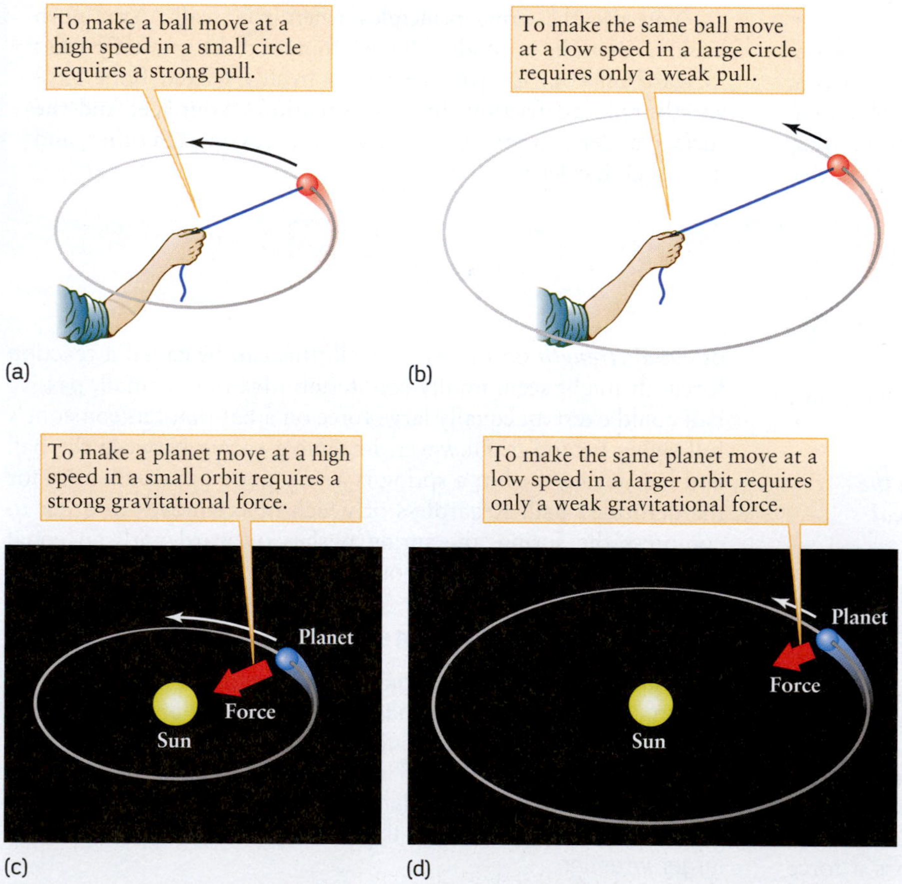
To make a ball move at a high speed in a small circle requires a strong pull.
To make the same ball move at a low speed in a large circle requires only a weak pull.
(a)
(b)
To make a planet move at a high speed in a small orbit requires a strong gravitational force.
To make the same planet move at a low speed in a larger orbit requires only a weak gravitational force.
Planet
Force
Sun
Planet
Force
Sun
(c)
(d)

FIGURE 4-19
An Orbit Analogy **(a)** To make a ball on a string move at high speed around a small circle, you have to exert a substantial pull on the string. **(b)** If you lengthen the string and make the same ball move at low speed around a large circle, much less pull is required. **(c)** Similarly, a planet that orbits close to the Sun moves at high speed and requires a substantial gravitational force from the Sun, while **(d)** a planet in a large orbit moves at low speed and requires less gravitational force to stay in orbit.

the string (Figure 4-19b). The orbits of the planets behave in the same way: The larger the size of the orbit, the slower the planet's speed (Figure 4-19c, d). By analogy to the force of the string on the orbiting ball, Newton concluded that the force that attracts a planet toward the Sun must decrease with increasing distance between the Sun and the planet.

Using his own three laws and Kepler's three laws, Newton succeeded in formulating a general statement that describes the nature of the gravitational force. Newton's **law of universal gravitation** is as follows:

Two objects attract each other with a force that is directly proportional to the mass of each object and inversely proportional to the square of the distance between them.

This law states that *any* two objects exert gravitational pulls on each other. Normally, you notice only the gravitational force that Earth exerts on you, otherwise known as your weight. In fact, you are gravitationally attracted to *all* the objects around you. For example, a book exerts a gravitational force on you as you read it. But because the force exerted on you by this book is proportional to the book's mass, which is very small compared to Earth's mass, the force is too small to notice. (It can actually be measured with sensitive equipment.)

The farther apart two objects are, the weaker the gravitational force between them. Since the gravitational force weakens by the square of the distance between two objects, doubling their distance reduces their attraction by

$$\frac{1}{2^2} = \frac{1}{4}$$

and tripling their distance reduces their attraction by

$$\frac{1}{3^2} = \frac{1}{9}$$

Newton's law of universal gravitation can be stated as an equation. If two objects have masses m_1 and m_2 and are separated by a distance r, then the gravitational force F between these two objects is given by the following equation:

Newton's law of universal gravitation

$$F = G\left(\frac{m_1 m_2}{r^2}\right)$$

F = gravitational force between two objects

m_1 = mass of first object

m_2 = mass of second object

r = distance between objects

G = universal constant of gravitation

If the masses are measured in kilograms and the distance between them in meters, then the force is measured in newtons. In this formula, G is a number called the **universal constant of gravitation.** Laboratory experiments have yielded a value for G of

$$G = 6.67 \times 10^{-11} \text{ newton} \cdot \text{m}^2/\text{kg}^2$$

We can use Newton's law of universal gravitation to calculate the force with which any two objects attract each other. For example, to compute the gravitational force that the Sun exerts on Earth, we substitute values for Earth's mass ($m_1 = 5.98 \times 10^{24}$ kg), the Sun's mass ($m_2 = 1.99 \times 10^{30}$ kg), the distance between them (r = 1 AU = 1.5×10^{11} m), and the value of G into Newton's equation. We get

$$F_{\text{Sun-Earth}} = 6.67 \times 10^{-11}\left[\frac{(5.98 \times 10^{24}) \times (1.99 \times 10^{30})}{(1.50 \times 10^{11})^2}\right]$$

$$= 3.53 \times 10^{22} \text{ newtons}$$

If we calculate the force that Earth exerts on the Sun, we get exactly the same result. (Mathematically, we just let m_1 be the Sun's mass and m_2 be Earth's mass instead of the other way around. The product of the two numbers is the same, so the force is the same.) This is in accordance with Newton's third law: Any two objects exert *equal* gravitational forces on each other.

Your weight is just the gravitational force that Earth exerts on you, so we can calculate it using Newton's law of universal gravitation. Earth's mass is $m_1 = 5.98 \times 10^{24}$ kg, and the distance r to use is the distance between the *centers* of Earth and you. This distance is just the radius of Earth, which is r = 6378 km = 6.378×10^6 m. If your mass is m_2 = 50 kg, your weight is

$$F_{\text{Earth-you}} = 6.67 \times 10^{-11}\left[\frac{(5.98 \times 10^{24}) \times (50)}{(6.378 \times 10^{6})^2}\right]$$

$$= 490 \text{ newtons}$$

This value is the same as the weight of a 50-kg person that we calculated in Section 4-6. This example shows that your weight would have a different value on a planet with a different mass m_1 and a different radius r.

CONCEPTCHECK 4-18

How much does the gravitational force of attraction change between two asteroids if the two asteroids drift 3 times closer together?

CALCULATIONCHECK 4-2

How much would a 75-kg astronaut, weighing about 165 pounds on Earth, weigh in newtons and in pounds if he were standing on Mars, which has a mass of 6.4×10^{23} kg and a radius of 3.4×10^6 m?

Answers appear at the end of the chapter.

4-7 Describing orbits with energy and gravity

Imagine a cannonball shot into the air; it arcs through the sky before it crashes back to the ground. If you want to shoot the cannonball into orbit around Earth (like a satellite), you might guess that the ball must be shot faster to give it more energy. This guess is correct, and next we'll look at how to understand orbits in terms of gravity and energy.

Different Forms of Energy

Energy comes in a variety of familiar forms. For example, consider a car: The faster a car moves, the greater its energy. In general, the greater an object's speed, the more **kinetic energy** it has (Figure 4-20). The word *kinetic* refers to motion, so kinetic energy is the energy of motion. However, the mass of an object also contributes to its kinetic energy, and a speeding car therefore has more kinetic energy than a speeding bullet.

Now let's look at some other forms of energy. Both food and batteries store chemical energy, although they each contain very different chemicals. When energy is stored, we say it has the potential to be used, and this stored energy is often called *potential energy.* Consider a girl on the diving board of a swimming pool (Figure 4-21); gravity has the potential to pull her into the water (whenever she steps off the board). This is an example of stored gravitational energy, or **gravitational potential energy.** The higher the diving board, the more gravitational potential energy she has. After she steps off the board, she'll have both kinetic energy and gravitational potential energy. As she falls towards the water, her kinetic energy increases and her gravitational potential energy decreases.

While the concept of energy is easier to introduce with a diving board, the same principles apply to orbiting objects: Planets, moons, and satellites also have both kinetic and gravitational potential energy. For example, the faster a satellite orbits Earth, the greater its kinetic energy, and if it falls back to Earth, it loses gravitational potential energy.

To understand another aspect of energy, let's think more about the diver in Figure 4-21. Standing high atop the board, the diver has more gravitational potential energy than she would have while standing on the ground. In midair during the dive, some of that gravitational potential energy has transformed into kinetic energy. This illustrates a law called the **conservation of energy:** While energy can change from one form to another, energy cannot be created or destroyed.

There's another way to express the law of conservation of energy. First consider a collection of objects that are *isolated,* which means no energy can leave or enter this collection of objects. Second, find the total energy by taking into account all the forms of energy the collection of objects has. This total energy is conserved, meaning that the total energy stays constant over time, even as objects interact with each other and the various forms of energy can change.

CONCEPTCHECK 4-19

Consider a cannonball shot vertically, straight up into the air. At the ball's highest point, it momentarily comes to a complete stop before beginning to fall back down. At its highest point, what form of energy does the ball have, and where did the energy come from?

Answer appears at the end of the chapter.

As a feather falls gently to the ground it loses gravitational potential energy without gaining much kinetic energy. This is an example of **air drag,** where the kinetic energy of an object is transferred to the molecules in the surrounding air and the air gets hotter. Energy is still conserved—as it always is—and the nearby air molecules actually speed up. Since the kinetic energy of individual particles gives a gas its temperature and thermal energy, we can see that air drag converts kinetic energy into thermal energy (we will learn more about thermal energy in Section 5-3). Air drag can be good or bad, depending on the situation. Astronauts returning to Earth rely on this transfer of energy, as they use air drag during atmospheric reentry to reduce their speed for a safe landing. As we will see shortly, air drag is undesirable when it transfers energy away from an orbiting satellite, causing it to fall back to Earth.

Gravitational Force, Energy, and Orbits

Because there is a gravitational force between any two objects, Newton concluded that gravity is also the force that keeps the Moon in orbit around Earth. It is also the force that keeps artificial satellites in orbit. But if the force of gravity attracts two objects to each other, why don't satellites immediately fall to

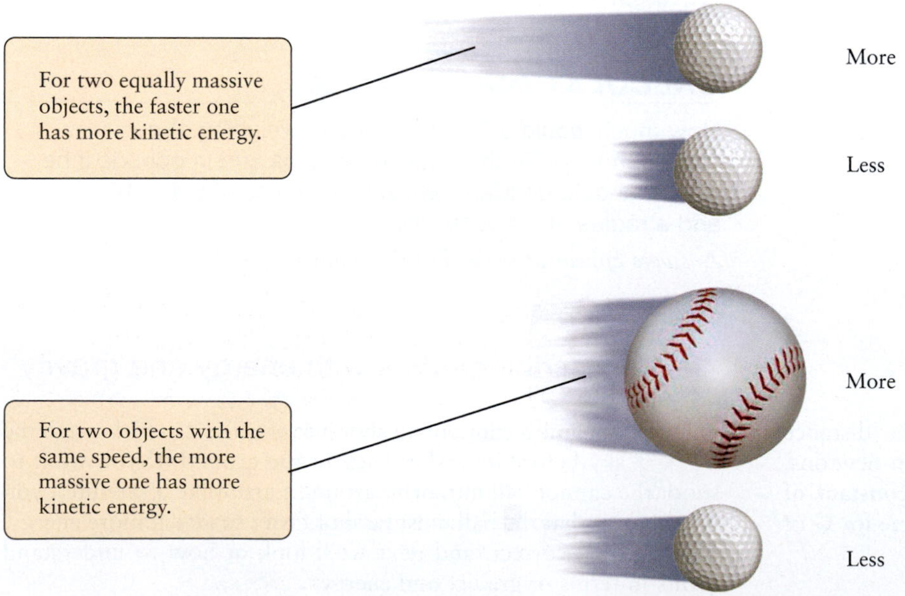
For two equally massive objects, the faster one has more kinetic energy.
More
Less
For two objects with the same speed, the more massive one has more kinetic energy.
More
Less

FIGURE 4-20
Kinetic Energy The kinetic energy of an object depends on both its speed and its mass.

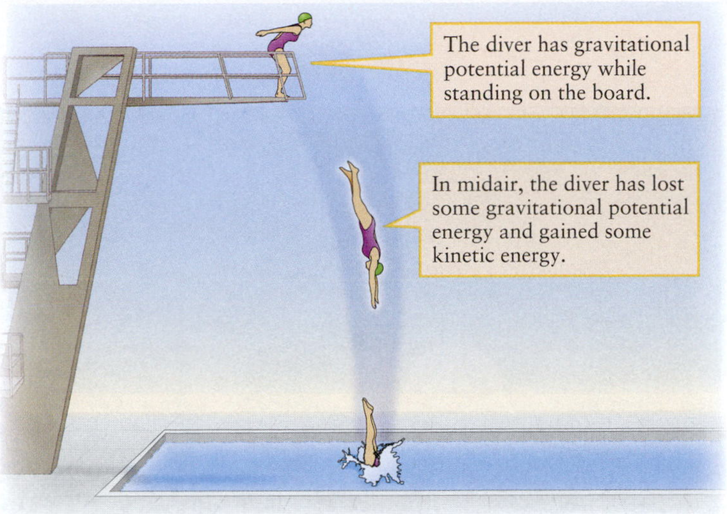

The diver has gravitational potential energy while standing on the board.
In midair, the diver has lost some gravitational potential energy and gained some kinetic energy.

FIGURE 4-21

Gravitational Potential Energy The higher the diver, the more gravitational potential energy she has. As the diver descends, some of her gravitational potential energy has transformed into kinetic energy.

Earth? Why doesn't the Moon fall into Earth? And, for that matter, why don't the planets fall into the Sun?

To see the answer, imagine (as Newton did) dropping a ball from a great height above Earth's surface, as in Figure 4-22. After you drop the ball, it, of course, falls straight down (path A in Figure 4-22). But if you *throw* the ball horizontally, it travels some distance across Earth's surface before hitting the ground (path B). If you throw the ball harder, it travels a greater distance (path C). If you could throw or shoot it at just the right speed, the curvature of the ball's path will exactly match the curvature of Earth's surface (path E). Although Earth's gravity is making the ball fall, Earth's surface is falling away under the ball at the same rate. Hence, the ball does not get any closer to the surface, and the ball is in circular orbit. So, the ball in path E is in fact falling, but it is falling *around* Earth rather than *toward* Earth.

As the hypothetical ball is launched faster and faster, you can also see that the paths (going from A to F) represent trajectories with more of both kinetic and gravitational potential energy; in other words, with increasing **orbital energy.** In general for an object orbiting the Sun or a planet:

The greater the orbital energy, the greater the average orbital distance (or the greater the semimajor axis, a).

A spacecraft is launched into orbit in just this way—by gaining enough speed and energy. Once the spacecraft is in orbit, no more rocket fuel is needed and, if not for air drag, the spacecraft could orbit indefinitely. However, air drag from the thin outer wisps of Earth's atmosphere slowly removes orbital energy from the closer satellites, slowing them down, and bringing them inward. Without occasional bursts of thrust, these satellites would fall back to Earth. The International Space Station is no exception, and air drag can decrease the station's altitude by a couple hundred feet each day!

CAUTION! An astronaut on board an orbiting spacecraft (like the one shown in the photograph that opens this chapter) feels "weightless." However, this is *not* because she is "beyond the pull of gravity." The astronaut is herself an independent satellite of Earth, and Earth's gravitational pull is what holds her in orbit. She feels "weightless" because she and her spacecraft are falling *together* around Earth, so there is nothing pushing her against any of the spacecraft walls. You feel the same "weightless" sensation whenever you are falling, such as when you jump off a diving board or ride the free-fall ride at an amusement park.

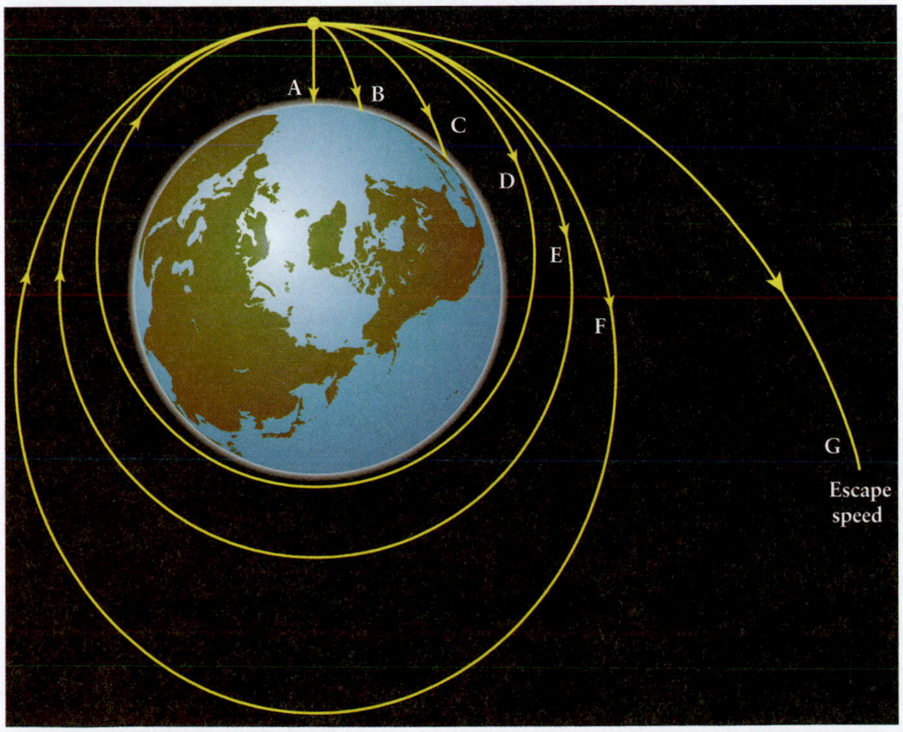

ANIMATION 4-7

FIGURE 4-22

Orbits and the Escape Speed If a ball is dropped from a great height above Earth's surface, it falls straight down (A). If the ball is thrown with some horizontal speed, it follows a curved path before hitting the ground (B, C). If thrown with just the right speed (E), the ball goes into circular orbit; the ball's path curves but it never gets any closer to Earth's surface. If the ball is thrown with a speed that is slightly less (D) or slightly more (F) than the speed for a circular orbit, the ball's orbit is an ellipse. If thrown faster than the escape speed (G), the ball will leave Earth and never return.

CONCEPTCHECK 4-20

Suppose a spacecraft is initially shot into a circular orbit with path E in Figure 4-22. Which new path will the spacecraft likely take if air drag is present? Which new path will likely result if a rocket thruster is fired?

CONCEPTCHECK 4-21

As Earth orbits the Sun, does Earth's orbit decay due to air drag from its own atmosphere?

Answers appear at the end of the chapter.

The Escape Speed

If an object is hurled with enough speed, it can escape a planet altogether. For example, when a very large asteroid smashes into Earth, most of the debris ejected from the crater falls back to Earth's surface, but some rocks have enough speed to escape and roam the solar system. In Figure 4-22, the object on path G is shot with the **escape speed,** has more energy than any of the "bound" orbits, and never returns to Earth's orbit. The speed needed for escape depends on the size and mass of the planet, but *not* on the mass of the escaping object:

Escape Speed

$$v_{\text{escape}} = \sqrt{\frac{2GM}{R}}$$

V_{escape} = escape speed
M = mass of the planet
R = radius of the planet
G = universal constant of gravitation

For Earth, the escape speed is 11.2 km/s. That's about 25,000 mi/h, or 33 times the speed of sound. This speed is so fast that, except for the largest asteroid impacts, very little debris would have escaped Earth. However, the escape speed for Mars is only about 5 km/s, so compared to Earth, it is easier for impact debris from Mars to get ejected into the solar system. (In fact, about 100 rocks found on Earth were originally ejected from Mars by asteroid impacts.)

What about spacecraft; are they launched at the escape speed? The escape speed assumes a simple scenario where no air drag is present and where there is no rocket fuel to add energy during flight. However, even with fuel, spacecraft going to the Moon or beyond often leave Earth near the escape speed.

CONCEPTCHECK 4-22

Two objects, a large rock and a small rock, are shot with just enough speed to escape the Moon. Can both rocks be shot with equal speed? Can both rocks be shot with the same kinetic energy?

CONCEPTCHECK 4-23

New planets around other stars are constantly being discovered. Would it be possible to find a planet with the same mass as Earth but with a lower escape speed from its surface?

Answers appear at the end of the chapter.

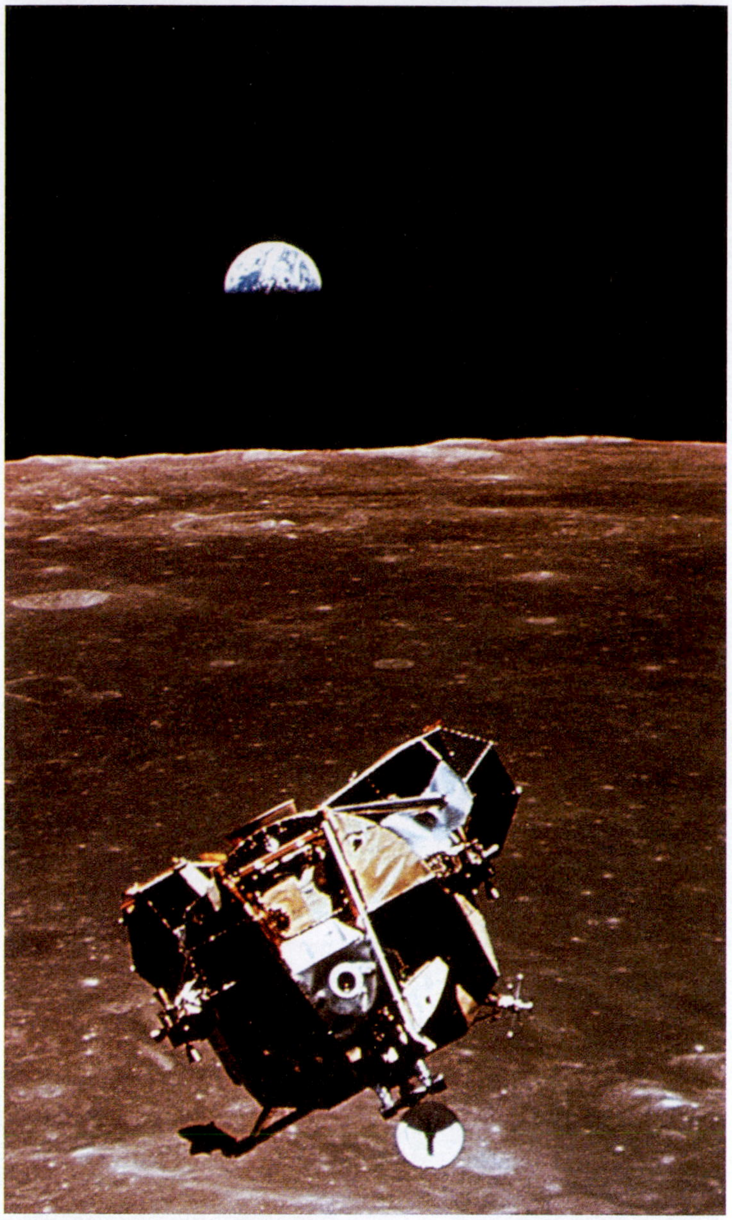

FIGURE 4-23 R I V U X G
In Orbit Around the Moon This photograph taken from the spacecraft *Columbia* shows the lunar lander *Eagle* after returning from the first human landing on the Moon in July 1969. Newton's form of Kepler's third law describes the orbit of a spacecraft around the Moon, as well as the orbit of the Moon around Earth (visible in the distance). (Michael Collins, *Apollo 11,* NASA)

Gravitation and Kepler's Laws

Using his three laws of motion and his law of gravity, Newton found that he could derive Kepler's three laws mathematically. (Newton also invented calculus, which helped in this endeavor!) While Kepler's laws describe the planets in the solar system specifically, Newton showed that similar behavior results from any orbital system, such as a planet and its moons. Kepler's first law, concerning the elliptical shape of planetary orbits, proved to be a direct consequence of the $1/r^2$ factor in the law of universal gravitation. The law of equal areas, or Kepler's second law, turns out to be a consequence of the Sun's gravitational force on a planet being directed straight toward the Sun.

Newton also demonstrated that Kepler's third law follows logically from his law of gravity. Specifically, he proved that if two objects with masses m_1 and m_2 orbit each other, the period P of their orbit and the semimajor axis a of their orbit (that is, the average distance between the two objects) are related by an equation that we call **Newton's form of Kepler's third law:**

$$P^2 = \left[\frac{4\pi^2}{G(m_1 + m_2)}\right]a^3$$

Newton's form of Kepler's third law is valid whenever two objects orbit each other because of their mutual gravitational attraction (Figure 4-23). It is invaluable in the study of binary star systems, in which two stars orbit each other. If the orbital period P and semimajor axis a of the two stars in a binary system are known, astronomers can use this formula to calculate the sum $m_1 + m_2$ of the masses of the two stars. Within our own solar system, Newton's form of Kepler's third law makes it possible to learn about the masses of planets. By measuring the period and semimajor axis for a satellite, astronomers can determine the sum of the masses of the planet and the satellite. (The satellite can be a moon of the planet or a spacecraft that we place in orbit around the planet. Newton's laws apply in either case.) Box 4-4 gives an example of using Newton's form of Kepler's third law.

Newton also discovered new features of orbits around the Sun. For example, his equations soon led him to conclude that the orbit of an object around the Sun need not be an ellipse. It could be any one of a family of curves called conic sections.

BOX 4-4 TOOLS OF THE ASTRONOMER'S TRADE

Newton's Form of Kepler's Third Law

Kepler's original statement of his third law, $P^2 = a^3$, is valid only for objects that orbit the Sun. (Box 4-2 shows how to use this equation.) But Newton's form of Kepler's third law is much more general: It can be used in *any* situation where two objects of masses m_1 and m_2 orbit each other. For example, Newton's form is the equation to use for a moon orbiting a planet or a satellite orbiting Earth. This equation is

Newton's form of Kepler's third law:

$$P^2 = \left[\frac{4\pi^2}{G(m_1 + m_2)}\right]a^3$$

P = sidereal period of orbit, in seconds

a = semimajor axis of orbit, in meters

m_1 = mass of first object, in kilograms

m_2 = mass of second object, in kilograms

G = universal constant of gravitation = 6.67×10^{-11}

Notice that P, a, m^1, and m^2 *must* be expressed in these particular units. If you fail to use the correct units, your answer will be incorrect.

EXAMPLE: Io (pronounced "eye-oh") is one of the four large moons of Jupiter discovered by Galileo and shown in Figure 4-16. It orbits at a distance of 421,600 km from the center of Jupiter and has an orbital period of 1.77 days. Determine the combined mass of Jupiter and Io.

Situation: We are given Io's orbital period P and semimajor axis a (the distance from Io to the center of its circular orbit, which is at the center of Jupiter). Our goal is to find the sum of the masses of Jupiter (m_1) and Io (m_2).

Tools: Because this orbit is not around the Sun, we must use Newton's form of Kepler's third law to relate P and a. This relationship also involves m_1 and m_2, whose sum $(m_1 + m_2)$ we are asked to find.

Answer: To solve for $m_1 + m_2$, we rewrite the equation in the form

$$m_1 + m_2 = \frac{4\pi^2 a^3}{GP^2}$$

To use this equation, we have to convert the distance a from kilometers to meters and convert the period P from days to seconds. There are 1000 meters in 1 kilometer and 86,400 seconds in 1 day, so

$$a = (421{,}600 \text{ km}) \times \frac{1000 \text{ m}}{1 \text{ km}} = 4.216 \times 10^8 \text{ m}$$

$$P = (1.77 \text{ days}) \times \frac{86{,}400 \text{ s}}{1 \text{ day}} = 1.529 \times 10^5 \text{ s}$$

We can now put these values and the value of G into the above equation:

$$m_1 + m_2 = \frac{4\pi^2(4.216 \times 10^8)^3}{(6.67 \times 10^{-11})(1.529 \times 10^5)^2} = 1.90 \times 10^{27} \text{ kg}$$

Review: Io is very much smaller than Jupiter, so its mass is only a small fraction of the mass of Jupiter. Thus, $m_1 + m_2$ is very nearly the mass of Jupiter alone. We conclude that Jupiter has a mass of 1.90×10^{27} kg, or about 300 times the mass of Earth. This technique can be used to determine the mass of any object that has a second, much smaller object orbiting it. Astronomers use this technique to find the masses of stars, black holes, and entire galaxies of stars.

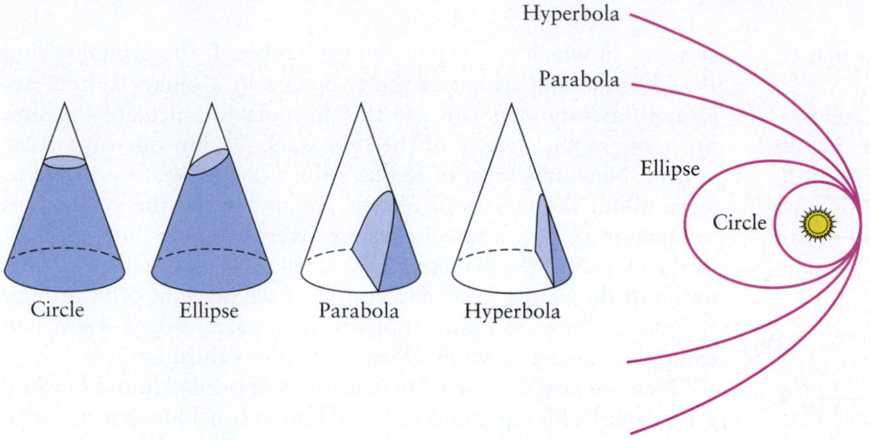

FIGURE 4-24
Conic Sections A conic section is any one of a family of curves obtained by slicing a cone with a plane. The orbit of one object about another can be any one of these curves: a circle, an ellipse, a parabola, or a hyperbola.

A **conic section** is any curve that you get by cutting a cone with a plane, as shown in Figure 4-24. You can get circles and ellipses by slicing all the way through the cone, and these represent the familiar circular and elliptical orbits. You can also get two types of open curves called **parabolas** and **hyperbolas.** An object traveling at the escape speed follows a parabolic curve, and for objects traveling even faster than the escape speed, the path is hyperbolic. Comets hurtling toward the Sun from the depths of space sometimes follow hyperbolic orbits and never return.

CONCEPTCHECK 4-24

A small celestial object on a hyperbolic orbit passes by a lonely moonless planet. Can the planet capture the object to acquire a moon?

Answer appears at the end of the chapter.

The Triumph of Newtonian Mechanics

Newton's ideas turned out to be applicable to an incredibly wide range of situations. Using his laws of motion, Newton himself proved that Earth's axis of rotation must precess because of the gravitational pull of the Moon and the Sun on Earth's equatorial bulge (see Figure 2-19). In fact, all the details of the orbits of the planets and their satellites could be explained mathematically with a body of knowledge built on Newton's work that is today called **Newtonian mechanics.**

Not only could Newtonian mechanics explain a variety of known phenomena in detail, but it could also predict new phenomena. For example, one of Newton's friends, Edmund Halley, was intrigued by three similar historical records of a comet that had been sighted at intervals of 76 years. Assuming these records to be accounts of the same comet, Halley used Newton's methods to work out the details of the comet's orbit and predicted its return in 1758. It was first sighted on Christmas night of 1757, a fitting memorial to Newton's birthday. To this day the comet bears Halley's name (Figure 4-25).

Another dramatic success of Newton's ideas was their role in the discovery of the eighth planet from the Sun. The seventh planet, Uranus, was discovered accidentally by William Herschel in 1781 during a telescopic survey of the sky. Fifty years later, however, it was clear that Uranus was not following its predicted orbit. John Couch Adams in England and Urbain Le Verrier in France independently calculated that the gravitational pull of a yet unknown, more distant planet could explain the deviations of Uranus from its orbit. Le Verrier predicted that the planet would be found at a certain location in the constellation of Aquarius. A brief telescopic search on September 23, 1846, revealed the planet Neptune within 1° of the calculated position. Before it was sighted with a telescope, Neptune was actually predicted with pencil and paper.

Because it has been so successful in explaining and predicting many important phenomena, Newtonian mechanics has become the cornerstone of modern physical science. Even today, as we send astronauts into Earth orbit and spacecraft to the outer planets, Newton's equations are used to calculate the orbits and trajectories of these spacecraft. The *Cosmic Connections* figure on the next page summarizes some of the ways that gravity plays an important role on scales from apples to galaxies.

FIGURE 4-25 R I V U X G
Comet Halley This most famous of all comets orbits the Sun with an average period of about 76 years. During the twentieth century, the comet passed near the Sun in 1910 and again in 1986 (when this photograph was taken). It will next be prominent in the sky in 2061. (Harvard College Observatory/Science Source)

COSMIC CONNECTIONS

Universal Gravitation

Gravity is one of the fundamental forces of nature. We can see its effects here on Earth as well as in the farthest regions of the observable universe.

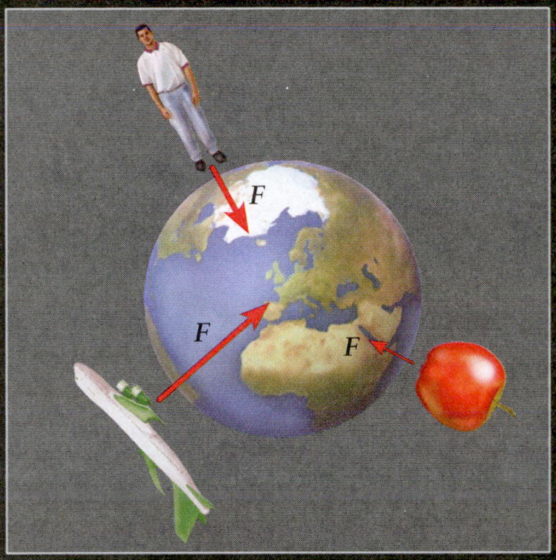

The weight of an ordinary object is just the gravitational force exerted on that object by Earth.

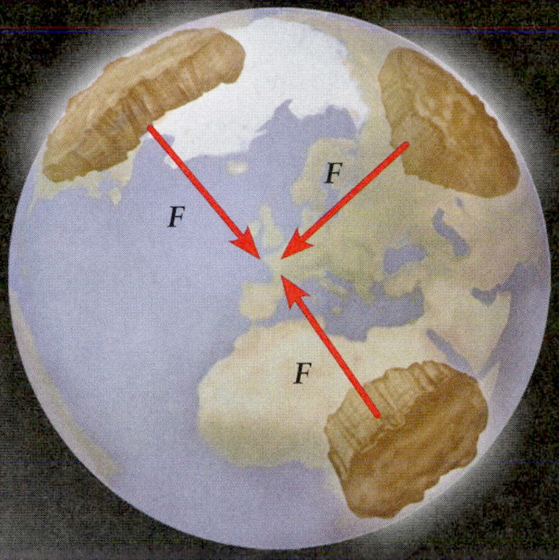

Earth is held together by the mutual gravitational attraction of its parts.

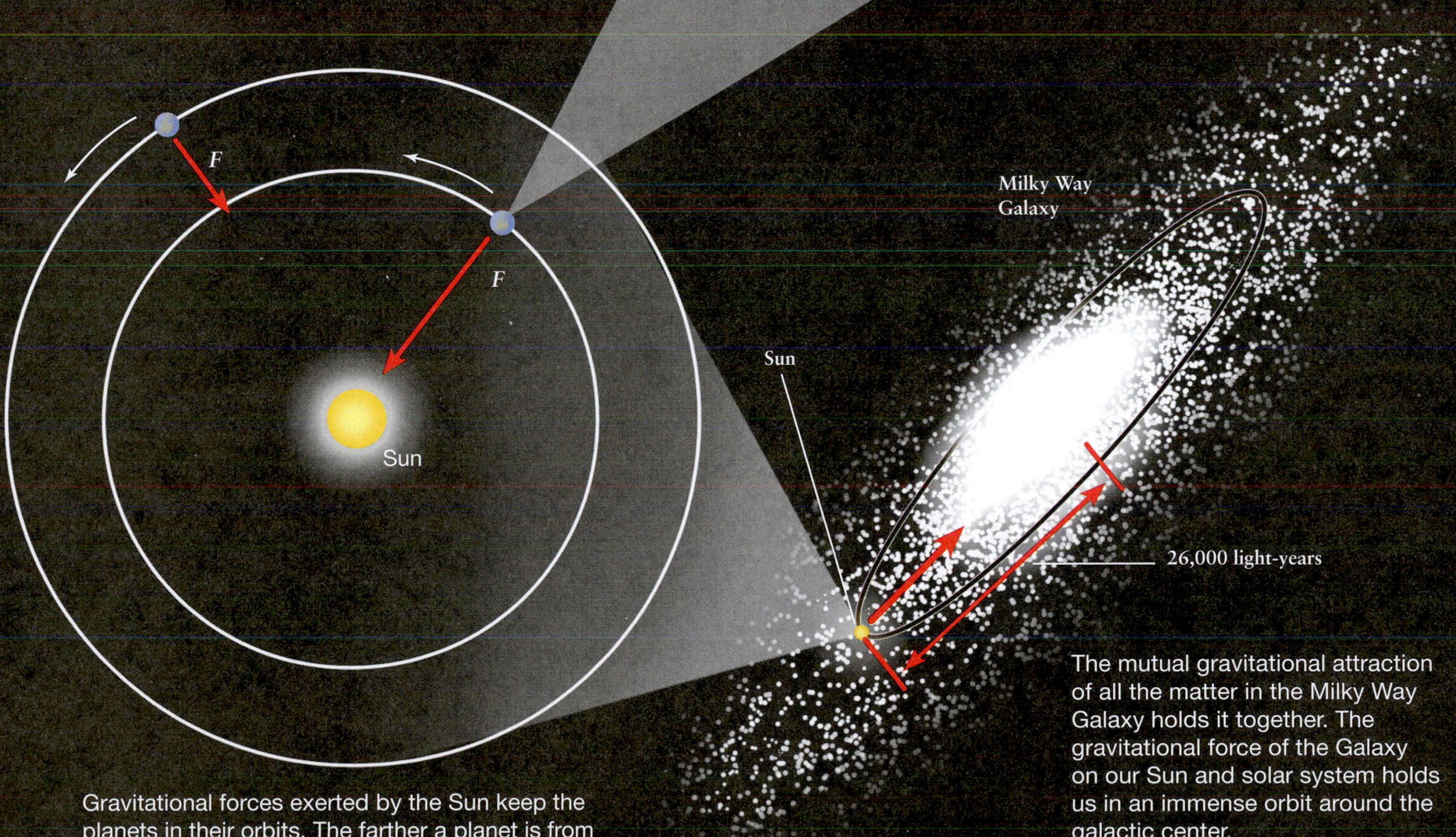

Gravitational forces exerted by the Sun keep the planets in their orbits. The farther a planet is from the Sun, the weaker the gravitational force that acts on the planet.

The mutual gravitational attraction of all the matter in the Milky Way Galaxy holds it together. The gravitational force of the Galaxy on our Sun and solar system holds us in an immense orbit around the galactic center.

In the twentieth century, scientists found that Newton's laws do not apply in all situations. A new theory called *quantum mechanics* had to be developed to explain the behavior of matter on the very smallest of scales, such as within the atom and within the atomic nucleus. Albert Einstein developed the *theory of relativity* to explain what happens at very high speeds approaching the speed of light and in places where gravitational forces are very strong. For many purposes in astronomy, however, Newton's laws are as useful today as when Newton formulated them more than three centuries ago.

4-8 Gravitational forces between Earth and the Moon produce tides

Tidal forces reveal how gravitation can pull objects apart rather than drawing them together

We have seen how Newtonian mechanics explains why the Moon stays in orbit around Earth. It also explains why there are ocean tides, as well as why the Moon always keeps the same face toward Earth. Both of these are the result of *tidal forces*—a consequence of gravity that deforms planets and reshapes galaxies.

Tidal forces are differences in the gravitational pull at different points in an object. As an illustration, imagine that three billiard balls are lined up in space at some distance from a planet, as in Figure 4-26a. According to Newton's law of universal gravitation, the force of attraction between two objects is greater the closer the two objects are to each other. Thus, the planet exerts more force on the 3-ball (in red) than on the 2-ball (in blue) and exerts more force on the 2-ball than on the 1-ball (in yellow). Now, imagine that the three balls are released and allowed to fall toward the planet. Figure 4-26b shows the situation a short time later. Because of the differences in gravitational pull, a short time later the 3-ball will have moved farther than the 2-ball, which will in turn have moved farther than the 1-ball. But now imagine that same motion from the perspective of the 2-ball. From this perspective, it appears as though the 3-ball is pulled toward the planet while the 1-ball is pushed away (Figure 4-26c). These apparent pushes and pulls are called tidal forces.

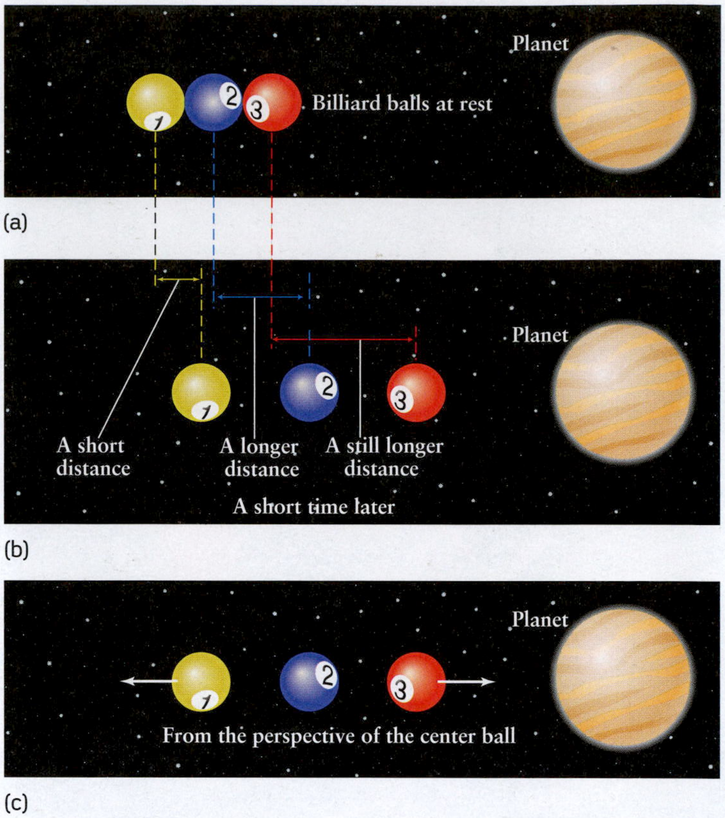

ANIMATION 4-8

FIGURE 4-26

The Origin of Tidal Forces **(a)** Imagine three identical billiard balls placed some distance from a planet and released. **(b)** The closer a ball is to the planet, the more gravitational force the planet exerts on it. Thus, a short time after the balls are released, the blue 2-ball has moved farther toward the planet than the yellow 1-ball, and the red 3-ball has moved farther still. **(c)** From the perspective of the 2-ball in the center, it appears that forces have pushed the 1-ball away from the planet and pulled the 3-ball toward the planet. These forces are called tidal forces.

Tidal Forces on Earth

The Moon has a similar effect on Earth as the planet in Figure 4-26 has on the three billiard balls. The arrows in Figure 4-27a indicate the strength and direction of the gravitational force of the Moon at several locations on Earth. The side of Earth closest to the Moon feels a greater gravitational pull than does Earth's center, and the side of Earth that faces away from the Moon feels less gravitational pull than does Earth's center. This means that just as for the billiard balls in Figure 4-26, there are tidal forces acting on Earth (Figure 4-27b). These tidal forces exerted by the Moon try to elongate Earth along a line connecting the centers of Earth and the Moon and try to squeeze Earth inward in the direction perpendicular to that line.

Because the body of Earth is largely rigid, it cannot deform very much in response to the tidal forces of the Moon. But the water in the oceans can and does deform into a football shape, as Figure 4-28a shows. As Earth rotates, a point on its surface goes from where the water is shallow to where the water is deep and back again. This is the origin of low and high ocean tides. (In this simplified description we have assumed that Earth is completely covered with water. The full story of the tides is much more complex, because the shapes of the continents and the effects of winds must also be taken into account.)

The Sun also exerts tidal forces on Earth's oceans. (The tidal effects of the Sun are about half as great as those of the Moon.) When the Sun, the Moon, and Earth are aligned, which happens at either new moon or full moon, the tidal effects of the Sun and Moon reinforce each other and the tidal distortion of the oceans is greatest. This produces large shifts in water level called **spring tides** (Figure 4-28b). At first quarter and last quarter, when the

Sun and Moon form a right angle with Earth, the tidal effects of the Sun and Moon partially cancel each other. Hence, the tidal distortion of the oceans is the least pronounced, producing smaller tidal shifts called **neap tides** (Figure 4-28c).

CAUTION! Note that spring tides have nothing to do with the season of the year called spring. Instead, the name refers to the way that the ocean level "springs up" to a greater than normal height. Spring tides occur whenever there is a new moon or full moon, no matter what the season of the year.

Tidal Forces on the Moon and Beyond

Just as the Moon exerts tidal forces on Earth, Earth exerts tidal forces on the Moon. Soon after the Moon formed some 4.56 billion years ago, it was molten throughout its volume. Earth's tidal forces deformed the molten Moon into a slightly elongated shape, with the long axis of the Moon pointed toward Earth. The Moon retained this shape and orientation when it cooled and solidified. To keep its long axis pointed toward Earth, the Moon spins once on its axis as it makes one orbit around Earth—that is, it is in synchronous rotation (Section 3-2). Hence, the same side of the Moon always faces Earth, and this is the side that we see. For the same reason, most of the satellites in the solar system are in synchronous rotation, and thus always keep the same side facing their planet.

Tidal forces are also important on scales much larger than the solar system. Figure 4-29 shows two spiral galaxies, like the one in Figure 1-9, undergoing a near-collision. During the millions of years that this close encounter has been taking place, the tidal forces of the larger galaxy have pulled an immense streamer of stars and interstellar gas out of the smaller galaxy.

Many galaxies, including our own Milky Way Galaxy, show signs of having been disturbed at some time by tidal interactions with other galaxies. By their effect on the interstellar gas from which stars are formed, tidal interactions can actually trigger the

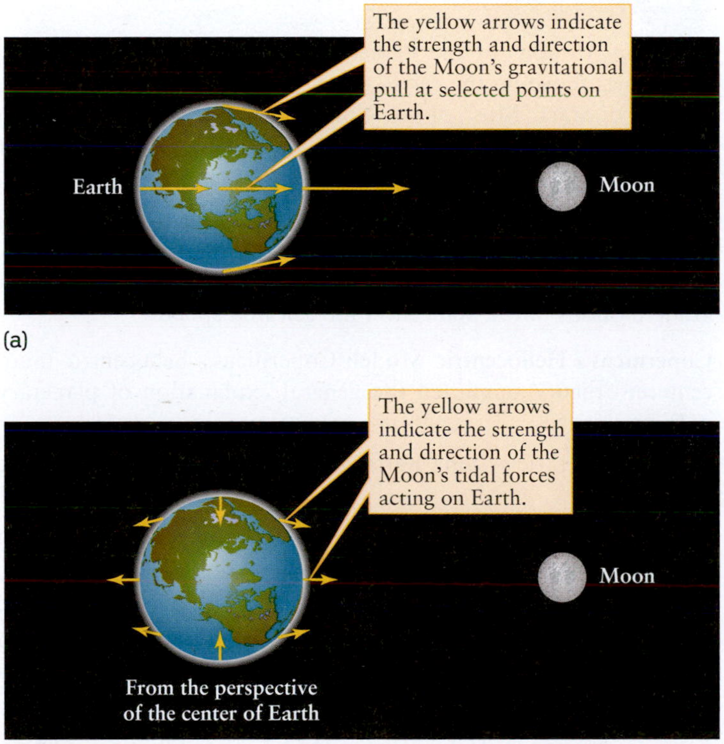

FIGURE 4-27

Tidal Forces on Earth **(a)** The Moon exerts different gravitational pulls at different locations on Earth. **(b)** At any location, the tidal force equals the Moon's gravitational pull at that location minus the gravitational pull of the Moon at the center of Earth. These tidal forces tend to deform Earth into a nonspherical shape.

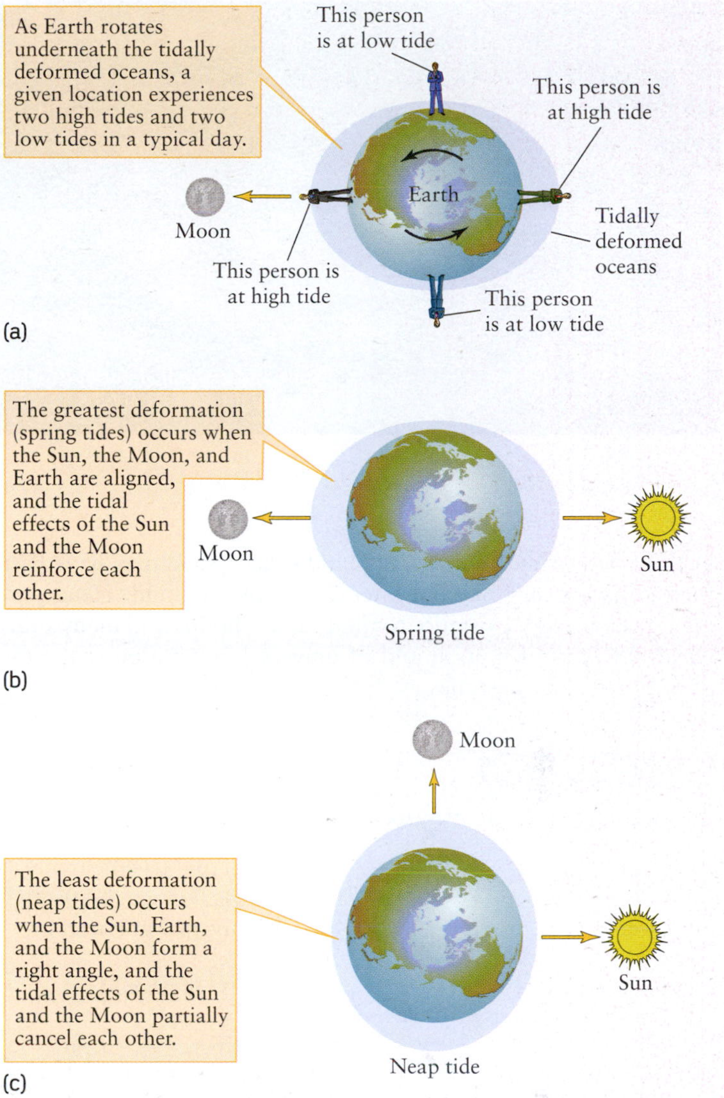

FIGURE 4-28

High and Low Tides **(a)** The gravitational forces of the Moon and the Sun deform Earth's oceans, giving rise to low and high tides. **(b), (c)** The strength of the tides depends on the relative positions of the Sun, the Moon, and Earth.

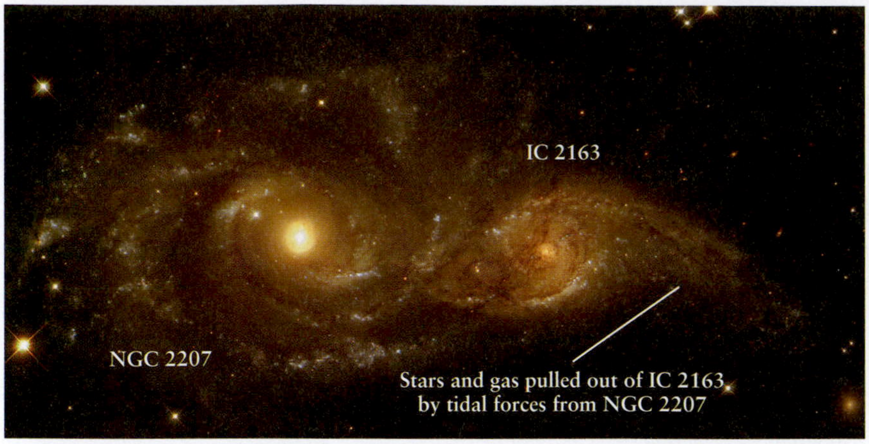

FIGURE 4-29 R I V U X G
Tidal Forces on a Galaxy For millions of years the galaxies NGC 2207 and IC 2163 have been moving ponderously past each other. The larger galaxy's tremendous tidal forces have drawn a streamer of material a hundred thousand light-years long out of IC 2163. If you lived on a planet orbiting a star within this streamer, you would have a magnificent view of both galaxies. NGC 2207 and IC 2163 are, respectively, 143,000 light-years and 101,000 light-years in diameter. Both galaxies are 114 million light-years away in the constellation Canis Major. (NASA and the Hubble Heritage Team, AURA/STScI)

birth of new stars. Our own Sun and solar system may have been formed as a result of tidal interactions of this kind. Hence, we may owe our very existence to tidal forces. In this and many other ways, the laws of motion and of universal gravitation shape our universe and our destinies.

KEY WORDS

Terms preceded by an asterisk (*) are discussed in the Boxes.

acceleration, p. 86
air drag, p. 90
aphelion, p. 80
conic section, p. 94
conjunction, p. 74
conservation of energy, p. 90
deferent, p. 72
direct motion, p. 71
eccentricity, p. 79
ellipse, p. 79
elongation, p. 74
epicycle, p. 72
escape speed, p. 92
focus (of an ellipse; *plural* foci), p. 79
force, p. 85
geocentric model, p. 70
gravitational force, p. 88
gravitational potential energy, p. 90
gravity, p. 88
greatest eastern elongation, p. 74
greatest western elongation, p. 74
heliocentric model, p. 73
hyperbola, p. 94
inferior conjunction, p. 74
inferior planet, p. 73
Kepler's first law, p. 79
Kepler's second law, p. 80
Kepler's third law, p. 80
kinetic energy, p. 90
law of equal areas, p. 80
law of universal gravitation, p. 89
major axis (of an ellipse), p. 79
mass, p. 86
neap tides, p. 97
Newtonian mechanics, p. 94
Newton's first law of motion, p. 85
Newton's form of Kepler's third law, p. 93
Newton's second law of motion, p. 86
Newton's third law of motion, p. 87
Occam's razor, p. 73
opposition, p. 74
orbital energy, p. 91
parabola, p. 94
parallax, p. 77
perihelion, p. 80
period (of a planet), p. 75
Ptolemaic system, p. 72
retrograde motion, p. 71
semimajor axis (of an ellipse), p. 79
sidereal period, p. 75
speed, p. 86
spring tides, p. 96
superior conjunction, p. 74
superior planet, p. 74
synodic period, p. 75
tidal forces, p. 96
universal constant of gravitation, p. 89
velocity, p. 86
weight, p. 86

KEY IDEAS

Apparent Motions of the Planets: Like the Sun and the Moon, the planets move on the celestial sphere with respect to the background of stars. Most of the time a planet moves eastward in direct motion, in the same direction as the Sun and the Moon, but from time to time it moves westward in retrograde motion.

The Ancient Geocentric Model: Ancient astronomers believed Earth to be at the center of the universe. They invented a complex system of epicycles and deferents to explain the direct and retrograde motions of the planets on the celestial sphere.

Copernicus's Heliocentric Model: Copernicus's heliocentric (Sun-centered) theory simplified the general explanation of planetary motions.

• In a heliocentric system, Earth is one of the planets orbiting the Sun.

• A planet undergoes retrograde motion as seen from Earth when Earth and the planet pass each other.

• The sidereal period of a planet, its true orbital period, is measured with respect to the stars. Its synodic period is measured with respect to Earth and the Sun (for example, from one opposition to the next).

Kepler's Improved Heliocentric Model and Elliptical Orbits: Copernicus thought that the orbits of the planets were combinations of circles. Using data collected by Tycho Brahe, Kepler deduced three laws of planetary motion: (1) the orbits are in fact ellipses; (2) a planet's speed varies as it moves around its elliptical orbit; and (3) the orbital period of a planet is related to the size of its orbit.

Evidence for the Heliocentric Model: The invention of the telescope led Galileo to new discoveries that supported a heliocentric model. These included his observations of the phases of Venus and of the motions of four moons around Jupiter.

Newton's Laws of Motion: Isaac Newton developed three principles, called the laws of motion, that apply to the motions of objects on Earth as well as in space. These are (1) the tendency of an object to maintain a constant velocity, (2) the relationship between the net outside force on an object and the object's acceleration, and (3) the principle of action and reaction. These laws and Newton's law of universal gravitation can be used to deduce Kepler's laws. They lead to extremely accurate descriptions of planetary motions.

• The mass of an object is a measure of the amount of matter in the object. Its weight is a measure of the force with which the gravity of some other object pulls on it.

• In general, the path of one object about another, such as that of a planet or comet about the Sun, is one of the curves called conic sections: circle, ellipse, parabola, or hyperbola.

Energy: There are different forms of energy an object can have. Kinetic and gravitational potential energy are the most important for orbiting objects.

• **Kinetic energy:** The greater an object's speed, the larger its kinetic energy. For two objects with equal speed, the more massive object has a larger kinetic energy.

• **Gravitational potential energy:** The farther an object is located from the surface of a planet, the greater its gravitational energy.

• **Conservation of energy:** Energy can change forms, and be transferred from one object to another, but energy cannot be created or destroyed.

Escape Speed: The escape speed of a planet is the ejection speed an object would need to have near the surface of the planet so that the object can "break free" from the planet and never return.

• The escape speed does not depend on the mass of the object, but it does depend on both the size and mass of the planet.

Tidal Forces: Tidal forces are caused by differences in the gravitational pull that one object exerts on different parts of a second object.

• The tidal forces of the Moon and the Sun produce tides in Earth's oceans.

• The tidal forces of Earth have locked the Moon into synchronous rotation.

QUESTIONS

Review Questions

1. How did the ancient Greeks explain why the Sun and the Moon slowly change their positions relative to the background stars?
2. In what direction does a planet move relative to the stars when it is in direct motion? When it is in retrograde motion? How do these compare with the direction in which we see the Sun move relative to the stars?
3. (**a**) In what direction does a planet move relative to the horizon over the course of one night? (**b**) The answer to (a) is the same whether the planet is in direct motion or retrograde motion. What does this tell you about the speed at which planets move on the celestial sphere?
4. What is an epicycle? How is it important in Ptolemy's explanation of the retrograde motions of the planets?
5. What is the significance of Occam's razor as a tool for analyzing theories?
6. How did the models of Aristarchus and Copernicus explain the retrograde motion of the planets?
7. How did Copernicus determine that the orbits of Mercury and Venus must be smaller than Earth's orbit? How did he determine that the orbits of Mars, Jupiter, and Saturn must be larger than Earth's orbit?
8. At what configuration (for example, superior conjunction, greatest eastern elongation, and so on) would it be best to observe Mercury or Venus with an Earth-based telescope? At what configuration would it be best to observe Mars, Jupiter, or Saturn? Explain your answers.
9. Is it ever possible to see Mercury at midnight? Explain your answer.
10. Which planets can never be seen at opposition? Which planets can never be seen at inferior conjunction? Explain your answers.
11. What is the difference between the synodic period and the sidereal period of a planet?
12. What is parallax? What did Tycho Brahe conclude from his attempt to measure the parallax of a supernova and a comet?
13. What observations did Tycho Brahe make in an attempt to test the heliocentric model? What were his results? Explain why modern astronomers get different results.
14. TUTORIAL 4-1 What are the foci of an ellipse? If the Sun is at one focus of a planet's orbit, what is at the other focus?
15. What are Kepler's three laws? Why are they important?
16. At what point in a planet's elliptical orbit does it move fastest? At what point does it move slowest? At what point does it sweep out an area at the fastest rate?
17. A line joining the Sun and an asteroid is found to sweep out an area of 6.3 AU^2 during 2010. How much area is swept out during 2011? Over a period of five years?
18. The orbit of a spacecraft about the Sun has a perihelion distance of 0.1 AU and an aphelion distance of 0.4 AU. What is the semimajor axis of the spacecraft's orbit? What is its orbital period?
19. A comet with a period of 125 years moves in a highly elongated orbit about the Sun. At perihelion, the comet comes very close to the Sun's surface. What is the comet's average distance from the Sun? What is the farthest it can get from the Sun?
20. What observations did Galileo make that reinforced the heliocentric model? Why did these observations contradict the older model of Ptolemy? Why could these observations not have been made before Galileo's time?

21. Why does Venus have its largest angular diameter when it is new and its smallest angular diameter when it is full?
22. What are Newton's three laws? Give an everyday example of each law.
23. How much force do you have to exert on a 3-kg brick to give it an acceleration of 2 m/s^2? If you double this force, what is the brick's acceleration? Explain your answer.
24. What is the difference between weight and mass?
25. What is your weight in pounds and in newtons? What is your mass in kilograms?
26. Suppose that Earth were moved to a distance of 3.0 AU from the Sun. How much stronger or weaker would the Sun's gravitational pull be on Earth? Explain your answer.
27. How far would you have to go from Earth to be completely beyond the pull of its gravity? Explain your answer.
28. A cannonball is shot horizontally from a barrel off a building. Name two forms of energy the cannonball has as it exits the barrel. How do these two forms of energy increase or decrease during the cannonball's flight? Before the cannonball was fired, where was the energy stored?
29. A satellite is in circular orbit. What two forms of energy are part of the satellite's orbital energy? Would its orbital energy need to increase or decrease in order to orbit at a larger distance from Earth?
30. If an object loses orbital energy through air drag, is energy still conserved? If so, where does the energy go?
31. Calculate the escape speed for Earth and show that it is 11.2 km/s. (Consult Appendix 2 for planetary data.)
32. Including the effects of Earth's atmosphere, would you expect the real escape speed for a hypothetical cannonball to be greater or less than 11.2 km/s? Why?
33. What are conic sections? In what way are they related to the orbits of planets in the solar system?
34. Why was the discovery of Neptune an important confirmation of Newton's law of universal gravitation?
35. What is a tidal force? How do tidal forces produce tides in Earth's oceans?
36. What is the difference between spring tides and neap tides?

Advanced Questions

Questions preceded by an asterisk (*) involve topics discussed in the Boxes.

Problem-solving tips and tools

Box 4-1 explains sidereal and synodic periods in detail. The semimajor axis of an ellipse is half the length of the long, or major, axis of the ellipse. For data about the planets and their satellites, see Appendices 1, 2, and 3 at the back of this book. If you want to calculate the gravitational force that you feel on the surface of a planet, the distance r to use is the planet's radius (the distance between you and the center of the planet). Boxes 4-2 and 4-4 show how to use Kepler's third law in its original form and in Newton's form.

37. Figure 4-2 shows the retrograde motion of Mars as seen from Earth. Sketch a similar figure that shows how Earth would appear to move against the background of stars during this same time period as seen by an observer on Mars.
*38. The synodic period of Mercury (an inferior planet) is 115.88 days. Calculate its sidereal period in days.
*39. Table 4-1 shows that the synodic period is *greater* than the sidereal period for Mercury, but the synodic period is *less* than the sidereal period for Jupiter. Draw diagrams like the one in Box 4-1 to explain why this is so.
*40. A general rule for superior planets is that the greater the average distance from the planet to the Sun, the more frequently that planet will be at opposition. Explain how this rule comes about.
41. In 2006, Mercury was at greatest western elongation on April 8, August 7, and November 25. It was at greatest eastern elongation on February 24, June 20, and October 17. Does Mercury take longer to go from eastern to western elongation, or vice versa? Explain why, using Figure 4-6.
42. Explain why the semimajor axis of a planet's orbit is equal to the average of the distance from the Sun to the planet at perihelion (the *perihelion distance*) and the distance from the Sun to the planet at aphelion (the *aphelion distance*).
43. A certain comet is 2 AU from the Sun at perihelion and 16 AU from the Sun at aphelion. (**a**) Find the semimajor axis of the comet's orbit. (**b**) Find the sidereal period of the orbit.
44. A comet orbits the Sun with a sidereal period of 64.0 years. (**a**) Find the semimajor axis of the orbit. (**b**) At aphelion, the comet is 31.5 AU from the Sun. How far is it from the Sun at perihelion?
45. One trajectory that can be used to send spacecraft from Earth to Mars is an elliptical orbit that has the Sun at one focus, its perihelion at Earth, and its aphelion at Mars. The spacecraft is launched from Earth and coasts along this ellipse until it reaches Mars, when a rocket is fired to either put the spacecraft into orbit around Mars or cause it to land on Mars. (**a**) Find the semimajor axis of the ellipse. (*Hint:* Draw a picture showing the Sun and the orbits of Earth, Mars, and the spacecraft. Treat the orbits of Earth and Mars as circles.) (**b**) Calculate how long (in days) such a one-way trip to Mars would take.
46. The mass of the Moon is 7.35×10^{22} kg, while that of Earth is 5.98×10^{24} kg. The average distance from the center of the Moon to the center of Earth is 384,400 km. What is the size of the gravitational force that Earth exerts on the Moon? What is the size of the gravitational force that the Moon exerts on Earth? How do your answers compare with the force between the Sun and Earth calculated in the text?
47. The mass of Saturn is approximately 100 times that of Earth, and the semimajor axis of Saturn's orbit is approximately 10 AU. To this approximation, how does the gravitational force that the Sun exerts on Saturn compare to the gravitational force that the Sun exerts on Earth? How do the accelerations of Saturn and Earth compare?

48. Suppose that you traveled to a planet with 4 times the mass and 4 times the diameter of Earth. Would you weigh more or less on that planet than on Earth? By what factor?

49. On Earth, a 50-kg astronaut weighs 490 newtons. What would she weigh if she landed on Jupiter's moon Callisto? What fraction is this of her weight on Earth? See Appendix 3 for relevant data about Callisto.

50. Except for some rare and exotic microbes, the Sun provides energy for life on Earth. Can you describe how energy in sunlight can end up in a kangaroo jumping through the air? In your description, can you name three forms of energy involved, other than sunlight?

51. You're talking with other students about how the energy of a satellite in circular orbit changes with altitude. One student says that lower orbits have more energy because Kepler's second law says that closer planets have higher speeds. Another student argues that orbits at larger distances must have more energy because rockets have to expend more energy to launch satellites farther into to space. Which student reaches the right conclusion?

52. Some argue that life might have started on either Mars or Earth and spread to the other planet by microbe-carrying debris ejected during very large asteroid impacts. Consult Appendix 2 for planetary data and calculate the escape speed for both Earth and Mars. Assuming all else being equal, for which planet is it easier for rocks to escape?

53. Imagine a planet like Earth orbiting a star with 4 times the mass of the Sun. If the semimajor axis of the planet's orbit (a) is 1 AU, what would be the planet's sidereal period? (*Hint:* Use Newton's form of Kepler's third law. Compared with the case of Earth orbiting the Sun, by what factor has the quantity $m^1 + m^2$ changed? Has a changed? By what factor must P^2 change?)

54. A satellite is said to be in a "geosynchronous" orbit if it appears always to remain over the exact same spot on rotating Earth. (**a**) What is the period of this orbit? (**b**) At what distance from the center of Earth must such a satellite be placed into orbit? (*Hint:* Use Newton's form of Kepler's third law.) (**c**) Explain why the orbit must be in the plane of Earth's equator.

55. Figure 4-23 shows the lunar module *Eagle* in orbit around the Moon after completing the first successful lunar landing in July 1969. (The photograph was taken from the command module *Columbia,* in which the astronauts returned to Earth.) The spacecraft orbited 111 km above the surface of the Moon. Calculate the period of the spacecraft's orbit. See Appendix 3 for relevant data about the Moon.

*56. In Box 4-4 we analyze the orbit of Jupiter's moon Io. Look up information about the orbits of Jupiter's three other large moons (Europa, Ganymede, and Callisto) in Appendix 3. Demonstrate that these data are in agreement with Newton's form of Kepler's third law.

*57. Suppose a newly discovered asteroid is in a circular orbit with synodic period 1.25 years. The asteroid lies between the orbits of Mars and Jupiter. (**a**) Find the sidereal period of the orbit. (**b**) Find the distance from the asteroid to the Sun.

58. The average distance from the Moon to the center of Earth is 384,400 km, and the diameter of Earth is 12,756 km. Calculate the gravitational force that the Moon exerts (**a**) on a 1-kg rock at the point on Earth's surface closest to the Moon, and (**b**) on a 1-kg rock at the point on Earth's surface farthest from the Moon. (**c**) Find the difference between the two forces you calculated in parts (a) and (b). This difference is the tidal force pulling these two rocks away from each other, like the 1-ball and 3-ball in Figure 4-26. Explain why tidal forces cause only a very small deformation of Earth.

Discussion Questions

59. Which planet would you expect to exhibit the greatest variation in apparent brightness as seen from Earth? Which planet would you expect to exhibit the greatest variation in angular diameter? Explain your answers.

60. Use two thumbtacks, a loop of string, and a pencil to draw several ellipses. Describe how the shape of an ellipse varies as the distance between the thumbtacks changes.

Web/eBook Questions

61. (**a**) Search the World Wide Web for information about Kepler. Before he realized that the planets move on elliptical paths, what other models of planetary motion did he consider? What was Kepler's idea of "the music of the spheres"? (**b**) Search the World Wide Web for information about Galileo. What were his contributions to physics? Which of Galileo's new ideas were later used by Newton to construct his laws of motion? (**c**) Search the World Wide Web for information about Newton. What were some of the contributions that he made to physics other than developing his laws of motion? What contributions did he make to mathematics?

62. **Monitoring the Retrograde Motion of Mars.** Watching Mars night after night reveals that it changes its position with respect to the background stars. To track its motion, access and view the animation "The Path of Mars in 2016" in Chapter 4 of the *Universe* Web site or eBook. (**a**) Through which constellations does Mars move? (**b**) On approximately what date does Mars stop its direct (west-to-east) motion and begin its retrograde motion? (*Hint:* Use the "Stop" function on your animation controls.) (**c**) Over how many days does Mars move retrograde?

ACTIVITIES

Observing Projects

63. It is quite probable that within a few weeks of your reading this chapter one of the planets will be near opposition or greatest eastern elongation, making it readily visible in the evening sky. Select a planet that is at or near such a configuration by searching the World Wide Web or by consulting a reference book, such as the current issue of the *Astronomical*

Almanac or the pamphlet entitled *Astronomical Phenomena* (both published by the U.S. government). At that configuration, would you expect the planet to be moving rapidly or slowly from night to night against the background stars? Verify your expectations by observing the planet once a week for a month, recording your observations on a star chart.

64. If Jupiter happens to be visible in the evening sky, observe the planet with a small telescope on five consecutive clear nights. Record the positions of the four Galilean satellites by making nightly drawings, just as the Jesuit priests did in 1620 (see Figure 4-17). From your drawings, can you tell which moon orbits closest to Jupiter and which orbits farthest? Was there a night when you could see only three of the moons? What do you suppose happened to the fourth moon on that night?

65. If Venus happens to be visible in the evening sky, observe the planet with a small telescope once a week for a month. On each night, make a drawing of the crescent that you see. From your drawings, can you determine if the planet is nearer or farther from Earth than the Sun is? Do your drawings show any changes in the shape of the crescent from one week to the next? If so, can you deduce if Venus is coming toward us or moving away from us?

66. Use the *Starry Night*™ program to observe retrograde motion. Select **Favourites > Explorations > Retrograde** from the menu. The view from Earth is centered on Mars against the background of stars and the framework of star patterns within the constellations. The **Time Flow Rate** is set to 1 day. Click **Play** and observe Mars as it moves against the background constellations. An orange line traces Mars's path in the sky from night to night. Watch the motion of Mars for at least 2 years of simulated time. Since the view is centered on and tracks Mars in the view, the sky appears to move but the relative motion of Mars against this sky is obvious. (**a**) For most of the time, does Mars move generally to the left (eastward) or to the right (westward) on the celestial sphere? Select **File > Revert** from the menu to return to the original view. Use the time controls in the toolbar (**Play, Step time forward,** and **Step time backward**) along with the **Zoom** controls (+ and – buttons at the right of the toolbar or the mouse wheel) to determine when Mars's usual *direct* motion ends, when it appears that Mars comes to a momentary halt in the west-east direction, and *retrograde* motion begins. On what date does retrograde motion end and direct motion resume? (**b**) You have been observing the motion of Mars as seen from Earth. To observe the motion of Earth as seen from Mars, locate yourself on the north pole of Mars by selecting **Favourites > Explorations > Retrograde Earth** from the menu. The view is centered on and will track Earth as seen from the north pole of Mars, beginning on **June 23, 2010**. Click the **Play** button. As before, watch the motion for 2 years of simulated time. In which direction does Earth appear to move for most of the time? On what date does its motion change from direct to retrograde? On what date does its motion change from retrograde back to direct? Are these roughly the same dates you found in part (a)? (**c**) To understand the motions of Mars as seen from Earth and vice versa, observe the motion of the planets from a point above the solar system. Select **Favourites > Explorations > Retrograde Overview** from the menu. This view, from a position 5 AU above the plane of the solar system, is centered on the Sun, and the orbits and positions of Mars and Earth on June 23, 2010, are shown. Click **Play** and watch the motions of the planets for 2 years of simulated time. Note that Earth catches up with and overtakes Mars as time proceeds. This relative motion of the two planets leads to our observation of retrograde motion. On what date during this 2-year period is Earth directly between Mars and the Sun? How does this date compare to the two dates you recorded in part (a) and the two dates you recorded in part (b)? Explain the significance of this.

67. Use *Starry Night*™ to observe the phases of Venus and of Mars as seen from Earth. Select **Favourites > Explorations > Phases of Venus** and click the **Now** button in the toolbar to see an image of Venus if you were to observe it through a telescope from Earth right at this moment. (**a**) Draw the current shape (phase) of Venus. With the **Time Flow Rate** set to **30 days**, step time forward, drawing Venus to scale at each step. Make a total of 20 time steps and drawings. (**b**) From your drawings, determine when the planet is nearer or farther from Earth than is the Sun. (**c**) Deduce from your drawings when Venus is coming toward us or is moving away from us. (**d**) Explain why Venus goes through this particular cycle of phases. Select **Favourites > Explorations > Phases of Mars** and click the **Now** button in the toolbar. With the **Time Flow Rate** set to **30 days**, step time forward, and observe the changing phase of Mars as seen from Earth. (**e**) Compare this with the phases that you observed for Venus. Why are the cycles of phases as seen from Earth different for the two planets?

68. Use *Starry Night*™ to observe the orbits of the planets of the inner solar system. Open **Favourites > Explorations > Kepler**. The view is centered on the Sun from a position in space 2.486 AU above the plane of the solar system and shows the Sun and the inner planets and their orbits, as well as many asteroids in the asteroid belt beyond the orbit of Mars. Click the **Play** button and observe the motions of the planets from this unique location. (**a**) Make a list of the planets visible in the view in the order of increasing distance from the Sun. (**b**) Make a list of the planets visible in the view in the order of increasing orbital period. (**c**) How do the lists compare? (**d**) What might you conclude from this observation? (**e**) Which of Kepler's laws accounts for this observation?

ANSWERS

ConceptChecks

ConceptCheck 4-1: When a planet is moving retrograde, the planet can be observed night after night to be slowly drifting from east to west compared to the very distant background stars (Figure 4-2). This is opposite to how a planet typically appears to move. However, regardless of this slow movement, all objects always appear to rise in the east and set in the west on a daily basis due to Earth's rotation.

ConceptCheck 4-2: The Greeks' ancient geocentric model used a nonspinning, stationary Earth where the stars, planets, and the Sun all moved around Earth.

ConceptCheck 4-3: In the Ptolemaic model, the planets are continuously orbiting in circles such that they appear to move backward for a brief time. However, the planets never actually stop and change their directions (Figure 4-3).

ConceptCheck 4-4: No. A planet only appears to move in retrograde motion if seen from another planet if the two planets move at different speeds and pass one another (Figure 4-5). An imaginary observer on the stationary Sun would only see planets moving in the same direction as they orbit the Sun.

ConceptCheck 4-5: Mars has an orbit around the Sun that is larger than Earth's orbit. As a result, Mars never moves to a position between Earth and the Sun, so Mars never is at inferior conjunction (Figure 4-6).

ConceptCheck 4-6: In Copernicus's heliocentric model, the more distant planets are moving slower than the planets closer to the Sun. As a faster-moving Earth moves past a slower-moving Mars, there is a brief time in which Mars appears to move backward through the sky. However, the planets never actually reverse their directions.

ConceptCheck 4-7: Slowly moving Jupiter does not move very far along its orbit in the length of time it takes for Earth to pass by Jupiter, move around the Sun, and pass by Jupiter again, giving Jupiter a synodic period similar to Earth's orbital period around the Sun. However, slow-moving Jupiter takes more than a decade to move around the Sun back to its original starting place as measured by the background stars, giving it a large sidereal period.

ConceptCheck 4-8: Initially, Copernicus's model was no more accurate at predicting the positions and motions of the planets than Ptolemy's model. However, Copernicus's model turned out to be more closely related to the actual motions of the planets around the Sun than was Ptolemy's model of planets orbiting Earth. With subsequent measurements, Ptolemy's model was proven wrong.

ConceptCheck 4-9: An ellipse with an eccentricity of zero is a perfect circle (see Figure 4-10b) and, compared to Mars's $e = 0.093$, the eccentricity of Venus' orbit is $e = 0.007$. Venus has a smaller eccentricity so its orbit is closer to a perfect circle in shape.

ConceptCheck 4-10: Kepler's second law says that objects are moving slowest when they are farthest from the object they are orbiting, so an Earth-orbiting satellite will move slowest when it is farthest from Earth. Just as the Sun is at one focus for a planet's elliptical orbit, Earth is at one focus for a satellite's elliptical orbit (Figure 4-11).

ConceptCheck 4-11: According to Kepler's third law, planets closer to the Sun move faster than planets farther from the Sun. For objects orbiting Earth, the object closer to Earth is also moving the fastest, which, in this case, is the space shuttle.

ConceptCheck 4-12: No. Kepler's laws of planetary motion apply to any objects in space that orbit around another object, including comets orbiting the Sun, man-made satellites and moons orbiting planets, and even stars orbiting other stars.

ConceptCheck 4-13: When Venus is on the opposite side of the Sun from Earth, it will be in a full or gibbous phase. The full phase can occasionally be observed because Earth and Venus do not orbit the Sun in the same exact plane.

ConceptCheck 4-14: Galileo was the first person to use and widely share what he learned from his telescope observations, and a telescope is necessary in order to observe Jupiter's tiny moons.

ConceptCheck 4-15: According to Newton's first law, the rock will continue to travel in the same direction, and with the same speed, as long as there is no net outside force. Note that this principle applies no matter how massive the rock is, or how fast it's traveling.

ConceptCheck 4-16: As described by Newton's third law, the forces the sumo wrestlers exert on each other are of equal strength, but in opposite directions. This is true even though they are different in size and even though one wrestler is pushed out of the ring. Frictional forces from the floor and their body masses influence which wrestler moves the most, but the forces they exert onto each other are of equal strength.

ConceptCheck 4-17: Expressing Newton's second law in terms of acceleration, the acceleration of an object is given by $a = F/m$. Since the gravitational forces on the diver and Earth are of equal strength, the smaller mass—the diver—will have the larger acceleration. Because Earth is so much more massive than the diver, Earth's acceleration in this case is nowhere near measurable.

ConceptCheck 4-18: According to Newton's universal law of gravitation, the gravitational attraction between two objects depends on the square of the distance between them. In this case, if the asteroids drift 3 times closer together, then the gravitational force of attraction between them increases 3^2 times, or, in other words, becomes 9 times greater.

ConceptCheck 4-19: At the cannonball's highest point, it is momentarily at rest and has gravitational potential energy, but no kinetic energy. Starting from the chemical energy in gunpowder, that chemical energy is transferred to the cannonball, giving it kinetic energy. As the ball rises, kinetic energy is transformed into gravitational potential energy until the ball reaches its peak.

ConceptCheck 4-20: Air drag *removes* orbital energy and could put the spacecraft on the elliptical path D. Firing the rocket thruster *adds* orbital energy and could put the spacecraft on the elliptical path F. A circular orbit requires special parameters, so it is not surprising that most moons and planets have at least slightly elliptical orbits.

ConceptCheck 4-21: In order for an object to feel air drag, it must pass through the air. Since Earth and our air orbit the Sun together, Earth feels no air drag and we can orbit the Sun indefinitely.

ConceptCheck 4-22: An object's escape speed from a planet does not depend on the object's mass, so both rocks can be shot at the same speed, which is the escape speed from the Moon. As illustrated in Figure 4-20, for two objects with the same speed, the more massive object has more kinetic energy, so the kinetic energy of the two rocks will not be equal, and more energy is required to shoot the larger rock off of the Moon.

ConceptCheck 4-23: Yes. The escape speed $v_{escape} = \sqrt{\frac{2GM}{R}}$ depends on both the mass and radius of the planet. If another Earth-mass planet had a larger radius (which is certainly possible), its escape speed would be less than that of Earth.

ConceptCheck 4-24: The object—on its high-energy hyperbolic orbit—would have to lose orbital energy in order to enter a lower-energy elliptical or circular orbit as a moon. Special circumstances are required to lose this much energy, which is why we do not think Earth simply captured our Moon as it passed by.

CalculationChecks

CalculationCheck 4-1: According to Kepler's third law, $P^2 = a^3$. So, if $P^2 = (39.5)^3$, then $P = 39.5^{3/2} = 248$ years.

CalculationCheck 4-2: Using Newton's universal law of gravitation, $F_{\text{Mars-astronaut}} = G(m_{\text{Mars}}) \times (m_{\text{astronaut}}) \div (\text{radius})^2 = 6.67 \times 10^{-11} \times 6.4 \times 10^{23} \times 75 \div (3.4 \times 10^6)^2 = 277$ newtons, which we can convert to pounds because 277 newtons $\times$ 0.255 lbs/N = 76 pounds.

CHAPTER 5

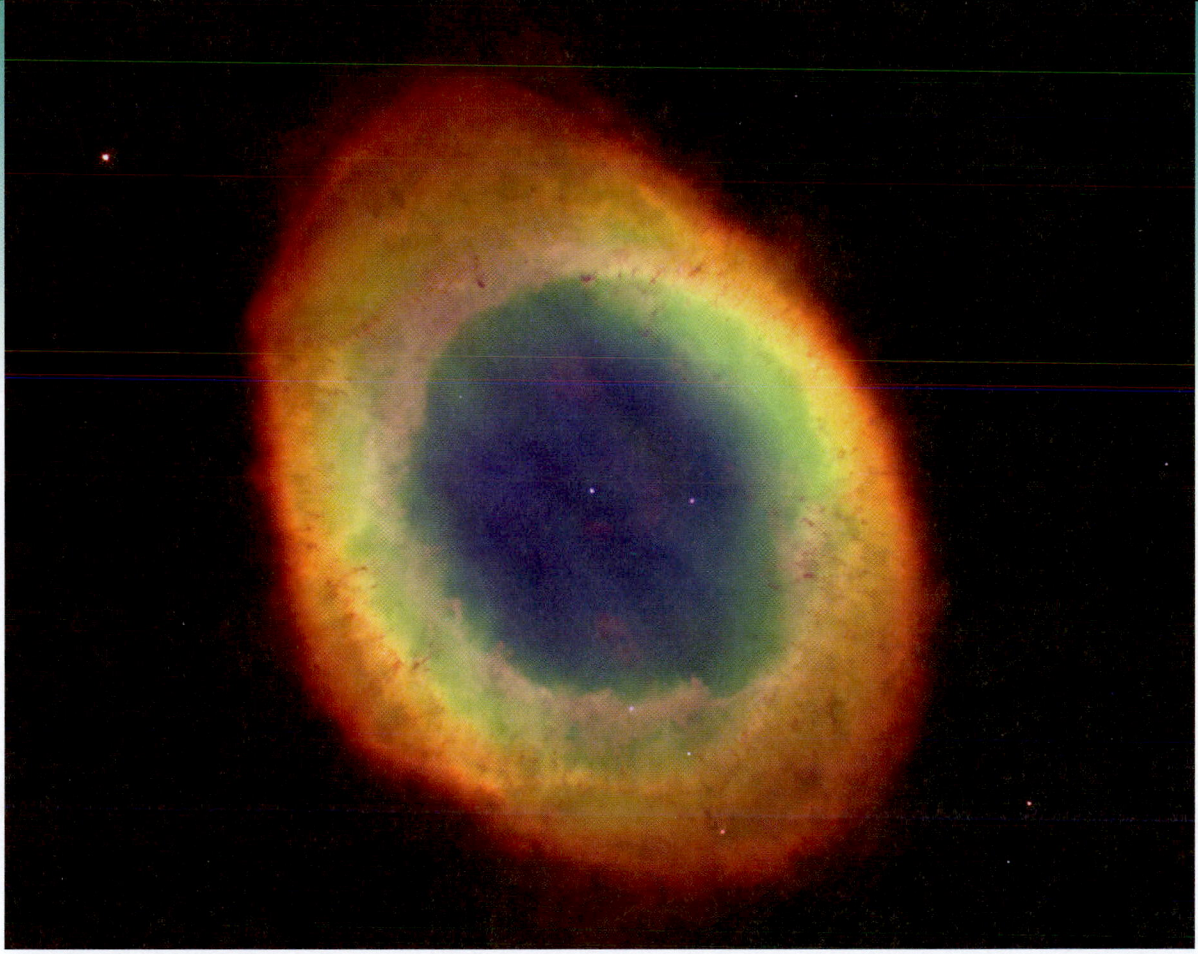

The Ring Nebula is a shell of glowing gases surrounding a dying star. The spectrum of the emitted light reveals which gases are present. (Hubble Heritage Team, AURA/STScI/NASA)) R I V U X G

The Nature of Light

LEARNING GOALS

By reading the sections of this chapter, you will learn

5-1 How we measure the speed of light

5-2 How we know that light is an electromagnetic wave

5-3 How an object's temperature is related to the radiation it emits

5-4 The relationship between an object's temperature and the amount of energy it emits

5-5 The evidence that light has both particle and wave aspects

5-6 How astronomers can detect an object's chemical composition by studying the light it emits

5-7 The quantum rules that govern the structure of an atom

5-8 The relationship between atomic structure and the light emitted by objects

5-9 How an object's motion affects the light we receive from that object

In the early 1800s, the French philosopher Auguste Comte argued that because the stars are so far away, humanity would never know their nature and composition. But the means to learn about the stars was already there for anyone to see—starlight. Just a few years after Comte's bold pronouncement, scientists began analyzing starlight to learn the very things that he had deemed unknowable.

We now know that atoms of each chemical element emit and absorb light at a unique set of wavelengths characteristic of that element alone. The red light in the accompanying image of a gas cloud in space is of a wavelength emitted by nitrogen and no other element; the particular green light in this image is unique to oxygen, and the particular blue light is unique to helium. The light from nearby planets, distant stars, and remote galaxies also has characteristic "fingerprints" that reveal the chemical composition of these celestial objects.

In this chapter we learn about the basic properties of light. Light has a dual nature: It has the properties of both waves and particles. The light emitted by an object depends upon the object's temperature; we can use this to determine the surface temperatures of stars. By studying the structure of atoms, we will learn why each element emits and absorbs light only at specific wavelengths and will see how astronomers determine what the atmospheres of planets and stars are made of. The motion of a light source also affects wavelengths, permitting us to deduce how fast stars and other objects are approaching or receding. These are but a few of the reasons why understanding light is a prerequisite to understanding the universe.

5-1 Light travels through empty space at a speed of 300,000 km/s

> The speed of light in a vacuum is a universal constant: It has the same value everywhere in the cosmos

Galileo Galilei and Isaac Newton were among the first scientific thinkers to ask basic questions about light. Does light travel instantaneously from one place to another, or does it move with a measurable speed? Whatever the nature of light, it does seem to travel swiftly from a source to our eyes. We see a distant event before we hear the accompanying sound. (For example, we see a flash of lightning before we hear the thunderclap.)

In the early 1600s, Galileo tried to measure the speed of light. He and an assistant stood at night on two hilltops a known distance apart, each holding a shuttered lantern. First, Galileo opened the shutter of his lantern; as soon as his assistant saw the flash of light, he opened his own. Galileo used his pulse as a timer to try to measure the time between opening his lantern and seeing the light from his assistant's lantern. From the distance and time, he hoped to compute the speed at which the light had traveled to the distant hilltop and back.

Galileo found that the measured time failed to increase noticeably, no matter how far away the assistant was stationed. Galileo therefore concluded that the speed of light is too high to be measured by slow human reactions. Thus, he was unable to tell whether or not light travels instantaneously.

The Speed of Light: Astronomical Measurements

The first evidence that light does *not* travel instantaneously was presented in 1676 by Olaus Rømer, a Danish astronomer. Rømer had been studying the orbits of the moons of Jupiter by carefully timing the moments when they passed into or out of Jupiter's shadow. To Rømer's surprise, the timing of these eclipses of Jupiter's moons seemed to depend on the relative positions of Jupiter and Earth. When Earth was far from Jupiter (that is, near conjunction; see Figure 4-6), the eclipses occurred several minutes later than when Earth was close to Jupiter (near opposition).

Rømer realized that this puzzling effect could be explained if light needs time to travel from Jupiter to Earth. When Earth is closest to Jupiter, the image of one of Jupiter's moons disappearing into Jupiter's shadow arrives at our telescopes a little sooner than it does when Jupiter and Earth are farther apart (Figure 5-1). The range of variation in the times at which such eclipses are observed is about 16.6 minutes, which Rømer interpreted as the length of time required for light to travel across the diameter of Earth's orbit (a distance of 2 AU). The size of Earth's orbit was not accurately known in Rømer's day, and he never actually calculated the speed of light. Today, using the modern value of 150 million kilometers for the astronomical unit, Rømer's method yields a value for the speed of light equal to roughly 300,000 km/s (186,000 mi/s).

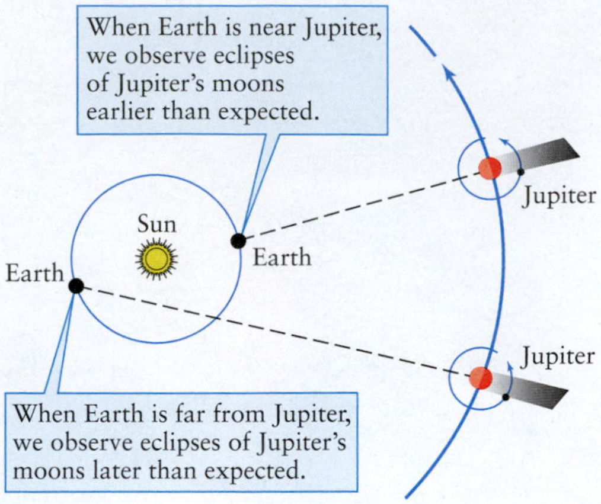
When Earth is near Jupiter, we observe eclipses of Jupiter's moons earlier than expected.
Sun
Earth
Earth
Jupiter
Jupiter
When Earth is far from Jupiter, we observe eclipses of Jupiter's moons later than expected.

FIGURE 5-1
Rømer's Evidence That Light Does Not Travel Instantaneously
The timing of eclipses of Jupiter's moons as seen from Earth depends on the Earth-Jupiter distance. Rømer correctly attributed this effect to variations in the time required for light to travel from Jupiter to Earth.

The Speed of Light: A Universal Speed Limit

Almost two centuries after Rømer, the speed of light was measured very precisely using an ingenious experiment. In 1850, the French physicists Armand-Hippolyte Fizeau and Jean Foucault built the apparatus sketched in Figure 5-2. Light from a light source reflects from a rotating mirror toward a stationary mirror 20 meters away. The rotating mirror moves slightly while the light is making the round trip, so the returning light ray is deflected away from the source by a small angle. By measuring this angle and knowing the dimensions of their apparatus, Fizeau and

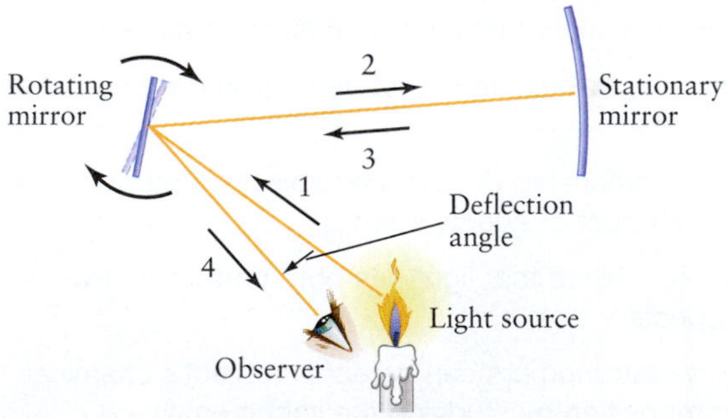
Rotating mirror
2
3
Stationary mirror
1
Deflection angle
4
Light source
Observer

FIGURE 5-2
The Fizeau-Foucault Method of Measuring the Speed of Light
Light from a light source (1) is reflected off a rotating mirror to a stationary mirror (2) and from there back to the rotating mirror (3). The ray that reaches the observer (4) is deflected away from the path of the initial beam because the rotating mirror has moved slightly while the light was making the round trip. The speed of light is calculated from the deflection angle and the dimensions of the apparatus.

Foucault could deduce the speed of light. Once again, the answer was very nearly 300,000 km/s. Based on modern measurements, whether measuring the speed of light on Earth, in the solar system, or amongst galaxies, the speed of light appears to be the same throughout the known universe.

The speed of light in a vacuum (a space devoid of matter) is usually designated by the letter c (from the Latin *celeritas*, meaning "speed"). The modern value is $c = 299{,}792.458$ km/s (186,282.397 mi/s). In most calculations you can use

$$c = 3.00 \times 10^5 \text{ km/s} = 3.00 \times 10^8 \text{ m/s}$$

The most convenient set of units to use for c is different in different situations. The value in kilometers per second (km/s) is often most useful when comparing c to the speeds of objects in space, while the value in meters per second (m/s) is preferred when doing calculations involving the wave nature of light (which we will discuss in Section 5-2).

The speed of light is extremely fast. In one second, light travels a distance equal to about 7½ times around Earth's equator! Since there are more than 31 million seconds in a year, one light-year—the distance light travels in one year—is about 6 trillion miles.

CAUTION! Note that the quantity c is the speed of light *in a vacuum*. Light travels more slowly through air, water, glass, or any other transparent substance than it does in a vacuum. In our study of astronomy, however, we will almost always consider light traveling through the vacuum (or near-vacuum) of space.

The speed of light in empty space is one of the most important numbers in modern physical science. This value appears in many equations that describe atoms, gravity, electricity, and magnetism. According to Einstein's special theory of relativity, there is a universal speed limit: **Nothing can travel faster than the speed of light.**

CONCEPTCHECK 5-1

Why has the speed of light historically been so difficult to measure?

Answer appears at the end of the chapter.

5-2 Light is electromagnetic radiation and is characterized by its wavelength

TUTORIAL 5-1

Light is energy. This fact is apparent to anyone who has felt the warmth of the sunshine on a summer's day. But what exactly is light? How is it produced? What is it made of? How does it move through space? Scholars have struggled with these questions throughout history.

> Visible light, radio waves, and X-rays are all the same type of wave: They differ only in their wavelength

Newton and the Nature of Color

The first major breakthrough in understanding light came from a simple experiment performed by Isaac Newton around 1670. Newton was familiar with what he called the "celebrated Phenomenon of Colours," in which a beam of sunlight passing through a glass prism spreads out into the colors of the rainbow (Figure 5-3). This rainbow is called a **spectrum** (plural **spectra**).

Until Newton's time, it was thought that a prism somehow added colors to white light. To test this idea, Newton placed a second prism so that just one color of the spectrum passed through it: In Figure 5-4, only yellow light can pass through a slit to the second prism. According to the old theory, this should have caused a further change in the color of the light. But Newton found that each color of the spectrum was unchanged by the second prism; yellow remained yellow, blue remained blue, and so on. He concluded that a prism merely separates colors and does not add color. Hence, the spectrum produced by the first prism shows that *sunlight is a mixture of all the colors of the rainbow*. We often call sunlight "white light," but keep in mind that white light is just our brain's perception of sunlight's mixed colors.

Newton suggested that light is made of particles too small to detect individually. In 1678, however, the Dutch physicist and astronomer Christiaan Huygens proposed a rival explanation. He suggested that light travels in the form of waves rather than particles.

CONCEPTCHECK 5-2

The rear brake lights on a car emit white light, but are covered with plastic that only allows red light to pass through. What color of plastic would the red light have to pass through in order to emerge as green light?

Answer appears at the end of the chapter.

Young and the Wave Nature of Light

Around 1801, the English scientist Thomas Young carried out an experiment that convincingly demonstrated the wavelike aspect of light. He passed a beam of light through two thin, parallel slits in a plate, as shown in Figure 5-5a. On a white screen some distance beyond the slits, the light formed a pattern of alternating bright and dark bands. Young reasoned that if a beam of light was a stream of particles (as Newton had suggested), the two beams of light from the slits would simply form bright images of the slits on the white surface. The pattern of bright and dark bands he observed, however, is just what would be expected if light had wavelike properties. An analogy with water waves demonstrates why (Figure 5-5b).

Maxwell and Light as an Electromagnetic Wave

The discovery of the wave nature of light posed some obvious questions. What exactly is "waving" in light? That is, what is it about light that goes up and down like water waves on the ocean? Because we can see light from the Sun, planets, and stars, light waves must be able to travel across empty space. Hence, whatever is "waving" cannot be any material substance. What, then, is it?

The answer came from a seemingly unlikely source—a comprehensive theory that described electricity and magnetism. Numerous experiments during the first half of the nineteenth century demonstrated an intimate connection between electric and magnetic forces. A central idea to emerge from these experiments is the concept of a *field*, an immaterial yet measurable disturbance

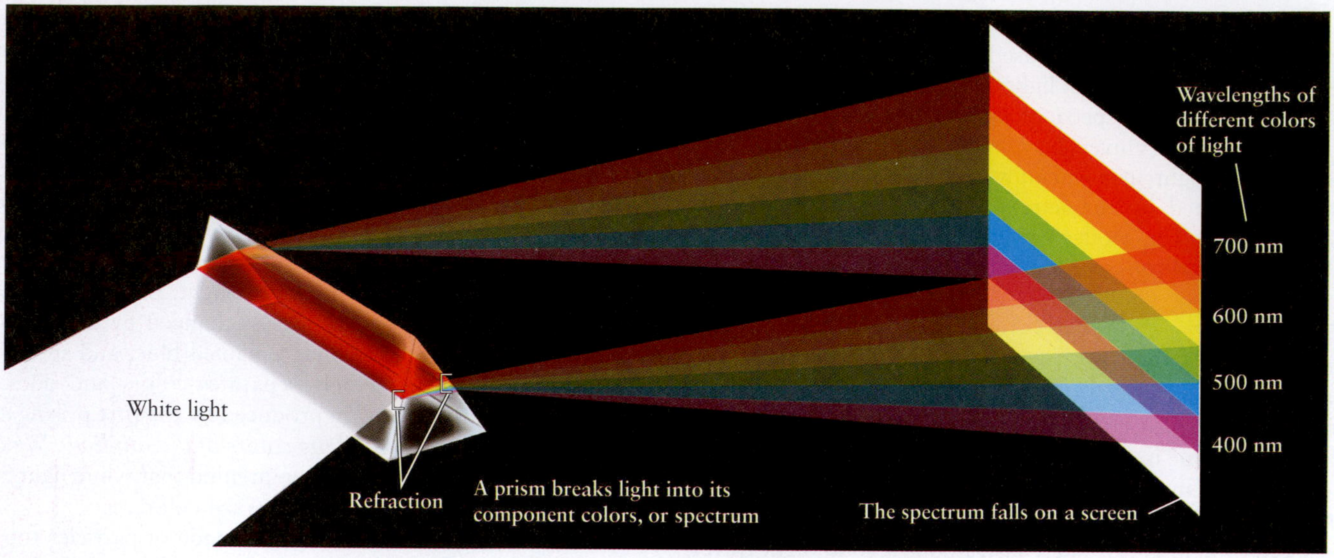

Wavelengths of different colors of light
700 nm
600 nm
500 nm
400 nm
White light
Refraction
A prism breaks light into its component colors, or spectrum
The spectrum falls on a screen

FIGURE 5-3

A Prism and a Spectrum When a beam of sunlight passes through a glass prism, the light is broken into a rainbow-colored band called a spectrum. The wavelengths of different colors of light are shown on the right (1 nm = 1 nanometer = 10^{-9} m).

of any region of space in which electric or magnetic forces are felt. Thus, an electric charge is surrounded by an electric field, and a magnet is surrounded by a magnetic field. Experiments in the early 1800s demonstrated that moving an electric charge produces a magnetic field; conversely, moving a magnet gives rise to an electric field.

In the 1860s, the Scottish mathematician and physicist James Clerk Maxwell succeeded in describing all the basic properties of electricity and magnetism in four equations. This mathematical achievement demonstrated that electric and magnetic forces are really two aspects of the same phenomenon, which we now call **electromagnetism.**

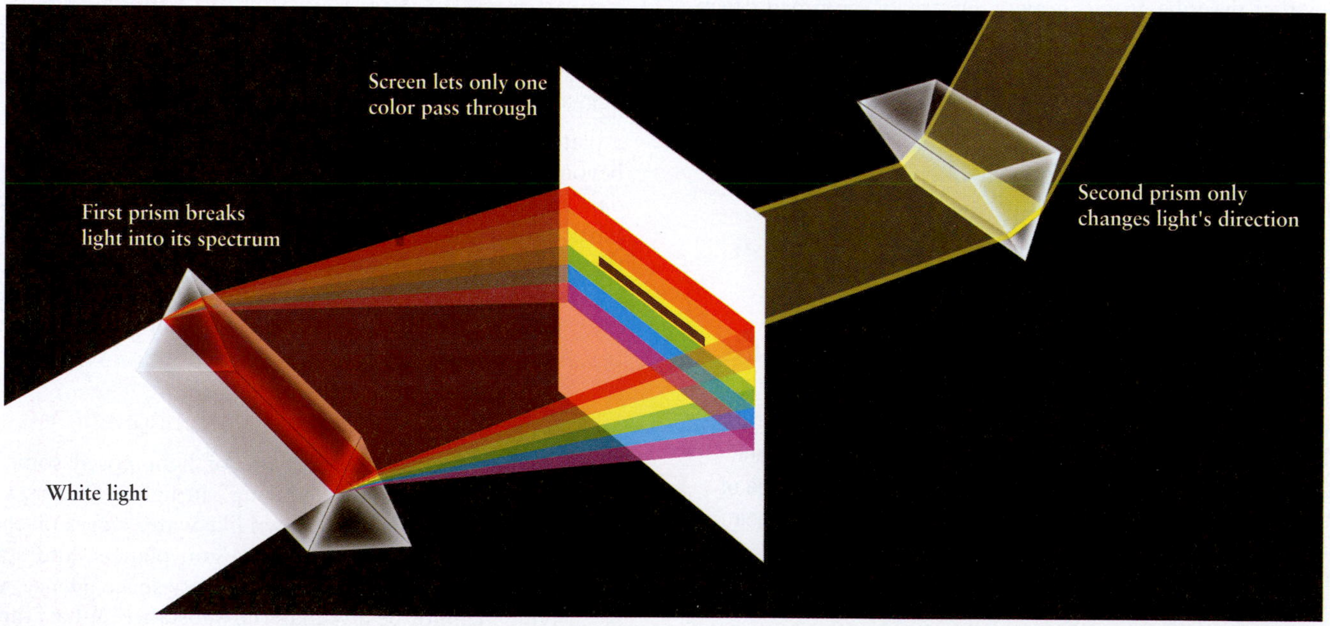

Screen lets only one color pass through
First prism breaks light into its spectrum
Second prism only changes light's direction
White light

FIGURE 5-4

Newton's Experiment on the Nature of Light In a crucial experiment, Newton took sunlight that had passed through a prism and sent it through a second prism. Between the two prisms was a screen with a hole in it that allowed only one color of the spectrum to pass through. This same color emerged from the second prism. Newton's experiment proved that prisms do not add color to light but merely bend different colors through different angles. It also showed that white light, such as sunlight, is actually a combination of all the colors that appear in its spectrum.

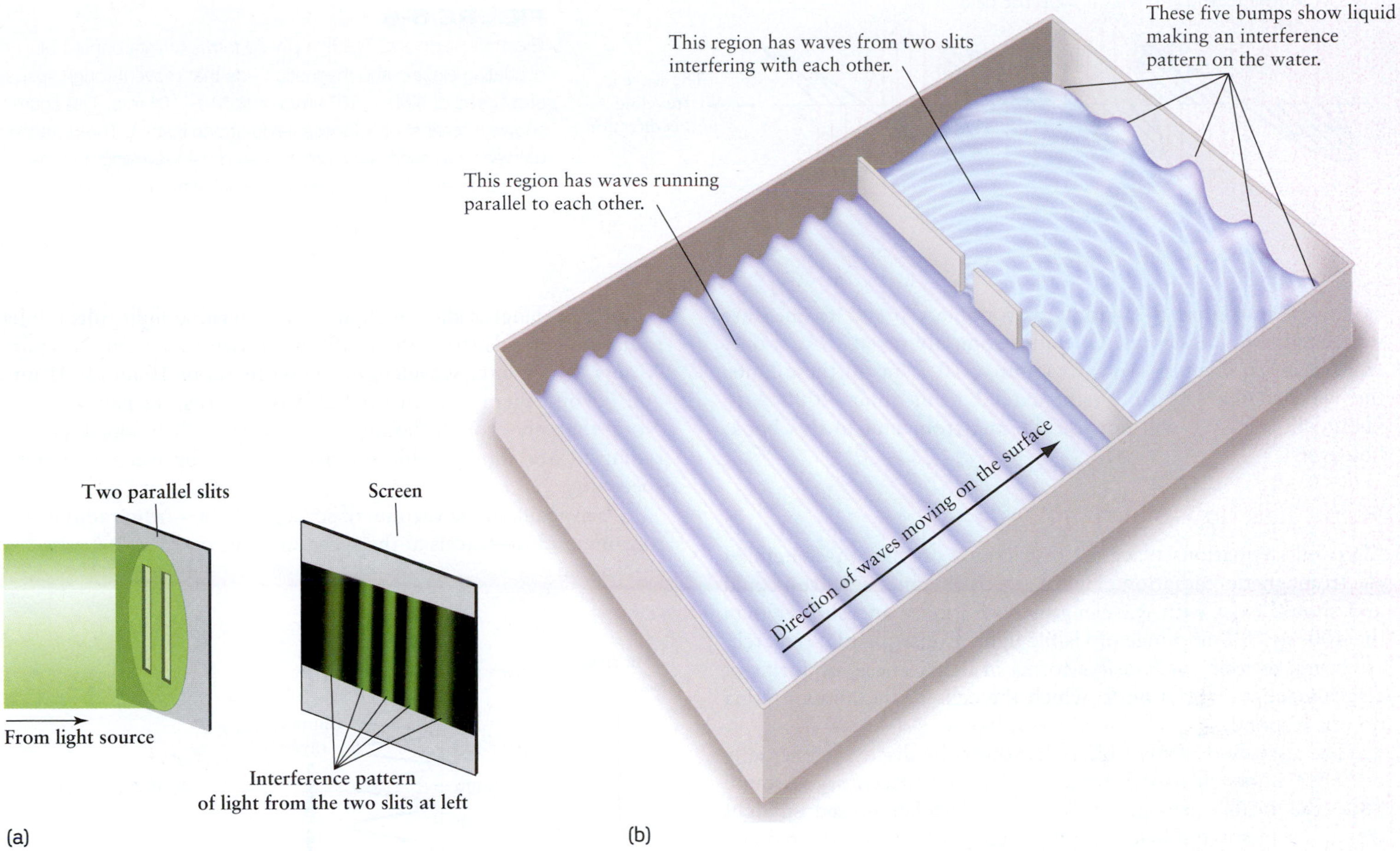
Two parallel slits
Screen
From light source
Interference pattern
of light from the two slits at left
(a)
This region has waves from two slits
interfering with each other.
These five bumps show liquid
making an interference
pattern on the water.
This region has waves running
parallel to each other.
Direction of waves moving on the surface
(b)

FIGURE 5-5

Wave Interference **(a)** Electromagnetic radiation travels as waves. Thomas Young's interference experiment shows that light of a single color passing through a barrier with two slits behaves as waves that create alternating light and dark patterns on a screen. **(b)** Water waves passing through two slits creates two overlapping waves on the right side of a ripple tank; the result is an interference pattern. Constructive interference occurs where the crests (high points) from waves combine together. Destructive interference occurs where the crest from one wave combines with the trough (low point) of another, leaving the water's surface undisturbed. In the experiment with light, the bright bands result from constructive interference, and the dark bands, from destructive interference, which shows that light is made of waves. (© University of Colorado, Center for Intergrated Plasma Studies, Boulder, CO)

By combining his four equations, Maxwell showed that electric and magnetic fields should travel through space in the form of waves at a speed of 3.0×10^5 km/s—a value equal to the best available value for the speed of light. Maxwell's suggestion that these waves do exist and are observed as light was soon confirmed by experiments. Because of its electric and magnetic properties, light is also called **electromagnetic radiation.**

CAUTION! You may associate the term *radiation* with radioactive materials like uranium, but this term refers to anything that radiates, or spreads away, from its source. For example, scientists sometimes refer to sound waves as "acoustic radiation." Radiation does not have to be related to radioactivity!

Electromagnetic radiation consists of oscillating electric and magnetic fields, as shown in Figure 5-6. The distance between two successive wave crests is called the **wavelength** of the light, usually designated by the Greek letter λ (lambda). It is essential to keep in mind that no matter what the wavelength, all electromagnetic radiation travels at the same speed in a vacuum: $c = 3.0 \times 10^5$ km/s $= 3.0 \times 10^8$ m/s.

More than a century elapsed between Newton's experiments with a prism and the confirmation of the wave nature of light. One reason for this delay is that **visible light,** the light to which the human eye is sensitive, has extremely short wavelengths—less than a thousandth of a millimeter—that are not easily detectable. To express such tiny distances conveniently, scientists use a unit of length called the **nanometer** (abbreviated nm), where 1 nm $= 10^{-9}$ m. Experiments demonstrated that visible light has wavelengths covering the range from about 400 nm for violet light to about 700 nm for red light. Intermediate colors of the rainbow like yellow (550 nm) have intermediate wavelengths,

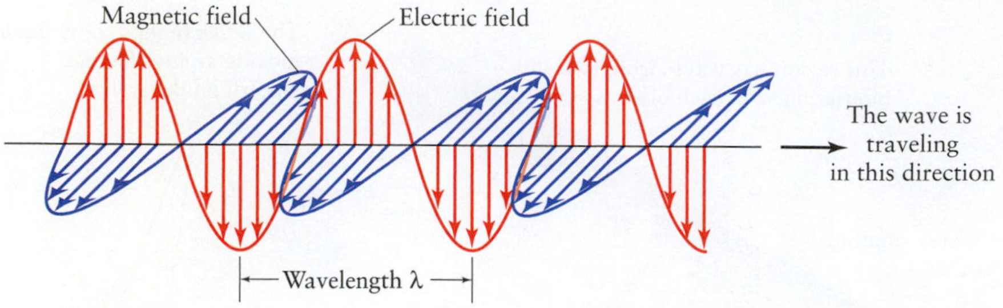

FIGURE 5-6
Electromagnetic Radiation All forms of light consist of oscillating electric and magnetic fields that move through space at a speed of 3.00 × 10^5 km/s = 3.00 × 10^8 m/s. This figure shows a "snapshot" of these fields at one instant. The distance between two successive crests, called the wavelength of the light, is usually designated by the Greek letter λ (lambda).

as shown in Figure 5-7. (Some astronomers prefer to measure wavelengths in *angstroms*. One angstrom, abbreviated Å, is one-tenth of a nanometer: 1 Å = 0.1 nm = 10^{-10} m. In these units, the wavelengths of visible light extend from about 4000 Å to about 7000 Å. We will not use the angstrom unit in this book, however.)

Visible and Nonvisible Light

Maxwell's equations place no restrictions on the wavelength of electromagnetic radiation. Hence, electromagnetic waves could and should exist with wavelengths both longer and shorter than the 400- to 700-nm range of visible light. Consequently, researchers began to look for *invisible* forms of light. These are forms of electromagnetic radiation to which the cells of the human retina do not respond.

The first kind of invisible radiation to be discovered actually preceded Maxwell's work by more than a half century. Around 1800 the British astronomer William Herschel passed sunlight through a prism and held a thermometer just beyond the red end of the visible spectrum. The thermometer registered a temperature increase, indicating that it was being exposed to an invisible form of energy. This invisible energy, now called **infrared radiation,** was later realized to be electromagnetic radiation with wavelengths somewhat longer than those of visible light.

In experiments with electric sparks in 1888, the German physicist Heinrich Hertz succeeded in producing electromagnetic radiation with even longer wavelengths of a few centimeters or more. These are now known as **radio waves.** In 1895 another German physicist, Wilhelm Röntgen, invented a machine that produces electromagnetic radiation with wavelengths shorter than 10 nm, now known as **X-rays.** The X-ray machines in modern medical and dental offices are direct descendants of Röntgen's invention. Over the years radiation has been discovered with many other wavelengths.

Thus, visible light occupies only a tiny fraction of the full range of possible wavelengths, collectively called the **electromagnetic spectrum.** As Figure 5-7 shows, the electromagnetic spectrum stretches from the longest-wavelength radio waves to the shortest-wavelength gamma rays. Figure 5-8 shows some applications of nonvisible light in modern technology.

On the long-wavelength side of the visible spectrum, infrared radiation covers the range from about 700 nm to 1 mm. Astronomers interested in infrared radiation often express wavelength in *micrometers* or *microns*, abbreviated μm, where 1 μm = 10^{-3} mm = 10^{-6} m. **Microwaves** have wavelengths from roughly 1 mm to 10 cm, while radio waves have even longer wavelengths.

At wavelengths shorter than those of visible light, **ultraviolet radiation** extends from about 400 nm down to 10 nm. Next are X-rays, which have wavelengths between about 10 and 0.01 nm, and beyond them at even shorter wavelengths are **gamma rays.** Note that the rough boundaries between different types of radiation are simply arbitrary divisions in the electromagnetic spectrum.

Once again, in the vacuum of space, all light—from radio waves to gamma rays—travels at the speed of light, $c = 3 \times 10^8$ m/s.

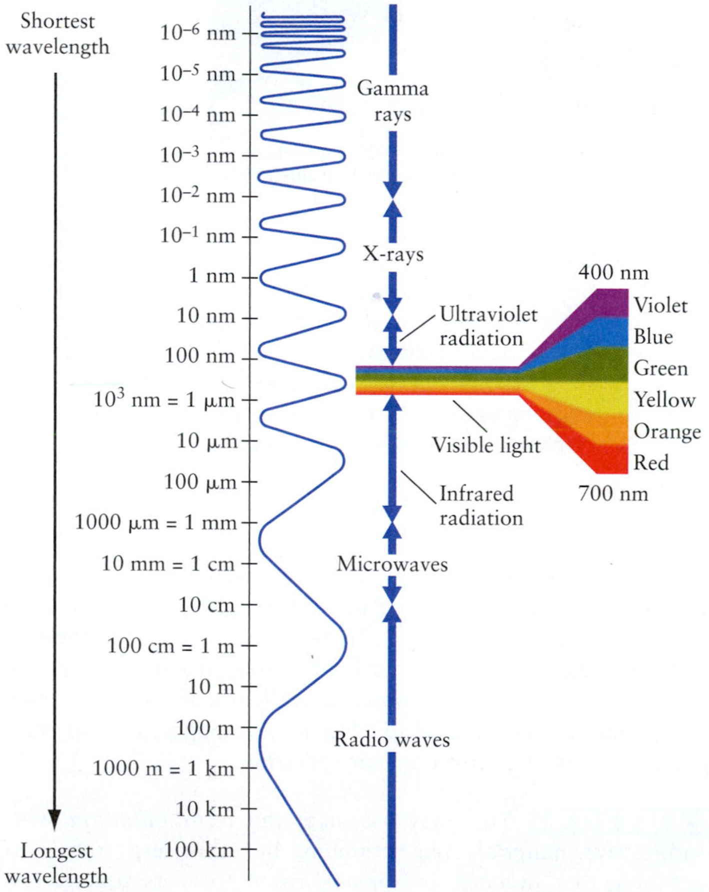

FIGURE 5-7
The Electromagnetic Spectrum The full array of all types of electromagnetic radiation is called the electromagnetic spectrum. It extends from the longest-wavelength radio waves to the shortest-wavelength gamma rays. Visible light occupies only a tiny portion of the full electromagnetic spectrum.

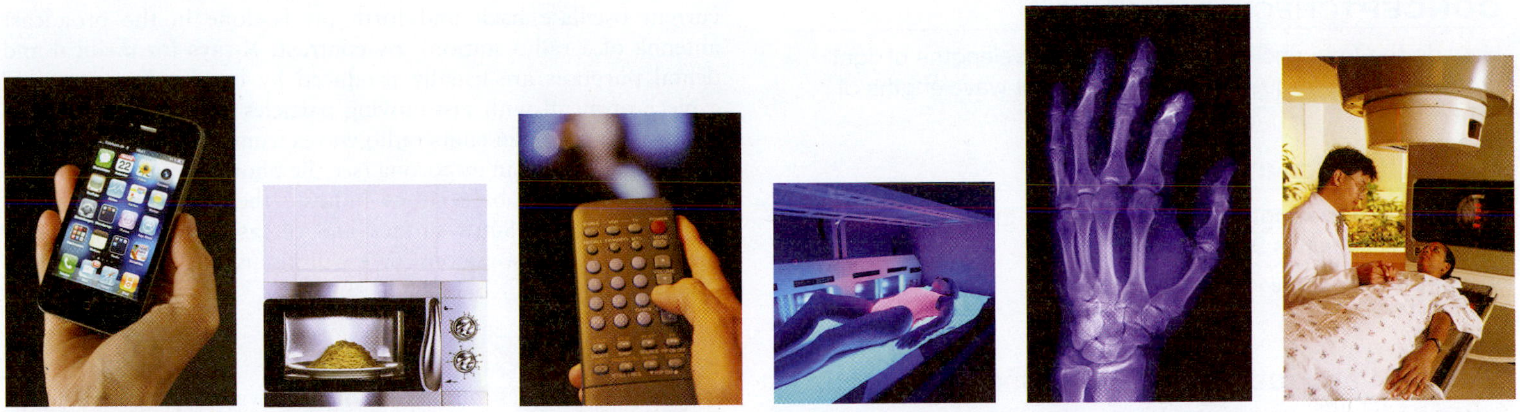

(a) Mobile phone: radio waves (b) Microwave oven: microwaves (c) TV remote: infrared light (d) Tanning booth: ultraviolet light (e) Medical imaging: X-rays (f) Cancer radiotherapy: gamma rays

FIGURE 5-8
Uses of Nonvisible Electromagnetic Radiation **(a)** A mobile phone is actually a radio transmitter and receiver. The wavelengths used are in the range 16 to 36 cm. **(b)** A microwave oven produces radiation with a wavelength near 10 cm. The water in food absorbs this radiation, thus heating the food. **(c)** A remote control sends commands to a television using a beam of infrared light. **(d)** Ultraviolet radiation in moderation gives you a suntan, but in excess can cause sunburn or skin cancer. **(e)** X-rays can penetrate through soft tissue but not through bone, which makes them useful for medical imaging. **(f)** Gamma rays destroy cancer cells by breaking their DNA molecules, making them unable to multiply. (a: Maurizio Gambarini/dpa/Corbis; b: Michael Haegele/Corbis; c: Bill Lush/Taxi/Getty; d: Neil McAllister/Alamy; e: Ted Kinsman/Photo Researchers, Inc.; f: Will and Deni McIntyre/Science Source)

CONCEPTCHECK 5-3

Which form of electromagnetic radiation has a wavelength similar to the diameter of your finger?

Answer appears at the end of the chapter.

Frequency and Wavelength

Astronomers who work with radio telescopes often prefer to speak of *frequency* rather than wavelength. The **frequency** of a wave is the number of wave crests that pass a given point in one second. Equivalently, it is the number of complete *cycles* of the wave that pass per second (a complete cycle is from one crest to the next). Frequency is usually denoted by the Greek letter ν (nu). The unit of frequency is the cycle per second, also called the *hertz* (abbreviated Hz) in honor of Heinrich Hertz, the physicist who first produced radio waves. For example, if 500 crests of a wave pass you in one second, the frequency of the wave is 500 cycles per second or 500 Hz.

In working with frequencies, it is often convenient to use the prefix *mega-* (meaning "million," or 10^6, and abbreviated M) or *kilo-* (meaning "thousand," or 10^3, and abbreviated k). For example, AM radio stations broadcast at frequencies between 535 and 1605 kHz (kilohertz), while FM radio stations broadcast at frequencies in the range from 88 to 108 MHz (megahertz).

The relationship between the frequency and wavelength of an electromagnetic wave is a simple one. Because light moves at a constant speed $c = 3 \times 10^8$ m/s, if the wavelength (distance from one crest to the next) is made shorter, the frequency must increase (more of those closely spaced crests pass you each second). Thus, the shorter the wavelength of light, the higher the frequency, and likewise, the longer the wavelength of light, the lower the frequency. The frequency ν of light is related to its wavelength λ by the equation

Frequency and wavelength of an electromagnetic wave

$$\nu = \frac{c}{\lambda}$$

ν = frequency of an electromagnetic wave (in Hz)

c = speed of light = 3×10^8 m/s

λ = wavelength of the wave (in meters)

That is, the frequency of a wave equals the wave speed divided by the wavelength.

For example, hydrogen atoms in space emit radio waves with a wavelength of 21.12 cm. To calculate the frequency of this radiation, we must first express the wavelength in meters rather than centimeters: $\lambda = 0.2112$ m. Then we can use the above formula to find the frequency ν:

$$\nu = \frac{c}{\lambda} = \frac{3 \times 10^8 \text{ m/s}}{0.2112 \text{ m}} = 1.42 \times 10^9 \text{ Hz} = 1420 \text{ MHz}$$

Visible light has a much shorter wavelength and higher frequency than radio waves. You can use the above formula to show that for yellow-orange light of wavelength 600 nm, the frequency is 5×10^{14} Hz or 500 *million* megahertz!

While Young's experiment (Figure 5-5) showed convincingly that light has wavelike aspects, it was discovered in the early 1900s that light *also* has some of the characteristics of a stream of particles and waves. We will explore light's dual nature in Section 5-5.

CONCEPTCHECK 5-4

How do the frequencies of the longest wavelengths of light compare to the frequencies of the shortest wavelengths of light?

CALCULATIONCHECK 5-1

What is the wavelength of radio waves for the FM radio station 99.9 KTYD (which is 99.9 MHz)?

Answers appear at the end of the chapter.

5-3 An opaque object emits electromagnetic radiation according to its temperature

> As an object is heated, it glows more brightly and its peak color shifts to shorter wavelengths

To learn about objects in the heavens, astronomers study the character of the electromagnetic radiation coming from those objects. Such studies can be very revealing because different kinds of electromagnetic radiation are typically produced in different ways. As an example, on Earth the most common way to generate radio waves is to make an electric current oscillate back and forth (as is done in the broadcast antenna of a radio station). By contrast, X-rays for medical and dental purposes are usually produced by bombarding atoms in a piece of metal with fast-moving particles extracted from other atoms. Our own Sun emits radio waves from near its glowing surface and X-rays from its corona (see the photo that opens Chapter 3). Hence, these observations indicate the presence of electric currents near the Sun's surface and of fast-moving particles in the Sun's outermost regions. (We will discuss the Sun at length in Chapter 18.)

Radiation from Heated Objects

TUTORIAL 5-2

The simplest and most common way to produce electromagnetic radiation, either on or off Earth, is to heat an object. The hot filament of wire inside an ordinary lightbulb emits white light, and a neon sign has a characteristic red glow because neon gas within the tube is heated by an electric current. In like fashion, almost all the visible light that we receive from space comes from hot objects like the Sun and the stars. The kind and amount of light emitted by a hot object tell us not only how hot it is but also about other properties of the object.

We can tell whether the hot object is made of relatively dense or relatively thin material. Consider the difference between a lightbulb

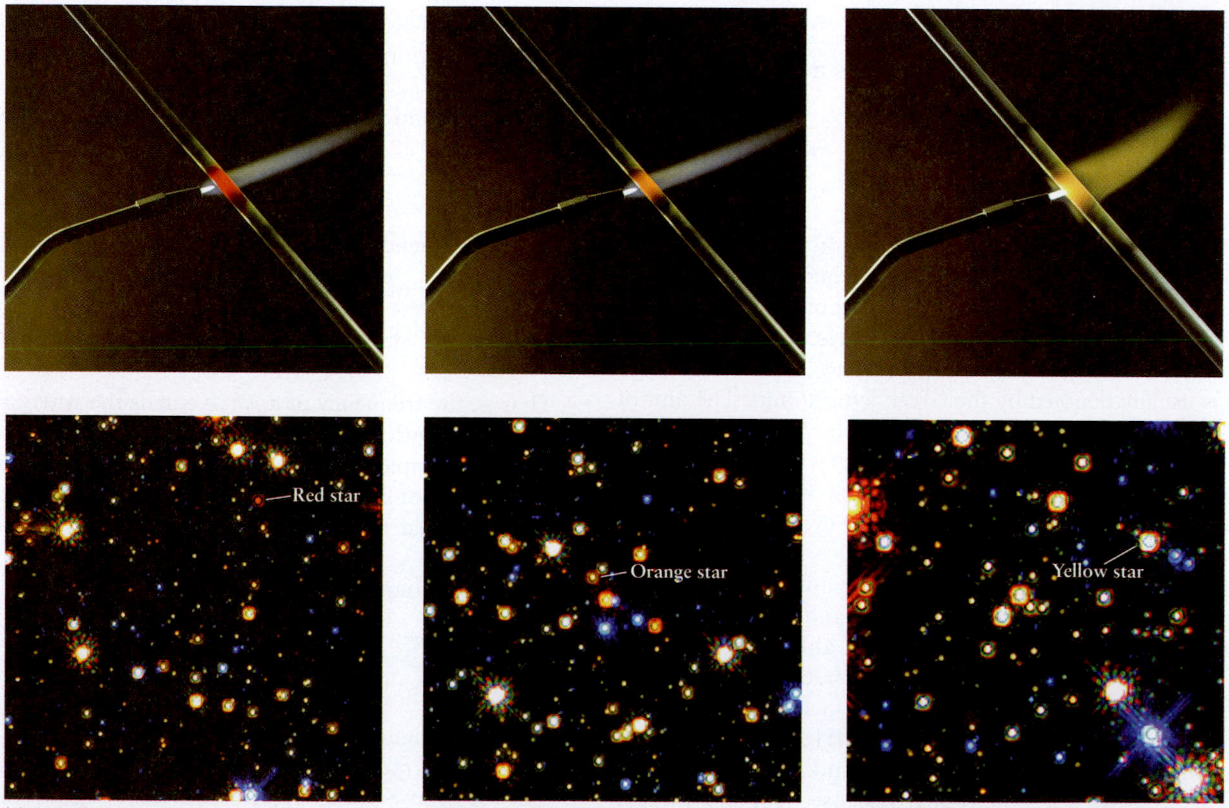

(a) Hot: glows deep red (b) Hotter: glows orange (c) Even hotter: glows yellow

FIGURE 5-9 R I V U X G

Objects at Different Temperatures Have Different Colors and Brightnesses This sequence of photographs shows the changing appearance of a piece of iron as it is heated. As the temperature increases, the amount of energy radiated by the bar increases, and so it appears brighter. The apparent color of the bar also changes because, as the temperature increases, the dominant or peak wavelength of light emitted by the bar decreases. The stars shown have roughly the same temperatures as the bars above them. (top row: © 1984 Richard Megna Fundamental Photographs; bottom row: NASA) and a neon sign. The dense, solid filament of a lightbulb makes white light, which is a mixture of all different visible wavelengths, while the thin, transparent neon gas produces light of a rather definite red color and, hence, a rather definite wavelength. For now we will concentrate our attention on the light produced by dense, *opaque* objects. An **opaque** object does not allow light to travel through it without being scattered or absorbed—opaque is the opposite of transparent. Earth's atmosphere absorbs and scatters only some of the incoming sunlight, and we say that it is partially opaque, or partially transparent. Even though the Sun and stars are gaseous, not solid, it turns out that because they are so large, light from the inside scatters many times before emerging from the surface. Therefore, stars are opaque and emit light with many of the same properties as light emitted by a hot, glowing, solid object.

Imagine a welder or blacksmith heating a bar of iron. As the bar becomes hot, it begins to glow deep red, as shown in Figure 5-9a. (You can see this same glow from the coils of a toaster or from an electric range turned on "high.") While the bar emits a range of wavelengths, the dominant wavelength is red in color. As the temperature rises further, the bar begins to give off a brighter orange light (Figure 5-9b). At still higher temperatures, it shines with a brilliant yellow color (Figure 5-9c); the dominant wavelength is yellow light. If the bar could be prevented from melting and vaporizing, at extremely high temperatures it would emit a dazzling blue-white light. As shown in Figure 5-9, the temperature of a star determines its color, and the hottest stars are blue.

As this example shows, the amount of energy emitted by the hot, dense object and the dominant wavelength of the emitted radiation both depend on the temperature of the object. The hotter the object, the more energy it emits and the shorter the wavelength at which most of the energy is emitted. Colder objects emit relatively little energy, and this emission is primarily at long wavelengths.

These observations explain why you cannot see in the dark. The temperatures of people, animals, and furniture are much less than even that of the iron bar in Figure 5-9a. So, while these objects emit radiation even in a darkened room, most of this emission is at wavelengths greater than those of red light, in the infrared part of the spectrum (see Figure 5-7). Your eye is not sensitive to infrared, and you thus cannot see ordinary objects in a darkened room. But you can detect this radiation by using a camera that is sensitive to infrared light (Figure 5-10).

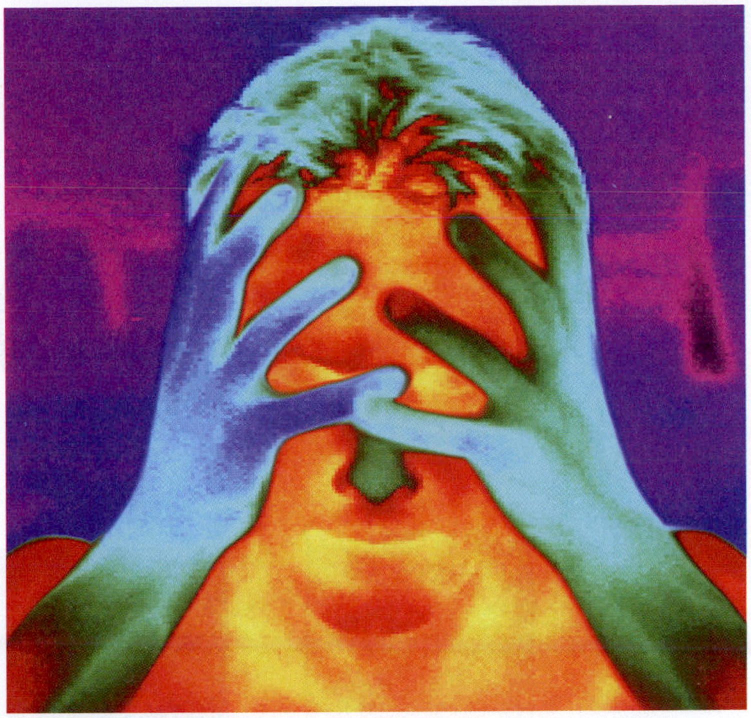

FIGURE 5-10 R **I** V U X G
An Infrared Portrait In this image made with a camera sensitive to infrared radiation, the different colors represent regions of different temperature. That this image is made from infrared radiation is indicated by the highlighted I in the wavelength tab. Red areas (like the man's face) are the warmest and emit the most infrared light, while blue-green areas (including the man's hands and hair) are at the lowest temperatures and emit the least radiation. (Dr. Arthur Tucker/Photo Researchers)

Temperature and Thermal Energy

To better understand the relationship between the temperature of a dense object and the radiation it emits, it is helpful to know just what "temperature" means. The temperature of a substance is directly related to the average speed of the tiny atoms or molecules that make up the substance. As shown in Figure 5-11, the hotter the substance, the faster its atoms or molecules are moving; the colder the substance, the slower the motion of its particles. The **thermal energy** of an object comes from the kinetic energy of its atoms and molecules. Therefore, the hotter a substance, the greater its thermal energy.

Scientists usually prefer to use the Kelvin temperature scale, on which temperature is measured in **kelvins** (K) upward from **absolute zero.** This is the coldest possible temperature, at which atoms move as slowly as possible (they can never quite stop completely). On the more familiar Celsius and Fahrenheit temperature scales, absolute zero (0 K) is 2273°C and 2460°F. Ordinary room temperature is 293 K, 20°C, or 68°F. Box 5-1 discusses the relationships among the Kelvin, Celsius, and Fahrenheit temperature scales.

Figure 5-12 depicts quantitatively how the radiation from a dense object depends on its Kelvin temperature. Each curve in this figure shows the intensity of light emitted at each wavelength by a dense object at a given temperature: 3000 K (the temperature of a "cool" star, and at which molten gold boils), 6000 K (around the temperature of our Sun, and of an iron-welding arc), and 12,000 K (the temperature of a "hot" star, and also found in special industrial furnaces). In other words, the curves show the spectrum of light emitted by such an object. At any temperature, a hot, dense object emits at all wavelengths, so its spectrum is a smooth, continuous curve with no gaps in it.

As we will see shortly, this is an example of *blackbody radiation*. Figure 5-12 shows that *the higher the temperature, the greater the intensity of light at all wavelengths*. One way to understand this response to temperature is that the emitted light results from the motion of the material's atoms and molecules, which, as we saw, increases with temperature (see Figure 5-11).

The temperature not only indicates the total intensity of the light, but also the shape of its spectrum. The most important feature in the spectrum—the dominant wavelength—is called the **wavelength of maximum emission,** at which the curve has its peak

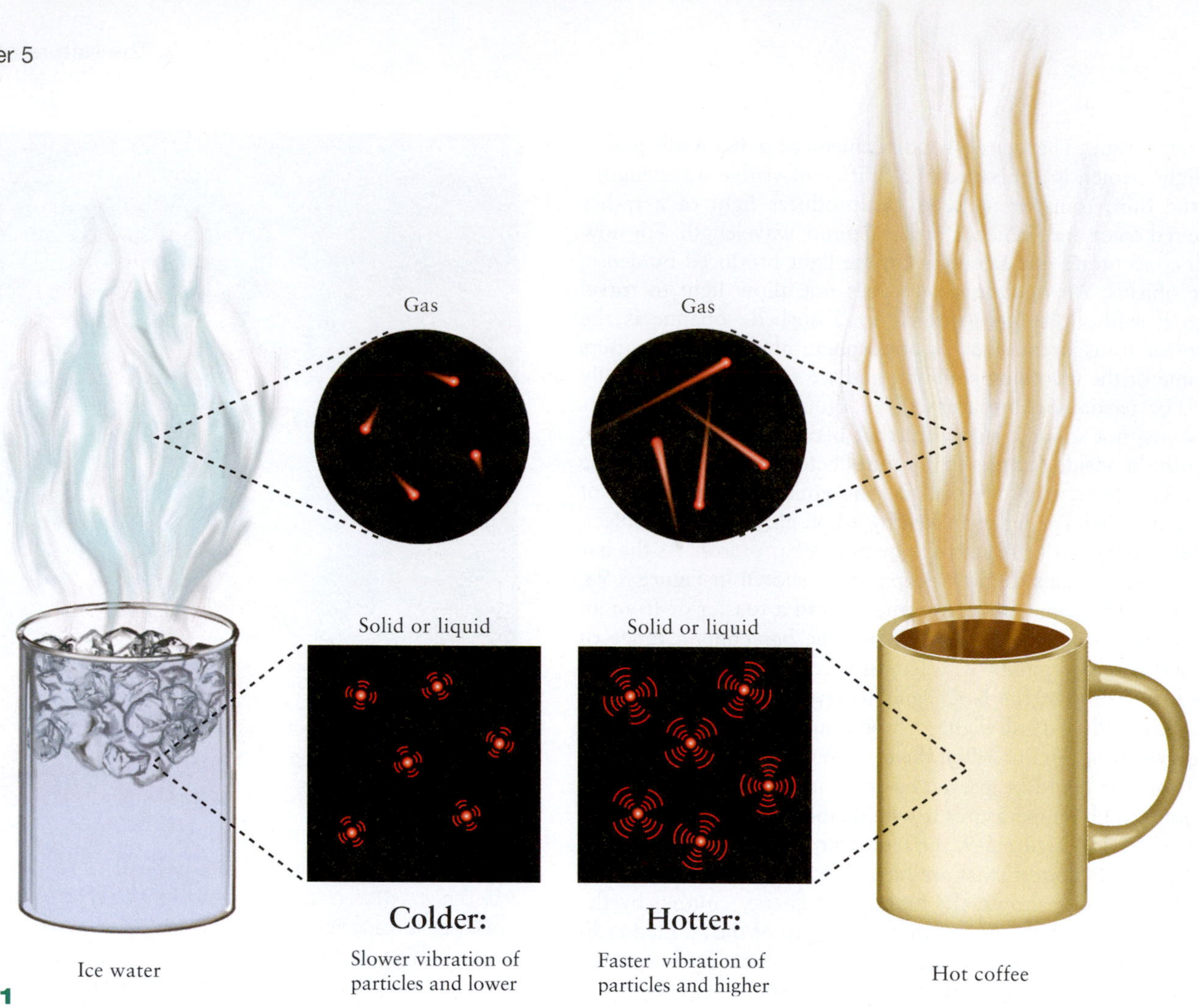

FIGURE 5-11
Thermal Energy The thermal energy of an object comes from the kinetic energy of its internal particles, and the hotter the material, the faster the particles move around. For hotter solids and liquids, particles vibrate faster and collide more intensely with their neighbors; for hotter gases the particles stream around at higher average speeds.

and the intensity of emitted energy is strongest (see Figure 5-12). The location of the peak also changes with temperature: *The higher the temperature, the shorter the wavelength of maximum emission.*

Figure 5-12 shows that for a dense object at a temperature of 3000 K, the wavelength of maximum emission is around 1000 nm (1 mm). Because this peak wavelength corresponds to the infrared range and is well outside the visible range, you might think that you cannot see the radiation from an object at this temperature. However, the glow from such an object *is* visible; the curve shows that this object emits light within the visible range, although about 90 percent of the emitted energy comes out at longer, invisible wavelengths (to the right of the visible spectrum). This invisible light is also emitted by lightbulbs. Incandescent lightbulbs create visible light by electrically heating a thin wire filament to about 3000 K, but because most of the energy is wasted at longer wavelengths, more efficient lights have become popular.

The 3000-K curve in Figure 5-12 is quite a bit higher at the red end of the visible spectrum than at the violet end, and a dense object at this temperature will appear yellowish in color. At even lower temperatures below 1000 K, the object would appear red (see Figure 5-9a). For higher temperatures, the 12,000-K curve has its wavelength of maximum emission in the ultraviolet part of the spectrum, at a wavelength shorter than visible light. But such a hot, dense object also emits copious amounts of visible light (much more than at 6000 K or 3000 K, for which the curves are lower) and thus will have a very visible glow. The curve for this temperature is higher for blue light than for red light, and so the color of a dense object at 12,000 K is a brilliant blue or blue-white. The same principles apply to stars of different temperatures, which accounts for the different colors of stars in Figure 5-9. A star that looks blue has a high surface temperature, while an orange star has a relatively cool surface.

To summarize these observations:

The higher an object's temperature, the more intensely the object emits electromagnetic radiation and the shorter the wavelength at which it emits most strongly.

We will make frequent use of this general rule to analyze the temperatures of celestial objects such as planets and stars.

The curves in Figure 5-12 are drawn for an idealized type of dense object called a **blackbody.** A perfect blackbody does not reflect any light at all; instead, it absorbs all radiation falling on it. Because it reflects no electromagnetic radiation, the radiation that it does emit is entirely the result of its temperature. Ordinary objects, like tables, textbooks, and people, are not perfect blackbodies; they

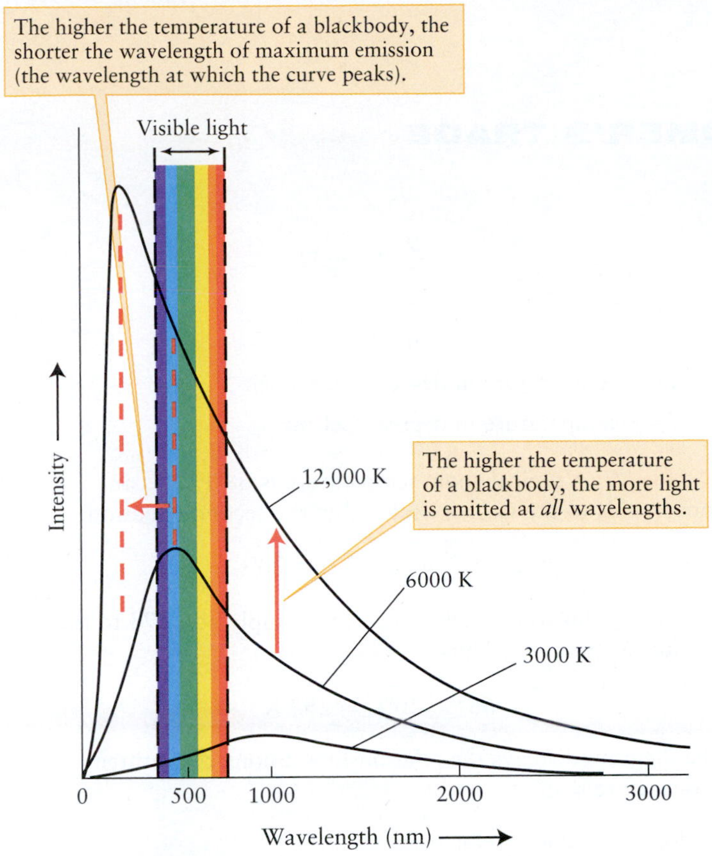

ANIMATION 5-1

FIGURE 5-12

Blackbody Curves Each of these curves shows the intensity of light at every wavelength that is emitted by a blackbody (an idealized case of a dense object) at a particular temperature. The rainbow-colored band shows the range of visible wavelengths. The vertical scale has been compressed so that all three curves can be seen; the peak intensity for the 12,000-K curve is actually about 1000 times greater than the peak intensity for the 3000-K curve.

reflect light, which is why they are visible. A star such as the Sun, however, behaves very much like a perfect blackbody, because it absorbs almost completely any radiation falling on it from outside. The light emitted by a blackbody is called **blackbody radiation,** and the curves in Figure 5-12 are often called **blackbody curves.**

CAUTION! Despite its name, a blackbody does not necessarily *look* black. The Sun, for instance, does not look black because its temperature is high (around 5800 K), and so it glows brightly. But a room-temperature (around 300 K) blackbody would appear very black indeed. Even if a 300-K blackbody were as large as the Sun, it would emit only about 1/100,000 as much energy: Its blackbody curve is far too low to graph in Figure 5-12. Furthermore, most of the radiation from a room-temperature object is emitted at infrared wavelengths that are too long for our eyes to perceive.

Figure 5-13 shows the blackbody curve for a temperature of 5800 K. It also shows the intensity curve for light from the Sun, as measured from above Earth's atmosphere. (This is necessary because Earth's atmosphere absorbs certain wavelengths.) The peak of both curves is at a wavelength of about 500 nm, near the middle of the visible spectrum. Note how closely the observed intensity curve for the Sun matches the blackbody curve. This is a strong indication that the temperature of the Sun's glowing surface is about 5800 K—a temperature that we can measure across a distance of 150 million kilometers! The close correlation between blackbody curves and the observed intensity curves for most stars is a key reason why astronomers are interested in the physics of blackbody radiation.

Blackbody radiation depends *only* on the temperature of the object emitting the radiation, not on the chemical composition of the object. The light emitted by molten gold at 2000 K is very nearly the same as that emitted by molten lead at 2000 K. Therefore, it might seem that analyzing the light from the Sun or from a star can tell astronomers the object's temperature but not what the star is made of. As Figure 5-13 shows, however, the intensity curve for the Sun (a typical star) is not precisely that of a blackbody. We will see later in this chapter that the *differences* between a star's spectrum and that of a blackbody allow us to determine the chemical composition of the star.

CONCEPTCHECK 5-5

An object is at 3000 K, emitting radiation that peaks at an infrared wavelength of around 1000 nm. Now the object is heated to 12,000 K and only emits a small fraction of its total energy at infrared wavelengths. At what temperature does the object emit more infrared radiation?

CONCEPTCHECK 5-6

Does a 0°C blackbody object give off any radiation?

Answers appear at the end of the chapter.

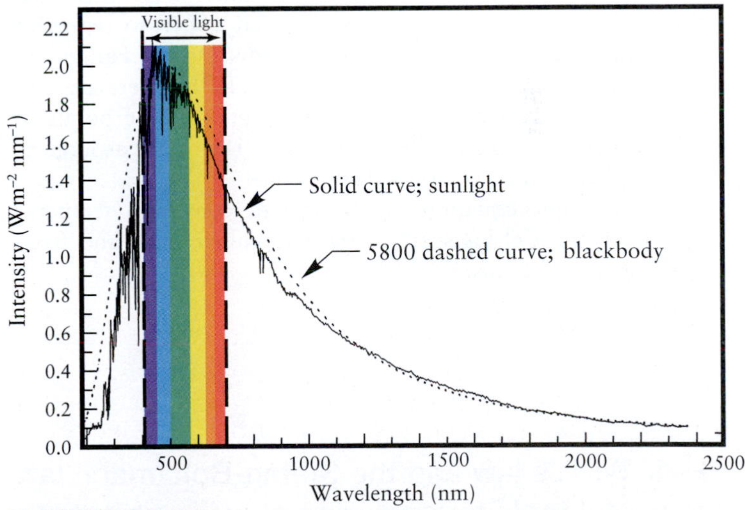

FIGURE 5-13

The Sun as a Blackbody This graph shows that the intensity of sunlight over a wide range of wavelengths (solid curve) is a remarkably close match to the intensity of radiation coming from a blackbody at a temperature of 5800 K (dashed curve). The measurements of the Sun's intensity were made above Earth's atmosphere (which absorbs and scatters certain wavelengths of sunlight). It is not surprising that the range of visible wavelengths includes the peak of the Sun's spectrum; the human eye evolved to take advantage of the most plentiful light available.

BOX 5-1 TOOLS OF THE ASTRONOMER'S TRADE

Temperatures and Temperature Scales

Three temperature scales are in common use. Throughout most of the world, temperatures are expressed in **degrees Celsius** (°C). The Celsius temperature scale is based on the behavior of water, which freezes at 0°C and boils at 100°C at sea level on Earth. This scale is named after the Swedish astronomer Anders Celsius, who proposed it in 1742.

Astronomers usually prefer the Kelvin temperature scale. This is named after the nineteenth-century British physicist Lord Kelvin, who made many important contributions to our understanding of heat and temperature. Absolute zero, the temperature at which atomic motion is at the absolute minimum, is –273°C in the Celsius scale but 0 K in the Kelvin scale. Atomic motion cannot be any less than the minimum, so nothing can be colder than 0 K; hence, there are no negative temperatures on the Kelvin scale. Note that we do *not* use degree (°) with the Kelvin temperature scale.

A temperature expressed in kelvins is always equal to the temperature in degrees Celsius plus 273. On the Kelvin scale, water freezes at 273 K and boils at 373 K. Water must be heated through a change of 100 K or 100°C to go from its freezing point to its boiling point. Thus, the "size" of a kelvin is the same as the "size" of a Celsius degree. When considering temperature changes, measurements in kelvins and Celsius degrees are the same. For extremely high temperatures the Kelvin and Celsius scales are essentially the same: For example, the Sun's core temperature is either 1.55×10^7 K or 1.55×10^7 °C.

The now-archaic Fahrenheit scale, which expresses temperature in **degrees Fahrenheit** (°F), is used only in the United States. When the German physicist Gabriel Fahrenheit introduced this scale in the early 1700s, he intended 100°F to represent the temperature of a healthy human body. On the Fahrenheit scale, water freezes at 32°F and boils at 212°F. There are 180 Fahrenheit degrees between the freezing and boiling points of water, so a degree Fahrenheit is only 100/180 = 5/9 as large as either a Celsius degree or a kelvin.

Two simple equations allow you to convert a temperature from the Celsius scale to the Fahrenheit scale and from Fahrenheit to Celsius:

$$T_F = \frac{9}{5}\, T_C + 32$$

$$T_C = \frac{5}{9}\, (T_F - 32)$$

T_F = temperature in degrees Fahrenheit

T_C = temperature in degrees Celsius

EXAMPLE: A typical room temperature is 68°F. We can convert this to the Celsius scale using the second equation:

$$T_C = \frac{5}{9}\,(68 - 32) = 20°\text{C}$$

To convert this to the Kelvin scale, we simply add 273 to the Celsius temperature. Thus,

$$68° = 20°\text{C} = 293\ \text{K}$$

The diagram displays the relationships among these three temperature scales.

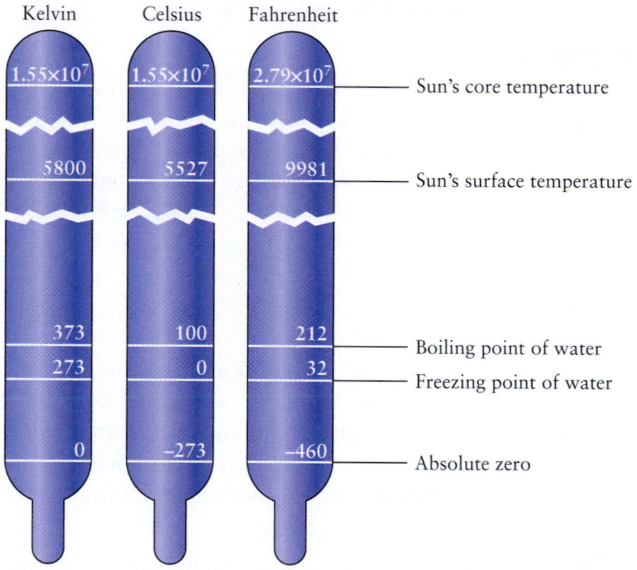

5-4 Wien's law and the Stefan-Boltzmann law are useful tools for analyzing glowing objects like stars

The mathematical formula that describes the blackbody curves in Figure 5-12 is a rather complicated one. But there are two simpler formulas for blackbody radiation that prove to be very useful in many branches of astronomy. They are used by astronomers who investigate the stars as well as by those who study the planets (which are dense, relatively cool objects that emit infrared radiation). One of these formulas relates the temperature of a blackbody to its wavelength of maximum emission, and the other relates the temperature to the amount of energy that the blackbody emits. These formulas, which we will use throughout this book, restate in precise mathematical terms the qualitative relationships that we described in Section 5-3.

Two simple mathematical formulas describing blackbodies are essential tools for studying the universe

Wien's Law

Figure 5-12 shows that the higher the temperature (T) of a blackbody, the shorter its wavelength of maximum emission (λ_{max}). In 1893 the German physicist Wilhelm Wien used ideas about both heat and electromagnetism to make this relationship quantitative. The formula that he derived, which today is called **Wien's law**, is

Wien's law for a blackbody

$$\lambda_{max} = \frac{0.0029 \text{ K m}}{T}$$

λ_{max} = wavelength of maximum emission of the object (in meters)

T = temperature of the object (in kelvins)

According to Wien's law, the wavelength of maximum emission of a blackbody is inversely proportional to its temperature in kelvins. In other words, if the temperature of the blackbody doubles, its wavelength of maximum emission is halved, and vice versa. For example, Figure 5-12 shows blackbody curves for temperatures of 3000 K, 6000 K, and 12,000 K. From Wien's law, a blackbody with a temperature of 6000 K has a wavelength of maximum emission $\lambda_{max} = (0.0029 \text{ K m})/(6000 \text{ K}) = 4.8 \times 10^{-7} \text{ m} = 480 \text{ nm}$, in the visible part of the electromagnetic spectrum. At 12,000 K, or twice the temperature, the blackbody has a wavelength of maximum emission half as great, or $\lambda_{max} = 240$ nm, which is in the ultraviolet. At 3000 K, just half our original temperature, the value of λ_{max} is twice the original value—960 nm, which is an infrared wavelength. You can see that these wavelengths agree with the peaks of the curves in Figure 5-12.

CAUTION! Remember that Wien's law involves the wavelength of maximum emission in *meters*. If you want to convert the wavelength to nanometers, you must multiply the wavelength in meters by $(10^9 \text{ nm})/(1 \text{ m})$.

Wien's law is very useful for determining the surface temperatures of stars. It is not necessary to know how far away the star is, how large it is, or how much energy it radiates into space. All we need to know is the dominant wavelength of the star's electromagnetic radiation.

CONCEPTCHECK 5-7

What single piece of information do astronomers need in order to determine if a star is hotter than our Sun?

CALCULATIONCHECK 5-2

Which wavelength of light would our Sun emit most if its temperature were twice its current temperature of 5800 K?

Answers appear at the end of the chapter.

The Stefan-Boltzmann Law

The other useful formula for the radiation from a blackbody involves the total amount of energy the blackbody radiates at all wavelengths. (By contrast, the curves in Figure 5-12 show how much energy a blackbody radiates at each individual wavelength.)

Energy is usually measured in **joules** (J), named after the nineteenth-century English physicist James Joule. A joule is the amount of kinetic energy contained in the motion of a 2-kilogram mass moving at a speed of 1 meter per second. The joule is a convenient unit of energy because it is closely related to the familiar **watt** (W): 1 watt is 1 joule per second, or $1 \text{ W} = 1 \text{ J/s} = 1 \text{ J s}^{-1}$. (The superscript –1 means you are dividing by that quantity.) For example, a 100-watt lightbulb uses energy at a rate of 100 joules per second, or 100 J/s. The energy content of food is also often measured in joules; in most of the world, diet soft drinks are labeled as "low joule" rather than "low calorie."

The amount of energy emitted by a blackbody depends both on its temperature and on its surface area. These characteristics make sense: A large burning log radiates much more heat than a burning match, even though the temperatures are the same. To consider the effects of temperature alone, it is convenient to look at the amount of energy emitted from each square meter of an object's surface in a second. This quantity is called the **energy flux** (F). Flux means "rate of flow," and thus F is a measure of how rapidly energy is flowing out of the object. It is measured in joules per square meter per second, usually written as $\text{J/m}^2\text{/s}$ or $\text{J m}^{-2} \text{ s}^{-1}$. Alternatively, because 1 watt equals 1 joule per second, we can express flux in watts per square meter (W/m^2, or W m^{-2}).

The nineteenth-century Irish physicist David Tyndall performed the first careful measurements of the amount of radiation emitted by a blackbody. (He studied the light from a heated platinum wire, which behaves approximately like a blackbody.) By analyzing Tyndall's results, the Slovenian physicist Josef Stefan deduced in 1879 that the flux from a blackbody is proportional to the fourth power of the object's temperature (measured in kelvins). Five years after Stefan announced his law, Austrian physicist Ludwig Boltzmann showed how it could be derived mathematically from basic assumptions about atoms and molecules. For this reason, Stefan's law is commonly known as the **Stefan-Boltzmann law.** Written as an equation, the Stefan-Boltzmann law is

Stefan-Boltzmann law for a blackbody

$$F = \sigma T^4$$

F = energy flux, in joules per square meter of surface per second

σ = a constant = $5.67 \times 10^{-8} \text{ W m}^{-2} \text{ K}^{-4}$

T = object's temperature, in kelvins

The value of the Stefan-Boltzmann constant σ (the Greek letter sigma) is known from laboratory experiments.

The Stefan-Boltzmann law says that if you double the temperature of an object (for example, from 300 K to 600 K), then the energy emitted from the object's surface each second increases by a factor of $2^4 = 16$. If you increase the temperature by a factor of 10 (for example, from 300 K to 3000 K), the rate of energy emission increases by a factor of $10^4 = 10{,}000$. Thus, a chunk of iron at room temperature (around 300 K) emits very little

BOX 5-2 TOOLS OF THE ASTRONOMER'S TRADE

Using the Laws of Blackbody Radiation

The Sun and stars behave like nearly perfect blackbodies. Wien's law and the Stefan-Boltzmann law can therefore be used to relate the surface temperature of the Sun or a distant star to the energy flux and wavelength of maximum emission of its radiation. The following examples show how to do this.

EXAMPLE: The maximum intensity of sunlight is at a wavelength of roughly 500 nm = 5.0×10^{-7} m. Use this information to determine the surface temperature of the Sun

Situation: We are given the Sun's wavelength of maximum emission λ_{max}, and our goal is to find the Sun's surface temperature, denoted by $T_\odot$. (The symbol $\odot$ is the standard astronomical symbol for the Sun.)

Tools: We use Wien's law to relate the values of λ_{max} and $T_\odot$.

Answer: As written, Wien's law tells how to find λ_{max} if we know the surface temperature. To find the surface temperature from λ_{max}, we first rearrange the formula, then substitute the value of λ_{max}:

$$T_\odot = \frac{0.0029 \text{ K m}}{\lambda_{max}} = \frac{0.0029 \text{ K m}}{5.0 \times 10^{-7} \text{ m}} = 5800 \text{ K}$$

Review: This temperature is very high by Earth standards, about the same as an iron-welding arc.

EXAMPLE: Using detectors above Earth's atmosphere, astronomers have measured the average flux of solar energy arriving at Earth. This value, called the **solar constant,** is equal to 1370 W m^{-2}. Use this information to calculate the Sun's surface temperature. (This calculation provides a check on our result from the preceding example.)

Situation: The solar constant is the flux of sunlight as measured at Earth. We want to use the value of the solar constant to calculate $T_\odot$.

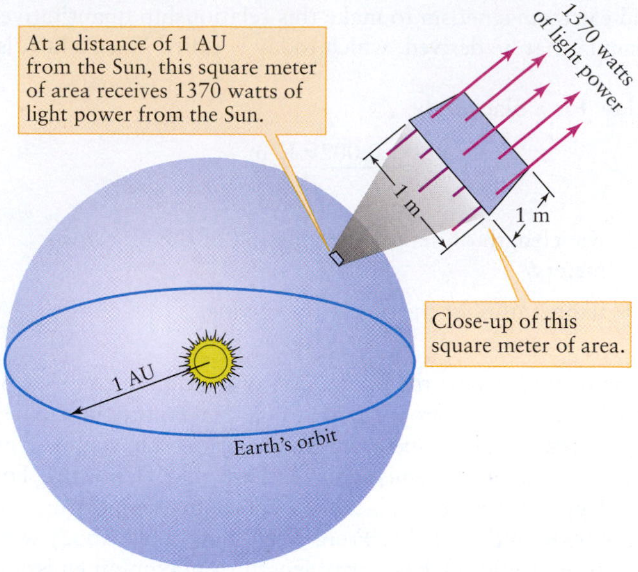

At a distance of 1 AU from the Sun, this square meter of area receives 1370 watts of light power from the Sun.
1370 watts of light power
1 m
1 m
Close-up of this square meter of area.
1 AU
Earth's orbit

Tools: It may seem that all we need is the Stefan-Boltzmann law, which relates flux to surface temperature. However, the quantity F in this law refers to the flux measured at the Sun's surface, *not* at Earth. Hence, we will first need to calculate F from the given information.

Answer: To determine the value of F, we first imagine a huge sphere of radius 1 AU with the Sun at its center, as shown in the figure. Each square meter of that sphere receives 1370 watts of power from the Sun, so the total energy radiated by the Sun per second is equal to the solar constant multiplied by the sphere's surface area. The result, called the **luminosity** of the Sun and denoted by the symbol $L_\odot$, is $L_\odot = 3.90 \times 10^{26}$ W. That is, in 1 second the Sun radiates 3.90×10^{26} joules of energy into space. Because we know the size of the Sun, we can compute the energy flux (energy emitted per square meter per second) at its surface. The radius of the Sun is

electromagnetic radiation (and essentially no visible light), but an iron bar heated to 3000 K glows quite intensely.

Box 5-2 gives several examples of applying Wien's law and the Stefan-Boltzmann law to typical astronomical problems.

CONCEPTCHECK 5-8

Is it ever possible for a cooler object to emit more energy than a warmer object?

Answer appears at the end of the chapter.

5-5 Light has properties of both waves and particles

At the end of the nineteenth century, physicists mounted a valiant effort to explain *how* a material can produce blackbody radiation. To this end they constructed theories based on Maxwell's description of light as electromagnetic waves. But while electromagnetic waves could explain many properties of light, it was not until the discovery of photons that all features of blackbody curves could be understood.

$R_\odot = 6.96 \times 10^8$ m, and the Sun's surface area is $4\pi R_\odot^2$. Therefore, its energy flux $F_\odot$ is the Sun's luminosity (total energy emitted by the Sun per second) divided by the Sun's surface area (the number of square meters of surface):

$$F_\odot = \frac{L_\odot}{4\pi R_\odot^2} = \frac{3.90 \times 10^{26}\ \text{W}}{4\pi(6.96 \times 10^8\ \text{m})^2} = 6.41 \times 10^7\ \text{W m}^{-2}$$

Once we have the Sun's energy flux $F_\odot$ we can use the Stefan-Boltzmann law to find the Sun's surface temperature $T_\odot$:

$$T_\odot^4 = F_\odot/\sigma = 1.13 \times 10^{15}\ \text{K}^4$$

Taking the fourth root (the square root of the square root) of this value, we find the surface temperature of the Sun to be $T_\odot = 5800$ K.

Review: Our result for $T_\odot$ agrees with the value we computed in the previous example using Wien's law. Notice that the solar constant of 1370 W m^{-2} is very much less than $F_\odot$ the flux at the Sun's surface. By the time the Sun's radiation reaches Earth, it is spread over a greatly increased area.

EXAMPLE: Sirius, the brightest star in the night sky, has a surface temperature of about 10,000 K. Find the wavelength at which Sirius emits most intensely.

Situation: Our goal is to calculate the wavelength of maximum emission of Sirius (λ_{max}) from its surface temperature T.

Tools: We use Wien's law to relate the values of λ_{max} and T.

Answer: Using Wien's law,

$$\lambda_{max} = \frac{0.0029\ \text{K m}}{T} = \frac{0.0029\ \text{K m}}{10{,}000\ \text{K}}$$

$$= 2.9 \times 10^{-7}\ \text{m} = 290\ \text{nm}$$

Review: Our result shows that Sirius emits light most intensely in the ultraviolet. In the visible part of the spectrum, it emits more blue light than red light (like the curve for 12,000 K in Figure 5-11), so Sirius has a distinct blue color.

EXAMPLE: How does the energy flux from Sirius compare to the Sun's energy flux?

Situation: To compare the energy fluxes from the two stars, we want to find the *ratio* of the flux from Sirius to the flux from the Sun.

Tools: We use the Stefan-Boltzmann law to find the flux from Sirius and from the Sun, which from the preceding examples have surface temperatures 10,000 K and 5800 K, respectively.

Answer: For the Sun, the Stefan-Boltzmann law is $F_\odot = \sigma T_\odot^4$, and for Sirius we can likewise write $F_* = \sigma T_*^4$, where the subscripts $\odot$ and $*$ refer to the Sun and Sirius, respectively. If we divide one equation by the other to find the ratio of fluxes, the Stefan-Boltzmann constants cancel out and we get

$$\frac{F_*}{F_\odot} = \frac{T_*^4}{T_\odot^4} = \frac{(10{,}000\ \text{K})^4}{(5800\ \text{K})^4} = \left(\frac{10{,}000}{5800}\right)^4 = 8.8$$

Review: Because Sirius has such a high surface temperature, each square meter of its surface emits 8.8 times more energy per second than a square meter of the Sun's relatively cool surface. Sirius is actually a larger star than the Sun, so it has more square meters of surface area and, hence, its *total* energy output is *more* than 8.8 times that of the Sun.

The Discovery of Photons

In 1905, the great physicist Albert Einstein developed a radically new model for the nature of light. Central to this new picture is that light is made out of particles! Each particle of light is called a **photon,** which is a distinct packet of electromagnetic energy.

Photons have a dual nature in that they are both particlelike and wavelike. They are particlelike in the sense that they are the small packets of energy that make up light. As expected with particles, you can count the number of photons in an electromagnetic wave. But, photons are also wavelike because each one is itself a wave, where each photon has the same wavelength as the electromagnetic wave of which it is a small part. As expected with waves, photons exhibit interference patterns when passed through a double-slit experiment like the one in Figure 5-5.

Light is made of smaller entities called photons, each one with a wavelength of its own

The energy of each photon is related to the wavelength of light: the longer the wavelength, the lower the energy. Thus, a photon of red light (wavelength $\lambda = 700$ nm) has less energy than a photon of violet light ($\lambda = 400$ nm). Conversely, the shorter a photon's wavelength, the higher its energy.

It was quickly realized that photons explain more than just blackbody curves, and now the study of photons plays a key role in almost every branch of science. While scientific equipment is usually needed to detect individual photons, there is a somewhat familiar effect that photons can help to explain. When it comes to tanning, your skin must be exposed to ultraviolet light. For example, a low intensity exposure to ultraviolet light causes tanning, but even a high intensity exposure of visible light will not cause the skin to darken. (Recall that you do not get tanned with regular indoor lighting, which has negligible UV light). The reason is that tanning, at the molecular level, results from a chemical reaction in the skin that involves a single photon. A single high-energy, short-wavelength ultraviolet photon can trigger a tanning reaction. However, even many lower-energy, longer-wavelength photons of visible light cannot trigger the reaction. In this situation, we see that it is not the total energy of light that matters, but how the energy is bundled into packets—and that is a photon.

The Energy of a Photon

The relationship between the energy E of a single photon and the wavelength of the electromagnetic radiation can be expressed in a simple equation:

Energy of a photon (in terms of wavelength)

$$E = \frac{hc}{\lambda}$$

E = energy of a photon

h = Planck's constant

c = speed of light

λ = wavelength of light

The value of the constant h in this equation, now called *Planck's constant,* has been shown in laboratory experiments to be

$$h = 6.625 \times 10^{-34} \text{ J s}$$

The units of h are joules multiplied by seconds, called "joule-seconds" and abbreviated J s.

Because the value of h is so tiny, a single photon carries a very small amount of energy. For example, a photon of red light with wavelength 633 nm has an energy of only 3.14×10^{-19} J (Box 5-3), which is why we ordinarily do not notice that light comes in the form of photons. Even a dim light source emits so many photons per second that it seems to be radiating a continuous stream of energy.

The energies of photons are sometimes expressed in terms of a small unit of energy called the **electron volt** (eV). One electron volt is equal to 1.602×10^{-19} J, so a 633-nm photon has an energy of 1.96 eV. If energy is expressed in electron volts, Planck's constant is best expressed in electron volts multiplied by seconds, abbreviated eV s:

$$h = 4.135 \times 10^{-15} \text{ eV s}$$

Because the frequency ν of light is related to the wavelength λ by $\nu = c/\lambda$, we can rewrite the equation for the energy of a photon as

Energy of a photon (in terms of frequency)

$$E = h\nu$$

E = energy of a photon

h = Planck's constant

ν = frequency of light

The equations $E = hc/\lambda$ and $E = h\nu$ are together called **Planck's law.** Both equations express a relationship between a particlelike property of light (the energy E of a photon) and a wavelike property (the wavelength λ or frequency ν).

BOX 5-3 ASTRONOMY DOWN TO EARTH

Photons at the Supermarket

A beam of light can be regarded as a stream of tiny packets of energy called photons. The Planck relationships $E = hc/\lambda$ and $E = h\nu$ can be used to relate the energy E carried by a photon to its wavelength λ and frequency ν.

As an example, the laser bar-code scanners used at stores and supermarkets emit orange-red light of wavelength 633 nm. To calculate the energy of a single photon of this light, we must first express the wavelength in meters. A nanometer (nm) is equal to 10^{-9} m, so the wavelength is

$$\lambda = (633 \text{ nm})\left(\frac{10^{-9} \text{ m}}{1 \text{ nm}}\right) = 633 \times 10^{-9} \text{ m} = 6.33 \times 10^{-7} \text{ m}$$

Then, using the Planck formula $E = hc/\lambda$, we find that the energy of a single photon is

$$E = \frac{hc}{\lambda} = \frac{(6.625 \times 10^{-34} \text{ J s})(3 \times 10^{8} \text{ m/s})}{6.33 \times 10^{-7} \text{ m}} = 3.14 \times 10^{-19} \text{ J}$$

This amount of energy is very small. The laser in a typical bar-code scanner emits 10^{-3} joule of light energy per second, so the number of photons emitted per second is

$$\frac{10^{-3} \text{ joule per second}}{3.14 \times 10^{-19} \text{ joule per photon}} = 3.2 \times 10^{15} \text{ photons per second}$$

This number is so large that the laser beam seems like a continuous flow of energy rather than a stream of little energy packets.

The photon picture of light is essential for understanding the detailed shapes of blackbody curves. As we will see, it also helps to explain how and why the spectra of the Sun and stars differ from those of perfect blackbodies.

CONCEPTCHECK 5-9

If a photon's wavelength is measured to be longer than the wavelength of a green photon, will it have a greater or lower energy than a green photon?

Answer appears at the end of the chapter.

5-6 Each chemical element produces its own unique set of spectral lines

Spectroscopy is the key to determining the chemical composition of planets and stars

In 1814 the German master optician Joseph von Fraunhofer repeated the classic experiment of shining a beam of sunlight through a prism (see Figure 5-3). But this time Fraunhofer subjected the resulting rainbow-colored spectrum to intense magnification. To his surprise, he discovered that the solar spectrum contains hundreds of fine, dark lines, now called **spectral lines** (Figure 5-14). A dark spectral line arises when light at a specific wavelength is at least partially *absorbed* so that the spectrum appears darker; it is also called an *absorption line*. By contrast, if the light from a perfect blackbody were sent through a prism, it would produce a smooth, continuous spectrum with no dark lines. Fraunhofer counted more than 600 dark lines in the Sun's spectrum; today we know of more than 30,000. The photograph of the Sun's spectrum in Figure 5-14 shows hundreds of these spectral lines.

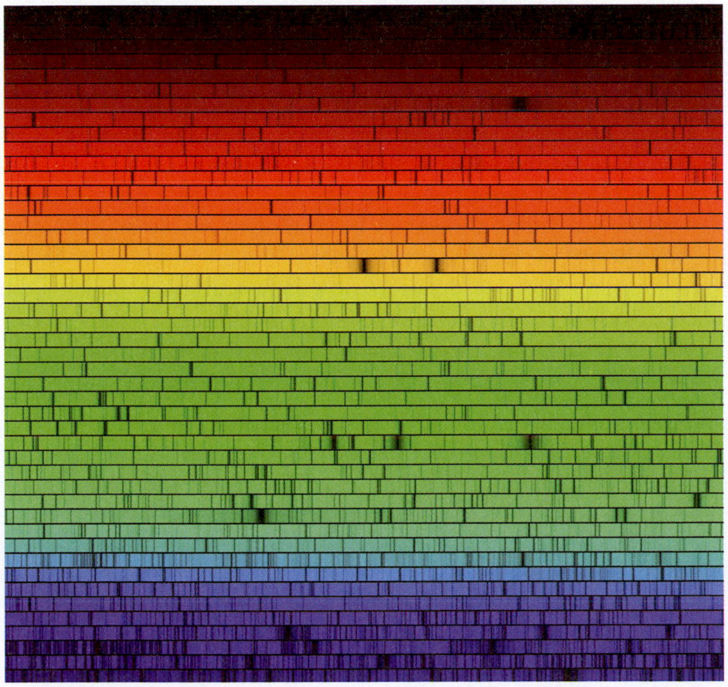

FIGURE 5-14 R I V U X G

The Sun's Spectrum Numerous dark spectral lines are seen in this image of the Sun's spectrum. The spectrum is spread out so much that it had to be cut into segments to fit on this page. (N. A. Sharp, NOAO/NSO/Kitt Peak FTS/AURA/NSF)

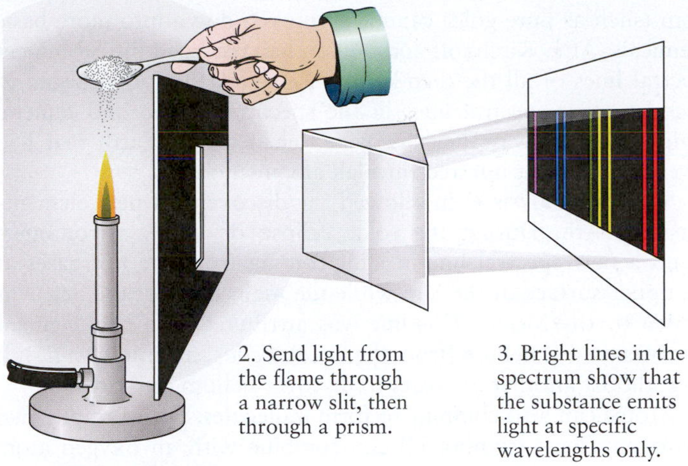

1. Add a chemical substance to a flame.
2. Send light from the flame through a narrow slit, then through a prism.
3. Bright lines in the spectrum show that the substance emits light at specific wavelengths only.

FIGURE 5-15

The Kirchhoff-Bunsen Experiment In the mid-1850s, Gustav Kirchhoff and Robert Bunsen discovered that when a chemical substance is heated and vaporized, the spectrum of the emitted light exhibits a series of bright spectral lines. They also found that each chemical element produces its own characteristic pattern of spectral lines. (In an actual laboratory experiment, lenses would be needed to focus the image of the slit onto the screen.)

Spectral Analysis

Half a century later, chemists discovered that they could produce spectral lines in the laboratory and use these spectral lines to analyze what kinds of atoms different substances are made of. Chemists had long known that many substances emit distinctive colors when sprinkled into a flame. To facilitate study of these colors, around 1857 the German chemist Robert Bunsen invented a gas burner (today called a Bunsen burner) that produces a clean flame with no color of its own. Bunsen's colleague, the Prussian-born physicist Gustav Kirchhoff, suggested that the colored light produced when substances were added to the flame might best be studied by passing the resulting light through a prism (Figure 5-15). The two scientists promptly discovered that the spectrum from the flame consists of a pattern of thin, bright spectral lines against a dark background. A bright spectral line arises because at least some additional light is being emitted at a specific wavelength; it is also called an *emission line*.

Kirchhoff and Bunsen then found that each chemical element produces its own unique pattern of spectral lines. Thus was born in 1859 the technique of **spectral analysis**: the identification of atoms and molecules by their unique patterns of spectral lines.

Atoms are the building blocks of substances on Earth and throughout the universe. Familiar examples are hydrogen, oxygen, carbon, iron, and gold. These atoms are also called chemical **elements** because a pure substance made from only one type of atom (such as pure gold) cannot be broken down into more basic chemicals. After Kirchhoff and Bunsen had recorded the prominent spectral lines of all the then-known elements, they soon began to discover other spectral lines in the spectra of vaporized mineral samples. In this way they discovered elements on Earth that had never been seen or inferred through any method.

Spectral analysis even allowed the discovery of new elements outside Earth. During the solar eclipse of 1868, astronomers found a new spectral line in light coming from the hot gases at the upper surface of the Sun while the main body of the Sun was hidden by the Moon. This line was attributed to a new element that was named helium (from the Greek *helios,* meaning "sun"). It took almost another 30 years to discover helium on Earth!

Atoms can also combine to form **molecules.** For example, two hydrogen atoms (symbol H) can combine with an oxygen atom (symbol O) to form a water molecule (symbol H_2O). Substances like water whose molecules include atoms of different elements are called **compounds.** Just as each type of atom has its own unique spectrum, so does each type of molecule. Figure 5-16 shows the spectra of several types of atoms and molecules. You can easily see in Figure 5-16 that each substance produces a unique pattern of spectral lines; each pattern can be thought of as a spectral "fingerprint" for identification. This is enormously important in astronomy because it allows us to determine the detailed composition of distant planets and stars by analyzing the spectra of their light that reaches Earth.

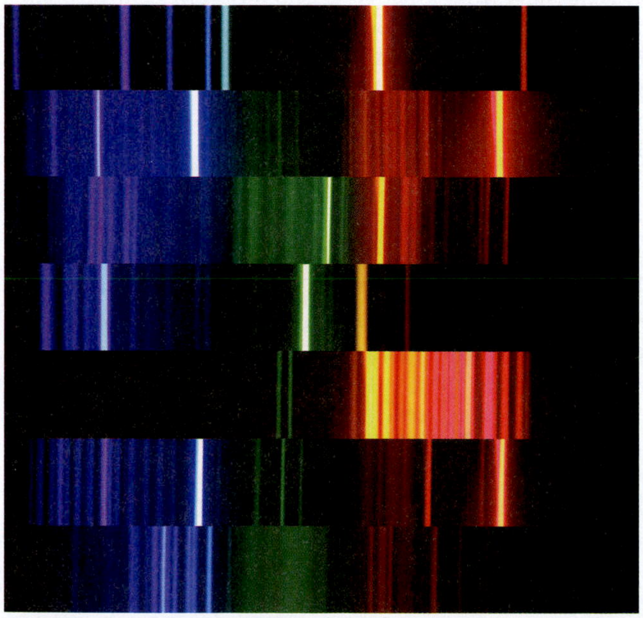

FIGURE 5-16 R I V U X G

Various Spectra These photographs show the spectra of different types of gases as measured in a laboratory on Earth. Each type of gas has a unique spectrum that is the same wherever in the universe the gas is found. Water vapor (H_2O) is a compound whose molecules are made up of hydrogen and oxygen atoms; the hydrogen molecule (H_2) is made up of two hydrogen atoms. For the hydrogen gas lamp, the bright lines are from single hydrogen atoms (H), and the fainter lines are emitted from hydrogen molecules (H_2). (Ted Kinsman/Science Photo Library)

Kirchhoff's Laws

TUTORIAL 5-3 The spectrum of the Sun, with its dark spectral lines superimposed on a bright background (see Figure 5-14), may seem to be unrelated to the spectra of bright lines against a dark background produced by substances in a flame (see Figure 5-15). But by the early 1860s, Kirchhoff's experiments had revealed a direct connection between these two types of spectra. His conclusions are summarized in three important statements about spectra that are today called **Kirchhoff's laws.** These laws, which are illustrated in Figure 5-17, are as follows:

Law 1 A hot opaque body, such as a perfect blackbody, or a hot, dense gas produces a **continuous spectrum**—a complete rainbow of colors without any spectral lines.

Law 2 A hot, transparent gas produces an **emission line spectrum**—a series of bright spectral lines against a dark background.

Law 3 A cool, transparent gas in front of a source of a continuous spectrum produces an **absorption line spectrum**—a series of dark spectral lines among the colors of the continuous spectrum. Furthermore, the dark lines in the absorption spectrum of a particular gas occur at exactly the *same* wavelengths as the bright lines in the emission spectrum of that same gas. As we will see in Section 5-8, absorption and emission lines correspond to distinct changes in an electron's orbit within an atom.

CAUTION! Figure 5-17 shows that light can either pass through a cloud of gas or be absorbed by the gas. But there is also a third possibility: The light can simply bounce off the atoms or molecules that make up the gas, a phenomenon called **light scattering.** In other words, photons passing through a gas cloud can miss the gas atoms altogether, be swallowed whole by the atoms (absorption), or bounce off the atoms like billiard balls colliding (scattering). Box 5-4 describes how light scattering explains the blue color of the sky and the red color of sunsets.

In general, a gas cloud can both emit and absorb light. Whether an emission line spectrum or an absorption line spectrum is observed from an intervening gas cloud depends on the relative temperatures of the gas cloud and its background. Absorption lines are seen if the background is hotter than the gas, and emission lines are seen if the background is cooler. This is why the gas cloud illustrating absorption (Law 3) in Figure 5-17 is cooler than the star.

Spectroscopy

Spectroscopy is the systematic study of spectra and spectral lines. As noted, spectral lines are tremendously important in astronomy, because they provide detailed evidence about the chemical composition of distant objects. As an example, the spectrum of the Sun shown in Figure 5-14 is an absorption line spectrum. The

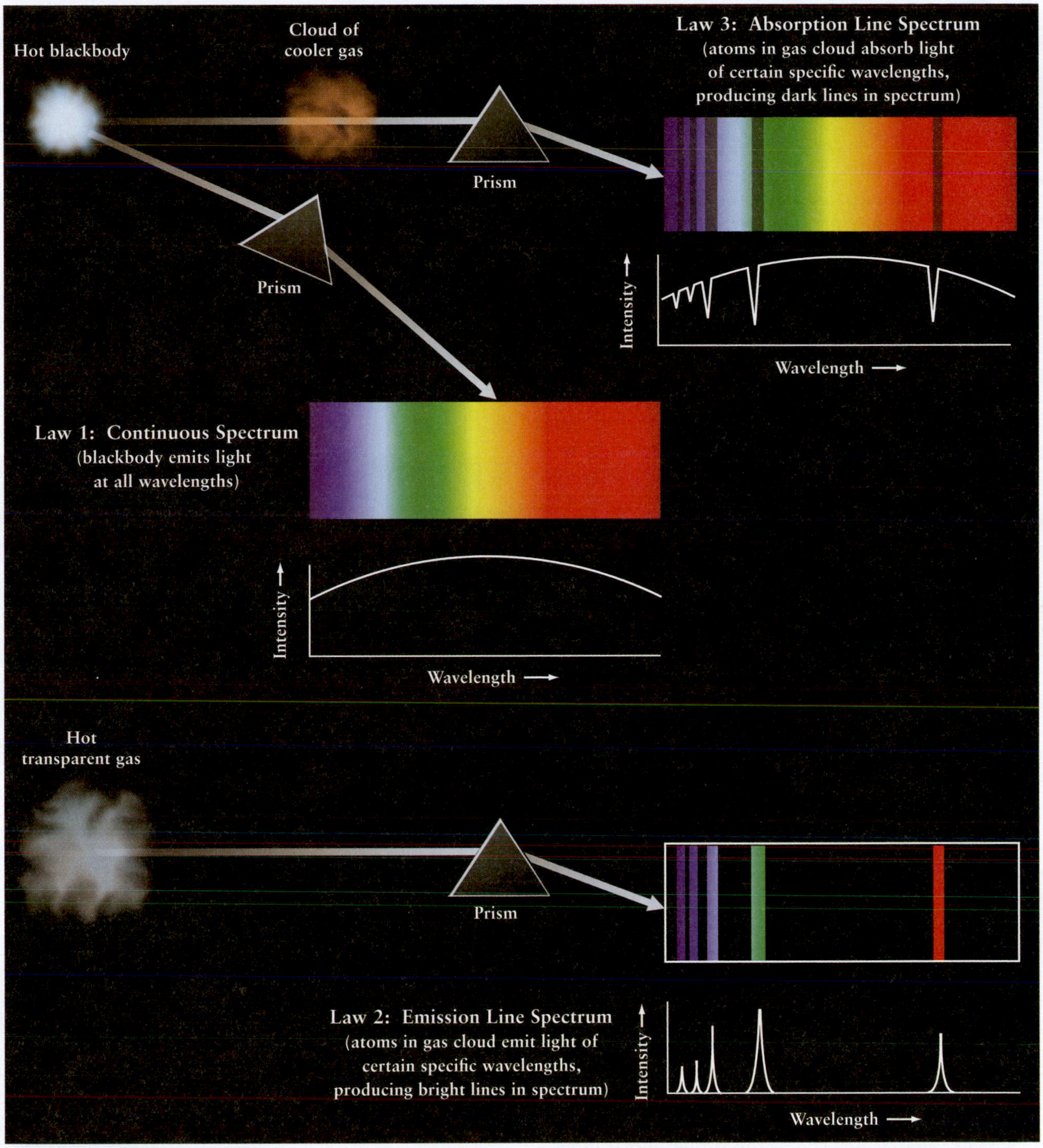

ANIMATION 5-2

FIGURE 5-17

Kirchhoff's Laws for Continuous, Absorption Line, and Emission Line Spectra **Law 1:** A hot, opaque body emits a continuous spectrum of light. **Law 3:** If this light is passed through a cloud of a cooler gas, the cloud absorbs light of certain specific wavelengths, leaving the light that passes directly through the cloud with an absorption line spectrum. **Law 2:** Atoms in a hot gas cloud collide energetically and emit light, creating an emission line spectrum that is unique to each type of atom. In each case the spectra is shown, as it would appear with a prism, and just below it, by a graph of intensity versus wavelength.

continuous spectrum comes from the hot surface of the Sun, which acts like a blackbody. The dark absorption lines are caused by this light passing through a cooler gas; this gas is the atmosphere that surrounds the Sun. Therefore, by identifying the spectral lines present in the solar spectrum, we can determine the chemical composition of the Sun's atmosphere.

BOX 5-4 ASTRONOMY DOWN TO EARTH

Why the Sky Is Blue

Light scattering is a process where photons bounce off particles, and change their direction of travel. The scattering particles can be atoms, molecules, or clumps of molecules. When you see regular objects like rocks or trees that do not produce their own intrinsic light, you see them because they scatter sunlight into your eyes.

An important fact about light scattering is that very small particles—ones that are smaller than a wavelength of visible light—are quite effective at scattering short-wavelength photons of blue light but less effective at scattering long-wavelength photons of red light. This fact explains a number of phenomena that you can see here on Earth.

The light that comes from the daytime sky is sunlight that has been scattered by the molecules that make up our atmosphere (see part a of the accompanying figure). Air molecules are less than 1 nm across, far smaller than the wavelength of visible light, so they scatter blue light more than red light—which is why the sky looks blue. In other words, the sky is blue because while incoming white sunlight contains all the colors, air molecules scatter blue light around more than red light, literally spreading blue light all throughout the atmosphere.

During the day, distant mountains often appear blue thanks to sunlight being scattered from the atmosphere between the mountains and your eyes. (The Blue Ridge Mountains, which extend from Pennsylvania to Georgia, and Australia's Blue Mountains derive their names from this effect.)

Light scattering also explains why sunsets are red. Again, sunlight contains all the colors, but as this light passes through our atmosphere the blue light is scattered away from the straight-line path from the Sun. On the other hand, red light undergoes relatively little scattering, so light coming directly from the Sun looks a bit redder once some blue has been removed. When you look toward the setting Sun (but, never look at the Sun!), the sunlight that reaches your eye has had to pass through a relatively thick layer of atmosphere (part b of the accompanying figure). Hence, a large fraction of the blue light from the Sun has been scattered, and the Sun appears quite red. If, at sunset, the sunlight shines onto clouds, the clouds also look red after much of the blue light has been scattered away.

The same effect also applies to sunrises, but sunrises seldom look as red as sunsets do. The reason is that dust is lifted into the atmosphere during the day by the wind (which, being driven by sunlight, typically blows stronger in the daytime), and dust particles in the atmosphere help to scatter even more blue light.

Finally, some scattered light appears white. If the small particles that scatter light are sufficiently concentrated, there will be almost as much scattering of red light as of blue light, and the scattered light will then appear white. This can occur when the scattering comes from dense ice crystals or water vapor and explains the white color of snow, clouds, and fog.

Light scattering has many applications to astronomy. For example, it explains why very distant stars in our Galaxy appear surprisingly red. The reason is that there are tiny dust particles throughout the space between the stars, and this dust scatters blue light. By studying how much scattering takes place, astronomers have learned about the tenuous material that fills interstellar space.

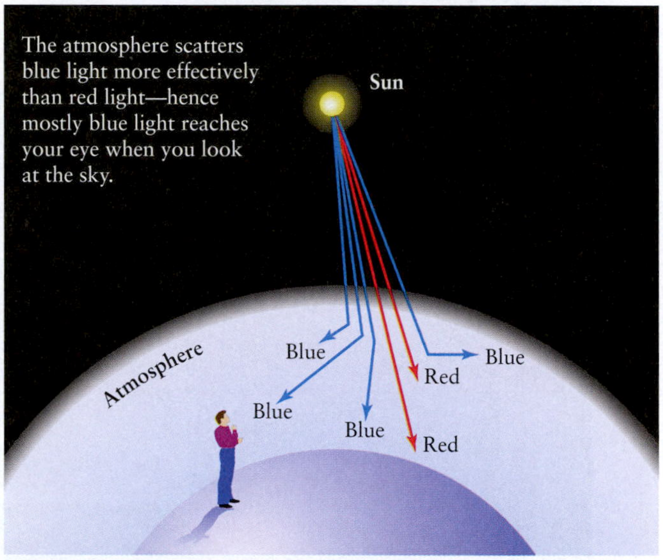

(a) Why the sky looks blue

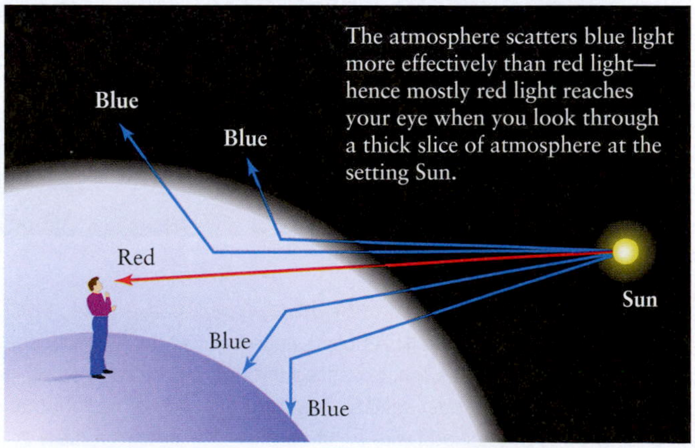

(b) Why the setting Sun looks red

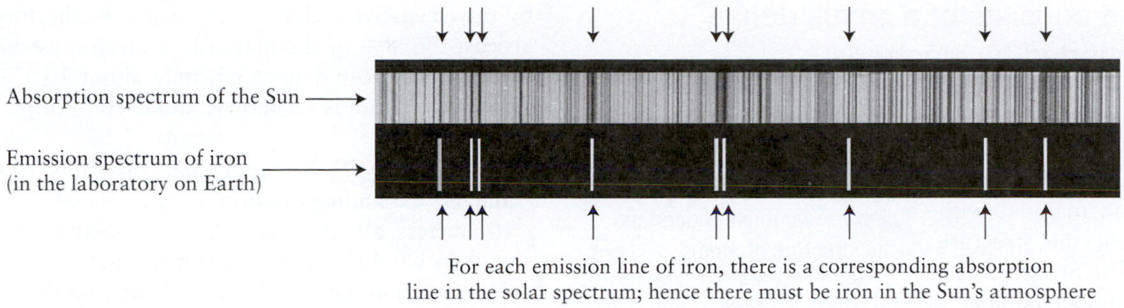
Absorption spectrum of the Sun
Emission spectrum of iron (in the laboratory on Earth)
For each emission line of iron, there is a corresponding absorption line in the solar spectrum; hence there must be iron in the Sun's atmosphere

FIGURE 5-18 R I V U X G
Iron in the Sun The upper part of this figure is a portion of the Sun's spectrum at violet wavelengths, showing numerous dark absorption lines. The lower part of the figure is a corresponding portion of the emission line spectrum of vaporized iron. The iron lines coincide with some of the solar lines, which proves that there is some iron (albeit a relatively small amount) in the Sun's atmosphere. (Carnegie Observatories)

Even when the spectral lines from more than one type of atom appear in the same spectrum, the different atoms can be identified. Figure 5-18 shows both a portion of the Sun's absorption line spectrum, which has numerous lines, and the emission line spectrum of iron vapor over the same wavelength range, measured on Earth. This pattern of bright spectral lines in the lower spectrum is iron's own distinctive "fingerprint," which no other substance can imitate. Because some absorption lines in the Sun's spectrum coincide with the iron lines, some vaporized iron must exist in the Sun's atmosphere.

Spectroscopy can also help us analyze gas clouds in space, such as the nebula surrounding the star cluster NGC 346 shown in Figure 5-19. The particular shade of red that dominates the color of this nebula is due to an emission line at a wavelength of 656 nm. This is one of the characteristic wavelengths emitted by hydrogen gas, so we can conclude that this nebula contains hydrogen. More detailed analyses of this kind show that hydrogen is the most common element in gaseous nebulae, and indeed in the universe as a whole. The spectra of other nebulae, such as the Ring Nebula shown in the image that opens this chapter, also reveal the presence of nitrogen, oxygen, helium, and other gases.

What is truly remarkable about spectroscopy is that it can determine chemical composition at any distance. The 656-nm red light produced by a sample of heated hydrogen gas on Earth (the bright red line in the hydrogen spectrum in Figure 5-16) is the same as that observed coming from the nebula shown in Figure 5-19, located about 210,000 light-years away. Throughout this book we will see many examples of how astronomers use spectra to determine the nature of celestial objects.

Why does an atom absorb light of only particular wavelengths? And why does it then emit light of only these same wavelengths? Maxwell's theory of electromagnetism (see Section 5-2) could not answer these questions. The answers did not come until scientists began to discover the structure and properties of atoms.

FIGURE 5-19 R I V U X G
Analyzing the Composition of a Distant Nebula The glowing gas cloud in this Hubble Space Telescope image lies 210,000 light-years away in the constellation Tucana (the Toucan). The red light emitted by the nebula has a wavelength of 656 nm, characteristic of an emission line from hydrogen gas. Hot stars within the nebula emit high energy, ultra violet photons, which ionize hydrogen (removing the electron). When the electron returns to the atom, it jumps towards the center and gives off emission lines. (NASA, ESA, and A. Nota, STScI/ESA)

CONCEPTCHECK 5-10

Which one of Kirchhoff's laws describes the Sun's spectrum in Figure 5-14? If helium gas is heated up enough to give the spectrum at the top of Figure 5-16, which of Kirchhoff's laws applies?

CONCEPTCHECK 5-11

What type of spectra would result from a glowing field of hot, dense lava as viewed by an orbiting satellite through Earth's atmosphere?

Answers appear at the end of the chapter.

5-7 An atom consists of a small, dense nucleus surrounded by electrons

> To decode the information in the light from immense objects like stars and galaxies, we must understand the structure of atoms

By the early twentieth century, it was known that **atoms** contain negatively charged particles called **electrons** and some material with positive charge, but the structure of atoms was still a mystery. The "plum pudding" model popular at that time proposed that the negative electrons were like plums uniformly mixed into a positively charged pudding. However, in 1910 this model was overturned by Ernest Rutherford, a gifted physicist from New Zealand.

The Nucleus of an Atom

Rutherford discovered a more accurate model for the structure of an atom, shown in Figure 5-20. According to this model, a massive, positively charged **nucleus** at the center of the atom is orbited by tiny, negatively charged electrons. Rutherford concluded that at least 99.98% of the mass of an atom must be concentrated in its nucleus, whose diameter is only about 10^{-14} m. (The diameter of a typical atom is far larger, about 10^{-10} m.)

ANALOGY To appreciate just how tiny the nucleus is, imagine expanding an atom by a factor of 10^{12} to a diameter of 100 meters, about the length of a football field. On this scale, the nucleus would be just a centimeter across (no larger than your thumbnail) in the middle of the field, and the electrons would be orbiting at a distance near the goal posts.

We know today that the nucleus of an atom contains two types of particles, **protons** and **neutrons.** A proton has a positive electric charge, equal in magnitude to that of the negatively charged electron. As its name suggests, a neutron has no electric charge—it is electrically neutral. As an example, the helium atom has two protons and two neutrons. Protons and neutrons are held together in a nucleus by the so-called *strong nuclear force,* whose great strength

BOX 5-5 TOOLS OF THE ASTRONOMER'S TRADE

Atoms, the Periodic Table, and Isotopes

Each different chemical element is made of a specific type of atom. Each specific atom has a characteristic number of protons in its nucleus. For example, a hydrogen atom has 1 proton in its nucleus, an oxygen atom has 8 protons in its nucleus, and so on.

The number of protons in an atom's nucleus is the **atomic number** for that particular element. The chemical elements are most conveniently listed in the form of a **periodic table** (shown in the figure). Elements are arranged in the periodic table in order of increasing atomic number. With only a few exceptions, this sequence also corresponds to increasing average mass of the atoms of the elements. Thus, hydrogen (symbol H), with atomic number 1, is the lightest element. Iron (symbol Fe) has atomic number 26 and is a relatively heavy element.

All the elements listed in a single vertical column of the periodic table have similar chemical properties. For example, the elements in the far right column are all gases under the conditions of temperature and pressure found at Earth's surface, and they are all very reluctant to react chemically with other elements.

In addition to nearly 100 naturally occurring elements, the periodic table includes a number of artificially produced elements. Most of these elements are heavier than uranium (symbol U) and are highly radioactive, which means that they decay into lighter elements within a short time of being created in laboratory experiments. Scientists have succeeded in creating only a few atoms of elements 104 and above.

The number of protons in the nucleus of an atom determines which element that atom is. Nevertheless, the same element may have different numbers of neutrons in its nucleus. For example, oxygen (O) has atomic number 8, so every oxygen nucleus has exactly 8 protons. But oxygen nuclei can have 8, 9, or 10 neutrons. These three slightly different kinds of oxygen are called **isotopes.** The isotope with 8 neutrons is by far the most abundant variety. It is written as ^{16}O, or oxygen-16. The rarer isotopes with 9 and 10 neutrons are designated as ^{17}O and ^{18}O, respectively.

The superscript that precedes the chemical symbol for an element equals the total number of protons and neutrons in a nucleus of that particular isotope. For example, a nucleus of the most common isotope of iron, ^{56}Fe or iron-56, contains a total of 56 protons and neutrons. From the periodic table, the atomic number of iron is 26, so every iron atom has 26 protons in its nucleus. Therefore, the number of neutrons in an iron-56 nucleus is $56 - 26 = 30$. (Most nuclei have more neutrons than protons, especially in the case of the heaviest elements.)

It is extremely difficult to distinguish chemically between the various isotopes of a particular element. Ordinary chemical reactions involve only the electrons that orbit the atom, never the neutrons buried in its nucleus. But there are small differences in the wavelengths of the spectral lines for different isotopes of the same element. For example, the spectral line wavelengths of the hydrogen isotope ^{2}H are about 0.03% greater than the wavelengths for the most common hydrogen isotope, ^{1}H. Thus, different isotopes can be distinguished by careful spectroscopic analysis.

Isotopes are important in astronomy for a variety of reasons. By measuring the relative amounts of different isotopes of a given element in a Moon rock or meteorite, the age of that sample can be determined. The mixture of isotopes left behind when a star explodes into a supernova (see Section 1-3) tells astronomers about the processes that led to the explosion. And knowing the properties of different isotopes of hydrogen and helium is crucial to understanding the nuclear reactions that make the Sun shine. Look for these and other applications of the idea of isotopes in later chapters.

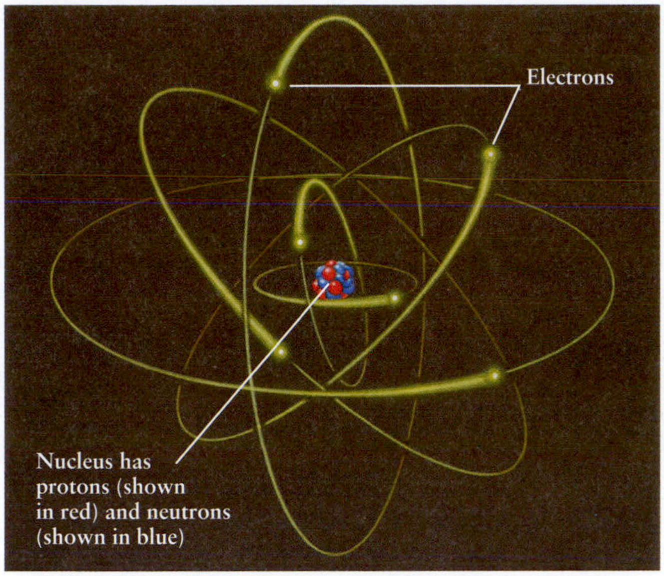
Electrons
Nucleus has protons (shown in red) and neutrons (shown in blue)

FIGURE 5-20
Rutherford's Model of the Atom In this model, electrons orbit the atom's nucleus, which contains most of the atom's mass. The nucleus contains two types of particles, protons and neutrons. The full behavior of electrons orbiting the nucleus is quite complicated and not represented in this simple illustration.

overcomes the electric repulsion between the positively charged protons. A proton and a neutron have almost the same mass, 1.7×10^{-27} kg, and each has about 2000 times as much mass as an electron (9.1×10^{-31} kg). In an ordinary atom there are as many positive protons as there are negative electrons, so the atom has no net electric charge. Because the mass of the electron is so small, the mass of an atom is not much greater than the mass of its nucleus.

While the solar system is held together by gravitational forces, electrons are held in atoms by electrical forces. Opposites attract: The negative charges on the orbiting electrons are attracted to the positive charges on the protons in the nucleus. Box 5-5 describes more about the connection between the structure of

Periodic Table of the Elements

1 H Hydrogen																	2 He Helium
3 Li Lithium	4 Be Beryllium											5 B Boron	6 C Carbon	7 N Nitrogen	8 O Oxygen	9 F Fluorine	10 Ne Neon
11 Na Sodium	12 Mg Magnesium											13 Al Aluminum	14 Si Silicon	15 P Phosphorus	16 S Sulfur	17 Cl Chlorine	18 Ar Argon
19 K Potassium	20 Ca Calcium	21 Sc Scandium	22 Ti Titanium	23 V Vanadium	24 Cr Chromium	25 Mn Manganese	26 Fe Iron	27 Co Cobalt	28 Ni Nickel	29 Cu Copper	30 Zn Zinc	31 Ga Gallium	32 Ge Germanium	33 As Arsenic	34 Se Selenium	35 Br Bromine	36 Kr Kryton
37 Rb Rubidium	38 Sr Strontium	39 Y Yttrium	40 Zr Zirconium	41 Nb Niobium	42 Mo Molybdenum	43 Tc Technetium	44 Ru Ruthenium	45 Rh Rhodium	46 Pd Palladium	47 Ag Silver	48 Cd Cadmium	49 In Indium	50 Sn Tin	51 Sb Antimony	52 Te Tellurium	53 I Iodine	54 Xe Xenon
55 Cs Cesium	56 Ba Barium	57 La Lanthanum	72 Hf Hafnium	73 Ta Tantaium	74 W Tungsten	75 Re Rhenium	76 Os Osmium	77 Ir Iridium	78 Pt Platinum	79 Au Gold	80 Hg Mercury	81 Tl Thallium	82 Pb Lead	83 Bi Bismuth	84 Po Polonium	85 At Astatine	86 Rn Radon
87 Fr Francium	88 Ra Radium	89 Ac Actinium	104 Rf Rutherfordium	105 Db Dubnium	106 Sg Seaborgium	107 Bh Bohrium	108 Hs Hassium	109 Mt Meitnerium	110 Ds Darmstadium	111 Rg Roentgenium							

58 Ce Cerium	59 Pr Praseodymium	60 Nd Neodymium	61 Pm Promethium	62 Sm Samarium	63 Eu Europium	64 Gd Gadolinium	65 Tb Terbium	66 Dy Dysprosium	67 Ho Holmium	68 Er Erbium	69 Tm Thulium	70 Yb Ytterbium	71 Lu Lutetium
90 Th Thorium	91 Pa Protactinium	92 U Uranium	93 Np Neptunium	94 Pu Plutonium	95 Am Americium	96 Cm Curium	97 Bk Berkelium	98 Cf Californium	99 Es Einsteinium	100 Fm Fermium	101 Md Mendelevium	102 No Nobelium	103 Lr Lawrencium

atoms and the chemical and physical properties of substances made of those atoms.

Rutherford's new model clarified the structure of the atom, but did not explain how these tiny particles within the atom give rise to spectral lines. The task of reconciling Rutherford's atomic model with Kirchhoff's laws of spectral analysis was undertaken by the young Danish physicist Niels Bohr, who joined Rutherford's laboratory in 1912.

5-8 Spectral lines are produced when an electron jumps from one energy level to another within an atom

Niels Bohr explained spectral lines with a radical new model of the atom

Niels Bohr began his study of the connection between atomic spectra and atomic structure by trying to understand the structure of hydrogen, the simplest and lightest of the elements. (As we discussed in Section 5-6, hydrogen is also the most common element in the universe.) When Bohr was done, he had not only found a way to explain this atom's spectrum but had also found a justification for Kirchhoff's laws in terms of atomic physics.

Hydrogen and the Balmer Series

The most common type of hydrogen atom consists of a single electron and a single proton. Hydrogen atoms have a simple visible-light spectrum consisting of a pattern of lines that begins at a wavelength of 656.3 nm and ends at 364.6 nm. The first spectral line is called H_α (H-alpha), the second spectral line is called H_β (H-beta), the third is H_γ (H-gamma), and so forth (Figure 5-21). (These are the bright lines in the spectrum of hydrogen shown in Figure 5-16. The fainter lines between these appear when hydrogen atoms form into molecules.) The closer you get to the short-wavelength end of the spectrum at 364.6 nm, the more spectral lines you see.

The regularity in this spectral pattern was described mathematically in 1885 by Johann Jakob Balmer, a Swiss schoolteacher. In Figure 5-21, the spectral lines of hydrogen at visible wavelengths are today called **Balmer lines,** and the entire pattern from H_α onward is called the **Balmer series.** Stars in general, including the Sun, have Balmer absorption lines in their spectra, which shows they have atmospheres that contain hydrogen. Notice that the H_α line in Figure 5-21 is in the red portion of the visible spectrum, and when seen as an emission line, it accounts for much of the red seen in beautiful nebulae (including Figure 5-19).

Using trial and error, Balmer discovered a formula from which the wavelengths (λ) of hydrogen's spectral lines can be calculated. Balmer's formula is usually written

$$\frac{1}{\lambda} = R\left(\frac{1}{4} - \frac{1}{n^2}\right)$$

In this formula R is the *Rydberg constant* ($R = 1.097 \times 10^7\ \text{m}^{-1}$), named in honor of the Swedish spectroscopist Johannes Rydberg, and n can be any integer (whole number) greater than 2. The meaning of n will become clear shortly when we discuss electron orbits in the hydrogen atom. To get the wavelength λ_α of the spectral line H_α, you first put $n = 3$ into Balmer's formula:

$$\frac{1}{\lambda_\alpha} = (1.097 \times 10^7\ \text{m}^{-1})\left(\frac{1}{4} - \frac{1}{3^2}\right) = 1.524 \times 10^6\ \text{m}^{-1}$$

Then take the reciprocal:

$$\lambda_\alpha = \frac{1}{1.524 \times 10^6\ \text{m}^{-1}} = 6.563 \times 10^{-7}\ \text{m} = 656.3\ \text{nm}$$

To get the wavelength of H_β, use $n = 4$, and to get the wavelength of H_γ use $n = 5$. If you use $n = \infty$ (the symbol ∞ stands for infinity), you get the short-wavelength end of the hydrogen spectrum at 364.6 nm. (Note that 1 divided by infinity equals zero.)

Bohr's Model of Hydrogen

Bohr realized that to fully understand the structure of the hydrogen atom, he had to be able to derive Balmer's formula using the laws of physics. He first made the rather wild assumption that the electron in a hydrogen atom can orbit the nucleus only in certain specific orbits. (This idea was a significant break with the ideas of Newton, in whose mechanics any orbit should be possible.) Figure 5-22 shows the four smallest of these **Bohr orbits,** labeled by the numbers $n = 1$, $n = 2$, $n = 3$, and so on.

Although confined to one of these allowed orbits while circling the nucleus, an electron can jump from one Bohr orbit to another. For an electron to jump between orbits, the hydrogen atom must gain or lose a specific amount of energy.

An atom must absorb energy for the electron to go from an inner to an outer orbit; an atom must release energy for the electron to go from an outer to an inner orbit.

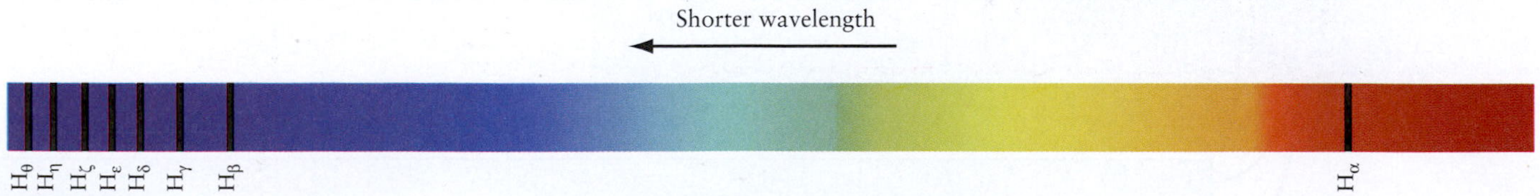

FIGURE 5-21 R I V U X G

Balmer Lines in the Spectrum of a Star This portion of the spectrum of the star Vega in the constellation Lyra (the Harp) shows eight Balmer lines, from H_α at 656.3 nm through H_θ (H-theta) at 388.9 nm. The series converges at 364.6 nm, slightly to the left of H_θ. Parts of this image were made using ultraviolet radiation, which is indicated by the highlighted U in the wavelength tab. (NOAO)

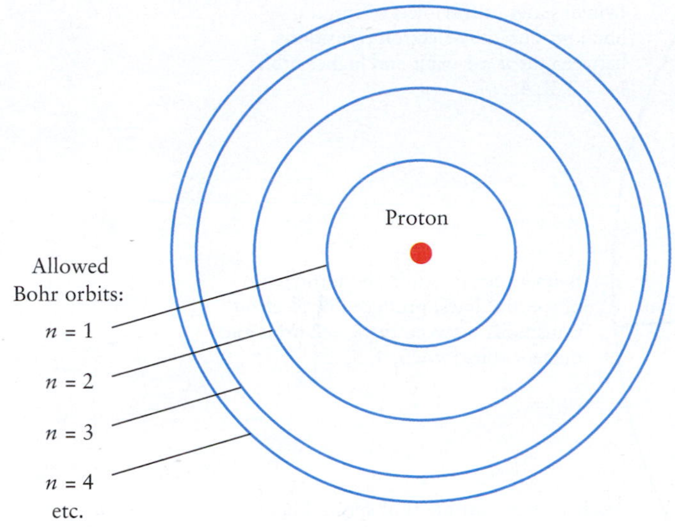

FIGURE 5-22
The Bohr Model of the Hydrogen Atom In this model, an electron circles the hydrogen nucleus (a proton) only in allowed orbits $n = 1, 2, 3$, and so forth. The first four Bohr orbits are shown here. This figure is not drawn to scale; in the Bohr model, the $n = 2$, 3, and 4 orbits are respectively 4, 9, and 16 times larger than the $n = 1$ orbit.

As an example, Figure 5-23 shows an electron jumping between the $n = 2$ and $n = 3$ orbits of a hydrogen atom as the atom absorbs or emits an H_α photon. Figure 5-23 shows absorption and emission of a single photon, but in a gas with many atoms, these same jumps produce absorption lines and emission lines.

Bohr's orbital model also explains Kirchhoff's observation (see Figure 5-17) that an atom can show the same spectral lines as either emission or absorption. Here's why: When the electron jumps from one orbit to another, the energy of the photon that is emitted or absorbed equals the difference in energy between these two orbits. This energy difference, and hence the photon energy, is the same whether the jump is from a low orbit to a high orbit (Figure 5-23a) or from the high orbit back to the low one (Figure 5-23b). If two photons have the same energy E, the relationship $E = hc/\lambda$ tells us that they must also have the same wavelength λ. Therefore, if an atom can emit photons of a given energy and wavelength, it can also absorb photons of precisely the same energy and wavelength.

The Bohr picture also helps us visualize how a hot, transparent gas produces an emission line spectrum (Kirchhoff's Law 2). When a gas is heated, its atoms move around rapidly and can collide forcefully with each other. These energetic collisions excite the atoms' electrons into various high orbits. The electrons then cascade back down to the innermost possible orbit, emitting photons whose energies are equal to the energy differences between different Bohr orbits. In this fashion, a hot gas produces an emission line spectrum with a variety of different wavelengths.

To produce an absorption line spectrum (Kirchhoff's Law 3), begin with a relatively cool gas, so that the electrons in most of the atoms are in inner, low-energy orbits. If a beam of light with a continuous spectrum is shone through the gas, most wavelengths will pass through undisturbed. *Only those photons will be absorbed whose energies are just right to excite an electron to an allowed outer orbit.* Hence, only certain wavelengths will be absorbed, and dark lines will appear in the spectrum at those wavelengths.

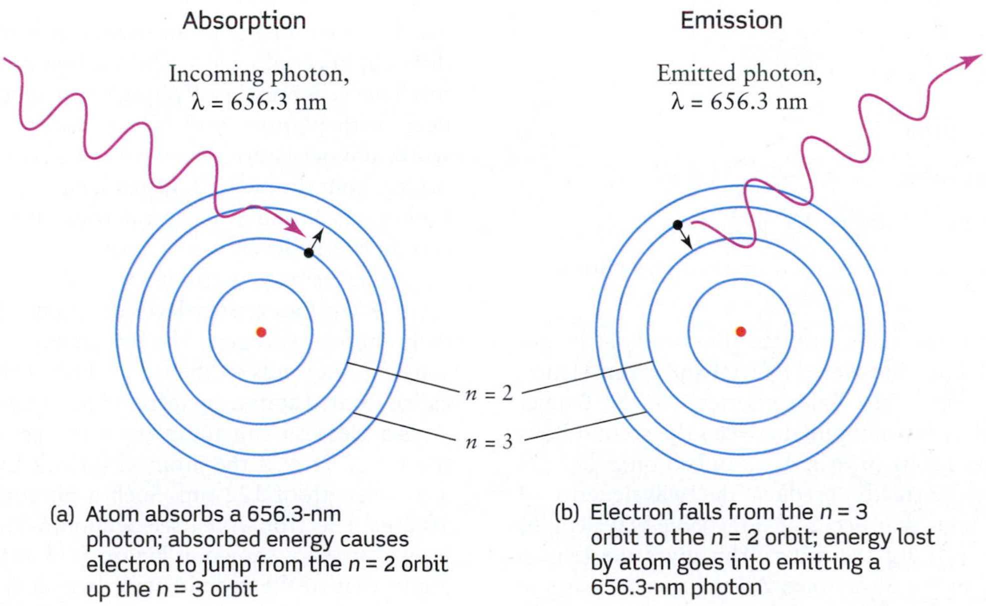

ANIMATION 5-3

FIGURE 5-23
The Absorption and Emission of an H_α Photon This schematic diagram, drawn according to the Bohr model, shows what happens when a hydrogen atom (a) absorbs or (b) emits a photon whose wavelength is 656.3 nm.

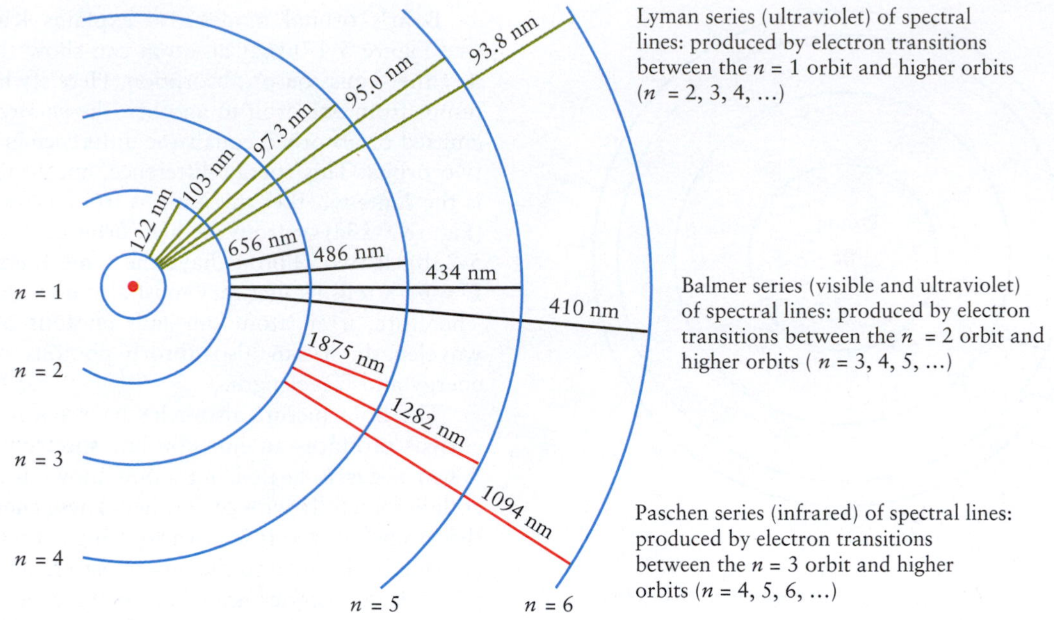
93.8 nm
95.0 nm
97.3 nm
103 nm
122 nm
656 nm
486 nm
434 nm
410 nm
1875 nm
1282 nm
1094 nm
n = 1
n = 2
n = 3
n = 4
n = 5
n = 6
Lyman series (ultraviolet) of spectral lines: produced by electron transitions between the n = 1 orbit and higher orbits (n = 2, 3, 4, ...)
Balmer series (visible and ultraviolet) of spectral lines: produced by electron transitions between the n = 2 orbit and higher orbits (n = 3, 4, 5, ...)
Paschen series (infrared) of spectral lines: produced by electron transitions between the n = 3 orbit and higher orbits (n = 4, 5, 6, ...)

FIGURE 5-24

Electron Transitions in the Hydrogen Atom This diagram shows the photon wavelengths associated with different electron transitions in hydrogen. In each case, the same wavelength occurs whether a photon is emitted (when the electron drops from a high orbit to a low one) or absorbed (when the electron jumps from a low orbit to a high one). The orbits are not shown to scale.

Using his picture of allowed orbits and the formula $E = hc/\lambda$, Bohr was able to prove mathematically that the wavelength λ of the photon emitted or absorbed as an electron jumps between an inner orbit N and an outer orbit n is

Bohr formula for hydrogen wavelengths

$$\frac{1}{\lambda} = R\left(\frac{1}{N^2} - \frac{1}{n^2}\right)$$

N = number of inner orbit

n = number of outer orbit

R = Rydberg constant = $1.097 \times 10^7 \text{ m}^{-1}$

λ = wavelength (in meters) of emitted or absorbed photon

If Bohr let $N = 2$ for the inner orbit in this formula, he got back the formula that Balmer discovered by trial and error. Hence, Bohr deduced the meaning of the Balmer series: All the Balmer lines are produced by electrons jumping between the second Bohr orbit ($N = 2$) and higher orbits ($n = 3, 4, 5$, and so on).

Bohr's formula also correctly predicts the wavelengths of other series of spectral lines that occur at nonvisible wavelengths (Figure 5-24). Using $N = 1$ for the inner orbit gives the **Lyman series,** which is entirely in the ultraviolet. All the spectral lines in this series involve electron transitions between the lowest Bohr orbit and all higher orbits ($n = 2, 3, 4$, and so on). This pattern of spectral lines begins with L_α (Lyman alpha) at 122 nm and converges on L_∞ at 91.2 nm. Using $N = 3$ gives a series of infrared wavelengths called the **Paschen series.** This series, which involves transitions between the third Bohr orbit and all higher orbits, begins with P_α (Paschen alpha) at 1875 nm and converges on P_∞ at 822 nm. Additional series exist at still longer wavelengths.

Atomic Energy Levels

Today's view of the atom owes much to the Bohr model, but is different in certain ways. The modern picture is based on **quantum mechanics,** a branch of physics developed during the 1920s that deals with photons and subatomic particles. As a result of this work, physicists no longer picture electrons as moving in specific orbits about the nucleus. Instead, electrons are now known to have both particle and wave properties and are said to occupy only certain **energy levels** in the atom.

An extremely useful way of displaying the structure of an atom is with an **energy-level diagram.** Figure 5-25 shows such a diagram for hydrogen. The lowest energy level, called the **ground state,** corresponds to the $n = 1$ Bohr orbit. Higher energy levels, called **excited states,** correspond to successively larger Bohr orbits.

An electron can jump from the ground state ($n = 1$) up to the $n = 2$ level if the atom absorbs a Lyman-alpha photon with a wavelength of 122 nm. Such a photon has energy $E = hc/\lambda$ = 10.2 eV (electron volts; see Section 5-5). That is why the energy level of $n = 2$ is shown in Figure 5-25 as having an energy 10.2 eV above that of the ground state (which is usually assigned a value of 0 eV). Similarly, the $n = 3$ level is 12.1 eV above the ground state, and so forth. Electrons can make transitions to higher energy levels by absorbing a photon or in a collision between atoms; they can make transitions to lower energy levels by emitting a photon.

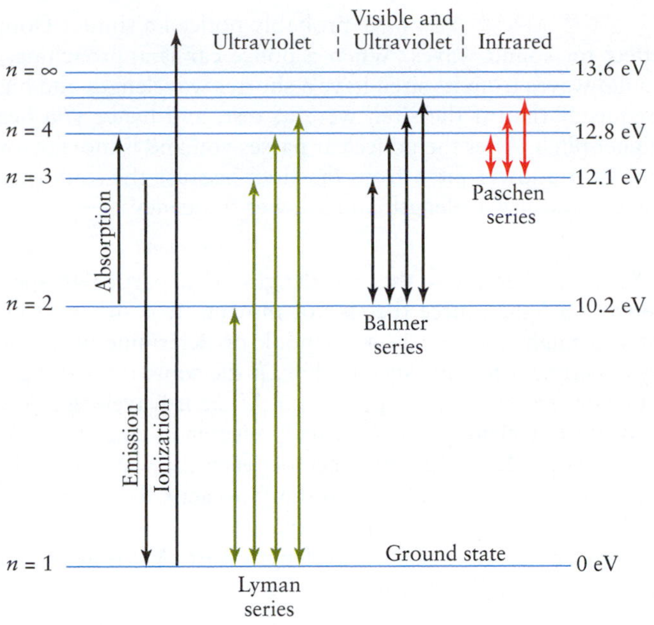

FIGURE 5-25

Energy-Level Diagram of Hydrogen A convenient way to display the structure of the hydrogen atom is in a diagram like this, which shows the allowed energy levels. The diagram shows a number of possible electron jumps, or transitions, between energy levels. An upward transition occurs when the atom absorbs a photon; a downward transition occurs when the atom emits a photon. (Compare with Figure 5-24.)

On the energy-level diagram for hydrogen, the $n = \infty$ level has an energy of 13.6 eV. (This corresponds to an infinitely large "orbit" in the Bohr model.) If the electron is initially in the ground state and the atom absorbs a photon of any energy greater than 13.6 eV, the electron will be removed completely from the atom. This process is called **ionization.** A 13.6-eV photon has a wavelength of 91.2 nm, equal to the shortest wavelength in the ultraviolet Lyman series (L_∞). So any photon with a wavelength of 91.2 nm or less can ionize hydrogen. (Recall that the Planck formula $E = hc/\lambda$ tells us that the shorter the wavelength, the higher the photon energy.)

As an example, the gaseous nebula shown in Figure 5-19 surrounds a cluster of hot stars that produce copious amounts of ultraviolet photons with wavelengths less than 91.2 nm. Hydrogen atoms in the nebula that absorb these photons become ionized and lose their electrons. When the electrons recombine with the nuclei, they cascade down the energy levels to the ground state and emit visible light in the process. This process is what makes the nebula glow. Note that this emission is due to the ionization of the atoms, not the temperature of the gas, and illustrates that Kirchhoff's Law 2 is not the only way to produce an emission line spectrum.

The Spectra of Other Elements

LOOKING DEEPER 5.1 The same basic principles that explain the hydrogen spectrum also apply to the atoms of other elements. Electrons in each kind of atom can only be in certain energy levels, so only photons of certain wavelengths can be emitted or absorbed. Because each kind of atom has its own unique arrangement of electron energy levels, the pattern of spectral lines is likewise unique to that particular type of atom. These patterns are in general much more complicated than for the hydrogen atom.

The properties of energy levels explain the emission line spectra and absorption line spectra of gases. But what about the continuous spectra produced by dense objects like the glowing coils of a toaster or other blackbody object? These objects are made of atoms, so why don't they emit light with an emission line spectrum characteristic of the particular atoms of which they are made?

The reason is directly related to the difference between a gas on the one hand and a liquid or solid on the other. In a gas, atoms are widely separated and can emit photons without interference from other atoms. But in a liquid or a solid, atoms are so close that they almost touch, and thus these atoms interact strongly with each other. In addition, each of the many possible interactions in a liquid or solid have their own energy levels. As a result, the pattern of distinctive bright spectral lines that the atoms would emit in isolation becomes crowded with additional lines forming a continuous spectrum.

ANALOGY Think of atoms as being like tuning forks. If you strike a single tuning fork, it produces a sound wave with a single clear frequency and wavelength, just as an isolated atom emits light of definite wavelengths. But if you shake a box packed full of tuning forks, you will hear a clanging noise that is a mixture of sounds of all different frequencies and wavelengths. This is directly analogous to the continuous spectrum of light emitted by a dense object with closely packed atoms.

With the work of such people as Planck, Einstein, Rutherford, and Bohr, the interchange between astronomy and physics came full circle. Modern physics was born when Newton set out to understand the motions of the planets. Two and a half centuries later, physicists in their laboratories probed the properties of light and the structure of atoms. Their labors had immediate applications in astronomy. Armed with this new understanding of light and matter, astronomers were able to investigate in detail the chemical and physical properties of planets, stars, and galaxies.

CONCEPTCHECK 5-12

Two photons are emitted from a hydrogen atom. One comes from the electron jumping down from the $n = 3$ level to the $n = 1$ level. The other photon comes from an electron transition from $n = 2$ to $n = 1$. Which photon has the highest frequency?

CONCEPTCHECK 5-13

Hydrogen is, by far, the most abundant element in the universe. Are any hydrogen emission lines visible to humans? If so, what color is most common?

Answers appear at the end of the chapter.

5-9 The wavelength of a spectral line is affected by the relative motion between the source and the observer

In addition to telling us about temperature and chemical composition, the spectrum of a planet, star, or galaxy can also reveal something about that object's motion through space. This idea dates from 1842, when Christian Doppler, a professor of mathematics in Prague, pointed out that the observed wavelength of light must be affected by motion.

The Doppler Effect

The Doppler effect makes it possible to tell whether astronomical objects are moving toward us or away from us

In Figure 5-26 a light source is moving from right to left; the circles represent the crests of waves emitted from the moving source at various positions. Each successive wave crest is emitted from a position slightly closer to the observer on the left, so she sees a shorter wavelength—the distance from one crest to the next—than she would if the source were stationary. What is the effect on a full spectrum of light? All the lines in the spectrum of an approaching source are shifted toward the short-wavelength (blue) end of the spectrum. This phenomenon is called a **blueshift.**

The source is receding from the observer on the right in Figure 5-26. The wave crests that reach him are stretched apart, so that he sees a longer wavelength than he would if the source were stationary. All the lines in the spectrum of a receding source are shifted toward the longer-wavelength (red) end of the spectrum, producing a **redshift.** In general, the effect of relative motion on wavelength is called the **Doppler effect.** Police radar guns use the Doppler effect to check for cars exceeding the speed limit: The radar gun sends a radio wave toward the car, and measures the wavelength shift of the reflected wave. During reflection of the wave, the car acts like a moving source so the Doppler shift measures the speed of the car.

ANALOGY You have probably noticed a similar Doppler effect for sound waves. When a police car is approaching, the sound waves from its siren have a shorter wavelength and higher frequency than if the siren were at rest, and hence you hear a higher pitch. After the police car passes you and is moving away, you hear a lower pitch from the siren because the sound waves have a longer wavelength and a lower frequency.

Suppose that λ_0 is the wavelength of a particular spectral line from a light source that is not moving. It is the wavelength that you might look up in a textbook or determine in a laboratory experiment for this spectral line. If the source is moving, this particular spectral line is shifted to a different wavelength λ. The size of the wavelength shift is usually written as $\Delta\lambda$, where $\Delta\lambda = \lambda - \lambda_0$. Thus, $\Delta\lambda$ is the difference between the wavelength listed in textbooks and the wavelength that you actually observe in the spectrum of a moving star or galaxy.

Doppler proved that the wavelength shift ($\Delta\lambda$) is governed by the following simple equation:

Doppler shift equation

$$\frac{\Delta\lambda}{\lambda_0} = \frac{v}{c}$$

$\Delta\lambda$ = wavelength shift

λ_0 = wavelength if source is not moving

v = velocity of the source measured along the line of sight

c = speed of light = 3.0×10^5 km/s

CAUTION! The capital Greek letter Δ (delta) is commonly used as a symbol to denote change in the value of a quantity. Thus, $\Delta\lambda$ is the change in the wavelength λ due to the Doppler effect. It is *not* equal to a quantity Δ multiplied by a second quantity λ.

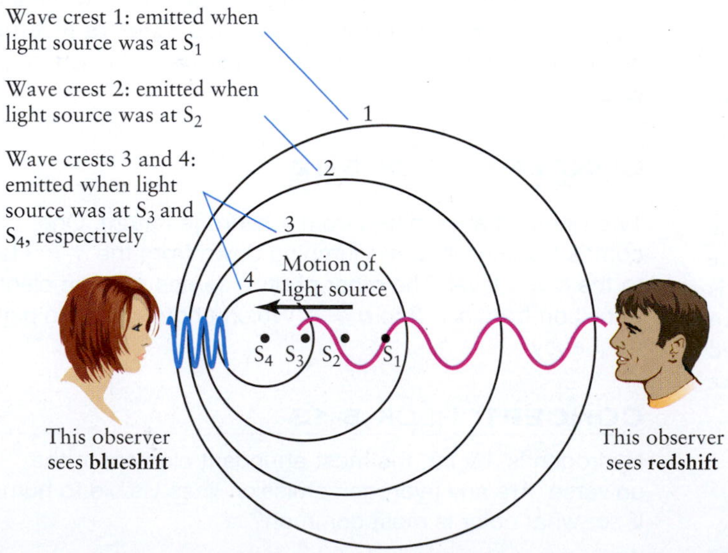

Wave crest 1: emitted when light source was at S_1
Wave crest 2: emitted when light source was at S_2
Wave crests 3 and 4: emitted when light source was at S_3 and S_4, respectively
1
2
3
4
Motion of light source
S_4 S_3 S_2 S_1
This observer sees **blueshift**
This observer sees **redshift**

FIGURE 5-26
The Doppler Effect The wavelength of light is affected by motion between the light source and an observer. The light source shown here is moving, so wave crests 1, 2, etc., emitted when the source was at points S_1, S_2, etc., are crowded together in front of the source but are spread out behind it. Consequently, wavelengths are shortened (blueshifted) if the source is moving toward the observer and lengthened (redshifted) if the source is moving away from the observer. Motion perpendicular to an observer's line of sight does not affect wavelength.

Interpreting the Doppler Effect

The velocity determined from the Doppler effect is called **radial velocity,** because v is the component of the star's motion parallel to our line of sight, or along a "radius" drawn from Earth to the star. Of course, a sizable fraction of a star's motion might be perpendicular to our line of sight, but this sideways or transverse movement across the sky does not cause Doppler shifts. Box 5-6 includes two examples of calculations with radial velocity using the Doppler formula.

CAUTION! The redshifts and blueshifts of stars visible to the naked eye, or even through a small telescope, are only a small fraction of a nanometer. These tiny wavelength changes are far too small to detect visually. (Astronomers were able to detect the tiny Doppler shifts of starlight only after they had developed highly sensitive equipment for measuring wavelengths. This was done around 1890, a half-century after Doppler's original proposal.) So, if you see a star with a red color, it means that the star really is red; it does *not* mean that it is moving rapidly away from us.

BOX 5-6 TOOLS OF THE ASTRONOMER'S TRADE

Applications of the Doppler Effect

Doppler's formula relates the radial velocity of an astronomical object to the wavelength shift of its spectral lines. Here are two examples that show how to use this remarkably powerful formula.

EXAMPLE: As measured in the laboratory, the prominent H_α spectral line of hydrogen has a wavelength $\lambda_0 = 656.285$ nm. But in the spectrum of the star Vega (Figure 5-21), this line has a wavelength $\lambda = 656.255$ nm. What can we conclude about the motion of Vega?

Situation: Our goal is to use the ideas of the Doppler effect to find the velocity of Vega toward or away from Earth.

Tools: We use the Doppler shift formula, $\Delta\lambda/\lambda_0 = v/c$, to determine Vega's velocity v.

Answer: The wavelength shift is

$$\Delta\lambda = \lambda - \lambda_0 = 656.255\text{ nm} - 656.285\text{ nm} = -0.030\text{ nm}$$

The negative value means that we see the light from Vega shifted to shorter wavelengths—that is, there is a blueshift. (Note that the shift is very tiny and can be measured only using specialized equipment.) From the Doppler shift formula, the star's radial velocity is

$$v = c\frac{\Delta\lambda}{\lambda_0} = (3.00 \times 10^5\text{ km/s})\left(\frac{-0.030\text{ nm}}{656.285\text{ nm}}\right) = -14\text{ km/s}$$

Review: The minus sign indicates that Vega is coming toward us at 14 km/s. *The star might also have some motion perpendicular to the line from Earth to Vega, but such motion produces no Doppler shift.*

By plotting the motions of stars such as Vega toward and away from us, astronomers have been able to learn how the Milky Way Galaxy (of which our Sun is a part) is rotating. From this knowledge, and aided by Newton's universal law of gravitation (see Section 4-6), they have made the surprising discovery that the Milky Way contains roughly 10 times more matter than had once been thought! The nature of this unseen *dark matter* is still one of the great unsolved mysteries in astronomy.

EXAMPLE: In the radio region of the electromagnetic spectrum, hydrogen atoms emit and absorb photons with a wavelength of 21.12 cm, giving rise to a spectral feature commonly called the *21-centimeter line.* The galaxy NGC 3840 in the constellation Leo (the Lion) is receding from us at a speed of 7370 km/s, or about 2.5% of the speed of light. At what wavelength do we expect to detect the 21-cm line from this galaxy?

Situation: Given the velocity of NGC 3840 away from us, our goal is to find the wavelength as measured on Earth of the 21-centimeter line from this galaxy.

Tools: We use the Doppler shift formula to calculate the wavelength shift $\Delta\lambda$, then use this to find the wavelength λ measured on Earth.

Answer: The wavelength shift is

$$\Delta\lambda = \lambda_0\left(\frac{v}{c}\right) = (21.12\text{ cm})\left(\frac{7370\text{ km/s}}{3.00 \times 10^5\text{ km/s}}\right) = 0.52\text{ cm}$$

Therefore, we will detect the 21-cm line of hydrogen from this galaxy at a wavelength of

$$\lambda = \lambda_0 + \Delta\lambda = 21.12\text{ cm} + 0.52\text{ cm} = 21.64\text{ cm}$$

Review: The 21-cm line has been redshifted to a longer wavelength because the galaxy is receding from us. In fact, most galaxies are receding from us. This observation is one of the key pieces of evidence that the universe is expanding, and has been doing so since the Big Bang that took place almost 14 billion years ago.

The Doppler effect is an important tool in astronomy because it uncovers basic information about the motion of planets, stars, and galaxies. For example, the rotation of the planet Venus was deduced from the Doppler shift of radar waves reflected from its surface. Small Doppler shifts in the spectrum of sunlight have shown that the entire Sun is vibrating like an immense gong. The back-and-forth Doppler shifting of the spectral lines of certain stars reveals that these stars are being orbited by unseen companions; from this astronomers have discovered planets around other stars and massive objects that may be black holes. Astronomers also use the Doppler effect along with Kepler's third law to measure a galaxy's mass. These are but a few examples of how Doppler's discovery has empowered astronomers in their quest to understand the universe.

In this chapter we have glimpsed how much can be learned by analyzing light from the heavens. To analyze this light, however, it is first necessary to collect as much of it as possible, because most light sources in space are very dim. Collecting the faint light from distant objects is the key purpose of telescopes. In the next chapter we will describe both how telescopes work and how they are used.

CONCEPTCHECK 5-14

How is the spectrum changed when looking at the absorption spectrum from an approaching star as compared to how the star's spectrum would look if the star were stationary?

CALCULATIONCHECK 5-3

How fast and in what direction is a star moving if it has a line that shifts from 486.2 nm to 486.3 nm?

Answers appear at the end of the chapter.

KEY WORDS

Terms preceded by an asterisk (*) are discussed in the Boxes.

absolute zero, p. 113
absorption line spectrum, p. 122
atom, p. 126
*atomic number, p. 126
Balmer line, p. 128
Balmer series, p. 128
blackbody, p. 114
blackbody curve, p. 115
blackbody radiation, p. 115
blueshift, p. 132
Bohr orbits, p. 128
compound, p. 122
continuous spectrum, p. 122
*degrees Celsius, p. 116
*degrees Fahrenheit, p. 116
Doppler effect, p. 132
electromagnetic radiation, p. 109
electromagnetic spectrum, p. 110
electromagnetism, p. 108
electron, p. 126
electron volt, p. 120
element, p. 122
emission line spectrum, p. 122
energy flux, p. 117
energy level, p. 130
energy-level diagram, p. 130
excited state, p. 130
frequency, p. 111
gamma rays, p. 110
ground state, p. 130
infrared radiation, p. 110
ionization, p. 131
*isotope, p. 126
joule, p. 117
kelvin, p. 113
Kirchhoff's laws, p. 122
light scattering, p. 122
*luminosity, p. 118
Lyman series, p. 130
microwaves, p. 110
molecule, p. 122
nanometer, p. 109
neutron, p. 126
nucleus, p. 126
opaque, p. 113
Paschen series, p. 130
*periodic table, p. 126
photon, p. 119
Planck's law, p. 120
proton, p. 126
quantum mechanics, p. 130
radial velocity, p. 133
radio waves, p. 110
redshift, p. 132
*solar constant, p. 118
spectral analysis, p. 121
spectral line, p. 121
spectroscopy, p. 122
spectrum (*plural* spectra), p. 107
Stefan-Boltzmann law, p. 117
thermal energy, p. 113
ultraviolet radiation, p. 110
visible light, p. 109
watt, p. 117
wavelength, p. 109
wavelength of maximum emission, p. 113
Wien's law, p. 117
X-rays, p. 110

KEY IDEAS

The Nature of Light: Light is electromagnetic radiation. It has wavelike properties described by its wavelength λ and frequency ν, and travels through empty space at the constant speed $c = 3.0 \times 10^8$ m/s $= 3.0 \times 10^5$ km/s.

Thermal energy: The thermal energy of a material comes from the kinetic energy of its microscopic particles (atoms and molecules). The hotter a material, the faster its particles move, and the greater its thermal energy.

Blackbody Radiation: A blackbody is a hypothetical object that emits a continuous spectrum; the hotter the object, the greater the emission. Stars closely approximate the behavior of blackbodies, as do other hot, dense objects.

• The intensities of radiation emitted at various wavelengths by a blackbody at a given temperature are shown by a blackbody curve.

• Wien's law states that the dominant wavelength at which a blackbody emits electromagnetic radiation is inversely proportional to the Kelvin temperature of the object: λ_{max} (in meters) $=$ (0.0029 K m)/T.

• The Stefan-Boltzmann law states that a blackbody radiates electromagnetic waves with a total energy flux F directly proportional to the fourth power of the Kelvin temperature T of the object: $F = \sigma T^4$.

Photons: Light is made of particles called photons. Each photon has a wavelength equal to the wavelength of the light that the photons make up.

• Planck's law relates the energy E of a photon to its frequency ν or wavelength λ: $E = h\nu = hc/\lambda$, where h is Planck's constant. The shorter the wavelength a photon has, the higher its frequency and the larger its energy.

Kirchhoff's Laws: Kirchhoff's three laws of spectral analysis describe conditions under which different kinds of spectra are produced.

• A hot, dense object such as a blackbody emits a continuous spectrum covering all wavelengths.

• A hot, transparent gas produces a spectrum that contains bright (emission) lines.

• A cool, transparent gas in front of a light source that itself has a continuous spectrum produces dark (absorption) lines in the continuous spectrum.

Atomic Structure: An atom has a small dense nucleus composed of protons and neutrons. The nucleus is surrounded by electrons that only occupy certain orbits or energy levels.

• When an electron jumps from one energy level to another, it emits or absorbs a photon of corresponding energy (and hence of a specific wavelength).

• The spectral lines of a particular atom correspond to various electron orbital transitions between energy levels in the atom. Each atom has a unique "spectral fingerprint."

The Doppler Shift: The Doppler shift enables us to determine the line-of-sight (radial) velocity of a light source from the displacement of its spectral lines.

• The spectral lines of an approaching light source are shifted toward short wavelengths (a blueshift); the spectral lines of a receding light source are shifted toward long wavelengths (a redshift).

• The size of a wavelength shift is proportional to the radial velocity of the light source relative to the observer.

QUESTIONS

Review Questions

1. When Jupiter is undergoing retrograde motion as seen from Earth, would you expect the eclipses of Jupiter's moons to occur several minutes early, several minutes late, or neither? Explain your answer.
2. Approximately how many times around Earth could a beam of light travel in one second?
3. How long does it take light to travel from the Sun to Earth, a distance of 1.50×10^8 km?
4. TUTORIAL 5-1 How did Newton show that a prism breaks white light into its component colors but does not add any color to the light?
5. For each of the following wavelengths, state whether it is in the radio, microwave, infrared, visible, ultraviolet, X-ray, or gamma-ray portion of the electromagnetic spectrum. Explain your reasoning. (a) 2.6 μm, (b) 34 m, (c) 0.54 nm, (d) 0.0032 nm, (e) 0.620 μm, (f) 310 nm, (g) 0.012 m
6. What is meant by the frequency of light? How is frequency related to wavelength?
7. A cellular phone is actually a radio transmitter and receiver. You receive an incoming call in the form of a radio wave of frequency 880.65 MHz. What is the wavelength (in meters) of this wave?
8. A light source emits infrared radiation at a wavelength of 1150 nm. What is the frequency of this radiation?
9. (a) What is a blackbody? (b) In what way is a blackbody black? (c) If a blackbody is black, how can it emit light? (d) If you were to shine a flashlight beam on a perfect blackbody, what would happen to the light?
10. What does it mean for a material to be opaque? Can water molecules form an opaque object in some forms and a transparent substance in others? (*Hint:* Think of clouds.)
11. Describe how the thermal energy changes at a microscopic level as a blacksmith heats a piece of iron. If the iron is then removed from the furnace and glows bright red, what effect does this have on the thermal energy?
12. Why do astronomers find it convenient to use the Kelvin temperature scale in their work rather than the Celsius or Fahrenheit scales?
13. Explain why astronomers are interested in blackbody radiation.
14. In what way is the Sun's spectrum similar to a blackbody spectrum? In what way does it differ from a blackbody spectrum?
15. TUTORIAL 5-2 Using Wien's law and the Stefan-Boltzmann law, explain the color and intensity changes that are observed as the temperature of a hot, glowing object increases.
16. If you double the Kelvin temperature of a hot piece of steel, how much more energy will it radiate per second?
17. The bright star Bellatrix in the constellation Orion has a surface temperature of 21,500 K. What is its wavelength of maximum emission in nanometers? What color is this star?
18. The bright star Antares in the constellation Scorpius (the Scorpion) emits the greatest intensity of radiation at a wavelength of 853 nm. What is the surface temperature of Antares? What color is this star?
19. (a) Describe an experiment in which light behaves like a wave. (b) Describe an experiment in which light behaves like a particle.
20. How is the energy of a photon related to its wavelength? What kind of photons carry the most energy? What kind of photons carry the least energy?
21. Which part of the electromagnetic spectrum contains light with a higher frequency: microwaves or radio waves?
22. To emit the same amount of light energy per second, which must emit more photons per second: a source of red light or a source of blue light? Explain your answer.
23. (a) Describe the spectrum of hydrogen at visible wavelengths. (b) Explain how Bohr's model of the atom accounts for the Balmer lines.
24. Why do different elements display different patterns of lines in their spectra?
25. What is the Doppler effect? Why is it important to astronomers?

26. If you see a blue star, what does its color tell you about how the star is moving through space? Explain your answer.

Advanced Questions

Questions preceded by an asterisk (*) involve topics discussed in the Boxes.

> **Problem-solving tips and tools**
>
> You can find formulas in Box 5-1 for converting between temperature scales. Box 5-2 discusses how a star's radius, luminosity, and surface temperature are related. Box 5-3 shows how to use Planck's law to calculate the energy of a photon. To learn how to do calculations using the Doppler effect, see Box 5-6.

27. Your normal body temperature is 98.6°F. What kind of radiation do you predominantly emit? At what wavelength (in nm) do you emit the most radiation?
28. What is the temperature of the Sun's surface in degrees Fahrenheit?
29. What wavelength of electromagnetic radiation is emitted with greatest intensity by this book? To what region of the electromagnetic spectrum does this wavelength correspond?
30. Can an object convert some of its orbital energy into thermal energy? If yes, describe a context in which this might occur.
31. Black holes are objects whose gravity is so strong that not even an object moving at the speed of light can escape from their surfaces. Hence, black holes do not themselves emit light. But it is possible to detect radiation from material falling *toward* a black hole. Calculations suggest that as this matter falls, it is compressed and heated to temperatures around 10^6 K. Calculate the wavelength of maximum emission for this temperature. In what part of the electromagnetic spectrum does this wavelength lie?

*32. Use the value of the solar constant given in Box 5-2 and the distance from Earth to the Sun to calculate the luminosity of the Sun.

*33. The star Alpha Lupi (the brightest in the constellation Lupus the Wolf) has a surface temperature of 21,600 K. How much more energy is emitted each second from each square meter of the surface of Alpha Lupi than from each square meter of the Sun's surface?

*34. Jupiter's moon Io has an active volcano named Pele whose temperature can be as high as 320°C. (a) What is the wavelength of maximum emission for the volcano at this temperature? In what part of the electromagnetic spectrum is this? (b) The average temperature of Io's surface is 2150°C. Compared with a square meter of surface at this temperature, how much more energy is emitted per second from each square meter of Pele's surface?

*35. The bright star Sirius in the constellation of Canis Major (the Large Dog) has a radius of 1.67 $R_{\odot}$ and a luminosity of 25 $L_{\odot}$. (a) Use this information to calculate the energy flux at the surface of Sirius. (b) Use your answer in part (a) to calculate the surface temperature of Sirius. How does your answer compare to the value given in Box 5-2?

36. Instruments on board balloons and spacecraft detect 511-keV photons coming from the direction of the center of our Galaxy. (The prefix k means *kilo,* or thousand, so 1 keV = 10^3 eV.) What is the wavelength of these photons? To what part of the electromagnetic spectrum do these photons belong?
37. (a) Calculate the wavelength of P_{Δ} (P-delta), the fourth wavelength in the Paschen series. (b) Draw a schematic diagram of the hydrogen atom and indicate the electron transition that gives rise to this spectral line. (c) In what part of the electromagnetic spectrum does this wavelength lie?
38. (a) Calculate the wavelength of H_{η} (H-eta), the spectral line for an electron transition between the $n = 7$ and $n = 2$ orbits of hydrogen. (b) In what part of the electromagnetic spectrum does this wavelength lie? Use this to explain why Figure 5-21 is labeled R I **V** **U** X G.
39. Certain interstellar clouds contain a very cold, very thin gas of hydrogen atoms. Ultraviolet radiation with any wavelength shorter than 91.2 nm cannot pass through this gas; instead, it is absorbed. Explain why.
40. (a) Can a hydrogen atom in the ground state absorb an H-alpha (H_{α}) photon? Explain why or why not. (b) Can a hydrogen atom in the $n = 2$ state absorb a Lyman-alpha (L_{α}) photon? Explain why or why not.
41. An imaginary atom has just three energy levels: 0 eV, 1 eV, and 3 eV. Draw an energy-level diagram for this atom. Show all possible transitions between these energy levels. For each transition, determine the photon energy and the photon wavelength. Which transitions involve the emission or absorption of visible light?
42. The star cluster NGC 346 and nebula shown in Figure 5-19 are located within the Small Magellanic Cloud (SMC), a small galaxy that orbits our Milky Way Galaxy. The SMC and the stars and gas within it are moving away from us at 158 km/s. At what wavelength does the red H_{α} line of hydrogen (which causes the color of the nebula) appear in the nebula's spectrum?
43. The wavelength of H_{β} in the spectrum of the star Megrez in the Big Dipper (part of the constellation Ursa Major the Great Bear) is 486.112 nm. Laboratory measurements demonstrate that the normal wavelength of this spectral line is 486.133 nm. Is the star coming toward us or moving away from us? At what speed?
44. You are given a traffic ticket for going through a red light (wavelength 700 nm). You tell the police officer that because you were approaching the light, the Doppler effect caused a blueshift that made the light appear green (wavelength 500 nm). How fast would you have had to be going for this to be true? Would the speeding ticket be justified? Explain your answer.

Discussion Questions

45. The equation that relates the frequency, wavelength, and speed of a light wave, $\nu = c/\lambda$, can be rewritten as $c = \nu\lambda$. A friend who has studied mathematics but not much astronomy or physics might look at this equation and say: "This equation tells me that the higher the frequency ν, the greater the wave speed c. Since visible light has a higher frequency than radio waves, this means that visible light goes faster than radio waves." How would you respond to your friend?
46. (**a**) If you could see ultraviolet radiation, how might the night sky appear different? Would ordinary objects appear different in the daytime? (**b**) What differences might there be in the appearance of the night sky and in the appearance of ordinary objects in the daytime if you could see infrared radiation?
47. The accompanying visible-light image shows the star cluster NGC 3293 in the constellation Carina (the Ship's Keel). What can you say about the surface temperatures of most of the bright stars in this cluster? In what part of the electromagnetic spectrum do these stars emit most intensely? Are your eyes sensitive to this type of radiation? If not, how is it possible to see these stars at all? There is at least one bright star in this cluster with a distinctly different color from the others; what can you conclude about its surface temperature?

R I V U X G (David Malin/Anglo-Australian Observatory)

48. The human eye is most sensitive over the same wavelength range at which the Sun emits the greatest intensity of radiation. Suppose creatures were to evolve on a planet orbiting a star somewhat hotter than the Sun. To what wavelengths would their vision most likely be sensitive?
49. Why do you suppose that ultraviolet light can cause skin cancer but ordinary visible light does not?

Web/eBook Question

50. Search the World Wide Web for information about rainbows. Why do rainbows form? Why do they appear as circular arcs? Why can you see different colors?

ACTIVITIES

Observing Projects

51. Turn on an electric stove or toaster oven and carefully observe the heating elements as they warm up. Relate your observations to Wien's law and the Stefan-Boltzmann law.
52. Use *Starry Night*™ to examine the celestial objects listed below. Select **Favourites > Explorations > Atlas** to show the whole sky as would be seen from the center of a transparent Earth. Ensure that deep space objects are displayed by selecting **View > Deep Space > Messier Objects** and **View > Deep Space > Bright NGC Objects** from the menu. Also, select **View > Deep Space > Hubble Images** and ensure that this option is turned **off**. To display each of the objects listed below, open the Find pane and then type the name of the selected object in the edit box and press the Enter (Return) key. The object will be centered in the view. Use the zoom controls to adjust your view until you can see the object in detail. For each object, decide whether you think it will have a continuous spectrum, an absorption line spectrum, or an emission line spectrum, and explain your reasoning. The objects to observe are (**a**) the Lagoon Nebula in Sagittarius (with a field of view of about 6° × 4°, you can compare and contrast the appearance of the Lagoon Nebula with the Trifid Nebula just to the north of it); (**b**) M31, the great galaxy in the constellation Andromeda (*Hint:* the light coming from this galaxy is the combined light of hundreds of billions of individual stars); (**c**) the Moon (*Hint:* moonlight is simply reflected sunlight).
53. Use the *Starry Night*™ program to compare the brightness of two similarly sized stars in the constellation Auriga. Select **Favourites > Explorations > Auriga**. The two stars Capella and Delta Aurigae are labeled in this view. Select **Preferences** from the **File** menu (Windows) or **Starry Night** menu (Mac) and set **Cursor Tracking (HUD)** options so that **Temperature** and **Radius** are shown in the HUD display. You will notice that these two stars have the same radius but differ in temperature. From these data, which of these stars is intrinsically brighter and by what proportion?
54. Use *Starry Night*™ to examine the temperatures of several relatively nearby stars. Select **Favourites > Explorations > Atlas**. Use the **File** menu (**Starry Night** menu on a Mac) and select **Preferences...** to open the **Preferences** dialog window. Click the box in the top left of this dialog window and choose **Cursor Tracking (HUD)**. Scroll through the **Show** list and click the checkbox next to **Temperature** to turn this option on. Then close the **Preferences** dialog window. Next, open the **Find** pane, click the magnifying glass icon in the edit box

at the top of this pane, and select **Star** from the dropdown menu. To locate each of the following stars—Altair; Procyon; Epsilon Indi; Tau Ceti; Epsilon Eridani; Lalande 21185—type the name of the star in the edit box and then press the **Enter** (**Return**) key. Use the **HUD** to find and record the star's temperature. (**a**) Which of the stars have a longer wavelength of maximum emission λ_{max} than the Sun? (**b**) Which of the stars have a shorter λ_{max} than the Sun? (**c**) Which of the stars will have a reddish color?

55. You can use the *Starry Night*™ program to measure the speed of light by observing a particular event, in this case one of Jupiter's moons emerging from the planet's shadow, from two locations separated by a known distance. These two locations are at the north poles of Earth and the planet Mercury, respectively. Open **Favourites > Explorations > Io from Earth.** The view shows Jupiter as seen from the north pole of Earth at 9:12:00 P.M. standard time on September 25, 2010. You will see the label for Io to the left of the planet. Keep the field of view about 11 arcminutes wide. Click the **Play** button and observe Io suddenly brighten as it emerges from Jupiter's shadow. Depending upon your computer monitor, you may need to **Zoom out** slightly so that Io's transition from being invisible to visible occurs instantaneously (at high zoom levels, Io will brighten gradually). Next, use the **Time** controls to **Step time backward** and **forward** in 1-second intervals to determine the time to the nearest second at which Io brightens. Open the **Status** pane and expand the **Time** layer. Record the **Universal Time** for this event. Then open the **Info** pane and be sure that **Io Info** appears at the top of the **Info** pane. Expand the **Position in Space** layer and record the value given for **Distance from Observer**. Now select **Favourites > Explorations > Io from Mercury**. This view once again shows Io labeled to the left of Jupiter but in this instance you are viewing the scene from the north pole of the planet Mercury. With the field of view set to 11 arcminutes wide, click the **Play** button. **Stop** time flow as soon as you see Io suddenly brighten as it emerges from eclipse. Again, it may be necessary to adjust the zoom level to make this event appear instantaneous rather than gradual. Use the time controls to find the time to the nearest second at which Io brightens. Open the **Status** pane and record the **Universal Time** for this event as seen from Mercury. Then open the contextual menu for Io and select **Show Info**. Record the **Distance from Observer** of Io from the **Position in Space** layer. Use your observations to calculate the speed of light. First, calculate the difference in the time between the two observations in seconds. Then calculate the difference in the **Distance from Observer** in AU for each of the locations. Divide the difference in the distance by the difference in time to calculate the speed of light in AU per second. Finally, convert this value to kilometers per second by multiplying the result by 1.496×10^8 (the number of kilometers in 1 AU). How does your result compare to the accepted value of the speed of light of 2.9979×10^5 kilometers per second? Explain the difference between your calculated value from these observations and the accepted value for the speed of light.

Collaborative Exercise

56. The Doppler effect describes how relative motion impacts wavelength. With a classmate, stand up and demonstrate each of the following: (**a**) a blueshifted source for a stationary observer; (**b**) a stationary source and an observer detecting a redshift; and (**c**) a source and an observer both moving in the same direction, but the observer is detecting a redshift. Create simple sketches to illustrate what you and your classmate did.

ANSWERS

ConceptChecks

ConceptCheck 5-1: The speed of light is incredibly fast, which made it very difficult to measure light's speed except over enormous distances.

ConceptCheck 5-2: As Newton found when passing sunlight through a series of prisms, when one color is isolated from white light, there are no longer any other colors present in the remaining light. As a result, you cannot turn pure red into any other color. White light contains all of the colors, but once any of those colors are absorbed, they cannot be recovered. The red plastic absorbs all the visible wavelengths other than red, so no green can be obtained.

ConceptCheck 5-3: The width of your finger is about 1 cm, which falls in the range of the wavelength of microwaves (1 mm–10 cm).

ConceptCheck 5-4: Because the relationship between wavelength and frequency is $c = \lambda \times f$, as one increases the other decreases (wavelength and frequency are inversely related). Thus, the longest wavelengths have the lowest frequencies and the shortest wavelengths have the highest frequencies.

ConceptCheck 5-5: As shown in Figure 5-12, at every wavelength, including infrared, the higher the temperature of a blackbody, the more energy it emits.

ConceptCheck 5-6: Yes. An object at 0°C still has a temperature of 237 K, and Kelvins are the temperature unit for working with blackbodies. At 273 K, the blackbody would peak in the infrared, although humans cannot see this light.

ConceptCheck 5-7: A star is hotter than our Sun if the star's peak wavelength of blackbody emission is a shorter wavelength than the peak from our Sun.

ConceptCheck 5-8: Yes. If the cooler object is much larger than the warmer object, the cooler object can emit a greater total energy. The energy flux F refers to the energy emitted per second for each square meter of surface area, so even for a cool object, the larger the object, the more energy it radiates. This is why some stars (called red giants) that are much larger than our Sun are actually brighter than our Sun, even though they are cooler.

ConceptCheck 5-9: Photons with longer wavelengths will have lower energy than those with shorter wavelengths because the greater the wavelength, the lower the energy of a photon associated with that wavelength.

ConceptCheck 5-10: While still hot, the Sun's outer atmosphere is cooler than deeper down where most of the Sun's continuous blackbody emission originates. Thus, Figure 5-14 is an absorption line spectrum described by Kirchhoff's third law. Kirchhoff's second law would describe the emission line spectrum from hot helium gas, as in Figure 5-16.

ConceptCheck 5-11: An absorption spectra results when the light from a hot, dense object passes through the cooler, transparent gas of our atmosphere. At visible wavelengths, the atmosphere does not produce a significant absorption spectrum, but at infrared wavelengths (which are also emitted by hot lava), the atmosphere has many absorption lines.

ConceptCheck 5-12: The greater a photon's energy, the higher its frequency. Since the change in energy for the $n = 3$ to $n = 1$ jump is the larger of the two transitions (see Figure 5-25), its emitted photon has the highest frequency. You can also compare the wavelengths of these two transitions in Figure 5-24, and the shorter the wavelength of light, the higher its frequency.

ConceptCheck 5-13: Yes. The most common visible-light hydrogen transition emits red photons of wavelength 656nm. It is an emission line (see Figure 5-24) with a transition from $n = 3$ to $n = 2$ and is referred to as H_α (H-alpha). This single transition accounts for most of the red seen in pretty nebulae such as Figure 5-19.

ConceptCheck 5-14: When the distance between an observer and a source is decreasing, the source's entire spectrum will be shifted toward shorter wavelengths; alternatively, when the distance between an object and a source is decreasing, the emissions lines will be shifted toward longer wavelengths. Thus, the star's absorption lines will be observed at shorter wavelengths.

CalculationCheck

CalculationCheck 5-1: This classic rock station in Santa Barbara, California, emits waves with a frequency of 99.9 MHz, but this can be rounded to 100 MHz. To calculate the wavelength of these radio waves, we rearrange the equation $c = \nu \div \lambda$ to get $\lambda = 3 \times 108$ m/s $\div\ 100 \times 106$ Hz $= 3.00$ m.

CalculationCheck 5-2: Wien's law can be rearranged to calculate the temperature of a star as $T = 0.0029$ K m $\div$ (5800 K $\times$ 2) $=$ 250 nm, which is ultraviolet.

CalculationCheck 5-3: $\nu = c \times \Delta\lambda \div \lambda_0 = 3 \times 10^5$ km/s $\times$ (486.3 nm $-$ 486.2 nm) $\div$ 486.2 nm $=$ 61.7 km/s, and because it is moving toward longer wavelengths, the distance between the observer and the star must be increasing.

CHAPTER

6

R I **V** U X G R **I** V U X G

Modern telescopes can view the universe in every range of electromagnetic radiation, although some must be placed above Earth's atmosphere. (NASA/JPL-Caltech/C. Engelbracht, University of Arizona)

Optics and Telescopes

LEARNING GOALS

By reading the sections of this chapter, you will learn

6-1 How a refracting telescope uses a lens to form an image

6-2 How a reflecting telescope uses a curved mirror to form an image

6-3 How a telescope's size and Earth's atmosphere limit the sharpness of a telescopic image

6-4 How electronic light detectors have revolutionized astronomy

6-5 How telescopes are used to obtain spectra of astronomical objects

6-6 The advantages of using telescopes that detect radio waves from space

6-7 The advantages of placing telescopes in Earth orbit

There is literally more to the universe than meets the eye. As seen through a small telescope, the galaxy M82 (shown in the accompanying image) appears as a bright patch that glows with the light of its billions of stars. But when observed with a telescope sensitive to infrared light—with wavelengths longer than your eye can detect—M82 displays an immense halo that extends for tens of thousands of light-years.

The spectrum of this halo reveals it to be composed of tiny dust particles, which are ejected by newly formed stars. The halo's tremendous size shows that new stars are forming in M82 at a far greater rate than within our own Galaxy.

These observations are just one example of the tremendous importance of telescopes to astronomy. Whether a telescope detects visible or nonvisible light, its fundamental purpose is the same—to gather more light than the unaided human eye. Telescopes are used to gather the feeble light from distant objects to make bright, sharp images. Telescopes also produce finely detailed spectra of objects in space. These spectra reveal the chemical compositions of nearby planets as well as of distant galaxies and help astronomers understand the nature and evolution of astronomical objects of all kinds.

As the accompanying images of the galaxy M82 show, telescopes for nonvisible light reveal otherwise hidden aspects of the universe. In addition to infrared telescopes, radio telescopes have mapped out the structure of our Milky Way Galaxy; ultraviolet telescopes have revealed the workings of the Sun's outer atmosphere; and gamma-ray telescopes have detected the most powerful explosions in the universe. The telescope, in all its variations, is by far astronomers' most useful tool for collecting data about the universe.

6-1 A refracting telescope uses a lens to concentrate incoming light at a focus

Refracting telescopes gave humans the first close-up views of the Moon and planets

The **optical telescope**—that is, a telescope designed for use with visible light—was invented in the Netherlands in the early seventeenth century. Soon after, Galileo used one of these new inventions for his groundbreaking astronomical observations (see Section 4-5). These first telescopes used carefully shaped pieces of glass, or **lenses,** to make distant objects appear larger and brighter. Telescopes of this same basic design are used today by many amateur astronomers. To understand telescopes of this kind, we need to understand how lenses work.

Refraction of Light

Lenses used in telescopes gather and focus light. Larger lenses can gather more light, which is why professional telescopes must be big. Focusing light requires that the light's path be bent. The ability of a lens to bend light is based on a universal fact: *Light travels at a slower speed in a dense substance.*

Thus, although the speed of light *in a vacuum* is 3.00×10^8 m/s, its speed in glass is less than 2×10^8 m/s. Just as a woman's walking pace slows suddenly when she walks from a boardwalk onto a sandy beach, so light slows abruptly as it enters a piece of glass. The same woman easily resumes her original pace when she steps back onto the boardwalk; in the same way, light resumes its original speed upon exiting the glass.

A material through which light travels is called a **medium** (*plural* **media**). As a beam of light passes from one transparent medium into another—say, from air into glass, or from glass back into air—the direction of the light can change. This path-bending phenomenon, called **refraction,** is caused by the change in the speed of light.

ANALOGY Imagine driving a car from a smooth pavement onto a sandy beach (Figure 6-1a). If the car approaches the beach head-on, it slows down when it enters the sand but keeps moving straight ahead. If the car approaches the beach at an angle, however, one of its front wheels will be slowed by the sand before the other is, and the car will veer from its original direction. In the same way, a beam of light changes direction when it enters a piece of glass at an angle (Figure 6-1b).

Figure 6-2a shows the refraction of a beam of light passing through a piece of flat glass. As the beam enters the upper surface of the glass, refraction takes place, and the beam is bent to a direction more nearly perpendicular to the surface of the glass. As the beam exits from the glass back into the surrounding air, a second refraction takes place, and the beam bends in the opposite sense.

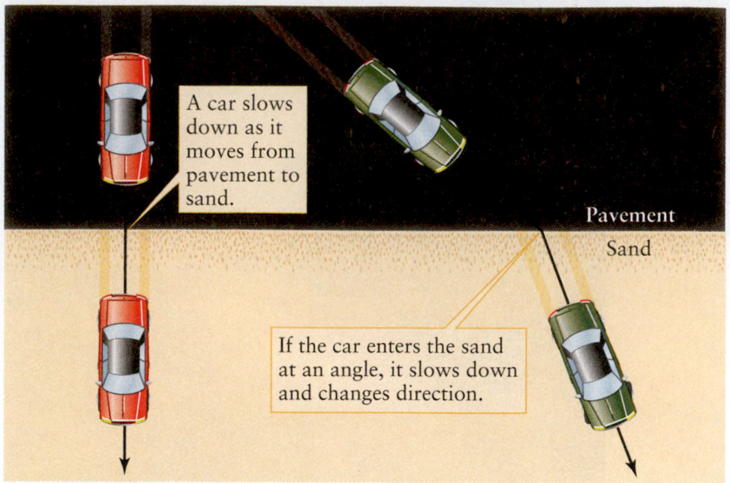

(a) How cars behave

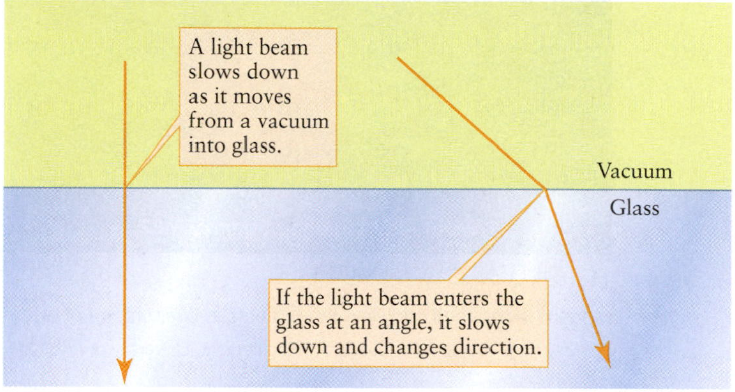

(b) How light beams behave

FIGURE 6-1
Refraction **(a)** When a car drives from smooth pavement into soft sand, it slows down. If it enters the sand at an angle, the front wheel on one side feels the drag of the sand before the other wheel, causing the car to veer to the side and change direction. **(b)** Similarly, light slows down when it passes from a vacuum into glass and changes direction if it enters the glass at an angle.

(The amount of bending depends on the speed of light in the glass, so different kinds of glass produce slightly different amounts of refraction.) Because the two surfaces of the glass are parallel, the beam emerges from the glass traveling in the same direction at which it entered.

Lenses and Refracting Telescopes

Something more useful happens if the glass is curved into a convex shape (one that is fatter in the middle than at the edges), like the lens in Figure 6-2b. When a beam of incoming light rays passes through the lens, refraction causes all the outgoing rays to converge at a point called the **focus.** If the light rays entering the lens are all parallel, the focus occurs at a special point called the **focal point.** The distance from the lens to the focal point is called the **focal length** of the lens.

The case of incoming parallel light rays, shown in Figure 6-2b, is not merely a theoretical ideal. The stars are so far away that light rays from them are essentially parallel, as Figure 6-3 shows. Consequently, a lens always focuses light from an astronomical object to the focal point. If the object has a very small angular

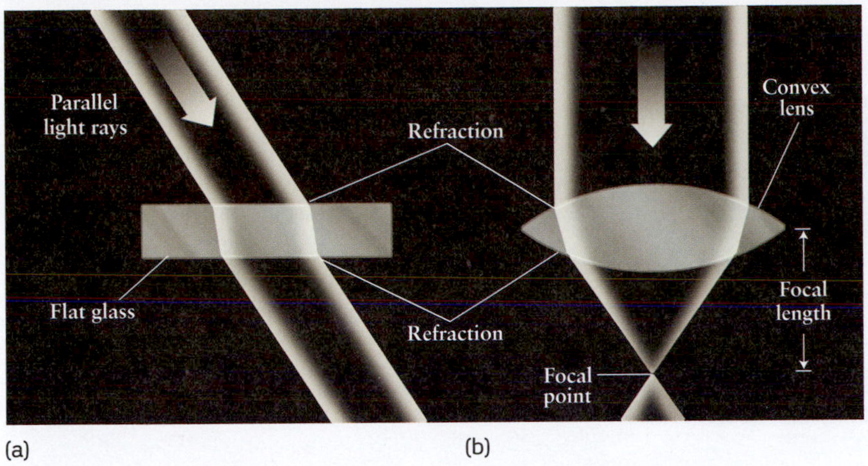

FIGURE 6-2
Refraction and Lenses **(a)** Refraction is the change in direction of a light ray when it passes into or out of a transparent medium such as glass. When light rays pass through a flat piece of glass, the two refractions bend the rays in opposite directions. There is no overall change in the direction in which the light travels. **(b)** If the glass is in the shape of a convex lens, parallel light rays converge to a focus at a special point called the focal point. The distance from the lens to the focal point is called the focal length of the lens.

size, like a distant star, all the light entering the lens from that object converges onto the focal point. The resulting image is just a single bright dot.

However, if the object is *extended*—that is, if it has a relatively large angular size, like the Moon or a nearby planet—then light coming from each point on the object is brought to a focus at its own individual point. The result is a clearly focused and extended image that lies in the **focal plane** of the lens (Figure 6-4), which is a plane that includes the focal point. You can use an ordinary magnifying glass in this way to make an image of the Sun on the ground.

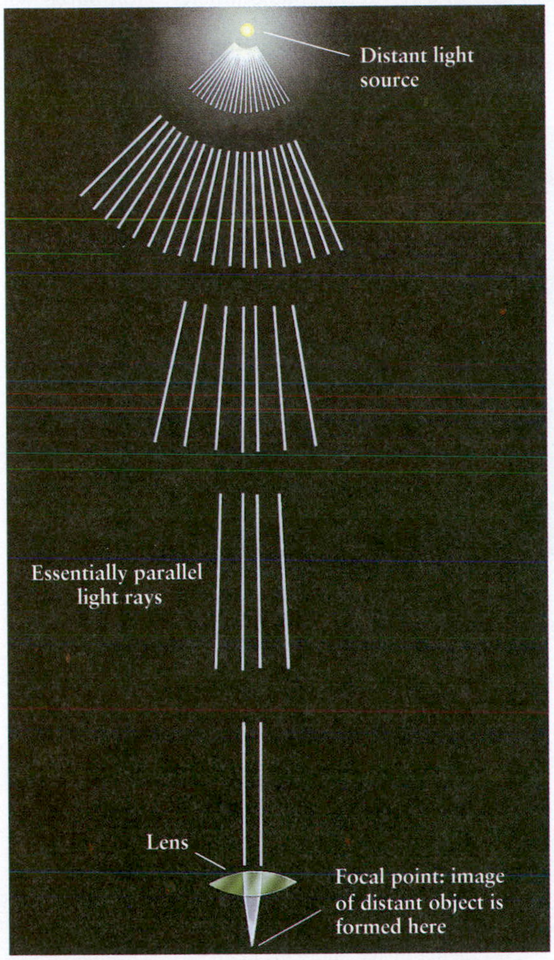

FIGURE 6-3
Light Rays from Distant Objects Are Parallel Light rays travel away in all directions from an ordinary light source. If a lens is located very far from the light source, only a few of the light rays will enter the lens, and these rays will be essentially parallel. This observation is why we drew parallel rays entering the lens in Figure 6-2b.

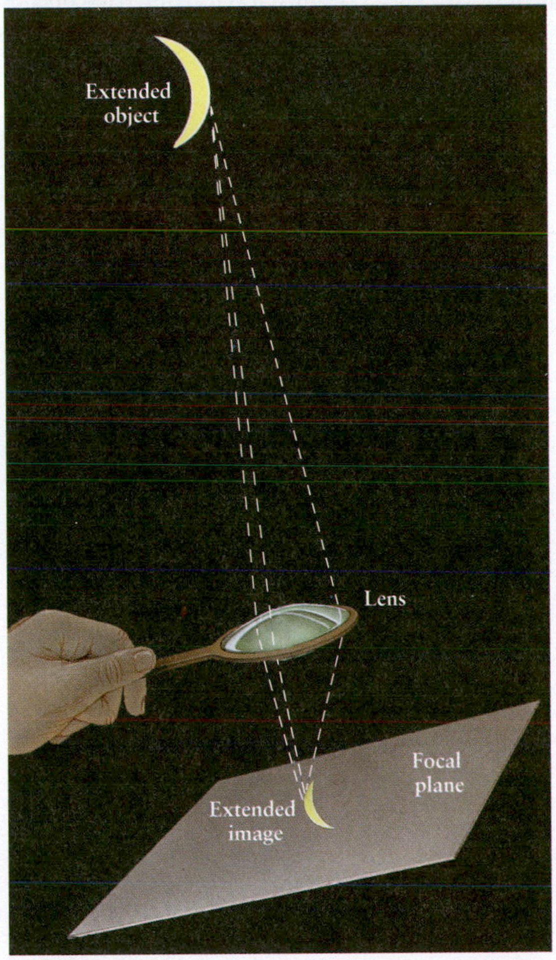

FIGURE 6-4
A Lens Creates an Extended Image of an Extended Object Light coming from each point on an extended object passes through a lens and produces an image of that point. All of these tiny images put together make an extended image of the entire object. The image of a very distant object is formed in a plane called the focal plane. The distance from the lens to the focal plane is called the focal length.

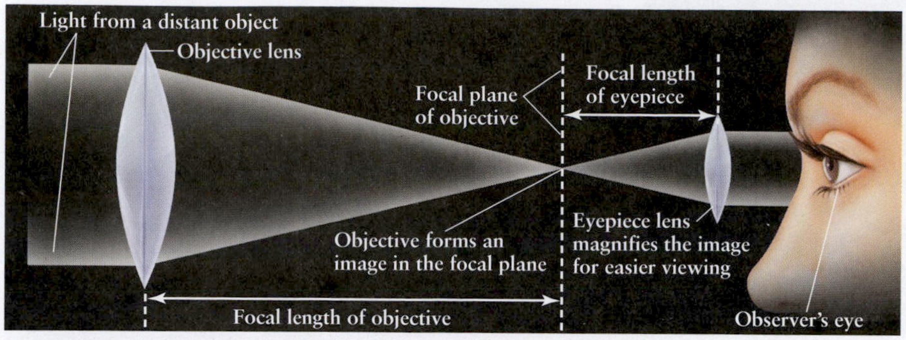

ANIMATION 6.1 **FIGURE 6-5**

A Refracting Telescope A refracting telescope consists of a large-diameter objective lens with a long focal length and a small eyepiece lens of short focal length. The eyepiece lens magnifies the image formed by the objective lens in its focal plane (shown as a dashed line). To take a photograph, the eyepiece is removed and an electronic detector is placed in the focal plane.

To use a lens to make a permanent picture of an astronomical object, you would place an electronic detector in the focal plane—the same type of detector used in digital cameras. In fact, an ordinary digital camera works in a very similar way for photographing objects here on Earth. However, many amateur astronomers want to view the image with their eye, not a camera, and so they add a second lens to magnify the image formed in the focal plane. Such an arrangement of two lenses is called a **refracting telescope,** or **refractor** (Figure 6-5). The large-diameter, long-focal-length lens at the front of the telescope, called the **objective lens,** forms the image; the smaller, shorter-focal-length lens at the rear of the telescope, called the **eyepiece lens,** magnifies the image for the observer.

Light-Gathering Power

In addition to the focal length, the other important dimension of the objective lens of a refractor is its diameter. Compared with a small-diameter lens, a large-diameter lens captures more light, produces

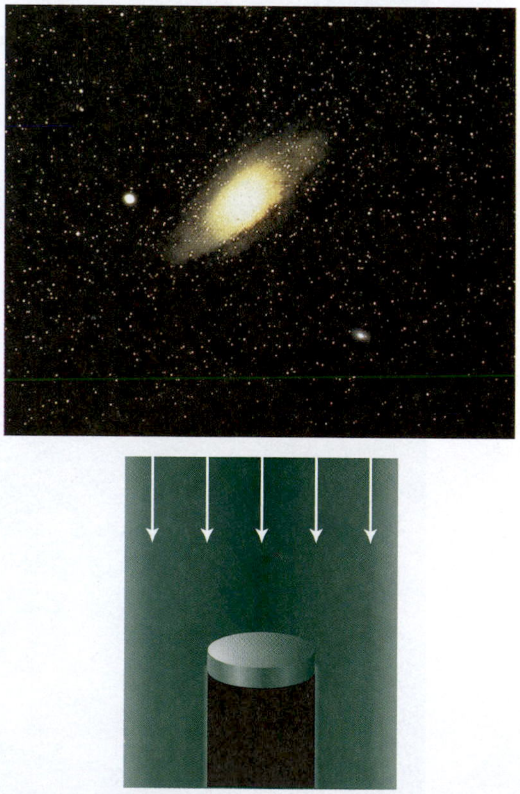

FIGURE 6-6 R I V U X G

Light-Gathering Power These two photographs of the galaxy M31 in Andromeda were taken using the same exposure time and at the same magnification, but with two different telescopes with objective lenses of different diameters. The right-hand photograph is brighter and shows more detail because it was made using the large-diameter lens, which intercepts more starlight than a small-diameter lens. This same principle applies to telescopes that use curved mirrors rather than lenses to collect light (see Section 6-2). (Association of Universities for Research in Astronomy)

BOX 6-1 TOOLS OF THE ASTRONOMER'S TRADE

Magnification and Light-Gathering Power

The magnification of a telescope is equal to the focal length of the objective divided by the focal length of the eyepiece. Telescopic eyepieces are usually interchangeable, so the magnification of a telescope can be changed by using eyepieces of different focal lengths.

EXAMPLE: A small refracting telescope has an objective of focal length 120 cm. If the eyepiece has a focal length of 4.0 cm, what is the magnification of the telescope?

Situation: We are given the focal lengths of the telescope's objective and eyepiece lenses. Our goal is to calculate the magnification provided by this combination of lenses.

Tools: We use the relationship that the magnification equals the focal length of the objective (120 cm) divided by the focal length of the eyepiece (4.0 cm).

Answer: Using this relationship,

$$\text{Magnification} = \frac{120\ \text{cm}}{4.0\ \text{cm}} = 30 \text{ (usually written as } 30\times\text{)}$$

Review: A magnification of 303 means that as viewed through this telescope, a large lunar crater that subtends an angle of 1 arcminute to the naked eye will appear to subtend an angle 30 times greater, or 30 arcminutes (one-half of a degree). This magnification makes the details of the crater much easier to see.

If a 2.0-cm-focal-length eyepiece is used instead, the magnification will be (120 cm)/(2.0 cm) = 60×. The shorter the focal length of the eyepiece, the greater the magnification.

The light-gathering power of a telescope depends on the diameter of the objective lens; it does not depend on the focal length. The light-gathering power is proportional to the square of the diameter. As an example, a fully dark adapted human eye has a pupil diameter of about 5 mm. By comparison, a small telescope whose objective lens is 5 cm in diameter has 10 times the diameter and $10^2 = 100$ times the light-gathering power of the eye. (Recall that there are 10 mm in 1 cm.) Hence, this telescope allows you to see objects 100 times fainter than you can see without a telescope.

EXAMPLE: The same relationships apply to reflecting telescopes, discussed in Section 6-2. Each of the two Keck telescopes on Mauna Kea in Hawaii (discussed in Section 6-3; see Figure 6-16) uses a concave mirror 10 m in diameter to bring starlight to a focus. How many times greater is the light-gathering power of either Keck telescope compared to that of the human eye?

Situation: We are given the diameters of the pupil of the human eye (5 mm) and of the mirror of either Keck telescope (10 m). Our goal is to compare the light-gathering powers of these two optical instruments.

Tools: We use the relationship that light-gathering power is proportional to the square of the diameter of the area that gathers light. Hence, the *ratio* of the light-gathering powers is equal to the square of the ratio of the diameters.

Answer: We first calculate the ratio of the diameter of the Keck mirror to the diameter of the pupil. To determine this ratio, we must first express both diameters in the same units. Because there are 1000 mm in 1 meter, the diameter of the Keck mirror can be expressed as

$$10\ \text{m} \times \frac{1000\ \text{mm}}{1\ \text{m}} = 10{,}000\ \text{mm}$$

Thus, the light-gathering power of either of the Keck telescopes is greater than that of the human eye by a factor of

$$\frac{(10{,}000\ \text{mm})^2}{(5\ \text{mm})^2} = (2000)^2 = 4 \times 10^6 = 4{,}000{,}000$$

Review: Either Keck telescope can gather *4 million* times as much light as a human eye. When it comes to light-gathering power, the bigger the telescope, the better!

brighter images, and allows astronomers to detect fainter objects. (For the same reason, the iris of your eye opens when you go into a darkened room to allow you to see dimly lit objects.)

The **light-gathering power** of a telescope is directly proportional to the area of the objective lens, which in turn is proportional to the square of the lens diameter (Figure 6-6). Thus, if you double the diameter of the lens, the light-gathering power increases by a factor of $2^2 = 2 \times 2 = 4$. Box 6-1 describes how to compare the light-gathering power of different telescopes.

Because light-gathering power is so important for seeing faint objects, the lens diameter is almost always given when describing a telescope. For example, the Lick telescope on Mount Hamilton in California is a 90-cm refractor, which means that it is a refracting telescope whose objective lens is 90 cm in diameter. By comparison, Galileo's telescope of 1610 was a 3-cm refractor. The Lick telescope has an objective lens 30 times larger in diameter, and so has 30^2 or $30 \times 30 = 900$ times the light-gathering power of Galileo's instrument.

CONCEPTCHECK 6-1

If someone says they are using an 8-inch telescope, which dimension of the telescope—length or tube diameter—are they most likely referring to?

CONCEPTCHECK 6-2

If a thick lens is able to bend light more than a thin lens, which lens has a greater focal length?

Answers appear at the end of the chapter.

Magnification

In addition to their light-gathering power, telescopes are useful because they magnify distant objects. As an example, the angular diameter of the Moon as viewed with the naked eye is about 0.5°. But when Galileo viewed the Moon through his telescope, its apparent angular diameter was 10°, large enough so that he could identify craters and mountain ranges. The **magnification,** or **magnifying power,** of a telescope is the ratio of an object's angular diameter seen through the telescope to its naked-eye angular diameter. Thus, the magnification of Galileo's telescope was 10°/0.5° = 20 times, usually written as 20×.

The magnification of a refracting telescope depends on the focal lengths of both of its lenses:

$$\text{Magnification} = \frac{\text{focal length of objective lens}}{\text{focal length of eyepiece lens}}$$

This formula shows that using a long-focal-length objective lens with a short-focal-length eyepiece gives a large magnification. Box 6-1 illustrates how this formula is used.

CAUTION! Many people think that the primary purpose of a telescope is to magnify images. But in fact magnification is not the most important aspect of a telescope. The reason is that there is a limit to how sharp any astronomical image can be, due either to the blurring caused by Earth's atmosphere or to fundamental limitations imposed by the nature of light itself. (We will describe these effects in more detail in Section 6-3.) Magnifying a blurred image may make it look bigger but will not make it any clearer. Thus, beyond a certain point, there is nothing to be gained by further magnification. Astronomers place more priority on the light-gathering power of a telescope than in its magnification. Greater light-gathering power means brighter images, which makes it easier to see faint details.

Disadvantages of Refracting Telescopes

If you were to build a telescope like that in Figure 6-5 using only the instructions given so far, you would probably be disappointed with the results. The problem is that a lens bends different colors of light through different angles, just as a prism does (recall Figure 5-3). As a result, different colors do not focus at the same point, and stars viewed through a telescope that uses a simple lens are surrounded by fuzzy, rainbow-colored halos. Figure 6-7a shows this optical defect, called **chromatic aberration.**

One way to correct for chromatic aberration is to use an objective lens that is not just a single piece of glass. Different types of glass can be manufactured by adding small amounts of chemicals to the glass when it is molten. Because of the different properties of these added chemicals, the speed of light varies slightly from one kind of glass to another, and the refractive properties vary as well. If a thin lens is mounted just behind the main objective lens of a telescope, as shown in Figure 6-7b, and if the telescope

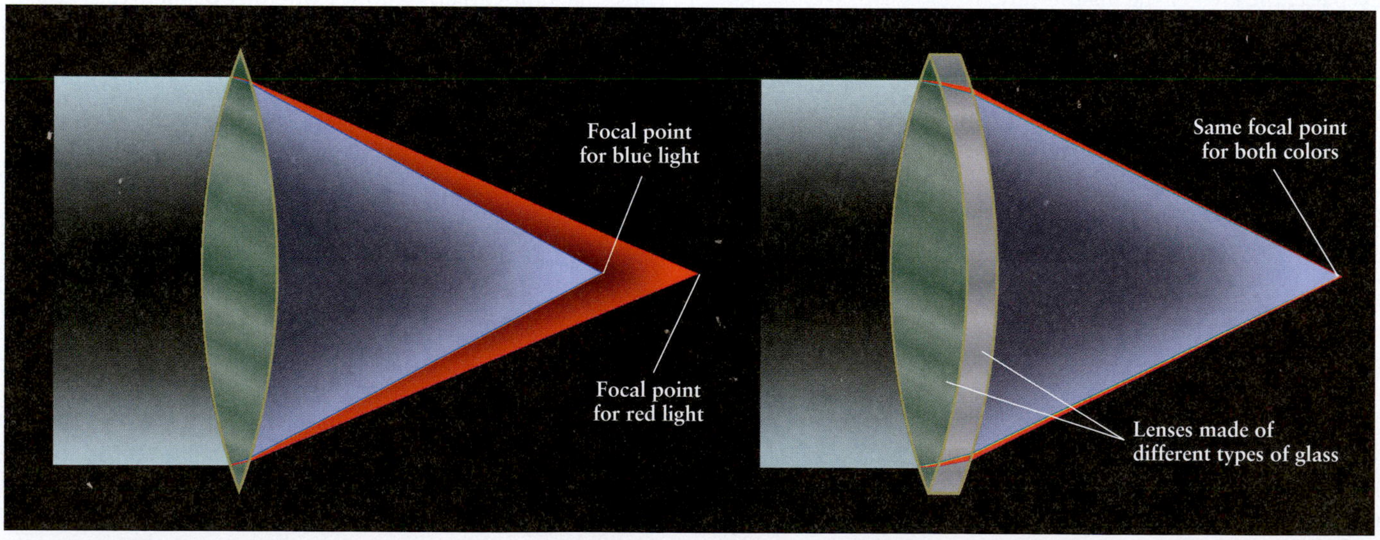

(a) The problem: chromatic aberration

(b) The solution: use two lenses

FIGURE 6-7
Chromatic Aberration **(a)** A single lens suffers from a defect called chromatic aberration, in which different colors of light are brought to a focus at different distances from the lens. While the aberration occurs for all colors, only red and blue are shown here. **(b)** This problem can be corrected by adding a second lens made from a different kind of glass.

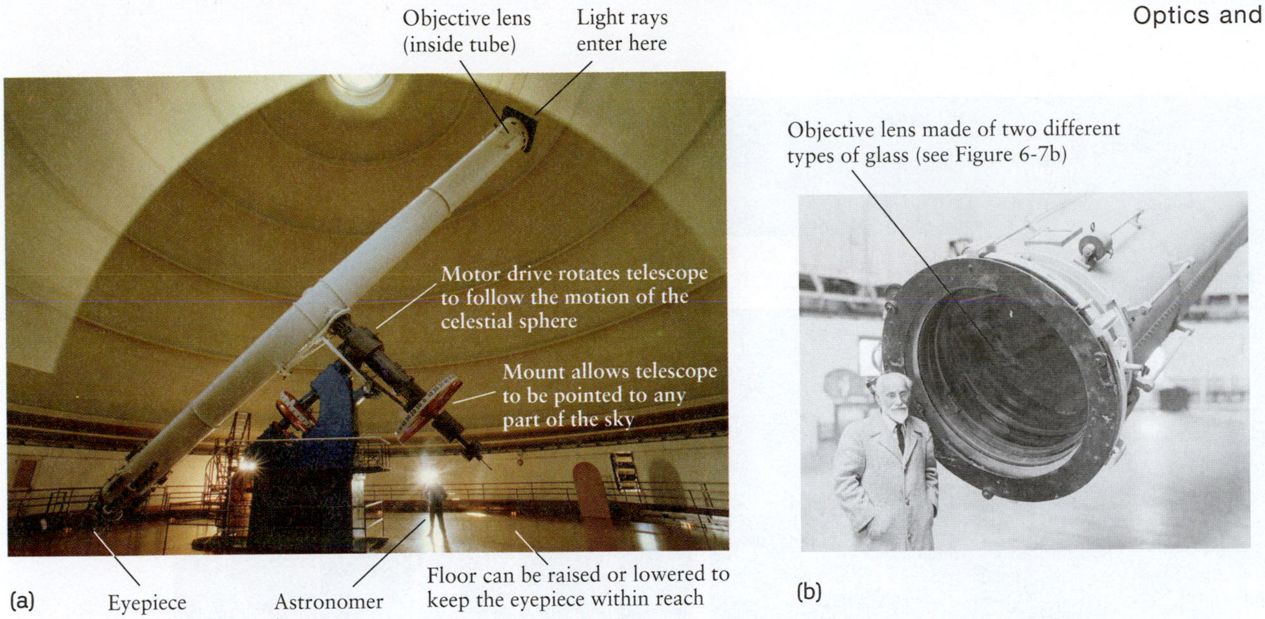
Objective lens (inside tube)
Light rays enter here
Motor drive rotates telescope to follow the motion of the celestial sphere
Mount allows telescope to be pointed to any part of the sky
(a)
Eyepiece
Astronomer
Floor can be raised or lowered to keep the eyepiece within reach
Objective lens made of two different types of glass (see Figure 6-7b)
(b)

FIGURE 6-8 R I V U X G

A Large Refracting Telescope **(a)** This giant refractor, built in 1897, is housed at Yerkes Observatory near Chicago. The telescope tube is 19.5 m (64 ft) long; it has to be this long because the focal length of the objective is just under 19.5 m (see Figure 6-5). As Earth rotates, the motor drive rotates the telescope in the opposite direction in order to keep the object being studied within the telescope's field of view. **(b)** This historical photograph shows the astronomer George van Biesbrock with the objective lens of the Yerkes refractor. This lens, the largest refracting lens still in use, is 102 cm (40 in.) in diameter. (a: Roger Ressmeyer/Corbis; b: Yerkes Observatory)

designer carefully chooses two different kinds of glass for these two lenses, different colors of light can be brought to a focus at the same point.

Chromatic aberration is only the most severe of a host of optical problems that must be solved in designing a high-quality refracting telescope. Master opticians of the nineteenth century devoted their careers to solving these problems, and several magnificent refractors were constructed in the late 1800s (Figure 6-8).

Unfortunately, there are several negative aspects of refractors that even the finest optician cannot overcome:

1. Because faint light must readily pass through the objective lens, the glass from which the lens is made must be totally free of defects, such as the bubbles that frequently form when molten glass is poured into a mold. Such defect-free glass is extremely expensive.

2. Glass is opaque to certain kinds of light. Ultraviolet light is absorbed almost completely, and even visible light is dimmed substantially as it passes through the thick slab of glass that makes up the objective lens.

3. It is impossible to produce a large lens that is entirely free of chromatic aberration.

4. Because a lens can be supported only around its edges, a large lens tends to sag and distort under its own weight as it tracks objects through the night. This distortion has negative effects on the image clarity.

For these reasons and more, few major refractors have been built since the beginning of the twentieth century. Instead, astronomers have avoided all of the limitations of refractors by building telescopes that use mirrors instead of lenses to form images.

CONCEPTCHECK 6-3

How do the eyepieces with the largest focal length affect a telescope's overall magnification?

Answer appears at the end of the chapter.

6-2 A reflecting telescope uses a mirror to concentrate incoming light at a focus

> All of the largest professional telescopes—and most amateur telescopes—are reflecting telescopes

Almost all modern telescopes form an image using the principle of **reflection.** To understand reflection, imagine drawing a dashed line perpendicular to the surface of a flat mirror at the point where a light ray strikes the mirror (Figure 6-9a). The angle i between the *incident* (arriving) light ray and the perpendicular is always equal to the angle r between the *reflected* ray and the perpendicular.

In 1663, the Scottish mathematician James Gregory first proposed a telescope using reflection from a concave mirror—one that is fatter at the edges than at the middle. Such a mirror makes parallel light rays converge to a focus (Figure 6-9b). The distance between the reflecting surface and the focus is the focal length of the mirror. A telescope that uses a curved mirror to make an image of a distant object is called a **reflecting telescope,** or **reflector.** Using terminology similar to that used for refractors, the mirror that forms the image is called the **objective mirror** or **primary mirror.**

To make a reflector mirror, an optician grinds and polishes a large slab of glass into the appropriate concave shape. The glass is then coated with silver, aluminum, or a similar highly reflective

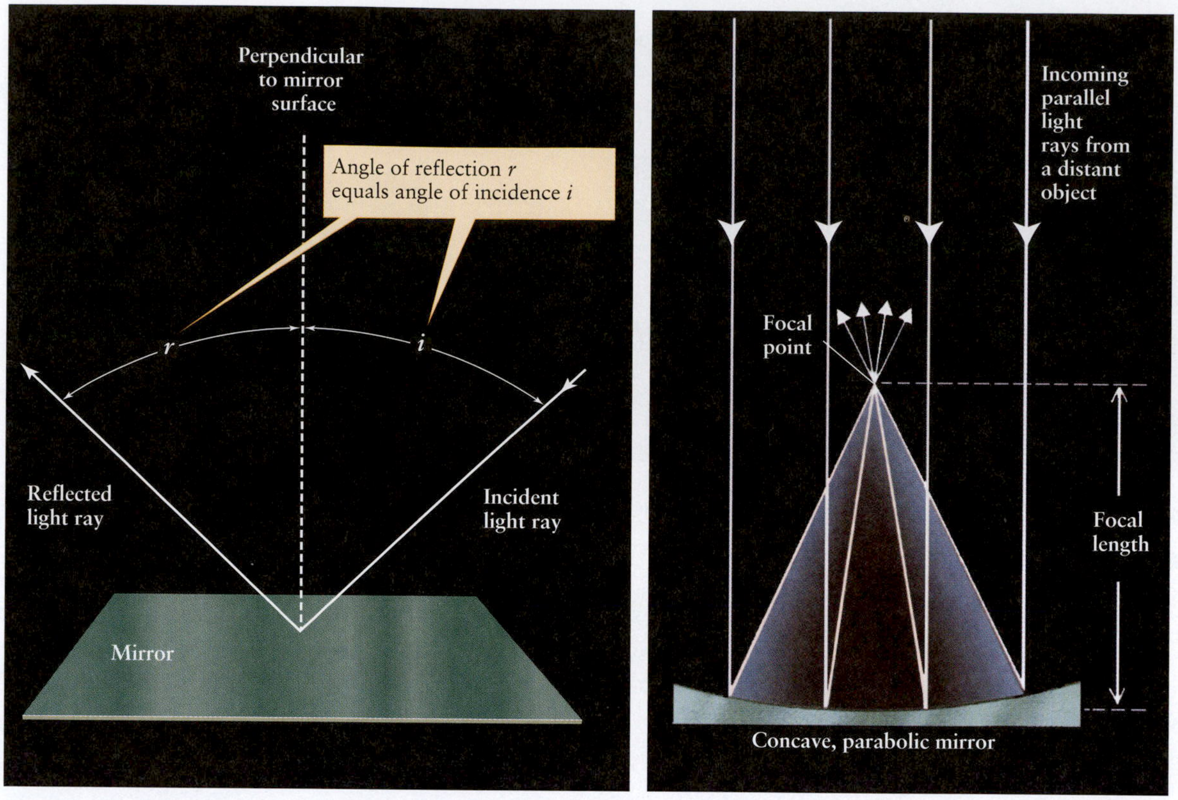

(a)

(b)

FIGURE 6-9

Reflection **(a)** The angle at which a beam of light approaches a mirror, called the angle of incidence (*i*), is always equal to the angle at which the beam is reflected from the mirror, called the angle of reflection (*r*). **(b)** A concave mirror causes parallel light rays to converge to a focus at the focal point. The distance between the mirror and the focal point is the focal length of the mirror.

substance. Because light reflects off the surface of the coated glass rather than passing through it, defects within the glass—which would have very negative consequences for the objective lens of a refracting telescope—have no effect on the optical quality of a reflecting telescope.

Another advantage of reflector mirrors is that they do not suffer from the chromatic aberration that plagues refractors. This is because reflection is not affected by the wavelength of the incoming light (only the angle it hits the mirror), so all wavelengths are reflected to the same focus. (A small amount of chromatic aberration may arise if the image is viewed using an eyepiece lens.) Furthermore, the mirror can be fully supported by braces on its back, so that a large, heavy mirror can be mounted and well-supported without its shape warping much from gravity.

Designs for Reflecting Telescopes

Although a reflecting telescope has many advantages over a refractor, the arrangement shown in Figure 6-9a is not ideal. One problem is that the focal point is in front of the objective mirror. If you try to view the image formed at the focal point, your head will block part or all of the light from reaching the mirror.

To get around this problem, in 1668 Isaac Newton simply placed a small, flat mirror at a 45° angle in front of the focal point, as sketched in Figure 6-10a. This secondary mirror deflects the light rays to one side, where Newton placed an eyepiece lens to magnify the image. A reflecting telescope with this optical design is appropriately called a **Newtonian reflector** (Figure 6-10b). The magnifying power of a Newtonian reflector is calculated in the same way as for a refractor: The focal length of the objective mirror is divided by the focal length of the eyepiece (see Box 6-1).

Later astronomers modified Newton's original design. The objective mirrors of some modern reflectors are so large that an astronomer could actually sit in an "observing cage" at the undeflected focal point directly in front of the objective mirror. (In practice, riding in this cage on a winter's night is a remarkably cold and uncomfortable experience.) This arrangement is called a **prime focus** (Figure 6-11a). It usually provides the highest-quality image, because there is no need for a secondary mirror (which might have imperfections).

Another popular optical design, called a **Cassegrain focus** after the French contemporary of Newton who first proposed it, also has a convenient, accessible focal point. A hole is drilled directly through the center of the primary mirror, and a convex secondary mirror placed in front of the original focal point reflects the light rays back through the hole (Figure 6-11b).

A fourth design is useful when there is optical equipment too heavy or bulky to mount directly on the telescope. Instead, a series of mirrors channels the light rays away from the telescope to a remote focal point where the equipment is located. This design is

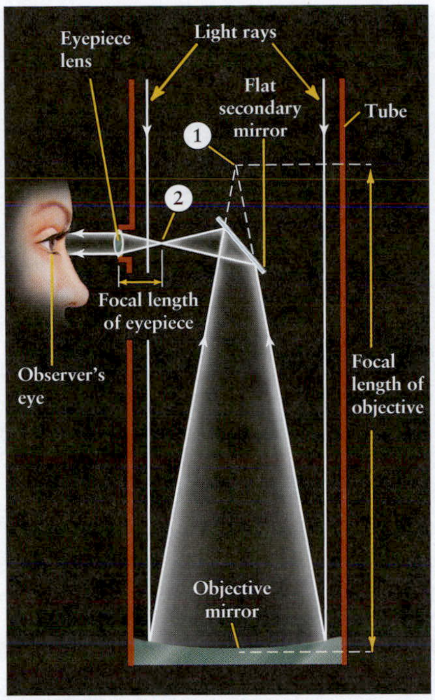

(a)

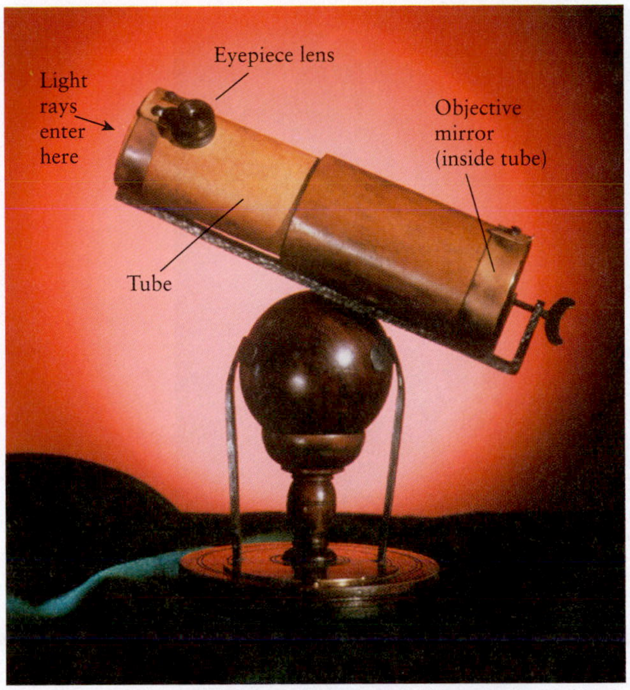

(b)

FIGURE 6-10 R I V U X G

A Newtonian Telescope **(a)** In a Newtonian telescope, the image made by the objective is moved from point 1 to point 2 by means of a flat mirror called the secondary. An eyepiece magnifies this image, just as for a refracting telescope (Figure 6-5). **(b)** This is a replica of a Newtonian telescope built by Isaac Newton in 1672. The objective mirror is 3 cm (1.3 inches) in diameter and the magnification is 403. (Royal Greenwich Observatory/Science Photo Library)

called a **coudé focus,** from a French word meaning "bent like an elbow" (Figure 6-11c).

CAUTION! You might think that the secondary mirror in the Newtonian design and the Cassegrain and coudé designs shown in Figure 6-11 would cause a black spot or hole in the center of the telescope image. But this spot does not form. The reason is that light from every part of the object lands on every part of the primary, objective mirror. Hence, any portion of the mirror can itself produce an image of the distant object, as Figure 6-12 shows. The only effect of the secondary mirror is that it prevents part of the light from reaching the objective mirror, which reduces somewhat the light-gathering power of the telescope.

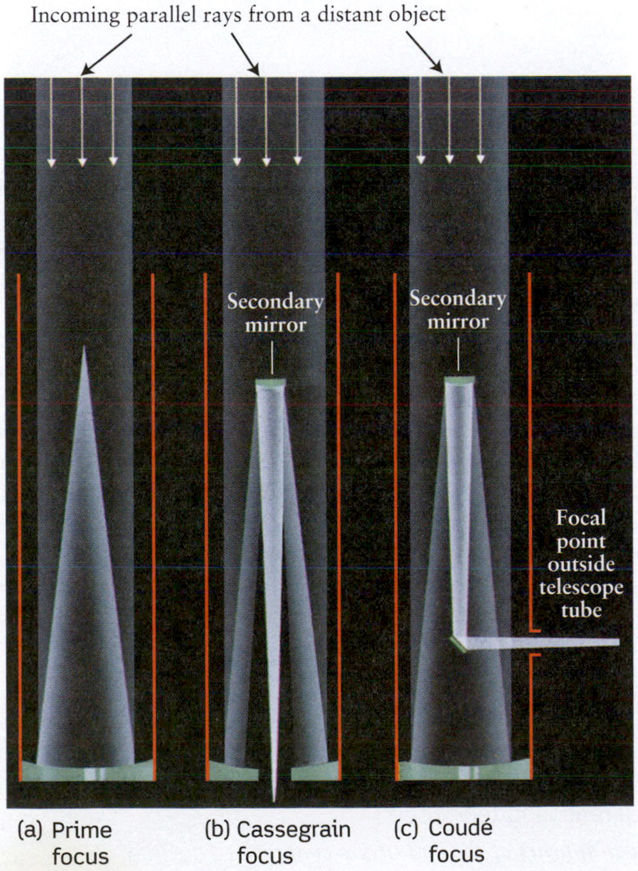

ANIMATION 6-2

FIGURE 6-11

Designs for Reflecting Telescopes Three common optical designs for reflecting telescopes are shown here. **(a)** Prime focus is used only on some large telescopes; an observer or instrument is placed directly at the focal point, within the barrel of the telescope. **(b)** The Cassegrain focus is used on reflecting telescopes of all sizes, from 90-mm (3.5-in.) reflectors used by amateur astronomers to giant research telescopes on Mauna Kea (see Figure 6-14). **(c)** The coudé focus is useful when large and heavy optical apparatus is to be used at the focal point. Light reflects off the objective mirror to a secondary mirror, then back down to a third, angled mirror. With this arrangement, the heavy apparatus at the focal point does not have to move when the telescope is repositioned.

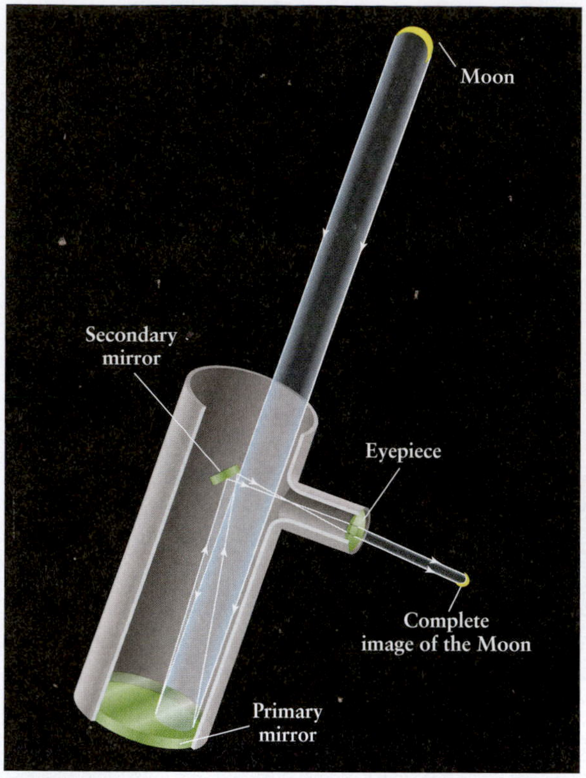

FIGURE 6-12

The Secondary Mirror Does Not Cause a Hole in the Image This illustration shows how even a small portion of the primary (objective) mirror of a reflecting telescope can make a complete image of the Moon. Thus, the secondary mirror does not cause a black spot or hole in the image. (It does, however, make the image a bit dimmer by reducing the total amount of light that reaches the primary mirror.)

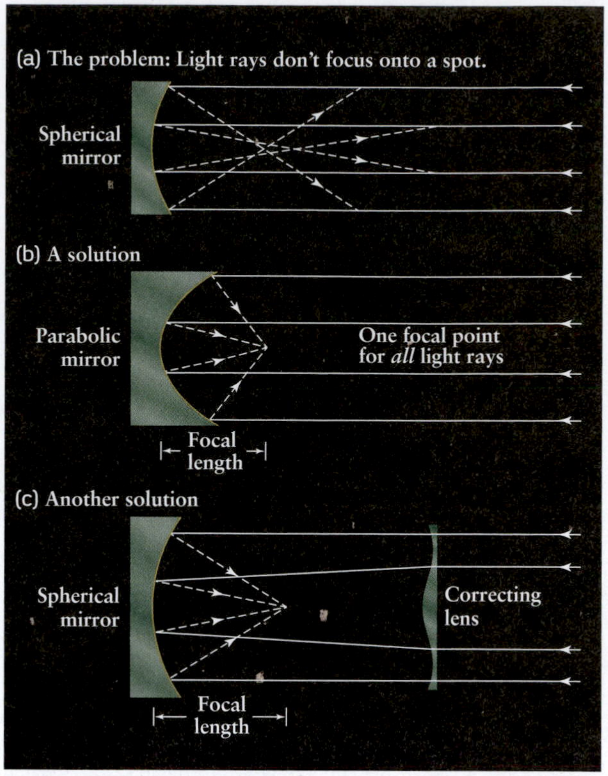

FIGURE 6-13

Spherical Aberration **(a)** Different parts of a spherically concave mirror reflect light to slightly different points. This effect, called spherical aberration, causes image blurring. This difficulty can be corrected by either **(b)** using a parabolic mirror or **(c)** using a correcting lens in front of the mirror.

Spherical Aberration

A reflecting telescope must be designed to minimize a defect called **spherical aberration** (Figure 6-13). At issue is the precise shape of a mirror's concave surface. A spherical surface is easy to grind and polish, but different parts of a spherical mirror focus light onto slightly different spots (Figure 6-13a). This results in a fuzzy image.

One common way to eliminate spherical aberration is to polish the mirror's surface to a parabolic shape, because a parabola reflects parallel light rays to a common focus (Figure 6-13b). Unfortunately, the astronomer then no longer has a wide-angle view. Furthermore, unlike spherical mirrors, parabolic mirrors suffer from a defect called **coma,** wherein star images far from the center of the field of view are elongated to look like tiny teardrops. A different approach is to use a spherical mirror, thus minimizing coma, and to place a thin correcting lens at the front of the telescope to eliminate spherical aberration (Figure 6-13c). This approach is only used on relatively small reflecting telescopes for amateur astronomers.

The Largest Reflectors

There are over a dozen optical reflectors in operation with primary mirrors between 8 meters (26.2 feet) and 11 meters (36.1 feet) in diameter. Figure 6-14a shows the objective mirror of one of the four Very Large Telescope (VLT) units in Chile, and Figure 6-14b shows the objective and secondary mirrors of the Gemini North telescope in Hawaii. A near-twin of Gemini North, called Gemini South, is in Cerro Pachón, Chile. These twins allow astronomers to observe both the northern and southern parts of the celestial sphere with essentially the same state-of-the-art instrument. Two other "twins" are the side-by-side 8.4-m objective mirrors of the Large Binocular Telescope in Arizona. Combining the light from these two mirrors gives double the light-gathering power, equivalent to a single 11.8-m mirror.

Several other reflectors around the world have objective mirrors between 3 and 6 meters in diameter, and dozens of smaller but still powerful telescopes have mirrors in the range of 1 to 3 meters. There are thousands of professional astronomers, each of whom has several ongoing research projects, and thus the demand for all of these telescopes is high. On any night of the year, nearly every research telescope in the world is being used to explore the universe.

CONCEPTCHECK 6-4

Does the angle of light reflected off a mirrored surface depend on wavelength? How does the answer to this question help or hurt a reflecting telescope?

Answer appears at the end of the chapter.

(a) A large objective mirror

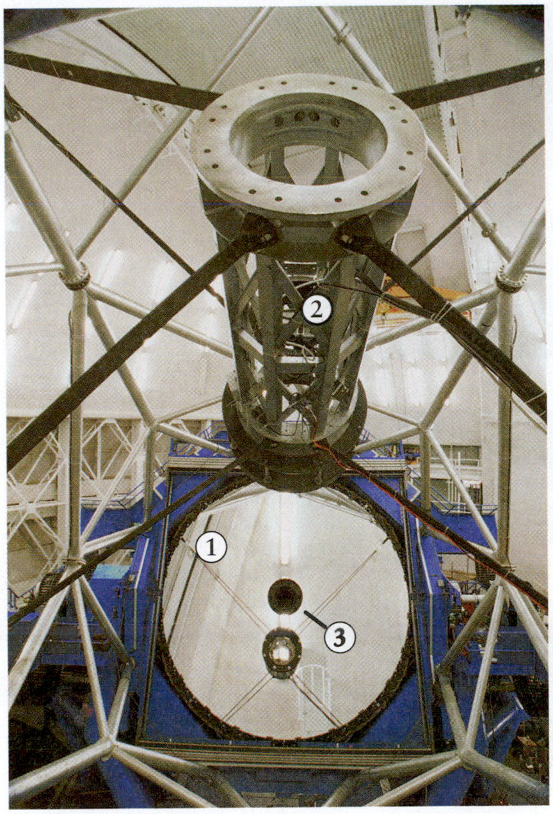

(b) A large Cassegrain telescope

FIGURE 6-14 R I V U X G

Reflecting Telescopes **(a)** This photograph shows technicians preparing an objective mirror 8.2 meters in diameter for the European Southern Observatory in Chile. The mirror was ground to a curved shape with a remarkable precision of 8.5 nanometers. **(b)** This view of the Gemini North telescope shows its 8.1-meter objective mirror (1). Light incident on this mirror is reflected toward the 1.0-meter secondary mirror (2), then through the hole in the objective mirror (3) to the Cassegrain focus (see Figure 6-11b). (a: SAGEM; b: NOAO/AURA/NSF)

6-3 Telescope images are degraded by the blurring effects of the atmosphere and by light pollution

TUTORIAL 6-1

In addition to providing a brighter image, a large telescope also helps achieve a second major goal: It produces star images that are sharp and crisp. A quantity called **angular resolution** gauges how well fine details can be seen. Poor angular resolution causes star images to be fuzzy and blurred together.

To determine the angular resolution of a telescope, pick out two adjacent stars whose separate images are just barely discernible (Figure 6-15). The angle θ (the Greek letter theta) between these stars is the telescope's angular resolution; the *smaller* that angle, the finer the details that can be seen and the sharper the image.

When you are asked to read the letters on an eye chart, what's being measured is the angular resolution of your eye. If you have 20/20 vision, the angular resolution θ of your eye is about 1 arcminute, or 60 arcseconds. (You may want to review the definitions of angles and their units in Section 1-5.) Hence, with the naked eye it is impossible to distinguish two stars less than 1 arcminute apart or to see details on the Moon with an angular size smaller than this. All the planets have angular sizes (as seen from Earth) of 1 arcminute or less, which is why they appear as featureless points of light to the naked eye.

Limits to Angular Resolution

One factor limiting angular resolution is **diffraction,** which is the tendency of light waves to spread out when they are confined to a small area like the lens or mirror of a telescope. (A rough analogy is the way water exiting a garden hose sprays out in a wider angle when you cover part of the end of the hose with your thumb.) As a result of diffraction, a narrow beam of light tends to spread out within a telescope's optics, thus blurring the image. If diffraction were the only limit, the angular resolution of a telescope would be given by the formula

Diffraction-limited angular resolution

$$\theta = 2.5 \times 10^5 \frac{\lambda}{D}$$

θ = diffraction-limited angular resolution of a telescope, in arcseconds

λ = wavelength of light, in meters

D = diameter of telescope objective, in meters

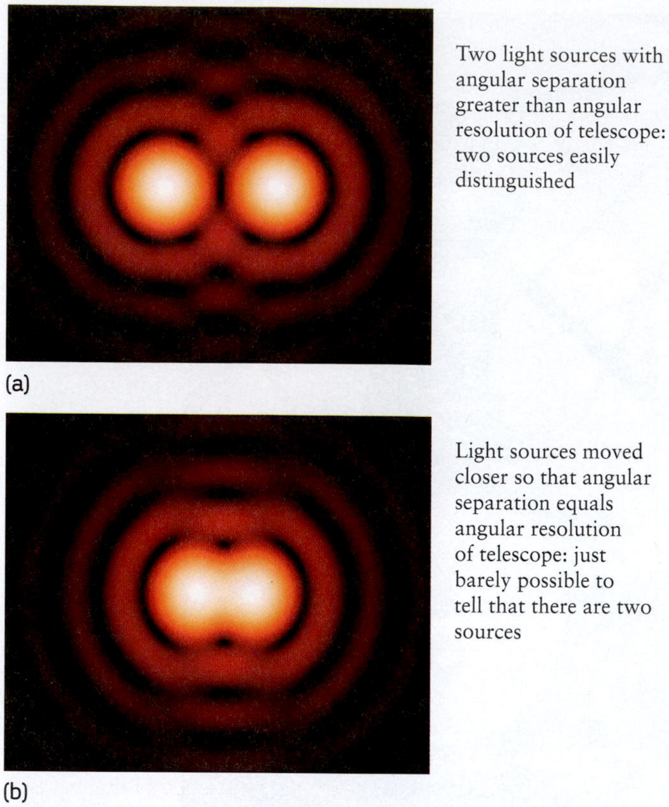

FIGURE 6-15 R I V U X G

Angular Resolution The angular resolution of a telescope indicates the sharpness of the telescope's images. **(a)** This telescope view shows two sources of light whose angular separation is greater than the angular resolution. **(b)** The light sources have been moved together so that their angular separation is equal to the angular resolution. If the sources were moved any closer together, the telescope image would show them as a single source. (Courtesy of John D. Monnier)

For a given wavelength of light, using a telescope with an objective of *larger* diameter *D reduces* the amount of diffraction and makes the angular resolution θ *smaller* (and hence better). For example, with red light with wavelength 640 nm, or 6.4×10^{-7} m, the diffraction-limited resolution of an 8-meter telescope (see Figure 6-15) would be

$$\theta = (2.5 \times 10^5)\,\frac{6.4 \times 10^{-7}\,\text{m}}{8\,\text{m}} = 0.02\ \text{arcsec}$$

In practice, however, ordinary optical telescopes cannot achieve such fine angular resolution. The problem is that turbulence in the air causes star images to jiggle around and twinkle. Even through the largest telescopes, a star still looks like a tiny blob rather than a pinpoint of light. A measure of the limit that atmospheric turbulence places on a telescope's resolution is called the **seeing disk.** This disk is the angular diameter of a star's image broadened by turbulence. The size of the seeing disk varies from one observatory site to another and from one night to another. At the observatories on Kitt Peak in Arizona and Cerro Tololo in Chile, the seeing disk is typically around 1 arcsec. Some of the very best conditions in the world can be found at the observatories atop Mauna Kea in Hawaii, where the seeing disk is often as small as 0.5 arcsec. These great conditions are one reason why so many telescopes have been built there (Figure 6-16).

CONCEPTCHECK 6-5

A small-diameter telescope collects less light than a larger-diameter telescope. If a telescope with a larger-diameter lens or mirror is used for observations of dim, barely illuminated features on Jupiter, does the larger diameter worsen or improve the image resolution compared to the smaller telescope?

Answer appears at the end of the chapter.

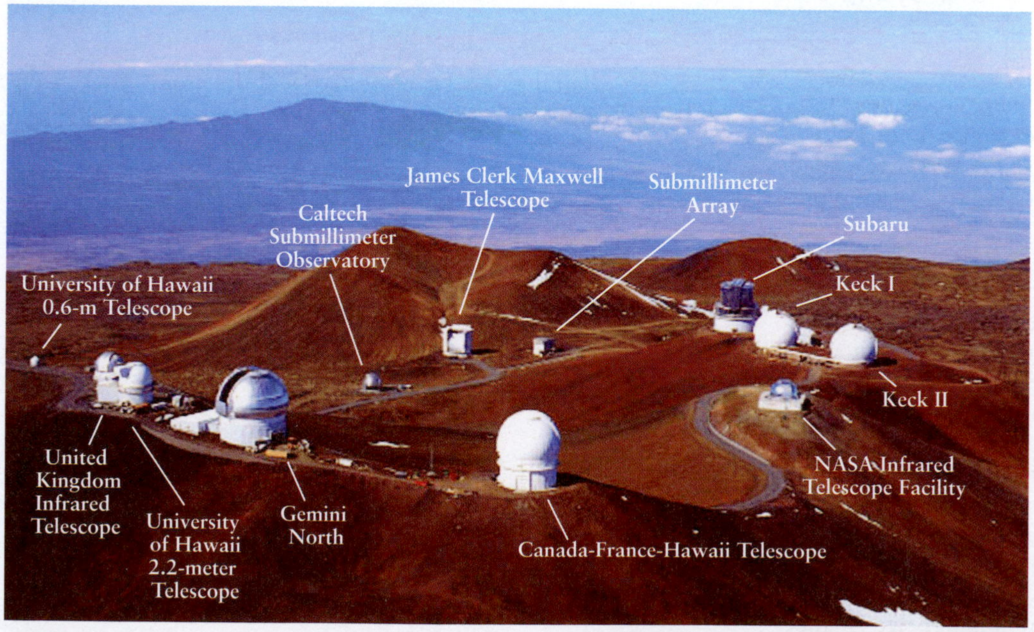

FIGURE 6-16 R I V U X G

The Telescopes of Mauna Kea The summit of Mauna Kea—an extinct Hawaiian volcano that reaches more than 4100 m (13,400 ft) above the waters of the Pacific—has nighttime skies that are unusually clear, still, and dark. To take advantage of these superb viewing conditions, Mauna Kea has become the home of many powerful telescopes. (AP Photo/University of Hawaii Institute of Astronomy, Richard Wainscoat)

Active Optics and Adaptive Optics

In many cases the angular resolution of a telescope is even worse than the limit imposed by the seeing disk. This occurs if the objective mirror deforms even slightly due to variations in air temperature or slight gravitational sagging of the telescope mount. To combat this, many large telescopes are equipped with an **active optics** system. Such a system slowly adjusts the mirror shape every few seconds to help keep the telescope in optimum focus and properly aimed at its target. These adjustments can be determined once during a testing phase for each orientation of the telescope and applied for all future observations. However, the adjustments discussed next are more challenging because they change rapidly throughout each observation.

FIGURE 6-17 R I **V** U X G

Creating an Artificial "Star" A laser beam shines upward from Yepun, an 8.2-meter telescope at the European Southern Observatory in the Atacama Desert of Chile. (Figure 6-14a shows the objective mirror for this telescope.) The beam strikes sodium atoms that lie about 90 km (56 miles) above Earth's surface, causing them to glow and make an artificial "star." Tracking the twinkling of this "star" makes it possible to undo the effects of atmospheric turbulence on telescope images. (European Southern Observatory)

Adaptive optics produces sharper images by "undoing" atmospheric turbulence

Changing the mirror shape is also at the heart of a more refined technique called **adaptive optics.** The goal of this technique is to compensate for atmospheric turbulence, so that the angular resolution can be smaller than the size of the seeing disk and can even approach the theoretical limit set by diffraction. Turbulence causes the image of a star to "dance" around erratically. While this turbulence produces twinkling stars seen with the naked eye, it produces blurry images in large telescopes. In an adaptive optics system, sensors monitor this dancing motion 10 to 100 times per second, and a powerful computer rapidly calculates the mirror shape needed to compensate. Fast-acting mechanical devices called *actuators* then deform the mirror accordingly, to produce a sharp, focused image. In some adaptive optics systems, the actuators deform a small secondary mirror rather than the large objective mirror.

One difficulty with adaptive optics is that a fairly bright star must be in or near the field of the telescope's view to serve as a "calibration target" for the sensors that track atmospheric turbulence. This is seldom the case, since the field of view of most telescopes is rather narrow, lowering the chances of viewing a sufficiently bright star. With no lack of creative problem-solving, astronomers get around this limitation by shining a laser beam toward a spot in the sky near the object to be observed (Figure 6-17). The laser beam causes atoms in Earth's upper atmosphere to glow, making an artificial "star." The light that comes down to Earth from this "star" travels through the same part of our atmosphere as the light from the object being observed, so its image in the telescope will "dance" around in the same erratic way as the image of a real star. By observing how the changing atmosphere distorts the artificial star, computers apply opposing adjustments to cancel the atmosphere's effects on the real star.

Figure 6-18 shows the dramatic improvement in angular resolution possible with adaptive optics. Images made with adaptive optics are nearly as sharp as if the telescope were in the vacuum of space, where there is no atmospheric distortion whatsoever and the only limit on angular resolution is diffraction. A number of large telescopes are now being used with adaptive optics systems.

CAUTION! The images in Figure 6-18 are **false color** images: They do not represent the true color of the stars shown. False color is often used when the image is made using wavelengths that the eye cannot detect, as with the infrared images in Figure 6-18. A different use of false color is to indicate the relative brightness of different parts of the image, as in the infrared image of a person in Figure 5-10. Throughout this book, we'll always point out when false color is used in an image.

Interferometry

Several large observatories are developing a technique called **interferometry** that promises to further improve the angular resolution of telescopes. The idea is to have two widely separated telescopes observe the same object simultaneously, then use fiber optic cables to "pipe" the light signals from each telescope to a central location where they "interfere" or blend together. This method makes the combined signal sharp and clear. The effective resolution of such

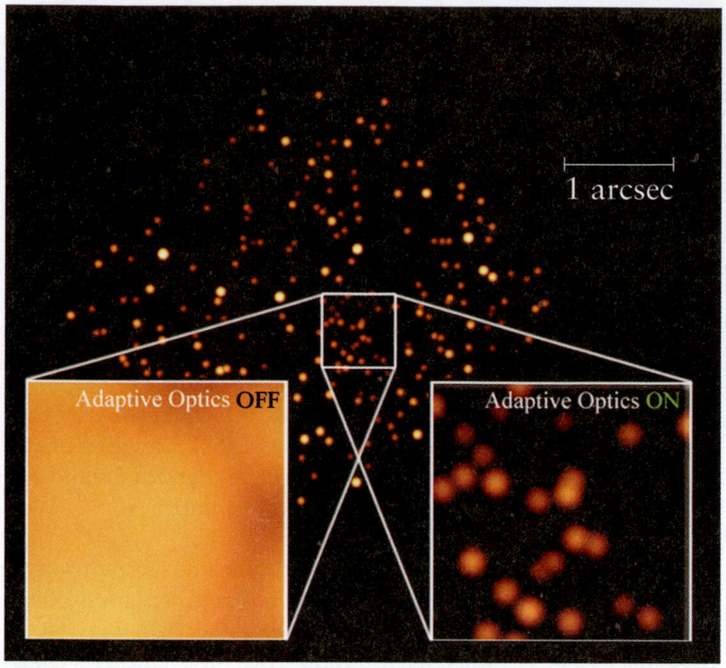

FIGURE 6-18 R I V U X G

Using Adaptive Optics to "Unblur" Telescope Images The two false-color, inset images show the same 1-arcsecond-wide region of the sky at infrared wavelengths as observed with the 10.0-m Keck II telescope on Mauna Kea (see Figure 6-16). Without adaptive optics, it is impossible to distinguish individual stars in this region. With adaptive optics turned on, more than two dozen stars can be distinguished. (UCLA Galactic Center Group)

a combination of telescopes is equivalent to that of one giant telescope with a diameter equal to the **baseline,** or distance between the two telescopes. For example, the Keck I and Keck II telescopes atop Mauna Kea (Figure 6-16) are 85 meters apart, so when used as an interferometer the angular resolution is the same as a single 85-meter telescope.

Interferometry has been used for many years with radio telescopes (which we will discuss in Section 6-6), but is still under development with telescopes for visible light or infrared wavelengths. Astronomers are devoting a great deal of effort to this development because the potential rewards are great. For example, the Keck I and II telescopes used together should give an angular resolution as small as 0.005 arcsec, which corresponds to being able to read the bottom row on an eye chart 36 km (22 miles) away!

Light Pollution

Light from city street lamps and from buildings also degrades telescope images. This **light pollution** illuminates the sky, making it more difficult to see the stars. You can appreciate the problem if you have ever looked at the night sky from a major city. Only a few of the very brightest stars can be seen, as against the thousands that can be seen with the naked eye from in the desert or the mountains. To avoid light pollution, observatories are built in remote locations far from any city lights.

Unfortunately for astronomers, the expansion of cities has brought light pollution to observatories that in former times had none. As an example, the growth of Tucson, Arizona, has had deleterious effects on observations at the nearby Kitt Peak National Observatory. Efforts have been made to have cities adopt light fixtures that provide safe illumination for their citizens but produce little light pollution. These efforts have met with only mixed success.

One factor over which astronomers have absolutely no control is the weather. Optical telescopes cannot see through clouds, so it is important to build observatories where the weather is usually clear. One advantage of mountaintop observatories such as Mauna Kea is that most clouds form at altitudes below the observatory, giving astronomers a better chance of having clear skies.

In many ways the best location for a telescope is in orbit around Earth, where it is unaffected by weather, light pollution, or atmospheric turbulence. We will discuss orbiting telescopes in Section 6-7.

CONCEPTCHECK 6-6

If astronomers are using an adaptive optics system on a night when the atmosphere is unusually turbulent, will the adaptive optics actuators deform the telescope's mirror more rapidly or less rapidly than on a typical night?

Answer appears at the end of the chapter.

6-4 A charge-coupled device is commonly used to record the image at a telescope's focus

TUTORIAL 6-2

Telescopes provide astronomers with detailed pictures of distant objects. The process of recording these pictures is called **imaging.**

Astronomical imaging really began in the nineteenth century with the invention of photography. It was soon realized that this new invention was a boon to astronomy. By mounting a camera at the focus of a telescope, long exposures can gather light from an object for an entire evening. Such long exposures can reveal details in galaxies, star clusters, and nebulae that would not be visible to an astronomer by simply looking through a telescope. Indeed, most large, modern telescopes do not have eyepieces at all.

> The same technology that makes digital cameras possible has revolutionized astronomy

Originally, cameras and telescopes used photographic film, but film has now been replaced by electronic detectors. Photographic film is not a very efficient light detector, with only 2% of the photons that strike photographic film actually triggering the chemical reaction needed to produce an image. And, of course, to analyze images on a computer we need them in some electronic form. For much greater efficiency, and to capture images digitally, we use charge-coupled devices (CCDs).

Charge-Coupled Devices

The most sensitive light detector currently available to astronomers is the **charge-coupled device (CCD).** At the heart of a CCD is a semiconductor wafer divided into an array of small,

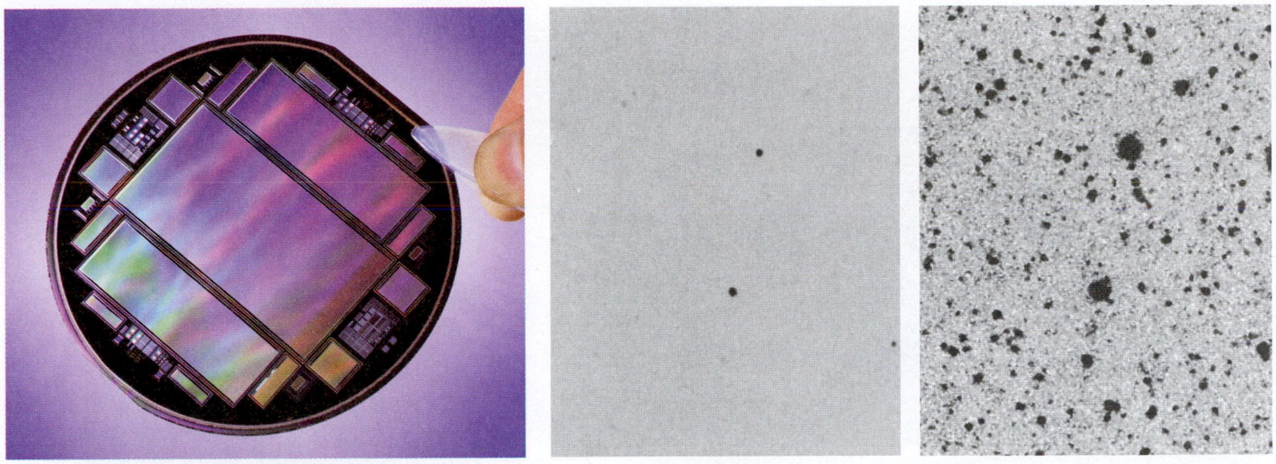

(a) Pan-STARRS detector CCDs

(b) An image made with photographic film

(c) An image of the same region of the sky made with a CCD

FIGURE 6-19 R I V U X G

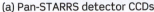

Charge-Coupled Devices (CCDs) and Imaging **(a)** Light is focused onto this CCD chip after passing through the Keck telescope on Mauna Kea, Hawaii. **(b)** This negative print (black stars and white sky) shows a portion of the sky as imaged with a 4-meter telescope and photographic film. **(c)** This negative image of the same region of the sky was made with the same telescope, but with the photographic film replaced by a CCD. Many more stars and galaxies are visible. (a: LBNL/Science Source; b, c: Patrick Seitzer, National Optical Astronomy Observatories)

light-sensitive squares called picture elements or, more commonly, **pixels.** A personal digital camera uses a single CCD chip and might have around 10 million pixels. A large astronomical camera combines many CCDs to capture a wide field of view at high resolution (Figure 6-19a). For example, the square highlighted in Figure 6-19a contains 64 CCDs; the full detector combines 60 of these squares for about 1.4 billion pixels.

When an image from a telescope is focused on the CCD, an electric charge builds up in each pixel in proportion to the number of photons falling on that pixel. When the exposure is finished, the amount of charge on each pixel is read by a computer, where the resulting image can be analyzed and stored in digital form. Compared with photographic film, CCDs are some 35 times more sensitive to light, can record much finer details, and respond more uniformly to light of different colors. Figures 6-19b and 6-19c show the dramatic difference between photographic and CCD images. The great sensitivity of CCDs also makes them useful for **photometry,** which measures the brightness of a star or other astronomical object.

In the modern world of CCD astronomy, astronomers no longer need to spend the night in the unheated dome of a telescope. Instead, they operate the telescope electronically from a separate control room, where the electronic CCD images can be viewed on a computer monitor. By viewing the CCD's electronic image during the evening's observations, an astronomer gets immediate feedback and often makes changes to get the best results. The control room need not even be adjacent to the telescope. Although the Keck I and II telescopes (see Figure 6-16) are at an altitude of 4100 m (13,500 feet), astronomers can now make observations from a facility elsewhere on the island of Hawaii that is much closer to sea level. This saves the laborious drive to the summit of Mauna Kea and eliminates the need for astronomers to acclimate to the high altitude.

Most of the images that you will see in this book were made with CCDs. Because of their extraordinary sensitivity and their ability to be used in conjunction with computers, CCDs have attained a role of central importance in astronomy.

CONCEPTCHECK 6-7

Why can CCDs more efficiently observe faint stars than photographic film or photographic plates?

Answer appears at the end of the chapter.

6-5 Spectrographs record the spectra of astronomical objects

The spectrum of a planet, star, or galaxy can reveal more about its nature than an image

We saw in Section 5-6 how the spectrum of an astronomical object provides a tremendous amount of information about that object, including its chemical composition and temperature. So, measuring spectra, or **spectroscopy,** is one of the most important uses of telescopes. Indeed, some telescopes are designed solely for measuring the spectra of distant, faint objects; they are never used for imaging.

Spectrographs and Diffraction Gratings

An essential tool of spectroscopy is the **spectrograph,** a device that records spectra. This optical device is mounted at the focus

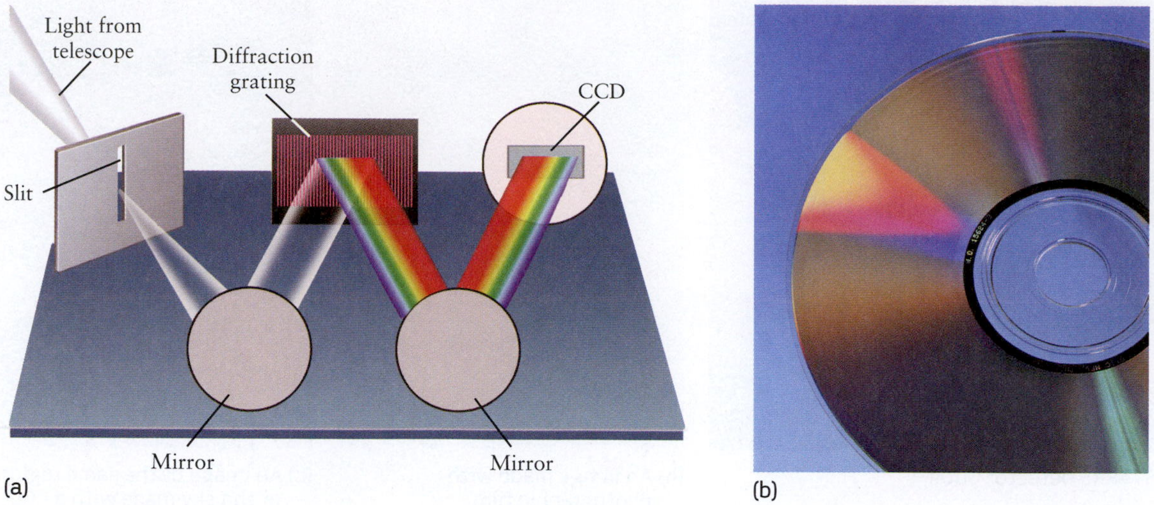

FIGURE 6-20 R I V U X G

A Grating Spectrograph **(a)** This optical device uses a diffraction grating to break up the light from a source into a spectrum. The spectrum is then recorded on a CCD. **(b)** A diffraction grating has a large number of parallel lines in its surface that reflect light of different colors in different direction. A compact disc, which stores information in a series of closely spaced pits, reflects light in a similar way. (Dale E. Boyer/Photo Researchers)

of a telescope. Figure 6-20a shows one design for a spectrograph, in which a **diffraction grating** is used to form the spectrum of a planet, star, or galaxy. A diffraction grating is a piece of glass on which thousands of very regularly spaced parallel lines have been cut. Some of the finest diffraction gratings have more than 10,000 lines per centimeter, which are usually cut by drawing a diamond back and forth across the glass. When light is shone on a diffraction grating, a spectrum is produced by the way in which light waves leaving different parts of the grating interfere with each other. (This same effect produces the rainbow of colors you see reflected from a compact disc or DVD, as shown in Figure 6-20b. Information is stored on the disc in a series of closely spaced pits, which acts as a diffraction grating.)

Older types of spectrographs used a prism rather than a diffraction grating to form a spectrum. Using a prism had several drawbacks. A prism does not disperse the colors of the rainbow evenly: Blue and violet portions of the spectrum are spread out more than the red portion. In addition, because the blue and violet wavelengths must pass through more of the prism's glass than do the red wavelengths (examine Figure 5-3), light is absorbed unevenly across the spectrum. Indeed, a glass prism is opaque to near-ultraviolet light. For these reasons, diffraction gratings are preferred in modern spectrographs.

In Figure 6-20a the spectrum of a planet, star, or galaxy formed by the diffraction grating is recorded on a CCD. Light from a hot gas (such as helium, neon, argon, iron, or a combination of these) is then focused on the spectrograph slit. The result is a *comparison spectrum* alongside the spectrum of the celestial object under study. The wavelengths of the bright spectral lines of the comparison spectrum are known from laboratory experiments and can therefore serve as reference markers. (See Figure 5-18, which shows the spectrum of the Sun and a comparison spectrum of iron.)

When the exposure is finished, electronic equipment measures the charge that has accumulated in each pixel. These data are used to graph light intensity versus wavelength. Dark absorption lines in the spectrum appear as depressions or valleys on the graph, while bright emission lines appear as peaks. Figure 6-21 compares two ways of exhibiting spectra with absorption lines and emission lines. Later in this book we shall see spectra presented in both these ways.

CONCEPTCHECK 6-8

A planet is viewed through a spectrograph like the one illustrated in Figure 6-20a. Does the CCD also record an image of the planet?

Answer appears at the end of the chapter.

6-6 A radio telescope uses a large concave dish to reflect radio waves to a focus

For thousands of years, all the information that astronomers gathered about the universe was based on ordinary visible light. In the twentieth century, however, astronomers first began to explore the nonvisible electromagnetic radiation coming from astronomical objects. In this way they have discovered aspects of the cosmos that are forever hidden to optical telescopes.

Observing at radio wavelengths reveals aspects of the universe hidden from ordinary telescopes

Astronomers have used ultraviolet light to map the outer regions of the Sun and the clouds of Venus, and used infrared radiation to see new stars and perhaps new planetary systems in the process of formation. By detecting radio waves from Jupiter and Saturn, they have mapped the intense magnetic fields that surround those giant planets; by detecting curious bursts of X-rays from space, they have learned about the utterly alien conditions in the vicinity of a black hole. It is no exaggeration to say that today's astronomers learn as much about the universe using telescopes for nonvisible wavelengths as they do using visible light.

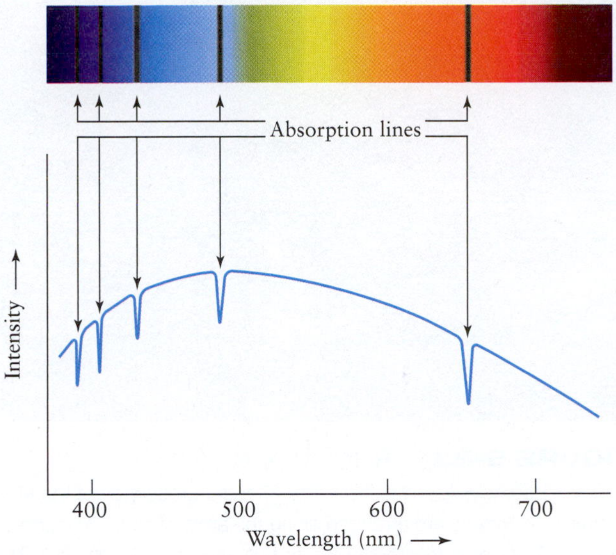

(a) Two representations of an absorption line spectrum

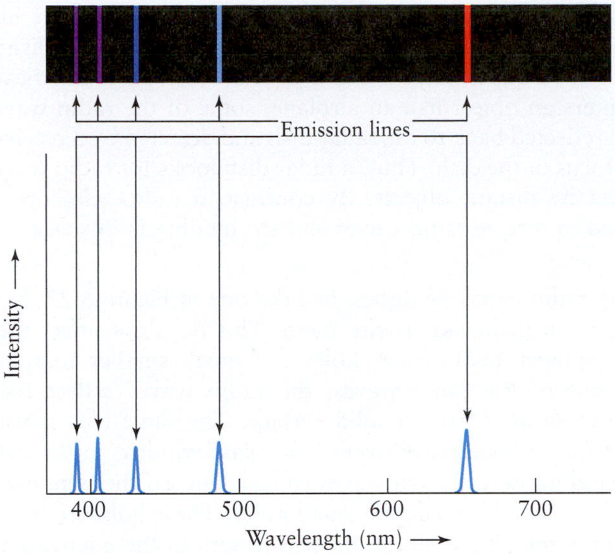

(b) Two representations of an emission line spectrum

FIGURE 6-21
Two Ways to Represent Spectra When a CCD is placed at the focus of a spectrograph, it records the rainbow-colored spectrum. A computer program can be used to convert the recorded data into a graph of intensity versus wavelength. **(a)** Absorption lines appear as dips on such a graph, while **(b)** emission lines appear as peaks. The dark absorption lines and bright emission lines in this example are the Balmer lines of hydrogen (see Section 5-8).

Radio Astronomy

Radio waves were the first part of the electromagnetic spectrum beyond the visible to be exploited for astronomy. The use of radio waves is a result of a research project seemingly unrelated to astronomy. In the early 1930s, Karl Jansky, a young electrical engineer at Bell Telephone Laboratories, was trying to locate the source of interference with the then-new transatlantic radio link. By 1932, he realized that one kind of radio noise is strongest when the constellation Sagittarius is high in the sky. The center of our Galaxy is located in the direction of Sagittarius, and Jansky concluded that he was detecting radio waves from an astronomical source.

At first, only Grote Reber, a radio engineer living in Illinois, took up Jansky's research. In 1936 Reber built in his backyard the first **radio telescope,** a radio-wave detector dedicated to astronomy. He modeled his design after an ordinary reflecting telescope, with a parabolic metal "dish" (reflecting antenna) measuring 10 m (31 ft) in diameter and a radio receiver at the focal point of the dish.

Reber spent the years from 1938 to 1944 mapping radio emissions from the sky at wavelengths of 1.9 m and 0.63 m. He found radio waves coming from the entire Milky Way, with the greatest emission from the center of the Galaxy. These results, together with the development of improved radio technology during World War II, encouraged the growth of radio astronomy and the construction of new radio telescopes around the world. Today, radio observatories are as common as major optical observatories.

Modern Radio Telescopes

Like Reber's prototype, a typical modern radio telescope has a large parabolic dish (Figure 6-22). An antenna tuned to the desired frequency is located at the focus (like the prime focus design for optical reflecting telescopes shown in Fig. 6-11a). The incoming signal is relayed from the antenna to amplifiers and recording instruments, typically located in a room at the base of the telescope's pier.

FIGURE 6-22 R I V U X G
A Radio Telescope The dish of the Parkes radio telescope in New South Wales, Australia, is 64 m (210 ft) in diameter. Radio waves reflected from the dish are brought to a focus and collected by an antenna at the focal point. (David Nunuk/Photo Researchers)

CAUTION! The radio telescope in Figure 6-22 looks like a radar dish but is used in a different way. In radar, the dish is used to send out a narrow beam of radio waves. If this beam encounters an object like an airplane, some of the radio waves will be reflected back to the radar dish and detected by a receiver at the focus of the dish. Thus, a radar dish looks for radio waves *reflected* by distant objects. By contrast, a radio telescope is designed to receive radio waves *emitted* by objects in space.

Many radio telescope dishes, like the one in Figure 6-22, have visible gaps in them like a wire mesh. This does not affect their reflecting power because the holes are much smaller than the wavelengths of the radio waves; the radio waves reflect from the wire mesh as if from a solid surface. The same idea is used in the design of microwave ovens. The glass window in the oven door would allow the microwaves to leak out, so the window is covered by a metal screen with small holes. These holes are much smaller than the 12.2-cm (4.8-in.) wavelength of the microwaves, so the screen reflects the microwaves back into the oven.

Radio Telescopes: Limits to Angular Resolution

One great drawback of early radio telescopes was their very poor angular resolution. Recall from Section 6-3 that angular resolution is the smallest angular separation between two stars that can just barely be distinguished as separate objects. Unlike visible light, radio waves are only slightly affected by turbulence in the atmosphere, so the limitation on the angular resolution of a radio telescope is diffraction. The problem is that diffraction-limited angular resolution is directly proportional to the wavelength being observed: The longer the wavelength, the larger (and hence worse) the angular resolution and the fuzzier the image. (See the formula for angular resolution θ in Section 6-3.) As an example, a 1-m radio telescope detecting radio waves of 5-cm wavelength has an angular resolution 100,000 times poorer than a 1-m optical telescope. Because radio radiation has very long wavelengths, *small* radio telescopes can produce only blurry, indistinct images. VIDEO 6.7 A very large radio telescope can produce a somewhat sharper radio image, because as the diameter of the telescope increases, the angular resolution decreases. In other words, the bigger the dish, the better the resolution. For this reason, most modern radio telescopes have dishes more than 30 m (100 ft) in diameter. A large dish is also useful for increasing light-gathering power, because radio signals from astronomical objects are typically very weak in comparison with the intensity of visible light emitted by the same objects. But even the largest single radio dish in existence, the 305-m (1000-ft) Arecibo radio telescope in Puerto Rico, cannot come close to the resolution of the best optical instruments.

To improve angular resolution at radio wavelengths, astronomers combine observations from two or more widely separated radio telescopes using the interferometry technique that we described in Section 6-3. Interferometry using radio waves is much easier than for visible or infrared light because radio signals can be carried over electrical wires. Consequently, two radio telescopes observing the same astronomical object can be hooked together, even if they are separated by a large distance, or baseline, of many kilometers. The connected radio dishes can act as a single large-diameter dish with vastly improved angular resolution.

FIGURE 6-23 R I **V** U X G
The Very Large Array (VLA) The 27 radio telescopes of the VLA in central New Mexico are arranged along the arms of a Y. The north arm of the array is 19 km long; the southwest and southeast arms are each 21 km long. By spreading the telescopes out along the legs and combining the signals received, the VLA can give the same angular resolution as a single dish many kilometers in radius. (Courtesy of NRAO/AUI)

One of the largest arrangements of radio telescopes for interferometry is the Very Large Array (VLA), located in the desert near Socorro, New Mexico (Figure 6-23). The VLA consists of 27 parabolic dishes, each 25 m (82 ft) in diameter. These 27 telescopes are arranged along the arms of a gigantic Y that covers an area 27 km (17 mi) in diameter. By pointing all 27 telescopes at the same object and combining the 27 radio signals, this system can produce radio views of the sky with an angular resolution as small as 0.05 arcsec, comparable to that of the very best optical telescopes.

Dramatically better angular resolution can be obtained by combining the signals from radio telescopes at different observatories thousands of kilometers apart. This technique is called **very-long-baseline interferometry (VLBI)**. VLBI is used by a system called the Very Long Baseline Array (VLBA), which consists of ten 25-meter dishes at different locations between Hawaii and the Caribbean. Although the 10 dishes are not physically connected, they are all used to observe the same object at the same time. The data from each telescope are recorded electronically and processed later. By carefully synchronizing the 10 recorded signals, they can be combined just as if the telescopes had been linked together during the observation. With the VLBA, features smaller than 0.001 arcsec can be distinguished at radio wavelengths. This angular resolution is 100 times better than a large optical telescope with adaptive optics.

Even better angular resolution can be obtained by adding radio telescopes in space; the baseline is then the distance from the VLBA to the orbiting telescope. The first such space radio telescope, the Japanese HALCA spacecraft, operated from 1997 to 2003. Orbiting at a maximum distance of 21,400 km (13,300 mi) above Earth's surface, HALCA gave a baseline 3 times longer—and thus an angular resolution 3 times better—than can be obtained with Earthbound telescopes alone. A Russian-led mission launched in 2011, Spektr-R, continues the effort for space-based interferometry.

Figure 6-24 shows how optical and radio images of the same object can give different and complementary kinds of information. The visible-light image of Saturn (Figure 6-24a) shows clouds in

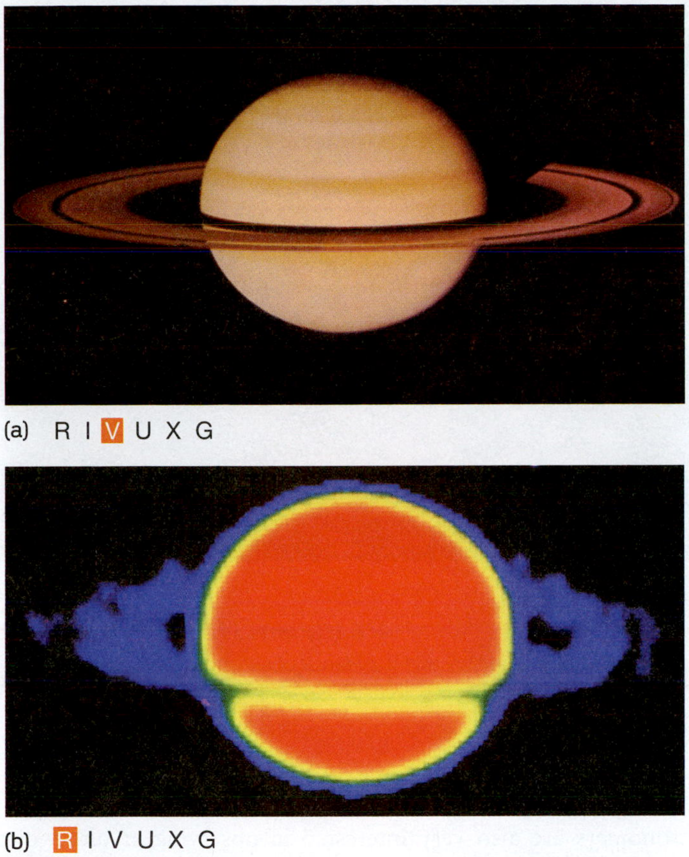

FIGURE 6-24

Optical and Radio Views of Saturn **(a)** This picture was taken by a spacecraft 18 million kilometers from Saturn. The view was produced by sunlight reflecting from the planet's cloudtops and rings. **(b)** The image is made by using radio waves, which is indicated by the highlighted R in the wavelength tab. This VLA image shows radio emission from Saturn at a wavelength of 2 cm. In this false-color image, the most intense radio emission is shown in red, the least intense in blue. Yellow and green represent intermediate levels of radio intensity; black indicates no detectable radio emission. Note the radio "shadow" caused by Saturn's rings where they lie in front of the planet. (NASA)

the planet's atmosphere and the structure of the rings. Like the visible light from the Moon, the light used to make this image is just reflected sunlight. By contrast, the false-color radio image of Saturn (Figure 6-24b) is a record of waves *emitted* by the planet and its rings. Analyzing such images provides information about the structure of Saturn's atmosphere and rings that could never be obtained from a visible-light image such as Figure 6-24a.

CONCEPTCHECK 6-9

What radio telescope would make an image with the better resolution: a single radio antenna dish with a diameter of 100 m, or two 25-m dishes connected together at a separation of 200 m?

Answer appears at the end of the chapter.

6-7 Telescopes in orbit around Earth detect radiation that does not penetrate the atmosphere

TUTORIAL 6-3

Space telescopes make it possible to study the universe across the entire electromagnetic spectrum

The many successes of radio astronomy show the value of observations at nonvisible wavelengths. But Earth's atmosphere is opaque to many wavelengths. Other than visible light and radio waves, very little radiation from space manages to penetrate the air we breathe. To overcome this, astronomers have placed a variety of telescopes in orbit around the planet.

Figure 6-25 shows the transparency of Earth's atmosphere to different wavelengths of electromagnetic radiation. The atmosphere is most transparent in two wavelength regions, the **optical window** (which includes the entire visible spectrum) and the **radio window** (which includes part, but not all, of the radio spectrum). There are also several relatively transparent regions at infrared wavelengths between 1 and 40 mm. Infrared radiation within these wavelength intervals can penetrate Earth's atmosphere somewhat and can be detected with ground-based telescopes. This wavelength range is

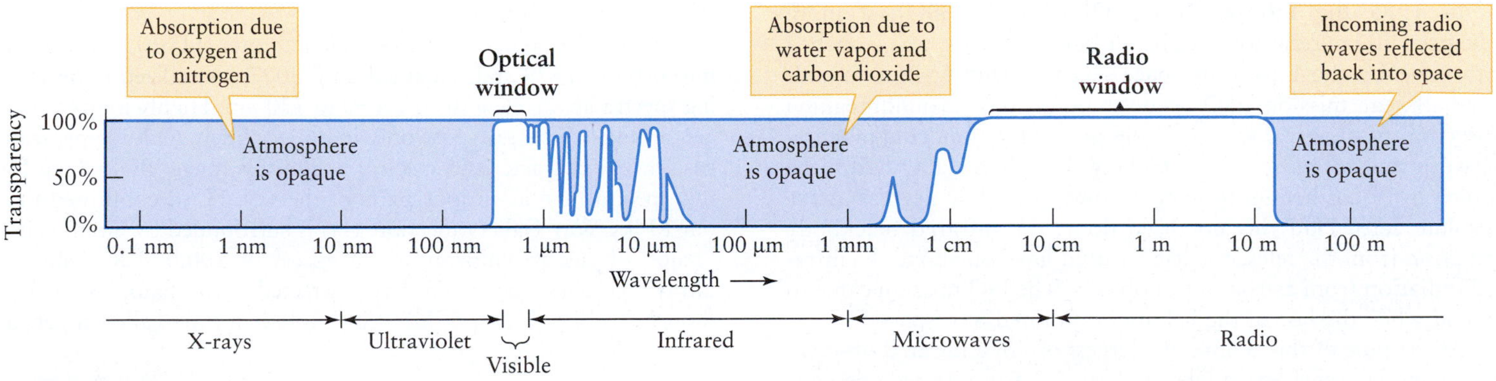

FIGURE 6-25

The Transparency of Earth's Atmosphere This graph shows the percentage of radiation that can penetrate Earth's atmosphere at different wavelengths. Regions in which the curve is high are called "windows," because the atmosphere is relatively transparent at those wavelengths. There are also three wavelength ranges in which the atmosphere is opaque and the curve is near zero. At wavelengths where the atmosphere is opaque, ground-based telescopes cannot receive light from space because it is either absorbed or reflected by the atmosphere.

called the *near-infrared,* because it lies just beyond the red end of the visible spectrum.

CONCEPTCHECK 6-10

Look at Figure 6-25, which shows the transparency of Earth's atmosphere. Would astronomers most prefer to have a new ground-based telescope constructed that is most sensitive in the X-ray region, the ultraviolet wavelength region, or in the microwave region?

Answer appears at the end of the chapter.

Infrared Astronomy

Water vapor is the main absorber of infrared radiation from space, which is why infrared observatories are located at sites with exceptionally low humidity. The site must also be at high altitude to get above as much of the atmosphere's water vapor as possible. One site that meets both criteria is the summit of Mauna Kea in Hawaii, shown in Figure 6-16. (The complete lack of vegetation on the summit attests to its extreme dryness.) Some of the telescopes on Mauna Kea are designed exclusively for detecting infrared radiation. Others, such as the Keck I and Keck II telescopes, are used for both visible and near-infrared observations (see Figure 6-18).

Even at the elevation of Mauna Kea, water vapor in the atmosphere restricts the kinds of infrared observations that astronomers can make. This situation can be improved by carrying telescopes on board high-altitude balloons or aircraft. But the ultimate solution is to place a telescope in Earth's orbit and radio its data back to astronomers on the ground. The first such orbiting infrared observatory, the Infrared Astronomical Satellite (IRAS), was launched in 1983. During its nine-month mission, IRAS used its 57-cm (22-in.) telescope to map almost the entire sky at wavelengths from 12 to 100 mm.

The IRAS data revealed the presence of dust disks around nearby stars. Planets are thought to coalesce from disks of this kind, so this was the first observational indication that there might be planets orbiting other stars. The dust that IRAS detected is warm enough to emit infrared radiation but too cold to emit much visible light, so it remained undetected by ordinary optical telescopes. IRAS also discovered distant, ultraluminous galaxies that emit almost all their radiation at infrared wavelengths.

In 1995 the Infrared Space Observatory (ISO), a more advanced 60-cm reflector with better light detectors, was launched into orbit by the European Space Agency. During its two-and-one-half-year mission, ISO made a number of groundbreaking observations of very distant galaxies and of the thin, cold material between the stars of our own Galaxy. Like IRAS, ISO had to be cooled by liquid helium to temperatures just a few degrees above absolute zero. Had this not been done, the infrared blackbody radiation from the telescope itself would have outshone the infrared radiation from astronomical objects. The ISO mission came to an end when the last of the helium evaporated into space.

At the time of this writing the largest orbiting infrared observatory is the Herschel Space Observatory, a 3.5-m infrared telescope designed to survey the infrared sky with unprecedented resolution (Figure 6-26). Placed in orbit in 2009, the Herschel Space Observatory is being used to study galaxy formation in the early universe, star formation, and the chemical composition of atmospheres around planets, moons, and comets in our solar system.

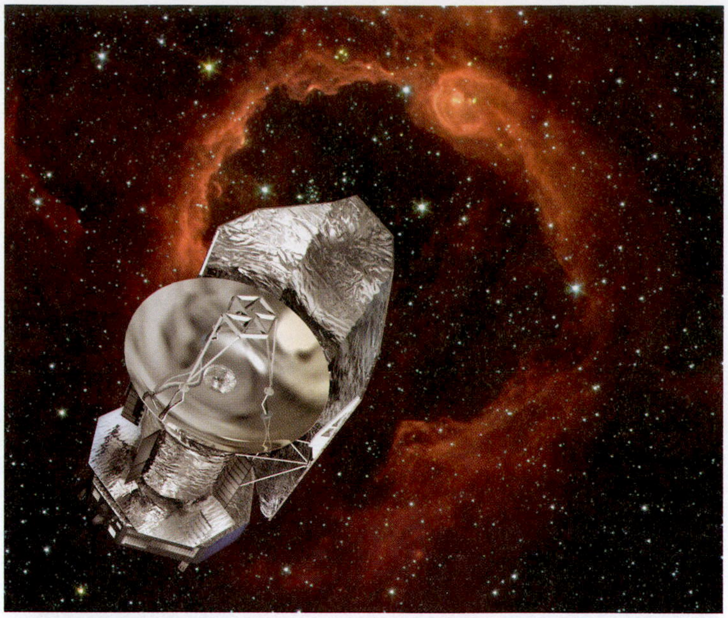

FIGURE 6-26
Herschel Space Observatory This is an artist's concept of the Herschel Space Observatory. Herschel and the Planck observatory lifted off together on May 14, 2009 aboard an Ariane ECA rocket. (ESA, D. Ducros, 2009/NASA)

Ultraviolet Astronomy

Astronomers are also very interested in observing at ultraviolet wavelengths. These observations can reveal a great deal about hot stars, ionized clouds of gas between the stars, and the Sun's high-temperature corona (see Section 3-5), all of which emit copious amounts of ultraviolet light. The spectrum of ultraviolet sunlight reflected from a planet can also reveal the composition of the planet's atmosphere. However, Earth's atmosphere is opaque to ultraviolet light except for the narrow *near-ultraviolet* range, which extends from about 400 nm (the violet end of the visible spectrum) down to 300 nm.

To see shorter-wavelength *far-ultraviolet* light, astronomers must again make their observations from space. The first ultraviolet telescope was placed in orbit in 1962, and several others have since followed it into space. Small rockets have also lifted ultraviolet cameras briefly above Earth's atmosphere. Figure 6-27 shows an ultraviolet view of the constellation Orion, along with infrared and visible views.

The Far Ultraviolet Spectroscopic Explorer (FUSE), which went into orbit in 1999 and remained until 2007, specialized in measuring spectra at wavelengths from 90 to 120 nm. Highly ionized oxygen atoms, which can exist only in an extremely high-temperature gas, have a characteristic spectral line in this range. By looking for this spectral line in various parts of the sky, FUSE confirmed that our Milky Way Galaxy (Section 1-4) is surrounded by an immense "halo" of gas at temperatures in excess of 200,000 K. Only an ultraviolet telescope could have detected this "halo," which is thought to have been produced by exploding stars called supernovae (Section 1-3).

The Hubble Space Telescope

Infrared and ultraviolet satellites give excellent views of the heavens at selected wavelengths. But since the 1940s, astronomers had dreamed of having one large telescope that could be operated at

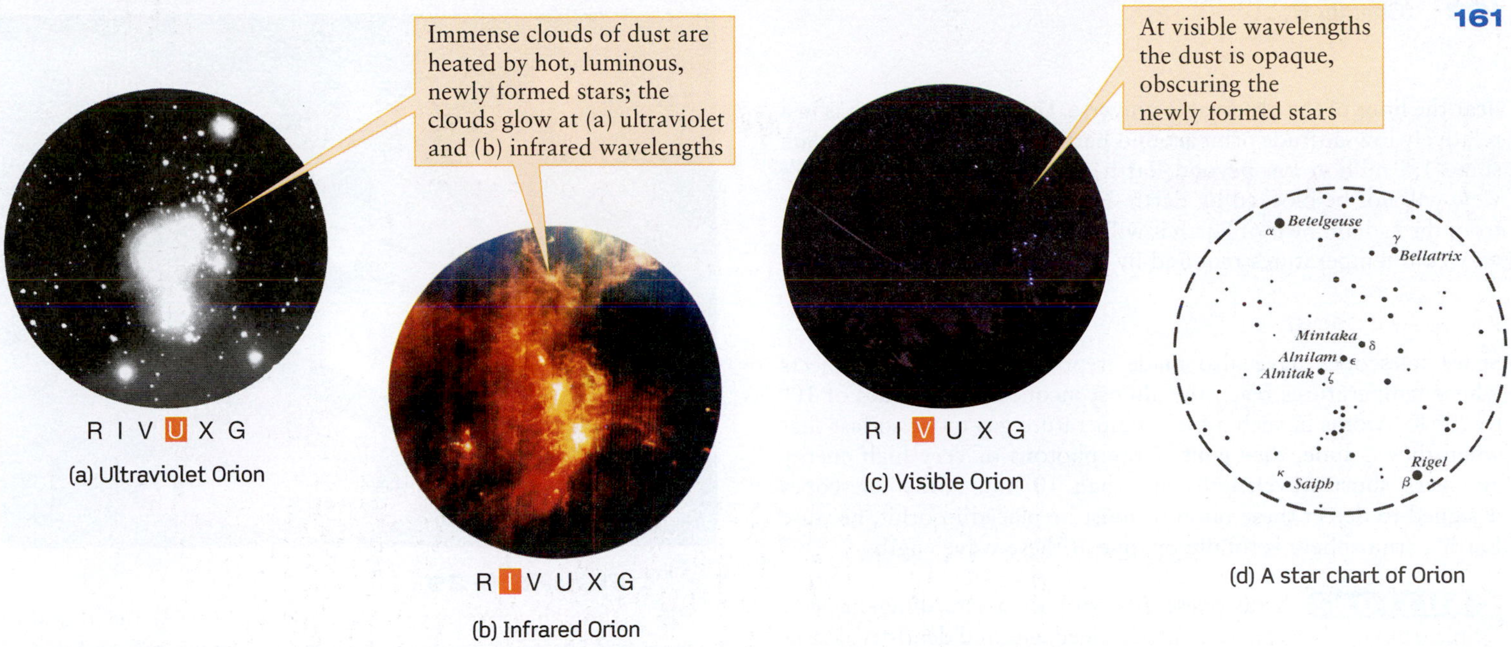

FIGURE 6-27

Orion Seen at Ultraviolet, Infrared, and Visible Wavelengths

(a) An ultraviolet view of the constellation Orion was obtained during a brief rocket flight in 1975. This 100-s exposure covers the wavelength range 125–200 nm. **(b)** The false-color view from the Infrared Astronomical Satellite displays emission at different wavelengths in different colors: red for 100-mm radiation, green for 60-mm radiation, and blue for 12-mm radiation. Compare these images with **(c)** an ordinary visible-light photograph and **(d)** a star chart of Orion. (a: G. R. Carruthers, Naval Research Laboratory; b: NASA; c: mike black photography/Flickr/Getty Images)

any wavelength from the near-infrared through the visible range and out into the ultraviolet. This is the mission of the Hubble Space Telescope (HST), which was placed in a 600-km-high orbit by the space shuttle *Discovery* in 1990 (Figure 6-28). HST has a 2.4-m (7.9-ft) objective mirror and was designed to observe at wavelengths from 115 nm to 1 mm. Like most ground-based telescopes, HST uses a CCD to record images. (In fact, the development of HST helped drive advances in CCD technology.) The images are then radioed back to Earth in digital form.

The great promise of HST was that from its vantage point high above the atmosphere, its angular resolution would be limited only by diffraction. But soon after HST was placed in orbit, astronomers discovered that a manufacturing error had caused the telescope's objective mirror to suffer from spherical aberration. The mirror should have been able to concentrate 70% of a star's light into an image with an angular diameter of 0.1 arcsec. Instead, only 20% of the light was focused into this small area. The remainder was smeared out over an area about 1 arcsec wide, giving images little better than those achieved at major ground-based observatories.

On an interim basis, astronomers used only the 20% of incoming starlight that was properly focused and, with computer processing, discarded the remaining poorly focused 80%. This was practical only for brighter objects on which astronomers could afford to waste light. But many of the observing projects scheduled for HST involved extremely dim galaxies and nebulae.

These problems were resolved by a second space shuttle mission in 1993. Astronauts installed a set of small secondary mirrors whose curvature exactly compensated for the error in curvature of the primary mirror. Once these were in place, HST was able to make truly sharp images of extremely faint objects. Astronomers have used the repaired HST to make discoveries about the nature of planets, the evolution of stars, the inner workings of galaxies, and the expansion of the universe. You will see many HST images in later chapters.

The success of HST has inspired plans for its larger successor, the James Webb Space Telescope, or JWST (Figure 6-29). Planned for a 2018 launch, JWST will observe at visible and infrared wavelengths from 600 nm to 28 mm. With its 6.5-m objective mirror—2.5 times the diameter of the HST objective mirror, with 6 times the light-gathering power—JWST will study faint objects such as planetary systems forming around other stars and galaxies

FIGURE 6-28 R I V U X G

The Hubble Space Telescope The largest telescope yet placed in orbit, HST is a joint project of NASA and the European Space Agency (ESA). HST has helped discover new moons of Pluto, probed the formation of stars, and found evidence that the universe is now expanding at a faster rate than several billion years ago. This photograph was taken from the space shuttle *Discovery* during a 1997 mission to service HST. (NASA) near the limit of the observable universe. Unlike HST, which is in a relatively low-altitude orbit around Earth, JWST will orbit the Sun some 1.5 million km beyond Earth. In this orbit the telescope's view will not be blocked by Earth. Furthermore, by remaining far from the radiant heat of Earth it will be easier to keep JWST at the very cold temperatures required by its infrared detectors.

X-ray Astronomy

Space telescopes have also made it possible to explore objects whose temperatures reach the almost inconceivable values of 10^6 to 10^8 K. Atoms in such a high-temperature gas move so fast that when they collide, they emit X-ray photons of very high energy and very short wavelengths less than 10 nm. X-ray telescopes designed to detect these photons must be placed in orbit, because Earth's atmosphere is totally opaque at these wavelengths.

CAUTION! X-ray telescopes work on a very different principle from the X-ray devices used in medicine and dentistry. If you have your foot "X- rayed" to check for a broken bone, a piece of photographic film (or an electronic detector) sensitive to X-rays is placed under your foot and an X-ray beam is directed at your foot from above. The radiation penetrates through soft tissue but not as much through bone, so the bones cast an "X-ray shadow" on the film. A fracture will show as a break in the shadow. X-ray telescopes, by contrast, do *not* send beams of X-rays toward astronomical objects in an attempt to see inside them. Rather, these telescopes detect X-rays that the objects emit on their own.

Astronomers got their first quick look at the X-ray sky from brief rocket flights during the late 1940s. These observations confirmed that the Sun's corona (see Section 3-5) is a source of X-rays, and must therefore be at a temperature of millions of kelvins. In 1962 a rocket experiment revealed that objects beyond the solar system also emit X-rays.

Since 1970, a series of increasingly sensitive and sophisticated X-ray observatories have been placed in orbit, including NASA's Einstein Observatory, the European Space Agency's Exosat, and the German-British-American ROSAT. These telescopes have shown that other stars also have high-temperature coronae and have found hot, X-ray–emitting gas clouds so immense that hundreds of galaxies fit inside them. They also discovered unusual stars that emit X-rays in erratic bursts. These bursts are now thought to be coming from heated gas swirling around a small but massive object—possibly a black hole.

X-ray astronomy took a quantum leap forward in 1999 with the launch of NASA's Chandra X-ray Observatory and the European Space Agency's XMM-Newton. Named for the Indian-American Nobel laureate astrophysicist Subrahmanyan Chandrasekhar, Chandra can view the X-ray sky with an angular resolution of 0.5 arcsec (Figure 6-30a). This is comparable to the best ground-based optical telescopes and more than a thousand times better than the resolution of the first orbiting X-ray telescope. Chandra can also measure X-ray spectra 100 times more precisely than any previous spacecraft and can detect variations in X-ray emissions on time scales as short as 16 microseconds. This latter capability is essential for understanding how X-ray bursts are produced around black holes.

XMM-Newton (for *X*-ray *M*ulti-mirror *M*ission) is actually three X-ray telescopes that all point in the same direction (Figure 6-30b). Their combined light-gathering power is 5 times greater than that of Chandra, which makes XMM-Newton able to observe fainter objects. (For reasons of economy, the mirrors were not ground as precisely as those on Chandra, so the angular resolution of XMM-Newton is only about 6 arcseconds.) It also carries a small but highly capable telescope for ultraviolet and visible observations. Hot X-ray sources are usually accompanied by cooler material that radiates at these longer wavelengths, so XMM-Newton can observe these hot and cool regions simultaneously.

The latest X-ray telescope in space is NuSTAR, with an incredible 10-m long mast (Figure 6-30c). NuSTAR will search for enormous black holes, study particles moving at almost the speed of light, and analyze how atoms are produced during supernova explosions.

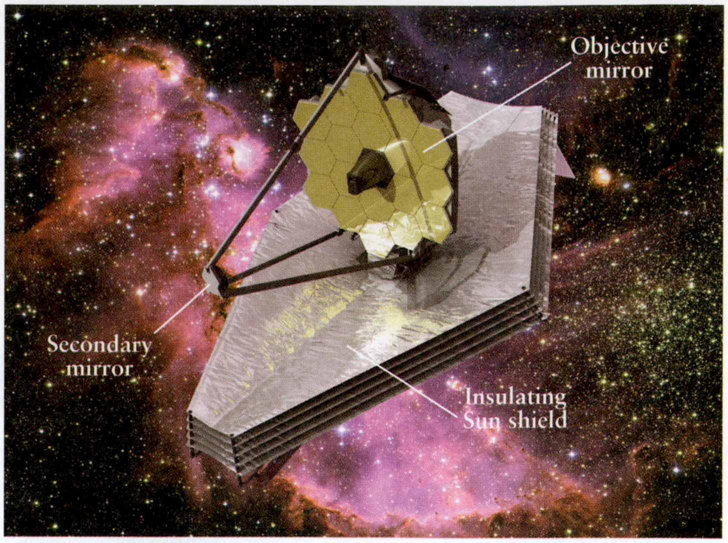

ANIMATION 6-3

FIGURE 6-29 **The James Webb Space Telescope (JWST)** The successor to the Hubble Space Telescope, JWST will have an objective mirror 6.5 m (21 ft) in diameter. Like many Earthbound telescopes, JWST will be of Cassegrain design (see Figures 6-11b and 6-14b). Rather than using liquid helium to keep the telescope and instruments at the low temperatures needed to observe at infrared wavelengths, JWST will keep cool using a multilayer sunshield the size of two tennis courts. (ESA, C. Carreau)

Gamma-Ray Astronomy

Gamma rays, the shortest-wavelength photons of all, help us to understand phenomena even more energetic than those that produce X-rays. As an example, when a massive star explodes into a supernova, it produces radioactive atomic nuclei that are strewn across interstellar space. Observing the gamma rays emitted by these nuclei helps astronomers understand the nature of supernova explosions.

Like X-rays, gamma rays do not penetrate Earth's atmosphere, so space telescopes are required. One of the first gamma-ray telescopes placed in orbit was the Compton Gamma Ray Observatory (CGRO). One particularly important task for CGRO was the study of gamma-ray bursts, which are brief, unpredictable, and very intense flashes of gamma rays that are found in all parts of the sky. By analyzing data from CGRO and other orbiting observatories, astronomers have shown that the sources of these gamma-ray bursts are billions of light-years away. For these bursts to be visible across such great distances, their sources must be among the most energetic objects in the universe. By combining these gamma-ray

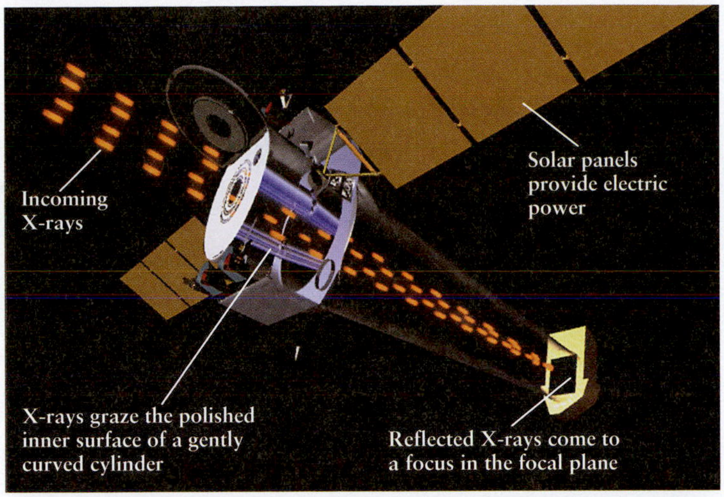
Incoming X-rays
Solar panels provide electric power
X-rays graze the polished inner surface of a gently curved cylinder
Reflected X-rays come to a focus in the focal plane

(a) Chandra X-ray Observatory

Apertures for the three X-ray telescopes

(b) XMM-Newton

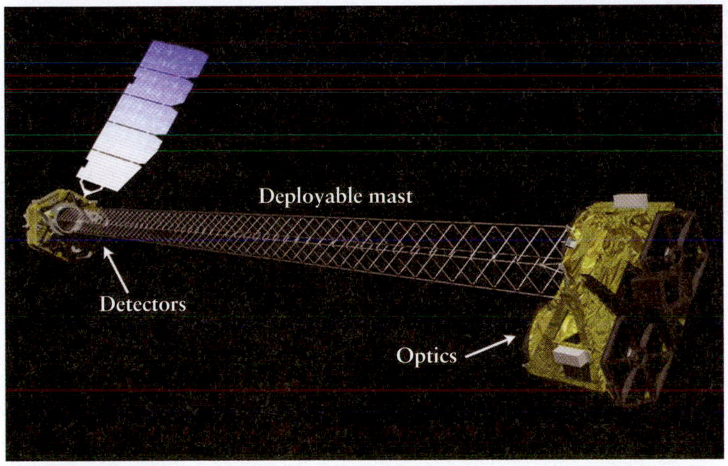
Deployable mast
Detectors
Optics

(c) NuSTAR

FIGURE 6-30

Three Orbiting X-Ray Observatories **(a)** X-rays are absorbed by ordinary mirrors like those used in optical reflectors, but they can be reflected if they graze the mirror surface at a very shallow angle. In the Chandra X-ray Observatory, X-rays are focused in this way onto a focal plane 10 m (33 ft) behind the mirror. **(b)** XMM-Newton is about the same size as Chandra, and its three X-ray telescopes form images in the same way. **(c)** NuSTAR has a 33-foot (10-meter) mast that deploys after launch to separate the optics modules (right) from the detectors in the focal plane (left) (a: NASA/Chandra X-ray Observatory Center/Smithsonian Astrophysical Observatory; b: D. Ducros/European Space Agency; c: NASA)

observations with images made by optical telescopes, astronomers have found that at least some of the gamma-ray bursts emanate from stars that explode catastrophically. Launched in 2008, the Fermi Gamma-ray Space Telescope (Figure 6-31) continues the study of gamma-ray bursts and other highly energetic phenomena.

The *Cosmic Connections* figure shows the wavelengths at which Earth-orbiting telescopes are particularly useful, as well as summarizing the design of refracting and reflecting Earth-based telescopes. The advantages and benefits of Earth-orbiting observatories cannot be overemphasized. We are no longer limited to the narrow ranges of whatever wavelengths manage to leak through our shimmering, hazy atmosphere (Figure 6-32). For the first time, we are really *seeing* the universe.

CONCEPTCHECK 6-11

What is the primary advantage of an orbiting space telescope, compared to a ground-based telescope?

Answer appears at the end of the chapter.

FIGURE 6-31 R I **V** U X G

The Fermi Gamma-Ray Space Telescope This photograph shows the telescope placed inside half of the launching rocket's nose cone. For scale: Several engineers in white clean-room gowns are at the base of the nose cone. (NASA)

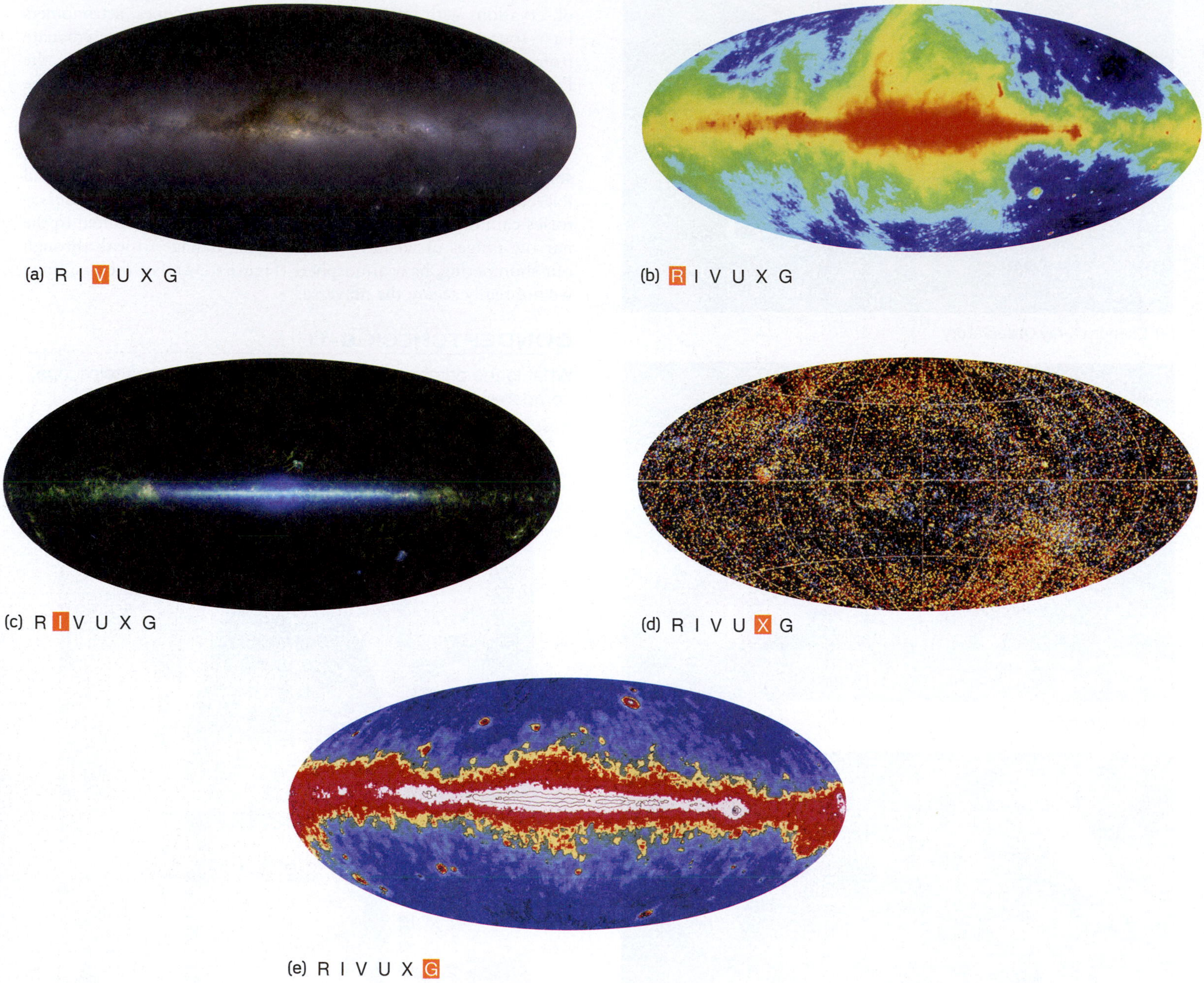

FIGURE 6-32

The Entire Sky at Five Wavelength Ranges These five views show the entire sky at visible, radio, infrared, X-ray, and gamma-ray wavelengths. The entire celestial sphere is mapped onto an oval, with the Milky Way stretching horizontally across the center. **(a)** In the visible view the constellation Orion is at the right, Sagittarius is in the middle, and Cygnus is toward the left. Many of the dark areas along the Milky Way are locations where interstellar dust is sufficiently thick to block visible light. **(b)** The radio view shows the sky at a wavelength of 21 cm. This wavelength is emitted by hydrogen atoms in interstellar space. The brightest regions (shown in red) are in the plane of the Milky Way, where the hydrogen is most concentrated. **(c)** This is a mosaic of the images covering the entire sky as observed by the Wide-field Infrared Survey Explorer (WISE). Most of the emission is from dust particles in the plane of the Milky Way that have been warmed by starlight. **(d)** The image is made by using X-rays, which is indicated by the highlighted X in the wavelength tab. The X-ray map from ROSAT shows about 50,000 X-ray sources whose intensity runs from red for the less intense sources through yellow and blue for the brightest ones. Extremely high temperature gas emits these X-rays. The white regions, which emit strongly at all X-ray wavelengths, are remnants of supernovae. **(e)** The image is made by using gamma rays, which is indicated by the highlighted G in the wavelength tab. The gamma-ray view from the Compton Gamma Ray Observatory includes all wavelengths less than about 1.2×10^{-5} nm (photon energies greater than 10^8 eV). The diffuse radiation from the Milky Way is emitted when fast-moving subatomic particles collide with the nuclei of atoms in interstellar gas clouds. The bright spots above and below the Milky Way are distant, extremely energetic galaxies. (a: Axel Mellinger; b: Max Planck Institute for Radio Astronomy/Science Photo Library/Science Source; c: NASA/JPL-Caltech/UCLA; d: Max Planck Institute for Extraterrestrial Physics/Science Photo Library/Science Source; e: NASA/Science Photo Library/Science Source)

COSMIC CONNECTIONS

Telescopes Across the EM Spectrum

Today, telescopes can view the universe in every range of electromagnetic radiation, although some must be above Earth's atmosphere to receive radiation without interference.

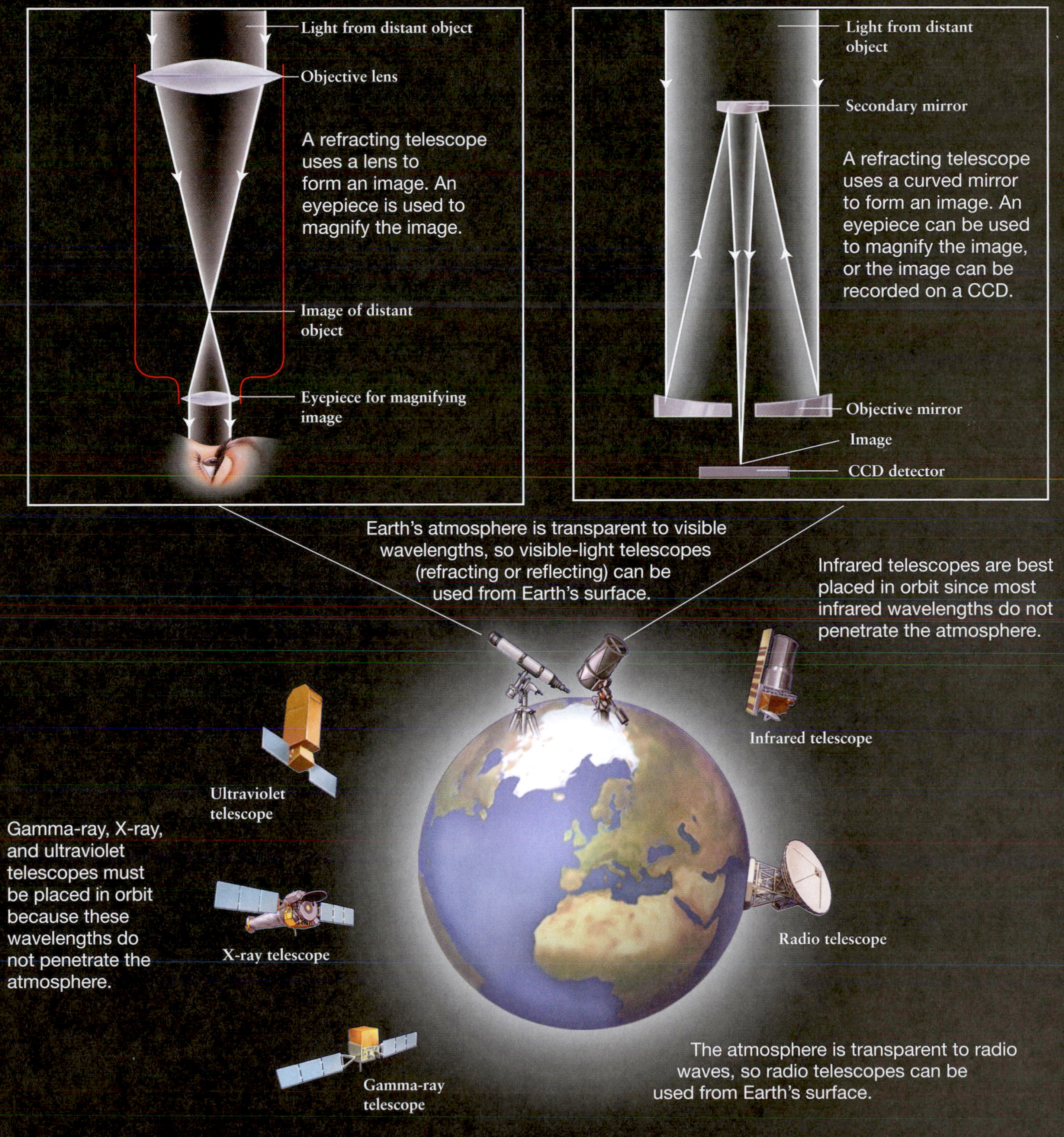

Light from distant object
Objective lens
A refracting telescope uses a lens to form an image. An eyepiece is used to magnify the image.
Image of distant object
Eyepiece for magnifying image
Light from distant object
Secondary mirror
A refracting telescope uses a curved mirror to form an image. An eyepiece can be used to magnify the image, or the image can be recorded on a CCD.
Objective mirror
Image
CCD detector
Earth's atmosphere is transparent to visible wavelengths, so visible-light telescopes (refracting or reflecting) can be used from Earth's surface.
Infrared telescopes are best placed in orbit since most infrared wavelengths do not penetrate the atmosphere.
Infrared telescope
Ultraviolet telescope
Gamma-ray, X-ray, and ultraviolet telescopes must be placed in orbit because these wavelengths do not penetrate the atmosphere.
X-ray telescope
Radio telescope
Gamma-ray telescope
The atmosphere is transparent to radio waves, so radio telescopes can be used from Earth's surface.

KEY WORDS

active optics, p. 153
adaptive optics, p. 153
angular resolution, p. 151
baseline, p. 154
Cassegrain focus, p. 148
charge-coupled device (CCD), p. 154
chromatic aberration, p. 146
coma, p. 150
coudé focus, p. 149
diffraction, p. 151
diffraction grating, p. 156
eyepiece lens, p. 144
false color, p. 153
focal length, p. 142
focal plane, p. 143
focal point, p. 142
focus (of a lens or mirror), p. 142
imaging, p. 154
interferometry, p. 153
lens, p. 142
light-gathering power, p. 145
light pollution, p. 154
magnification (magnifying power), p. 146
medium (*plural* media), p. 142
Newtonian reflector, p. 148
objective lens, p. 144
objective mirror (primary mirror), p. 147
optical telescope, p. 142
optical window (in Earth's atmosphere), p. 159
photometry, p. 155
pixel, p. 154
prime focus, p. 148
radio telescope, p. 157
radio window (in Earth's atmosphere), p. 159
reflecting telescope (reflector), p. 147
reflection, p. 147
refracting telescope (refractor), p. 144
refraction, p. 142
seeing disk, p. 152
spectrograph, p. 155
spectroscopy, p. 155
spherical aberration, p. 150
very-long-baseline interferometry (VLBI), p. 158

KEY IDEAS

Refracting Telescopes: Refracting telescopes, or refractors, produce images by bending light rays as they pass through glass lenses.

• Chromatic aberration is an optical defect whereby light of different wavelengths is bent in different amounts by a lens.

• Glass impurities, chromatic aberration, opacity to certain wavelengths, and structural difficulties make it inadvisable to build extremely large refractors.

Reflecting Telescopes: Reflecting telescopes, or reflectors, produce images by reflecting light rays to a focus point from curved mirrors.

• Reflectors are not subject to most of the problems that limit the useful size of refractors.

Angular Resolution: A telescope's angular resolution, which indicates ability to see fine details, is limited by two key factors.

• Diffraction is an intrinsic property of light waves. Its effects can be minimized by using a larger objective lens or mirror.

• The blurring effects of atmospheric turbulence can be minimized by placing the telescope atop a tall mountain with very smooth air. They can be dramatically reduced by the use of adaptive optics and can be eliminated entirely by placing the telescope in orbit.

Charge-Coupled Devices: Sensitive electronic light detectors called charge-coupled devices (CCDs) are often used at a telescope's focus to digitally record faint images.

Spectrographs: A spectrograph uses a diffraction grating to form the spectrum of an astronomical object.

Radio Telescopes: Radio telescopes use large reflecting dishes to focus radio waves onto a detector.

• Very large dishes provide reasonably sharp radio images. Higher resolution is achieved with interferometry techniques that link smaller dishes together.

Transparency of Earth's Atmosphere: Earth's atmosphere absorbs much of the radiation that arrives from space.

• The atmosphere is transparent chiefly in two wavelength ranges known as the optical window and the radio window. A few wavelengths in the near-infrared also reach the ground.

Telescopes in Space: For observations at wavelengths to which Earth's atmosphere is opaque, astronomers depend on telescopes carried above the atmosphere by rockets or spacecraft.

• Satellite-based observatories provide new information about the universe and permit coordinated observation of the sky at all wavelengths.

QUESTIONS

Review Questions

1. Describe refraction and reflection. Explain how these processes enable astronomers to build telescopes.
2. Explain why a flat piece of glass does not bring light to a focus while a curved piece of glass can.
3. Explain why the light rays that enter a telescope from an astronomical object are essentially parallel.
4. With the aid of a diagram, describe a refracting telescope. Which dimensions of the telescope determine its light-gathering power? Which dimensions determine the magnification?
5. What is the purpose of a telescope eyepiece? What aspect of the eyepiece determines the magnification of the image? In what circumstances would the eyepiece not be used?
6. Do most professional astronomers actually look through their telescopes? Why or why not?
7. Quite often advertisements appear for telescopes that extol their magnifying power. Is this a good criterion for evaluating telescopes? Explain your answer.
8. What is chromatic aberration? For what kinds of telescopes does it occur? How can it be corrected?
9. With the aid of a diagram, describe a reflecting telescope. Describe four different ways in which an astronomer can access the focal plane.
10. Explain some of the disadvantages of refracting telescopes compared to reflecting telescopes.
11. What kind of telescope would you use if you wanted to take a color photograph entirely free of chromatic aberration? Explain your answer.

12. Explain why a Cassegrain reflector can be substantially shorter than a refractor of the same focal length.

13. No major observatory has a Newtonian reflector as its primary instrument, whereas Newtonian reflectors are extremely popular among amateur astronomers. Explain why this is so.

14. What is spherical aberration? How can it be corrected?

15. What is diffraction? Why does it limit the angular resolution of a telescope? What other physical phenomenon is often a more important restriction on angular resolution?

16. What is active optics? What is adaptive optics? Why are they useful? Would either of these be a good feature to include on a telescope to be placed in orbit?

17. Explain why combining the light from two or more optical telescopes can give dramatically improved angular resolution.

18. What is light pollution? What effects does it have on the operation of telescopes? What can be done to minimize these effects?

19. What is a charge-coupled device (CCD)? Why have CCDs replaced photographic film for recording astronomical images?

20. What is a spectrograph? Why do many astronomers regard it as the most important device that can be attached to a telescope?

21. What are the advantages of using a diffraction grating rather than a prism in a spectrograph?

22. Compare an optical reflecting telescope and a radio telescope. What do they have in common? How are they different?

23. Why can radio astronomers make observations at any time during the day, whereas optical astronomers are mostly limited to observing at night? (*Hint:* Does your radio work any better or worse in the daytime than at night?)

24. Why are radio telescopes so large? Why does a single radio telescope have poorer angular resolution than a large optical telescope? How can the resolution be improved by making simultaneous observations with several radio telescopes?

25. What are the optical window and the radio window? Why isn't there an X-ray window or an ultraviolet window?

26. Why is it necessary to keep an infrared telescope at a very low temperature?

27. How are the images made by an X-ray telescope different from those made by a medical X-ray machine?

28. Why must astronomers use satellites and Earth-orbiting observatories to study the heavens at X-ray and gamma-ray wavelengths?

Advanced Questions

Problem-solving tips and tools

You may find it useful to review the small-angle formula discussed in Box 1-1. The area of a circle is proportional to the square of its diameter. Data on the planets can be found in the appendices at the end of this book. Section 5-2 discusses the relationship between frequency and wavelength. Box 6-1 gives examples of how to calculate magnifying power and light-gathering power.

29. Show by means of a diagram why the image formed by a simple refracting telescope is upside down.

30. Ordinary photographs made with a telephoto lens make distant objects appear close. How does the focal length of a telephoto lens compare with that of a normal lens? Explain your reasoning.

31. The observing cage in which an astronomer can sit at the prime focus of the 5-m telescope on Palomar Mountain is about 1 m in diameter. Calculate what fraction of the incoming starlight is blocked by the cage.

32. (**a**) Compare the light-gathering power of the Keck I 10.0-m telescope with that of the Hubble Space Telescope (HST), which has a 2.4-m objective mirror. (**b**) What advantages does Keck I have over HST? What advantages does HST have over Keck I?

33. Suppose your Newtonian reflector has an objective mirror 20 cm (8 in.) in diameter with a focal length of 2 m. What magnification do you get with eyepieces whose focal lengths are (**a**) 9 mm, (**b**) 20 mm, and (**c**) 55 mm? (**d**) What is the telescope's diffraction-limited angular resolution when used with orange light of wavelength 600 nm? (**e**) Would it be possible to achieve this angular resolution if you took the telescope to the summit of Mauna Kea? Why or why not?

34. Several groups of astronomers are making plans for large ground-based telescopes. (**a**) What would be the diffraction-limited angular resolution of a telescope with a 40-meter objective mirror? Assume that yellow light with wavelength 550 nm is used. (**b**) Suppose this telescope is placed atop Mauna Kea. How will the actual angular resolution of the telescope compare to that of the 10-meter Keck I telescope? Assume that adaptive optics is not used.

35. The Hobby-Eberly Telescope (HET) at the McDonald Observatory in Texas has a spherical mirror, which is the least expensive shape to grind. Consequently, the telescope has spherical aberration. Explain why this does not affect the usefulness of HET for spectroscopy. (The telescope is not used for imaging.)

36. The four largest moons of Jupiter are roughly the same size as our Moon and are about 628 million (6.28×10^8) kilometers from Earth at opposition. What is the size in kilometers of the smallest surface features that the Hubble Space Telescope

(resolution of 0.1 arcsec) can detect? How does this compare with the smallest features that can be seen on the Moon with the unaided human eye (resolution of 1 arcmin)?

37. The Hubble Space Telescope (HST) has been used to observe the galaxy M100, some 70 million light-years from Earth. (a) If the angular resolution of the HST image is 0.1 arcsec, what is the diameter in light-years of the smallest detail that can be discerned in the image? (b) At what distance would a U.S. dime (diameter 1.8 cm) have an angular size of 0.1 arcsec? Give your answer in kilometers.

38. At its closest to Earth, Pluto is 28.6 AU from Earth. Can the Hubble Space Telescope distinguish any features on Pluto? Justify your answer using calculations.

39. The Institute of Space and Astronautical Science in Japan proposes to place a radio telescope into an even higher orbit than the HALCA telescope. Using this telescope in concert with ground-based radio-telescopes, baselines as long as 25,000 km may be obtainable. Astronomers want to use this combination to study radio emission at a frequency of 43 GHz from the molecule silicon monoxide, which is found in the interstellar clouds from which stars form. (1 GHz = 1 gigahertz = 10^9 Hz.) (a) What is the wavelength of this emission? (b) Taking the baseline to be the effective diameter of this radio-telescope array, what angular resolution can be achieved?

40. The mission of the Submillimeter Wave Astronomy Satellite (SWAS), launched in 1998, was to investigate interstellar clouds within which stars form. One of the frequencies at which it observed these clouds is 557 GHz (1 GHz = 1 gigahertz = 10^9 Hz), characteristic of the emission from interstellar water molecules. (a) What is the wavelength (in meters) of this emission? In what part of the electromagnetic spectrum is this? (b) Why was it necessary to use a satellite for these observations? (c) SWAS had an angular resolution of 4 arcminutes. What was the diameter of its primary mirror?

41. To search for ionized oxygen gas surrounding our Milky Way Galaxy, astronomers aimed the ultraviolet telescope of the FUSE spacecraft at a distant galaxy far beyond the Milky Way. They then looked for an ultraviolet spectral line of ionized oxygen in that galaxy's spectrum. Were they looking for an emission line or an absorption line? Explain.

42. A sufficiently thick interstellar cloud of cool gas can absorb low-energy X-rays but is transparent to high-energy X-rays and gamma rays. Explain why both Figure 6-32b and Figure 6-32d reveal the presence of cool gas in the Milky Way. Could you infer the presence of this gas from the visible-light image in Figure 6-32a? Explain.

Discussion Questions

43. If you were in charge of selecting a site for a new observatory, what factors would you consider important?

44. Discuss the advantages and disadvantages of using a small telescope in Earth's orbit versus a large telescope on a mountaintop.

Web/eBook Questions

45. Several telescope manufacturers build telescopes with a design called a Schmidt-Cassegrain. These use a correcting lens in an arrangement like that shown in Figure 6-13c. Consult advertisements on the World Wide Web to see the appearance of these telescopes and find out their cost. Why do you suppose they are very popular among amateur astronomers?

46. The Large Zenith Telescope (LZT) in British Columbia, Canada, uses a 6.0-m *liquid* mirror made of mercury. Use the World Wide Web to investigate this technology. How can a liquid metal be formed into the necessary shape for a telescope mirror? What are the advantages of a liquid mirror? What are the disadvantages?

47. Three of the telescopes shown in Figure 6-16—the James Clerk Maxwell Telescope (JCMT), the Caltech Submillimeter Observatory (CSO), and the Submillimeter Array (SMA)—are designed to detect radiation with wavelengths close to 1 mm. Search for current information about JCMT, CSO, and SMA on the World Wide Web. What kinds of celestial objects emit radiation at these wavelengths? What can astronomers see using JCMT, CSO, and SMA that cannot be observed at other wavelengths? Why is it important that they be at high altitude? How large are the primary mirrors used in JCMT, CSO, and SMA? What are the differences among the three telescopes? Which can be used in the daytime? What recent discoveries have been made using JCMT, CSO, or SMA?

48. In 2003 an ultraviolet telescope called GALEX (Galaxy Evolution Explorer) was placed into orbit. Use the World Wide Web to learn about GALEX and its mission. What aspects of galaxies was GALEX designed to investigate? Why is it important to make these observations using ultraviolet wavelengths?

49. At the time of this writing, NASA's plans for the end of the Hubble Space Telescope's mission were uncertain. Consult the Space Telescope Science Institute Web site to learn about plans for HST's final years of operation. Are future space shuttle missions planned to service HST? If so, what changes will be made to HST on such missions? What will become of HST at the end of its mission lifetime?

ACTIVITIES

Observing Projects

50. Obtain a telescope during the daytime along with several eyepieces of various focal lengths. If you can determine the telescope's focal length, calculate the magnifying powers of the eyepieces. Focus the telescope on some familiar object, such as a distant lamppost or tree. **DO NOT FOCUS ON THE SUN! Looking directly at the Sun can cause blindness.** Describe the image you see through the telescope. Is it upside down? How does the image move as you slowly and gently shift the telescope left and right or up and down? Examine the eyepieces, noting their focal lengths. By changing the eyepieces, examine the distant object under different

magnifications. How do the field of view and the quality of the image change as you go from low power to high power?

51. On a clear night, view the Moon, a planet, and a star through a telescope using eyepieces of various focal lengths and known magnifying powers. (To determine the locations in the sky of the Moon and planets, you may want to use the *Starry Night*™ program if you have access. You may also want to consult such magazines as *Sky & Telescope* and *Astronomy* or their Web sites.) In what way does the image seem to degrade as you view with increasingly higher magnification? Do you see any chromatic aberration? If so, with which object and which eyepiece is it most noticeable?

52. Many towns and cities have amateur astronomy clubs. If you are so inclined, attend a "star party" hosted by your local club. People who bring their telescopes to such gatherings are delighted to show you their instruments and take you on a telescopic tour of the heavens. Such an experience can lead to a very enjoyable, lifelong hobby.

53. Use the *Starry Night*™ program to compare the field of view, magnification, and quality of image provided by different optical instruments when observing various celestial objects. Open **Favourites > Explorations > Field of View** to examine several solar system objects. Click the **Find** tab on the left side of the **View** window. Remove any text from the edit box at the top of the **Find** pane, click on the magnifying glass icon at the left side of the edit box, and select the **Orbiting Objects** item from the dropdown menu that appears to bring up a list of solar system objects. Select each of the following objects in turn: the Moon, Jupiter, and Saturn. Double-click on the name of the selected object in the **Find** pane to center it in the view. Click the down arrow to the right of the **Zoom** panel in the toolbar and select first the **7 × 50 Binoculars** from the dropdown menu. Note the change in the quality of the view of the object under observation. Open the **Zoom** dropdown menu again and select the **25mm Plossl** eyepiece on a **Sample 4″ refractor**, followed by the **25mm Plossl** eyepiece on a **Sample 8″ Schmidt-Cassegrain** telescope, followed by the **10mm Plossl** eyepiece on an **8″ Schmidt-Cassegrain.** Before finding the next object, use the **Zoom** dropdown menu to return the field of view to **120°**. (**a**) Describe the change in the level of detail visible in the observed object as the field of view decreases. Open the **Find** pane again. Click on the magnifying glass icon on the left-hand side of the edit box at the top of the **Find** pane and select **Messier Objects** from the dropdown menu that appears. From the Messier Objects listed in the **Find** pane, double-click the entry for **Ring Nebula** to center this object in the view. If nothing appears in the view, open the **Options** side pane and click the checkbox to the left of **Messier Objects** under the **Deep Space** layer. Once more, use the **Zoom** dropdown menu to examine this object through the various instruments. (**b**) Assuming that the 8″ Schmidt-Cassegrain telescope has a focal length of 2000mm, what is the magnification produced by the 25mm and the 10mm eyepieces? (**c**) What is the ratio of the magnification of the 10mm eyepiece to that of the 25mm eyepiece on the 8″ Schmidt-Cassegrain telescope? (**d**) What is the nominal field of view of the 10mm and 25mm eyepieces on the Schmidt-Cassegrain telescope (shown next to the eyepiece name in the **Zoom** dropdown menu)? (**e**) What is the ratio of the field of view of the 10mm eyepiece to the 25mm eyepiece? (Note that this is the inverse ratio to that which you calculated for the magnification yield of the two eyepieces.) (**f**) Describe the relationship between magnification and field of view for a particular telescope.

54. Use *Starry Night*™ to explore the difference between magnification and resolution. Select **Favourites > Explorations > Resolution.** The view is centered upon the full Moon as it might appear to the naked eye from Earth. Right-click the image of the Moon and select **Magnify** from the contextual menu to see the Moon as it might appear in good binoculars or a small telescope. (**a**) Describe the difference between the naked-eye image of the Moon and the magnified image. (**b**) What happens to the field of view when the image is magnified? (**c**) **Zoom** in to a field of view about **10′** wide (this is equivalent to increasing the magnification of the telescope). How does this further magnification affect the quality of the image, that is, the ability to distinguish close details on the lunar surface? (**d**) **Zoom** in to a field of view about **3′** wide. Does the clarity of the image improve, deteriorate, or remain unchanged with this further magnification? (**e**) Slowly **Zoom** in further to a field of view about **1′** wide. What happens to the quality of the image with this increase in magnification (zoom)? (**e**) Explain your observations based on the concept of resolution.

55. Use *Starry Night*™ to explore the effect of light pollution on the night sky. This exercise will also help you to determine the brightness of the faintest stars that are likely to be visible to the naked eye from your location under your present sky conditions and to estimate the fraction of possible stars that you can see with the unaided eye. This exercise is best done outdoors on a dark, clear night with a laptop computer and *Starry Night*™ set to function in night vision mode (select **Options > Night Vision**). Click the **Home** button to ensure that the view is correct for your present location and time. Open the **Options** side pane and expand the **Local View** layer. Turn the **Daylight** option off and then place the cursor over the words **Local Light Pollution** and click the **Local Light Pollution Options…** button that appears. This will open the **Local View Options** dialog window. Move this dialog window to the side of the view. In the **Local View Options** dialog window, click the checkbox to the left of the **Local Light Pollution** option to turn this feature on. You can now use the slide bar to adjust the **Local Light Pollution** level until the view matches your night sky. When you are satisfied that the view matches your sky, click on **OK** to dismiss the **Local View Options** dialog window. Be sure not to change the **Zoom** from the standard field of view, which is 100° wide. Move the cursor over some of the stars that appear on your screen to display their properties in the **HUD** (Heads-up Display). (If necessary, open the **Preferences** dialog from the **File** menu in Windows or **Starry Night** menu on a Mac and add the **apparent magnitude** option to the **Cursor Tracking (HUD)** options.) (**a**) Make a note of the apparent magnitudes of some of the faintest stars in this view and compare these values to the faintest apparent magnitude of about +6 that can be seen under ideal conditions by the human eye. (*Reminder:* Magnitude values increase as stars become fainter.) (**b**) The second goal is to estimate what fraction of the visible stars you can see under these conditions, compared to the total number of stars you might see under ideal conditions. Open the **FOV** pane and click the **Add...** button in the **Other (This Chart)** layer. Select **Rectangular...** from the popup menu and use the **FOV Indicator** dialog window to create a rectangular indicator with a width and height of 10° and click **OK**. If necessary, expand the **Other (This Chart)** layer and click the checkbox for the 10° indicator you created. This will display the indicator in the center of the screen. Without changing the **Zoom** from the standard 100°–wide field of view, use the hand tool to drag the view so that a region of sky with a reasonable number of stars lies within the central 10° indicator. Count and record the number of stars in this square with your present setting. Now open the **Local View Options** dialog window again and set the **Local Light Pollution** to **less** (which essentially gives you ideal conditions) and repeat the count of visible stars. Divide the first number you count by the second. What fraction of the stars that would be visible under ideal conditions were you seeing? (**c**) To see how much light pollution occurs in large cities, adjust the slide bar for **Local Light Pollution** in the **Local View Options** dialog window to the far right for maximum light pollution and repeat the star-counting within the same limited sky region. Compared to viewing under ideal conditions, what fraction of the stars are observers in large urban centers seeing?

Collaborative Exercises

56. Stand up and have everyone in your group join hands, making as large a circle as possible. If a telescope mirror were built as big as your circle, what would be its diameter? What would be your telescope's diffraction-limited angular resolution for blue light? Would atmospheric turbulence have a noticeable effect on the angular resolution?

57. Are there enough students in your class to stand and join hands and make two large circles that recreate the sizes of the two Keck telescopes? Explain how you determined your answer.

ANSWERS

ConceptChecks

ConceptCheck 6-1: The diameter, because it is the width of the opening that determines how much light can be captured by the telescope.

ConceptCheck 6-2: The thin lens bends light less, so, according to Figure 6-5, it must focus light at a more distant location. Thus, the thinner lens has the greater focal length.

ConceptCheck 6-3: The larger the eyepiece focal length, the smaller the magnification.

ConceptCheck 6-4: Reflection off a mirror does not depend on wavelength. This is a key advantage for a reflecting telescope, because it prevents the chromatic aberration that blurs images in telescopes made from lenses.

ConceptCheck 6-5: The angular resolution is decreased for larger telescope diameters. This improves the sharpness of the image, by reducing the angle that light is spread out and blurred by the telescope. Therefore, a larger telescope is better in two ways: sharper images of even dimmer features.

ConceptCheck 6-6: Adaptive optics actuators slightly deform the telescope's mirror to match the apparent movement of a star due to distortions in Earth's atmosphere. The actuators must deform the mirror even more on nights when the atmosphere is fluctuating more rapidly.

ConceptCheck 6-7: Photographic film only captures 2% of the light it receives. For a faint star, there might be very little light captured by the film. CCDs, on the other hand, are 35 times more sensitive, capturing $35 \times 2\% = 70\%$ of the light they receive.

ConceptCheck 6-8: No. Instead of the CCD capturing a focused image of the planet, the spectrograph device spreads light from the planet out in wavelength to form a spectrum. A choice must be made to either look at an image using imaging equipment, or to look at a spectrum using a spectrograph.

ConceptCheck 6-9: If more than one radio dish is connected together, the distance between the two dishes—not the size of the dishes—determines the telescope's overall diameter when considering the angular resolution. Thus, two dishes at 200 m produce sharper images than a single 100-m dish.

ConceptCheck 6-10: Given these three choices, astronomers would much prefer to have a new telescope in the microwave region because X-rays and ultraviolet wavelengths rarely pass through Earth's atmosphere to the ground. While the atmosphere absorbs some portion of microwaves, not all microwaves are absorbed.

ConceptCheck 6-11: Orbiting space telescopes are placed far above most of Earth's atmosphere. Without Earth's atmosphere to absorb infrared, ultraviolet, X-ray, and gamma-ray light, astronomers can study the universe at these wavelengths.

March 10, 2006: The *Mars Reconnaissance Orbiter* spacecraft arrives at Mars (artist's impression). (JPL/NASA)

CHAPTER 7

Comparative Planetology I: Our Solar System

LEARNING GOALS

By reading the sections of this chapter, you will learn

7-1 The important differences between the two broad categories of planets: terrestrial and Jovian

7-2 The similarities and differences among the large planetary satellites, including Earth's Moon

7-3 How the spectrum of sunlight reflected from a planet reveals the composition of its atmosphere and surface

7-4 Why some planets have atmospheres and others do not

7-5 The categories of the many small bodies that also orbit the Sun

7-6 How craters on a planet or satellite reveal the age of its surface and the nature of its interior

7-7 Why a planet's magnetic field indicates a fluid interior in motion

7-8 How the diversity of the solar system is a result of its origin and evolution

As recently as the mid-1900s, astronomers knew precious little about the other worlds that orbit the Sun. Even the best telescopes provided images of the planets that were frustratingly hazy and indistinct. Of asteroids, comets, and the satellites (or moons) of the planets, we knew even less.

Today, our knowledge of the solar system has grown exponentially, due almost entirely to robotic spacecraft. (The illustration shows an artist's impression of one such robotic explorer, the *Mars Reconnaissance Orbiter* spacecraft, as it approached Mars in 2006.) Spacecraft have been sent to fly past all the planets at close range, revealing details unimagined by astronomers of an earlier generation. We have landed spacecraft on the Moon, Venus, and Mars and dropped a probe into the immense atmosphere of Jupiter. This is truly the golden age of solar system exploration.

In this chapter we paint in broad outline our present understanding of the solar system. We will see that the planets come in a variety of sizes and chemical compositions. A rich variety also exists among the moons (or satellites) of the planets and among smaller bodies we call asteroids, comets, and trans-Neptunian objects. We will investigate the nature of craters on the Moon and other worlds of the solar system. And by exploring the magnetic fields of planets, we will be able to peer inside Earth and other worlds and learn about their interior compositions.

An important reason to study the solar system is to search for our own origins. In Chapter 8 we will see how astronomers have used evidence from the present-day solar system to understand how the Sun, Earth, and the other planets formed some four and a half billion years ago, and how they have evolved since then. But for now, we invite you to join us on a guided tour of the worlds that orbit our Sun.

CONCEPTCHECK 7-1

How many stars are in our solar system?

Answer appears at the end of the chapter.

7-1 The solar system has two broad categories of planets: Earthlike and Jupiterlike

Each of the planets that orbit the Sun is unique. Only Earth has liquid water and an atmosphere that humans can breathe; only Venus has a perpetual cloud layer made of sulfuric acid droplets; and only Jupiter has immense storm systems that persist for centuries. But there are also striking similarities among planets. Volcanoes are found not only on Earth but also on Venus and Mars; rings encircle Jupiter, Saturn, Uranus, and Neptune; and impact craters dot the surfaces of Mercury, Venus, Earth, and Mars, showing that all of these planets have been bombarded by interplanetary debris.

How can we make sense of the many similarities and differences among the planets? An important first step is to organize our knowledge of the planets in a systematic way. We can organize this information in two ways. First, we can contrast the orbits of different planets around the Sun; and second, we can compare the planets' physical properties such as size, mass, average density, and chemical composition.

We can understand the most important similarities and differences among the planets by comparing their orbits, masses, and diameters

Comparing the Planets: Orbits

TUTORIAL 7-1 A planet falls naturally into one of two categories according to the size of its orbit. As Figure 7-1 shows, the orbits of the four inner planets (Mercury, Venus, Earth, and Mars) are crowded in close to the Sun. In contrast, the orbits of the next four planets (Jupiter, Saturn, Uranus, and Neptune) are widely spaced at great distances from the Sun. Table 7-1 lists the orbital characteristics of these eight planets.

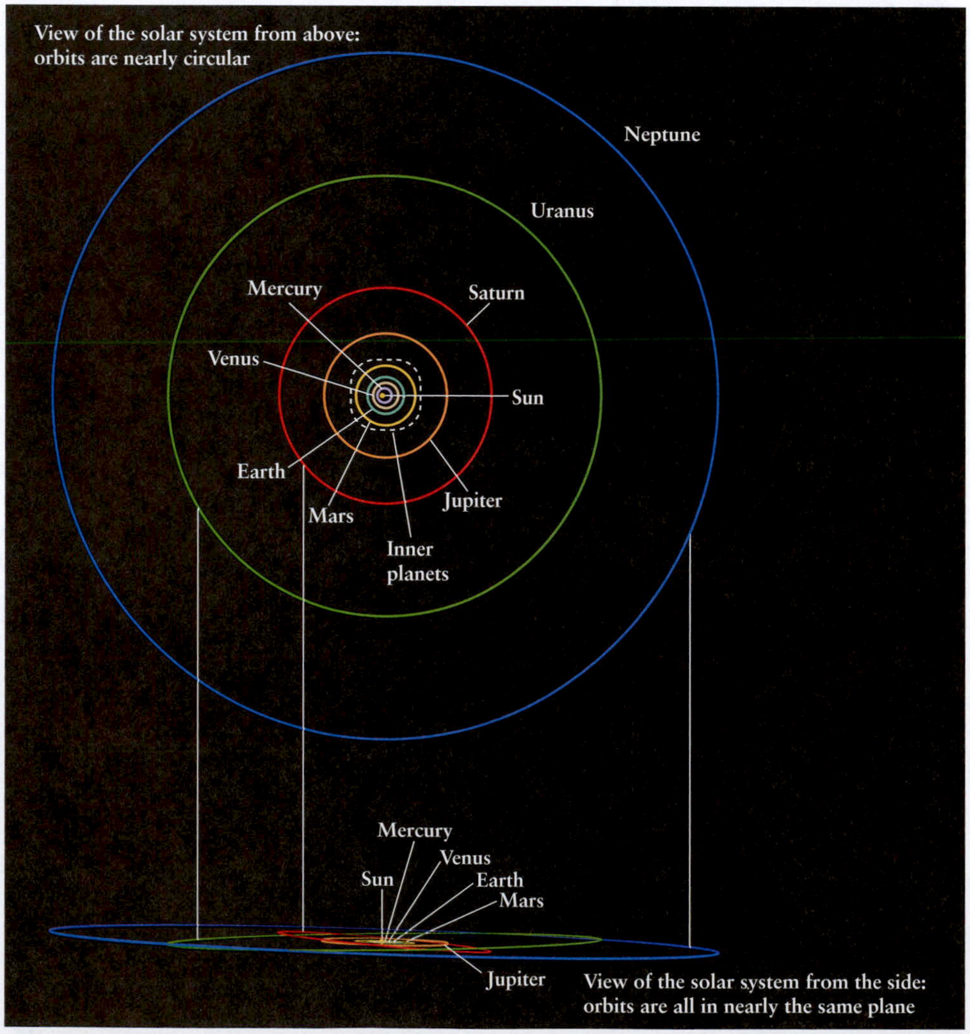

ANIMATION 7-1 **FIGURE 7-1** **The Solar System to Scale** This scale drawing shows the orbits of the planets around the Sun. The four inner planets are crowded in close to the Sun, while the four outer planets orbit the Sun at much greater distances. On the scale of this drawing, the planets themselves would be much smaller than the diameter of a human hair and too small to see.

TABLE 7-1 Characteristics of the Planets

	The Inner (Terrestrial) Planets			
	Mercury	Venus	Earth	Mars
Average distance from the Sun (10^6 km)	57.9	108.2	149.6	227.9
Average distance from the Sun (AU)	0.387	0.723	1.000	1.524
Orbital period (years)	0.241	0.615	1.000	1.88
Orbital eccentricity	0.206	0.007	0.017	0.093
Inclination of orbit to the ecliptic	7.00°	3.39°	0.00°	1.85°
Equatorial diameter (km)	4880	12,104	12,756	6794
Equatorial diameter (Earth = 1)	0.383	0.949	1.000	0.533
Mass (kg)	3.302×10^{23}	4.868×10^{24}	5.974×10^{24}	6.418×10^{23}
Mass (Earth = 1)	0.0553	0.8150	1.0000	0.1074
Average density (kg/m3)	5430	5243	5515	3934

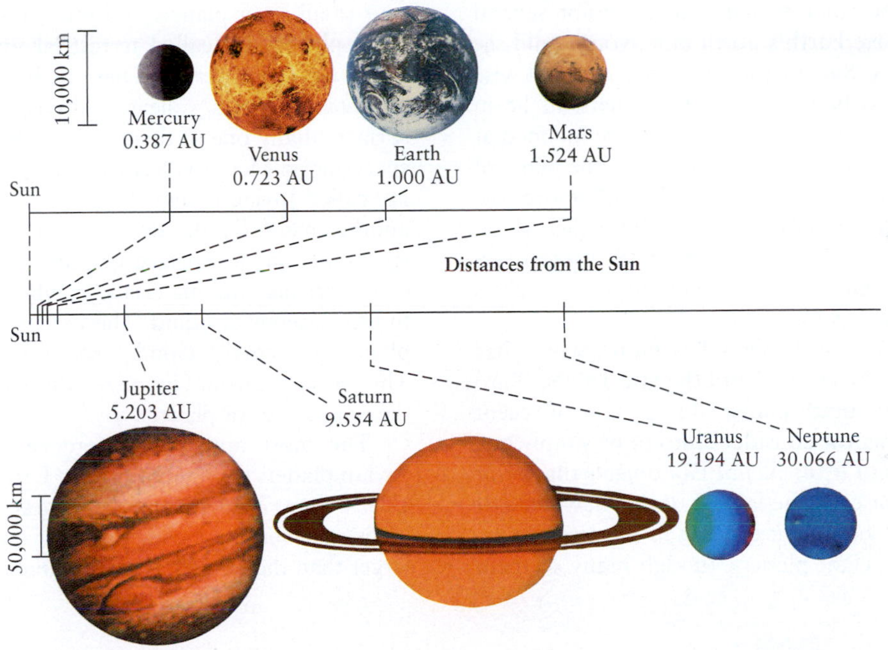

R I V U X G
(NASA)

	The Outer (Jovian) Planets			
	Jupiter	Saturn	Uranus	Neptune
Average distance from the Sun (10^6 km)	778.3	1429	2871	4498
Average distance from the Sun (AU)	5.203	9.554	19.194	30.066
Orbital period (years)	11.86	29.46	84.10	164.86
Orbital eccentricity	0.048	0.053	0.043	0.010
Inclination of orbit to the ecliptic	1.30°	2.48°	0.77°	1.77°
Equatorial diameter (km)	142,984	120,536	51,118	49,528
Equatorial diameter (Earth = 1)	11.209	9.449	4.007	3.883
Mass (kg)	1.899×10^{27}	5.685×10^{26}	8.682×10^{25}	1.024×10^{26}
Mass (Earth = 1)	317.8	95.16	14.53	17.15
Average density (kg/m^3)	1326	687	1318	1638

CAUTION! While Figure 7-1 shows the orbits of the planets, it does not show the planets themselves. The reason is simple: If Jupiter, the largest of the planets, were to be drawn to the same scale as the rest of this figure, it would be a dot just 0.0002 cm across—about $^1/_{300}$ of the width of a human hair and far too small to be seen without a microscope. The planets themselves are *very* small compared to the distances between them. Indeed, while an airplane traveling at 1000 km/h (620 mi/h) can fly around Earth in less than two days, at this speed it would take 17 *years* to fly from Earth to the Sun. The solar system is a very large and very empty place!

Most of the planets have orbits that are nearly circular. As we learned in Section 4-4, Kepler discovered in the seventeenth century that these orbits are actually ellipses. Astronomers denote the elongation of an ellipse by its *eccentricity* (see Figure 4-10b). The eccentricity of a circle is zero, and indeed most of the eight planets (with the notable exception of Mercury) have orbital eccentricities that are very close to zero.

If you could observe the solar system from a point several astronomical units (AU) above Earth's north pole, you would see that all the planets orbit the Sun in the same counterclockwise direction. Furthermore, the orbits of the eight planets all lie in nearly the same plane. In other words, these orbits are inclined at only slight angles to the plane of the ecliptic, which is the plane of Earth's orbit around the Sun (see Section 2-5). What's more, the plane of the Sun's equator is very closely aligned with the orbital planes of the planets. As we will see in Chapter 8, these near-alignments are not a coincidence. They provide important clues about the origin of the solar system.

Not included in Figure 7-1 or Table 7-1 is Pluto, which has an orbit that reaches beyond Neptune. Until the late 1990s, Pluto was generally regarded as the ninth planet. But in light of recent discoveries most astronomers now consider Pluto to be simply one member of a large collection of *trans-Neptunian objects* that orbit far from the Sun. Pluto is not even the largest of this new class of objects! Trans-Neptunian objects orbit the Sun in the same counterclockwise direction as the eight planets, though many of them have orbits that are steeply inclined to the plane of the ecliptic and have high eccentricities (that is, the orbits are quite elongated and noncircular). We will discuss trans-Neptunian objects, along with other small bodies that orbit the Sun, in Section 7-5.

CONCEPTCHECK 7-2

Is Mars classified as an inner planet or an outer planet? Is Mars a terrestrial planet?

CALCULATIONCHECK 7-1

Using Table 7-1, which of the planets has an orbital path that is most nearly a perfect circle in shape?

Answers appear at the end of the chapter.

Comparing the Planets: Physical Properties

TUTORIAL 7-2

When we compare the physical properties of the planets, we again find that they fall naturally into two classes—four small inner planets and four large outer ones. The four small inner planets are called **terrestrial planets** because they resemble Earth (in Latin, *terra*). They all have hard, rocky surfaces with mountains, craters, valleys, and volcanoes. You could stand on the surface of any one of them, although you would need a protective spacesuit on Mercury, Venus, or Mars. The four large outer planets are called **Jovian planets** because they resemble Jupiter. (Jove was another name for the Roman god Jupiter.) An attempt to land a spacecraft on the surface of any of the Jovian planets would be futile, because the materials of which these planets are made are mostly gaseous or liquid. The visible "surface" features of a Jovian planet are actually cloud formations in the planet's atmosphere. The photographs in Figure 7-2 show the distinctive appearances of the two classes of planets.

The most apparent difference between the terrestrial and Jovian planets is their *diameters*. Earth, with its diameter of about 12,756 km (7926 mi), is the largest of the four inner, terrestrial planets. In sharp contrast, the four outer, Jovian planets are much larger than the terrestrial planets. First place goes to Jupiter, whose

FIGURE 7-2 R I V U X G

The Planets to Scale This figure shows the planets from Mercury to Neptune to the same scale. The four terrestrial planets have orbits nearest the Sun, and the Jovian planets are the next four planets from the Sun. (Calvin J. Hamilton and NASA/JPL)

BOX 7-1 ASTRONOMY DOWN TO EARTH

Average Density

Average density—the mass of an object divided by that object's volume—is a useful quantity for describing the differences between planets in our solar system. This same quantity has many applications here on Earth.

A rock tossed into a lake sinks to the bottom, while an air bubble produced at the bottom of a lake (for example, by the air tanks of a scuba diver) rises to the top. These are examples of a general principle: An object sinks in a fluid if its average density is greater than that of the fluid, but rises if its average density is less than that of the fluid. The average density of water is 1000 kg/m^3, which is why a typical rock (with an average density of about 3000 kg/m^3) sinks, while an air bubble (average density of about 1.2 kg/m^3) rises.

At many summer barbecues, cans of soft drinks are kept cold by putting them in a container full of ice. When the ice melts, the cans of diet soda always rise to the top, while the cans of regular soda sink to the bottom. Why is this? The average density of a can of diet soda—which includes water, flavoring, artificial sweetener, and the trapped gas that makes the drink fizzy—is slightly less than the density of water, and so the can floats. A can of regular soda contains sugar instead of artificial sweetener, and the sugar is a bit heavier than the sweetener. The extra weight is just enough to make the average density of a can of regular soda slightly more than that of water, making the can sink. (You can test these statements for yourself by putting unopened cans of diet soda and regular soda in a sink or bathtub full of water.)

The concept of average density provides geologists with important clues about the early history of Earth. The average density of surface rocks on Earth, about 3000 kg/m^3, is less than Earth's average density of 5515 kg/m^3. The simplest explanation is that in the ancient past, Earth was completely molten throughout its volume, so that low-density materials rose to the surface and high-density materials sank deep into Earth's interior in a process called *chemical differentiation.* This series of events also suggests that Earth's core must be made of relatively dense materials, such as iron and nickel. A tremendous amount of other geological evidence has convinced scientists that this picture is correct.

equatorial diameter is more than 11 times that of Earth. On the other end of the scale, Mercury's diameter is less than two-fifths that of Earth. Figure 7-2 shows the Sun and the planets drawn to the same scale. The diameters of the planets are given in Table 7-1.

The *masses* of the terrestrial and Jovian planets are also dramatically different. If a planet has a moon, you can calculate the planet's mass from the moon's period and semimajor axis by using Newton's form of Kepler's third law (see Section 4-7 and Box 4-4). Astronomers have also measured the mass of each planet by sending a spacecraft to pass near the planet. The planet's gravitational pull (which is proportional to its mass) deflects the spacecraft's path, and the amount of deflection tells us the planet's mass. Using these techniques, astronomers have found that the four Jovian planets have masses that range from tens to hundreds of times greater than the mass of any of the terrestrial planets. Again, first place goes to Jupiter, whose mass is 318 times greater than Earth's.

Once we know the diameter and mass of a planet, we can learn something about what that planet is made of. The trick is to calculate the planet's **average density,** or mass divided by volume, measured in kilograms per cubic meter (kg/m^3). The average density of any substance depends in part on that substance's composition. For example, air near sea level on Earth has an average density of 1.2 kg/m^3, water's average density is 1000 kg/m^3, and a piece of concrete has an average density of 2000 kg/m^3. Box 7-1 describes some applications of the idea of average density to everyday phenomena on Earth.

The four inner, terrestrial planets have very high average densities (see Table 7-1); the average density of Earth, for example, is 5515 kg/m^3. By contrast, a typical rock found on Earth's surface has a lower average density, about 3000 kg/m^3. Thus, Earth must contain a large amount of material that is denser than rock. This information provides our first clue that terrestrial planets have dense iron cores.

In sharp contrast, the outer, Jovian planets have quite low densities. Saturn has an average density less than that of water. This information strongly suggests that the giant outer planets are composed primarily of light elements such as hydrogen and helium. All four Jovian planets probably have large cores of mixed rock and highly compressed water that are buried beneath low-density outer layers tens of thousands of kilometers thick.

We can conclude that the following general rule applies to the planets:

The terrestrial planets are made of rocky materials and have dense iron cores. These planets have solid surfaces and high average densities. The Jovian planets are composed primarily of light elements such as hydrogen and helium, resulting in low average densities. These planets have interiors made mostly of gas and liquid, and have no solid surface.

CONCEPTCHECK 7-3

A planet's average density can be estimated by measuring its size and how much the planet's gravity deflects a nearby spacecraft's path. If the density of rocks recovered from a planet's surface is lower than the planet's average density, what can one infer about the density of the planet's core?

CALCULATIONCHECK 7-2

If Earth's diameter is 12,756 km and Saturn's diameter is 120,536 km, how many Earths could fit across the diameter of Saturn?

Answers appear at the end of the chapter.

7-2 Seven large satellites are almost as big as the terrestrial planets

All the planets except Mercury and Venus have moons (also called satellites). At least 170 satellites are known: Earth has 1 (the Moon), Mars has 2, Jupiter has at least 66, Saturn at least 62, Uranus at least 27, and Neptune at least 13. Dozens of other small satellites probably remain to be discovered as our telescope technology continues to improve. Like the terrestrial planets, all of the satellites of the planets have solid surfaces.

You can see that there is a striking difference between the terrestrial planets, with few or no satellites, and the Jovian planets, each of which has so many moons that it resembles a miniature solar system. In Chapter 8 we will explore this evidence (as well as other evidence) that the Jovian planets formed in a manner similar to the solar system as a whole, but on a smaller scale.

> The various moons of the planets are not simply copies of Earth's Moon

Seven of the Jovian satellites are roughly as big as the planet Mercury! Table 7-2 lists these satellites and shows them to the same scale. Note that Earth's Moon and Jupiter's satellites Io and Europa have relatively high average densities, indicating that these moons are made primarily of rocky materials. By contrast, the average densities of Ganymede, Callisto, Titan, and Triton are all relatively low. Planetary scientists conclude that the interiors of these four moons also contain substantial amounts of water ice, which is less dense than rock. (In Section 7-4 we will learn about types of frozen "ice" made of substances other than water.)

CAUTION! Water ice may seem like a poor material for building a satellite, since the ice you find in your freezer can easily be cracked or crushed. But under high pressure, such as is found in the interior of a large satellite, water ice becomes as rigid as rock. (It also becomes denser than the ice found in ice cubes, although not as dense as rock.) Note that water ice is an important constituent only for satellites in the outer solar system, where the Sun is far away and temperatures are very low. For example, the surface temperature of Titan is a frigid 95 K (2178°C = 2288°F). In Section 7-4 we will learn more about the importance of temperature in determining the composition of a planet or satellite.

The satellites listed in Table 7-2 are actually unusually large. Most of the known satellites have diameters less than 2000 km, and many are irregularly shaped and just a few kilometers across.

Interplanetary spacecraft have made many surprising and fascinating discoveries about the satellites of the solar system. We now know that Jupiter's satellite Io is the most geologically active world in the solar system, with numerous volcanoes that continually belch forth sulfur-rich compounds. The fractured surface of Europa, another of Jupiter's large satellites, suggests that a worldwide ocean of liquid water may lie beneath its icy surface. On Saturn's moon Titan, lakes of hydrocarbons, the primary source of energy on Earth, have been found near the poles—some the size of Lake Superior! In 2008, the space probe *Cassini* flew right through the plumes of geysers on Saturn's moon Enceladus as they ejected water-ice, dust, and gas. While Jupiter and Saturn lack a solid surface, their many moons contain a rich variety of features.

TABLE 7-2 The Seven Giant Satellites

	Moon	Io	Europa	Ganymede	Callisto	Titan	Triton
Parent planet	Earth	Jupiter	Jupiter	Jupiter	Jupiter	Saturn	Neptune
Diameter (km)	3476	3642	3130	5268	4806	5150	2706
Mass (kg)	7.35×10^{22}	8.93×10^{22}	4.80×10^{22}	1.48×10^{23}	1.08×10^{23}	1.34×10^{23}	2.15×10^{22}
Average density (kg/m^3)	3340	3530	2970	1940	1850	1880	2050
Substantial atmosphere?	No	No	No	No	No	Yes	No

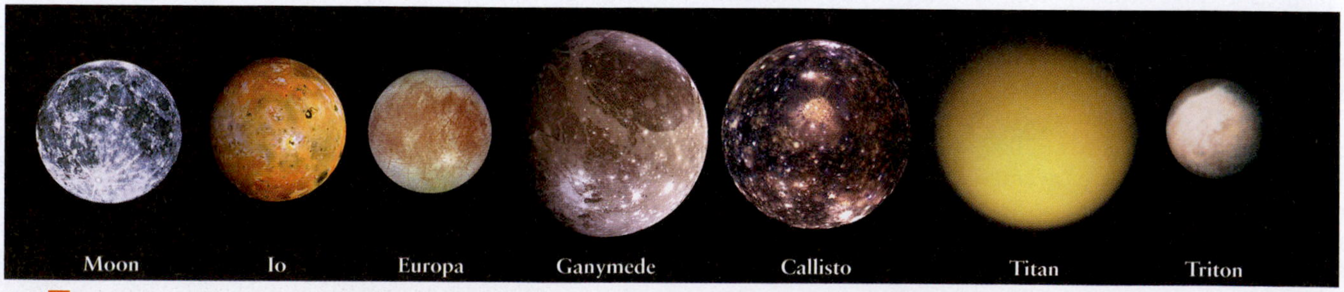

R I V U X G (NASA/JPL/Space Science Institute)

CONCEPTCHECK 7-4

How many moons in the solar system are larger than Earth's Moon (see Table 7-2)?

Answer appears at the end of the chapter.

7-3 Spectroscopy reveals the chemical composition of the planets

The light we receive from a planet or satellite is reflected sunlight—but with revealing differences in its spectrum

As we have seen, the average densities of the planets and satellites give us a crude measure of their **chemical compositions**—that is, what substances they are made of. For example, the low average density of the Moon (3340 kg/m^3) compared with Earth (5515 kg/m^3) tells us that the Moon contains relatively little iron or other dense metals. But to truly understand the nature of the planets and satellites, we need to know their chemical compositions in much greater detail than we can learn from average density alone.

The most accurate way to determine chemical composition is by directly analyzing samples taken from a planet's atmosphere and soil. Unfortunately, of all the planets and satellites, we have such direct information only for Earth and the four worlds on which spacecraft have landed—Venus, the Moon, Mars, and Titan. In all other cases, astronomers must analyze sunlight reflected from the distant planets and their satellites. To do that, astronomers bring to bear one of their most powerful tools, **spectroscopy,** the systematic study of spectra and spectral lines. (We discussed spectroscopy in Sections 5-6 and 6-5.)

Determining Atmospheric Composition

Spectroscopy is a sensitive probe of the composition of a planet's *atmosphere*. If a planet has an atmosphere, then sunlight reflected from that planet must have passed through its atmosphere before coming back out. During this passage, some of the wavelengths of sunlight will have been absorbed. Hence, the spectrum of this reflected sunlight will have dark absorption lines. Astronomers look at the particular wavelengths absorbed and the amount of light absorbed at those wavelengths. Both of these depend on the kinds of chemicals present in the planet's atmosphere and the abundance of those chemicals.

For example, astronomers have used spectroscopy to analyze the atmosphere of Saturn's largest satellite, Titan (see Figure 7-3a and Table 7-2). The graph in Figure 7-3b shows the spectrum of visible sunlight reflected from Titan. (We first saw this method

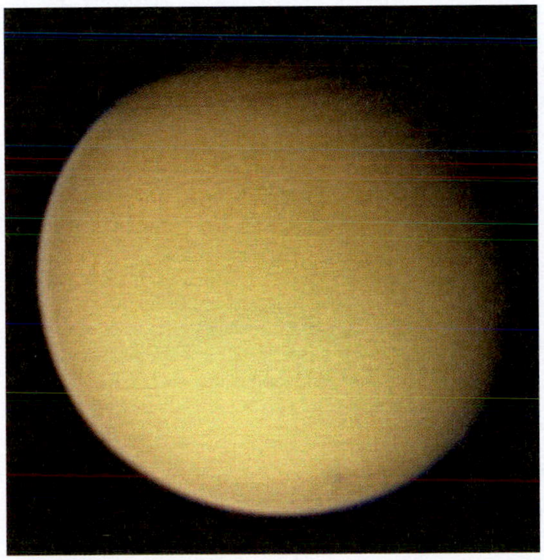

(a) Saturn's satellite Titan

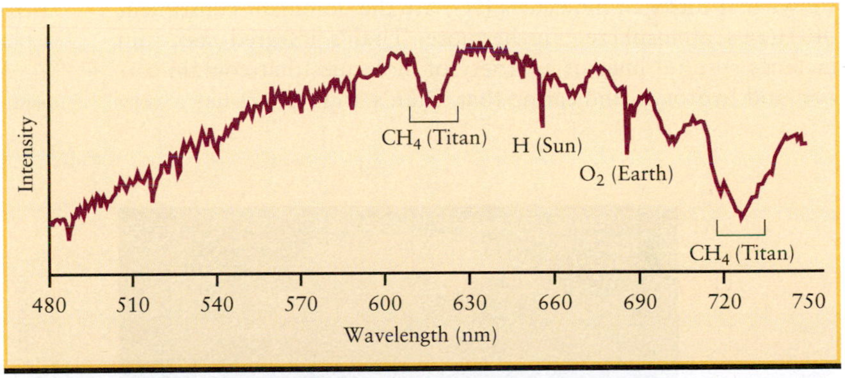

(b) The spectrum of sunlight reflected from Titan

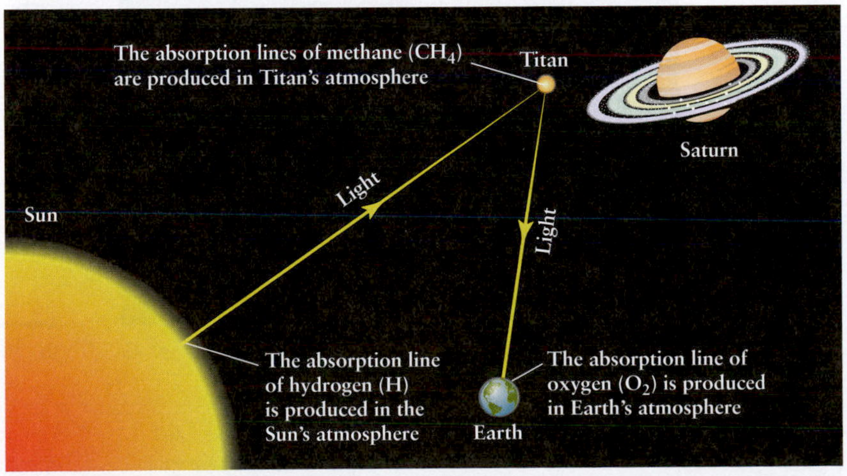

(c) Interpreting Titan's spectrum

FIGURE 7-3 R I V U X G

Analyzing a Satellite's Atmosphere through Its Spectrum

(a) Titan is the only satellite in the solar system with a substantial atmosphere. **(b)** The dips in the spectrum of sunlight reflected from Titan are due to absorption by hydrogen atoms (H), oxygen molecules (O_2), and methane molecules (CH_4). Of these, only methane is actually present in Titan's atmosphere. **(c)** This illustration shows the path of the light that reaches us from Titan. To interpret the spectrum of this light as shown in (b), astronomers must account for the absorption that takes place in the atmospheres of the Sun and Earth. (a: NASA/JPL/Space Science Institute) of displaying spectra in Figure 6-21.) The dips in this curve of intensity versus wavelength represent absorption lines. However, not all of these absorption lines are produced in the atmosphere of Titan (Figure 7-3c). Before reaching Titan, light from the Sun's glowing surface must pass through the Sun's own hydrogen-rich atmosphere. This produces the hydrogen absorption line in Figure 7-3b at a wavelength of 656 nm. After being reflected from Titan, the light must pass through Earth's atmosphere before reaching the observing telescope; this is where the oxygen absorption line in Figure 7-3b is produced. Only the two dips near 620 nm and 730 nm are caused by gases in Titan's atmosphere.

These two absorption lines are caused not by individual atoms in the atmosphere of Titan but by atoms combined to form molecules. (We introduced the idea of molecules in Section 5-6.) Molecules, like atoms, also produce unique patterns of lines in the spectra of astronomical objects. The absorption lines in Figure 7-3b indicate the presence in Titan's atmosphere of methane molecules (CH_4, a molecule made of one carbon atom and four hydrogen atoms). Titan must be a curious place indeed, because on Earth, methane is a rather rare substance that is the primary ingredient in natural gas! When we examine other planets and satellites with atmospheres, we find that all of their spectra have absorption lines of molecules of various types.

In addition to visible-light measurements such as those shown in Figure 7-3b, it is very useful to study the *infrared* and *ultraviolet* spectra of planetary atmospheres. Many molecules have much stronger spectral lines in these nonvisible wavelength bands than in the visible. As an example, the ultraviolet spectrum of Titan shows that nitrogen molecules (N_2) are the dominant constituent of Titan's atmosphere. Furthermore, Titan's infrared spectrum includes spectral lines of a variety of molecules that contain carbon and hydrogen, indicating that Titan's atmosphere has a very complex chemistry. None of these molecules could have been detected by visible light alone. Because Earth's atmosphere is largely opaque to infrared and ultraviolet wavelengths (see Section 6-7), telescopes in space are important tools for these spectroscopic studies of the solar system.

It is a testament to the power of spectroscopy that when the robotic spacecraft *Huygens* landed on Titan in 2005, its onboard instruments confirmed the presence of methane and nitrogen in Titan's atmosphere—just as had been predicted years before by spectroscopic observations.

CONCEPTCHECK 7-5

If planets reflect some of the Sun's light rather than emitting visible light of their own, how can spectroscopy reveal information about a planet's atmosphere?

Answer appears at the end of the chapter.

Determining Surface Composition

Spectroscopy can also provide useful information about the *solid surfaces* of planets and satellites without atmospheres. When light shines on a solid surface, some wavelengths are absorbed while others are reflected. (For example, a plant leaf absorbs red and violet light but reflects green light—which is why leaves look green.) Unlike a gas, a solid illuminated by sunlight does not produce sharp, definite spectral lines. Instead, only broad absorption features appear in the spectrum. By comparing such a spectrum with the spectra of samples of different substances on Earth, astronomers can infer the chemical composition of the surface of a planet or satellite.

As an example, Figure 7-4a shows Jupiter's satellite Europa (see Table 7-2), and Figure 7-4b shows the infrared spectrum of

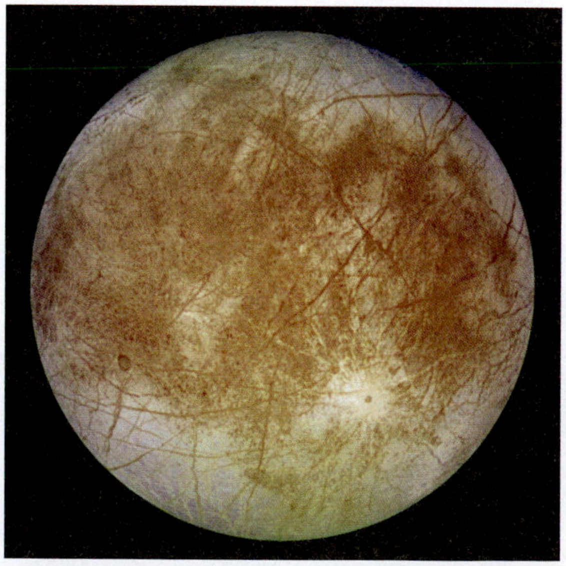

(a) Jupiter's moon Europa

R I V U X G

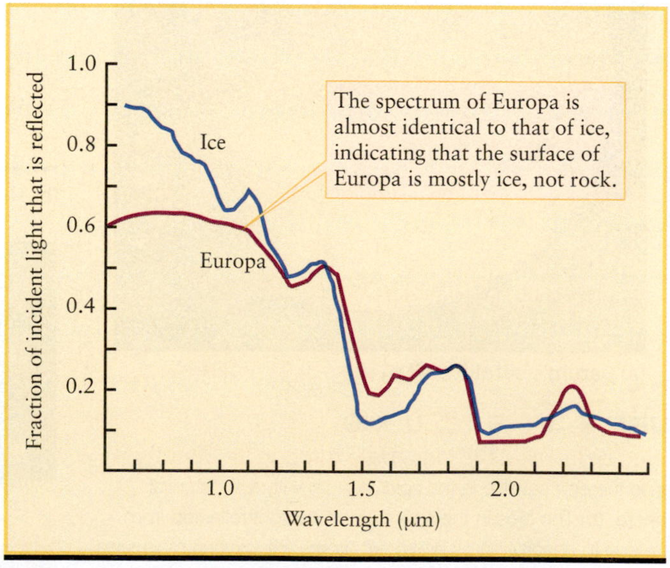

Fraction of incident light that is reflected
1.0
0.8
0.6
0.4
0.2
Ice
Europa
The spectrum of Europa is almost identical to that of ice, indicating that the surface of Europa is mostly ice, not rock.
1.0
1.5
2.0
Wavelength (μm)

(b) The spectum of light reflected from Europa

FIGURE 7-4

Analyzing a Satellite's Surface from Its Spectrum (a) Unlike Titan (Figure 7-3a), Jupiter's satellite Europa has no atmosphere. (b) Infrared light from the Sun that is reflected from the surface of Europa has nearly the same spectrum as sunlight reflected from ordinary water ice.

light reflected from the surface of Europa. Because this spectrum is so close to that of water ice—that is, frozen water—astronomers conclude that water ice is the dominant constituent of Europa's surface. (We saw in Section 7-2 that water ice cannot be the dominant constituent of Europa's *interior,* because this satellite's density is too high.)

Unfortunately, spectroscopy tells us little about what the material is like just below the surface of a satellite or planet. For this purpose, there is simply no substitute for sending a spacecraft to a planet and examining its surface directly.

CONCEPTCHECK 7-6

How is the spectrum of reflected sunlight from a solid planetary surface different from the spectrum produced when sunlight passes through a planet's gaseous atmosphere?

Answer appears at the end of the chapter.

7-4 The Jovian planets are made of lighter elements than the terrestrial planets

Spectroscopic observations from Earth and spacecraft show that the outer layers of the Jovian planets are composed primarily of the lightest gases, hydrogen and helium (see Box 5-5). In contrast, chemical analysis of soil samples from Venus, Earth, and Mars demonstrate that the terrestrial planets are made mostly of heavier elements, such as iron, oxygen, silicon, magnesium, nickel, and sulfur. Spacecraft images such as Figure 7-5 and Figure 7-6 only hint at these striking differences in chemical composition, which are summarized in Table 7-3.

> Differences in distance from the Sun, and hence in temperature, explain many distinctions between terrestrial and Jovian planets

Temperature plays a major role in determining whether the materials of which planets are made exist as solids, liquids, or gases. Hydrogen (H_2) and helium (He) are gaseous except at

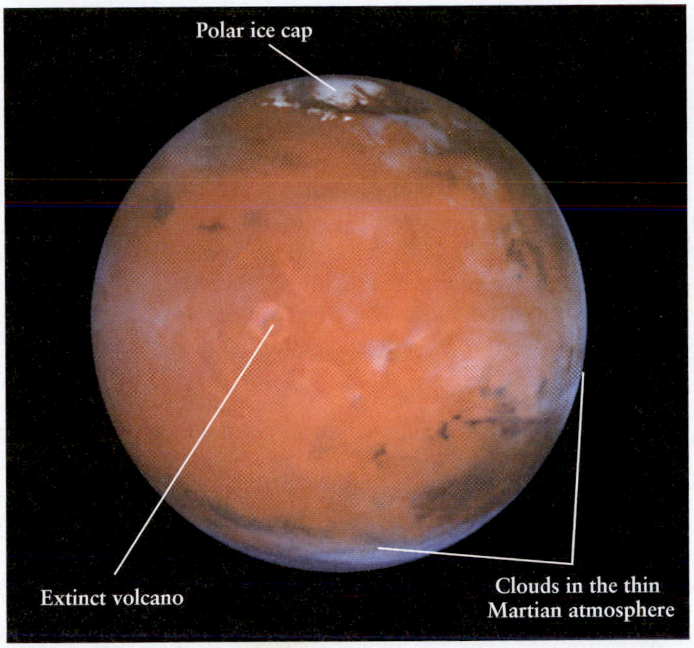

FIGURE 7-6 R I V U X G
A Terrestrial Planet Mars is composed mostly of heavy elements such as iron, oxygen, silicon, magnesium, nickel, and sulfur. The planet's red surface can be seen clearly in this Hubble Space Telescope image because the Martian atmosphere is thin and nearly cloudless. Olympus Mons, the extinct volcano to the left of center, is nearly 3 times the height of Mount Everest. (Space Telescope Science Institute/JPL/NASA)

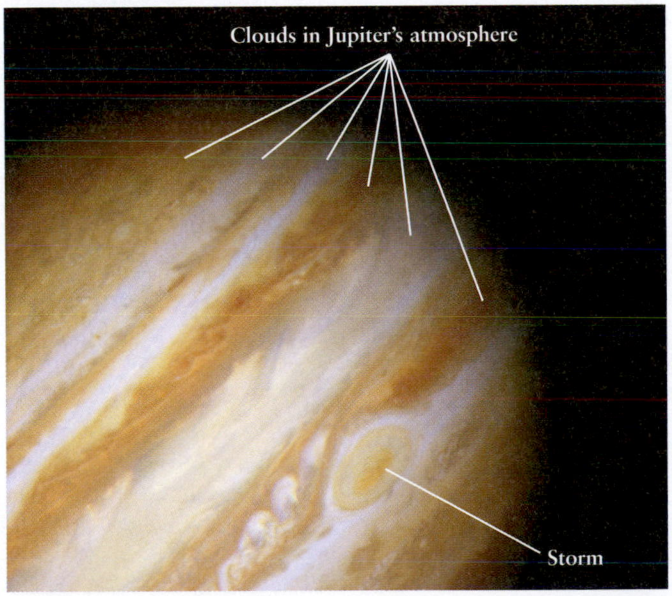

FIGURE 7-5 R I V U X G
A Jovian Planet This Hubble Space Telescope image gives a detailed view of Jupiter's cloudtops. Jupiter is composed mostly of the lightest elements, hydrogen and helium, which are colorless; the colors in the atmosphere are caused by trace amounts of other substances. The giant storm at lower right, called the Great Red Spot, has been raging for more than 300 years. (Space Telescope Science Institute/JPL/NASA)

TABLE 7-3 Comparing Terrestrial and Jovian Planets

	Terrestrial Planets	Jovian Planets
Distance from the Sun	Less than 2 AU	More than 5 AU
Size	Small	Large
Composition	Mostly rocky materials containing iron, oxygen, silicon, magnesium, nickel, and sulfur	Mostly light elements such as hydrogen and helium
Density	High	Low

extremely low temperatures and extraordinarily high pressures. By contrast, rock-forming substances such as iron and silicon are solids except at temperatures well above 1000 K. (You may want to review the discussion of temperature scales in Box 5-1.) Between these two extremes are substances such as water (H_2O), carbon dioxide (CO_2), methane (CH_4), and ammonia (NH_3), which solidify at low temperatures (from below 100 to 300 K) into solids called **ices.** (In astronomy, frozen water is just one kind of "ice.") At somewhat higher temperatures, they can exist as liquids or gases. For example, clouds of ammonia ice crystals are found in the cold upper atmosphere of Jupiter, but within Jupiter's warmer interior, ammonia exists primarily as a liquid.

CAUTION! The Jovian planets are sometimes called "gas giants." It is true that their primary constituents, including hydrogen, helium, ammonia, and methane, are gases under normal conditions on Earth. But in the interiors of these planets, pressures are so high that these substances are *liquids,* not gases. The Jovian planets might be better described as "liquid giants"!

As you might expect, a planet's surface temperature is related to its distance from the Sun. The four inner planets are quite warm. For example, midday temperatures on Mercury may climb to 700 K (= 427°C = 801°F), and during midsummer on Mars, it is sometimes as warm as 290 K (= 17°C = 63°F). The outer planets, which receive much less solar radiation, are cooler. Typical temperatures range from about 125 K (= −148°C = −234°F) in Jupiter's upper atmosphere to about 55 K (= −218°C = −360°F) at the tops of Neptune's clouds.

How Temperature Affects Atmospheres

The higher surface temperatures of the terrestrial planets help to explain the following observation: The atmospheres of the terrestrial planets contain virtually *no* hydrogen molecules or helium atoms. Instead, the atmospheres of Venus, Earth, and Mars are composed of heavier molecules such as nitrogen (N_2, 14 times more massive than a hydrogen molecule), oxygen (O_2, 16 times more massive), and carbon dioxide (22 times more massive). To understand the connection between surface temperature and the absence of hydrogen and helium, we need to know a few basic facts about gases.

The temperature of a gas is directly related to the speeds at which the atoms or molecules of the gas move: The higher the gas temperature, the greater the speed of its atoms or molecules. Furthermore, for a given temperature, lightweight atoms and molecules move more rapidly than heavy ones. On the four inner, terrestrial planets, where atmospheric temperatures are high, low-mass hydrogen molecules and helium atoms move so swiftly that they can escape from the relatively weak gravity of these planets. Hence, the atmospheres that surround the terrestrial planets are composed primarily of more massive, slower-moving molecules such as CO_2, N_2, O_2, and water vapor (H_2O). On the four Jovian planets, low temperatures and relatively strong gravity prevent even lightweight hydrogen and helium gases from escaping into space, and so their atmospheres are much more extensive. The combined mass of Jupiter's atmosphere, for example, is about a million (10^6) times greater than that of Earth's atmosphere. This is comparable to the entire mass of Earth! Box 7-2 describes more about the ability of a planet's gravity to retain gases.

CONCEPTCHECK 7-7

If the average speed of hydrogen molecules in Earth's atmosphere is below Earth's escape speed, why do Earth's atmospheric hydrogen molecules slowly leak into space?

Answer appears at the end of the chapter.

7-5 Small chunks of rock and ice also orbit the Sun

The smaller bodies of the solar system contain important clues about its origin and evolution

In addition to the eight planets, many smaller objects orbit the Sun. These objects fall into three broad categories: asteroids, which are rocky objects found in the inner solar system; trans-Neptunian objects, which are found beyond Neptune in the outer solar system and contain both rock and ice; and comets, which are mixtures of rock and ice that originate in the outer solar system but can venture close to the Sun.

Asteroids

Within the orbit of Jupiter are hundreds of thousands of rocky objects called **asteroids.** There is no sharp dividing line between planets and asteroids, which is why asteroids are also called **minor planets.** The largest asteroid, Ceres, has a diameter of about 900 km (around one-quarter our Moon's diameter). The next largest, Pallas and Vesta, are each about 500 km in diameter. Still smaller ones, like the asteroid shown in close-up in Figure 7-7, are more numerous. Hundreds of thousands of kilometer-sized

FIGURE 7-7 R I V U X G

An Asteroid The asteroid shown in this image, 433 Eros, is only 33 km (21 mi) long and 13 km (8 mi) wide—about the same size as the island of Manhattan. Because Eros is so small, its gravity is too weak to have pulled it into a spherical shape. This image was taken in 2000 by NEAR Shoemaker, the first spacecraft to orbit around and land on an asteroid. (NEAR Project, NLR, JHUAPL, Goddard SVS, NASA) asteroids are known, and there are probably hundreds of thousands more that are boulder-sized or smaller. All of these objects orbit the Sun in the same direction as the planets.

Most (although not all) asteroids orbit the Sun at distances of 2 to 3.5 AU. This region of the solar system between the orbits of Mars and Jupiter is called the **asteroid belt.**

CAUTION! One common misconception about asteroids is that they are the remnants of an ancient planet that somehow broke apart or exploded, like the fictional planet Krypton in the comic book adventures of Superman. In fact, the combined mass of all the asteroids (including Ceres, Pallas, and Vesta) is much less than the mass of the Moon, and they were probably never part of any planet-sized body. The early solar system is thought to have been filled with asteroidlike objects, most of which were incorporated into the planets. The "leftover" objects that missed out on this process make up our present-day population of asteroids.

CONCEPTCHECK 7-8

Is the largest asteroid, Ceres, about the same size as a large city, a large U.S. state, a large country, or Earth itself?

Answer appears at the end of the chapter.

Trans-Neptunian Objects

While asteroids are the most important small bodies in the inner solar system, the outer solar system is the realm of the **trans-Neptunian objects.** As the name suggests, these are small bodies whose orbits lie beyond the orbit of Neptune. The first of these to be discovered (in 1930) was Pluto, with a diameter of only 2274 km. This is larger than any asteroid, but smaller than any planet or any of the moons listed in Table 7-2. The orbit of Pluto has a greater semimajor axis (39.54 AU), is more steeply inclined to the ecliptic (17.15°), and has a greater eccentricity (0.250) than that of any of the planets (Figure 7-8). In fact, Pluto's noncircular orbit sometimes takes it nearer the Sun than Neptune. (Happily, the orbits of Neptune and Pluto are such that they will never collide.) Pluto's density is only 2000 kg/m^3, about the same as Neptune's moon Triton, shown in Table 7-2. Hence, its composition is thought to be a mixture of about 70% rock and 30% ice.

Since 1992 astronomers have discovered more than 900 other trans-Neptunian objects, and are discovering more each year. Like asteroids, the vast majority of trans-Neptunian objects orbit the Sun in the same direction as the planets. A handful of trans-Neptunian objects are comparable in size to Pluto; at least one, Eris, is even larger than Pluto, as well as being in an orbit that is much larger, more steeply inclined, and more eccentric (Figure 7-8).

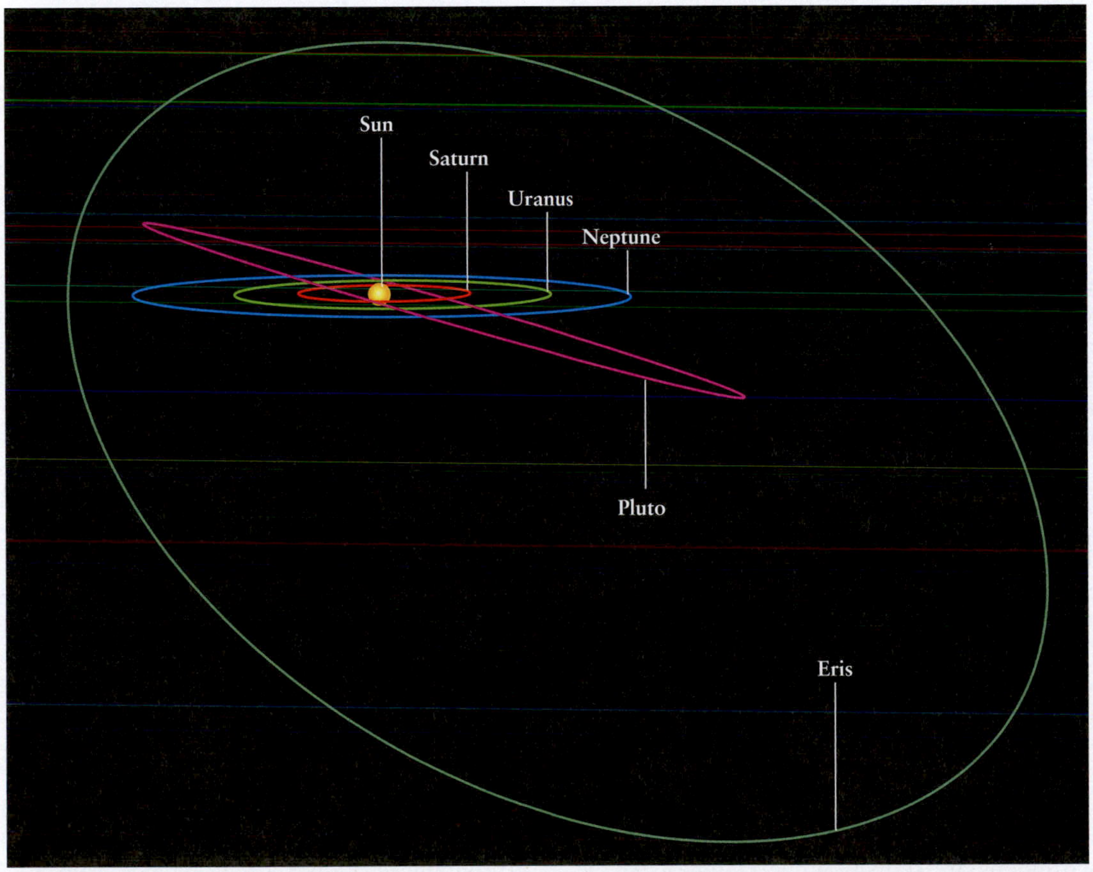

FIGURE 7-8
Trans-Neptunian Objects Pluto and Eris are the two largest trans-Neptunian objects, small worlds of rock and ice that orbit beyond Neptune. Unlike the orbits of the planets, the orbits of these two objects are steeply inclined to the ecliptic: Pluto's orbit is tilted by about 17°, and that of Eris is tilted by 44°.

TUTORIAL 7-3

BOX 7-2 TOOLS OF THE ASTRONOMER'S TRADE

Kinetic Energy, Temperature, and Whether Planets Have Atmospheres

A moving object possesses energy as a result of its motion. The faster it moves, the more energy it has (see Figure 4-20). Energy of this type is called **kinetic energy.** If an object of mass m is moving with a speed v its kinetic energy E_k is given by

Kinetic energy of an object

$$E_k = \frac{1}{2} mv^2$$

E_k = kinetic energy of an object

m = mass of object

v = speed of object

This expression for kinetic energy is valid for all objects, both big and small, from atoms and molecules to planets and stars, as long as their speeds are slow in relation to the speed of light. If the mass is expressed in kilograms and the speed in meters per second, the kinetic energy is expressed in joules (J).

EXAMPLE: An automobile of mass 1000 kg driving at a typical freeway speed of 30 m/s (= 108 km/h = 67 mi/h) has a kinetic energy of

$$E_k = \frac{1}{2}(1000 \times 30^2) = 450{,}000 \text{ J} = 4.5 \times 10^5 \text{ J}$$

Consider a gas, such as the atmosphere of a star or planet. Some of the gas atoms or molecules will be moving slowly, with little kinetic energy, while others will be moving faster and have more kinetic energy. The temperature of the gas is a direct measure of the *average* amount of kinetic energy per atom or molecule. The hotter the gas, the faster atoms or molecules move, on average, and the greater the average kinetic energy of an atom or molecule (see Figure 5-11).

If the gas temperature is sufficiently high, typically several thousand kelvins, molecules move so fast that when they collide with one another, the energy of the collision can break the molecules apart into their constituent atoms. Thus, the Sun's atmosphere, where the temperature is 5800 K, consists primarily of individual hydrogen atoms rather than hydrogen molecules. By contrast, the hydrogen atoms in Earth's atmosphere (temperature 290 K) are combined with oxygen atoms into molecules of water vapor (H_2O).

The physics of gases tells us that in a gas of temperature T (in kelvins), the average kinetic energy of an atom or molecule is

Kinetic energy of a gas atom or molecule

$$E_k = \frac{3}{2} kT$$

E_k = average kinetic energy of a gas atom or molecule, in joules

k = 1.38×10^{-23} J/K

T = temperature of gas, in kelvins

The quantity k is called the Boltzmann constant. Note that the higher the gas temperature, the greater the average kinetic energy of an atom or molecule of the gas. This average kinetic energy becomes zero at absolute zero, or $T = 0$, the temperature at which molecular motion is at a minimum.

At a given temperature, all kinds of atoms and molecules will have the same average kinetic energy. But the average *speed* of a given kind of atom or molecule depends on the particle's mass. To see this dependence on mass, note that the average kinetic energy of a gas atom or molecule can be written in two equivalent ways:

$$E_k = \frac{1}{2} mv^2 = \frac{3}{2} kT$$

where v represents the average speed of an atom or molecule in a gas with temperature T. Rearranging this equation, we obtain

Average speed of a gas atom or molecule

$$v = \sqrt{\frac{3kT}{m}}$$

v = average speed of a gas atom or molecule, in m/s

k = 1.38×10^{-23} J/K

T = temperature of gas, in kelvins

m = mass of the atom or molecule, in kilograms

For a given gas temperature, the greater the mass of a given type of gas atom or molecule, the slower its average speed. (The value of v given by this equation is actually slightly higher than

Table 7-4 lists the seven largest trans-Neptunian objects known as of this writing. (Note that Charon is actually a satellite of Pluto.) Haumea and its two moons are shown. While Haumea's shape has not been observed directly, modeling of its fuzzy image suggests that it is unusually elongated, possibly from a collision.

Just as most asteroids lie in the asteroid belt, most trans-Neptunian objects orbit within a band called the **Kuiper belt** (pronounced "ki-per") that extends from 30 AU to 50 AU from the Sun and is centered on the plane of the ecliptic. Like asteroids, many more trans-Neptunian objects remain to be discovered:

the average speed of the atoms or molecules in the gas, but it is close enough for our purposes here. If you are studying physics, you may know that v is actually the root-mean-square speed.)

EXAMPLE: What is the average speed of the oxygen molecules that you breathe at a room temperature of 20°C (= 68°F)?

Situation: We are given the temperature of a gas and are asked to find the average speed of the gas molecules.

Tools: We use the relationship $v = \sqrt{3kT/m}$ where T is the gas temperature in kelvins and m is the mass of a single oxygen molecule in kilograms.

Answer: To use the equation to calculate the average speed v, we must express the temperature T in kelvins (K) rather than degrees Celsius (°C). As we learned in Box 5-1, we do this by adding 273 to the Celsius temperature, so 20°C becomes (20 + 273) = 293 K. The mass m of an oxygen molecule is not given, but you can easily find that the mass of an oxygen *atom* is 2.66×10^{-26} kg. The mass of an oxygen molecule (O_2) is twice the mass of an oxygen atom, or $2(2.66 \times 10^{-26}\ \text{kg}) = 5.32 \times 10^{-26}$ kg. Thus, the average speed of an oxygen molecule in 20°C air is

$$v = \sqrt{\frac{3(1.38 \times 10^{-23})(293)}{5.32 \times 10^{-26}}} = 478\ \text{m/s} = 0.478\ \text{km/s}$$

Review: This speed is about 1700 kilometers per hour, or about 1100 miles per hour. Hence, atoms and molecules move rapidly in even a moderate-temperature gas.

In some situations, atoms and molecules in a gas may be moving so fast that they can overcome the attractive force of a planet's gravity and escape into interplanetary space. The minimum speed that an object at a planet's surface must have in order to permanently leave the planet is called the planet's **escape speed** (see Figure 4-22). The escape speed for a planet of mass M and radius R is given by

$$v_{\text{escape}} = \sqrt{\frac{2GM}{R}}$$

where $G = 6.67 \times 10^{-11}\ \text{N m}^2/\text{kg}^2$ is the universal constant of gravitation.

The accompanying table gives the escape speed for the Sun, the planets, and the Moon. For example, to escape Earth, a cannon ball would have to be shot with an implausible speed greater than 11.2 km/s (25,100 mi/h).

A good rule of thumb is that a planet can retain a gas if the escape speed is at least 6 times greater than the average speed of the molecules in the gas. (Some molecules are moving slower than average, and others are moving faster, but very few are moving more than 6 times faster than average.) In such a case, very few molecules will be moving fast enough to escape from the planet's gravity.

EXAMPLE: Consider Earth's atmosphere. We saw that the average speed of oxygen molecules is 0.478 km/s at room temperature. The escape speed from Earth (11.2 km/s) is much more than 6 times the average speed of the oxygen molecules, so Earth has no trouble keeping oxygen in its atmosphere.

A similar calculation for hydrogen molecules (H_2) gives a different result, however. At 293 K, the average speed of a hydrogen molecule is 1.9 km/s. Six times this speed is 11.4 km/s, which is about the escape speed from Earth. Thus, Earth does not retain hydrogen in its atmosphere. Any hydrogen released into the air slowly leaks away into space. On Jupiter, by contrast, the escape speed is so high that even the lightest gases such as hydrogen are retained in its atmosphere. But on Mercury the escape speed is low and the temperature high (so that gas molecules move faster), and Mercury cannot retain any significant atmosphere at all.

Object	Escape speed (km/s)
Sun	618.0
Mercury	4.3
Venus	10.4
Earth	11.2
Moon	2.4
Mars	5.0
Jupiter	59.5
Saturn	35.5
Uranus	21.3
Neptune	23.5

Astronomers estimate that there are 35,000 or more such objects with diameters greater than 100 km. If so, the combined mass of all trans-Neptunian objects is comparable to the mass of Jupiter, and is several hundred times greater than the combined mass of all the asteroids found in the inner solar system.

Like asteroids, trans-Neptunian objects are thought to be debris left over from the formation of the solar system. In the inner regions of the solar system, rocky fragments have been able to endure continuous exposure to the Sun's heat, but any ice originally present would have evaporated. Far from the Sun, ice has survived for

TABLE 7-4 Seven Large Trans-Neptunian Objects

	Quaoar	Haumea	Sedna	Makemake	Pluto	Charon (satellite of Pluto)	Eris
Average distance from the Sun (AU)	43.54	43.34	489	45.71	39.54	39.54	67.67
Orbital period (years)	287	285	10,800	309	248.6	248.6	557
Orbital eccentricity	0.035	0.189	0.844	0.155	0.250	0.250	0.442
Inclination of orbit to the ecliptic	8.0°	28.2°	11.9°	29.0°	17.15°	17.15°	44.2°
Approximate diameter (km)	1250	1500	1600	1800	2274	1190	2900

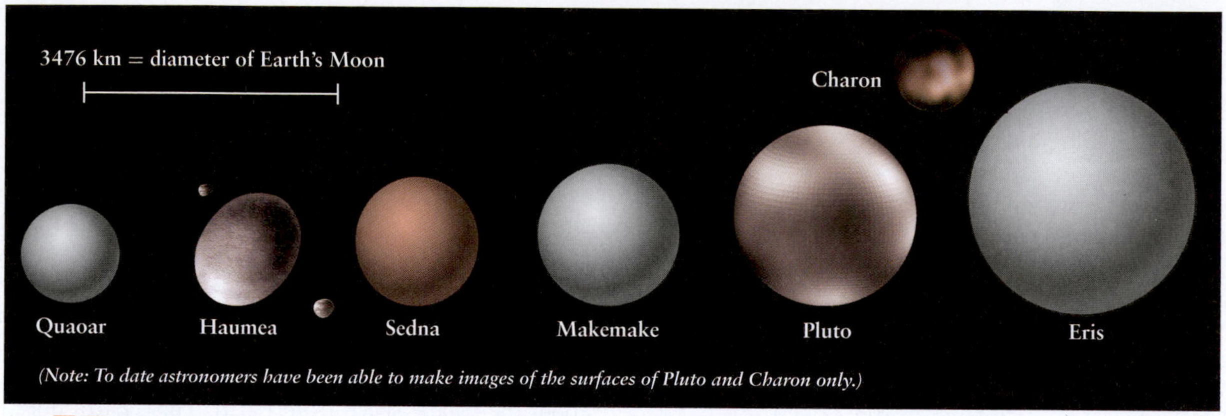

R I V U X G

(Haumea: A. Field [STScI]/NASA; Charon: Lanthanum-138; all others: Alan Stern [Southwest Research Institute]/Marc Buie [Lowell Observatory]/ NASA/ESA)

billions of years. Thus, *debris* in the solar system naturally divides into two families—rocky asteroids and partially icy trans-Neptunian objects—which can be arranged according to distance from the Sun, just like the two categories of planets (terrestrial and Jovian).

CONCEPTCHECK 7-9

Is Pluto an asteroid, planet, Kuiper belt object, or trans-Neptunian object?

CALCULATIONCHECK 7-3

Which has a larger average orbital radius, the asteroid belt or the Kuiper belt?

Answers appear at the end of the chapter.

Comets

Two objects in the Kuiper belt can collide if their orbits cross each other. When this happens, a fragment a few kilometers across can be knocked off one of the colliding objects and be diverted into an elongated orbit that brings it close to the Sun. Such small objects, each a combination of rock and ice, are called **comets**. When a comet comes close enough to the Sun, the Sun's radiation vaporizes some of the comet's ices, producing long flowing tails of gas and dust particles (Figure 7-9). Astronomers deduce the composition of comets by studying the spectra of these tails.

FIGURE 7-9 R I V U X G

A Comet This photograph shows Comet Hale-Bopp as it appeared in April 1997. The solid part of a comet like this is a chunk of dirty ice a few tens of kilometers in diameter. When a comet passes near the Sun, solar radiation vaporizes some of the icy material, forming a bluish tail of gas and a white tail of dust. Both tails can extend for tens of millions of kilometers. (Agencia el Universal/AP Images)

CAUTION! Science-fiction movies and television programs sometimes show comets tearing across the night sky like a rocket, which would be a pretty impressive sight. However, comets do not zoom across the sky. Like the planets, comets orbit the Sun. And like the planets, comets move hardly at all against the background of stars over the course of a single night (see Section 4-1). If you are lucky enough to see a bright comet, it will not zoom dramatically from horizon to horizon. Instead, it will seem to hang majestically among the stars, so you can admire it at your leisure.

Some comets appear to originate from locations far beyond the Kuiper belt. The source of these is thought to be a swarm of comets that forms a spherical "halo" around the solar system called the **Oort comet cloud** (also known as Oort cloud). This hypothesized "halo" extends to 50,000 AU from the Sun (about one-fifth of the way to the nearest other star). Because the Oort cloud is so distant, it has not yet been possible to detect objects in the Oort cloud directly.

CONCEPTCHECK 7-10

Consider a comet that has an orbital plane that is perpendicular to the plane of Earth's orbit around the Sun. Did this comet most likely originate from the Kuiper belt or the Oort cloud?

Answer appears at the end of the chapter.

7-6 Craters on planets and satellites are the result of impacts from interplanetary debris

One of the great challenges in studying planets and satellites is determining their internal structures. Are they solid or liquid inside? If there is liquid in the interiors, is the liquid calm or in agitated motion? Because planets are opaque, we cannot see directly into their interiors to answer these questions. But we can gather important clues about the interiors of terrestrial planets and satellites by studying the extent to which their surfaces are covered with craters (Figure 7-10). To see how information about the interiors is gathered, we first need to understand where craters come from.

Scientists study craters on a planet or satellite to learn the age and geologic history of the surface

The Origin of Craters

The planets orbit the Sun in roughly circular orbits. But many asteroids and comets are in more elongated orbits. Such an elongated orbit can put these small objects on a collision course with a planet or satellite. If the object collides with a Jovian planet, it is swallowed up by the planet's thick atmosphere. (Astronomers actually saw an event of this kind in 1994, when a comet crashed into Jupiter.) But if the object collides with the solid surface of a terrestrial planet or a satellite, the result is an **impact crater** (see Figure 7-10). Such impact craters, found throughout the solar system, offer stark evidence of these violent collisions.

The easiest way to view impact craters is to examine the Moon through a telescope or binoculars. Some 30,000 lunar craters are visible, with diameters ranging from 1 km to several hundred km. Close-up photographs from lunar orbit have revealed millions of craters too small to be seen from Earth (Figure 7-10a). These smaller craters are thought to have been caused by impacts of relatively small objects called **meteoroids,** which range in size from a few hundred meters across to the size of a pebble or smaller. Meteoroids are the result of collisions between asteroids, whose

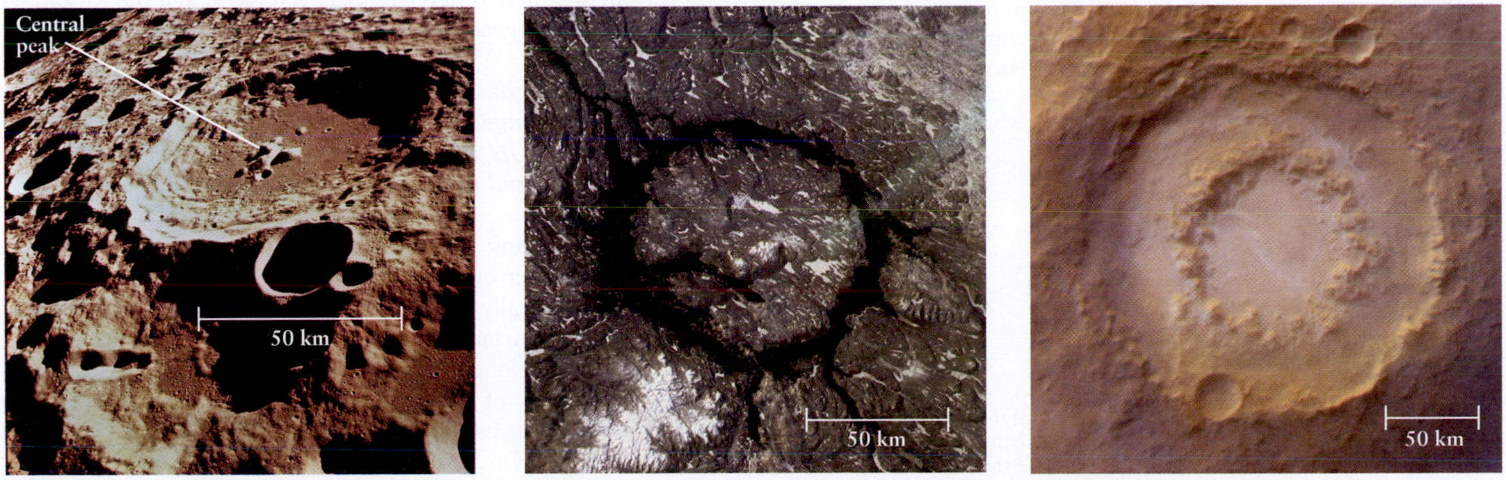

(a) A crater on the Moon (b) A crater on Earth (c) A crater on Mars

FIGURE 7-10 R I V U X G

Impact Craters These images, all taken from spacecraft, show impact craters on three different worlds. **(a)** The Moon's surface has craters of all sizes. The large crater near the middle of this image is about 80 km (50 mi) in diameter, equal to the length of San Francisco Bay. **(b)** Manicouagan Reservoir in Quebec is the relic of a crater formed by an impact more than 200 million years ago. The crater was eroded over the ages by the advance and retreat of glaciers, leaving a ring lake 100 km (60 mi) across. **(c)** Lowell Crater in the southern highlands of Mars is 201 km (125 mi) across. Like the image of the Moon in part (a), there are craters on top of craters. Note the light-colored frost formed by condensation of carbon dioxide from the Martian atmosphere. (a: NASA; b: JSC/NASA; c: NASA/JPL/MSSS) orbits sometimes cross. The chunks of rock that result from these collisions go into independent orbits around the Sun, which can lead them to collide with the Moon or another world.

When German astronomer Franz Gruithuisen proposed in 1824 that lunar craters were the result of impacts, a major sticking point was the observation that nearly all craters are circular. If craters were merely gouged out by high-speed rocks, a rock striking the Moon in any direction except straight downward would have created a noncircular crater. A century after Gruithuisen, it was realized that a meteoroid colliding with the Moon generates a shock wave in the lunar surface that spreads out from the point of impact. Such a shock wave produces a circular crater no matter what direction the meteoroid was moving. (In a similar way, the craters made by artillery shells are almost always circular.) Many of the larger lunar craters also have a central peak, which is characteristic of a high-speed impact (see Figure 7-10a). Craters made by other processes, such as volcanic action, would not have central peaks of this sort.

Comparing Cratering on Different Worlds

Not all planets and satellites show the same amount of cratering. The Moon is heavily cratered over its entire surface, with craters on top of craters, as shown in Figure 7-10a. On Earth, by contrast, craters are very rare. Geologists have identified fewer than 200 impact craters on our planet (Figure 7-10b). Our understanding is that both Earth and the Moon formed at nearly the same time and have been bombarded at comparable rates over their histories. Why, then, are craters so much rarer on Earth than on the Moon?

The answer is that Earth is a *geologically active* planet: Through **plate tectonics**—the motion of rocky plates over Earth's surface—the continents slowly change their positions over eons, new material flows onto the surface from the interior (as occurs in a volcanic eruption), and old surface material is pushed back into the interior (as occurs off the coast of Chile, where the ocean bottom is slowly being pushed beneath the South American continent). These processes, coupled with erosion from wind and water, cause craters on Earth to be erased over time. The few craters found on Earth today must be relatively recent, since there has not yet been time to erase them.

The Moon, by contrast, is geologically *inactive*. There are no volcanoes and no motion of continents (and, indeed, no continents). Furthermore, the Moon has neither oceans nor an atmosphere, so there is no erosion as we know it on Earth. With none of the processes that tend to erase craters on Earth, the Moon's surface remains pockmarked with the scars of billions of years of impacts.

In order for a planet to be geologically active, its interior must be at least partially molten. Even if the surface does not undergo plate tectonics—as in the case of Jupiter's moon Io—a partially molten interior is required to generate volcanoes with molten lava. Therefore, we would expect geologically inactive (and hence heavily cratered) worlds like the Moon to have less molten material in their interiors than does Earth. Investigation of these inactive worlds bears this out. But *why* is the Moon's interior less molten than Earth's?

ANALOGY To see one simple answer to this question, notice that a large turkey or roast taken from the oven will stay warm inside for hours, but a single meatball will cool off much more rapidly. The reason is that the meatball has more surface area relative to its volume, and so it can more easily lose heat to its surroundings. A planet or satellite also tends to cool down as it emits electromagnetic radiation into space (see Section 5-3); the smaller the planet or satellite, the greater its surface area relative to its volume, and the more readily it can radiate away heat (Figure 7-11). Both Earth and the Moon were probably completely molten when they first formed, but because the Moon (diameter 3476 km) is so much smaller than Earth (diameter 12,756 km), it has lost much of its internal heat and has a much more solid interior. There are two main sources for a planet's heat, which we will discuss in Chapter 8—gravity and radioactivity—and both of these sources result in more heat for larger planets.

Cratering Measures Geologic Activity

By considering these differences between Earth and the Moon, we have uncovered a general rule for worlds with solid surfaces:

The smaller the terrestrial world, the less internal heat it is likely to have retained, and, thus, the less geologic activity it will display on its surface. The less geologically active the world, the older and hence more heavily cratered its surface.

This rule means that we can use the amount of cratering visible on a planet or satellite to estimate the age of its surface and how geologically active it is. As an example, Mercury has a heavily cratered surface, which means that the surface is very old. This geologically inactive surface is in agreement with Mercury being the smallest of the terrestrial planets (see Table 7-1). Due to its small size, it has lost the internal heat required to sustain geologic activity. On Venus, by comparison, there are only about

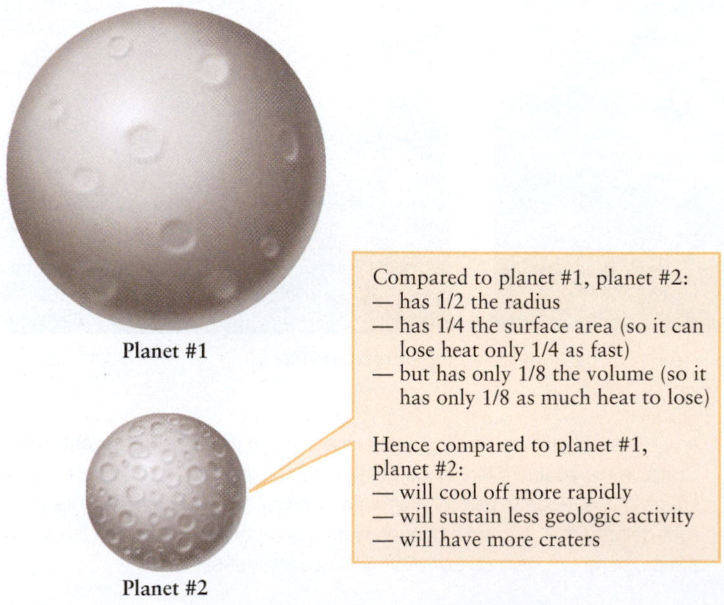
Planet #1
Planet #2
Compared to planet #1, planet #2:
— has 1/2 the radius
— has 1/4 the surface area (so it can lose heat only 1/4 as fast)
— but has only 1/8 the volume (so it has only 1/8 as much heat to lose)
Hence compared to planet #1, planet #2:
— will cool off more rapidly
— will sustain less geologic activity
— will have more craters

FIGURE 7-11
Planet Size and Cratering Of these two hypothetical planets, the smaller one (#2) has less volume and less internal heat, as well as less surface area from which to radiate heat into space. But the ratio of surface area to volume is greater for the smaller planet. Hence, the smaller planet will lose heat faster, have a colder interior, and be less geologically active. It will also have a more heavily cratered surface, since it takes geologic activity to erase craters.

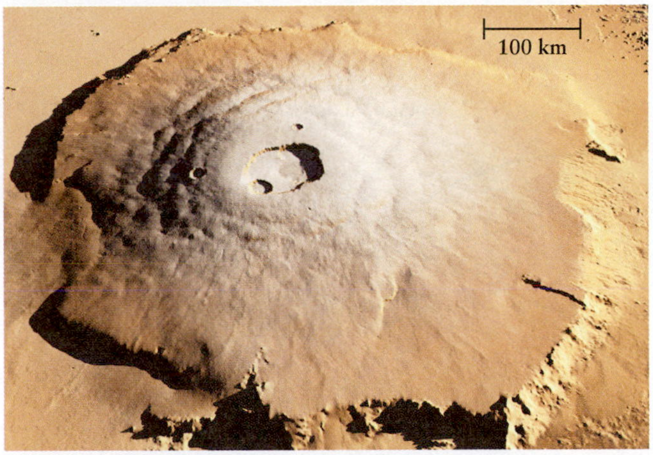

FIGURE 7-12 R I V U X G

A Martian Volcano Olympus Mons is the largest of the inactive volcanoes of Mars and the largest volcano in the solar system. The base of Olympus Mons measures 600 km (370 mi) in diameter, and the scarps (cliffs) that surround the base are 6 km (4 mi) high. The caldera, or volcanic crater, at the summit is approximately 70 km across, large enough to contain the state of Rhode Island. (Kees Veenenbos/Science Source)

a thousand craters larger than a few kilometers in diameter, many more than have been found on Earth but only a small fraction of the number on the Moon or Mercury. Venus is only slightly smaller than Earth, and it has enough internal heat to power the geologic activity required to erase most of its impact craters.

Mars is an unusual case, in that extensive cratering (Figure 7-10c) is found only in the higher terrain; the lowlands of Mars are remarkably smooth and free of craters. Thus, it follows that the Martian highlands are quite old, while the lowlands have a younger surface from which most craters have been erased. Considering the planet as a whole, the amount of cratering on Mars is intermediate between that on Mercury and Earth. This agrees with our general rule, because Mars is intermediate in size between Mercury and Earth. The interior of Mars was once hotter and more molten than it is now, so that geologic processes were able to erase some of the impact craters. A key piece of evidence that supports this picture is that Mars has a number of immense volcanoes (Figure 7-12). These volcanoes were active when Mars was young, but as this relatively small planet cooled down and its interior solidified, the supply of molten material to the volcanoes from the Martian interior was cut off. As a result, all of the volcanoes of Mars are now inactive.

As for all rules, there are limitations and exceptions to the rule relating a world's size to its geologic activity. One limitation is that the four terrestrial planets all have slightly different compositions, which affects the types and extent of geologic activity that can take place on their surfaces. The different compositions also complicate the relationship between the number of craters and the age of the surface. An important exception to our rule is Jupiter's satellite Io, which, despite its small size, is the most volcanic world in the solar system (see Section 7-2). Something must be supplying Io with energy to keep its interior hot; this energy comes from Jupiter, which exerts powerful tidal forces on Io as it moves in a relatively small orbit around its planet. These tidal forces cause Io to flex like a ball of clay being kneaded between your fingers, and this flexing heats up the satellite's interior. But despite these limitations and exceptions, the relationships between a world's size, internal heat, geologic activity, and amount of cratering are powerful tools for understanding the terrestrial planets and satellites.

CONCEPTCHECK 7-11

Io, which is a *moon,* is an exception to the rule that smaller worlds should have less geologic activity. Do you expect that some small *planets* might break this rule as well?

Answer appears at the end of the chapter.

7-7 A planet with a magnetic field indicates a fluid interior in motion

By studying the magnetic field of a planet or satellite, scientists can learn about that world's interior

The amount of impact cratering on a terrestrial planet or satellite provides indirect evidence about whether the planet or satellite has a molten interior. But another, more direct tool for probing the interior of *any* planet or satellite is an ordinary compass, which senses the magnetic field outside the planet or satellite. Magnetic field measurements prove to be an extremely powerful way to investigate the internal structure of a world without having to actually dig into its interior. To illustrate how this works, consider the behavior of a compass on Earth.

The needle of a compass on Earth points north because it aligns with Earth's *magnetic field.* Such magnetic fields arise whenever electrically charged particles are in motion. For example, a loop of wire carrying an electric current generates a magnetic field in the space around it. The source of the magnetic field that surrounds an ordinary bar magnet (Figure 7-13a) is complicated, but part of it is created by the motion of negatively charged electrons within the iron atoms of which the magnet is made. Earth's magnetic field is similar to that of a bar magnet, as Figure 7-13b shows. The consensus among geologists is that this magnetic field is caused by the motion of the liquid portions of Earth's interior. Because this molten material (mostly iron) conducts electricity, these motions give rise to electric currents, which in turn produce Earth's magnetic field. Our planet's rotation helps to sustain these motions and hence the magnetic field. This process for producing a magnetic field is called a **dynamo.**

CAUTION! While Earth's magnetic field is similar to that of a giant bar magnet, you should not take this picture too literally. Earth is *not* simply a magnetized ball of iron. In an iron bar magnet, the electrons of different atoms orbit their nuclei in the same general direction, so that the magnetic fields generated by individual atoms add together to form a single, strong field. But at temperatures above 770°C (1418°F = 1043 K), the orientations of the electron orbits become randomized. The fields of individual atoms tend to cancel each other out, and the iron loses its magnetism. Geological evidence shows that almost all of Earth's interior is hotter than 770°C, so the iron there cannot be extensively magnetized. The correct picture is that Earth acts as a dynamo: The liquid iron carries electric currents, and these currents create Earth's magnetic field.

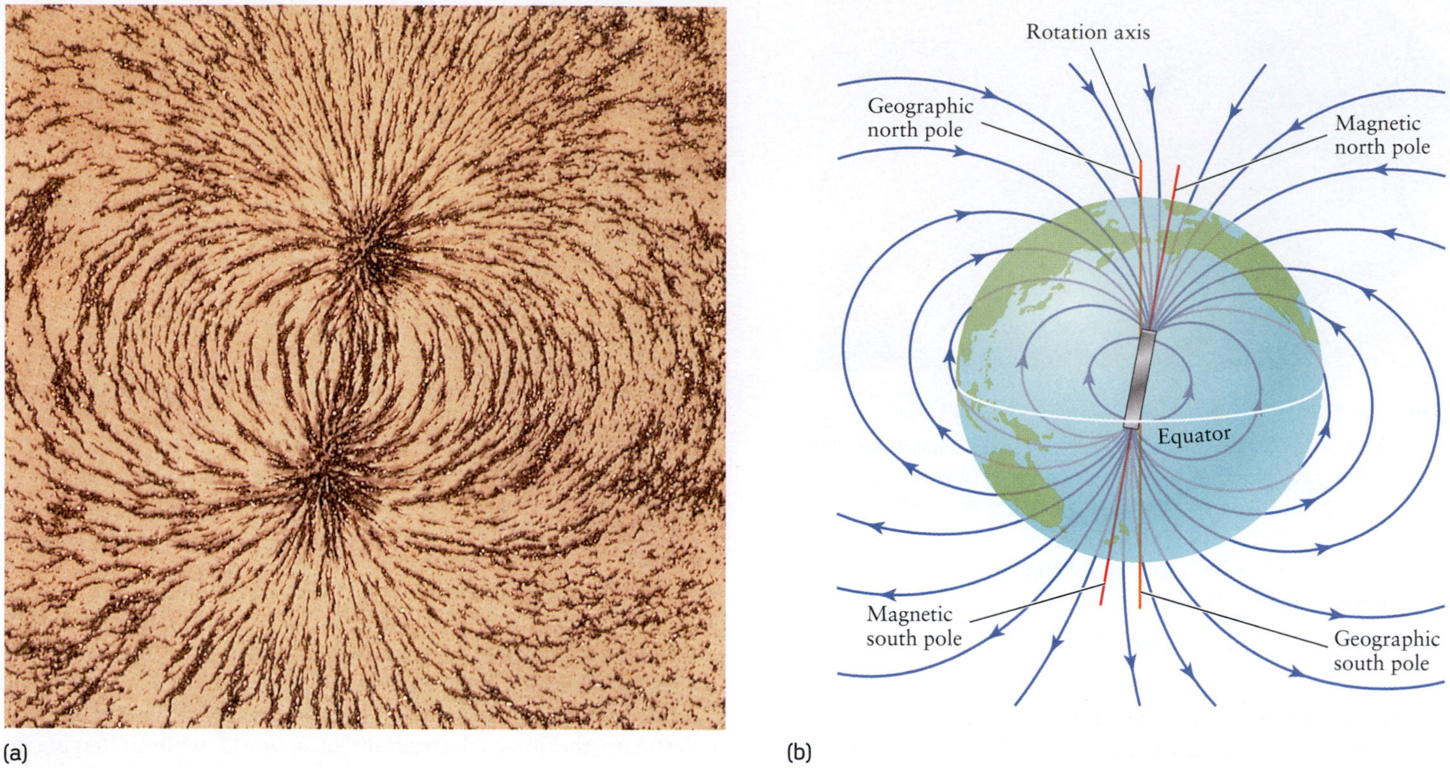

FIGURE 7-13

The Magnetic Fields of a Bar Magnet and of Earth **(a)** This picture was made by placing a piece of paper on top of a bar magnet, then spreading thin iron filings on the paper. Each elongated iron filing acts like a compass needle and aligns with the magnetic field at its location. The pattern of the filings show the magnetic field lines, which appear to stream from one of the magnet's poles to the other. **(b)** Earth's magnetic field lines have a similar pattern. Although Earth's field is produced in a different way—by electric currents in the liquid portion of our planet's interior—the field is much the same as if there were a giant bar magnet inside Earth. This "bar magnet" is not exactly aligned with Earth's rotation axis, which is why the magnetic north and south poles are not at the same locations as the true, or geographic, poles. A compass needle points toward the north magnetic pole, not the true north pole. (a: Jules Bucher/Photo Researchers)

If a planet or satellite has a mostly solid interior, then the dynamo mechanism cannot work: Material in the interior cannot flow, there are no electric currents, and the planet or satellite does not generate a magnetic field. One example of this is the Moon. As we saw in Section 7-6, the extensive cratering of the lunar surface indicates that the Moon has no geologic activity and must therefore have a mostly solid interior. Measurements made during the *Apollo* missions, in which 12 humans visited the lunar surface between 1969 and 1972, showed that the present-day Moon indeed has no global magnetic field. However, careful magnetic measurements of lunar rocks returned by the *Apollo* astronauts indicate that the Moon *did* have a weak magnetic field when the rocks solidified. These rocks, like the rest of the lunar surface, are very old. Hence, in the distant past the Moon may have had a small amount of molten iron in its interior that acted as a dynamo. This material presumably solidified at least partially as the Moon cooled, so that the lunar magnetic field disappeared.

We have now identified another general rule about planets and satellites:

A planet or satellite with a global magnetic field has liquid material in its interior that conducts electricity and is in motion, generating the magnetic field.

Thus, by studying the magnetic field of a planet or satellite, we can learn about that world's interior. So, many spacecraft carry devices called **magnetometers** to measure magnetic fields. Magnetometers are often placed on a long boom extending outward from the body of the spacecraft (Figure 7-14). The boom isolates them from the magnetic fields produced by electric currents in the spacecraft's own circuitry.

Measurements made with magnetometers on spacecraft have led to a number of striking discoveries. For example, it has been found that Mercury has a planetwide magnetic field like Earth's, although it's only about 1% as strong as Earth's field. If Mercury's magnetic field was simply due to a magnetized crust, variations in the field due to crust-altering craters would have been detected, yet no such variations were observed. Instead, Mercury's smooth magnetic field—created beneath the surface—suggests that at least *some* of the planet's interior is in a liquid state to act as a magnetic dynamo. A liquid interior is quite surprising because Mercury's heavily cratered surface indicates a lack of geologic activity, which is usually associated with a solid interior. By contrast, Venus has no measurable planetwide magnetic field, even though the paucity of craters on its surface indicates the presence of geologic activity and hence a hot interior of the planet. One possible reason for the lack of a magnetic field on Venus is that the planet turns on its

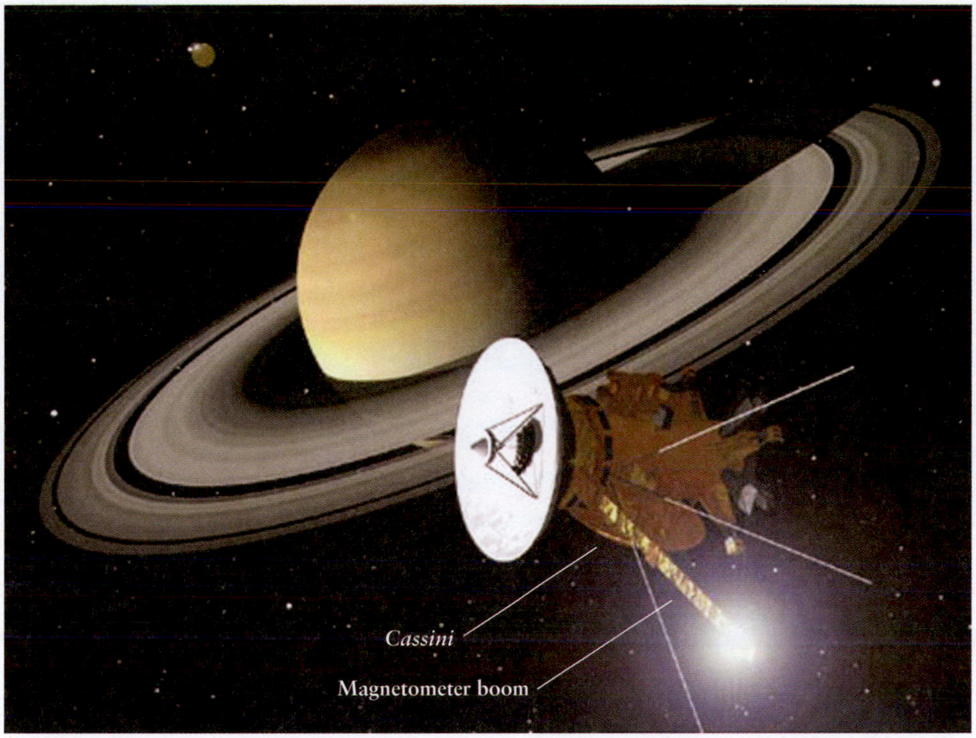

FIGURE 7-14

Probing the Magnetic Field of Saturn This illustration depicts the *Cassini* spacecraft as it entered orbit around Saturn on July 1, 2004. In addition to telescopes for observing Saturn and its satellites, *Cassini* carries a magnetometer for exploring Saturn's magnetic field. The magnetometer is located on the long boom that extends down and to the right from the body of the spacecraft. (The glow at the end of the boom is the reflection of the Sun.) (JPL/NASA)

axis very slowly, taking 243 days for a complete rotation. Because of this slow rotation, the fluid material within the planet is hardly agitated at all, and so may not move in the fashion that generates a magnetic field.

Mars has a very interesting magnetic field that sheds light on its early history. The magnetometer aboard the *Mars Global Surveyor* spacecraft, which went into orbit around Mars in 1997, found that most of the planet's crust is magnetized (Figure 7-15): Mars has no planetwide magnetic field like Earth, but portions of its rocky surface are magnetized. This information actually tells us a lot about the Martian past. During the planet's first billion years or so, the interior was still hot and molten. Electric currents in the flowing molten material could then produce a planetwide magnetic field. As surface material cooled and solidified, the crust became magnetized by the planetwide field, and remained magnetized even after the internally generated planetwide field shut down.

The alternating stripes of magnetism on the Martian crust—seen most clearly in red and blue in the middle of Figure 7-15—tell us even more about the Martian past. Far from being "alien" features, these familiar stripes have been heavily studied in the places they occur on Earth, where they result from plate tectonics. Plate tectonics—the motion of rocky plates over a planet's surface—are driven by a planet's internal heat. Furthermore, on Earth plate tectonics stabilizes our atmosphere; Mars may have lost its atmosphere after its interior cooled too much to power further plate motion.

The most intense planetary magnetic field in the solar system is that of Jupiter: At the tops of Jupiter's clouds, the magnetic field is about 14 times stronger than the field at Earth's surface. It is thought that Jupiter's field, like Earth's, is produced by a dynamo acting deep within the planet's interior. Unlike Earth, however, Jupiter is composed primarily of hydrogen and helium, not substances like iron that conduct electricity. How, then, can Jupiter have a dynamo that generates such strong magnetic fields?

To answer this question, recall from Section 5-8 that a hydrogen atom consists of a single proton orbited by a single electron. Deep inside Jupiter, the pressure is so great and hydrogen atoms are squeezed so close together that electrons can hop from one atom to another. This hopping motion creates an electric current, just as the ordered movement of electrons in the copper wires of a flashlight constitutes an electric current. In other words, the highly compressed hydrogen deep inside Jupiter behaves like an electrically conducting metal; thus, it is called **liquid metallic hydrogen.**

Laboratory experiments show that hydrogen becomes a liquid metal when the pressure is more than about 1.4 million times ordinary atmospheric pressure on Earth. Recent calculations suggest that this transition occurs about 7000 km below Jupiter's cloudtops. Most of the planet's enormous bulk lies below this level, so there is a tremendous amount of liquid metallic hydrogen within Jupiter. Since Jupiter rotates rapidly—a "day" on Jupiter is just less than 10 hours long—this liquid metal moves rapidly, generating the planet's powerful magnetic field. Saturn also has

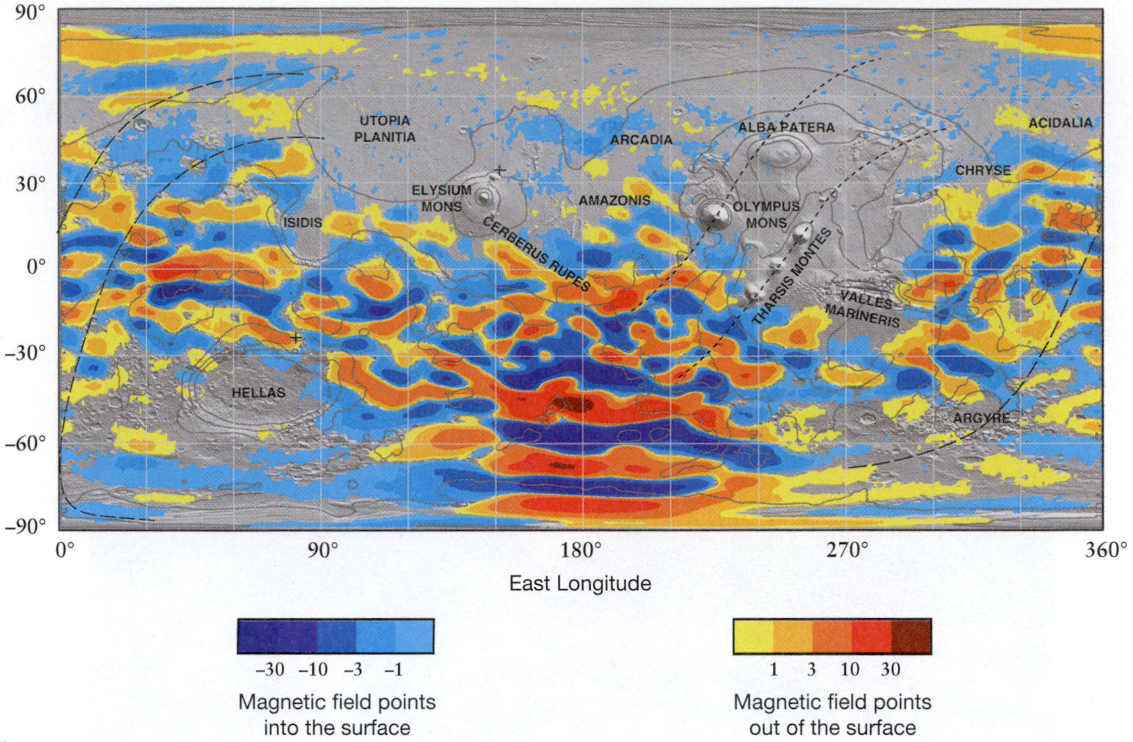

FIGURE 7-15

Relic Magnetism on Mars For almost a billion years after Mars formed, molten material with electric currents generated a planetwide magnetic field. Most of the surface was magnetized during this time, which is the crustal field we measure today. Regions without magnetism probably experienced higher temperatures that can remove magnetism, or were formed by lava flows after the planetwide field shut down. Some magnetic stripes are visible in the lower middle of the image, where the field alternately points up and down. These stripes suggest that Mars went through a period with plate tectonics. (NASA)

a magnetic field produced by dynamo action in liquid metallic hydrogen. (The field is weaker than Jupiter's because Saturn is a smaller planet with less internal pressure, so there is less of the liquid metal available.)

Uranus and Neptune also have magnetic fields, but they cannot be produced in the same way: Because these planets are relatively small, the internal pressure is not great enough to turn liquid hydrogen into a metal. Instead, it is thought that both Uranus and Neptune have large amounts of liquid water in their interiors and that this water has molecules of ammonia and other substances dissolved in it. (The fluid used for washing windows has a similar chemical composition.) Under the pressures found in this interior water, the dissolved molecules lose one or more electrons and become electrically charged (that is, they become ionized; see Section 5-8). Water is a good conductor of electricity when it has such electrically charged molecules dissolved in it, and electric currents in this fluid would be the source of the magnetic fields of Uranus and Neptune. However, there is no direct evidence of currents in liquid water for these planets, and an alternative model proposes an electrically conductive mixture of hydrogen and rocky material.

CONCEPTCHECK 7-12

By comparing parts (a) and (b) of Figure 7-13, can you conclude that there is a large bar magnet inside Earth?

CONCEPTCHECK 7-13

Does every planet with a molten interior produce a magnetic field? Do any of the Jovian planets create their magnetic fields through electrically conductive molten iron?

Answers appear at the end of the chapter.

7-8 The diversity of the solar system is a result of its origin and evolution

> The similarities and differences among the planets can be logically explained by a model of the solar system's origin and evolution

Our brief tour of the solar system has revealed its almost dizzying variety. No two planets are alike, satellites come in all sizes, the extent of cratering varies from one terrestrial planet to another, and the magnetic fields of different planets vary dramatically in their strength and in how they are produced. (The *Cosmic Connections* that closes this chapter summarizes these properties of the planets.) All of this variety leads us to a simple yet profound question: *Why* are the planets and satellites of the solar system so different from each other?

Among humans, the differences from one individual to another result from heredity (the genetic traits passed on from an individual's parents) and environment (the circumstances under which the individual matures to an adult). As we will find in the following chapter, much the same is true for the worlds of the solar system.

In Chapter 8 we will see evidence that the entire solar system shares a common "heredity," in that the planets, satellites, comets, asteroids, and the Sun itself formed from the same cloud of interstellar gas and dust. The composition of this cloud was shaped by cosmic processes, including nuclear reactions that took place within stars that died long before our solar system was formed. We will see how different planets formed in different environments depending on their distance from the Sun and will discover how these environmental variations gave rise to the planets and satellites of our present-day solar system. And we will see how we can test these ideas of solar system origin and evolution by studying planetary systems orbiting other stars.

Our journey through the solar system is just beginning. In this chapter we have explored space to examine the variety of the present-day solar system; in Chapter 8 we will journey through time to see how our solar system came to be.

CALCULATIONCHECK 7-4

Using the *Cosmic Connections* figure, *Characteristics of the Planets,* how many Earth masses does it take to equal Jupiter's mass? How many Saturn masses does it take to equal Jupiter's mass?

Answer appears at the end of the chapter.

KEY WORDS

Terms preceded by an asterisk (*) are discussed in the Boxes.

asteroid, p. 180
asteroid belt, p. 181
average density, p. 175
chemical composition, p. 177
comet, p. 184
dynamo, p. 187
*escape speed, p. 183
ices, p. 180
impact crater, p. 185
Jovian planet, p. 174
*kinetic energy, p. 182
Kuiper belt, p. 182
liquid metallic hydrogen, p. 189
magnetometer, p. 188
meteoroid, p. 185
minor planet, p. 180
Oort comet cloud (Oort cloud), p. 185
plate tectonics, p. 186
spectroscopy, p. 177
terrestrial planet, p. 174
trans-Neptunian object, p. 181

KEY IDEAS

Properties of the Planets: All of the planets orbit the Sun in the same direction and in almost the same plane. Most of the planets have nearly circular orbits.

• The four inner planets are called terrestrial planets. They are relatively small (with diameters of 5000 to 13,000 km), have high average densities (4000 to 5500 kg/m^3), and are composed primarily of rocky materials.

• The four giant outer planets are called Jovian planets. They have large diameters (50,000 to 143,000 km) and low average densities (700 to 1700 kg/m^3) and are composed primarily of light elements such as hydrogen and helium.

Satellites and Small Bodies in the Solar System: Besides the planets, the solar system includes satellites of the planets, asteroids, comets, and trans-Neptunian objects.

• Seven large planetary satellites (one of which is the Moon) are comparable in size to the planet Mercury. The remaining satellites of the solar system are much smaller.

• Asteroids are small, rocky objects, while comets and trans-Neptunian objects are made of ice and rock. All are remnants left over from the formation of the planets.

• Most asteroids are found in the asteroid belt between the orbits of Mars and Jupiter, and most trans-Neptunian objects lie in the Kuiper belt outside the orbit of Neptune. Pluto is one of the largest members of the Kuiper belt.

Spectroscopy and the Composition of the Planets: Spectroscopy, the study of spectra, provides information about the chemical composition of objects in the solar system.

• The spectrum of a planet or satellite with an atmosphere reveals the atmosphere's composition. If there is no atmosphere, the spectrum indicates the composition of the surface.

• The substances that make up the planets can be classified as gases, ices, or rock, depending on the temperatures at which they solidify.

Impact Craters: When an asteroid, comet, or meteoroid collides with the surface of a terrestrial planet or satellite, the result is an impact crater.

• Geologic activity renews the surface and erases craters, so a terrestrial world with extensive cratering has an old surface and little or no geologic activity.

• Because geologic activity is powered by internal heat, and smaller worlds lose heat more rapidly, as a general rule smaller terrestrial worlds are more extensively cratered.

Magnetic Fields and Planetary Interiors: Planetary magnetic fields are produced by the motion of electrically conducting liquids inside the planet. This mechanism is called a dynamo. If a planet has no magnetic field, that is evidence that there is little such liquid material in the planet's interior or that the liquid is not in a state of motion.

• The magnetic fields of terrestrial planets are produced by metals such as iron in the liquid state. The stronger fields of the Jovian planets are generated by liquid metallic hydrogen or by water with ionized molecules dissolved in it.

QUESTIONS

Review Questions

1. Do all the planets orbit the Sun in the same direction? Are all of the orbits circular?

COSMIC CONNECTIONS

Characteristics of the Planets

The Inner (Terrestrial) Planets — Close to the Sun – Small diameter, small mass – High density

	Mercury	Venus	Earth	Mars
Average distance from the Sun (AU)	0.387	0.723	1.000	1.524
Equatorial diameter (Earth = 1)	0.383	0.949	1.000	0.533
Mass (Earth = 1)	0.0553	0.8150	1.0000	0.1074
Average density (kg/m^3)	5430	5243	5515	3934

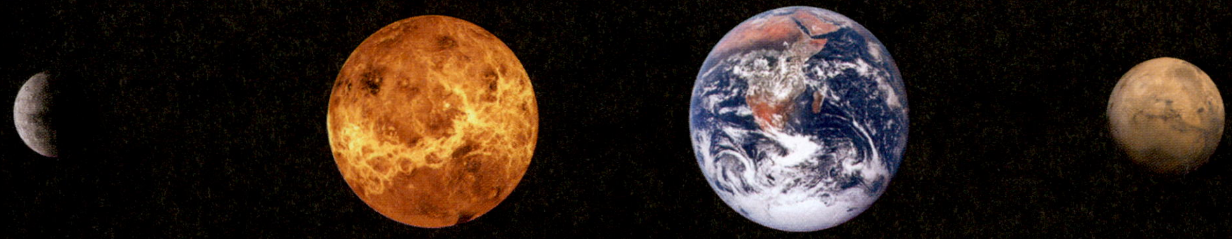

Mercury | Venus | Earth | Mars

Atmosphere None | **Atmosphere** Carbon dioxide | **Atmosphere** Nitrogen, oxygen | **Atmosphere** Carbon dioxide

Magnetic field Weak | **Magnetic field** None | **Magnetic field** Moderate, due to liquid iron core | **Magnetic field** Weak, due to magnetized crust

II Planets and Moons

The polar region in the north of Mars holds truly alien landscapes. Wind-blown dunes of dust form the mounds appearing somewhat pink in this color-enhanced image. In the Martian winter, the dunes are covered in carbon-dioxide snow. As summer approaches, the carbon dioxide evaporates, exposing the underlying dust. The black streaks occur on the steepest slopes where the exposed dust forms miniature avalanches. (NASA/JPL/University of Arizona)

R I V U X G

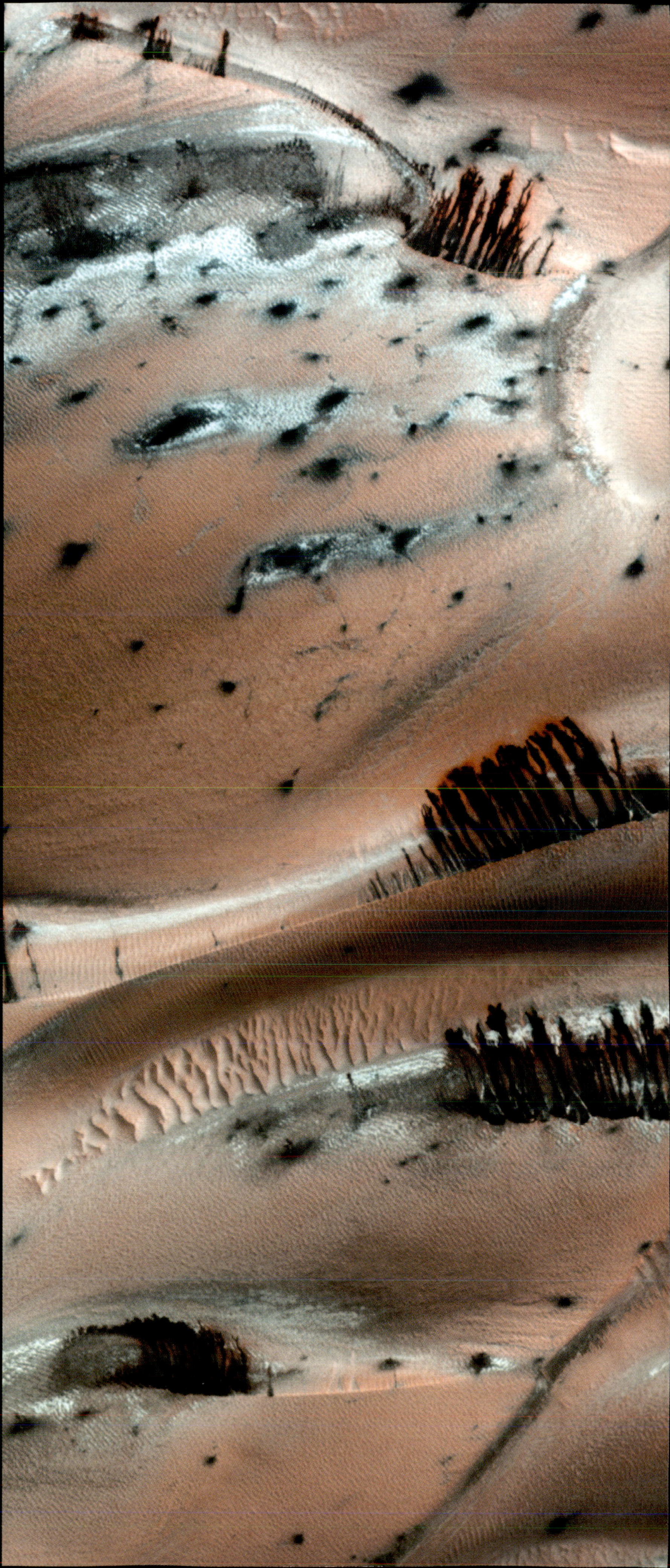

To attain this stunning view, the *Cassini* spacecraft was carefully lined up to watch Saturn eclipse the Sun. With the Sun directly behind Saturn, otherwise faint and diffuse ring particles appear bright in this color-enhanced image. The outermost ring contains Saturn's moon Enceladus, and the diffuse ring itself is produced by ejecta from that moon's ice-volcanoes. At about the "10 o'clock" position in the upper left, Earth can be seen just above the bright inner rings. (NASA/JPL/Space Science Institute)

R I V U X G

The Outer (Jovian) Planets

Far from the Sun – Large diameter, large mass – Low density

	Jupiter	Saturn	Uranus	Neptune
Average distance from the Sun (AU)	5.203	9.554	19.194	30.066
Equatorial diameter (Earth = 1)	11.209	9.449	4.007	3.883
Mass (Earth = 1)	317.8	95.16	14.53	17.15
Average density (kg/m^3)	1326	687	1318	1638

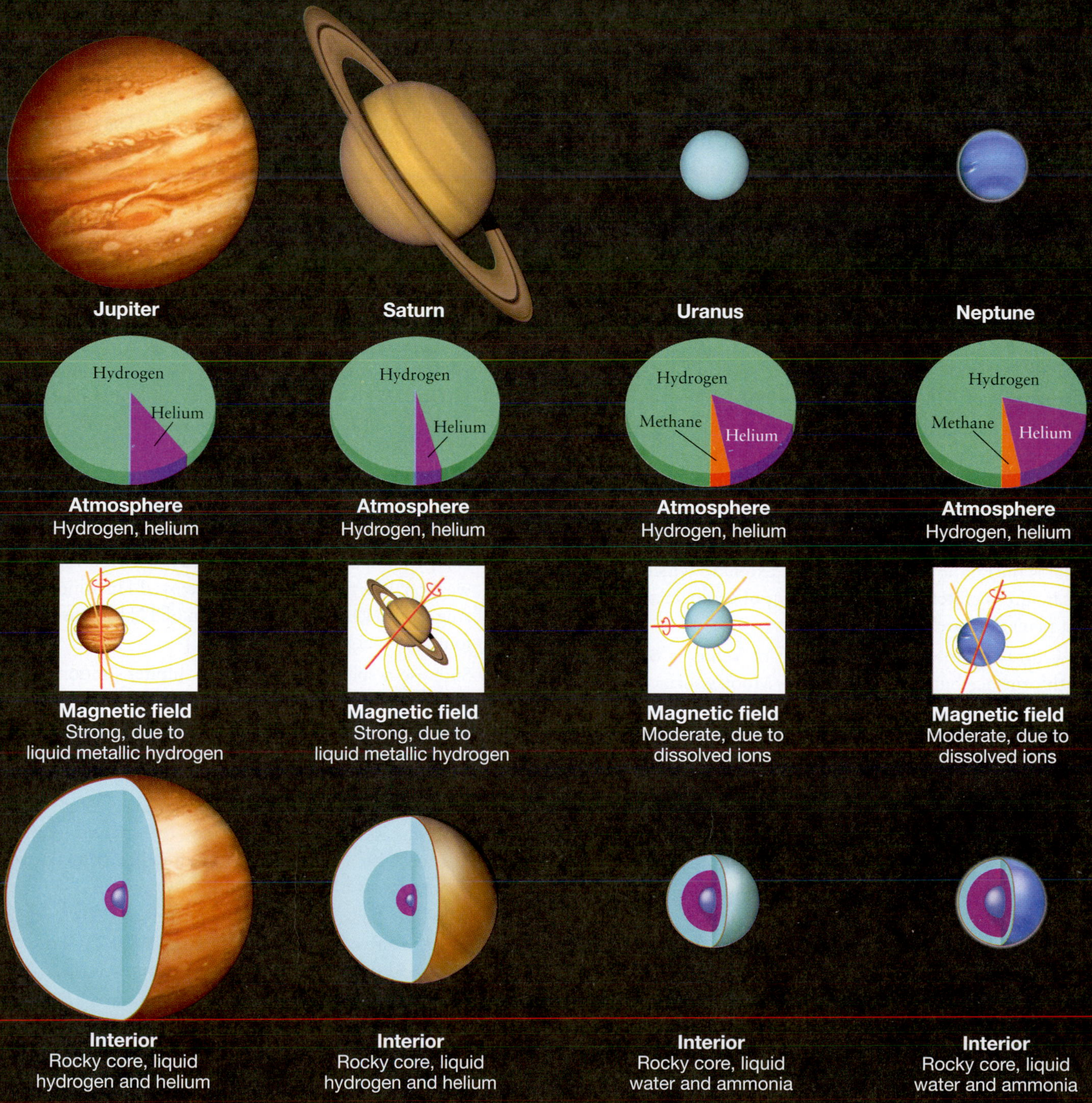

2. What are the characteristics of a terrestrial planet?
3. What are the characteristics of a Jovian planet?
4. In what ways are the largest satellites similar to the terrestrial planets? In what ways are they different? Which satellites are largest?
5. On March 16, 2007, Venus was 1.97×10^8 km from Earth and had an angular diameter of 12.7 arcsec. Using the small-angle formula from Box 1-1, calculate the diameter of Venus.
6. What is meant by the average density of a planet? What does the average density of a planet tell us?
7. What are the differences in chemical composition between the terrestrial and Jovian planets?
8. The absorption lines in the spectrum of a planet or satellite do not necessarily indicate the composition of the planet or satellite's atmosphere. Why not?
9. Why are hydrogen and helium abundant in the atmospheres of the Jovian planets but present in only small amounts in Earth's atmosphere?
10. What is an asteroid? What is a trans-Neptunian object? In what ways are these minor members of the solar system like or unlike the planets?
11. What are the asteroid belt, the Kuiper belt, and the Oort cloud? Where are they located? How do the objects found in these three regions compare?
12. In what ways is Pluto similar to a terrestrial planet? In what ways is it different?
13. What is the connection between comets and the Kuiper belt? Between comets and the Oort cloud?
14. What is one piece of evidence that impact craters are actually caused by impacts?
15. What is the relationship between the extent to which a planet or satellite is cratered and the amount of geologic activity on that planet or satellite?
16. How do we know that the surface of Venus is older than Earth's surface but younger than the Moon's surface?
17. Why do smaller worlds retain less of their internal heat?
18. How does the size of a terrestrial planet influence the amount of cratering on the planet's surface?
19. How is the magnetic field of a planet different from that of a bar magnet? Why is a large planet more likely to have a magnetic field than a small planet?
20. Could you use a compass to find your way around Venus? Why or why not?
21. If Mars has no planetwide magnetic field, why does it have magnetized regions on its surface?
22. What is liquid metallic hydrogen? Why is it found only in the interiors of certain planets?

Advanced Questions

Questions preceded by an asterisk (*) involve topics discussed in the Boxes.

Problem-solving tips and tools

The volume of a sphere of radius r is $4\pi r^3/3$, and the surface area of a sphere of radius r is $4\pi r^2$. The surface area of a circle of radius r is πr^2. The average density of an object is its mass divided by its volume. To calculate escape speeds, you will need to review Box 7-2. Be sure to use the same system of units (meters, seconds, kilograms) in all your calculations involving escape speeds, orbital speeds, and masses. Appendix 6 gives conversion factors between different sets of units, and Box 5-1 has formulas relating various temperature scales.

23. Mars has two small satellites, Phobos and Deimos. Phobos circles Mars once every 0.31891 day at an average altitude of 5980 km above the planet's surface. The diameter of Mars is 6794 km. Using this information, calculate the mass and average density of Mars.
24. Figure 7-3 shows the spectrum of Saturn's largest satellite, Titan. Can you think of a way that astronomers can tell which absorption lines are due to Titan's atmosphere and which are due to the atmospheres of the Sun and Earth? Explain.

*25. (a) Find the mass of a hypothetical spherical asteroid 2 km in diameter and composed of rock with average density 2500 kg/m^3. (b) Find the speed required to escape from the surface of this asteroid. (c) A typical jogging speed is 3 m/s. What would happen to an astronaut who decided to go for a jog on this asteroid?

*26. The hypothetical asteroid described in Question 25 strikes Earth with a speed of 25 km/s. (a) What is the kinetic energy of the asteroid at the moment of impact? (b) How does this energy compare with that released by a 20-kiloton nuclear weapon, like the device that destroyed Hiroshima, Japan, on August 6, 1945? (*Hint:* 1 kiloton of TNT releases 4.2×10^{12} joules of energy.)

*27. Suppose a spacecraft landed on Jupiter's moon Europa (see Table 7-2), which moves around Jupiter in an orbit of radius 670,900 km. After collecting samples from the satellite's surface, the spacecraft prepares to return to Earth. (a) Calculate the escape speed from Europa. (b) Calculate the escape speed from Jupiter at the distance of Europa's orbit. (c) In order to begin its homeward journey, the spacecraft must leave Europa with a speed greater than either your answer to (a) or your answer to (b). Explain why.

*28. A hydrogen atom has a mass of 1.673×10^{-27} kg, and the temperature of the Sun's surface is 5800 K. What is the average speed of hydrogen atoms at the Sun's surface?

*29. The Sun's mass is 1.989×10^{30} kg, and its radius is 6.96×10^8 m. (a) Calculate the escape speed from the Sun's surface

(b) Using your answer to Question 28, explain why the Sun has lost very little hydrogen over its entire 4.56-billion-year history.

*30. Saturn's satellite Titan has an appreciable atmosphere, yet Jupiter's satellite Ganymede—which is about the same size and mass as Titan—has no atmosphere. Explain why there is a difference.

31. The distance from the asteroid 433 Eros (Figure 7-7) to the Sun varies between 1.13 and 1.78 AU. (a) Find the period of Eros's orbit. (b) Does Eros lie in the asteroid belt? How can you tell?

32. Imagine a trans-Neptunian object with roughly the same mass as Earth but located 50 AU from the Sun. (a) What do you think this object would be made of? Explain your reasoning. (b) On the basis of this speculation, assume a reasonable density for this object and calculate its diameter. How many times bigger or smaller than Earth would it be?

33. Consider a hypothetical trans-Neptunian object located 100 AU from the Sun. (a) What would be the orbital period (in years) of this object? (b) There are 360 degrees in a circle, and 60 arcminutes in a degree. How long would it take this object to move 1 arcminute across the sky? (c) Trans-Neptunian objects are discovered by looking for "stars" that move on the celestial sphere. Use your answer from part (b) to explain why these discoveries require patience. (d) Discovering trans-Neptunian objects also requires large telescopes equipped with sensitive detectors. Explain why.

34. The surfaces of Mercury, the Moon, and Mars are riddled with craters formed by the impact of space debris. Many of these craters are billions of years old. By contrast, there are only a few conspicuous craters on Earth's surface, and these are generally less than 500 million years old. What do you suppose explains the difference?

35. During the period of most intense bombardment by space debris, a new 1-km-radius crater formed somewhere on the Moon about once per century. During this same period, what was the probability that such a crater would be created within 1 km of a certain location on the Moon during a 100-year period? During a 10^6-year period? (*Hint:* If you drop a coin onto a checkerboard, the probability that the coin will land on any particular one of the board's 64 squares is $^1/_{64}$.)

36. When an impact crater is formed, material (called *ejecta*) is sprayed outward from the impact. (The accompanying photograph of the Moon shows light-colored ejecta extending outward from the crater Copernicus.) While ejecta are found surrounding the craters on Mercury, they do not extend as far from the craters as do ejecta on the Moon. Explain why, using the difference in surface gravity between the Moon (surface gravity = 0.17 that on Earth) and Mercury (surface gravity = 0.38 that on Earth).

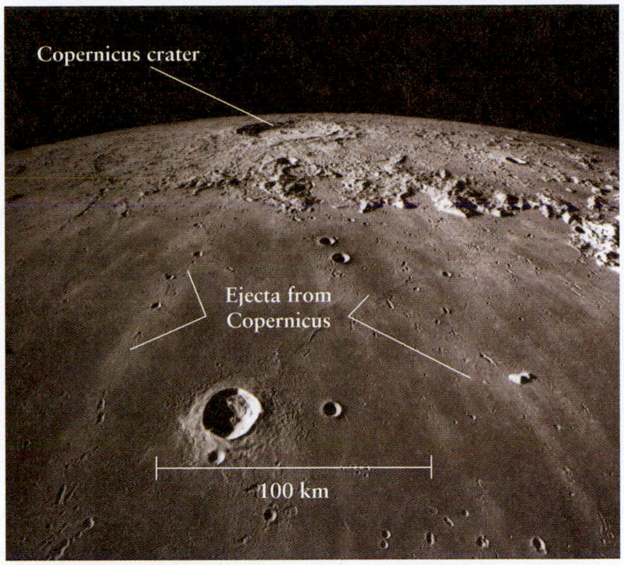

R I V U X G (NASA)

37. Mercury rotates once on its axis every 58.646 days, compared to 1 day for Earth. Use this information to argue why Mercury's magnetic field should be much smaller than Earth's.

38. As you can see in Figure 7-15, *Mars Global Surveyor* did not find a significant magnetic field in the northern region of Utopia Planitia, which is a large lava field. Based on this observation, would you expect the lava field to have formed before or after Mars ceased to have a molten core?

39. Liquid metallic hydrogen is the source of the magnetic fields of Jupiter and Saturn. Explain why liquid metallic hydrogen cannot be the source of Earth's magnetic field.

Discussion Questions

*40. There are no asteroids with an atmosphere. Discuss why not.

41. The *Galileo* spacecraft that orbited Jupiter from 1995 to 2003 discovered that Ganymede (Table 7-2) has a magnetic field twice as strong as that of Mercury. Does this discovery surprise you? Why or why not?

Web/eBook Question

42. ANIMATION 7-1 Search the World Wide Web for information about impact craters on Earth. Where is the largest crater located? How old is it estimated to be? Which crater is closest to where you live?

43. **Determining Terrestrial Planet Orbital Periods.** Access the animation "Planetary Orbits" in Chapter 7 of the *Universe* Web site or eBook. Focus on the motions of the inner planets at the last half of the animation. Using the stop and start buttons, determine how many days it takes Mars, Venus, and Mercury to orbit the Sun once if Earth takes approximately 365 days.

ACTIVITIES

Observing Projects

44. Use a telescope or binoculars to observe craters on the Moon. Make a drawing of the Moon, indicating the smallest and largest craters that you can see. Can you estimate their sizes? For comparison, the Moon as a whole has a diameter of 3476 km. *Hint:* You can see craters most distinctly when the Moon is near first quarter or third quarter (see Figure 3-2). At these phases, the Sun casts long shadows across the portion of the Moon in the center of your field of view, making the variations in elevation between the rims and centers of craters easy to identify. You can determine the phase of the Moon by looking at a calendar or the weather page of the newspaper, by using the *Starry Night*™ program, or on the World Wide Web.

45. Use *Starry Night*™ to examine magnified images of the terrestrial major planets Mercury, Venus, Earth, and Mars, and the dwarf planet Ceres. Select each of these planets from **Favourites > Explorations** in turn. Use the location scroller cursor to rotate the image to see different views of the planet. (**a**) Describe each planet's appearance. From what you observe in each case, is there any way of knowing whether you are looking at a planet's surface or at complete cloud cover over the planet? (**b**) Which planet or planets have clouds? If a planet has clouds, open its contextual menu and choose **Surface Image/Model > Default** and use the location scroller to examine the planet's surface. (**c**) Which major planet shows the heaviest cratering? (**d**) Which of these terrestrial planets show evidence of liquid water? (**e**) What do you notice about Venus's rotation compared to the other planets?

46. Use *Starry Night*™ to examine the Jovian planets Jupiter, Saturn, Uranus, and Neptune. Select each of these planets from **Favourites > Explorations**. Use the location scroller cursor to examine each planet from different views. (**a**) Describe each planet's appearance. Which has the greatest color contrast in its cloud tops? (**b**) Which planet has the least color contrast in its cloud tops? (**c**) What can you say about the thickness of Saturn's rings compared to their diameter?

ANSWERS

ConceptChecks

ConceptCheck 7-1: Our solar system contains only one star, the Sun. The other stars are very far away.

ConceptCheck 7-2: Mars is classified as an inner planet, and all four inner planets are also terrestrial planets with hard surfaces (these are Mercury, Venus, Earth, and Mars). The inner planets orbit within 2 AU of the Sun, while the outer planets (Jupiter, Saturn, Uranus, and Neptune) all orbit farther out, more than 5 AU from the Sun.

ConceptCheck 7-3: If the rocks on the surface have a density lower than the planet's average density, then the planet's core has a density greater than the planet's average. This is not surprising, since gravity causes material of greater density to sink towards the center of a planet. (The concept of average density is discussed in Box 7-1.)

ConceptCheck 7-4: Table 7-2 lists four having diameters greater than our Moon's 3476 km: Io, Ganymede, Callisto, and Titan.

ConceptCheck 7-5: As sunlight passes through a planet's atmosphere (just before and after reflection off of the planet's surface), the atoms and molecules in the atmosphere absorb specific wavelengths of light unique to these atoms and molecules. By looking at this absorption spectrum in the reflected sunlight, astronomers can infer the composition of the atmosphere as illustrated in Figure 7-3.

ConceptCheck 7-6: The reflected spectrum from a solid surface shows broad absorption features, whereas the spectrum observed from light passing through a gaseous atmosphere shows sharper spectral lines.

ConceptCheck 7-7: Hydrogen molecules in a gaseous atmosphere have a range of speeds based on the temperature of the gas, with some moving about six times the average speed. These faster molecules exceed Earth's escape speed and leave Earth entirely as described in Box 7-2. Continuously, some of the remaining slower hydrogen molecules speed up through molecular collisions and steadily escape.

ConceptCheck 7-8: Ceres has a diameter of about 900 km. This is about the same size as a large U.S. state, such as the length of California.

ConceptCheck 7-9: Pluto orbits beyond Neptune. All such objects are called trans-Neptunian objects. One group of trans-Neptunian objects—including Pluto—orbits in the Kuiper belt, so Pluto is both a trans-Neptunian object and a Kuiper belt object. Pluto is neither a planet nor an asteroid.

ConceptCheck 7-10: Whereas the Kuiper belt lies in the same plane as Earth's orbit around the Sun, the Oort cloud is a spherical distribution of comets that completely surrounds the solar system. If a comet has an orbit that is considerably different from that of the flat plane of the solar system, it most likely came from the spherical Oort cloud that exists in all directions around our solar system.

ConceptCheck 7-11: No. Io's heat comes from tidal forces exerted by a very massive Jupiter. A planet, on the other hand, is not likely to experience strong interior-melting tidal forces from its smaller, orbiting moons.

ConceptCheck 7-12: No. Figure 7-13a shows the field created by a bar magnet. Figure 7-13b shows that these fields are similar in structure, but the source of the magnetic field is not a bar magnet. Instead, Earth's magnetic field is due to electric currents flowing in a molten iron interior.

ConceptCheck 7-13: The answer to both questions is no. Based on its similar size to Earth, Venus probably has a molten interior. However, it exhibits no magnetic field, probably due to its very slow rotation: Earth rotates about 243 times for each single rotation of Venus. As in the cases of Jupiter, Saturn, Uranus, and Neptune, some electrically conducting material other than iron is needed to explain the observed magnetic fields.

CalculationChecks

CalculationCheck 7-1: The shape of a planet's orbit is given by the value of its eccentricity. The closer this value is to zero, the closer the orbit's shape is to that of a perfect circle. According to the table, the orbit of Venus has the eccentricity closest to zero (0.007), making it the most circlelike of all planetary orbits.

CalculationCheck 7-2: If we divide Saturn's 120,536-km diameter by Earth's 12,756-km diameter—we find that 120,536 km ÷ 12,756 km = 9.449, so about 9½ Earth's would fit across Saturn's diameter.

CalculationCheck 7-3: The asteroid belt is located between Mars and Jupiter at about 3 AU from the Sun, whereas the much larger Kuiper belt is beyond the orbit of Neptune and is located between about 30 and 50 AU from the Sun.

CalculationCheck 7-4: The masses in this figure are all given as multiples of Earth's mass, so it would take just over 317 Earth masses to equal Jupiter's mass. The number of Saturn masses that would equal Jupiter's mass is 317.8/95.16 = 3.33.

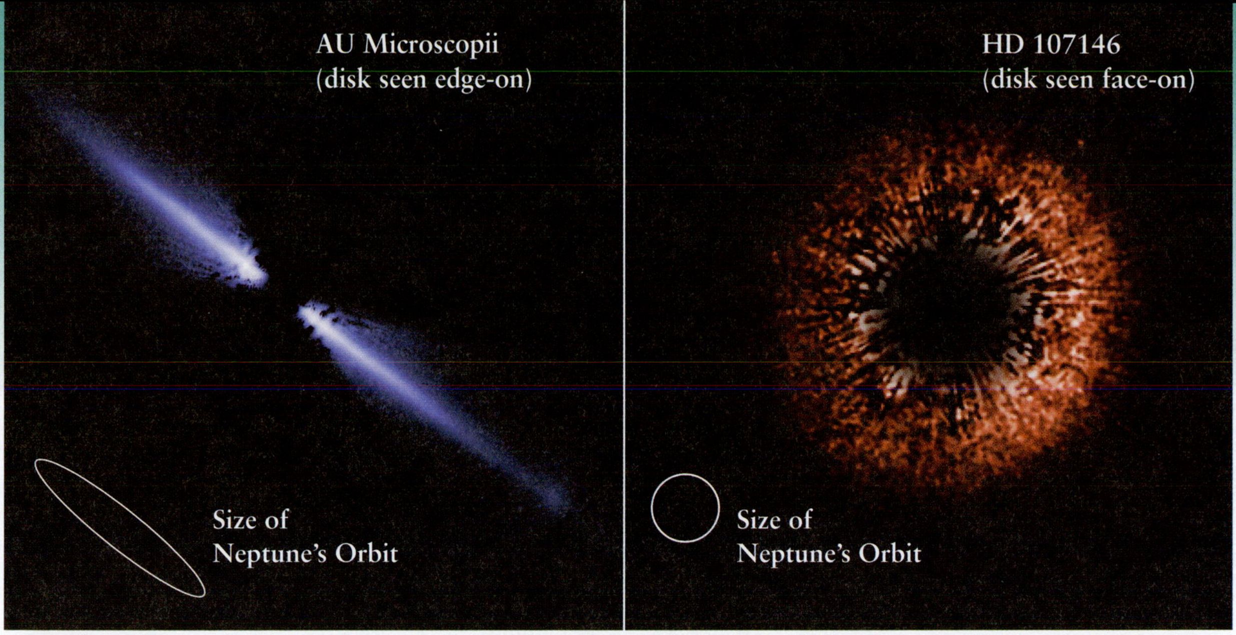

R I V U X G

Planets are thought to form within the disks surrounding young stars such as these. Neptune's orbit, shown for scale, is about 60 AU across. (NASA, ESA, D. R. Ardila (JHU), D. A. Golimowski (JHU), J. E. Krist (STScI/JPL), M. Clampin (NASA/GSFC), J. P. Williams (UH/IfA), J. P. Blakeslee (JHU), H. C. Ford (JHU), G. F. Hartig (STScI), G. D. Illingworth (UCO-Lick) and the ACS Science Team)

CHAPTER 8

Comparative Planetology II: The Origin of Our Solar System

LEARNING GOALS

By reading the sections of this chapter, you will learn

8-1 The key characteristics of the solar system that must be explained by any theory of its origins

8-2 How the abundances of chemical elements in the solar system and beyond explain the sizes of the planets

8-3 How we can determine the age of the solar system by measuring abundances of radioactive elements

8-4 Why scientists think the Sun and planets all formed from a cloud called the solar nebula

8-5 How the solar nebula model explains the formation of the terrestrial planets

8-6 How the Jovian planets formed and migrated in the early solar system

8-7 How astronomers search for planets around other stars

What did our solar system look like before the planets were fully formed? The answer may lie in these remarkable images from the Hubble Space Telescope. Each image shows an immense disk of gas and dust surrounding a young star. (In each image the light from the star itself was blocked out within the telescope to make the rather faint disk more visible.) Astronomers strongly suspect that in its infancy our own Sun was surrounded by a similar disk from which the planets of our solar system eventually coalesced.

In this chapter we will examine the evidence that led astronomers to this model of the origin of the planets. We will see how the abundances of different chemical elements in the solar system indicate that the Sun and planets formed from a thin cloud of interstellar matter some 4.54 billion years ago, an age determined by measuring the radioactivity of meteorites. We will learn how the nature of planetary orbits gives important clues to what happened as this cloud contracted and how evidence from meteorites reveals the chaotic conditions that existed within the cloud. And we will see how this cloud eventually evolved into the solar system that we see today.

In the past decade, astronomers have been able to test this picture of planetary formation by examining disks around young stars (like the ones in the accompanying images). Most remarkably, they have discovered planets in orbit around dozens of other stars. These recent observations provide valuable information about how our own system of planets came to be.

8-1 Any model of solar system origins must explain the present-day Sun and planets

How did the Sun and planets form? In other words, where did the solar system come from? This question has tantalized astronomers for centuries. Our goal in this chapter is to examine our current understanding of how the solar system came to be—that is, our current best *theory* of the origin of the solar system.

Recall from Section 1-1 that a theory is not merely a set of wild speculations, but a self-consistent collection of ideas that must pass the test of providing an accurate description of the real world. Since no humans were present to witness the formation of the planets, scientists must base their theories of solar system origins on their observations of the present-day solar system. (In an analogous way, paleontologists base their understanding of the lives of dinosaurs on the evidence provided by fossils that have survived to the present day.) In so doing, they are following the steps of the scientific method that we described in Section 1-1.

What key attributes of the solar system should guide us in building a theory of solar system origins? Among the many properties of the planets that we discussed in Chapter 7, three of the most important are listed in Table 8-1. Any theory that attempts to describe the origin of the solar system must be able to explain how these attributes came to be. We begin by considering what Property 1 tells us; we will return to Properties 2 and 3 and the orbits of the planets later in this chapter.

CONCEPTCHECK 8-1

Is Earth an exception to any of the three key properties of our solar system in Table 8-1?

Answer appears at the end of the chapter.

8-2 The cosmic abundances of the chemical elements are the result of how stars evolve

The small sizes of the terrestrial planets compared to the Jovian planets (Property 1 in Table 8-1) suggest that some chemical elements are quite common in our solar system, while others are quite rare. The tremendous masses of the Jovian planets—Jupiter alone has more mass than all of the other planets combined—means that the elements of which they are made, primarily hydrogen and helium, are very abundant. The Sun, too, is made almost entirely of hydrogen and helium. Its average density of 1410 kg/m^3 is in the same range as the densities of the Jovian planets (see Table 7-1), and its absorption spectrum (see Figure 5-14) shows the dominance of hydrogen and helium in the Sun's atmosphere. Hydrogen, the most abundant element, makes up nearly three-quarters of the combined mass of the Sun and planets. Helium is the second most abundant element. Together, hydrogen and helium account for about 98% of the mass of all the material in the solar system. All of the other chemical elements are relatively rare; combined, they make up the remaining 2% (Figure 8-1).

> The terrestrial planets are small because they are made of less abundant elements

The dominance of hydrogen and helium is not merely a characteristic of our local part of the universe. By analyzing the spectra of stars and galaxies, astronomers have found essentially the same pattern of chemical abundances out to the farthest distance attainable by the most powerful telescopes. Hence, the vast majority of the atoms in the universe are hydrogen and helium atoms. The elements that make up the bulk of Earth—mostly iron, oxygen, and silicon—are relatively rare in the universe as a whole, as are the elements of which living organisms are made—carbon, oxygen, nitrogen, and phosphorus, among others. (You may find it useful to review the periodic table of the elements, described in Box 5-5.)

The Origin of the Elements and Cosmic "Recycling"

There is a good reason for this overwhelming abundance of hydrogen and helium. A wealth of evidence has led astronomers to conclude that the universe began some 13.7 billion years ago with a violent event called the Big Bang (Chapter 26). Only the lightest elements—hydrogen and helium, as well as tiny amounts of lithium and perhaps beryllium—emerged from the enormously high temperatures following this cosmic event. All the heavier elements were later manufactured by stars, either by thermonuclear fusion reactions deep in their interiors or by the violent explosions that mark the end of massive stars. Were it not for these processes that take place only in stars, there would be no heavy elements in the universe, no planet like our Earth, and no humans to contemplate the nature of the cosmos.

TABLE 8-1 Three Key Properties of Our Solar System

Any theory of the origin of the solar system must be able to account for these properties of the planets.

Property 1: Sizes and compositions of terrestrial planets versus Jovian planets	The terrestrial planets, which are composed primarily of rocky substances, are relatively small, while the Jovian planets, which are composed primarily of hydrogen and helium, are relatively large (see Sections 7-1 and 7-4).
Property 2: Directions and orientations of planetary orbits	All of the planets orbit the Sun in the same direction, and all of their orbits are in nearly the same plane (see Section 7-1).
Property 3: Sizes of terrestrial planet orbits versus Jovian planet orbits	The terrestrial planets orbit close to the Sun, while the Jovian planets orbit far from the Sun orbits versus Jovian planet orbits (see Section 7-1).

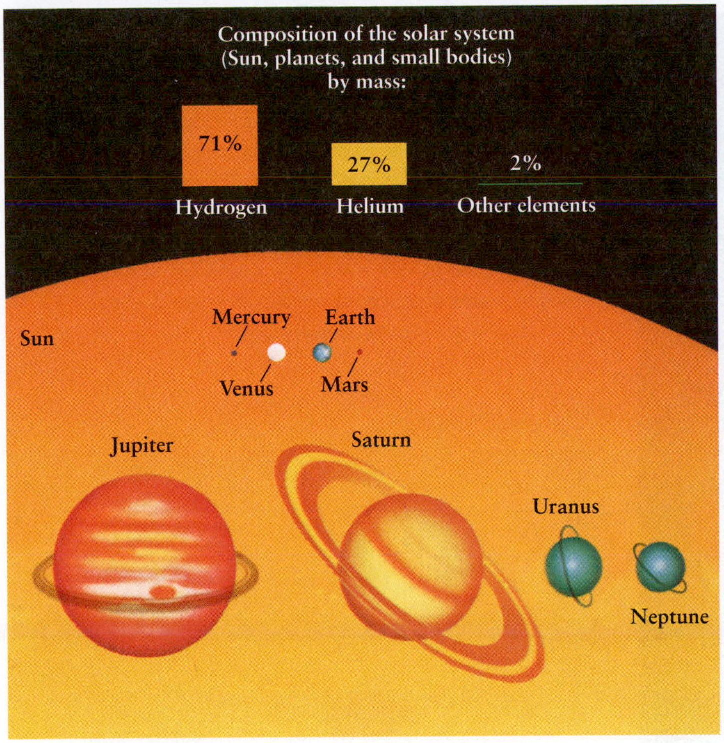

FIGURE 8-1

Composition of the Solar System Hydrogen and helium make up almost all of the mass of our solar system. Other elements such as carbon, oxygen, nitrogen, iron, gold, and uranium constitute only 2% of the total mass.

Because our solar system contains heavy elements, it must be that at least some of its material was once inside other stars. But how did this material become available to help build our solar system? The answer is that near the ends of their lives, stars cast much of their matter back out into space. For most stars this process is a comparatively gentle one, in which a star's outer layers are gradually expelled. Figure 8-2 shows a star losing material in this fashion. This ejected material appears as the cloudy region, or **nebulosity** (from *nubes,* Latin for "cloud"), that surrounds the star and is illuminated by it. A few stars eject matter much more dramatically at the very end of their lives, in a spectacular detonation called a *supernova,* which blows the star apart (see Figure 1-8).

No matter how it escapes, the ejected material contains heavy elements dredged up from the star's interior, where they were formed. This material becomes part of the **interstellar medium,** a tenuous collection of gas and dust that pervades the spaces between the stars. As different stars die, they increasingly enrich the interstellar medium with heavy elements. Observations show that new stars form as condensations in the interstellar medium (Figure 8-3). Thus, these new stars have an adequate supply of heavy elements from which to develop a system of planets, satellites, comets, and asteroids. Our own solar system must have formed from enriched material in just this way. Thus, our solar system contains "recycled" material that was produced long ago inside a now-dead star. This "recycled" material includes all of the carbon in your body, all of the oxygen that you breathe, and all of the iron and silicon in the soil beneath your feet. Scientifically accurate, a rock song from 1970 proclaimed: "We are stardust."

The Abundances of the Elements

Stars create different heavy elements in different amounts. For example, oxygen (as well as carbon, silicon, and iron) is readily produced in the interiors of massive stars, whereas gold (as well as silver, platinum, and uranium) is created only under special circumstances. Consequently, gold is rare in our solar system and in the universe as a whole, while oxygen is relatively abundant (although still much less abundant than hydrogen or helium).

A convenient way to express the relative abundances of the various elements is to say how many atoms of a particular element are found for every trillion (10^{12}) hydrogen atoms. For example, for every 10^{12} hydrogen atoms in space, there are about 100 billion

FIGURE 8-2 R I **V** U X G

A Mature Star Ejecting Gas and Dust The star Antares is shedding material from its outer layers, forming a thin cloud around the star. We can see the cloud because some of the ejected material has condensed into tiny grains of dust that reflect the star's light. (Dust particles in the air around you reflect light in the same way, which is why you can see them within a shaft of sunlight in a darkened room). Antares lies some 600 light-years from Earth in the constellation Scorpio. (David Malin/Anglo-Australian Observatory)

Gas and dust ejected from earlier generations of stars have coalesced to form new stars.
The dust reflects light emitted by the newly formed stars.

FIGURE 8-3 R I V U X G

New Stars Forming from Gas and Dust Unlike Figure 8-2, which depicts an old star that is ejecting material into space, this image shows young stars in the constellation Orion (the Hunter) that have only recently formed from a cloud of gas and dust. The bluish, wispy appearance of the cloud (called NGC 1973-1975-1977) is caused by starlight reflecting off interstellar dust grains within the cloud (see Box 5-4). The grains are made of heavy elements produced by earlier generations of stars. (David Malin/Anglo-Australian Observatory)

(10^{11}) helium atoms. From spectral analysis of stars and chemical analysis of Earth rocks, Moon rocks, and bits of interplanetary debris called meteorites, scientists have determined the relative abundances of the elements in our part of the Milky Way Galaxy today. Figure 8-4 shows the relative abundances of the 30 lightest elements, arranged in order of their **atomic numbers.** An element's atomic number is the number of protons in the nucleus of an atom of that element. It is also equal to the number of electrons orbiting

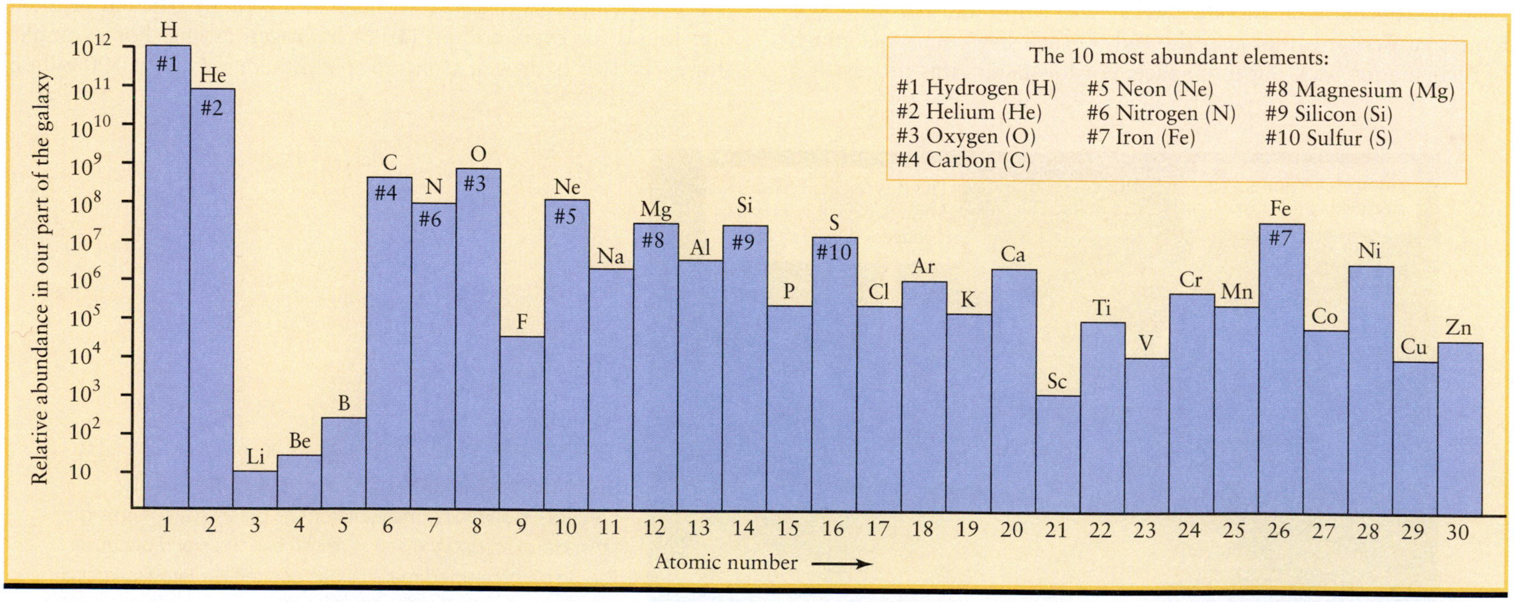

FIGURE 8-4

Abundances of the Lighter Elements This graph shows the abundances in our part of the Galaxy of the 30 lightest elements (listed in order of increasing atomic number) compared to a value of 10^{12} for hydrogen. The inset lists the 10 most abundant of these elements, which are also indicated in the graph. Notice that the vertical scale is not linear; each division on the scale corresponds to a tenfold increase in abundance. All elements heavier than zinc (Zn) have abundances of fewer than 1000 atoms per 10^{12} atoms of hydrogen.

the nucleus (see Box 5-5). In general, the greater the atomic number of an atom, the greater its mass.

CAUTION! Figure 8-4 shows that there is about 10 times more hydrogen than helium when comparing the *number of atoms*. Figure 8-1 shows that our solar system is made of 71% hydrogen versus 27% helium when comparing *mass*. The explanation of this seeming inconsistency lies in the difference between comparing the mass of atoms versus their numbers: Each helium atom has about 4 times the mass of a hydrogen atom, which makes helium's contribution to total mass larger than its contribution to the total number of atoms.

The box inset in Figure 8-4 lists the 10 most abundant elements. Note that even oxygen (chemical symbol O), the third most abundant element, is quite rare relative to hydrogen (H) and helium (He): There are only 8.5×10^8 oxygen atoms for each 10^{12} hydrogen atoms and each 10^{11} helium atoms. Expressed another way, for each oxygen atom in our region of the Milky Way Galaxy, there are about 1200 hydrogen atoms and 120 helium atoms.

In addition to the 10 most abundant elements listed in Figure 8-4, five elements are moderately abundant: sodium (Na), aluminum (Al), argon (Ar), calcium (Ca), and nickel (Ni). These elements have abundances in the range of 10^6 to 10^7 relative to the standard 10^{12} hydrogen atoms. Most of the other elements are much rarer. For example, for every 10^{12} hydrogen atoms in the solar system, there are only 6 atoms of gold.

The small cosmic abundances of elements other than hydrogen and helium help to explain why the terrestrial planets are so small (Property 1 in Table 8-1). Because the heavier elements required to make a terrestrial planet are rare, only relatively small planets can form out of them. By contrast, hydrogen and helium are so abundant that it was possible for these elements to form large Jovian planets.

CONCEPTCHECK 8-2

What is meant by the phrase "we are stardust"?

CONCEPTCHECK 8-3

From the abundances in Figure 8-4, do you expect that water (H_20) might be a common substance in the galaxy?

Answers appear at the end of the chapter.

8-3 The abundances of radioactive elements reveal the solar system's age

Our solar system, which formed 9 billion years after the Big Bang, is a relative newcomer to the universe

The heavy elements can tell us even more about the solar system: They also help us determine its age. The particular heavy elements that provide us with this information are *radioactive*. Their atomic nuclei are unstable because they contain too many protons or too many neutrons. A radioactive nucleus therefore ejects particles until it becomes stable. In doing so, a nucleus may change from one element to another. Physicists refer to this transmutation as **radioactive decay.** For example, a radioactive form of the element rubidium (atomic number 37) decays into the element strontium (atomic number 38) when one of the neutrons in the rubidium nucleus decays into a proton and an electron (which is ejected from the nucleus).

Experiment shows that each type of radioactive nucleus decays at its own characteristic rate, which can be measured in the laboratory. Furthermore, the older a solid rock is, the less of its original radioactive nuclei remains. This behavior is the key to a technique called **radioactive dating,** which is used to determine how many years ago a rock cooled and solidified, or simply, to determine the "ages" of rocks. For example, if a rock contained a certain amount of radioactive rubidium when it first solidified, over time more and more of the atoms of rubidium within the rock will decay into strontium atoms. The ratio of the number of strontium atoms the rock contains to the number of rubidium atoms it contains then gives a measure of the age of the rock. Box 8-1 describes radioactive dating in more detail.

Dating the Solar System

Scientists have applied techniques of radioactive dating to rocks taken from all over Earth. The results show that most rocks are tens or hundreds of millions of years old, but that some rocks are as much as 4 billion (4×10^9) years old. These results confirm that geologic processes—for example, lava flows—have produced new surface material over Earth's history, as we concluded from the small number of impact craters found on Earth (see Section 7-6). They also show that Earth must be at least 4×10^9 years old.

Radioactive dating has also been applied to rock samples brought back from the Moon by the *Apollo* astronauts. The oldest *Apollo* specimen, collected from one of the most heavily cratered and hence most ancient regions of the Moon, is 4.5×10^9 years old. But the oldest rocks found anywhere in the solar system are **meteorites,** bits of interplanetary debris that survive passing through Earth's atmosphere and land on our planet's surface (Figure 8-5). Radioactive dating of meteorites reveals that they are all nearly

FIGURE 8-5 R I V U X G

A Meteorite Although it resembles an ordinary Earth rock, this is actually a meteorite that fell from space. The proof of its extraterrestrial origin is the meteorite's composition and its surface. Searing heat melted the surface as the rock slammed into our atmosphere. Meteorites are the oldest objects in the solar system. (Ted Kinsman/Photo Researchers, Inc.)

BOX 8-1 TOOLS OF THE ASTRONOMER'S TRADE

Radioactive Dating

How old are the rocks found on Earth and other planets? Are rocks found at different locations the same age or different ages? How old are meteorites? Questions like these are important to scientists who wish to reconstruct the history of our solar system. The age of a rock is how long ago it solidified, but simply looking at a rock cannot tell us whether it was formed a million years, or a billion years ago. Fortunately, most rocks contain trace amounts of radioactive elements such as uranium. By measuring the relative abundances of various radioactive isotopes and their decay products within a rock, scientists can determine the rock's age.

As we saw in Box 5-5, every atom of a particular element has the same number of protons in its nucleus. However, different isotopes of the same element have different numbers of neutrons in their nuclei. For example, the common isotopes of uranium are ^{235}U and ^{238}U. Each isotope of uranium has 92 protons in its nucleus (correspondingly, uranium is element 92 in the periodic table; see Box 5-5). However, a ^{235}U nucleus contains 143 neutrons, whereas a ^{238}U nucleus has 146 neutrons.

A radioactive nucleus with too many protons or too many neutrons is unstable; to become stable, it *decays* by ejecting particles until it becomes stable. If the number of protons (the atomic number) changes in this process, the nucleus changes from one element to another.

Some radioactive isotopes decay rapidly, while others decay slowly. Physicists find it convenient to talk about the decay rate in terms of an isotope's **half-life.** The half-life of an isotope is the time interval in which one-half of the nuclei decay. For example, the half-life of ^{238}U is 4.5 billion (4.5×10^9) years. Uranium's half-life means that if you start out with 1 kg of ^{238}U, after 4.5 billion years, you will have only ½ kg of ^{238}U remaining; the other ½ kg will have turned into other elements. If you wait another half-life, so that a total of 9.0 billion years has elapsed, only 4¼ kg of ^{238}U—one-half of one-half of the original amount—will remain. Several isotopes useful for determining the ages of rocks are listed in the accompanying table.

To see how geologists date rocks, consider the slow conversion of radioactive rubidium (^{87}Rb) into strontium (^{87}Sr). (The periodic table in Box 5-5 shows that the atomic numbers for these elements are 37 for rubidium and 38 for strontium, so in the decay a neutron is transformed into a proton. In this process an electron is ejected from the nucleus.) Over the years, the amount of ^{87}Rb in a rock decreases, while the amount of ^{87}Sr increases. Because the ^{87}Sr appears in the rock due to radioactive decay, this isotope is called *radiogenic.* Dating the rock is not simply a matter of measuring its ratio of rubidium to strontium, however, because the rock already had some strontium in it when it was formed. Geologists must therefore determine how much fresh strontium came from the decay of rubidium after the rock's formation.

To make this determination, geologists use as a reference another isotope of strontium whose concentration has remained constant. In this case, they use ^{86}Sr, which is stable and is not created by radioactive decay; it is said to be *nonradiogenic.* Dating a rock thus entails comparing the ratio of radiogenic and nonradiogenic strontium ($^{87}Sr/^{86}Sr$) in the rock to the ratio of radioactive rubidium to nonradiogenic strontium ($^{87}Rb/^{86}Sr$). Because the half-life for converting ^{87}Rb into ^{87}Sr is known, the rock's age can then be calculated from these ratios (see the table).

Radioactive isotopes decay with the same half-life no matter where in the universe they are found. Hence, scientists have used the same techniques to determine the ages of rocks from the Moon and of meteorites.

Original Radioactive Isotope	Final Stable Isotope	Half-Life (Years)	Range of Ages that Can Be Determined (Years)
Rubidium (^{87}Rb)	Strontium (^{87}Sr)	47.0 billion	10 million–4.54 billion
Uranium (^{238}U)	Lead (^{206}Pb)	4.5 billion	10 million–4.54 billion
Potassium (^{40}K)	Argon (^{40}Ar)	1.3 billion	50,000–4.54 billion
Carbon (^{14}C)	Nitrogen (^{14}N)	5730	100–70,000

the *same* age, about 4.54 billion years old. The absence of any younger or older meteorites indicates that these are all remnants of objects that formed around the same time when rocky material in the early solar system—which was initially hot—first cooled and solidified. We conclude that the age of the oldest meteorites, about 4.54×10^9 years, is the age of the solar system itself. Note that this almost inconceivably long span of time is only about one-third of the current age of the universe, 13.7×10^9 years.

Thus, by studying the abundances of radioactive elements, we are led to a remarkable insight: Some 4.54 billion years ago, a collection of hydrogen, helium, and a much smaller amount of heavy elements came together to form the Sun and all of the objects that orbit around it. All of those heavy elements, including the carbon atoms in your body and the oxygen atoms that you breathe, were created and cast off by stars that lived and died long before our solar system formed, during the first 9 billion years of the universe's existence. We are literally made of old star dust, and our solar system is relatively young.

CONCEPTCHECK 8-4

What is meant by the "age" of a rock? Is it the age of the rock's atoms?

Answer appears at the end of the chapter.

8-4 The Sun and planets formed from a solar nebula

Astronomers see young stars that may be forming planets today in the same way that our solar system did billions of years ago

We have seen how processes in the Big Bang and within ancient stars produced the raw ingredients of our solar system. But given these ingredients, how did they combine to make the Sun and planets? Astronomers have developed a variety of models for the origin of the solar system. The test of these models is whether they explain the properties of the present-day system of Sun and planets.

The Failed Tidal Hypothesis

Any model of the origin of the solar system must explain why all the planets orbit the Sun in the same direction and in nearly the same plane (Property 2 in Table 8-1). One model that was devised explicitly to address this issue was the *tidal hypothesis,* proposed in the early 1900s. As we saw in Section 4-8, two nearby planets, stars, or galaxies exert tidal forces on each other that cause the objects to elongate. In the tidal hypothesis, another star happened to pass close by the Sun, and the star's tidal forces drew a long filament out of the Sun. The filament material would then go into orbit around the Sun, and all of it would naturally orbit in the same direction and in the same plane. From this filament the planets would condense. However, it was shown in the 1930s that the same tidal forces strong enough to pull a filament out of the Sun would also cause the filament to disperse before it could condense into planets. Hence, the tidal hypothesis cannot be correct.

The Successful Nebular Hypothesis

An entirely different model is now thought to describe the most likely series of events that led to our present solar system (Figure 8-6). The central idea of this model dates to the late 1700s, when the German philosopher Immanuel Kant and the French scientist Pierre-Simon de Laplace turned their attention to the manner in which the planets orbit the Sun. Both concluded that the arrangement of the orbits—all in the same direction and in nearly the same plane—could not be mere coincidence. To explain the orbits, Kant and Laplace independently proposed that our entire solar system, including the Sun as well as all of its planets and satellites, formed from a vast, rotating cloud of gas and dust called the **solar nebula** (Figure 8-6a). This model is called the **nebular hypothesis.**

The consensus among today's astronomers is that Kant and Laplace were exactly right. In the modern version of the nebular hypothesis, at the outset the solar nebula was similar in character to the nebulosity shown in Figure 8-3 and had a mass somewhat greater than that of our present-day Sun.

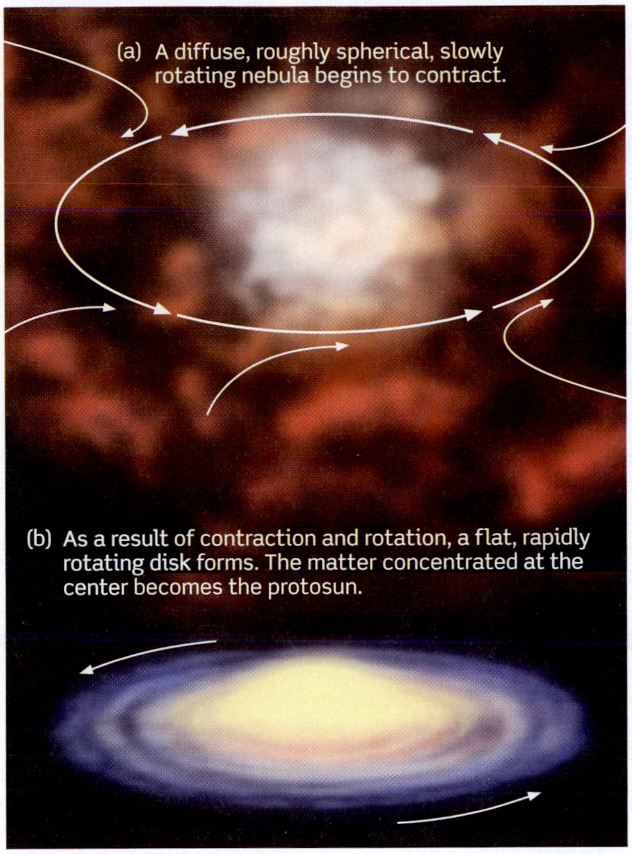

(a) A diffuse, roughly spherical, slowly rotating nebula begins to contract.
(b) As a result of contraction and rotation, a flat, rapidly rotating disk forms. The matter concentrated at the center becomes the protosun.

FIGURE 8-6

The Birth of the Solar System **(a)** A cloud of interstellar gas and dust begins to contract because of its own gravity. **(b)** As the cloud flattens and spins more rapidly around its rotation axis, a central condensation develops that evolves into a glowing protosun. The planets will form out of the surrounding disk of gas and dust.

Each part of the nebula exerted a gravitational attraction on the other parts, and these mutual gravitational pulls tended to make the nebula contract. As it contracted, the greatest concentration of matter occurred at the center of the nebula, forming a relatively dense region called the **protosun.** As its name suggests, this part of the solar nebula eventually developed into the Sun. The planets formed from the much sparser material in the outer regions of the solar nebula. Indeed, the mass of all the planets together is only 0.1% of the Sun's mass.

Evolution of the Protosun

When you drop a ball, the gravitational attraction of Earth makes the ball fall faster and faster as it falls; in the same way, material falling inward toward the protosun would have gained speed as it approached the center of the solar nebula. As this fast-moving material ran into the protosun, the kinetic energy of the collision was converted into thermal energy, causing the temperature deep inside the solar nebula to climb. This process, in which the gravitational energy of a contracting gas cloud is converted into thermal energy, is called **Kelvin-Helmholtz contraction,** after the nineteenth-century physicists who first described it.

As the newly created protosun continued to contract and become denser, its temperature continued to climb as well. After about 10^5 (100,000) years, the protosun's surface temperature stabilized at about 6000 K, but the temperature in its interior kept increasing to ever higher values as the central regions of the protosun became denser and denser. Eventually, after perhaps 10^7 (10 million) years had passed since the solar nebula first began to contract, the gas at the center of the protosun reached a density of about 10^5 kg/m^3 (about 13 times denser than typical iron) and a temperature of a few million kelvins (that is, a few times 10^6 K). Under these extreme conditions, nuclear reactions that convert hydrogen into helium began in the protosun's interior. These nuclear reactions released energy that significantly increased the pressure in the protosun's core. When the pressure built up enough, it stopped further contraction of the protosun and a true star was born. In fact, the onset of nuclear reactions defines the end of a protostar and beginning of a star. Nuclear reactions continue to the present day in the interior of the Sun and are the source of all the energy that the Sun radiates into space.

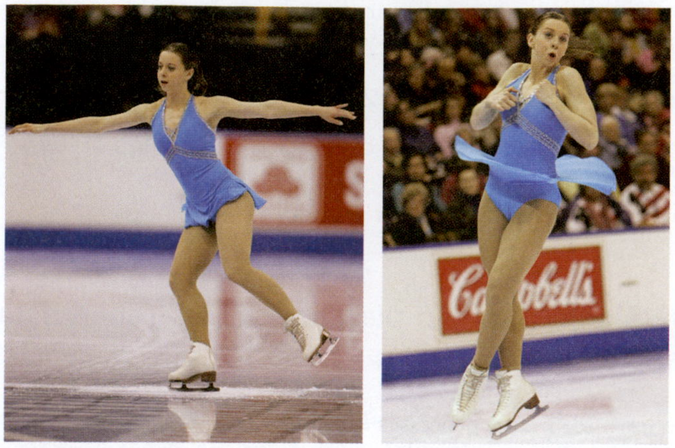

(a) (b)

FIGURE 8-7 R I V U X G

Conservation of Angular Momentum A figure skater who **(a)** spins slowly with her limbs extended will naturally speed up when **(b)** she pulls her limbs in. In the same way, the solar nebula spun more rapidly as its material contracted toward the center of the nebula. (AP Photo/Amy Sancetta)

The Protoplanetary Disk

If the solar nebula had not been rotating at all, everything would have fallen directly into the protosun, leaving nothing behind to form the planets. Instead, the solar nebula must have had an overall slight rotation, which caused its evolution to follow a different path. As the slowly rotating nebula collapsed inward, it would naturally have tended to rotate faster. This relationship between the size of an object and its rotation speed is an example of a general principle called the **conservation of angular momentum.**

ANALOGY Figure skaters make use of the conservation of angular momentum. When a spinning skater pulls her arms and legs in close to her body, the rate at which she spins automatically increases (Figure 8-7). Even if you are not a figure skater, you can demonstrate this by sitting on a rotating office chair. Sit with your arms outstretched and hold a weight, like a brick or a full water bottle, in either hand. Now use your feet to start your body and the chair rotating, lift your feet off the ground, and then pull your arms inward. Your rotation will speed up quite noticeably.

Astronomers see young stars that may be forming planets today in the same way that our solar system did billions of years ago. As the solar nebula began to rotate more rapidly, it also tended to flatten out (Figure 8-6b)—but why? From the perspective of a particle rotating along with the nebula, it felt as though there were a force pushing the particle away from the nebula's axis of rotation. (Likewise, passengers on a merry-go-round or spinning carnival ride seem to feel a force pushing them outward and away from the ride's axis of rotation.) This apparent force was directed opposite to the inward pull of gravity, and so it tended to slow the contraction of material toward the nebula's rotation axis. But there was no such effect opposing contraction in a direction parallel to the rotation axis. Some 10^5 (100,000) years after the solar nebula first began to contract, it had developed the structure shown in Figure 8-6b, with a rotating, flattened disk surrounding what will become the protosun. This disk is called the **protoplanetary disk,** since planets formed from its material. This model explains why their orbits all lie in essentially the same plane and why they all orbit the Sun in the same direction.

There were no humans to observe these processes taking place during the formation of the solar system. But Earth astronomers have seen disks of material surrounding other stars that formed only recently. These, too, are called protoplanetary disks, because it is thought that planets can eventually form from these disks around other stars. Hence, these disks are planetary systems that are still "under construction." By studying these disks around other stars, astronomers are able to examine what our solar nebula may have been like some 4.5×10^9 years ago.

Figure 8-8 shows a number of protoplanetary disks in the Orion Nebula, a region of active star formation. A star is visible at the center of each disk, which reinforces the idea that our Sun began to shine before the planets were fully formed. (The images that open this chapter show even more detailed views of disks surrounding young stars.) A study of 110 young stars in the Orion Nebula detected protoplanetary disks around 56 of them, which suggests that systems of planets may form around a substantial fraction of stars. Later in this chapter we will see direct evidence for planets that have formed around stars other than the Sun.

CONCEPTCHECK 8-5

If the nebular hypothesis is correct, what must it explain about the planetary orbits?

CONCEPTCHECK 8-6

Was the solar nebula cold until nuclear reactions began powering the Sun?

Answers appear at the end of the chapter.

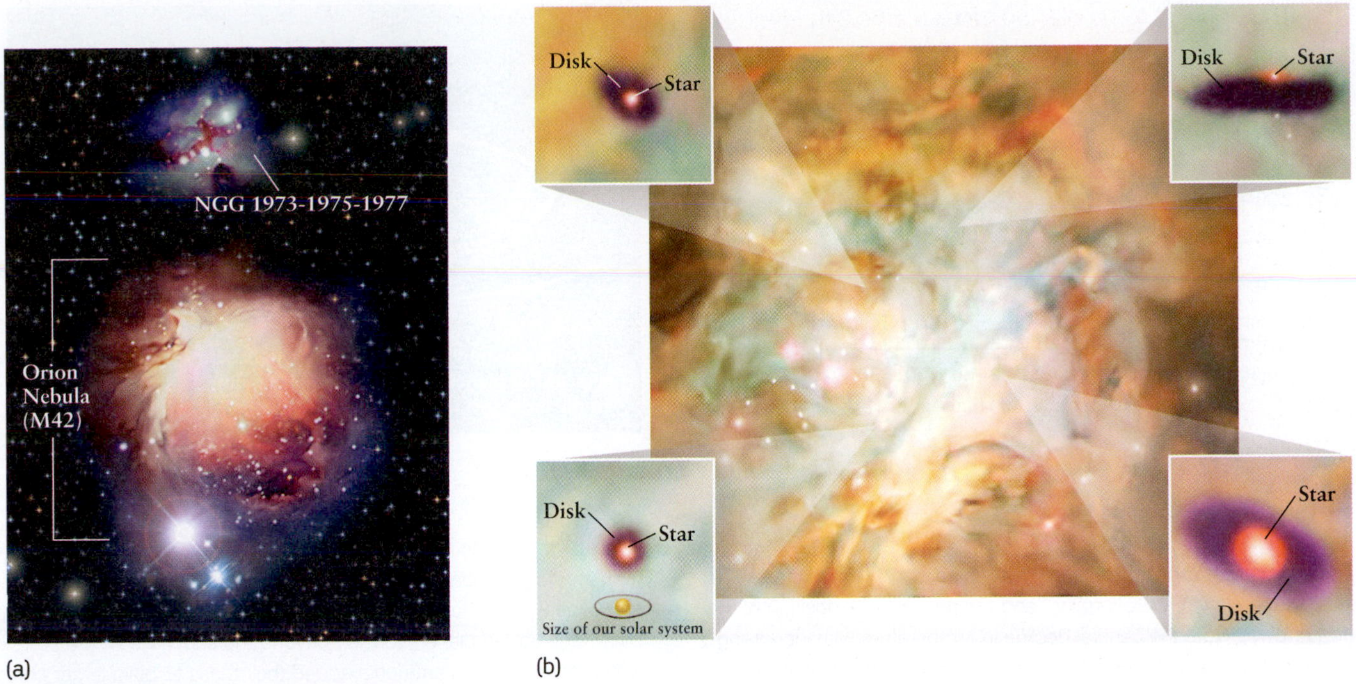

VIDEO 8-1

FIGURE 8-8 R I V U X G

Protoplanetary Disks **(a)** The Orion Nebula is a star-forming region located some 1500 light-years from Earth. It is the middle "star" in Orion's "sword" (see Figure 2-2a). The smaller, bluish nebula is the object shown in Figure 8-3. **(b)** This view of the center of the Orion Nebula is a mosaic of Hubble Space Telescope images. The four insets are false-color close-ups of four protoplanetary disks that lie within the nebula. A young, recently formed star is at the center of each disk. (The disk at upper right is seen nearly edge-on.) The inset at the lower left shows the size of our own solar system for comparison. (a: Anglo-Australian Observatory image by David Malin; b: C. R. O'Dell and S. K. Wong, Rice University; NASA)

8-5 The terrestrial planets formed by the accretion of planetesimals

> Rocky planets formed in the inner solar nebula as a consequence of the high temperatures close to the protosun

We have seen how the solar nebula would have contracted to form a young Sun with a protoplanetary disk rotating around it. But how did the material in this disk form into planets? Why are the small terrestrial planets located in the inner solar system (Mercury, Venus, Earth, and Mars), while the giant Jovian planets are in the outer solar system (Jupiter, Saturn, Uranus, and Neptune)? In this section and the next we will see how the nebular hypothesis provides answers to these questions.

The Condensation Temperature and the Snow Line

To understand how the planets, asteroids, and comets formed, we start by considering a cold and low-pressure solar nebula, before it was warmed by an emerging protosun. At the low pressures that prevailed, a substance does not form into a liquid state, but must exist as either a solid or a gas. An example of such a solid is in Figure 8-9, which shows a dust grain of the sort that would have been present throughout the solar nebula. Other substances

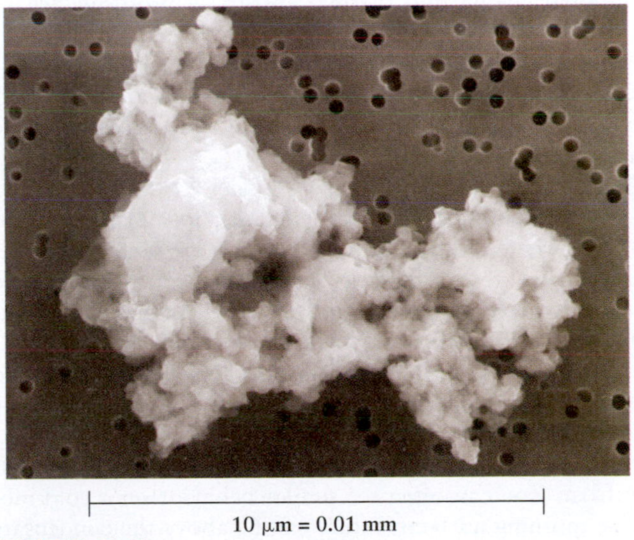

FIGURE 8-9

A Grain of Cosmic Dust This highly magnified image shows a microscopic dust grain that came from interplanetary space. It entered Earth's upper atmosphere and was collected by a high-flying aircraft. Dust grains of this sort are abundant in star-forming regions like that shown in Figure 8-3. These tiny grains were also abundant in the solar nebula and served as the building blocks of the planets. (NASA)

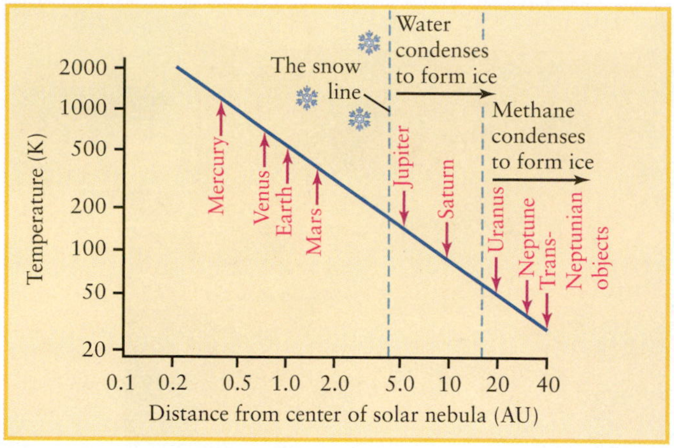

(a)

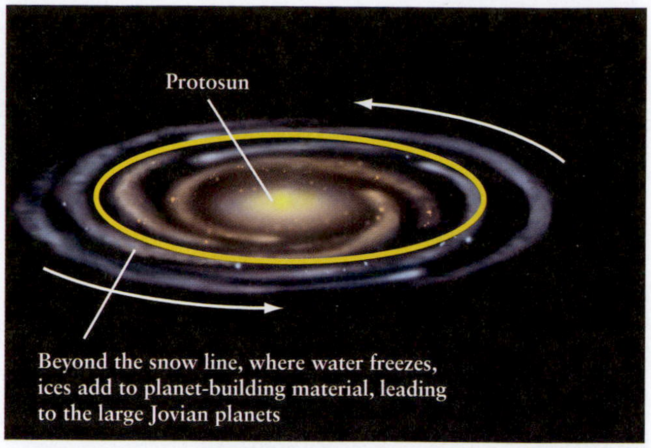

Protosun
Beyond the snow line, where water freezes, ices add to planet-building material, leading to the large Jovian planets

(b)

ANIMATION 8-2

FIGURE 8-10

Temperature Distribution and Snow Line in the Solar Nebula **(a)** This graph shows how temperatures probably varied across the solar nebula as the planets were forming, and the present-day position of the planets (red arrows). Note the general decline in temperature with increasing distance from the center of the nebula. Between the present-day distances of Mars and Jupiter, the snow line marked where temperatures were low enough for water to condense and form ice; beyond about 16 AU, methane (CH_4) could also condense into ice. **(b)** Terrestrial planets formed inside the snow line, where the low abundance of solid dust grains kept these planets small. The Jovian planets formed beyond the snow line where solid ices of water, methane, and ammonia added their mass to build larger cores, and attract surrounding gas.

in the early solar nebula would have been in the form of small ice crystals (like snow), although at higher temperatures, these ices would evaporate to form a gas. Together, these solids—referred to as ices, dust grains, and ice-coated dust grains—were mixed with gaseous hydrogen and helium. But, then things began to heat up.

Due to Kelvin-Helmholtz contraction—the conversion of gravitational energy into thermal energy—a protosun began to form in the center of the nebula, and it was actually quite a bit more luminous than the present-day Sun. The temperature in the nebula varied significantly, rising above 2000 K closer to the hot protosun, and dropping below 50 K in the outermost regions (Figure 8-10a).

Our first goal in determining a model for the origin of the solar system is to understand why the inner four planets are small and rocky, whereas the outer Jovian planets are large and made of lighter elements (recall Properties 1 and 3 in Table 8-1). To answer this, we first need to understand the **condensation temperature,** which determines when a substance forms a solid or a gas.

If the temperature of a substance in the solar nebula is above its condensation temperature, the substance is a gas. On the other hand, if the temperature is below the condensation temperature, the substance solidifies (or condenses) into tiny specks of dust or icy frost. You can often see similar behavior on a cold morning. The morning air temperature can be above the condensation temperature of water, while the cold windows of parked cars may have temperatures below the condensation temperature. Thus, water molecules in the warmer air remain as a gas (water vapor) but form solid ice particles (frost) on the colder car windows.

With heat from the emerging protosun, the solar nebula was radically changed into two distinct regions. These regions—inner and outer—had very different properties and eventually formed very different planets:

- **Inner Region: rock and metal.** In the hot inner region of the nebula, only substances with high condensation temperatures could have remained solid. These rocky and metallic materials, in the form of solid dust grains (Figure 8-9), eventually formed much of the rock and metal of the terrestrial planets. Hydrogen compounds—such as water (H_2O), methane (CH_4), and ammonia (NH_3)—remained as gas; they could not condense into solids in the warm inner solar system. (Hydrogen and helium could not condense into solids anywhere in the solar system.) The terrestrial planets that formed in this inner region are Mercury, Venus, Earth, and Mars.

- **Outer Region: ices beyond the snow line.** In the area between the present-day orbits of Mars and Jupiter, the temperature dropped below around 170 K. In the low pressures of the solar nebula, this is the condensation temperature at which water vapor (H_2O) forms ice (see Figure 8-10a). One aspect of this **snow line**—the distance from the Sun at which water vapor solidifies into ice or frost—is that beyond this line, the solid ice particles can join with rock and metal grains resulting in more mass to build planets (Figure 8-10b). Farther out, at even cooler temperatures, methane (CH_4), and ammonia (NH_3) also form ices. (Recall from Section 7-4 that "ice" can refer to frozen carbon dioxide, methane, or ammonia as well as to frozen water.) To summarize: Beyond the snow line, rocky and metallic dust grains were coated with frost, which means that additional icy mass went into building even larger planets. The Jovian planets that formed in this outer region are Jupiter, Saturn, Uranus, and Neptune.

The abundances of material in the solar nebula provide the final insight about planet sizes. With a mixture similar to the Sun, the solar nebula was composed of about 98% hydrogen and helium, about 1.4% hydrogen compounds (H_2O, CH_4, and NH_3), and the remaining 0.6% of rocks and metals. Inside the snow line, the only solid material available to build planets consisted of rock and metal. Thus, with rock and metal having the lowest abundances, the inner terrestrial planets that formed from these dust grains are the smallest. On the other hand, beyond the snow line there was more solid mass available in the form of ice-coated dust grains to help build larger planets.

The phenomena described so far help us understand why smaller terrestrial planets formed close to the Sun, and larger Jovian planets formed farther out. However, there is more to this story, and some big mysteries remain. Next, we consider the process of building planets in more detail, beginning with the terrestrial planets.

Planetesimals, Protoplanets, and Terrestrial Planets

ANIMATION 8-1 In the inner part of the solar nebula, the grains of rocks and metals would have collided and merged into small chunks. Initially, electric forces—that is, chemical bonds—held these chunks together, in the same way that chemical bonds hold an ordinary rock together. Over a few million years, these chunks coalesced into roughly a billion asteroidlike objects called **planetesimals**, with diameters of a kilometer or so. These larger planetesimals were massive enough to be held together by their own gravity, resulting from the mutual gravitational attraction of all the material within the planetesimal.

During the next stage, gravitational attraction between different planetesimals caused them to collide and accumulate into around a hundred still-larger objects called **protoplanets** (also called planetary embryos), each of which was roughly the size and mass of our Moon. This accumulation of material to form larger and larger objects is called **accretion.** For about a hundred million years, these Moon-sized protoplanets collided to form the inner planets. This final episode must have involved some truly spectacular, world-shattering collisions. In fact, detailed modeling suggests that one of these protoplanet collisions with the forming Earth probably created our Moon! Snapshots from a computer simulation of accreting planetesimals are shown in Figure 8-11.

In the inner solar nebula only materials with high condensation temperatures could form dust grains and hence protoplanets, so the result was a set of planets made predominantly of materials such as iron, silicon, magnesium, and nickel. In strong support for this model of solar system formation, these rocky and metallic materials match the composition of the present-day terrestrial planets. There is also supporting evidence from meteorites (dated from their radioactivity) that are thought to have come from smashed-up planetesimals during this early period. As expected for planetesimals from the inner solar system, these meteorites contain metallic grains mixed with rocky material (Figure 8-12).

At first the material that coalesced to form protoplanets in the inner solar nebula remained largely in solid form, despite the high temperatures close to the protosun. But as the protoplanets grew, they were heated by violent impacts as they collided with other planetesimals, as well as by the energy released from the decay of radioactive elements, and all this heat caused melting. Thus, the terrestrial planets began their existence as spheres of at least partially molten rocky materials. Material was free to move within these molten spheres, so the denser, iron-rich minerals sank

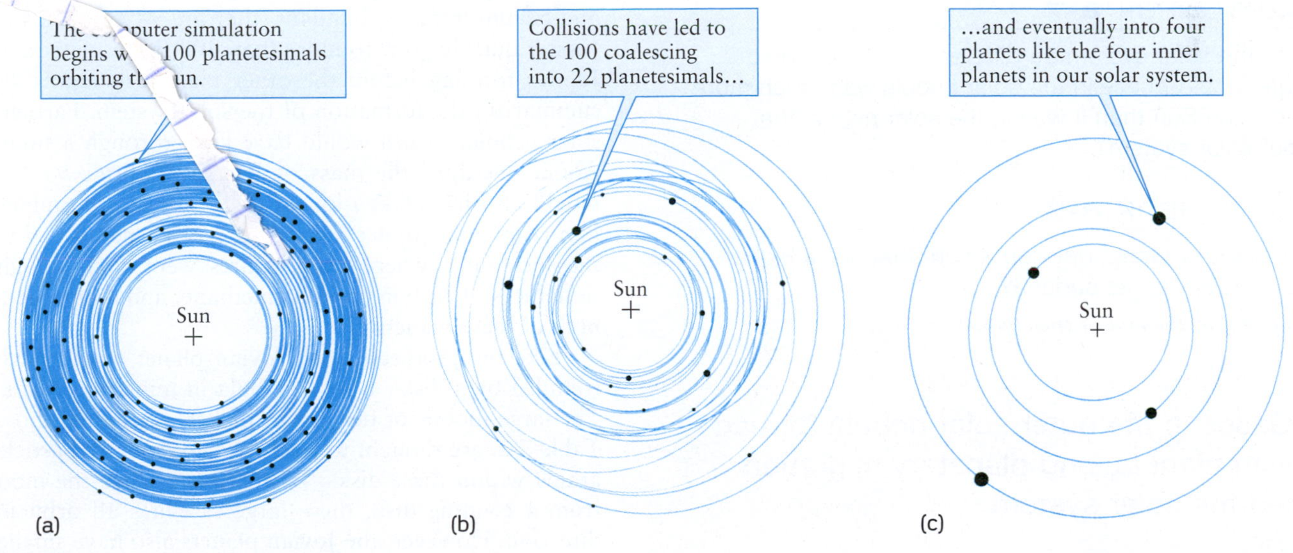

puter simulation
100 planetesimals
Sun
Collisions have led to the 100 coalescing into 22 planetesimals...
Sun
...and eventually into four planets like the four inner planets in our solar system.
Sun
(a)
(b)
(c)

FIGURE 8-11
Accretion of the Inner Planets This computer simulation shows the formation of the inner planets over time. (Adapted from George W. Wetherill)

FIGURE 8-12 R I V U X G

Primitive Meteorite This is a cross-section from a fragment of the Allende meteorite that landed in Chihuahua, Mexico, in 1969. A fireball was observed at 1:05 A.M. local time, and over the next 25 years about 3 tons of meteorite fragments were collected over an 8 km by 50 km area. Radioactive dating yields an age of 4.57 billion years, corresponding to the very early solar system. The meteorite contains rocks and metals that are consistent with the inner solar system and is thought to come from a smashed-up planetesimal. (ChinellatoPhoto/Shutterstock)

to the centers of the planets while the less dense silicon-rich rocky minerals floated to their surfaces. This process is called **chemical differentiation** (see Box 7-1). In this way the terrestrial planets developed their dense iron cores.

CONCEPTCHECK 8-7

How might the solar system be different if the location at which water could freeze in the solar nebula was much more distant from the Sun than it was in the solar nebula that formed our solar system?

CONCEPTCHECK 8-8

How does a rocky planet develop a core that has a higher density than rocks on its surface?

Answers appear at the end of the chapter.

8-6 Gases in the outer solar nebula formed the Jovian planets, and planetary migration reshaped the solar system

Low temperatures in the outer solar nebula made it possible for planets to grow to titanic size

We have seen how the low abundance of solid material in the inner solar nebula led to the formation of the small, rocky terrestrial planets. To explain the very different properties of the Jovian planets, we need to consider the conditions that prevailed in the relatively cool outer regions of the solar nebula.

The Core Accretion Model

The large Jovian planets *initially* formed through a similar process as the terrestrial planets—through the accretion of planetesimals. As discussed in Section 8-5, a key difference is that ices—in addition to rock and metal grains—were able to survive in the cooler outer regions of the solar nebula (see Figure 8-10). The elements of which ices are made are much more abundant than those that form rocky grains. Thus, more solid material would have been available to form planetesimals in the outer solar nebula than in the inner part. As a result, solid objects larger than any of the terrestrial planets could have formed in the outer solar nebula. Each such object could have become the core of a Jovian planet and served as a "seed" around which the rest of the planet eventually grew. For example, the mass of Jupiter's rock and metal core is estimated to equal about 10 Earth masses. However, while additional solid material beyond the snow line plays a role in forming the larger cores of Jovian planets, astronomers do not fully understand how their cores get as large as they do, and this is an active area of research.

For Jupiter, a large seed mass of rock, metal, and ice is only the beginning: Most of Jupiter's mass is hydrogen and helium. Recall that for a planet, retaining a gaseous atmosphere depends on both the planet's mass and on the gas temperature (see Box 7-2). Accordingly, due to the lower temperatures in the outer solar system, Jupiter's large seed mass could capture and retain hydrogen and helium gas. This picture—where a Jovian protoplanet core captures gas and grows by accretion—is called the **core accretion model.**

As Jupiter grew, its gravitational pull increased, allowing it to capture more gases and grow even larger until most of the available gas in its region had been captured. Because hydrogen and helium were so abundant (they are 98% of the solar nebula), Jupiter quickly grew to more than 300 Earth masses. You can see Jupiter forming before the other planets in Figure 8-13, which summarizes the formation of the solar system. Farther out in the solar nebula, Saturn would have gone through a similar process. About one-third the mass of Jupiter, Saturn's 95 Earth masses would also have taken longer to accumulate, forming a few million years after Jupiter. Uranus and Neptune formed well beyond the snow line, where temperatures were cold enough for additional ices of carbon dioxide, methane, and ammonia to form the bulk of these planets.

Like the protosun, each Jovian planet would have been surrounded by a disk—a solar nebula in miniature (see Figure 8-6). The large moons of the Jovian planets, including those shown in Table 7-2, are thought to have formed from ice particles and dust grains within these disks. Furthermore, since the moons formed from a rotating disk, these large satellites all orbit in the same direction. However, the Jovian planets also have smaller bodies—called irregular satellites—that orbit in the *opposite* direction, and these were probably captured after the planets formed.

Observations of protoplanetary disks around other protostars (such as those in Figure 8-8b) suggest that high-energy photons

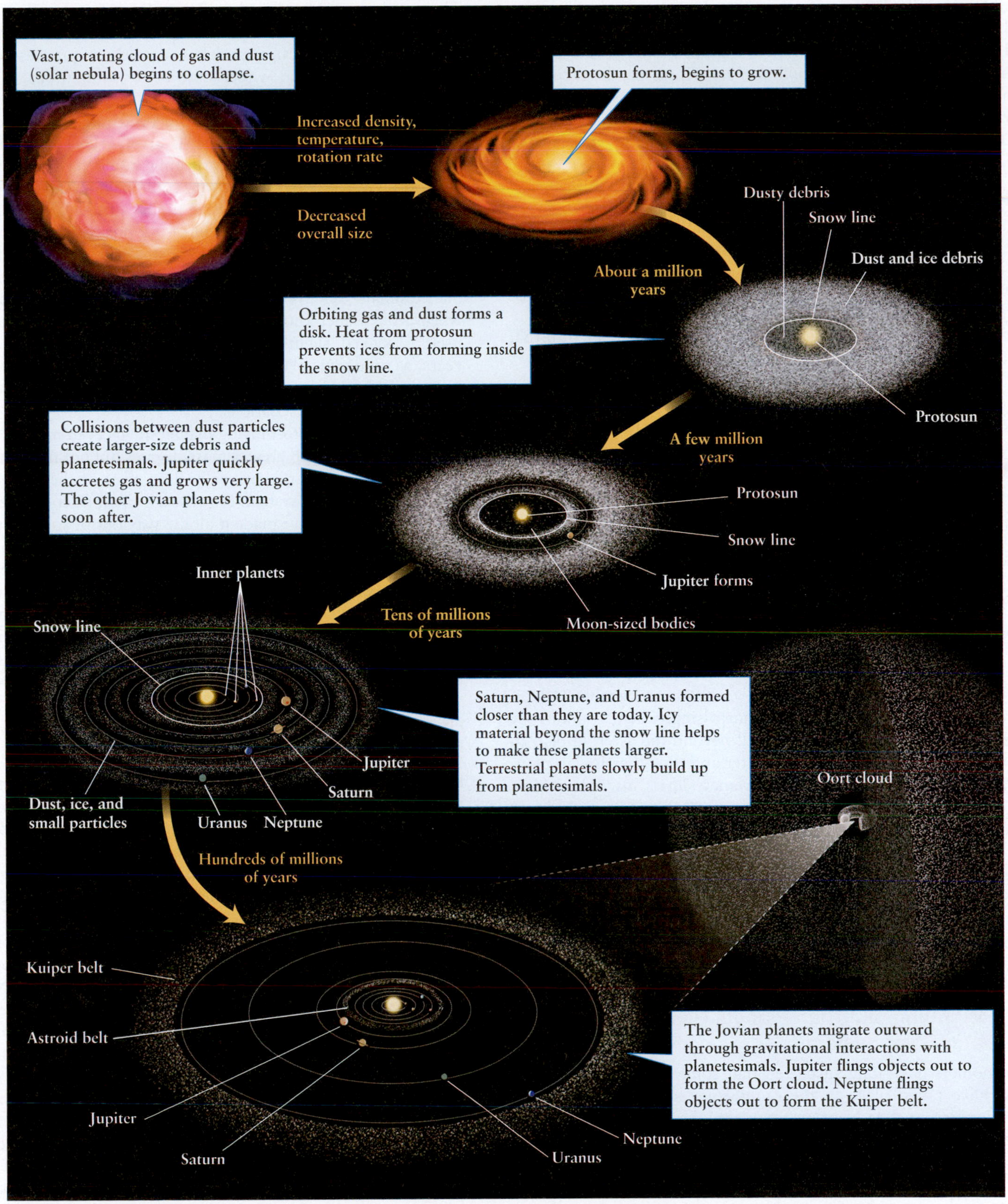
Vast, rotating cloud of gas and dust (solar nebula) begins to collapse.
Increased density, temperature, rotation rate
Decreased overall size
Protosun forms, begins to grow.
About a million years
Dusty debris
Snow line
Dust and ice debris
Orbiting gas and dust forms a disk. Heat from protosun prevents ices from forming inside the snow line.
Protosun
A few million years
Collisions between dust particles create larger-size debris and planetesimals. Jupiter quickly accretes gas and grows very large. The other Jovian planets form soon after.
Protosun
Snow line
Jupiter forms
Moon-sized bodies
Tens of millions of years
Inner planets
Snow line
Saturn, Neptune, and Uranus formed closer than they are today. Icy material beyond the snow line helps to make these planets larger. Terrestrial planets slowly build up from planetesimals.
Jupiter
Saturn
Uranus
Neptune
Dust, ice, and small particles
Oort cloud
Hundreds of millions of years
Kuiper belt
Astroid belt
Jupiter
Saturn
Uranus
Neptune
The Jovian planets migrate outward through gravitational interactions with planetesimals. Jupiter flings objects out to form the Oort cloud. Neptune flings objects out to form the Kuiper belt.

FIGURE 8-13

The Formation of the Solar System This sequence of drawings shows stages in the formation of the solar system.

would eventually disperse any remaining hydrogen and helium gas, and this gas ends up in the space between the stars. With hydrogen and helium gone from the disk, formation of the Jovian planets came to an end after a few million years. However, the accumulation of planetesimals that built the terrestrial planets did not stop when gas was expelled from the protoplanetary disk, and the terrestrial planets took much longer to form than the Jovian planets.

Early Migration of Jovian Planets Shapes the Inner Solar System

We now look at **migration,** which refers to changes in orbital distances of the planets. Migration of the planets results from interactions—gravitational tugs and pulls—between planets and different parts of the forming solar system. The details of planetary migration are far from certain, but the broad picture that follows is emerging from computer simulations.

Within the first few hundred thousand years, long before the terrestrial planets formed, Jupiter's interaction with the gaseous disk of hydrogen and helium led to an inward migration of Jupiter, followed by an outward migration. At its closest distance to the protosun, Jupiter migrated inward to about 1.5 AU, which is near the current orbit of Mars. With its large mass, Jupiter gravitationally deflected many of the planetesimals near the current Martian orbit. As a result, when Mars eventually formed in this region, it ended up with a low mass of about one-tenth of Earth's mass. In fact, it was the mysteriously low mass of Mars that prompted this analysis of Jupiter's inward migration. Even more, inward-then-outward migration of all the Jovian planets—called the **Grand Tack model**—also solves a long-standing mystery involving the asteroid belt.

The asteroid belt (Section 7-5) contains rocky objects typical of the inner solar system, but, surprisingly, also contains icy objects expected to have formed well beyond the asteroid belt. In the Grand Tack model, as both Jupiter and the other Jovian planets migrate outward, they deflect planetesimals inward to form the asteroid belt. Some of these planetesimals come from the inner solar system, but some also come from much farther out beyond the snow line, providing a very natural explanation for the icy objects in the asteroid belt.

This early migration, along with Jovian planet formation, was over within the solar system's first few million years or so. Another type of migration occurred over the next few hundred million years, and reshaped the outer solar system.

Late Migration of Jovian Planets Reshapes the Outer Solar System

Astronomers have long suspected that some planetary migration must have taken place in the outer solar system. To see why, consider Neptune. At Neptune's current large orbital distance, the timescale necessary to build a planet of Neptune's large size by core accretion is much longer than the time that the protoplanetary disk was around. However, planets can grow much faster closer to the protosun, where the protoplanetary disk is denser. Therefore, astronomers suspect that Neptune formed closer in, and then migrated outward.

The leading theory described here for late migration of the Jovian planets is based on the **Nice** (pronounced "niece") **model.** This model was developed in Nice, France, around 2005 and results from detailed computer simulations. It is important to emphasize that the Nice model is a work in progress and has already evolved with new ideas and better simulations.

In the Nice model, the Jovian planets would have all originally formed within about 20 AU of the protosun, even though the outermost planet today, Neptune, is at 30 AU. Furthermore, the outermost planet during this early time could have been Uranus, at 20 AU. Gravitational interactions could have switched the orbital ordering of Neptune and Uranus; in the Nice model, both initial scenarios can lead to the arrangement of our present-day solar system.

There was also a disk of planetesimals beyond the outermost Jovian planet at 20 AU, and through gravitational encounters, these planetesimals, on average, were scattered inward. Furthermore, when a big planet knocks a small planetesimal inward, the planet itself is kicked slightly outward. Over a few hundred million years, as numerous planetesimals were knocked inward by Saturn, Neptune, and Uranus, these Jovian planets slowly migrated outward to their current locations.

Most of the inward-moving planetesimals even made it close to Jupiter. With its much greater mass, Jupiter did not migrate inward very much, but did gravitationally fling most of the planetesimals clear out of the solar system. However, a small fraction of these icy objects would not have made it all the way out of solar system and are thought to currently orbit at about 50,000 AU from the Sun—about 0.8 light-year away and one-fifth the distance to the nearest star. These planetesimals would be very loosely bound to our Sun and gravitational deflections from passing stars would spread many of their orbits into a spherical "halo." We call this hypothesized distribution of icy planetesimals the Oort cloud (Figure 8-14a), which formed beyond the objects of our next topic—the Kuiper belt.

Unstable Orbits, the Kuiper Belt, and the Late Heavy Bombardment

During the slow outward migration of the Jovian planets, their orbits remain mostly circular; in other words, they have very low eccentricity. The slowly expanding Jovian orbits are also quite stable and allow the migration to continue for about 600 million years or so. However, gravitational interactions with the planetesimals eventually destabilize planetary orbits, leading to elongated, or eccentric, orbits.

With elongated orbits, the planets gravitationally interact with each other more strongly. In many computer simulations, this nearly *doubles* the orbital distance of Neptune, sending Neptune out to its current distance of about 30 AU. As Neptune moves outward, its gravity flings nearby planetesimals to greater distances as well, creating an orbiting collection of icy objects called the Kuiper belt (Figure 8-14b). (The images that open this chapter show two young stars surrounded by dusty disks that resemble the Kuiper belt, one seen edge-on and the other seen face-on.) Taken together, the Oort cloud and Kuiper belt contain icy planetesimals that could not have formed at these great distances where matter was sparse, but instead were gravitationally deflected outward by Jupiter and then by Neptune.

The Nice model seems to solve another mystery of the solar system. Many astronomers think there was a cataclysmic period when the planets and moons of the inner solar system were subjected to a short but intense period of large impacts; this

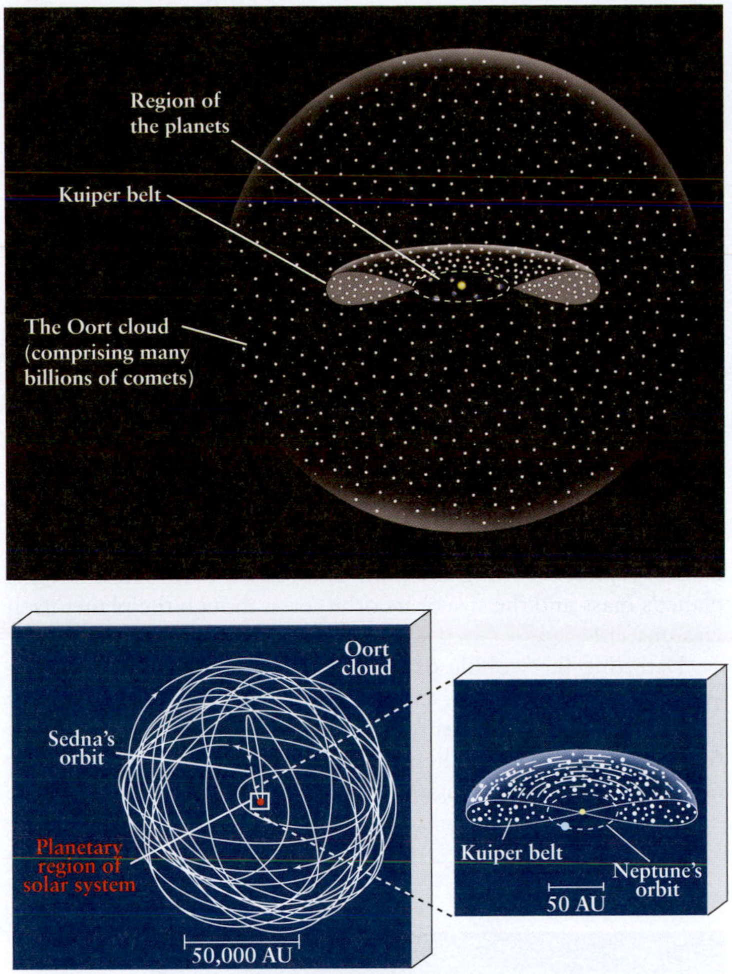

(a) Orbits of some Oort and Kuiper belt objects

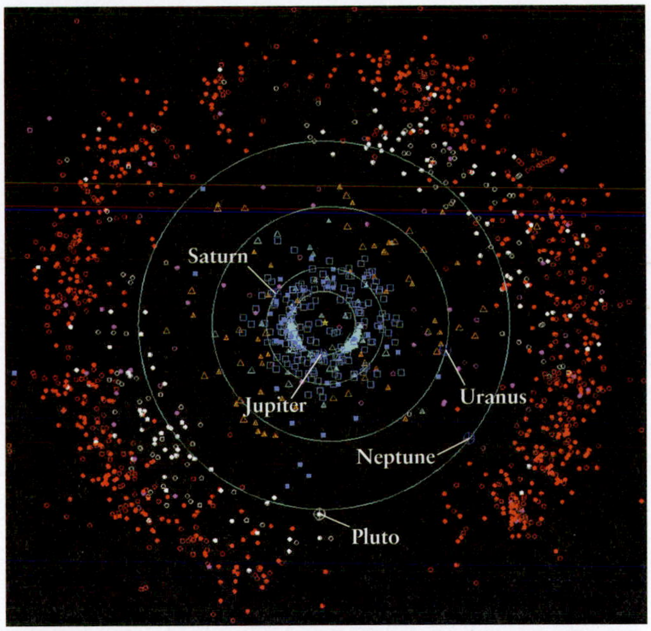

(b)

ANIMATION 8-4 **FIGURE 8-14**
The Kuiper Belt and Oort Cloud (a) The classical Kuiper belt of comets spreads from Neptune out to 50 AU from the Sun. Most of the estimated 200 million belt comets are believed to orbit in or near the plane of the ecliptic. The spherical Oort cloud extends from beyond the Kuiper belt. (b) Orbits of bodies in the Kuiper belt. (b: NASA and A. Field/Space Telescope Science Institute)

hypothesis is referred to as the **Late Heavy Bombardment.** The evidence comes from radioactive dating of craters on the Moon, but more lunar samples are needed to firmly establish a spike in the timing of impact events. This bombardment is considered "late" because the frequent planetesimal collisions of planet-building had long since passed.

So, what could cause a big jump in impacts? There are several ideas, but here is the most favored scenario. In the Nice model, the destabilized and elongated planetary orbits also led to deflections of planetesimals throughout the inner solar system. Very significantly, it takes about 600 million years before widespread deflection of the planetesimals, which is consistent with the surprisingly late timing of the Late Heavy Bombardment.

ANIMATION 8-3 To a *much* lesser extent, gravitational deflections continue today. Planetesimals from either the Kuiper belt or the Oort cloud are occasionally knocked into new orbits. If one of these icy objects is deflected on a path through the inner solar system, it begins to evaporate, producing a visible tail, and is seen as a comet (see Figure 7-9).

Prior to 1995 the only fully formed planetary system to which we could apply this model was our own. As we will see in the next section, astronomers can now further test this model on an ever-growing number of planets known to orbit other stars.

CONCEPTCHECK 8-9

Why is it unlikely that Neptune formed by core accretion at its current location?

CONCEPTCHECK 8-10

In the Nice model, how did Neptune get to its present location?

CONCEPTCHECK 8-11

Does accretion refer to the accumulation of matter by gravitational attraction or the formation of chemical bonds?

Answers appear at the end of the chapter.

8-7 A variety of observational techniques reveal planets around other stars

Many planets have been discovered around other stars, and some are Earth-size

If planets formed around our Sun, have they formed around other stars? The answer is yes, and they are called **extrasolar planets,** or **exoplanets.** The discovery of exoplanets brings up several questions that we will consider next. What are the properties of these exoplanets—are they similar to the planets of our solar system? What do they tell us about the formation of other solar systems? And perhaps the biggest question of all: Are there other Earthlike planets?

Since exoplanets were first detected in the 1990s, this field of astronomy has exploded with new discoveries. At the time of this writing, 834 exoplanets have been discovered with a wide range of properties, and some big surprises. (This number refers to confirmed exoplanets, and several thousand candidates are awaiting confirmation.) To appreciate how remarkable these discoveries are, we must look at the process that astronomers go through to search for exoplanets.

Searching for Exoplanets

It is very difficult to make direct observations of planets orbiting other stars. The problem is that planets are small and dim compared with stars; for example, the Sun is a billion times brighter than Jupiter at visible wavelengths. Exoplanets can be imaged directly, as in Figure 8-15a, but a star's glare makes this method the least productive method of detection. By the time the exoplanets in Figure 8-15b were imaged in 2010, hundreds had been detected through other means. One very powerful method of exoplanet detection is to search for stars that appear to "wobble" (Figure 8-16). If a star has a planet, it is not quite correct to say that the planet orbits the star. Rather, both the planet and the star move in elliptical orbits around a point called the **center of mass** (Figure 8-16a). Imagine the planet and the star as sitting at opposite ends of a very long seesaw; the center of mass is where you would have to place the support-point (or fulcrum) in order to balance the seesaw. Because of the star's much greater mass, the center of mass is much closer to the star than to the planet. Thus, while the planet may move in a very large orbit, the star will move in a much smaller orbit.

As an example, the Sun and Jupiter both orbit their common center of mass with an orbital period of 11.86 years. (Jupiter has more mass than the other seven planets put together, so it is a reasonable approximation to consider the Sun's wobble as being due to Jupiter alone.) The Sun's orbit around the common center of mass has a semimajor axis only slightly greater than the Sun's radius, so the Sun slowly wobbles around a point not far outside its surface. If astronomers elsewhere in the Galaxy could detect the Sun's wobbling motion, they could tell that there was a large planet (Jupiter) orbiting the Sun. They could even determine the planet's mass and the size of its orbit, even though the planet itself was unseen!

Detecting the wobble of other stars is not an easy task. One approach to the problem, called the **astrometric method,** involves making very precise measurements of a star's position in the sky relative to other stars. The goal is to find stars whose positions change in a cyclic way (Figure 8-16b). The measurements must be

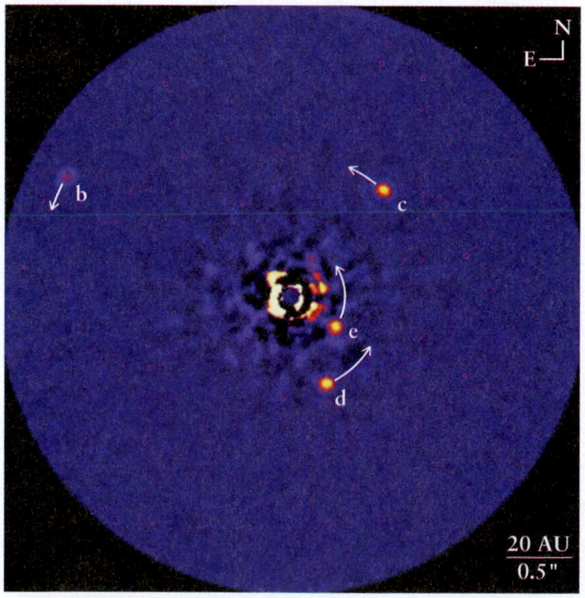

(a)

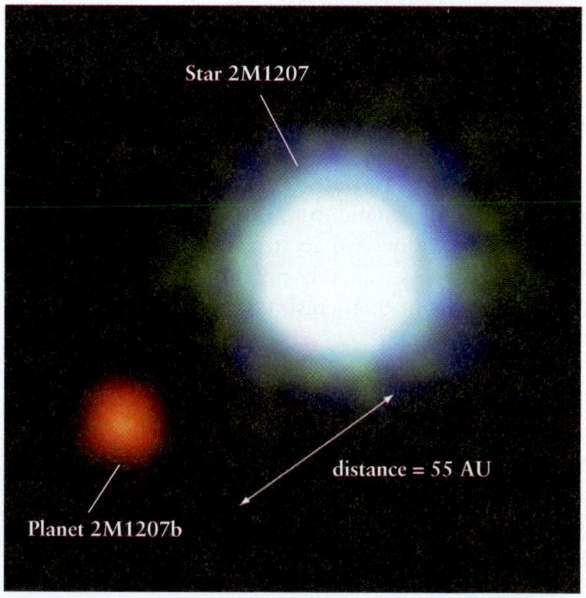

(b)

FIGURE 8-15 R I V U X G

Imaging Extrasolar Planets **(a)** Four planets can be seen orbiting the star HR 8799. The star is about 129 light-years from Earth and can be seen with the naked eye on a clear night. In capturing this image, special techniques make the star less visible to reduce its glare and reveal its planets. The white arrows indicate possible trajectories that each star might take over the next ten years. **(b)** About 170 light-years away, the star 2M1207 and a planet with about 1.5 times the diameter of Jupiter are captured in this infrared image. First observed in 2004, this extrasolar planet was the first to be visible in a telescopic image. At infrared wavelengths a star outshines a Jupiter-sized planet by only about 100 to 1, compared to 10^9 to 1 at visible wavelengths, making imaging possible in the infrared. (a: NASA/W. M. Keck Observatory; b: ESO/VLT/NACO)

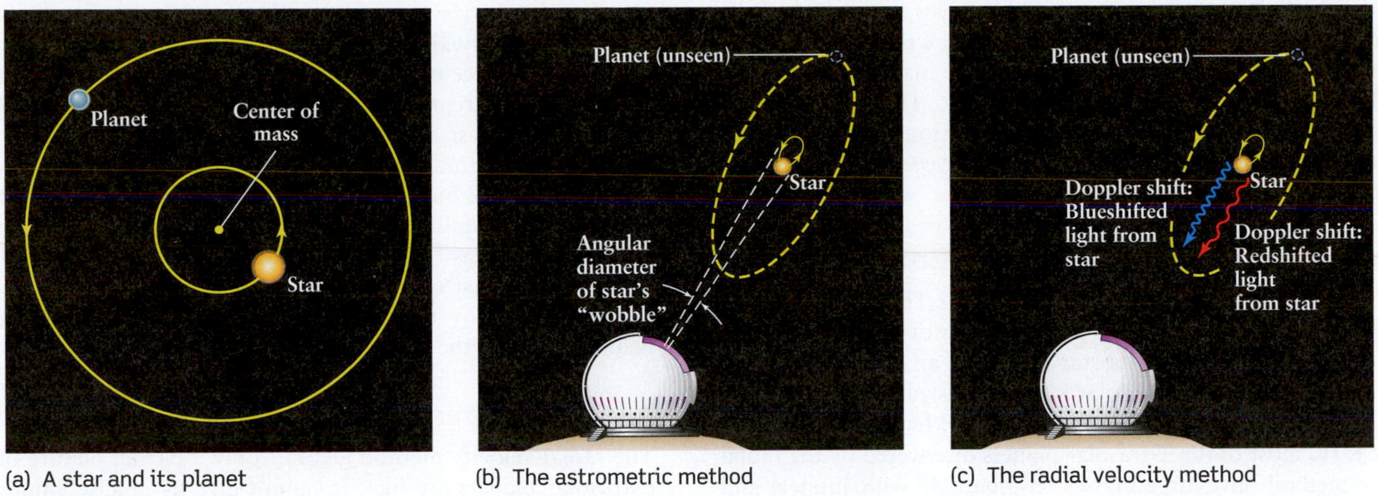

FIGURE 8-16

Detecting a Planet by Measuring Its Parent Star's Motion

(a) A planet and its star both orbit around their common center of mass, always staying on *opposite sides* of this point. Even if the planet cannot be seen, its presence can be inferred if the star's motion can be detected. **(b)** The astrometric method of detecting the unseen planet involves making direct measurements of the star's orbital motion. **(c)** In the radial velocity method, astronomers measure the Doppler shift of the star's spectrum as it moves alternately toward and away from Earth. Analyzing the Doppler shift of the star's light leads to the orbital period, orbital size (semimajor axis), and mass of the unseen planet. The radial velocity method works for circular and elliptical orbits and can even determine the orbital eccentricity.

made with very high accuracy (0.001 arcsec or better) and, ideally, over a long enough time to span an entire orbital period of the star's motion. The direct observation of a star's changing position has not yet played a significant role in detecting exoplanets, but as we will see, that will change in the near future.

ANIMATION 8-5 A star's wobble can also be detected by the **radial velocity method** (Figure 8-16c). This method is based on the Doppler effect, which we described in Section 5-9. During repeated orbits, a wobbling star will alternately move toward and away from Earth. Only this motion along a radius or line between Earth and the star—in other words, the star's radial velocity—contributes to the Doppler shift. The star's movement causes the wavelengths of its entire spectrum to shift in a periodic fashion, and this can be measured in the star's dark absorption lines (see Figure 5-14). When the star is moving away from us, its spectrum will undergo a redshift to longer wavelengths. When the star is approaching, there will be a blueshift of the spectrum to shorter wavelengths (see Figure 5-26). These wavelength shifts are very small because the star's motion around its orbit is quite slow. As an example, the Sun moves around its small orbit at only 12.5 m/s (45 km/h, or 28 mi/h)—slow enough that someone could beat the Sun on a fast bike ride! If the Sun was moving directly toward an observer at this speed, the hydrogen absorption line at a wavelength of 656 nm in the Sun's spectrum would be shifted by only 2.6×10^{-5} nm, or about 1 part in 25 million. Detecting these tiny shifts requires extraordinarily careful measurements and painstaking data analysis.

In 1995, the radial velocity method was used to discover a planet orbiting the star 51 Pegasi, which is 47.9 light-years away. For the first time, solid evidence was found for a planet orbiting a star like our own Sun. Astronomers have since used the radial velocity method to discover about 500 exoplanets.

From radial velocity plots such as Figure 8-17, a planet's orbital distance and mass can be determined. To determine these quantities, the period of the radial velocity—this is measured

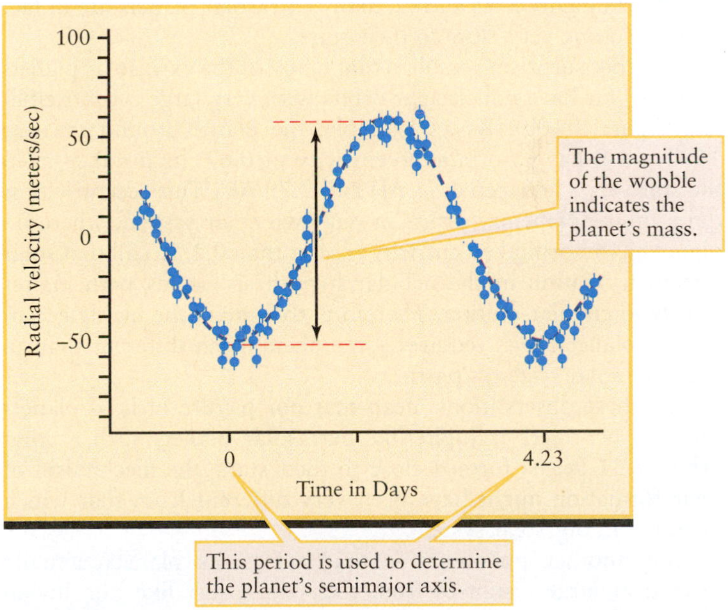

FIGURE 8-17

Discovering Planets from a Star's Wobble: 51 Pegasi As a star wobbles, it moves toward us and away from us, with a varying radial velocity that is measured through the Doppler shift. The shape of the curve indicates the planet's semimajor axis and also the planet's mass. However, the planetary mass is actually a minimum mass because an orbital plane tilted out of our line of sight would require an even larger mass to produce the observed Doppler shift.

from the star's Doppler shifting wobble—gives the period of the planet's orbit. Using Newton's form of Kepler's third law (see Box 4-4) and standard methods for estimating the mass of the star, we find the semimajor axis of the planet's orbit. Then, by knowing the planet's orbital distance and the gravitational force required to cause the star's observed radial velocity, we can determine the planet's mass.

Exoplanets: Surprising Masses and Orbits

What do we find in these other solar systems? The planet orbiting 51 Pegasi has a mass of at least 0.47 times that of Jupiter but orbits only 0.052 AU away from its star and with an orbital period of just 4.23 days. In our own solar system, this planet would orbit about *7 times closer to the Sun than Mercury!* In fact, as shown in Figure 8-18, most of the extrasolar planets discovered by the radial velocity method have large masses (comparable with Jupiter) and also orbit very close to their stars. However, this is an example of a **selection effect,** where an observational technique is more likely to detect some systems, and not others. Since a larger planet closer to its star creates a greater Doppler shift, the radial velocity method is more likely to detect this type of extrasolar system.

The difficulty of the radial velocity method to detect Earthlike planets in orbits similar to our own does not lessen the mystery of the so-called **hot Jupiters** that have been found. Orbiting so close to their stars, these planets would, in fact, be very hot; the temperatures at Mercury's distance can melt lead! Furthermore, according to the picture of our own solar system formation (Section 8-5), such Jovian planets would be expected to orbit relatively far from their stars, where temperatures were low enough to allow the buildup of a massive envelope of hydrogen and helium gas. But as Figure 8-18 shows, many extrasolar planets are in fact found orbiting very *close* to their stars.

Another surprising result is that many of the extrasolar planets found so far have noncircular orbits with very large eccentricities (see Figure 4-10b). As an example, the planet around the star 16 Cygni B has an orbital eccentricity of 0.67; its distance from the star varies between 0.55 AU and 2.79 AU. This eccentricity is quite unlike planetary orbits in our own solar system, where no planet has an orbital eccentricity greater than 0.2. Recall that solar system formation in the nebular hypothesis begins with matter mostly in circular motion. Therefore, to explain the highly eccentric exoplanet orbits requires some phenomena different than in our own solar system's past.

Do these observations mean that our picture of how planets form is incorrect? If Jupiterlike extrasolar planets such as that orbiting 51 Pegasi formed close to their stars, the mechanism of their formation might have been very different from that which operated in our solar system.

But another possibility is that extrasolar planets actually formed at large distances from their stars, just like our Jovian planets, and then migrated inward since their formation. Recall that in the Grand Tack model of our solar system formation (Section 8-6), Jupiter migrates inward, but to a much lesser extent than the extreme migration implied by these hot Jupiters. However, computer modeling indicates that Jupiter-size planets can migrate very close in to their stars due to interactions with a gaseous disk, if the migration occurs before the gas is dispersed. Simulations even suggest that planets can migrate *all the way* inward and get swallowed by the central star.

One piece of evidence that young planets may spiral into their parent stars is the spectrum of the star HD 82943, which has at least two planets orbiting it. The spectrum shows that this star's atmosphere contains a rare form of lithium that is found in planets, but which is destroyed in stars by nuclear reactions within 30 million years. The presence of this exotic form of lithium, known as ^{6}Li, indicates that in the past HD 82943 was orbited by at least one other planet. It is thought that this planet was consumed by the star when it spiraled too close, leaving behind its ^{6}Li in the star's atmosphere.

Analyzing Extrasolar Planets with the Transit Method

The radial velocity method gives us only a partial picture of what extrasolar planets are like. It cannot give us precise values for a planet's mass, only a lower limit. (An exact determination of the mass would require knowing how the plane of the planet's orbit is inclined to our line of sight. Unfortunately, this angle is not known because the planet itself is unseen.) Furthermore, the radial velocity method by itself cannot tell us the diameter of a planet or what the planet is made of.

A technique called the **transit method** makes it possible to fill in these blanks about the properties of certain extrasolar planets. This method looks for the rare situation in which a planet comes between us and its parent star, in an event called a **transit** (Figure 8-19). As in a partial solar eclipse (Section 3-5), this causes a small but measurable dimming of the star's light. If a transit is seen, the orbit must be nearly edge-on to our line of sight. By knowing the orientation of the orbit due to the transit, and then using information obtained by additional radial velocity measurements of the star, we learn the true mass (not just a lower limit) of the orbiting planet. But learning the planet's mass is only the beginning.

There are three other benefits of the transit method. One, the amount by which the star is dimmed during the transit depends on how large the planet is, and so tells us the planet's diameter (Figure 8-19a). Two, during the transit the star's light passes through the planet's atmosphere, where certain wavelengths are absorbed by the atmospheric gases. This absorption affects the spectrum of starlight that we measure and therefore allows us to determine the composition of the planet's atmosphere (Figure 8-19b). Three, with an infrared telescope it is possible to detect a slight dimming when the planet goes *behind* the star (Figure 8-19c). That is because the planet emits infrared radiation due to its own temperature, and this radiation is blocked when the planet is behind the star. Measuring the amount of infrared dimming tells us the amount of radiation emitted by the planet, which in turn tells us the planet's surface temperature.

One example of what the transit method can tell us is the planet that orbits the star HD 209458 in Figure 8-19, which is 153 light-years from the Sun. The mass of this planet is 0.69 that of Jupiter, but its diameter is 1.32 times larger than that of Jupiter. This lower-mass, larger-volume planet has only one-quarter the density of Jupiter. It is also very hot: Because the planet orbits just 0.047 AU from the star, its surface temperature is a torrid 1130 K (860°C, or 1570°F), about twice the temperature needed to melt

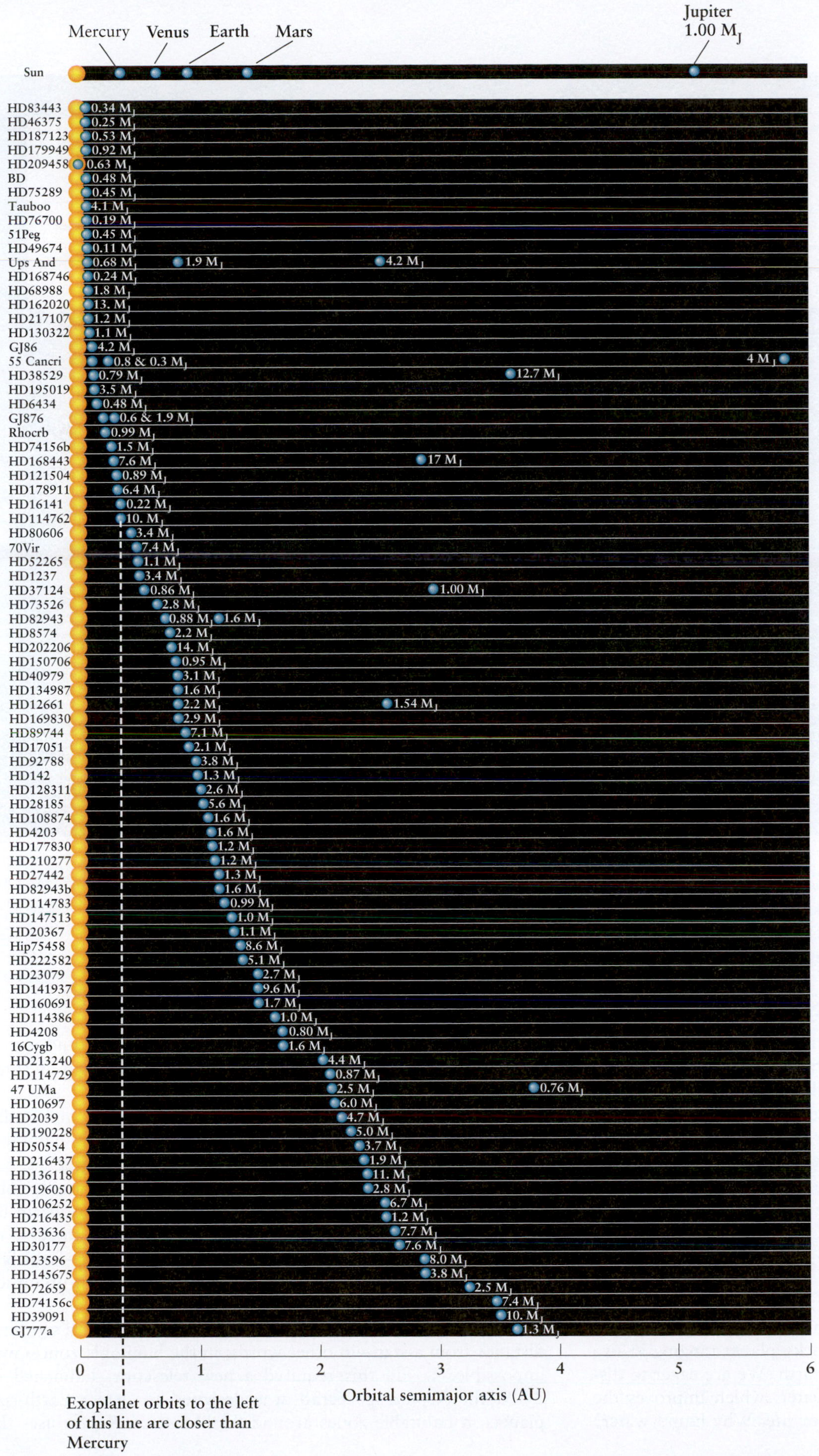

FIGURE 8-18

A Selection of Extrasolar Planets This figure summarizes what we know about planets orbiting a number of other stars. The star name is given at the left of each line. Each planet is shown at its average distance from its star (equal to the semimajor axis of its orbit), and some stars are found with multiple planets. The mass of each planet—actually a lower limit—is given as a multiple of Jupiter's mass (M_J), equal to 318 Earth masses. Comparison with our own solar system (at the top of the figure) shows how closely many of these extrasolar planets orbit their stars—many planets about the size of Jupiter, orbit closer to their star than Mercury does to our Sun! Due to higher temperatures closer to the Sun, many of these exoplanets are called "hot Jupiters." However, the radial velocity method detects large close-in planets most easily, and if all planets could be detected, the majority would be smaller and farther away. (Adapted from the California and Carnegie Planet Search)

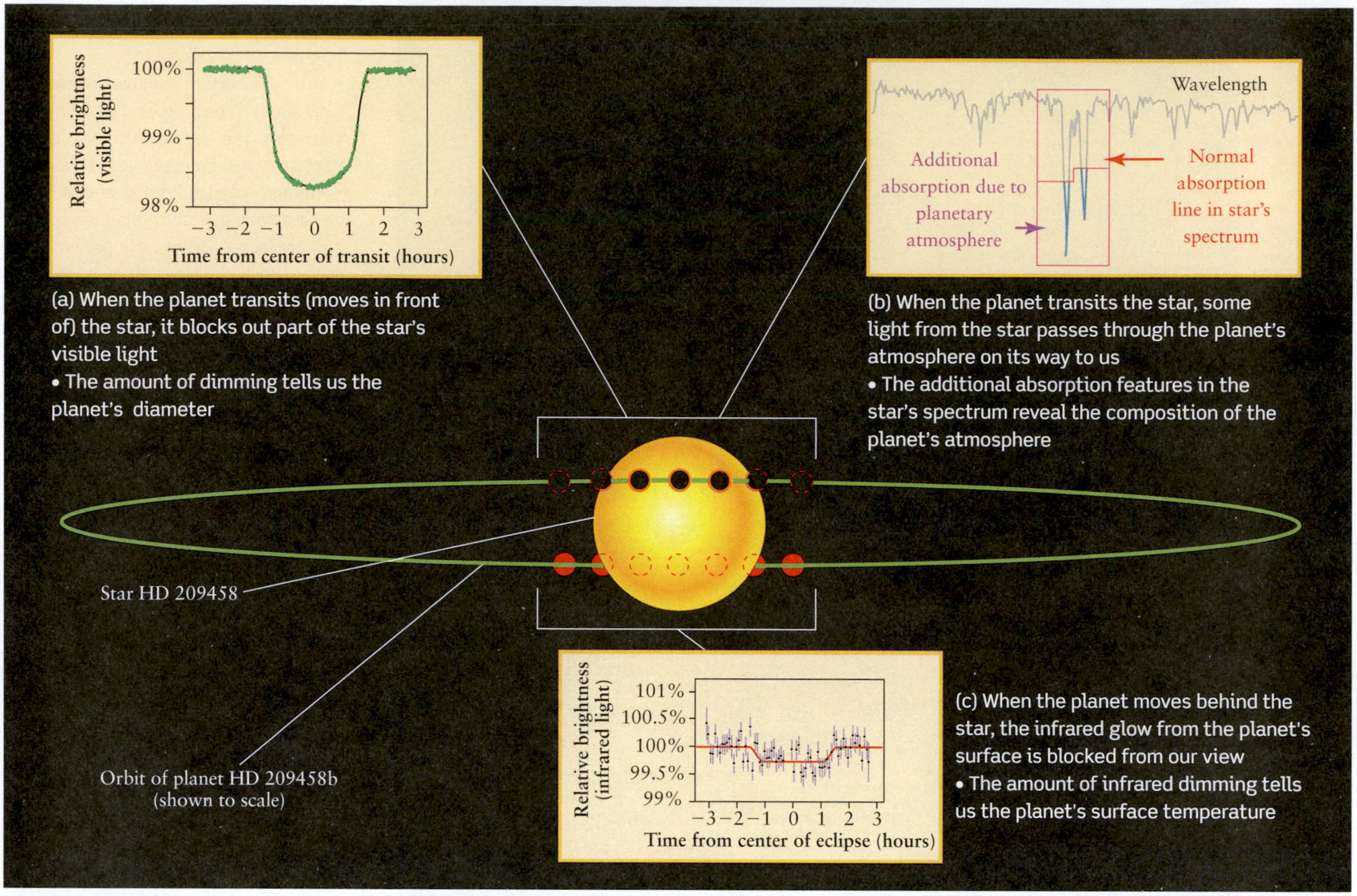

Relative brightness (visible light)
100%
99%
98%
−3 −2 −1 0 1 2 3
Time from center of transit (hours)
(a) When the planet transits (moves in front of) the star, it blocks out part of the star's visible light
• The amount of dimming tells us the planet's diameter
Wavelength
Additional absorption due to planetary atmosphere
Normal absorption line in star's spectrum
(b) When the planet transits the star, some light from the star passes through the planet's atmosphere on its way to us
• The additional absorption features in the star's spectrum reveal the composition of the planet's atmosphere
Star HD 209458
Orbit of planet HD 209458b (shown to scale)
Relative brightness (infrared light)
101%
100.5%
100%
99.5%
99%
−3 −2 −1 0 1 2 3
Time from center of eclipse (hours)
(c) When the planet moves behind the star, the infrared glow from the planet's surface is blocked from our view
• The amount of infrared dimming tells us the planet's surface temperature

FIGURE 8-19

A Transiting Extrasolar Planet If the orbit of an extrasolar planet is nearly edge-on to our line of sight, like the planet that orbits the star HD 209458, we can learn about the planet's **(a)** diameter, **(b)** atmospheric composition, and **(c)** surface temperature. (S. Seager and C. Reed, *Sky and Telescope;* H. Knutson, D. Charbonneau, R. W. Noyes (Harvard-Smithsonian CfA), T. M. Brown (HAO/NCAR), and R. L. Gilliland (STScI); A. Feild (STScI); NASA/JPL-Caltech/D. Charbonneau, Harvard-Smithsonian CfA)

lead! Observations of the spectrum during a transit of this hot Jupiter reveal the presence of hydrogen, carbon, oxygen, and sodium in a planetary atmosphere that is literally evaporating away in the intense starlight. Discoveries such as these give insight into the exotic circumstances that can be found in other planetary systems. Yet, we are also very interested in less exotic systems: the transit method can also discover terrestrial Earth-sized planets farther away from their star where temperatures might be just right for liquid water to form.

The Search for Earth-size Planets in the Habitable Zone

The ultimate goal of planetary searches is to find planets like Earth. An "Earth-size" planet is any rocky planet ranging in size from about half to double the size of Earth. We are eager to discover Earth-size planets with liquid water, which improves the chances that such a planet might harbor life. Why liquid water? Even the most exotic forms of life on Earth require liquid water, and many biochemists argue that an alien life-form based on some alternative biochemistry might still require water.

A rocky planet in the so-called **habitable zone** of a solar system might have liquid water on its surface: In the habitable zone a planet is close enough to its star to melt water ice, yet not close enough to boil the water away. But the habitable zone is only a rough guide: Planetary greenhouse gases can warm up a cold planet or even boil off a planet's water (as in the case of Venus). Furthermore, as in the case of Jupiter's Europa—which is far beyond our habitable zone—tidal forces can warm up an icy moon to create liquid water.

Until recently, detecting Earth-size planets at Earth's orbital distance from a star—in other words, in the habitable zone—was impossible. To do this required a new telescope. Launched in 2009, the *Kepler* spacecraft was designed to detect Earth-size planets in habitable zones around Sunlike stars. *Kepler* uses the

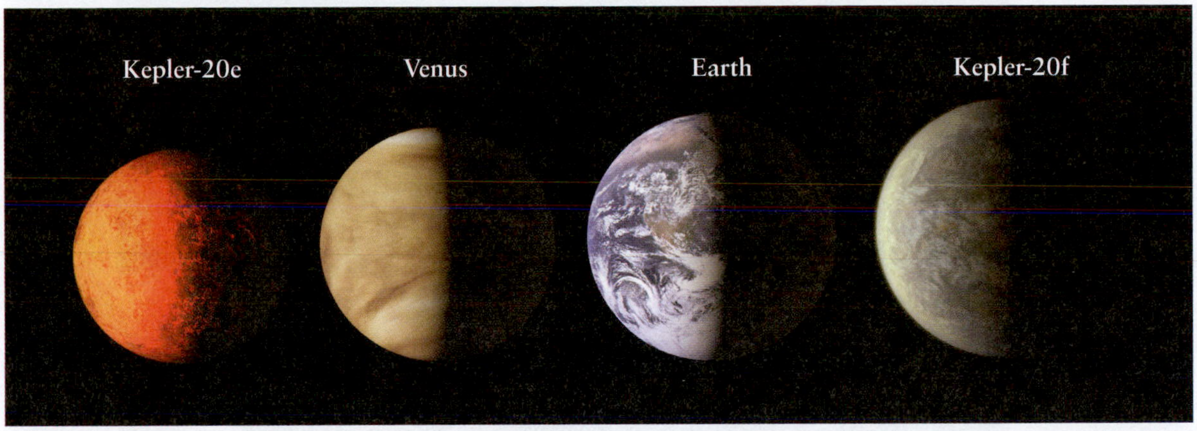

FIGURE 8-20

Earth-Size Exoplanets Kepler-20 is a star with five known exoplanets. The exoplanets Kepler-20e and Kepler-20f are illustrated in comparison to Earth and Venus. Kepler-20e (at 0.05 AU) has a radius just 0.87 times smaller than Earth, while Kepler-20f (at 0.1 AU) is 1.03 times larger than Earth. However, these objects orbit extremely close to their star and are much too hot for liquid water on their surfaces. (Ames/JPL-Caltech/NASA)

transit method by continuously monitoring the brightness of 145,000 stars in our Galaxy to search for a transit's dimming effect. As of this writing, *Kepler* has found 2321 candidates, and more than 90% of these will probably be confirmed as exoplanets. Of these, 207 candidates are roughly Earth-size (Figure 8-20) and 48 planets of various sizes were found in their stars' habitable zones. The *Kepler* team also estimates from this data set that about 5.4% of stars in our Galaxy host an Earth-size planet, and 17% of stars host multiple-planet systems.

Kepler is getting quite close to finding an Earth-size planet that is also in the habitable zone. One exoplanet only 2.4 times larger than Earth, Kepler-22b, is within the habitable zone, as illustrated in Figure 8-21. With more data on the way, habitable exoplanets might be just over the horizon.

Additional Searches for Extrasolar Planets

In 2004 astronomers began to use a property of space discovered by Albert Einstein as a tool for detecting extrasolar planets. As predicted by Einstein's general theory of relativity, a star's gravity can deflect the path of a light beam just as it deflects the path of a planet or spacecraft. If a star drifts through the line of sight between Earth and a more distant star, the closer star's gravity acts like a lens that focuses the more distant star's light (Figure 8-22). Such **microlensing** causes the distant star's image as seen in a telescope to become brighter. If the closer star has a planet, the planet's gravity will cause a secondary brightening whose magnitude depends on the planet's mass. Using this technique astronomers have detected a handful of extrasolar planets as far as 24,000 light-years away, and expect to find many more.

The most extensive search for extrasolar planets will be carried out by the European Space Agency's *Gaia* mission, scheduled for launch in 2013. *Gaia* will survey a billion (10^9) stars—1% of all the stars in the Milky Way Galaxy—and will be able to detect tiny stellar wobbles. To detect these wobbles, *Gaia* will measure minute changes in a star's position, as small as 2×10^{-6} arcsec, which is enough to see something move 1 cm on the Moon all the way from Earth! With such high-precision data, the astrometric method (Figure 8-16b) will finally come into its own. *Gaia* will also be able to follow up with radial velocity measurements on the

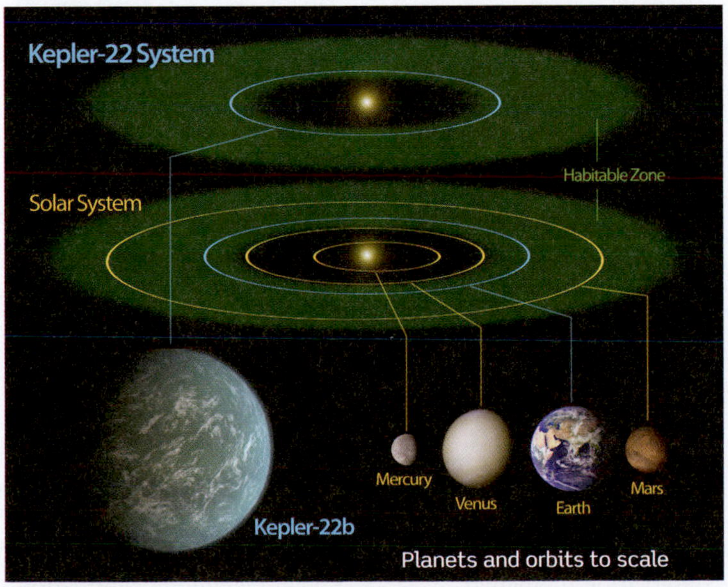

FIGURE 8-21

Exoplanets in the Habitable Zone The exoplanet Kepler-22b lies inside the habitable zone and is compared to our solar system in this illustration. This planet is 2.4 times the size of Earth, which raises hopes for a hard surface with at least some liquid water. Its mass is not yet known, but it could be around 10 times the mass of Earth. Each star determines the size of its surrounding habitable zone, and the two zones in this image are similar because the star, Kepler-22, is Sunlike. (NASA/Ames/JPL-Caltech)

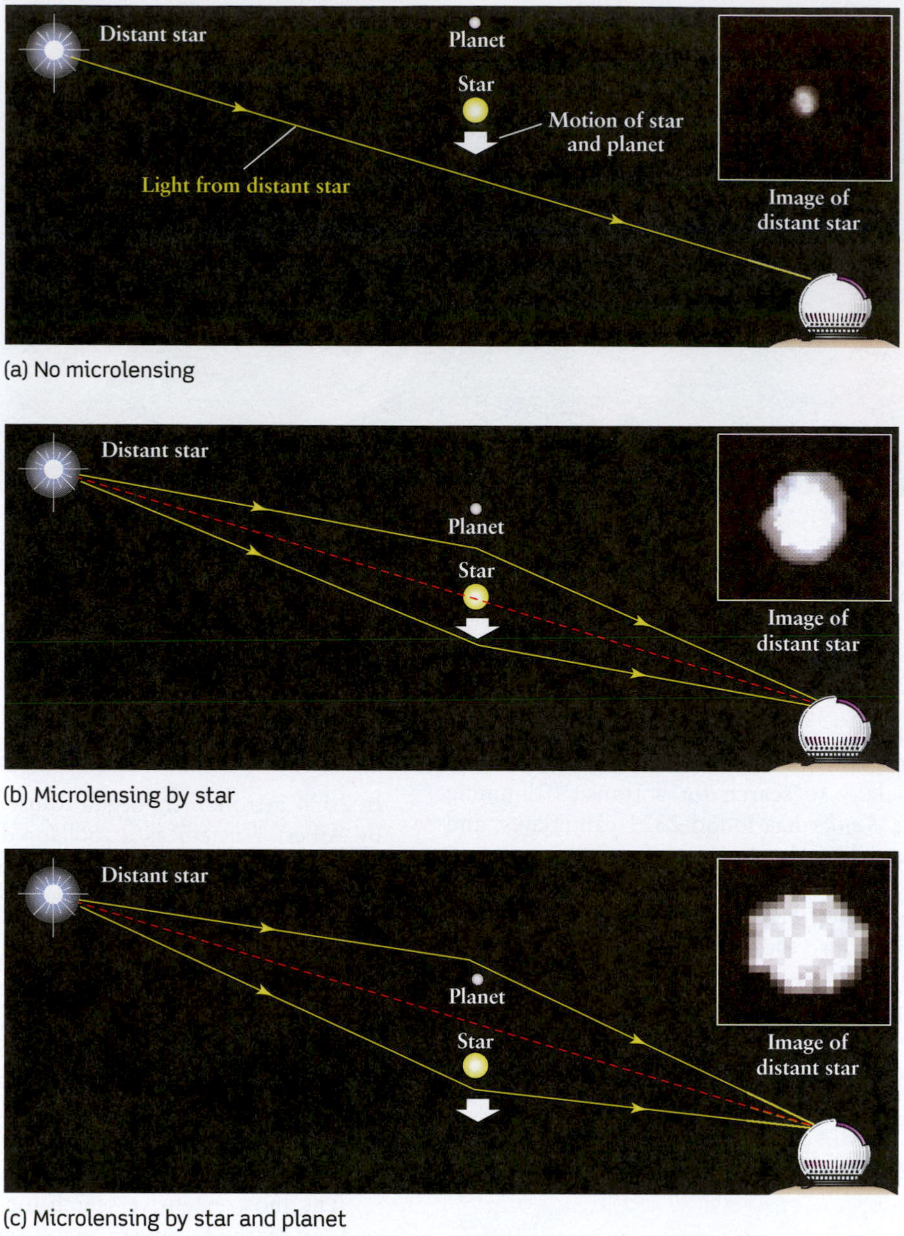

ANIMATION 8-9

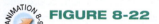

FIGURE 8-22

Microlensing Reveals an Extrasolar Planet **(a)** A star with a planet drifts across the line of sight between a more distant star and a telescope on Earth. **(b)** The gravity of the closer star bends the light rays from the distant star, focusing the distant star's light and making it appear brighter. **(c)** The gravity of the planet causes a second increase in the distant star's brightness.

most promising candidates, as well as on transits. Astronomers estimate that *Gaia* will find about 10 exoplanets a day, leading to about 15,000 discoveries over the life of its mission. Over the next several years, the data from *Kepler* and *Gaia* will greatly advance our understanding of the formation and evolution of planetary systems, as well as find planets similar to Earth.

CONCEPTCHECK 8-12

Why will the radial velocity method fail to discover an extrasolar planet if the plane of the extrasolar planet's orbit is oriented perpendicular to the line between Earth and the wobbling star?

Answer appears at the end of the chapter.

KEY WORDS

Term preceded by an asterisk (*) is discussed in Box 8-1.

accretion, p. 209
astrometric method (for detecting extrasolar planets), p. 214
atomic number, p. 202
center of mass, p. 214
chemical differentiation, p. 210
condensation temperature, p. 208
conservation of angular momentum, p. 206
core accretion model, p. 210
exoplanet, p. 213
extrasolar planet, p. 213
Grand Tack model, p. 212
habitable zone, p. 218
*half-life, p. 204
hot Jupiter, p. 216
interstellar medium, p. 201
Kelvin-Helmholtz contraction, p. 206

KEY IDEAS

The Nebular Hypothesis: The most successful model of the origin of the solar system is called the nebular hypothesis. According to this hypothesis, the solar system formed from a cloud of interstellar material called the solar nebula. This occurred 4.54 billion years ago (as determined by radioactive dating).

The Solar Nebula and Its Evolution: The chemical composition of the solar nebula, by mass, was 98% hydrogen and helium (elements that formed shortly after the beginning of the universe) and 2% heavier elements (produced much later in the centers of stars and cast into space when the stars died). The heavier elements were in the form of ice and dust particles.

• The nebula flattened into a disk in which all the material orbited the center in the same direction, just as do the present-day planets.

Formation of the Planets and Sun: The terrestrial planets, the Jovian planets, and the Sun followed different pathways to formation.

• The four terrestrial planets formed through the accretion of rock and metal dust grains into planetesimals, then into larger protoplanets.

• Closer to the Sun—inside the snow line—water could not freeze into ice to help build larger planets. With rocks and metals making up only a small fraction of mass throughout the nebula, the terrestrial planets made from these materials in the inner solar system are small.

• Jovian planets form beyond the snow line, where ices of water and methane contribute to make larger cores as these planets form. These larger cores can then attract copious amounts of hydrogen and helium gas, becoming large Jovian planets.

• Gravitational interactions between the Jovian planets and a gaseous disk, and much later with a disk of planetesimals, leads to planetary migration. These migrations can explain the small size of Mars, the composition of the asteroid belt, the locations of the Jovian planets, aspects of the Kuiper belt and Oort cloud, and even account for a Late Heavy Bombardment.

• The Sun formed by gravitational contraction of the center of the nebula. After about 10^8 years, temperatures at the protosun's center became high enough to ignite nuclear reactions that convert hydrogen into helium, thus forming a true star.

Extrasolar Planets: Astronomers have discovered many planets orbiting other stars.

• Many of these planets are detected by the "wobble" of the stars around which they orbit. The radial velocity method detects this wobble through Doppler shifts. The astrometric method directly observes a star's change in position as it wobbles.

• Many extrasolar planets have been discovered by the transit method, and a lesser number by microlensing, and direct imaging.

• Most of the extrasolar planets discovered to date are quite massive and have orbits that are very different from planets in our solar system. This is expected because big planets close to stars are easier to detect. We are just beginning to detect Earth-size planets at distances that might allow liquid water on their surfaces.

QUESTIONS

Review Questions

1. Describe three properties of the solar system that are thought to be a result of how the solar system formed.
2. The graphite in your pencil is a form of carbon. Where were these carbon atoms formed?
3. What is the interstellar medium? How does it become enriched over time with heavy elements?
4. What is the evidence that other stars existed before our Sun was formed?
5. Why are terrestrial planets smaller than Jovian planets?
6. How do radioactive elements make it possible to determine the age of the solar system? What are the oldest objects that have been found in the solar system?
7. Consider a 4.54×10^9 year old meteorite. Are the atoms in that meteorite—which were created in stars—4.54×10^9 years old? Explain your reasoning.
8. What is the tidal hypothesis? What aspect of the solar system was it designed to explain? Why was this hypothesis rejected?
9. The half-life for uranium-238 in Box 8-1 is 4.5 billion years. There is plenty of uranium-238 in Earth, and much of Earth's heat comes from the radioactive decay of this isotope. Compared to today, about how much more uranium-238 was there when Earth formed? (*Hint:* See Box 8-1.)
10. What is the nebular hypothesis? Why is this hypothesis accepted?
11. What was the protosun? What caused it to shine? Into what did it evolve?
12. Why is it thought that a disk appeared in the solar nebula?
13. What are protoplanetary disks? What do they tell us about the plausibility of our model of the solar system's origin?
14. What is meant by a substance's condensation temperature? What role did condensation temperatures play in the formation of the planets?
15. At distances within the snow line, what is the state of water (solid, liquid, or gas)? How does this affect the formation of terrestrial planets?
16. Why are terrestrial planets smaller than Jovian planets?
17. What is a planetesimal? How did planetesimals give rise to the terrestrial planets?

18. (a) What is meant by accretion? (b) Why are the terrestrial planets denser at their centers than at their surfaces?
19. If hydrogen and helium account for 98% of the mass of all the atoms in the universe, why aren't Earth and the Moon composed primarily of these two gases?
20. Why did the terrestrial planets form close to the Sun while the Jovian planets formed far from the Sun?
21. How did the Jovian planets form?
22. Explain how our current understanding of the formation of the solar system can account for the following characteristics of the solar system: (a) All planetary orbits lie in nearly the same plane. (b) All planetary orbits are nearly circular. (c) The planets orbit the Sun in the same direction in which the Sun itself rotates.
23. In the Grand Tack model, why is Mars much smaller than Earth or Venus?
24. How does the Grand tack model help us understand the composition of the asteroid belt?
25. Why do we think that Neptune must have formed closer to the Sun, and later migrated outward to its present position?
26. In the Nice model of the Jovian planets, which planet migrates more, Jupiter or Neptune?
27. In the Nice model, how does the Oort cloud and the Kuiper belt form?
28. What is the Late Heavy Bombardment? How does the Nice model explain this event?
29. Explain why most of the satellites of Jupiter orbit that planet in the same direction that Jupiter rotates.
30. What is the radial velocity method used to detect planets orbiting other stars? Why is it difficult to use this method to detect planets like Earth?
31. What type of planets and orbits are easiest to detect in the radial velocity method? What are "hot Jupiters"? Explain your answer.
32. Summarize the differences between the planets of our solar system and those found orbiting other stars.
33. Is there evidence that planets have fallen into their parent stars? Explain your answer.
34. What does it mean for a planet to transit a star? What can we learn from such events?
35. What combination of methods are required to determine the average density of an exoplanet? Explain your reasoning. (*Hint:* The average density is the planet's mass divided by its volume.)
36. What is so special about the habitable zone? Is every planet in this zone like Earth?
37. What is microlensing? How does it enable astronomers to discover extrasolar planets?
38. A 1999 news story about the discovery of three planets orbiting the star Upsilon Andromedae ("Ups And" is near the top of stars listed in Figure 8-18) stated that "the newly discovered galaxy, with three large planets orbiting a star known as Upsilon Andromedae, is 44 light-years away from Earth." What is wrong with this statement?

Advanced Questions

Questions preceded by an asterisk (*) involve topics discussed in Box 4-4 or Box 8-1.

Problem-solving tips and tools

The volume of a disk of radius r and thickness t is $\pi r^2 t$. Box 1-1 explains the relationship between the angular size of an object and its actual size. An object moving at speed v for a time t travels a distance $d = vt$; Appendix 6 includes conversion factors between different units of length and time. To calculate the mass of 70 Virginis or the orbital period of the planet around 2M1207, review Box 4-4. Section 5-4 describes the properties of blackbody radiation.

39. Figure 8-4 shows that carbon, nitrogen, and oxygen are among the most abundant elements (after hydrogen and helium). In our solar system, the atoms of these elements are found primarily in the molecules CH_4 (methane), NH_3 (ammonia), and H_2O (water). Explain why you suppose this is.
40. (a) If Earth had retained hydrogen and helium in the same proportion to the heavier elements that exist elsewhere in the universe, what would its mass be? Give your answer as a multiple of Earth's actual mass. Explain your reasoning. (b) How does your answer to (a) compare with the mass of Jupiter, which is 318 Earth masses? (c) Based on your answer to (b), would you expect Jupiter's rocky core to be larger, smaller, or the same size as Earth? Explain your reasoning.

*41. If you start with 0.80 kg of radioactive potassium (^{40}K), how much will remain after 1.3 billion years? After 2.6 billion years? After 3.9 billion years? How long would you have to wait until there was *no* ^{40}K remaining?

*42. Three-quarters of the radioactive potassium (^{40}K) originally contained in a certain volcanic rock has decayed into argon (^{40}Ar). How long ago did this rock form?

43. Suppose you were to use the Hubble Space Telescope to monitor one of the protoplanetary disks shown in Figure 8-8b. Over the course of 10 years, would you expect to see planets forming within the disk? Why or why not?
44. The protoplanetary disk at the upper right of Figure 8-8b is seen edge-on. The diameter of the disk is about 700 AU. (a) Make measurements on this image to determine the thickness of the disk in AU. (b) Explain why the disk will continue to flatten as time goes by.
45. The accompanying infrared image shows IRAS 0430212247, a young star that is still surrounded by a disk of gas and dust. The scale bar at the lower left of the image shows that at the distance of IRAS 0430212247, an angular size of 2 arcseconds corresponds to a linear size of 280 AU. Use this information to find the distance to IRAS 0430212247.

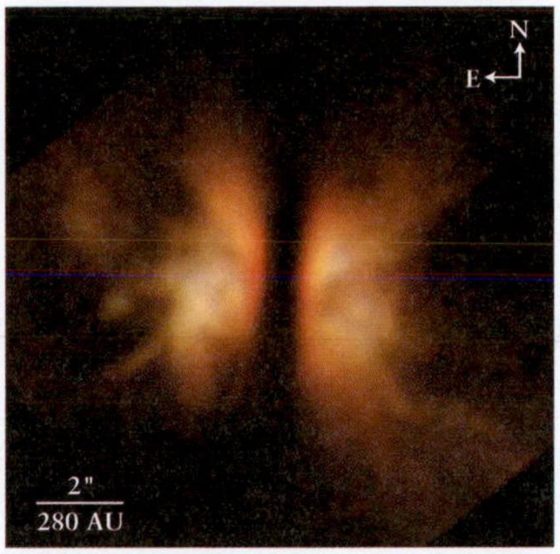

(Courtesy of D. Padgett and W. Brandner, IPAC/Caltech; K. Stapelfeldt, JPL; and NASA)

R I V U X G

46. The image accompanying Question 45 shows a dark, opaque disk of material surrounding the young star IRAS 0430212247. The disk is edge-on to our line of sight, so it appears as a dark band running vertically across this image. The material to the left and right of this band is still falling onto the disk. (**a**) Make measurements on this image to determine the diameter of the disk in AU. Use the scale bar at the lower left of this image. (**b**) If the thickness of the disk is 50 AU, find its volume in cubic meters. (**c**) The total mass of the disk is perhaps 2×10^{28} kg (0.01 of the mass of the Sun). How many atoms are in the disk? Assume that the disk is all hydrogen. A single hydrogen atom has a mass of 1.673×10^{-27} kg. (**d**) Find the number of atoms per cubic meter in the disk. Is the disk material thick or thin compared to the air that you breathe, which contains about 5.4×10^{25} atoms per cubic meter?

*47. The planet discovered orbiting the star 70 Virginis (70Vir is near the middle of the stars listed in Figure 8-18), 59 light-years from Earth, moves in an orbit with semimajor axis 0.48 AU and eccentricity 0.40. The period of the orbit is 116.7 days. Find the mass of 70 Virginis. Compare your answer with the mass of the Sun. (*Hint:* The planet has far less mass than the star.)

*48. Because of the presence of Jupiter, the Sun moves in a small orbit of radius 742,000 km with a period of 11.86 years. (**a**) Calculate the Sun's orbital speed in meters per second. (**b**) An astronomer on a hypothetical planet orbiting the star Vega, 25 light-years from the Sun, wants to use the astrometric method to search for planets orbiting the Sun. What would be the angular diameter of the Sun's orbit as seen by this alien astronomer? Would the Sun's motion be discernible if the alien astronomer could measure positions to an accuracy of 0.001 arcsec? (**c**) Repeat part (b), but now let the astronomer be located on a hypothetical planet in the Pleiades star cluster, 360 light-years from the Sun. Would the Sun's motion be discernible to this astronomer?

49. (**a**) Figure 8-19c shows how astronomers determine that the planet of HD 209458 has a surface temperature of 1130 K. Treating the planet as a blackbody, calculate the wavelength at which it emits most strongly. (**b**) The star HD 209458 itself has a surface temperature of 6030 K. Calculate its wavelength of maximum emission, assuming it to be a blackbody. (**c**) If a high-resolution telescope were to be used in an attempt to record an image of the planet orbiting HD 209458, would it be better for the telescope to use visible or infrared light? Explain your reasoning.

*50. (**a**) The star 2M1207 shown in Figure 8-15b is 170 light-years from Earth. Find the angular distance between this star and its planet as seen from Earth. Express your answer in arcseconds. (**b**) The mass of 2M1207 is 0.025 that of the Sun; the mass of the planet is very much smaller. Calculate the orbital period of the planet, assuming that the distance between the star and planet shown in Figure 8-15b is the semimajor axis of the orbit. Is it possible that an astronomer could observe a complete orbit in one lifetime?

Discussion Questions

51. Propose an explanation why the Jovian planets are orbited by terrestrial-like satellites.

52. Suppose that a planetary system is now forming around some protostar in the sky. In what ways might this planetary system turn out to be similar to or different from our own solar system? Explain your reasoning.

53. Suppose astronomers discovered a planetary system in which the planets orbit a star along randomly inclined orbits. How might a theory for the formation of that planetary system differ from that for our own?

Web/eBook Question

54. Search the World Wide Web for information about recent observations of protoplanetary disks. What insights have astronomers gained from these observations? Is there any evidence that planets have formed within these disks?

55. In 2000, extrasolar planets with masses comparable to that of Saturn were first detected around the stars HD 16141 (also called 79 Ceti) and HD 46375. Search the World Wide Web for information about these "lightweight" planets. Do these planets move around their stars in the same kind of orbit as Saturn follows around the Sun? Why do you suppose this is? How does the discovery of these planets reinforce the model of planet formation described in this chapter?

56. In 2006 a planet called XO-1b was discovered using the transit method. Search the World Wide Web for information about this planet and how it was discovered. What unusual kind of telescope was used to make this discovery? Have other extrasolar planets been discovered using the same kind of telescope?

ACTIVITIES

Observing Projects

57. Use *Starry Night*™ to make observations of the solar system. Select **Favourites > Explorations > Solar System**. The view shows the names and orbits of the major planets of the solar system against the backdrop of the stars of the Milky Way Galaxy, from a location hovering 64 AU from the Sun. You may also see many smaller objects moving in the asteroid

belt between the orbits of Mars and Jupiter. (If not, select **View > Solar System** and click on the **Asteroids** box.) (**a**) Use the location scroller to look at the solar system from different angles, and observe the general distribution and motion of the major planets. Make a list of your observations. (**b**) How does the nebular hypothesis of solar system formation account for your observations?

58. Use *Starry Night*™ to examine stars that have planets. Select **Favourites > Explorations > Extrasolar Planets.** This is the view of nearby space looking back toward the location of Earth from a distance of about 90 light-years. Use the location scroller to look around. Stars marked with a light blue halo show evidence of having at least one planet. Right-click (Ctrl-click on a Mac) on a sample of these circled stars and select **Show Info** from the contextual menu to learn more about the star's properties. Under the **Other Data** layer, note the star's apparent magnitude, which is a measure of how bright each star appears as seen from Earth. Apparent magnitude uses an inverse scale: The greater the apparent magnitude, the dimmer the star. Most of the brighter stars you can see with the naked eye from Earth have apparent magnitudes between 0 and 1, while the dimmest star you can see from a dark location has apparent magnitude 6. Also note, under the **Other Data** layer info for your sample stars, the extrasolar mass (the mass of the extrasolar planet) and extrasolar semimajor axis (a measure of the planet's distance from its parent star). (**a**) Are most of the circled stars in your sample visible to the naked eye from Earth? List at least two stars that have extrasolar planets that are visible to the naked eye from Earth and include their apparent magnitudes. Expand your sample if necessary. (**b**) Are most of the planets in your sample larger than Jupiter? Offer a hypothesis to account for this. (**c**) You may have noted that a surprising number of extrasolar planets in your sample are both very massive and also orbit very close to their parent star. What hypothesis might account for this?

59. Use *Starry Night*™ to investigate stars that have planets orbiting them. Click the **Home** button in the toolbar. Open the **Options** pane and use the checkboxes in the **Local View** layer to turn off **Daylight** and the **Local Horizon**. Expand the **Stars** layer in the **Options** pane and then expand the **Stars** item and check the **Mark stars with extrasolar planets** option. Then use the **Find** pane to find and center each of the stars listed below. To do this, click the magnifying glass icon on the side of the edit box at the top of the **Find** pane and select **Star** from the dropdown menu; then type the name of the star in the edit box and press the **Enter** or **Return** key on the keyboard. Click on the **Info** tab for full information about the star. Expand the **Other Data** layer and note the luminosity of each of these four stars: 47 Ursae Majoris (3 known planets); 51 Pegasi (1 known planet); 70 Virginis (1 known planet); Rho Coronae Borealis (1 known planet). (**a**) Which stars are more luminous than the Sun? (**b**) Which are less luminous? (**c**) How do you think these differences would have affected temperatures in the nebula in which each star's planets formed?

ANSWERS

ConceptChecks

ConceptCheck 8-1: No. Earth is not an exception to any of the three properties in Table 8-1, and in fact, all the planets are consistent with these properties.

ConceptCheck 8-2: The carbon atoms that make up much of our human bodies and the very oxygen atoms we breathe were made inside of stars. Stars have made most of the atoms other than hydrogen.

ConceptCheck 8-3: Yes. Hydrogen (H) is the first most abundant element, and oxygen (O) is the third most abundant element. Water (H_20) is, in fact, quite common even though it is usually frozen.

ConceptCheck 8-4: No. The age of a rock refers to how much time has passed since the rock cooled and solidified. It is not the age of the rock's atoms, which were made in stars, and possibly different stars at different times.

ConceptCheck 8-5: To be correct, the nebular hypothesis must explain why all the planets orbit the Sun in the same direction and in nearly the same plane. This arrangement is unlikely to have arisen purely by chance.

ConceptCheck 8-6: No. Through Kelvin-Helmholtz contraction, gravitational energy is converted into heat—a lot of heat. The protosun, which formed before nuclear reactions began, had a surface temperature of 6000 K, slightly higher than our present-day Sun. Thus, the solar nebula was warmed by the protosun.

ConceptCheck 8-7: Frozen water helped to make Jupiter large enough to hold onto hydrogen and helium. The closest that a large, Jupiterlike planet could form would be much farther away in a much hotter solar nebula.

ConceptCheck 8-8: Called chemical differentiation, denser, iron-rich materials migrate to the center of a planet while the less dense silicon-rich minerals float to the outside surface before the initially molten planet solidifies in the early solar system.

ConceptCheck 8-9: At Neptune's present distance, the time to build a planet the size of Neptune is much longer than the time that the protoplanetary disk is around.

ConceptCheck 8-10: In the Nice model, once planetary orbits became unstable, Neptune was deflected to about twice its original orbital distance.

ConceptCheck 8-11: Accretion refers to the gravitational accumulation of matter. Chemical bonds might, or might not, be formed once matter has accumulated.

ConceptCheck 8-12: An extrasolar planet with an orbit that causes a star to wobble side to side will not exhibit any Doppler shifted spectra as seen from Earth because the star will not be moving alternately toward and away from Earth, and therefore the radial velocity method will fail to detect such an orbiting extrasolar planet.

III Stars and Stellar Evolution

Spanning more than 300 light-years across, the Carina Nebula shown here is one of the largest and brightest nebulae in our Milky Way Galaxy. In fact, if aliens were to gaze upon us from another galaxy, this nebula would be one of the prominent features they would see. Within this nebula shines one of our Galaxy's brightest stars, Eta Carinae, at more than 5 million times the luminosity of the Sun. Ultraviolet light from the brightest stars excite the nebula's gas and the enhanced coloring in this image tracks emission from hydrogen, oxygen, nitrogen, and sulfur. In the darkest regions, dense gas is forming a new generation of stars. (NASA)

R I V U X G

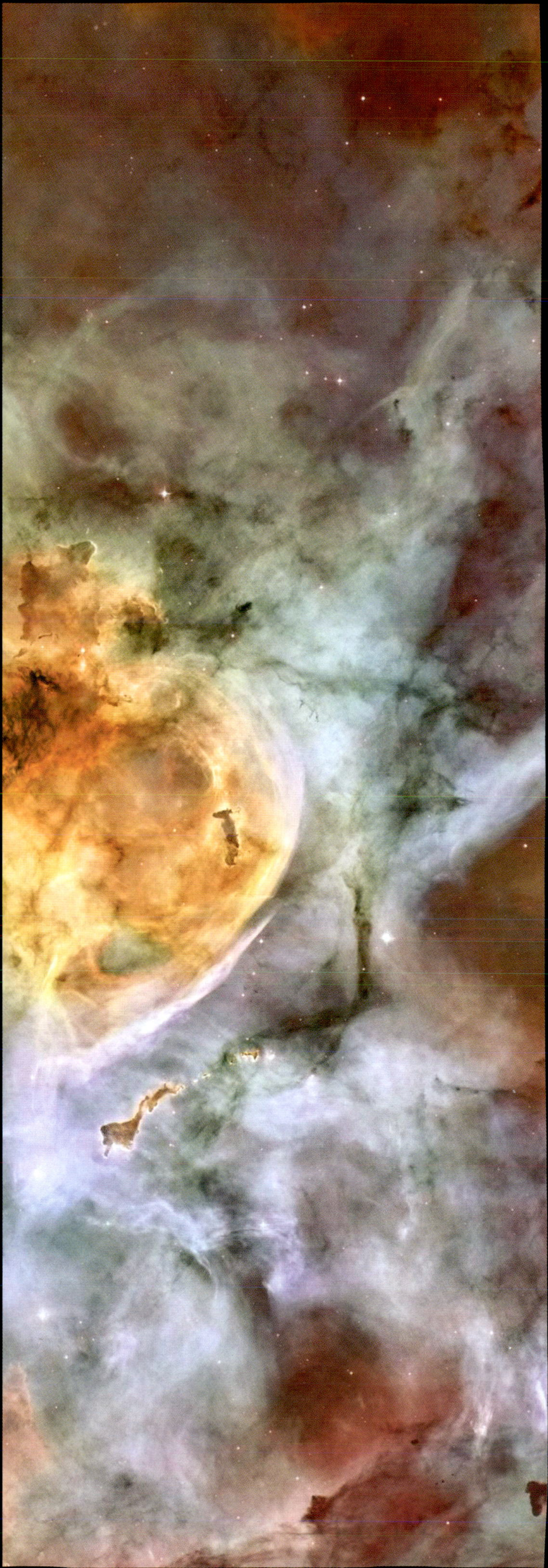

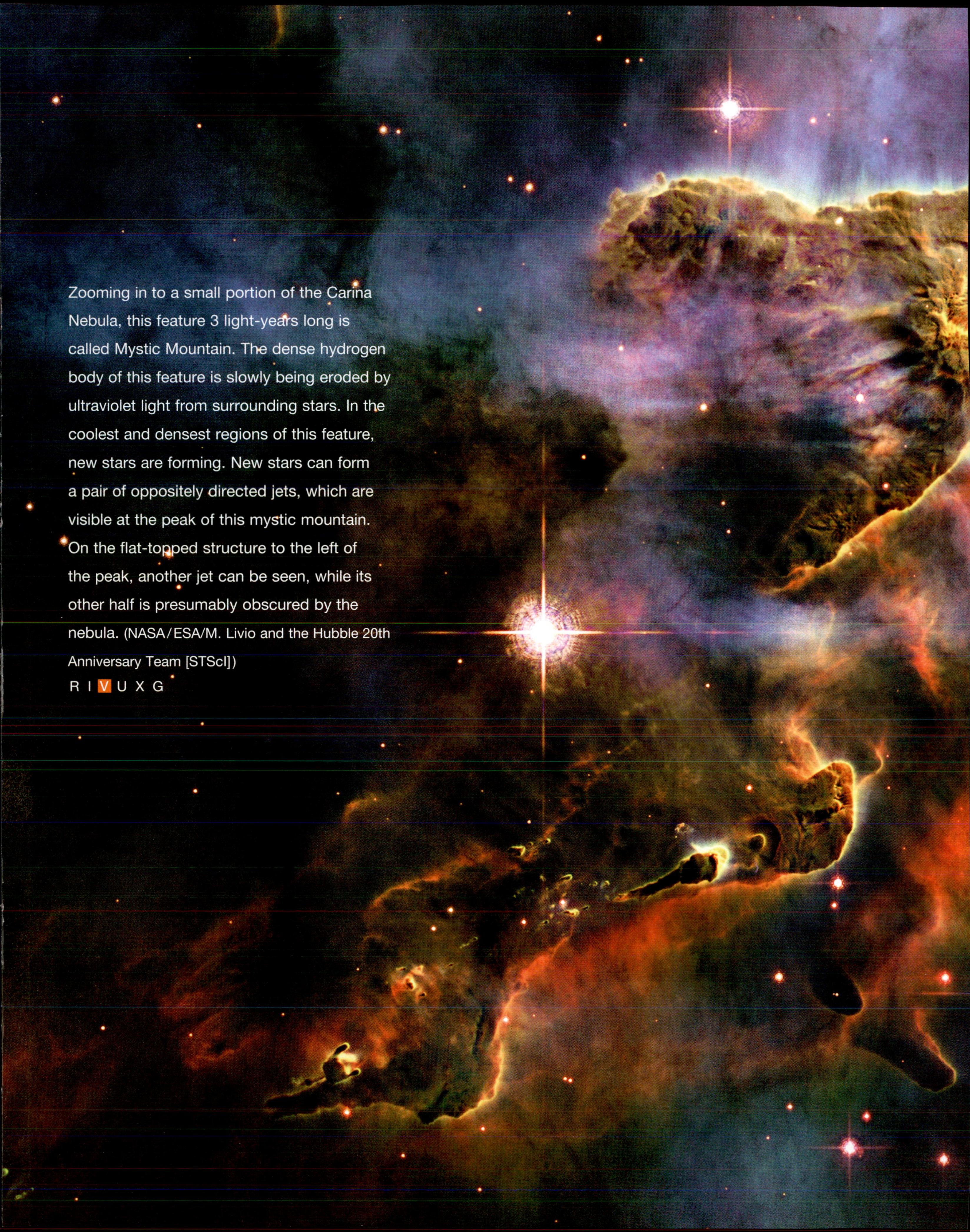

Zooming in to a small portion of the Carina Nebula, this feature 3 light-years long is called Mystic Mountain. The dense hydrogen body of this feature is slowly being eroded by ultraviolet light from surrounding stars. In the coolest and densest regions of this feature, new stars are forming. New stars can form a pair of oppositely directed jets, which are visible at the peak of this mystic mountain. On the flat-topped structure to the left of the peak, another jet can be seen, while its other half is presumably obscured by the nebula. (NASA/ESA/M. Livio and the Hubble 20th Anniversary Team [STScI])

R I V U X G

CHAPTER

16

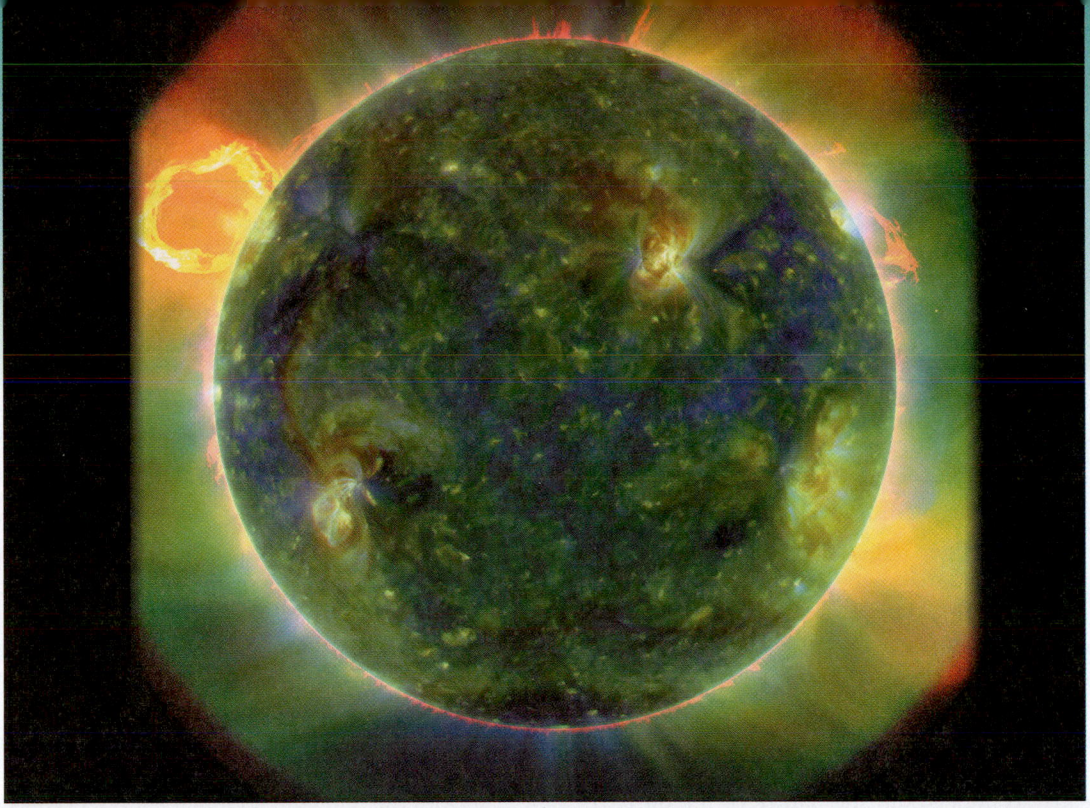

This ultraviolet image of the Sun, taken in 2010 by the Solar Dynamics Observatory, shows a prominence (upper left) in which magnetic fields carry gases above the Sun's surface in a loop that can extend upward hundreds of thousands of kilometers. (NASA/GSFC/AIA)

R I V **U** X G

Our Star, the Sun

LEARNING GOALS

By reading the sections of this chapter, you will learn

16-1 The source of the Sun's heat and light

16-2 How scientists model the Sun's internal structure

16-3 How the Sun's vibrations reveal what lies beneath its glowing surface

16-4 How scientists are able to probe the Sun's energy-generating core

16-5 Why the gaseous Sun appears to have a sharp outer edge

16-6 Why the upper regions of the solar atmosphere have an emission spectrum

16-7 The relationship between the Sun's corona and the solar wind

16-8 The nature of sunspots

16-9 The connection between sunspots and the Sun's magnetic field

16-10 How magnetic reconnection can power immense solar eruptions

The Sun is by far the brightest object in the sky. By earthly standards, the temperature of its glowing surface is remarkably high, about 5800 K, but there are regions of the Sun that reach far higher temperatures of tens of thousands or even millions of kelvins. Gases at such temperatures emit ultraviolet light or X-rays, which makes the Sun's outer regions prominent in the accompanying image from an ultraviolet telescope in space. Far from being smooth and placid, the Sun is a dynamic object with many features. Some of the hottest and most energetic regions on the Sun spawn immense disturbances and can propel solar material across space to reach Earth and other planets.

In recent decades, we have learned that the Sun shines because at its core hundreds of millions of tons of hydrogen are converted to helium every second. We have confirmed this picture by detecting the by-products of this transmutation—strange, ethereal particles called neutrinos—streaming outward from the Sun into space. We have discovered that the Sun has a surprisingly violent atmosphere, with a host of features such as sunspots whose numbers rise and fall on a predictable 11-year cycle. By studying the Sun's vibrations, we have begun to probe beneath its surface into hitherto unexplored realms. And we have just begun to investigate how changes in the Sun's activity can affect Earth's environment as well as our technological society.

16-1 The Sun's energy is generated by thermonuclear reactions in its core

The Sun is the largest member of the solar system. It has almost a thousand times more mass than all the planets, moons, asteroids, comets, and meteoroids put together. But the Sun is also a star. In fact, it is a remarkably typical star, with a mass, size, surface temperature, and chemical composition that are roughly midway between the extremes exhibited by the myriad other stars in the heavens. Table 16-1 lists essential data about the Sun.

Solar Energy

For most people, what matters most about the Sun is the energy that it radiates into space. Without the Sun's warming rays, our oceans would freeze. Furthermore, since most life is part of a food chain dependent on sunlight for energy, an Earth without sunlight would be nearly dead. To understand the nature of life on Earth, we must understand the Sun.

The Sun's surface is hot enough to emit visible light

Why is the Sun such a strong source of energy? Recall that the hotter an object, the more light it emits (see Figure 5-9). The Sun's spectrum is close to that of an idealized blackbody with a temperature of 5800 K (see Sections 5-3 and 5-4, especially Figure 5-13). Thanks to this high temperature, each square meter of the Sun's surface emits a tremendous amount of radiation, principally at visible wavelengths. Indeed, the Sun is the only object in the solar system that *emits* substantial amounts of visible light. The surface temperatures of the Moon and planets are very low compared to the Sun; the visible light that we see from bodies in the solar system is actually sunlight that struck those worlds and was *reflected* toward Earth.

The bright Sun emits a tremendous amount of energy. The total amount of energy emitted by the Sun each second, called its **luminosity,** is about 3.9×10^{26} watts (3.9×10^{26} joules of energy emitted per second). (We discussed the relationship between the Sun's surface temperature, radius, and luminosity in Box 5-2.) Astronomers denote the Sun's luminosity by the symbol $L_\odot$. A circle with a dot in the center is the astronomical symbol for the Sun and was also used by ancient astrologers.

Does the Sun emit so much energy because it is hot or because it is big? To put the Sun's temperature and size into perspective,

TABLE 16-1 Sun Data

Distance from Earth:	Mean: 1 AU = 149,598,000 km Maximum: 152,000,000 km Minimum: 147,000,000 km
Light travel time to Earth:	8.32 min
Mean angular diameter:	32 arcmin
Radius:	696,000 km = 109 Earth radii
Mass:	1.9891×10^{30} kg = 3.33×10^5 Earth masses
Composition (by mass):	74% hydrogen, 25% helium, 1% other elements
Composition (by number of atoms):	92.1% hydrogen, 7.8% helium, 0.1% other elements
Mean density:	1410 kg/m^3
Mean temperature:	Surface: 5800 K; Center: 1.55×10^7 K
Luminosity:	3.9×10^{26} W
Distance from center of Galaxy:	8000 pc = 26,000 ly
Orbital period around center of Galaxy:	220 million years
Orbital speed around center of Galaxy:	220 km/s

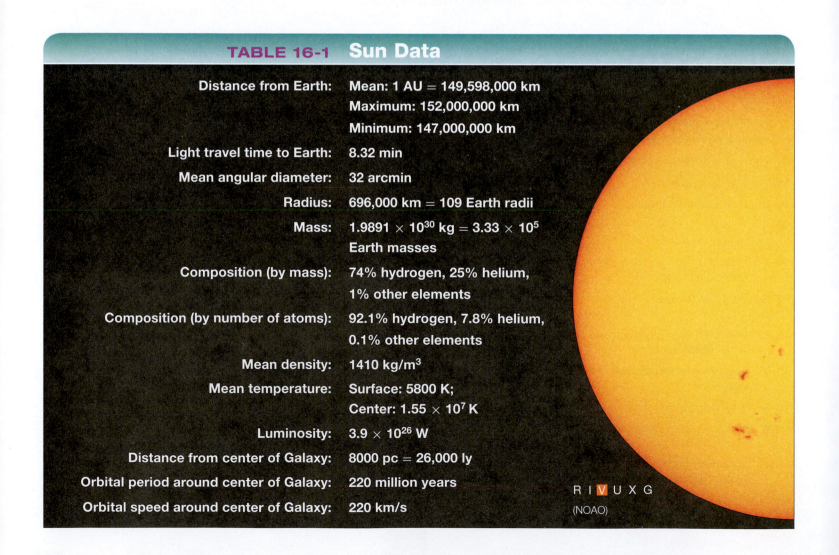

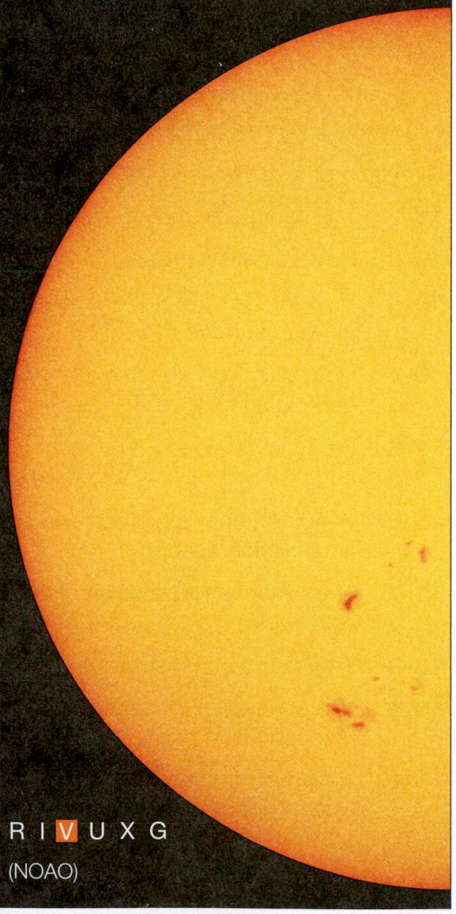

R I V U X G
(NOAO)

we can make a comparison with ovens and lightbulbs. The solar surface is about 10 times hotter than the maximum temperature of a household oven. A factor of 10 might not sound like a great amount between the Sun and an oven, but we saw in Section 5-4, for the Stefan-Boltzmann law, F (energy flux) $= \sigma T^4$, that raising an object's temperature 10 times will increase the energy it emits per second by a factor of 10,000! But the Sun is also much bigger than an oven, and to appreciate the effect of size, let us consider lightbulbs.

Suppose you stood in a room that contained an object at the same temperature as the Sun but with only a surface area of 1 square millimeter. Would it burn you up? No. Such an object would emit around 100 watts, similar to a household lightbulb. Just like a lightbulb, you would have to put your hand quite close to feel any warmth. What helps make the Sun so strong is that each square meter patch of the solar surface emits about as much as a million 100-watt lightbulbs, and the Sun's surface area has more than a million trillion of these 1-square-meter patches. Taken together, it is the high temperature *and* large size of the Sun that make it so bright.

CONCEPTCHECK 16-1

The Sun emits light at nearly all possible wavelengths. At which range of wavelengths is most of the Sun's energy emitted?

Answer appears at the end of the chapter.

The Source of the Sun's Energy: Early Ideas

The bright Sun leads us to a more fundamental question: What keeps the Sun's visible surface so hot? Or, put another way, what is the fundamental source of the tremendous energy that the Sun radiates into space? For centuries, this question was one of the greatest mysteries in science. The mystery deepened in the nineteenth century, when geologists and biologists found convincing evidence that life has existed on Earth for at least several hundred million years. Since life as we know it depends crucially on sunlight, the conclusion was that the Sun was at least several hundred million years old. (We now know that Earth is 4.54 billion years old and that life has existed on it for most of its history.) The source of the Sun's energy posed a severe problem for physicists. What source of energy could have kept the Sun shining for so long (Figure 16-1)?

One attempt to explain solar energy was made in the mid-1800s by the English physicist Lord Kelvin (for whom the temperature scale is named) and the German scientist Hermann von Helmholtz. They argued that the tremendous weight of the Sun's outer layers should cause the Sun to contract gradually, compressing its interior gases. Whenever a gas is compressed, its temperature rises. (You can demonstrate this with a bicycle pump: As you pump air into a tire, the temperature of the air increases and the pump becomes warm to the touch.) Kelvin and Helmholtz thus suggested that gravitational contraction could cause the Sun's gases to become hot enough to radiate energy out into space.

This process, called *Kelvin-Helmholtz contraction,* actually does occur during the earliest stages of the birth of a star like the Sun (see Section 8-4). But Kelvin-Helmholtz contraction cannot be the major source of the Sun's energy today. If it were, the Sun would have had to be much larger in the relatively recent past.

FIGURE 16-1 R I **V** U X G
The Sun The Sun's visible surface has a temperature of about 5800 K. At this temperature, all solids and liquids vaporize to form gases. It was only in the twentieth century that scientists discovered what has kept the Sun so hot for billions of years—the thermonuclear fusion of hydrogen nuclei in the Sun's core. (Jeremy Woodhouse/PhotoDisc)

Helmholtz's own calculations showed that the Sun could have started its initial collapse from the solar nebula no more than about 25 million years ago. But the geological and fossil record shows that Earth is far older than that, and so the Sun must be as well. Hence, this model of a Sun that shines because it shrinks cannot be correct.

On Earth, a common way to produce heat and light is by burning fuel, such as a log in a fireplace or charcoal in a barbeque grill. Is it possible that a similar process explains the energy released by the Sun? The answer is no, because this process could not continue for a long enough time to explain the age of Earth. It might seem difficult to estimate how long the Sun could last if its energy came from burning, but the calculation only requires a few steps. To begin with, the chemical reactions involved in burning release roughly 10^{-19} joule of energy per atom. Therefore, the number of atoms that would have to undergo chemical reactions each second to generate the Sun's luminosity of 3.9×10^{26} joules per second is approximately

$$\frac{3.9 \times 10^{26} \text{ joules per second}}{10^{-19} \text{ joule per atom}} = 3.9 \times 10^{45} \text{ atoms per second}$$

From its mass and chemical composition, we know that the Sun contains about 10^{57} atoms. Thus, the length of time that would be required to consume the entire Sun by burning is

$$\frac{10^{57} \text{ atoms}}{3.9 \times 10^{45} \text{ atoms per second}} = 3 \times 10^{11} \text{ seconds}$$

There are about 3×10^7 seconds in a year. Hence, in this model, the Sun would burn itself out in a mere 10,000 (10^4) years! This period of time is far shorter than the known age of Earth, so chemical reactions cannot explain how the Sun shines. This is a very powerful result because even though there are numerous types of chemical reactions that release energy (anything from fire to dynamite), the maximum possible chemical energy in the Sun's total mass could not keep it burning for very long.

The Source of the Sun's Energy: Discovering Thermonuclear Fusion

The source of the Sun's luminosity could be explained if there were a process that was like burning but released much more energy per atom. Then the rate at which atoms would have to be consumed would be far less, and the lifetime of the Sun could be long enough to be consistent with the known age of Earth. Albert Einstein discovered the key to such a process in 1905. According to his *special theory of relativity,* a quantity m of mass can in principle be converted into an amount of energy E according to a now-famous equation:

Einstein's mass-energy equation

$$E = mc^2$$

m = quantity of mass, in kg

c = speed of light = 3×10^8 m/s

E = amount of energy into which the mass can be converted, in joules

The speed of light c is a large number, so c^2 is huge. Therefore, a small amount of matter can release an awesome amount of energy. For example, if the mass in just one penny (about 2.5 grams) were completely converted into energy, it would release about 5 times the energy of the nuclear bomb detonated over Hiroshima, Japan, during World War II. (While the energy in a penny's worth of mass is impressive, the largest nuclear bombs release the energy contained in the mass of 10 rolls of quarters; see Figure 1-6.)

Inspired by Einstein's ideas, astronomers began to wonder if the Sun's energy output might come from the conversion of matter into energy. The Sun's low density of 1410 kg/m^3 indicates that it must be made of the very lightest atoms, primarily hydrogen and helium. In the 1920s, the British astronomer Arthur Eddington showed that temperatures near the center of the Sun must be so high that atoms become completely ionized. Hence, at the Sun's center we expect to find hydrogen nuclei and electrons flying around independent of each other.

Ideas from relativity and nuclear physics led to an understanding of how the Sun shines

Another British astronomer, Robert Atkinson, suggested that under these conditions hydrogen nuclei could fuse together to produce helium nuclei in a *nuclear reaction* that transforms a tiny amount of mass into a large amount of energy. Experiments in the laboratory using individual nuclei show that such reactions can indeed take place. The process of converting hydrogen into helium is called **hydrogen fusion.** (It is also sometimes called *hydrogen burning,* even though nothing is actually burned in the conventional sense. Ordinary burning involves chemical reactions that rearrange the outer electrons of atoms but have no effect on the atoms' nuclei.)

The fusing together of nuclei is also called **thermonuclear fusion,** because it can take place only at extremely high temperatures. This is because all nuclei have a positive electric charge and so tend to repel one another. But in the extreme heat and pressure at the Sun's center, hydrogen nuclei (protons) are moving so fast that they can overcome their electric repulsion and actually touch one another. When that happens, thermonuclear fusion can take place.

ANALOGY You can think of protons as tiny electrically charged spheres that are coated with a very powerful glue. If the spheres are not touching, the repulsion between their charges pushes them apart. But if the spheres are forced into contact, the strength of the glue "fuses" them together. In this fusion, a little bit of mass is lost, which gets converted into energy.

CAUTION! Be careful not to confuse thermonuclear fusion with the similar-sounding process of *nuclear fission.* In nuclear fusion, energy is released by joining or fusing together nuclei of lightweight atoms such as hydrogen. In nuclear fission, by contrast, the nuclei of very massive atoms such as uranium or plutonium release energy by fragmenting into smaller nuclei. Nuclear power plants produce energy using fission, not fusion. (Generating power using fusion has been a goal of researchers for decades, but no one has yet devised a commercially viable way to do this.)

CONCEPTCHECK 16-2

If hydrogen nuclei are positively charged, under what conditions can two hydrogen nuclei overcome the repulsive force of their electric charges and fuse together, thus releasing energy according to Einstein's equation, $E = mc^2$?

CALCULATIONCHECK 16-1

How much energy is released when just 5 kg of mass is converted into energy?

Answers appear at the end of the chapter.

Converting Hydrogen to Helium

We learned in Section 5-8 that the nucleus of a hydrogen atom (H) consists of a single proton. The nucleus of a helium atom (He) consists of two protons and two neutrons. In the nuclear process that Atkinson described, four hydrogen nuclei combine to form one helium nucleus, with a concurrent release of energy:

$$4\ \text{H} \rightarrow \text{He} + \text{energy}$$

In several separate reactions, two of the four protons are changed into neutrons, and eventually combine with the remaining protons to produce a helium nucleus. This sequence of reactions is called the **proton-proton chain.** The *Cosmic Connections* figure depicts the proton-proton chain in detail.

Each time this process takes place, a small fraction (0.7%) of the initial combined mass of the hydrogen nuclei does not show up in the final mass of the helium nucleus. This "lost" mass is converted into energy. Box 16-1 describes how to use Einstein's mass-energy equation to calculate the amount of energy released.

The Proton–Proton Chain

The most common form of hydrogen fusion in the Sun involves three steps, each of which releases energy.

Hydrogen fusion in the Sun usually takes place in a sequence of steps called the proton-proton chain. Each of these steps releases energy that heats the Sun and gives it its luminosity.

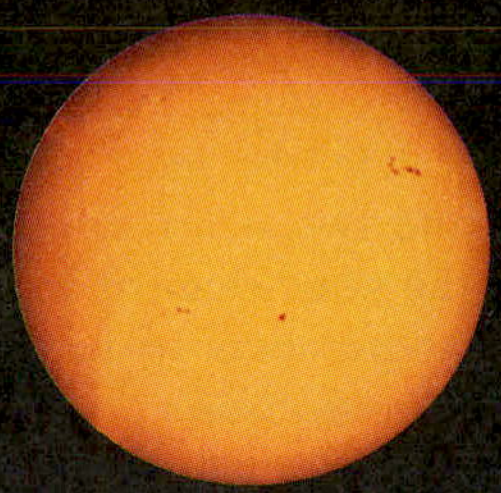

STEP 1

(a) Two protons (hydrogen nuclei, ^1H) collide.

(b) One of the protons changes into a neutron (shown in blue). The proton and neutron form a hydrogen isotope (^2H).

(c) One byproduct of converting a proton to a neutron is a neutral, nearly massless neutrino (ν), which escapes from the Sun.

(d) The other byproduct of converting a proton to a neutron is a positively charged electron, or positron (e^+). This encounters an ordinary electron (e^-), annihilating both particles and converting them into gamma-ray photons (γ). The energy of these photons goes into sustaining the Sun's internal heat.

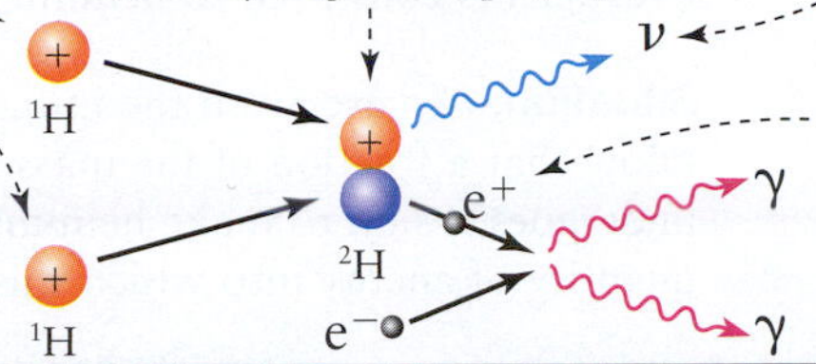

STEP 2

(a) The ^2H nucleus produced in Step 1 collides with a third proton (^1H).

(b) The result of the collision is a helium isotope (^3He) with two protons and one neutron.

(c) This nuclear reaction releases another gamma-ray photon (γ). Compared to other sources in these three steps, these gamma rays contribute the most energy to the Sun's internal heat.

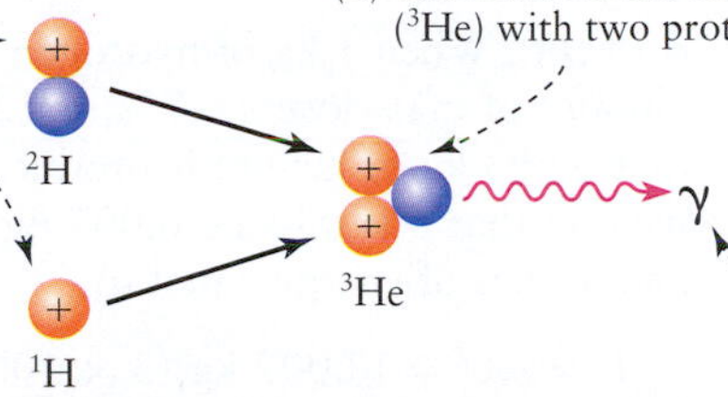

STEP 3

(a) The ^3He nucleus produced in Step 2 collides with another ^3He nucleus produced from three other protons.

(b) Two protons and two neutrons from the two ^3He nuclei rearrange themselves into a different helium isotope (^4He).

(c) The two remaining protons are released. The energy of their motion contributes to the Sun's internal heat.

(d) Six ^1H nuclei went into producing the two ^3He nuclei, which combine to make one ^4He nucleus. Since two of the original ^1H nuclei are returned to their original state, we can summarize the three steps as:

$$4\ ^1H \longrightarrow\ ^4He + \text{energy}$$

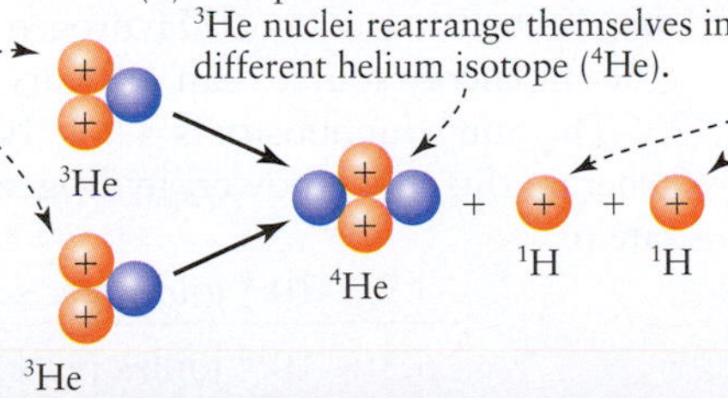

ANIMATION 16-1

Hydrogen fusion also takes place in all of the stars visible to the naked eye. (Fusion follows a different sequence of steps in the most massive stars, but the net result is the same.)

BOX 16-1 TOOLS OF THE ASTRONOMER'S TRADE

Converting Mass into Energy

The *Cosmic Connections: The Proton-Proton Chain* figure shows the steps involved in the thermonuclear fusion of hydrogen at the Sun's center. In these steps, four protons are converted into a single nucleus of ^{4}He, the most common isotope of helium with two protons and two neutrons. (As we saw in Box 5-5, different isotopes of the same element have the same number of protons but different numbers of neutrons.) The reaction depicted in the *Cosmic Connections* Step 1 also produces a neutral, nearly massless particle called the *neutrino*. Neutrinos respond hardly at all to ordinary matter, so they travel almost unimpeded through the Sun's massive bulk. Hence, the energy that neutrinos carry is quickly lost into space. This loss is not great, however, because the neutrinos carry relatively little energy. (See Section 16-4 for more about these curious particles.)

Most of the energy released by thermonuclear fusion appears in the form of gamma-ray photons. The energy of these photons remains trapped within the Sun for a long time, thus maintaining the Sun's intense internal heat. Some gamma-ray photons are produced by the reaction shown as Step 2 in the *Cosmic Connections* figure. Others appear when an electron in the Sun's interior annihilates a positively charged electron, or **positron,** which is a by-product of the reaction shown in Step 1 in the *Cosmic Connections* figure. An electron and a positron are respectively matter and antimatter, and they convert their mass entirely into energy when they meet. (You may have thought that "antimatter" was pure science fiction. In fact, tremendous amounts of antimatter are being created and annihilated in the Sun as you read these words.)

We can summarize the thermonuclear fusion of hydrogen as follows:

$$4\,^1\text{H} \rightarrow {}^4\text{He} + \text{neutrinos} + \text{gamma-ray photons}$$

To calculate how much energy is released in this process, we use Einstein's mass-energy formula: The energy released is equal to the amount of mass consumed multiplied by c^2, where c is the speed of light. To see how much mass is consumed, we compare the combined mass of four hydrogen atoms (the ingredients) to the mass of one helium atom (the product):

$$\begin{array}{r} 4 \text{ hydrogen atoms} = 6.693 \times 10^{-27} \text{ kg} \\ -1 \text{ helium atom} = 6.645 \times 10^{-27} \text{ kg} \\ \hline \text{Mass lost} = 0.048 \times 10^{-27} \text{ kg} \end{array}$$

Thus, a small fraction (0.7%) of the mass of the hydrogen going into the nuclear reaction does not show up in the mass of the helium. This lost mass is converted into an amount of energy $E = mc^2$:

$$\begin{aligned} E &= mc^2 = (0.048 \times 10^{-27} \text{ kg})(3 \times 10^8 \text{ m/s})^2 \\ &= 4.3 \times 10^{-12} \text{ joule} \end{aligned}$$

This amount of energy is released by the formation of a single helium atom. It would light a 10-watt lightbulb for almost one-half of a trillionth of a second.

EXAMPLE: How much energy is released when 1 kg of hydrogen is converted to helium?

Situation: We are given the initial mass of hydrogen. We know that a fraction of the mass is lost when the hydrogen undergoes fusion to make helium; our goal is to find the quantity of energy into which this lost mass is transformed.

Tools: We use the equation $E = mc^2$ and the result that 0.7% of the mass is lost when hydrogen is converted into helium.

Answer: When 1 kg of hydrogen is converted to helium, the amount of mass lost is 0.7% of 1 kg, or 0.007 kg. (This means that 0.993 kg of helium is produced.) Using Einstein's equation, we find that this missing 0.007 kg of matter is transformed into an amount of energy equal to

$$E = mc^2 = (0.007 \text{ kg})(3 \times 10^8 \text{ m/s})^2 = 6.3 \times 10^{14} \text{ joules}$$

Review: The energy released by the fusion of 1 kg of hydrogen is the same as that released by burning *20,000 metric tons* (2×10^7 kg) of coal! Hydrogen fusion is a *much* more efficient energy source than ordinary burning.

The Sun's luminosity is 3.9×10^{26} joules per second. To generate this much power, hydrogen must be consumed at a rate of

$$\frac{3.9 \times 10^{26} \text{ joules per second}}{6.3 \times 10^{14} \text{ joules per kilogram}}$$

$$= 6 \times 10^{11} \text{ kilograms per second}$$

That is, the Sun converts 600 million metric tons of hydrogen into helium every second.

CAUTION! You may have heard the idea that mass is always conserved (that is, it is neither created nor destroyed), or that energy is always conserved in a reaction. Einstein's ideas show that neither of these statements is quite correct, because mass can be converted into energy and vice versa. A more accurate statement is that the total amount of mass *plus* energy is conserved. Hence, the destruction of mass in the Sun does not violate any laws of nature.

For every four hydrogen nuclei converted into a helium nucleus, 4.3×10^{-12} joule of energy is released—only enough energy to power a small Christmas light for a trillionth of a second. This amount of energy may seem tiny, but it is about 10^7 times larger than the amount of energy released in a typical chemical reaction, such as occurs in ordinary burning. Thus, thermonuclear fusion explains how the Sun could have been shining for billions of years.

To produce the Sun's luminosity of 3.9×10^{26} joules per second, 6×10^{11} kg (600 million metric tons) of hydrogen must be converted into helium each second—this is about 100 times the mass of Egypt's largest pyramid. This rate is prodigious, but there is, literally, an astronomical amount of hydrogen in the Sun. In particular, the Sun's core contains enough hydrogen to have been giving off energy at the present rate for as long as the solar system has existed, about 4.56 billion years, and to continue doing so for more than 6 billion years into the future.

The proton-proton chain is also the energy source for many of the stars in the sky. In stars with central temperatures that are much hotter than that of the Sun, however, hydrogen fusion proceeds according to a different set of nuclear reactions, called the **CNO cycle,** in which carbon, nitrogen, and oxygen nuclei absorb protons to produce helium nuclei. Still other thermonuclear reactions, such as helium fusion, carbon fusion, and oxygen fusion, occur late in the lives of many stars.

CONCEPTCHECK 16-3

If 1 kg of hydrogen combines to form helium in the proton-proton chain, why is only 0.007 kg (0.7%) available to be converted into energy?

CONCEPTCHECK 16-4

How do astronomers estimate that our Sun can continue to shine for more than 6 billion years?

Answers appear at the end of the chapter.

16-2 A theoretical model of the Sun shows how energy gets from its center to its surface

While thermonuclear fusion is the source of the Sun's energy, this process cannot take place everywhere within the Sun. As we have seen, extremely high temperatures—in excess of 10^7 K—are required for atomic nuclei to fuse together to form larger nuclei. The temperature of the Sun's visible surface, about 5800 K, is too low for these reactions to occur there. Hence, thermonuclear fusion can be taking place only within the Sun's interior. But precisely where does it take place? And how does the energy produced by fusion make its way to the surface, where it is emitted into space in the form of photons?

To answer these questions, we must understand conditions in the Sun's interior. Ideally, we would send an exploratory spacecraft to probe deep into the Sun; in practice, the Sun's intense heat would vaporize even the sturdiest spacecraft. Instead, astronomers use the laws of physics to construct a theoretical model of the Sun, and then compare predictions of the model to a variety of observations. (We discussed the use of models in science in Section 1-1.) Let's see what ingredients go into building a model of this kind.

Hydrostatic Equilibrium

Note first that the Sun is not undergoing any dramatic changes. The Sun is not exploding or collapsing, nor is it significantly heating or cooling. The Sun is thus said to be in both *hydrostatic equilibrium* and *thermal equilibrium.*

To understand what is meant by **hydrostatic equilibrium,** imagine a slab of material in the solar interior (Figure 16-2a). In equilibrium, the slab on average will move neither up nor down. (In fact, there are upward and downward motions of material inside the Sun, but these motions average out.) Equilibrium is maintained by a balance among three forces that act on this slab:

1. The downward pressure of the layers of solar material above the slab.
2. The upward pressure of the hot gases beneath the slab.
3. The slab's weight—that is, the downward gravitational pull it feels from the rest of the Sun.

The pressure from below must balance both the slab's weight and the pressure from above. Hence, the pressure below the slab

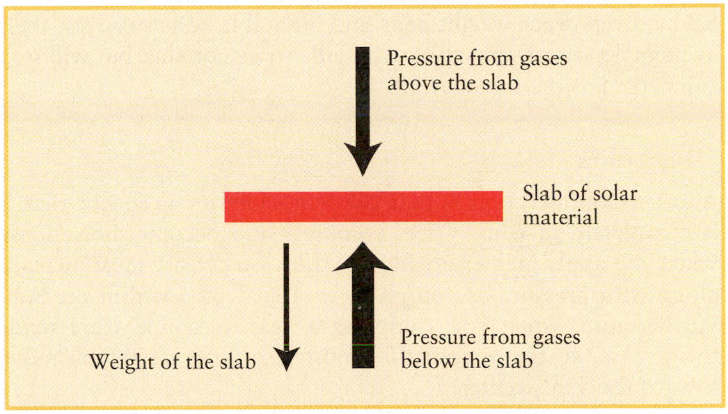

(a) Material inside the sun is in hydrostatic equilibrium, so forces balance

(b) A fish floating in water is in hydrostatic equilibrium, so forces balance

FIGURE 16-2

Hydrostatic Equilibrium **(a)** Material in the Sun's interior tends to move neither up nor down. The upward forces on a slab of solar material (due to pressure of gases below the slab) must balance the downward forces (due to the slab's weight and the pressure of gases above the slab). Hence, the pressure must increase with increasing depth. **(b)** The same principle applies to a fish floating in water. In equilibrium, the forces balance and the fish neither rises nor sinks. (Ken Usami/PhotoDisc) must be greater than that above the slab. In other words, pressure has to increase with increasing depth. For the same reason, pressure increases as you dive deeper into the ocean (Figure 16-2b) or as you move toward lower altitudes in our atmosphere.

ANALOGY If you had to join a vertical stack of 10 students piled on top of each other, you would naturally prefer to be on the top. Each student lower in the stack must bear the weight of all the students above them, so the pressure on each body increases with depth.

Hydrostatic equilibrium also tells us about the density of the slab of solar material. If the slab is too dense, its weight will be too large and it will sink; if the density is too low, the slab will rise. To prevent this, the density of solar material must have a certain value at each depth within the solar interior. (The same principle applies to objects that float beneath the surface of the ocean. Scuba divers wear weight belts and inflatable vests to adjust their average density so that they will neither rise nor sink but will stay submerged at the same level.)

Thermal Equilibrium

Another consideration is that the Sun's interior is so hot that it is completely gaseous. Gases compress and become more dense when you apply greater pressure to them, so density must increase along with pressure as you go to greater depths within the Sun. Furthermore, when you compress a gas, its temperature tends to increase, so the temperature must also increase as you move toward the Sun's center.

While the temperature in the solar interior is different at different depths, the temperature at each depth remains constant in time. This principle is called **thermal equilibrium.** For the Sun to be in thermal equilibrium, all the energy generated by thermonuclear reactions in the Sun's core must be transported to the Sun's glowing surface, where it can be radiated into space. If too much energy flowed from the core to the surface to be radiated away, the Sun's interior would cool down; alternatively, the Sun's interior would heat up if too little energy flowed to the surface.

CONCEPTCHECK 16-5

If our Sun were much less massive, how would the pressure at the Sun's center be different from what it actually is?

Answer appears at the end of the chapter.

Transporting Energy Outward from the Sun's Core

But exactly how is energy transported from the Sun's center to its surface? There are three methods of energy transport: *conduction, convection,* and *radiative diffusion.* Only the last two are important inside the Sun.

If you heat one end of a metal bar with a blowtorch, energy flows to the other end of the bar so that it too becomes warm. The efficiency of this method of energy transport, called **conduction,** varies significantly from one substance to another. For example, metal is a good conductor of heat, but plastic is not (which is why metal cooking pots often have plastic handles). Conduction is *not* an efficient means of energy transport in substances with low average densities, including the gases inside stars like the Sun.

Inside stars like our Sun, energy moves from center to surface by two other means: convection and radiative diffusion. **Convection** is the circulation of fluids—gases or liquids—between hot and cool regions. Hot gases (with lower density) rise toward a star's surface, while cool gases (with higher density) sink back down toward the star's center. This physical movement of gases transports heat energy outward in a star, just as the physical movement of water boiling in a pot transports energy from the bottom of the pot (where the heat is applied) to the cooler water at the surface.

In **radiative diffusion,** photons created in the thermonuclear inferno at a star's center diffuse outward toward the star's surface. Individual photons are absorbed and reemitted by atoms and electrons inside the star. The overall result is an outward migration from the hot core, where photons are constantly created, toward the cooler surface, where they escape into space.

CONCEPTCHECK 16-6

Why is the energy transport process of conduction relatively unimportant when studying how energy moves toward the Sun's surface?

Answer appears at the end of the chapter.

Modeling the Sun

To construct a model of a star like the Sun, astrophysicists express the ideas of hydrostatic equilibrium, thermal equilibrium, and energy transport as a set of equations. To ensure that the model applies to the particular star under study, they also make use of astronomical observations of the star's surface. (For example, to construct a model of the Sun, they use the data that the Sun's surface temperature is 5800 K, its luminosity is 3.9×10^{26} watts, and the gas pressure and density at the surface are almost zero.) The astrophysicists then use a computer to solve their set of equations and calculate conditions layer by layer in toward the star's center. The result is a model of how temperature, pressure, and density increase with increasing depth below the star's surface (see Figure 9-5).

Table 16-2 and Figure 16-3 show a theoretical model of the Sun that was calculated in just this way. Different models of the Sun use slightly different assumptions, but all models give essentially the same results as those shown here. From such computer models we have learned that at the Sun's center the density is 160,000 kg/m^3 (14 times the density of lead!), the temperature is 1.55×10^7 K, and the pressure is 3.4×10^{11} atm. (One atmosphere, or 1 atm, is the average atmospheric pressure at sea level on Earth.)

Table 16-2 and Figure 16-3 show that the solar luminosity rises to 100% at about one-quarter of the way from the Sun's center to its surface. In other words, the Sun's energy production occurs within a volume that extends out only to 0.25 $R_\odot$. (The symbol $R_\odot$ denotes the solar radius, or radius of the Sun as a whole, equal to 696,000 km.) Outside 0.25 $R_\odot$, the density and temperature are too low for thermonuclear reactions to take place. Also note that 94% of the total mass of the Sun is found within the inner 0.5 $R_\odot$. Hence, the outer 0.5 $R_\odot$ contains only a relatively small amount of material.

How energy flows from the Sun's center toward its surface depends on how easily photons move through the gas. If the solar gases are relatively transparent, photons can travel moderate distances before being scattered or absorbed, and energy is thus transported by radiative diffusion. You can think of radiative

TABLE 16-2 A Theoretical Model of the Sun

Distance from the Sun's center (solar radii)	Fraction of luminosity	Fraction of mass	Temperature ($\times 10^6$ K)	Density (kg/m^3)	Pressure relative to pressure at center
0.0	0.00	0.00	15.5	160,000	1.00
0.1	0.42	0.07	13.0	90,000	0.46
0.2	0.94	0.35	9.5	40,000	0.15
0.3	1.00	0.64	6.7	13,000	0.04
0.4	1.00	0.85	4.8	4,000	0.007
0.5	1.00	0.94	3.4	1,000	0.001
0.6	1.00	0.98	2.2	400	0.0003
0.7	1.00	0.99	1.2	80	4×10^{-5}
0.8	1.00	1.00	0.7	20	5×10^{-6}
0.9	1.00	1.00	0.3	2	3×10^{-7}
1.0	1.00	1.00	0.006	0.00030	4×10^{-13}

Note: The distance from the Sun's center is expressed as a fraction of the Sun's radius ($R_\odot$). Thus, 0.0 is at the center of the Sun and 1.0 is at the surface. The fraction of luminosity is that portion of the Sun's total luminosity produced within each distance from the center; this is equal to 1.00 for distances of 0.25 $R_\odot$ or more, which means that all of the Sun's nuclear reactions occur within 0.25 solar radius from the Sun's center. The fraction of mass is that portion of the Sun's total mass lying within each distance from the Sun's center. The pressure is expressed as a fraction of the pressure at the center of the Sun.

diffusion as the "free flight" of photons, and the energy is transported in these photons. If the gases are relatively *opaque*, photons are frequently scattered or absorbed and cannot easily get through the gas. In an opaque gas, heat builds up and convection then becomes the most efficient means of energy transport. In convection, the energy is in the form of heat; the movement of hot gas transports the energy. The gases start to churn, with hot lower-density gas moving outward and cooler gas sinking inward.

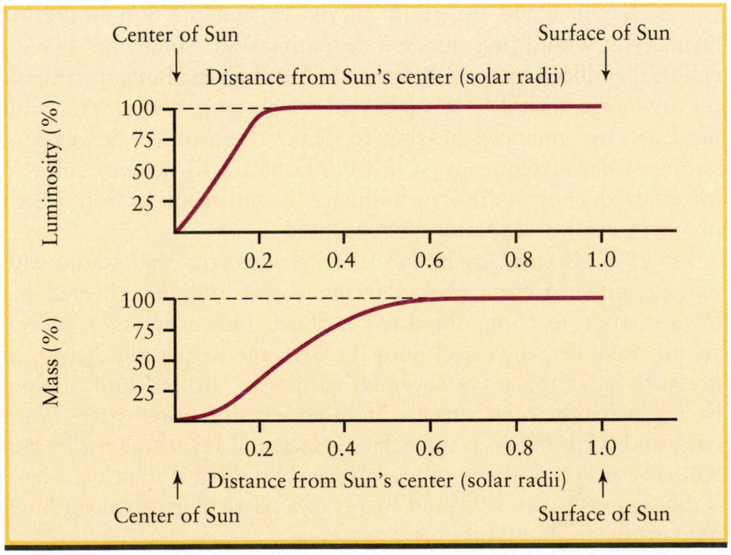

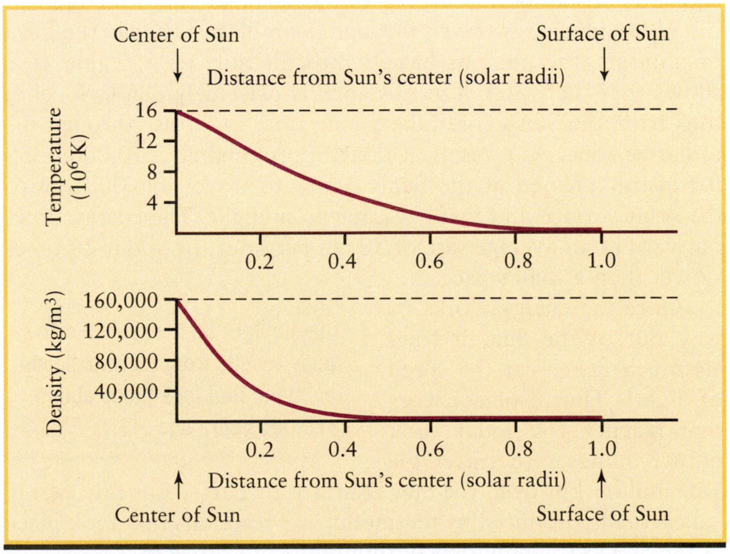

FIGURE 16-3

A Theoretical Model of the Sun's Interior These graphs depict what percentage of the Sun's total luminosity is produced within each distance from the center (upper left), what percentage of the total mass lies within each distance from the center (lower left), the temperature at each distance (upper right), and the density at each distance (lower right). (See Table 16-2 for a numerical version of this model.)

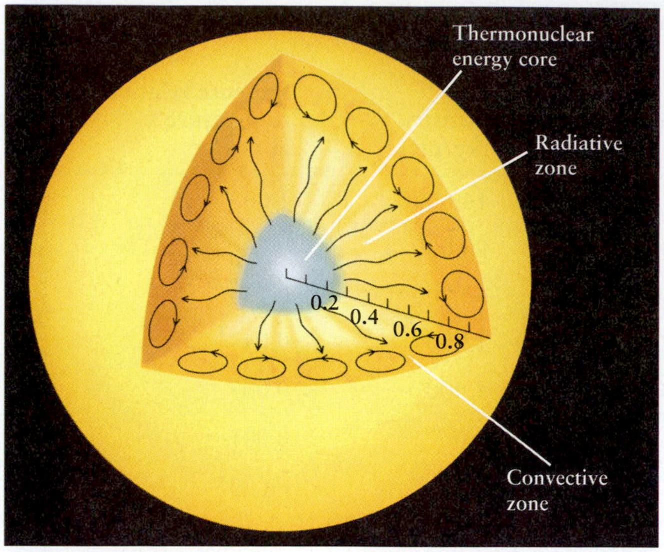

FIGURE 16-4

The Sun's Internal Structure Thermonuclear reactions occur in the Sun's core, which extends out to a distance of 0.25 $R_\odot$ from the center. Energy is transported outward, via radiative diffusion, to a distance of about 0.71 $R_\odot$. In the outer layers between 0.71 $R_\odot$ and 1.00 $R_\odot$, energy flows outward by convection.

From the center of the Sun out to about 0.71 $R_\odot$, energy is transported by radiative diffusion. Hence, this region is called the **radiative zone.** Beyond about 0.71 $R_\odot$, the temperature is low enough (a mere 2×10^6 K or so) for electrons and hydrogen nuclei to join into hydrogen atoms. These atoms are very effective at absorbing photons, much more so than free electrons or nuclei, and this absorption chokes off the outward flow of photons. Therefore, beyond about 0.71 $R_\odot$, radiative diffusion is not an effective way to transport energy. Instead, convection dominates the energy flow in this outer region, which is why it is called the **convective zone.** Figure 16-4 shows these aspects of the Sun's internal structure.

Although energy travels through the radiative zone in the form of photons, the photons have a difficult time of it. Table 16-2 shows that the material in this zone is extremely dense, so photons from the Sun's core take a long time to diffuse through the radiative zone. As a result, it takes approximately 170,000 years for energy created at the Sun's center to travel 696,000 km to the solar surface and finally escape as sunlight. The energy flows outward at an average rate of 50 cm per hour, or about 20 times slower than a snail's pace.

> The sunlight that reaches Earth today results from thermonuclear reactions that took place about 170,000 years ago

Once the energy works its way out of the Sun, it travels much faster—at the speed of light! Thus, solar energy that reaches you today took only 8 minutes to travel the 150 million km from the Sun's surface to Earth. But this energy was actually produced by thermonuclear reactions that took place in the Sun's core hundreds of thousands of years ago.

CONCEPTCHECK 16-7

Which of the following decreases when we move from the Sun's central core toward its surface: temperature or luminosity?

CALCULATIONCHECK 16-2

By how much (in percent) does the Sun's temperature drop from its temperature at the central core to its temperature at 0.5 $R_\odot$?

Answers appear at the end of the chapter.

16-3 Astronomers probe the solar interior using the Sun's own vibration

We have described how astrophysicists construct models of the Sun. But since we cannot see into the Sun's opaque interior, how can we check these models to see if they are accurate? What is needed is a technique for probing the Sun's interior. A very powerful technique of just this kind involves measuring vibrations of the Sun as a whole. This field of solar research is called **helioseismology.**

Vibrations are a useful tool for examining the hidden interiors of all kinds of objects. Food shoppers test whether melons are ripe by tapping on them and listening to the sound made by the resulting vibrations. Since the vibrations of a melon depend on its density profile, the sounds from tapping provide clues about the melon's interior. Similarly, geologists can determine the structure of Earth's interior by using seismographs to record vibrations during earthquakes.

Although there are no true "sunquakes," the Sun does vibrate at a variety of frequencies, somewhat like a ringing bell. These vibrations were first noticed in 1960 by Robert Leighton of the California Institute of Technology, who made high-precision Doppler shift observations of the solar surface. These measurements revealed that parts of the Sun's surface move up and down about 10 m every 5 minutes. Since the mid-1970s, several astronomers have reported slower vibrations, having periods ranging from 20 to 160 minutes. The detection of extremely slow vibrations has inspired astronomers to organize networks of telescopes around and in orbit above Earth to monitor the Sun's vibrations on a continuous basis.

The vibrations of the Sun's surface can be compared with sound waves. If you could somehow survive within the Sun's outermost layers, you would first notice a deafening roar, somewhat like a jet engine, produced by turbulence in the Sun's gases. But superimposed on this noise would be a variety of nearly pure tones. You would need greatly enhanced hearing to detect these tones, however; the strongest has a frequency of just 0.003 hertz, 13 octaves below the lowest frequency audible to humans. (Recall from Section 5-2 that one hertz is one oscillation per second.)

In 1970, Roger Ulrich, at UCLA, pointed out that sound waves moving upward from the solar interior would be reflected back inward after reaching the solar surface. However, as a reflected sound wave descends back into the Sun, the increasing density and pressure bend the wave severely, turning it around and aiming it back toward the solar surface. In other words, sound waves bounce back and forth between the solar surface and layers deep within the Sun. These sound waves can reinforce each other if their wavelength is the right size, just as sound waves of a particular wavelength resonate inside an organ pipe.

> The Sun's oscillations resemble those inside an organ pipe but are much more complex

The Sun oscillates in millions of ways as a result of waves resonating in its interior. Figure 16-5 is a computer-generated illustration of one such mode of vibration.

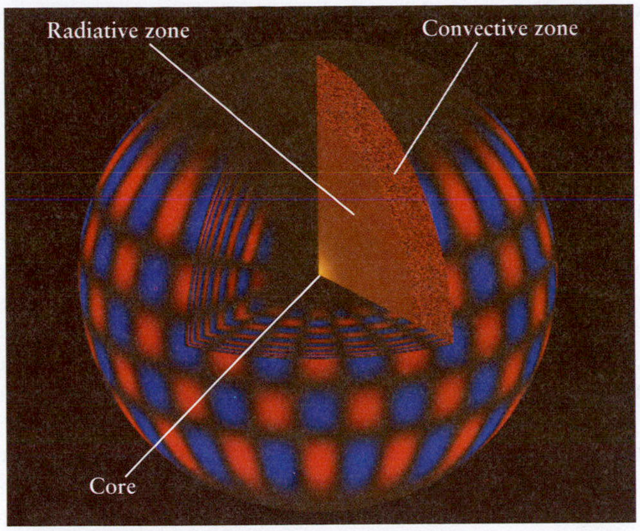

FIGURE 16-5
A Sound Wave Resonating in the Sun This computer-generated image shows one of the millions of ways in which the Sun's interior vibrates. The regions that are moving outward are colored blue, those moving inward, red. As the cutaway shows, these oscillations are thought to extend into the Sun's radiative zone (compare Figure 16-4). (National Solar Observatory)

Helioseismologists can deduce information about the solar interior from measurements of these oscillations. In addition to providing information about interior temperatures and densities, helioseismology sets limits on the amount of helium in the Sun's core and convective zone and determines the thickness of the transition region between the radiative zone and convective zone.

CONCEPTCHECK 16-8

Do solar sound waves travel exclusively on the surface of the Sun—like ripples across a pond—or do they travel through the Sun's interior as well?

CONCEPTCHECK 16-9

What can be determined from carefully monitoring the Sun's vibrations?

Answers appear at the end of the chapter.

16-4 Neutrinos reveal information about the Sun's core—and have surprises of their own

We have seen circumstantial evidence that thermonuclear fusion is the source of the Sun's power. To be certain, however, we need more definitive evidence. How can we show that thermonuclear fusion really is taking place in the Sun's core?

What Sunlight Cannot Tell Us

Although the light energy that we receive from the Sun originates in the core, it provides few clues about conditions there. The problem is that this energy has changed form repeatedly during its passage from the core: It appeared first as photons diffusing through the radiative zone, then as heat transported through the outer layers by convection, and then again as photons emitted from the Sun's glowing surface. As a result of these transformations, much of the information that the Sun's radiated energy once carried about conditions in the core has been lost.

ANALOGY If you make a photocopy of a photocopy of a photocopy of an original document, the final result may be so blurred as to be unreadable. In an analogous way, because solar energy is transformed many times while en route through the Sun, the story it could tell us about the Sun's core is hopelessly blurred.

Solar Neutrinos

Scientists use the most ethereal of subatomic particles to learn about the Sun, and vice versa

Happily, there is a way for scientists to learn about conditions in the Sun's core and to get direct evidence that thermonuclear fusion really does happen there. The trick is to detect the subatomic by-products of thermonuclear fusion reactions.

As part of the process of hydrogen fusion, protons change into neutrons and release **neutrinos** (see the *Cosmic Connections* figure in Section 16-1 as well as Box 16-1). Like photons, neutrinos are particles that have no electric charge. Unlike photons, however, neutrinos interact only very weakly with matter. Even the vast bulk of the Sun offers little impediment to their passage, so neutrinos must be streaming out of the core and into space. Indeed, the conversion of hydrogen into helium at the Sun's center produces 10^{38} neutrinos each second. Some of these neutrinos hit Earth, and about 10^{14} neutrinos from the Sun—that is, **solar neutrinos**—must pass through each square meter of Earth every second. Perhaps the exceedingly weak interaction between neutrinos and matter is a good thing, since about a trillion neutrinos pass through our heads each second.

If it were possible to detect these solar neutrinos, we would have direct evidence that thermonuclear reactions really do take place in the Sun's core. Beginning in the 1960s, scientists began to build neutrino detectors for precisely this purpose.

The challenge is that neutrinos are exceedingly difficult to detect. Just as neutrinos pass unimpeded through the Sun, they also pass through Earth almost as if it were not there. We stress the word "almost," because neutrinos can and do interact with matter, albeit infrequently.

On rare occasions a neutrino will strike a neutron and convert it into a proton. This effect was the basis of the original solar neutrino detector, designed and built by Raymond Davis of the Brookhaven National Laboratory in the 1960s. This device used 100,000 gallons of perchloroethylene (C_2Cl_4), a fluid used in dry cleaning, in a huge tank buried deep underground. Most of the solar neutrinos that entered Davis's tank passed right through it with no effect whatsoever. But occasionally a neutrino struck the nucleus of one of the chlorine atoms (^{37}Cl) in the cleaning fluid and converted one of its neutrons into a proton, creating a radioactive atom of argon (^{37}Ar). Fortunately for experimenters, what individual neutrinos lack in barely interacting with matter, the Sun partially compensates for by producing neutrinos in extraordinarily large numbers.

The rate at which argon is produced is related to the neutrino flux—that is, the number of neutrinos from the Sun arriving at Earth per square meter per second. By counting the number of newly created argon atoms, Davis was able to determine the neutrino flux from the Sun. (Other subatomic particles besides neutrinos can also induce reactions that create radioactive atoms. By placing the experiment deep underground, however, the overlying Earth absorbs essentially all such particles—with the exception of neutrinos.)

The Solar Neutrino Problem

Davis and his collaborators found that solar neutrinos created one radioactive argon atom in the tank every three days. But this rate corresponded to only one-third of the neutrino flux predicted from standard models of the Sun. This troubling discrepancy between theory and observation, called the **solar neutrino problem,** motivated scientists around the world to conduct further experiments to measure solar neutrinos.

One key question was whether the neutrinos that Davis had detected had really come from the Sun. (The Davis experiment had no way to determine the direction from which neutrinos had entered the tank of cleaning fluid.) The direction was resolved by an experiment in Japan called Kamiokande, which was designed by the physicist Masatoshi Koshiba. A large underground tank containing 3000 tons of water was surrounded by 1100 light detectors. From time to time, a high-energy solar neutrino struck an electron in one of the water molecules, dislodging it and sending it flying like a pin hit by a bowling ball. The recoiling electron produced a streaking flash of light, which was sensed by the detectors. By analyzing the flashes, scientists could tell the direction from which the neutrinos were coming and confirmed that they emanated from the Sun. These results in the late 1980s gave direct evidence that thermonuclear fusion is indeed occurring in the Sun's core. (Davis and Koshiba both received the 2002 Nobel Prize in Physics for their pioneering research on solar neutrinos.)

Like Davis's experiment, however, Koshiba and his colleagues at Kamiokande detected only a fraction of the expected flux of neutrinos. Where, then, were the missing solar neutrinos?

LOOKING DEEPER 16-1

One proposed solution had to do with the energy of the detected neutrinos. The vast majority of neutrinos from the Sun are created during the first step in the proton-proton chain, in which two protons combine to form a heavy isotope of hydrogen (see the *Cosmic Connections* figure in Section 16-1). But these neutrinos have too little energy to convert chlorine into argon. Both Davis's and Koshiba's experiments responded only to high-energy neutrinos produced by reactions that occur only part of the time near the end of the proton-proton chain. (The *Cosmic Connections* figure does not show these reactions.) Could it be that the discrepancy between theory and observation would go away if the flux of low-energy neutrinos could be measured?

To test this idea, two teams of physicists constructed neutrino detectors that used several tons of gallium (a liquid metal) rather than cleaning fluid. Low-energy neutrinos convert gallium (^{71}Ga) into a radioactive isotope of germanium (^{71}Ge). By chemically separating the germanium from the gallium and counting the radioactive atoms, the physicists were able to measure the flux of low-energy solar neutrinos. These experiments—GALLEX in Italy and SAGE (Soviet-American Gallium Experiment) in Russia—detected only 50% to 60% of the expected neutrino flux. Hence, the solar neutrino problem was a discrepancy between theory and observation for neutrinos of all energies.

Another proposed solution to the neutrino problem was that the Sun's core is cooler than predicted by solar models. If the Sun's central temperature were only 10% less than the current estimate, fewer neutrinos would be produced and the neutrino flux would agree with experiments. However, a lower central temperature would cause other obvious features, such as the Sun's size and surface temperature, to be different from what we observe.

Finding the Missing Neutrinos

Only very recently has the solution to the neutrino problem been found. The answer lies not in how neutrinos are produced, but rather in what happens to them between the Sun's core and detectors on Earth. Motivated by the lack of observed solar neutrinos, physicists have discovered that there are actually three types of neutrinos. Only one of these types is produced in the Sun, and it is only this type that can be detected by the experiments we have described. But if some of the solar neutrinos change *in flight* into a different type of neutrino, the detectors in these experiments would record only a fraction of the total neutrino flux. This effect, where one type of neutrino can spontaneously transform into another type, is called **neutrino oscillation.** In June 1998, scientists at the Super-Kamiokande neutrino observatory (a larger and more sensitive device than Kamiokande) revealed evidence that neutrino oscillation does indeed take place.

The best confirmation of this idea has come from the Sudbury Neutrino Observatory (SNO) in Canada. Like Kamiokande and Super-Kamiokande, SNO uses a large tank of water placed deep underground (Figure 16-6). But unlike those earlier experiments, SNO can detect all three types of neutrinos. It detects neutrinos by using *heavy water*. In ordinary, or "light," water, each hydrogen atom in the H_2O molecule has a solitary proton as its nucleus. In heavy water, by contrast, each of the hydrogen nuclei has a nucleus made up of a proton and a neutron. (This isotope ^{2}H is shown in the *Cosmic Connections* figure in Section 16-1.) If a high-energy solar neutrino of any type passes through SNO's tank of heavy water, it can knock the neutron out of one of the ^{2}H nuclei. The ejected neutron can then be captured by another nucleus, and this capture releases energy that manifests itself as a tiny burst of light. As in Kamiokande, detectors around SNO's water tank record these light flashes.

Scientists using SNO have found that the combined flux of all three types of neutrinos coming from the Sun is *equal* to the theoretical prediction. Together with the results from earlier neutrino experiments, this result strongly suggests that the Sun is indeed producing neutrinos at the predicted rate as a by-product of thermonuclear reactions. But before these neutrinos can reach Earth, about two-thirds of them undergo spontaneous oscillation and change their type. Thus, there is really no solar neutrino problem (no lack of neutrinos)—scientists merely needed the right kind of detectors to observe all the neutrinos, including the ones that transformed in flight.

The story of the solar neutrino problem illustrates how two different branches of science—in this case, studies of the solar interior and investigations of subatomic particles—can sometimes interact, to the mutual benefit of both. While there is still much we do not understand about the Sun and about neutrinos, a new generation of neutrino detectors in Japan, Canada, and elsewhere promises to further our knowledge of these exotic realms of astronomy and physics.

CONCEPTCHECK 16-10

Why did the SNO experiment find three times as many neutrinos as the Kamiokande experiment?

Answer appears at the end of the chapter.

FIGURE 16-6 R I V U X G
A Solar Neutrino Experiment Located 2073 m (6800 ft) underground in the Creighton nickel mine in Sudbury, Canada, the Sudbury Neutrino Observatory is centered around a tank that contains 1000 tons of water. Occasionally, a neutrino entering the tank interacts with one or another of the particles already there. Such interactions create flashes of light, called Cherenkov radiation. Some 9600 light detectors sense this light. The numerous silver protrusions are the back sides of the light detectors prior to their being wired and connected to electronics in the lab (seen at the bottom of the photograph). (Science Source)

16-5 The photosphere is the lowest of three main layers in the Sun's atmosphere

Although the Sun's core is hidden from our direct view, we can easily see sunlight coming from the high-temperature gases that make up the Sun's atmosphere. These outermost layers of the Sun prove to be the sites of truly dramatic activity, much of which has a direct impact on our planet. By studying these layers, we gain further insight into the character of the Sun as a whole.

Observing the Photosphere

A visible-light photograph like Figure 16-7 makes it appear that the Sun has a definite surface. This surface is actually an illusion; the Sun is gaseous throughout its volume because of its high internal temperature, and the gases simply become less and less dense as you move farther away from the Sun's center.

Why, then, does the Sun appear to have a sharp, well-defined surface? The reason is that essentially all of the Sun's visible light emanates from a single, thin layer of gas called the **photosphere** ("sphere of light"). Just as you can see only a certain distance through Earth's atmosphere before objects vanish in the haze, we can see only about 400 km into the photosphere. This distance is so small compared with the Sun's radius of 696,000 km that the photosphere appears to be a definite surface.

The photosphere is actually the lowest of the three layers that together constitute the solar atmosphere. Above it are the *chromosphere* and the *corona*, both of which are transparent to visible light. We can see them only using special techniques, which we discuss later in this chapter. Everything below the photosphere is called the *solar interior*.

The Lesson of Limb Darkening

The photosphere is heated from below by energy streaming outward from the solar interior. Hence, temperature should decrease as you go upward in the photosphere, just as in the solar interior (see Table 16-2 and Figure 16-3). We know this is the case because the photosphere appears darker around the edge, or *limb*, of the Sun than it does toward the center of the solar disk, an effect called **limb darkening** (examine Figure 16-7). This happens because when we look near the Sun's limb, we do not see as deeply into the photosphere as we do when we look near the center of the disk (Figure 16-8). The high-altitude gas we observe at the limb is not as hot and thus does not glow as brightly as the deeper, hotter gas seen near the disk center.

The spectrum of the Sun's photosphere confirms how its temperature varies with altitude. As we saw in Section 16-1, the photosphere shines like a nearly perfect blackbody with an average temperature of about 5800 K. However, superimposed on this spectrum are many dark absorption lines (see Figure 5-12). As discussed in Section 5-6, we see an *absorption line spectrum* of this sort whenever we view a hot, glowing object through a relatively cool gas. In this case, the hot object is the lower part of the photosphere; the cooler gas is in the upper part of the photosphere, where the temperature declines to about 4400 K. All the absorption lines in the Sun's spectrum are produced in this relatively cool layer, as atoms selectively absorb photons of various wavelengths streaming outward from the hotter layers below.

CAUTION! You may find it hard to think of 4400 K as "cool." But keep in mind that the ratio of 4400 K to 5800 K, the temperature in the lower photosphere, is the same as the ratio of the temperature on a Siberian winter night to that of a typical day in Hawaii.

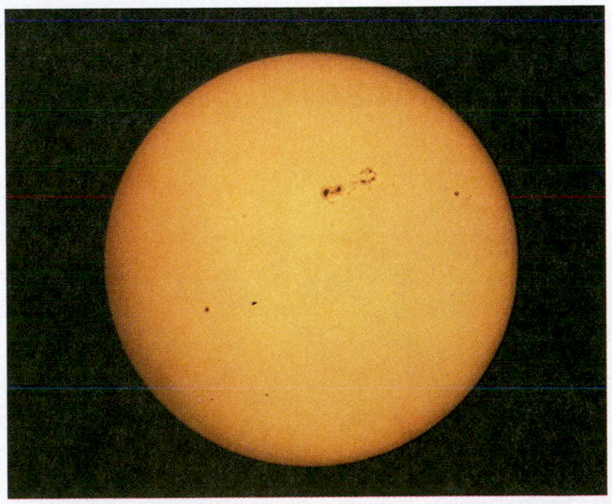

FIGURE 16-7 R I V U X G
The Photosphere The photosphere is the layer in the solar atmosphere from which the Sun's visible light is emitted. Note that the Sun appears darker around its limb, or edge; here we are seeing the upper photosphere, which is relatively cool and thus glows less brightly. (The dark sunspots, which we discuss in Section 16-8, are also relatively cool regions.) (Celestron International)

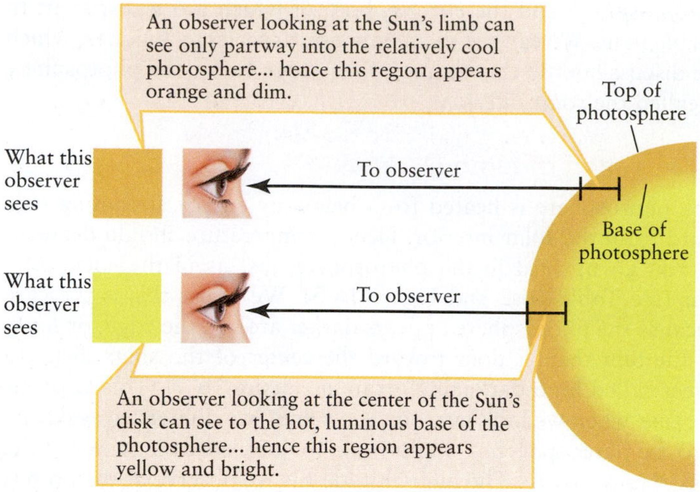

An observer looking at the Sun's limb can see only partway into the relatively cool photosphere... hence this region appears orange and dim.
Top of photosphere
What this observer sees
To observer
Base of photosphere
What this observer sees
To observer
An observer looking at the center of the Sun's disk can see to the hot, luminous base of the photosphere... hence this region appears yellow and bright.

FIGURE 16-8

The Origin of Limb Darkening Light from the Sun's limb and light from the center of its disk both travel about the same straight-line distance through the photosphere to reach us. Because of the Sun's curved shape, light from the limb comes from a greater height within the photosphere, where the temperature is lower and the gases glow less brightly. Hence, the limb appears darker and more orange.

Granules and Supergranules in the Photosphere

We can learn still more about the photosphere by examining it with a telescope—but only when using special dark filters to prevent eye damage. ***Looking directly at the Sun without the correct filter, whether with the naked eye or with a telescope, can cause permanent blindness!*** Under good observing conditions, astronomers using such filter-equipped telescopes can often see a blotchy pattern in the photosphere, called **granulation** (Figure 16-9). Each light-colored **granule** measures about 1000 km (600 mi) across—equal in size to the areas of Texas and Oklahoma combined—and is surrounded by a darkish boundary. The difference in brightness between the center and the edge of a granule corresponds to a temperature drop of about 300 K.

Granulation is caused by convection of the gas in the photosphere. The inset in Figure 16-9 shows how hot gas from lower levels rises upward in granules, cools off, spills over the edges of the granules, and then plunges back down into the Sun. Convection can occur only if the gas is heated from below, like a pot of water being heated on a stove (see Section 16-2). Along with limb darkening and the Sun's absorption line spectrum, granulation shows that the upper part of the photosphere must be cooler than the lower part.

Like water boiling on a stove, the photosphere bubbles with convection cells

Time-lapse photography reveals more of the photosphere's dynamic activity. Granules form, disappear, and reform in cycles lasting only a few minutes. At any one time, about 4 million granules cover the solar surface.

Superimposed on the pattern of granulation are even larger convection cells called **supergranules** (Figure 16-10). As in granules, gases rise upward in the middle of a supergranule, move horizontally outward toward its edge, and descend back into the Sun. The difference is that a typical supergranule is about 35,000 km in diameter, large enough to enclose several hundred granules. This large-scale convection moves at only about 0.4 km/s (1400 km/h, or 900 mi/h), about one-tenth the speed of gases churning in a regular-sized granule. A given supergranule lasts about a day.

ANALOGY Similar patterns of large-scale and small-scale convection can be found in Earth's atmosphere. On the large scale, air rises gradually at a low-pressure area, then sinks gradually at a high-pressure area, which might be hundreds of kilometers away. This pattern is analogous to the flow in a supergranule. Thunderstorms in our atmosphere are small but intense convection cells within which air moves rapidly up and down. Like granules, they last only a relatively short time before they dissipate.

The Photosphere: Hot, Thin, and Opaque

Although the photosphere is a very active place, it actually contains relatively little material. Careful examination of the spectrum shows that it has a density of only about 10^{-24} kg/m^3, roughly 0.01% the density of Earth's atmosphere at sea level. The photosphere is made primarily of hydrogen and helium, the most abundant elements in the solar system (see Figure 8-4).

Despite being such a thin gas, the photosphere is surprisingly opaque to visible light. (This means that visible light is scattered numerous times as it passes through the photosphere.) If it were

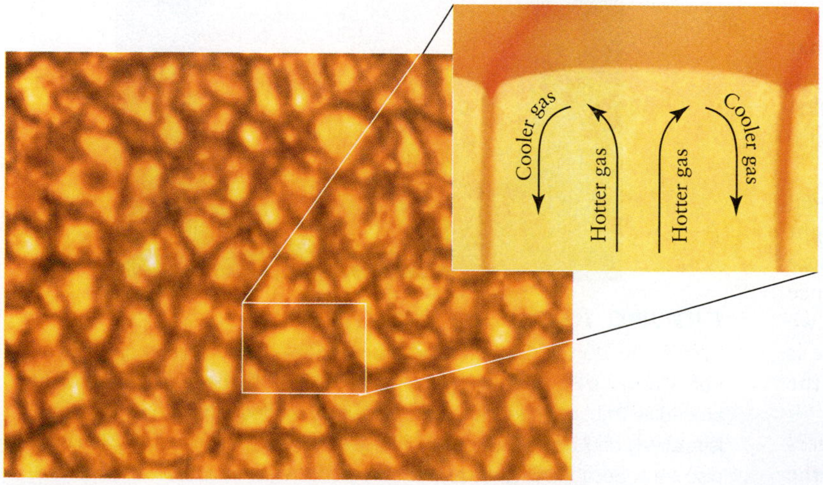

VIDEO 16-1 VIDEO 16-2 **FIGURE 16-9** R I V U X G

Solar Granulation High-resolution photographs of the Sun's surface reveal a blotchy pattern called granulation. Granules are convection cells about 1000 km (600 mi) wide in the Sun's photosphere. **Inset:** Rising hot gas produces bright granules. Cooler gas sinks downward along the boundaries between granules; this gas glows less brightly, giving the boundaries their dark appearance. This convective motion transports heat from the Sun's interior outward to the solar atmosphere. (MSFC/NASA; inset: Goran Scharmer, Lund Observatory)

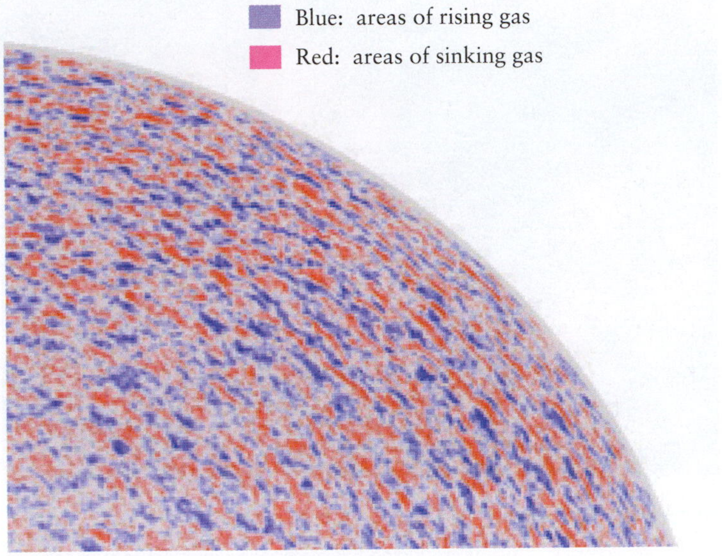

FIGURE 16-10 R I V U X G
Supergranules and Large-Scale Convection Supergranules display relatively little contrast between their center and edges, so they are hard to observe in ordinary images. But they can be seen in a false-color Doppler image like this one. Light from gas that is approaching us (that is, rising) is shifted toward shorter "bluer" wavelengths, while light from receding gas (that is, descending) is shifted toward longer "redder" wavelengths (see Section 5-9). (David Hathaway, MSFC/NASA)

not so opaque, we could see into the Sun's interior to a depth of hundreds of thousands of kilometers, instead of the mere 400 km that we can see down into the photosphere. What makes the photosphere so opaque is that its hydrogen atoms sometimes acquire an extra electron, becoming **negative hydrogen ions.** The scattering of photons can be pictured in two steps. In the first step, the extra electron of a negative hydrogen ion is only loosely attached and can be dislodged if it absorbs a photon of any visible wavelength. In the second step, a similar photon can be emitted when the dislodged electron once again combines with hydrogen to form another negative hydrogen ion, but the emitted photon *heads in a random direction each time.* Hence, negative hydrogen ions absorb light very efficiently, and there are enough of these light-absorbing ions in the photosphere to make it quite opaque. Because it is so opaque, the photosphere's spectrum is close to that of an ideal blackbody (blackbodies are discussed in Section 5-3).

CONCEPTCHECK 16-11

What causes the photosphere to bubble like water boiling on the stove?

Answer appears at the end of the chapter.

16-6 Spikes of rising gas extend through the Sun's chromosphere

An ordinary visible-light image such as Figure 16-7 gives the impression that the Sun ends at the top of the photosphere. But during a total solar eclipse, the Moon blocks the photosphere from our view, revealing a glowing, pinkish layer of gas above the photosphere (Figure 16-11). This layer is the tenuous **chromosphere** ("sphere of color"), the second of the three major levels in the Sun's atmosphere. The chromosphere is only about one ten-thousandth (10^{-4}) as dense as the photosphere, or about 10^{-8} as dense as our own atmosphere. No wonder it is normally invisible!

Comparing the Chromosphere and Photosphere

Unlike the photosphere, which has an absorption line spectrum, the chromosphere has a spectrum dominated by emission lines. An emission line spectrum is produced by the atoms of a hot, thin gas (see Section 5-6 and Section 5-8). As their electrons fall from higher to lower energy levels, the atoms emit photons.

One of the strongest emission lines in the chromosphere's spectrum is the H_α line at 656.3 nm, which is emitted by a hydrogen atom when its single electron falls from the $n = 3$ level to the $n = 2$ level (recall Figure 5-23b). This wavelength is in the red part of the spectrum, which gives the chromosphere its characteristic pinkish color. The spectrum also contains emission lines of singly ionized calcium, as well as lines due to ionized helium and ionized metals. In fact, helium—named after Helios, a Greek reference to the Sun—was originally discovered through detection in the chromospheric spectrum in 1868, almost 30 years before helium was discovered on Earth.

Analysis of the chromospheric spectrum shows that temperature *increases* with increasing height in the chromosphere. This trend is just the opposite of the situation in the photosphere, where

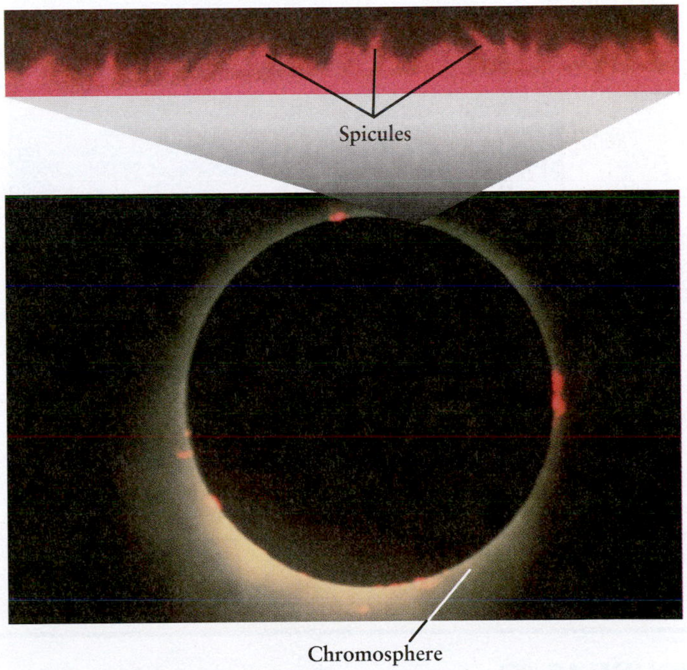

FIGURE 16-11 R I V U X G
The Chromosphere During a total solar eclipse, the Sun's glowing chromosphere can be seen around the edge of the Moon. It appears pinkish because its hot gases emit light at only certain discrete wavelengths, principally the H_α emission of hydrogen at a red wavelength of 656.3 nm. The expanded area above shows spicules, jets of chromospheric gas that surge upward into the Sun's outer atmosphere. (NOAO)

temperature decreases with increasing height. The temperature is about 4400 K at the top of the photosphere; 2000 km higher, at the top of the chromosphere, the temperature is nearly 25,000 K. This temperature difference is very surprising, since temperatures should decrease as you move away from the Sun's interior. In Section 16-10 we will see how solar scientists explain this seeming paradox.

The photosphere's spectrum is dominated by absorption lines at certain wavelengths, while the spectrum of the chromosphere has emission lines at these same wavelengths. In other words, the photosphere appears dark at the specific wavelengths at which the chromosphere emits most strongly, such as the H_α wavelength of 656.3 nm. By viewing the Sun through a special filter that is transparent to light only at the wavelength of H_α, astronomers can screen out light from the photosphere and make the chromosphere visible. Filters make it possible to see the chromosphere at any time, not just during a solar eclipse.

Spicules

Jets of gas thousands of kilometers in height rise through the chromosphere

The top photograph in Figure 16-11 is a high-resolution image of the Sun's chromosphere taken through an H_α filter. This image shows numerous vertical spikes, which are actually jets of rising gas called **spicules.** A typical spicule lasts just 15 minutes or so: It rises at the rate of about 20 km/s (72,000 km/h, or 45,000 mi/h), can reach a height of several thousand kilometers, and then collapses and fades away (Figure 16-12). Approximately 300,000 spicules exist at any one time, covering about 1% of the Sun's surface. When viewed from above, spicules can surround a sunspot, as in Figure 16-13 (we will learn more about sunspots in Section 16-8). While the Sun's magnetic field appears to be involved in the production of spicules, the details are not understood.

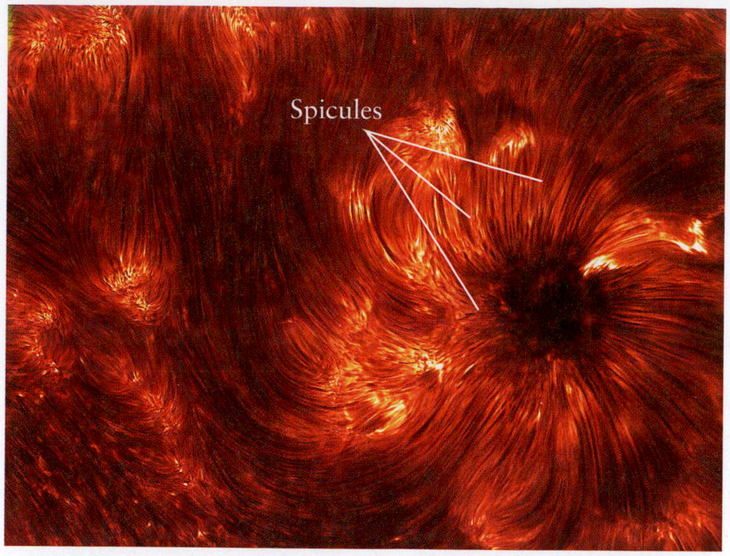

FIGURE 16-13 R I V U X G

Spicules from Above Surrounding this sunspot are numerous spicules, also called "fibrils" because of their long and thin fiberlike shape. These spicules are about as long as Earth's diameter and contain very hot gas confined by magnetic fields. This is one of the highest-resolution images of the Sun ever made. (K. Reardon [Osservatorio Astrofisico di Arcetri, INAF], IBIS, DST, NSO)

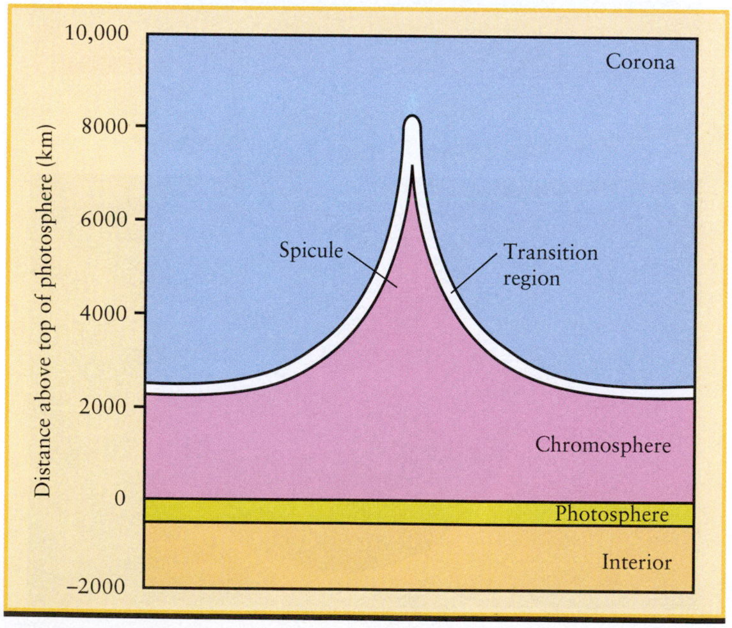

FIGURE 16-12

The Solar Atmosphere This schematic diagram shows the three layers of the solar atmosphere. The lowest, the photosphere, is about 400 km thick. The chromosphere extends about 2000 km higher, with spicules jutting up to nearly 10,000 km above the photosphere. Above a transition region is the Sun's outermost layer, the corona, which we discuss in Section 16-7. It extends many millions of kilometers out into space. (Adapted from J. A. Eddy)

CONCEPTCHECK 16-12

Why do the spicules in Figure 16-11 appear red? Do spicules emit light at other wavelengths?

Answer appears at the end of the chapter.

16-7 The corona ejects mass into space to form the solar wind

The **corona,** or outermost region of the Sun's atmosphere, begins at the top of the chromosphere. It extends out to a distance of several million kilometers. Despite its tremendous extent, the corona is only about one-millionth (10^{-6}) as bright as the photosphere—no brighter than the full moon. Hence, the corona can be viewed only when the light from the photosphere is blocked out, either by use of a specially designed telescope or during a total solar eclipse.

Figure 16-14 is an exceptionally detailed photograph of the Sun's corona taken during a solar eclipse. It shows that the corona is not merely a spherical shell of gas surrounding the Sun. Rather, numerous streamers extend in different directions far above the solar surface. The shapes of these streamers vary on timescales of days or weeks. (For another view of the corona during a solar eclipse, see Figure 3-10b.)

Comparing the Corona, Chromosphere, and Photosphere

Like the chromosphere that lies below it, the corona has an emission line spectrum characteristic of a hot, thin gas. When the

FIGURE 16-14 R I V U X G
The Solar Corona This striking photograph of the corona was taken during the total solar eclipse of July 11, 1991. Numerous streamers extend for millions of kilometers above the solar surface. The unearthly light of the corona is one of the most extraordinary aspects of experiencing a solar eclipse. (Courtesy of R. Christen and M. Christen, Astro-Physics, Inc.)

spectrum of the corona was first measured in the nineteenth century, astronomers found a number of emission lines at wavelengths that had never been seen in the laboratory. Their explanation was that the corona contained elements that had not yet been detected on Earth. However, laboratory experiments in the 1930s revealed that these unusual emission lines were in fact caused by the same atoms found elsewhere in the universe—but in highly ionized states. For example, a prominent green line at 530.3 nm is caused by highly ionized iron atoms, each of which has been stripped of 13 of its 26 electrons. In order to strip that many electrons from atoms, temperatures in the corona must reach 2 million kelvins (2×10^6 K) or even higher—far greater than the temperatures in the chromosphere. Figure 16-15 shows how temperature varies in the chromosphere and corona.

CAUTION! The corona is actually not very "hot"—that is, it contains very little thermal energy. The reason is that the corona is nearly a vacuum. In the corona there are only about 10^{11} atoms per cubic meter, compared with about 10^{23} atoms per cubic meter in the Sun's photosphere and about 10^{25} atoms per cubic meter in the air that we breathe. Because of the corona's high temperature, the atoms there are moving at very high speeds. But because there are so few atoms in the corona, the total amount of energy in these moving atoms (a measure of how "hot" the gas is) is rather low. If you flew a spaceship into the corona, you would have to worry about becoming overheated by the intense light coming from the photosphere, but you would notice hardly any heating from the corona's ultrathin gas.

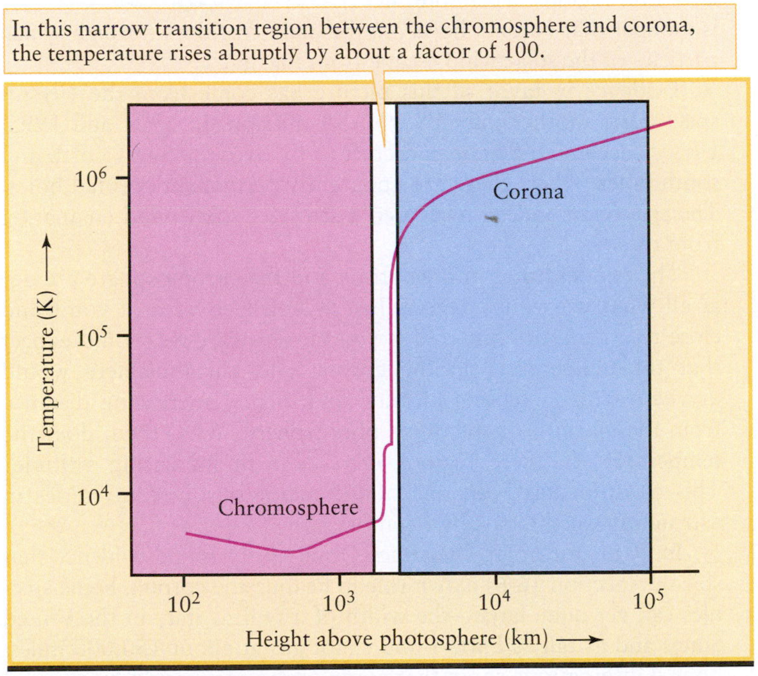

(a)

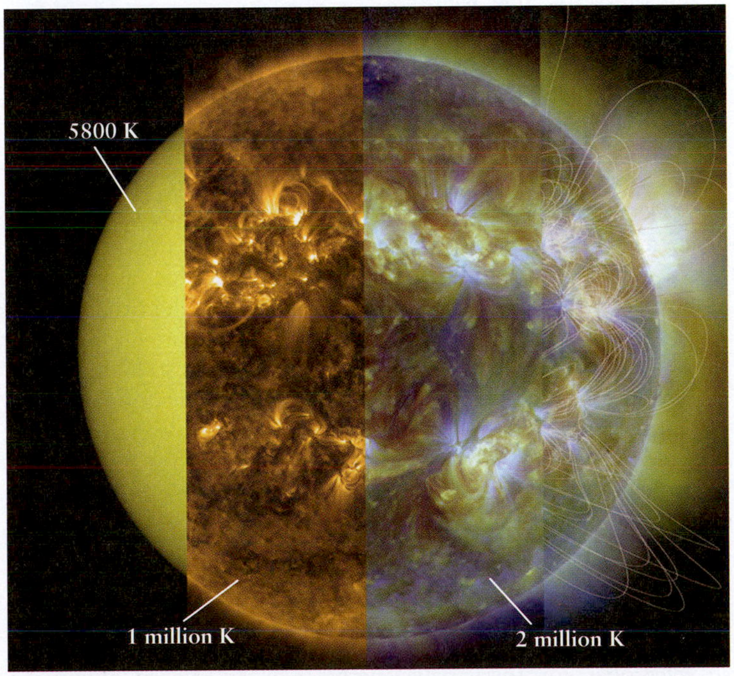

(b)

FIGURE 16-15 R I V U X G
Temperatures in the Sun's Upper Atmosphere **(a)** This graph shows how temperature varies with altitude in the Sun's chromosphere and corona and in the narrow transition region between them. In order to show a large range of values, both the vertical and horizontal scales are nonlinear. **(b)** This composite portrait shows the different features that appear as wavelengths corresponding to higher temperatures are imaged. At visible wavelengths, the 5800 K photosphere is visible, while ultraviolet light images features at millions of degrees. On the far right, modeling has been used to draw lines of the Sun's magnetic field in the corona. (a: Adapted from A. Gabriel; b: NASA)

ANALOGY The situation in the corona is similar to that inside a conventional oven that is being used for baking. Both the walls of the oven and the air inside the oven are at the same high temperature, but the air contains very few atoms and thus carries little energy. If you put your hand in the oven momentarily, the lion's share of the heat you feel is radiation from the oven walls.

The low density of the corona explains why it is so dim compared with the photosphere. In general, the higher the temperature of a gas, the brighter it glows. But because there are so few atoms in the corona, the net amount of light that it emits is very feeble compared with the light from the much cooler, but also much denser, photosphere.

CONCEPTCHECK 16-13

Why is the corona so difficult to see if it is so much hotter than the photosphere?

Answer appears at the end of the chapter.

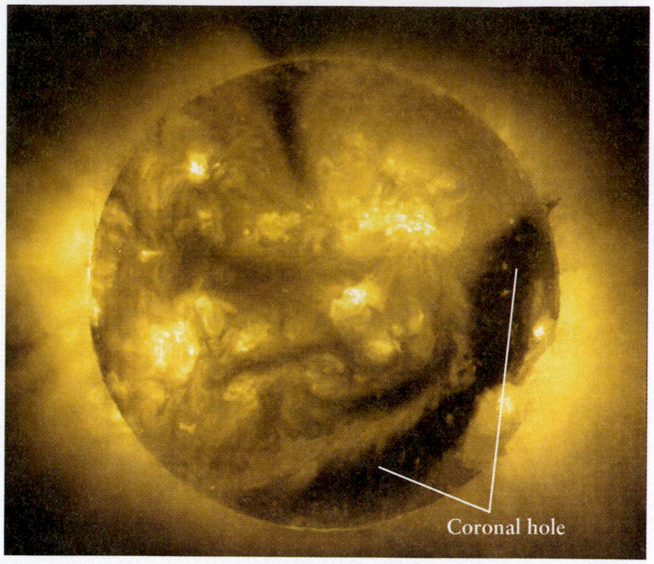

VIDEO 16.3 **FIGURE 16-16** R I V U X G

The Ultraviolet Corona The *SOHO* spacecraft recorded this false-color ultraviolet view of the solar corona. The dark feature running across the Sun's disk from the top is a coronal hole, a region where the coronal gases are thinner than elsewhere. Such holes are often the source of strong gusts in the solar wind. (SOHO/EIT/ESA/NASA)

The Solar Wind and Coronal Holes

Earth's gravity keeps our atmosphere from escaping into space. In the same way, the Sun's powerful gravitational attraction keeps most of the gases of the photosphere, chromosphere, and corona from escaping. But the corona's high temperature means that its atoms and ions are moving at very high speeds, around a million km per hour. As a result, some of the coronal gas can and does escape. This outflow of gas, which we first encountered in Section 8-5, is called the **solar wind.**

Each second the Sun ejects about a million tons (10^9 kg) of material into the solar wind. But the Sun is so massive that, even over its entire lifetime, it will eject only a few tenths of a percent of its total mass. The solar wind is composed almost entirely of electrons and nuclei of hydrogen and helium. About 0.1% of the solar wind is made up of ions of more massive atoms, such as silicon, sulfur, calcium, chromium, nickel, iron, and argon. The aurorae seen at far northern or southern latitudes on Earth are produced when electrons and ions from the solar wind enter our upper atmosphere.

Special telescopes enable astronomers to see the origin of the solar wind. To appreciate what sort of telescopes are needed, note that because the temperature of the coronal gas is so high, ions in the corona are moving very fast (see Box 7-2). When ions collide, the energy of the impact is so great that the ion's electrons are boosted to very high energy levels. As the electrons fall back to lower levels, they emit high-energy photons in the ultraviolet and X-ray portions of the spectrum—wavelengths at which the photosphere and chromosphere are relatively dim. Hence, telescopes sensitive to these short wavelengths are ideal for studying the corona and the flow of the solar wind.

Earth's atmosphere absorbs most ultraviolet light and X-rays, so telescopes for these wavelengths must be placed above the atmosphere on spacecraft (see Section 6-7, especially Figure 6-25). Figure 16-16 shows an ultraviolet view of the corona from the *SOHO* spacecraft (*Solar and Heliospheric Observatory*), a joint project of the European Space Agency (ESA) and NASA.

Figure 16-16 reveals that the corona is not uniform in temperature or density. The densest, highest-temperature regions appear bright, while the thinner, lower-temperature regions are dark. Note the large dark area, called a **coronal hole** because it is almost devoid of luminous gas. Particles streaming away from the Sun can most easily flow outward through these particularly thin regions. Therefore, it is thought that coronal holes are the main corridors through which particles of the solar wind escape from the Sun.

Unlike the lower levels of the Sun's atmosphere, the corona has immense holes that shift and reshape

Evidence in favor of this picture has come from the *Ulysses* spacecraft, another joint ESA/NASA mission. In 1994 and 1995, *Ulysses* became the first spacecraft to fly over the Sun's north and south poles, where there are apparently permanent coronal holes. The spacecraft indeed measured a stronger solar wind emanating from these holes.

The temperatures in the corona and the chromosphere are not at all what we would expect. Just as you feel warm if you stand close to a campfire but cold if you move away, we would expect that the temperature in the corona and chromosphere would *decrease* with increasing altitude and, hence, increasing distance from the warmth of the Sun's photosphere. Why, then, does the temperature in these regions *increase* with increasing altitude? This question has been one of the major unsolved mysteries in astronomy for the past half-century.

In 2011, the Solar Dynamics Observatory found evidence that spicules (Section 16-6) play a role in heating the corona. Some spicules can get quite large—the width of a typical state in the United States and as tall as Earth—and shoot gas at about 150,000 miles/hour. Observations show that some of the gas ejected by spicules is heated to millions of degrees as it is shot into the corona. With so many spicules present at any given time (about 300,000), estimates indicate this ejected gas makes a significant contribution to coronal heating.

However, spicules are only one contribution to coronal heating, and as astronomers have tried to resolve this dilemma, they have found important clues in one of the Sun's most familiar features—sunspots.

CONCEPTCHECK 16-14

What is the connection between the solar wind and coronal holes?

Answer appears at the end of the chapter.

16-8 Sunspots are low-temperature regions in the photosphere

Granules, supergranules, spicules, and the solar wind occur continuously. These features are said to be aspects of the *quiet* Sun. But other, more dramatic features appear periodically, including massive eruptions and regions of concentrated magnetic fields. When these features are present, astronomers refer to the *active* Sun. The features of the active Sun that can most easily be seen with even a small telescope are sunspots (although only with a safety filter attached).

Observing Sunspots

Sunspots are irregularly shaped dark regions in the photosphere. Sometimes sunspots appear in isolation (Figure 16-17a), but frequently they are found in sunspot groups (Figure 16-17b; see also Figure 16-7). Although sunspots vary greatly in size, typical ones measure a few tens of thousands of kilometers across—comparable to the diameter of Earth. Sunspots are not permanent features of the photosphere but last between a few hours and a few months.

Each sunspot has a dark central core, called the *umbra,* and a brighter border called the *penumbra.* We used these same terms in Section 3-4 to refer to different parts of Earth's shadow or the Moon's shadow. But a sunspot is not a shadow: It is a region in the photosphere where the temperature is relatively low, which makes it appear darker than its surroundings. As we saw in Section 5-4, Wien's law relates the color of a blackbody, such as the photosphere, to the blackbody's temperature; the cooler the blackbody, the longer the wavelength at which it emits the most light. If the surrounding hot photosphere is blocked from view, a sunspot's cooler umbra appears red and the warmer penumbra appears orange. The colors of a sunspot indicate that the temperature of the umbra is typically 4300 K and that of the penumbra is typically 5000 K. While high by earthly standards, these temperatures are quite a bit lower than the average photospheric temperature of 5800 K.

The Stefan-Boltzmann law (see Section 5-4) tells us that the energy flux from a blackbody is proportional to the fourth power of its temperature. This law lets us compare the amounts of light energy emitted by a square meter of a sunspot's umbra and by a square meter of undisturbed photosphere. The ratio is

$$\frac{\text{flux from umbra}}{\text{flux from photosphere}} = \left(\frac{4300\text{ K}}{5800\text{ K}}\right)^4 = 0.30$$

That is, the umbra emits only 30% as much light as an equally large patch of undisturbed photosphere, which is why sunspots appear so dark.

Sunspots and the Sun's Rotation

Occasionally, a sunspot group is large enough to be seen without a telescope. Chinese astronomers recorded such sightings 2000 years ago, and huge sunspot groups visible to the naked eye (with an appropriate filter) were seen in 1989 and 2003. But it was not until Galileo introduced the telescope into astronomy (see Section 4-5) that anyone was able to examine sunspots in detail.

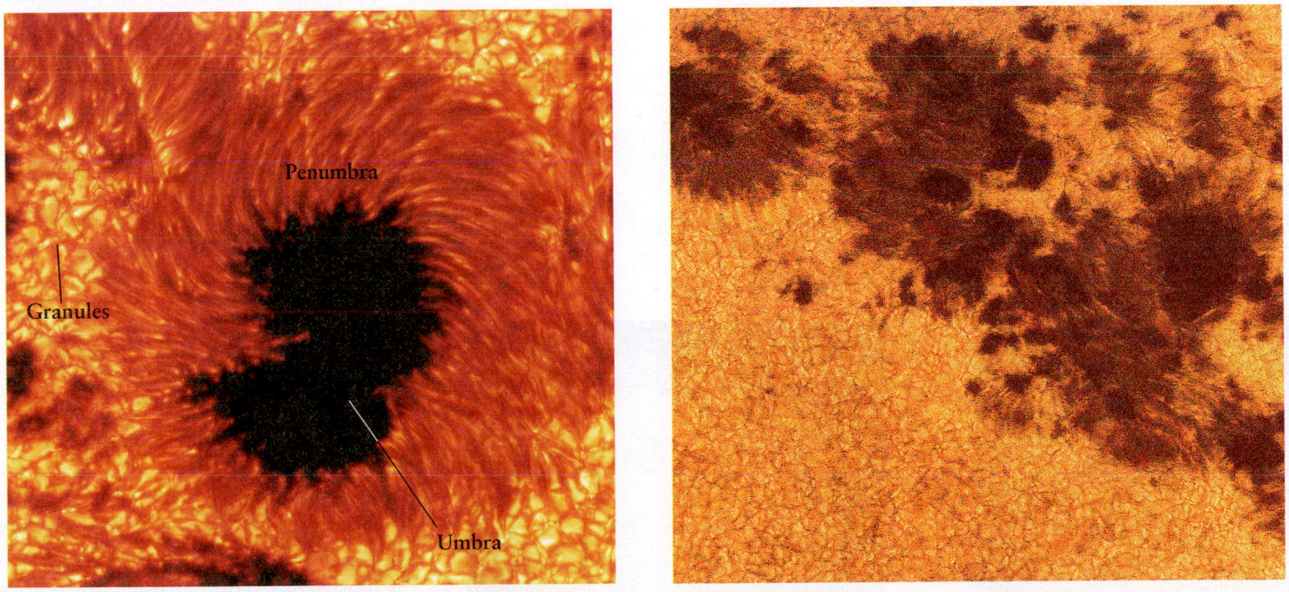

(a) (b)

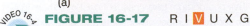

VIDEO 16.4 **FIGURE 16-17** R I V U X G

Sunspots (a) This high-resolution photograph of the photosphere shows a mature sunspot. The dark center of the spot is called the umbra and has the lowest temperature. It is bordered by the penumbra, which is less dark, is hotter than the umbra, and has a featherlike appearance. (Even the umbra emits visible light, but not enough to capture in an image against the glare of the Sun.) **(b)** In this view of a typical sunspot group, several sunspots are close enough to overlap. In both images you can see granulation in the surrounding, undisturbed photosphere. (a: Scharmer et al., Royal Swedish Academy of Sciences/Science Source; b: NOAO)

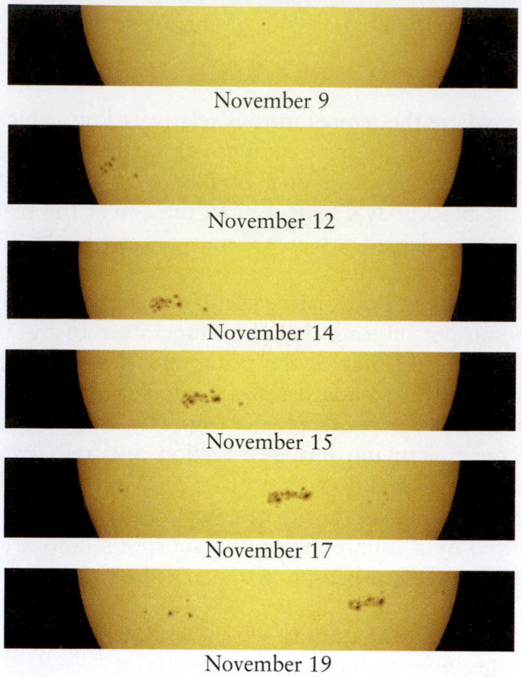

VIDEO 16-5

FIGURE 16-18 R I V U X G

Tracking the Sun's Rotation with Sunspots This series of photographs taken in 1999 shows the rotation of the Sun. By observing the same group of sunspots from one day to the next, Galileo found that the Sun rotates once in about four weeks. (The equatorial regions of the Sun actually rotate somewhat faster than the polar regions.) Notice how the sunspot group shown here changed its shape. (The Carnegie Observatories)

Galileo discovered that he could determine the Sun's rotation rate by tracking sunspots as they moved across the solar disk (Figure 16-18). He found that the Sun rotates once in about four weeks. A typical sunspot group lasts about two months, so a specific one can be followed for two solar rotations.

Further observations by the British astronomer Richard Carrington in 1859 demonstrated that the Sun does not rotate as a rigid body. Instead, the equatorial regions rotate more rapidly than the polar regions. This phenomenon is known as **differential rotation.** Thus, while a sunspot near the solar equator takes only 25 days to go once around the Sun, a sunspot at 30° north or south of the equator takes 27½ days. The rotation period at 75° north or south is about 33 days, while near the poles it may be as long as 35 days.

CONCEPTCHECK 16-15

If the center of a sunspot has a temperature of about 4300 K, why does it appear dark?

Answer appears at the end of the chapter.

The Sunspot Cycle

The number of sunspots increases and decreases on an 11-year cycle

The average number of sunspots on the Sun is not constant, but varies in a predictable **sunspot cycle** (Figure 16-19a). This phenomenon was first reported by the German astronomer Heinrich Schwabe in 1843 after many years of observing. As Figure 16-19a shows, the average number of sunspots varies with a period of about 11 years. A period of exceptionally many sunspots is a **sunspot maximum** (Figure 16-19b), as occurred in 1979, 1989, and 2000. Conversely, the Sun is almost devoid of sunspots at a **sunspot minimum** (Figure 16-19c), as occurred in 1976, 1986, 1996, and 2008. During the 2008 minimum in sunspot activity, there were fewer sunspots observed than in any other year since 1913.

The locations of sunspots also vary with the same 11-year sunspot cycle. At the beginning of a cycle, just after a sunspot minimum, sunspots first appear at latitudes around 30° north and south of the solar equator (Figure 16-20). Over the succeeding years, the sunspots occur closer and closer to the equator.

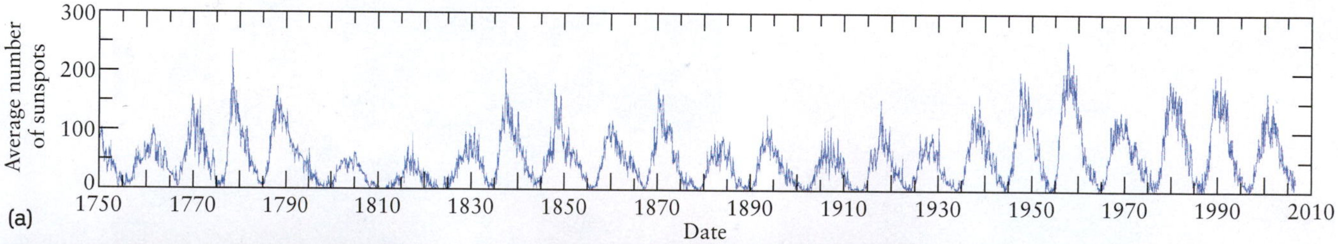

(a)

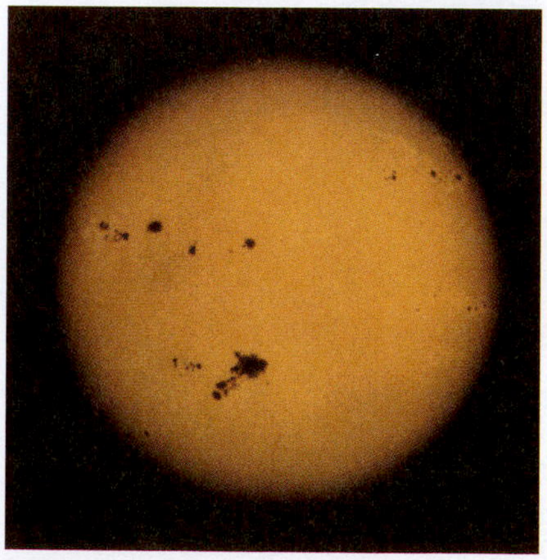

(b) Near sunspot maximum

(c) Near sunspot minimum

FIGURE 16-19 R I V U X G

The Sunspot Cycle **(a)** The number of sunspots on the Sun varies with a period of about 11 years. The most recent sunspot maximum occurred in 2000. **(b)** This photograph, taken near sunspot maximum in 1989, shows a number of sunspots and large sunspot groups. The sunspot group visible near the bottom of the Sun's disk has about the same diameter as the planet Jupiter. **(c)** Near sunspot minimum, as in this 1986 photograph, essentially no sunspots are visible. (NOAO)

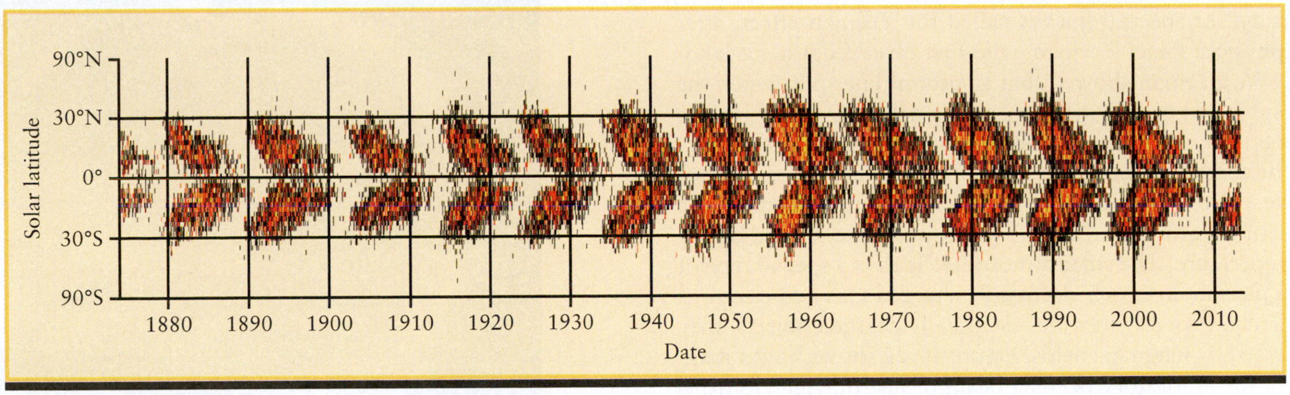

FIGURE 16-20

Variations in the Average Latitude of Sunspots The dots in this graph (sometimes called a "butterfly diagram") record how far north or south of the Sun's equator sunspots were observed. At the beginning of each sunspot cycle, most sunspots are found near latitudes 30° north or south. As the cycle goes on, sunspots typically form closer to the equator. (NASA Marshall Space Flight Center)

16-9 Sunspots are produced by a 22-year cycle in the Sun's magnetic field

TUTORIAL 16-3

Why should the number of sunspots vary with an 11-year cycle? Why should their average latitude vary over the course of a cycle? And why should sunspots exist at all? The first step toward answering these questions came in 1908, when the American astronomer George Ellery Hale discovered that sunspots are associated with intense magnetic fields on the Sun.

Probing Solar Magnetism

LOOKING DEEPER 16-2

When Hale focused a spectroscope on sunlight coming from a sunspot, he found that many spectral lines appear to be split into several closely spaced lines (**Figure 16-21**).

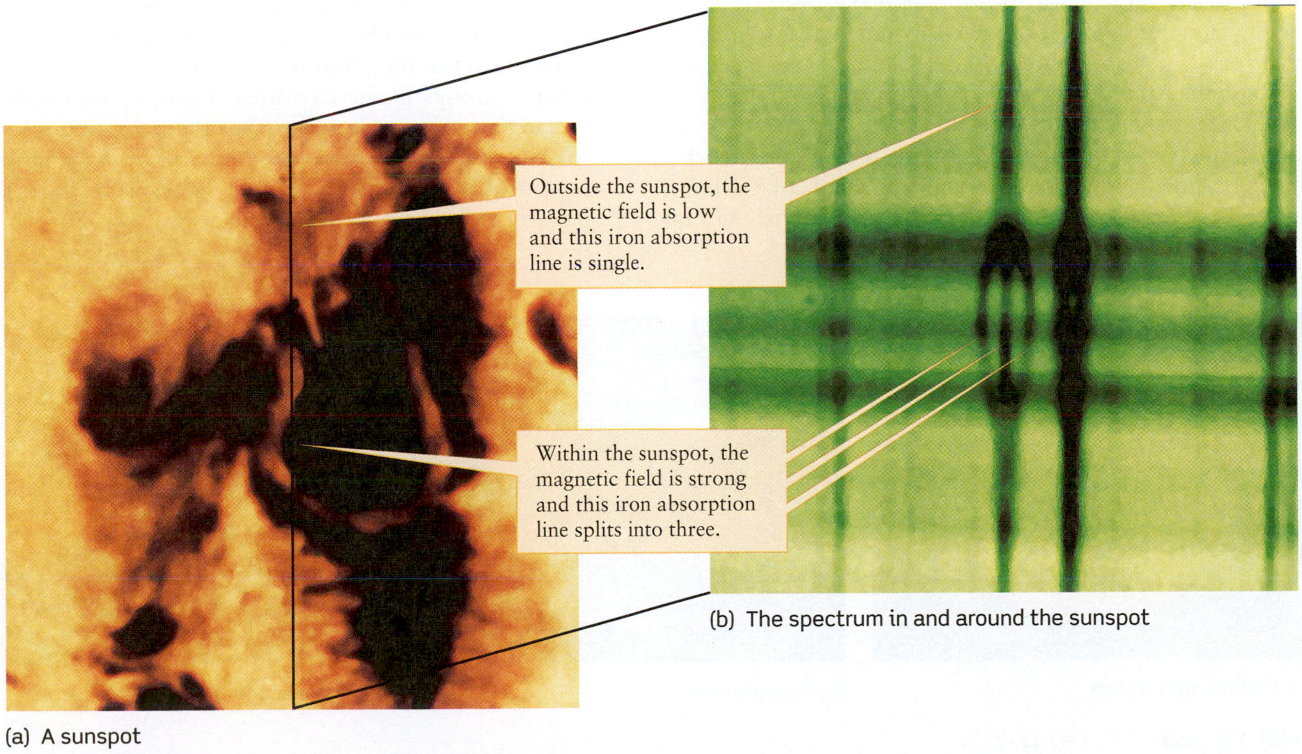

Outside the sunspot, the magnetic field is low and this iron absorption line is single.
Within the sunspot, the magnetic field is strong and this iron absorption line splits into three.
(a) A sunspot
(b) The spectrum in and around the sunspot

FIGURE 16-21 R I V U X G

Sunspots Have Strong Magnetic Fields **(a)** A black line in this image of a sunspot shows where the slit of a spectrograph was aimed. **(b)** This is a portion of the resulting spectrum, including a dark absorption line caused by iron atoms in the photosphere. The splitting of this line by the sunspot's magnetic field can be used to calculate the field strength. Typical sunspot magnetic fields are over 5000 times stronger than Earth's field at its north and south poles. (NOAO)

This "splitting" of spectral lines is called the **Zeeman effect,** after the Dutch physicist Pieter Zeeman, who first observed it in his laboratory in 1896. Zeeman showed that a spectral line splits when the atoms are subjected to an intense magnetic field. The more intense the magnetic field, the wider the separation of the split lines.

Hale's discovery showed that sunspots are places where the hot gases of the photosphere are bathed in a concentrated magnetic field. Many of the atoms of the Sun's atmosphere are ionized due to the high temperature. The solar atmosphere is thus a special type of gas called a **plasma,** in which electrically charged ions and electrons can move freely. Like any moving, electrically charged objects, they can be deflected by magnetic fields. Figure 16-22 shows how a magnetic field in the laboratory bends a beam of fast-moving electrons into a curved trajectory. Similarly, the paths of moving ions and electrons in the photosphere are deflected by the Sun's magnetic field. In particular, magnetic forces act on the hot plasma that rises from the Sun's interior due to convection. Where the magnetic field is particularly strong, these forces push the hot plasma away. The result is a localized region where the gas is relatively cool and thus glows less brightly—in other words, a sunspot.

To get a fuller picture of the Sun's magnetic fields, astronomers take images of the Sun at two wavelengths, one just less than and one just greater than the wavelength of a magnetically split spectral line. From the difference between these two images, they can construct a picture called a **magnetogram,** which displays the magnetic fields in the solar atmosphere. Figure 16-23a is an ordinary white-light photograph of the Sun taken at the same time as the magnetogram in Figure 16-23b. In the magnetogram, dark blue indicates areas of the photosphere with one magnetic polarity (north), and yellow indicates areas with the opposite (south) magnetic polarity. This image shows that many sunspot groups have roughly comparable areas covered by north and south magnetic polarities (see also Figure 16-23c). Thus, a sunspot group resembles a giant bar magnet, with a north magnetic pole at one end and a south magnetic pole at the other.

If different sunspot groups were unrelated to one another, their magnetic poles would be randomly oriented, like a bunch of compass needles all pointing in random directions. As Hale discovered, however, there is a striking regularity in the magnetization of sunspot groups. As a given sunspot group moves with the Sun's rotation, the sunspots in front are called the "preceding members" of the group. The spots that follow behind are referred to as the "following members." Hale compared the sunspot groups in the two solar hemispheres, north or south of the Sun's equator. He found that the preceding members in one solar hemisphere all have the same magnetic polarity, while the preceding members in the other hemisphere have the opposite polarity. Furthermore, in the hemisphere where the Sun has its

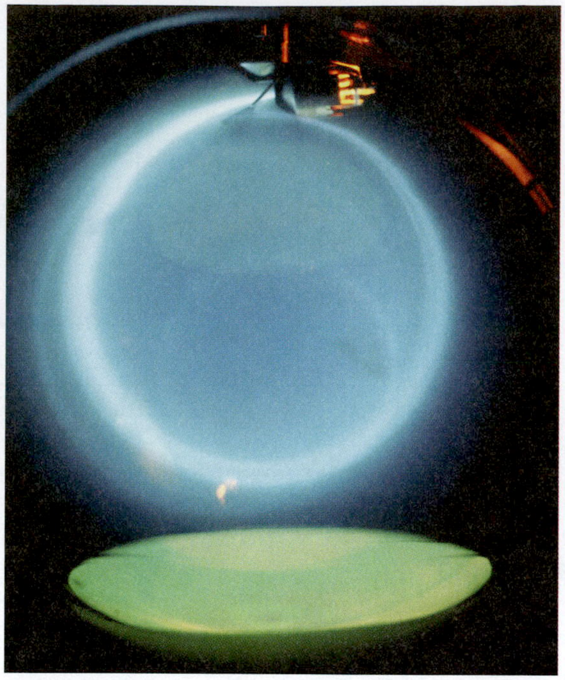

FIGURE 16-22 R I V U X G
Magnetic Fields Deflect Moving, Electrically Charged Objects
In this laboratory experiment, a beam of negatively charged electrons (shown by a blue arc) is aimed straight upward from the center of the apparatus. The entire apparatus is inside a large magnet, and the magnetic field deflects the beam into a curved path. (Andrew Lambert Photography/Science Source)

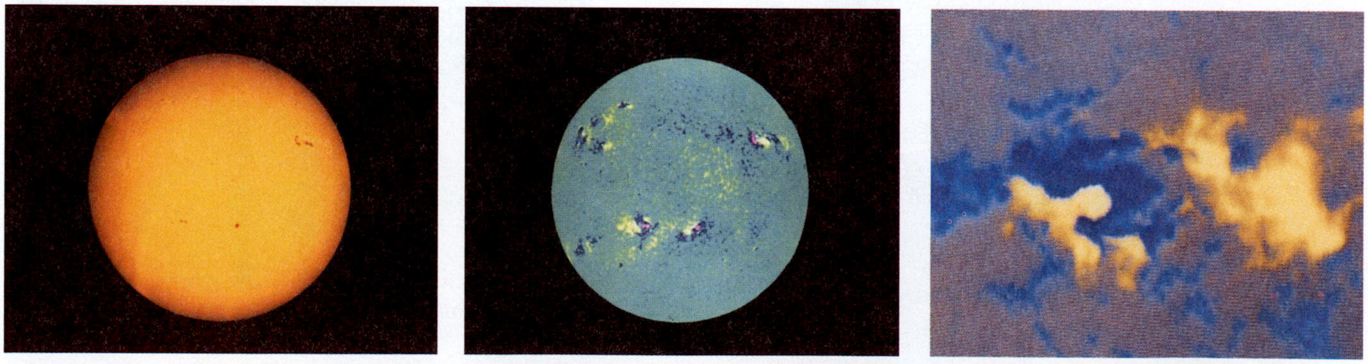

(a) Visible-light image

(b) Magnetogram

(c) Magnetogram of a sunspot group

FIGURE 16-23 R I V U X G
Mapping the Sun's Magnetic Field **(a)** This visible-light image and **(b)** this false-color magnetogram were recorded at the same time. Dark blue and yellow areas in the magnetogram have north and south magnetic polarity, respectively; blue-green regions have weak magnetic fields. The highly magnetized regions in (b) correlate with the sunspots in (a). **(c)** The two ends of this large sunspot group have opposite magnetic polarities (colored blue and yellow), like the ends of a giant bar magnet. (NOAO)

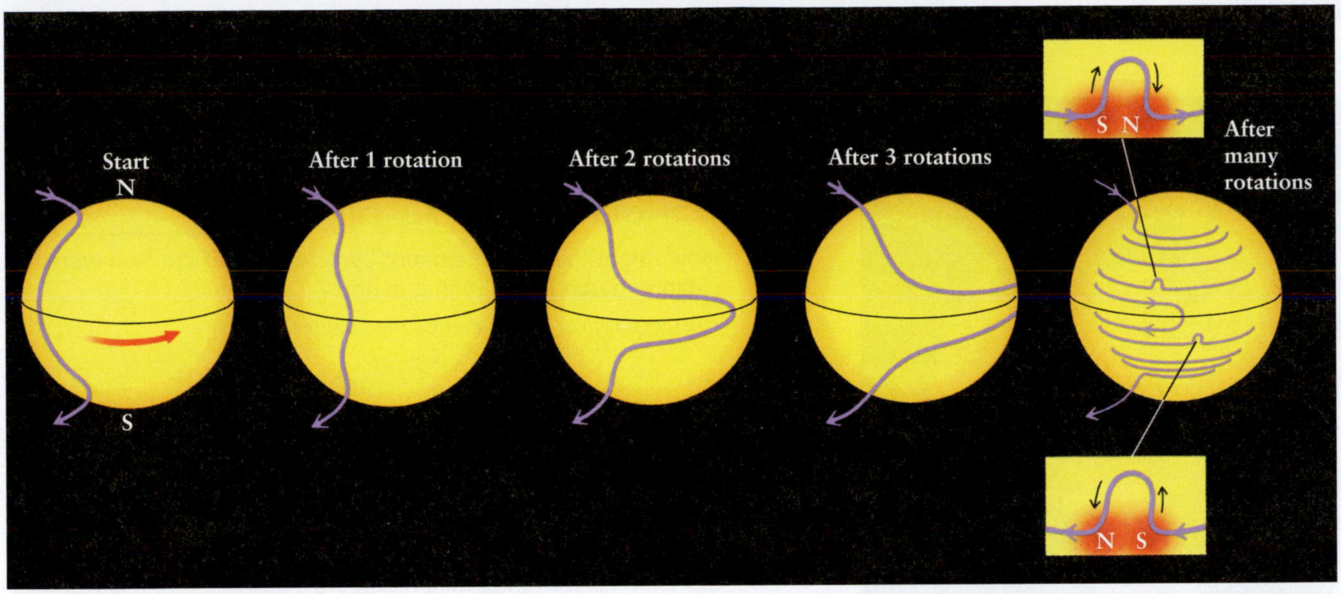

FIGURE 16-24

Babcock's Magnetic Dynamo Model Magnetic field lines tend to move along with the plasma in the Sun's outer layers. Because the Sun rotates faster at the equator than near the poles, a field line that starts off running from the Sun's north magnetic pole (N) to its south magnetic pole (S) ends up wrapped around the Sun like twine wrapped around a ball. The insets on the far right show how sunspot groups appear where the concentrated magnetic field rises through the photosphere.

north magnetic pole, the preceding members of all sunspot groups have north magnetic polarity. In the opposite hemisphere, where the Sun has its south magnetic pole, the preceding members all have south magnetic polarity.

Along with his colleague Seth B. Nicholson, Hale also discovered that the Sun's polarity pattern completely reverses itself every 11 years—the same interval as the time from one solar maximum to the next. The hemisphere that has preceding north magnetic poles during one 11-year sunspot cycle will have preceding south magnetic poles during the next 11-year cycle, and vice versa. The north and south magnetic poles of the Sun itself also reverse every 11 years. Thus, the Sun's magnetic pattern repeats itself only after two sunspot cycles, which is why astronomers speak of a **22-year solar cycle.**

The Magnetic-Dynamo Model

In 1960, the American astronomer Horace Babcock proposed a description that seems to account for many features of this 22-year solar cycle. Babcock's scenario, called a **magnetic-dynamo model,** makes use of two basic properties of the Sun's photosphere—differential rotation and convection. Differential rotation causes the magnetic field in the photosphere to become wrapped around the Sun (Figure 16-24). As a result, the magnetic field becomes concentrated at certain latitudes on either side of the solar equator. Convection in the photosphere creates tangles in the concentrated magnetic field, and "kinks" erupt through the solar surface. Sunspots appear where the magnetic field protrudes through the photosphere. The theory suggests that sunspots should appear first at northern and southern latitudes and later form nearer to the equator, which is just what is observed (see Figure 16-20). Note also that as shown on the far right in Figure 16-24, the preceding member of a sunspot group has the same polarity (N or S) as the Sun's magnetic pole in that hemisphere, which is just as Hale observed.

Differential rotation eventually undoes the twisted magnetic field. The preceding members of sunspot groups move toward the Sun's equator, while the following members migrate toward the poles. Because the preceding members from the two hemispheres have opposite magnetic polarities, their magnetic fields cancel each other out when they meet at the equator. The following members in each hemisphere have the opposite polarity to the Sun's pole in that hemisphere; hence, when they converge on the pole, the following members first cancel out and then reverse the Sun's overall magnetic field. The fields are now completely relaxed. Once again, differential rotation begins to twist the Sun's magnetic field, but now with all magnetic polarities reversed. In this way, Babcock's model helps to explain the change in field direction every 11 years.

The Sun's differential rotation makes the magnetic field twist like a rubber band

Recent discoveries in helioseismology (Section 16-3) offer new insights into the Sun's magnetic field. By comparing the speeds of sound waves that travel with and against the Sun's rotation, helioseismologists have been able to determine the Sun's rotation rate at different depths and latitudes. As shown in Figure 16-25, the Sun's surface pattern of differential rotation persists through the convective zone. Farther in, within the radiative zone, the Sun seems to rotate like a rigid object with a period of 27 days at all latitudes. Astronomers suspect that the Sun's magnetic field originates in a relatively thin layer where the radiative and convective zones meet and slide past each other due to their different rotation rates.

One dilemma about sunspots is that compressed magnetic fields tend to push themselves apart, which means that sunspots should dissipate rather quickly. Yet observations show that sunspots can persist for many weeks. The resolution of this paradox may have been found using helioseismology (see Section 16-3). Analysis of the vibrations of the Sun around sunspots shows that beneath the surface of the photosphere, the gases surrounding each sunspot are circulating at high speed—rather like a hurricane as large as Earth. The circulation of charged gases around the magnetic field holds the fields in place, thus stabilizing the sunspot.

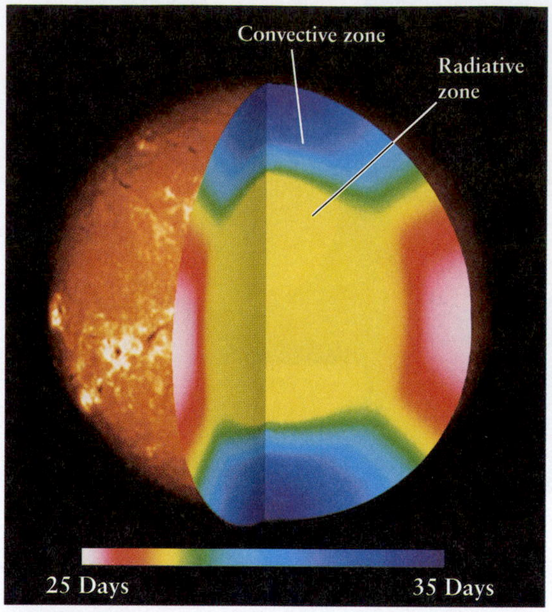

FIGURE 16-25

Rotation of the Solar Interior This cutaway picture of the Sun shows how the solar rotation period (shown by different colors) varies with depth and latitude. The surface and the convective zone have differential rotation (a short period at the equator and longer periods near the poles). Deeper within the Sun, the radiative zone seems to rotate like a rigid sphere. (Courtesy of K. Libbrecht, Big Bear Solar Observatory)

Helioseismology—the analysis of solar sound waves—can also take advantage of the relationship between sunspots and the solar magnetic field. Compared to typical solar material, sunspots absorb solar sound waves more strongly. When sunspots occur on the *opposite* side of the Sun, their increased absorption means that interior sound waves do not reflect back as strongly through the body of the Sun. Amazingly, this effect is used to estimate sunspot activity on the *opposite side* of the Sun. Knowing the number of sunspots facing away from Earth is actually useful: Added to the sunspots facing Earth, the *total* sunspot activity is continuously monitored to help forecast **space weather**—these are variations in the solar wind and magnetic field that can affect satellites and astronauts.

Much about sunspots and solar activity remains mysterious. There are perplexing irregularities in the solar cycle. For example, the overall reversal of the Sun's magnetic field is often piecemeal and haphazard. One pole may reverse polarity long before the other. For several weeks the Sun's surface may have two north magnetic poles and no south magnetic pole at all.

Furthermore, there seem to be times when all traces of sunspots and the sunspot cycle vanish for many years. For example, virtually no sunspots were seen from 1645 through 1715. Curiously, during these same years Europe experienced record low temperatures, often referred to as the Little Ice Age, whereas the western United States was subjected to severe drought. By contrast, there was apparently a period of increased sunspot activity during the eleventh and twelfth centuries, during which Earth was warmer than it is today. Thus, variations in solar activity appear to affect climates on Earth. The origin of this Sun-Earth connection is a topic of ongoing research.

CONCEPTCHECK 16-16

Is the sunspot cycle an 11-year cycle or a 22-year cycle?

CONCEPTCHECK 16-17

How might the Sun's sunspot cycle change if the Sun were rotating much faster than it is now?

Answers appear at the end of the chapter.

16-10 The Sun's magnetic field heats the corona, produces flares, and causes massive eruptions

Astronomers now understand that the Sun's magnetic field does more than just explain the presence of sunspots. It is also responsible for the existence of spicules, as well as a host of other dramatic phenomena in the chromosphere and corona.

Magnetic Reconnection

In a plasma, magnetic field lines and the material of the plasma tend to move together. Their moving together means that as convection pushes material toward the edge of a supergranule, it pushes magnetic field lines as well. The result is that vertical magnetic field lines pile up around a supergranule. Plasma that "sticks" to these magnetic field lines thus ends up lifted upward, forming a spicule (see Figures 16-12 and 16-13).

The tendency of plasma to follow the Sun's magnetic field can also explain why the temperature of the chromosphere and corona is so high. Spacecraft observations show magnetic field arches extending tens of thousands of kilometers into the corona, with streamers of electrically charged particles moving along each arch (Figure 16-26a). If the magnetic fields of two arches come into proximity, their magnetic fields can rearrange in a process called **magnetic reconnection** (Figure 16-26b). The tremendous amount of energy stored in the magnetic field is then released into the solar atmosphere. (A single arch contains as much energy as a hydroelectric power plant would generate in a million years.) The amount of energy released in this way is thought to help maintain the temperatures of the chromosphere and corona.

ANALOGY The idea that a magnetic field can heat gases has applications on Earth as well as on the Sun. In an automobile engine's ignition system, an electric current is set up in a coil of wire, which produces a magnetic field. When the current is shut off, the magnetic field collapses and its energy is directed to a spark plug in one of the engine's cylinders. The released energy heats the mixture of air and gasoline around the plug, causing the mixture to ignite. This drives the piston in that cylinder and makes the automobile go.

VIDEO 16.6 Magnetic heating can also explain why the parts of the corona that lie on top of sunspots are often the most prominent in ultraviolet images. (Some examples are the bright regions in Figure 16-16.) The intense magnetic field of the sunspots helps trap and compress hot coronal gas, giving it such a high temperature that it emits copious amounts of high-energy ultraviolet photons and even more energetic X-ray photons.

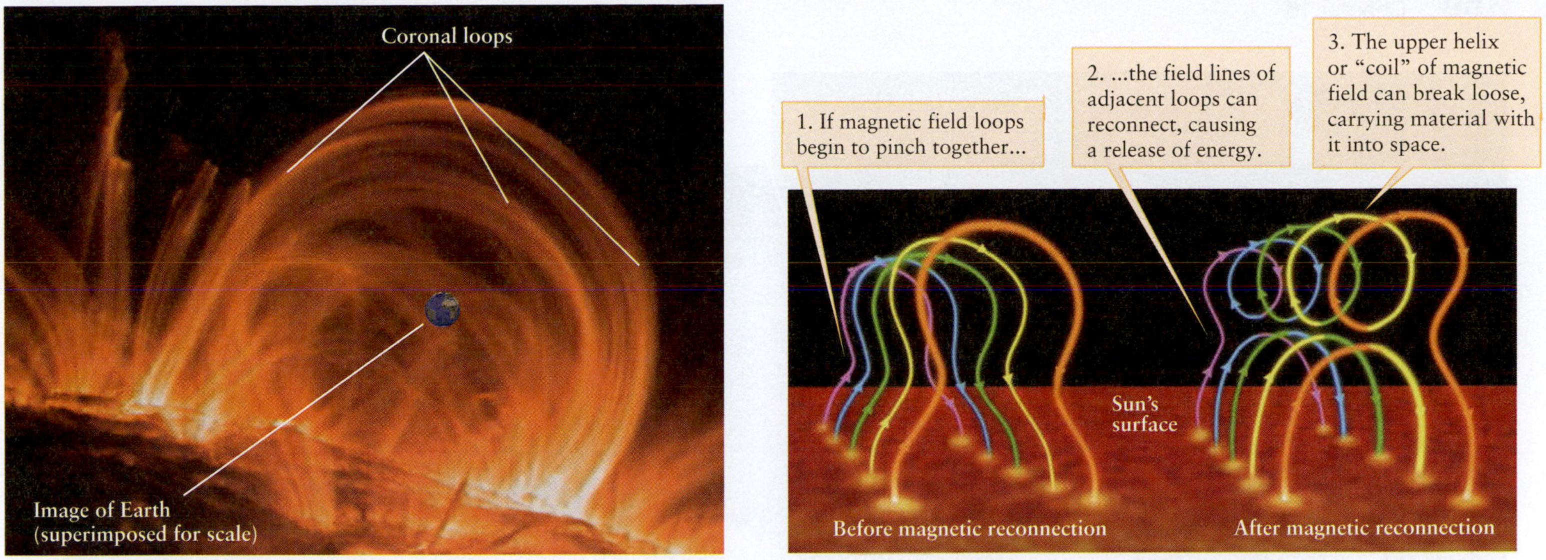

Coronal loops
Image of Earth (superimposed for scale)
1. If magnetic field loops begin to pinch together...
2. ...the field lines of adjacent loops can reconnect, causing a release of energy.
3. The upper helix or "coil" of magnetic field can break loose, carrying material with it into space.
Sun's surface
Before magnetic reconnection
After magnetic reconnection

(a) (b)

FIGURE 16-26 R I V **U** X G

Magnetic Arches and Magnetic Reconnection **(a)** This false-color ultraviolet image from the *TRACE* spacecraft *(Transition Region and Coronal Explorer*) shows magnetic field loops suspended high above the solar surface. The loops are made visible by the glowing gases trapped within them. **(b)** When the magnetic fields in these loops change their arrangement, a tremendous amount of energy is released and solar material can be ejected upward. (a: Stanford-Lockheed Institute for Space Research; *TRACE;* and NASA)

CONCEPTCHECK 16-18

Why does glowing plasma on the Sun appear to arch up above the Sun's photosphere?

Answer appears at the end of the chapter.

Prominences, Flares, and Coronal Mass Ejections

Spicules and coronal heating occur even when the Sun is quiet. But magnetic fields can also explain many aspects of the active Sun in addition to sunspots. Figure 16-27 is an image of the chromosphere made with an H_α filter during a sunspot maximum. The bright areas are called **plages** (from the French word for "beach"). These plages are bright, hot regions in the chromosphere that tend to form just before the appearance of new sunspots. They are probably created by magnetic fields that push upward from the Sun's interior, compressing and heating a portion of the chromosphere. The dark streaks, called **filaments,** are relatively cool and dense parts of the chromosphere that have been pulled along with magnetic field lines as they arch to high altitudes. Images with an H_α filter are made from visible light, and images using ultraviolet light can highlight filaments and sunspots (Figure 16-28).

When seen from the side, so that they are viewed against the dark background of space, filaments appear as bright, arching columns of gas called **prominences** (Figure 16-29). They can extend for tens of thousands of kilometers above the photosphere. Some prominences last for only a few hours, while others persist for many months. The most energetic prominences break free of the magnetic fields that confined them and burst into space.

VIDEO 16.7 Magnetic reconnection also causes violent, eruptive events on the Sun, called **solar flares.** Rooted to the same magnetic phenomena, the solar flares occur in complex sunspot groups. Within only a few minutes, temperatures in a compact region may soar to 5×10^6 K and vast quantities of particles and radiation—including as much material as is in the prominence shown in Figure 16-29—are blasted out into space. These eruptions can also cause disturbances that spread outward in the solar atmosphere, like the ripples that appear when you drop a rock into a pond.

While the external effects of the Sun's magnetism are on full display with solar flares, the advancing magnetic fields can also be discerned below the surface. Observations from NASA's Solar Dynamics Observatory employed helioseismology to detect magnetic fields about 65,000 km below the surface (that distance is

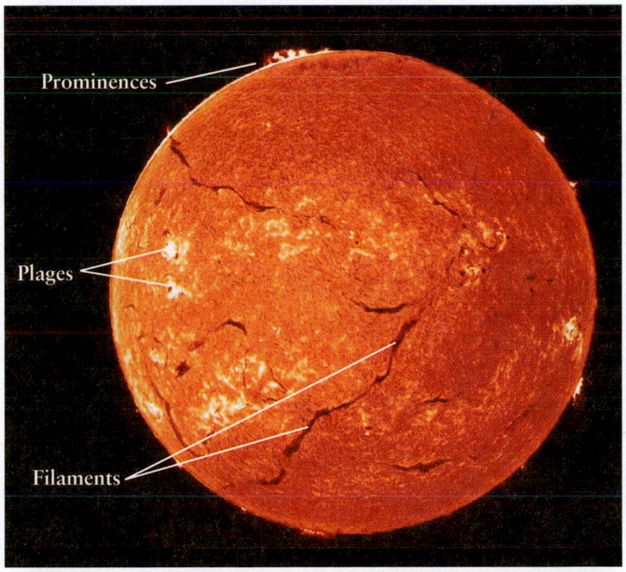

FIGURE 16-27 R I **V** U X G

The Active Sun Seen through an H_α Filter This image was made using a red filter that only passes light at a wavelength of 656 nm. The spectrum of the photosphere has an absorption line at this wavelength and so appears dark. Hence, this filter reveals the photosphere and corona. Prominences, plages, and filaments are associated with strong magnetic fields. (NASA)

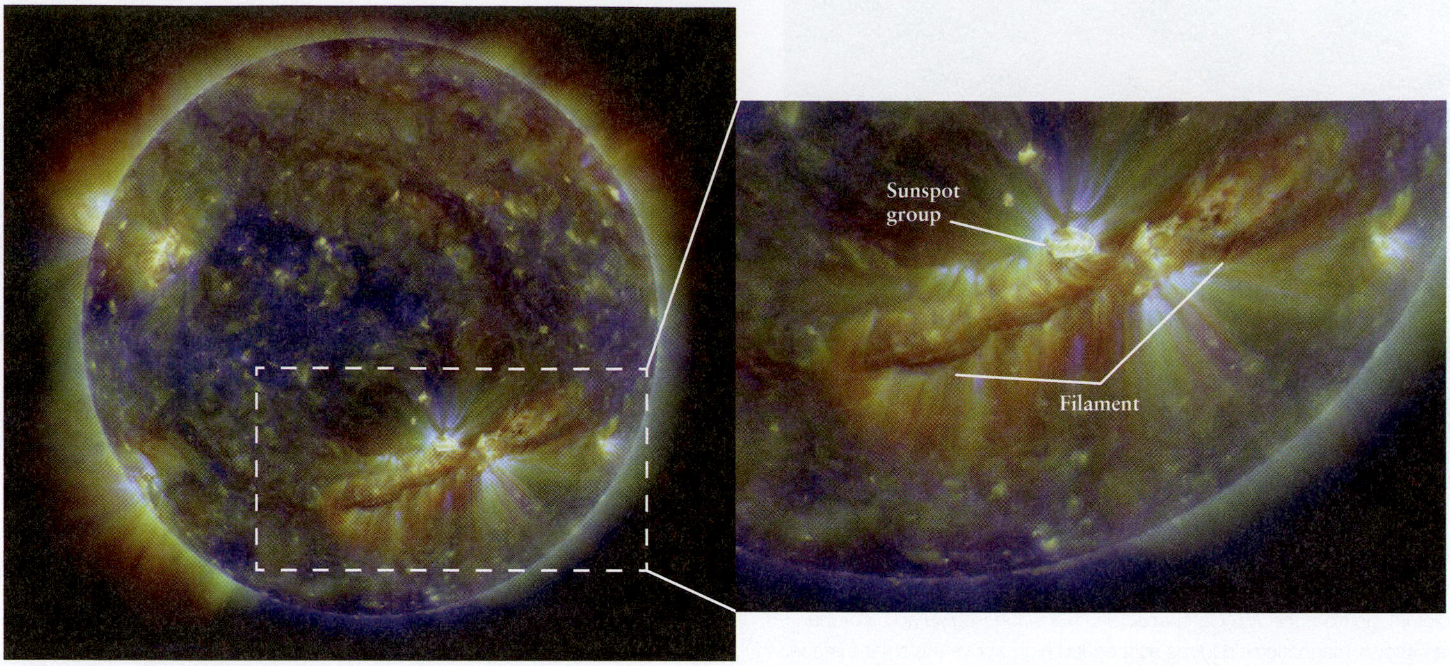

FIGURE 16-28 R I V U X G

The Sun in Ultraviolet Light This false-color image was made in ultraviolet light by the Solar Dynamics Observatory. A very long filament is seen cutting across the Sun's southern hemisphere; it is longer than the distance from Earth to the Moon! The filament is confined by a strong magnetic field, and above the filament is a bright sunspot group. (NASA)

about 2½ times Earth's circumference). As predicted, the effects of these magnetic fields emerged several days later as sunspots.

The most energetic flares carry as much as 10^{30} joules of energy, equivalent to 10^{14} one-megaton nuclear weapons being exploded at once! However, the energy of a solar flare does not come from thermonuclear fusion in the solar atmosphere; instead, it appears to be released from the intense magnetic field around a sunspot group.

> A solar flare can have as much energy as 100 trillion nuclear bombs

As energetic as solar flares are, they are dwarfed by **coronal mass ejections.** One such event is shown in Figure 16-30. If directed toward Earth, these events are large enough to affect us directly (Figure 16-31). In a coronal mass ejection, more than 10^{12} kilograms of high-temperature coronal gas is blasted into space at speeds of hundreds of kilometers per second. That is about 200 times the mass of the largest Egyptian pyramid, and a typical coronal mass ejection lasts a few hours. These explosive events seem to be related to large-scale alterations in the Sun's magnetic field, like the magnetic reconnection shown in Figure 16-26b. Coronal mass ejections occur every few months; smaller eruptions may occur almost daily.

If a solar flare or coronal mass ejection happens to be aimed toward Earth, a stream of high-energy electrons and nuclei reaches us a few days later (Figure 16-31b); this is part of space weather (Section 16-9). When this plasma arrives, it can interfere with satellites, pose a health hazard to astronauts in orbit, and disrupt electrical and communications equipment on Earth's surface. Telescopes on Earth and on board spacecraft now monitor the Sun continuously to provide warnings of dangerous levels of solar particles. By design, the walls of the International Space Station are thick enough to protect astronauts from material ejected during regular solar activity, but extra precautions are needed during a stronger solar event.

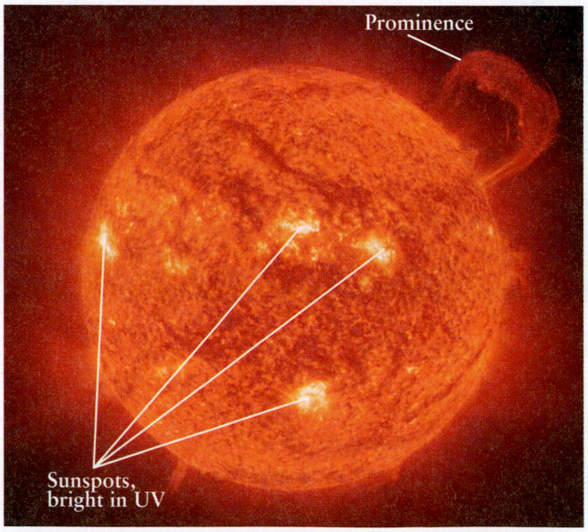

FIGURE 16-29 R I V U X G

A Solar Prominence A huge prominence arches above the solar surface in this ultraviolet image from the *SOHO* spacecraft. The image was recorded using ultraviolet light at a wavelength of 30.4 nm, emitted by singly ionized helium atoms at a temperature of about 60,000 K. By comparison, the material within the arches in Figure 16-25 reaches temperatures in excess of 2×10^6 K. (NASA)

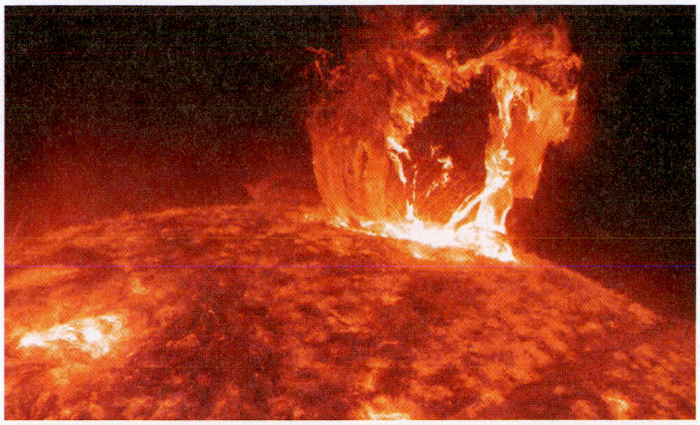

FIGURE 16-30 R I V U X G

Coronal Mass Ejection The Sun produced a huge coronal mass ejection over several hours on April 16, 2012. Not all of the gas was shot into the solar system, and some material from this event was observed crashing back onto the solar surface. (This ejection was not directed toward Earth.) (NASA)

The numbers of plages, filaments, solar flares, and coronal mass ejections all vary with the same 11-year cycle as sunspots. But unlike sunspots, coronal mass ejections never completely cease, even when the Sun is at its quietest. Astronomers are devoting substantial effort to understanding these and other aspects of our dynamic Sun.

CONCEPTCHECK 16-19

Which of the following phenomena is the most energetic: prominences, solar flares, or coronal mass ejections?

Answer appears at the end of the chapter.

KEY WORDS

The term preceded by an asterisk (*) is discussed in Box 16-1.

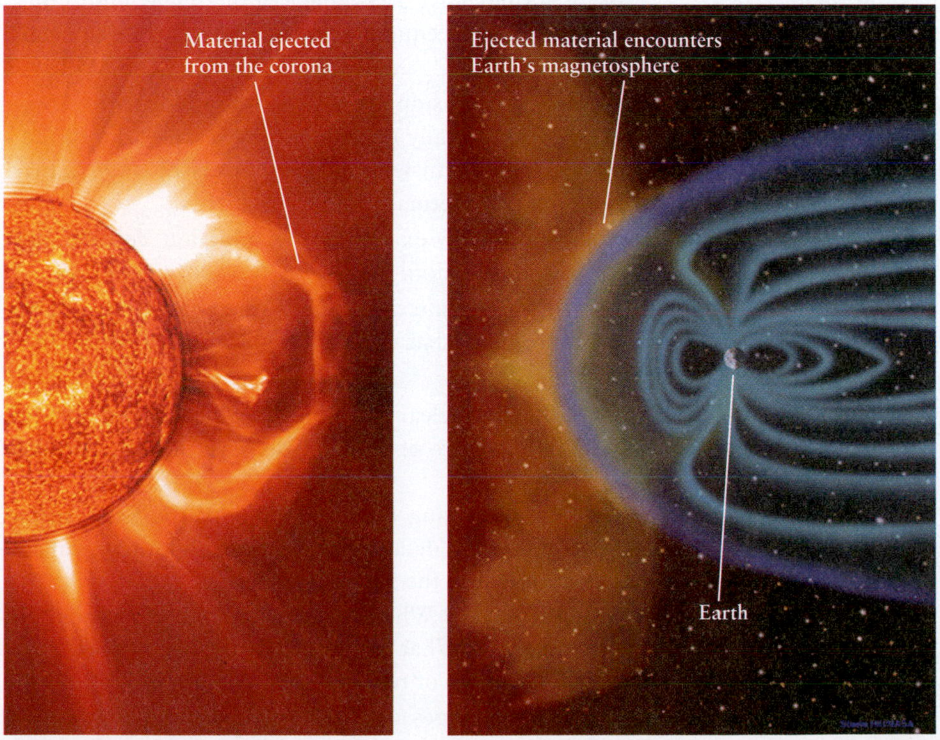

(a) A coronal mass ejection (b) Two to four days later

FIGURE 16-31 R I V U X G

A Sun-Earth Connection **(a)** *SOHO* recorded this coronal mass ejection in an X-ray image. (The image of the Sun itself was made at ultraviolet wavelengths.) **(b)** In this artist's illustration we see that within two to four days the fastest-moving ejected material reaches a distance of 1 AU from the Sun. Most particles are deflected by Earth's magnetosphere, but some are able to reach Earth where they can damage satellites and our electrical power grid. (The ejection shown in (a) was not aimed toward Earth and did not affect us.) (SOHO/EIT/LASCO/ESA/NASA)

KEY IDEAS

Hydrogen Fusion in the Sun's Core: The Sun's energy is produced by hydrogen fusion, a sequence of thermonuclear reactions in which four hydrogen nuclei combine to produce a single helium nucleus.

• The energy released in a nuclear reaction corresponds to a slight reduction of mass according to Einstein's equation $E = mc^2$.

• Thermonuclear fusion occurs only at very high temperatures; for example, hydrogen fusion occurs only at temperatures in excess of about 10^7 K. In the Sun, fusion occurs only in the dense, hot core.

Models of the Sun's Interior: A theoretical description of a star's interior can be calculated using the laws of physics.

• The standard model of the Sun suggests that hydrogen fusion takes place in a core extending from the Sun's center to about 0.25 solar radius.

• The core is surrounded by a radiative zone extending to about 0.71 solar radius. In this zone, energy travels outward through radiative diffusion.

• The radiative zone is surrounded by a rather opaque convective zone of gas at relatively low temperature and pressure. In this zone, energy travels outward primarily through convection.

Solar Neutrinos and Helioseismology: Conditions in the solar interior can be inferred from measurements of solar neutrinos and of solar vibrations.

• Neutrinos emitted in thermonuclear reactions in the Sun's core have been detected, but in smaller numbers than expected. Recent neutrino experiments explain why this is so.

• Helioseismology is the study of how the Sun vibrates. These vibrations have been used to infer pressures, densities, chemical compositions, and rotation rates within the Sun.

The Sun's Atmosphere: The Sun's atmosphere has three main layers: the photosphere, the chromosphere, and the corona. Everything below the solar atmosphere is called the solar interior.

• The visible surface of the Sun, the photosphere, is the lowest layer in the solar atmosphere. Its spectrum is similar to that of a blackbody at a temperature of 5800 K. Convection in the photosphere produces granules.

• Above the photosphere is a layer of less dense but higher-temperature gases called the chromosphere. Spicules extend upward from the chromosphere into the corona.

• The outermost layer of the solar atmosphere, the corona, is made of very high-temperature gases at extremely low density. Activity in the corona includes coronal mass ejections and coronal holes. The solar corona blends into the solar wind at great distances from the Sun.

The Active Sun: The Sun's surface features vary in an 11-year cycle. This is related to a 22-year cycle in which the surface magnetic field increases, decreases, and then increases again with the opposite polarity.

• Sunspots are relatively cool regions produced by local concentrations of the Sun's magnetic field. The average number of sunspots increases and decreases in a regular cycle of approximately 11 years, with reversed magnetic polarities from one 11-year cycle to the next. Two such cycles make up the 22-year solar cycle.

• The magnetic-dynamo model suggests that many features of the solar cycle are due to changes in the Sun's magnetic field. These changes are caused by convection and the Sun's differential rotation.

• A solar flare is a brief eruption of hot, ionized gases from a sunspot group. A coronal mass ejection is a much larger eruption that involves immense amounts of gas from the corona.

QUESTIONS

Review Questions

1. What is meant by the luminosity of the Sun?
2. What is Kelvin-Helmholtz contraction? Why is it ruled out as a source of the present-day Sun's energy?
3. TUTORIAL 16-1 Why is it impossible for the burning of substances like coal to be the source of the Sun's energy?
4. What is hydrogen fusion? Why is hydrogen fusion fundamentally unlike the burning of a log in a fireplace?
5. If the electric force between protons were somehow made stronger, what effect would this have on the temperature required for thermonuclear fusion to take place?
6. Why do thermonuclear reactions occur only in the Sun's core, not in its outer regions?
7. Describe how the net result of the reactions shown in the *Cosmic Connections* figure (Section 16-1) is the conversion of four protons into a single helium nucleus. What other particles are produced in this process? How many of each particle are produced?
8. TUTORIAL 16-2 Give an everyday example of hydrostatic equilibrium. Give an example of thermal equilibrium. Explain how these equilibrium conditions apply to each example.
9. If thermonuclear fusion in the Sun were suddenly to stop, what would eventually happen to the overall radius of the Sun? Justify your answer using the ideas of hydrostatic equilibrium and thermal equilibrium.
10. Give some everyday examples of conduction, convection, and radiative diffusion.
11. Describe the Sun's interior. Include references to the main physical processes that occur at various depths within the Sun.
12. Suppose thermonuclear fusion in the Sun stopped abruptly. Would the intensity of sunlight decrease just as abruptly? Why or why not?
13. Explain how studying the oscillations of the Sun's surface can give important, detailed information about physical conditions deep within the Sun.
14. What is a neutrino? Why is it useful to study neutrinos coming from the Sun? What do they tell us that cannot be learned from other avenues of research?
15. Unlike all other types of telescopes, neutrino detectors are placed deep underground. Why?

16. What was the solar neutrino problem? What solution to this problem was suggested by the results from the Sudbury Neutrino Observatory?
17. Describe the dangers in attempting to observe the Sun. How have astronomers learned to circumvent these observational problems?
18. Briefly describe the three layers that make up the Sun's atmosphere. In what ways do they differ from one another?
19. What is solar granulation? Describe how convection gives rise to granules.
20. High-resolution spectroscopy of the photosphere reveals that absorption lines are blueshifted in the spectrum of the central, bright regions of granules but are redshifted in the spectrum of the dark boundaries between granules. Explain how these observations show that granulation is due to convection.
21. What is the difference between granules and supergranules?
22. What are spicules? Where are they found? How can you observe them? What causes them?
23. How do astronomers know that the temperature of the corona is so high?
24. TUTORIAL 16-3 How do astronomers know when the next sunspot maximum and minimum will occur?
25. Why do astronomers say that the solar cycle is really 22 years long, even though the number of sunspots varies over an 11-year period?
26. Explain how the magnetic-dynamo model accounts for the solar cycle.
27. Describe one explanation for why the corona has a higher temperature than the chromosphere.
28. Why should solar flares and coronal mass ejections be a concern for businesses that use telecommunication satellites?

Advanced Questions

Questions preceded by an asterisk (*) involve the topic discussed in Box 16-1.

Problem-solving tips and tools

You may have to review Wien's law and the Stefan-Boltzmann law, which are the subjects of Section 5-4. Section 5-5 discusses the properties of photons. As we described in Box 5-2, you can simplify calculations by taking ratios, such as the ratio of the flux from a sunspot to the flux from the undisturbed photosphere. When you do this, all the cumbersome constants cancel out. Figure 5-7 shows the various parts of the electromagnetic spectrum. We introduced the Doppler effect in Section 5-9 and Box 5-6. For information about the planets, see Table 7-1.

29. Calculate how much energy would be released if each of the following masses were converted *entirely* into their equivalent energy: (**a**) a carbon atom with a mass of 2×10^{-26} kg, (**b**) 1 kg, and (**c**) a planet as massive as Earth (6×10^{24} kg).
30. Use the luminosity of the Sun (given in Table 16-1) and the answers to the previous question to calculate how long the Sun must shine in order to release an amount of energy equal to that produced by the complete mass-to-energy conversion of (**a**) a carbon atom, (**b**) 1 km, and (**c**) Earth.
31. Assuming that the current rate of hydrogen fusion in the Sun remains constant, what fraction of the Sun's mass will be converted into helium over the next 5 billion years? How will this affect the overall chemical composition of the Sun?
32. (**a**) Estimate how many kilograms of hydrogen the Sun has consumed over the past 4.56 billion years, and estimate the amount of mass that the Sun has lost as a result. Assume that the Sun's luminosity has remained constant during that time. (**b**) In fact, the Sun's luminosity when it first formed was only about 70% of its present value. With this in mind, explain whether your answers to part (a) are an overestimate or an underestimate.

*33. To convert 1 kg of hydrogen (^{1}H) into helium (^{4}He) as described in Box 16-1, you must start with 1.5 kg of hydrogen. Explain why, and explain what happens to the other 0.5 kg. (*Hint:* How many ^{1}H nuclei are used to make the two ^{3}He nuclei shown in the *Cosmic Connections* figure in Section 16-1? How many of these ^{1}H nuclei end up being incorporated into the ^{4}He nucleus shown in the figure?)

*34. (**a**) A positron has the same mass as an electron (see Appendix 7). Calculate the amount of energy released by the annihilation of an electron and positron. (**b**) The products of this annihilation are two photons, each of equal energy. Calculate the wavelength of each photon, and confirm from Figure 5-7 that this wavelength is the gamma-ray range.

35. Sirius is the brightest star in the night sky. It has a luminosity of 23.5 $L_\odot$, that is, it is 23.5 times as luminous as the Sun and burns hydrogen at a rate 23.5 times greater than the Sun. How many kilograms of hydrogen does Sirius convert into helium each second?
36. (Refer to the preceding question.) Sirius has 2.3 times the mass of the Sun. Do you expect that the lifetime of Sirius will be longer, shorter, or the same length as that of the Sun? Explain your reasoning.
37. (**a**) If the Sun were not in a state of hydrostatic equilibrium, would its diameter remain the same? Explain your reasoning. (**b**) If the Sun were not in a state of thermal equilibrium, would its luminosity remain the same? What about its surface temperature? Explain your reasoning.
38. Using the mass and size of the Sun given in Table 16-1, verify that the average density of the Sun is 1410 kg/m^3. Compare your answer with the average densities of the Jovian planets.
39. Use the data in Table 16-2 to calculate the average density of material within 0.1 $R_\odot$ of the center of the Sun. (You will need to use the mass and radius of the Sun as a whole, given in Table 16-1.) Explain why your answer is not the same as the density at 0.1 $R_\odot$ given in Table 16-2.
40. In a typical solar oscillation, the Sun's surface moves up or down at a maximum speed of 0.1 m/s. An astronomer sets out to measure this speed by detecting the Doppler shift of an

absorption line of iron with wavelength 557.6099 nm. What is the maximum wavelength shift that she will observe?

41. Explain why the results from the Sudbury Neutrino Observatory (SNO) only provide an answer to the solar neutrino problem for relatively high-energy neutrinos. (*Hint:* Can SNO detect solar neutrinos of all energies?)

42. The amount of energy required to dislodge the extra electron from a negative hydrogen ion is 1.2×10^{-19} J. (**a**) The extra electron can be dislodged if the ion absorbs a photon of sufficiently short wavelength. (Recall from Section 5-5 that the higher the energy of a photon, the shorter its wavelength.) Find the longest wavelength (in nm) that can accomplish this. (**b**) In what part of the electromagnetic spectrum does this wavelength lie? (**c**) Would a photon of visible light be able to dislodge the extra electron? Explain. (**d**) Explain why the photosphere, which contains negative hydrogen ions, is quite opaque to visible light but is less opaque to light with wavelengths longer than the value you calculated in (a).

43. Astronomers often use an H_α filter to view the chromosphere. Explain why this can also be accomplished with filters that are transparent only to the wavelengths of the H and K lines of ionized calcium. (*Hint:* The H and K lines are dark lines in the spectrum of the photosphere.)

44. Calculate the wavelengths at which the photosphere, chromosphere, and corona emit the most radiation. Explain how the results of your calculations suggest the best way to observe these regions of the solar atmosphere. (*Hint:* Treat each part of the atmosphere as a perfect blackbody. Assume average temperatures of 50,000 K and 1.5×10^6 K for the chromosphere and corona, respectively.)

45. On November 15, 1999, the planet Mercury passed in front of the Sun as seen from Earth. The *TRACE* spacecraft made these time-lapse images of this event using ultraviolet light (top) and visible light (bottom). (Mercury moved from left to right in these images. The time between successive views of Mercury is 6 to 9 minutes.) Explain why the Sun appears somewhat larger in the ultraviolet image than in the visible-light image.

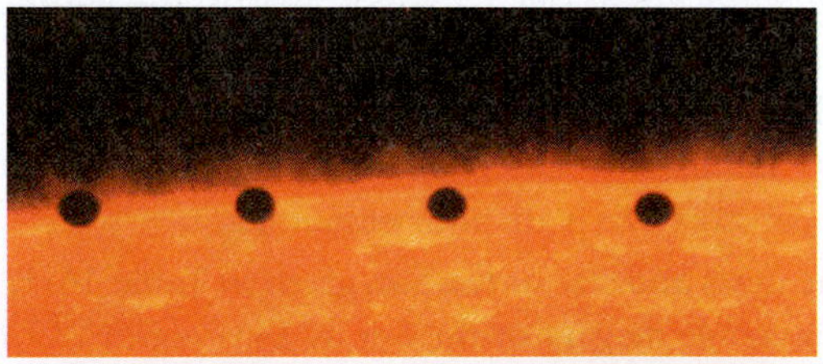

R I V **U** X G

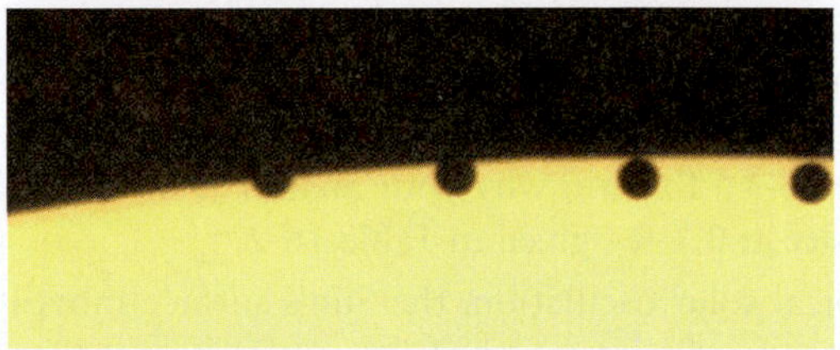

R I **V** U X G

(K. Schrijver, Stanford-Lockheed Institute for Space Research, TRACE, and NASA)

46. The moving images on a television set that uses a tube (as opposed to a LCD or plasma flat-screen TV) are made by a fast-moving electron beam that sweeps over the back of the screen. Explain why placing a strong magnet next to the screen distorts the picture. (*Caution:* Do not try this with your television; it can cause permanent damage.)

47. Find the wavelength of maximum emission of the umbra of a sunspot and the wavelength of maximum emission of a sunspot's penumbra. In what part of the electromagnetic spectrum do these wavelengths lie?

48. (**a**) Find the ratio of the energy flux from a patch of a sunspot's penumbra to the energy flux from an equally large patch of undisturbed photosphere. Which patch is brighter? (**b**) Find the ratio of the energy flux from a patch of a sunspot's penumbra to the energy flux from an equally large patch of umbra. Again, which patch is brighter?

49. Suppose that you want to determine the Sun's rotation rate by observing its sunspots. Is it necessary to take Earth's orbital motion into account? Why or why not?

50. (**a**) Using a ruler to make measurements of Figure 16-26, determine how far the arches in that figure extend above the Sun's surface. The diameter of Earth is 12,756 km. (**b**) In Figure 16-26 (an ultraviolet image) the photosphere appears dark compared to the arches. Explain why.

51. The amount of visible light emitted by the Sun varies only a little over the 11-year sunspot cycle. But the number of X-rays emitted by the Sun can be 10 times greater at solar maximum than at solar minimum. Explain why these two types of radiation should be so different in their variability.

Discussion Questions

52. Discuss the extent to which cultures around the world have worshipped the Sun as a deity throughout history. Why do you suppose there has been such widespread veneration?

53. In the movie *Star Trek IV: The Voyage Home,* the starship *Enterprise* flies on a trajectory that passes close to the Sun's surface. What features should a real spaceship have to survive such a flight? Why?

54. Discuss some of the difficulties in correlating solar activity with changes in Earth's climate.

55. Describe some of the advantages and disadvantages of observing the Sun (**a**) from space and (**b**) from Earth's south pole. What kinds of phenomena and issues might solar astronomers want to explore from these locations?

Web/eBook Questions

56. Search the World Wide Web for information about features in the solar atmosphere called *sigmoids.* What are they? What causes them? How might they provide a way to predict coronal mass ejections?

57. VIDEO 16.1 **Determining the Lifetime of a Solar Granule.** Access and view the video "Granules on the Sun's Surface" in Chapter 16 of the *Universe* Web site or eBook. Your task is to determine the approximate lifetime of a solar granule on the photosphere. Select an area, then slowly and rhythmically repeat "start, stop, start, stop" until you can consistently predict the appearance and disappearance of granules. While keeping your rhythm, move to a different area of the video and continue monitoring the appearance and disappearance of granules. When you are confident you have the timing right, move your eyes (or use a partner) to look at the clock shown in the video. Determine the length of time between the appearance and disappearance of the granules and record your answer.

ACTIVITIES

Observing Projects

Observing tips and tools

At the risk of repeating ourselves, we remind you to ***never look directly at the Sun, because it can easily cause permanent blindness.*** You can view the Sun safely without a telescope just by using two pieces of white cardboard. First, use a pin to poke a small hole in one piece of cardboard; this will be your "lens," and the other piece of cardboard will be your "viewing screen." Hold the "lens" piece of cardboard so that it is face-on to the Sun and sunlight can pass through the hole. With your other hand, hold the "viewing screen" so that the sunlight from the "lens" falls on it. Adjust the distance between the two pieces of cardboard so that you see a sharp image of the Sun on the "viewing screen." This image is perfectly safe to view. It is actually possible to see sunspots with this low-tech apparatus.

For a better view, use a telescope with a solar filter that fits on the front of the telescope. A standard solar filter is a piece of glass coated with a thin layer of metal to give it a mirror-like appearance. This coating reflects almost all the sunlight that falls on it, so that only a tiny, safe amount of sunlight enters the telescope. An H_α filter, which looks like a red piece of glass, keeps the light at a safe level by admitting only a very narrow range of wavelengths. (Filters that fit on the back of the telescope are *not* recommended. The telescope focuses concentrated sunlight on such a filter, heating it and making it susceptible to cracking—and if the filter cracks when you are looking through it, your eye will be ruined instantly and permanently.)

To use a telescope with a solar filter, first aim the telescope away from the Sun, then put on the filter. Keep the lens cap on the telescope's secondary wide-angle "finder scope" (if it has one), because the heat of sunlight can fry the finder scope's optics. Next, aim the telescope toward the Sun, using the telescope's shadow to judge when you are pointed in the right direction. You can then safely look through the telescope's eyepiece. When you are done, make sure you point the telescope away from the Sun before removing the filter and storing the telescope.

Note that the amount of solar activity that you can see (sunspots, filaments, flares, prominences, and so on) will depend on where the Sun is in its 11-year sunspot cycle.

58. Use a telescope with a solar filter to observe the surface of the Sun. Do you see any sunspots? Sketch their appearance. Can you distinguish between the umbrae and penumbrae of the sunspots? Can you see limb darkening? Can you see any granulation?

59. If you have access to an H_α filter attached to a telescope especially designed for viewing the Sun safely, use this instrument to examine the solar surface. How does the appearance of the Sun differ from that in white light? What do sunspots look like in H_α Can you see any prominences? Can you see any filaments? Are the filaments in the H_α image near any sunspots seen in white light? (Note that the amount of activity that you see will be much greater at some times during the solar cycle than at others.)

60. Use the *Starry Night*™ program to examine simulations of various features that appear on the surface of the Sun. Select **Favourites > Explorations > Sun** to show a simulated view of the visible surface of the Sun as it might appear from a spacecraft. **Stop** time flow and use the **Location Scroller** to examine this surface. (**a**) Which layer of the Sun's atmosphere is shown in this part of the simulation? (**b**) List the different features that are visible in this view of the Sun's surface. (**c**) Click and hold the **Decrease current elevation** button in the toolbar to move to a location on the surface of the Sun, from which you can look out into the chromosphere. (The **Viewing Location** panel will indicate the location on the Sun's surface.) This simulated view of the chromosphere is at the color of the wavelength of hydrogen light. The opacity of the gas at this wavelength means that you can see the structure of the hot chromosphere that lies above the visible surface. Use the hand tool or cursor keys to change the gaze direction to view different features of the Sun, zooming in when necessary for a closer look at features on the horizon. (**d**) Provide a detailed description of the various features visible in this simulation of the Sun's surface. You can see current solar images from both ground and space-based solar telescopes by opening the **LiveSky** pane if you have an Internet connection on your computer.

61. Use *Starry Night*™ to measure the Sun's rotation. Select **Favourites > Explorations > Solar Rotation** to display the Sun as seen from about 0.008 AU above its surface, well inside the orbit of Mercury. Use the time controls to stop the Sun's rotation at a time when a line of longitude on the Sun makes a straight line between the solar poles, preferably a line crossing a recognizable solar feature. Note the date and time. **Run Time Forward** and adjust the date and time to place this selected meridian in this position again. (**a**) What is the rotation rate of the Sun as shown in *Starry Night*™? This demonstration does not show one important feature of the Sun, namely its differential rotation, where the equator of this fluid body rotates faster than the polar regions. (**b**) To which region of the Sun does your measured rotation rate refer? It is this differential rotation that is thought to generate the magnetic fields and active regions that make the Sun an active star. Occasional emission of high-energy particles from these active regions can disturb Earth's environment and disrupt electrical transmission systems. Examine this image of the Sun and compare it to real images seen in textbooks, the

Internet, or from the links in the **Solar Images** layer in the **LiveSky** pane. (c) In particular, how does the distribution of sunspots and active regions on this image compare to the distribution of these regions on the real Sun?

Collaborative Exercises

62. Figure 16-20 shows variations in the average latitude of sunspots. Estimate the average latitude of sunspots in the year you were born and estimate the average latitude on your twenty-first birthday. Make rough sketches of the Sun during those years to illustrate your answers.
63. Create a diagram showing a sketch of how limb darkening on the Sun would look different if the Sun had either a thicker or thinner photosphere. Be sure to include a caption explaining your diagram.
64. Solar granules, shown in Figure 16-9, are about 1000 km across. What city is about that distance away from where you are right now? What city is that distance from the birthplace of each group member?
65. Magnetic arches in the corona are shown in Figure 16-26a. How many Earths high are these arches, and how many Earths could fit inside one arch?

ANSWERS

ConceptChecks

ConceptCheck 16-1: The Sun emits most of its energy in the form of visible light.

ConceptCheck 16-2: At the extremely high temperatures and pressures existing in the Sun's core, hydrogen nuclei can move fast enough to overcome the repulsive force from their positive electric charges and fuse together.

ConceptCheck 16-3: When 1 kg of hydrogen combines to form helium, the vast majority of the mass ends up in helium atoms, with only 0.7% of the original mass converted into energy.

ConceptCheck 16-4: Astronomers use the current energy output of the Sun to estimate how fast the Sun is consuming its usable fuel (hydrogen). Then, the amount of hydrogen fuel remaining in the Sun's core indicates how much longer the Sun will shine.

ConceptCheck 16-5: Because pressure in the Sun's core is due to the downward pushing weight of the overlying mass of material, having less mass pressing down would result in a lower pressure at the core.

ConceptCheck 16-6: The energy transport process of conduction occurs when energy moves through a relatively dense material by hot material transferring its kinetic energy to cooler material through direct contact. The Sun's density is simply too low for conduction to be an important process in transferring energy from one part of the Sun to another part.

ConceptCheck 16-7: Only the temperature decreases with increasing distance from the Sun's central core.

ConceptCheck 16-8: As shown in Figure 16-5, sound waves do penetrate into the Sun, which allows helioseismologists to study the solar interior.

ConceptCheck 16-9: By carefully monitoring how sound waves move through the Sun, astronomers are able to deduce the amount of helium in the Sun's core and convective zone, and the thickness of various zones.

ConceptCheck 16-10: Kamiokande could only detect one of three possible types of neutrinos, and this was the type emitted by the Sun. However, along the way, two-thirds of these neutrinos transformed into the other types and were undetectable by Kamiokande. SNO, on the other hand, was able to detect all three types of neutrinos.

ConceptCheck 16-11: The energy transport process of convection causes warmer material to rise to the surface and cooler material to sink back into the photosphere, giving the photosphere a granule-like appearance.

ConceptCheck 16-12: Figure 16-11 is taken with an H_α filter that only allows red light to pass through and form the image. This method is used to block out the Sun's other light that would overwhelm the camera. Since the H_α line is red (at 656.3 nm), the spicules appear red, but they emit other wavelengths as well.

ConceptCheck 16-13: The corona has an extremely low density and can only be observed when the more dominant photosphere is blocked, such as during a solar eclipse.

ConceptCheck 16-14: The solar wind must escape through the corona, and this occurs in greater abundance through holes in the corona that contain less gas.

ConceptCheck 16-15: The photosphere surrounding a sunspot has a temperature of about 5800 K, which far outshines the relatively cooler sunspot region, resulting in the sunspots appearing to be quite dark in comparison.

ConceptCheck 16-16: The length of time between large numbers of sunspots to few numbers of sunspots and back to large numbers of sunspots averages about 11 years. However, the magnetic character of sunspots flips every 11-year cycle, suggesting an overarching 22-year sunspot cycle.

ConceptCheck 16-17: According to the magnetic-dynamo model, the sunspot cycle is a result of twisting magnetic fields. In the event the Sun was turning faster, the twisting would occur more quickly and the length of the sunspot cycle would decrease.

ConceptCheck 16-18: The glowing plasma tends to follow the pathways created by the Sun's invisible magnetic field lines, which form curved arches above the Sun's surface.

ConceptCheck 16-19: Coronal mass ejections are many times more energetic than any other event on the Sun and, when directed at Earth, can cause severe problems with telecommunications, electrical power distribution, and radiation health hazards for astronauts working in space.

CalculationChecks

CalculationCheck 16-1: According to Einstein's equation that $E = mc^2$, a mass of 5 kg is equivalent to 5 kg $\times$ $(3 \times 10^8 \text{ m/s})^2 =$ 15×10^8 joules, which is equivalent to burning about 100,000 metric tons of coal!

CalculationCheck 16-2: According to Figure 16-3, the Sun's core temperature is about 16 million Kelvins. At a distance of 50% of the Sun's radius, the Sun's temperature has dropped to 4 million Kelvins, which is a drop of about 75%.

Stars come in various colors from red to blue. Different temperatures lead to different colors, and stars come in many sizes and brightness. This is the central region of the star cluster Omega Centauri, which is about 15,800 ly from Earth. (NASA, ESA, and the Hubble SM4 ERO Team) R I **V** U X G

CHAPTER 17

The Nature of the Stars

LEARNING GOALS

By reading the sections of this chapter, you will learn

17-1 How we can measure the distances to the stars

17-2 How we measure a star's brightness and luminosity

17-3 The magnitude scale for brightness and luminosity

17-4 How a star's color indicates its temperature

17-5 How a star's spectrum reveals its chemical composition

17-6 How we can determine the sizes of stars

17-7 How H-R diagrams summarize our knowledge of the stars

17-8 How we can deduce a star's size from its spectrum

17-9 How we can use binary stars to measure the masses of stars

17-10 How we can learn about binary stars in very close orbits

17-11 What eclipsing binaries are and what they tell us about the sizes of stars

To the unaided eye, the night sky is spangled with several thousand stars, each appearing as a bright pinpoint of light. With a pair of binoculars, you can see some 10,000 other, fainter stars; with a 15-cm (6-in.) telescope, the total rises to more than 2 million. Astronomers now know that there are more than 100 billion (10^{11}) stars in our Milky Way Galaxy alone.

But what are these distant pinpoints? To the great thinkers of ancient Greece, the stars were bits of light embedded in a vast sphere with Earth at the center. They thought the stars were composed of a mysterious "fifth element," quite unlike anything found on Earth.

Today, we know that the stars are made of the same chemical elements found on Earth. We know their sizes, their temperatures, their masses, and something of their internal structures. We understand, too, why the stars in the accompanying image come in a range of beautiful colors: Blue stars have high surface temperatures, while the surface temperatures of red and yellow stars are relatively low.

How have we learned these things? How can we know the nature of the stars, objects so distant that their light takes years or centuries to reach us? In this chapter, we will learn about the measurements and calculations that astronomers make to determine the properties of stars. We will also take a first look at the Hertzsprung-Russell diagram, an important tool that helps astronomers systematize the wealth of available information about the stars. In later chapters, we will use this diagram to help us understand how stars are born, evolve, and eventually die.

17-1 Careful measurements of the parallaxes of stars reveal their distances

The vast majority of stars are objects very much like the Sun. This understanding followed from the discovery that the stars are tremendously far from us, at distances so great that their light takes years to reach us. Because the stars at night are clearly visible to the naked eye despite these huge distances, it must be that the **luminosity** of the stars—that is, how much light energy they emit into space per second—is comparable to or greater than that of the Sun. Just as for the Sun, the only explanation for such tremendous luminosities is that thermonuclear reactions are occurring within the stars (see Section 16-1).

Clearly, then, it is important to know how distant the stars are. But how do we measure these distances? You might think these distances are determined by comparing how bright different stars appear. Perhaps the star Betelgeuse in the constellation Orion appears bright because it is relatively close, while the dimmer and less conspicuous star Polaris (the North Star, in the constellation Ursa Minor) is farther away.

But this line of reasoning is incorrect: Polaris is actually closer to us than Betelgeuse! How bright a star appears is *not* a good indicator of its distance. If you see a light on a darkened road, it could be a motorcycle headlight a kilometer away or a person holding a flashlight just a few meters away. In the same way, a bright star might be extremely far away but have an unusually high luminosity, and a dim star might be relatively close but have a rather low luminosity. Astronomers must use other techniques to determine the distances to the stars.

CONCEPTCHECK 17-1

Which appears brighter, a small handheld flashlight 10 cm away or a large spotlight atop a lighthouse 10 km away?

Answer appears at the end of the chapter.

When you are at position B, the tree appears to be in front of this mountain.
When you are at position A, the tree appears to be in front of this mountain.
Position A
Position B

FIGURE 17-1

Parallax Imagine looking at some nearby object (a tree) against a distant background (mountains). When you move from one location to another, the nearby object appears to shift with respect to the distant background scenery. This familiar phenomenon is called parallax.

Parallax and the Distances to the Stars

TUTORIAL 17-1

The most straightforward way of measuring stellar distances uses an effect called **parallax,** which is the apparent displacement of an object because of a change in the observer's point of view (Figure 17-1). To see how parallax works, hold your arm out straight in front of you. Now look at the hand on your outstretched arm, first with your left eye closed, then with your right eye closed. When you close one eye and open the other, your hand appears to shift back and forth against the background of more distant objects.

The closer the object you are viewing, the greater the parallax shift. To see this increased shift, repeat the experiment with your hand held closer to your face. Your brain analyzes such parallax shifts constantly as it compares the images from your left and right eyes, and in this way determines the distances to objects around you. This analysis is the basis for depth perception.

You measure distances around you by comparing the images from your left and right eyes—we find the distances to stars using the same principle

To measure the distance to a star, astronomers measure the parallax shift of the star using two points of view that are as far apart as possible—at opposite sides of Earth's orbit. The direction from Earth to a nearby star changes as our planet orbits the Sun, and the nearby star appears to move back and forth against the background of more distant stars (Figure 17-2). This motion is called **stellar parallax.** The parallax (p) of a star is equal to half the angle through which the star's apparent position shifts as Earth moves from one side of its orbit to the other. The larger the parallax p, the smaller the distance d to the star (compare Figure 17-2a with Figure 17-2b).

It is convenient to measure the distance d in **parsecs.** A star with a parallax angle of 1 second of arc ($p = 1$ arcsec) is at a distance of 1 parsec ($d = 1$ pc). (The word "parsec" is a contraction of the phrase "the distance at which a star has a *par*allax of one arc*sec*ond." Recall from Section 1-7 that 1 parsec equals 3.26 ly, 3.09×10^{13} km, or 206,265 AU; see Figure 1-14.) If the angle p is measured in arcseconds, then the distance d to the star in parsecs is given by the following equation:

Relation between a star's distance and its parallax

$$d = \frac{1}{p}$$

d = distance to a star, in parsecs

p = parallax angle of that star, in arcseconds

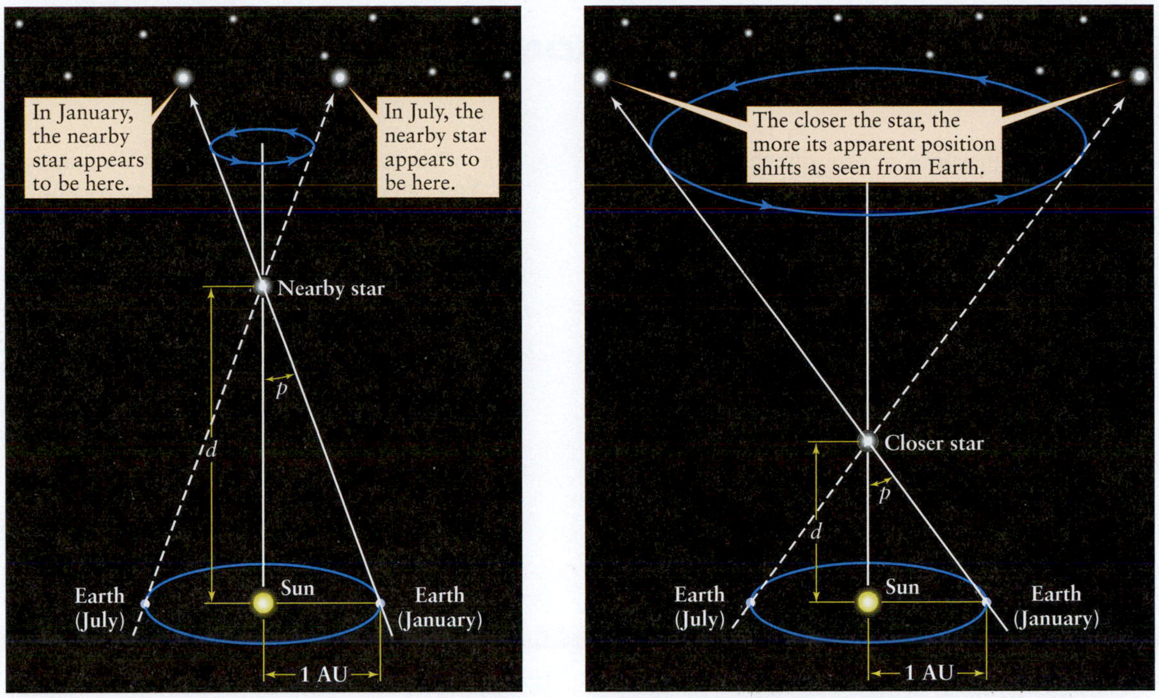

(a) Parallax of a nearby star

(b) Parallax of an even closer star

ANIMATION 17-1

FIGURE 17-2

Stellar Parallax **(a)** As Earth orbits the Sun, a nearby star appears to shift its position against the background of distant stars. The parallax (p) of the star is equal to half the angle that the star appears to shift as Earth moves from one side of its orbit to the other. **(b)** The closer the star is to us, the greater the parallax angle p. The distance d to the star (in parsecs) is equal to the reciprocal of the parallax angle p (in arcseconds): $d = 1/p$.

This simple relationship between parallax and distance in parsecs is one of the main reasons that astronomers usually measure cosmic distances in parsecs rather than light-years. For example, a star whose parallax is $p = 0.1$ arcsec is at a distance $d = 1/(0.1) = 10$ parsecs from Earth. Barnard's star, named for the American astronomer Edward E. Barnard, has a parallax of 0.547 arcsec. Hence, the distance to this star is

$$d = \frac{1}{p} = \frac{1}{0.547} = 1.83 \text{ pc}$$

Because 1 parsec is 3.26 ly, this distance can also be expressed as

$$d = 1.83 \text{ pc} \times \frac{3.26 \text{ ly}}{1 \text{ pc}} = 5.97 \text{ ly}$$

All known stars have parallax angles of less than 1 arcsecond. In other words, the closest star is more than 1 parsec away. Such small parallax angles are difficult to detect, so it was not until 1838 that the first successful parallax measurements were made by the German astronomer and mathematician Friedrich Wilhelm Bessel. He found the parallax angle of the star 61 Cygni to be just $1/3$ arcsec—equal to the angular diameter of a dime at a distance of 11 km, or 7 mi. He thus determined that this star is about 3 pc from Earth. (Modern measurements give a slightly smaller parallax angle, which means that 61 Cygni is actually more than 3 pc away.) The star Proxima Centauri has the largest known parallax angle, 0.772 arcsec, and hence is the closest known star (other than the Sun); its distance is $1/(0.772) = 1.30$ pc.

Appendix 4 at the back of this book lists all the stars within 4 pc of the Sun, as determined by parallax measurements. Most of these stars are far too dim to be seen with the naked eye, which is why their names are probably unfamiliar to you. By contrast, the majority of the familiar, bright stars in the nighttime sky (listed in Appendix 5) are so far away that their parallaxes cannot be measured from Earth's surface. They appear bright not because they are close, but because they are far more luminous than the Sun. The brightest stars in the sky are *not* necessarily the nearest stars!

CONCEPTCHECK 17-2

If an astronomer erroneously measures a parallax angle for a star to be smaller than it actually is, does the measurement suggest the star is farther away or closer to Earth than it actually is?

CALCULATIONCHECK 17-1

How many light-years away is Alpha Centauri if it has a parallax angle of 0.772 arcseconds?

Answers appear at the end of the chapter.

Measuring Parallax from Space

Parallax angles smaller than about 0.01 arcsec are extremely difficult to measure from Earth, in part because of the blurring effects of the atmosphere. Therefore, the parallax method used with ground-based telescopes can give fairly reliable distances only for stars nearer than

BOX 17-1 TOOLS OF THE ASTRONOMER'S TRADE

Stellar Motions

Stars can move through space in any direction. The **space velocity** of a star describes how fast and in what direction it is moving. As the accompanying figure shows, a star's space velocity v can be broken into components parallel and perpendicular to our line of sight.

The component of velocity perpendicular to our line of sight—that is, across the plane of the sky—is called the star's **tangential velocity** (v_t). To determine it, astronomers must know the distance to a star (d) and its **proper motion** (μ, the Greek letter mu), which is the number of arcseconds that the star appears to move per year on the celestial sphere. Proper motion does not repeat itself yearly, so it can be distinguished from the apparent back-and-forth motion due to parallax. In terms of a star's distance and proper motion, its tangential velocity (in km/s) is

$$v_t = 4.74\ \mu d$$

where μ is in arcseconds per year and d is in parsecs. For example, Barnard's star (Figure 17-3) has a proper motion of 10.358 arcseconds per year and a distance of 1.83 pc. Hence, its tangential velocity is

$$v_t = 4.74(10.358)(1.83) = 89.8 \text{ km/s}$$

The component of a star's motion parallel to our line of sight—that is, either directly toward us or directly away from us—is its **radial velocity** (v_r). It can be determined from measurements of the Doppler shifts of the star's spectral lines (see Section 5-9 and Box 5-6). If a star is approaching us, the wavelengths of all of its spectral lines are decreased (blueshifted); if the star is receding from us, the wavelengths are increased (redshifted). The radial velocity v_r is related to the wavelength shift by the equation

$$\frac{\lambda - \lambda_0}{\lambda_0} = \frac{v_r}{c}$$

In this equation, λ is the wavelength of light coming from the star, λ_0 is what the wavelength would be if the star were not moving, and c is the speed of light. As an illustration, a particular spectral line of iron in the spectrum of Barnard's star has a wavelength (λ) of 516.445 nm. As measured in a laboratory on Earth, the same spectral line has a wavelength (λ_0) of 516.629 nm. Thus, for Barnard's star, our equation becomes

$$\frac{516.445 \text{ nm} - 516.629 \text{ nm}}{516.629 \text{ nm}} = -0.000356 = \frac{v_r}{c}$$

Solving this equation for the radial velocity v_r, we find

$$v_r = (-0.000356)\ c = (-0.000356)(3.00 \times 10^5 \text{ km/s}) = -107 \text{ km/s}$$

The minus sign means that Barnard's star is moving toward us. You can check this interpretation by noting that the wavelength $\lambda = 516.445$ nm received from Barnard's star is less than the laboratory wavelength $\lambda_0 = 516.629$ nm; hence, the light from the star is blueshifted, which indeed means that the star is approaching. If the star were receding, its light would be redshifted, and its radial velocity would be positive.

The illustration shows that the tangential velocity and radial velocity form two sides of a right triangle. The long side (hypotenuse) of this triangle is the space velocity (v). From the Pythagorean theorem, the space velocity is

$$v = \sqrt{v_t^2 + v_r^2}$$

For Barnard's star, the space velocity is

$$v = \sqrt{(89.4 \text{ km/s})^2 + (-107 \text{ km/s})^2} = 140 \text{ km/s}$$

Therefore, Barnard's star is moving through space at a speed of 140 km/s (503,000 km/h, or 312,000 mi/h) relative to the Sun.

Determining the space velocities of stars is essential for understanding the structure of the Galaxy. Studies show that the stars in our local neighborhood are moving in wide orbits around the center of the Galaxy, which lies some 8000 pc (26,000 ly) away in the direction of the constellation Sagittarius (the Archer). While many of the orbits are roughly circular and lie in nearly the same plane, others are highly elliptical or steeply inclined to the galactic plane. We will see in Chapter 22 how the orbits of stars and gas clouds reveal the Galaxy's spiral structure.

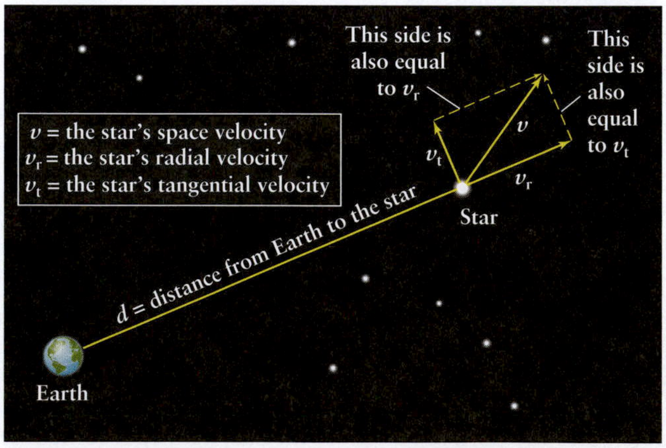

about 1/0.01 = 100 pc. But an observatory in space is unhampered by the atmosphere. Observations made from spacecraft therefore permit astronomers to measure even smaller parallax angles and thus determine the distances to more remote stars.

In 1989 the European Space Agency (ESA) launched the satellite *Hipparcos*, an acronym for *h*igh *p*recision *par*allax *co*llecting *s*atellite (and a commemoration of the ancient Greek astronomer Hipparchus, who created one of the first star charts). Over more than three years of observations, the telescope aboard *Hipparcos* was used to measure the parallaxes of 118,000 stars, some with an accuracy of 0.001 arcsecond. This telescope has enabled astronomers to determine stellar distances out to several hundred parsecs, and with much greater precision than has been possible with ground-based observations. In the years to come, astronomers will increasingly turn to space-based observations to determine stellar distances.

Unfortunately, most of the stars in the Galaxy are so far away that their parallax angles are too small to measure even with an orbiting telescope. Later in this chapter, we will discuss a technique that can be used to find the distances to these more remote stars. In Chapter 23 we will learn about other techniques that astronomers use to determine the much larger distances to galaxies beyond the Milky Way. These techniques also help us understand the overall size, age, and structure of the universe.

The Importance of Parallax Measurements

Because it can be used only on relatively close stars, stellar parallax might seem to be of limited usefulness. But parallax measurements are the cornerstone for all other methods of finding the distances to remote objects. These other methods require a precise and accurate knowledge of the distances to nearby stars, as determined by stellar parallax. Hence, any inaccuracies in the parallax angles for nearby stars can translate into substantial errors in measurement for the whole universe. To minimize these errors astronomers are continually trying to perfect their parallax-measuring techniques.

Stellar parallax is an *apparent* motion of stars caused by Earth's orbital motion around the Sun. But stars are not fixed objects, and they actually do move through space. As a result, stars change their positions on the celestial sphere (Figure 17-3), and they move either toward or away from the Sun. These motions are sufficiently slow, however, that changes in the positions of the stars are hardly noticeable over a human lifetime. Box 17-1 describes how astronomers study these motions and what insights they gain from these studies.

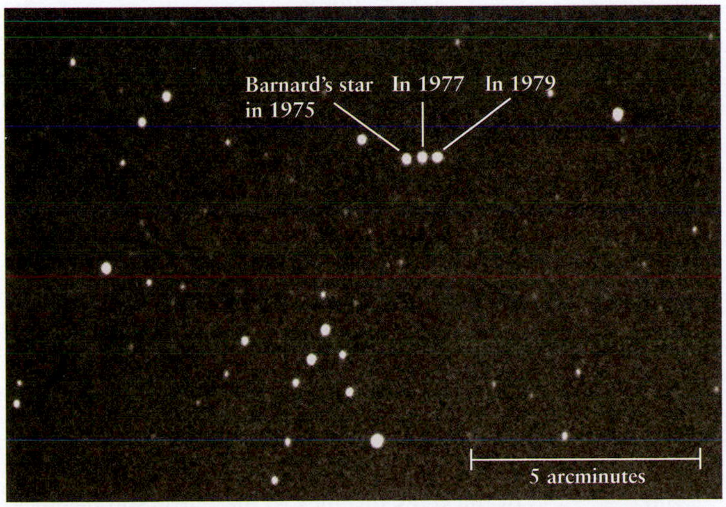

FIGURE 17-3 R I V U X G

The Motion of Barnard's Star Three photographs taken over a four-year period were combined to show the motion of Barnard's star, which lies 1.82 pc away in the constellation Ophiuchus. Over this time interval, Barnard's star moved more than 41 arcseconds on the celestial sphere (about 0.69 arcminutes, or 0.012°), more than any other star. (John Sanford/Science Source)

CONCEPTCHECK 17-3

Why can't the distances to most stars in our Galaxy be measured using parallax?

Answer appears at the end of the chapter.

17-2 If a star's distance is known, its luminosity can be determined from its apparent brightness

TUTORIAL 17-2

All the stars you can see in the nighttime sky shine by thermonuclear fusion, just as the Sun does (see Section 16-1). But they are by no means merely identical copies of the Sun. Stars differ in their luminosity (L), the amount of light energy they emit each second. Luminosity is usually measured either in watts (1 watt, or 1 W, is 1 joule per second) or as a multiple of the Sun's luminosity ($L_{\odot}$, equal to 3.90×10^{26} W). Most stars are less luminous than the Sun, but some blaze forth with a million times the Sun's luminosity. Knowing a star's luminosity is essential for determining the star's history, its present-day internal structure, and its future evolution.

Luminosity, Apparent Brightness, and the Inverse-Square Law

Imagine a star radiating light into space. As the light energy moves away from the star, it spreads out over increasingly larger regions of space—the energy thins out. Now imagine a sphere of radius d centered on the star, as in Figure 17-4. The amount of energy that passes each second through a square meter of the sphere's surface area is called the **brightness** (b). The brightness is the total luminosity of the source (L) divided by the total surface area of the sphere (equal to $4\pi d^2$). This quantity is also called the **apparent brightness** (b), because how bright a light source appears depends on how far away it is being observed. Apparent brightness is measured in watts per square meter (W/m^2).

Apparent brightness is a measure of how faint a star looks to us; luminosity is a measure of the star's total light output

Written in the form of an equation, the relationship between apparent brightness and luminosity is

Inverse-square law relating apparent brightness and luminosity

$$b = \frac{L}{4\pi d^2}$$

b = apparent brightness of a star's light, in W/m^2

L = star's luminosity, in watts

d = distance to star, in meters

This relationship is called the **inverse-square law**, because the apparent brightness of light that an observer can see or measure is inversely proportional to the square of the observer's distance (d) from the source. If you double your distance from a light source, its

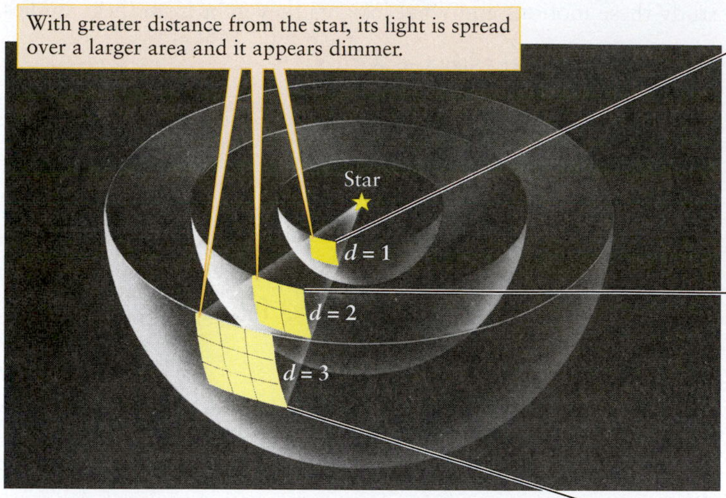

(a)

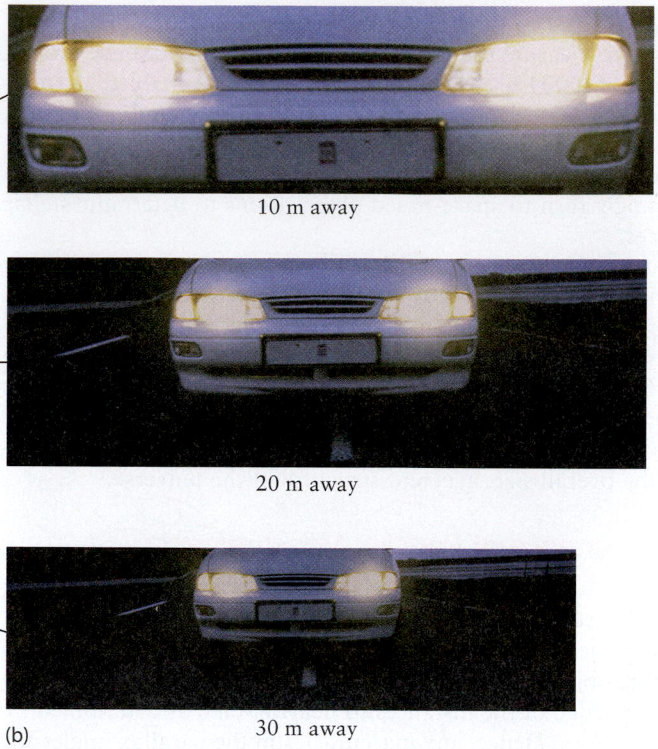

(b)

ANIMATION 17-2

FIGURE 17-4

The Inverse-Square Law Radiation from a light source illuminates an area that increases as the square of the distance from the source. Hence, the apparent brightness decreases as the square of the distance. The brightness at $d = 2$ is $1/(2^2) = 1/4$ of the brightness at $d = 1$, and the brightness at $d = 3$ is $1/(3^2) = 1/9$ of that at $d = 1$. (b: Corbis)

radiation is spread out over an area 4 times larger, so the apparent brightness you see is decreased by a factor of 4. Similarly, at triple the distance, the apparent brightness is 1/9 as great (see Figure 17-4).

We can apply the inverse-square law to the Sun, which is 1.50×10^{11} m from Earth. Its apparent brightness ($b_\odot$) is

$$b_\odot = \frac{3.90 \times 10^{26}\ \text{W}}{4\pi(1.50 \times 10^{11}\ \text{m})^2} = 1370\ \text{W/m}^2$$

That is, a solar panel with an area of 1 square meter receives 1370 watts of power from the Sun.

When astronomers measure the apparent brightness of a star, the process is called **photometry.**

CAUTION! No matter what your distance is from a star, the star's luminosity (L) remains unchanged. Luminosity is the intrinsic energy output of the star. Only the apparent brightness (b) changes with distance as the star's energy thins out. Furthermore, a big telescope and a small telescope on Earth will both measure the same apparent brightness; while a larger telescope collects more light, its larger lens or mirror is taken into account when computing b.

CALCULATIONCHECK 17-2

How many times less light falls on a book illuminated by a lightbulb if the book is moved a distance of 3 times farther away?

Answer appears at the end of the chapter.

Calculating a Star's Luminosity

The inverse-square law says that we can find a star's luminosity if we know its distance and its apparent brightness. For convenience, this law can be expressed in a somewhat different form. We first rearrange the above equation:

$$L = 4\pi d^2 b$$

We then apply this equation to the Sun. That is, we write a similar equation relating the Sun's luminosity ($L_\odot$), the distance from Earth to the Sun ($d_\odot$, equal to 1 AU), and the Sun's apparent brightness ($b_\odot$):

$$L_\odot = 4\pi d_\odot^2 b_\odot$$

If we take the ratio of these two equations, the unpleasant factor of 4π drops out and we are left with the following:

Determining a star's luminosity from its apparent brightness

$$\frac{L}{L_\odot} = \left(\frac{d}{d_\odot}\right)^2 \frac{b}{b_\odot}$$

$L/L_\odot$ = ratio of the star's luminosity to the Sun's luminosity

$d/d_\odot$ = ratio of the star's distance to the Earth-Sun distance

$b/b_\odot$ = ratio of the star's apparent brightness to the Sun's apparent brightness

We need to know just two things to find a star's luminosity: the distance to a star as compared to the Earth-Sun distance (the ratio $d/d_\odot$), and how that star's apparent brightness compares to that of the Sun (the ratio $b/b_\odot$). Then we can use the above equation to determine how luminous that star is compared to the Sun (the ratio $L/L_\odot$).

In other words, this equation gives us a general rule relating the luminosity, distance, and apparent brightness of a star:

We can determine the luminosity of a star from its distance and apparent brightness. For a given distance, the brighter the star, the more luminous that star must be. For a given apparent brightness, the more distant the star, the more luminous it must be to be seen at that distance.

Box 17-2 shows how to use the above equation to determine the luminosity of the nearby star ε (epsilon) Eridani, the fifth brightest star in the constellation Eridanus (named for a river in Greek mythology). Parallax measurements indicate that ε Eridani is 3.23 pc away, and photometry shows that the star appears only 6.73×10^{-13} as bright as the Sun. Using the above equation, we find that ε Eridani has only 0.30 times the luminosity of the Sun.

CALCULATIONCHECK 17-3

The star Pleione in the constellation Taurus is 190 times as luminous as the Sun but appears only 3.19×10^{-13} as bright as the Sun. How many times farther is Pleione from Earth compared to the Earth-Sun distance?

Answer appears at the end of the chapter.

BOX 17-2 TOOLS OF THE ASTRONOMER'S TRADE

Luminosity, Distance, and Apparent Brightness

The inverse-square law (Section 17-2) relates a star's luminosity, distance, and apparent brightness to the corresponding quantities for the Sun:

$$\frac{L}{L_\odot} = \left(\frac{d}{d_\odot}\right)^2 \frac{b}{b_\odot}$$

We can use a similar equation to relate the luminosities, distances, and apparent brightnesses of any two stars, which we call star 1 and star 2:

$$\frac{L_1}{L_2} = \left(\frac{d_1}{d_2}\right)^2 \frac{b_1}{b_2}$$

EXAMPLE: The star ε (epsilon) Eridani is 3.23 pc from Earth. As seen from Earth, this star appears only 6.73×10^{-13} as bright as the Sun. What is the luminosity of ε Eridani compared with that of the Sun?

Situation: We are given the distance to ε Eridani ($d = 3.23$ pc) and this star's brightness compared to that of the Sun ($b/b_\odot = 6.73 \times 10^{-13}$). Our goal is to find the ratio of the luminosity of ε Eridani to that of the Sun, that is, the quantity $L/L_\odot$.

Tools: Since we are asked to compare this star to the Sun, we use the first of the two equations given above, $L/L_\odot = (d/d_\odot)^2(b/b_\odot)$, to solve for $L/L_\odot$.

Answer: Our equation requires the ratio of the star's distance to the Sun's distance, $d/d_\odot$. The distance from Earth to the Sun is $d_\odot = 1$ AU. To calculate the ratio $d/d_\odot$, we must express both distances in the same units. There are 206,265 AU in 1 pc, so we can write the distance to ε Eridani as $d = (3.23 \text{ pc})(206{,}265 \text{ AU/pc}) = 6.66 \times 10^5$ AU. Hence, the ratio of distances is $d/d_\odot = (6.66 \times 105 \text{ AU})/(1 \text{ AU}) = 6.66 \times 10^5$. Then we find that the ratio of the luminosity of ε Eridani (L) to the Sun's luminosity ($L_\odot$) is

$$\frac{L}{L_\odot} = \left(\frac{d}{d_\odot}\right)^2 \frac{b}{b_\odot} = (6.66 \times 10^5)^2 \times (6.73 \times 10^{-13}) = 0.30$$

Review: This result means that ε Eridani is only 0.30 as luminous as the Sun; that is, its power output is only 30% as great.

EXAMPLE: Suppose star 1 is at half the distance of star 2 (that is, $d_1/d_2 = ½$) and that star 1 appears twice as bright as star 2 (that is, $b_1/b_2 = 2$). How do the luminosities of these two stars compare?

Situation: For these two stars, we are given the ratio of distances (d_1/d_2) and the ratio of apparent brightnesses (b_1/b_2). Our goal is to find the ratio of their luminosities (L_1/L_2).

Tools: Since we are comparing two stars, neither of which is the Sun, we use the second of the two equations above: $L_1/L_2 = (d_1/d_2)^2(b_1/b_2)$.

Answer: Plugging values into our equation, we find

$$\frac{L_1}{L_2} = \left(\frac{d_1}{d_2}\right)^2 \frac{b_1}{b_2} = \left(\frac{1}{2}\right)^2 \times 2 = \frac{1}{2}$$

Review: This result says that star 1 has only one-half the luminosity of star 2. Despite this, star 1 appears brighter than star 2 because it is closer to us.

The two equations above are also useful in the method of *spectroscopic parallax*, which we discuss in Section 17-8. It turns out that a star's luminosity can be determined simply by analyzing the star's spectrum. If the star's apparent brightness

(continued on the next page)

is also known, the star's distance can be calculated. The inverse-square law can be rewritten as an expression for the ratio of the star's distance from Earth (d) to the Earth-Sun distance ($d_{\odot}$):

$$\frac{d}{d_{\odot}} = \sqrt{\frac{(L/L_{\odot})}{(b/b_{\odot})}}$$

We can also use this formula as a relation between the properties of any two stars, 1 and 2:

$$\frac{d_1}{d_2} = \sqrt{\frac{(L_1/L_2)}{(b_1/b_2)}}$$

EXAMPLE: The star Pleione in the constellation Taurus is 190 times as luminous as the Sun but appears only 3.19×10^{-13} as bright as the Sun. How far is Pleione from Earth?

Situation: We are told the ratio of Pleione's luminosity to that of the Sun ($L/L_{\odot} = 190$) and the ratio of their apparent brightnesses ($b/b_{\odot} = 3.19 \times 10^{-13}$). Our goal is to find the distance d from Earth to Pleione.

Tools: Since we are comparing Pleione to the Sun, we use the first of the two equations above.

Answer: Our equation tells us the ratio of the Earth-Pleione distance to the Earth-Sun distance:

$$\frac{d}{d_{\odot}} = \sqrt{\frac{(L/L_{\odot})}{(b/b_{\odot})}} = \sqrt{\frac{190}{3.19 \times 10^{-13}}}$$

$$= \sqrt{5.95 \times 10^{14}} \times 2.44 \times 10^{7}$$

Hence, the distance from Earth to Pleione is 2.44×10^7 times greater than the distance from Earth to the Sun. The Sun-Earth distance is $d_{\odot} = 1$ AU and 206,265 AU = 1 pc, so we can express the star's distance as $d = (2.44 \times 10^7 \text{ AU}) \times (1 \text{ pc}/206{,}265 \text{ AU}) = 118$ pc.

Review: We can check our result by comparing it with the above example about the star ε Eridani. Pleione has a much greater luminosity than ε Eridani (190 times the Sun's luminosity versus 0.30 times), but Pleione appears dimmer than ε Eridani (3.19×10^{-13} times as bright as the Sun compared to 6.73×10^{-13} times). For this to be true, Pleione must be much farther away from Earth than is ε Eridani. This is just what our results show: $d = 118$ pc for Pleione compared to $d = 3.23$ pc for ε Eridani.

EXAMPLE: The star δ (delta) Cephei, which lies 300 pc from Earth, is thousands of times more luminous than the Sun. Thanks to this great luminosity, stars like δ Cephei can be seen in galaxies millions of parsecs away. As an example, the Hubble Space Telescope has detected stars like δ Cephei within the galaxy NGC 3351, which lies in the direction of the constellation Leo. These stars appear only 9×10^{-10} as bright as δ Cephei. What is the distance to NGC 3351?

Situation: To determine the distance we want, we need to find the distance to a star within NGC 3351. We are told that certain stars within this galaxy have the same luminosity as δ Cephei but appear only 9×10^{-10} as bright.

Tools: We use the equation $d_1/d_2 = \sqrt{(L_1/L_2)/(b_1/b_2)}$ to relate two stars, one within NGC 3351 (call this star 1) and the identical star δ Cephei (star 2). Our goal is to find d_1.

Answer: Since the two stars are identical, they have the same luminosity ($L_1 = L_2$, or $L_1/L_2 = 1$). The brightness ratio is $b_1/b_2 = 9 \times 10^{-10}$, so our equation tells us that

$$\frac{d_1}{d_2} = \sqrt{\frac{(L_1/L_2)}{(b_1/b_2)}} = \sqrt{\frac{1}{9 \times 10^{-10}}} = \sqrt{1.1 \times 10^{9}} = 33{,}000$$

Hence, NGC 3351 is 33,000 times farther away than δ Cephei, which is 300 pc from Earth. The distance from Earth to NGC 3351 is therefore (33,000)(300 pc) = 10^7 pc, or 10 megaparsecs (10 Mpc).

Review: This example illustrates one technique that astronomers use to measure extremely large distances. We will learn more about stars like δ Cephei in Chapter 19, and in Chapter 23 we will explore further how they are used to determine the distances to remote galaxies.

The Stellar Population

Calculations of this kind show that stars come in a wide variety of different luminosities, with values that range from about 10^6 $L_{\odot}$ (a million times the Sun's luminosity) to only about 10^{-4} $L_{\odot}$ (a mere ten-thousandth of the Sun's light output). The most luminous star emits roughly 10^{10} times more energy each second than the least luminous! (To put this number in perspective, about 10^{10} human beings have lived on Earth since our species first evolved.)

As stars go, our Sun is neither extremely luminous nor extremely dim; it is a rather ordinary, garden-variety star. It is somewhat more luminous than most stars, however. Of more than 30 stars within 4 pc of the Sun (see Appendix 4), only 3 (α Centauri, Sirius, and Procyon) have a greater luminosity than the Sun.

To better characterize a typical population of stars, astronomers count the stars out to a certain distance from the Sun and plot the number of stars that have different luminosities. The

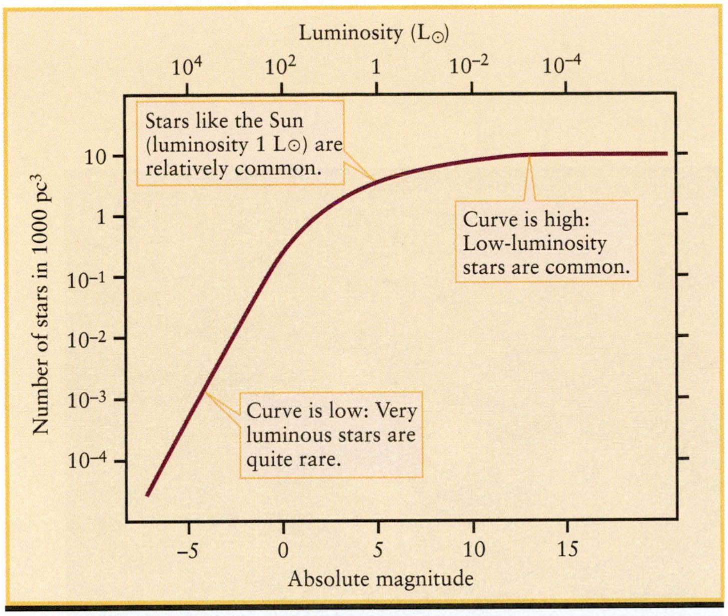

FIGURE 17-5

The Luminosity Function This graph shows how many stars of a given luminosity lie within a representative 1000–cubic-parsec volume. The scale at the bottom of the graph shows absolute magnitude, an alternative measure of a star's luminosity (described in Section 17-3). (Adapted from J. Bahcall and R. Soneira)

resulting graph is called the **luminosity function.** Figure 17-5 shows the luminosity function for stars in our part of the Milky Way Galaxy. The curve declines very steeply for the most luminous stars toward the left side of the graph, indicating that they are quite rare. For example, this graph shows that stars like the Sun are about 10,000 times more common than stars like Spica (which has a luminosity of 2100 $L_\odot$).

The exact shape of the curve in Figure 17-5 applies only to the vicinity of the Sun and similar regions in our Milky Way Galaxy. Other locations have somewhat different luminosity functions. In stellar populations in general, however, low-luminosity stars are much more common than high-luminosity ones.

17-3 Astronomers often use the magnitude scale to denote brightness

Because astronomy is among the most ancient of sciences, some of the tools used by modern astronomers are actually many centuries old. One such tool is the **magnitude scale,** which astronomers frequently use to denote the brightness of stars. This scale was introduced in the second century B.C.E. by the Greek astronomer Hipparchus, who called the brightest stars first-magnitude stars. Stars about half as bright as first-magnitude stars were called second-magnitude stars, and so forth, down to sixth-magnitude stars, the dimmest ones he could see. After telescopes came into use, astronomers extended Hipparchus's magnitude scale to include even dimmer stars.

Apparent Magnitudes

The magnitudes in Hipparchus's scale are properly called **apparent magnitudes,** because they describe how bright an object *appears* to an Earth-based observer. Apparent magnitude is directly related to apparent brightness.

CAUTION! The magnitude scale can be confusing because it works "backward." Keep in mind that the *greater* the apparent magnitude, the *dimmer* the star. A star of apparent magnitude 13 (a third-magnitude star) is dimmer than a star of apparent magnitude 12 (a second-magnitude star).

In the nineteenth century, astronomers developed better techniques for measuring the light energy arriving from a star. These measurements showed that a first-magnitude star is about 100 times brighter than a sixth-magnitude star. In other words, it would take 100 stars of magnitude 16 to provide as much light energy as we receive from a single star of magnitude 11. To make computations easier, the magnitude scale was redefined so that a magnitude difference of 5 corresponds exactly to a factor of 100 in brightness. A magnitude difference of 1 corresponds to a factor of 2.512 in brightness, because

$$2.512 \times 2.512 \times 2.512 \times 2.512 \times 2.512 = (2.512)^5 = 100$$

Thus, it takes 2.512 third-magnitude stars to provide as much light as we receive from a single second-magnitude star. Stated another way, a second-magnitude star is 2.512 times brighter than a third-magnitude star.

Figure 17-6 illustrates the modern apparent magnitude scale. The dimmest stars visible through a pair of binoculars have an apparent magnitude of 110, and the dimmest stars that can be photographed in a one-hour exposure with the Keck telescopes (see Section 6-2) or the Hubble Space Telescope have apparent magnitude 130. Modern astronomers also use negative numbers to extend Hipparchus's scale to include very bright objects. For example, Sirius, the brightest star in the sky, has an apparent magnitude of –1.43. The Sun, the brightest object in the sky, has an apparent magnitude of –26.7.

Absolute Magnitudes

Apparent magnitude measures a star's brightness; absolute magnitude measures its luminosity

Apparent magnitude is a measure of a star's apparent brightness as seen from Earth. A related quantity that measures a star's true energy output—that is, its luminosity—is called **absolute magnitude.** This is the apparent magnitude a star would have if it were located exactly 10 parsecs from Earth.

ANALOGY If you wanted to compare the light output of two different lightbulbs, you would naturally place them side by side so that both bulbs were the same distance from you. In the absolute magnitude scale, we imagine doing the same thing with stars at a distance of 10 pc.

If the Sun were moved to a distance of 10 pc from Earth, it would have an apparent magnitude of 14.8. The absolute magnitude of the Sun is thus 14.8. The absolute magnitudes of the stars

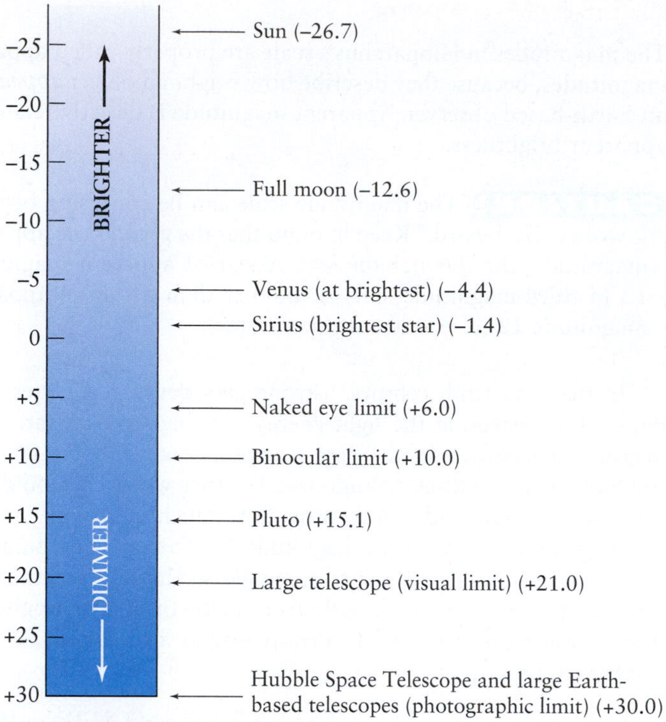

(a) Some apparent magnitudes

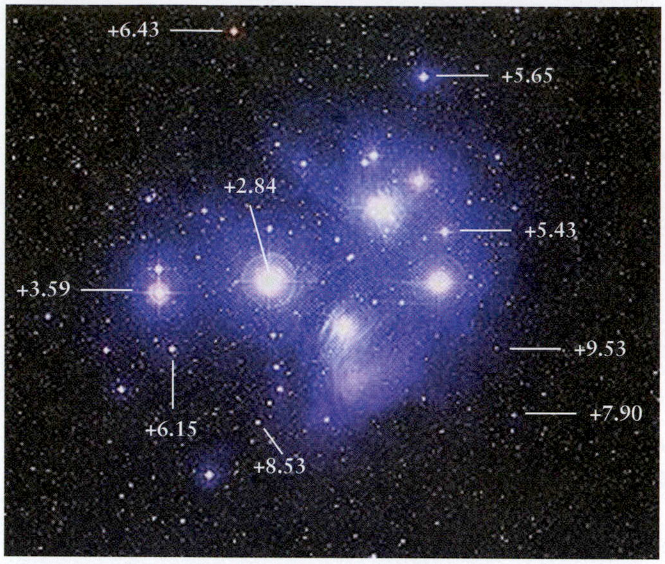

(b) Apparent magnitudes of stars in the Pleiades

FIGURE 17-6 R I V U X G

The Apparent Magnitude Scale **(a)** Astronomers denote the apparent brightness of objects in the sky by their apparent magnitudes. The greater the apparent magnitude, the dimmer the object. **(b)** This photograph of the Pleiades cluster, located about 120 pc away in the constellation Taurus, shows the apparent magnitudes of some of its stars. Most are too faint to be seen by the naked eye. (b: © Australian Astronomical Observatory/David Malin Images)

range from approximately 115 for the least luminous to 210 for the most luminous. (*Note:* Like apparent magnitudes, absolute magnitudes work "backward": The *greater* the absolute magnitude, the *less luminous* the star.) The Sun's absolute magnitude is about in the middle of this range.

We saw in Section 17-2 that we can calculate the luminosity of a star if we know its distance and apparent brightness. There is a mathematical relationship between absolute magnitude and luminosity, which astronomers use to convert one to the other as they see fit. It is also possible to rewrite the inverse-square law, which we introduced in Section 17-2, as a mathematical relationship that allows you to calculate a star's absolute magnitude (a measure of its luminosity) from its distance and apparent magnitude (a measure of its apparent brightness). Box 17-3 describes these relationships and how to use them.

Because the backward magnitude scales can be confusing, we will use them only occasionally in this book. We will usually speak of a star's luminosity rather than its absolute magnitude and will describe a star's appearance in terms of apparent brightness rather than apparent magnitude. But if you go on to study more about astronomy, you will undoubtedly make frequent use of apparent magnitude and absolute magnitude.

CONCEPTCHECK 17-4

If we were observing our Sun from a distance of 10 pc, what would be its apparent and absolute magnitudes?

CONCEPTCHECK 17-5

The star Tau Ceti has an apparent magnitude of about +3 and an absolute magnitude of about +6. Is it much closer or much farther from Earth than 10 pc?

Answers appear at the end of the chapter.

17-4 A star's color depends on its surface temperature

The image that opens this chapter shows that stars come in different colors. You can see these colors even with the naked eye. For example, you can easily see the reddish color of Betelgeuse (the star in the "armpit" of the constellation Orion) and the blue tint of Bellatrix at Orion's other "shoulder" (see Figure 2-2). Colors are most evident for the brightest stars, because human color vision works poorly at low light levels.

CAUTION! It is true that the light from a star will appear redshifted if the star is moving away from you and blueshifted if it is moving toward you. But for even the fastest stars, these color shifts are so tiny that it takes sensitive instruments to measure them. The red color of Betelgeuse and the blue color of Bellatrix are not due to their motions; they are the actual colors of the stars.

BOX 17-3 TOOLS OF THE ASTRONOMER'S TRADE

Apparent Magnitude and Absolute Magnitude

Astronomers commonly express a star's apparent brightness in terms of apparent magnitude (denoted by a lowercase m), and the star's luminosity in terms of absolute magnitude (denoted by a capital M). While we do not use these quantities extensively in this book, it is useful to know a few simple relationships involving them.

Consider two stars, labeled 1 and 2, with apparent magnitudes m_1 and m_2 and brightnesses b_1 and b_2, respectively. The *ratio* of their apparent brightnesses (b_1/b_2) corresponds to a *difference* in their apparent magnitudes ($m_2 - m_1$). As we learned in Section 17-3, each step in magnitude corresponds to a factor of 2.512 in brightness; we receive 2.512 times more energy per square meter per second from a third-magnitude star than from a fourth-magnitude star. This idea was used to construct the following table:

Apparent magnitude difference ($m_2 - m_1$)	Ratio of apparent brightness (b_1/b_2)
1	2.512
2	$(2.512)^2 = 6.31$
3	$(2.512)^3 = 15.85$
4	$(2.512)^4 = 39.82$
5	$(2.512)^5 = 100$
10	$(2.512)^{10} = 10^4$
15	$(2.512)^{15} = 10^6$
20	$(2.512)^{20} = 10^8$

A simple equation relates the difference between two stars' apparent magnitudes to the ratio of their brightnesses:

Magnitude difference related to brightness ratio

$$m_2 - m_1 = 2.5 \log\left(\frac{b_1}{b_2}\right)$$

m_1, m_2 = apparent magnitudes of stars 1 and 2

b_1, b_2 = apparent brightnesses of stars 1 and 2

In this equation, $\log (b_1/b_2)$ is the logarithm of the brightness ratio. The logarithm of $1000 = 10^3$ is 3, the logarithm of $10 = 10^1$ is 1, and the logarithm of $1 = 10^0$ is 0.

EXAMPLE: At their most brilliant, Venus has a magnitude of about −4 and Mercury has a magnitude of about −2. How many times brighter are these planets than the dimmest stars visible to the naked eye, with a magnitude of +6?

Situation: In each case we want to find a ratio of two apparent brightnesses (the brightness of Venus or Mercury compared to that of the dimmest naked-eye stars).

Tools: In each case we will convert a *difference* in apparent magnitude between the planet and the naked-eye star into a *ratio* of their brightnesses.

Answer: The magnitude difference between Venus and the dimmest stars visible to the naked eye is $+6 - (-4) = 10$. From the table, this difference corresponds to a brightness ratio of $(2.512)^{10} = 10^4 = 10{,}000$, so Venus at its most brilliant is 10,000 times brighter than the dimmest naked-eye stars.

The magnitude difference between Mercury and the dimmest naked-eye stars is $+4 - (-4) = 8$. While this value is not in the table, you can see that the corresponding ratio of brightnesses is $(2.512)^8 = (2.512)^{5+3} = (2.512)^5 \times (2.512)^3$. From the table, $(2.512)^5 = 100$ and $(2.512)^3 = 15.85$, so the ratio of brightnesses is $100 \times 15.85 = 1585$. Hence, Mercury at its most brilliant is 1585 times brighter than the dimmest stars visible to the naked eye.

Review: Can you show that when at their most brilliant, Venus is 6.31 times brighter than Mercury? (*Hint:* No multiplication or division is required—just notice the difference in apparent magnitude between Venus and Mercury, and consult the table.)

EXAMPLE: The variable star RR Lyrae in the constellation Lyra (the Harp) periodically doubles its light output. By how much does its apparent magnitude change?

Situation: We are given a ratio of two brightnesses (the star at its maximum is twice as bright as at its minimum). Our goal is to find the corresponding difference in apparent magnitude.

Tools: We let 1 denote the star at its maximum brightness and 2 denote the same star at its dimmest, so the ratio of brightnesses is $b_1/b_2 = 2$. We then use the equation $m_2 - m_1 = 2.5 \log (b_1/b_2)$ to solve for the apparent magnitude difference $m_2 - m_1$.

Answer: Using a calculator, we find $m_2 - m_1 = 2.5 \log (2) = 2.5 \times 0.30 = 0.75$. RR Lyrae therefore varies periodically in brightness by 0.75 magnitude.

Review: Our answer means that at its dimmest, RR Lyrae has an apparent magnitude m_2 that is 0.75 *greater* than its apparent magnitude m_1 when it is brightest. (Remember that a greater value of apparent magnitude means the star is dimmer, not brighter!)

The inverse-square law relating a star's apparent brightness and luminosity can be rewritten in terms of the

(continued on the next page)

star's apparent magnitude (m), absolute magnitude (M), and distance from Earth (d). This can be expressed as an equation:

Relation between a star's apparent magnitude and absolute magnitude

$$m - M = 5 \log d - 5$$

m = star's apparent magnitude

M = star's absolute magnitude

d = distance from Earth to the star in parsecs

In this expression $m - M$ is called the **distance modulus,** and log d means the logarithm of the distance d in parsecs. For convenience, the following table gives the values of the distance d corresponding to different values of $m - M$.

Distance modulus $m - M$	Distance d (pc)
−4	1.6
−3	2.5
−2	4.0
−1	6.3
0	10
1	16
2	25
3	40
4	63
5	100
10	10^3
15	10^4
20	10^5

This table shows that if a star is less than 10 pc away, its distance modulus $m - M$ is negative. That is, its apparent magnitude (m) is less than its absolute magnitude (M). If the star is more than 10 pc away, $m - M$ is positive and m is greater than M. As an example, the star ε (epsilon) Indi, which is in the direction of the southern constellation Indus, has apparent magnitude $m = 14.7$. It is 3.6 pc away, which is less than 10 pc, so its apparent magnitude is less than its absolute magnitude.

EXAMPLE: Find the absolute magnitude of ε Indi.

Situation: We are given the distance to ε Indi ($d = 3.6$ pc) and its apparent magnitude ($m = 14.7$). Our goal is to find the star's absolute magnitude M.

Tools: We use the formula $m - M = 5 \log d - 5$ to solve for M.

Answer: Since $d = 3.6$ pc, we use a calculator to find log $d = \log 3.6 = 0.56$. Therefore the star's distance modulus is $m - M = 5(0.56) - 5 = -2.2$, and the star's absolute magnitude is $M = m - (-2.2) = +4.7 + 2.2 = +6.9$.

Review: As a check on our calculations, note that this star's distance modulus $m - M = -2.2$ is less than zero, as it should be for a star less than 10 pc away. Note that our Sun has absolute magnitude +4.8; ε Indi has a greater absolute magnitude, so it is less luminous than the Sun.

EXAMPLE: Suppose you were viewing the Sun from a planet orbiting another star 100 pc away. Could you see it without using a telescope?

Situation: We learned in the preceding examples that the Sun has absolute magnitude $M = +4.8$ and that the dimmest stars visible to the naked eye have apparent magnitude $m = +6$. Our goal is to determine whether the Sun would be visible to the naked eye at a distance of 100 pc.

Tools: We use the relationship $m - M = 5 \log d - 5$ to find the Sun's apparent magnitude at $d = 100$ pc. If this is greater than +6, the Sun would not be visible at that distance. (Remember that the greater the apparent magnitude, the dimmer the star.)

Answer: From the table, at $d = 100$ pc the distance modulus is $m - M = 5$. So, as seen from this distant planet, the Sun's apparent magnitude would be $m = M + 5 = +4.8 + 5 = +9.8$. This is greater than the naked-eye limit $m = +6$, so the Sun could not be seen.

Review: The Sun is by far the brightest object in Earth's sky. But our result tells us that to an inhabitant of a planetary system 100 pc away—a rather small distance in a galaxy that is thousands of parsecs across—our own Sun would be just another insignificant star, visible only through binoculars or a telescope.

The magnitude system is also used by astronomers to express the colors of stars as seen through different filters, as we describe in Section 17-4. For example, rather than quantifying a star's color by the color ratio b_V/b_B (a star's apparent brightness as seen through a V filter divided by the brightness through a B filter), astronomers commonly use the *color index* B–V, which is the difference in the star's apparent magnitude as measured with these two filters. We will not use this system in this book, however (but see Advanced Questions 53 and 54).

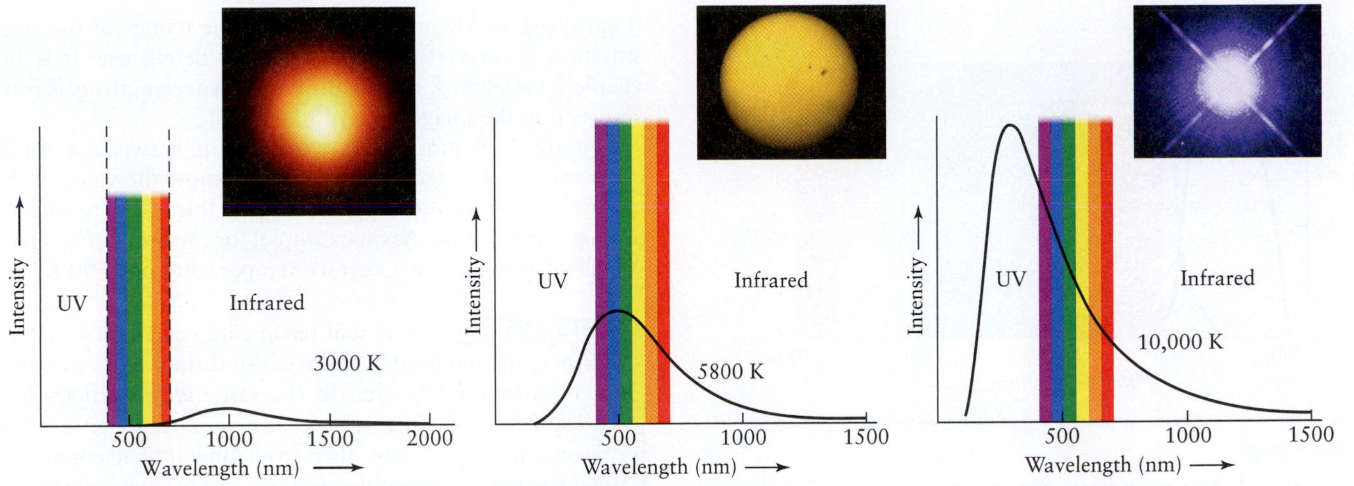
Intensity
UV
Infrared
3000 K
500
1000
1500
2000
Wavelength (nm)
(a) A **cool star** with surface temperature 3000 K emits much more red light than blue light, and so appears red.
Intensity
UV
Infrared
5800 K
500
1000
1500
Wavelength (nm)
(b) A **warmer star** with surface temperature 5800 K (like the Sun) emits similar amounts of all visible wavelengths, and so appears yellow-white.
Intensity
UV
Infrared
10,000 K
500
1000
1500
Wavelength (nm)
(c) A **hot star** with surface temperature 10,000 K emits much more blue light than red light, and so appears blue.

FIGURE 17-7

Temperature and Color These graphs show the intensity of light emitted by three hypothetical stars plotted against wavelength (compare with Figure 5-11). The rainbow band indicates the range of visible wavelengths. The star's apparent color depends on whether the intensity curve has larger values at the short-wavelength (blue) or long-wavelength (red) end of the visible spectrum. However, the human brain combines all of the colors in a spectrum, so where the intensity peaks is just one factor shaping the color perceived by the eye. The image insets show stars of about these surface temperatures. UV stands for ultraviolet, which extends from 10 to 400 nm. See Figure 5-12 for more on wavelengths of the spectrum. (inset a: Andrea Dupree [Harvard-Smithsonian CfA], Ronald Gilliland [STScI, NASA, and ESA]; inset b: NSO/AURA/NSF; inset c: NASA, H. E. Bond, and E. Nelan [STScI]; M. Barstow and M. Burleigh [U. of Leicester, U.K.]; and J. B. Holberg [U. of Arizona])

Color and Temperature

We saw in Section 5-3 that a star's color is directly related to its surface temperature. The intensity of light from a relatively cool star peaks at long wavelengths, making the star look red (Figure 17-7a). A hot star's intensity curve peaks at shorter wavelengths, so the star looks blue (Figure 17-7c). For a star with an intermediate temperature, such as the Sun, the intensity peak is near the middle of the visible spectrum. The human visual system interprets an object with this spectrum of wavelengths as yellowish in color (Figure 17-7b). This leads to an important general rule about star colors and surface temperatures:

Red stars are relatively cold, with low surface temperatures; blue stars are relatively hot, with high surface temperatures.

Astronomers use a set of filters in their telescopes to measure the surface temperatures of stars

Figure 17-7 shows that astronomers can accurately determine the surface temperature of a star by carefully measuring its color. To measure color, the star's light is collected by a telescope and passed through various color filters. For example, a red filter passes red light while blocking other wavelengths. The filtered light is then collected by a light-sensitive device such as a CCD (see Section 6-4). The process is then repeated with each of the filters in the set. The star's image will have a different brightness through each colored filter, and by comparing these brightnesses, astronomers can find the wavelength at which the star's intensity curve has its peak—and hence the star's temperature.

UBV Photometry

Let's look at this procedure in more detail. The most commonly used filters are called U, B, and V, and the technique that uses them is called **UBV photometry.** Each filter is transparent to a different band of wavelengths: the ultraviolet (U), the blue (B), and the yellow-green (V, for visual) region in and around the visible spectrum (Figure 17-8). The transparency of the V filter mimics the sensitivity of the human eye.

To determine a star's temperature using UBV photometry, the astronomer first measures the star's brightness through each of the filters individually. This gives three apparent brightnesses for the star, designated b_U, b_B, and b_V. The astronomer then compares the intensity of starlight in neighboring wavelength bands by taking the ratios of these brightnesses: b_V/b_B and b_B/b_U. Table 17-1 gives values for these **color ratios** for several stars with different surface temperatures.

If a star is very hot, its radiation is skewed toward short, ultraviolet wavelengths as in Figure 17-7c. This makes the star dim through the V filter, brighter through the B filter, and brightest through the U filter. Hence, for a hot star b_V is less than b_B, which in turn is less than b_U, and the ratios b_V/b_B and b_B/b_U are both less than 1. One such star is Bellatrix (see Table 17-1), which has a surface temperature of 21,500 K.

In contrast, if a star is cool, its radiation peaks at long wavelengths as in Figure 17-7a. Such a star appears brightest through

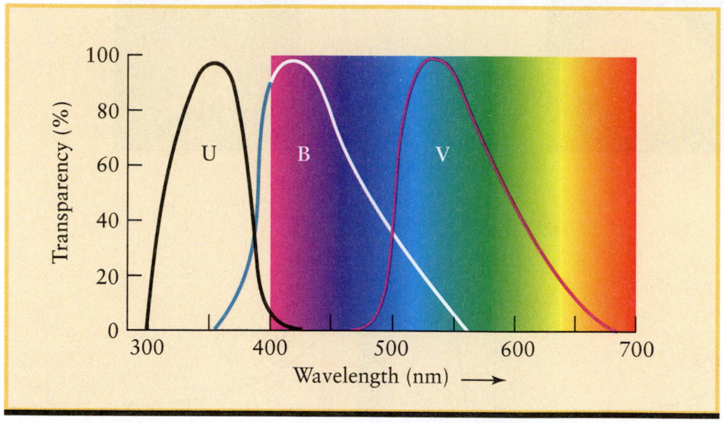

FIGURE 17-8

U, B, and V Filters This graph shows the wavelengths to which the standard filters are transparent. The U filter is transparent to near-ultraviolet light. The B filter is transparent to violet, blue, and green light, while the V filter is transparent to green and yellow light. By measuring the apparent brightness of a star with each of these filters and comparing the results, an astronomer can determine the star's surface temperature.

the V filter, dimmer through the B filter, and dimmest through the U filter (see Figure 17-8). In other words, for a cool star b_V is greater than b_B, which in turn is greater than b_U. Hence, the ratios b_V/b_B and b_B/b_U will both be greater than 1. The star Betelgeuse (surface temperature 3500 K) is an example.

You can see these differences between hot and cool stars in parts a and c of Figure 6-27, which show the constellation Orion at ultraviolet wavelengths (a bit shorter than those transmitted by the U filter) and at visible wavelengths that approximate the transmission of a V filter. The hot star Bellatrix is brighter in the ultraviolet image (Figure 6-27a) than at visible wavelengths (Figure 6-27c). (Figure 6-27d shows the names of the stars.) The situation is reversed for the cool star Betelgeuse: It is bright at visible wavelengths, but at ultraviolet wavelengths it is too dim to show up in the image.

Figure 17-9 graphs the relationship between a star's b_V/b_B color ratio and its temperature. If you know the value of the b_V/b_B color ratio for a given star, you can use this graph to find the star's surface temperature. As an example, for the Sun b_V/b_B equals 1.87, which corresponds to a surface temperature of 5800 K.

CAUTION! As we will see in Chapter 18, tiny dust particles that pervade interstellar space cause distant stars to appear redder than they really are. (In the same way, particles in Earth's atmosphere make the setting Sun look redder; see Box 5-4.) Astronomers must take this reddening into account whenever they attempt to determine a star's surface temperature from its color ratios. A star's spectrum provides a more precise measure of a star's surface temperature, as we will see next. But it is quicker and easier to observe a star's colors with a set of U, B, and V filters than it is to take the star's spectrum with a spectrograph.

CONCEPTCHECK 17-6

Which is hotter: a star with an orange color or one with a more blue color?

Answer appears at the end of the chapter.

17-5 The spectra of stars reveal their chemical compositions as well as their surface temperatures

We have seen how the color of a star's light helps astronomers determine its surface temperature. To determine the other

TABLE 17-1 Colors of Selected Stars

Star	Surface temperature (K)	b_V/b_B	b_B/b_U	Apparent color
Bellatrix (γ Orionis)	21,500	0.81	0.45	Blue
Regulus (α Leonis)	12,000	0.90	0.72	Blue-white
Sirius (α Canis Majoris)	9400	1.00	0.96	Blue-white
Megrez (δ Ursae Majoris)	8630	1.07	1.07	White
Altair (α Aquilae)	7800	1.23	1.08	Yellow-white
Sun	5800	1.87	1.17	Yellow-white
Aldebaran (α Tauri)	4000	4.12	5.76	Orange
Betelgeuse (α Orionis)	3500	5.55	6.66	Red

Source: J.-C Mermilliod, B. Hauck, and M. Mermilliod, University of Lausanne

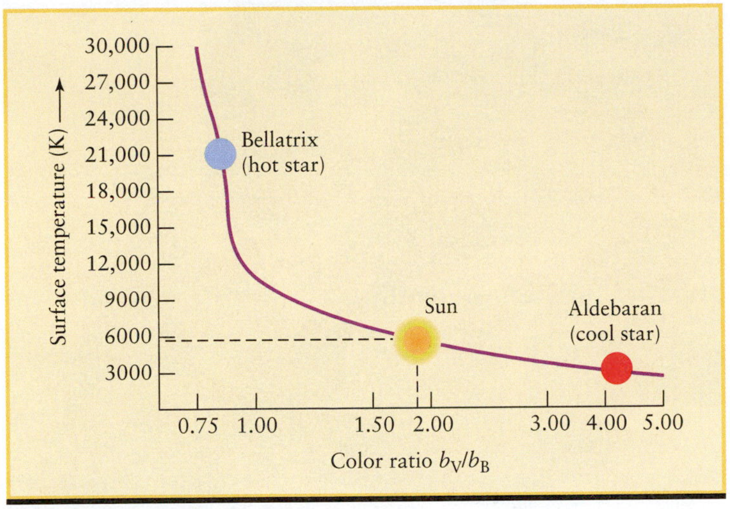

FIGURE 17-9

Temperature, Color, and Color Ratio The b_V/b_B color ratio is the ratio of a star's apparent brightnesses through a V filter and through a B filter. This ratio is small for hot, blue stars but large for cool, red stars. After measuring a star's brightness with the B and V filters, an astronomer can estimate the star's surface temperature from a graph like this one.

properties of a star, astronomers must analyze the spectrum of its light in more detail. This technique of *stellar spectroscopy* began in 1817 when Joseph Fraunhofer, a German instrument maker, attached a spectroscope to a telescope and pointed it toward the stars. Fraunhofer had earlier observed that the Sun has an absorption line spectrum—that is, a continuous spectrum with dark absorption lines (see Section 5-6). He found that stars have the same kind of spectra, which reinforces the idea that our Sun is a rather typical star. But Fraunhofer also found that the pattern of absorption lines is different for different stars.

Classifying Stars: Absorption Line Spectra and Spectral Classes

We see an absorption line spectrum when a cool gas lies between us and a hot, glowing object (recall Figure 5-17). The light from the hot, glowing object itself has a continuous spectrum. In the case of a star, light with a continuous spectrum is produced at low-lying levels of the star's atmosphere where the gases are hot and dense. The absorption lines are created when this light flows outward through the upper layers of the star's atmosphere. Atoms in these cooler, less dense layers absorb radiation at specific wavelengths, which depend on the specific kinds of atoms present—hydrogen, helium, or other elements—and on whether or not the atoms are ionized. Absorption lines in the Sun's spectrum are produced in this same way (see Section 16-5).

Some stars have spectra in which the Balmer absorption lines of hydrogen are prominent. But in the spectra of other stars, including the Sun, the Balmer lines are nearly absent and the dominant absorption lines are those of heavier elements such as calcium, iron, and sodium. Still other stellar spectra are dominated by broad absorption lines caused by molecules, such as titanium oxide, rather than single atoms. To cope with this diversity, astronomers group similar-appearing stellar spectra into **spectral classes.** In a popular classification scheme that emerged in the late 1890s, a star was assigned a letter from A through O according to the strength or weakness of the hydrogen Balmer lines in the star's spectrum.

> Deciphering the information in starlight took the painstaking work of generations of astronomers

Nineteenth-century science could not explain why or how the spectral lines of a particular chemical are affected by the temperature and density of the gas. Nevertheless, a team of astronomers at the Harvard College Observatory forged ahead with a monumental project of examining the spectra of hundreds of thousands of stars. Their goal was to develop a system of spectral classification in which all spectral features, not just Balmer lines, change smoothly from one spectral class to the next.

The Harvard project was financed by the estate of Henry Draper, a wealthy New York physician and amateur astronomer who in 1872 became the first person to photograph stellar absorption lines. Researchers on the project included Edward C. Pickering, Williamina Fleming, Antonia Maury, and Annie Jump Cannon (Figure 17-10). As a result of their efforts, many of the original A-through-O classes were dropped and others were consolidated. The remaining spectral classes were reordered in the sequence **OBAFGKM.** You can remember this sequence with the mnemonic "*O*h, *B*e *A* *F*ine *G*irl (or *G*uy), *K*iss *M*e!"

CONCEPTCHECK 17-7

What principal characteristic of a star's spectrum most dominates which spectral class letter of the alphabet it is assigned?

Answer appears at the end of the chapter.

Refining the Classification: Spectral Types

Cannon refined the original OBAFGKM sequence into smaller steps called **spectral types.** These steps are indicated by attaching an integer from 0 through 9 to the original letter. For example, the spectral class F includes spectral types F0, F1, F2, . . . , F8, F9, which are followed by the spectral types G0, G1, G2, . . . , G8, G9, and so on.

Figure 17-11 shows representative spectra of several spectral types. The strengths of spectral lines change gradually from one spectral type to the next. For example, the Balmer absorption lines of hydrogen become increasingly prominent as you go from spectral type B0 to A0. From A0 onward through the F and G classes, the hydrogen lines weaken and almost fade from view. The Sun, whose spectrum is dominated by calcium and iron, is a G2 star.

The Harvard project culminated in the *Henry Draper Catalogue,* published between 1918 and 1924. It listed 225,300 stars, each of which Cannon had personally classified. Meanwhile, physicists had been making important discoveries about the structure of atoms. Ernest Rutherford had shown that atoms have nuclei (recall Figure 5-20), and Niels Bohr made the remarkable hypothesis that electrons circle atomic nuclei along discrete orbits (see Figure 5-22).

(a)

(b)

FIGURE 17-10 R I V U X G

Classifying the Spectra of the Stars The modern classification scheme for stars, based on their spectra, was developed at the Harvard College Observatory in the late nineteenth century. Female astronomers, initially led by Edward C. Pickering (not shown) and **(a)** Williamina Fleming, standing, and then by **(b)** Annie Jump Cannon, analyzed hundreds of thousands of spectra. Social conventions of the time prevented most female astronomers from using research telescopes or receiving salaries comparable to those of men. (a: Harvard College Observatory; b: Bettmann/Corbis)

These advances gave scientists the conceptual and mathematical tools needed to understand stellar spectra.

In the 1920s, the Harvard astronomer Cecilia Payne and the Indian physicist Meghnad Saha demonstrated that the OBAFGKM spectral sequence is actually a sequence in temperature. The hottest stars are O stars. Their absorption lines can occur only if these stars have surface temperatures above 25,000 K. M stars are the coolest stars. The spectral features of M stars are consistent with stellar surface temperatures of about 3000 K.

CONCEPTCHECK 17-8

Is a spectral class F2 star more similar to an A-spectral class star or a G-spectral class star?

CONCEPTCHECK 17-9

As spectral type numbers increase within the G-spectral class of stars, do the larger numbers represent higher temperature stars?

Answers appear at the end of the chapter.

Why Surface Temperature Affects Stellar Spectra

To see why the appearance of a star's spectrum is profoundly affected by the star's surface temperature, consider the Balmer lines of hydrogen. Hydrogen is by far the most abundant element in the universe, accounting for about three-quarters of the mass of a typical star. Yet the Balmer lines do not necessarily show up in a star's spectrum. As we saw in Section 5-8, Balmer absorption lines are produced when an electron in the $n = 2$ orbit of hydrogen is lifted into a higher orbit by absorbing a photon with the right amount of energy (see Figure 5-24). If the star is much hotter than 10,000 K, the photons pouring out of the star's interior have such high energy that they easily knock electrons out of hydrogen atoms in the star's atmosphere. This process ionizes the gas. With its only electron torn away, a hydrogen atom cannot produce absorption lines. Hence, the Balmer lines will be relatively weak in the spectra of such hot stars, such as the hot O and B2 stars in Figure 17-11.

Conversely, if the star's atmosphere is much cooler than 10,000 K, almost all the hydrogen atoms are in the lowest ($n = 1$) energy state. Most of the photons passing through the star's atmosphere possess too little energy to boost electrons up from the $n = 1$ to the $n = 2$ orbit of the hydrogen atoms. Hence, very few of these atoms will have electrons in the $n = 2$ orbit, and only these few can absorb the photons characteristic of the Balmer lines. As a result, these lines are nearly absent from the spectrum of a cool star. (You can see this in the spectra of the cool M0 and M2 stars in Figure 17-11.)

For the Balmer lines to be prominent in a star's spectrum, the star must be hot enough to excite the electrons out of the ground state but not so hot that all the hydrogen atoms become ionized. A stellar surface temperature of about 9000 K produces the strongest hydrogen lines; this is the case for the stars of spectral types A0 and A5 in Figure 17-11.

LOOKING DEEPER 17.1 Every other type of atom or molecule also has a characteristic temperature range in which it produces prominent absorption lines in the observable part of the spectrum. Figure 17-12 shows the relative strengths of absorption lines produced by different chemicals. By measuring the details of these lines in a given star's spectrum, astronomers can accurately determine that star's surface temperature.

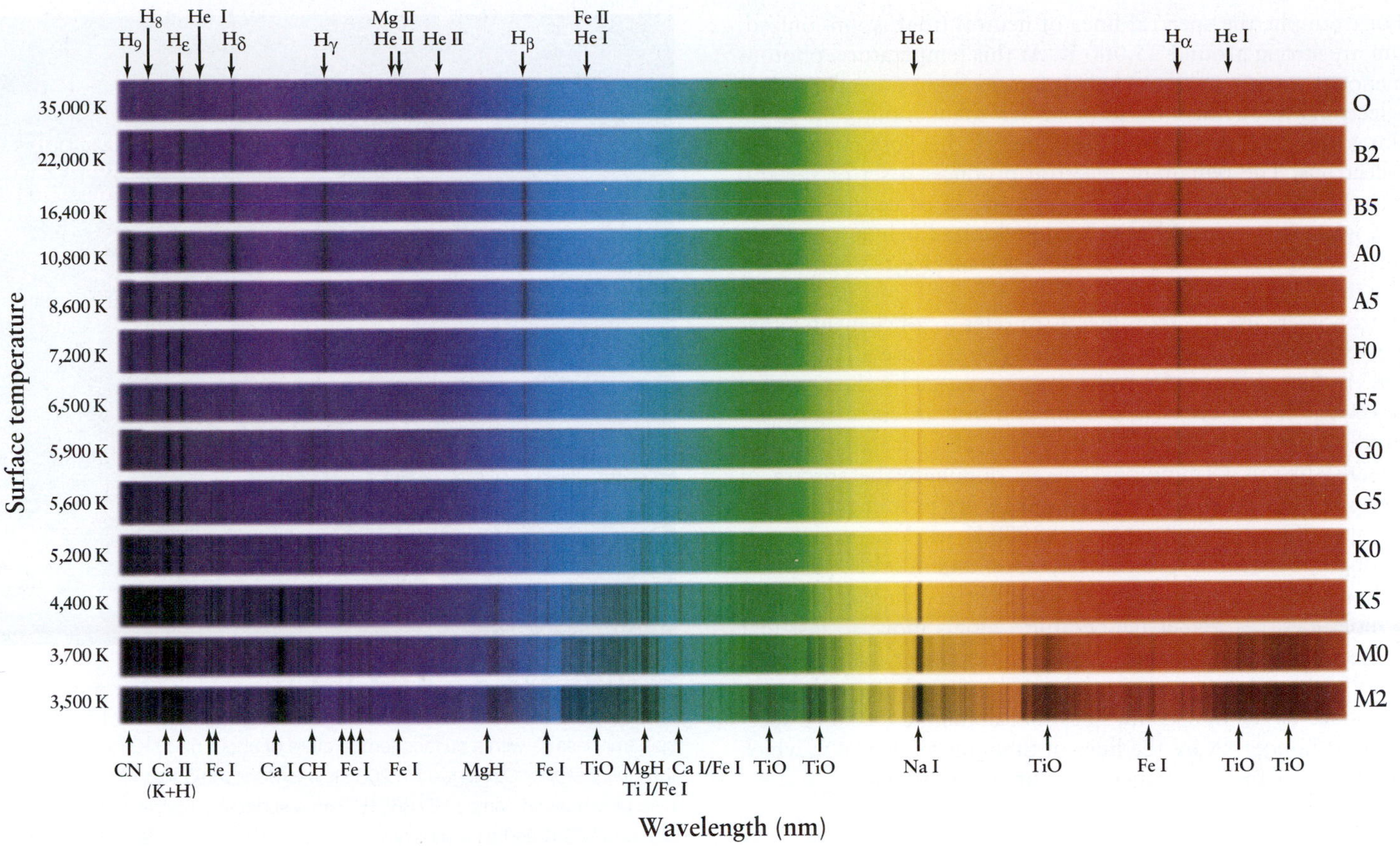

FIGURE 17-11 R I V U X G

Principal Types of Stellar Spectra Stars of different spectral classes and different surface temperatures have spectra dominated by different absorption lines. Notice how the Balmer lines of hydrogen (H_α, H_β, H_γ, and H_δ) are strongest for hot stars of spectral class A, while absorption lines due to calcium (Ca) are strongest in cooler K and M stars. The spectra of M stars also have broad, dark bands caused by molecules of titanium oxide (TiO), which can only exist at relatively low temperatures. A roman numeral after a chemical symbol shows whether the absorption line is caused by un-ionized atoms (roman numeral I) or by atoms that have lost one electron (roman numeral II). The Sun, a G2 star, has a spectrum between G0 and G5. (R. Bell, University of Maryland, and M. Briley, University of Wisconsin at Oshkosh)

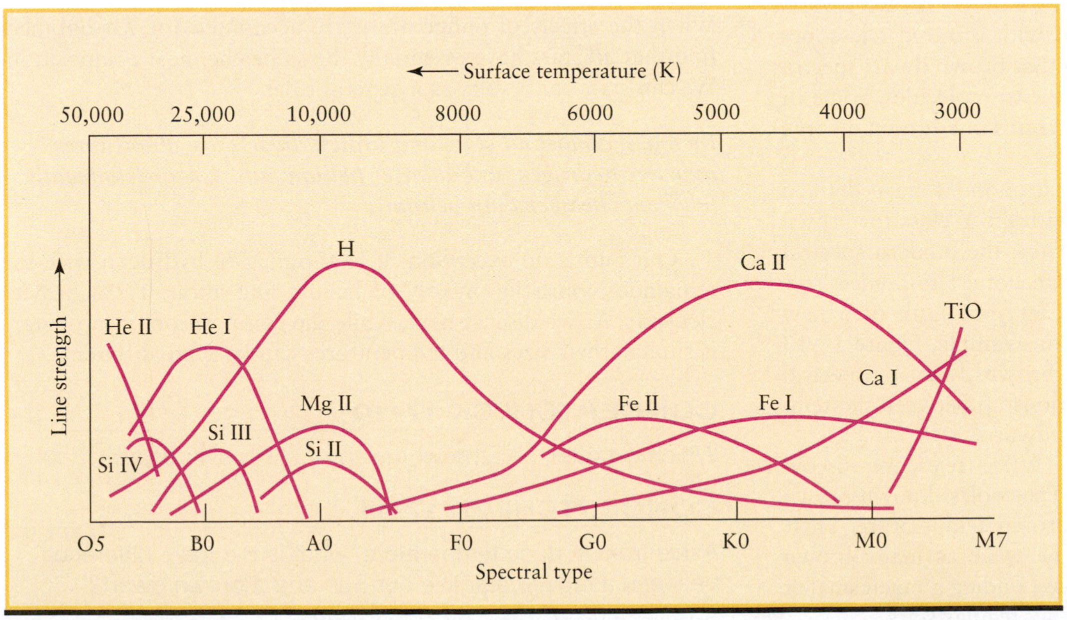

FIGURE 17-12

The Strengths of Absorption Lines Each curve in this graph peaks at the stellar surface temperature for which that chemical's absorption line is strongest. For example, hydrogen (H) absorption lines are strongest in A stars with surface temperatures near 10,000 K. Roman numeral I denotes neutral, un-ionized atoms; II, III, and IV denote atoms that are singly, doubly, or triply ionized (that is, have lost one, two, or three electrons).

For example, the spectral lines of neutral (that is, un-ionized) helium are strong around 25,000 K. At this temperature, photons have enough energy to excite helium atoms without tearing away the electrons altogether. In stars hotter than about 30,000 K, helium atoms become singly ionized, that is, they lose one of their two electrons. The remaining electron produces a set of spectral lines that is recognizably different from those of neutral helium. Hence, when the spectral lines of singly ionized helium appear in a star's spectrum, we know that the star's surface temperature is greater than 30,000 K.

Astronomers use the term **metals** to refer to all elements other than hydrogen and helium. This idiosyncratic use of the term "metal" is quite different from the definition used by chemists and other scientists. To a chemist, sodium and iron are metals but carbon and oxygen are not; to an astronomer, all of these substances are metals. In this terminology, metals dominate the spectra of stars cooler than 10,000 K. Ionized metals are prominent for surface temperatures between 6000 and 8000 K, while neutral metals are strongest between approximately 5500 and 4000 K.

Below 4000 K, certain atoms in a star's atmosphere combine to form molecules. (At higher temperatures atoms move so fast that when they collide, they bounce off each other rather than "sticking together" to form molecules.) As these molecules vibrate and rotate, they produce bands of spectral lines that dominate the star's spectrum. Most noticeable are the lines of titanium oxide (TiO), which are strongest for surface temperatures of about 3000 K.

Spectral Classes for Brown Dwarfs

Since 1995 astronomers have found a number of stars with surface temperatures even lower than those of spectral class M. Strictly speaking, these are not stars but **brown dwarfs**. Brown dwarfs are too small to sustain thermonuclear fusion in their cores; they have masses between 13 M_J and 80 M_J (where M_J stands for mass of Jupiter). Instead, these "substars" glow primarily from the heat released by Kelvin-Helmholtz contraction. This is the process where gravitational contraction compresses and heats up gas, which we described in Section 16-1. (They do undergo fusion reactions for a brief period during their evolution.) Brown dwarfs are so cold that they are best observed with infrared telescopes (Figure 17-13). Such observations reveal that brown dwarf spectra have a rich variety of absorption lines due to molecules. Some of these molecules actually form into solid grains in a brown dwarf's atmosphere.

To describe brown dwarf spectra, astronomers have defined three new spectral classes, L, T, and Y, with the Y class for brown dwarfs below temperatures of 600 K. Thus, the modern spectral sequence of stars and brown dwarfs from hottest to coldest surface temperature is OBAFGKMLTY. (Can you think of a new mnemonic that includes L, T, and Y?) For example, Figure 17-13 shows a star of spectral class K and a brown dwarf of spectral class T. Table 17-2 summarizes the relationship between the temperature and spectra of stars and brown dwarfs.

NASA's Wide-field Infrared Satellite (WISE) telescope discovered more than 100 new brown dwarfs. The coolest known brown dwarf is a mere 80°F, which is about room temperature. Early predictions suggested that there might be twice as many brown dwarfs as regular stars, but observations are finding a much smaller ratio—about one brown dwarf for every six regular stars.

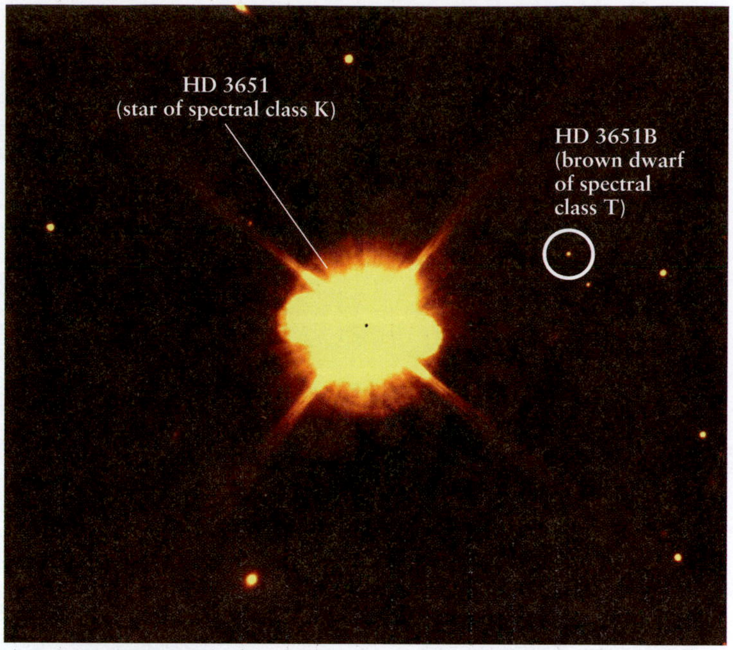

FIGURE 17-13 R **I** V U X G

An Infrared Image of Brown Dwarf HD 3651B The star HD 3561 is of spectral class K, with a surface temperature of about 5200 K. ("HD" refers to the *Henry Draper Catalogue*.) In 2006 it was discovered that HD 3651 is orbited by a brown dwarf named HD 3651B with a surface temperature between 800 and 900 K and a luminosity just 1/300,000 that of the Sun. The brown dwarf emits most of its light at infrared wavelengths, so an infrared telescope was used to record this image. The hotter and more luminous star HD 3651 is greatly overexposed in this image and appears much larger than its true size. HD 3651 and HD 3651B are both 11 pc (36 ly) from Earth in the constellation Pisces (the Fish); the other stars in this image are much farther away. (M. Mugrauer and R. Neuhäuser [U. of Jena], A. Seifahrt [ESO], and T. Mazeh [Tel Aviv U.])

All Stars Have Similar Composition

When the effects of temperature are accounted for, astronomers find that *all* stars have essentially the same chemical composition. We can state the results as a general rule:

By mass, almost all stars and brown dwarfs are about three-quarters hydrogen, one-quarter helium, and 1% or less metals (elements heavier than helium).

Our Sun is no exception: It is about 75% hydrogen with the remainder consisting mostly of helium and about 1.7% heavier elements. As we discuss next, while the composition of most stars is similar, their sizes and temperatures can be quite different.

CONCEPTCHECK 17-10

Which is hotter, an L brown dwarf or a T brown dwarf?

CONCEPTCHECK 17-11

Aside from surface temperature, what is the main difference between a regular star like our Sun and a brown dwarf?

Answers appear at the end of the chapter.

TABLE 17-2 The Spectral Sequence

Spectral class	Color	Temperature (K)	Spectral lines	Examples
O	Blue-violet	30,000–50,000	Ionized atoms, especially helium	Naos (ζ Puppis), Mintaka (δ Orionis)
B	Blue-white	11,000–30,000	Neutral helium, some hydrogen	Spica (α Virginis), Rigel (β Orionis)
A	White	7500–11,000	Strong hydrogen, some ionized metals	Sirius (α Canis Majoris), Vega (α Lyrae)
F	Yellow-white	5900–7500	Hydrogen and ionized metals such as calcium and iron	Canopus (α Carinae), Procyon (α Canis Minoris)
G	Yellow	5200–5900	Both neutral and ionized metals, especially ionized calcium	Sun, Capella (α Aurigae)
K	Orange	3900–5200	Neutral metals	Arcturus (α Boötis), Aldebaran (α Tauri)
M	Red-orange	2500–3900	Strong titanium oxide and some neutral calcium	Antares (α Scorpii), Betelgeuse (α Orionis)
L	Red	1300–2500	Neutral potassium, rubidium, and cesium, and metal hydrides	Brown dwarf Teide 1
T	Red	700–1300	Methane, strong neutral potassium and some water (H_2O)	Brown dwarfs Gliese 229B, HD 3651B
Y	Red	below 600	Possibly ammonia	Brown dwarfs CFBDS J005910.90–011401.3, WISE 1828+2650

17-6 Stars come in a wide variety of sizes and temperatures

A star's radius can be calculated if we know its luminosity and surface temperature

With even the best telescopes, stars appear as nothing more than bright points of light. On a photograph or CCD image, brighter stars appear larger than dim ones (see Figures 17-3, 17-6b, and 17-13), but these apparent sizes give no indication of the star's actual size. To determine the size of a star, astronomers combine information about its luminosity (determined from its distance and apparent brightness) and its surface temperature (determined from its spectral type). In this way, they find that some stars are quite a bit smaller than the Sun, while others are a thousand times larger.

Calculating the Radii of Stars

The key to finding a star's radius from its luminosity and surface temperature is the Stefan-Boltzmann law (see Section 5-4). This law says that the amount of energy radiated per second from a square meter of a blackbody's surface—that is, the energy flux (F)—is proportional to the fourth power of the temperature of that surface (T), as given by the equation $F = \sigma T^4$. This equation applies very well to stars, whose spectra are quite similar to that of a perfect blackbody. (Absorption lines, while important for determining the star's chemical composition and surface temperature, make only relatively small modifications to a star's blackbody spectrum.)

A star's luminosity is the amount of energy emitted per second from its entire surface. This quantity equals the energy flux F multiplied by the total number of square meters on the star's surface (that is, the star's surface area). We expect that most stars are nearly spherical, like the Sun, so we can use the formula for the surface area of a sphere. The formula is $4\pi R^2$, where R is the star's radius (the distance from its center to its surface). Multiplying together the formulas for energy flux and surface area, we can write the star's luminosity as follows:

Relationship between a star's luminosity, radius, and surface temperature

$$L = 4\pi R^2 \sigma T^4$$

L = star's luminosity, in watts

R = star's radius, in meters

σ = Stefan-Boltzmann constant = 5.67×10^{-8} W m^{-2} K^{-4}

T = star's surface temperature, in kelvins

This equation says that a relatively cool star (low surface temperature T), for which the energy flux is quite low, can nonetheless be very luminous if it has a large enough radius R. Alternatively, a relatively hot star (large T) can have a very low luminosity if the star has only a little surface area (small R).

Box 17-4 describes how to use the above equation to calculate a star's radius if its luminosity and surface temperature are known.

We can express the idea behind these calculations in terms of the following general rule:

We can determine the radius of a star from its luminosity and surface temperature.

For a given luminosity: The greater the surface temperature, the smaller the radius must be.

For a given surface temperature: The greater the luminosity, the larger the radius must be.

ANALOGY In a similar way, a roaring campfire can emit more light than a welder's torch. The campfire is at a lower temperature than the torch, but has a much larger surface area from which it emits light.

The Range of Stellar Radii

Using this general rule as shown in Box 17-4, astronomers find that stars come in a wide range of sizes. The smallest stars visible through ordinary telescopes, called *white dwarfs,* are about the same size as Earth. Although their surface temperatures can be very high (25,000 K or more), white dwarfs have so little surface area that their luminosities are very low (less than 0.01 $L_{\odot}$). The largest stars, called *supergiants,* are a thousand times larger in radius than the Sun and 10^5 times larger than white dwarfs. If our own Sun were replaced by one of these supergiants, Earth's orbit would lie completely inside the star!

Figure 17-14 summarizes how astronomers determine the distance from Earth, luminosity, surface temperature, chemical composition, and radius of a star close enough to us so that its parallax can be measured. Remarkably, all of these properties can be deduced from just a few measured quantities: the star's parallax angle, apparent brightness, and spectrum.

CONCEPTCHECK 17-12

What makes lighting a candle using a large, roaring bonfire much more difficult than using a handheld lighter of the same temperature?

CALCULATIONCHECK 17-4

If two stars are at the same temperature, but one is 3 times larger, how many times more luminous is it?

Answers appear at the end of the chapter.

17-7 Hertzsprung-Russell (H-R) diagrams reveal different kinds of stars

Astronomers have collected a wealth of data about the stars, but merely having tables of numerical data is not enough. Like all scientists, astronomers want to analyze their data to look for trends and underlying principles. One of the best ways to look for trends in any set of data, whether it comes from astronomy, finance, medicine, or meteorology, is to create a graph showing how one quantity depends on another. For example, investors consult graphs of stock market values versus dates, and weather forecasters make graphs of temperature versus altitude to determine whether thunderstorms are likely to form. Astronomers have found that a particular graph of stellar properties shows that stars fall naturally into just a few categories. This graph, one of

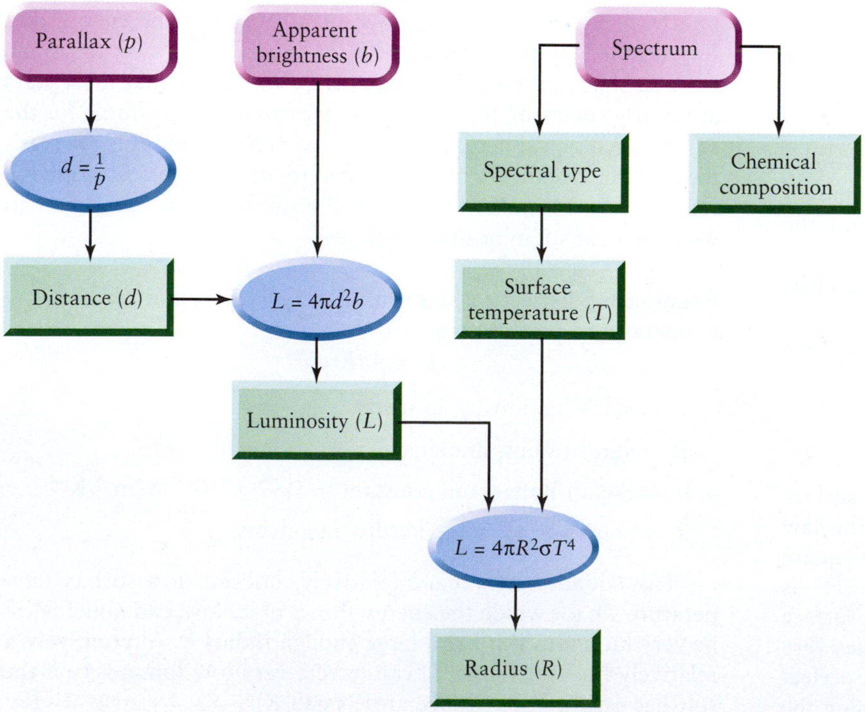

FIGURE 17-14
Finding Key Properties of a Nearby Star This flowchart shows how astronomers determine the properties of a relatively nearby star (one close enough that its parallax can be measured). The rounded purple boxes show the measurements that must be made of the star, the blue ovals show the key equations that are used (from Sections 17-2, 17-5, and 17-6), and the green rectangles show the inferred properties of the stars. A different procedure is followed for more distant stars (see Section 17-8, especially Figure 17-17).

BOX 17-4 TOOLS OF THE ASTRONOMER'S TRADE

Stellar Radii, Luminosities, and Surface Temperatures

Because stars emit light in almost exactly the same fashion as blackbodies, we can use the Stefan-Boltzmann law to relate a star's luminosity (L), surface temperature (T), and radius (R). The relevant equation is

$$L = 4\pi R^2 \sigma T^4$$

As written, this equation involves the Stefan-Boltzmann constant σ, which is equal to 5.67×10^{-8} W m^{-2} K^{-4}. In many calculations, it is more convenient to relate everything to the Sun, which is a typical star. Specifically, for the Sun we have $L_\odot = 4\pi R_\odot^2 \sigma T_\odot^4$, where $L_\odot$ is the Sun's luminosity, $R_\odot$ is the Sun's radius, and $T_\odot$ is the Sun's surface temperature (equal to 5800 K). Dividing the general equation for L by this specific equation for the Sun, we obtain

$$\frac{L}{L_\odot} = \left(\frac{R}{R_\odot}\right)^2 \left(\frac{T}{T_\odot}\right)^4$$

This formula is easier to use because the constant σ has cancelled out. We can also rearrange terms to arrive at a useful equation for the radius (R) of a star:

Radius of a star related to its luminosity and surface temperature

$$\frac{R}{R_\odot} = \left(\frac{T_\odot}{T}\right)^2 \sqrt{\frac{L}{L_\odot}}$$

$R/R_\odot$ = ratio of the star's radius to the Sun's radius

$T_\odot/T$ = ratio of the Sun's surface temperature to the star's surface temperature

$L/L_\odot$ = ratio of the star's luminosity to the Sun's luminosity

EXAMPLE: The bright reddish star Betelgeuse in the constellation Orion (see Figure 2-2) is 60,000 times more luminous than the Sun and has a surface temperature of 3500 K. What is its radius?

Situation: We are given the star's luminosity $L = 60{,}000\ L_\odot$ and its surface temperature $T = 3500$ K. Our goal is to find the star's radius R.

Tools: We use the above equation to find the ratio of the star's radius to the radius of the Sun, $R/R_\odot$. Note that we also know the Sun's surface temperature, $T_\odot = 5800$ K.

Answer: Substituting these data into the above equation, we get

$$\frac{R}{R_\odot} = \left(\frac{5800\text{ K}}{3500\text{ K}}\right)^2 \sqrt{6 \times 10^4} = 670$$

Review: Our result tells us that Betelgeuse's radius is 670 times larger than that of the Sun. The Sun's radius is 6.96×10^5 km, so we can also express the radius of Betelgeuse as $(670)(6.96 \times 10^5 \text{ km}) = 4.7 \times 10^8$ km, which is more than 3 AU. If Betelgeuse were located at the center of our solar system, it would extend beyond the orbit of Mars!

EXAMPLE: Sirius, the brightest star in the sky, is actually two stars orbiting each other (a binary star). The less luminous star, Sirius B, is a white dwarf that is too dim to see with the naked eye. Its luminosity is $0.0025\ L_\odot$ and its surface temperature is 10,000 K. How large is Sirius B compared to Earth?

Situation: Again we are asked to find a star's radius from its luminosity and surface temperature.

Tools: We use the same equation as in the preceding example.

Answer: The ratio of the radius of Sirius B to the Sun's radius is

$$\frac{R}{R_\odot} = \left(\frac{5800\text{ K}}{10{,}000\text{ K}}\right)^2 \sqrt{0.0025} = 0.017$$

Since the Sun's radius is $R_\odot = 6.96 \times 10^5$ km, the radius of Sirius B is $(0.017)(6.96 \times 10^5 \text{ km}) = 12{,}000$ km. From Table 7-1, Earth's radius (half its diameter) is 6378 km. Hence, this star is only about twice the radius of Earth.

Review: Sirius B's radius would be large for a terrestrial planet, but it is minuscule for a star. The name *dwarf* is well deserved!

The radii of some stars have been measured with other techniques (see Section 17-11). These other methods yield values consistent with those calculated by the methods just described.

---

the most important in all astronomy, will in later chapters help us understand how stars form, evolve, and eventually die.

H-R Diagrams

Which properties of stars should we include in a graph? Most stars have about the same chemical composition, but two properties of stars—their luminosities and surface temperatures—differ substantially from one star to another. Stars also come in a wide range of radii, but a star's radius is a secondary property that can be found from the luminosity and surface temperature (as we saw in Section 17-6 and Box 17-4). We also relegate the positions and space velocities of stars to secondary importance. (In a similar way, a physician is more interested in your weight and blood pressure than in where you live or how fast you drive.) We can then ask the

> **A graph of luminosity versus surface temperature reveals that stars fall into a few basic categories**

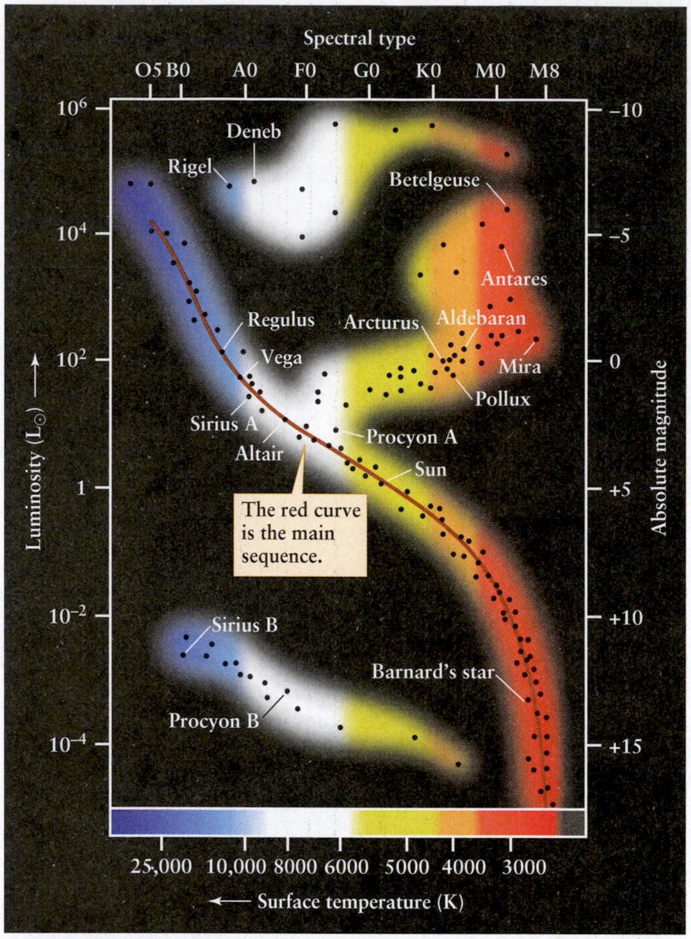

(a) A Hertzsprung-Russell (H-R) diagram

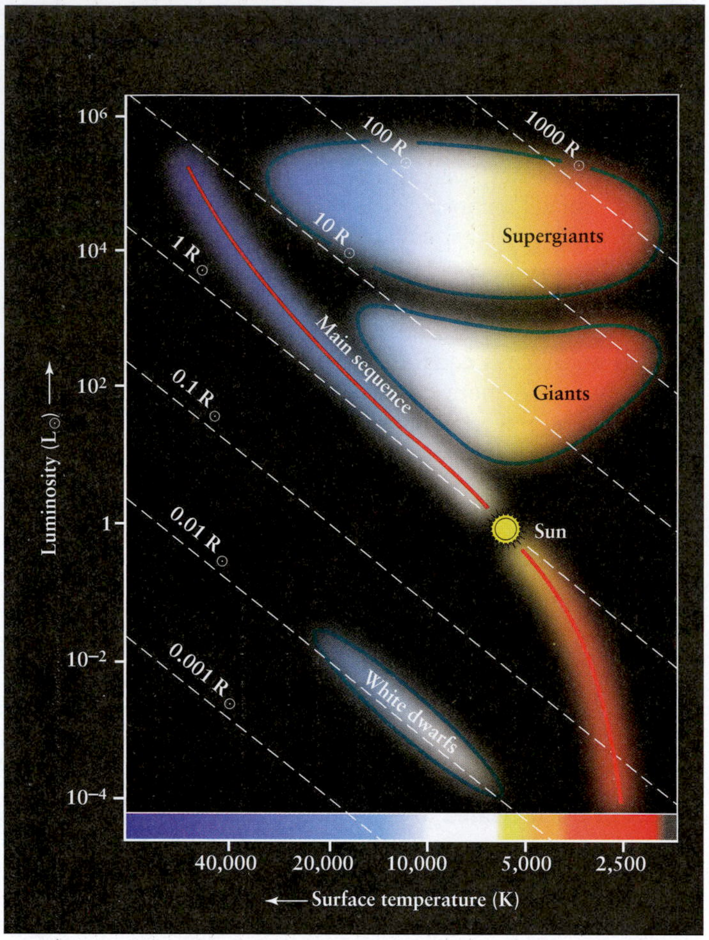

(b) The sizes of stars on an H-R diagram

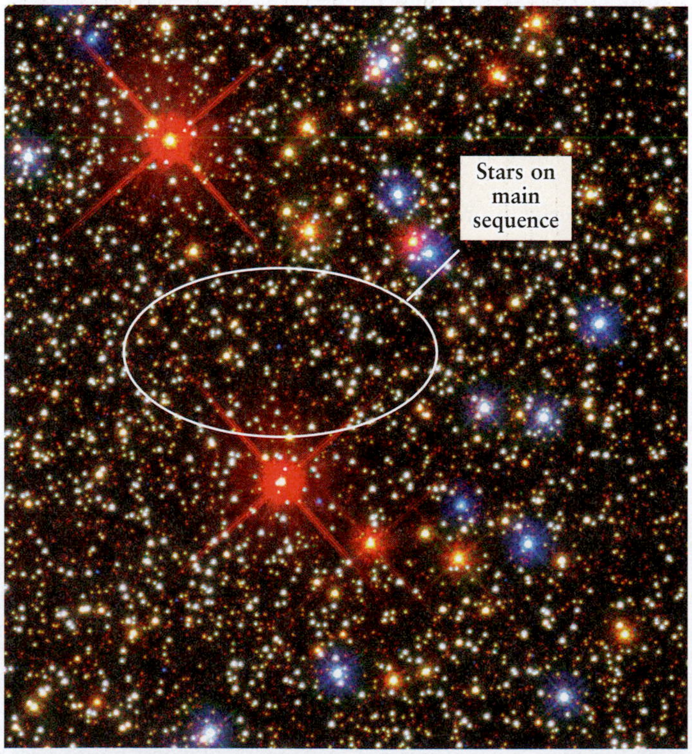

(c) A portion of the star cluster Omega Centauri

FIGURE 17-15 R I **V** U X G

Hertzsprung-Russell (H-R) Diagrams On an H-R diagram, the luminosities (or absolute magnitudes) of stars are plotted against their spectral types (or surface temperatures). **(a)** The data points are grouped in just a few regions on the graph, showing that luminosity and spectral type are correlated. Most stars lie along the red curve called the main sequence. Giants like Arcturus as well as supergiants like Rigel and Betelgeuse are above the main sequence, and white dwarfs like Sirius B are below it. **(b)** The blue curves on this H-R diagram enclose the regions of the diagram in which different types of stars are found. The dashed diagonal lines indicate different stellar radii. For a given stellar radius, as the surface temperature increases (that is, moving from right to left in the diagram), the star glows more intensely and the luminosity increases (that is, moving upward in the diagram). Note that the Sun is intermediate in luminosity, surface temperature, and radius. **(c)** Some stars in Omega Centauri are burning hydrogen in their cores and would appear on the main sequence in an H-R diagram. The orange- and red-colored stars are red giants, and the blue stars have begun burning helium in their cores. (c: NASA, ESA, and the Hubble SM4 ERO Team) following question: What do we learn when we graph the luminosities of stars versus their surface temperatures?

The first answer to this question was given in 1911 by the Danish astronomer Ejnar Hertzsprung. He pointed out that a regular pattern appears when the absolute magnitudes of stars (which measure their luminosities) are plotted against their colors (which measure their surface temperatures). Two years later, the American astronomer Henry Norris Russell independently discovered a similar regularity in a graph using spectral types (another measure of surface temperature) instead of colors. In recognition of their originators, graphs of this kind are today known as **Hertzsprung-Russell diagrams,** or **H-R diagrams** (Figure 17-15).

Figure 17-15a is a typical Hertzsprung-Russell diagram. Each dot represents a star whose spectral type and luminosity have been determined. The most luminous stars are near the top of the diagram, the least luminous stars near the bottom. Hot stars of spectral classes O and B are toward the left side of the graph and cool stars of spectral class M are toward the right.

CAUTION! You are probably accustomed to graphs in which the numbers on the horizontal axis increase as you move to the right. (For example, the business section of a newspaper includes a graph of stock market values versus dates, with later dates to the right of earlier ones.) But on an H-R diagram the temperature scale on the horizontal axis increases toward the *left*. This practice stems from the original diagrams of Hertzsprung and Russell, who placed hot O stars on the left and cool M stars on the right. This arrangement is a tradition that no one has seriously tried to change.

CONCEPTCHECK 17-13

Where on the H-R diagram are the stars with the greatest luminosity and highest temperatures?

Answer appears at the end of the chapter.

Star Varieties: Main-Sequence Stars, Giants, Supergiants, White Dwarfs, and Brown Dwarfs

The most striking feature of the H-R diagram is that the data points are not scattered randomly over the graph but are grouped in a few distinct regions. The luminosities and surface temperatures of stars do *not* have random values; instead, these two quantities are related!

The band stretching diagonally across the H-R diagram includes about 90% of the stars in the night sky. This band, called the **main sequence,** extends from the hot, luminous, blue stars in the upper left corner of the diagram to the cool, dim, red stars in the lower right corner. A star whose properties place it in this region of an H-R diagram is called a **main-sequence star.** The Sun (spectral type G2, luminosity 1 $L_{\odot}$, absolute magnitude 14.8) is such a star. We will find that all main-sequence stars are like the Sun in that *hydrogen fusion*—thermonuclear reactions that convert hydrogen into helium (see Section 16-1)—is taking place in their cores:

Main-sequence stars gain their energy through conversion of hydrogen to helium in their cores.

The upper right side of the H-R diagram shows a second major grouping of data points. Stars represented by these points are both luminous and cool. From the Stefan-Boltzmann law, we know that a cool star radiates much less light per unit of surface area than a hot star. In order for these stars to be as luminous as they are, they must be huge (see Section 17-6), and so they are called **giants.** These stars are about 10 to 100 times larger than the Sun. You can see this size difference in Figure 17-15b, which is an H-R diagram to which dashed lines have been added to represent stellar radii. Most giant stars are around 100 to 1000 times more luminous than the Sun and have surface temperatures of about 3000 to 6000 K. Cooler members of this class of stars (those with surface temperatures from about 3000 to 4000 K) are often called **red giants** because they appear reddish. In the collection of stars in Figure 17-15c the brightest orange and red stars are red giants. However, while these giant stars also appear larger in this image, their true size is much too small to see; it is really their brightness that overexposes the camera and makes them appear large. (A number of red giants can easily be seen with the naked eye, including Aldebaran in the constellation Taurus and Arcturus in Boötes.)

A few rare stars in the night sky are considerably bigger and brighter than typical red giants, with radii up to 1000 $R_{\odot}$. Appropriately enough, these superluminous stars are called **supergiants.** Betelgeuse in Orion (Figure 17-16) and Antares in Scorpius are two supergiants you can find in the nighttime sky. Together, giants and supergiants make up about 1% of the stars in the sky.

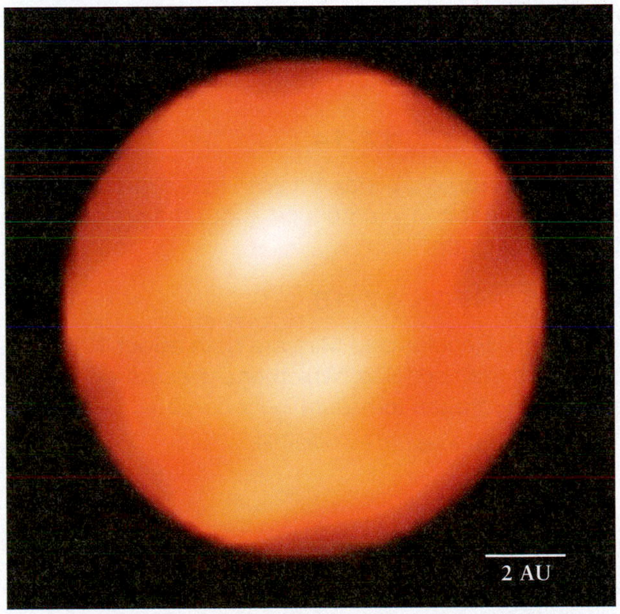

FIGURE 17-16 R I V U X G

Supergiant Betelgeuse Betelgeuse is just close enough and big enough that its surface features can be reconstructed using special techniques (interferometry at infrared wavelengths). The two brighter regions near the center are thought to be hotter gas rising by convection. The size of Betelgeuse is uncertain because its precise distance from Earth is not known, and its radius changes as the star expands and contracts. In this image, the radius is about 5.5 AU—slightly larger than Jupiter's orbital distance. (Xavier Haubois [Observatoire de Paris] et al.)

Both giants and supergiants have thermonuclear reactions occurring in their interiors, but the character of those reactions and where in the star they occur can be quite different than for a main-sequence star like the Sun. We will study these stars in more detail in Chapter 21.

The remaining 9% of stars in the H-R diagram of Figure 17-15a form a distinct grouping of data points toward the lower left corner. Although these stars are hot, their luminosities are quite low; hence, they must be small. They are appropriately called **white dwarfs.** These stars, which are so dim that they can be seen only with a telescope, are approximately the same size as Earth. As we will learn in Chapter 20, no thermonuclear reactions take place within white dwarf stars. Rather, like embers left from a fire, they are the still-glowing remnants of what were once giant stars.

By contrast, *brown* dwarfs (which lie at the extreme lower right of the main sequence, off the bottom and right-hand edge of Figure 17-15a or Figure 17-15b) are objects that will never become stars. They are not massive enough for significant fusing of hydrogen in their cores.

ANALOGY You can think of white dwarfs as "has-been" stars whose days of glory have passed. In this analogy, a brown dwarf is a "never-will-be" and is sometimes called a "failed star."

The existence of fundamentally different types of stars is the first important lesson to come from the H-R diagram. In later chapters we will find that these different types represent various stages in the lives of stars. For example, stars can move through the H-R diagram as they leave the main sequence and become red giants. We will use the H-R diagram as an essential tool for understanding how stars evolve.

CONCEPTCHECK 17-14

Betelgeuse and Barnard's star are both in the red portion of Figure 17-15. What can you deduce about their relative sizes from the H-R diagram?

CONCEPTCHECK 17-15

In Figure 17-15, Regulus and Mira are both about 100 times more luminous than our Sun. Which one is powered through conversion of hydrogen to helium in its core?

Answers appear at the end of the chapter.

17-8 Details of a star's spectrum reveal whether it is a giant, a white dwarf, or a main-sequence star

A star's surface temperature largely determines which lines are prominent in its spectrum. Therefore, classifying stars by spectral type is essentially the same as categorizing them by surface temperature. But as the H-R diagram in Figure 17-15b shows, stars of the same surface temperature can have very different luminosities. As an example, a star with surface temperature 5800 K could be a white dwarf, a main-sequence star, a giant, or a supergiant, depending on its luminosity. By examining the details of a star's spectrum, however, astronomers can determine to which of these categories a star belongs. As we will see, this gives astronomers a tool to determine the distances to stars millions of parsecs away, far beyond the maximum distance that can be measured using stellar parallax.

Determining a Star's Size from Its Spectrum

The smaller a star and the denser its atmosphere, the broader the absorption lines in its spectrum

Figure 17-17 compares the spectra of two stars of the same spectral type but different luminosity (and hence different size): a B8 supergiant and a B8 main-sequence star. Note that the Balmer lines of hydrogen are narrow in the spectrum of the very large, very luminous supergiant but are quite broad in the spectrum of the small, less luminous main-sequence star. In general, for stars of spectral types B through F, the larger and more luminous the star, the narrower its hydrogen lines.

Fundamentally, these differences between stars of different luminosity are due to differences between the stars' atmospheres, where absorption lines are produced. Hydrogen lines in particular are affected by the density and pressure of the gas in a star's atmosphere. The higher the density and pressure, the more frequently hydrogen atoms collide and interact with other atoms and ions in the atmosphere. These collisions shift the energy levels in the hydrogen atoms and thus broaden the hydrogen spectral lines.

In the atmosphere of a luminous giant star, the density and pressure are quite low because the star's mass is spread over a huge volume. Atoms and ions in the atmosphere are relatively far apart; hence, collisions between them are sufficiently infrequent

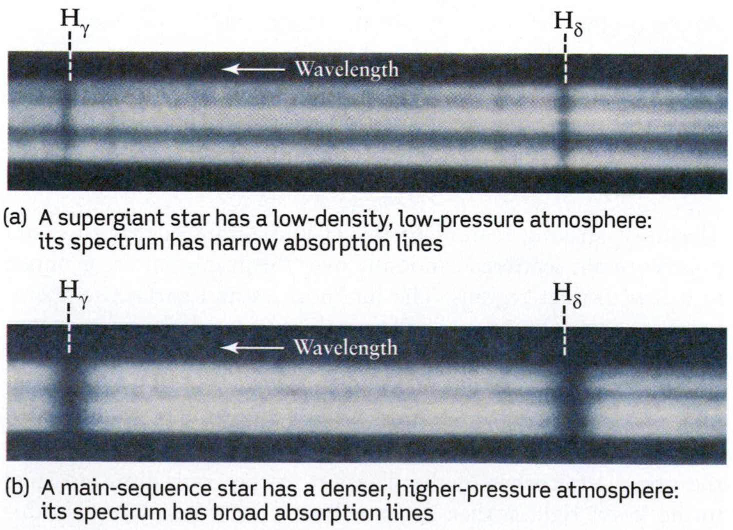

H_γ
H_δ
Wavelength
(a) A supergiant star has a low-density, low-pressure atmosphere: its spectrum has narrow absorption lines
H_γ
H_δ
Wavelength
(b) A main-sequence star has a denser, higher-pressure atmosphere: its spectrum has broad absorption lines

FIGURE 17-17 R I V U X G
How a Star's Size Affects Its Spectrum These are the spectra of two stars of the same spectral type (B8) and surface temperature (13,400 K) but different radii and luminosities: **(a)** the B8 supergiant Rigel (luminosity 58,000 $L_\odot$) in Orion, and **(b)** the B8 main-sequence star Algol (luminosity 100 $L_\odot$) in Perseus. (From *An Atlas of Stellar Spectra,* W. W. Morgan, P. C. Keenan, and E. Kellman, University of Chicago Press, © 1943)

that hydrogen atoms can produce narrow Balmer lines. A main-sequence star, however, is much more compact than a giant or supergiant. In the denser atmosphere of a main-sequence star, frequent interatomic collisions perturb the energy levels in the hydrogen atoms, thereby producing broader Balmer lines.

Luminosity Classes

In the 1930s, W. W. Morgan and P. C. Keenan of the Yerkes Observatory of the University of Chicago developed a system of **luminosity classes** based upon the subtle differences in spectral lines. When these luminosity classes are plotted on an H-R diagram (Figure 17-18), they provide a useful subdivision of the star types in the upper right of the diagram. Luminosity classes Ia and Ib are composed of supergiants; luminosity class V includes all the main-sequence stars. The intermediate classes distinguish giant stars of various luminosities. Note that for stars of a given surface temperature (that is, a given spectral type), the *higher* the number of the luminosity class, the *lower* the star's luminosity.

As we will see in Chapters 19 and 20, different luminosity classes represent different stages in the evolution of a star. White dwarfs are not always given a luminosity class of their own; as we mentioned in Section 17-7, they represent a final stage in stellar evolution in which no thermonuclear reactions take place.

Astronomers commonly use a shorthand description that combines a star's spectral type and its luminosity class. For example, the Sun is said to be a G2 V star. The spectral type indicates the star's surface temperature, and the luminosity class indicates its luminosity. Thus, an astronomer knows immediately that any G2 V star is a main-sequence star with a luminosity of about 1 $L_\odot$ and a surface temperature of about 5800 K. Similarly, a description of Aldebaran as a K5 III star tells an astronomer that it is a red giant with a luminosity of around 370 $L_\odot$ and a surface temperature of about 4000 K.

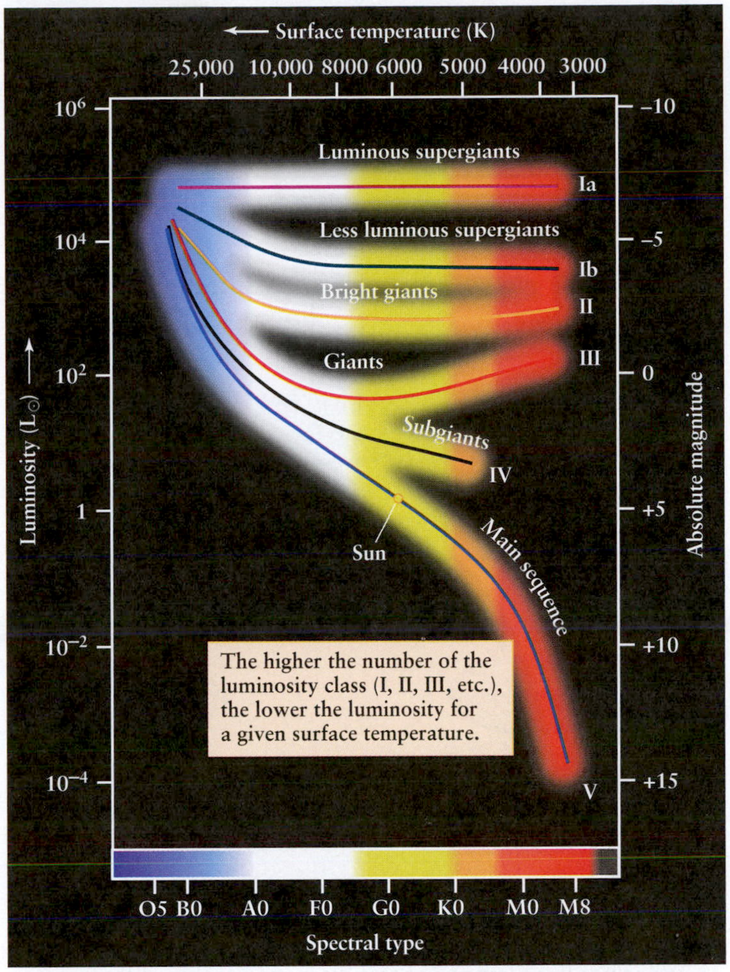

FIGURE 17-18

Luminosity Classes The H-R diagram is divided into regions corresponding to stars of different luminosity classes. (White dwarfs do not have their own luminosity class.) A star's spectrum reveals both its spectral type and its luminosity class; from these, the star's luminosity can be determined.

Spectroscopic Parallax

A star's spectral type and luminosity class, combined with the information on the H-R diagram, enable astronomers to estimate the star's distance from Earth. As an example, consider the star Pleione in the constellation Taurus. Its spectrum reveals Pleione to be a B8 V star (a hot, blue, main-sequence star, like the one in Figure 17-16b). Using Figure 17-17, we can read off that such a star's luminosity is 190 $L_\odot$. Given the star's luminosity and its apparent brightness—in the case of Pleione, 3.9×10^{-13} of the apparent brightness of the Sun—we can use the inverse-square law to determine its distance from Earth. The mathematical details are worked out in Box 17-2.

This method for determining distance, in which the luminosity of a star is found using spectroscopy, is called **spectroscopic parallax.** Figure 17-19 summarizes the method of spectroscopic parallax.

CAUTION! The term "spectroscopic parallax" is a bit misleading, because no parallax angle is involved! The idea is that measuring the star's spectrum takes the place of measuring its parallax as a way to find the star's distance. A better name for this method, although not the one used by astronomers, would be "spectroscopic distance determination."

Spectroscopic parallax is an incredibly powerful technique. No matter how remote a star is, this technique allows astronomers to determine its distance, provided only that its spectrum and apparent brightness can be measured. Box 17-2 gives an example of how spectroscopic parallax has been used to find the distance to stars in other galaxies tens of millions of parsecs away. By contrast, we saw in Section 17-1 that "real" stellar parallaxes can be measured only for stars within a few hundred parsecs.

Unfortunately, spectroscopic parallax has its limitations; distances to individual stars determined using this method often have errors greater than 10%. The reason is that the luminosity classes shown in Figure 17-17 are not thin lines on the H-R diagram but are moderately broad bands. Hence, even if a star's spectral type and luminosity class are known, there is still some uncertainty in the luminosity that we read off an H-R diagram. Nonetheless, spectroscopic parallax is often the only means that an astronomer has to estimate the distance to remote stars.

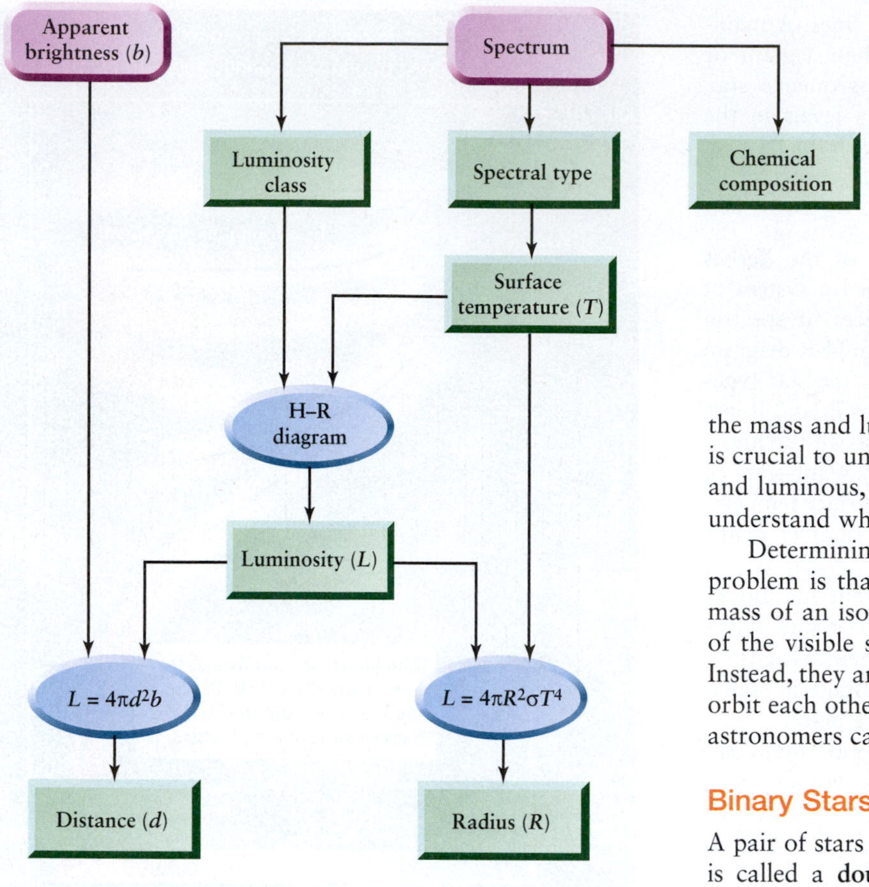

FIGURE 17-19
The Method of Spectroscopic Parallax If a star is too far away, its parallax angle is too small to allow a direct determination of its distance. This flowchart shows how astronomers deduce the properties of such a distant star. Note that the H-R diagram plays a central role in determining the star's luminosity from its spectral type and luminosity class. Just as for nearby stars (see Figure 17-14), the star's chemical composition is determined from its spectrum, and the star's radius is calculated from the luminosity and surface temperature.

What has been left out of this discussion is *why* different stars have different spectral types and luminosities. One key factor, as we shall see, turns out to be the mass of the star.

CONCEPTCHECK 17-16

What is the luminosity class of the main sequence (see Figure 17-18)? Are all stars in this luminosity class less luminous than the stars in luminosity class III?

CONCEPTCHECK 17-17

How does a K5 V star compare to a K5 II star in terms of temperature, luminosity, and radius?

Answers appear at the end of the chapter.

17-9 Observing binary star systems reveals the masses of stars

TUTORIAL 17-3

For main-sequence stars, there is a direct correlation between mass and luminosity

We now know something about the sizes, temperatures, and luminosities of stars. To complete our picture of the physical properties of stars, we need to know their masses. In this section, we will see that stars come in a wide range of masses. We will also discover an important relationship between the mass and luminosity of main-sequence stars. This relationship is crucial to understanding why some main-sequence stars are hot and luminous, while others are cool and dim. It will also help us understand what happens to a star as it ages and evolves.

Determining the masses of stars is not trivial, however. The problem is that there is no practical, direct way to measure the mass of an isolated star. Fortunately for astronomers, about half of the visible stars in the night sky are not isolated individuals. Instead, they are *multiple-star systems,* in which two or more stars orbit each other. By carefully observing the motions of these stars, astronomers can glean important information about their masses.

Binary Stars

A pair of stars located at nearly the same position in the night sky is called a **double star.** The Anglo-German astronomer William Herschel made the first organized search for such pairs. Between 1782 and 1821, he published three catalogs listing more than 800 double stars. Late in the nineteenth century, his son, John Herschel, discovered 10,000 more doubles. Some of these double stars are **optical double stars,** which are two stars that lie along nearly the same line of sight but are actually at very different distances from us. But many double stars are true **binary stars,** or **binaries**—pairs of stars that actually orbit each other. Figure 17-20 shows an example of this orbital motion.

When astronomers can actually see the two stars orbiting each other, a binary is called a **visual binary.** By observing the binary over an extended period, astronomers can plot the orbit that one star appears to describe around the other, as shown in the center diagram in Figure 17-20.

In fact, *both* stars in a binary system are in motion. They orbit each other because of their mutual gravitational attraction, and their orbital motions obey Kepler's third law as formulated by Isaac Newton (see Section 4-7 and Box 4-4). This law can be written as follows:

Kepler's third law for binary star systems

$$M_1 + M_2 = \frac{a^3}{P^2}$$

M_1, M_2 = masses of two stars in binary system, in solar masses

a = semimajor axis of one star's orbit around the other, in AU

P = orbital period, in years

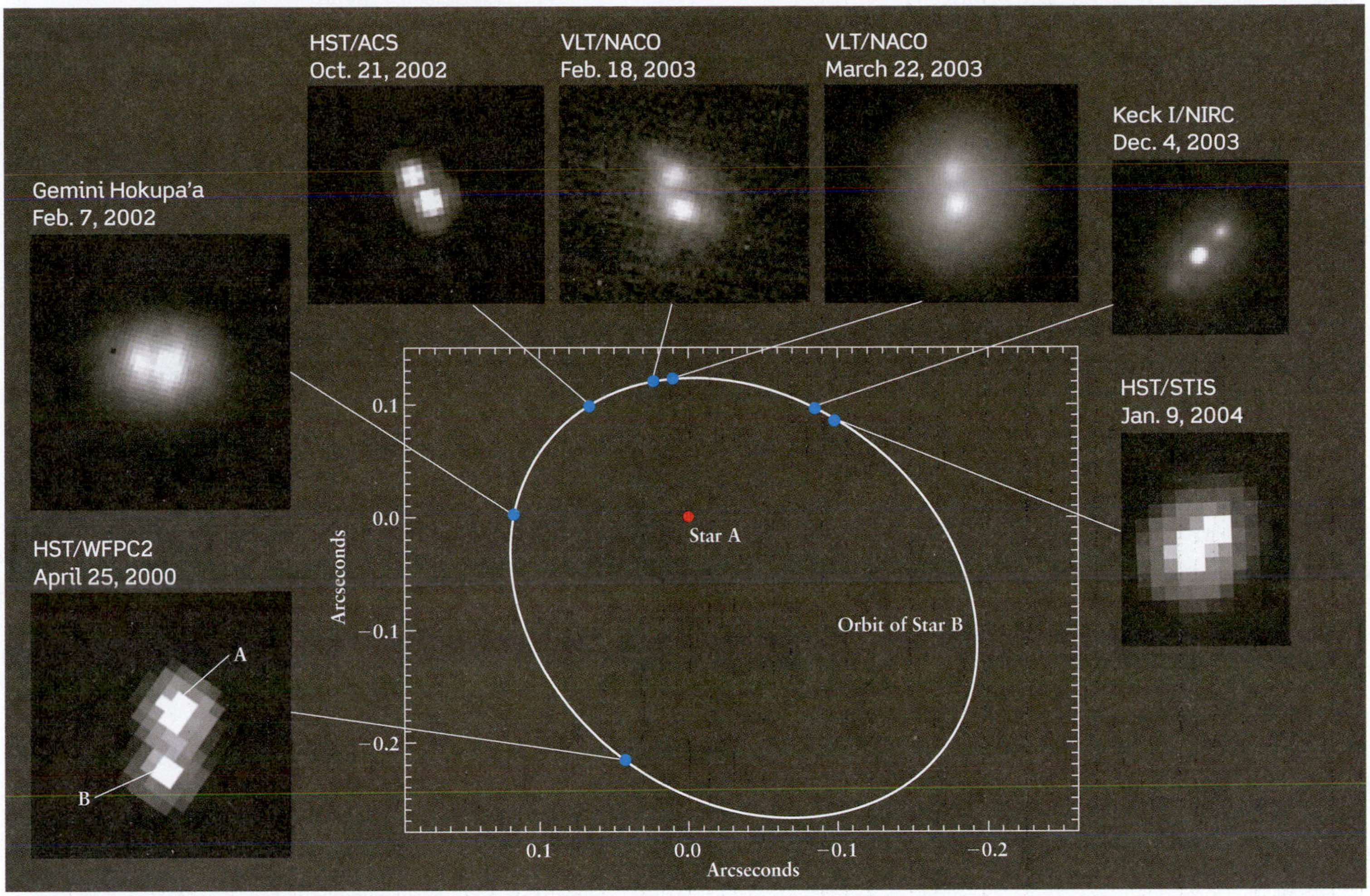

FIGURE 17-20 R I V U X G

A Binary Star System As seen from Earth, the two stars that make up the binary system called 2MASSW J074642512000321 are separated by less than 1/3 arcsecond. The images surrounding the center diagram show the relative positions of the two stars over a four-year period. These images were made by the Hubble Space Telescope (HST), the European Southern Observatory's Very Large Telescope (VLT), and Keck I and Gemini North in Hawaii (see Figure 6-16). For simplicity, the diagram shows one star as remaining stationary; in reality, both stars move around their common center of mass. (H. Bouy et al., MPE and ESO)

Here a is the semimajor axis of the elliptical orbit that one star appears to describe around the other, plotted as in the center diagram in Figure 17-20. As this equation indicates, if we can measure this semimajor axis (a) and the orbital period (P), we can learn something about the masses of the two stars.

In principle, the orbital period of a visual binary is easy to determine. All you have to do is see how long it takes for the two stars to revolve once about each other. The two stars shown in Figure 17-20 are relatively close, about 2.5 AU on average, and their orbital period is only 10 years. Many binary systems have much larger separations, however, and the period may be so long that more than one astronomer's lifetime is needed to complete the observations.

Determining the semimajor axis of an orbit can also be a challenge. The *angular* separation between the stars can be determined by observation. To convert this angle into a physical distance between the stars, we need to know the distance between the binary and Earth. This distance can be found from parallax measurements or by using spectroscopic parallax. The astronomer must also take into account how the orbit is tilted to our line of sight.

Once both P and a have been determined, Kepler's third law can be used to calculate $M_1 + M_2$, the sum of the masses of the two stars in the binary system. But this analysis tells us nothing about the *individual* masses of the two stars. To obtain these masses, more information about the motions of the two stars is needed.

Each of the two stars in a binary system actually moves in an elliptical orbit about the **center of mass** of the system. Imagine two children sitting on opposite ends of a seesaw (Figure 17-21a). For the seesaw to balance properly, they must position themselves so that their center of mass—an imaginary point that lies along a line connecting their two bodies—is at the fulcrum, or pivot point of the seesaw. If the two children have the same mass, the center

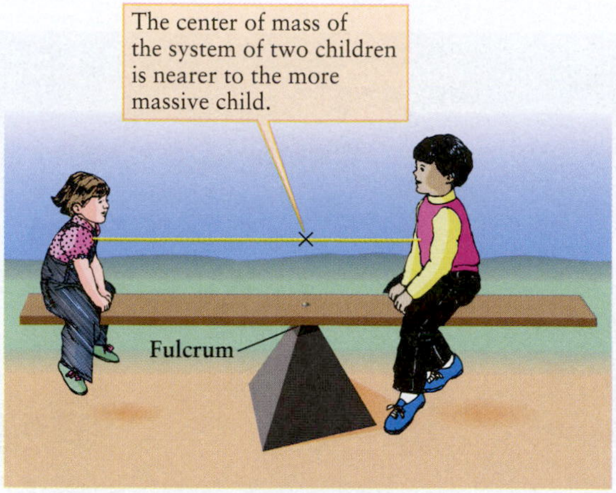
The center of mass of the system of two children is nearer to the more massive child.
Fulcrum

(a) A "binary system" of two children

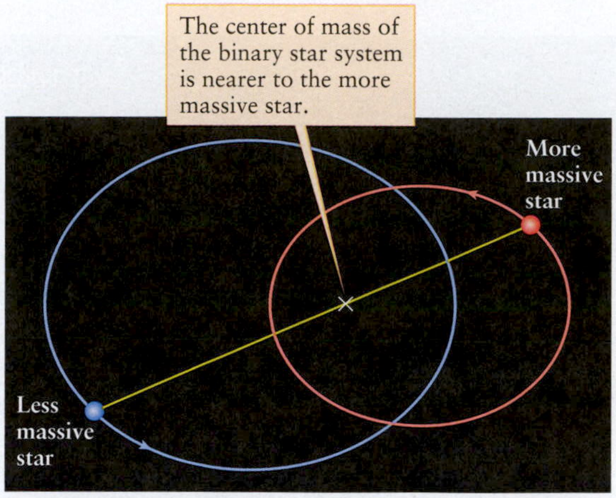
The center of mass of the binary star system is nearer to the more massive star.
More massive star
Less massive star

(b) A binary star system

ANIMATION 17-3

FIGURE 17-21
Center of Mass in a Binary Star System **(a)** A seesaw balances if the fulcrum is at the center of mass of the two children. **(b)** The members of a binary star system orbit around the center of mass of the two stars. Although their elliptical orbits cross each other, the two stars are always on opposite sides of the center of mass and thus never collide.

of mass lies midway between them, and they should sit equal distances from the fulcrum. If their masses are different, the center of mass is closer to the heavier child.

Just as the seesaw naturally balances at its center of mass, the two stars that make up a binary system naturally orbit around their center of mass (Figure 17-21b). The center of mass always lies along the line connecting the two stars and is closer to the more massive star.

The center of mass of a visual binary is located by plotting the separate orbits of the two stars, as in Figure 17-21b, using the background stars as reference points. The center of mass lies at the common focus of the two elliptical orbits. Comparing the relative sizes of the two orbits around the center of mass yields the ratio of the two stars' masses, M_1/M_2. With the sum $M_1 + M_2$ already known from Kepler's third law, the individual masses of the two stars can then be determined using algebra.

CONCEPTCHECK 17-18

If two stars in a binary system were moved farther apart, how would their masses and orbital periods change?

Answer appears at the end of the chapter.

Main-Sequence Masses and the Mass-Luminosity Relation

Years of careful, patient observations of binaries have slowly yielded the masses of many stars. As the data accumulated, an important trend began to emerge: For main-sequence stars, there is a direct correlation between mass and luminosity. The more massive a main-sequence star, the more luminous it is. Figure 17-22 depicts this **mass-luminosity relation** as a graph. The range of stellar masses extends from less than 0.1 of a solar mass to more than 50 solar masses. The Sun's mass lies between these extremes.

The *Cosmic Connections* figure depicts the mass-luminosity relation for main-sequence stars on an H-R diagram. This figure shows that the main sequence on an H-R diagram is a progression in mass as well as in luminosity and surface temperature. The hot, bright, bluish stars in the upper left corner of an H-R diagram are the most massive main-sequence stars. Likewise, the dim, cool, reddish stars in the lower right corner of an H-R diagram are the

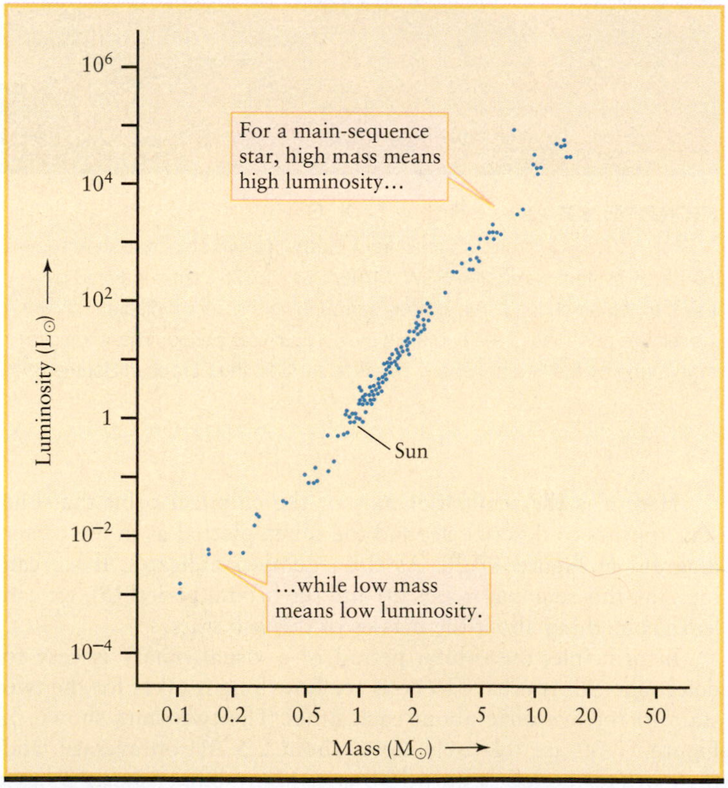
Luminosity (L⊙) →
10^6
10^4
10^2
1
10^{-2}
10^{-4}
For a main-sequence star, high mass means high luminosity...
Sun
...while low mass means low luminosity.
0.1 0.2 0.5 1 2 5 10 20 50
Mass (M⊙) →

FIGURE 17-22
The Mass-Luminosity Relation For main-sequence stars, there is a direct correlation between mass and luminosity—the more massive a star, the more luminous it is. A main-sequence star of mass 10 M⊙ (that is, 10 times the Sun's mass) has roughly 3000 times the Sun's luminosity (3000 L⊙); one with 0.1 M⊙ has a luminosity of only about 0.001 L⊙.

COSMIC CONNECTIONS

The Main Sequence and Masses

The main sequence is an arrangement of stars according to their mass. The most massive main-sequence stars have the greatest luminosity, greatest radius, and greatest surface temperature. These characteristics are consequences of the behavior of thermonuclear reactions at the core of a main-sequence star.

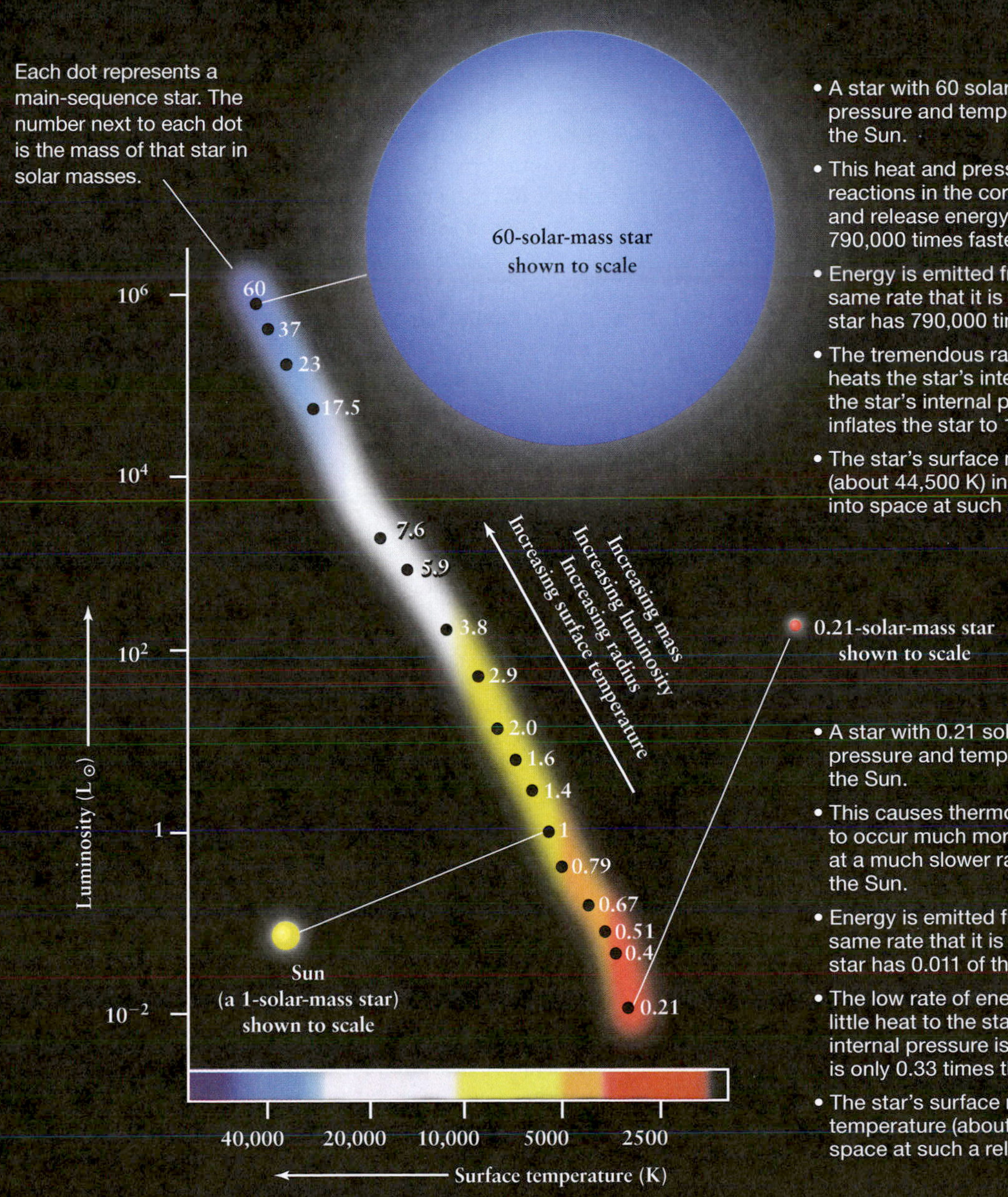

- A star with 60 solar masses has much higher pressure and temperature at its core than does the Sun.
- This heat and pressure cause thermonuclear reactions in the core to occur much more rapidly and release energy at a much faster rate — 790,000 times faster than in the Sun.
- Energy is emitted from the star's surface at the same rate that it is released in the core, so the star has 790,000 times the Sun's luminosity.
- The tremendous rate of energy release also heats the star's interior tremendously, increasing the star's internal pressure. This pressure inflates the star to 15 times the Sun's radius.
- The star's surface must be at a high temperature (about 44,500 K) in order for it to emit energy into space at such a rapid rate.

- A star with 0.21 solar mass has much lower pressure and temperature at its core than does the Sun.
- This causes thermonuclear reactions in the core to occur much more slowly and release energy at a much slower rate — 0.011 times as fast as in the Sun.
- Energy is emitted from the star's surface at the same rate that it is released in the core, so the star has 0.011 of the Sun's luminosity.
- The low rate of energy release supplies relatively little heat to the star's interior, so the star's internal pressure is low. Hence the star's radius is only 0.33 times the Sun's radius.
- The star's surface need be at only a low temperature (about 3200 K) to emit energy into space at such a relatively slow rate.

least massive. Main-sequence stars of intermediate temperature and luminosity also have intermediate masses.

The mass of a main-sequence star also helps determine its radius. Referring back to Figure 17-15b, we see that if we move along the main sequence from low luminosity to high luminosity, the radius of the star increases. Thus, we have the following general rule for main-sequence stars:

The greater the mass of a main-sequence star, the greater its luminosity, its surface temperature, and its radius.

Mass and Main-Sequence Stars

Why is mass the controlling factor in determining the properties of a main-sequence star? The answer is that all main-sequence stars are objects like the Sun, with essentially the same chemical composition as the Sun but with different masses. Like the Sun, all main-sequence stars shine because thermonuclear reactions at their cores convert hydrogen to helium and release energy. The greater the total mass of the star, the greater the pressure and temperature at the core, the more rapidly thermonuclear reactions take place in the core, and the greater the energy output—that is, the luminosity—of the star. In other words, the greater the mass of a main-sequence star, the greater its luminosity. This statement is just the mass-luminosity relation, which we can now recognize as a natural consequence of the nature of main-sequence stars.

Like the Sun, main-sequence stars are in a state of both hydrostatic equilibrium and thermal equilibrium. Calculations using models of a main-sequence star's interior (like the solar models we discussed in Section 16-2) show that as a star settles into equilibrium, a more massive star must have a larger radius and a higher surface temperature. This result is just what we see when we plot the curve of the main sequence on an H-R diagram (see Figure 17-15b). As you move up the main sequence from less massive stars (at the lower right in the H-R diagram) to more massive stars (at the upper left), the radius and surface temperature both increase.

Calculations using hydrostatic and thermal equilibrium also show that if a star's mass is less than about $0.08M_{\odot}$, the core pressure and temperature are too low for thermonuclear reactions to take place. The "star" (or so-called failed star) is then a brown dwarf. Brown dwarfs are still luminous; as they slowly contract, gravitational energy is converted into heat.

CAUTION! The mass-luminosity relation we have discussed applies to main-sequence stars only. There are *no* simple mass-luminosity relations for giant, supergiant, or white dwarf stars. Why these stars lie where they do on an H-R diagram will become apparent when we study the evolution of stars in Chapters 19 and 20. We will find that main-sequence stars evolve into giant and supergiant stars, and that some of these eventually end their lives as white dwarfs.

CONCEPTCHECK 17-19

From the mass-luminosity relation in the Cosmic Connections figure, what can you say about the masses, temperatures, and radii for the most luminous stars?

Answer appears at the end of the chapter.

17-10 Spectroscopy makes it possible to study binary systems in which the two stars are close together

We have described how the masses of stars can be determined from observations of visual binaries, in which the two stars can be distinguished from each other. But if the two stars in a binary system are too close together, the images of the two stars can blend to produce the semblance of a single star. Happily, in many cases we can use spectroscopy to decide whether a seemingly single star is in fact a binary system. Spectroscopic observations of binaries provide additional useful information about star masses.

Some binaries are discovered when the spectrum of a star shows incongruous spectral lines. For example, the spectrum of what appears to be a single star may include both strong hydrogen lines (characteristic of a type A star) and strong absorption bands of titanium oxide (typical of a type M star). Because a single star cannot have the differing physical properties of these two spectral types, such a star must actually be a binary system that is too far away for us to resolve its individual stars. A binary system detected in this way is called a **spectrum binary.**

Other binary systems can be detected using the Doppler effect. If a star is moving toward Earth, its spectral lines are displaced toward the short-wavelength (blue) end of the spectrum. Conversely, the spectral lines of a star moving away from us are shifted toward the long-wavelength (red) end of the spectrum. The upper portion of Figure 17-23 applies these ideas to a hypothetical binary star system with an orbital plane that is edge-on to our line of sight.

As the two stars move around their orbits, they periodically approach and recede from us. Hence, the spectral lines of the two stars are alternately blueshifted and redshifted. The two stars in this hypothetical system are so close together that they appear through a telescope as a single star with a single spectrum. Because one star shows a blueshift while the other is showing a redshift, the spectral lines of the binary system appear to split apart and rejoin periodically. Stars whose binary character is revealed by such shifting spectral lines are called **spectroscopic binaries.**

Exploring Spectroscopic Binary Stars

Stars in close binary systems move so rapidly that we can detect their motion using the Doppler effect

To analyze a spectroscopic binary, astronomers measure the wavelength shift of each star's spectral lines and use the Doppler shift formula (introduced in Section 5-9 and Box 5-6) to determine the *radial velocity* of each star—that is, how fast and in what direction it is moving along our line of sight. The lower portion of Figure 17-22 shows a graph of the radial velocity versus time, called a **radial velocity curve,** for the binary system HD 171978. Each of the two stars alternately approaches and recedes as it orbits around the center of mass. The pattern of the curves repeats every 15 days, which is the orbital period of the binary.

It is important to emphasize that the Doppler effect applies only to motion along the line of sight. Motion perpendicular to the line of sight does not affect the observed wavelengths of spectral lines. Hence, the ideal orientation for a spectroscopic binary is to

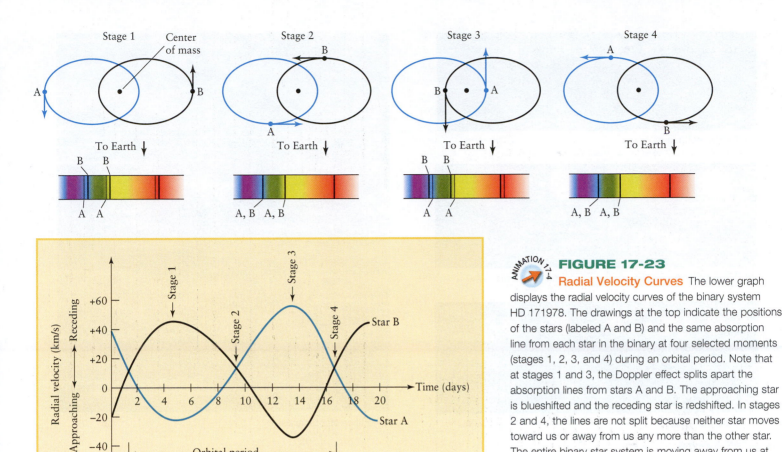

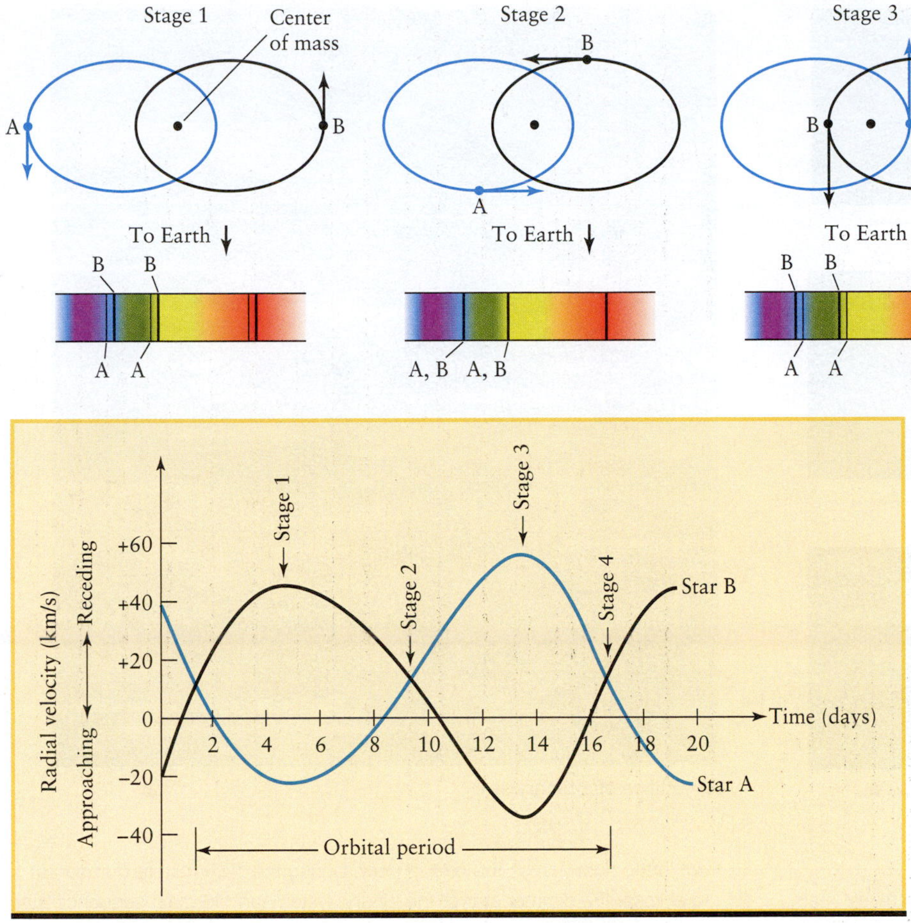

ANIMATION 17-4

FIGURE 17-23

Radial Velocity Curves The lower graph displays the radial velocity curves of the binary system HD 171978. The drawings at the top indicate the positions of the stars (labeled A and B) and the same absorption line from each star in the binary at four selected moments (stages 1, 2, 3, and 4) during an orbital period. Note that at stages 1 and 3, the Doppler effect splits apart the absorption lines from stars A and B. The approaching star is blueshifted and the receding star is redshifted. In stages 2 and 4, the lines are not split because neither star moves toward us or away from us any more than the other star. The entire binary star system is moving away from us at 12 km/s, which is why the entire pattern of radial velocity curves is displaced upward from the zero-velocity line.

have the stars orbit in a plane that is edge-on to our line of sight. (By contrast, a *visual* binary is best observed if the orbital plane is face-on to our line of sight.) For the Doppler shifts to be noticeable, the orbital speeds of the two stars should be at least a few kilometers per second.

As for visual binaries, spectroscopic binaries allow astronomers to learn about stellar masses. From a radial velocity curve, one can find the *ratio* of the masses of the two stars in a binary. The *sum* of the masses is related to the orbital speeds of the two stars by Kepler's laws and Newtonian mechanics. If both the ratio of the masses and their sum are known, the individual masses can be determined using algebra. However, determining the sum of the masses requires that we know how the binary orbits are tilted from our line of sight. This is because the Doppler shifts reveal only the radial velocities of the stars rather than their true orbital speeds. This tilt is often impossible to determine, because we cannot see the individual stars in the binary. Thus, the masses of stars in spectroscopic binaries tend to be uncertain.

There is one important case in which we can determine the orbital tilt of a spectroscopic binary. If the two stars are observed to eclipse each other periodically, then we must be viewing the orbit nearly edge-on. As we will see next, individual stellar masses—as well as other useful data—can be determined if a spectroscopic binary also happens to be such an *eclipsing* binary.

CONCEPTCHECK 17-20

In Figure 17-23, describe the Doppler shift observed at Stage 2 of the binary orbit.

Answer appears at the end of the chapter.

17-11 Light curves of eclipsing binaries provide detailed information about the two stars

Some binary systems are oriented so that the two stars periodically eclipse each other as seen from Earth. These **eclipsing binaries** can be detected even when the two stars cannot be resolved visually as two distinct images in the telescope. The apparent brightness of the image of the binary dims briefly each time one star blocks the light from the other.

Using a sensitive detector at the focus of a telescope, an astronomer can measure the incoming light intensity quite accurately and create a **light curve** (Figure 17-24). The shape of the light curve for an eclipsing binary reveals at a glance whether the eclipse is partial or total (compare Figures 17-24a and 17-24b). Figure 17-24d shows an observation of a binary system undergoing a total eclipse.

In fact, the light curve of an eclipsing binary can yield a surprising amount of information. For example, the ratio of the surface

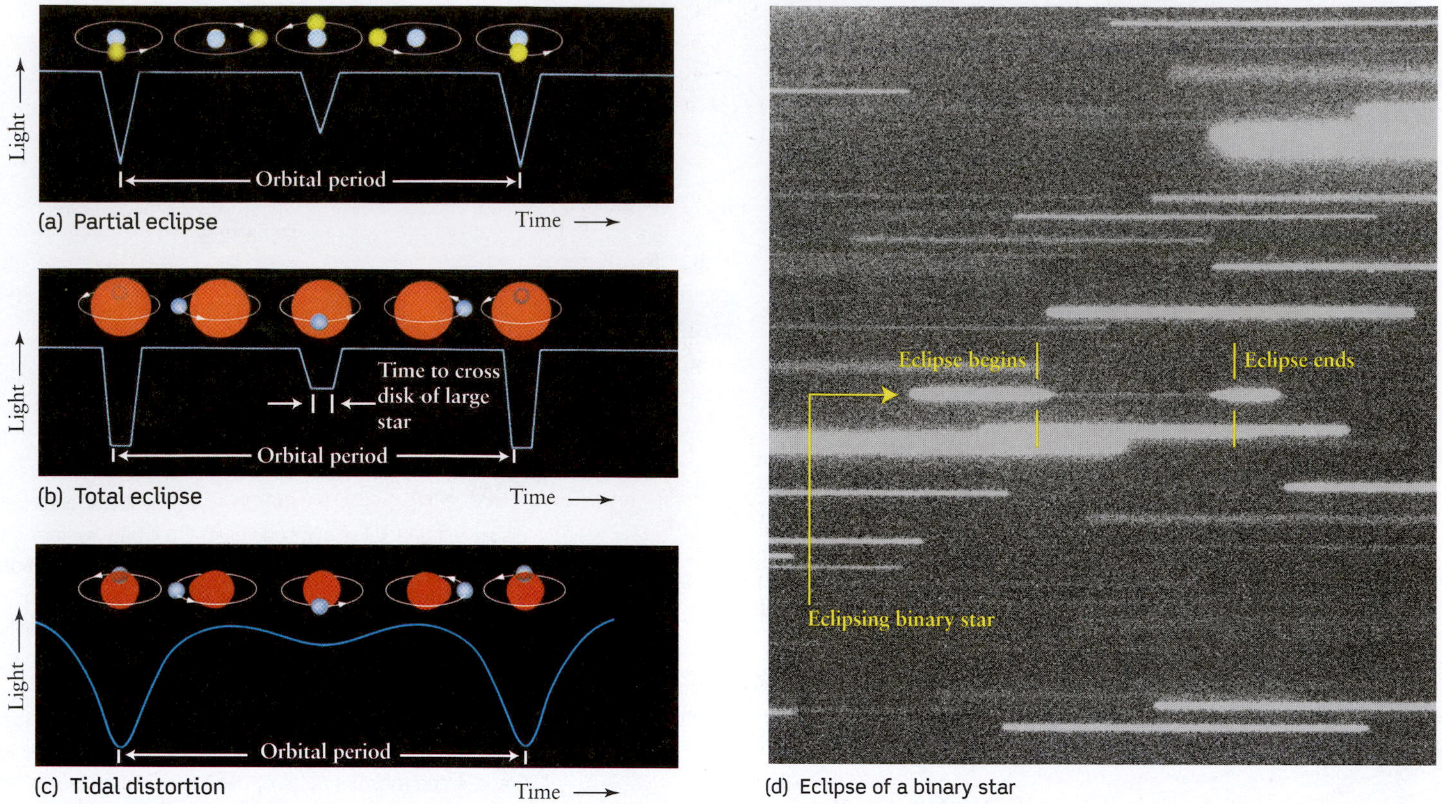

FIGURE 17-24 R I V U X G

Representative Light Curves of Eclipsing Binaries

(a), (b), (c) The shape of the light curve of an eclipsing binary can reveal many details about the two stars that make up the binary. **(d)** This image shows the binary star NN Serpens (indicated by the arrow) undergoing a total eclipse. The telescope was moved during the exposure so that the sky drifted slowly from left to right across the field of view. During the 10.5-minute duration of the eclipse, the dimmer star of the binary system (an M6 main-sequence star) passed in front of the other, more luminous star (a white dwarf). The binary became so dim that it almost disappeared. (European Southern Observatory)

temperatures can be determined from how much their combined light is diminished when the stars eclipse each other. Also, the duration of a mutual eclipse tells astronomers about the relative sizes of the stars and their orbits.

Eclipsing binaries can reveal the sizes and shapes of stars

If the eclipsing binary is also a double-line spectroscopic binary, an astronomer can calculate the mass and radius of each star from the light curves and the velocity curves. Unfortunately, very few binary stars are of this ideal type. Stellar radii determined in this way agree well with the values found using the Stefan-Boltzmann law, as described in Section 17-6.

The shape of a light curve can reveal many additional details about a binary system. In some binaries, for example, the gravitational pull of one star distorts the other, much as the Moon distorts Earth's oceans in producing tides (see Figure 4-26). Figure 17-24c shows how such tidal distortion gives the light curve a different shape than in Figure 17-24b.

Information about stellar atmospheres can also be derived from light curves. Suppose that one star of a binary is a luminous main-sequence star and the other is a bloated red giant. By observing exactly how the light from the bright main-sequence star is gradually cut off as it moves behind the edge of the red giant during the beginning of an eclipse, astronomers can infer the pressure and density in the upper atmosphere of the red giant.

Binary systems are tremendously important because they enable astronomers to measure stellar masses as well as other key properties of stars. In the next several chapters, we will use this information to help us piece together the story of *stellar evolution*—how stars are born, evolve, and eventually die.

CONCEPTCHECK 17-21

Consider the total eclipse in Figure 17-24d. Why is there still a thin line of light during the total eclipse?

Answer appears at the end of the chapter.

KEY WORDS

Terms preceded by an asterisk (*) are discussed in the Boxes.

absolute magnitude, p. 473
apparent brightness (brightness), p. 469
apparent magnitude, p. 473
binary star (binary), p. 490
brown dwarf, p. 482
center of mass, p. 491
color ratio, p. 477
*distance modulus, p. 476
double star, p. 490
eclipsing binary, p. 495
giant, p. 487
Hertzsprung-Russell diagram (H-R diagram), p. 487
inverse-square law, p. 469
light curve, p. 495
luminosity, p. 466
luminosity class, p. 489

luminosity function, p. 473
magnitude scale, p. 473
main sequence, p. 487
main-sequence star, p. 487
mass-luminosity relation, p. 492
metals, p. 482
OBAFGKM, p. 479
optical double star, p. 490
parallax, p. 466
parsec, p. 466
photometry, p. 470
*proper motion, p. 468
*radial velocity, p. 468
radial velocity curve, p. 494
red giant, p. 487
*space velocity, p. 468
spectral classes, p. 479
spectral types, p. 479
spectroscopic binary, p. 494
spectroscopic parallax, p. 489
spectrum binary, p. 494
stellar parallax, p. 466
supergiant, p. 487
*tangential velocity, p. 468
UBV photometry, p. 477
visual binary, p. 490
white dwarf, p. 488

KEY IDEAS

Measuring Distances to Nearby Stars: Distances to the nearer stars can be determined by parallax, the apparent shift of a star against the background stars observed as Earth moves along its orbit.

• Parallax measurements made from orbit, above the blurring effects of the atmosphere, are much more accurate than those made with Earth-based telescopes.

• Stellar parallaxes can only be measured for stars within a few hundred parsecs.

The Inverse-Square Law: A star's luminosity (total light output), apparent brightness, and distance from Earth are related by the inverse-square law. If any two of these quantities are known, the third can be calculated.

The Population of Stars: Stars of relatively low luminosity are more common than more luminous stars. Our own Sun is a rather average star of intermediate luminosity.

The Magnitude Scale: The apparent magnitude scale is an alternative way to measure a star's apparent brightness.

• The absolute magnitude of a star is the apparent magnitude it would have if viewed from a distance of 10 pc. A version of the inverse-square law relates a star's absolute magnitude, apparent magnitude, and distance.

Photometry and Color Ratios: Photometry measures the apparent brightness of a star. The color ratios of a star are the ratios of brightness values obtained through different standard filters, such as the U, B, and V filters. These ratios are a measure of the star's surface temperature.

Spectral Types: Stars are classified into spectral types (subdivisions of the spectral classes O, B, A, F, G, K, and M), based on the major patterns of spectral lines in their spectra. The spectral class and type of a star is directly related to its surface temperature: O stars are the hottest and M stars are the coolest.

• Most brown dwarfs are in even cooler spectral classes called L, T, and Y. Unlike true stars, brown dwarfs are too small to sustain thermonuclear fusion.

Hertzsprung-Russell Diagram: The Hertzsprung-Russell (H-R) diagram is a graph plotting the absolute magnitudes of stars against their spectral types—or, equivalently, their luminosities against surface temperatures.

• The positions on the H-R diagram of most stars are along the main sequence, a band that extends from high luminosity and high surface temperature to low luminosity and low surface temperature.

• On the H-R diagram, giant and supergiant stars lie above the main sequence, while white dwarfs are below the main sequence.

• By carefully examining a star's spectral lines, astronomers can determine whether that star is a main-sequence star, giant, supergiant, or white dwarf. Using the H-R diagram and the inverse-square law, the star's luminosity and distance can be found without measuring its stellar parallax.

Binary Stars: Binary stars, in which two stars are held in orbit around each other by their mutual gravitational attraction, are surprisingly common. Those that can be resolved into two distinct star images by a telescope are called visual binaries.

• Each of the two stars in a binary system moves in an elliptical orbit about the center of mass of the system.

• Binary stars are important because they allow astronomers to determine the masses of the two stars in a binary system. The masses can be computed from measurements of the orbital period and orbital dimensions of the system.

Mass-Luminosity Relation for Main-Sequence Stars: Main-sequence stars are stars like the Sun but with different masses.

• The mass-luminosity relation expresses a direct correlation between mass and luminosity for main-sequence stars. The greater the mass of a main-sequence star, the greater its luminosity (and also the greater its radius and surface temperature).

Spectroscopic Observations of Binary Stars: Some binaries can be detected and analyzed, even though the system may be so distant or the two stars so close together that the two star images cannot be resolved.

• A spectrum binary appears to be a single star but has a spectrum with the absorption lines for two distinctly different spectral types.

• A spectroscopic binary has spectral lines that shift back and forth in wavelength. This is caused by the Doppler effect, as the orbits of the stars carry them first toward and then away from Earth.

• An eclipsing binary is a system whose orbits are viewed nearly edge-on from Earth, so that one star periodically eclipses the other. Detailed information about the stars in an eclipsing binary can be obtained from a study of the binary's radial velocity curve and its light curve.

QUESTIONS

Review Questions

1. Explain the difference between a star's apparent brightness and its luminosity.
2. TUTORIAL 17-1 Describe how the parallax method of finding a star's distance is similar to binocular (two-eye) vision in humans.
3. Why does it take at least six months to make a measurement of a star's parallax?
4. Why are measurements of stellar parallax difficult to make? What are the advantages of making these measurements from orbit?

5. TUTORIAL 17-2 What is the inverse-square law? Use it to explain why an ordinary lightbulb can appear brighter than a star, even though the lightbulb emits far less light energy per second.
6. Briefly describe how you would determine the luminosity of a nearby star. Of what value is knowing the luminosity of various stars?
7. Which are more common, stars more luminous than the Sun or stars less luminous than the Sun?
8. Why is the magnitude scale called a "backward" scale? What is the difference between apparent magnitude and absolute magnitude?
9. The star Zubenelgenubi (from the Arabic for "scorpion's southern claw") has apparent magnitude 12.75, while the star Sulafat (Arabic for "tortoise") has apparent magnitude 13.25. Which star appears brighter? From this information alone, what can you conclude about the luminosities of these stars? Explain your answer.
10. Explain why the color ratios of a star are related to the star's surface temperature.
11. Would it be possible for a star to appear bright when viewed through a U filter or a V filter, but dim when viewed through a B filter? Explain your answer.
12. Which gives a more accurate measure of a star's surface temperature, its color ratios or its spectral lines? Explain.
13. Menkalinan (Arabic for "shoulder of the rein-holder") is an A2 star in the constellation Auriga (the Charioteer). What is its spectral class? What is its spectral type? Which gives a more precise description of the spectrum of Menkalinan?
14. What are the most prominent absorption lines you would expect to find in the spectrum of a star with a surface temperature of (**a**) 35,000 K, (**b**) 2800 K, and (**c**) 5800 K (like the Sun)? Briefly describe why these stars have such different spectra even though they have essentially the same chemical composition.
15. A fellow student expresses the opinion that since the Sun's spectrum has only weak absorption lines of hydrogen, this element cannot be a major constituent of the Sun. How would you enlighten this person?
16. If a red star and a blue star both have the same radius and both are the same distance from Earth, which one looks brighter in the night sky? Explain why.
17. If a red star and a blue star both appear equally bright and both are the same distance from Earth, which one has the larger radius? Explain why.
18. If a red star and a blue star both have the same radius and both appear equally bright, which one is farther from Earth? Explain why.
19. Sketch a Hertzsprung-Russell diagram. Indicate the regions on your diagram occupied by (**a**) main-sequence stars, (**b**) red giants, (**c**) supergiants, (**d**) white dwarfs, and (**e**) the Sun.
20. Most of the bright stars in the night sky (see Appendix 5) are giants and supergiants. How can this be, if giants and supergiants make up only 1% of the population of stars?
21. Explain why the dashed lines in Figure 17-15b slope down and to the right.
22. Some giant and supergiant stars are of the same spectral type (G2) as the Sun. What aspects of the spectrum of a G2 star would you concentrate on to determine the star's luminosity class? Explain what you would look for.
23. Briefly describe how you would determine the distance to a star whose parallax is too small to measure.
24. What information about stars do astronomers learn from binary systems that cannot be learned in any other way? What measurements do they make of binary systems to garner this information?
25. Suppose that you want to determine the temperature, diameter, and luminosity of an isolated star (not a member of a binary system). Which of these physical quantities require you to know the distance to the star? Explain your answer.
26. What is the mass-luminosity relation? Does it apply to stars of all kinds?
27. Use Figure 17-22 to (**a**) estimate the mass of a main-sequence star that is 1000 times as luminous as the Sun, and (**b**) estimate the luminosity of a main-sequence star whose mass is one-fifth that of the Sun. Explain your answers.
28. Which is more massive, a red main-sequence star or a blue main-sequence star? Which has the greater radius? Explain your answers.
29. How do white dwarfs differ from brown dwarfs? Which are more massive? Which are larger in radius? Which are denser?
30. TUTORIAL 17-3 Sketch the radial velocity curves of a binary consisting of two identical stars moving in circular orbits that are (**a**) perpendicular to our line of sight and (**b**) parallel to our line of sight.
31. Give two reasons why a visual binary star is unlikely to also be a spectroscopic binary star.
32. Sketch the light curve of an eclipsing binary consisting of two identical stars in highly elongated orbits oriented so that (**a**) their major axes are pointed toward Earth and (**b**) their major axes are perpendicular to our line of sight.

Advanced Questions

Questions preceded by an asterisk (*) involve the topic discussed in the Boxes.

Problem-solving tips and tools

Look carefully at the worked examples in Boxes 17-1, 17-2, 17-3, and 17-4 before attempting these exercises. For data on the planets, see Table 7-1 or Appendices 1 and 2 at the back of this book. Remember that a telescope's light-gathering power is proportional to the area of its objective or primary mirror. The volume of a sphere of radius r is $4\pi r^3/3$. Make use of the H-R diagrams in this chapter to answer questions involving spectroscopic parallax. As Box 17-3 shows, some of the problems concerning magnitudes may require facility with logarithms.

33. Find the average distance from the Sun to Neptune in parsecs. Compared to Neptune, how many times farther away from the Sun is Proxima Centauri?

34. Suppose that a dim star were located 2 million AU from the Sun. Find (**a**) the distance to the star in parsecs and (**b**) the parallax angle of the star. Would this angle be measurable with present-day techniques?

35. The star GJ 1156 has a parallax angle of 0.153 arcsec. How far away is the star?

*36. Kapteyn's star (named after the Dutch astronomer who found it) has a parallax of 0.255 arcsec, a proper motion of 8.67 arcsec per year, and a radial velocity of 1246 km/s. (**a**) What is the star's tangential velocity? (**b**) What is the star's actual speed relative to the Sun? (**c**) Is Kapteyn's star moving toward the Sun or away from the Sun? Explain.

*37. How far away is a star that has a proper motion of 0.08 arcseconds per year and a tangential velocity of 40 km/s? For a star at this distance, what would its tangential velocity have to be in order for it to exhibit the same proper motion as Barnard's star (see Box 17-1)?

*38. The space velocity of a certain star is 120 km/s and its radial velocity is 72 km/s. Find the star's tangential velocity.

*39. In the spectrum of a particular star, the Balmer line H_α has a wavelength of 656.15 nm. The laboratory value for the wavelength of H_α is 656.28 nm. (**a**) Find the star's radial velocity. (**b**) Is this star approaching us or moving away? Explain your answer. (**c**) Find the wavelength at which you would expect to find H_α in the spectrum of this star, given that the laboratory wavelength of H_α is 486.13 nm. (**d**) Do your answers depend on the distance from the Sun to this star? Why or why not?

*40. Derive the equation given in Box 17-1 relating proper motion and tangential velocity. (*Hint:* See Box 1-1.)

41. How much dimmer does the Sun appear from Neptune than from Earth? (*Hint:* The average distance between a planet and the Sun equals the semimajor axis of the planet's orbit.)

42. Stars *A* and *B* are both equally bright as seen from Earth, but *A* is 120 pc away while *B* is 24 pc away. Which star has the greater luminosity? How many times greater is it?

43. Stars C and *D* both have the same luminosity, but *C* is 32 pc from Earth while *D* is 128 pc from Earth. Which star appears brighter as seen from Earth? How many times brighter is it?

44. Suppose two stars have the same apparent brightness, but one star is 8 times farther away than the other. What is the ratio of their luminosities? Which one is more luminous, the closer star or the farther star?

45. The *solar constant,* equal to 1370 W/m^2, is the amount of light energy from the Sun that falls on 1 square meter of Earth's surface in 1 second (see Section 17-2). What would the distance between Earth and the Sun have to be in order for the solar constant to be 1 watt per square meter (1 W/m^2)?

46. The star Procyon in Canis Minor (the Small Dog) is a prominent star in the winter sky, with an apparent brightness 1.3×10^{-11} that of the Sun. It is also one of the nearest stars, being only 3.50 pc from Earth. What is the luminosity of Procyon? Express your answer as a multiple of the Sun's luminosity.

47. The star HIP 92403 (also called Ross 154) is only 2.97 pc from Earth but can be seen only with a telescope, because it is 60 times dimmer than the dimmest star visible to the unaided eye. How close to us would this star have to be in order for it to be visible without a telescope? Give your answer in parsecs and in AU. Compare with the semimajor axis of Pluto's orbit around the Sun.

*48. The star HIP 72509 has an apparent magnitude of +12.1 and a parallax angle of 0.222 arcsecond. (**a**) Determine its absolute magnitude. (**b**) Find the approximate ratio of the luminosity of HIP 72509 to the Sun's luminosity.

*49. Suppose you can just barely see a twelfth-magnitude star through an amateur's 6-inch telescope. What is the magnitude of the dimmest star you could see through a 60-inch telescope?

*50. A certain type of variable star is known to have an average absolute magnitude of 0.0. Such stars are observed in a particular star cluster to have an average apparent magnitude of +14.0. What is the distance to that star cluster?

*51. (**a**) Find the absolute magnitudes of the brightest and dimmest of the labeled stars in Figure 17-6b. Assume that all of these stars are 110 pc from Earth. (**b**) If a star in the Pleiades cluster is just bright enough to be seen from Earth with the naked eye, what is its absolute magnitude? Is such a star more or less luminous than the Sun? Explain.

52. (**a**) On a copy of Figure 17-8, sketch the intensity curve for a blackbody at a temperature of 3000 K. Note that this figure shows a smaller wavelength range than Figure 17-7a. (**b**) Repeat part (a) for a blackbody at 12,000 K (see Figure 17-7c). (**c**) Use your sketches from parts (a) and (b) to explain why the color ratios b_V/b_B and b_B/b_U are less than 1 for very hot stars but greater than 1 for very cool stars.

*53. Astronomers usually express a star's color using apparent magnitudes. The star's apparent magnitude as viewed through a B filter is called m_B, and its apparent magnitude as viewed through a V filter is m_V. The difference $m_B - m_V$ is called the *B–V color index* ("B minus V"). Is the B–V color index positive or negative for very hot stars? What about very cool stars? Explain your answers.

*54. (See Question 53.) The B–V color index is related to the color ratio b_V/b_B by the equation

$$m_B = m_V = 2.5 \log\left(\frac{b_V}{b_B}\right)$$

(**a**) Explain why this equation is correct. (**b**) Use the data in Table 17-1 to calculate the B–V color indices for Bellatrix, the Sun, and Betelgeuse. From your results, describe a simple rule that relates the value of the B–V color index to a star's color.

55. The bright star Rigel in the constellation Orion has a surface temperature about 1.6 times that of the Sun. Its luminosity is about 64,000 $L_\odot$. What is Rigel's radius compared to the radius of the Sun?

56. (See Figure 17-12.) What temperature and spectral classification would you give to a star with equal line strengths of hydrogen (H) and neutral helium (He I)? Explain your answer.

57. The Sun's surface temperature is 5800 K. Using Figure 17-12, arrange the following absorption lines in the Sun's spectrum from the strongest to the weakest, and explain your reasoning: (i) neutral calcium; (ii) singly ionized calcium; (iii) neutral iron; (iv) singly ionized iron.

58. Star *P* has one-half the radius of star *Q*. Stars *P* and *Q* have surface temperatures 4000 K and 8000 K, respectively. Which star has the greater luminosity? How many times greater is it?

59. Star *X* has 12 times the luminosity of star *Y*. Stars *X* and *Y* have surface temperatures 3500 K and 7800 K, respectively. Which star has the larger radius? How many times larger is it?

60. Suppose a star experiences an outburst in which its surface temperature doubles but its average density (its mass divided by its volume) decreases by a factor of 8. The mass of the star stays the same. By what factors do the star's radius and luminosity change?

61. The Sun experiences solar flares (see Section 16-10). The amount of energy radiated by even the strongest solar flare is not enough to have an appreciable effect on the Sun's luminosity. But when a flare of the same size occurs on a main-sequence star of spectral class M, the star's brightness can increase by as much as a factor of 2. Why should there be an appreciable increase in brightness for a main-sequence M star but not for the Sun?

62. The bright star Zubeneschmali (β Librae) is of spectral type B8 and has a luminosity of 130 $L_{\odot}$. What is the star's approximate surface temperature? How does its radius compare to that of the Sun?

63. Castor (α Geminorum) is an A1 V star with an apparent brightness of 4.4×10^{-12} that of the Sun. Determine the approximate distance from Earth to Castor (in parsecs).

64. A brown dwarf called CoD–33°7795 B has a luminosity of $0.0025 L_{\odot}$. It has a relatively high surface temperature of 2550 K, which suggests that it is very young and has not yet had time to cool down by emitting radiation. (**a**) What is this brown dwarf's spectral class? (**b**) Find the radius of CoD–33°7795 B. Express your answer in terms of the Sun's radius and in kilometers. How does this compare to the radius of Jupiter? Is the name "dwarf" justified?

65. The star HD 3651 shown in Figure 17-13 has a mass of 0.79 $M_{\odot}$. Its brown dwarf companion, HD 3651B, has about 40 times the mass of Jupiter. The average distance between the two stars is about 480 AU. How long does it take the two stars to complete one orbit around each other?

66. The visual binary 70 Ophiuchi (see the accompanying figure) has a period of 87.7 years. The parallax of 70 Ophiuchi is 0.2 arcsec, and the apparent length of the semimajor axis as seen through a telescope is 4.5 arcsec. (**a**) What is the distance to 70 Ophiuchi in parsecs? (**b**) What is the actual length of the semimajor axis in AU? (**c**) What is the sum of the masses of the two stars? Give your answer in solar masses.

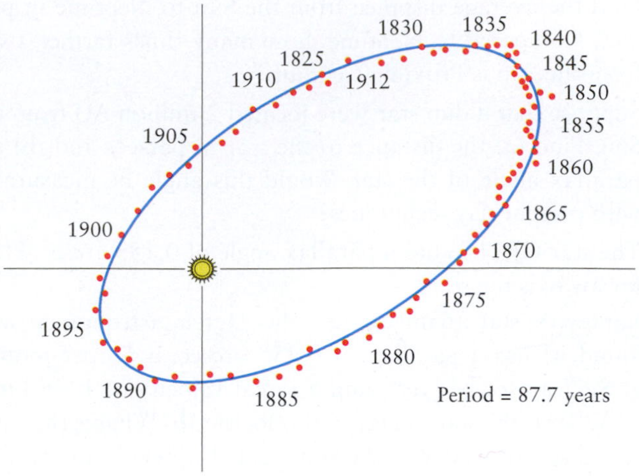

67. An astronomer observing a binary star finds that one of the stars orbits the other once every 5 years at a distance of 10 AU. (**a**) Find the sum of the masses of the two stars. (**b**) If the mass ratio of the system is $M_1/M_2 = 0.25$, find the individual masses of the stars. Give your answers in terms of the mass of the Sun.

Discussion Questions

68. From its orbit around Earth, the *Hipparcos* satellite could measure stellar parallax angles with acceptable accuracy only if the angles were larger than about 0.002 arcsec. Discuss the advantages or disadvantages of making parallax measurements from a satellite in a large solar orbit, say at the distance of Jupiter from the Sun. If this satellite can also measure parallax angles of 0.002 arcsec, what is the distance of the most remote stars that can be accurately determined? How much bigger a volume of space would be covered compared to Earth-based observations? How many more stars would you expect to be contained in that volume?

*69. As seen from the starship *Enterprise* in the *Star Trek* television series and movies, stars appear to move across the sky due to the starship's motion. How fast would the *Enterprise* have to move in order for a star 1 pc away to appear to move 1° per second? (*Hint:* The speed of the star as seen from the *Enterprise* is the same as the speed of the *Enterprise* relative to the star.) How does this compare with the speed of light? Do you think the stars appear to move as seen from an orbiting space shuttle, which moves at about 8 km/s?

70. It is desirable to be able to measure the radial velocity of stars (using the Doppler effect) to an accuracy of 1 km/s or better. One complication is that radial velocities refer to the motion of the star relative to the Sun, while the observations are made using a telescope on Earth. Is it important to take into account the motion of Earth around the Sun? Is it important to take into account Earth's rotational motion? To answer this question, you will have to calculate Earth's orbital speed and the speed of a point on Earth's equator (the part of Earth's surface that moves at the greatest speed because of the planet's rotation). If one or both of these effects are of importance, how do you suppose astronomers compensate for them?

Web/eBook Questions

71. Search the World Wide Web for information about *Gaia*, a European Space Agency (ESA) spacecraft planned to extend the work carried out by *Hipparcos*. When is the spacecraft planned to be launched? How will *Gaia* compare to *Hipparcos*? For how many more stars will it be able to measure parallaxes? What other types of research will it carry out?

72. Search the World Wide Web for recent discoveries about brown dwarfs. Are all brown dwarfs found orbiting normal stars, or are they also found orbiting other brown dwarfs? Are any found in isolation (that is, not part of a binary system)? The Sun experiences flares (see Section 16-10), as do other normal stars; is there any evidence that brown dwarfs also experience flares? If so, is there anything unusual about these flares? What is the lowest temperature for a brown dwarf found so far?

ACTIVITIES

Observing Projects

Observing tips and tools

Even through a telescope, the colors of stars are sometimes subtle and difficult to see. To give your eye the best chance of seeing color, use the "averted vision" trick: When looking through the telescope eyepiece, direct your vision a little to one side of the star you are most interested in. This places the light from that star on a more sensitive part of your eye's retina.

73. The table below lists five well-known red stars. It includes their right ascension and declination (celestial coordinates described in Box 2-1), apparent magnitudes, and color ratios. As their apparent magnitudes indicate, all these stars are somewhat variable. Observe at least two of these stars both by eye and through a small telescope. Is the reddish color of the stars readily apparent, especially in contrast to neighboring stars? (The Jesuit priest and astronomer Angelo Secchi named Y Canum Venaticorum "La Superba," and μ Cephei is often called William Herschel's "Garnet Star.")

Star	Right ascension	Declination	Apparent magnitude	b_V/b_B
Betelgeuse	$5^h\ 55.2^m$	+7° 24′	0.4–1.3	5.5
Y Canum Venaticorum	12 45.1	+45 26	5.5–6.0	10.4
Antares	16 29.4	−26 26	0.9–1.8	5.4
μ Cephei	21 43.5	+58 47	3.6–5.1	8.7
TX Piscium	23 46.4	+3 29	5.3–5.8	11.0

Note: The right ascensions and declinations are given for epoch 2000.

74. The table of double stars shown below includes vivid examples of contrasting star colors. The table lists the angular separation between the stars of each double. Observe at least four of these double stars through a telescope. Use the spectral types listed to estimate the difference in surface temperature of the stars in each pair you observe. Does the double with the greatest difference in temperature seem to present the greatest color contrast? From what you see through the telescope and on what you know about the H-R diagram, explain why all the cool stars (spectral types K and M) listed are probably giants or supergiants.

75. Observe the eclipsing binary Algol (β Persei), using nearby stars to judge its brightness during the course of an eclipse. Algol has an orbital period of 2.87 days, and, with the onset of primary eclipse, its apparent magnitude drops from 2.1 to 3.4. It remains this faint for about 2 hours. The entire eclipse, from start to finish, takes about 10 hours. Consult the "Celestial Calendar" section of the current issue of *Sky & Telescope* for the predicted dates and times of the minima of Algol. Note that the schedule is given in Universal Time (the same as Greenwich Mean Time), so you will have to convert the time to that of your own time zone. Algol is normally the second brightest star in the constellation of Perseus. Because of its position on the celestial sphere (R.A. = $3^h\ 08.2^m$, Decl. = 40° 57′), Algol is readily visible from northern latitudes during the fall and winter months.

Star	Right ascension	Declination	Apparent magnitude	Angular separation (arcseconds)	Spectral types
55 Piscium	$0^h\ 39.9^m$	+21° 26′	5.4 and 8.7	6.5	K0 and F3
γ Andromedae	2 03.9	+42 20	2.3 and 4.8	9.8	K3 and A0
32 Eridani	3 54.3	−2 57	4.8 and 6.1	6.8	G5 and A
ι Cancri	8 46.7	+28 46	4.2 and 6.6	30.5	G5 and A5
γ Leonis	10 20.0	+19 51	2.2 and 3.5	4.4	K0 and G7
24 Coma Berenicis	12 35.1	+18 23	5.2 and 6.7	20.3	K0 and A3
ν Boötis	14 45.0	+27 04	2.5 and 4.9	2.8	K0 and A0
α Herculis	17 14.6	+14 23	3.5 and 5.4	4.7	M5 and G5
59 Serpentis	18 27.2	+0 12	5.3 and 7.6	3.8	G0 and A6
β Cygni	19 30.7	+27 58	3.1 and 5.1	34.3	K3 and B8
δ Cephei	22 29.2	+58 25	4* and 7.5	20.4	F5 and A0

Note: The right ascensions and declinations are given for epoch 2000.

**The brighter star in the δ Cephei binar system is a variable star of approximately the fourth magnitude.*

76. Use *Starry Night*™ to examine the 10 brightest stars in Earth's night sky. Select **Favourites > Explorations > Atlas.** Use the **View > Constellations** menu command to display constellation **Boundaries, Labels,** and **Astronomical** stick figures. Use the **File** (Windows) or **Starry Night** (Mac) menu command to open the **Preferences** dialog window. Ensure that the **Cursor Tracking (HUD)** preferences include **Apparent Magnitude, Distance from observer, Luminosity,** and **Temperature** in the Show list. Before closing the **Preferences** dialog window, it might be helpful to increase the saturation for **Star color** under the **Brightness/Contrast** preferences. Click on the **Lists** side pane tab, expand the **Observing Lists** and click the **10 Brightest Stars** option. Then expand the **List Viewer** layer and select **All Targets** from the **Show** dropdown menu to see a list of the 10 brightest stars in Earth's night sky. Double-click on each of the stars in this list in turn to center the star in the view. Use the HUD to compile a table of these stars that includes each star's apparent magnitude, distance, luminosity, and temperature. You may also wish to sketch the star's position within its constellation. Alternatively, you may find it helpful to print out relevant star charts around these stars, using *Starry Night*™. (**a**) Which is the brightest star in Earth's night sky? What features of this star make it so bright in our sky? (**b**) Which of these brightest stars has the highest temperature? What would you expect to be the color of this star compared to others in the list? (**c**) Which of these stars is intrinsically the most luminous? (**d**) Use *Starry Night*™ to determine which of these stars is visible from your location. Click the **Home** button, then the **Stop** button, and finally the **Sunset** button to show the view from your home location today at sunset. Again, it may be helpful to display the constellation **Boundaries, Labels,** and **Astronomical** stick figures in the view. Open the **Lists** side pane and double-click each entry in the list of the 10 brightest stars. If the star is visible in your sky, the program will center it in the view or alternately suggest a **Best Time** for observing this star. For those stars in the list that are visible from your home location, go outside if possible and observe them in the real sky. See if you can tell which of these stars has the highest temperature on the basis of your conclusion regarding the star's color and check your estimate against the table you compiled in part (a). (*Hint:* The colors of stars are not very distinct and a dark sky background is needed in order to distinguish differences in stellar colors.)

77. Use the *Starry Night*™ program to investigate the Hertzsprung-Russell (H-R) diagram. Select **Favourites > Explorations > Denver.** Open the **Status** pane, expand the **H-R Options** layer, and choose the following options: **Use absolute magnitudes** and **Labels.** In the expanded **Labels** panel, click on the **Gridlines, Regions, Main Sequence,** and **Spectral class** options. Now, expand the **Hertzsprung-Russell** layer to show the H-R diagram that plots all of the stars that are currently in the main view. This graphical representation shows the absolute magnitudes of stars as a function of their spectral class. The sequence of spectral class, from O, B, A, F, G, K and M, represents the star's surface temperatures, plotted in an inverse direction, hottest stars appearing to the left of the diagram. Absolute magnitude is related to the star's luminosity, the smaller the absolute value of absolute magnitude, the larger the luminosity. (**a**) Use the hand tool to scroll around the sky. Watch the H-R diagram change as different stars enter and leave the main window. Right-click (Ctrl-click on a Mac) on a blank part of the sky and select **Hide Horizon** from the contextual menu so that you can survey the entire sky. Does the distribution of stars in the H-R diagram change drastically from one part of the sky to another, or are all types of stars approximately equally represented in all directions from the Earth? (**b**) If you place the cursor over a star, a red dot appears in the H-R diagram at the position for this star. Use this facility to estimate and make a note of the spectral-luminosity classification and the absolute magnitude, M_V, of the following stars that are labeled in the main window: Altair, Deneb, Enif, 74 Ophiuchi, and 51 Pegasi. (Remember that each spectral class is divided into 10 subclasses from 0 to 9.) (**c**) Based on its spectral-luminosity classification, which of these five stars is most similar to the Sun? What is the name of the region of the H-R diagram occupied by this star? If these two physical properties are similar for the Sun and this star, which other two parameters will necessarily be similar? Explain. (**d**) The apparent magnitude, m_V, of Altair is +0.8 and that of Enif is +2.4. For both stars, calculate the quantity $m_V - M_V$ (that is, apparent magnitude minus absolute magnitude.) Based on your results, would you expect Altair to be closer to us than 10 pc or farther away than 10 pc? What about Enif? Explain your reasoning. (**e**) Open the **Info** pane and expand **Other Data.** Open the contextual menu for Enif in the main view and select **Show Info** to display relevant data for this star. Make a note of the radius, temperature, and luminosity of Enif. Use the given temperature and luminosity to calculate the radius of Enif, and compare your result to the value from the **Info** pane. The temperature of the Sun is 5780 K.

78. Use *Starry Night*™ to examine the effect of distance upon the apparent brightness of the Sun. Prepare a table with five columns labeled Distance, Apparent Magnitude, Difference, Apparent Brightness, and Reciprocal of Distance squared. Open **Favourites > Explorations > Sun from 1 AU.** The view is centered upon the Sun from a location in space 1 AU from the Sun. Use the HUD or the Info pane to determine the apparent magnitude of the Sun as seen from this distance. Next, select **Options > Viewing Location...** from the menu and, in the **Viewing Location** dialog window, choose **stationary location** in the View from selection box. Then enter "2 AU" for the **Radius** under the Spherical coordinates column and click the **Go To Location** button to view the Sun from a distance of 2 AU. Use the HUD of **Info** pane to obtain the Sun's apparent magnitude from this distance. Use the **Options > Viewing Location...** command to make one

more observation of the apparent magnitude of the Sun from a distance of 3 AU. Now you need to convert the values for apparent magnitude into values that represent the apparent brightness of the Sun when viewed from these different distances. To do this, first you must calculate the difference in apparent magnitude of the Sun as seen from more distant locations to the apparent magnitude of the Sun as seen from 1 AU. In other words, subtract the apparent magnitude of the Sun from each of the larger distances from the apparent magnitude of the Sun as seen from 1 AU and record these values in the Differences column of your data table. Since each magnitude difference of 1 AU corresponds to a factor of 2.512 in apparent brightness, use a calculator or spreadsheet to raise the number 2.512 to the power of the magnitude difference to obtain a value for the relative brightness of the Sun from each distance and record these values in the Apparent Brightness column of your data table. Finally, in column 5 of your table, calculate the reciprocal of the square of the distance for each observation. Limit your answers to 2 significant figures. Interpret the meaning of your results.

ANSWERS

ConceptChecks

ConceptCheck 17-1: A very nearby flashlight will *appear* much brighter than a very distant spotlight. Although a large spotlight emits more light than a small flashlight, the light that appears brighter depends on their distances away.

ConceptCheck 17-2: The parallax angle is smallest for the most distant stars, so if a parallax angle was measured to be too small, then the astronomer would assume the star is farther away than it actually is.

ConceptCheck 17-3: Parallax only works for stars that are relatively nearby, and the majority of stars in our Galaxy are much too far away to exhibit any apparent shift in position as Earth orbits the Sun.

ConceptCheck 17-4: Because a star's apparent magnitude when viewed from a distance of 10 pc is the same as its absolute magnitude, both magnitudes would be the same, +4.8.

ConceptCheck 17-5: From Earth, Tau Ceti appears to be relatively bright at +3. However, if it were viewed from a distance of 10 pc, it would appear much dimmer, at +6. The only way this can be is if Tau Ceti is much closer to Earth than 10 pc.

ConceptCheck 17-6: The shorter the peak wavelength, the higher the temperature. The blue star emits light that peaks at blue wavelengths, and these are shorter wavelengths than the other star's peak at orange wavelengths. Thus, the blue star is hotter. (Figure 17-7 illustrates these effects.)

ConceptCheck 17-7: The prominence of certain hydrogen absorption lines visible in the star, with A class being the greatest and O class being the least.

ConceptCheck 17-8: In the sequence OBAFGKM, F-spectral class stars with designations F1, F2, F3, and F4 are more similar to A-spectral class stars, and F6, F7, F8, and F9 are more similar to G-spectral class stars.

ConceptCheck 17-9: No. Within a particular spectral class, the larger numbers correspond to cooler stars. For example, a G2 star is hotter than a G8 star because a G2 is closer to the hotter F-spectral class stars in the sequence OBAFGKM.

ConceptCheck 17-10: Given that the modern spectral sequence of stars and brown dwarfs from hottest to coolest is OBAFGKMLT, L-spectral class brown dwarfs are hotter than T-spectral class brown dwarfs.

ConceptCheck 17-11: Stars like our Sun derive their energy from nuclear reactions in the core, whereas brown dwarfs derive their energy from gravitational contraction. Brown dwarfs are not true stars.

ConceptCheck 17-12: A bonfire often extends over a large area so that it has an enormously high luminosity, even at the same temperature of a match or a handheld lighter.

ConceptCheck 17-13: The H-R diagram is plotted as increasing luminosity on the vertical axis and decreasing temperature along the horizontal axis. The stars with the greatest luminosity and the highest temperatures are found in the upper left-hand corner of the H-R diagram.

ConceptCheck 17-14: You can read from the vertical axis of the H-R diagram that Betelgeuse is much more luminous than Barnard's star. Since they both have similar surface temperatures (around 3500 K), Betelgeuse must be much larger to emit its larger luminosity.

ConceptCheck 17-15: Regulus. All main-sequence stars, and only these stars (such as Regulus, but not Mira), convert hydrogen to helium in their cores.

ConceptCheck 17-16: The main sequence and all its stars have luminosity class V. At higher temperatures, the main-sequence stars can have a greater luminosity than the lower-temperature stars in luminosity class III.

ConceptCheck 17-17: Main-sequence stars are luminosity class "V" whereas luminosity class "II" are giant stars. In this instance, the two stars have the same temperature because they are in the same spectral class, but the luminosity class II star is larger and has a greater luminosity than the much smaller main-sequence star.

ConceptCheck 17-18: If two stars orbiting a common center of mass were moved farther apart, their masses would not change, but the period would increase following Kepler's third law.

ConceptCheck 17-19: According to the mass-luminosity relation, the most luminous main-sequence stars are also the most massive, have the largest radii, and have the greatest temperatures.

ConceptCheck 17-20: At Stage 2, neither of the stars moves toward or away from Earth. The only radial velocity Doppler shift arises from the pair of stars as they drift together through space, which in this case, can be read near the bottom of Figure 17-23 for Stage 2 at about +12 km/s away from Earth. Since both stars have this same radial velocity at Stage 2, their absorption lines are not split.

ConceptCheck 17-21: When the bright white dwarf passes behind the larger and dimmer star, the light coming from the dimmer star is still visible.

CalculationChecks

CalculationCheck 17-1: Given that distance in parsecs is the inverse of the parallax angle, $d = 1/0.7772$, then $d = 1.35$ pc. Because 1 pc is 3.26 ly, we find that 1.35 pc × (3.26 ly/1pc) = 4.4 ly.

CalculationCheck 17-2: According to the inverse-square law nature of light, the intensity of the light received decreases with the square of the distance. In this instance, increasing the distance 3 times means the newspaper receives only $1/(3)^2$ or one-ninth the amount of light it originally received.

CalculationCheck 17-3: Pleione's luminosity relative to that of the Sun is $L/L_\odot = 190$. The ratio of their apparent brightnesses is $b/b_\odot = 3.19 \times 10^{-13}$. Rearranging the equation relating luminosity, distance, and brightness yields the distance from Earth to Pleione as 2.44×10^7 times greater than the distance from Earth to the Sun:

$$\frac{d}{d_\odot} = \sqrt{\frac{(L/L_\odot)}{(b/b_\odot)}} = \sqrt{\frac{190}{3.19 \times 10^{-13}}}$$

$$= \sqrt{5.95 \times 10^{14}} \times 2.44 \times 10^7$$

CalculationCheck 17-4: If luminosity is given by the Stefan-Boltzmann law where $L = 4\pi R^2 \times T^4$, then if radius, R, is tripled, then L must increase by the R^2, which is, in this instance, 3^2 or 9 times, for a star 3 times larger but at the same temperature.

A region of star formation about 1400 pc (4000 ly) from Earth in the southern constellation Ara (the Altar). (European Southern Observatory) R I **V** U X G

CHAPTER 18

The Birth of Stars

LEARNING GOALS

By reading the sections of this chapter, you will learn

18-1 How astronomers have pieced together the story of stellar evolution

18-2 What interstellar nebulae are and what they are made of

18-3 What happens as a star begins to form

18-4 The stages of growth from young protostars to main-sequence stars

18-5 How stars gain and lose mass during their growth

18-6 What insights star clusters add to our understanding of stellar evolution

18-7 Where new stars form within galaxies

18-8 How the death of old stars can trigger the birth of new stars

The stars that illuminate our nights seem eternal and unchanging. But this permanence is an illusion. Each of the stars visible to the naked eye shines due to thermonuclear reactions and has only a finite amount of fuel available for these reactions. Hence, stars cannot last forever: They form from material in interstellar space, evolve over millions or billions of years, and eventually die. In this chapter our concern is with how stars are born and become part of the main sequence.

Stars form within cold, dark clouds of gas and dust that are scattered abundantly throughout our Galaxy. One such cloud appears as a dark area on the far right-hand side of the photograph on this page. Perhaps a dark cloud like this encounters one of the Galaxy's spiral arms, or perhaps a supernova detonates nearby. From the shock of events like these, the cloud begins to contract under the pull of gravity, forming protostars—the fragments that will one day become stars. As a protostar develops, its internal pressure builds and its temperature rises. In time, hydrogen fusion begins, and a star is born. The hottest, bluest, and brightest young stars, like those in the accompanying image, emit ultraviolet radiation that excites atoms in the surrounding interstellar gas. The result is a beautiful glowing nebula, which typically has the red color characteristic of excited hydrogen (as shown in the photograph).

In Chapters 19 and 20 we will see how stars mature and grow old. Some even blow themselves apart in death throes that enrich interstellar space with the material for future generations of stars. Thus, like the mythical phoenix, new stars arise from the ashes of the old.

18-1 Understanding how stars evolve requires observation as well as ideas from physics

Stars consume the material of which they are made, and so cannot last forever

Over the past several decades, astronomers have labored to develop an understanding of **stellar evolution,** that is, how stars are born, live their lives, and finally die. Our own Sun provides evidence that stars are not permanent. The energy radiated by the Sun comes from thermonuclear reactions in its core, which consume 6×10^{11} kg of hydrogen each second and convert it into helium (see Section 16-1). While the amount of hydrogen in the Sun's core is vast, it is not infinite; therefore, the Sun cannot always have been shining, nor can it continue to shine forever. The same is true for all other main-sequence stars, which are fundamentally the same kinds of objects as the Sun but with different masses (see Section 17-9). Thus, stars must have a beginning as well as an end.

Stars last very much longer than the lifetime of any astronomer—indeed, far longer than the entire history of human civilization. Thus, it is impossible to watch a single star go through its formation, evolution, and eventual demise. Rather, astronomers have to piece together the evolutionary history of stars by studying different stars at different stages in their life cycles.

ANALOGY To see the magnitude of this task, imagine that you are a biologist from another planet who sets out to understand the life cycles of human beings. You send a spacecraft to fly above Earth and photograph humans in action. Unfortunately, the spacecraft fails after collecting only 20 seconds of data, but during that time its sophisticated equipment sends back observations of thousands of different humans. From this brief snapshot of life on Earth—only 10^{-8} (a hundred-millionth) of a typical human lifetime—how would you decide which were the young humans and which were the older ones? Without a look inside our bodies to see the biological processes that shape our lives, could you tell how humans are born and how they die? And how could you deduce the various biological changes that humans undergo as they age?

Astronomers, too, have data spanning only a tiny fraction of any star's lifetime. A star like the Sun can last for about 10^{10} years, whereas astronomers have been observing stars in detail for only about a century—as in our analogy, roughly 10^{-8} of the life span of a typical star. Astronomers are also frustrated by being unable to see the interiors of stars. For example, we cannot see the thermonuclear reactions that convert hydrogen into helium. But astronomers have an advantage over the biologist in our story. Unlike humans, stars are made of relatively simple substances, primarily hydrogen and helium, that are found almost exclusively in the form of gases. Of the three phases of matter—gas, liquid, and solid—gases are by far the simplest to understand.

Astronomers use our understanding of gases to build theoretical models of the interiors of stars, like the model of the Sun we saw in Section 16-2. Models help to complete the story of stellar evolution. In fact, like all great dramas, the story of stellar evolution can be regarded as a struggle between two opposing and unyielding forces: Gravity continually tries to make a star shrink, while the star's internal pressure tends to make the star expand. When these two opposing forces are in balance, the star is in a state of hydrostatic equilibrium (see Figure 16-2).

But what happens when changes within the star cause either pressure or gravity to predominate? The star must then either expand or contract until it reaches a new equilibrium. In the process, it will change not only in size but also in luminosity and color.

In the following chapters, we will find that giant and supergiant stars are the result of pressure gaining the upper hand over gravity. Both giants and supergiants turn out to be aging stars that have become tremendously luminous and ballooned to hundreds or thousands of times their previous size. White dwarfs, by contrast, are the result of the balance tipping in gravity's favor. These dwarfs are even older stars that have collapsed to a fraction of the size they had while on the main sequence. In this chapter, however, we will see how the opposing influences of gravity and pressure explain the birth of stars. We start our journey within the diffuse clouds of gas and dust that permeate our Galaxy.

18-2 Interstellar gas and dust pervade the galaxy

Different types of nebulae emit, absorb, or reflect light

Where do stars come from? As we saw in Section 8-4, our Sun condensed from a solar nebula, a collection of gas and dust in interstellar space. Observations suggest that other stars originate in a similar way (see Figure 8-8). To understand the formation of stars, we must first understand the nature of the interstellar matter from which the stars form.

Nebulae and the Interstellar Medium

At first glance, the space between the stars seems to be empty. On closer inspection, we find that it is filled with a thin gas laced with microscopic dust particles. This combination of gas and dust is called the **interstellar medium.** Evidence we will discuss for this medium includes interstellar clouds of various types, curious lines in the spectra of binary star systems, and an apparent dimming and reddening of distant stars.

You can see evidence for the interstellar medium with the naked eye. Look carefully at the constellation Orion (Figure 18-1a), visible on winter nights in the northern hemisphere and summer nights in the southern hemisphere. While most of the stars in the constellation appear as sharply defined points of light, the middle "star" in Orion's sword has a fuzzy appearance. This fuzziness becomes more obvious with binoculars or a telescope. As Figure 18-1b shows, this "star" is actually not a star at all, but the Orion Nebula—a cloud in interstellar space. Any interstellar cloud is called a **nebula** (plural **nebulae**) or **nebulosity.**

Emission Nebulae: Clouds of Excited Gas

The Orion Nebula emits its own light, with the characteristic emission line spectrum of a hot, thin gas. For this reason it is called an **emission nebula.** Many emission nebulae can be seen with a small telescope. Figure 18-2 shows some of these nebulae in a different part of the constellation Orion. Emission nebulae are direct evidence of gas atoms in the interstellar medium.

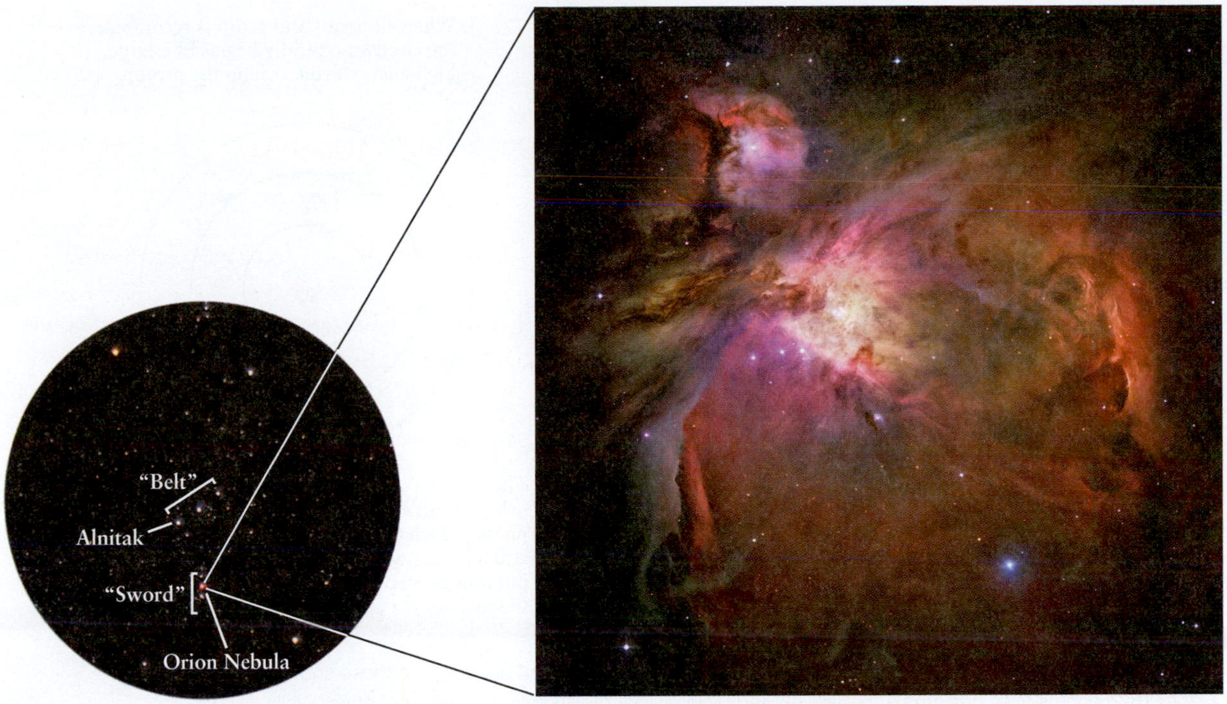

(a) A wide-angle view of Orion (b) A closeup of the Orion Nebula

FIGURE 18-1 R I V U X G

The Orion Nebula **(a)** The middle "star" of the three that make up Orion's sword is actually an interstellar cloud called the Orion Nebula. **(b)** The nebula is about 450 pc (1500 ly) from Earth and contains about 300 solar masses of material. Most of the ultraviolet light that makes the nebula glow comes from just five hot massive stars. (a: Australian Astronomical Observatory/David Malin Images; b: NASA,ESA, M. Robberto [Space Telescope Science Institute/ESA] and the Hubble Space Telescope Orion Treasury Project Team)

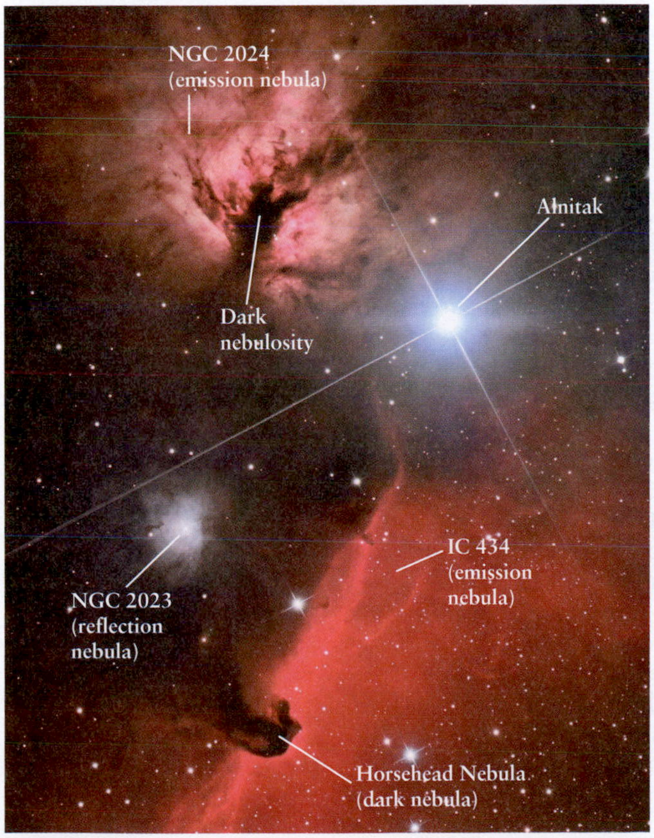

Typical emission nebulae have masses that range from about 100 to about 10,000 solar masses. Because this mass is spread over a huge volume that is light-years across, the density is quite low by Earth standards, only a few thousand hydrogen atoms per cubic centimeter. (By comparison, the air you are breathing contains more than 10^{19} atoms per cm^3.)

Emission nebulae are found near hot, luminous stars of spectral types O and B. Such stars emit copious amounts of ultraviolet radiation. When atoms in the nearby interstellar gas absorb these energetic ultraviolet photons, the atoms become ionized. Indeed, emission nebulae are composed primarily of ionized hydrogen atoms, that is, free protons (hydrogen nuclei) and electrons. Astronomers use the notation H I for neutral, un-ionized hydrogen atoms and H II for ionized hydrogen atoms, which is why emission nebulae are also called **H II regions.**

H II regions emit red visible light when some of the free protons and electrons get back together to form hydrogen atoms, a process

FIGURE 18-2 R I V U X G

Emission, Reflection, and Dark Nebulae in Orion A variety of different nebulae appear in the sky around Alnitak, the easternmost star in Orion's belt (see Figure 18-1a). All the nebulae lie approximately 500 pc (1600 ly) from Earth. They are actually nowhere near Alnitak, which is only 250 pc (820 ly) distant. This photograph shows an area of the sky about 1.5° across. (Stocktrek Images/Roth Ritter/Stocktrek Images/Corbis)

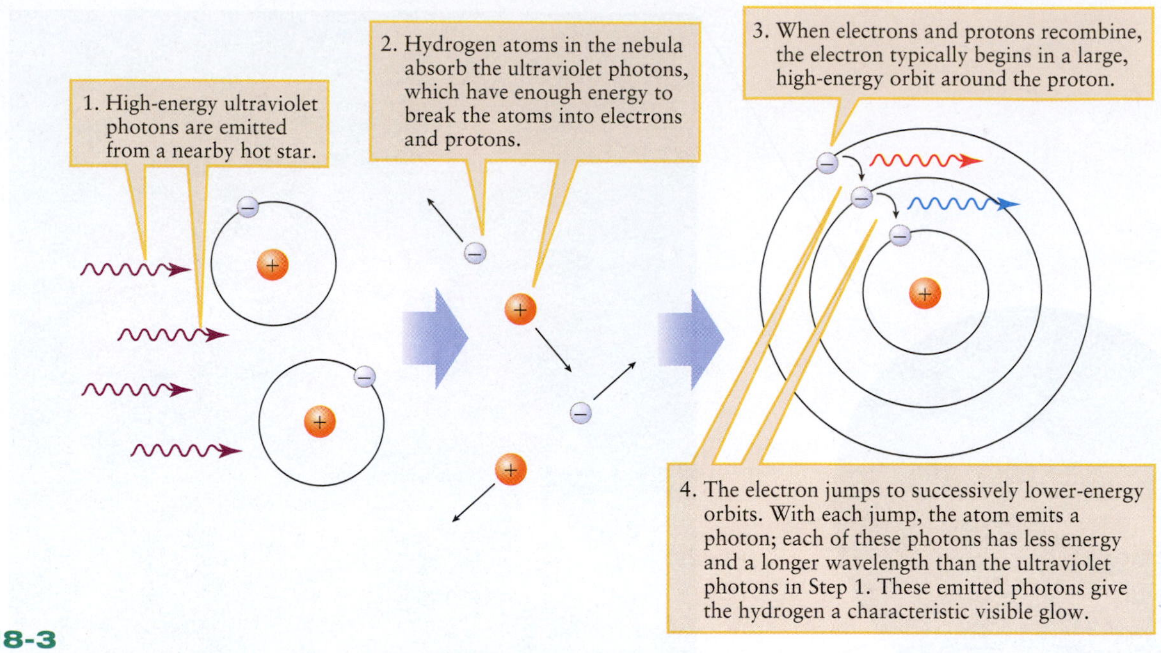

1. High-energy ultraviolet photons are emitted from a nearby hot star.
2. Hydrogen atoms in the nebula absorb the ultraviolet photons, which have enough energy to break the atoms into electrons and protons.
3. When electrons and protons recombine, the electron typically begins in a large, high-energy orbit around the proton.
4. The electron jumps to successively lower-energy orbits. With each jump, the atom emits a photon; each of these photons has less energy and a longer wavelength than the ultraviolet photons in Step 1. These emitted photons give the hydrogen a characteristic visible glow.

FIGURE 18-3

Ionization and Recombination In an H II region, the characteristic red glow of emission nebulae (like those shown in Figure 18-1 and Figure 18-2) comes from gas atoms that are excited by ultraviolet radiation from nearby hot stars. While many photons are emitted during recombination, only some are visible, and red is the most common.

called **recombination** (Figure 18-3). When an atom forms by recombination, the electron is typically captured into a high-energy orbit. As the electron cascades downward through the atom's energy levels toward the ground state, the atom emits photons with lower energies and longer wavelengths than the photons that originally caused the ionization. Particularly important is the transition from $n = 3$ to $n = 2$. It produces H_α photons with a wavelength of 656 nm, in the red portion of the visible spectrum (see Section 5-8, especially Figure 5-23b). These photons give H II regions their distinctive reddish color.

For each high-energy, ultraviolet photon absorbed by a hydrogen atom to ionize it, several photons of lower energy are emitted when a proton and electron recombine. As Box 18-1 describes, a

BOX 18-1 ASTRONOMY DOWN TO EARTH

Fluorescent Lights

The light that comes from glowing interstellar clouds is, quite literally, otherworldly. But the same principles that explain how such clouds emit light are also at the heart of light phenomena that we see here on Earth.

A fluorescent lamp produces light in a manner not too different from an emission nebula (H II region). In both cases, the physical effect is called **fluorescence:** High-energy ultraviolet photons are absorbed, and the absorbed energy is reradiated as lower-energy photons of visible light.

Within the glass tube of a fluorescent lamp is a small amount of the element mercury. When you turn on the lamp, an electric current passes through the tube, vaporizing the mercury and exciting its atoms. This excited mercury vapor radiates light with an emission-line spectrum, including lines in the ultraviolet. The white fluorescent coating on the inside of the glass tube absorbs these ultraviolet photons, exciting electrons in the coating's molecules to high energy levels.

The electrons in the molecules then cascade down through a number of lower levels before reaching the ground state. During this cascade, visible-light photons of many different wavelengths are emitted, giving an essentially continuous spectrum and a very white light. (By comparison, the hydrogen atoms in an H II region emit at only certain discrete wavelengths, because the spectrum of hydrogen is much simpler than that of the fluorescent coating's molecules. Another difference is that the molecules in the fluorescent tube never become ionized.)

Many common materials display fluorescence. Among them are teeth, fingernails, and certain minerals. When illuminated with ultraviolet light, these materials glow with a blue or green color. (Most natural history museums and science museums have an exhibit showing fluorescent minerals.) Laundry detergent also contains fluorescent material. After washing, your laundry absorbs ultraviolet light from the Sun and glows faintly, making it appear "whiter than white."

similar effect takes place in a fluorescent lightbulb. In this sense, H II regions are immense fluorescent light fixtures!

Further evidence of interstellar gas comes from the spectra of binary star systems. As the two stars that make up a binary system orbit their common center of mass, their spectral lines shift back and forth (see Figure 17-23). But certain calcium and sodium lines are found to remain at fixed wavelengths and these lines have captured the curiosity of observers. These **stationary absorption lines** are therefore not associated with the binary star. Instead, they are caused by interstellar gas between us and the binary system.

CONCEPTCHECK 18-1

Would it be correct to describe a reddish H II region as "red hot," the way filaments in an electric toaster glow red?

Answer appears at the end of the chapter.

Dark Nebulae and Reflection Nebulae: Abodes of Dust

In addition to the presence of gas atoms in red H II regions, Figure 18-2 also shows two kinds of evidence for larger bits of matter, called **dust grains,** in the interstellar medium. These dust grains make their appearance in *dark nebulae* and *reflection nebulae.*

A **dark nebula** is so opaque that it blocks any visible light coming from stars that lie behind it. One such dark nebula, called the Horsehead Nebula for its shape, protrudes in front of one of the H II regions in Figure 18-2. Figure 18-4 shows another dark nebula, called Barnard 86. These nebulae have a relatively dense concentration of microscopic dust grains, which scatter and absorb light much more efficiently than single atoms. Typical dark nebulae have temperatures between 10 to 100 K, which is low enough for hydrogen atoms to form molecules. Such nebulae contain from 10^4 to 10^9 particles (atoms, molecules, and dust grains) per cubic centimeter. Although tenuous by Earth standards, dark nebulae are large enough—typically many light-years deep—that they block the passage of light. In the same way, a sufficient depth of haze or smoke in our atmosphere can make it impossible to see distant mountains.

The other evidence for dust in Figure 18-2 is the bluish haze, labeled NGC 2023, surrounding the star immediately above and to the left of the Horsehead Nebula. Figure 18-5 shows a similar haze around a different set of stars. A haze of this kind, called a **reflection nebula,** is caused by fine grains of dust in a lower concentration than that found in dark nebulae. The light we see coming from the nebula is starlight that has been scattered and reflected by these dust grains. The grains are only about 500 nm across, no larger than a typical wavelength of visible light, and they scatter short-wavelength blue light more efficiently than long-wavelength red light. Hence, reflection nebulae have a characteristic blue color. Box 5-4 explains how a similar process—the scattering of sunlight in our atmosphere—gives rise to the blue color of the sky.

CONCEPTCHECK 18-2

Does the blue color in Figure 18-5 come from hot gas similar to the surfaces of massive hot blue stars?

Answer appears at the end of the chapter.

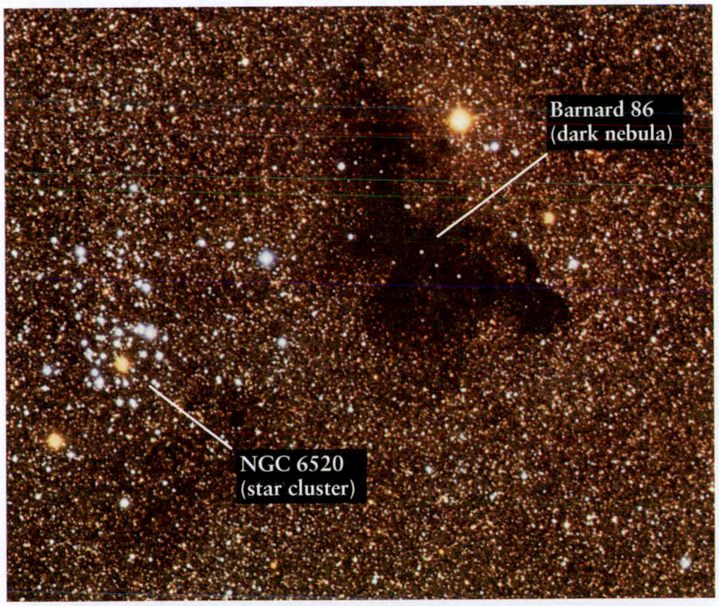

FIGURE 18-4 R I V U X G

A Dark Nebula When first discovered in the late 1700s, dark nebulae were thought to be "holes in the heavens" where very few stars are present. In fact, they are opaque regions that block out light from the stars beyond them. The few stars that appear to be within Barnard 86 lie between us and the nebula. Barnard 86 is in the constellation Sagittarius and has an angular diameter of 4 arcminutes, about 1/7 the angular diameter of the full moon. (Australian Astronomical Observatory/David Malin Images)

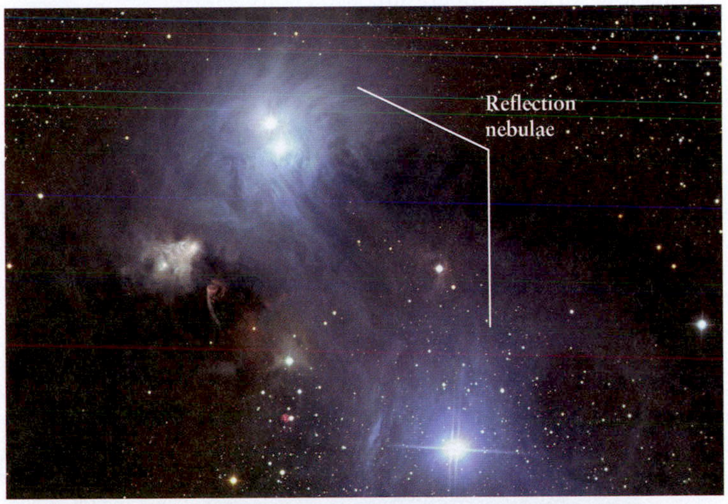

FIGURE 18-5 R I V U X G

Reflection Nebulae Wispy reflection nebulae called NGC 6726-27-29 surround several stars in the constellation Corona Australis (the Southern Crown). Unlike emission nebulae, reflection nebulae do not emit their own light, but scatter and reflect light from the stars that they surround. This scattered starlight is quite blue in color. The region shown here is about 23 arcminutes across. (Dr. Stefan Binnewies and Josef Popsel/www.capella-observatory.com)

Interstellar Extinction and Reddening

In the 1930s, the American astronomer Robert Trumpler discovered two other convincing pieces of evidence for the existence of interstellar matter—*interstellar extinction* and *interstellar reddening*. While studying the brightness and distances of certain star clusters, Trumpler noticed that remote clusters seem to be dimmer than would be expected from their distance alone. His observations demonstrated that the intensity of light from remote stars is reduced as the light passes through material in interstellar space. This process is called **interstellar extinction.** (In the same way, the headlights of oncoming cars appear dimmer when there is smoke or fog in the air.) The light from remote stars is also reddened as it passes through the interstellar medium, because the blue component of their starlight is scattered and absorbed by interstellar dust. This effect, shown in Figure 18-6, is called **interstellar reddening.** The same effect makes the setting Sun look red (see Box 5-4).

CAUTION! It is important to understand the distinction between interstellar reddening and the Doppler effect. If an object is moving away from us, the Doppler effect *shifts* all of that object's light toward longer wavelengths (a redshift). Interstellar reddening, by contrast, makes objects appear red not by shifting wavelengths but by filtering out shorter blue wavelengths. The effect is the same as if we looked at an object through red-colored glasses, which let red light pass but block out blue light. The one similarity between Doppler shifts and interstellar reddening is that neither has any discernible effect on what you see with the naked eye. Both of these effects cause only subtle color changes that your eye (which does a poor job of seeing colors in faint objects like stars) cannot detect. If you see a star with a red color, the reason is *not* interstellar reddening or the Doppler effect; it is because the star really is red due to its low surface temperature (see Figure 17-7a).

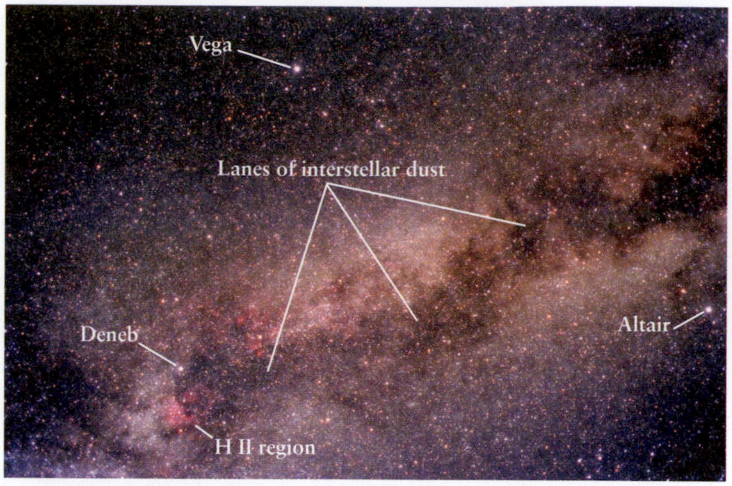

FIGURE 18-7 R I V U X G
Gas and Dust in the Milky Way There are so many stars that they aren't easy to see individually, resulting in the diffuse whitish band that gives the Milky Way its name. Red glowing gas clouds (emission nebulae or H II regions) can be seen in the foreground, and dark, dusty regions that block starlight are concentrated close to the midplane of the Milky Way Galaxy. This wide-angle photograph also shows the three bright stars that make up the "summer triangle" (see Figure 2-8). (Jerry Lodriguss/Science Source)

Interstellar Gas and Dust in Spiral Galaxies

Observations indicate that interstellar gas and dust are largely confined to the plane of the Milky Way—that faint, hazy band of stars you can see stretching across the sky on a dark, moonless night. Figure 18-7 shows glowing emission nebulae and dark lanes of dust along the Milky Way.

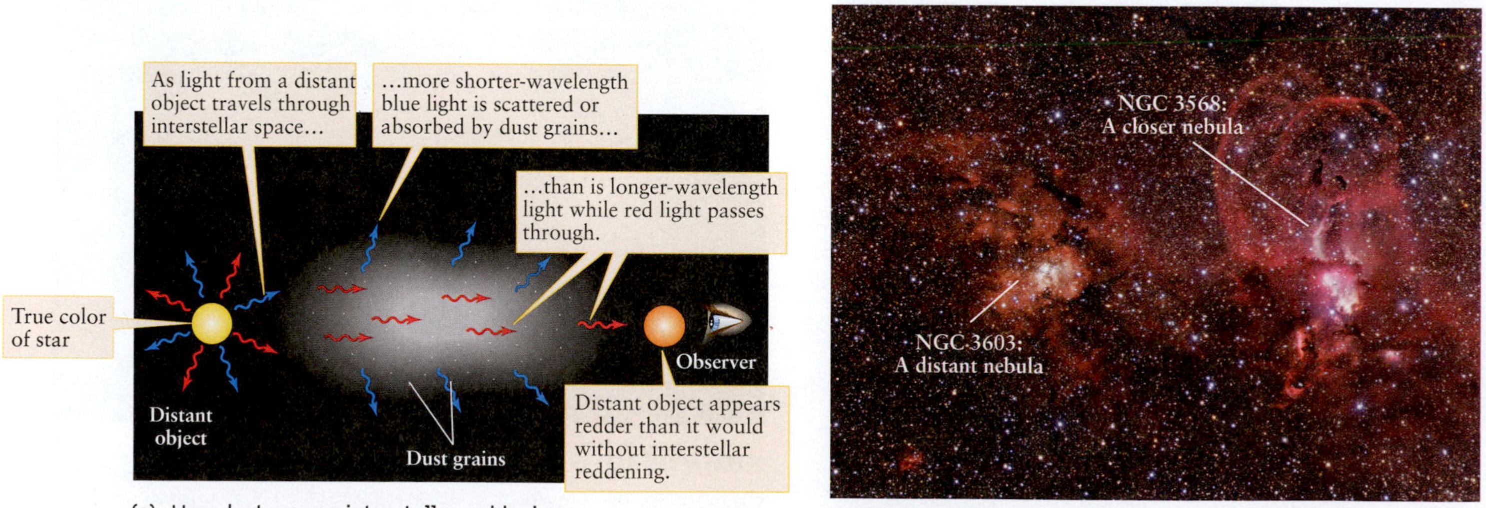

(a) How dust causes interstellar reddening

(b) Reddening depends on distance

FIGURE 18-6 R I V U X G
Interstellar Reddening **(a)** Dust grains in interstellar space scatter or absorb blue light more than red light. Thus, light from a distant object appears redder than it really is. **(b)** The emission nebulae NGC 3603 and NGC 3576 are different distances from Earth. Light from the more distant nebula must pass through more interstellar dust to reach us, so more interstellar reddening occurs and NGC 3603 is a deeper shade of red. The two nebulae are about 1° apart in the sky. (b: Stocktrek Images/Corbis)

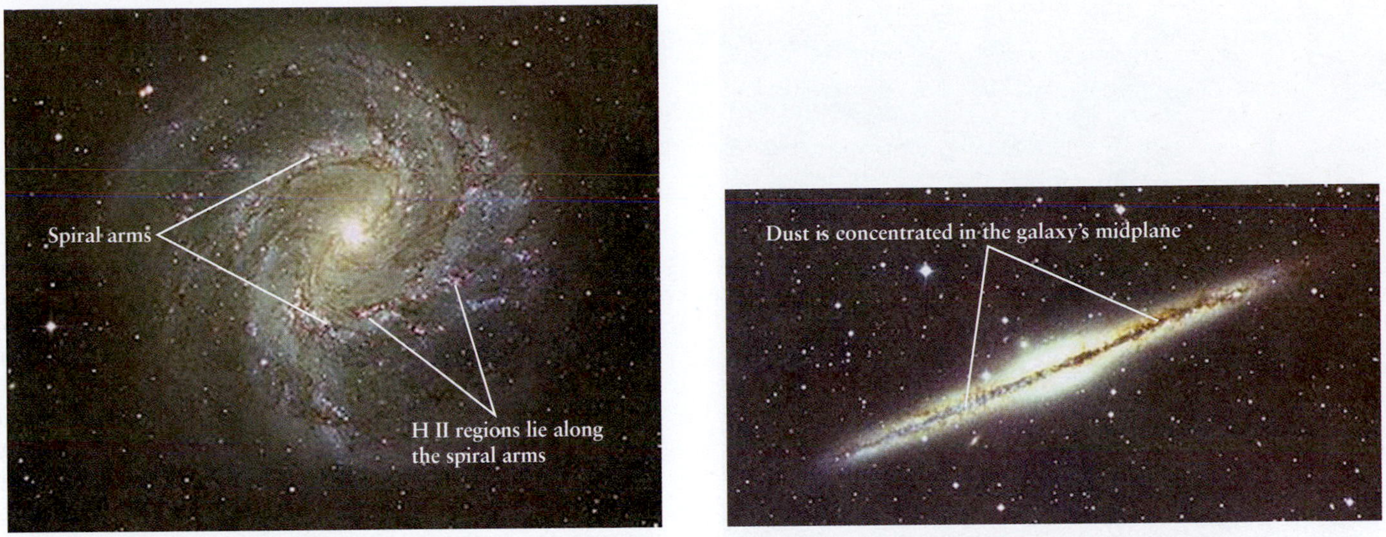

(a) We see spiral galaxy M83 nearly face-on

(b) We see spiral galaxy NGC 891 nearly edge-on

FIGURE 18-8 R I V U X G

Spiral Galaxies Spiral galaxies, like our own Milky Way Galaxy, consist of stars, gas, and dust that are largely confined to a flattened, rotating disk. **(a)** This face-on view of M83 shows luminous stars and H II regions along the spiral arms. **(b)** This edge-on view of NGC 891 shows a dark band caused by dust in this galaxy's interstellar medium. Although in different parts of the sky, both galaxies are about 7 million pc (23 million ly) from Earth and have angular diameters of about 13 arcminutes. (a: Australian Astronomical Observatory/David Malin Images; b: Instituto de Astrofísica de Canarias/Royal Greenwich Observatory/David Malin)

As we will see in Chapter 23, the band of the Milky Way is actually our inside view of our Galaxy, which is a flat, pancake-shaped collection of several hundred billion stars about 50,000 pc (160,000 ly) in diameter. The Sun is located inside this pancake shape and is about 8000 pc (26,000 ly) from our Galaxy's center. Astronomers know these dimensions because they can map our Galaxy from the locations of bright stars and nebulae and also by using radio telescopes. Observations indicate that bright stars, gas, and glowing nebulae are mostly located within a few hundred parsecs of the midplane of our Galaxy, form somewhat of a "pancake" shape, and are concentrated along arching spiral arms that wind outward from the Galaxy's center. If we could view our Galaxy from a great distance, it would look somewhat like one of the galaxies shown in Figure 18-8.

Interstellar gas and dust are the raw material from which stars are made. The disk of our Galaxy, where most of this matter is concentrated, is therefore the site of ongoing star formation. Our next goal is to examine the steps by which stars are formed.

18-3 Protostars form in cold, dark nebulae

TUTORIAL 18-1

How do stars form in the interstellar medium? In this section, we discuss how a contracting cloud gives birth to a clump called a *protostar*—a future main-sequence star.

Protostars: Initial Gravitational Collapse

In order for interstellar material to condense and form a star, the force of gravity—which tends to draw interstellar material together—must overwhelm the internal pressure pushing the material apart. This means that stars will most easily form in regions where the interstellar material is relatively dense, so that atoms and dust grains are close together and gravitational attraction is enhanced.

To assist star formation, the pressure of the interstellar medium should be relatively low. This means that the star-forming region of the interstellar medium should be as cold as possible, because the pressure of a gas goes down as the gas temperature decreases. (Conversely, increasing the temperature of a gas makes the pressure increase. For example, the air pressure increases inside a covered pot of boiling water, which often causes steam to shoot out.)

The only parts of the interstellar medium with high enough density and low enough temperature for stars to form are the dark nebulae. Many of these nebulae, such as the Horsehead Nebula (see Figure 18-2) and Barnard 86 (see Figure 18-4), were discovered and catalogued around 1900 by Edward Barnard and are known as **Barnard objects.** (The Horsehead Nebula is also known as Barnard 33.) Other relatively small, nearly spherical dark nebulae are known as **Bok globules,** after the Dutch-American astronomer Bart Bok, who first called attention to them during the 1940s (Figure 18-9). A Bok globule resembles the inner core of a Barnard object with the outer, less dense portions stripped away. The density of the gas and dust within a Barnard object or a Bok globule is indeed quite high by cosmic standards, in the range from 100 to 10,000 particles per cm^3; by comparison, most of the interstellar medium contains only 0.1 to 20 particles per cm^3. Radio emissions from molecules within these clouds indicate that their internal temperatures are very low, only about 10 K.

A typical Barnard object contains a few thousand solar masses of gas and dust spread over a volume roughly 10 pc (30 ly) across;

FIGURE 18-9 R I V U X G

Bok Globules The dark blobs in this photograph of a glowing H II region are clouds of dust and gas called Bok globules. A typical Bok globule is a parsec or less in size and contains from one to a thousand solar masses of material. The Bok globules and H II region in this image are part of a much larger star-forming region called NGC 281, which lies about 2900 pc (9500 ly) from Earth in the constellation Cassiopeia. The image shows an area about 2.7 pc (8.8 ly) across. (NASA, ESA, and The Hubble Heritage Team [STScI/AURA])

a typical Bok globule is about one-tenth as large. The chemical composition of this material is the standard "cosmic abundance" of about 74% (by mass) hydrogen, 25% helium, and 1% heavier elements (review Figure 8-4). Within these clouds, the densest portions can contract under their own mutual gravitational attraction and form clumps called **protostars.** Each protostar will eventually evolve into a main-sequence star. Because dark nebulae contain many solar masses of material, it is possible for a large number of protostars to form out of a single such nebula. Thus, we can think of dark nebulae as "stellar nurseries."

In the first stage of star formation, a protostar coalesces and contracts due to the mutual gravitational attraction of its parts

The Evolution of a Protostar

In the 1950s, astrophysicists such as Louis Henyey in the United States and C. Hayashi in Japan performed calculations that enabled them to describe the earliest stages of a protostar. At first, a protostar is merely a cool blob of gas several times larger than our solar system. The pressure inside the protostar is too low to support all this cool gas against the mutual gravitational attraction of its parts, and so the protostar collapses. As the protostar collapses, gravitational energy is converted into thermal energy, making the gases heat up and start glowing. (We discussed this process, called Kelvin-Helmholtz contraction, in Section 8-4 and Section 16-1.) Energy from the interior of the protostar is transported outward by convection, warming its surface.

After only a few thousand years of gravitational contraction, the protostar's surface temperature reaches 2000 to 3000 K. At this point the protostar is still quite large, so its glowing gases produce substantial luminosity. (The greater a star's radius and surface temperature, the greater its luminosity. You can review the details in Section 17-6 and Box 17-4.) For example, after only 1000 years of contraction, a protostar of 1 solar mass (1 $M_\odot$) is 20 times larger in radius than the Sun and has 100 times the Sun's luminosity. Unlike the Sun, however, the luminosity of a young protostar is not the result of thermonuclear fusion, because the protostar's core is not yet hot enough for fusion reactions to begin. Instead, the radiated energy comes exclusively from the heating of the protostar as it contracts.

In order to determine the conditions inside a contracting protostar, astrophysicists use computers to solve equations similar to those used for calculating the structure of the Sun (see Section 16-2). The results tell how the protostar's luminosity and surface temperature change at various stages in its contraction. This information, when plotted on a Hertzsprung-Russell diagram, provides a protostar's **evolutionary track.** The track shows us how the protostar's appearance changes because of changes in its interior. These theoretical tracks agree quite well with actual observations of protostars.

CAUTION! When astronomers refer to a star "following an evolutionary track" or "moving on the H-R diagram," they mean that the star's luminosity, surface temperature, or both change. Hence, the point that represents the star on an H-R diagram changes its position. This change is completely unrelated to how the star physically moves through space!

Figure 18-10 shows evolutionary tracks for protostars of seven different masses, ranging from 0.5 $M_\odot$ to 15 $M_\odot$. Because protostars are relatively cool when they begin to shine at visible wavelengths, these tracks all begin near the right (low-temperature) side of the H-R diagram. The subsequent evolution is somewhat different, however, depending on the protostar's mass. Note that the greater a star's mass, the more rapidly it moves along its evolutionary track: a 15-$M_\odot$ protostar takes only about 10^5 years to reach the main sequence, while a 1-$M_\odot$ protostar takes over a hundred times longer (more than 10^7 years).

CONCEPTCHECK 18-3

Is it possible for a protostar to shine brighter than the main-sequence star it becomes?

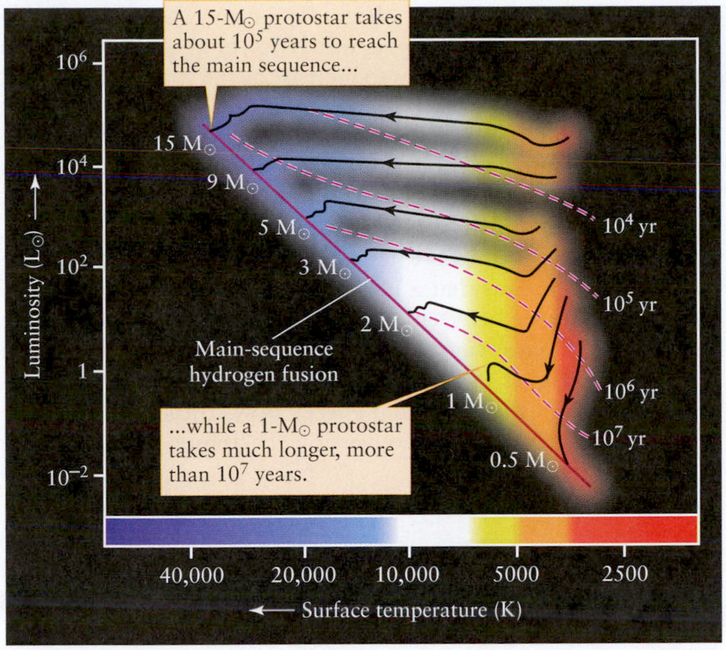

FIGURE 18-10

Pre–Main-Sequence Evolutionary Tracks on an H-R Diagram As a protostar evolves, its luminosity and surface temperature both change. The tracks shown here depict these changes for protostars of seven different masses. Each dashed red line shows the age of a protostar when its evolutionary track crosses that line. (We will see in Section 18-5 that protostars lose quite a bit of mass as they evolve: The mass shown for each track is the value when the protostar finally settles down as a main-sequence star.)

CONCEPTCHECK 18-4

About how long does it take for a 2-$M_\odot$ protostar to reach the main sequence?

Answers appear at the end of the chapter.

Observing Protostars

Observing the evolution of a protostar can be quite a challenge. Indeed, while Figure 18-10 shows that young protostars are much more luminous than the Sun, it is quite unlikely that you have ever seen a protostar shining in the night sky. The reason is that protostars form within clouds that contain substantial amounts of interstellar dust. The dust in a protostar's immediate surroundings, called its **cocoon nebula,** absorbs the vast amounts of visible light emitted by the protostar and makes it very hard to detect using visible wavelengths.

Protostars can be seen, however, using infrared wavelengths. Because a cocoon nebula absorbs so much energy from the protostar that it surrounds, the nebula becomes heated to a few hundred kelvins. The warmed dust then reradiates its thermal energy at infrared wavelengths, to which the cocoon nebula is relatively transparent. So, by using infrared telescopes, astronomers can see protostars within the "stellar nursery" of a dark nebula.

Figure 18-11 shows visible-light and infrared views of one such stellar nursery. The visible-light view (Figure 18-11a) shows a dark, dusty nebula (or Bok globule) that appears completely opaque. The infrared image (Figure 18-11b) allows us to see through the dust, revealing a newly formed protostar within the dark nebula. The properties of this protostar agree well with the theoretical ideas outlined in this section.

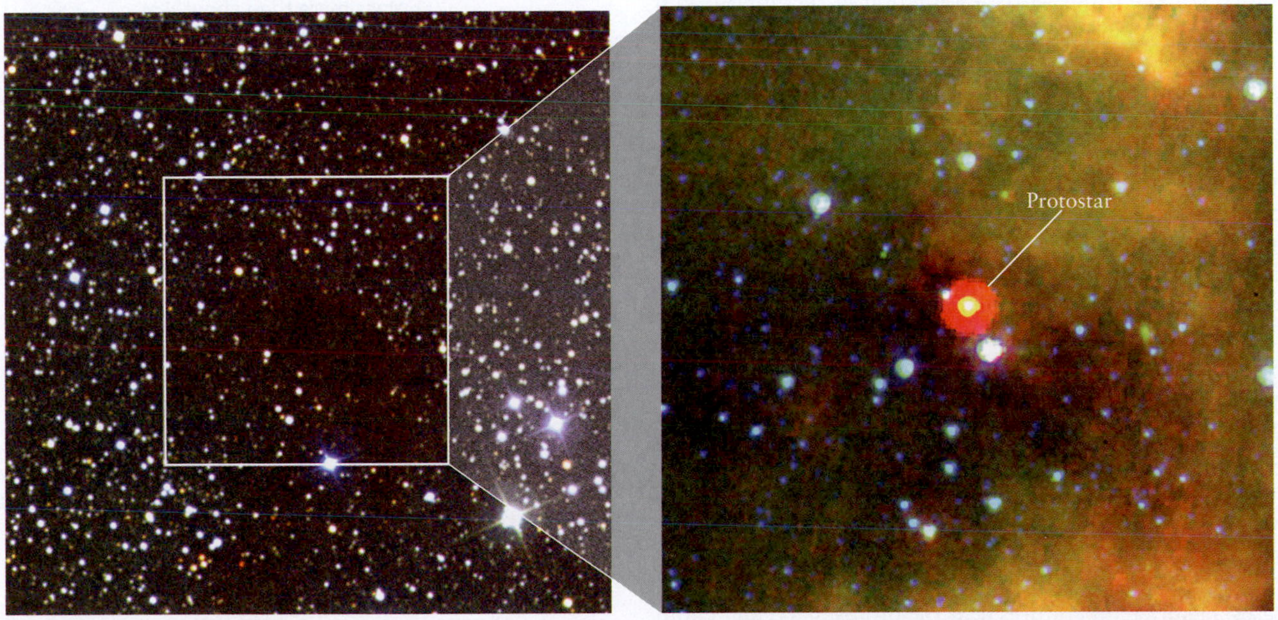

(a) A dark nebula R I V U X G

(b) A hidden protostar within the dark nebula R I V U X G

FIGURE 18-11

Revealing a Hidden Protostar (a) This visible-light view shows a dark nebula (or Bok globule) called L1014 in the constellation Cygnus (the Swan). No stars are visible within the nebula. **(b)** The Spitzer Space Telescope was used to make this false-color infrared image of the outlined area in (a). The bright red-yellow spot is a protostar within the dark nebula. (NASA/JPL-Caltech/N. Evans [University of Texas at Austin])

18-4 Protostars evolve into main-sequence stars

> The course of a protostar's evolution depends on its mass

Evolutionary tracks in Figure 18-10 show how a protostar matures into a star as its gases contract. The details of this evolution depend on the star's mass. To follow these details, you should keep in mind the basic principle that a star's luminosity, radius, and surface temperature are intimately related. The luminosity, which is the total energy output per second, depends on how hot and how large the star is. Specifically, the luminosity is proportional to the square of a star's radius and to the fourth power of its surface temperature (see Section 17-6).

A One-Solar-Mass Protostar

For a protostar with the same mass as our Sun (1 $M_\odot$), the outer layers are relatively cool and quite opaque (for the same reasons that the Sun's photosphere is opaque; see the discussion at the end of Section 16-5). Due to frequent scattering or absorption, light does not pass easily through an opaque material. This means that energy released from the shrinking inner layers in the form of radiation cannot easily reach the protostar's surface. Instead, energy flows outward by the slower and less effective method of convection. The result is that for a contracting 1-$M_\odot$ protostar, the surface temperature stays roughly constant, the luminosity decreases as the radius decreases, and the protostar's evolutionary track initially moves downward on the H-R diagram in Figure 18-10.

Although its surface temperature changes relatively little, the internal temperature of the shrinking protostar increases. After a time, the interior becomes more ionized, which makes it less opaque. Energy is then conveyed outward by radiation in the interior and by convection in the opaque outer layers, just as in the present-day Sun (see Section 16-2, especially Figure 16-4). This makes it easier for energy to escape from the protostar, so the luminosity—the rate at which energy is emitted from the protostar's surface—increases. As a result, the evolutionary track for a 1-$M_\odot$ protostar in Figure 18-10 bends upward (higher luminosity) and to the left (higher surface temperature, caused by the increased energy flow).

In time, the 1-$M_\odot$ protostar's interior temperature reaches a few million kelvins, hot enough for thermonuclear reactions to begin converting hydrogen into helium. As we saw in Section 16-1 and Box 16-1, these reactions release enormous amounts of energy. Eventually, these reactions provide enough heat and internal pressure to stop the star's gravitational contraction, and hydrostatic equilibrium is reached. The protostar's evolutionary track has now led it to the main sequence, and the protostar has become a full-fledged main-sequence star.

In summary, energy transport is initially dominated by convection as a 1-$M_\odot$ protostar shrinks and dims. Next, the protostar transports more energy through radiation and its surface temperature increases. Finally, temperatures near the protostar's center climb high enough to cause thermonuclear reactions, and a star is born.

High-Mass and Low-Mass Protostars

More massive protostars evolve a bit differently. If its mass is more than about 4 $M_\odot$, a protostar contracts and heats more rapidly, and hydrogen fusion begins quite early. As a result, the luminosity quickly stabilizes at nearly its final value, while the surface temperature continues to increase as the star shrinks. Thus, the evolutionary tracks of massive protostars traverse the H-R diagram roughly horizontally (signifying approximately constant luminosity) in the direction from right to left (from low to high surface temperature). You can see this most easily for the 9-$M_\odot$ and 15-$M_\odot$ evolutionary tracks in Figure 18-10.

Greater mass means greater pressure and temperature in the interior, which means that a massive star has an even larger temperature difference between its core and its outer layers than the Sun. This difference causes convection deep in the interior of a massive star (Figure 18-12). By contrast, a massive star's outer layers are of such low density that energy flows through them more easily by radiation than by convection. Therefore, main-sequence

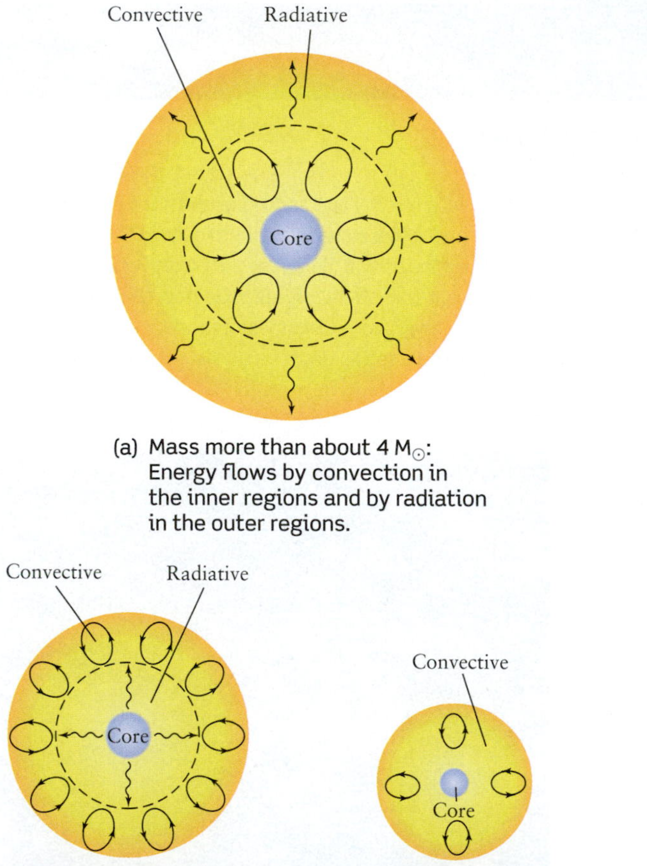

(a) Mass more than about 4 $M_\odot$: Energy flows by convection in the inner regions and by radiation in the outer regions.

(b) Mass between about 4 $M_\odot$ and 0.4 $M_\odot$: Energy flows by radiation in the inner regions and by convection in the outer regions.

(c) Mass less than 0.4 $M_\odot$: Energy flows by convection throughout the star's interior.

FIGURE 18-12
Main-Sequence Stars of Different Masses Stellar models show that when a protostar evolves into a main-sequence star, its internal structure depends on its mass. *Note:* The three stars shown here are not drawn to scale. Compared with a 1-$M_\odot$ main-sequence star like that shown in (b), a 6-$M_\odot$ main-sequence star like that in (a) has more than 4 times the radius, and a 0.2-$M_\odot$ main-sequence star like that in (c) has only one-third the radius.

stars with masses more than about 4 $M_\odot$ have convective interiors but radiative outer layers (Figure 18-12a). By contrast, less massive main-sequence stars such as the Sun have radiative interiors and convective outer layers (Figure 18-12b).

The internal structure is also different for main-sequence stars of very low mass. When such a star forms from a protostar, the interior temperature is never high enough to fully ionize the interior. The interior remains too opaque for radiation to flow efficiently, so energy is transported by convection throughout the volume of the star (Figure 18-12c).

Arriving on the Main Sequence

The main sequence represents stars in which thermonuclear reactions are converting hydrogen into helium, and all of the protostar evolutionary tracks in Figure 18-10 end on the main sequence. For most stars, these thermonuclear reactions are part of a stable situation. For example, our Sun will remain on or very near the main sequence with a roughly constant temperature and luminosity, quietly fusing hydrogen into helium at its core, for a total of some 10^{10} years. The point along the main sequence where each evolutionary track ends depends on the star's mass. The most massive stars are the most luminous and their evolutionary tracks end at the upper left of the main sequence, while the least massive stars are the least luminous and their evolutionary tracks end at the lower right of the main sequence. That a main-sequence star's luminosity is greater for increasing mass should be familiar because this relationship is just the mass-luminosity relation (recall the *Cosmic Connections* figure in Section 17-9).

The theory of how protostars evolve helps explain why the main sequence has both an upper mass limit and a lower mass limit. As we saw in Section 17-5, protostars less massive than about 0.08 $M_\odot$ can never develop the necessary pressure and temperature to start hydrogen fusion in their cores. Instead, such "failed stars" end up as *brown dwarfs,* which shine faintly by Kelvin-Helmholtz contraction (see Figure 17-13).

Protostars with masses greater than about 200 solar masses also do not become main-sequence stars. Such a protostar rapidly becomes very luminous, resulting in tremendous internal pressures. This pressure is so great that it overwhelms the effects of gravity, expelling the outer layers into space and disrupting the star. Main-sequence stars therefore have masses between about 0.08 and 200 $M_\odot$, although the high-mass stars are extremely rare.

CAUTION! A few words of caution are in order here. First, while the evolutionary tracks of protostars begin in the red giant region of the H-R diagram (the upper right), protostars are *not* red giants. As we will see in Chapter 19, red giant stars represent a stage in the evolution of stars that comes *after* being a main-sequence star. Second, it is worth remembering that stars live out most of their lives on the main sequence, after only a relatively brief period as protostars. A 15-$M_\odot$ protostar takes only 20,000 years to become a main-sequence star, and a 1-$M_\odot$ protostar takes about 2×10^7 years. By contrast, the Sun has been a main-sequence star for about 4.56×10^9 years. By astronomical standards, pre–main-sequence stars are quite transitory. Furthermore, during the last 4.56×10^9 years, the Sun has moved very little from its spot on the main sequence.

CONCEPTCHECK 18-5

A 9-$M_\odot$ protostar can have a somewhat constant luminosity, and this holds up even when it reaches the main sequence (see Figure 18-10). Does that mean it has the same energy source while it is a protostar as when it is a main-sequence star?

Answer appears at the end of the chapter.

18-5 During the birth process, stars both gain and lose mass

Before settling down, protostars eject more than half of their initial mass into space

After reading the previous section, you may think that a main-sequence star forms simply by collapsing inward. In fact, much of the material of a cold, dark nebula is ejected into space and never incorporated into stars. As it is ejected, this material may help sweep away the dust surrounding a young star, making the star observable at visible wavelengths.

T Tauri Stars

Mass ejection into space is a hallmark of **T Tauri stars.** These objects are protostars with emission lines as well as absorption lines in their spectra and whose luminosity can change irregularly on timescales of a few days. The namesake of this class of stars, T Tauri, is a protostar in the constellation Taurus (the Bull).

T Tauri stars have final masses less than about 3 $M_\odot$ and ages around 10^6 years, so on an H-R diagram such as Figure 18-10 they appear above the right-hand end of the main sequence. (The stellar masses shown in Figure 18-10 are those of the final, main-sequence stars.) The emission lines show that these protostars are surrounded by a thin, hot gas. The Doppler shifts of these emission lines suggest that the protostars eject gas at speeds around 80 km/s (300,000 km/h, or 180,000 mi/h).

On average, T Tauri stars eject about 10^{-8} to 10^{-7} solar masses of material per year. This amount may seem small, but by comparison the present-day Sun loses only about 10^{-14} $M_\odot$ per year. The T Tauri phase of a protostar may last 10^7 years or so, during which time the protostar may eject roughly a solar mass of material. Thus, the mass of the final main-sequence star is quite a bit less than that of the cloud of gas and dust from which the star originated.

Young protostars that are more massive than about 3 $M_\odot$ do not vary in luminosity like T Tauri stars. They do lose mass, however, because the pressure of radiation at their surfaces is so strong that it blows gas into space (Figure 18-13).

Bipolar Outflows and Herbig-Haro Objects

ANIMATION 18.1 In the early 1980s, it was discovered that many young stars, including T Tauri stars, lose mass by ejecting gas along two narrow, oppositely directed jets—a phenomenon called **bipolar outflow.** As this material is ejected into space at speeds of several hundred kilometers per second, it collides with the surrounding interstellar medium and produces knots of hot, ionized gas that glow with an emission-line spectrum. These glowing knots are called **Herbig-Haro objects** after the two astronomers, George Herbig in

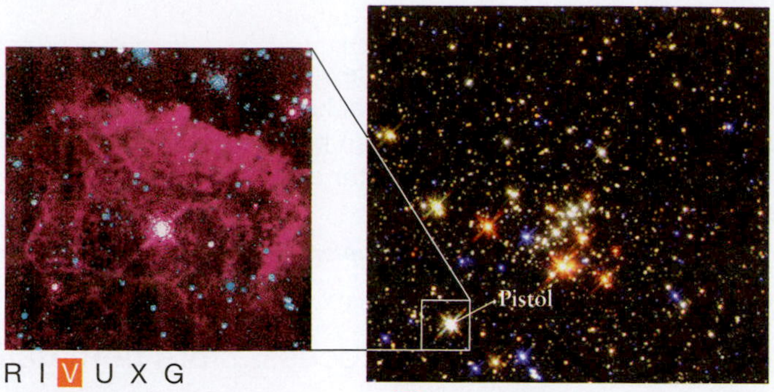

R I V U X G

R I V U X G

FIGURE 18-13

Mass Loss from a Supermassive Star The Quintuplet Cluster is 25,000 ly from Earth. **Inset:** Within the cluster is one of the brightest known stars, called the Pistol. Astronomers calculate that the Pistol formed nearly 3 million years ago and originally had 100–200 $M_\odot$. The structure of the gas cloud suggests the star ejected the gas we see in two episodes, 6000 and 4000 years ago. The gas from any previous ejections is so thinly spread now that we cannot see it. The nebula shown in the inset is more than 1.25 pc (4 ly) across—it would stretch from the Sun nearly to the closest star, Proxima Centauri. (2 MASS/UMass/IPAC-Caltech/NASA/NSF; inset: D. Filger, NASA)

the United States and Guillermo Haro in Mexico, who discovered them independently. Thus, a Herbig-Haro object is the observational manifestation of bipolar outflow around a star. Figure 18-14 is a Hubble Space Telescope image of the Herbig-Haro objects HH 1 and HH 2, which are produced by the two jets from a single young star in the constellation Orion. Herbig-Haro objects like these change noticeably in position, size, shape, and brightness from year to year, indicating the dynamic character of bipolar outflows.

Observations suggest that most protostars eject material in the form of jets at some point during their evolution. These bipolar outflows are very short-lived by astronomical standards, a mere 10^4 to 10^5 years, but they are so energetic that they typically eject into space more mass than ends up in the final protostar.

In one object, water has been observed flowing out in the jets along with other material. The amount of water ejected each second is about 100 million times the amount of water flowing out of the Amazon River each second. The water in the jets moves at about 80 times the speed of a machine gun's bullet and eventually slams into surrounding material about 5000 AU away.

Accretion Disks

Protostars slowly add mass to themselves at the same time that they rapidly eject it into space. In fact, the two processes are related. As a protostar's nebula contracts, it spins faster and flattens into a disk with the protostar itself at the center. The same flattening took place in the solar nebula from which the Sun and planets formed (see Section 8-4). Particles orbiting the protostar within this disk collide with each other, causing them to lose energy, spiral inward onto the protostar, and add to the protostar's mass. This process is called **accretion,** and the disk of material being added to the protostar in this way is called a **circumstellar accretion disk.** Figure 18-15 is an edge-on view of a circumstellar accretion disk, showing two oppositely directed jets emanating from a point at or near the center of the disk (where the protostar is located).

What causes some of the material in the disk to be blasted outward in a pair of jets? One model involves the magnetic field of the dark nebula in which the star forms (Figure 18-16). As material in the circumstellar accretion disk falls inward, it drags the magnetic field lines along with it. (We saw in Section 16-9 how a similar mechanism may explain the Sun's 22-year cycle.) Parts of the disk at different distances from the central protostar orbit at different speeds, and this can twist the magnetic field lines into two helix shapes, one on each side of the disk. The helices then act as channels that guide initially infalling material *away* from the protostar, forming two opposing jets. While it's clear that jets do form, the details of their formation are still uncertain. The star-forming region in Figure 18-17 shows two spectacular outflows from jets.

Protoplanetary Disks

In the 1990s, astronomers using the Hubble Space Telescope discovered many examples of disks around newly formed stars in the Orion Nebula (see Figure 18-1), one of the most prominent star-forming regions in the northern sky. Figure 8-8 shows a number of these **protoplanetary disks,** or **proplyds,** that surround young stars within the nebula. As the name suggests, protoplanetary disks are thought to contain the material from which planets form around stars. They are what remains of a circumstellar accretion disk after much of the material has either fallen onto the star or been ejected by bipolar outflows.

Not all stars are thought to form protoplanetary disks; the exceptions probably include stars with masses in excess of about 3 $M_\odot$, as well as many stars in binary systems. But surveys of the Orion Nebula show that disks are found around most young,

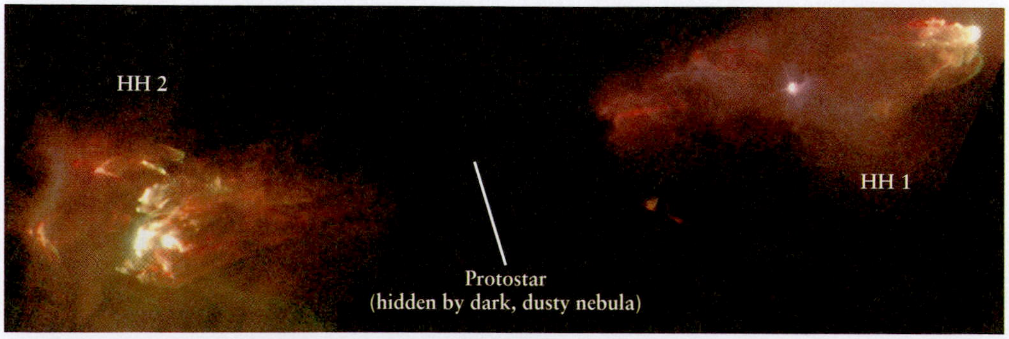

FIGURE 18-14 R I V U X G

Bipolar Outflow and Herbig-Haro Objects The two bright knots of glowing, ionized gas called HH 1 and HH 2 are Herbig-Haro objects. They are created when fast-moving gas ejected from a protostar slams into the surrounding interstellar medium, heating the gas to high temperature. HH 1 and HH 2 are 0.34 pc (1.1 ly) apart and lie 470 pc (1500 ly) from Earth in the constellation Orion. (J. Hester [Arizona State University], the WFPC-2 Investigation Team, and NASA)

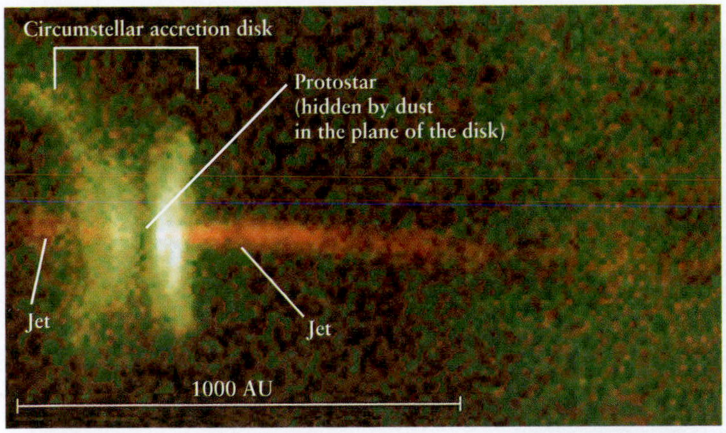

FIGURE 18-15 R I V U X G

A Circumstellar Accretion Disk and Jets This false-color image shows a T Tauri star surrounded by an accretion disk, which we see nearly edge-on. Red denotes emission from ionized gas, while green denotes starlight scattered from dust particles in the disk. The midplane of the accretion disk is so dusty and opaque that it appears dark. Two oppositely directed jets flow away from the star, perpendicular to the disk and along the disk's rotation axis. This star, Herbig-Haro object HH 30, lies 140 pc (460 ly) from Earth. (C. Burrows, the WFPC-2 Investigation Definition Team, and NASA)

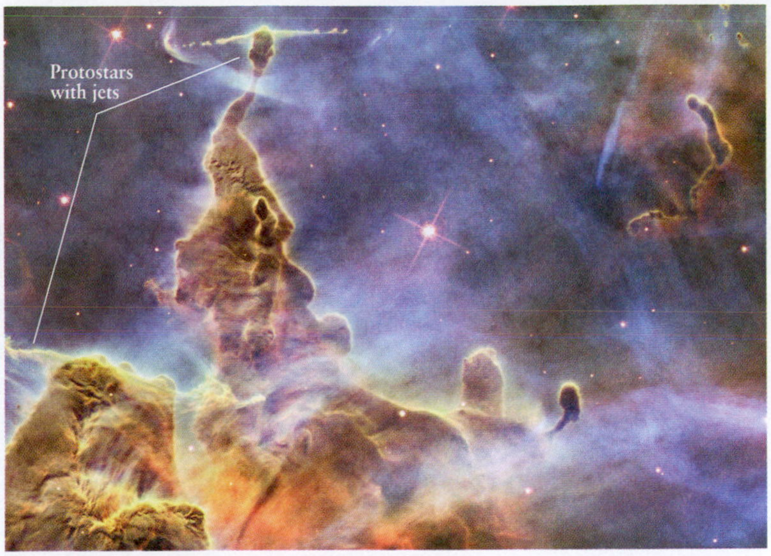

FIGURE 18-17 R I V U X G

Star Formation and Jets While the protostars aren't visible, some of their jets can be seen emanating from the thick gas and dust from which the stars are forming. As new stars form, their strong winds and intense ultraviolet radiation blow away lower-density material, leaving behind distinct regions with greater density. The denser material makes an ideal star-forming region to produce stars; the first few bright stars are already visible. This is only part of the 30 ly–wide Carina Nebula, which is about 7500 ly away. (NASA, ESA, and M. Livio and the Hubble 20th Anniversary Team [STScI])

low-mass stars. Thus, disk formation may be a natural stage in the birth of many stars.

CONCEPTCHECK 18-6

Protostars gain and lose mass on their way to the main sequence, but which effect is dominant in the formation of main-sequence stars with less than 3 $M_\odot$?

Answer appears at the end of the chapter.

18-6 Young star clusters give insight into star formation and evolution

TUTORIAL 18-2

Dark nebulae contain tens or hundreds of solar masses of gas and dust, enough to form many stars. As a consequence, these nebulae tend to form groups or **clusters** of young stars. One cluster still in the act of star formation is M16, shown in Figure 18-18. As a cluster ages, intense radiation

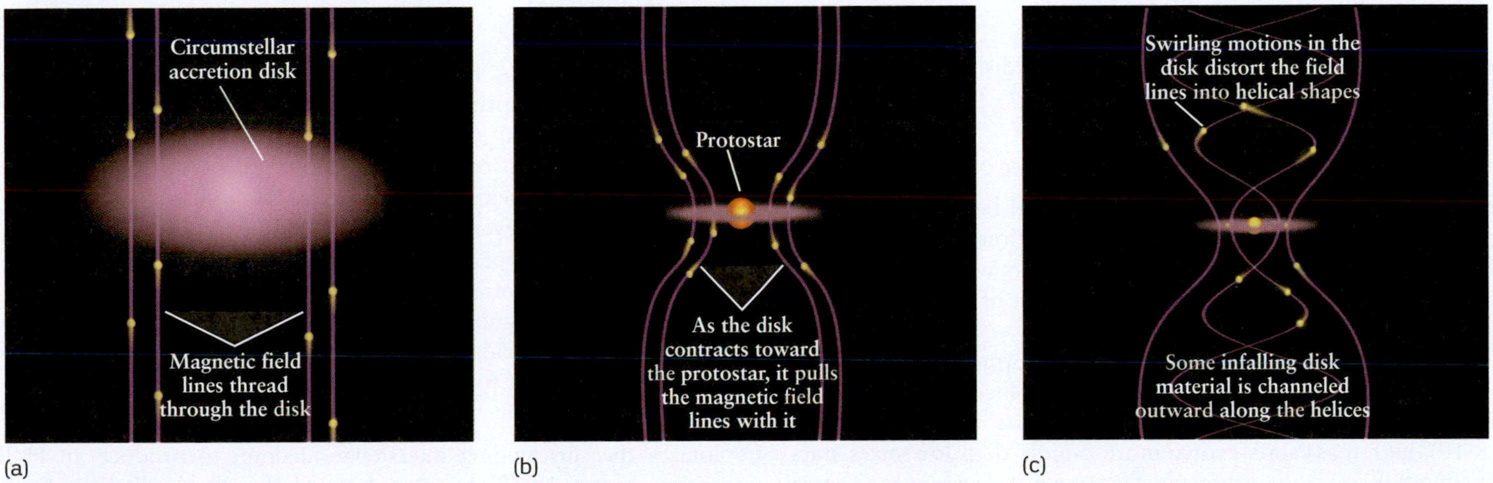

FIGURE 18-16

A Magnetic Model for Bipolar Outflow **(a)** Observations suggest that circumstellar accretion disks are threaded by magnetic field lines, as shown here. **(b)** Magnetic field lines move with the material they thread. **(c)** The contraction and rotation of the disk make the magnetic field lines distort and twist into helices. These helices steer some of the disk material into jets that stream perpendicular to the plane of the disk, as in Figure 18-15. (Adapted from Alfred T. Kamajian/Thomas P. Ray, "Fountain of Youth: Early Days in the Life of a Star," *Scientific American,* August 2000)

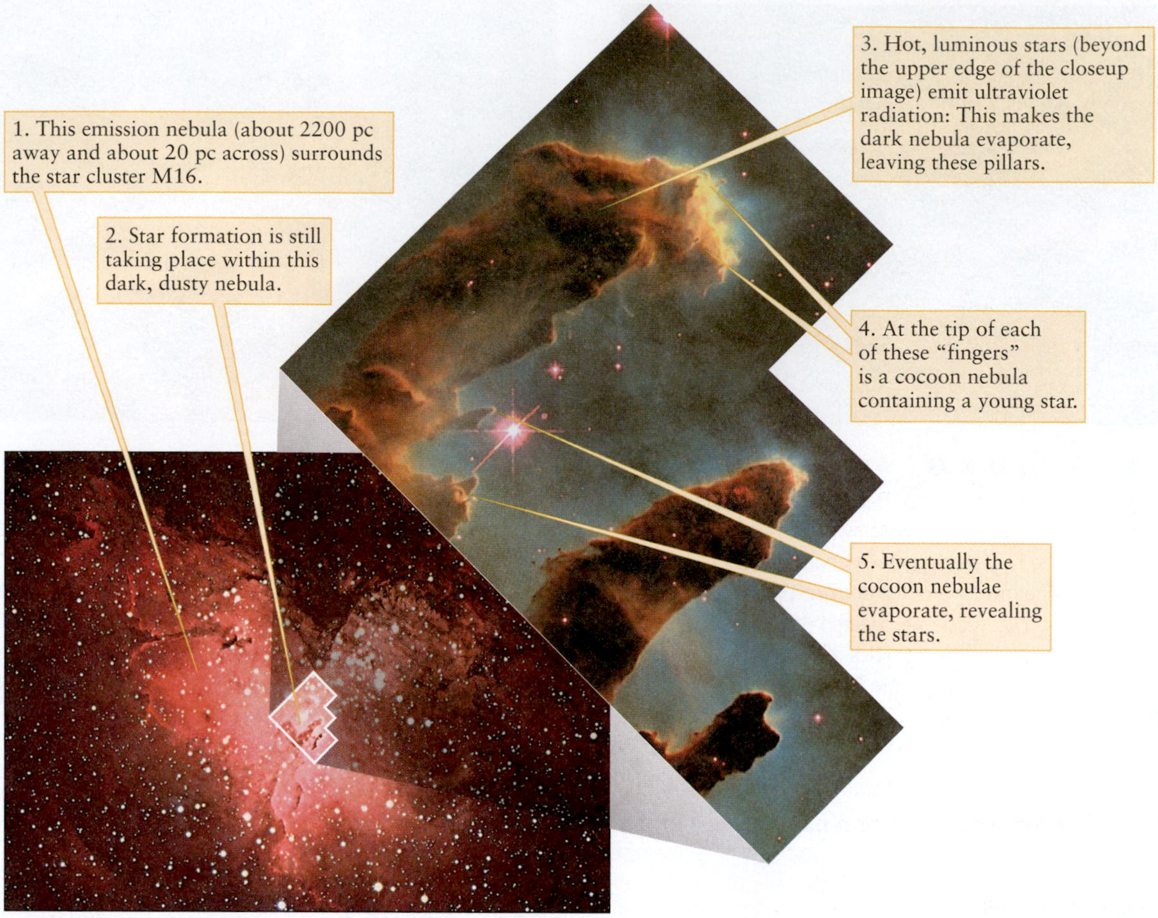

1. This emission nebula (about 2200 pc away and about 20 pc across) surrounds the star cluster M16.
2. Star formation is still taking place within this dark, dusty nebula.
3. Hot, luminous stars (beyond the upper edge of the closeup image) emit ultraviolet radiation: This makes the dark nebula evaporate, leaving these pillars.
4. At the tip of each of these "fingers" is a cocoon nebula containing a young star.
5. Eventually the cocoon nebulae evaporate, revealing the stars.

ANIMATION 18-3 **FIGURE 18-18** R I V U X G
Formation of a Star Cluster The star cluster M16 is thought to be no more than 800,000 years old, and star formation is still taking place within adjacent dark, dusty globules. Massive young stars in the cluster power red emission lines of the H II region (called the Eagle Nebula for its shape). The inset zooms in on three dense, cold pillars of gas with ongoing star formation. The pillar at the upper left extends about 0.3 pc (1 ly) from base to tip, and each of its "fingers" is somewhat broader than our entire solar system. (Australian Astronomical Observatory; J. Hester and P. Scowen, Arizona State University; NASA)

transforms the dark nebula, producing a reddish H II region as in Figure 18-19.

Star Clusters as Evolutionary Laboratories

In addition to being objects of great natural beauty, star clusters give us a unique way to compare the evolution of different stars. That is because clusters typically include stars with a range of different masses, all of which began to form out of the parent nebula at roughly the same time.

All the stars in a cluster may begin to form nearly simultaneously, but they do not all become main-sequence stars at the same time. As you can see from their evolutionary tracks (see Figure 18-10), high-mass stars evolve more rapidly than low-mass stars. The more massive the protostar, the sooner it develops the central pressures and temperatures needed for steady hydrogen fusion to begin, thus joining the main sequence.

Upon reaching the main sequence, *high-mass* protostars become hot, ultraluminous stars of spectral types O and B. As we saw in Section 18-2, these types of stars have ultraviolet radiation that ionizes the surrounding interstellar medium to produce an H II region. Figure 18-18 shows such an H II region, called the Eagle Nebula, surrounding the young star cluster M16. A few hundred thousand years ago, this region of space would have had a far less dramatic appearance. It was then a dark nebula, with protostars just beginning to form. Over the intervening millennia, mass ejection from these evolving protostars swept away the obscuring dust.

ANALOGY A foot race is a useful way to compare the performance of sprinters because all the competitors start the race simultaneously. A young star cluster gives us the same kind of opportunity to compare the evolution of stars of different masses that all began to form roughly simultaneously. Unlike a foot race, however, the entire "race" of stellar evolution in a single cluster happens too slowly for us to observe; as Figure 18-10 shows, protostars take many thousands or millions of years to evolve significantly. Instead, we must compare different star clusters at various stages in their evolution to piece together the history of star formation in a cluster.

FIGURE 18-19 R I **V** U X G

A Mature Star Cluster NGC 3603 is one the closest star clusters at only 20,000 ly away. Star formation has ended and ultraviolet light from its stars power the largest visible reddish H II region in our Galaxy. (NASA/Hubble Space Telescope Collection)

The exposed young, hot stars heated the relatively thin remnants of the original dark nebula, creating the H II region that we see today.

When the most massive protostars to form out of a dark nebula have reached the main sequence, other *low-mass* protostars are still evolving nearby within their dusty cocoons. The evolution of these low-mass stars can be disturbed by their more massive neighbors. As an example, the inset in Figure 18-18 is a close-up of part of the Eagle Nebula. Within these opaque pillars of cold gas and dust, protostars are still forming. At the same time, however, the pillars are being eroded by intense ultraviolet light from hot, massive stars that have already shed their cocoons. As each pillar evaporates, the embryonic stars within have their surrounding material stripped away prematurely, limiting the total mass that these stars can accrete.

Analyzing Young Clusters Using H-R Diagrams

Star clusters tell us still more about how high-mass and low-mass stars evolve. Figure 18-20a shows the young star cluster NGC 2264 and its associated emission nebula. Astronomers have measured each star's apparent brightness and color ratio. Knowing the distance to the cluster, they have deduced the luminosities and

(a) The star cluster NGC 2264

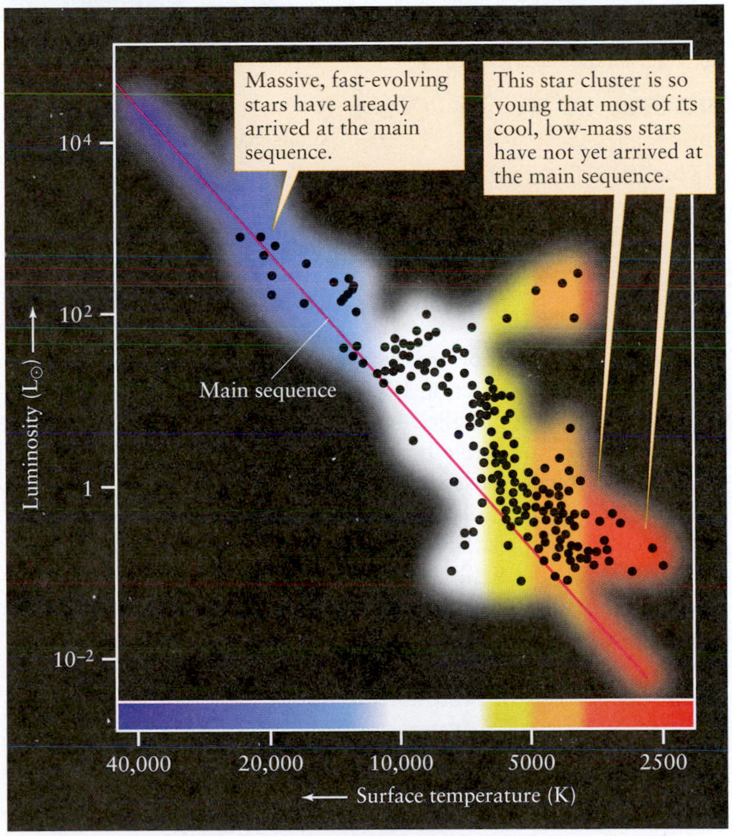

(b) An H-R diagram of the stars in NGC 2264

FIGURE 18-20 R I **V** U X G

A Young Star Cluster and Its H-R Diagram **(a)** This image shows the young star cluster NGC 2264 in a reddish H II region known as the Fox Fur Nebula. Stars within the cluster provide the ultraviolet light that powers this emission nebula. The cluster lies about 800 pc (2600 ly) from Earth. **(b)** Each dot plotted on this H-R diagram represents a star in NGC 2264 whose luminosity and surface temperature have been determined. This star cluster probably started forming only 2 million years ago, and its lower-mass stars have not yet reached the main sequence. (a: © 2004–2013 R. Jay GaBany, Cosmotography.com)

(a) The Pleiades star cluster

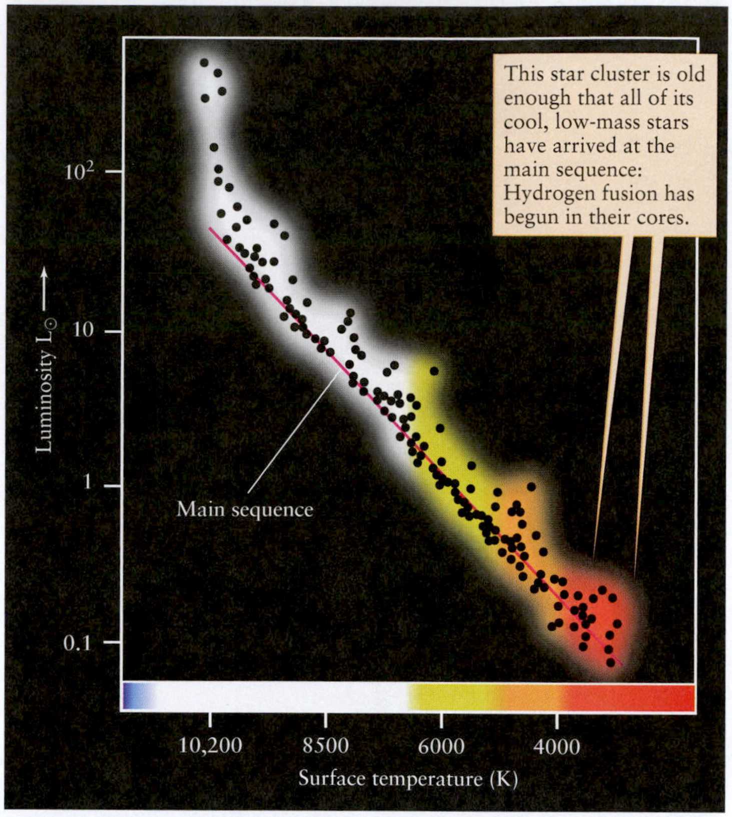

(b) An H-R diagram of the stars in the Pleiades

FIGURE 18-21 R I **V** U X G
The Pleiades and Its H-R Diagram **(a)** The Pleiades star cluster is 117 pc (380 ly) from Earth in the constellation Taurus and can be seen with the naked eye. **(b)** Each dot plotted on this H-R diagram represents a star in the Pleiades whose luminosity and surface temperature have been measured. (*Note:* The scales on this H-R diagram are different from those in Figure 18-18b.) The Pleiades is about 50 million (5×10^7) years old. (Australian Astronomical Observatory/David Malin Images)

The H-R diagram of a young cluster reveals how much time has elapsed since its stars began to form

surface temperatures of the stars (see Section 17-2 and Section 17-4). Figure 18-20b shows all these stars on an H-R diagram. Note that the hottest and most massive stars, with surface temperatures around 20,000 K, are on the main sequence. Stars cooler than about 10,000 K, however, have not yet quite arrived at the main sequence. These are less massive stars in the final stages of pre–main-sequence contraction and are just now beginning to ignite thermonuclear reactions at their centers. To find the ages of these stars, we can compare Figure 18-20b with the theoretical calculations of protostar evolution in Figure 18-10. It turns out that this particular cluster is probably about 2 million years old.

Figure 18-21a shows another young star cluster called the Pleiades. The photograph shows gas that must once have formed an H II region around this cluster and has dissipated into interstellar space, leaving only traces of dusty material that forms reflection nebulae around the cluster's stars. This implies that the Pleiades must be older than NGC 2264, the cluster in Figure 18-20a, which is still surrounded by an H II region. The H-R diagram for the Pleiades in Figure 18-21b bears out this idea. In contrast to the H-R diagram for NGC 2264, nearly all the stars in the Pleiades are on the main sequence. The cluster's age is about 50 million years, which is how long it takes for the least massive stars to finally begin hydrogen fusion in their cores.

CAUTION! Note that the data points for the most massive stars in the Pleiades (at the upper left of the H-R diagram in Figure 18-21b) lie above the main sequence. This is *not* because these stars have yet to arrive at the main sequence. Rather, these stars were the first members of the cluster to arrive at the main sequence some time ago and are now the first members to leave it. They have used up the hydrogen in their cores, so the steady process of core hydrogen fusion that characterizes main-sequence stars cannot continue. In Chapter 19 we will see why massive stars spend a rather short time as main-sequence stars and will study what happens to stars after the main-sequence phase of their lives.

There are two ways to get the ages of clusters with H-R diagrams. What we have seen in Figure 18-20 and Figure 18-21 is for *young* clusters. The age is estimated by figuring out how long it takes the less massive, slower-evolving protostars to *arrive* at the main sequence. In the next chapter (Section 19-4), we will estimate the ages of *old* clusters by determining how long it has taken more massive stars to *leave* the main sequence.

A loose collection of stars such as NGC 2264 or the Pleiades is referred to as an **open cluster** (or *galactic cluster,* since such clusters are usually found in the plane of the Milky Way Galaxy). Open clusters possess barely enough mass to hold themselves together by gravitation. Occasionally, a star moving faster than average will escape, or "evaporate," from an open cluster. Indeed, by the time the stars are a few billion years old, they may be so widely separated that a cluster no longer exists.

If a group of stars is gravitationally unbound from the very beginning—that is, if the stars are moving away from one another so rapidly that gravitational forces cannot keep them together—then the group is called a **stellar association.** Because young stellar associations are typically dominated by luminous O and B main-sequence stars, they are also called **OB associations.** The image that opens this chapter shows part of an OB association in the southern constellation Ara (the Altar).

CONCEPTCHECK 18-7

Consider the pillars shown in the inset of Figure 18-18. Are the pillars eroded away by the stars forming in their cocoon nebulae within the pillars?

Answer appears at the end of the chapter.

18-7 Star birth can begin in giant molecular clouds

We have seen that star formation takes place within dark nebulae. But where within our Galaxy are these dark nebulae found? Does star formation take place everywhere within the Milky Way, or only in certain special locations? The answers to such questions can enhance our understanding of star formation and of the nature of our home Galaxy.

Exploring the Interstellar Medium at Millimeter Wavelengths

Dark nebulae are a challenge to locate simply because they *are* dark—they do not emit visible light. Nearby dark nebulae can be seen silhouetted against background stars or reddish H II regions (see Figure 18-2), but sufficiently distant dark nebulae are impossible to see in contrast with background visible light because of interstellar extinction from dust grains. They can, however, be detected using longer-wavelength radiation that can pass unaffected through interstellar dust. In fact, dark nebulae actually emit radiation at millimeter wavelengths.

> **Observing the Galaxy at millimeter wavelengths reveals the cold gas that spawns new stars**

Such emission takes place because in the cold depths of interstellar space, atoms combine to form molecules. The laws of quantum mechanics predict that just as electrons within atoms can occupy only certain specific energy levels (see Section 5-8), molecules can vibrate and rotate only at certain specific rates. When a molecule goes from one vibrational state or rotational state to another, it either emits or absorbs a photon. (In the same way, an atom emits or absorbs a photon as an electron jumps from one energy level to another.) Most molecules are strong emitters of radiation with wavelengths of around 1 to 10 millimeters (mm). Consequently, observations with radio telescopes tuned to millimeter wavelengths make it possible to detect interstellar molecules of different types. More than 100 different kinds of molecules have so far been discovered in interstellar space, and the list is constantly growing.

Hydrogen is by far the most abundant element in the universe. Unfortunately, in cold nebulae much of it is in a molecular form (H_2) that is difficult to detect. The reason is that the hydrogen molecule is symmetric, with two atoms of equal mass joined together, and such molecules do not emit many photons at radio frequencies. In contrast, asymmetric molecules that consist of two atoms of unequal mass joined together, such as carbon monoxide (CO), are easily detectable at radio frequencies. When a carbon monoxide molecule makes a transition from one rate of rotation to another, it emits a photon at a wavelength of 2.6 mm or shorter.

The ratio of carbon monoxide to hydrogen in interstellar space is reasonably constant: For every CO molecule, there are about 10,000 H_2 molecules. As a result, carbon monoxide is an excellent "tracer" for molecular hydrogen gas. Wherever astronomers detect strong emission from CO, they know molecular hydrogen gas must be abundant.

Giant Molecular Clouds

The first systematic surveys of our Galaxy looking for 2.6-mm CO radiation were undertaken in 1974 by the American astronomers Philip Solomon and Nicholas Scoville. In mapping the locations of CO emission, they discovered huge clouds, now called **giant molecular clouds,** that must contain enormous amounts of hydrogen. These clouds have masses in the range of 10^5 to 2×10^6 solar masses and diameters that range from about 15 to 100 pc (50 to 300 ly). Inside one of these clouds, there are about 200 hydrogen molecules per cubic centimeter. This density is several thousand times greater than the average density of matter in the disk of our Galaxy, yet only 10^{-17} as dense as the air we breathe. Astronomers now estimate that our Galaxy contains about 5000 of these enormous clouds.

Figure 18-22 is a map of radio emissions from carbon monoxide in the constellations Orion and Monoceros. Note the extensive areas of the sky covered by giant molecular clouds. This part of the sky is of particular interest because it includes several star-forming regions. By comparing the radio map with the star chart overlay, you can see that the areas where CO emission is strongest, and, thus, where giant molecular clouds are densest, are sites of star formation. Therefore, giant molecular clouds are associated with the formation of stars. Particularly dense regions within these clouds form dark nebulae, and within these stars are born.

By using CO emissions to map out giant molecular clouds, astronomers can find the locations in our Galaxy where star formation occurs. These investigations reveal that molecular clouds clearly outline our Galaxy's spiral arms, as **Figure 18-23** shows. These clouds lie roughly 1000 pc (3000 ly) apart and are strung along the spiral arms like beads on a string. This arrangement

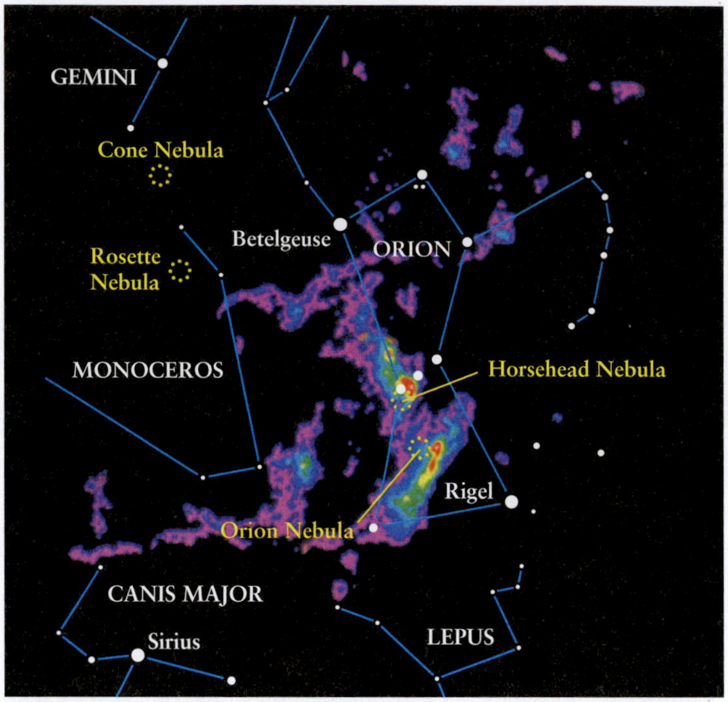

FIGURE 18-22 R I V U X G

Mapping Molecular Clouds A radio telescope was tuned to a wavelength of 2.6 mm to detect emissions from carbon monoxide (CO) molecules in the constellations Orion and Monoceros. The result was this false-color map, which shows a 35° × 40° section of the sky. The Orion and Horsehead star-forming nebulae are located at sites of intense CO emission (shown in red and yellow), indicating the presence of a particularly dense molecular cloud at these sites of star formation. The molecular cloud is much thinner at the positions of the Cone and Rosette nebulae, where star formation is less intense. (Courtesy of R. Maddalena, M. Morris, J. Moscowitz, and P. Thaddeus)

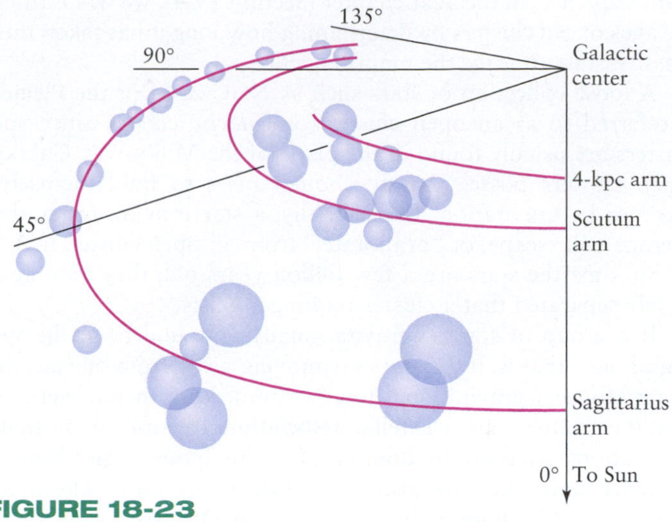

FIGURE 18-23

Giant Molecular Clouds in the Milky Way This perspective drawing shows the locations of giant molecular clouds in an inner part of our Galaxy as seen from a vantage point above the Sun. These clouds lie primarily along the Galaxy's spiral arms, shown by red arcs. The distance from the Sun to the galactic center is about 8000 pc (26,000 ly). (Adapted from T. M. Dame and colleagues)

resembles the spacing of H II regions along the arms of other spiral galaxies, such as the galaxy shown in Figure 18-8a. The presence of both molecular clouds and H II regions shows that spiral arms are sites of ongoing star formation.

Star Formation in Spiral Arms

In Chapter 23 we will learn that spiral arms are locations where matter "piles up" temporarily as it orbits the center of the Galaxy. You can think of matter in a spiral arm as analogous to a freeway traffic jam. Just as cars are squeezed close together when they enter a traffic jam, a giant molecular cloud is compressed when it passes through a spiral arm. When this happens, vigorous star formation begins in the cloud's densest regions.

As soon as massive O and B stars form, they emit ultraviolet light that ionizes the surrounding hydrogen, and an H II region is born. An H II region is thus a small, bright "hot spot" in a giant molecular cloud (but keep in mind that light from an H II region comes from emission lines, and not blackbody radiation from hot matter). An example is the Orion Nebula, shown in Figure 18-1b. Four hot, luminous O and B stars at the heart of the nebula produce the ionizing radiation that makes the surrounding gases glow. The

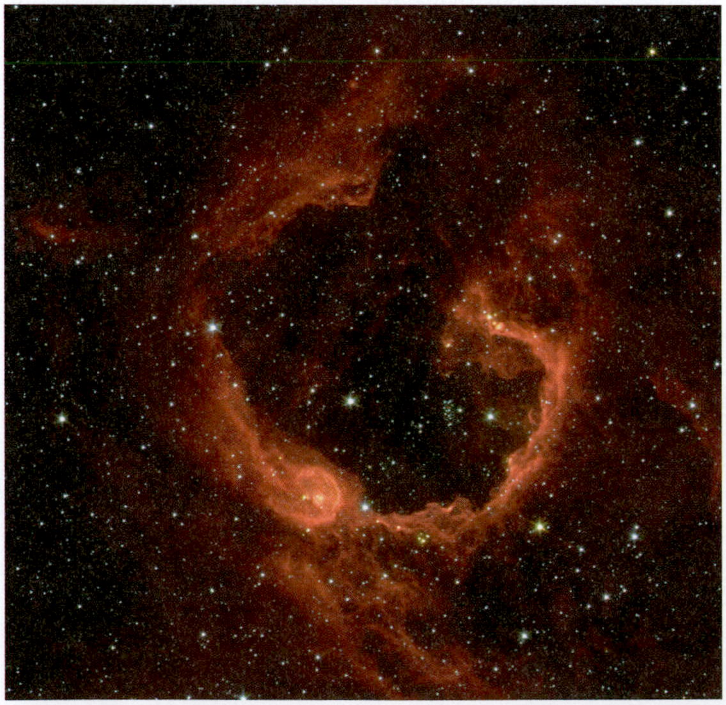

FIGURE 18-24 R I V U X G

A Star-Forming Bubble Radiation and winds from the hot, young O and B stars at the center of this Spitzer Space Telescope image have carved out a bubble about 20 pc (70 ly) in diameter in the surrounding gas and dust. The material around the surface of the bubble has been compressed and heated, making the dust glow at the infrared wavelengths used to record this image. The compressed material is so dense that new stars have formed within that material. This glowing cloud, called RCW 79, lies about 5300 pc (17,200 ly) from Earth in the constellation Centaurus. (NASA; JPL-Caltech; and E. Churchwell, University of Wisconsin-Madison)

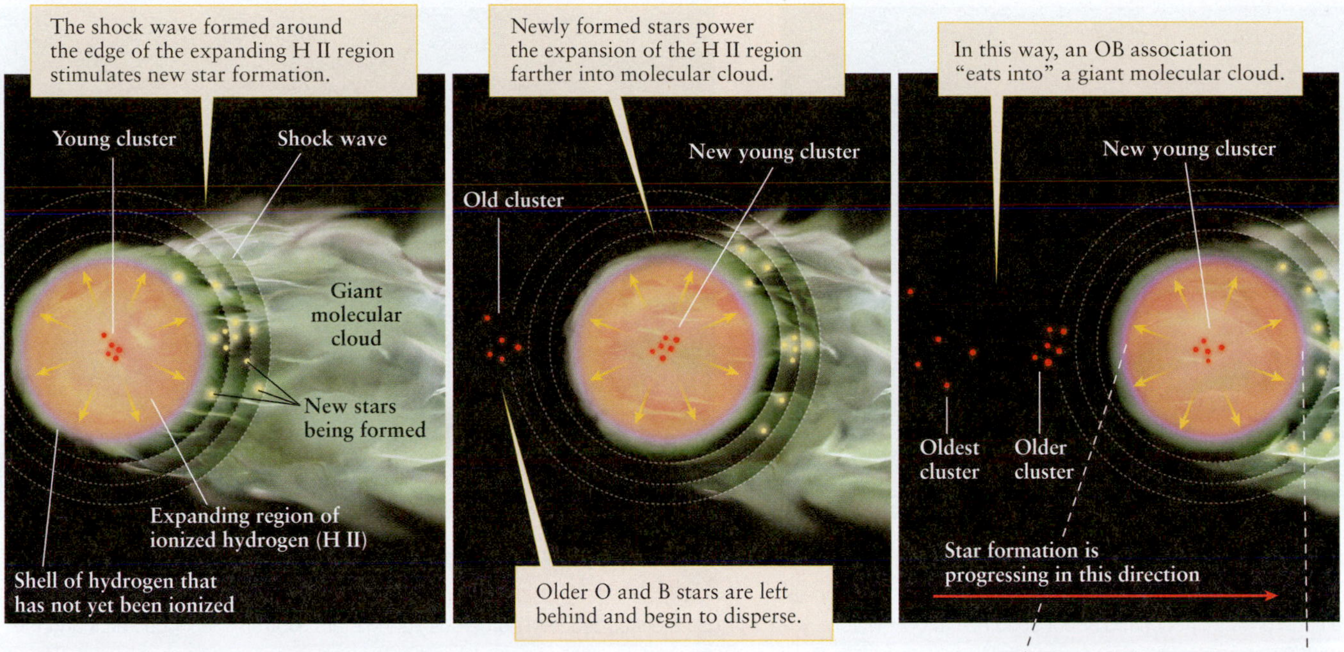
The shock wave formed around the edge of the expanding H II region stimulates new star formation.
Young cluster
Shock wave
Giant molecular cloud
New stars being formed
Expanding region of ionized hydrogen (H II)
Shell of hydrogen that has not yet been ionized
Newly formed stars power the expansion of the H II region farther into molecular cloud.
New young cluster
Old cluster
Older O and B stars are left behind and begin to disperse.
In this way, an OB association "eats into" a giant molecular cloud.
New young cluster
Oldest cluster
Older cluster
Star formation is progressing in this direction

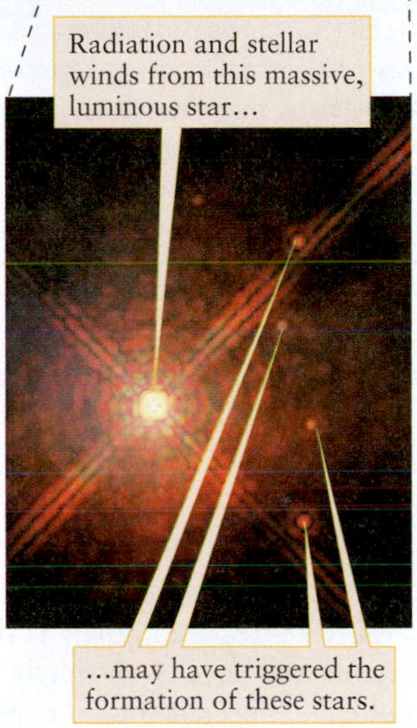
Radiation and stellar winds from this massive, luminous star…
…may have triggered the formation of these stars.

FIGURE 18-25 R I V U X G

How O and B Stars Trigger Star Formation Stellar winds and ultraviolet radiation from young O and B stars produce a shock wave that compresses gas farther into the giant molecular cloud. This stimulates star formation, producing more O and B stars, which stimulate still more star formation, and so on. Meanwhile, older stars are left behind. This figure only illustrates new star formation in the rightward direction, but it could be spherical in a real cloud. The inset shows a massive star that has spawned other, smaller stars in this way. These stars are about 770 pc (2500 ly) from Earth in the Cone Nebula, a star-forming region in the constellation Monoceros. The younger stars are just 0.04 to 0.08 ly (2500 to 5000 AU) from the central star. (Adapted from C. Lada, L. Blitz, and B. Elmegreen; inset: R. Thompson, M. Rieke, G. Schneider, and NASA)

Orion Nebula is embedded on the edge of a giant molecular cloud whose mass is estimated at 500,000 $M_{\odot}$. The H II regions in Figure 18-2 are located at a different point on the edge of the same molecular cloud, some 25 pc (80 ly) from the Orion Nebula.

Once star formation has begun and an H II region has formed, the massive O and B stars at the core of the H II region induce star formation in the rest of the giant molecular cloud. Ultraviolet radiation and vigorous stellar winds from the O and B stars carve out a cavity in the cloud, and the reddish H II region, powered by the stars, expands into it. These winds travel faster than the speed of sound in the gas—that is, they are **supersonic.** Just as an airplane creates a shock wave (a sonic boom) if it flies faster than sound waves in our atmosphere, a shock wave forms where the expanding H II region pushes at supersonic speed into the rest of the giant molecular cloud. This shock wave compresses the gas through which it passes, stimulating more star birth (Figure 18-24).

Newborn O and B stars further expand the H II region into the giant molecular cloud. Meanwhile, the older O and B stars, which were left behind, begin to disperse (Figure 18-25). In this way, an OB association "eats into" a giant molecular cloud, leaving stars in its wake.

CONCEPTCHECK 18-8

How can O and B stars initiate star formation in a giant molecular cloud?

Answer appears at the end of the chapter.

18-8 Supernovae compress the interstellar medium and can trigger star birth

Spiral arms are not the only mechanism for triggering the birth of stars. Presumably, anything that compresses interstellar clouds will do the job. The most dramatic is a *supernova,* caused by the violent death of a massive star after it has left the main sequence. As we will see in Chapter 20, the core of the doomed star collapses suddenly, releasing vast quantities of particles and energy that

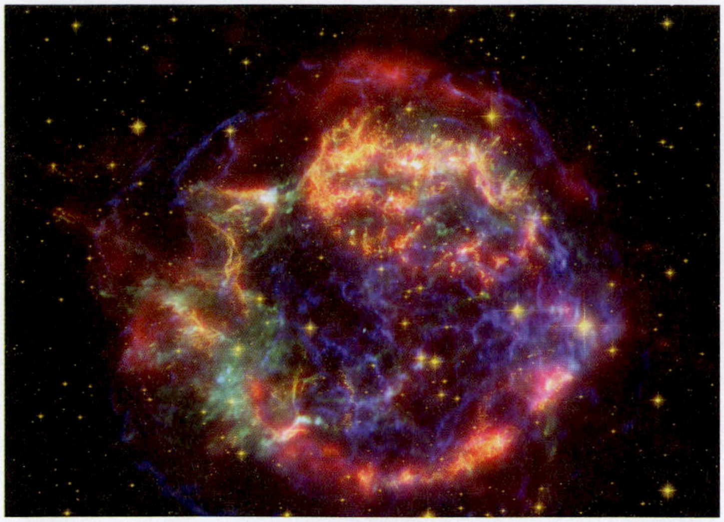

ANIMATION 18-4 **FIGURE 18-26** R I V U X G

A Supernova Remnant This composite image shows Cassiopeia A, the remnant of a supernova that occurred about 3000 pc (10,000 ly) from Earth. In the roughly 300 years since the supernova explosion, a shock wave has expanded about 3 pc (10 ly) outward in all directions from the explosion site. The shock wave has warmed interstellar dust to a temperature of about 300 K (Spitzer Space Telescope infrared image in red), and has heated interstellar gases to temperatures that range from 10^4 K (Hubble Space Telescope visible-light image in yellow) to 10^7 K (Chandra X-ray Observatory X-ray image in green and blue). (NASA; JPL-Caltech; and O. Krause, Steward Observatory)

ANIMATION 18-5 **FIGURE 18-27** R I V U X G

The Canis Major R1 Association These luminous arcs of gas are studded with numerous young stars. Both the luminous arcs and the young stars within can be traced to the same source—a supernova explosion. The shock wave from the supernova explosion is exciting the gas and making it glow; the same shock wave also compresses the interstellar medium through which it passes, triggering star formation. (Davide De Martin)

blow the star apart. The star's outer layers are blasted into space at speeds of several thousand kilometers per second.

Supernova Remnants and Star Formation

Astronomers have found many nebulae across the sky that are the shredded funeral shrouds of these dead stars. Such nebulae, like the one shown in Figure 18-26, are known as **supernova remnants.** Many supernova remnants have a distinctly circular or arched appearance, as would be expected for an expanding shell of gas. This wall of gas is typically moving away from the dead star faster than sound waves can travel through the interstellar medium. As we saw in Section 18-7, such supersonic motion produces a shock wave that abruptly compresses the medium through which it passes. When a gas is compressed rapidly, its temperature rises, and this temperature rise causes the gas to glow as shown in Figure 18-26.

When the expanding shell of a supernova remnant slams into an interstellar cloud, it compresses the cloud, stimulating star birth. This kind of star birth is taking place in the stellar association seen in Figure 18-27. This stellar nursery is located along reddish luminous arcs of gas extending about 30 pc (100 ly) in length that is presumably the remnant of a supernova explosion. (In fact, these arcs are part of an almost complete ring of glowing gas). Since a supernova remnant does not last very long, this interpretation requires that the stars triggered by the supernova are young. But are the stars young? Spectroscopic observations of stars along this arc reveal substantial T Tauri activity. This activity results from young protostars undergoing mass loss in their final stages of contraction before they become main-sequence stars.

A supernova a few light-years away may have triggered the formation of our own Sun. Supernovae produce a variety of radioactive atomic nuclei, including some that are not produced in any other way. These nuclei are dispersed into space by the explosion, and act as telltale signs of their supernova origin. Traces of these telltale nuclei have been discovered in meteorites that have fallen to Earth. Since meteorites formed very early in the history of our solar system (see Section 8-3), this suggests that a supernova occurred nearby when our solar system was very young. Detailed modeling suggests this interpretation is plausible, with a supernova about 15 ly away triggering the collapse of a cloud that became our solar system.

> Our Sun may have formed in association with a number of other stars

Many other processes can also trigger star formation. For example, a collision between two interstellar clouds can create new stars. Compression occurs at the interface between the two colliding clouds and vigorous star formation follows. Similarly, stellar winds from a group of O and B stars may exert strong enough pressure on interstellar clouds to cause compression, followed by star formation. (This process is similar to the one depicted in Figure 18-25.)

Star Birth in Perspective

Our understanding of star birth has improved dramatically in recent years, primarily through infrared- and millimeter-wavelength observations. The *Cosmic Connections* figure summarizes our present state of knowledge about the formation of stars.

How Stars Are Born

f a clump of interstellar matter is cold and dense enough, it will begin to collapse thanks to the mutual gravitational attraction of its parts. If the clump is massive enough, it will evolve into a main-sequence star through the sequence of events shown here.

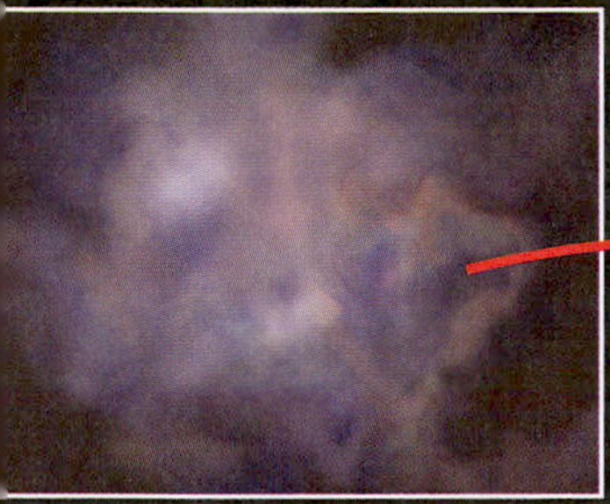

n this cold, dark nebula, gas atoms and dust particles move so slowly that gravity can draw them together.

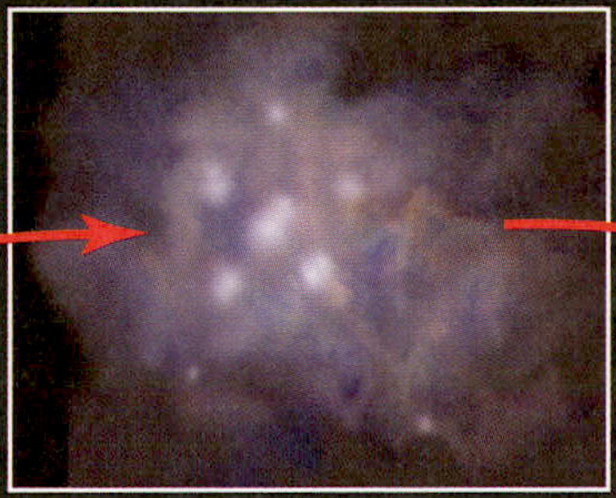

Gas and dust begin to condense into clumps, forming the cores of protostars.

As the cores condense, their density and temperature both increase.

As the protostars continue to heat up and accrete matter from the nebula, they begin to glow due to their increasing temperature.

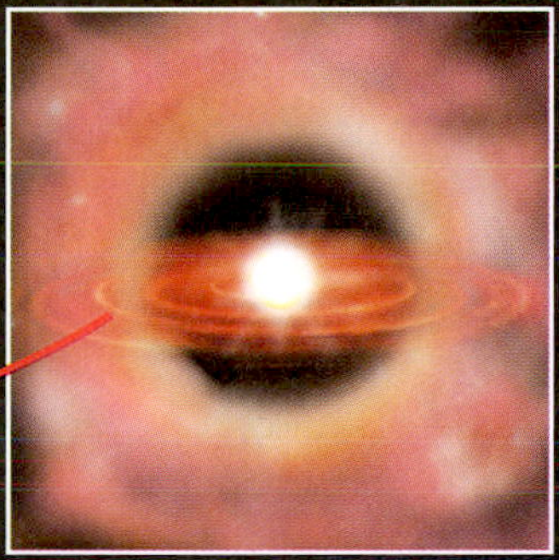

Once the temperature at the center of a protostar becomes sufficiently high, thermonuclear fusion of hydrogen into helium begins. The mass that is continuing to fall onto the star forms an accretion disk.

In the T Tauri stage, the young star ejects mass into space in a bipolar outflow. A stellar wind blows away the remaining parts of the nebula that surround the star, exposing the star to space.

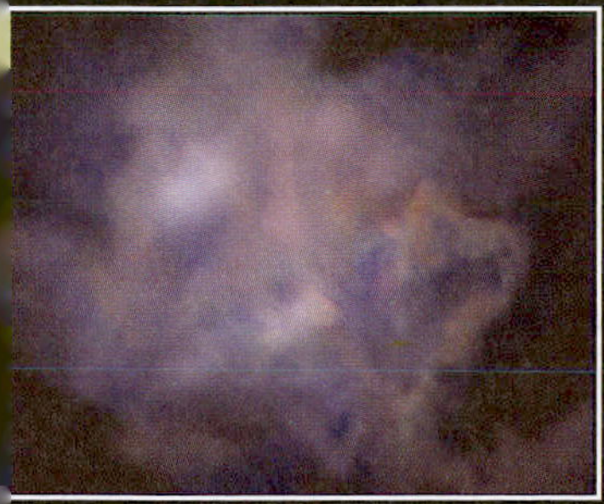

The ejected mass can induce a shock wave in the surrounding interstellar material, triggering the formation of additional stars.

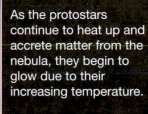

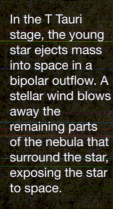

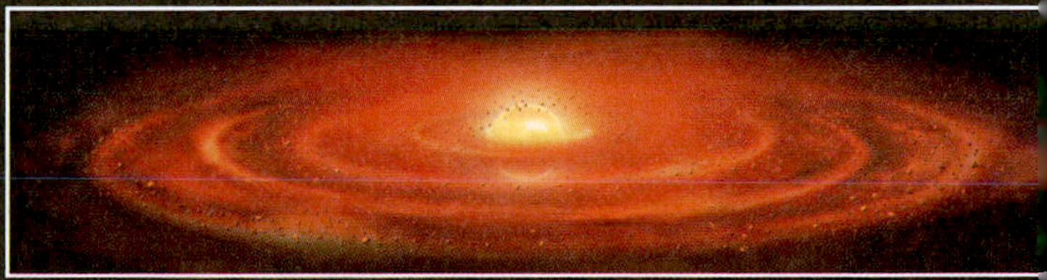

Processes that cause the star to lose or gain mass come to an end, and the star stabilizes as a main-sequence star in hydrostatic equilibrium. The remnants of the accretion disk may remain as a protoplanetary disk, from which a system of planets may form around the star.

Nevertheless, many puzzles and mysteries remain. One problem is that different modes of star birth tend to produce different percentages of different kinds of stars. For example, the passage of a spiral arm through a giant molecular cloud tends to produce an abundance of massive O and B stars. In contrast, the shock wave from a supernova seems to produce fewer O and B stars but many more of the less massive A, F, G, and K stars. We do not yet know why this is so.

Despite these unanswered questions, it is now clear that star birth involves mechanisms on a colossal scale, from the deaths of massive stars to the rotation of an entire galaxy. In many respects, we have just begun to appreciate these cosmic processes. The study of cold, dark stellar nurseries will be an active and exciting area of astronomical research for many years to come.

In the chapters that follow, we will learn that the interstellar medium is both the birthplace of new stars and a dumping ground for dying stars. At the end of its life, a star can shed most of its mass in an outburst that enriches interstellar space with new chemical elements. The interstellar medium is therefore both nursery and graveyard. Because of this intimate relationship with stars, the interstellar medium evolves as successive generations of stars live out their lives. Understanding the details of this cosmic symbiosis is one of the challenges of modern astronomy.

CONCEPTCHECK 18-9

Several processes have been considered in this chapter that initiate star formation in clouds. What do they have in common?

Answer appears at the end of the chapter.

KEY WORDS

The term preceded by an asterisk (*) is discussed in Box 18-1.

accretion, p. 516
Barnard object, p. 511
bipolar outflow, p. 515
Bok globule, p. 511
circumstellar accretion disk, p. 516
cluster (of stars), p. 517
cocoon nebula, p. 513
dark nebula, p. 509
dust grains, p. 509
emission nebula, p. 506
evolutionary track, p. 512
*fluorescence, p. 508
giant molecular cloud, p. 521
H II region, p. 507
Herbig-Haro object, p. 515
interstellar extinction, p. 510
interstellar medium, p. 506
interstellar reddening, p. 510
nebula (*plural* nebulae), p. 506
nebulosity, p. 506
OB association, p. 521
open cluster, p. 521
protoplanetary disk (proplyd), p. 516
protostar, p. 512
recombination, p. 508
reflection nebula, p. 509
stationary absorption lines, p. 509
stellar association, p. 521
stellar evolution, p. 506
supernova remnant, p. 524
supersonic, p. 523
T Tauri star, p. 515

KEY IDEAS

Stellar Evolution: Because stars shine by thermonuclear reactions, they have a finite life span. The theory of stellar evolution describes how stars form and change during that life span.

The Interstellar Medium: Interstellar gas and dust, which make up the interstellar medium, are concentrated in the disk of the Galaxy. Clouds within the interstellar medium are called nebulae.

• Dark nebulae are so dense that they are opaque. They appear as dark blots against a background of distant stars.

• Emission nebulae, or H II regions, are reddish glowing, ionized clouds of gas. Emission nebulae are powered by ultraviolet light that they absorb from nearby hot stars.

• Reflection nebulae are produced when starlight is reflected from dust grains in the interstellar medium, producing a characteristic bluish glow.

Protostars: Star formation begins in dense, cold nebulae, where gravitational attraction causes a clump of material to condense into a protostar.

• As a protostar grows by the gravitational accretion of gases, Kelvin-Helmholtz contraction causes it to heat and begin glowing. Its relatively low temperature and high luminosity place it in the upper right region on an H-R diagram.

• Further evolution of a protostar causes it to move toward the main sequence on the H-R diagram. When its core temperatures become high enough to ignite steady hydrogen burning, it becomes a main-sequence star.

• The more massive the protostar, the more rapidly it evolves.

Mass Loss by Protostars: In the final stages of pre–main-sequence contraction, when thermonuclear reactions are about to begin in its core, a protostar may eject large amounts of gas into space.

• Low-mass stars that vigorously eject gas are called T Tauri stars.

• A circumstellar accretion disk provides material that a young star ejects as jets. Clumps of glowing gas called Herbig-Haro objects are sometimes found along these jets and at their ends.

Star Clusters: Newborn stars may form an open or galactic cluster. Stars are held together in such a cluster by gravity. Occasionally a star moving more rapidly than average will escape, or "evaporate," from such a cluster.

• A stellar association is a group of newborn stars that are moving apart so rapidly that their gravitational attraction for one another cannot pull them into orbit about one another.

O and B Stars and Their Relation to H II Regions: The most massive protostars to form out of a dark nebula rapidly become main-sequence O and B stars. They emit strong ultraviolet radiation that ionizes hydrogen in the surrounding cloud, thus creating the reddish emission nebulae called H II regions.

• Ultraviolet radiation and stellar winds from the O and B stars at the core of an H II region create shock waves that move outward through the gas cloud, compressing the gas and triggering the formation of more protostars.

Giant Molecular Clouds: The spiral arms of our Galaxy are laced with giant molecular clouds, immense nebulae so cold that their constituent atoms can form into molecules.

• Star-forming regions appear when a giant molecular cloud is compressed. This can be caused by the cloud's passage through one of the spiral arms of our Galaxy, by a supernova explosion, or by other mechanisms.

QUESTIONS

Review Questions

1. If no one has ever seen a star go through the complete formation process, how are we able to understand how stars form?
2. Why is it more difficult to observe the life cycles of stars than the life cycles of planets or animals?
3. If an interstellar medium fills the space between the stars, how is that we are able to see the stars at all?
4. Summarize the evidence that interstellar space contains (**a**) gas and (**b**) dust.
5. What are H II regions? Near what kinds of stars are they found? Why do only these stars give rise to H II regions?
6. What are stationary absorption lines? In what sort of spectra are they seen? How do they give evidence for the existence of the interstellar medium?
7. In Figure 18-2, what makes the Horsehead Nebula dark? What makes IC 434 glow?
8. Why is the daytime sky blue? Why are distant mountains purple? Why is the Sun red when seen near the horizon at sunrise or sunset? In what ways are your answers analogous to the explanations for the bluish color of reflection nebulae and the process of interstellar reddening?
9. To see the constellation Coma Berenices (Berenice's Hair) you must look perpendicular to the plane of the Milky Way. By contrast, the Milky Way passes through the constellation Cassiopeia (named for a mythical queen). Would you expect H II regions to be more abundant in Coma Berenices or in Cassiopeia? Explain your reasoning.
10. The interior of a dark nebula is billions of times less dense than the air that you breathe. How, then, are dark nebulae able to block out starlight?
11. TUTORIAL 18-1 Why are low temperatures necessary in order for protostars to form inside dark nebulae?
12. Compare and contrast Barnard objects and Bok globules. How many Sun-sized stars could you make out of a Barnard object? Out of a Bok globule?
13. Describe the energy source that causes a protostar to shine. How does this source differ from the energy source inside a main-sequence star?
14. What is an evolutionary track? How can evolutionary tracks help us interpret the H-R diagram?
15. What happens inside a protostar to slow and eventually halt its gravitational contraction?
16. Why are the evolutionary tracks of high-mass stars different from those of low-mass stars? For which kind of star is the evolution more rapid? Why?
17. Why are protostars more easily seen with an infrared telescope than with a visible-light telescope?
18. In what ways is the internal structure of a 1-$M_\odot$ main-sequence star different from that of a 5-$M_\odot$ main-sequence star? From that of a 0.5-$M_\odot$ main-sequence star? What features are common to all these stars?
19. What sets the limits on the maximum and minimum masses of a main-sequence star?
20. What are T Tauri stars? How do we know that they eject matter at high speed? How does their rate of mass loss compare to that of the Sun?
21. What are Herbig-Haro objects? Why are they often found in pairs?
22. Why do disks form around contracting protostars? What is the connection between disks and bipolar outflows?
23. TUTORIAL 18-2 Young open clusters like those shown in Figures 18-20a and 18-21 are found only in the plane of the Galaxy. Explain why this should be.
24. Why are observations at millimeter wavelengths so much more useful in exploring interstellar clouds than observations at visible wavelengths?
25. What are giant molecular clouds? What role do these clouds play in the birth of stars?
26. Giant molecular clouds are among the largest objects in our Galaxy. Why, then, were they discovered only relatively recently?
27. Consider the following stages in the evolution of a young star cluster: (i) H II region; (ii) dark nebula; (iii) formation of O and B stars; (iv) giant molecular cloud. Put these stages in the correct chronological order and discuss how they are related.
28. Briefly describe four mechanisms that compress the interstellar medium and trigger star formation.

Advanced Questions

Problem-solving tips and tools

You may find it helpful to review Box 17-4, which describes the relationship among a star's luminosity, radius, and surface temperature. The small-angle formula is described in Box 1-1. Orbital periods are described by Kepler's third law, which we discussed in Box 4-2 and Box 4-4. Remember that the Stefan-Boltzmann law (Box 5-2) relates the temperature of a blackbody to its energy flux. Remember, too, that the volume of a sphere of radius r is $4\pi r^3/3$.

29. If you looked at the spectrum of a reflection nebula, would you see absorption lines, emission lines, or no lines? Explain your answer. As part of your explanation, describe how the spectrum demonstrates that the light was reflected from nearby stars.
30. In the direction of a particular star cluster, interstellar extinction allows only 15% of a star's light to pass through each kiloparsec (1000 pc) of the interstellar medium. If the star

cluster is 3.0 kiloparsecs away, what percentage of its photons survive the trip to Earth?

31. The visible-light photograph below shows the Trifid Nebula in the constellation Sagittarius. Label the following features on this photograph: (**a**) reflection nebulae (and the star or stars whose light is being reflected); (**b**) dark nebulae; (**c**) H II regions; (**d**) regions where star formation may be occurring. Explain how you identified each feature.

R I V U X G
(Australian Astronomical Observatory/David Malin Images)

32. Find the density (in atoms per cubic centimeter) of a Bok globule having a radius of 1 ly and a mass of 100 $M_{\odot}$. How does your result compare with the density of a typical H II region, between 80 and 600 atoms per cm^3? (Assume that the globule is made purely of hydrogen atoms.)

33. The *Becklin-Neugebauer object* is a newly formed star within the Orion Nebula. It is substantially more luminous than the other newly formed stars in that nebula. Assuming that all these stars began the process of formation of the same time, what can you conclude about the mass of the Becklin-Neugebauer object compared with those of the other newly formed stars? Does your conclusion depend on whether or not the stars have reached the main sequence? Explain your reasoning.

34. The two false-color images opposite show a portion of the Trifid Nebula (see Question 31). The reddish-orange view is a false-color infrared image, while the bluish picture (shown to the same scale) was made with visible light. Explain why the dark streaks in the visible-light image appear bright in the infrared image.

35. At one stage during its birth, the protosun had a luminosity of 1000 $L_{\odot}$ and a surface temperature of about 1000 K. At

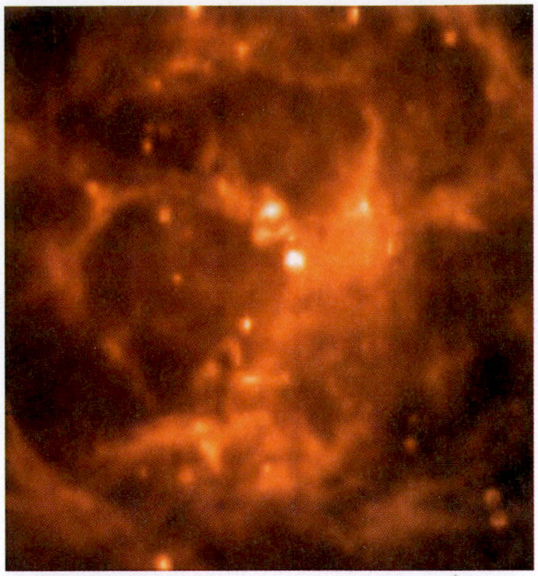

R I V U X G

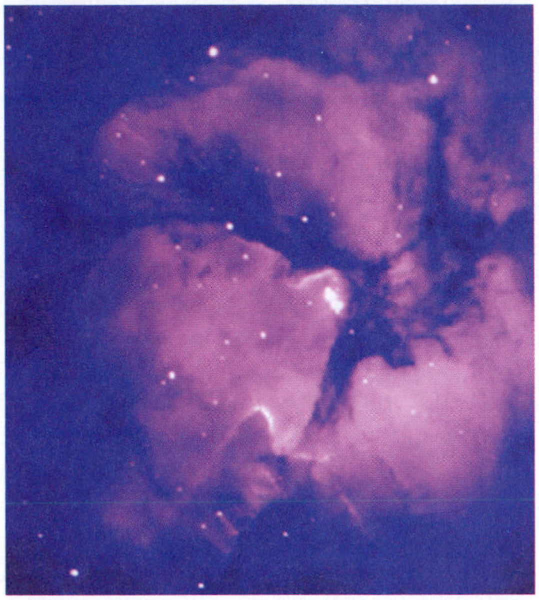

R I V U X G
(ESA/ISO, ISOCAM, and J. Cernicharo et. al.; IAC, Observatorio del Teide, Tenerife)

this time, what was its radius? Express your answer in three ways: as a multiple of the Sun's present-day radius, in kilometers, and in astronomical units.

36. A newly formed protostar and a red giant are both located in the same region on the H-R diagram. Explain how you could distinguish between these two.

37. (**a**) Determine the radius of the circumstellar accretion disk in Figure 18-15. (You will need to measure this image with a ruler. Note the scale bar in this figure.) Give your answer in astronomical units and in kilometers. (**b**) Assume that the young star at the center of this disk has a mass of 1 $M_{\odot}$. What is the orbital period (in years) of a particle at the outer edge of the disk? (**c**) Using your ruler again, determine the length of

the jet that extends to the right of the circumstellar disk in Figure 18-15. At a speed of 200 km/s, how long does it take gas to traverse the entire visible length of the jet?

38. The star cluster NGC 2264 (Figure 18-20) contains numerous T Tauri stars, while the Pleiades (Figure 18-21) contains none. Explain why there is a difference.

39. The concentration or abundance of ethyl alcohol in a typical molecular cloud is about 1 molecule per 10^8 cubic meters. What volume of such a cloud would contain enough alcohol to make a martini (about 10 grams of alcohol)? A molecule of ethyl alcohol has 46 times the mass of a hydrogen atom (that is, ethyl alcohol has a molecular weight of 46).

40. From the information given in the caption to Figure 18-26, calculate the angular diameter in arcminutes of Cassiopeia A as seen from Earth.

41. From the information given in the caption for Figure 18-26, calculate the average speed at which the shock wave has spread away from the site of the supernova explosion. Give your answer in kilometers per second and as a fraction of the speed of light. (*Hint:* There are 3.16×10^7 seconds in a year and the speed of light is 3.00×10^5 km/s.)

Discussion Questions

42. Some science-fiction movies show stars suddenly becoming dramatically brighter when they are "born" (that is, when thermonuclear fusion reactions begin in their cores). Discuss whether this is a reasonable depiction.

43. Suppose that the electrons in hydrogen atoms were not as strongly attracted to the nuclei of those atoms, so that these atoms were easier to ionize. What consequences might this have for the internal structure of main-sequence stars? Explain your reasoning.

44. What do you think would happen if our solar system were to pass through a giant molecular cloud? Do you think Earth has ever passed through such clouds?

45. Many of the molecules found in giant molecular clouds are *organic* molecules (that is, they contain carbon). Speculate about the possibility of life-forms and biological processes occurring in giant molecular clouds. In what ways might the conditions existing in giant molecular clouds favor or hinder biological evolution?

46. Speculate on why a shock wave from a supernova seems to produce relatively few high-mass O and B stars, compared to the lower-mass A, F, G, and K stars.

Web/eBook Questions

47. In recent years astronomers have been able to learn about the character of the interstellar medium in the vicinity of the Sun. Search the World Wide Web for information about aspects of the nearby interstellar medium, including features called the Local Interstellar Cloud and the Local Bubble. How do astronomers study the nearby interstellar medium? What makes these studies difficult? Is the interstellar medium relatively uniform in our neighborhood, or is it clumpy? If the latter, is our solar system in a relatively thin or thick part of the interstellar medium? How is our solar system moving through the interstellar medium?

48. Search the World Wide Web for recent discoveries about how brown dwarfs form. Do they tend to form in the same locations as "real" stars? Do they form in relatively small or relatively large numbers compared to "real" stars? What techniques are used to make these discoveries?

49. ANIMATION 18-1 **Measuring a Stellar Jet.** Access the animation "A Stellar Jet in the Trifid Nebula" in Chapter 18 of the *Universe* Web site or eBook. (**a**) The Trifid Nebula as a whole has an angular diameter of 28 arcmin. By stepping through the animation, estimate the angular size of the stellar jet shown at the end of the animation. (**b**) The Trifid Nebula is about 2800 pc (9000 ly) from Earth. Estimate the length of the jet in light-years. (**c**) If gas in the jet travels at 200 km/s, how long does it take to traverse the length of the jet? Give your answer in years. (*Hint:* There are 3.16×10^7 seconds in a year.)

ACTIVITIES

Observing Projects

Observing tips and tools

After looking at the beautiful color photographs of nebulae in this chapter, you may find the view through a telescope a bit disappointing at first, but fear not. You can see a great deal with even a small telescope. To get the best view of a dim nebula using a telescope, use the same "averted vision" trick we described in "Observing tips and tools" in Chapter 17: If you direct your vision a little to one side of the object that you are looking at, the light from that object will go onto a more sensitive part of the retina.

50. Use a telescope to observe at least two of the H II regions listed in the following table. In each case, can you guess which stars are probably responsible for the ionizing radiation that causes the nebula to glow? Can you see any obscuration or silhouetted features that suggest the presence of interstellar dust? Draw a picture of what you see through the telescope and compare it with a photograph of the object. Take note of which portions of the nebula were not visible through your telescope.

Nebula	Right ascension	Declination
M42 (Orion)	$5^h\ 35.4^m$	−5° 27′
M43	$5^h\ 35.6^m$	−5° 16′
M20 (Trifid)	$18^h\ 02.6^m$	−23° 02′
M8 (Lagoon)	$18^h\ 03.8^m$	−24° 23′
M17 (Omega)	$18^h\ 20.8^m$	−16° 11′

Note: The right ascensions and declinations are given for epoch 2000.

51. On an exceptionally clear, moonless night, use a telescope to observe at least one of the dark nebulae listed in the following

table. These nebulae are very difficult to find, because they are recognizable only by the absence of stars in an otherwise starry part of the sky. Are you confident that you actually saw the dark nebula? Does the pattern of background stars suggest a particular shape to the nebula?

Nebula	Right ascension	Declination
Barnard 72 (the Snake)	$17^h\ 23.5^m$	−23° 38′
Barnard 86	$18^h\ 02.7^m$	−27° 50′
Barnard 133	$19^h\ 06.1^m$	−6° 50′
Barnard 142 and 143	$19^h\ 40.7^m$	−10° 57′

Note: The right ascensions and declinations are given for epoch 2000.

52. A few fine objects cover such large regions of the sky that they are best seen with binoculars. If you have access to a high-quality pair of binoculars, observe the North American Nebula in Cygnus and the Pipe Nebula in Ophiuchus. Both nebulae are quite faint, so you should attempt to observe them only on an exceptionally dark, clear, moonless night. The North America Nebula is a cloud of glowing hydrogen gas located about 3° east of Deneb, the brightest star in Cygnus. While searching for the North America Nebula, you may glimpse another diffuse H II region, the Pelican Nebula, located about 2° southeast of Deneb. The Pipe Nebula is a 7°-long, meandering, dark nebula to the south and to the east of the star θ (theta) Ophiuchi, which is in a section of Ophiuchus that extends southward between the constellations of Scorpius and Sagittarius. Located about 12° east of the bright red star Antares, you can locate θ Ophiuchi using the *Starry Night*™ software.

53. Use the *Starry Night*™ program to examine the Milky Way Galaxy. Open **Favourites > Explorations > Star forming regions.** The view looks toward the center of the Milky Way from the center of a transparent Earth. The view also displays the galactic equator and the constellations. Observe the mottled appearance of the Milky Way caused by regions of opaque dust and gas. Click on the **Options** tab, expand the **Deep Space** layer and click the checkbox to turn **Nebulae On.** Click on the + sign to the left of **Nebulae** to expand this layer. Click **Off** all the checkboxes for the Nebula options except that for **Dark Nebula.** Numerous dark nebulae visible in Earth's night sky are shown with green outlines or markers. (**a**) Use the hand tool and zoom controls to look around the view and describe the distribution of these dark nebulae in the sky. (**b**) Now click **Off** the **Dark Nebula** option in the Options pane and click **On** the **Emission Nebula.** Again use the hand tool and zoom controls to look around the sky and observe the distribution of these nebulae and describe your observations. (**c**) Use the **Find** tool to locate and identify several of the more prominent emission nebulae: M8, M20, NGC 3372, M42, and NGC 7000 (North American Nebula). Magnify each of these nebulae in turn and describe your observations of the details of some of them, such as color, shape, and structure. (**d**) Zoom out again to return to the wide-field view and use the hand tool to find the tight knot of emission nebulae between the constellations of Ursa Major and Draco. Zoom in on this region to a field of view about 1° wide. You will notice that the outlined emission nebulae belong to the galaxy M101 (the Pinwheel Galaxy), which is about 27 million ly from Earth. Describe the distribution of these emission nebulae within this galaxy and compare it to your observations of the distribution of emission nebulae in the Milky Way. (**e**) Given the distance to M101, what does the fact that these emission nebulae in this galaxy are visible from Earth suggest about their properties?

54. Use *Starry Night*™ to examine the H-R diagram of the Pleiades star cluster. This group of stars was formed relatively recently in astronomical time. Select **Favourites > Explorations > Pleiades** to display this cluster of young stars in the view. Click on the **Status** tab to display the H-R diagram of all stars in this field of view around the Pleiades. Note that this appears to be similar to that of stars in our local neighborhood. However, if you restrict the distance to display only the stars within this localized cluster, a different pattern emerges. Click on the **Distance cut-off** checkbox in the H-R Options layer of the **Status** pane to restrict the distance to a range between 320 and 420 ly. (**a**) Where on the H-R diagram do you now find the majority of the stars of this cluster? (**b**) In view of the existence within this cluster of very hot stars with high output of energy, what does this tell you about the age of this cluster compared to the general population of stars?

Collaborative Exercises

55. Imagine that your group walks into a store that specializes in selling antique clothing. Prepare a list of observable characteristics that you would look for to distinguish which items were from the early, middle, and late twentieth century. Also, write a paragraph that specifically describes how this task is similar to how astronomers understand the evolution of stars.

56. Consider advertisement signs visible at night in your community and provide specific examples of ones that are examples of the three different types of nebulae that astronomers observe and study. If an example doesn't exist in your community, creatively design an advertisement sign that could serve as an example.

57. The pre–main-sequence evolutionary tracks shown in Figure 18-10 describe the tracks of seven protostars of different masses. Imagine a new sort of H-R diagram that plots a human male's increasing age versus decreasing hair density on the head instead of increasing luminosity versus decreasing temperature. Create and carefully label a sketch of this imaginary H-R diagram showing both the majority of the U.S. male population and a few oddities. Finally, draw a line that clearly labels your sketch to show how a typical male undergoing male-pattern baldness might slowly change position on the H-R diagram over the course of a human life span.

ANSWERS

ConceptChecks

ConceptCheck 18-1: No. Toaster filaments glow red because they are hot blackbodies. The red light from an H II region comes from a single red emission line. The gas in an H II region is too thin to emit as a blackbody.

ConceptCheck 18-2: No. The blue gas is simply reflecting starlight, but with more reflection at blue wavelengths.

ConceptCheck 18-3: Yes. The evolutionary track for a 1 $M_{\odot}$ solar star like our Sun in Figure 18-10 indicates that such a protostar is almost 100 times more luminous when it forms then when it reaches the main sequence.

ConceptCheck 18-4: The red dashed lines indicate how long it takes for protostars to cross those lines on their way to the main sequence. A 2 $M_{\odot}$ star reaches the main sequence after about 10^7 (10 ten million) years.

ConceptCheck 18-5: No. The energy source of a protostar is gravitational contraction compressing and heating the protostar's interior. The energy source for all main-sequence stars are nuclear reactions converting hydrogen to helium in the star's core.

ConceptCheck 18-6: For a main-sequence star that ends up with less than 3 $M_{\odot}$ it was more massive as a protostar; these protostars lose more mass than they gain. The masses shown in Figure 18-10 are for the final main-sequence stars.

ConceptCheck 18-7: No. The pillars are eroded away by hot luminous stars outside of the pillars that emit ultraviolet radiation.

ConceptCheck 18-8: Ultraviolet light from hot O and B stars creates an H II region that expands into the surrounding molecular cloud. This compresses the cloud, causing star formation.

ConceptCheck 18-9: Protostars begin to form when gravity overwhelms a gas cloud's internal pressure. This is initiated when an external process compresses the cloud.

CHAPTER

19

The red stars in this image of open cluster NGC 290 are red giants, a late stage in stellar evolution. (ESA/NASA/Edward W. Olszewski, U. of Arizona) R I V U X G

Stellar Evolution: On and After the Main Sequence

LEARNING GOALS

By reading the sections of this chapter, you will learn

19-1 How a main-sequence star changes as it converts hydrogen to helium

19-2 What happens to a star when it runs out of hydrogen fuel

19-3 How aging stars can initiate a second stage of thermonuclear fusion

19-4 How H-R diagrams for star clusters reveal the later stages in the evolution of stars

19-5 The two kinds of stellar populations and their significance

19-6 Why some aging stars pulsate and vary in luminosity

19-7 How stars in a binary system can evolve very differently from single, isolated stars

Imagine a world like Earth, but orbiting a star more than 100 times larger and 2000 times more luminous than our Sun. Bathed in the star's intense light, the surface of this world is utterly dry, airless, and hot enough to melt iron. If you could somehow survive on the daytime side of this world, you would see the star filling almost the entire sky.

This bizarre planet is not a creation of science fiction—it is our own Earth some 7.6 billion years from now. The bloated star is our own Sun, which in that remote era will have become a red giant.

In this chapter, we will learn how a main-sequence star evolves into a red giant when all the hydrogen in its core is consumed. The star's core contracts and heats up, but its outer layers expand and cool. In the hot, compressed core, helium fusion becomes a new energy source. The more massive a star, the more rapidly it consumes its core's hydrogen and the sooner it evolves into a giant.

The interiors of stars are hidden from our direct view, so much of the story in this chapter is based on theory. We back up those calculations with observations of star clusters, which contain stars of different masses but roughly the same age. (An example is the cluster shown here, many of whose stars have evolved into luminous red giants.) Other observations show that some red giant stars actually pulsate, and that stars can evolve along very different paths if they are part of a binary star system.

19-1 During a star's main-sequence lifetime, it expands and becomes more luminous

> Over the past 4.56 billion years, thermonuclear reactions have caused an accumulation of helium in our Sun's core

In their cores, main-sequence stars are all fundamentally alike. As we saw in Section 18-4, it is in their cores that all such stars convert hydrogen into helium by thermonuclear reactions. This process is called **core hydrogen fusion.** The total time that a star will spend fusing hydrogen into helium in its core, and thus the total time that it will spend as a main-sequence star, is called its **main-sequence lifetime.** For a star of a given mass, its main-sequence lifetime can be estimated by modeling the nuclear reactions within its core. For our Sun, the main-sequence lifetime is about 12 billion (1.2×10^{10}) years. Hydrogen fusion has been going on in the Sun's core for the past 4.56 billion (4.56×10^9) years, so our Sun is less than halfway through its main-sequence lifetime.

What happens to a star like the Sun after the core hydrogen has been used up, so that it is no longer a main-sequence star? As we will see, it expands dramatically to become a red giant. To understand why this happens, it is useful to first look at how a star evolves *during* its main-sequence lifetime. The nature of that evolution depends on whether its mass is less than or greater than about 0.4 $M_\odot$.

Main-Sequence Stars of 0.4 $M_\odot$ or Greater: Consuming Core Hydrogen

A protostar becomes a main-sequence star when steady hydrogen fusion begins in its core and it achieves *hydrostatic equilibrium*—a balance between the inward force of gravity and the outward pressure produced by hydrogen fusion (see Section 16-2 and Section 18-4). Such a freshly formed main-sequence star is called a **zero-age main-sequence star.**

We make the distinction between "main-sequence" and "zero-age main-sequence" because a star undergoes noticeable changes in luminosity, surface temperature, and radius during its main-sequence lifetime. These changes are a result of core hydrogen fusion, which alters the chemical composition of the core. As an example, when our Sun first formed, its composition was the same at all points throughout its volume: by mass, about 74% hydrogen, 25% helium, and 1% heavy elements. But as Figure 19-1 shows, the Sun's core now contains a greater mass of helium than of hydrogen. (There is still enough hydrogen in the Sun's core for another 7 billion years or so of core hydrogen fusion.)

CAUTION! Although the outer layers of the Sun are also predominantly hydrogen, there are two reasons why this hydrogen cannot undergo fusion. The first reason is that while the temperature and pressure in the core are high enough for thermonuclear reactions to take place, the temperatures and pressure in the outer layers are not. The second reason is that there is no flow of material between the Sun's core and outer layers, so the hydrogen in the outer layers cannot move into the hot, high-pressure core to undergo fusion. The same is true for main-sequence stars with masses of about 0.4 $M_\odot$ or greater. (We will see below that the outer layers *can* undergo fusion in main-sequence stars with a mass less than about 0.4 $M_\odot$.)

Thanks to core hydrogen fusion, the total number of atomic nuclei in a star's core decreases with time: In each reaction, four hydrogen nuclei are converted to a single helium nucleus (see the *Cosmic Connections* figure in Section 16-1, as well as Box 16-1). With fewer particles bouncing around to provide the core's

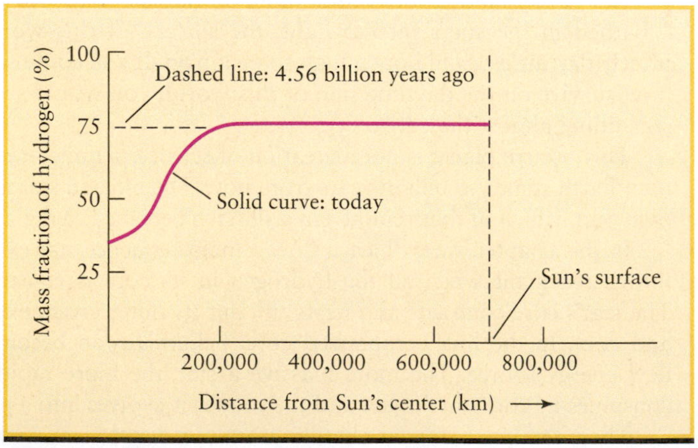

(a) Hydrogen in the Sun's interior

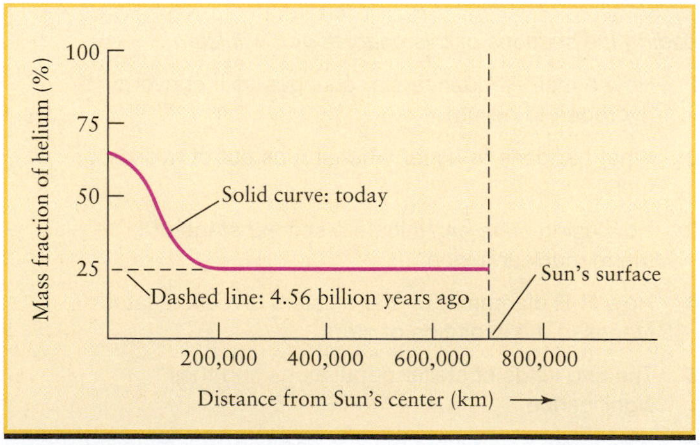

(b) Helium in the Sun's interior

FIGURE 19-1

Changes in the Sun's Chemical Composition These graphs show the percentage by mass of **(a)** hydrogen and **(b)** helium at different points within the Sun's interior. The dashed horizontal lines show that these percentages were the same throughout the Sun's volume when it first formed. As the solid curves show, over the past 4.56×10^9 years, thermonuclear reactions at the core have depleted hydrogen in the core and increased the amount of helium in the core.

internal pressure, the core contracts slightly under the weight of the star's outer layers. Compression makes the core denser and increases its temperature. (Box 19-1 gives some everyday examples of how the temperature of a gas changes when it compresses or expands.) As a result of these changes in density and temperature, the pressure in the compressed core is actually higher than before.

As the star's core shrinks, its outer layers expand and shine more brightly. Here's why: As the core's density and temperature increase, hydrogen nuclei in the core collide with one another more frequently, and the rate of core hydrogen fusion increases. Hence, the star's release of energy increases, which increases the star's luminosity. The radius of the star as a whole also increases slightly, because increased core pressure pushes outward on the star's outer layers. The star's surface temperature changes as well, because it is related to the luminosity and radius (see Section 17-6 and Box 17-4). As an example, theoretical calculations indicate that over the past 4.56×10^9 years, our Sun has become 40% more luminous, grown in radius by 6%, and increased in surface temperature by 300 K (Figure 19-2).

Main-Sequence Stars of Less than 0.4 $M_\odot$: Consuming All Their Hydrogen

The story is somewhat different for the least massive main-sequence stars, with masses between 0.08 $M_\odot$ (the minimum mass for sustained thermonuclear reactions to take place in a star's core) and about 0.4 $M_\odot$. These stars, of spectral class M, are called **red dwarfs** because they are small in size and have a red color due to their low surface temperature. They are also very numerous; about 85% of all stars in the Milky Way Galaxy are red dwarfs.

In a red dwarf, helium does *not* accumulate in the core to the same extent as in the Sun's core. The reason is that in a red dwarf there are convection cells of rising and falling gas that extend throughout the star's volume and penetrate into the core (see Figure 18-12c). These convection cells drag helium outward

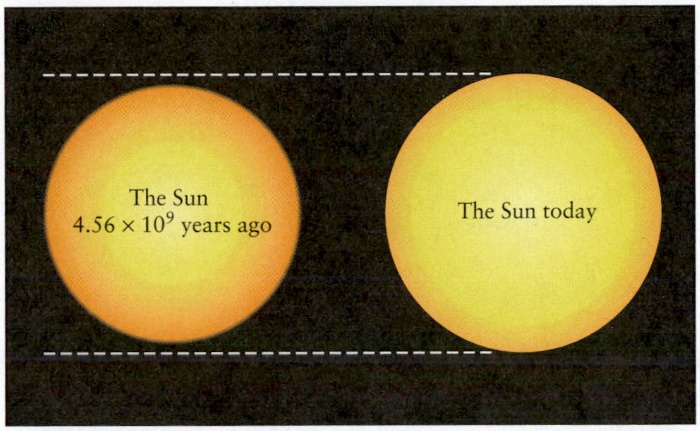

FIGURE 19-2
The Zero-Age Sun and Today's Sun Over the past 4.56×10^9 years, about half of the hydrogen in the Sun's core has been converted into helium, the core has contracted a bit, and the Sun's luminosity has gone up by about 40%. These changes in the core have made the Sun's outer layers expand in radius by 6% and increased the surface temperature from 5500 K to 5800 K.

BOX 19-1 ASTRONOMY DOWN TO EARTH

Compressing and Expanding Gases

As a star evolves, various parts of the star either contract or expand. When this happens, the gases behave in much the same way as gases here on Earth when they are forced to compress or allowed to expand.

When a gas is compressed, its temperature rises. You know this by personal experience if you have ever had to inflate a bicycle tire with a hand pump. As you pump, the compressed air gets warm and makes the pump warm to the touch. The same effect happens on a larger scale in southern California during Santa Ana winds or downwind from the Rocky Mountains when there are Chinook winds. Both of these strong winds blow from the mountains down to the lowlands. Even though the mountain air is cold, the winds that reach low elevations can be very hot. (Chinook winds have been known to raise the temperature by as much as 27°C, or 49°F, in only 2 minutes!) The explanation is compression. Air blown downhill by the winds is compressed by the greater air pressure at lower altitudes, and this compression raises the temperature of the air.

On the other hand, expanding gases tend to drop in temperature. When you open a bottle of carbonated beverage, the gases trapped in the bottle expand and cool down. The cooling can be so great that a little cloud forms within the neck of the bottle. Clouds form in the atmosphere in the same way. Rising air cools as it goes to higher altitudes, where the pressure is lower, and the cooling makes water in the air condense into droplets.

Here's an experiment you can do to feel the cooling of expanding gases. Your breath is actually quite warm, as you can feel if you open your mouth wide, hold the back of your hand next to your mouth, and exhale. But if you bring your lips together to form a small "o" and again blow on your hand, your breath feels cool. In this second case, your exhaled breath has to expand as it passes between your lips to the outside, which makes its temperature drop.

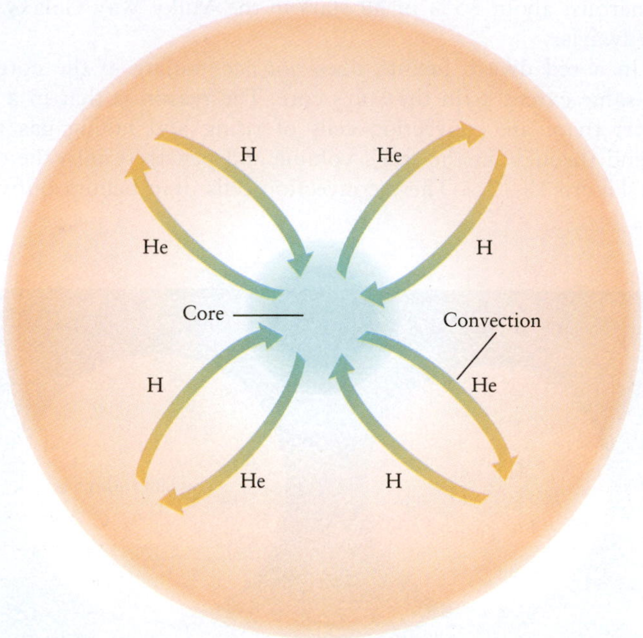

FIGURE 19-3

A Fully Convective Red Dwarf In a red dwarf—a main-sequence star with less than about 0.4 solar masses—helium (He) created in the core by thermonuclear reactions is carried to the star's outer layers by convection. Convection also brings fresh hydrogen (H) from the outer layers into the core. This process continues until the entire star is helium.

from the core and replace it with hydrogen from the outer layers (Figure 19-3). The fresh hydrogen can undergo thermonuclear fusion that releases energy and makes additional helium. This helium is then dragged out of the core by convection and replaced by even more hydrogen from the red dwarf's outer layers.

As a consequence, over a red dwarf's main-sequence lifetime essentially all of the star's hydrogen can be consumed and converted to helium. The core temperature and pressure in a red dwarf is less than in the Sun, so thermonuclear reactions happen more slowly than in our Sun. Does it take a long time for a red dwarf to burn all its hydrogen? Indeed it does! Calculations indicate that it takes hundreds of billions of years for a red dwarf to convert all of its hydrogen completely to helium. The present age of the universe is only 13.7 billion years, so there has not yet been time for *any* red dwarfs—not one—to become pure helium.

A Star's Mass Determines Its Main-Sequence Lifetime

The main-sequence lifetime of a star depends critically on its mass. As Table 19-1 shows, ***massive stars have short main-sequence lifetimes.*** The greater core pressure created by a larger total mass causes nuclear reactions to occur more rapidly, and burn through the available hydrogen fuel more quickly. Hence, even though a massive main-sequence star contains much more hydrogen fuel in its core than is in the entire volume of a red dwarf, a massive star exhausts its hydrogen much sooner. More rapid burning also produces a greater luminosity so that ***massive stars are more luminous*** (see Section 17-9, and particularly the *Cosmic Connections* figure for Chapter 17). Thus, a main-sequence star's mass determines not only its luminosity, but also how long it can remain a main-sequence star (see Box 19-2 for details).

We saw in Section 18-4 how more-massive stars evolve more quickly through the protostar phase to become main-sequence stars (see Figure 18-10). In general, the more massive the star, the more rapidly it goes through *all* the phases of its life. Still, most of the stars we are able to detect are in their main-sequence phase, because this phase lasts so much longer than other luminous phases.

In the remainder of this chapter we will look at the luminous phases that can take place after the end of a star's main-sequence lifetime. (In Chapters 20 and 21 we will explore the final phases of a star's existence, when it ceases to have an appreciable luminosity.)

CONCEPTCHECK 19-1

Why are there no red dwarfs that have already left the main sequence?

Answer appears at the end of the chapter.

TABLE 19-1 Approximate Main-Sequence Lifetimes

Mass ($M_\odot$)	Surface temperature (K)	Spectral class	Luminosity ($L_\odot$)	Main-sequence lifetime (10^6 years)
25	35,000	O	80,000	4
15	30,000	B	10,000	15
3	11,000	A	60	800
1.5	7000	F	5	4500
1.0	6000	G	1	12,000
0.75	5000	K	0.5	25,000
0.50	4000	M	0.03	700,000

The main-sequence lifetimes were estimated using the relationship $t \propto 1/M^{2.5}$ (see Box 19-2).

BOX 19-2 TOOLS OF THE ASTRONOMER'S TRADE

Main-Sequence Lifetimes

Hydrogen fusion converts a portion of a star's mass into energy. We can use Einstein's famous equation relating mass and energy to calculate how long a star will remain on the main sequence.

Suppose that M is the mass of a star and f is the fraction of the star's mass that is converted into energy by hydrogen fusion. During its main-sequence lifetime, the total energy E supplied by the hydrogen fusion can be expressed as

$$E = fMc^2$$

In this equation c is the speed of light.

This energy from hydrogen fusion is released gradually over millions or billions of years. If L is the star's luminosity (energy released per unit time) and t is the star's main-sequence lifetime (the total time over which the hydrogen fusion occurs), then we can write

$$L = \frac{E}{t}$$

(Actually, this equation is only an approximation. A star's luminosity is not quite constant over its entire main-sequence lifetime. But the variations are not important for our purposes.) We can rewrite this equation as

$$E = Lt$$

From this equation and $E = fMc^2$, we see that

$$Lt = fMc^2$$

We can rearrange this equation as

$$t = \frac{fMc^2}{L}$$

Thus, a star's lifetime on the main sequence is proportional to its mass (M) divided by its luminosity (L). Using the symbol $\propto$ to denote "is proportional to," we write

$$t \propto \frac{M}{L}$$

We can carry this analysis further by recalling that main-sequence stars obey the mass-luminosity relation (see Section 17-9, especially the *Cosmic Connections* figure). The distribution of data on the graph in the *Cosmic Connections* figure in Section 17-9 tells us that a star's luminosity is roughly proportional to the 3.5 power of its mass:

$$L \propto M^{3.5}$$

Substituting this relationship into the previous proportionality, we find that

$$t \propto \frac{M}{M^{3.5}} = \frac{1}{M^{2.5}} = \frac{1}{M^2\sqrt{M}}$$

This approximate relationship can be used to obtain rough estimates of how long a star will remain on the main sequence. It is often convenient to relate these estimates to the Sun (a typical 1-$M_\odot$ star), which will spend 1.2×10^{10} years on the main sequence.

EXAMPLE: How long will a star whose mass is 4 $M_\odot$ remain on the main sequence?

Situation: Given the mass of a star, we are asked to determine its main-sequence lifetime.

Tools: We use the relationship $t \propto 1/M^{2.5}$.

Answer: The star has 4 times the mass of the Sun, so it will be on the main sequence for approximately

$$\frac{1}{4^{2.5}} = \frac{1}{4^2\sqrt{4}} = \frac{1}{32}$$

times the Sun's main-sequence lifetime

Thus, a 4-$M_\odot$ main-sequence star will fuse hydrogen in its core for about $(1/32) \times 1.2 \times 10^{10}$ years, or about 4×10^8 (400 million) years.

Review: Our result makes sense: A star more massive than the Sun must have a shorter main-sequence lifetime.

19-2 When core hydrogen fusion ceases, a main-sequence star like the Sun becomes a red giant

TUTORIAL 19-1 Like so many properties of stars, what happens at the end of a star's main-sequence lifetime depends on its mass. If the star is a red dwarf of less than about 0.4 $M_\odot$, after hundreds of billions of years the star has converted all of its hydrogen to helium. It is possible for helium to undergo thermonuclear fusion, but this requires temperatures and pressures far higher than those found within a red dwarf. Thus, this red dwarf will end its life as a ball of helium, which has no further nuclear reactions but still glows due to its internal heat. As it radiates energy into space, it slowly cools and shrinks. This slow, quiet demise is the ultimate fate of the red dwarfs, which are 85% of the stars in the Milky Way. (As we have seen, there has not yet been time in the history of the universe for any red dwarf to reach this final stage in its evolution.)

What is the fate of stars more massive than about 0.4 $M_\odot$, including the Sun? As we will see, the late stages of their evolution are far more dramatic. Studying these stages will give us insight into the fate of our solar system and of life on Earth.

Stars of 0.4 $M_\odot$ or Greater: From Main-Sequence Star to Red Giant

When a star of at least 0.4 solar masses reaches the end of its main-sequence lifetime, all of the hydrogen in its core has been used up and hydrogen fusion ceases there. In this new stage, hydrogen fusion continues only in the hydrogen-rich material just outside the core, a situation called **shell hydrogen fusion.** At first, this process occurs only in the hottest region just outside the core, where the hydrogen fuel has not yet been exhausted. Outside this region, no fusion reactions take place.

In the transition from main-sequence star to red giant, the star's core contracts while its outer layers expand

Strangely enough, the end of the core hydrogen fusion process leads to an *increase* in the core's temperature. Here's why: When thermonuclear reactions first cease in the core, nothing remains to generate heat there. Hence, the core starts to cool and the pressure in the core starts to decrease. This pressure decrease allows the star's core to again compress under the weight of the outer layers. As the core contracts, its temperature again increases, and heat begins to flow outward from the core even though no nuclear reactions are taking place there. (Technically, gravitational energy is converted into thermal energy, as in Kelvin-Helmholtz contraction; see Section 8-4 and Section 16-1).

This new flow of heat warms the gases around the core, increasing the rate of shell hydrogen fusion and causing the shell of fusion to "eat" further outward into the surrounding matter. Helium produced by reactions in the shell falls down onto the core, which continues to contract and heat up as it gains mass. Over the course of hundreds of millions of years, the core of a 1-$M_\odot$ star compresses to about one-third of its original radius, while its central temperature increases from about 15 million (1.5×10^7) K to about 100 million (10^8) K.

During this post–main-sequence phase, the star's outer layers expand just as dramatically as the core contracts. As the hydrogen-fusing shell works its way outward, egged on by heat from the contracting core, the star's luminosity increases substantially. This increases the star's internal pressure and makes the star's outer layers expand to many times its original radius. This tremendous expansion causes those outer layers to cool down, and the star's surface temperature drops (see Box 19-1). Once the temperature of the star's bloated surface falls to about 3500 K, the gases glow with a reddish hue, in accordance with Wien's law (see Figure 17-7a). The star is then appropriately called a **red giant** (Figure 19-4). Thus, we see that red giant stars are former main-sequence stars that have evolved into a new stage of existence. We can summarize these observations as a general rule:

Stars join the main sequence when they begin hydrogen fusion in their cores. They leave the main sequence and become giant stars when the core hydrogen is depleted.

Red giant stars undergo substantial **mass loss** because of their large diameters and correspondingly weak surface gravity. This makes it relatively easy for gases to escape from the red giant into space. Mass loss can be detected in a star's spectrum, because gas escaping from a red giant toward a telescope on Earth produces

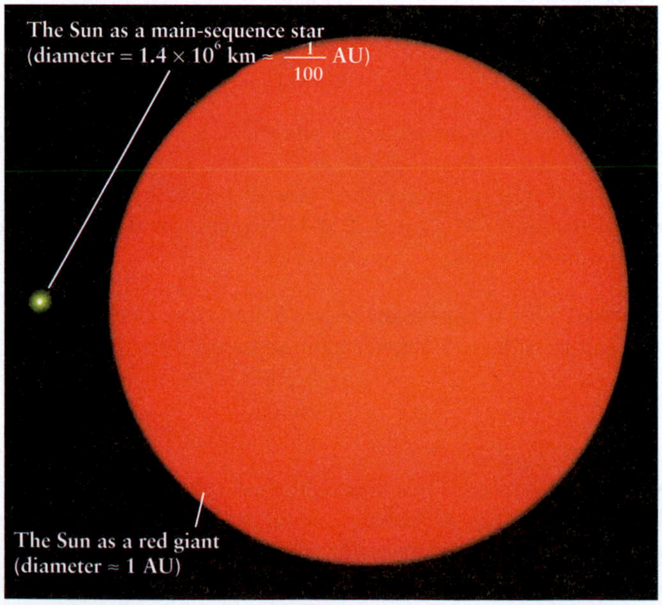

(a) The Sun today and as a red giant

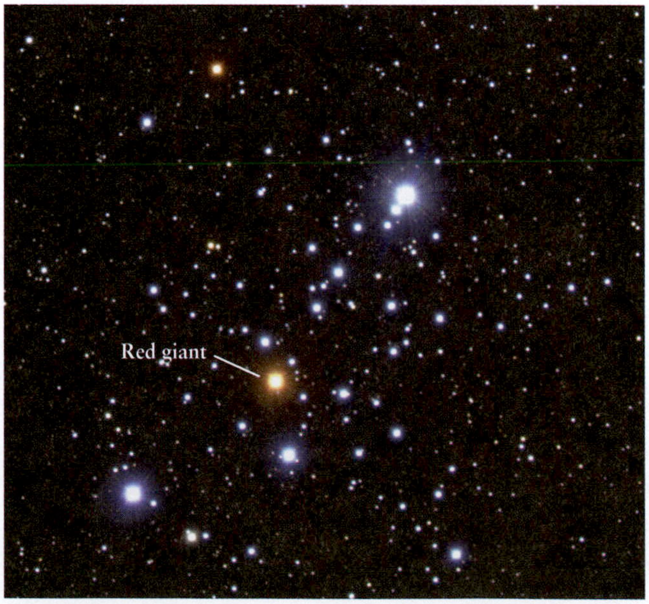

(b) A red giant star in the star cluster M103 R I V U X G

FIGURE 19-4

Red Giants (a) The present-day Sun produces energy in a hydrogen-fusing core about 100,000 km in diameter. Some 7.6 billion years from now, when the Sun becomes a red giant, its energy source will be a shell only about 30,000 km in diameter within which hydrogen fusion will take place at a furious rate. The Sun's luminosity will be about 2000 times greater than today, and the increased luminosity will make the Sun's outer layers expand to approximately 100 times their present size. **(b)** The cluster M103 contains a red giant star. M103 is 14 ly across and lies around 8000 ly from Earth in the constellation Cassiopeia. (NOAO/Science Source)

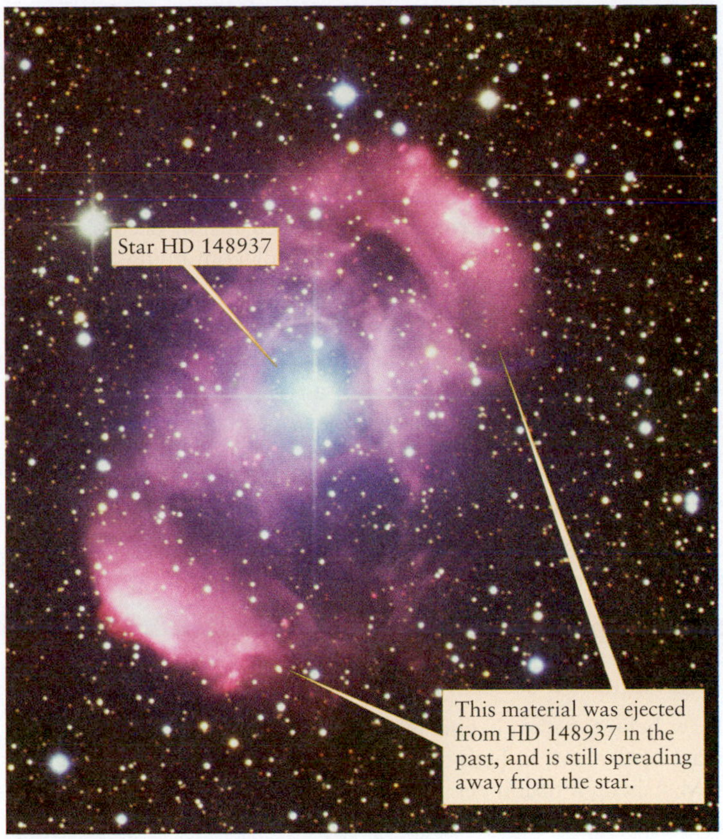

FIGURE 19-5 R I V U X G

A Mass-Loss Star As stars age and become giant stars, they expand tremendously and shed matter into space. This star, HD 148937, is losing matter at a high rate. Other strong outbursts in the past ejected the clouds that surround HD 148937. These clouds absorb ultraviolet radiation from the star, which excites the atoms in the clouds and causes them to glow. The characteristic red color of the clouds reveals the presence of hydrogen (see Section 5-6) that was ejected from the star's outer layers. (Australian Astronomical Observatory/David Malin Images)

narrow absorption lines that are slightly blueshifted by the Doppler effect (review Figure 5-26). Typical observed blueshifts correspond to a speed of about 10 km/s. A typical red giant loses roughly 10^{-7} $M_\odot$ of matter per year. For comparison, the Sun's present-day mass loss rate is only 10^{-14} $M_\odot$ per year. Hence, an evolving star loses a substantial amount of mass as it becomes a red giant. Figure 19-5 shows a star losing mass in this way.

The Distant Future of Our Solar System

We can use these ideas to peer into the future of our planet and our solar system. The Sun's luminosity will continue to increase as it goes through its main-sequence lifetime, and the temperature of Earth will increase with it. One and a half billion years from now Earth's average surface temperature will be 50°C (122°F). By 3½ billion years from now the surface temperature of Earth will exceed the boiling temperature of water. All the oceans will boil away, and Earth will become utterly incapable of supporting life. These increasingly hostile conditions will pose the ultimate challenge to whatever intelligent beings might inhabit Earth in the distant future.

About 7 billion years from now, our Sun will finish converting hydrogen into helium at its core. As the Sun's core contracts, its atmosphere will expand to envelop Mercury and perhaps reach to the orbit of Venus. Roughly 700 million years after leaving the main sequence, our red giant Sun will have swollen to a diameter of about 1 AU—roughly 100 times larger than its present size—and its surface temperature will have dropped to about 3500 K. Shell hydrogen fusion will proceed at such a furious rate that our star will shine with the brightness of 2000 present-day Suns. Finding themselves *inside* the bloated Sun, some of the inner planets, possibly including Earth, will be vaporized. Even the thick atmospheres of the outer planets will evaporate away to reveal tiny, rocky cores. Thus, in its later years, the aging Sun may destroy the planets that have accompanied it since its birth.

CONCEPTCHECK 19-2

When core hydrogen burning ceases in a red giant, why does the core's temperature increase?

CONCEPTCHECK 19-3

Why are red giants so large?

Answers appear at the end of the chapter.

19-3 Fusion of helium into carbon and oxygen begins at the center of a red giant

When a star with a mass greater than 0.4 $M_\odot$ first changes from a main-sequence star (Figure 19-6a) to a red giant (Figure 19-6b), its hydrogen-fusing shell surrounds a small, compact core of

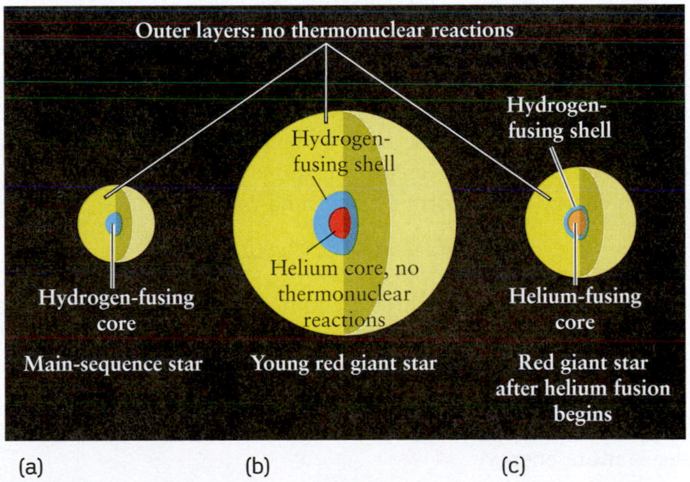

FIGURE 19-6

Stages in the Evolution of a Star with More than 0.4 Solar Masses **(a)** During the star's main-sequence lifetime, hydrogen is converted into helium in the star's core. **(b)** When the core hydrogen is exhausted, hydrogen fusion continues in a shell, and the star expands to become a red giant. **(c)** When the temperature in the red giant's core becomes high enough because of contraction, core helium fusion begins. (These three pictures are not drawn to scale. The star is about 100 times larger in its red-giant phase than in its main-sequence phase, then shrinks somewhat when core helium fusion begins.)

almost pure helium. In a red giant of moderately low mass, which the Sun will become 7 billion years from now, the dense helium core is about twice the diameter of Earth. Most of this core helium was *produced* by thermonuclear reactions during the star's main-sequence lifetime; during the red-giant era, this helium will *undergo* thermonuclear reactions.

Core Helium Fusion

Helium, the "ash" left over from hydrogen fusion, is a potential nuclear fuel: **Helium fusion,** the thermonuclear fusion of helium nuclei to make heavier nuclei, releases energy. But this reaction cannot take place within the core of our present-day Sun because the temperature there is too low. Each helium nucleus contains two protons, so it has twice the positive electric charge of a hydrogen nucleus, and there is a much stronger electric repulsion between two helium nuclei than between two hydrogen nuclei. For helium nuclei to overcome this repulsion and get close enough to fuse together, they must be moving at very high speeds, which means that the temperature of the helium gas must be very high. (For more on the relationship between the temperature of a gas and the speed of atoms in the gas, see Box 7-2.)

When a star first becomes a red giant, the temperature of its contracted helium core is still too low for helium nuclei to fuse. But as the hydrogen-fusing shell adds mass to the helium core, the core contracts even more, further increasing the star's central temperature. When the central temperature finally reaches 100 million (10^8) K, **core helium fusion**—that is, thermonuclear fusion of helium in the core—begins. As a result, the aging star again has a central energy source for the first time since leaving the main sequence (see Figure 19-6c).

Helium fusion occurs in two steps. First, two helium nuclei combine to form a beryllium nucleus:

$$^{4}\mathrm{He} + {}^{4}\mathrm{He} \rightarrow {}^{8}\mathrm{Be}$$

This particular beryllium isotope, which has four protons and four neutrons, is very unstable and breaks into two helium nuclei soon after it forms. However, in the star's dense core a third helium nucleus may strike the ^{8}Be nucleus before it has a chance to fall apart. Such a collision creates a stable isotope of carbon and releases energy in the form of a gamma-ray photon (γ):

$$^{8}\mathrm{Be} + {}^{4}\mathrm{He} \rightarrow {}^{12}\mathrm{C} + \gamma$$

This process of fusing three helium nuclei to form a carbon nucleus is called the **triple alpha process,** because helium nuclei (^{4}He) are also called **alpha particles** by nuclear physicists. Some of the carbon nuclei created in this process can fuse with an additional helium nucleus to produce a stable isotope of oxygen and release more energy:

$$^{12}\mathrm{C} + {}^{4}\mathrm{He} \rightarrow {}^{16}\mathrm{O} + \gamma$$

Thus, both carbon and oxygen make up the "ash" of helium fusion. The second step in the triple alpha process, and the process of oxygen formation, both release energy. The pressure resulting from this energy prevents any further gravitational contraction of the star's core. The *Cosmic Connections* figure summarizes the reactions involved in helium fusion.

It is interesting to note that ^{12}C and ^{16}O are the most common isotopes of carbon and oxygen, respectively; the vast majority of the carbon atoms in your body are ^{12}C, and almost all the oxygen you breathe is ^{16}O. We will discuss the significance of this in Section 19-5.

A mature red giant fuses helium in its core for a much shorter time than it spent fusing hydrogen in its core as a main-sequence star. For example, in the distant future the Sun will sustain helium core fusion for only about 100 million years—this period is only about 1% of the time that hydrogen fusion occurs. (While this is going on, hydrogen fusion is still continuing in a shell around the core.)

CONCEPTCHECK 19-4

What is different about the core of a red giant that allows helium to fuse even though this does not occur in the core of our present-day Sun?

CONCEPTCHECK 19-5

Where is a red giant hotter than it was during its main-sequence phase? Where is it cooler?

Answers appear at the end of the chapter.

The Helium Flash and Electron Degeneracy

How helium fusion begins at a red giant's center depends on the mass of the star. In high-mass red giants (greater than about 2 to 3 $M_\odot$), helium fusion begins gradually as temperatures in the star's core approach 10^8 K. In red giants with a mass less than about 2 to 3 $M_\odot$, helium fusion begins explosively and suddenly, in what is called the **helium flash.** Table 19-2 summarizes these differences.

The helium flash occurs because of unusual conditions that develop in the core of a moderately low-mass star as it becomes a red giant. To appreciate these conditions we must first understand how an ordinary gas behaves. Then we can explore how the densely packed electrons at the star's center alter this behavior.

When a gas is compressed, it usually becomes denser and warmer. To describe this process, scientists use the convenient concept of an **ideal gas,** which has a simple relationship between pressure, temperature, and density. Specifically, the pressure exerted by an ideal gas is directly proportional to both the density and the temperature of the gas. Many real gases actually behave like an ideal gas over a wide range of temperatures and densities.

Under most circumstances, the gases inside a star act like an ideal gas. If the gas expands, it cools down, and if it is compressed, it heats up (see Box 19-1). This behavior—like a safety valve—has

TABLE 19-2 How Helium Core Fusion Begins in Different Red Giants

Mass of star	Onset of helium fusion in core
More than about 0.4 but less than 2–3 solar masses	Explosive (helium flash)
More than 2–3 solar masses	Gradual

Stars with less than about 0.4 solar masses do not become red giants (see Section 19-2).

Helium Fusion in a Red Giant

A star becomes a red giant after the fusion of hydrogen into helium in its core has come to an end. As the red giant's core shrinks and heats up, a new cycle of reactions can occur that create the even heavier elements carbon and oxygen.

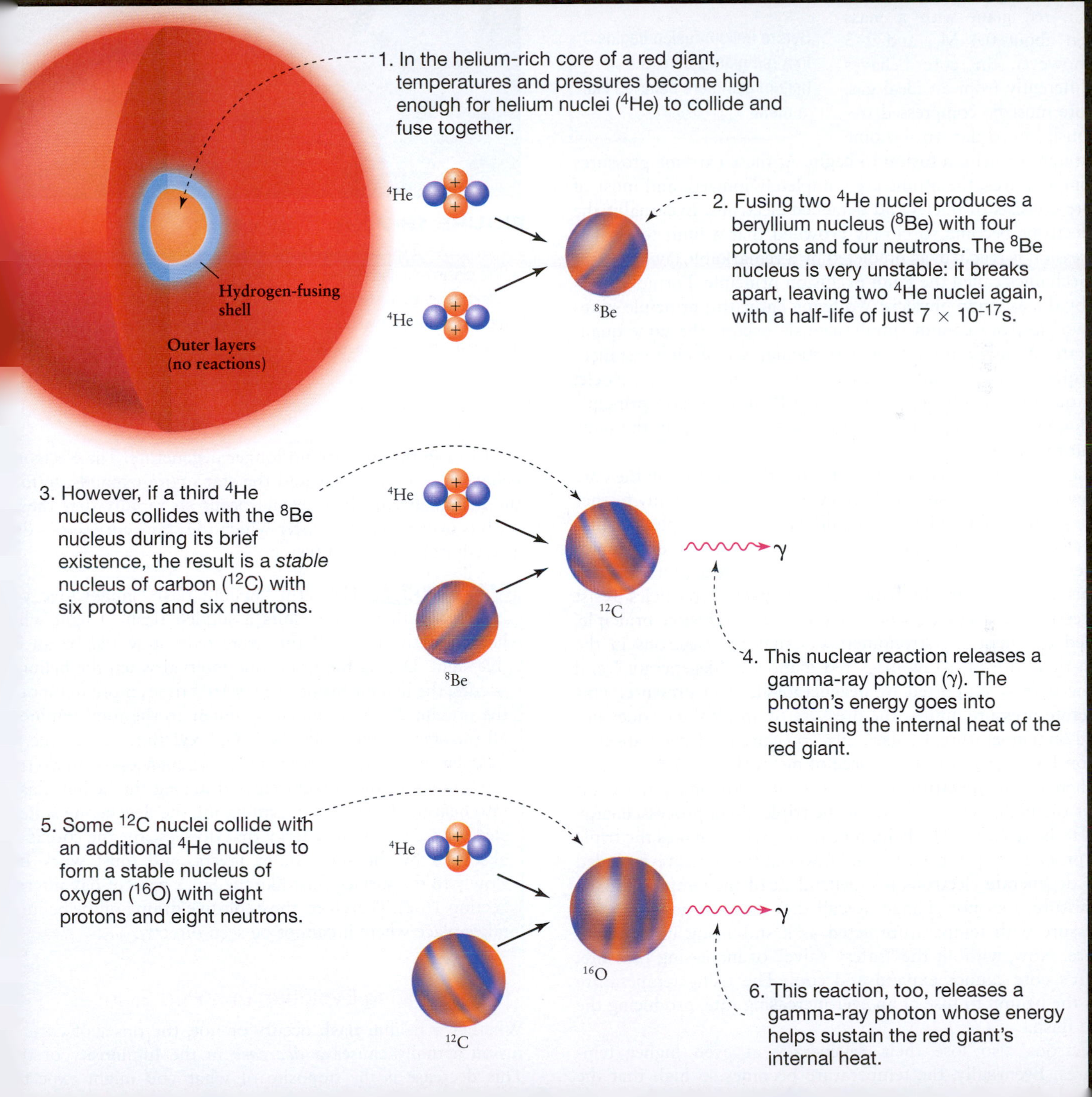
1. In the helium-rich core of a red giant, temperatures and pressures become high enough for helium nuclei (^{4}He) to collide and fuse together.
Hydrogen-fusing shell
Outer layers (no reactions)
^{4}He
^{4}He
^{8}Be
2. Fusing two ^{4}He nuclei produces a beryllium nucleus (^{8}Be) with four protons and four neutrons. The ^{8}Be nucleus is very unstable: it breaks apart, leaving two ^{4}He nuclei again, with a half-life of just 7×10^{-17}s.
3. However, if a third ^{4}He nucleus collides with the ^{8}Be nucleus during its brief existence, the result is a *stable* nucleus of carbon (^{12}C) with six protons and six neutrons.
^{4}He
^{8}Be
^{12}C
γ
4. This nuclear reaction releases a gamma-ray photon (γ). The photon's energy goes into sustaining the internal heat of the red giant.
5. Some ^{12}C nuclei collide with an additional ^{4}He nucleus to form a stable nucleus of oxygen (^{16}O) with eight protons and eight neutrons.
^{4}He
^{12}C
^{16}O
γ
6. This reaction, too, releases a gamma-ray photon whose energy helps sustain the red giant's internal heat.

a stabilizing influence, ensuring that the star neither collapses nor explodes (see Section 16-2). For example, if the rate of thermonuclear reactions in the star's core should increase, the additional energy releases heat and expands the core. This expansion cools the core's gases and slows the rate of thermonuclear reactions back to the original value. Conversely, if the rate of thermonuclear reactions should decrease, the core will cool down and compress under the pressure of the overlying layers. The compression of the core will make its temperature increase, thus speeding up the thermonuclear reactions and returning them to their original rate.

Before helium fusion begins in a red giant's core, the helium gas there behaves like a metal

In a red giant with a mass between about 0.4 $M_\odot$ and 2–3 $M_\odot$, however, the core behaves very differently from an ideal gas. The core must be compressed tremendously in order to become hot enough for helium fusion to begin. At these extreme pressures and temperatures, the atoms are completely ionized, and most of the core consists of nuclei and detached electrons. Eventually, the free electrons become so closely crowded that a limit to further compression is reached, as predicted by a remarkable law of quantum mechanics called the **Pauli exclusion principle.** Formulated in 1925 by the Austrian physicist Wolfgang Pauli, this principle states that two electrons cannot simultaneously occupy the same quantum state. A quantum state is a particular set of circumstances concerning locations and speeds that are available to a particle. In the quantum world of electrons, the Pauli exclusion principle is analogous to saying that you can't have two things in the same place at the same time.

Just before the onset of helium fusion, the electrons in the core of a low-mass star are so closely crowded together that any further compression would violate the Pauli exclusion principle. Because the electrons cannot be squeezed any closer together, they produce a powerful pressure that resists further core contraction.

This phenomenon, in which closely packed particles resist compression as a consequence of the Pauli exclusion principle, is called **degeneracy.** Astronomers say that the electrons in the helium-rich core of a low-mass red giant are "degenerate," and that the core is supported by **degenerate-electron pressure.** This degenerate pressure, unlike the pressure of an ideal gas, does not depend on temperature. Remarkably, you can find degenerate electrons on Earth in an ordinary piece of metal (Figure 19-7).

When the temperature in the core of a low-mass red giant reaches the high level required for the triple alpha process, energy begins to be released. The helium heats up, which makes the triple alpha process happen even faster. However, the pressure provided by the degenerate electrons is independent of the temperature, so the pressure does not change. Recall that an ideal gas's increase in pressure with temperature acted as a stabilizing influence in the core. Now, without the "safety valve" of increasing pressure, the star's core cannot expand and cool. The rising temperature causes the helium to fuse at an ever-increasing rate, producing the helium flash.

Electrons also lose their degeneracy at even higher temperatures. Eventually, the temperature becomes so high that the electrons in the core are no longer degenerate. The electrons then behave like an ideal gas and the star's core expands, terminating the helium flash. These events occur so rapidly that the helium flash is over in seconds, after which the star's core settles down to a steady rate of helium fusion.

FIGURE 19-7 R I **V** U X G
Degenerate Electrons In an ordinary piece of metal, like the chrome grille on this classic car, the electrons are affected by the Pauli exclusion principle. The resulting degenerate electron pressure helps make metals strong and difficult to compress. A more powerful version of this same effect happens inside the cores of low-mass red giant stars. (Santokh Kochar/PhotoDisc/Getty Images)

CAUTION! The term "helium flash" might give you the impression that a star emits a sudden flash of light when the helium flash occurs. If this were true, it would be an incredible sight. During the brief time interval when the helium flash occurs, the helium-fusing core is 10^{11} times more luminous than the present-day Sun, which is similar to the total luminosity of all the stars in the Milky Way Galaxy! But, in fact, the helium flash has no immediately visible consequences—for two reasons. First, much of the energy released during the helium flash goes into heating the core and terminating the degenerate state of the electrons. Second, the energy that does escape the core is largely absorbed by the star's outer layers, and must work its way slowly to the surface (just like the Sun's present-day interior; see Section 16-2). Therefore, the explosive drama of the helium flash takes place where it cannot be seen directly.

The Continuing Evolution of a Red Giant

Whether a helium flash occurs or not, the onset of core helium fusion actually causes a *decrease* in the luminosity of the star. This decrease is the opposite of what you might expect—after all, turning on a new energy source should make the luminosity greater, not less. What happens is that after the onset of core helium fusion, a star's superheated core expands like an ideal gas. (If the star is of sufficiently low mass to have had a degenerate core, the increased temperature after the helium flash makes the core too hot to remain degenerate. Hence, these stars also end up with cores that behave like ideal gases.) Temperatures drop around the expanding core, so the hydrogen-fusing shell reduces its energy output and the star's luminosity decreases. This temperature decrease allows the star's outer layers to contract and heat up. Consequently, a post–helium-flash star is less luminous, hotter at the surface, and smaller than a red giant.

Core helium fusion lasts for only a relatively short time. Calculations suggest that a 1-$M_\odot$ star like the Sun sustains core hydrogen fusion for about 12 billion (1.2×10^{10}) years, followed by about 250 million (2.5×10^8) years of shell hydrogen fusion leading up to the helium flash. After the helium flash, such a star can fuse helium in its core (while simultaneously fusing hydrogen in a shell around the core) for only 100 million (10^8) years, a mere 1% of its main-sequence lifetime. **Figure 19-8** summarizes these evolutionary stages in the life of a 1-$M_\odot$ star. In Chapter 20 we will take up the story of what happens after a star has consumed all the helium in its core.

Here is the story of post–main-sequence evolution in its briefest form: Before the beginning of core helium fusion, the star's core compresses and the outer layers expand, and just after core helium fusion begins, the core expands and the outer layers compress. We will see in Chapter 20 that this behavior, in which the inner and outer regions of the star change in opposite ways, occurs again and again in the final stages of a star's evolution.

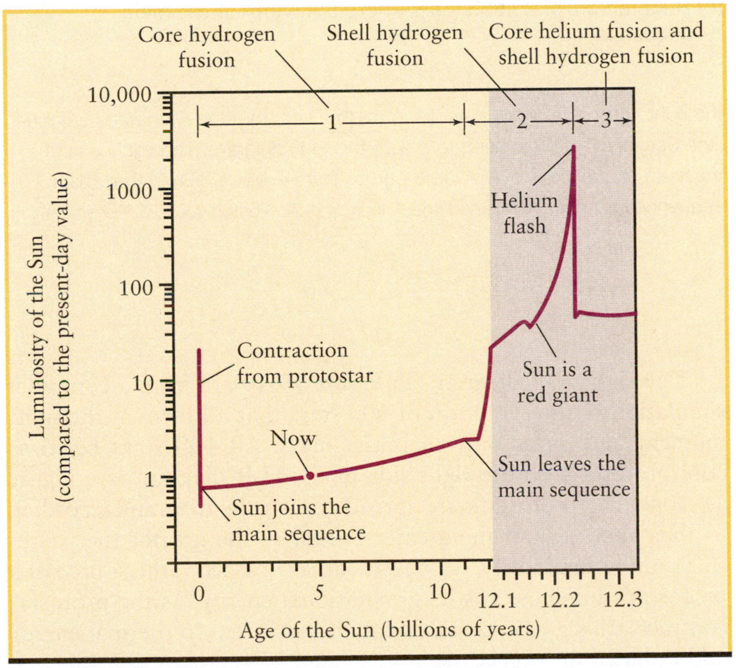

FIGURE 19-8

Stages in the Evolution of the Sun This diagram shows how the luminosity of the Sun (a 1-$M_\odot$ star) changes over time. The Sun began as a protostar whose luminosity decreased rapidly as the protostar contracted. Once established as a main-sequence star with core hydrogen fusion, the Sun's luminosity increases slowly over billions of years. The post–main-sequence evolution is much more rapid, so a different timescale is used in the right-hand portion of the graph. (Adapted from Mark A. Garlick, based on calculations by I.-Juliana Sackmann and Kathleen E. Kramer)

CONCEPTCHECK 19-6

What is the "safety valve" in a main-sequence star's nuclear reactions, and how is it not in place for a red giant with a core supported by degenerate-electron pressure?

CONCEPTCHECK 19-7

What happens to the luminosity when the helium-fusing core is no longer supported by degenerate-electron pressure?

Answers appear at the end of the chapter.

19-4 H-R diagrams and observations of star clusters reveal how red giants evolve

TUTORIAL 19-2 To see how stars evolve during and after their main-sequence lifetimes, it is helpful to follow them on a Hertzsprung-Russell (H-R) diagram. On such a diagram, zero-age main-sequence stars lie along a line called the **zero-age main sequence**, or **ZAMS** (**Figure 19-9a**). These stars have just emerged from their protostar stage, are steadily fusing hydrogen into helium in their cores, and have attained hydrostatic equilibrium. With the passage of time, hydrogen in a main-sequence star's core is converted to helium, the luminosity slowly increases, the star slowly expands, and the star's position on the H-R diagram inches away from the ZAMS. As a result, the main sequence on an H-R diagram is a fairly broad band rather than a narrow line (Figure 19-9b).

Post–Main-Sequence Evolution on an H-R Diagram

The dashed line in Figure 19-9a denotes stars whose cores have been exhausted of hydrogen and in which core hydrogen fusion has ceased. These stars have reached the ends of their main-sequence lifetimes. From there, the points representing high-mass stars (3 $M_\odot$, 5 $M_\odot$, and 9 $M_\odot$) move rapidly from left to right across the H-R diagram. This means that, although the star's surface temperature is decreasing, its surface area is increasing at a rate that keeps its overall luminosity roughly constant. During this transition, the star's core contracts and its outer layers expand as energy flows outward from the hydrogen-fusing shell.

Just before core helium fusion begins, the evolutionary tracks of high-mass stars turn upward in the red-giant region of the H-R diagram (to the upper right of the main sequence). After core helium fusion begins, however, the cores of these stars expand, the

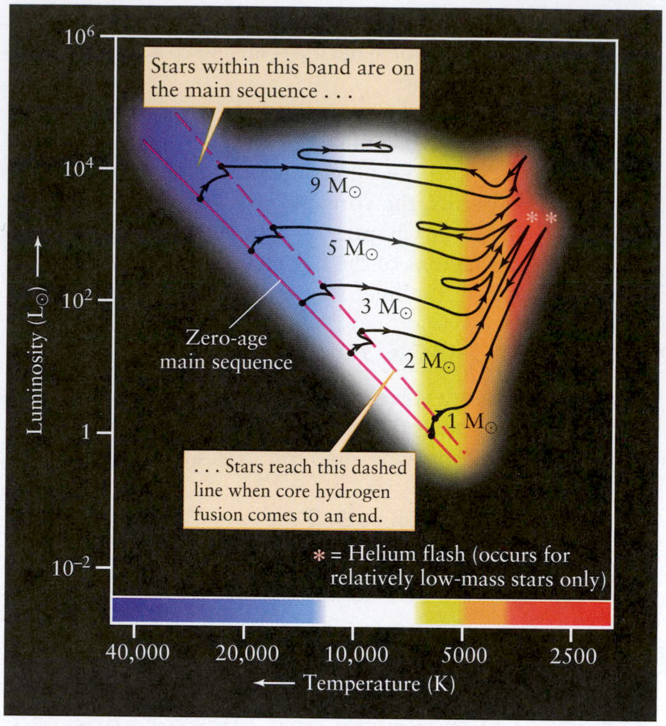

(a) Post-main-sequence evolutionary tracks of five stars with different mass

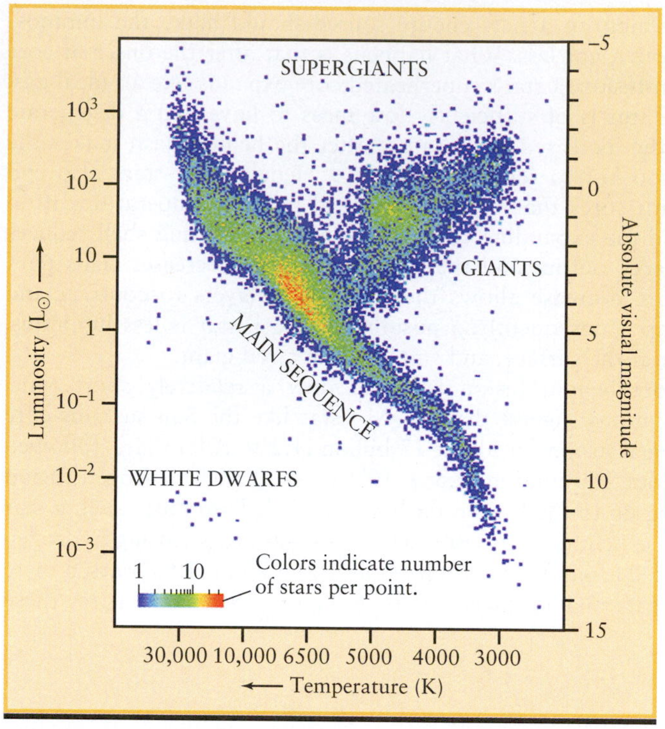

(b) H-R diagram of 20,853 stars—note the width of the main sequence

FIGURE 19-9

H-R Diagrams of Stellar Evolution On and Off the Main Sequence **(a)** The two lowest-mass stars shown here (1 $M_\odot$ and 2 $M_\odot$) undergo a helium flash at their centers, as shown by the asterisks. In the high-mass stars, core helium fusion ignites more gradually where the evolutionary tracks make a sharp downward turn in the red-giant region on the right hand side of the H-R diagram. **(b)** Data from the *Hipparcos* satellite (see Section 17-1) was used to create this H-R diagram. The thickness of the main sequence is due in large part to stars evolving during their main-sequence lifetimes. (a: Adapted from I. Iben; b: Adapted from M. A. C. Perryman)

outer layers contract, and the evolutionary tracks back away from these temporary peak luminosities. The tracks then wander back and forth in the red-giant region while the stars readjust to their new energy sources.

Figure 19-9a also shows the evolutionary tracks of two stars of moderately low mass (1 $M_\odot$ and 2 $M_\odot$). The onset of core helium fusion in these stars occurs with a helium flash, indicated by the red asterisks in the figure. As we saw in the previous section, after the helium flash, these stars shrink and become less luminous. The decrease in size is proportionately greater than the decrease in luminosity, and so the surface temperatures increase. Hence, after the helium flash, the evolutionary tracks for the 1-$M_\odot$ and 2-$M_\odot$ stars move down and to the left.

A Simulated Star Cluster: Tracking 4½ Billion Years of Stellar Evolution

We can summarize our understanding of stellar evolution from birth through the onset of helium fusion by following the evolution of a hypothetical cluster of stars. We saw in Section 18-6 that the stars that make up a cluster all begin to form at essentially the same time but have different initial masses. Hence, studying star clusters allows us to compare how stars of different masses evolve.

The eight H-R diagrams in Figure 19-10 are from a computer simulation of the evolution of 100 stars that all form at the same moment and differ only in initial mass. All 100 stars begin as cool protostars on the right side of the H-R diagram (see Figure 19-10a). The protostars are spread out on the diagram according to their masses, and the greater the mass, the greater the protostar's initial luminosity. As we saw in Section 18-3, the source of a protostar's luminosity is its gravitational energy. As the protostar contracts, this gravitational energy is converted to thermal energy and radiated into space.

The most massive protostars contract and heat up very rapidly. After only 5000 years, they have already moved across the H-R diagram toward the main sequence (see Figure 19-10b). After 100,000 years, these massive stars have ignited hydrogen fusion in their cores and have settled down on the main sequence as O stars (see Figure 19-10c). After 3 million years, stars of moderate mass have also ignited core hydrogen fusion and become main-sequence stars of spectral classes B and A (see Figure 19-10d). Meanwhile, low-mass protostars continue to inch their way toward the main sequence as they leisurely contract and heat up.

After 30 million years (see Figure 19-10e), the most massive stars have depleted the hydrogen in their cores and become red

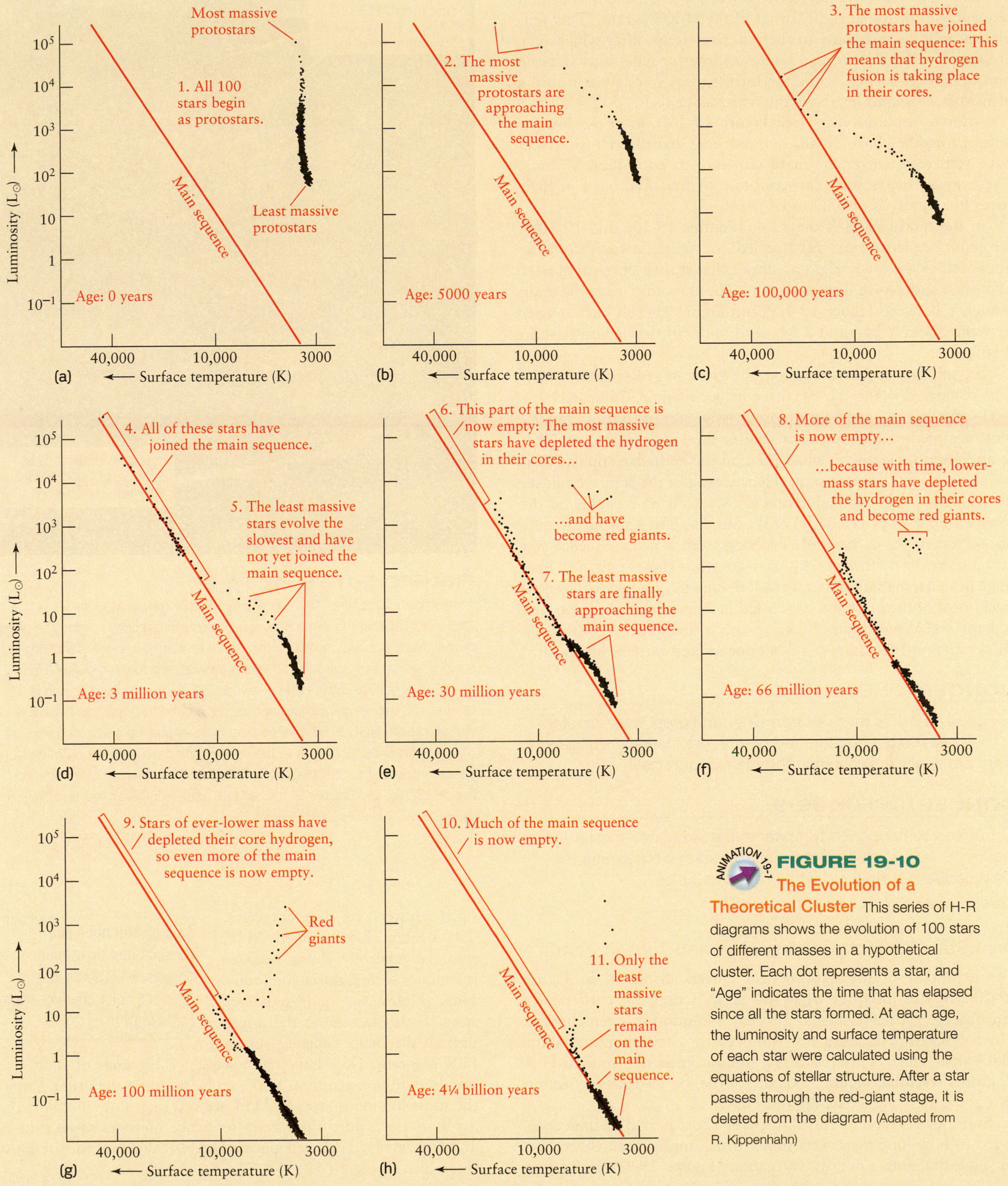

FIGURE 19-10 The Evolution of a Theoretical Cluster This series of H-R diagrams shows the evolution of 100 stars of different masses in a hypothetical cluster. Each dot represents a star, and "Age" indicates the time that has elapsed since all the stars formed. At each age, the luminosity and surface temperature of each star were calculated using the equations of stellar structure. After a star passes through the red-giant stage, it is deleted from the diagram (Adapted from R. Kippenhahn)

giants. These stars have moved from the upper left end of the main sequence to the upper right corner of the H-R diagram. (This simulation follows stars only to the red-giant stage, after which they are simply deleted from the diagram.) Intermediate-mass stars lie on the main sequence, while the lowest-mass stars are still in the protostar stage and lie above the main sequence.

After 66 million years (see Figure 19-10f), even the lowest-mass protostars have finally ignited core hydrogen fusion and have settled down on the main sequence as cool, dim, M stars. These lowest-mass stars can continue to fuse hydrogen in their cores for hundreds of billions of years.

In the final two H-R diagrams (Figures 19-10g and 19-10h), the main sequence stars get "peeled" or "eaten away" from the upper left to the lower right as stars exhaust their core supplies of hydrogen and evolve into red giants. The stars that leave the main sequence between Figure 19-10g and Figure 19-10h have masses between about 1 $M_\odot$ and 3 $M_\odot$ and undergo the helium flash in their cores.

For all stars in this simulation, the giant stage lasts only a brief time compared to the star's main-sequence lifetime. Compared to a 1-$M_\odot$ star (see Figure 19-8), a more massive star has a shorter main-sequence lifetime *and* spends a shorter time as a giant star. Thus, at any given time, only a small fraction of the stellar population is passing through the giant stage. Hence, most of the stars we can see through telescopes are main-sequence stars. As an example, of the stars within 4.00 pc (13.05 ly) of the Sun listed in Appendix 4, only one—Procyon A—is presently evolving from a main-sequence star into a giant. Two other nearby stars are white dwarfs, an even later stage in stellar evolution that we will discuss in Chapter 20. (By contrast, most of the *brightest* stars listed in Appendix 5 are giants and supergiants. Although they make up only a small fraction of the stellar population, these stars stand out due to their extreme luminosity.)

CONCEPTCHECK 19-8

In Figure 19-10, is there an age when the most massive stars in the cluster have become red giants, yet the least massive stars are finally approaching the main sequence?

CONCEPTCHECK 19-9

Suppose you happened to observe the simulated cluster in Figure 19-10 just as its most massive stars were joining the main sequence. How old would the cluster be?

Answers appear at the end of the chapter.

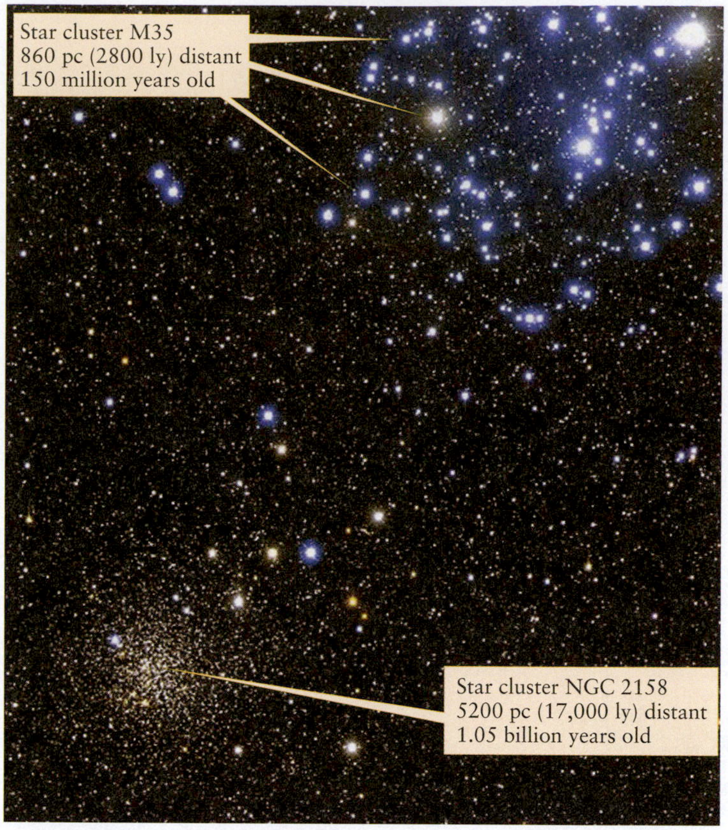

FIGURE 19-11 R I V U X G

Two Open Clusters The two clusters in this image, M35 and NGC 2158, lie in almost the same direction in the constellation Gemini. The nearer cluster, M35, has a number of luminous blue main-sequence stars with surface temperatures around 10,000 K as well as a few red giants. Hence, its H-R diagram resembles that shown in Figure 19-10g, and its age is around 100 million years (more accurately, around 150 million). The more distant cluster, NGC 2158, has no blue main-sequence stars; long ago, all of these massive stars came to the end of their main-sequence lifetimes and became giants. The H-R diagram for NGC 2158 is intermediate between Figure 19-10g and Figure 19-10h, and its age (1.05 billion years) is as well. (Jean-Charles Cuillandre (CFHT), © 2001 CFHT)

Real Star Clusters: Cluster Ages and Turnoff Points

The evolution of the hypothetical cluster displayed in Figure 19-10 helps us interpret what we see in actual star clusters. We can observe the early stages of stellar evolution in **open clusters,** which typically contain a few hundred to a few thousand stars. Many open clusters are just a few million years old, so their H-R diagrams resemble Figure 19-10d, 19-10e, or 19-10f (see Section 18-6, especially Figure 18-20 and Figure 18-21).

Star clusters tend to form in a very short period of time compared to how long their stars shine, which means each cluster has an age—the time since its formation. Figure 19-11 shows two open clusters of different ages. The nearer cluster, called M35, must be relatively young because it contains several dozen luminous, blue, high-mass main-sequence stars. These stars lie in the upper part of the main sequence on an H-R diagram. They have main-sequence lifetimes of only a few hundred million years, so M35 can be no older than that. Some of the most luminous stars in M35 are red or yellow in color; these are stars that ended their main-sequence lifetimes some time ago and have evolved into red giants. The H-R diagram for this cluster resembles Figure 19-10g.

As a cluster ages and its stars end their main-sequence lifetimes, the cluster's color changes from blue to red

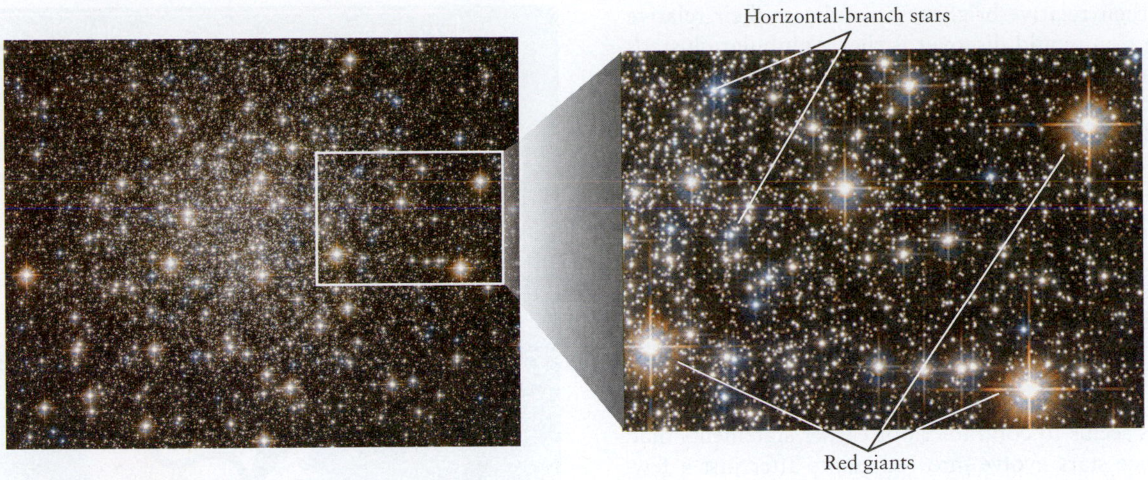

FIGURE 19-12 R I V U X G

A Globular Cluster This cluster, called M10, contains a few hundred thousand stars within a diameter of only 20 pc (70 ly). It lies approximately 5000 pc (16,000 ly) from Earth in the constellation Ophiuchus (the Serpent Holder). Most of the stars in this image are either red giants or blue, horizontal-branch stars with both core helium fusion and shell hydrogen fusion. (ESA/Hubble & NASA)

There are no high-mass blue main-sequence stars at all in NGC 2158, the more distant cluster shown in Figure 19-11. Any such stars that were once in NGC 2158 have long since come to the end of their main-sequence lifetimes. As a result, the main sequence in this cluster has been "eaten away" more than that of M35, leaving only stars that are yellow or red in color. This tells us that NGC 2158 must be older than M35 (compare Figures 19-10g and 19-10h). This example shows that as a cluster ages, it generally becomes redder in its average color.

We can see even later stages in stellar evolution by studying **globular clusters,** so called because of their spherical shape. A typical globular cluster contains up to 1 million stars in a volume less than 100 pc across (Figure 19-12). Among these are many highly evolved post–main-sequence stars.

Globular clusters must be old, because they contain no high-mass stars on the main-sequence. To determine that these clusters are old, you would measure the apparent magnitude (a measure of apparent brightness, which we introduced in Section 17-3) and color ratio of many stars in a globular cluster, then plot the data as shown in Figure 19-13. Such a **color-magnitude diagram** for a cluster is equivalent to an H-R diagram. The color ratio of a star tells you its surface temperature (as described in Section 17-4), and because all the stars in the cluster are at essentially the same

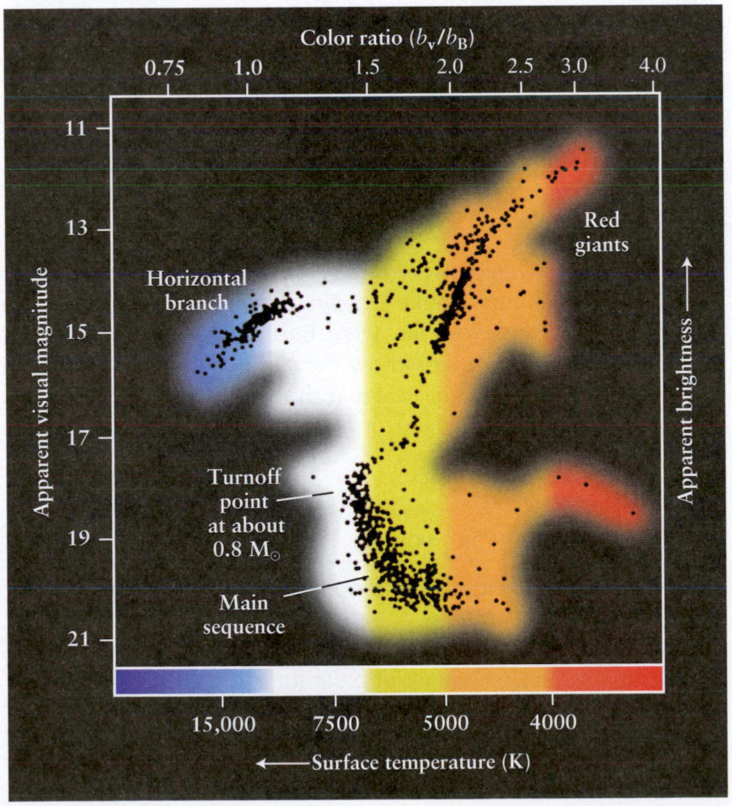

FIGURE 19-13

Age of a Globular Cluster This color-magnitude diagram is equivalent to an H-R diagram and reveals the age of the cluster. Below the turnoff point, stars are still on the main sequence. Above the turnoff point, higher mass stars have already moved off the main sequence. The turnoff point occurs for stars around 0.8 $M_\odot$; since stars of this mass are only on the main sequence for about 12 billion years, this tells us the age of the cluster. Each dot in this diagram represents the apparent visual magnitude (a measure of the brightness as seen through a V filter) and surface temperature (as measured by the color ratio b_V/b_B) of a star in the globular cluster M55 in Sagittarius. (Adapted from D. Schade, D. VandenBerg, and F. Hartwick) distance from us, their relative brightnesses indicate their relative luminosities. What you would discover is that a globular cluster's main sequence has been "peeled" or "eaten away" even more extensively than the open cluster NGC 2158 in Figure 19-11. Hence, globular clusters must be even older than NGC 2158. In a typical globular cluster, all the main-sequence stars with masses more than about 1 $M_\odot$ or 2 $M_\odot$ evolved long ago into red giants. Only low-mass, slowly evolving stars still have core hydrogen fusion. (Compare Figure 19-13 with Figure 19-10h.) As we will see next, not only can we tell that globular clusters are old, we can also figure out just how old they are.

CAUTION! The inset in Figure 19-12 shows something surprising: There are luminous *blue* stars in the ancient globular cluster M10. This seems to contradict our earlier statements that blue main-sequence stars evolve into red giants after just a few hundred million years. The explanation is that these are *not* main-sequence stars, but rather **horizontal-branch stars.** These stars get their name because in the color-magnitude diagram of a globular cluster, they form a horizontal grouping in the left-of-center portion of the diagram (see Figure 19-13). Horizontal-branch stars are relatively low-mass stars that have already become red giants and undergone a helium flash, so there is both core helium fusion and shell hydrogen fusion taking place in their interiors. After the helium flash their luminosity decreased to about 50 $L_\odot$ (compared to about 1000 $L_\odot$ before the flash) and their outer layers contracted and heated, giving these stars their blue color. In years to come, these stars will move back toward the red-giant region as their fuel is devoured. Our own Sun will go through a horizontal-branch phase in the distant future; this is the phase labeled by the number 3 at the far right of Figure 19-8.

Measuring the Ages of Star Clusters

The idea that a cluster's main sequence is progressively "eaten away" is the key to determining the age of a cluster. In the H-R diagram for a very young cluster, all the stars are on or near the main sequence. (An example is the open cluster NGC 2264, shown in Figure 18-20.) As a cluster gets older, however, stars begin to leave the main sequence. The high-mass, high-luminosity stars are the first to consume their core hydrogen and become red giants. As time passes, fewer and fewer stars remain on the main sequence (see parts d through h of Figure 19-10).

The age of a cluster can be found from the **turnoff point,** which is at the peak of the surviving portion of the main-sequence stars on the cluster's H-R diagram (see Figure 19-13). The stars at the turnoff point are just at the stage of exhausting the hydrogen in their cores, so their main-sequence lifetime is equal to the age of the cluster. Since we know the main-sequence lifetimes of stars with differing masses through modeling, we are able to determine how long ago the cluster formed. For example, in the case of the globular cluster M55 plotted in Figure 19-13, 0.8-$M_\odot$ stars have just left the main sequence, indicating that the cluster's age is more than 12 billion (1.2×10^{10}) years (see Table 19-1).

Figure 19-14 shows data for several star clusters plotted on a single H-R diagram. This graph also shows turnoff-point times from which the ages of the clusters can be estimated. Notice that the lower down the turnoff point occurs (which corresponds to lower masses moving off the main sequence), the older the cluster.

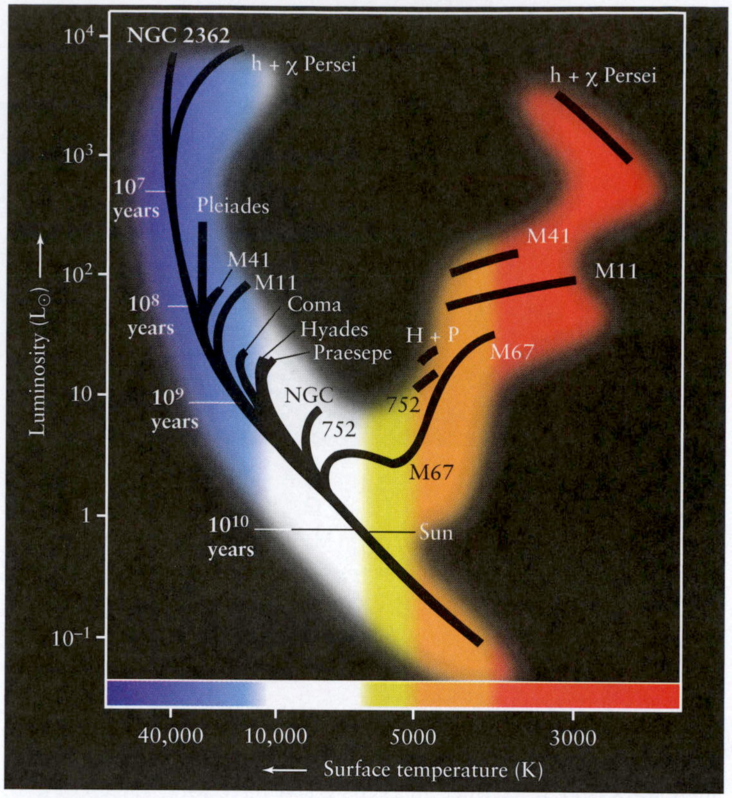

FIGURE 19-14
An H-R Diagram for Open Star Clusters The black bands indicate where stars from various open clusters fall on the H-R diagram. The age of a cluster can be estimated from the location of the cluster's turnoff point, where the cluster's most massive stars are just now leaving the main sequence. The times for these turnoff points are listed alongside the main sequence. For example, the Pleiades cluster turnoff point is near the 10^8-year point, so this cluster is about 10^8 years old. (Adapted from A. Sandage)

ANALOGY Suppose you knew that certain birthday candles took 10 minutes to burn down. If you walked into a room and saw that these candles were burned halfway down on a birthday cake, you would know they were lit 5 minutes ago. In this analogy, knowing the rate that the candle burns is analogous to knowing how long a star of a given mass will remain on the main sequence.

CONCEPTCHECK 19-10

The turnoff point in Figure 19-13 corresponds to main-sequence stars that are about 7500 K. Suppose that the turnoff point was instead for stars of around 10,000 K. For a 10,000-K turnoff point, would the cluster be younger or older than 12 billion years?

CONCEPTCHECK 19-11

Using Figure 19-14, explain why turnoff points at lower luminosities correspond to older clusters.

Answers appear at the end of the chapter.

19-5 Stellar evolution has produced two distinct populations of stars

Studies of star clusters reveal a curious difference between the youngest and oldest stars in our Galaxy. Stars in the youngest clusters (those with most of their main sequences still intact) are said to be **metal rich,** because their spectra contain many prominent spectral lines of heavy elements. (Recall from Section 17-5 that astronomers use the term "metal" to denote any element other than hydrogen and helium, which are the two lightest elements.) Such stars are also called **Population I stars.** The Sun is a relatively young, metal-rich, Population I star.

By contrast, the spectra of stars in the oldest clusters show only weak lines of heavy elements. These ancient stars are thus said to be **metal poor,** because heavy elements are only about 3% as abundant in these stars as in the Sun. They are also called **Population II stars.** The stars in globular clusters are metal-poor, Population II stars. Figure 19-15 shows the difference in spectra between a metal-poor, Population II star and the Sun (a metal-rich, Population I star).

CAUTION! Note that "metal rich" and "metal poor" are relative terms. In even the most metal-rich star known, metals make up just a few percent of the total mass of the star.

Stellar Populations and the Origin of Heavy Elements

To explain why there are two distinct populations of stars, we must go back to the Big Bang, the explosive origin of the universe that took place some 13.7 billion years ago. As we will discuss in Chapter 26, the early universe consisted almost exclusively of hydrogen and helium, with almost no heavy elements (metals). The first stars to form were likewise metal poor. The least massive of these stars have survived to the present day and are now the ancient stars of Population II.

Stars like the Sun contain material that was processed through an earlier generation of stars

The more massive of the original stars evolved more rapidly and no longer shine. But as these stars evolved, helium fusion in their cores produced metals—carbon and oxygen. In the most massive stars, as we will learn in Chapter 20, further thermonuclear reactions produced even heavier elements. As these massive original stars aged and died, they expelled their metal-enriched gases into space. (The star shown in Figure 19-5 is going through such a mass-loss phase late in its life.) This expelled material joined the interstellar medium and was eventually incorporated into a second generation of stars that have a higher concentration of heavy elements. These metal-rich members of the second stellar generation are the Population I stars, of which our Sun is an example.

CAUTION! Be careful not to let the designations of the two stellar populations confuse you. Population *I* stars are members of a *second* stellar generation, while Population *II* stars belong to an older *first* generation. Unfortunately, the names were assigned in the order that these populations of stars were discovered, and not the order that they occur in a galaxy.

The relatively high concentration of heavy elements in the Sun means that the solar nebula, from which both the Sun and planets formed (see Section 8-4), must likewise have been metal rich. Earth is composed almost entirely of heavy elements, as are our bodies. Thus, our very existence is intimately linked to the Sun's being a Population I star. A planet like Earth probably could not have formed from the metal-poor gases that went into making Population II stars.

The concept of two stellar populations provides insight into our own origins. Recall from Section 19-3 that helium fusion in red giant stars produces the same isotopes of carbon (^{12}C) and oxygen (^{16}O) that are found most commonly on Earth. The reason is that

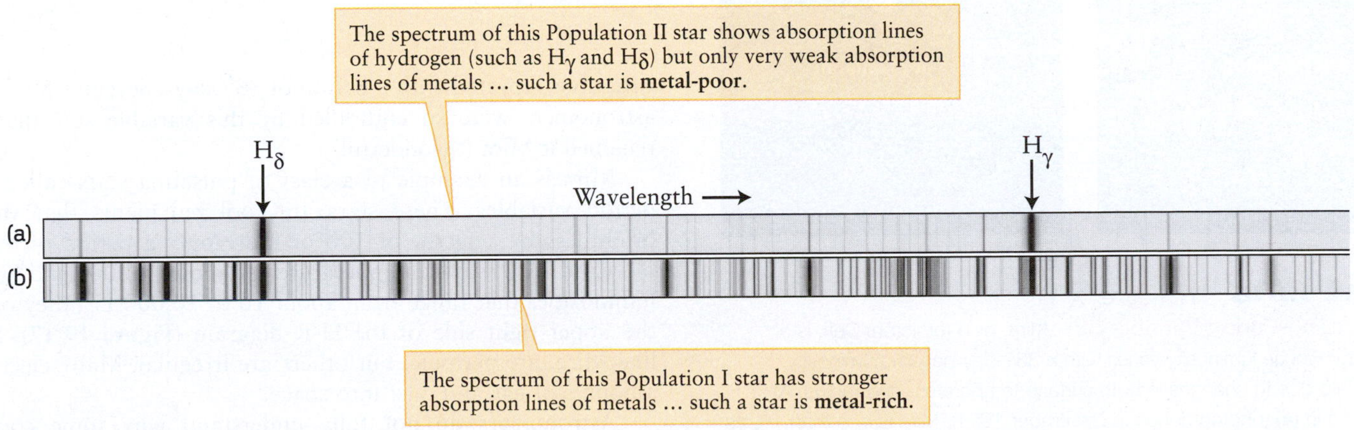

The spectrum of this Population II star shows absorption lines of hydrogen (such as Hγ and Hδ) but only very weak absorption lines of metals ... such a star is metal-poor.
Hδ
Hγ
Wavelength →
(a)
(b)
The spectrum of this Population I star has stronger absorption lines of metals ... such a star is metal-rich.

FIGURE 19-15 R I V U X G
Spectra of a Metal-Poor Star and a Metal-Rich Star The abundance of metals (elements heavier than hydrogen and helium) in a star can be inferred from its spectrum. These spectra compare (a) a metal-poor, Population II star and (b) a metal-rich, Population I star (the Sun) of the same surface temperature. We described the hydrogen absorption lines Hγ (wavelength 434 nm) and Hδ (wavelength 410 nm) in Section 5-8.

Earth's carbon and oxygen atoms, including all of those in your body, actually *were* produced by helium fusion. These reactions occurred billions of years ago within an earlier generation of stars that died and gave up their atoms to the interstellar medium—the same atoms that later became part of our solar system, our planet, and our bodies. We are literally children of the stars.

CONCEPTCHECK 19-12

How do we know that Population II stars had to form before Population I stars?

Answer appears at the end of the chapter.

19-6 Many mature stars pulsate

ANIMATION 19-2

We saw in Section 16-3 that the surface of our Sun vibrates in and out, although by only a small amount. But other stars undergo substantial changes in size, alternately swelling and shrinking. As these stars pulsate, their surface temperatures also vary. As a result, pulsating stars can vary dramatically in brightness. We now understand that these **pulsating variable stars** are actually evolved, post–main-sequence stars.

Long-Period Variables

Pulsating variable stars were first discovered in 1595 by David Fabricius, a Dutch minister and amateur astronomer. He noticed that the star o (omicron) Ceti is sometimes bright enough to be easily seen with the naked eye but at other times fades to invisibility (Figure 19-16). By 1660, astronomers realized that these brightness variations repeated with a period of 332 days. Seventeenth-century astronomers were so enthralled by this variable star that they renamed it Mira ("wonderful").

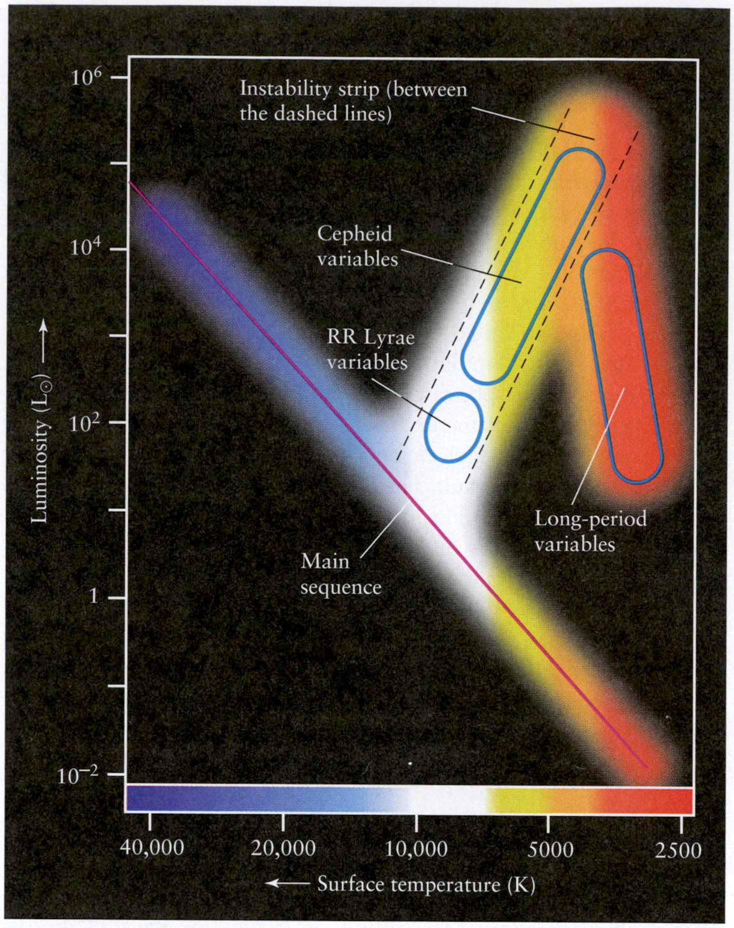

FIGURE 19-17

Variable Stars on the H-R Diagram Pulsating variable stars are found in the upper right of the H-R diagram. Long-period variables like Mira are cool red giant stars that pulsate slowly, changing their brightness in a semiregular fashion over months or years. Cepheid variables and RR Lyrae variables are located in the instability strip, which lies between the main sequence and the red giant region. A star passing through this strip along its evolutionary track becomes unstable and pulsates.

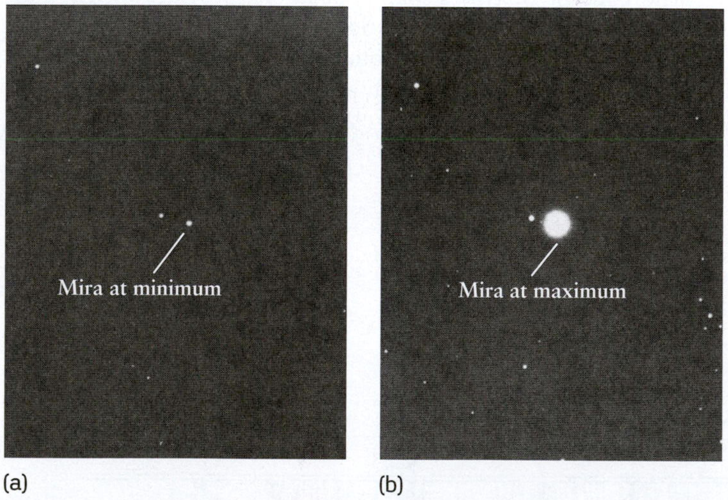

FIGURE 19-16 R I **V** U X G

Mira—A Long-Period Variable Star Mira, or o (omicron) Ceti, is a variable star whose luminosity varies with a 332-day period. Luminosity variations are due to changes in both surface temperature and radius. At its dimmest, as in **(a)** (photographed in December 1961), Mira is less than 1% as bright as when it is at maximum, as in **(b)** (January 1965). The physical size of the star is too small to discern with the telescope used, but variations in brightness change how wide it appears in the image. (Lowell Observatory)

Mira is an example of a class of pulsating stars called **long-period variables.** These stars are cool red giants that vary in brightness by a factor of 100 or more over a period of months or years. With surface temperatures of about 3500 K and average luminosities that range from about 10 to 10,000 $L_\odot$, they occupy the upper right side of the H-R diagram (Figure 19-17). Some, like Mira, are periodic, but others are irregular. Many eject large amounts of gas and dust into space.

Astronomers do not fully understand why some cool red giants become long-period variables. It is difficult to calculate accurate stellar models to describe such huge stars with extended, tenuous atmospheres.

Cepheid Variables

Astronomers have a much better understanding of other pulsating stars, called **Cepheid variables,** or simply Cepheids. As we will see, Cepheid stars play a crucial role in measuring the large-scale behavior of the universe and were some of the first objects targeted by the Hubble Space Telescope.

A Cepheid variable is recognized by the characteristic way in which its light output varies—rapid brightening followed by gradual dimming. They are named for δ (delta) Cephei, an example of this type of star discovered in 1784 by John Goodricke, a deaf, mute, 19-year-old English amateur astronomer. He found that at its most brilliant, δ Cephei is 2.3 times as bright as at its dimmest. The cycle of brightness variations repeats every 5.4 days. (Sadly, Goodricke paid for his discoveries with his life; he caught pneumonia while making his nightly observations and died before his twenty-second birthday.) The surface temperatures and luminosities of the Cepheid variables place them in the upper middle of the H-R diagram (see Figure 19-17).

By studying variable stars, astronomers gain insight into late stages of stellar evolution

After core helium fusion begins, mature stars move across the middle of the H-R diagram. Figure 19-9a shows the evolutionary tracks of high-mass stars crisscrossing the H-R diagram. Post–helium-flash stars of moderate mass also cross the middle of the H-R diagram between the red-giant region and the horizontal branch.

During these transitions across the H-R diagram, a star can become unstable and pulsate. In fact, there is a region on the H-R diagram between the upper main sequence and the red-giant branch called the **instability strip** (see Figure 19-17). When an evolving star passes through this region, the star pulsates and its brightness varies periodically. Figure 19-18a shows the brightness variations of δ Cephei, which lies within the instability strip.

A Cepheid variable brightens and fades because the star's outer envelope cyclically expands and contracts. The first to observe this was the Russian astronomer Aristarkh Belopolsky, who noticed in 1894 that spectral lines in the spectrum of δ Cephei shift back and forth with the same 5.4-day period as that of the magnitude variations. From the Doppler effect, we can translate these wavelength shifts into radial velocities and draw a velocity curve (Figure 19-18b). Negative velocities mean that the star's surface is expanding toward us; positive velocities mean that the star's surface is

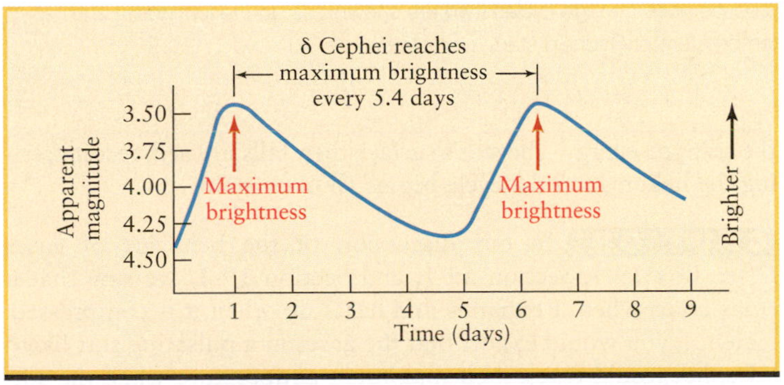

(a) The light curve of δ Cephei (a graph of brightness versus time)

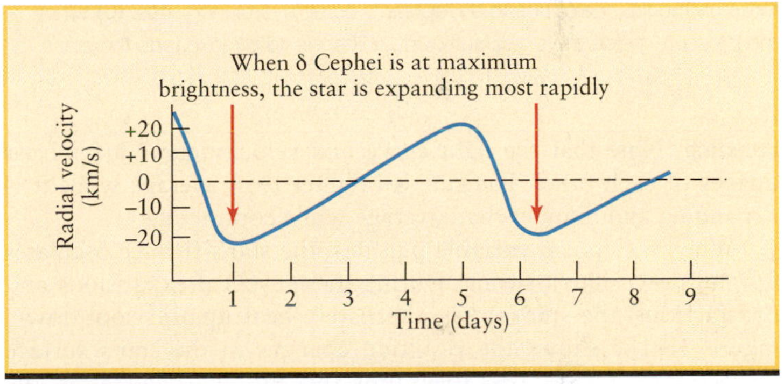

(b) Radial velocity versus time for δ Cephei (positive: star is contracting; negative: star is expanding)

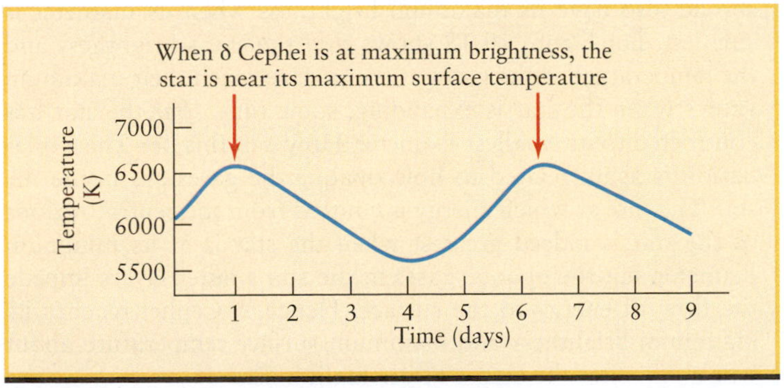

(c) Surface temperature versus time for δ Cephei

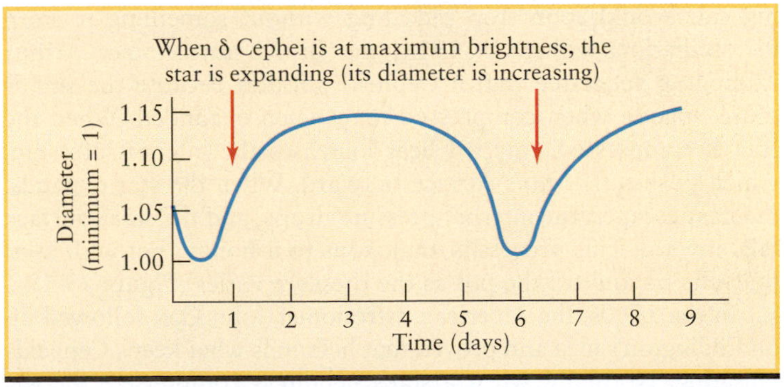

(d) Diameter versus time for δ Cephei

FIGURE 19-18

δ Cephei—A Pulsating Star **(a)** As δ Cephei pulsates, it brightens quickly (the light curve moves upward sharply) but fades more slowly (the curve declines more gently). The increases and decreases in brightness are nearly in step with variations in **(b)**, the star's radial velocity (positive when the star contracts and the surface moves away from us, negative when the star expands and the surface approaches us), as well as in **(c)**, the star's surface temperature. **(d)** The star is still expanding when it is at its brightest and hottest (compare with parts a and b).

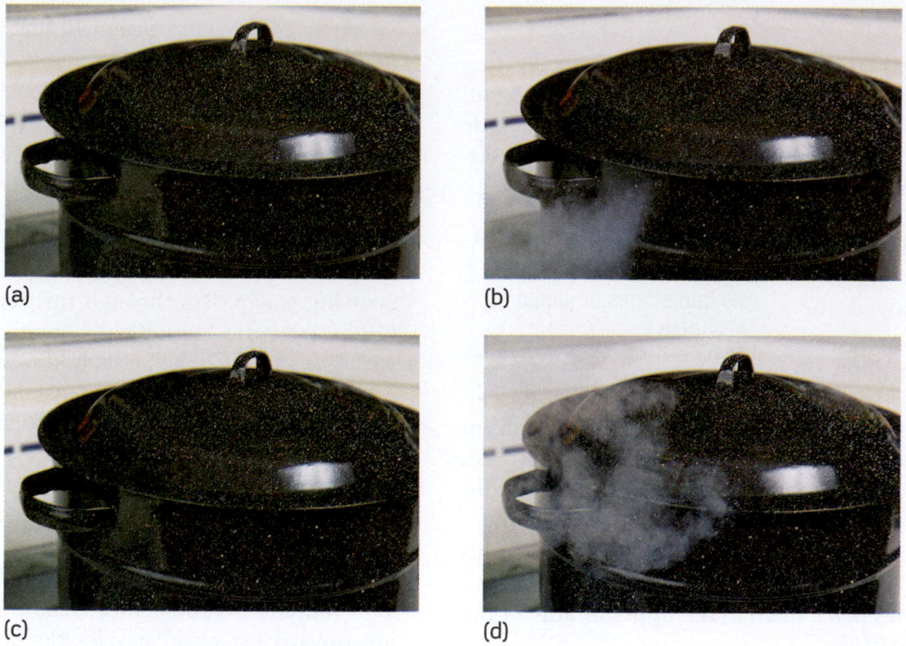

FIGURE 19-19 R I V U X G
Analogy for Cepheid Variability **(a)** As pressure builds up in this pot, the force on the lid (analogous to a Cepheid's outer layers) increases. **(b)** When the pressure inside the pot is sufficient, it lifts the lid off (expands the star's outer layers) and thereby allows some of the energy inside to escape. This process goes through cycles (two are shown), as do the luminosity and temperature of Cepheid stars. (Janet Horton)

receding. Note that the light curve and velocity curve are mirror images of each other. The star is brighter than average while it is expanding and dimmer than average while contracting.

When a Cepheid variable pulsates, the star's surface oscillates up and down like a spring. During these cyclical expansions and contractions, the star's gases alternately heat up and cool down. Figure 19-18c shows the resulting changes in the star's surface temperature. Figure 19-18d graphs the periodic changes in the star's diameter.

Just as a bouncing ball eventually comes to rest, a pulsating star would soon stop pulsating without something to keep its oscillations going. In 1914, the British astronomer Arthur Eddington suggested that a Cepheid pulsates because the star is more opaque when compressed than when expanded. When the star is compressed, trapped heat increases the internal pressure, which pushes the star's surface outward. When the star expands, the heat escapes, the internal pressure drops, and the star's surface falls inward. This process is analogous to a boiling pot with a lid that lifts on and off the pot as the pressure varies (Figure 19-19).

In the 1960s, the American astronomer John Cox followed up on Eddington's idea and proved that helium is what keeps Cepheids pulsating. Normally, when a star's helium is compressed, the gas increases in temperature and becomes more transparent. But in certain layers near the star's surface, compression may ionize helium (remove one of its electrons) instead of raising its temperature. Ionized helium gas is quite opaque, so these layers effectively trap heat and make the star expand, as Eddington suggested. This expansion cools the outer layers and makes the helium ions recombine with electrons, which makes the gas more transparent and releases the trapped energy. The star's surface then falls inward, recompressing the helium, and the cycle begins all over again.

CAUTION! In our discussion of the behavior of gases (see Box 19-1, Section 19-1, and Section 19-3) we saw that a gas cools when it expands and heats up when it is compressed. Hence, you would expect that the gases in a pulsating star like δ Cephei would reach their maximum temperature when the star is at its smallest diameter, so that the gases are most compressed. The hotter the gas, the more brightly it glows, so δ Cephei should also have its maximum brightness when its diameter is smallest. But Figure 19-18 shows that the star's brightness and the temperature of the gases at the surface reach their maximum values when the star is expanding, some time *after* the star has contracted to its smallest diameter. How can this be? The explanation is again related to how opaque the gases are inside the star. The rate at which energy is emitted from the central regions of the star is indeed greatest when the star is at its minimum diameter, but the opaque gases in the star's outer layers impede the flow of energy to the surface. Hence, δ Cephei reaches its maximum brightness and maximum surface temperature about half a day after the star is at its smallest size.

CONCEPTCHECK 19-13

What accounts for most of the brightening of a Cepheid variable: an increase in temperature, or an increase in size? (Consult Figure 19-18.)

Answer appears at the end of the chapter.

Determining the Distance of a Cepheid

Cepheid variables are important because they allow astronomers to determine the distances to these stars. This method works so well that it can even be used to measure the distances to other galaxies containing Cepheids. Why is this so important? As we will see in Chapter 23, mapping out distances to other galaxies has lead to profound discoveries about the very structure and nature of the universe.

Determining the distance to a Cepheid is based on two properties. First, Cepheids can be seen even at distances of millions of parsecs, because they are very luminous, ranging from a few hundred times solar luminosity to more than 10^4 $L_\odot$. Second, there is a direct relationship between a Cepheid's period and its average luminosity: The dimmest Cepheid variables pulsate rapidly, with periods of 1 to 2 days, while the most luminous Cepheids pulsate with much slower periods of about 100 days.

Figure 19-20 shows this **period-luminosity relation,** discovered by the American astronomer Henrietta Swan Leavitt, a "computer" at the observatory at Harvard College. Harvard did not let women use its telescope in the early 1900s, but in 1912 Leavitt analyzed photographic plates containing thousands of nearby Cepheids to establish a relationship between their period and luminosity. By first measuring the period of a distant Cepheid's brightness variations, and then using a graph like Figure 19-20, an astronomer can determine the star's intrinsic luminosity. By also measuring the star's apparent brightness, the distance to the Cepheid can then be found by using the inverse-square law (see Section 17-2). By applying the period-luminosity relation in this way to Cepheids in other galaxies, astronomers have been able to calculate the distances to those galaxies with great accuracy. (Box 17-2 gives an example of such a calculation.) As we will see in Chapters 23 and 25, such measurements play an important role in determining the overall size and structure of the universe.

The evolutionary tracks of mature, high-mass stars pass back and forth through the upper end of the instability strip on the H-R diagram. These stars become Cepheids when helium ionization occurs at just the right depth to drive the pulsations. For stars on the high-temperature (left) side of the instability strip, helium ionization occurs too close to the surface and involves only an insignificant fraction of the star's mass. For stars on the cool (right) side of the instability strip, convection in the star's outer layers prevents the storage of the energy needed to drive the pulsations. Thus, Cepheids exist only in a narrow temperature range on the H-R diagram.

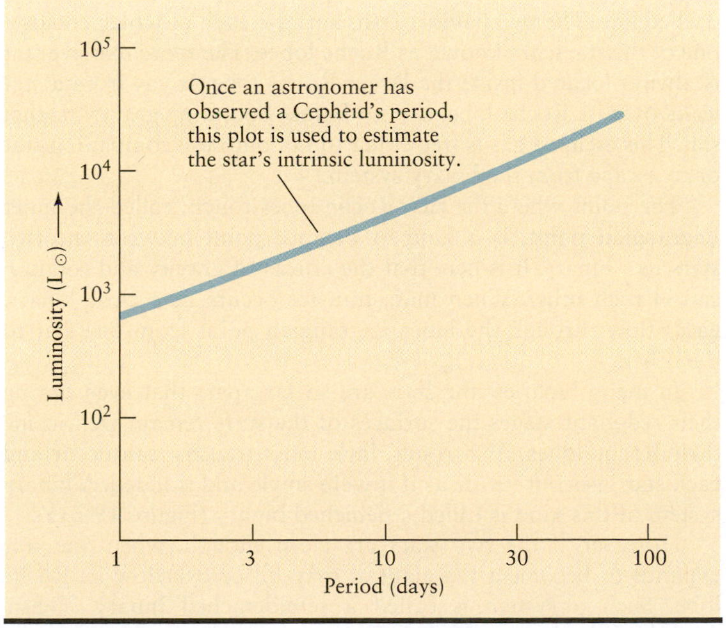
10^5
10^4
10^3
10^2
Luminosity ($L_\odot$) ⟶
Once an astronomer has observed a Cepheid's period, this plot is used to estimate the star's intrinsic luminosity.
1
3
10
30
100
Period (days)

FIGURE 19-20
Period-Luminosity Relation for Cepheids The greater the average luminosity of a Cepheid variable, the longer its period and the slower its pulsations. (Adapted from H. C. Arp)

RR Lyrae Variables

Stars of lower mass do not become Cepheids. Instead, after leaving the main sequence, becoming red giants, and undergoing the helium flash, their evolutionary tracks pass through the lower end of the instability strip as they move along the horizontal branch. Some of these stars become **RR Lyrae variables,** named for their prototype in the constellation Lyra (the Harp). RR Lyrae variables all have periods shorter than one day and roughly the same average luminosity as horizontal-branch stars, about 100 $L_\odot$. In fact, the RR Lyrae region of the instability strip (see Figure 19-17) is actually a segment of the horizontal branch. RR Lyrae stars are all metal-poor, Population II stars. Many have been found in globular clusters, and they have been used to determine the distances to those clusters in the same way that Cepheids are used to find the distances to other galaxies. In Chapter 22 we will see how RR Lyrae stars helped astronomers determine the size of the Milky Way Galaxy.

In some cases the expansion speed of a pulsating star exceeds the star's escape speed. When this happens, the star's outer layers are ejected completely. We will see in Chapter 20 that dying stars eject significant amounts of mass in this way, renewing and enriching the interstellar medium for future generations of stars.

CONCEPTCHECK 19-14

If a Cepheid variable star was observed to vary in brightness with a period of 30 days, how would you use Figure 19-20 to determine the distance to this star?

Answer appears at the end of the chapter.

19-7 Mass transfer can affect the evolution of stars in a close binary system

We have outlined what happens when a main-sequence star evolves into a red giant. What we have ignored is that more than half of all stars are members of multiple-star systems, including binaries. If the stars in such a system are widely separated, the individual stars follow the same course of evolution as if they were isolated. In a **close binary,** however, when one star expands to become a red giant, its outer layers can be gravitationally captured by the nearby companion star. In other words, a bloated red

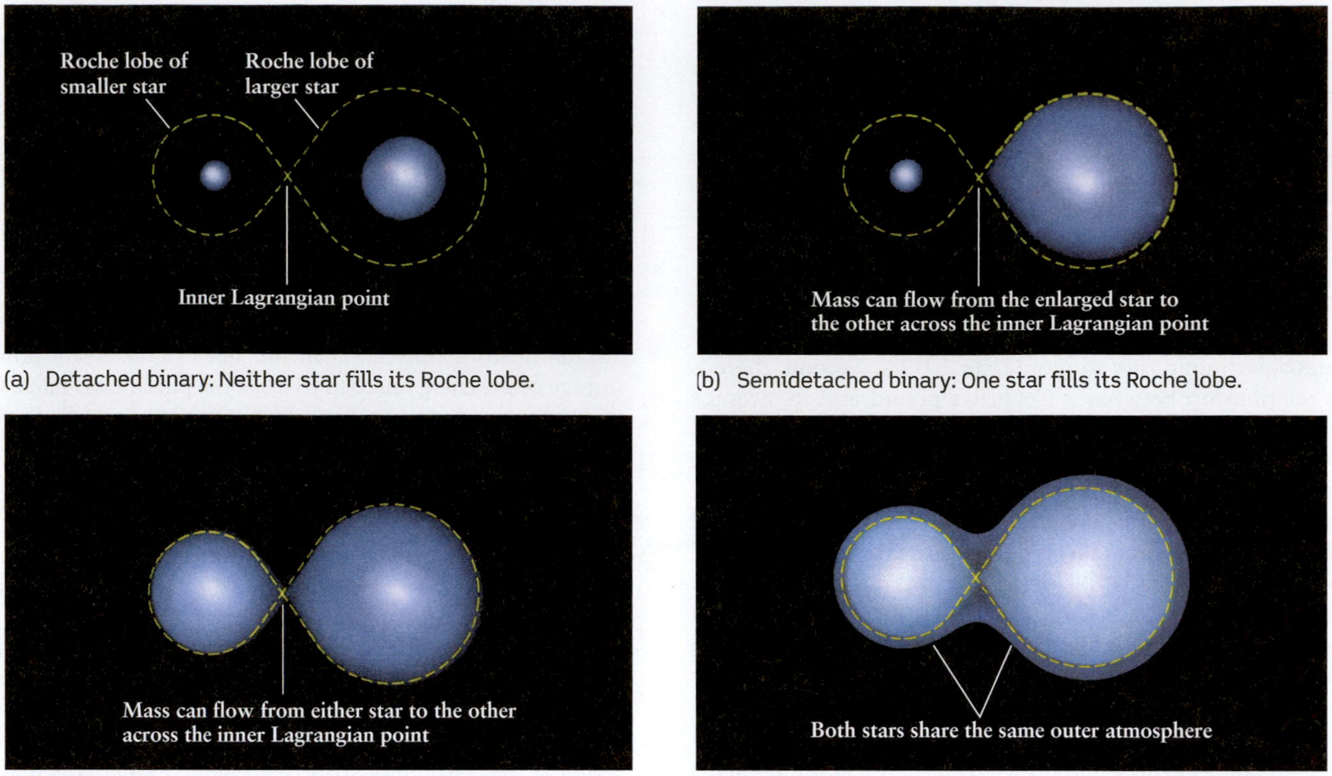

(a) Detached binary: Neither star fills its Roche lobe.

(b) Semidetached binary: One star fills its Roche lobe.

(c) Contact binary: Both stars fill their Roche lobes.

(d) Overcontact binary: Both stars overfill their Roche lobes.

FIGURE 19-21
Close Binary Star Systems The gravitational domain of a star in a close binary system is called its Roche lobe. The two Roche lobes meet at the inner Lagrangian point. The sizes of the stars relative to their Roche lobes determine whether the system is **(a)** a detached binary, **(b)** a semidetached binary, **(c)** a contact binary, or **(d)** an overcontact binary.

giant in a close binary system can dump gas onto its companion, a process called **mass transfer.**

Roche Lobes and Lagrangian Points

Our modern understanding of mass transfer in close binaries is based on the work of the French mathematician Edouard Roche. In the mid-1800s, Roche studied how rotation and mutual tidal interaction affect the stars in a binary system. Tidal forces cause the two stars in a close binary to keep the same sides facing each other, just as our Moon keeps its same side facing Earth (see Section 4-8). But because stars are gaseous, not solid, rotation and tidal forces can have significant effects on their shapes.

> If the stars in a binary system are sufficiently close, tidal forces can pull gases off one star and onto the other

In widely separated binaries, the stars are so far apart that tidal effects are small, and, therefore, the stars are nearly perfect spheres. In close binaries, where the separation between the stars is not much greater than their sizes, tidal effects are strong, causing the stars to be somewhat egg-shaped.

Roche discovered a mathematical surface that marks the gravitational domain of each star in a close binary. (This surface is not a real physical one, like the surface of a balloon, but a mathematical construct.) Figure 19-21a shows the outline of this surface as a dashed line. The two halves of this surface, each of which encloses one of the stars, are known as **Roche lobes.** The more massive star is always located inside the larger Roche lobe. If gas from a star leaks over its Roche lobe, it is no longer bound by gravity to that star. This escaped gas is free either to fall onto the companion star or to escape from the binary system.

The point where the two Roche lobes touch, called the **inner Lagrangian point,** is a kind of balance point between the two stars in a binary. It is here that the effects of gravity and rotation cancel each other. When mass transfer occurs in a close binary, gases flow through the inner Lagrangian point from one star to the other.

In many binaries, the stars are so far apart that even during their red-giant stages the surfaces of the stars remain well inside their Roche lobes. As a result, little mass transfer can occur and each star lives out its life as if it were single and isolated. A binary system of this kind is called a **detached binary** (Figure 19-21a).

However, if the two stars are close enough, when one star expands to become a red giant, it may fill or overflow its Roche lobe. Such a system is called a **semidetached binary** (Figure 19-21b). If both stars fill their Roche lobes, the two stars actually touch and the system is called a **contact binary** (Figure 19-21c). It is quite unlikely, however, that both stars exactly fill their Roche lobes at the same time. (This would only be the case if the two stars had identical masses, so that they both evolved at exactly the

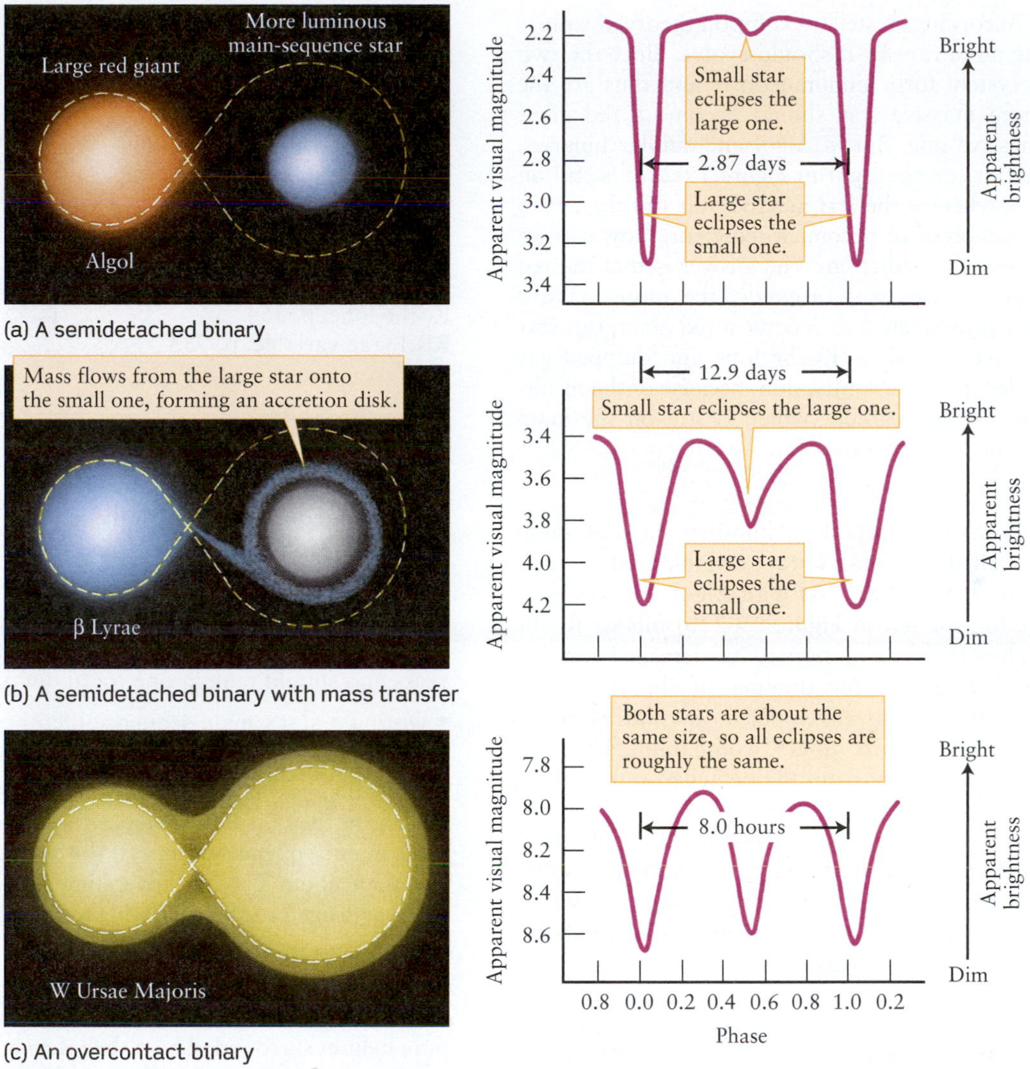

FIGURE 19-22

Three Eclipsing Binaries Compare the light curves of these three eclipsing binaries to those in Figure 17-24. **(a)** Algol is a semidetached binary. The deep eclipse occurs when the large red giant star blocks the light from the smaller but more luminous main-sequence star. (Compare with Figure 17-24b.) **(b)** β Lyrae's light curve is also at its lowest when the larger star completely eclipses the smaller one. Half an orbital period later, the smaller star partially eclipses the larger one, making a shallower dip in the light curve. **(c)** W Ursae Majoris is an overcontact binary in which both stars overfill their Roche lobes. The extremely short period of this binary indicates that the two stars are very close to each other.

same rate.) It is more likely that they overflow their Roche lobes, giving rise to a common envelope of gas. Such a system is called an **overcontact binary** (Figure 19-21d).

CONCEPTCHECK 19-15

What happens if the mass from one star enters the Roche lobe of another star?

Answer appears at the end of the chapter.

Observations of Mass Transfer

The binary star system Algol (from an Arabic term for "demon") provided the first clear evidence of mass transfer in close binaries. Also called β (beta) Persei, Algol can easily be seen with the naked eye. Ancient astronomers knew that Algol varies periodically in brightness by a factor of more than 2. In 1782, John Goodricke (the discoverer of δ Cephei's variability) first suggested that these brightness variations take place because Algol is an *eclipsing* binary. (We discussed this type of binary in Section 17-11.) The orbital plane of the two stars that make up the binary system happens to be nearly edge-on to our line of sight, so one star periodically eclipses the other. Algol's light curve (Figure 19-22a) and spectrum show that Goodricke's brilliant hypothesis is correct, and that Algol is a semidetached binary. The detached star (on the right in Figure 19-22a) is a luminous blue main-sequence star, while its less massive companion is a dimmer red giant that fills its Roche lobe.

CAUTION! According to stellar evolution theory, the more massive a star, the more rapidly it should evolve. Since the two stars in a binary system form simultaneously and thus are the same age, the more massive star should become a red giant before the less massive one. But in Algol and similar binaries, the *more* massive star (on the right in Figure 19-22a) is still on the main sequence, whereas the *less* massive star (on the left in Figure 19-22a) has evolved to become a red giant. How can we explain this apparent contradiction? The answer is that the red giant in Algol-type binaries was *originally* the more massive star. As it left the main sequence to become a red giant, this star expanded until it overflowed its Roche lobe and dumped gas onto its originally less massive companion. Because of the resulting mass transfer, that companion (which is still on the main sequence) became the more massive star.

Mass transfer is also important in another class of semidetached binaries, called β (beta) Lyrae variables, after their prototype in the constellation Lyra. As with Algol, the less massive star in β Lyrae (on the left in Figure 19-22b) fills its Roche lobe. Unlike Algol, however, the more massive detached star (on the right in Figure 19-22b) is the dimmer of the two stars. Apparently, this detached star is enveloped in a rotating *accretion disk* of gas captured from its bloated companion. This disk partially blocks the light coming from the detached star, making it appear dimmer.

What is the fate of an Algol or β Lyrae system? If the detached star is massive enough, it will evolve rapidly, expanding to also fill its Roche lobe. The result will be an overcontact binary in which the two stars share the gases of their outer layers. Such binaries are sometimes called W Ursae Majoris stars, after the prototype of this class (Figure 19-22c).

Mass transfer can also continue even after nuclear reactions cease in one of the stars in a close binary. In Chapters 20 and 21 we will see how mass transfer onto different types of dead stars called *white dwarfs, neutron stars,* and *black holes* can produce some of the most unusual and dramatic objects in the sky.

KEY WORDS

alpha particle, p. 541
Cepheid variable, p. 550
close binary, p. 553
color-magnitude diagram, p. 547
contact binary, p. 554
core helium fusion, p. 540
core hydrogen fusion, p. 534
degeneracy, p. 542
degenerate-electron pressure, p. 542
detached binary, p. 554
globular cluster, p. 547
helium flash, p. 540
helium fusion, p. 540
horizontal-branch star, p. 548
ideal gas, p. 540
inner Lagrangian point, p. 554
instability strip, p. 551
long-period variable, p. 550
main-sequence lifetime, p. 534
mass loss, p. 538
mass transfer, p. 554
metal-poor star, p. 549
metal-rich star, p. 549
open cluster, p. 546
overcontact binary, p. 555
Pauli exclusion principle, p. 542
period-luminosity relation, p. 553
Population I and Population II stars, p. 549
pulsating variable star, p. 550
red dwarf, p. 535
red giant, p. 538
Roche lobe, p. 554
RR Lyrae variable, p. 553
semidetached binary, p. 554
shell hydrogen fusion, p. 538
triple alpha process, p. 540
turnoff point, p. 548
zero-age main sequence (ZAMS), p. 543
zero-age main-sequence star, p. 534

KEY IDEAS

The Main-Sequence Lifetime: The duration of a star's main-sequence lifetime depends on the amount of hydrogen available to be consumed in the star's core and the rate at which this hydrogen is consumed.

- The more massive a star, the shorter is its main-sequence lifetime. The Sun has been a main-sequence star for about 4.56 billion years and should remain one for about another 7 billion years.
- During a star's main-sequence lifetime, the star expands somewhat and undergoes a modest increase in luminosity.
- If a star's mass is greater than about 0.4 $M_\odot$, only the hydrogen present in the core can undergo thermonuclear fusion during the star's main-sequence lifetime. If the star is a red dwarf with a mass less than about 0.4 $M_\odot$, over time convection brings all of the star's hydrogen to the core where it can undergo fusion.

Becoming a Red Giant: Core hydrogen fusion ceases when the hydrogen has been exhausted in the core of a main-sequence star with mass greater than about 0.4 $M_\odot$. This leaves a core of nearly pure helium surrounded by a shell through which hydrogen fusion works its way outward in the star. The core shrinks and becomes hotter, while the star's outer layers expand and cool. The result is a red giant star.

- As a star becomes a red giant, its evolutionary track moves rapidly from the main sequence to the red-giant region of the H-R diagram. The more massive the star, the more rapidly this evolution takes place.

Helium Fusion: When the central temperature of a red giant reaches about 100 million K, helium fusion begins in the core. This process, also called the triple alpha process, converts helium to carbon and oxygen.

- In a more massive red giant, helium fusion begins gradually; in a less massive red giant, it begins suddenly, in a process called the helium flash.
- After the helium flash, a low-mass star moves quickly from the red-giant region of the H-R diagram to the horizontal branch.

Star Clusters and Stellar Populations: The age of a star cluster can be estimated by plotting its stars on an H-R diagram.

- The cluster's age is equal to the age of the main-sequence stars at the turnoff point (the upper end of the remaining main sequence).

• As a cluster ages, the main sequence is "eaten away" from the upper left as stars of progressively smaller mass evolve into red giants.

• Relatively young Population I stars are metal rich; ancient Population II stars are metal poor. The metals (heavy elements) in Population I stars were manufactured by thermonuclear reactions in an earlier generation of Population II stars, then ejected into space and incorporated into a later stellar generation.

Pulsating Variable Stars: When a star's evolutionary track carries it through a region in the H-R diagram called the instability strip, the star becomes unstable and begins to pulsate.

• Cepheid variables are high-mass pulsating variables. There is a direct relationship between their periods of pulsation and their luminosities.

• RR Lyrae variables are low-mass, metal-poor pulsating variables with short periods.

• Long-period variable stars also pulsate but in a fashion that is less well understood.

Close Binary Systems: Mass transfer in a close binary system occurs when one star in a close binary overflows its Roche lobe. Gas flowing from one star to the other passes across the inner Lagrangian point. This mass transfer can affect the evolutionary history of the stars that make up the binary system.

QUESTIONS

Review Questions

1. How does the chemical composition of the present-day Sun's core compare to the core's composition when the Sun formed? What caused the change?
2. Which regions of the Sun are denser today than they were a billion years ago? Which regions are less dense? What has caused these changes?
3. What is a red dwarf? How are thermonuclear reactions in the core of a red dwarf able to consume hydrogen from the star's outer layers?
4. Why do high-mass main-sequence stars have shorter lifetimes than those of lower mass?
5. On what grounds are astronomers able to say that the Sun has about 7×10^9 years remaining in its main-sequence stage?
6. What will happen inside the Sun 7 billion years from now, when it begins to mature into a red giant?
7. TUTORIAL 19-1 Explain why Earth is expected to become inhospitable to life long before the Sun becomes a red giant.
8. Explain how it is possible for the core of a red giant to contract at the same time that its outer layers expand.
9. Why does helium fusion require much higher temperatures than hydrogen fusion?
10. How is a degenerate gas different from ordinary gases?
11. What is the helium flash? Why does it happen in some stars but not in others?
12. Why does a star's luminosity decrease after helium fusion begins in its core?
13. TUTORIAL 19-2 What does it mean when an astronomer says that a star "moves" from one place to another on an H-R diagram?
14. Explain why the majority of the stars visible through telescopes are main-sequence stars.
15. On an H-R diagram, main-sequence stars do not lie along a single narrow line but are spread out over a band (see Figure 19-9b). On the basis of how stars evolve during their main-sequence lifetimes, explain why this should be so.
16. Explain how and why the turnoff point on the H-R diagram of a cluster is related to the cluster's age.
17. There is a good deal of evidence that our universe is about 13.7 billion years old. Explain why no main-sequence stars of spectral class M have yet evolved into red giant stars.
18. How do astronomers know that globular clusters are made of old stars?
19. Red giant stars appear more pronounced in composites of infrared images and visible-light images, like those in Figure 19-4b and Figure 19-12. Explain why.
20. The horizontal-branch stars in Figure 19-12 appear blue. (**a**) Explain why this is consistent with the color-magnitude diagram shown in Figure 19-13. (**b**) All horizontal-branch stars were once red giants. Explain what happened to these stars to change their color.
21. What is the difference between Population I and Population II stars? In what sense can the stars of one population be regarded as the "children" of the other population?
22. Both diamonds and graphite (the material used in pencils to make marks on paper) are crystalline forms of carbon. Most of the carbon atoms in these substances have nuclei with 6 protons and 6 neutrons (^{12}C). Where did these nuclei come from?
23. Why do astronomers attribute the observed Doppler shifts of a Cepheid variable to pulsation, rather than to some other cause, such as orbital motion?
24. Why do Cepheid stars pulsate? Why are these stars important to astronomers who study galaxies beyond the Milky Way?
25. What is a Roche lobe? What is the inner Lagrangian point? Why are Roche lobes important in close binary star systems?
26. What is the difference between a detached binary, a semidetached binary, a contact binary, and an overcontact binary?
27. Massive main-sequence stars turn into red giants before less massive stars. Why, then, is the more massive star in Algol a main-sequence star and the less massive star a red giant?

Advanced Questions

Questions preceded by an asterisk (*) involve topics discussed in Box 7-2, the Boxes in Chapter 17, or the Boxes in this chapter.

Problem-solving tips and tools

Recall from Section 16-1 that 6×10^{11} kg of hydrogen is converted into helium each second at the Sun's center. Recall also that you must use absolute (Kelvin) temperatures when using the Stefan-Boltzmann law. You may find it helpful to review the discussion of apparent magnitude, absolute magnitude, and luminosity in Box 17-3. Section 17-4 discusses the connection between the surface temperatures and colors of stars. Newton's form of Kepler's third law (see Section 17-9) describes the orbits of stars in binary systems. Box 7-2 gives the formula for escape speed and for the average speeds of gas molecules.

28. The radius of the Sun has increased over the past several billion years. Over the same time period, the size of the Moon's orbit around Earth has also increased. A few billion years ago, were annular eclipses of the Sun (see Figure 3-12) more or less common than they are today? Explain your answer.

29. The Sun has increased in radius by 6% over the past 4.56 billion years. Its present-day radius is 696,000 km. What was its radius 4.56 billion years ago? (*Hint:* The answer is *not* 654,000 km.)

***30.** Calculate the escape speed from (**a**) the surface of the present-day Sun and (**b**) the surface of the Sun when it becomes a red giant, with essentially the same mass as today but with a radius that is 100 times larger. (**c**) Explain how your results show that a red giant star can lose mass more easily than a main-sequence star.

***31.** Calculate the average speed of a hydrogen atom (mass 1.67×10^{-27} kg) (**a**) in the atmosphere of the present-day Sun, with temperature 5800 K, and (**b**) in the atmosphere of a 1-$M_\odot$ red giant, with temperature 3500 K. (**c**) Compare your results with the escape speeds that you calculated in Question 30. Use this comparison to discuss how well the present-day Sun and a 1-$M_\odot$ red giant can retain hydrogen in their atmospheres.

32. Use the value of the Sun's luminosity (3.90×10^{26} watts, or 3.90×10^{26} joules per second) to calculate what mass of hydrogen the Sun will convert into helium during its entire main-sequence lifetime of 1.2×10^{10} years. (Assume that the Sun's luminosity remains nearly constant during the entire 1.2×10^{10} years.) What fraction does this represent of the total mass of hydrogen that was originally in the Sun?

33. (**a**) The main-sequence stars Sirius (spectral type A1), Vega (A0), Spica (B1), Fomalhaut (A3), and Regulus (B7) are among the 20 brightest stars in the sky. Explain how you can tell that all these stars are younger than the Sun. (**b**) The third-brightest star in the sky, although it can be seen only south of 29° north latitude, is α (alpha) Centauri A. It is a main-sequence star of spectral type G2, the same as the Sun. Can you tell from this whether α Centauri A is younger than the Sun, the same age, or older? Explain your reasoning.

34. Using the same horizontal and vertical scales as in Figure 19-9a, make points on an H-R diagram for each of the stars listed in Table 19-1. Label each point with the star's mass and its main-sequence lifetime. Which of these stars will remain on the main sequence after 10^9 years? After 10^{11} years?

***35.** Explain why the quantity f in Box 19-2 has a different value for stars with masses less than 0.4 $M_\odot$ than for stars with masses greater than 0.4 $M_\odot$. In which case does f have a greater value?

***36.** Calculate the main-sequence lifetimes of (**a**) a 9-$M_\odot$ star and (**b**) a 0.25-$M_\odot$ star. Compare these lifetimes with that of the Sun.

***37.** The earliest fossil records indicate that life appeared on Earth about a billion years after the formation of the solar system. What is the most mass that a star could have in order that its lifetime on the main sequence is long enough to permit life to form on one or more of its planets? Assume that the evolutionary processes would be similar to those that occurred on Earth.

38. As a red giant, the Sun's luminosity will be about 2000 times greater than it is now, so the amount of solar energy falling on Earth will increase to 2000 times its present-day value. Hence, to maintain thermal equilibrium, each square meter of Earth's surface will have to radiate 2000 times as much energy into space as it does now. Use the Stefan-Boltzmann law to determine what Earth's surface temperature will be under these conditions. (*Hint:* The present-day Earth has an average surface temperature of 14°C.)

39. When the Sun becomes a red giant, its luminosity will be about 2000 times greater than it is today. Assuming that this luminosity is caused *only* by fusion of the Sun's remaining hydrogen, calculate how long our star will be a red giant. (In fact, only a fraction of the remaining hydrogen will be consumed, and the luminosity will vary over time as shown in Figure 19-8.)

40. What observations would you make of a star to determine whether its primary source of energy is hydrogen fusion or helium fusion?

41. The star whose spectrum is shown in Figure 19-15a has a lower percentage of heavy elements than the Sun, whose spectrum is shown in Figure 19-15b. Hence, the star in Figure 19-15a has a higher percentage of hydrogen. Why, then, isn't the H_δ absorption line of hydrogen noticeably darker for the star in Figure 19-15a?

42. Would you expect the color of a Cepheid variable star (consider Figure 19-18) to change during the star's oscillation period? If not, why not? If so, describe why the color should change, and describe the color changes you would expect to see during an oscillation period.

43. The brightness of a certain Cepheid variable star changes from maximum brightness to minimum brightness in 26 days. In 4 more days, the star returns to maximum brightness. (**a**) What is this star's period? (**b**) What is this star's approximate luminosity?

44. The star X Arietis is an RR Lyrae variable. Its apparent brightness varies between 2.0×10^{-15} and 4.9×10^{-15} that of the Sun with a period of 0.65 day. Interstellar extinction dims the star by 37%. Approximately how far away is the star?

45. The apparent brightness of δ Cephei (a Cepheid variable star) varies with a period of 5.4 days. Its average apparent brightness is 5.1×10^{-13} that of the Sun. Approximately how far away is δ Cephei? (Ignore interstellar extinction.)

46. Suppose you find a binary star system in which the more massive star is a red giant and the less massive star is a main-sequence star. Would you expect that mass transfer between the stars has played an important role in the evolution of these stars? Explain your reasoning.

47. The larger star in the Algol binary system (see Figure 19-22a) is of spectral class K, while the smaller star is of spectral class B. Discuss how the color of Algol changes as seen through a small telescope (through which Algol appears as a single star). What is the color during a deep eclipse, when the large star eclipses the small one? What is the color when the small star eclipses the large one?

48. Suppose the detached star in β Lyrae (Figure 19-22b) did not have an accretion disk. Would the deeper dips in the light curve be deeper, shallower, or about the same? What about the shallower dip? Explain your answers.

49. The two stars that make up the overcontact binary W Ursae Majoris (Figure 19-22c) have estimated masses of 0.99 $M_\odot$ and 0.62 $M_\odot$. (**a**) Find the average separation between the two stars. Give your answer in kilometers. (**b**) The radii of the two stars are estimated to be 1.14 $R_\odot$ and 0.83 $R_\odot$. Show that these values and your result in part (a) are consistent with the statement that this is an overcontact binary.

50. The stars that make up the binary system W Ursae Majoris (see Figure 19-22c) have particularly strong magnetic fields. Explain how astronomers could have discovered this. (*Hint:* See Section 16-9.)

51. Consult recent issues of *Sky & Telescope* and *Astronomy* to find out when Mira will next reach maximum brightness. Look up the star's location in the sky using the *Starry Night*™ program if you have access. (Use the **Find...** command in the **Edit** menu to search for Omicron Ceti.) Why is it unlikely that you will be able to observe Mira at maximum brightness?

Discussion Questions

52. Eventually the Sun's luminosity will increase to the point where Earth can no longer sustain life. Discuss what measures a future civilization might take to preserve itself from such a calamity.

53. The half-life of the ^{8}Be nucleus, 2.6×10^{-16} second, is the average time that elapses before this unstable nucleus decays into two alpha particles. How would the universe be different if instead the ^{8}Be half-life were zero? How would the universe be different if the ^{8}Be nucleus were stable and did not decay?

54. Discuss how H-R diagrams of star clusters could be used to set limits on the age of the universe. Could they be used to set lower limits on the age? Could they be used to set upper limits? Explain your reasoning.

Web/eBook Questions

55. Suppose that an oxygen nucleus (^{16}O) were fused with a helium nucleus (^{4}He). What element would be formed? Look up the relative abundance of this element in, for example, the *Handbook of Chemistry and Physics* or on the World Wide Web. Based on the abundance, comment on whether such a process is likely. (*Hint:* See Figure 8-4.)

56. Although Polaris, the North Star, is a Cepheid variable, it pulsates in a somewhat different way than other Cepheids. Search the World Wide Web for information about this star's pulsations and how they have been measured by astronomers at the U.S. Naval Observatory. How does Polaris pulsate? How does this differ from other Cepheids?

57. ANIMATION 19-1 **Observing Stellar Evolution.** Step through the animation "The Hertzsprung-Russell Diagram and Stellar Evolution" in Chapter 19 of the *Universe* Web site or eBook. Use this animation to answer the following questions. (**a**) How does a 1-$M_\odot$ star move on the H-R diagram during its first 4.56 billion (4560 million) years of existence? Compare this with the discussion in Section 19-1 of how the Sun has evolved over the past 4.56 billion years. (**b**) What is the zero-age spectral class of a 2-$M_\odot$ star? At what age does such a star evolve into a red giant of spectral class K? (**c**) What is the approximate zero-age luminosity of a 1.3-$M_\odot$ star? What is its approximate luminosity when it becomes a red giant? (**d**) Suppose a star cluster has no main-sequence stars of spectral classes O or B. What is the approximate age of the cluster? (**e**) Approximately how long do the most massive stars of spectral class B live before leaving the main sequence? What about the most massive stars of spectral class F?

ACTIVITIES

Observing Projects

Observing tips and tools

An excellent resource for learning how to observe variable stars is the Web site of the American Association of Variable Star Observers. A wealth of data about specific variable stars can be found on the *Sky & Telescope* Web site and in the three volumes of *Burnham's Celestial Handbook: An Observer's Guide to the Universe Beyond the Solar System* (Dover, 1978). This book also provides useful information about observing star clusters.

58. Use *Starry Night*™ to demonstrate your knowledge of the evolution of a main sequence star. Open **Favourites > Explorations > Stellar Evolution.** The view shows the sky as seen from the center of a transparent Earth with the constellations displayed and five labeled stars. Use the **File** menu (Windows) or **Starry Night** menu (Mac) to access **Preferences…**, open the dropdown menu and set the **Cursor Tracking** (**HUD**) options to show **Radius, Luminosity,** and **Spectral class.** Open the **Status** pane and expand the **Hertzsprung-Russell** layer to show the H-R diagram for all of the stars within this view. Now position the cursor over each of the labeled stars and form a table of values of the above parameters for each star. Using this information and the position of the star in the H-R diagram (appearing as a red dot in the diagram when the cursor is over the star), sort and list the labeled stars in increasing evolutionary age.

59. Use the *Starry Night*™ program to look for signs of stellar evolution in M101, the Pinwheel Galaxy. Select **Favourites > Explorations > Atlas** and open the **Find** pane. Type **M101** in the search box and then right-click (Ctrl-click on Mac) on M101 in the list and select **Magnify** from the contextual menu. (a) What is the color of the central part of this Galaxy? (b) What is the color of the outer regions of this galaxy? (c) Based on your observations, what type of stars would you expect to predominate in each of these two regions of this galaxy?

Collaborative Exercises

60. The inverse relationship between a star's mass and its main-sequence lifetime is sometimes likened to automobiles in that the more massive vehicles, such as commercial semi–tractor-trailer trucks, need to consume significantly more fuel to travel at highway speeds than more lightweight and economical vehicles. As a group, create a table called "Maximum Vehicle Driving Distances," much like Table 19-1, "Main-Sequence Lifetimes," by making estimates for any five vehicles of your groups' choosing. The table's column headings should be (1) vehicle make and model; (2) estimated gas tank size; (3) cost to fill tank; (4) estimated mileage (in miles per gallon); and (5) number of miles driven on a single fill-up.

61. Consider Figure 19-20, showing the period-luminosity relation for Cepheids. If a certain Cepheid star has an average luminosity of 4310^{30} W, what is the variable star's approximate period?

62. Figure 19-18a shows a light curve of apparent magnitude versus time in days for δ Cephei—a pulsating star that reaches maximum brightness every 5.4 days. Create a new sketch of apparent magnitude versus time in days showing three different stars: (1) δ Cephei; (2) a slightly smaller pulsating star; and (3) a slightly larger pulsating star, all of which have about the same total change in apparent magnitude.

ANSWERS

ConceptChecks

ConceptCheck 19-1: To leave the main sequence, a red dwarf has to finish converting all of its hydrogen to helium. This takes hundreds of billions of years for a red dwarf—longer than the age of the universe.

ConceptCheck 19-2: When main-sequence hydrogen burning stops, core pressure decreases, and the core contracts. As the core contracts, gravitational energy is converted into heat, and the core's temperature increases well beyond the temperature it had during its main-sequence burning.

ConceptCheck 19-3: The higher temperatures in a red giant's core lead to an increased rate of shell hydrogen fusion. The additional energy released by this fusion produces a greater pressure outside the core than before, which greatly expands the star.

ConceptCheck 19-4: In a red giant, the core is at a significantly higher temperature and density that allows helium atoms to combine and fuse together, releasing energy.

ConceptCheck 19-5: Compared to when it was on the main sequence, a red giant is hotter in the core and cooler on the surface.

ConceptCheck 19-6: If fusion in a main-sequence star increases its energy output, the higher temperatures increase the internal pressure. This in turn expands the core's gas, lowers its temperature, and reduces the rate of nuclear reactions. The "safety valve" aspect is that increased nuclear reactions end up reducing further reactions so that the rate of reactions remains stable. When a red giant's core is supported by degenerate-electron pressure, an increase in temperature from an increase in nuclear reactions does not increase the pressure to expand the core, and does not reduce further reaction rates.

ConceptCheck 19-7: At high enough temperatures the core is no longer supported by degenerate-electron pressure and the core begins to expand. By expanding, the core cools somewhat, nuclear reaction rates slow down, and luminosity of the star decreases. This is shown on the right of Figure 19-8.

ConceptCheck 19-8: Yes. After 30 million years you can see in Figure 19-10e a few red giants just as low-mass stars are finally approaching the main sequence.

ConceptCheck 19-9: In Figure 19-10c, showing the cluster after 100,000 years, the most massive stars are just joining the main sequence.

ConceptCheck 19-10: The turnoff point in Figure 19-13 corresponds to $0.8\text{-}M_\odot$ stars that stay on the main sequence for about 12 billion years. If the turnoff point were instead at around 10,000 K, these stars would be more massive and spend *less* time on the main sequence, and the cluster would be younger.

ConceptCheck 19-11: Recall that lower-mass stars are less luminous. Therefore, the lower the luminosity of the turnoff point, the lower the mass of the main-sequence star at the turnoff point. Lower-mass stars remain on the main sequence longer, indicating an older cluster when these stars finally leave the main sequence.

ConceptCheck 19-12: Population I stars are enriched with heavier atomic elements (they are metal rich). The heavier elements can only be made by an earlier generation of stars through supernova explosions. The earlier generation lacks the heavier elements (they are metal poor) and are called Population II stars.

ConceptCheck 19-13: You can see that the Cepheid in Figure 19-18a peaks in luminosity on day 1. This is about when the temperature peaks as well in Figure 19-18c. However, when the star reaches its maximum size in Figure 19-18d, the luminosity is already declining. Therefore, surface temperature contributes more to the peak luminosity.

ConceptCheck 19-14: Figure 19-20 indicates that a variable star with a 30-day period has an intrinsic luminosity (L) about 10,000 times brighter than our Sun. Direct observations of the star also tell you its apparent brightness (b). Section 17-2 describes how these two quantities are directly related to the distance (d) of the star, so the distance is easily determined.

ConceptCheck 19-15: If a star expands enough that some of its mass enters the Roche lobe of its binary companion, the matter becomes gravitationally bound to the binary companion. Over time, this can transfer a significant amount of mass to the companion.

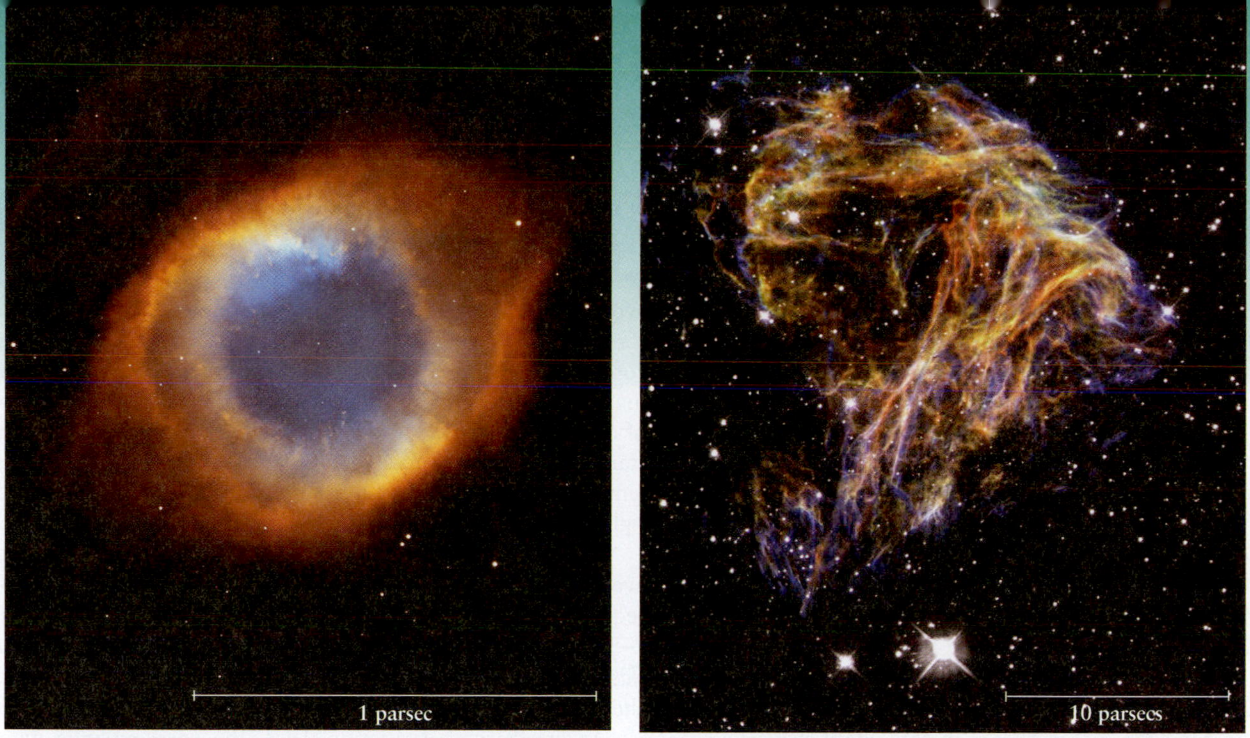

Left: The planetary nebula NGC 7293 (the Helix Nebula). Right: The supernova remnant LMC N49. (NASA, NOAO, ESA, the Hubble Helix Nebula Team, M. Meixner/STScI, and T. A. Rector/NRAO; NASA and the Hubble Heritage Team, STScI/AURA) R I V U X G

CHAPTER 20

Stellar Evolution: The Deaths of Stars

LEARNING GOALS

By reading the sections of this chapter, you will learn

20-1 What kinds of nuclear reactions occur inside a star of moderately low mass as it ages

20-2 How evolving stars disperse carbon into the interstellar medium

20-3 How stars of moderately low mass eventually die

20-4 The nature of white dwarfs and how they are formed

20-5 What kinds of reactions occur inside a high-mass star as it ages

20-6 How high-mass stars end with a supernova explosion

20-7 Why supernova SN 1987A was both important and unusual

20-8 What role neutrinos play in the death of a massive star

20-9 How white dwarfs in close binary systems can explode

20-10 What remains after a supernova explosion

20-11 How neutron stars and pulsars are related

20-12 How novae and X-ray bursts come from binary systems

When a star of 0.4 solar mass or more reaches the end of its main-sequence lifetime and becomes a red giant, it has a compressed core and a bloated atmosphere. Finally, the star devours its remaining nuclear fuel and begins to die. As we will learn in this chapter, the character of the star's death depends crucially on the value of its mass.

A star of relatively low mass—such as our own Sun—ends its evolution by gently expelling its outer layers into space. These ejected gases form a glowing cloud called a *planetary nebula* such as the one shown here in the left-hand image. The burned-out core that remains is called a *white dwarf.*

In contrast, a high-mass star ends its life in almost inconceivable violence. At the end of its short life, the core of such a star collapses suddenly, which triggers a powerful *supernova* explosion that can be as luminous as an entire galaxy of stars. A white dwarf, too, can become a supernova if it accretes gas from a companion star in a close binary system.

In supernovae, nuclear reactions produce a wide variety of heavy elements, which are ejected into the interstellar medium. (The supernova remnant shown above in the right-hand image is rich in these elements.) Such heavy elements are essential building blocks for terrestrial worlds like our Earth. Thus, the deaths of massive stars can provide the seeds for planets orbiting succeeding generations of stars.

20-1 Stars of between 0.4 and 4 solar masses go through two distinct red giant stages

TUTORIAL 20-1

All main-sequence stars convert hydrogen to helium in their cores in a series of energy-releasing nuclear reactions. As we saw in Section 19-1, convection within a low-mass main-sequence star—a so-called *red dwarf* with a mass between 0.08 and 0.4 $M_{\odot}$—will eventually bring all of the star's hydrogen into the core. Thus, over hundreds of billions of years a red dwarf will end up as a ball of helium. Convection is less important in main-sequence stars with masses greater than 0.4 $M_{\odot}$, so these stars are able to consume only the hydrogen that is present within the core. These stars of greater mass then leave the main sequence. Let's examine what happens next for a star of moderately low mass, between 0.4 and 4 $M_{\odot}$. One example for such a star is our own Sun, with a mass of 1 $M_{\odot}$. We will begin by reviewing what we learned in Chapter 19 about the first stages of post–main-sequence evolution for such a star. (Later in this chapter, we will study the evolution of more massive stars.)

Note: Technically, the nuclear reactions referred to above are **thermonuclear.** The "thermo" in thermonuclear indicates that high-speed collisions, due to very high temperatures, help fuse atoms together despite electric repulsion between protons. In this chapter, we will often refer to these simply as nuclear reactions.

The Red Giant and Horizontal-Branch Stages: A Review

We can describe a star's post–main-sequence evolution using an evolutionary track on a H-R diagram. Figure 20-1 shows the track for a 1-$M_{\odot}$ star like the Sun. Once core hydrogen fusion ceases, the core shrinks, heating the nearby surrounding hydrogen and triggering shell hydrogen fusion (see Section 19-2). The new outpouring of energy causes the star's outer layers to expand and cool, and the star becomes a red giant. As the luminosity increases and the surface temperature drops, the post–main-sequence star moves up and to the right along the **red giant branch** on an H-R diagram (Figure 20-1a).

Next, the helium-rich core of the star shrinks and heats until eventually **core helium fusion** begins. This second post–main-sequence stage begins gradually in stars more massive than about 2–3 $M_{\odot}$, but for less massive stars it comes suddenly—in a *helium flash*. During core helium fusion, the surrounding hydrogen-fusing shell still provides most of the red giant's luminosity.

As we learned in Section 19-3, the core expands when core helium fusion begins, which makes the core cool down a bit. (We saw in Box 19-1 that letting a gas expand tends to lower its temperature, while compressing a gas tends to increase its temperature.) The cooling of the core also cools the surrounding hydrogen-fusing shell, so that the shell releases energy more slowly. Hence, the luminosity goes down a bit after core helium fusion begins.

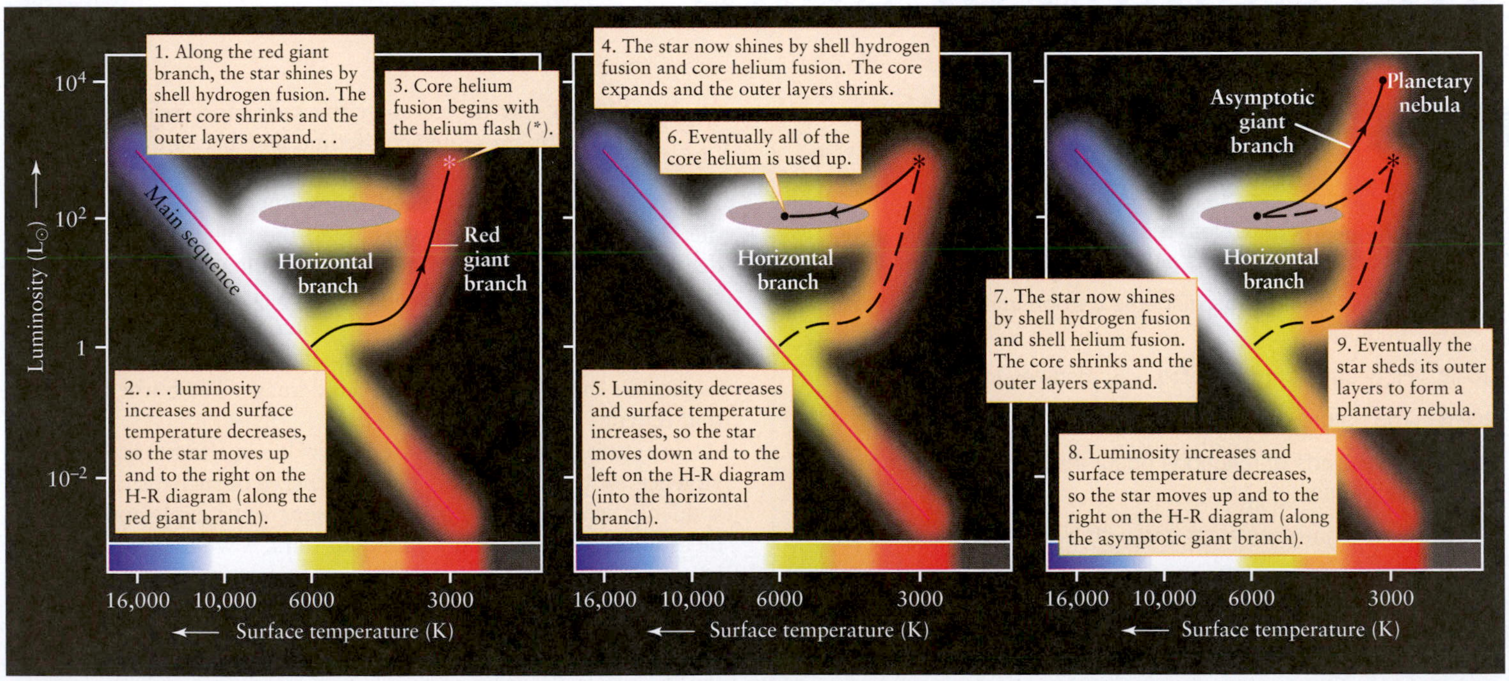

ANIMATION 20-1

FIGURE 20-1
The Post–Main-Sequence Evolution of a 1-$M_{\odot}$ Star These H-R diagrams show the evolutionary track of a star like the Sun as it goes through the stages of being **(a)** a red giant star, **(b)** a horizontal-branch star, and **(c)** an asymptotic giant branch (AGB) star. The star eventually evolves into a planetary nebula (described in Section 20-3).

The slower rate of energy release also lets the star's outer layers contract. As they contract, they heat up, so the star's surface temperature increases and its evolutionary track moves to the left on the H-R diagram in Figure 20-1b. The luminosity changes relatively little during this stage, so the evolutionary track moves almost horizontally, along a path called the **horizontal branch.** Horizontal-branch stars have helium-fusing cores surrounded by hydrogen-fusing shells. Figure 19-12 shows horizontal-branch stars in a globular cluster, and Figure 19-8 shows the evolution of the luminosity of a 1-$M_\odot$ star up to this point in its history. (Technically, a star ceases to be a red giant as soon as it begins core helium fusion even though it is still red and giant for a while; we did not make this distinction in Chapter 19.)

CAUTION! It is easy to get confused about the relationship between fusion inside and outside the core of a star versus the star's overall size, temperature, and luminosity. Compared to fusion in the core, shell fusion releases energy at a greater rate. This increased energy puffs up the star's outer layers. Shell fusion results in a larger star with a larger luminosity, even though its surface temperature decreases.

AGB Stars: The Second Red Giant Stage

Stars like the Sun go through a second red giant phase, during which helium fusion takes place in a shell around an inert core

Helium fusion produces nuclei of carbon and oxygen. After about a hundred million (10^8) years of core helium fusion, essentially all the helium in the core of a 1-$M_\odot$ star has been converted into carbon and oxygen, and the fusion of helium in the core ceases. (This corresponds to the right-hand end of the graph in Figure 19-8.) Without nuclear reactions to maintain the core's internal pressure, the core again contracts, until it is stopped by degenerate electron pressure (described in Section 19-3). This contraction releases heat into the surrounding helium-rich gases, and a new stage of helium fusion begins in a thin shell around the core. This process is called **shell helium fusion.**

History now repeats itself—the star enters a *second* red giant phase. A star first becomes a red giant at the end of its main-sequence lifetime, when the outpouring of energy from shell hydrogen fusion makes the star's outer layers expand and cool. In the same way, the outpouring of energy from shell *helium* fusion causes the outer layers to expand again. The low-mass star ascends into the red giant region of the H-R diagram for a second time (Figure 20-1c), but now with even greater luminosity than during its first red giant phase.

Stars in this second red giant phase are commonly called **asymptotic giant branch stars,** or **AGB stars,** and their evolutionary tracks follow what is called the **asymptotic giant branch.** (*Asymptotic* means "approaching"; the name means that a star on the asymptotic giant branch approaches the red giant branch from the left on an H-R diagram.)

When a low-mass star first becomes an AGB star, it consists of a carbon-oxygen core (supported by degenerate electron pressure) with no nuclear reactions, and a helium-fusing shell, both inside a hydrogen-fusing shell, all within a volume not much larger than Earth. This small, dense central region is surrounded by an enormous hydrogen-rich envelope about as big as Earth's orbit around the Sun. After a while, the expansion of the star's outer layers causes the hydrogen-fusing shell to also expand and cool, and nuclear reactions in this shell temporarily cease. This leaves the aging star's structure as shown in Figure 20-2.

We saw in Section 19-1 that the more massive a star, the shorter the amount of time it remains on the main sequence. Similarly, the greater the star's mass, the more rapidly it goes through the stages of post–main-sequence evolution. Hence, we

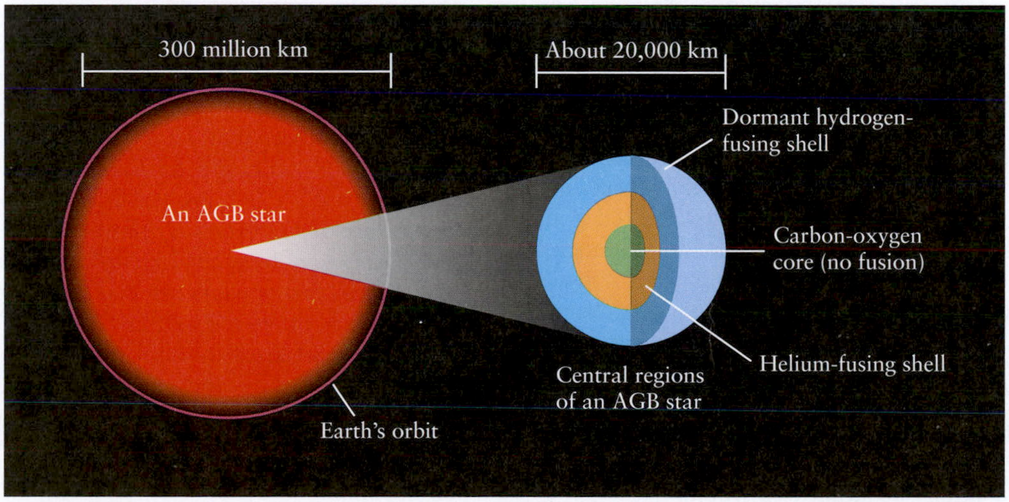

FIGURE 20-2

The Structure of an Old, Moderately Low-Mass AGB Star Near the end of its life, a star like the Sun becomes an immense, red, asymptotic giant branch (AGB) star. The star's inert core, active helium-fusing shell, and dormant hydrogen-fusing shell are all contained within a volume roughly the size of Earth. Nuclear reactions in the helium-fusing shell are so rapid that the star's luminosity is thousands of times that of the present-day Sun. (The relative sizes of the shells in the star's interior are not shown to scale.)

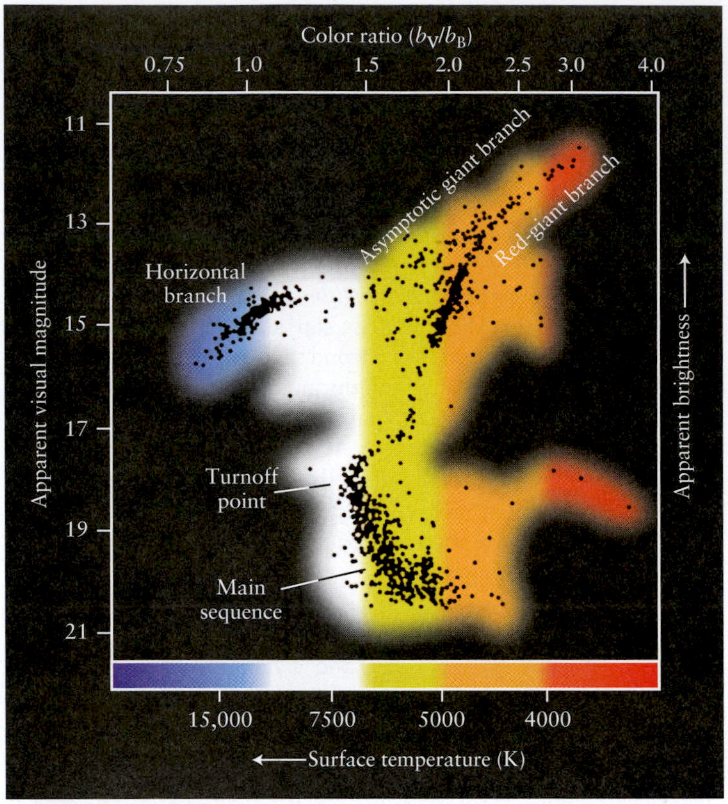

FIGURE 20-3

Stellar Evolution in a Globular Cluster In the old globular cluster M55, stars with masses less than about 0.8 $M_\odot$ are still on the main sequence, converting hydrogen into helium in their cores. Slightly more massive stars have consumed their core hydrogen and are ascending the red giant branch; even more massive stars have begun core helium fusion and are found on the horizontal branch. The most massive stars (which still have less than 4 $M_\odot$) have consumed all the helium in their cores and are ascending the asymptotic giant branch. (Adapted from D. Schade, D. VandenBerg, and F. Hartwick)

can see all of these stages by studying star clusters, which contain stars that are all the same age but that have a range of masses (see Section 19-4). Figure 20-3 shows a color-magnitude diagram for the globular cluster M55, which is at least 13 billion years old. The least massive stars in this cluster are still on the main sequence. Progressively more massive stars have evolved to the red giant branch, the horizontal branch, and the asymptotic giant branch.

A 1-$M_\odot$ AGB star can reach a maximum luminosity of nearly 10^4 $L_\odot$, as compared with approximately 10^3 $L_\odot$ when it reached the helium flash and a relatively paltry 1 $L_\odot$ during its main-sequence lifetime. When the Sun becomes an AGB star some 7.8 billion years from now, this tremendous increase in luminosity will cause Mars and the Jovian planets to largely evaporate away. The Sun's bloated outer layers will reach to Earth's orbit. Mercury and perhaps Venus will simply be swallowed whole.

CONCEPTCHECK 20-1

Why does helium fusion in a star's core, which releases energy, actually lead to a decrease in its luminosity? What are these stars called on the H-R diagram?

Answer appears at the end of the chapter.

20-2 Dredge-ups bring the products of nuclear fusion to a giant star's surface

As we saw in Section 16-2, energy is transported outward from a star's core by one of two processes—radiative diffusion or convection. The first is the passage of energy in the form of electromagnetic radiation, and it dominates only when a star's gases are relatively transparent. The second involves up-and-down movement of the star's gases. Convection plays a very important role in giant stars, and it helps supply the cosmos with the elements essential to life.

Convection, Dredge-ups, and Carbon Stars

ANIMATION 20-2 In the Sun, convection dominates only the outer layers, from around 0.71 solar radius (measured from the center of the Sun) up to the photosphere (recall Figure 16-4). During the final stages of a star's life, however, the convective zone can become so broad that it extends down to the star's core. At these times, convection can "dredge up" the heavy elements produced in and around the core by nuclear fusion, transporting them all the way to the star's surface.

LOOKING DEEPER 20-1 The *first* **dredge-up** takes place after core hydrogen fusion stops, when the star becomes a red giant for the first time. Convection dips so deeply into the star that material processed by the CNO cycle of hydrogen fusion (see Section 16-1) is carried up to the star's surface, changing the relative abundances of carbon, nitrogen, and oxygen. A *second* dredge-up occurs after core helium fusion ceases, further altering the abundances of carbon, nitrogen, and oxygen. Still later, during the AGB stage, a *third* dredge-up can occur if the star has a mass greater than about 2 $M_\odot$. This third dredge-up transports large amounts of freshly synthesized carbon to the star's surface, and the star's spectrum thus exhibits prominent absorption bands of carbon-rich molecules like C_2, CH, and CN. For this reason, an AGB star that has undergone a third dredge-up is called a **carbon star.**

> The carbon that forms the basis of all life on Earth was ejected billions of years ago from giant stars

All AGB stars have very strong stellar winds that cause them to lose mass at very high rates, up to 10^{-4} $M_\odot$ per year (a thousand times greater than that of a red giant, and 10^{10} times greater than the rate at which our present-day Sun loses mass). The surface temperature of AGB stars is relatively low, around 3000 K, so any ejected carbon-rich molecules can condense to form tiny grains of soot. Indeed, carbon stars are commonly found to be obscured in sooty cocoons of ejected matter (Figure 20-4).

Carbon stars are important because they enrich the interstellar medium with carbon and some nitrogen and oxygen. The triple alpha process that occurs in helium fusion is the *only* way that carbon can be made, and carbon stars are the primary avenue by which this element is dispersed into interstellar space. Indeed, most of the carbon in your body was produced long ago inside a star by the triple alpha process (see Section 19-3). This carbon was later dredged up to the star's surface and ejected into space. Some 4.56 billion years ago a clump of the interstellar medium that contained this carbon coalesced into the solar nebula from which our Earth—and all of the life on it—eventually formed. In this sense you can think of your body as containing "recycled"

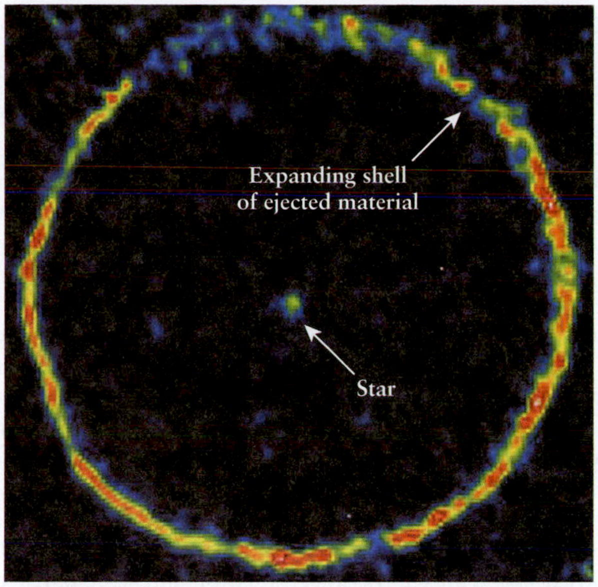

FIGURE 20-4 R I V U X G

A Carbon Star TT Cygni is an AGB star in the constellation Cygnus that ejects some of its carbon-rich outer layers into space. Some of the ejected carbon combines with oxygen to form molecules of carbon monoxide (CO), whose emissions can be detected with a radio telescope. This radio image shows the CO emissions from a shell of material that TT Cygni ejected some 7000 years ago. Over that time, the shell has expanded to a diameter of about ½ ly. (H. Olofsson, Stockholm Observatory, et al./NASA)

material—substances that were once in the heart of a star that formed and evolved long before our solar system existed.

CONCEPTCHECK 20-2

How did most of the carbon in your body leave the stars where these atoms formed? Through a supernova explosion, or a stellar wind?

Answer appears at the end of the chapter.

20-3 Stars of moderately low mass die by gently ejecting their outer layers, creating planetary nebulae

An aging AGB star casts off much of the mass that it possesses and makes a beautiful glowing nebula

For a star, like our Sun, that began with a moderately low mass (between about 0.4 and 4 $M_\odot$), the AGB stage in its evolution is a dramatic turning point. Before this stage, a star loses mass only gradually through steady stellar winds. But as it evolves during its AGB stage, a star divests itself completely of its outer layers. The aging star undergoes a series of bursts in luminosity, and in each burst it ejects a shell of material into space. (The shell around the AGB star TT Cygni, shown in Figure 20-4, was probably created in this way.) Eventually, all that remains of a low-mass star is a fiercely hot, exposed core, surrounded by glowing shells of ejected gas. This late stage in the life of a star is called a **planetary nebula;** see Figure 20-5. (The left-hand image on the opening page of this chapter also shows a planetary nebula called the Helix Nebula.)

Making a Planetary Nebula

To understand how an AGB star can eject its outer layers in shells, consider the internal structure of such a star as shown in Figure 20-2. As the helium in the helium-fusing shell is used up, the pressure that holds up the dormant hydrogen-fusing shell decreases. Hence, the dormant hydrogen shell contracts and heats up, and hydrogen fusion begins anew. This revitalized hydrogen fusion creates helium, which rains downward onto the temporarily dormant helium-fusing shell. As the helium shell gains mass, it shrinks and heats up. When the temperature of the helium shell reaches a certain critical value, it reignites in a **helium shell flash** that is similar to (but less intense than) the helium flash that occurred earlier in the evolution of a low-mass star (see Section 19-3). The released energy pushes the hydrogen-fusing shell outward, making it cool off, so that hydrogen fusion ceases and this shell again becomes dormant. The process then starts over again.

When a helium shell flash occurs, the luminosity of an AGB star increases substantially in a relatively short-lived burst called a **thermal pulse.** Figure 20-6, which is based on a theoretical calculation of the evolution of a 1-$M_\odot$ star, shows that thermal pulses begin when the star is about 12.365 billion years old. The

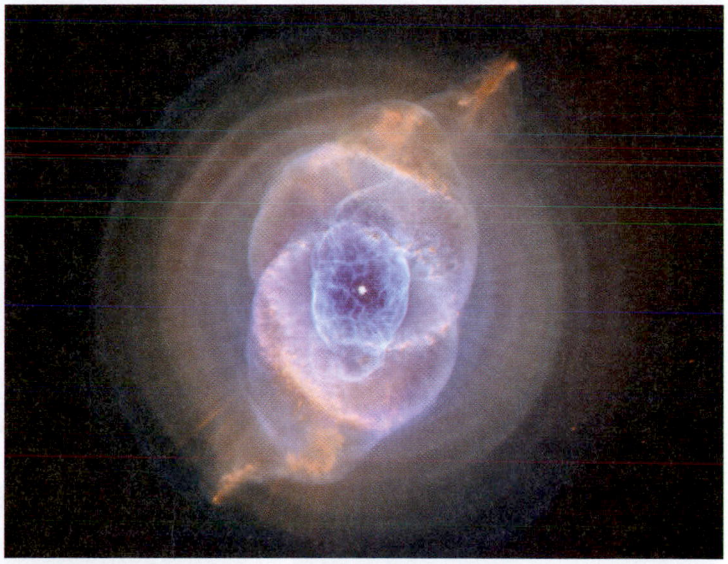

FIGURE 20-5 R I V U X G

The Cat's Eye Nebula Named after its shape, this nebula was first discovered in 1786. The exposed core of the central star—clearly visible in the image—is about 10,000 times more luminous than the Sun, and its intense ultraviolet radiation powers the nebula by ionizing its gas. While the nebula is hotter than the Sun, the gas is too diffuse to shine by thermal radiation as the Sun does. Instead, like all planetary nebula, the visible light comes from emission lines: After electrons are excited to higher energies, photons are emitted as the electrons cascade down to lower energies in the gas's atoms. As with many planetary nebulae, the mechanisms shaping the intricate visible structures are not well understood. (NASA, ESA, HEIC, The Hubble Heritage Team STScI/AURA)

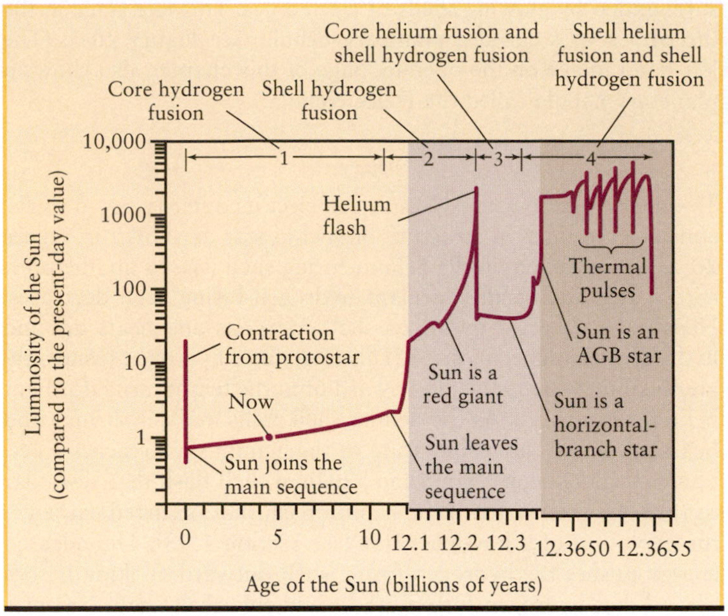

FIGURE 20-6

Further Stages in the Evolution of the Sun This diagram, which shows how the luminosity of the Sun (a 1-$M_\odot$ star) changes over time, is an extension of Figure 19-8. We use different scales for the final stages because the evolution is so rapid. During the AGB stage there are brief periods of runaway helium fusion, causing spikes in luminosity called thermal pulses. (Adapted from Mark A. Garlick, based on calculations by I.-Juliana Sackmann and Kathleen E. Kramer)

calculations predict that thermal pulses occur at ever-shorter intervals of about 100,000 years.

During these thermal pulses, the dying star's outer layers can separate completely from its carbon-oxygen core. As the ejected material expands into space, dust grains condense out of the cooling gases. Radiation pressure from the star's hot, burned-out core acts on the specks of dust, propelling them further outward, and the star sheds its outer layers altogether. In this way an aging 1-$M_\odot$ star loses as much as 40% of its mass. More massive stars eject even greater fractions of their original mass.

As a dying star ejects its outer layers, the star's hot core becomes exposed. With a surface temperature of about 100,000 K, this exposed core emits ultraviolet radiation intense enough to ionize and excite atoms in the expanding shell of ejected gases. These gases therefore glow and emit visible light through the process of fluorescence (see Figure 18-3 and Box 18-1). In fluorescence, visible photons are emitted after electrons knocked up to higher energy levels fall back down to lower energy levels. This mechanism produces the light seen in Figure 20-5, as well as more colorful emission in planetary nebula like those shown in Figure 20-7.

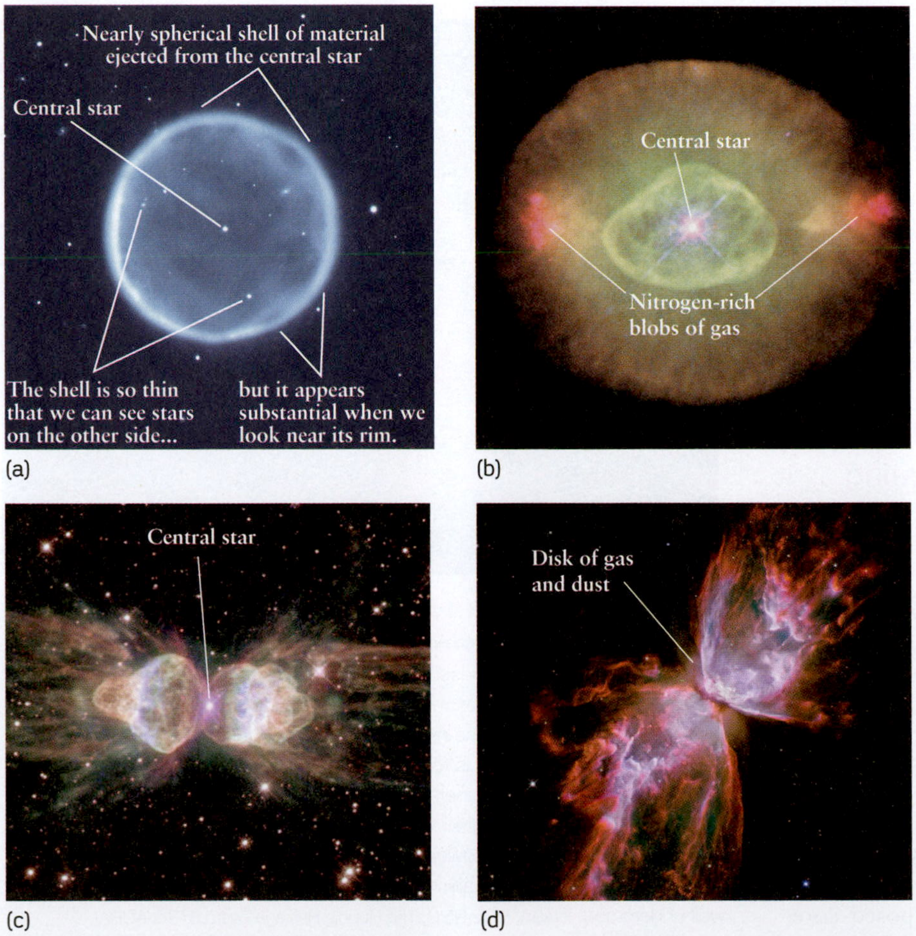

FIGURE 20-7 R I V U X G

Some Shapes of Planetary Nebulae The outer shells of dying low-mass stars are ejected in a wonderful variety of patterns. **(a)** This shell of expanding gas, in the globular cluster M15, is about 2150 pc (7000 ly) away from Earth. The almost perfectly spherical shell that comprises the nebula is about 1.5 pc (5 ly) in diameter; the thickness of the shell is only about 0.1 pc (0.3 ly). **(b)** NGC 6826 shows, among other features, lobes of nitrogen-rich gas (red). The process by which they were ejected is as yet unknown. **(c)** Mz 3 (Menzel 3) is 900 pc (3000 ly) from Earth. The dying star, creating these bubbles of gas, may be part of a binary system. **(d)** The spectacular Butterfly Nebula is thought to have been channeled by a doughnut-shaped cloud at the center of the disk that separates the two "wings." (a: WIYN/NOAO/NSF; b: AURA/STScI/NASA; c: R. Sahai and J. Trauser, JPL, the WFPC-2 Science Team, and NASA; d: NASA, ESA and the Hubble SM4 ERO Team)

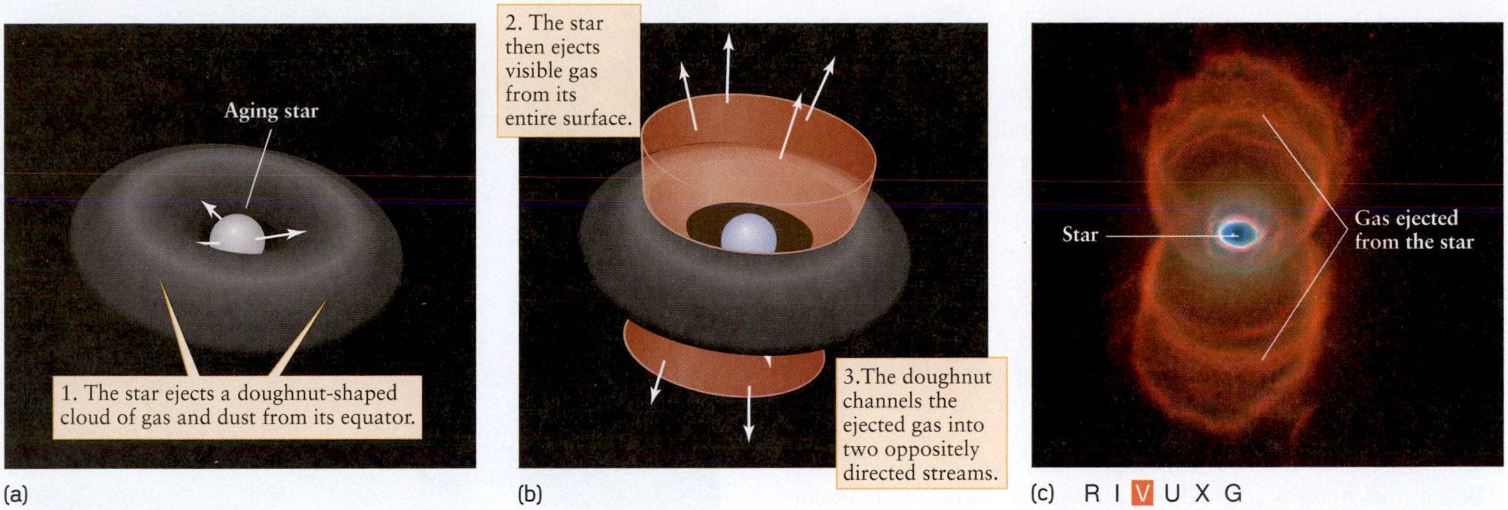

Aging star
1. The star ejects a doughnut-shaped cloud of gas and dust from its equator.
2. The star then ejects visible gas from its entire surface.
3.The doughnut channels the ejected gas into two oppositely directed streams.
Star
Gas ejected from the star

(a) (b) (c) R I V U X G

FIGURE 20-8

Making an Elongated Planetary Nebula **(a), (b)** These illustrations show one proposed explanation for why many planetary nebulae have an elongated shape. **(c)** The planetary nebula MyCn18, shown here in false color, may have acquired its elongated shape in this way. It lies some 2500 pc (8000 ly) from Earth in the constellation Musca (R. Sahai and J. Trauger, Jet Propulsion Laboratory; the WFPC-2 Science Team; and NASA)

CAUTION! Despite their name, planetary nebulae have nothing to do with planets. This misleading term was introduced in the nineteenth century because these glowing objects looked like distant Jovian planets—a small colored blur—when viewed through the small telescopes then available. The difference between planets and planetary nebulae first became obvious with the advent of spectroscopy: Planets have *absorption* line spectra (see Section 7-3), but the excited gases of planetary nebulae have *emission* line spectra.

The Properties of Planetary Nebulae

Planetary nebulae are quite common. Astronomers estimate that there are 20,000 to 50,000 planetary nebulae in our Galaxy alone. Many planetary nebulae, such as those in Figure 20-7, are more or less spherical in shape. This shape is a result of the symmetrical way in which the gases were ejected. But if the rate of expansion is not the same in all directions, the resulting nebula can take on an hourglass or dumbbell appearance (Figure 20-8).

Spectroscopic observations of planetary nebulae show emission lines of ionized hydrogen, oxygen, and nitrogen. From the Doppler shifts of these lines, astronomers have concluded that the expanding shell of gas moves outward from a dying star at speeds from 10 to 30 km/s. For a shell expanding at such speeds to have attained the typical diameter of a planetary nebula, about 1 ly, it must have begun expanding about 10,000 years ago. Thus, by astronomical standards, the planetary nebulae we see today were created only very recently.

We do not observe planetary nebulae that are more than about 50,000 years old. After this length of time, the shell has spread out so far from the cooling central star that its gases cease to glow and simply fade from view. The nebula's gases then mix with the surrounding interstellar medium.

Astronomers estimate that all the planetary nebulae in the Galaxy return a total of about 5 $M_\odot$ of matter to the interstellar medium each year. This amount is about 15% of all the matter expelled by all the various sorts of stars in the Galaxy each year. Because this contribution is so significant, and because the ejected material includes heavier elements (metals) manufactured within a nebula's central star, planetary nebulae play an important role in the chemical evolution of the Galaxy as a whole.

CONCEPTCHECK 20-3

What provides the energy for a planetary nebula's emission? Does a planetary nebula fade from view when its gases cool?

Answer appears at the end of the chapter.

20-4 The burned-out core of a moderately low-mass star cools and contracts until it becomes a white dwarf

We have seen that after a moderately low-mass star (from about 0.4 to about 4 solar masses) consumes all the hydrogen in its core, it is able to ignite nuclear reactions that convert helium to carbon and oxygen. Given sufficiently high temperature and pressure, carbon and oxygen can also undergo fusion reactions that release energy. But for such a moderately low-mass star, the core temperature and pressure never reach the extremely high values needed for these reactions to take place. Instead, as we have seen, the process of mass ejection just strips away the star's outer layers and leaves behind the hot carbon-oxygen core. With no nuclear reactions taking place, the core simply cools down like a dying ember. This burned-out relic of a star's former glory is called a **white dwarf.**

Such white dwarfs prove to have exotic physical properties that are wholly unlike any object found on Earth.

CAUTION! Unfortunately, the word *dwarf* is used in astronomy for several very different kinds of small objects. Here's a review of the three kinds that we have encountered so far in this book. A *white* dwarf is the relic that remains at the very end of the evolution of a star of initial mass between about 0.4 $M_\odot$ and 4 $M_\odot$. Nuclear reactions are no longer taking place in its interior; it emits light simply because it is still hot. A *red* dwarf, discussed in Section 19-1, is a cool main-sequence star with a mass between about 0.08 $M_\odot$ and 0.4 $M_\odot$. The energy emitted by a red dwarf in the form of light comes from its core, where fusion reactions convert hydrogen into helium. Finally, a *brown* dwarf (see Section 8-6 and Section 17-5) is an object like a main-sequence star but with a mass less than about 0.08 $M_\odot$. Because its mass is so small, its internal pressure and temperature are too low to sustain nuclear reactions. Instead, a brown dwarf emits light because it is slowly contracting, a process that releases energy (see Section 16-1). White dwarfs are comparable in size to Earth (see Section 17-7); by contrast, brown dwarfs are larger than the planet Jupiter, and red dwarfs are even larger.

Properties of White Dwarfs

A white dwarf is kept from collapsing by the pressure of its degenerate electrons

You might think that without nuclear reactions to provide internal heat and pressure, a white dwarf should keep on shrinking under the influence of its own gravity as it cools. Actually, a cooling white dwarf maintains its size, because the burned-out stellar core is so dense that most of its electrons are degenerate (see Section 19-3). Thus, degenerate electron pressure supports the star against further collapse. This pressure does not depend on temperature, so it continues to hold up the star even as the white dwarf cools and its temperature drops.

Many white dwarfs are found in the solar neighborhood, but all are too faint to be seen with the naked eye. One of the first white dwarfs to be discovered is a companion to Sirius, the brightest star in the night sky. In 1844 the German astronomer Friedrich Bessel noticed that Sirius was moving back and forth slightly, as if it was being orbited by an unseen object. This companion, designated Sirius B (Figure 20-9), was first glimpsed in 1862 by the American astronomer Alvan Clark. Recent Hubble Space Telescope observations at ultraviolet wavelengths, where hot white dwarfs emit most of their light, show that the surface temperature of Sirius B is 25,200 K. (By contrast, the main-sequence star Sirius A has a surface temperature of 10,500 K, while the Sun's surface temperature is a relatively frosty 5800 K.)

The Density of White Dwarfs

Observations of white dwarfs in binary systems like Sirius allow astronomers to determine the mass, radius, and density of these stars (see Sections 17-9, 17-10, and 17-11). Such observations show that the density of the degenerate matter in a white dwarf is typically 10^9 kg/m^3 (a million times denser than water). A teaspoonful of white dwarf matter brought to Earth would weigh nearly 5.5 tons—as much as an elephant!

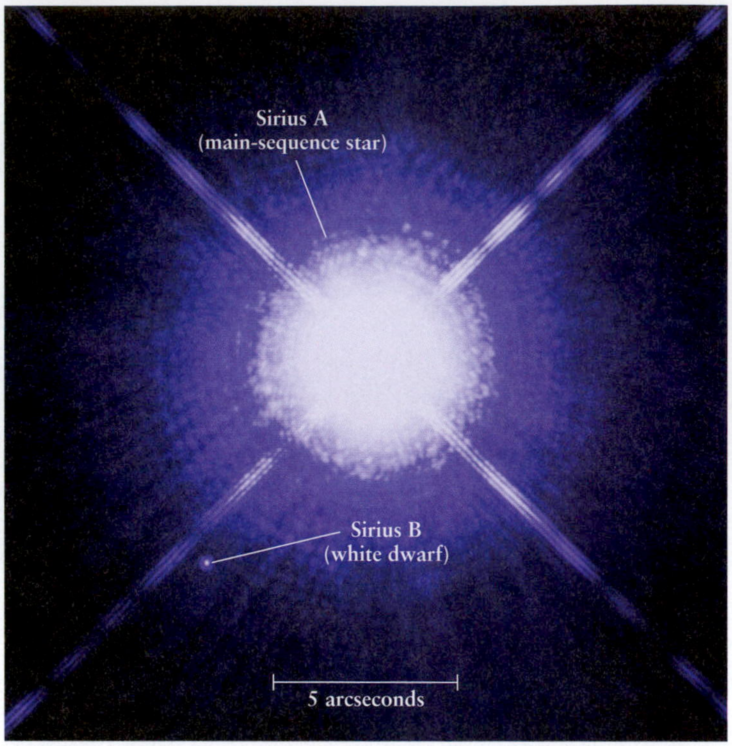

FIGURE 20-9 R I V U X G
Sirius A and Its White Dwarf Companion Sirius, the brightest-appearing star in the sky, is actually a binary star: The secondary star, called Sirius B, is a white dwarf. In this Hubble Space Telescope image, Sirius B is almost obscured by the glare of the overexposed primary star, Sirius A, which is about 10^4 times more luminous than Sirius B. The halo and rays around Sirius A are the result of optical effects within the telescope. (NASA; H. E. Bond and E. Nelan, STScI; M. Barstow and M. Burleigh, U. of Leicester; and J. B. Holberg, U. of Arizona)

Compressing matter so that an elephant could fit into a teaspoon might seem impossible until you remember that most of an atom is empty space. We mentioned in Section 5-7 that if you imagine a typical atom expanded to the size of a football field, the nucleus at the 50-yard line would only be 1 cm wide. That leaves a lot of space between the nucleus and its electrons that would be out near the goal posts; even farther out would be the nucleus of the next nearest atom. The incredible density of a white dwarf leaves much less empty space between its atoms, but as we will see, objects called neutron stars are even denser than white dwarfs.

CONCEPTCHECK 20-4

Do all atoms consist mostly of empty space, or is empty space a property of only the carbon and oxygen atoms that lead to a white dwarf?

Answer appears at the end of the chapter.

The Mass-Radius Relation for White Dwarfs

As we learned in Section 17-3, degenerate matter has a very different relationship between its pressure, density, and temperature than that of ordinary gases. Consequently, white dwarf stars have an unusual **mass-radius relation:** The more massive a white dwarf star, the *smaller* it is. Figure 20-10 displays the mass-radius relation for white dwarfs. Note that the more degenerate matter you

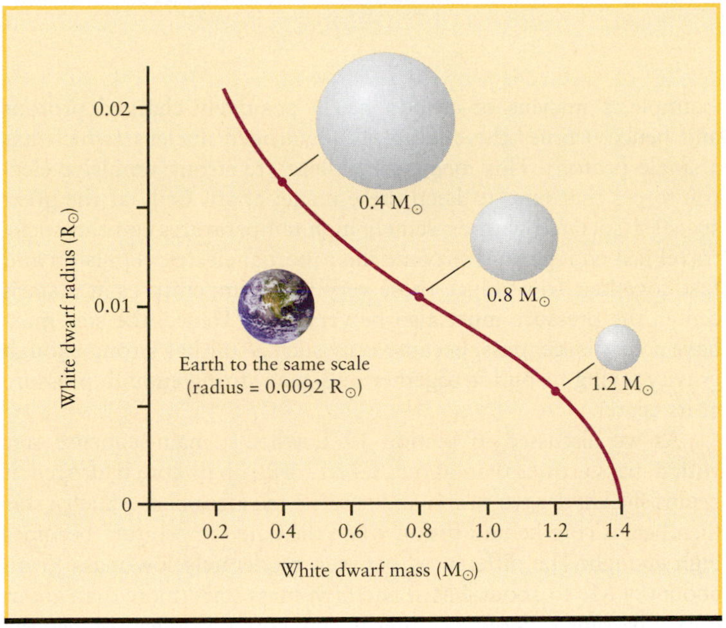

FIGURE 20-10

The Mass-Radius Relationship for White Dwarfs The more massive a white dwarf is, the smaller its radius. (The drawings of white dwarfs of different mass are drawn to the same scale as the image of Earth.) This unusual relationship is a result of the degenerate-electron pressure that supports the star. The maximum mass of a white dwarf, called the Chandrasekhar limit, is 1.4 $M_\odot$.

pile onto a white dwarf, the smaller it becomes. However, there is a limit to how much pressure degenerate electrons can produce. As a result, there is an upper limit to the mass that a white dwarf can have. This maximum mass is called the **Chandrasekhar limit,** after the Indian-American scientist Subrahmanyan Chandrasekhar, who pioneered theoretical studies of white dwarfs in the 1930s. (The orbiting Chandra X-ray Observatory, described in Section 6-7, is named in his honor.) The Chandrasekhar limit is equal to 1.4 $M_\odot$, meaning that all white dwarfs must have masses less than 1.4 $M_\odot$. As we will see in Section 20-6, when the Chandrasekhar limit is exceeded, the star is no longer a white dwarf, and the results are explosive.

It is worth pointing out that the Chandrasekhar limit represents a huge scientific success. Electron degeneracy pressure is based on the quantum physics of electrons on the smallest of imaginable scales, yet the Chandrasekhar limit applies this physics to a solar mass of material about the size of Earth. Observations verify that white dwarfs do not exceed the Chandrasekhar limit: Numerous white dwarfs are in binary systems (that allows their masses to be measured; see Section 4-7), and amazingly, none are above 1.4 $M_\odot$!

The material inside a white dwarf consists mostly of ionized carbon and oxygen atoms floating in a sea of degenerate electrons. As the dead star cools, the carbon and oxygen ions slow down, and electric forces between the ions begin to prevail over the random thermal motions. About 5×10^9 years after the star first becomes a white dwarf, when its luminosity has dropped to about 10^{-4} $L_\odot$ and its surface temperature is a mere 4000 K, the ions no longer move freely. Instead, they arrange themselves in orderly rows, like an immense crystal lattice. From this time on, you could say that the star is "solid." The degenerate electrons move around freely in this crystalline material, just as electrons move freely through an electrically conducting metal like copper or silver. A diamond is also crystalline (with carbon ordered in a crystal lattice), so a cool carbon-oxygen white dwarf has some similarities to a giant diamond in the sky!

CONCEPTCHECK 20-5

Could electron degeneracy pressure support a 1.2-$M_\odot$ white dwarf? If so, how would the white dwarf compare in size to Earth?

Answer appears at the end of the chapter.

From Red Giant to Planetary Nebula to White Dwarf

Figure 20-11 shows the evolutionary tracks followed by three burned-out stellar cores as they pass through the planetary nebula stage and become white dwarfs. When these three stars were red giants, they had masses of 0.8, 1.5, and 3.0 $M_\odot$. Mass ejection

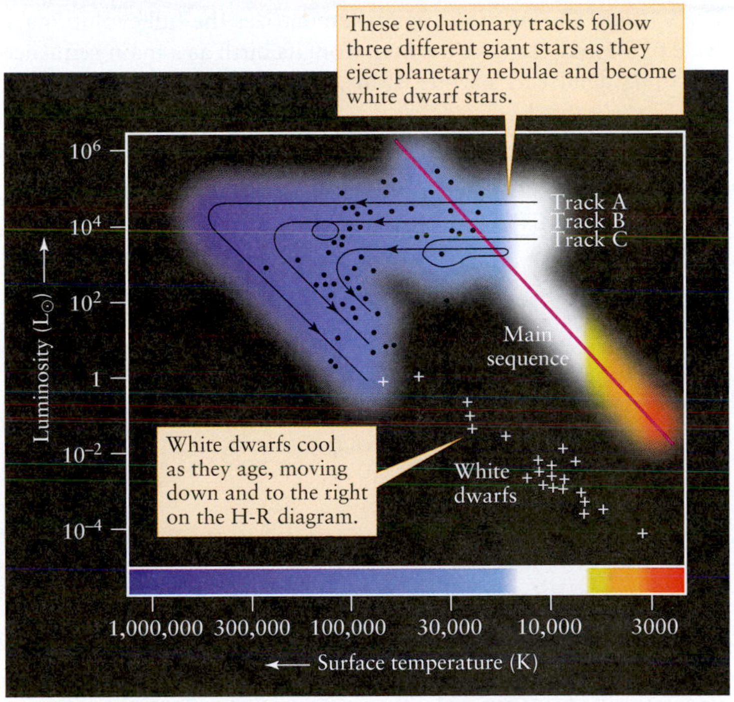

Evolutionary track	Mass ($M_\odot$) Giant star	Ejected nebula	White dwarf
A	3.0	1.8	1.2
B	1.5	0.7	0.8
C	0.8	0.2	0.6

FIGURE 20-11

Evolution from Giants to White Dwarfs This H-R diagram shows the evolutionary tracks of three low-mass giant stars as they eject planetary nebulae. The table gives the extent of mass loss in each case. The dots represent the central stars of planetary nebulae whose surface temperatures and luminosities have been determined; the crosses represent white dwarfs of known temperature and luminosity. (Adapted from B. Paczynski) strips these dying stars of up to 60% of their matter. During their final spasms, the luminosity and surface temperature of these stars change quite rapidly. The points representing these stars on an H-R diagram race along their evolutionary tracks, sometimes executing loops corresponding to thermal pulses (see Track B and Track C in Figure 20-11). Finally, as the ejected nebulae fade and the stellar cores cool, the evolutionary tracks of these dying stars take a sharp turn toward the white dwarf region of the H-R diagram. As the table accompanying Figure 20-11 shows, the final white dwarf has only a fraction of the mass of the giant star from which it evolved.

Although a white dwarf maintains the same size as it cools, its luminosity and surface temperature both decrease with time. Consequently, the evolutionary tracks of aging white dwarfs point toward the lower right corner of the H-R diagram. You can see this in the lower right portion of Figure 20-11. After ejecting much of its mass into space, our own Sun will eventually evolve into a white dwarf star about the size of Earth and with perhaps one-tenth of its present luminosity. It will become even dimmer as it cools. After 5 billion years as a white dwarf, the Sun will radiate with no more than one ten-thousandth of its present brilliance. With the passage of eons, our Sun will simply fade into obscurity. The *Cosmic Connections* figure summarizes the full evolutionary cycle of a 1-$M_{\odot}$ star like the Sun, from its birth as a main-sequence star to its demise as a white dwarf.

CONCEPTCHECK 20-6

Consult the *Cosmic Connections* figure. When the Sun becomes a red giant, is that the largest it will ever be? The most luminous?

CONCEPTCHECK 20-7

Consult the *Cosmic Connections* figure. Why does the formation of a planetary nebula around the Sun precipitate the end of nuclear reactions in the core?

Answers appear at the end of the chapter.

20-5 High-mass stars create heavy elements in their cores

During the entire lifetime of a low-mass red dwarf star (with an initial mass less than about 0.4 $M_{\odot}$), the only nuclear reaction that takes place is the fusion of hydrogen nuclei to form helium nuclei. In stars with initial masses from about 0.4 $M_{\odot}$ to about 4 $M_{\odot}$, a second kind of nuclear reaction takes place—helium fusion. The heaviest elements manufactured by helium fusion are carbon and oxygen.

The life story of a *high-mass* star (with a main-sequence mass greater than about 4 $M_{\odot}$) begins with these same reactions. But calculations show that high-mass stars can also go through several additional stages of nuclear reactions involving the fusion of carbon, oxygen, and other heavy nuclei. As a result, high-mass stars end their lives quite differently from low-mass stars.

Heavy-Element Fusion in Massive Stars

Why is fusion of heavy nuclei possible only in a high-mass star? The reason is that heavy nuclei have large electric charges: For example, a nucleus of carbon has 6 positively charged protons and hence 6 times the charge of a hydrogen nucleus (which has a single proton). This means that there are strong repulsive electric forces that tend to keep these nuclei apart. Only at the great speeds associated with extremely high temperatures can the nuclei travel fast enough to overcome their mutual electric repulsion and fuse together. To produce these very high temperatures at a star's center, the pressure must also be very high. Hence, the star must have a very large mass, because only such a star has strong enough gravity trying to pull it together and thus strong enough pressure at its center.

As we discussed in Section 19-2, when a main-sequence star with a mass greater than about 0.4 $M_{\odot}$ uses up its core hydrogen, it begins shell hydrogen fusion and enters a red giant phase. Such a star then begins core helium fusion when the core temperature becomes high enough. The differences between moderately low-mass (from about 0.4 $M_{\odot}$ to about 4 $M_{\odot}$) and high-mass stars (more than about 4 $M_{\odot}$) become pronounced after helium core fusion ends, when the core is composed primarily of carbon and oxygen.

Let us consider how the late stages in the evolution of a high-mass star differ from those of a low-mass star. In low-mass stars, as we saw in Section 20-4, fusion stops in the carbon-oxygen core; it eventually becomes a white dwarf. But in stars whose overall mass is more than about 4 $M_{\odot}$, the temperature and pressure of the carbon-oxygen core is enough to trigger further fusion reactions. The first of the new nuclear reactions is **carbon fusion,** which consumes carbon nuclei and produces oxygen, neon, sodium, and magnesium.

If a star has an even larger main-sequence mass of about 8 $M_{\odot}$ or so (before mass ejection), even more nuclear reactions can take place. After the cessation of carbon fusion, the core will again contract, and the star's central temperature exceeds 1 billion kelvins (10^9 K). Around this temperature **neon fusion** begins. This uses up the neon accumulated from carbon fusion and further increases the amount of oxygen and magnesium in the star's core.

After neon fusion ends, the core will again contract, and **oxygen fusion** begins. The principal product of oxygen fusion is silicon. Once oxygen fusion is over, the core will contract yet again. With this contraction, the central temperature rises and **silicon fusion** begins, producing a variety of nuclei from sulfur to iron and nickel. While all of this fusion is going on in the star's interior, at the surface the star is losing mass at a rapid rate (Figure 20-12).

As a high-mass star consumes increasingly heavier nuclei, the nuclear reactions produce a wider variety of products. For example, oxygen fusion produces not only silicon but also magnesium, phosphorus, and sulfur. Some nuclear reactions that create heavy elements also release neutrons. A neutron is like a proton except that it carries no electric charge. Therefore, neutrons are not repelled by positively charged nuclei, and so can easily collide and combine with them. This absorption of neutrons by nuclei, called **neutron capture,** creates many elements and isotopes that are not produced directly in fusion reactions.

Each stage of nuclear reactions in a high-mass star helps to trigger the succeeding stage. In each stage, when the star exhausts a given type of nuclear fuel in its core, gravitational contraction takes the core to ever-higher densities and temperatures, thereby igniting the "ash" of the previous fusion stage—and possibly the outlying shell of unburned fuel as well.

COSMIC CONNECTIONS

Our Sun: The Next Eight Billion Years

The Sun is presently less than halfway through its lifetime as a main-sequence star. The H-R diagram and cross-sections on this page summarize the dramatic changes that will take place when the Sun's main-sequence lifetime comes to an end.

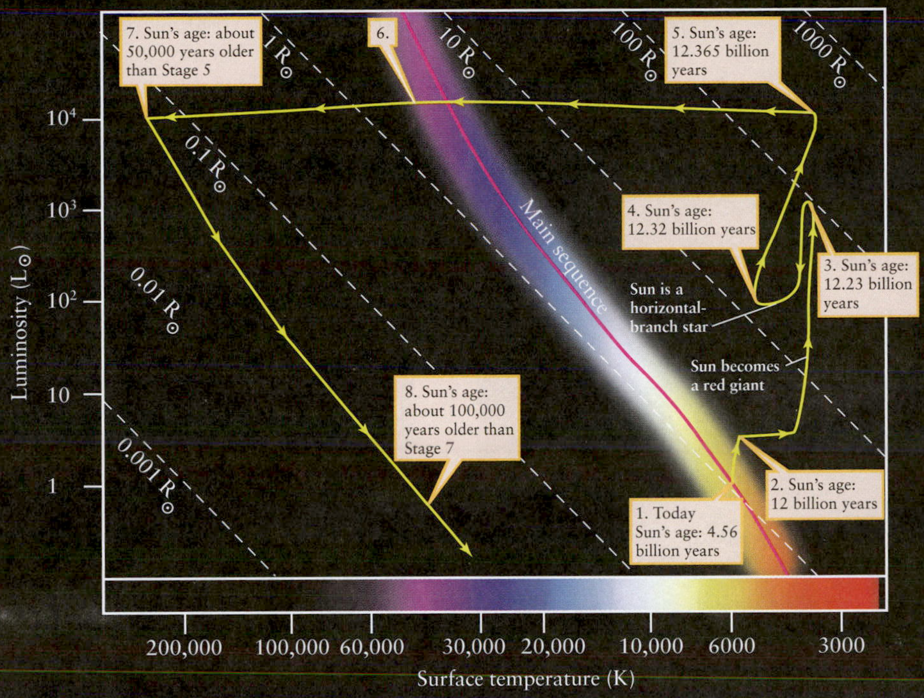

NOTE: The illustrations below do *not* show the dramatic changes in the Sun's radius as it evolves. The sizes of the various layers are not shown to scale.

1. On the main sequence

The present-day Sun is a main-sequence star – in its core, hydrogen fuses to produce helium.

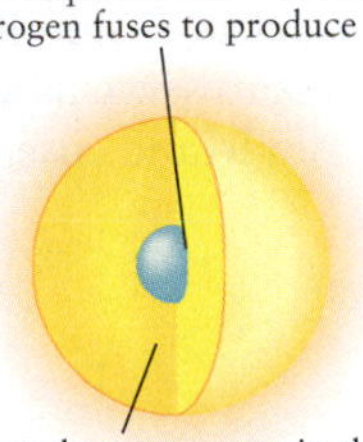

Fusion does not occur in the outer layers (which contain predominantly hydrogen and helium).

2. Becoming a red giant

At the end of the Sun's main-sequence lifetime, fusion stops in the core (which has been converted to helium).

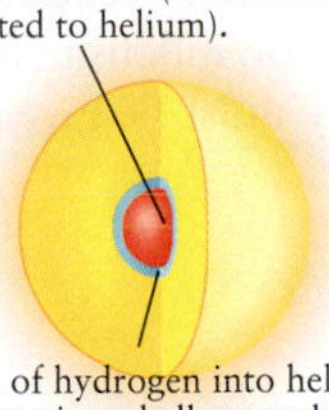

Fusion of hydrogen into helium continues in a shell around the core. The core shrinks, accelerating the fusion reactions in the shell and making the outer layers expand and cool.

3. The helium flash

As the core contracts and heats, the core helium begins to fuse to make carbon and oxygen. The core expands and the rate of energy release slows.

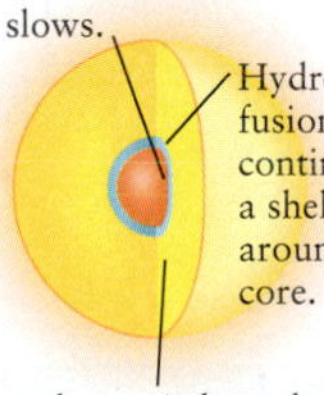

The outer layers (where there are still no fusion reactions) contract and get hotter due to the slower rate of energy release.

4. Beginning the second red giant phase

Once the core helium is consumed, what remains is an inert core of carbon and oxygen. The core again shrinks and gets hotter.

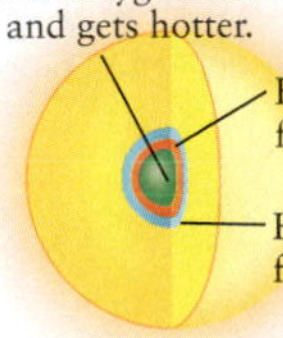

The shrinkage of the core again accelerates fusion reactions in the shells, making the inert outer layers expand and cool.

5. The Sun reaches its maximum size

Inert carbon-oxygen core

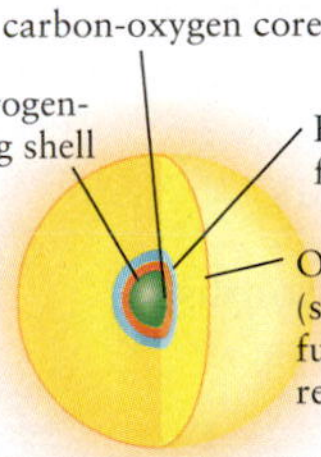

The Sun is more than 100 times larger in radius than when it was a main-sequence star. Part of the outer layers escapes into space in a stellar wind.

6. A planetary nebula

Thermal pulses cause spikes in luminosity that eject the star's outer layers.

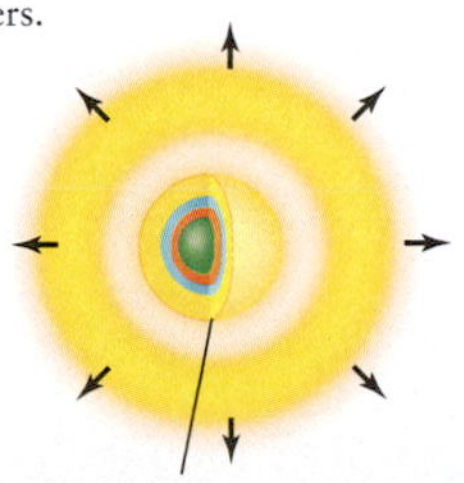

As the hot interior of the star is exposed, we observe an increase in the star's surface temperature.

7. The end of nuclear reactions

With the outer layers gone, the pressure on the shells around the core is too little to sustain nuclear reactions.

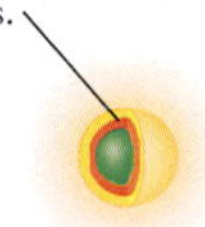

The star still glows intensely because of its high temperature. As energy is lost in the form of electromagnetic radiation, the star slowly cools.

8. A white dwarf

The core is now a white dwarf star, and the former shells around the core become its thin atmosphere.

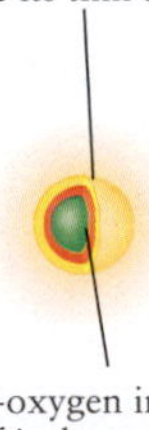

The carbon-oxygen interior of the white dwarf is degenerate, so it does not contract as it cools. Hence the white dwarf's radius no longer changes.

Hydrogen and helium, no fusion | Hydrogen fusion producing helium | Helium, no fusion | Helium fusion producing carbon and oxygen | Carbon and oxygen, no fusion

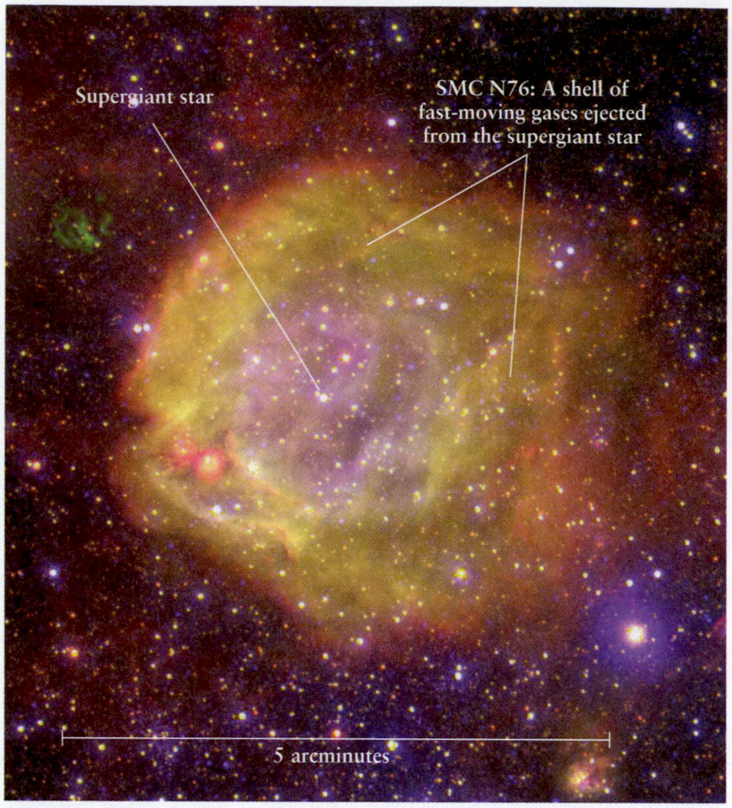

FIGURE 20-12 R I **V** U X G

Mass Loss from a Supergiant Star At the heart of this nebulosity, called SMC N76, lies a supergiant star with a mass of at least 18 $M_\odot$. This star is losing mass at a rapid rate in a strong stellar wind. As this wind collides with the surrounding interstellar gas and dust, it creates the "bubble" shown here. SMC N76, which has an angular diameter of 130 arcsec, lies within the Small Magellanic Cloud, a small galaxy that orbits our Milky Way. It is about 60,600 pc (198,000 ly) distant. (European Southern Observatory)

Supergiant Stars and Their Evolution

The increasing density and temperature of the core make each successive nuclear reaction more rapid than the one that preceded it. As an example, Table 20-1 shows a theoretical calculation of the evolutionary stages for a star with an initial main-sequence mass of 25 $M_\odot$. This calculation indicates that carbon fusion in such a star lasts for a 1000 years, neon fusion for 3 years, and oxygen fusion for only 3.6 months. The last, and briefest, stage of nuclear reactions is silicon fusion. The entire core supply of silicon in a 25-$M_\odot$ star is used up in only 5 days!

Each stage of core fusion in a high-mass star generates a new shell of material around the core. After several such stages, the internal structure of a truly massive star—say, 25 to 30 $M_\odot$ or greater—resembles that of an onion (Figure 20-13). Notice the order of the layers in Figure 20-13—the earlier the fusion reaction begins, the farther it is from the center. Because nuclear reactions can take place simultaneously in several shells, energy is released at such a rapid rate that the star's outer layers expand tremendously. The result is a **supergiant** star, whose luminosity and radius are much larger than those of a giant (see Section 17-7).

A star of 8 or more solar masses evolves into a supergiant 100 times (or more) larger than the Sun

Several of the brightest stars in the sky are supergiants, including Betelgeuse and Rigel in the constellation Orion and Antares in the constellation Scorpius. (Figure 17-15 shows the locations of these stars on an H-R diagram.) They appear bright not because they are particularly close, but because they are extraordinarily luminous.

The End of Nuclear Fusion

A supergiant star cannot keep adding shells to its "onion" structure forever, because the sequence of nuclear fusion reactions cannot go on indefinitely. In order for atoms to serve as a nuclear fuel, energy must be given off when they collide and fuse.

TABLE 20-1 Evolutionary Stages of a 25-$M_\odot$ Star

Stage	Core temperature (K)	Core density (kg/m^3)	Duration of stage
Hydrogen fusion	7×10^7	10×10^3	10^7 years
Helium fusion	2×10^8	2×10^6	10^6 years
Carbon fusion	8×10^8	10^9	1000 years
Neon fusion	1.6×10^9	10^{10}	3 years
Oxygen fusion	1.8×10^9	10^{10}	3.6 months
Silicon fusion	2.5×10^9	10^{11}	5 days
Core collapse	10^{10}	10^{13}	¼ second
Explosive (supernova)	about 10^9	varies	10 seconds

Based on calculations by Stanford Woosley (University of California, Santa Cruz)and H.-T. Janka.

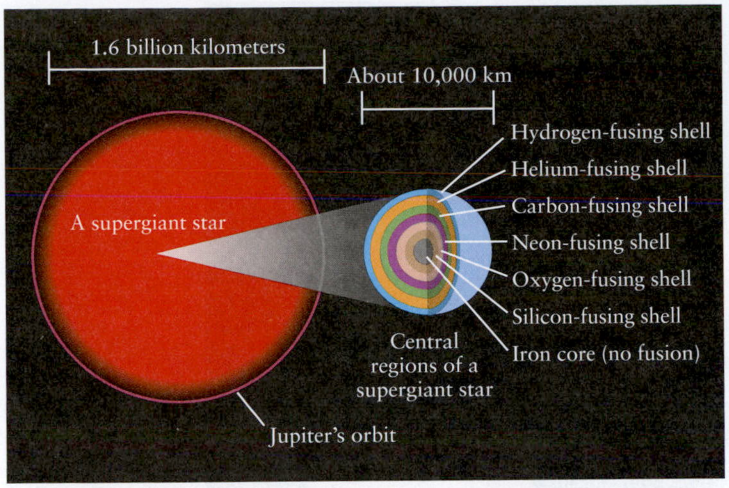

FIGURE 20-13

The Structure of an Old High-Mass Star Near the end of its life, a star with an initial mass greater than about 8 $M_\odot$ becomes a red supergiant. The star's overall size can be as large as Jupiter's orbit around the Sun. The stars energy comes from a series of concentric fusing shells, all combined within a volume roughly the same size as Earth. Nuclear reactions do not occur within the iron core, because fusion reactions that involve iron require an input of energy rather than release it.

Let's consider a common case in which a helium nucleus fuses with a larger atomic nucleus, although the results apply beyond helium. When they are close enough (due to high temperatures and pressures), the helium atom and the larger atom are pulled together by strong nuclear forces and energy is released. However, to get close enough, the atoms must overcome a repulsive force because all protons repel each other. As a result, fusion with larger atomic nuclei becomes increasingly difficult as the larger atom's greater number of protons leads to a greater *repulsive* electric force against the protons in the smaller helium nucleus. Very large atoms can be fused with helium, but they require an *input* of energy, rather than releasing energy. Everything in a star changes when a fusion reaction requires an input of energy in order to occur.

To act as a fuel, nuclear fusion reactions must release energy; this is the energy that powers a star. As larger and larger atoms are built up inside a star's core, they will eventually approach a size where they can no longer release energy through fusion and would instead require an input of energy to fuse. At this point, the star has run out of nuclear fuel in its core: **Iron is the first atom produced for which further fusion does not release energy.** As silicon fusion produces iron, the iron atoms in the core steadily grow in number without fusing into even larger atoms. Hence, the sequence of energy-producing fusion stages ends with silicon fusion, and no new layers in the fusing "onion" shells of Figure 20-13 will be produced. Shell fusion in the layers surrounding the iron-rich core consumes the star's remaining reserves of fuel. At this stage the entire energy-producing region of the star is contained in a volume no bigger than Earth, some 10^6 times smaller in radius than the overall size of the star. This state of affairs will soon come to an end, because the buildup of the iron-rich core signals the impending violent death of a massive supergiant star.

CONCEPTCHECK 20-8

Is it possible for helium nuclei to fuse together with atoms larger than iron to make even larger atoms?

Answer appears at the end of the chapter.

20-6 High-mass stars violently blow apart in core-collapse supernova explosions

Our present understanding is that stars of about 8 $M_\odot$ or less shed most of their mass in the form of planetary nebulae. The burned-out core that remains settles down to become a white dwarf star. But the truly massive stars—stellar heavyweights that begin their lives with more than 8 solar masses of material—do not pass through a planetary nebula phase. Instead, they die in spectacular *core-collapse supernova explosions*.

The Violent End of a High-Mass Star

To understand what happens in a core-collapse supernova explosion, we rely on theoretical models based on the known behavior of gases and atomic nuclei. The following model accounts for the observed properties of supernovae fairly well, which lends support to the overall picture. And, as we will see in Section 20-8, a special kind of "telescope" has allowed us to glimpse the interior of at least one relatively nearby supernova.

> By forming a dense core of iron, a massive star sows the seeds of its own destruction

The core of an aging, massive star gets progressively hotter as it contracts and ignites successive stages of nuclear fusion (see Stage 1 in Figure 20-14). But even as more of the star's core is converted into iron, this iron is unable to undergo further fusion. Without this heat-producing nuclear energy, the usual pressure that supports a star's core from collapsing is absent. Nonetheless, the core is held up briefly by degenerate electron pressure, which is the same pressure that keeps a white dwarf from collapsing. However, degenerate electron pressure can only hold up a 1.4 $M_\odot$ core against its own tremendous inward gravitational pull (this is the Chandrasekhar limit, see Section 20-4). When enough iron builds up to cross this threshold of mass, the result is a cataclysmic core collapse (Stage 2 in Figure 20-14). The collapse is so sudden and gravity is so strong that the outer core comes rushing in at over 20% the speed of light!

Even though steady fusion reactions in the core end with iron, the collapsing core soars in temperature due to the gravitational implosion. Temperatures skyrocket up to about 10 billion K within a fraction of a second. As a result, gamma-ray photons emitted by the intensely hot core have so much energy that when they collide with iron nuclei, they begin to break the iron nuclei down into much smaller helium nuclei. This process is called **photodisintegration.** As Table 20-1 shows, it takes a high-mass star millions of years and several stages of nuclear reactions to build up an iron core, but, within a fraction of a second, photodisintegration blasts apart much of that core.

Within another fraction of a second, the core becomes so dense that the negatively charged electrons within it are forced to combine with the positively charged protons to produce neutrons.

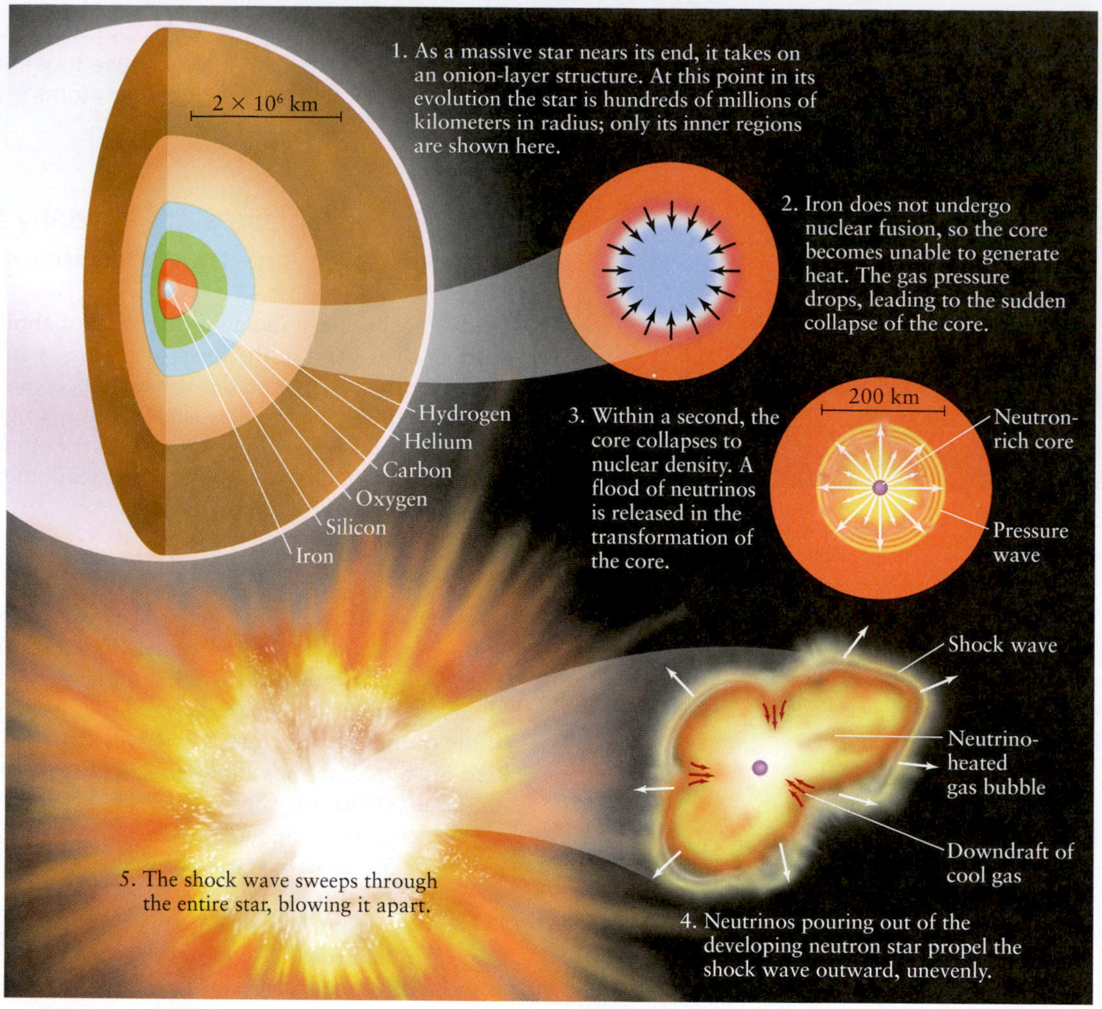

1. As a massive star nears its end, it takes on an onion-layer structure. At this point in its evolution the star is hundreds of millions of kilometers in radius; only its inner regions are shown here.
2 × 10⁶ km
Hydrogen
Helium
Carbon
Oxygen
Silicon
Iron
2. Iron does not undergo nuclear fusion, so the core becomes unable to generate heat. The gas pressure drops, leading to the sudden collapse of the core.
3. Within a second, the core collapses to nuclear density. A flood of neutrinos is released in the transformation of the core.
200 km
Neutron-rich core
Pressure wave
Shock wave
Neutrino-heated gas bubble
Downdraft of cool gas
4. Neutrinos pouring out of the developing neutron star propel the shock wave outward, unevenly.
5. The shock wave sweeps through the entire star, blowing it apart.

FIGURE 20-14

A Core-Collapse Supernova This series of illustrations depicts our understanding of the last day in the life of a star of more than about 8 $M_{\odot}$. (Illustration by Don Dixon, adapted from Wolfgang Hillebrandt, Hans-Thomas Janka, and Ewald Müller, "How to Blow Up a Star," *Scientific American,* October 2006)

In fact, most of the core is converted into neutrons. This process also releases a flood of *neutrinos* (see Section 16-4), denoted by the Greek letter υ (nu):

$$e^- + p \rightarrow n + \upsilon$$

There are other processes that produce neutrinos as well, and because these neutrinos carry away a substantial amount of energy from the core, the core cools down and condenses even further.

At about 0.25 seconds after its rapid contraction begins, the core is less than 20 km in diameter and its density is in excess of 4×10^{17} kg/m^3. This is **nuclear density,** the density with which neutrons and protons are packed together inside nuclei. (If Earth were compressed to this density, it would be only 300 meters, or 1000 feet, in diameter!)

Matter at nuclear density or higher is extraordinarily difficult to compress. Just as compressed electrons experience a degenerate electron pressure, these neutrons in the core attain a **degenerate neutron pressure.** Thus, when the density of the neutron-rich core begins to exceed nuclear density, the core suddenly becomes very stiff and rigid. The core's contraction comes to a sudden halt, and the inner part of the core actually bounces back and expands somewhat. This *core bounce* sends a powerful wave of pressure, like an unimaginably intense sound wave, outward into the outer core (see Stage 3 in Figure 20-14).

Initially, it was thought that the core bounce was sufficient to generate a shock wave that propelled a star's matter outward during a supernova. While there is evidence for a shock wave, more energy is needed than can be provided by the core bounce alone. One of the mechanisms that can provide additional energy to the shock wave appears to involve neutrinos.

At first glance, you might not expect neutrinos to have an explosive effect because they interact so weakly with matter. Recall that after a trillion or so neutrinos pass through your head each second—which is our Sun's normal output—most of them pass right through Earth without getting absorbed. In a collapsing core, however, the density of matter is so high during the collapse that enough neutrinos are absorbed to significantly heat matter around the core.

These neutrinos are essential to what follows. The absorbed neutrinos are thought to carry enough energy to produce hot inflating bubbles (Stage 4 in Figure 20-14). As the outer boundary of these bubbles pushes on matter around the core, the resulting shock wave ejects material from the star altogether. **The result is an explosive supernova.**

When the shock wave breaks out of the star, a portion of the explosive energy escapes in a torrent of light (see Stage 5 in Figure 20-14). At this point, the star has become a **supernova** (plural **supernovae**). Specifically, what we have described is the formation of a **core-collapse supernova.** We use this term because, as we will see in Section 20-9, it is possible for a different type of supernova explosion to occur that does not involve the collapse of the core in a massive star.

The details of the shock wave are not fully understood, neither the mechanism that drives it nor all the forms of energy that power it. Supernova explosions are still an active area of investigation, and as one researcher put it: "Astrophysicists still don't know how to blow up a star."

CAUTION! The energy released in a core-collapse supernova is an incomprehensibly large 10^{46} joules—a hundred times more energy than the Sun has emitted due to nuclear reactions over its *entire* 4.56-billion year history. However, it is important to recognize that the source of the supernova's energy release is *not* nuclear reactions. Rather, it is the *gravitational* energy released by the collapse of the core. (You release gravitational energy when you step off a diving board, and this released energy goes into making a big splash in the swimming pool.) The nuclear reactions that occur in a supernova do not occur unless the atoms are pushed very close together, and this is powered by gravity. In a similar way, gravitational energy can power a bouncing ball: While the elastic (or springlike) energy pushes the ball back up when it is compressed against the ground, it was gravitational energy that powered the compression.

CONCEPTCHECK 20-9

Our Sun cannot create the pressures necessary to fuse silicon into iron. However, would a 1-$M_\odot$ core of iron collapse and produce a supernova?

CONCEPTCHECK 20-10

What can provide pressure to halt the collapse of an iron core that exceeds the Chandrasekhar limit?

CONCEPTCHECK 20-11

Hot, expanding bubbles might produce a shock wave during a supernova explosion. What delivers energy to heat up these bubbles?

Answers appear at the end of the chapter.

The Big Picture of Core-Collapse

One way to think about the evolution of a massive star's core is that it undergoes an unrelenting gravitational contraction. Before the star arrives on the main sequence, hydrogen contracts until hydrogen fusion produces enough heat and pressure to temporarily halt further contraction. But as the resulting helium builds up, the core contracts until helium fusion turns on and produces a temporary halt to further gravitational contraction. This pattern continues all the way through to the production of iron in the core, with each stage of fusion—each "onion" layer in Figure 20-13—providing only a "pause" in gravity's unrelenting contraction.

But these pauses in gravitational contraction required pressure that was produced by the release of nuclear energy. While that was possible when fusing elements lighter than iron, the iron core does not release nuclear energy, and offers nothing to support the surrounding layers against gravity.

With nothing to hold off gravity, the surrounding layers rush inward, gaining tremendous speed and kinetic energy. The core is also compressed further as a variety of processes unfold. However, more than anything else, the kinetic energy of the infalling matter is ultimately converted into neutrinos. In this sense, **a supernova can be seen as a gravity-powered neutrino explosion.**

For a few seconds during the collapse, the neutrino energy output is unimaginably intense: While a supernova can outshine an entire galaxy with its visible light, the energy in neutrinos is a hundred times greater. By depositing only a tiny fraction of this energy in surrounding layers of the star, the neutrinos produce hot inflating bubbles that propel most of the star's matter into space.

The Wreckage of a Core-Collapse Supernova

Supercomputer simulations provide many insights into the violent, complex, and rapidly changing conditions deep inside a star as it is torn apart by a supernova explosion (Figure 20-15). For example, Figure 20-15a shows a computer simulation of a high-mass star several hours after the core bounce that initiates the supernova. The simulation produces turbulent swirls and eddies. Evidence in favor of such turbulence comes from images of the remnants of long-ago supernovae (Figure 20-15b). Such images show that material is ejected from the supernova not in uniform shells but in irregular clumps. These clumps are just what would be expected from a turbulent explosion. (We discuss supernova remnants further in Section 20-10.)

Detailed computer calculations suggest that a 25-$M_\odot$ star ejects about 96% of its material to the interstellar medium for use in producing future generations of stars. Less massive stars eject a smaller percentage of their mass into space when they become core-collapse supernovae.

Before this material is ejected into space, it is compressed so much that **a new wave of nuclear reactions sets in.** These reactions produce many of the atoms on the periodic table of the elements (see Box 5-5) that are heavier than iron. Reactions producing the heavier elements require a tremendous *input* of energy (recall Section 20-5), and thus cannot take place during the star's presupernova lifetime. Hence, much of the heavy elemental material that makes up our solar system, Earth, and even our bodies was produced in violent, explosive supernovae.

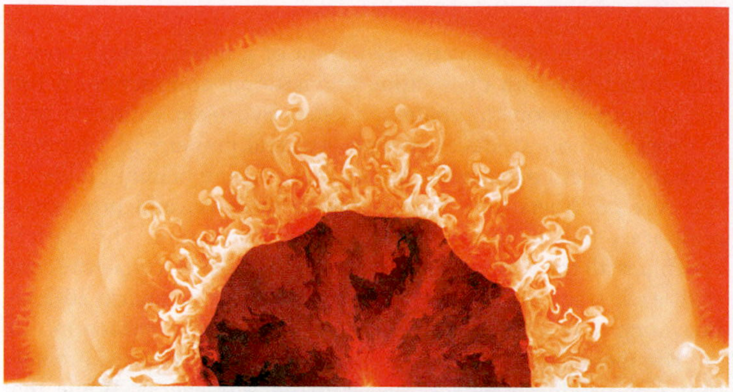

(a) A simulated supernova 2.8 hours after the core bounce

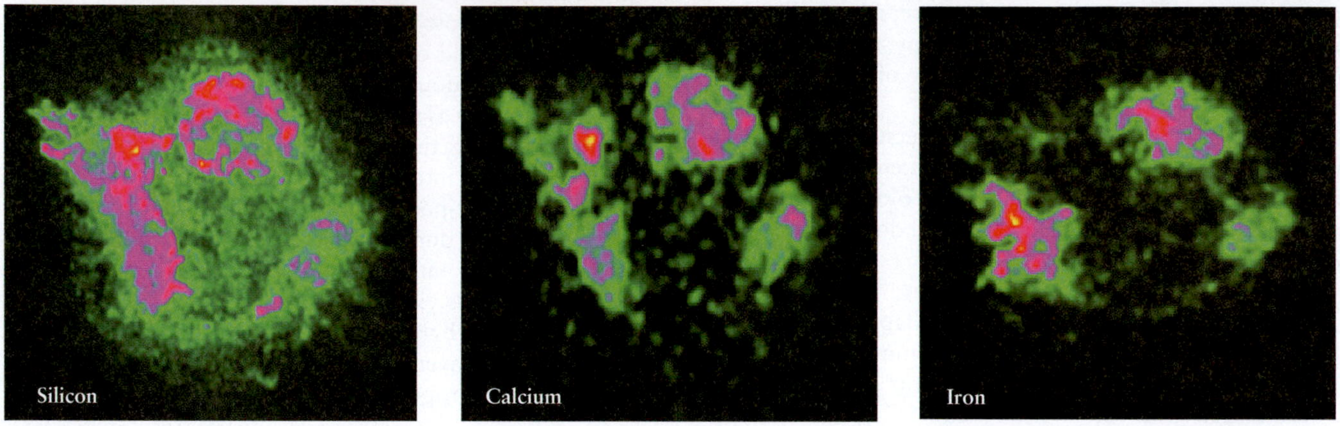

(b) Material was ejected in "blobs" from the supernova that produced the Cassiopeia A supernova remnant R I V U X G

ANIMATION 20-4 **FIGURE 20-15**

Turbulence in a Core-Collapse Supernova **(a)** This image from a supercomputer simulation shows a cross section of a massive star 2.8 hours into the supernova explosion. Density increases from black through red and orange to white. The darker inner region of the core is lower density because its heavier elements have been ejected into the outer layers. These dense metals appear almost as white flames as they undergo turbulence on the outermost shells of helium and hydrogen. The simulated supernova is SN 1987A. **(b)** Turbulence causes material to be ejected from the supernova in irregular "blobs," as shown by these images of the supernova remnant Cassiopeia A. Each image was made using an X-ray wavelength emitted by a particular element. (Figure 18-26 shows a false-color image of Cassiopeia A made using visible, infrared, and X-ray wavelengths.) (a: Konstantinos Kifonidis/Science Photo Library/Science Source; b: U. Hwang et al., NASA/GSFC)

CONCEPTCHECK 20-12

Uranium atoms are much heavier than iron atoms. During what stage of a star's life are uranium atoms produced? Do they release energy when produced?

Answer appears at the end of the chapter.

20-7 In 1987 a nearby supernova gave us a close-up look at the death of a massive star

Typical supernovae have peak luminosities of visible light as great as 10^9 $L_\odot$, rivaling the light output of an entire galaxy. These large luminosities make it possible to see supernovae in galaxies far beyond our own Milky Way Galaxy, and indeed hundreds of these distant supernovae are observed each year. In fact, the vast majority of supernovae occur so far beyond our Galaxy that astronomers cannot tell from earlier images the type of star that exploded. However, in a handful of nearby cases, images made before the explosion have allowed astronomers to identify the star that subsequently "went supernova," called the **progenitor star.** One significant case occurred in 1987.

A Supernova in the Galaxy Next Door

On February 23, 1987, a supernova was discovered in the Large Magellanic Cloud (LMC), a companion galaxy to our Milky Way some 51,500 pc (168,000 ly) from Earth. The supernova, designated SN 1987A (because it was the first discovered that year) was so bright that observers in the southern hemisphere could see it without a telescope (Figure 20-16).

Supernova 1987A was the first nearby supernova to be seen since the invention of the telescope

Such a bright supernova is a rare event. In the past thousand years, only five other supernovae—in 1006, 1054, 1181, 1572, and 1604—have been bright enough to be seen with the naked eye, and all of these occurred in the Milky Way Galaxy. As the brightest supernova with modern telescopes in place, SN 1987A has given astronomers an unprecedented view of the violent death of a massive star.

The light from a supernova such as SN 1987A does not all come in a single brief flash; the outer layers continue to glow as they expand into space. For the first 20 days after the detonation of SN 1987A, its glow was powered primarily by the explosion's tremendous heat. As the expanding gases cooled, the light energy

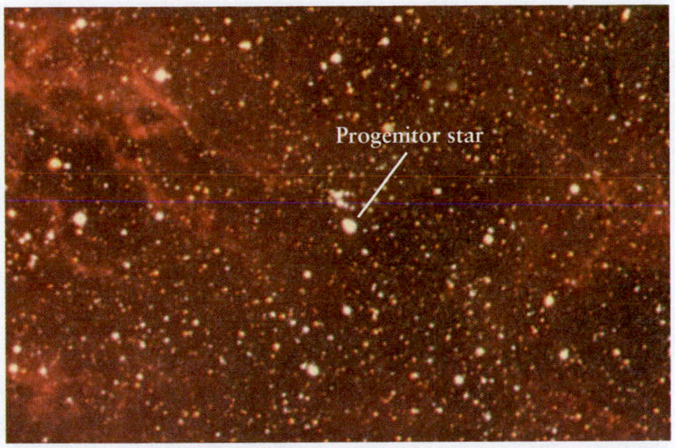

(a) Before the star exploded

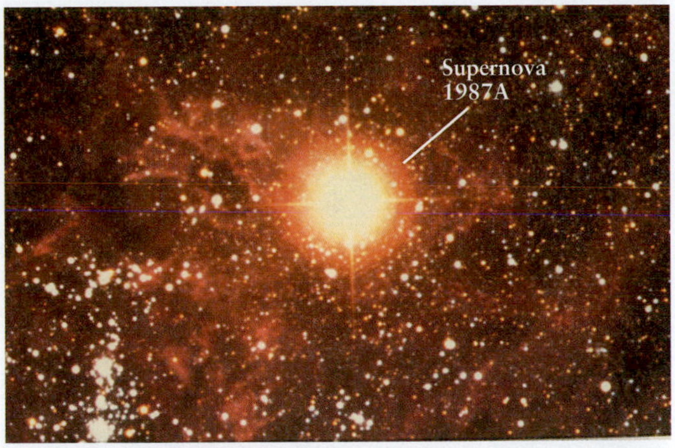

(b) After the star exploded

FIGURE 20-16 R I V U X G
SN 1987A—Before and After **(a)** This photograph shows a small section of the Large Magellanic Cloud as it appeared before the explosion of SN 1987A. The supernova's progenitor star was a B3 blue supergiant. **(b)** This image shows a somewhat larger region of the sky a few days after the supernova exploded into brightness. (Australian Astronomical Observatory/David Malin Images)

began to be provided by a different source—the decay of radioactive atoms that were produced during the supernova explosion.

Why SN 1987A Was Unusual

Ideally, SN 1987A would have confirmed the theories of astronomers about typical supernovae. But SN 1987A was *not* typical. Its luminosity peaked at roughly 10^8 $L_\odot$, only one-tenth of the maximum luminosity observed for typical supernovae. Fortunately, the doomed star was close enough to be observed prior to becoming a supernova, and these observations helped explain why SN 1987A was an exceptional case.

Figure 20-13 indicates that a massive star is usually in a red supergiant stage when the iron core collapses and the star becomes a supernova. The progenitor star of SN 1987A was indeed a high-mass star: Its estimated main-sequence mass was about 20 $M_\odot$, although by the time it exploded—some 10^7 years after it first formed—it probably had shed a few solar masses. However, this progenitor star was identified not as a red supergiant, but as a *blue* B3 I supergiant.

The explanation of this seeming contradiction is that stars in the Large Magellanic Cloud, including the progenitor of SN 1987A, are Population II stars with a very low percentage of metals—that is, elements heavier than hydrogen and helium (see Section 19-5). A small difference in the amount of metals present can cause a star to follow a somewhat different evolutionary track on the H-R diagram. Apparently, when the progenitor star of SN 1987A developed an iron core and became a supernova, it was in the blue supergiant stage.

When in a blue supergiant phase, a star's radius may be less than one-tenth as large as when it is in a red supergiant phase. Hence, for its large mass, the progenitor star was relatively small when its core collapsed (though its radius was still more than 10 times that of our Sun). This means that the star's outer layers were held more strongly by the core's gravitational attraction. When the detonation occurred, a relatively large fraction of the supernova's energy had to be used against this gravitational attraction to push the outer layers into space. Hence, the amount of energy available to be converted into light was smaller than for most supernovae. This explains why SN 1987A was only one-tenth as bright as an exploding red supergiant would have been.

The Aftermath of SN 1987A

Three and a half years after SN 1987A exploded, astronomers used the newly launched Hubble Space Telescope to obtain a picture of the supernova. To their surprise, the image showed a ring of glowing gas around the exploded star. After the optics of the Hubble Space Telescope were repaired in 1994 (see Section 6-7), SN 1987A was observed again and a set of *three* glowing rings was revealed (Figure 20-17a).

These rings are relics of a hydrogen-rich outer atmosphere that was ejected by gentle stellar winds from the progenitor star when it was a red supergiant, about 20,000 years ago. This diffuse gas expanded in a hourglass shape (Figure 20-17b), because it was blocked from expanding around the star's equator either by a preexisting ring of gas or by the orbit of an as yet unseen companion star. (Figure 20-8 shows a similar model used to explain the shapes of certain planetary nebulae.) The outer rings in Figure 20-17a are parts of the hourglass that were ionized by the initial flash of ultraviolet radiation from the supernova; as electrons recombine with the ions, the rings emit visible-light photons. (We described this process of *recombination* in Section 18-2.)

ANIMATION 20-5 By the early years of the twenty-first century, the explosion debris from the supernova was beginning to collide with the "waist" of the hourglass shown in Figure 20-17b. This collision is making the hourglass glow more brightly in visible wavelengths—though not enough, unfortunately, to make the

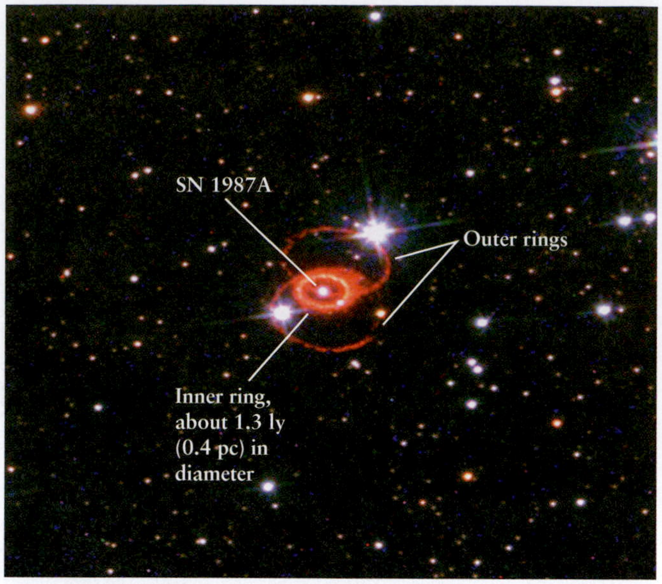

(a) Supernova 1987A seen in 1996

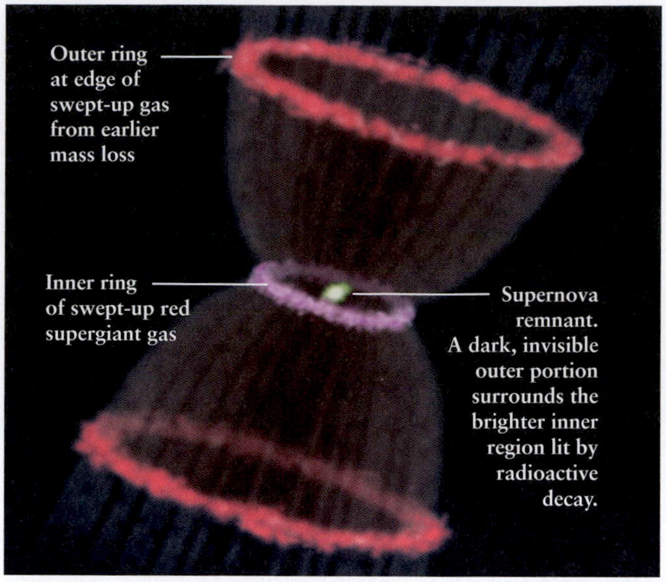

(b) An explanation of the rings

ANIMATION 20-4

FIGURE 20-17 R I V U X G **SN 1987A and Its "Three-Ring Circus"** **(a)** This true-color view from the Hubble Space Telescope shows three bright rings around SN 1987A. **(b)** This drawing shows the probable origin of the rings. A wind from the progenitor star, shown in Figure 20-16a, formed an hourglass-shaped shell surrounding the star. (Compare with Figure 20-8.) Ultraviolet light from the supernova explosion ionized ring-shaped regions in the shell, causing them to glow. (Robert Kirshner and Peter Challis, Harvard-Smithsonian Center for Astrophysics; STScI)

supernova again visible to the naked eye—and emit copious radiation at X-ray and ultraviolet wavelengths. Because SN 1987A provides a unique laboratory for studying the evolution of a supernova, astronomers will monitor it carefully for decades to come.

CONCEPTCHECK 20-13

Why was SN 1987A about one-tenth as bright than was expected?

Answer appears at the end of the chapter.

20-8 Neutrinos emanate from supernovae like SN 1987A

> Observations of SN 1987A confirm that a core-collapse supernova emits most of its energy in the form of neutrinos

In addition to electromagnetic radiation, supernovae also emit a brief but intense burst of neutrinos from their collapsing cores. In fact, theory suggests that *most* of the energy released by the exploding star is in the form of neutrinos. If it were possible to detect the flood of neutrinos from an exploding star, astronomers would have direct evidence of the nuclear processes that occur within the star during its final seconds before becoming a supernova.

Unfortunately, detecting neutrinos is difficult because under most conditions matter is transparent to neutrinos (see Section 16-4). Consequently, when supernova neutrinos encounter Earth, almost all of them pass completely through the planet as if it were not there. The challenge to scientists is to detect the tiny fraction of neutrinos that *do* interact with the matter through which they pass.

Neutrino Telescopes

During the 1980s, two "neutrino telescopes" designed uniquely for detecting supernova neutrinos went into operation—the Kamiokande detector in Japan (a joint project of the University of Tokyo and the University of Pennsylvania), and the IMB detector (a collaboration of the University of California at Irvine, the University of Michigan, and Brookhaven National Laboratory). Both detectors consisted of large tanks containing thousands of tons of water. On the rare occasion when a neutrino collided with one of the water molecules, it produced a brief flash of light. Any light flash was recorded by photomultiplier tubes lining the walls of the tank; photomultiplier tubes simply convert any detected light into an electronic signal that can be recorded. Because other types of subatomic particles besides neutrinos could also produce similar light flashes, the detectors were placed deep underground so that hundreds of meters of Earth would screen out almost all particles except neutrinos. (The more recent Sudbury Neutrino Observatory, shown in Figure 16-6, has a similar design.)

What causes the light flashes? And how do we know whether a neutrino comes from a supernova? The key is that a single supernova neutrino carries a relatively large amount of energy, typically 20 MeV or more. (We introduced the unit of energy called the electron volt, or eV, in Section 5-5. One MeV is equal to 10^6 electron volts. Only the most energetic nuclear reactions produce particles with energies of more than 1 MeV.) If such a high-energy neutrino hits a proton in the water-filled tank of a neutrino telescope, the collision produces a positron (see Box 16-1). The positron is emitted at a speed greater than the speed of light in water, which is 2.3×10^5 km/s. (Such motion does not violate the ultimate speed limit in the universe, 3×10^5 km/s, which is the speed of light in a *vacuum*.)

Just as an airplane that flies faster than sound produces a shock wave (a sonic boom), a positron that moves through a substance such as water faster than the speed of light in that substance produces a shock wave of light. This shock wave is called **Cherenkov radiation,** after the Russian physicist Pavel A. Cherenkov who first observed it in 1934. It is this radiation that is detected by the photomultiplier tubes that line the detector walls.

By measuring the properties of the Cherenkov radiation from an emitted positron, scientists can determine the positron's energy and, therefore, the energy of the neutrino that created the positron. Determining the energy allows them to tell the difference between the high-energy neutrinos from supernovae and neutrinos from the Sun, which typically have energies of 1 MeV or less.

Neutrinos from SN 1987A

Both the Kamiokande and IMB detectors were operational on February 23, 1987, when SN 1987A was first observed in the Large Magellanic Cloud. Soon afterward, the physicists working with these detectors excitedly reported that they had detected Cherenkov flashes from a 12-second burst of neutrinos that reached Earth 3 hours before astronomers saw the light from the exploding star. Only a few neutrinos were seen: The Kamiokande detector saw flashes from 11 neutrinos at about the same time that 8 were recorded by the IMB detector. But when the physicists factored in the sensitivity of their detectors, they calculated that Kamiokande and IMB had actually been exposed to a torrent of more than 10^{16} neutrinos.

Given the flux of neutrinos measured by the detectors and the distance of 168,000 ly from SN 1987A to Earth, physicists used the inverse-square law to determine the total number of neutrinos that had been emitted from the supernova. (This law applies to neutrinos just as it does to electromagnetic radiation; see Section 17-2.) They found that over a 10-second period, SN 1987A emitted 10^{58} neutrinos with a total energy of 10^{46} joules. This energy is more than 100 times as much energy as the Sun has emitted in its entire history and more than 100 times the amount of energy that the supernova emitted in the form of electromagnetic radiation. Indeed, for a few seconds the supernova's neutrino luminosity—that is, the *rate* at which it emitted energy in the form of neutrinos—was 10 times greater than the total luminosity in electromagnetic radiation of all of the stars in the observable universe! Such comparisons give a hint of the incomprehensible violence with which a supernova explodes.

Why did the neutrinos from SN 1987A arrive 3 hours *before* the first light was seen? As we saw in Section 20-6, neutrinos are produced when nuclear reactions cease in the core of a massive star and the core collapses. The neutrinos that are not absorbed around the core pass easily through the volume of the star. However, the tremendous increase in the star's light output occurs only when the shock wave reaches the star's outermost layers (see Figure 20-15a). For SN 1987A, it took 3 hours for this shock wave to travel outward from the star's core to its surface, by which time the neutrino burst was already billions of kilometers beyond the dying star. Traveling toward Earth for the next 168,000 years, the neutrinos that would eventually produce light flashes in Kamiokande and IMB remained in front of the photons emitted from the star's surface, and so the neutrinos were detected before the supernova's light. Thus, the neutrino data from SN 1987A gave astronomers direct confirmation of theoretical ideas about how supernova explosions take place.

Kamiokande and IMB have both been replaced by a new generation of neutrino telescopes (see Section 16-4). While these new detectors are intended primarily to observe neutrinos from the Sun, they are also fully capable of measuring neutrino bursts from nearby supernovae. Astronomers have identified a number of supergiant stars in our Galaxy that are likely to explode into supernovae, among them the bright red supergiant Betelgeuse in the constellation Orion (see Figure 2-2 and Figure 6-27).

Unfortunately, astronomers do not yet know how to predict precisely when such stars will explode into supernovae. It may be many thousands of years before Betelgeuse explodes. Then again, it could happen tomorrow. If it does, neutrino telescopes will be ready to record the collapse of its massive core. However, you will not need a telescope to observe this spectacular event, as it would be about ten times as bright as the full Moon!

CONCEPTCHECK 20-14

How do high-energy neutrinos produce detectable light?

Answer appears at the end of the chapter.

20-9 White dwarfs in close binary systems can also become supernovae

In just a few seconds, a thermonuclear supernova completely destroys an entire white dwarf star

Astronomers discover dozens of supernovae in distant galaxies every year, but not all of these are the result of massive stars dying violently. A totally different type of supernova occurs when a white dwarf star in a binary system (possibly with a second white dwarf) blows itself completely apart. We now look at the different types of supernovae.

Types of Supernovae

There are two main mechanisms for a supernova explosion. We have already discussed one mechanism, which is a core-collapse supernova. A second mechanism involves at least one white dwarf in a binary system. Had astronomers understood these processes all along, they might have named the different supernovae according to their underlying mechanisms. However, the first indication that there were different types of supernova came from their *differing spectral lines*. As a result, supernovae are labeled not by their underlying mechanism, but by their spectra.

Supernovae with hydrogen emission lines are called **Type II supernovae;** these are core-collapse supernovae of the sort we described in Section 20-6. They are caused by the deaths of highly evolved massive stars that still have ample hydrogen in their atmospheres when they explode. When the star explodes, the hydrogen atoms are excited and glow prominently, producing hydrogen emission lines. SN 1987A (the topic of Sections 20-7 and 20-8) was a Type II supernova.

(a) Type Ia supernova
- The spectrum has no hydrogen or helium lines, but does have a strong absorption line of ionized silicon (Si II).
- Produced by runaway carbon fusion in a white dwarf in a close binary system (the ionized silicon is a by-product of carbon fusion).

(b) Type Ib supernova
- The spectrum has no hydrogen lines, but does have a strong absorption line of un-ionized helium (He I).
- Produced by core collapse in a massive star that lost the hydrogen from its outer layers.

(c) Type Ic supernova
- The spectrum has no hydrogen lines or helium lines.
- Produced by core collapse in a massive star that lost the hydrogen and the helium from its outer layers.

(d) Type II supernova
- The spectrum has prominent hydrogen lines such as H_α.
- Produced by core collapse in a massive star whose outer layers were largely intact.

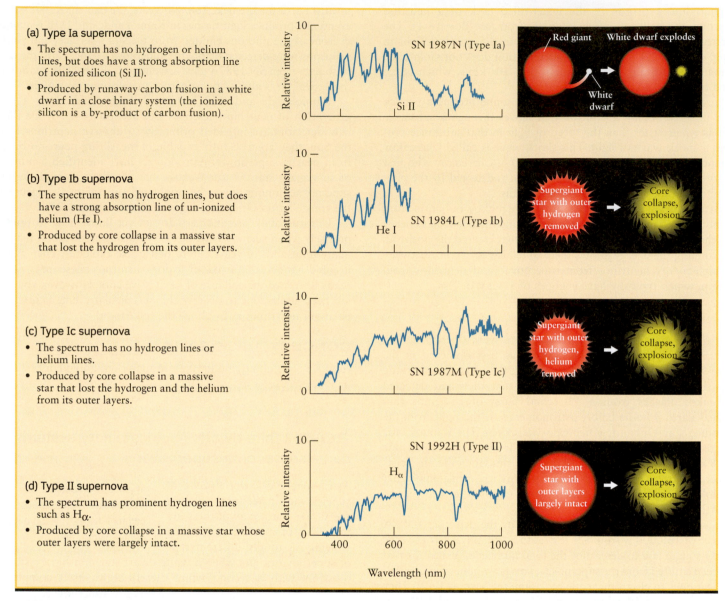

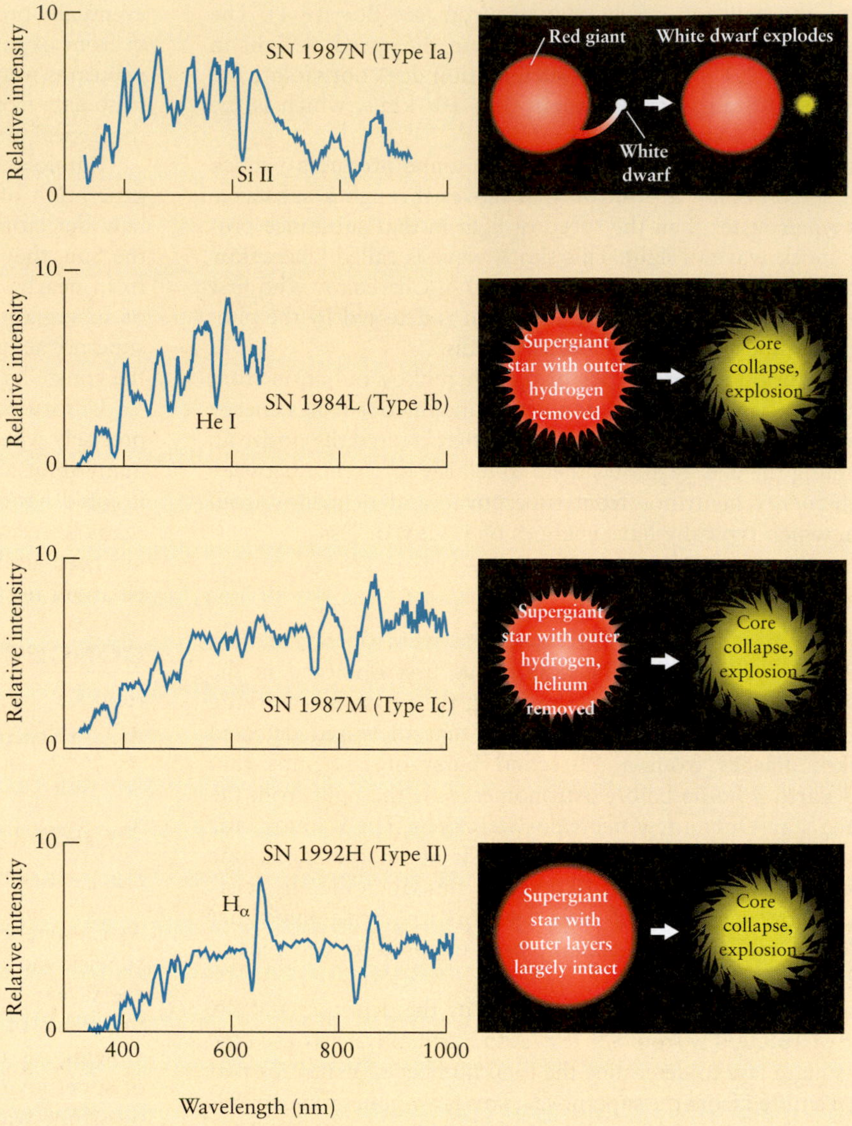

FIGURE 20-18

Supernova Types These illustrations show the characteristic spectra and the probable origins of supernovae of **(a)** Type Ia, **(b)** Type Ib, **(c)** Type Ic, and **(d)** Type II. (Spectra courtesy of Alexei V. Filippenko, University of California, Berkeley)

Hydrogen lines are missing in the spectrum of a **Type I supernova** (Figure 20-18), which tells us that little or no hydrogen is left in the debris from the explosion. Type I supernovae are further divided into three important subclasses. **Type Ia supernovae** have spectra that include a strong absorption line of ionized silicon. **Type Ib** and **Type Ic supernovae** both lack the ionized silicon line. The difference between them is that the spectra of Type Ib supernovae have a strong helium absorption line, while those of Type Ic supernovae do not.

Astronomers suspect that Type Ib and Ic supernovae are caused by core-collapse in dying massive stars, just like Type II supernovae. The difference is that the progenitor stars of Type Ib and Ic supernovae have been stripped of their outer layers before they explode. A star can lose its outer layers to a strong stellar wind (see Figure 20-12) or, if it is part of a close binary system, by transferring mass to its companion star (see Figure 19-21b). If enough mass remains for the star's core to collapse, the star dies as a Type Ib supernova. Because the outer layers of hydrogen are absent, the supernova's spectrum exhibits no hydrogen lines but many helium lines (Figure 20-18b). Type Ic supernovae have apparently undergone even more mass loss prior to their explosion; their spectra show that they have lost much of their helium as well as their hydrogen (Figure 20-18c).

As Figure 20-18 shows, dividing supernovae by their underlying explosive mechanism leads to those involving white dwarfs

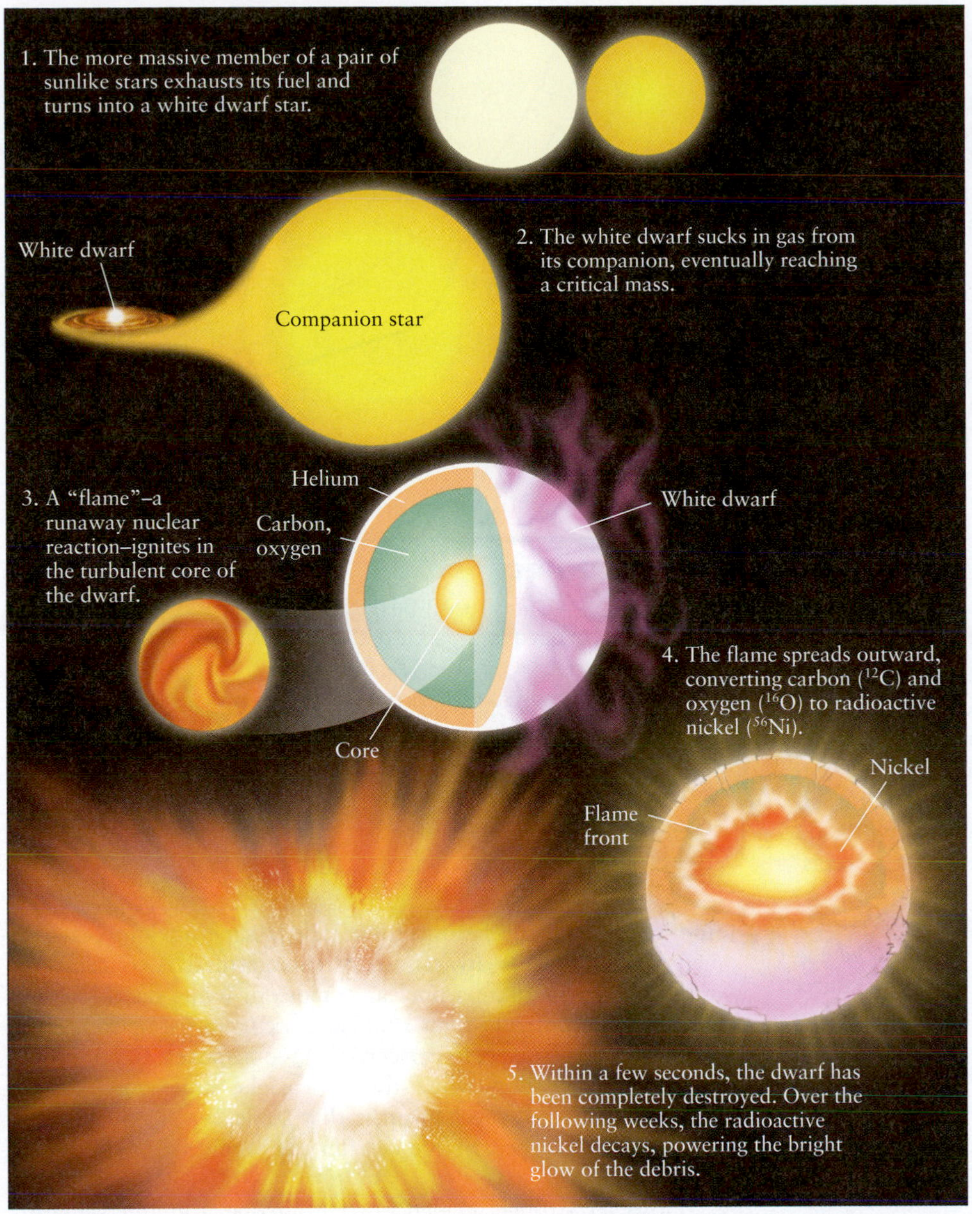

1. The more massive member of a pair of sunlike stars exhausts its fuel and turns into a white dwarf star.
White dwarf
2. The white dwarf sucks in gas from its companion, eventually reaching a critical mass.
Companion star
Helium
3. A "flame"–a runaway nuclear reaction–ignites in the turbulent core of the dwarf.
Carbon, oxygen
White dwarf
Core
4. The flame spreads outward, converting carbon (^{12}C) and oxygen (^{16}O) to radioactive nickel (^{56}Ni).
Nickel
Flame front
5. Within a few seconds, the dwarf has been completely destroyed. Over the following weeks, the radioactive nickel decays, powering the bright glow of the debris.

ANIMATION 20-6 ANIMATION 20-7

FIGURE 20-19 **A Type Ia Supernova** This series of illustrations depicts our understanding of how a white dwarf in a close binary system can undergo a sudden nuclear detonation that destroys it completely. Such a cataclysmic event is called a Type Ia supernova or thermonuclear supernova. (Illustration by Don Dixon, adapted from Wolfgang Hillebrandt, Hans-Thomas Janka, and Ewald Müller, "How to Blow Up a Star," *Scientific American,* October 2006)

(Type Ia), and core-collapse (Type II, Type Ib, and Type Ic). We now look at how to detonate a white dwarf.

Type Ia Supernovae: Detonating a White Dwarf

Type Ia supernovae are *not* the death throes of massive supergiant stars. Instead, Type Ia supernovae are thought to result from the thermonuclear explosion of a white dwarf star. Recall that the "thermo" in thermonuclear indicates that high-speed collisions at very high temperatures fuse the atoms together. Furthermore, a supernova resulting from this rapid release of energy is called a **thermonuclear supernova.** As we will see, a thermonuclear supernova begins with a white dwarf.

This may seem contradictory, because we saw in Section 20-4 that white dwarf stars have no nuclear reactions going on in their interiors. But these reactions *can* occur if a carbon-oxygen–rich white dwarf gains mass in a close, semidetached binary system with a red giant star (see Figure 19-20b and Figure 20-18a).

Figure 20-19 shows the likely series of events that lead to a Type Ia supernova involving only one white dwarf. Stage 1 in this figure shows a close binary system in which both stars have less than 4 $M_{\odot}$. The more massive star on the left evolves more rapidly than its less massive companion and eventually becomes a white dwarf. As the companion evolves and its outer layers expand, it overflows its Roche lobe and dumps gas from its outer layers onto the white dwarf (see Stage 2 in Figure 20-19). When the total mass of the white dwarf approaches the Chandrasekhar limit (1.4 $M_{\odot}$), the increased pressure applied to the white dwarf's interior causes carbon fusion to begin there (Stage 3 in Figure 20-19). Hence, the interior temperature of the white dwarf increases.

If the white dwarf were made of ordinary matter, the temperature increase would cause a further increase in pressure, the white dwarf would expand and cool, and the carbon-fusing reactions would abate. But because the white dwarf is composed of degenerate matter, this "safety valve" between temperature and pressure does not operate. Instead, the increased temperature just makes the reactions proceed at an ever-increasing rate, in a catastrophic runaway process reminiscent of the helium flash in low-mass stars (Stage 4 in Figure 20-19). The reaction spreads rapidly outward from the white dwarf's *center,* with its leading edge (called the *flame front*) being propelled by convection and turbulence, in a manner analogous to what happens to the shock wave in a core-collapse supernova. Within seconds the white dwarf blows apart, dispersing 100% of its mass into space (Stage 5 in Figure 20-19).

Before exploding, the white dwarf contained primarily carbon and oxygen and almost no hydrogen or helium, which explains the absence of hydrogen and helium lines in the spectrum of the resulting supernova. Silicon is a by-product of the carbon-fusing reaction and gives rise to the silicon absorption line characteristic of Type Ia supernovae.

CAUTION! Different types of supernovae have fundamentally different energy sources. We saw in Section 20-6 that core-collapse supernovae (which we have now sorted into Types II, Ib, and Ic) are powered by *gravitational* energy released as the star's iron-rich core and outer layers fall inward. The core undergoes nuclear reactions during the supernova, but these reactions *consume* energy provided by gravity. Therefore, the nuclear reactions that occur are gravity-powered; gravity "fuels" the explosion. Type Ia supernovae, by contrast, are powered by *nuclear* energy released in the explosive thermonuclear fusion of a white dwarf star. In these supernovae, gravity only acts as a *trigger*—just as a spark can light a candle—but, the source of energy is thermonuclear carbon-fusion reactions; nuclear energy contained in the carbon atoms is the "fuel." To highlight the source of the energy, we also use the term thermonuclear supernova to refer to a Type Ia supernova. The different types of supernova also differ in their energy output. While Type Ia supernovae typically emit more energy in the form of visible light than supernovae of other types, they do not emit copious numbers of neutrinos because there is no core collapse. If we include the energy emitted in the form of neutrinos, the most energetic supernovae by far are those of Type II. A Type Ia supernova consumes its nuclear fuel in much the same way as a fusing star, but instead of lasting a stellar lifetime, the thermonuclear reactions are over within seconds.

The Decay of a Supernova: Light Curves

In addition to the differences in their spectra, different types of supernovae can be distinguished by their light curves (Figure 20-20). All supernovae begin with a sudden rise in brightness that occurs in less than a day. After reaching peak luminosity, Type Ia, Ib, and Ic supernovae settle into a steady, gradual decline in luminosity. By contrast, most Type II light curves have a long plateau, where they remain at a similar brightness for about 100 days, before rapidly fading. For all supernova types, during the period of declining brightness—after the peak—energy comes from the decay of radioactive isotopes. These decays heat the surrounding material enough to emit visible light.

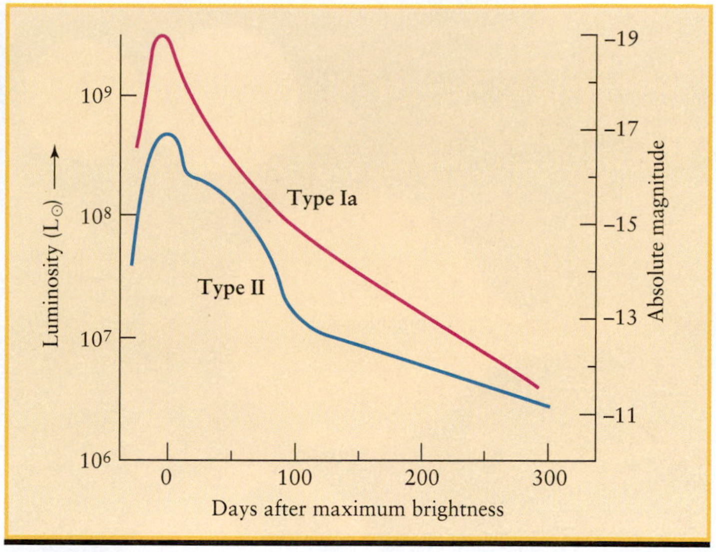

FIGURE 20-20

Supernova Light Curves A Type Ia supernova reaches maximum brightness in about a day, followed by a gradual decline in brightness. A Type II supernova reaches a maximum brightness only about one-fourth that of a Type Ia supernova and usually has alternating intervals of steep and gradual declines.

Merging White Dwarfs

Another possible mechanism for a Type Ia supernova involves the merger of two white dwarfs, where enough mass is concentrated in the merger to ignite a nuclear explosion. While the initial scenario is quite different than a binary system consisting of a white dwarf and a red giant, under certain circumstances, the predicted outcome of merging white dwarfs can be quite similar. One motivation for this mechanism is that some Type Ia supernovae fail to show evidence, before or after the event, for any red giant companion.

However, stronger evidence would involve a unique signature predicted by the merger of two white dwarfs. In 2011 the Type Ia supernova SN 2011fe exploded, and it is close enough that observations rule out a red giant companion; this supernova generates a lot of excitement because it is our best case for a nearby white dwarf merger. One signature that might indicate a white dwarf merger is that such a merger is expected to produce a greater proportion of radioactive cobalt atoms that power its light curve (unique from those in Figure 20-20). For SN 2011fe, it is predicted that after several years of monitoring the light curve, this higher abundance of cobalt atoms might be revealed and tell us if the event was most likely due to merging white dwarfs.

Type Ia Supernova Are Used to Measure Distances to Other Galaxies

A number of astronomers are now measuring the distances to remote galaxies by looking for Type Ia supernovae in those galaxies. This is possible because there is a simple relationship between the rate at which a Type Ia supernova fades away and its peak luminosity: The slower it fades, the greater its luminosity. Hence, by observing how rapidly a distant Type Ia supernova fades, astronomers can determine its peak luminosity—not just its apparent brightness, but its actual intrinsic luminosity at its peak.

Then, a measurement of the supernova's peak apparent brightness tells us (through the inverse-square law) the distance to the supernova, and, therefore, the distance to the supernova's host galaxy. In this way, the Type Ia supernova is being used as a "standard candle" of known luminosity, and by comparison to its apparent brightness, its distance is determined.

The tremendous luminosity of Type Ia supernovae allows this method to be used for galaxies more than 10^9 ly distant. In Chapter 25 we will learn what such studies tell us about the size and evolution of the universe as a whole. In fact, we will see that distances determined from Type Ia supernovae underlie such discoveries as "dark energy."

CONCEPTCHECK 20-15

If a supernova is observed with no lines in emission or absorption of hydrogen, helium, or silicon, did it result from core-collapse, or involve a white dwarf (consult Figure 20-18)? What "type" is it?

Answer appears at the end of the chapter.

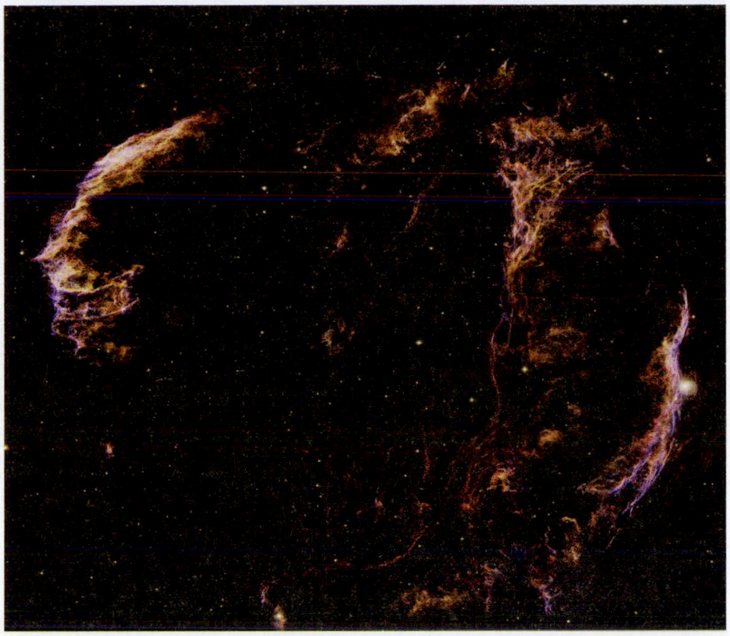

FIGURE 20-21 R I V U X G
The Veil Nebula—A Supernova Remnant The Cygnus Loop is a roughly spherical remnant of a supernova that exploded about 8,000 years ago. The distance to the nebula is about 1,500 ly and the overall diameter of the loop is about 100,000 ly. This supernova remnant is so large that it covers an area of about 45 full moons! The image contains more than 1.6 Gb of data and is one of the largest astronomical images ever made. (National Optical Astronomy Observatory [NOAO] and WIYN partners)

20-10 A supernova remnant can be detected at many wavelengths for centuries after the explosion

Many supernovae in our Galaxy are hidden from us by the obscuring interstellar medium

Astronomers find the debris of supernova explosions, called **supernova remnants,** scattered across the sky. A beautiful example of a supernova remnant is the Veil Nebula, shown in Figure 20-21. The doomed star's outer layers were blasted into space with such violence that they are still traveling through the interstellar medium at supersonic speeds 8000 years later. As this expanding shell of gas plows through space, it collides with atoms in the interstellar medium, exciting the gas and making it glow. We saw in Section 18-8 that the passage of a supernova remnant through the interstellar medium can trigger the formation of new stars, so the death of a single massive star (in a core-collapse supernova) or white dwarf (in a thermonuclear supernova) can cause a host of new stars to be born.

A few nearby supernova remnants cover sizable areas of the sky. The largest is the Gum Nebula, named after the astronomer Colin Gum, who first noticed its faint glowing wisps on photographs of the southern sky. Its astonishingly wide 40° angular diameter is centered on the constellation Vela (the Ship's Sail).

The Gum Nebula looks big because it is quite close to us and has had a long time to expand—its center is only about 1000 ly from Earth, and it originated from a supernova explosion about a million years ago. Embedded in the Gum Nebula is another large supernova remnant called the Vela Nebula (Figure 20-22), which is about 16° wide. Studies of the nebula's expansion rate suggest

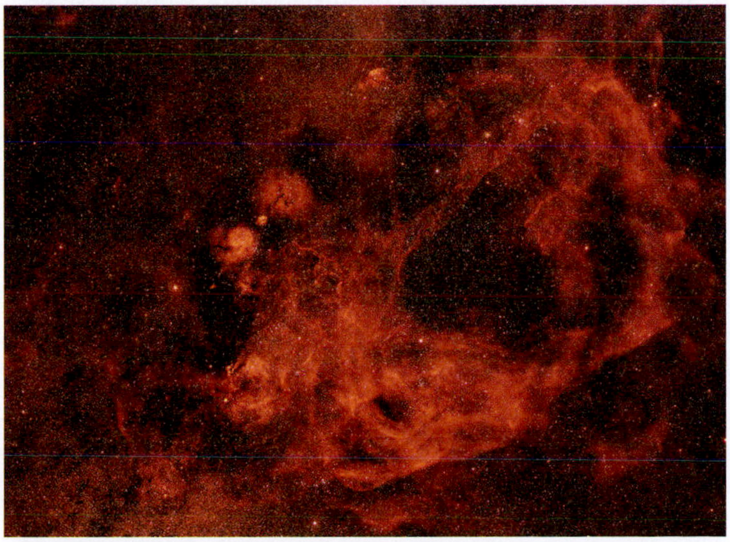

FIGURE 20-22 R I V U X G
The Vela Nebula—A Supernova Remnant The Vela Nebula is embedded in an even larger remnant called the Gum Nebula. The Vela supernova explosion occurred about 11,000 years ago, and the remnant now has a diameter of about 700 pc (2300 ly). (© Axel Mellinger)

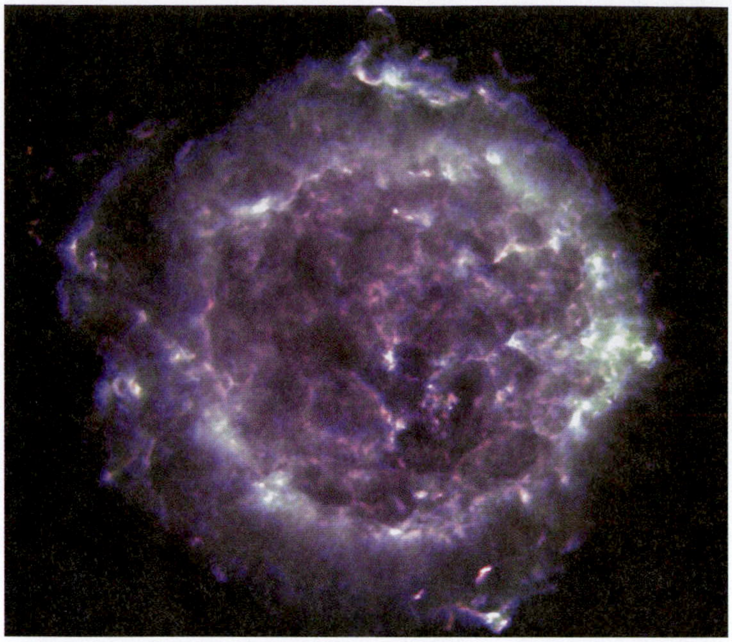

ANIMATION 20-8

FIGURE 20-23 R I V U X G

Cassiopeia A—A Supernova Remnant This false-color radio image of Cassiopeia A was produced by the Very Large Array (see Figure 6-23). The impact of supernova material on the interstellar medium causes ionization, and the liberated electrons generate radio waves as they move. Cassiopeia A is roughly 3300 pc (11,000 ly) from Earth. (NRAO/AUI)

that this supernova exploded around 9000 B.C.E. At maximum brilliance, the exploding star probably was as bright as the Moon at first quarter. Like the first quarter moon, it would have been visible in the daytime!

Many supernova remnants are nearly invisible at optical wavelengths. However, when the expanding gases collide with the interstellar medium, they radiate energy at a wide range of wavelengths, from X-rays through radio waves. For example, Figure 20-23 shows a radio image of the supernova remnant Cassiopeia A. (Compare to Figure 18-26, which is a composite of observations of Cassiopeia A at X-ray, visible, and infrared wavelengths.) As a rule, radio searches for supernova remnants are more fruitful than optical searches. Only two dozen supernova remnants have been found in visible-light images, but more than 100 remnants have been discovered by radio astronomers.

From the expansion rate of Cassiopeia A, astronomers conclude that this supernova explosion occurred about 300 years ago. Although telescopes were in wide use by the late 1600s, no one saw the outburst (and no one today knows why). The last supernova seen in our Galaxy, which occurred in 1604, was observed by Johannes Kepler. In 1572, Tycho Brahe also recorded the sudden appearance of an exceptionally bright star in the sky. To find any other accounts of nearby bright supernovae, we must delve into astronomical records that are almost 1000 years old.

At first glance, this apparent lack of nearby supernovae may seem puzzling. From the frequency with which supernovae occur in distant galaxies, it is expected that we should have about two supernovae per century. Where have they been?

As we will learn when we study galaxies in Chapters 22 and 23, the plane of our Galaxy is where massive stars are born and supernovae explode. This disklike region is so rich in interstellar dust, however, that we simply cannot see very far into space when looking in the plane of the disk, which is in the directions occupied by the Milky Way (see Section 20-2). In other words, supernovae probably do in fact erupt every few decades in remote parts of our Galaxy, but their detonations are hidden from our view by intervening interstellar matter.

Relics of the Fall: White Dwarfs, Neutron Stars, and Black Holes

A supernova remnant may be all that is left after some supernovae explode. But for core-collapse supernovae of Types II, Ib, and Ic, the core itself may also remain. If there is a relic of the core, it may be either a *neutron star* or a *black hole*, depending on the mass of the core and the conditions within it during the collapse. Neutron stars, as the name suggests, are made primarily of neutrons. Wholly unlike anything we have studied so far, these exotic objects are the subject of the next section. We will study black holes, which are far stranger even than neutron stars, in Chapter 21.

CAUTION! Although neutron stars and black holes can be part of the debris from a supernova explosion, they are *not* called "supernova remnants." That term is applied exclusively to the gas and dust that spreads away from the site of the supernova explosion.

We have seen in this chapter that only the most massive stars end their lives as core-collapse supernovae. We learned earlier that mass plays a central role in determining the speed with which a star forms and joins the main sequence (see Figure 18-10), the star's luminosity and surface temperature while on the main sequence (see Figure 17-22 and the *Cosmic Connections* figure in Chapter 17), and how long a star can remain on the main sequence (see Table 19-1). Now we see that a star's initial mass also determines its eventual fate (Figure 20-24a). We also see that a star's fate is connected to the formation of future stars and the matter from which even we are made of (Figure 20-24b).

CONCEPTCHECK 20-16

How can astronomers estimate how long ago a supernova exploded by its remnant?

Answer appears at the end of the chapter.

20-11 Neutron stars

On the morning of July 4, 1054, Yang Wei-T'e, the imperial astronomer to the Chinese court, made a startling discovery. Just a few minutes before sunrise, a new and dazzling object ascended above the eastern horizon. This "guest star," as Yang called it, was far brighter than Venus and more resplendent than any star he had ever seen.

Yang's records show that the "guest star" was so brilliant that it could easily be seen during broad daylight for the rest of

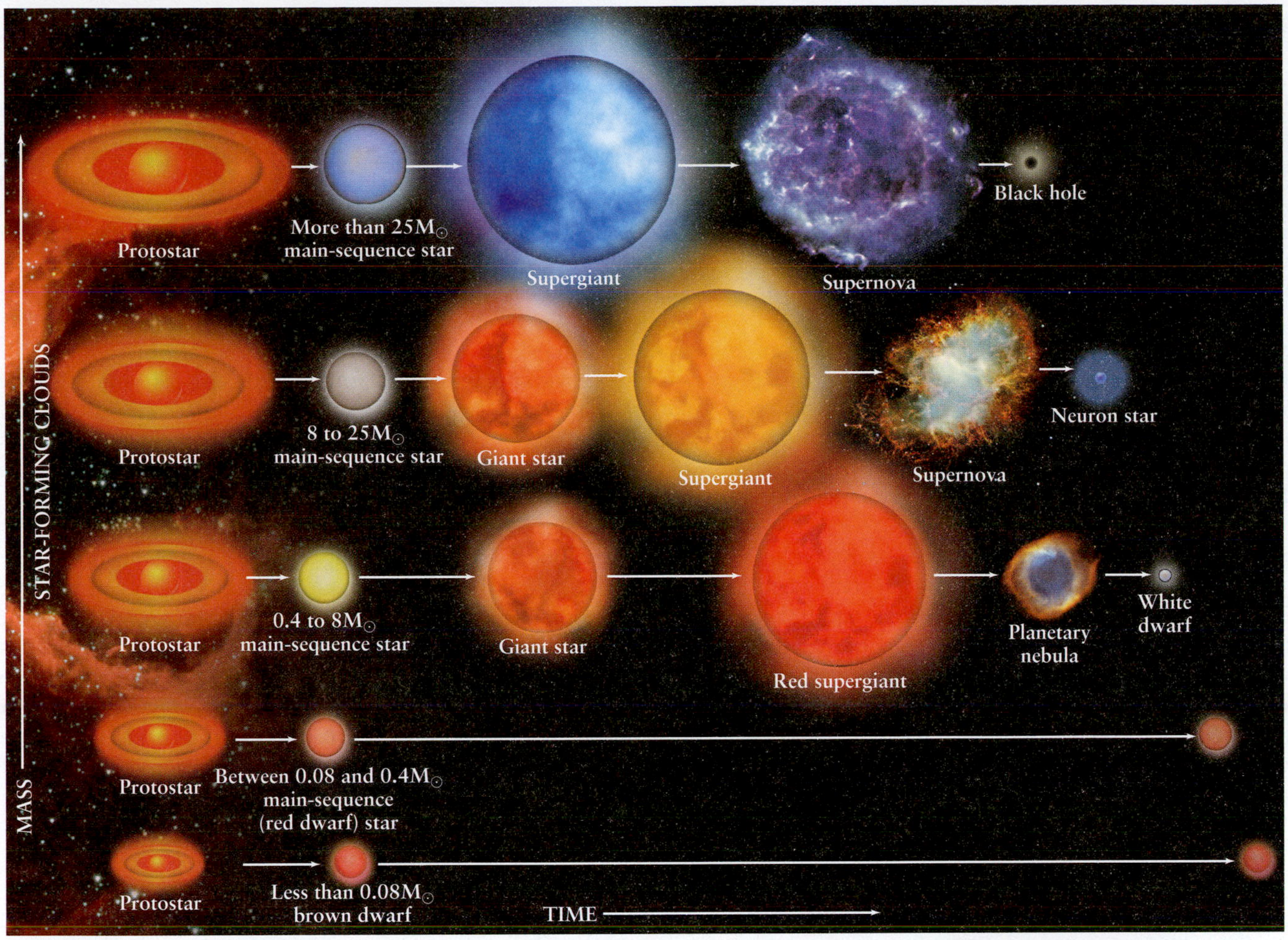

(a)

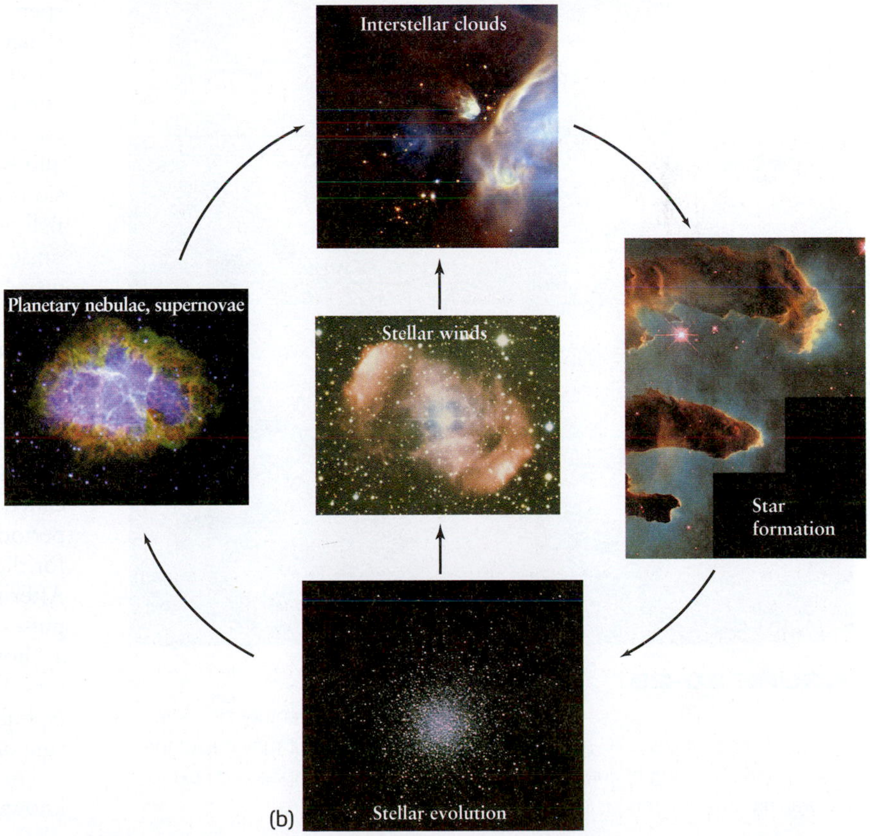

(b)

R I V U X G

FIGURE 20-24

A Summary of Stellar Evolution **(a)** The evolution of an isolated star (one that is not part of a close multiple-star system) depends on the star's mass. The more massive the star, the more rapid its evolution. If the star's initial mass is less than about 0.4 $M_\odot$, it evolves slowly over the eons into an inert ball of helium. If the initial mass is in the range from about 0.4 $M_\odot$ to about 8 $M_\odot$, it ejects enough mass over its lifetime so that what remains is a white dwarf with a mass less than the Chandrasekhar limit of 1.4 $M_\odot$. If the star's initial mass is more than about 8 $M_\odot$, it ends as a core-collapse supernova, leaving behind a neutron star or black hole. **(b)** These images summarize the key stages in the cycle of stellar evolution.

(b: *top:* Infrared Space Observatory, NASA; *right:* Australian Astronomical Observatory; *bottom:* NASA; *left:* NASA; *middle:* Australian Astronomical Observatory/David Malin Images)

July. Records from Constantinople (now Istanbul, Turkey) also describe this object, and works of art made by the Chacoans in the American Southwest suggest that they may have seen it as well (Figure 20-25). Over the next 21 months, however, the "guest star" faded to invisibility.

We now know that the "guest star" of 1054 was actually a remarkable stellar transformation: A massive star some 6500 ly away perished in a supernova explosion, leaving behind both a supernova remnant and a bizarre object called a **neutron star**—an incredibly dense sphere composed primarily of neutrons. (We learned in Section 5-7 that a neutron has about the same mass as a proton, but has no electric charge.) Today, the remnant of this supernova that was seen nearly a thousand years ago is called the Crab Nebula; the neutron star at its center is known as the Crab pulsar for reasons we will see shortly (Figure 20-26a).

We saw in Section 20-6 that under the very high pressures within a core-collapse supernova, a proton and an electron can combine to form a neutron (as well as a neutrino). Far from being a rare transformation, **a core-collapse can convert over a solar mass of material into almost pure neutrons.** Recall that white dwarfs are supported by degenerate electron pressure, and by a similar effect, neutron stars are supported against further gravitational collapse by degenerate neutron pressure. (Like electrons, neutrons obey the Pauli exclusion principle that we described in Section 19-3.)

FIGURE 20-25

A Supernova Pictograph This drawing in an eleventh-century structure in New Mexico shows a 10-pointed star next to a crescent. It may depict the scene on the morning of July 5, 1054, when a "guest star" appeared next to the waning crescent moon. (Courtesy of National Parks Service)

Although the possibility for neutron stars to form was proposed in 1934, most scientists politely ignored the idea for years. After all, a neutron star must be a rather weird object. If brought to Earth's surface, a single teaspoonful of neutron star matter would weigh the same as about 20 million elephants (about 100 million tons)!

Neutron stars can only exist for a narrow range of masses. Supported by degenerate electron pressure, stellar cores below the Chandrasekhar limit of 1.4 $M_\odot$ form white dwarfs. Above the Chandrasekhar limit, a neutron star forms, but the maximum possible neutron star mass is only around 2 to 3 $M_\odot$. Beyond a stellar core mass of 2 to 3 $M_\odot$, gravity overwhelms internal pressure, and a black hole is formed. Keep in mind that these are only the masses of the stellar cores. Stars with total masses between 8 to 25 $M_\odot$ are expected to develop neutron stars, and above 25 $M_\odot$, black holes.

A couple of solar masses is a lot of material, but at extreme densities, these objects are surprisingly small. A 1.4-$M_\odot$ neutron star would have a diameter of only 20 km (12 mi). Thus, a neutron star is like a giant atomic nucleus made only of neutrons, and the size of a city! Its surface gravity would be so strong that its escape speed—the speed an object would need to have in order to escape into space (Section 4-7) would be one-half the speed of light. Even climbing a 1-millimeter "mountain" on a neutron star would require about the same energy as a person on Earth jumping straight to the Moon! These conditions seemed outrageous until the late 1960s, when astronomers discovered pulsating radio sources.

The Discovery of Pulsars

The rapid flashing of radiation from a pulsar gave evidence that white dwarfs are not the only endpoint of stellar evolution

As a young graduate student at Cambridge University, Jocelyn Bell spent many months helping construct an array of radio antennas covering 4½ acres of the English countryside. The instrument was completed by the summer of 1967, and Bell and her colleagues began using it to scrutinize radio emissions from the sky. While searching for something quite different, Bell noticed that the antennas had detected regular pulses of radio noise from one particular location in the sky. These radio pulses were arriving at regular intervals of 1.3373011 seconds—much more rapid than those of any other astronomical object known at that time. Indeed, they were so rapid and regular that the Cambridge team first wondered if they might be signals from an advanced alien civilization.

That possibility had to be discarded within a few months after several more of these pulsating radio sources, which came to be called **pulsars**, were discovered across the sky. In all cases, the periods were extremely regular, ranging from about 0.25 second for the fastest to about 1.5 seconds for the slowest (Figure 20-27). After ruling out other mechanisms to generate such rapid periodic pulses (such as rapidly orbiting binary systems), it appeared that a "hot spot" on a rotating white dwarf might account for pulsars. However, the discovery of a pulsar at the center of the Crab Nebula—called the Crab pulsar (Figure 20-26b)—was about to rule out white dwarfs and fundamentally change astronomy.

At the time of its discovery, the Crab pulsar was the fastest pulsar known to astronomers. Its period is 0.0333 seconds per rotation,

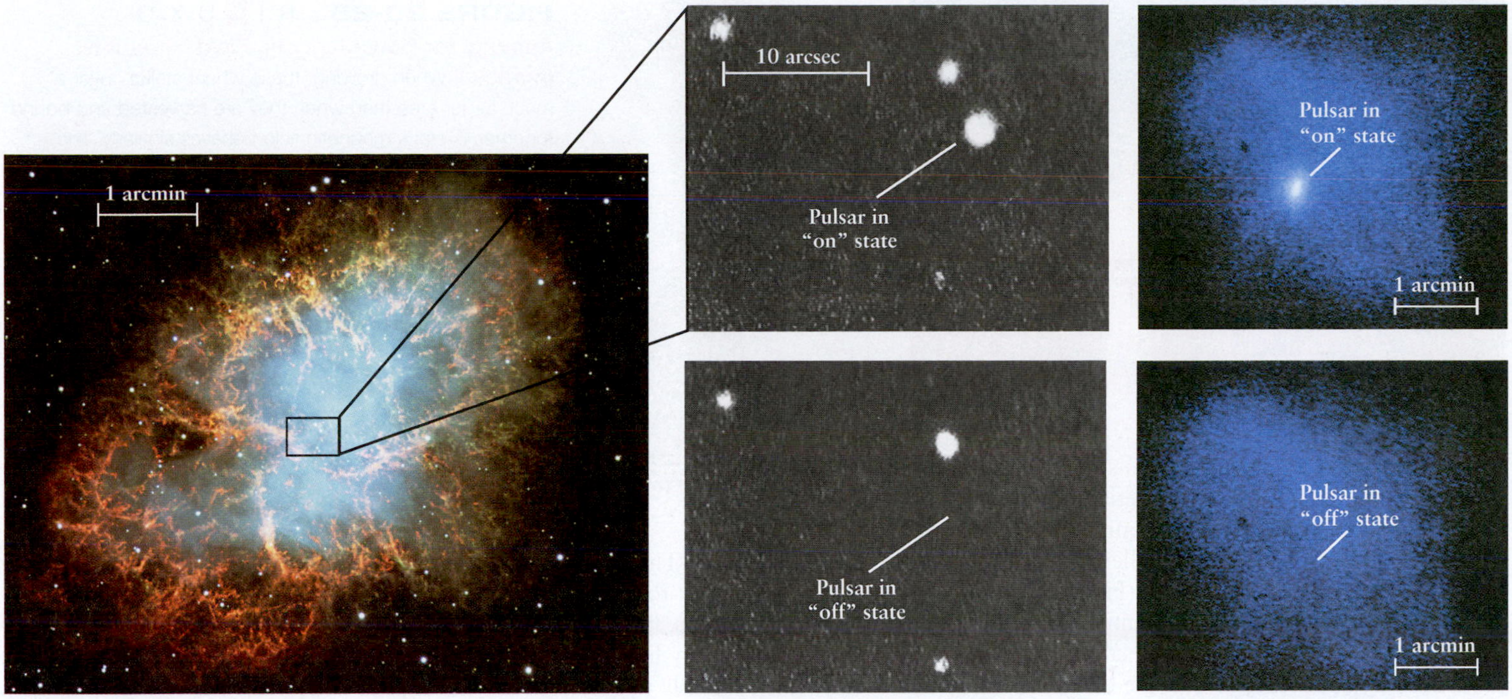

(a) The Crab Nebula R I V U X G
(b) The Crab pulsar in visible light R I V U X G
(c) The Crab pulsar in X-rays R I V U X G

FIGURE 20-26

The Crab Pulsar **(a)** A pulsar is located at the center of the Crab Nebula, which is about 2000 pc (6500 ly) from Earth and about 3 pc (10 ly) across. The boxed area is shown in close-up in part (b). **(b)** We see a flash of visible light when the rotating pulsar's beam is directed toward us (the "on" state). The pulsar fades (the "off" state) when the beam is aimed elsewhere. **(c)** The Crab pulsar also pulses "on" and "off" at X-ray wavelengths. The radio pulses, visual flashes, and X-ray flashes all have a period of 0.033 s. (a, b: The FORS Team, VLT, European Southern Observatory; c: Harvard-Smithsonian Center for Astrophysics)

which means that it flashes or rotates about 1/0.0333 = 30 times each second. It was immediately apparent to astrophysicists that white dwarfs are too big and bulky to generate 30 signals per second; calculations demonstrated that a white dwarf could not rotate that fast without tearing itself apart! Hence, the Crab pulsar indicated that pulsars had to be much smaller and more compact than a white dwarf. We now understand that pulsars arise from rapidly rotating neutron stars, which were created in supernova explosions. This was truly remarkable, because most astronomers in the mid-1960s thought that *all* stellar corpses are white dwarfs.

The neutron star model of pulsars had to pass several stringent tests to be accepted by astronomers

Because neutron stars are very small, they should also rotate rapidly. A typical star, such as our Sun, takes nearly a full month to rotate once about its axis. But just as a spinning ice skater speeds up when she pulls in her arms, a collapsing star also speeds up as its size shrinks. (We introduced this principle, called the *conservation of angular momentum,* in Section 8-4; see Figure 8-7.) If our Sun were compressed to the size of a neutron star, it would spin about 1000 times per second! Because neutron stars are so small and dense, they can spin this rapidly without flying apart.

The small size of neutron stars also leads to intense magnetic fields. The magnetic field of a main-sequence star is spread out over billions of square kilometers of the star's surface. However, if such

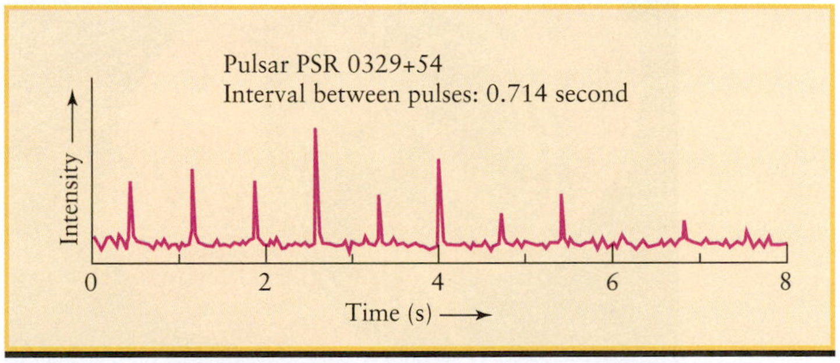

FIGURE 20-27

A Recording of a Pulsar This chart recording shows the intensity of radio emission from one of the first pulsars to be discovered. (The designation PSR 0329+54 means pulsar at a right ascension of $03^h\ 29^m$ and a declination of +54°.) Note that the interval between pulses is very regular, even though some pulses are weak and others are strong. (Adapted from R. N. Manchester and J. H. Taylor)

FIGURE 20-28 R I V U X G
Analogy for How Magnetic Field Strengths Increase When growing, these wheat stalks cover a much larger area than when they are harvested and bound together. A star's magnetic field behaves similarly. The collapsing star carries the field inward, thereby increasing its strength. (a: Corbis; b: Oscar Burriel/Photo Researchers, Inc.)

a star collapses down to a neutron star, its surface area shrinks by a factor of about 10^{10}. The magnetic field, which weaves through and is connected to the star's ionized gases (see Section 16-9), becomes concentrated over a smaller area than before the collapse and increases by a factor of 10^{10} (Figure 20-28)

How can astronomers determine the strength of a neutron star's magnetic field? Certain spectral lines can be split apart—after initially appearing as a single line—and the greater the magnetic field, the farther apart the lines move. This method reveals that magnetic fields on neutron stars are over a million times stronger than the strongest fields produced in labs on Earth.

The magnetic field of a neutron star makes it possible for the star to radiate pulses of energy toward our telescopes. As a neutron star rotates, the magnetic axis of a neutron star—the imaginary line passing through the north and south magnetic poles—is likely to be inclined at an angle to the rotation axis (Figure 20-29). After all, there is no fundamental reason for these two axes to coincide. (Indeed, these two axes do not coincide for any of the planets of our solar system.) Furthermore, nearby electrically

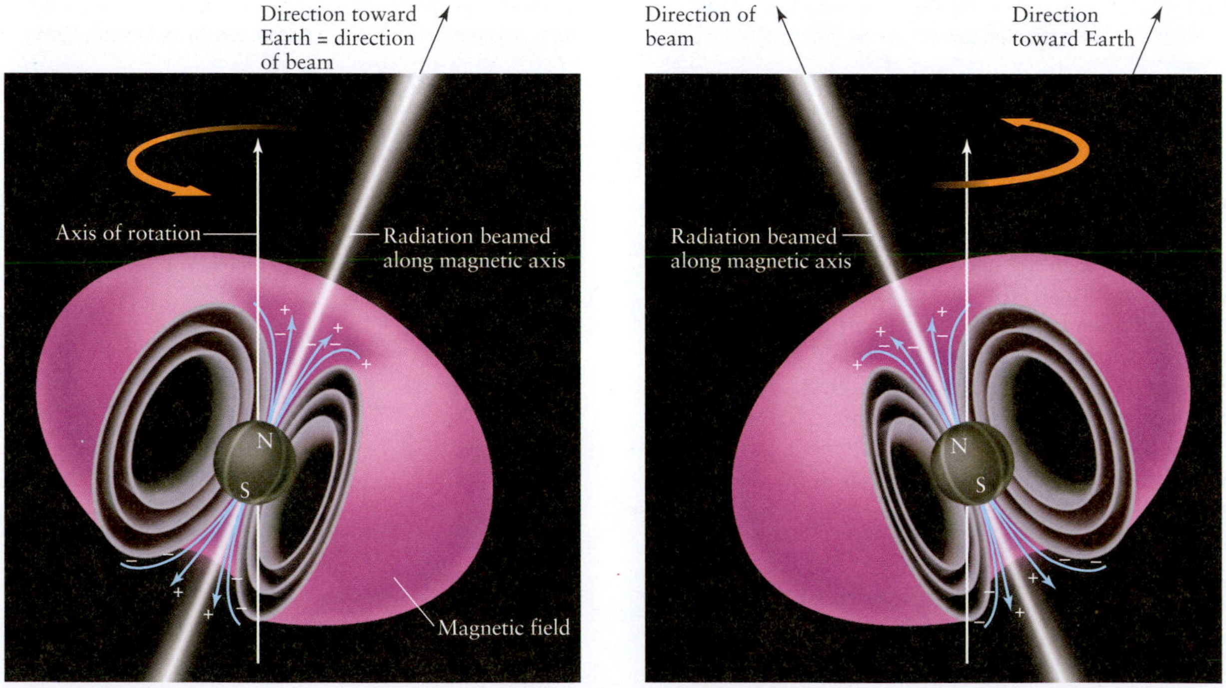

(a) One of the beams from the rotating neutron star is aimed toward Earth: We detect a pulse of radiation.

(b) Half a rotation later, neither beam is aimed toward Earth: We detect that the radiation is "off."

FIGURE 20-29
A Rotating, Magnetized Neutron Star Charged particles are accelerated near a magnetized neutron star's magnetic poles (labeled N and S), producing two oppositely directed beams of radiation. If the star's magnetic axis (a line that connects the north and south magnetic poles) is tilted at an angle from the axis of rotation, as shown here, the beams sweep around the sky as the star rotates. If Earth happens to lie in the path of one of the beams, we detect radiation that appears to pulse **(a)** on and **(b)** off.

charged particles are accelerated by the magnetic fields, and this motion produces electromagnetic radiation. The result is that two narrow beams of radiation pour out of the neutron star's north and south magnetic polar regions. While pulsars were originally discovered at radio wavelengths, some also pulse in visible light and X-rays (Figure 20-26b,c). Note that a neutron *star* is not a star in the usual sense: it isn't powered by nuclear reactions.

ANALOGY A rotating, magnetized neutron star is somewhat like a lighthouse beacon. As the star rotates, the beams of radiation sweep around the sky. If at some point during the rotation one of those beams happens to point toward Earth, as shown in Figure 20-29a, we will detect a brief flash as the beam sweeps over us. At other points during the rotation, the beam will be pointed away from Earth, and the radiation from the neutron star will appear to have turned off (Figure 20-29b). Hence, a radio telescope will detect regular pulses of radiation, with one pulse being received for each rotation of the neutron star.

CAUTION! The name *pulsar* may lead you to think that the source of radio waves is actually pulsing. But in the model just described, this is not the case at all. Instead, beams of radiation are emitted continuously from the magnetic poles of the neutron star. The pulsing that astronomers detect here on Earth is simply a result of the rapid rotation of the neutron star, which brings one of the beams periodically into our line of sight, as Figure 20-29 shows. In this sense, the analogy between a pulsar and a lighthouse beacon is a very close one.

CONCEPTCHECK 20-17

What prevents a neutron star from collapsing under its own tremendous gravitational attraction?

CONCEPTCHECK 20-18

How can a rotating neutron star lead to radio pulses?

Answers appear at the end of the chapter.

20-12 Explosive nuclear processes on white dwarfs and neutron stars produce novae and bursters

Still other exotic phenomena occur when a stellar corpse is part of a close binary system. One example is a **nova** (plural **novae**), in which a faint star suddenly brightens by a factor of 10^4 to 10^8 over a few days or hours, reaching a peak luminosity of about 10^5 $L_\odot$. By contrast, a *supernova* has a peak luminosity of about 10^9 $L_\odot$.

Novae and White Dwarfs

In the 1950s, painstaking observations led to the conclusion that all novae are members of close binary systems containing a white dwarf. Gradual mass transfer from the ordinary companion star deposits fresh hydrogen onto the white dwarf (see Figure 19-21b for a schematic diagram of this sort of mass transfer).

Because of the white dwarf's strong gravity, this hydrogen is compressed into a dense layer covering the hot surface of the white dwarf. As more gas is deposited and compressed, the temperature in the hydrogen layer increases. When the temperature reaches about 10^7 K, hydrogen fusion ignites throughout the gas layer, embroiling the white dwarf's surface in a nuclear detonation that we see as a nova (Figure 20-30).

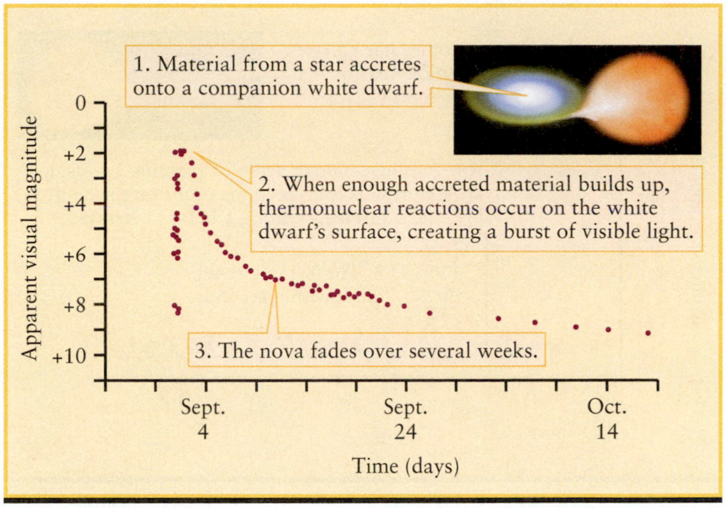
1. Material from a star accretes onto a companion white dwarf.
2. When enough accreted material builds up, thermonuclear reactions occur on the white dwarf's surface, creating a burst of visible light.
3. The nova fades over several weeks.
Apparent visual magnitude
0
+2
+4
+6
+8
+10
Sept. 4
Sept. 24
Oct. 14
Time (days)

FIGURE 20-30

The Light Curve of a Nova This illustration and graph show the history of Nova Cygni 1975, a typical nova. Its rapid rise and gradual decline in apparent brightness are characteristic of all novae. This nova, also designated V1500 Cyg, was easily visible to the naked eye (that is, was brighter than an apparent magnitude of +6) for nearly a week. (Illustration courtesy CXC/M. Weiss)

CAUTION! It is important to understand the similarities and differences between novae and the Type Ia supernovae that we described in Section 20-9 (see Figure 20-19). Both kinds of celestial explosions are thought to occur in close binary systems where one or both of the stars is a white dwarf. But, as befits their name, supernovae are much more energetic. A Type Ia supernova explosion radiates 10^{44} joules of energy into space, while the corresponding figure for a typical nova is 10^{37} joules. (To be fair to novae, this relatively paltry figure is as much energy as our Sun emits in 1000 years.) The difference is thought to be that in a Type Ia supernova, the white dwarf accretes much more mass from its companion, or a merger produces a larger mass. This added mass causes so much compression that nuclear reactions can take place *inside* the white dwarf and blow the white dwarf completely apart. In a nova, by contrast, nuclear reactions occur only within the accreted material on the surface.

Novae and thermonuclear supernovae both occur in close binary systems with a white dwarf, but a nova can recur while a supernova is a one-shot event

Because the white dwarf itself survives a nova explosion, it is possible for the same star to undergo more than one nova. As an example, the star RS Ophiuchi erupted as a nova in 1898, then put

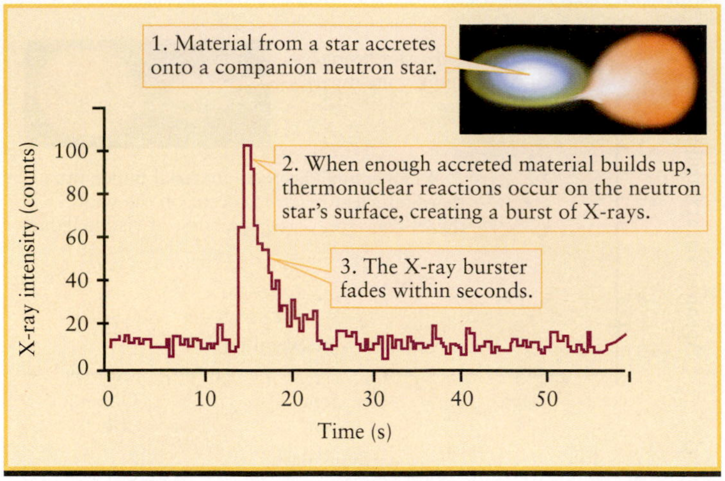
1. Material from a star accretes onto a companion neutron star.
2. When enough accreted material builds up, thermonuclear reactions occur on the neutron star's surface, creating a burst of X-rays.
3. The X-ray burster fades within seconds.
X-ray intensity (counts)
100
80
60
40
20
0
0
10
20
30
40
50
Time (s)

FIGURE 20-31

The Light Curve of an X-ray Burster This illustration and graph show the history of a typical X-ray burster. A burster emits a constant low intensity of X-rays interspersed with occasional powerful X-ray bursts. This burst was recorded on September 28, 1975, by an Earth-orbiting X-ray telescope. Contrast this figure with Figure 20-30, which shows a typical nova. (Data adapted from W. H. G. Lewin; illustration courtesy CXC/M. Weiss)

in repeat performances in 1933, 1958, 1967, 1985, and 2006. By contrast, a given star can only be a supernova once.

X-ray Bursters and Neutron Stars

A surface explosion similar to a nova also occurs with neutron stars. In 1975 it was discovered that some objects in the sky emit sudden, powerful bursts of X-rays. Figure 20-31 shows the record of a typical burst. The source emits X-rays at a constant low level until suddenly, without warning, there is an abrupt increase in X-rays, followed by a gradual decline. An entire burst typically lasts for only 20 seconds, and the same object can repeat these bursts, although at irregular intervals. Sources that behave in this fashion are known as **X-ray bursters.** Several dozen X-ray bursters have been discovered in our Galaxy.

X-ray bursters, like novae, are thought to involve close binaries whose stars are engaged in mass transfer. With a burster, however, the stellar corpse is a neutron star rather than a white dwarf. Gases escaping from the ordinary companion star fall onto the neutron star. The energy released as these gases crash down onto the neutron star's surface produces the low-level X-rays that are continuously emitted by the burster.

Most of the gas falling onto the neutron star is hydrogen, which the star's powerful gravity compresses against its hot surface. In fact, temperatures and pressures in this accreting layer become so high that the arriving hydrogen is converted into helium by hydrogen fusion. As a result, the accreted gases develop a layered structure that covers the entire neutron star, with a few tens of centimeters of hydrogen lying atop a similar thickness of helium. The structure is reminiscent of the layers within an evolved giant star (see Figure 20-2), although the layers atop a neutron star are much more compressed, thanks to the star's tremendous surface gravity.

When the helium layer grows to about 1 m thick, helium fusion ignites explosively and heats the neutron star's surface to about 3×10^7 K. At this temperature the surface predominantly emits X-rays, but the emission ceases within a few seconds as the surface cools. Hence, we observe a sudden burst of X-rays only a few seconds in duration. New hydrogen then flows onto the neutron star, and the whole process starts over. Indeed, X-ray bursters typically emit a burst every few hours or days.

Whereas explosive *hydrogen* fusion on a white dwarf produces a nova, explosive *helium* fusion on a neutron star produces an X-ray burster. In both cases, the process is explosive, because the fuel is compressed so tightly against the star's surface that it becomes degenerate, like the star itself. As with the helium flash inside red giants (described in Section 19-3), the ignition of a degenerate nuclear fuel involves a sudden thermal runaway. This is because an increase in temperature does not produce a corresponding increase in pressure that would otherwise relieve compression of the gases and slow the nuclear reactions.

Pulsars and X-ray bursters are just some of the intriguing phenomena associated with neutron stars. However, some of the most interesting objects of all—black holes—are created when there is so much matter that a neutron star *cannot* form. The gravity associated with a neutron star is so strong that the escape speed from it is roughly one-half the speed of light. But if a stellar corpse has a mass greater than 3 $M_\odot$, so much matter is crushed into such a small volume that the escape speed actually *exceeds* the speed of light. Because nothing can travel faster than light, nothing—not even light—can leave this dead star. Its gravity is so powerful that it leaves a hole in the fabric of space and time. We take up this story of black holes in the next chapter.

CONCEPTCHECK 20-19

Why is a nova less luminous than a Type Ia supernova (which both involve white dwarfs)?

CONCEPTCHECK 20-20

Why does a Type Ia supernova only occur once for a given object, but novae and X-ray bursters can occur repeatedly from the same object?

Answers appear at the end of the chapter.

KEY WORDS

asymptotic giant branch, p. 565
asymptotic giant branch star (AGB star), p. 565
carbon fusion, p. 572
carbon star, p. 566
Cherenkov radiation, p. 581
Chandrasekhar limit, p. 571
core-collapse supernova, p. 577
core helium fusion, p. 564
degenerate neutron pressure, p. 576
dredge-up, p. 566
helium shell flash, p. 567
horizontal branch, p. 565
mass-radius relation, p. 570
neon fusion, p. 572
neutron capture, p. 572
neutron star, p. 588
nova (*plural* novae), p. 591
nuclear density, p. 576
oxygen fusion, p. 572
photodisintegration, p. 575
planetary nebula, p. 567
progenitor star, p. 578
pulsar, p. 588
red giant branch, p. 564

shell helium fusion, p. 565
silicon fusion, p. 572
supergiant, p. 574
supernova (*plural* supernovae), p. 577
supernova remnant, p. 585
thermal pulse, p. 567
thermonuclear, p. 564
thermonuclear supernova, p. 583
Type I supernova, p. 582
Type Ia supernova, p. 582
Type Ib supernova, p. 582
Type Ic supernova, p. 582
Type II supernova, p. 581
white dwarf, p. 569
X-ray burster, p. 592

KEY IDEAS

Late Evolution of Low-Mass Stars: A star of moderately low mass (about 0.4 $M_\odot$ to about 4 $M_\odot$) becomes a red giant when shell hydrogen fusion begins, a horizontal-branch star when core helium fusion begins, and an asymptotic giant branch (AGB) star when the helium in the core is exhausted and shell helium fusion begins.

• As a moderately low-mass star ages, convection occurs over a larger portion of its volume. This takes heavy elements formed in the star's interior and distributes them throughout the star.

Planetary Nebulae and White Dwarfs: Helium shell flashes in an old, moderately low-mass star produce thermal pulses during which more than half the star's mass may be ejected into space. This exposes the hot carbon-oxygen core of the star.

• Ultraviolet radiation from the exposed core ionizes and excites the ejected gases, producing a planetary nebula.

• No further nuclear reactions take place within the exposed core. Instead, it becomes a degenerate, dense sphere about the size of Earth and is called a white dwarf. It glows from thermal radiation; as a white dwarf cools, it becomes dimmer.

Late Evolution of High-Mass Stars: Unlike a moderately low-mass star, a high-mass star (initial mass more than about 4 $M_\odot$) undergoes an extended sequence of nuclear reactions in its core and shells. These include carbon fusion, neon fusion, oxygen fusion, and silicon fusion.

• In the last stages of its life, a high-mass star has an iron-rich core surrounded by concentric shells hosting the various nuclear reactions. The sequence of nuclear reactions stops here, because the formation of elements heavier than iron requires an input of energy rather than causing energy to be released.

Core-Collapse Supernovae: A star with an initial mass greater than 8 $M_\odot$ dies in a violent cataclysm in which its core collapses and most of its matter is ejected into space at high speeds. The luminosity of the star increases suddenly by a factor of around 10^8 during this explosion, producing a supernova.

• More than 99% of the energy from a core-collapse supernova is emitted in the form of neutrinos from the collapsing core.

• The matter ejected from the supernova, moving at supersonic speeds through interstellar gases and dust, glows as a nebula called a supernova remnant.

• A Type II supernova is the result of the collapse of the core of a massive star, as are supernovae of Type Ib and Type Ic.

White Dwarf Supernovae: An accreting white dwarf in a close binary system can also become a supernova when carbon fusion ignites explosively throughout such a degenerate star. This is called a thermonuclear supernova.

• One scenario for Type Ia supernovae is an accreting white dwarf in a close binary with a star; another scenario involves the merger of two white dwarfs.

Neutron Stars: A neutron star is a dense stellar corpse consisting primarily of closely packed degenerate neutrons.

• Neutron stars form in a Type II core-collapse supernova. They are held up from further collapse by degenerate neutron pressure.

• A neutron star typically has a diameter of about 20 km, a mass less than 3 $M_\odot$, a magnetic field 10^{12} times stronger than that of the Sun, and a rotation period of roughly 1 second.

Pulsars: A pulsar is a source of periodic pulses of radio emission. These pulses are produced as beams of radio waves from a neutron star's magnetic poles sweep past Earth.

Novae and Bursters: Material from an ordinary star in a close binary can fall onto the surface of the companion white dwarf or neutron star to produce a thin surface layer in which nuclear reactions can explosively ignite.

• Explosive hydrogen fusion may occur in the surface layer of a companion white dwarf, producing the sudden increase in luminosity that we call a nova.

• Explosive helium fusion may occur in the surface layer of a companion neutron star. This produces a sudden increase in X-ray radiation, which we call a burster.

QUESTIONS

Review Questions

1. TUTORIAL 20.1 What is the horizontal branch? Where is it located on an H-R diagram? How do stars on the horizontal branch differ from red giants or main-sequence stars?
2. Horizontal-branch stars are sometimes referred to as "helium main-sequence stars." In what sense is this true?
3. What is the asymptotic giant branch? Where is it located on an H-R diagram? How do asymptotic giant branch stars differ from red giants or main-sequence stars?
4. Is a carbon star a star that is made of carbon? Explain your answer.
5. What is the connection between dredge-ups in old stars and life on Earth?
6. What are thermal pulses in AGB stars? What causes them? What effect do they have on the luminosity of the star?
7. How is a planetary nebula formed?
8. How can an astronomer tell the difference between a planetary nebula and a planet?
9. What is the evidence that typical planetary nebulae are only a few thousand years old?

10. Why do we not observe planetary nebulae that are more than about 50,000 years old?
11. What is a white dwarf? Does it produce light in the same way as a star like the Sun?
12. What is nuclear density? Why is it significant when a star's core reaches this density?
13. How does the radius of a white dwarf depend on its mass? How is this different from other types of stars?
14. What is the significance of the Chandrasekhar limit?
15. On an H-R diagram, sketch the evolutionary track that the Sun will follow from when it leaves the main sequence to when it becomes a white dwarf. Approximately how much mass will the Sun have when it becomes a white dwarf? Where will the rest of the mass go?
16. What prevents nuclear reactions from occurring at the center of a white dwarf? If no nuclear reactions are occurring in its core, why doesn't the star collapse?
17. A white dwarf has a greater mass than either a red dwarf or a brown dwarf. Yet a white dwarf has a smaller radius than either a red dwarf or a brown dwarf. Explain why, in terms of the types of pressure that keep the different kinds of dwarfs from collapsing under their own gravity.
18. Why do you suppose that all the white dwarfs known to astronomers are relatively close to the Sun?
19. Why does the mass of a star play such an important role in determining the star's evolution?
20. Why is the temperature in a star's core so important in determining which nuclear reactions can occur there?
21. What is the difference between a red giant and a red supergiant?
22. Why does the evolutionary track of a high-mass star move from left to right and back again in the H-R diagram?
23. In what way does the structure of an aging supergiant resemble that of an onion?
24. Can stars fuse iron as nuclear fuel? How does this affect the core of a supergiant?
25. What happens if the iron core of a star exceeds the Chandrasekhar limit? About how large must the star be for this to occur?
26. Supernovae can produce uranium that, billions of years later, release nuclear energy through radioactive decay. In what form was this energy during the supernova event?
27. Why is SN 1987A so interesting to astronomers? In what ways was it not a typical supernova?
28. Why are neutrinos emitted by core-collapse supernovae? How can these neutrinos be detected? How can they be distinguished from solar neutrinos?
29. What causes a thermonuclear supernova? How does a thermonuclear supernova compare with a core-collapse supernova?
30. What are the differences among Type Ia, Type Ib, Type Ic, and Type II supernovae? Which type is most unlike the other three, and why?
31. How can a supernova continue to shine for many years after it explodes?
32. How do supernova remnants produce radiation at nonvisible wavelengths?
33. There may have been recent supernovae in our Galaxy that have not been observable even though they are incredibly luminous. How is it this possible?
34. Is our own Sun likely to become a supernova? Why or why not?
35. What are neutron stars? Where do the neutrons come from?
36. In what type of supernova can a neutron star form? Why?
37. What is degenerate neutron pressure? How is it related to degenerate electron pressure?
38. What does the radio emission from a pulsar look like?
39. During the weeks immediately following the discovery of the first pulsar, one suggested explanation was that the pulses might be signals from an extraterrestrial civilization. Why did astronomers soon discard this idea?
40. How are rotating neutron stars able to produce pulses of radiation as seen by an observer on Earth?
41. Why do neutron stars rotate so much more rapidly than ordinary stars? Why do they have such strong magnetic fields?
42. Is our Sun likely to end up as a neutron star? Why or why not?
43. Why does a neutron star need to have a mass above 1.4 $M_\odot$? What effects contribute to the pressure as a neutron star approaches its maximum mass?
44. What are the similarities between a nova and a Type Ia supernova? What are the differences?
45. What are the similarities between novae and X-ray bursters? What are the differences?

Advanced Questions

Questions preceded by an asterisk (*) involve topics discussed in the Boxes in Chapter 1, Chapter 7, or Chapter 17.

Problem-solving tips and tools

The small-angle formula is given in Box 1-1. You may find it useful to review Box 17-4, which discusses stellar radii and their relationship to temperature and luminosity. Section 4-7 explains the formula for gravitational force, and Box 7-2 explains the concept of escape speed. Sections 5-2 and 5-4 describe some key properties of light, especially blackbody radiation. We discussed the relationship among luminosity, apparent brightness, and distance in Box 17-2. The relationship among absolute magnitude, apparent magnitude, and distance was the topic of Box 17-3. Section 17-6 gives the formula relating a star's luminosity, surface temperature, and radius (see Box 17-4 for worked examples). In our discussion of binary stars in Section 17-9 we saw how the masses of the stars, the orbital period, and the average distance between the two stars are all related. The volume of a sphere of radius r is $4\pi r^3/3$. Appendix 6 gives the conversion between seconds and years as well as the radius of the Sun.

46. Some blue main-sequence stars in our region of the Galaxy have the same luminosity and surface temperature as the horizontal-branch stars in the globular cluster M55 (see Figure 20-3). How do we know that the horizontal-branch stars in M55 are not main-sequence stars?

47. Stellar winds from an AGB star can cause it to lose mass at a rate of up to 10^{-4} $M_\odot$ per year. (**a**) Express this rate in metric tons per second. (One metric ton equals 1000 kilograms.) (**b**) At this rate, how long would it take an AGB star to eject an amount of mass equal to the mass of Earth? Express your answer in days.

48. The globular cluster M15 depicted in Figure 20-7a contains 30,000 old stars, but only one of these stars is presently in the planetary nebula stage of its evolution. Explain why planetary nebulae are not more prevalent in M15.

49. The central star in a newly formed planetary nebula has a luminosity of 1000 $L_\odot$ and a surface temperature of 100,000 K. What is the star's radius? Give your answer as a multiple of the Sun's radius.

50. You want to determine the age of a planetary nebula. What observations should you make, and how would you use the resulting data?

51. Figure 20-7a shows the planetary nebula Abell 39. Use the information given in the figure caption to calculate the angular diameter of the nebula as seen from Earth.

*52. The Ring Nebula is a planetary nebula in the constellation Lyra. It has an angular size of 1.4 arcmin × 1.0 arcmin and is expanding at the rate of about 20 km/s. Approximately how long ago did the central star shed its outer layers? Assume that the nebula is 2,700 ly from Earth.

53. The accompanying image shows the planetary nebula IC 418 in the constellation Lepus (the Hare). (**a**) The image shows a small shell of glowing gas (shown in blue) within a larger glowing gas shell (shown in orange). Discuss how IC 418 could have acquired this pair of gas shells. (**b**) Explain why the outer shell looks thicker around the edges than near the middle.

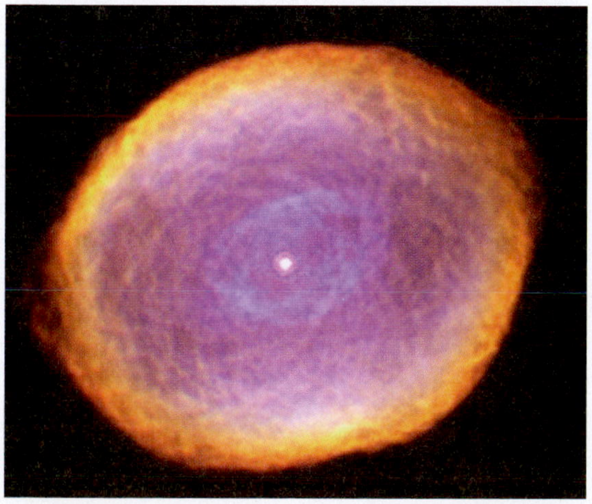

R I V U X G
(NASA and Hubble Heritage Team, STScI/AURA)

54. (**a**) Calculate the wavelength of maximum emission of the white dwarf Sirius B. In what part of the electromagnetic spectrum does this wavelength lie? (**b**) In a visible-light photograph such as Figure 20-9, Sirius B appears much fainter than its primary star. But in an image made with an X-ray telescope, Sirius B is the brighter star. Explain the difference.

*55. Sirius is 2.63 pc from Earth. By making measurements on Figure 20-9, calculate the distance between the centers of Sirius A and Sirius B at the time that this image was made in October 2003. Give your answer in astronomical units (AU). (Note that your result is the true distance only if Sirius A and Sirius B were exactly the same distance from Earth at the time the image was made. If one of the stars was closer to us than the other, the actual distance between them is greater than what you calculate.)

56. (**a**) Find the average density of a 1-$M_\odot$ white dwarf having the same diameter as Earth. (**b**) What speed is required to eject gas from the white dwarf's surface? (This is also the speed with which interstellar gas falling from a great distance would strike the star's surface.)

57. In the classic 1960s science-fiction comic book *The Atom*, a physicist discovers a basketball-sized meteorite (about 10 cm in radius) that is actually a fragment of a white dwarf star. With some difficulty, he manages to hand-carry the meteorite back to his laboratory. Estimate the mass of such a fragment, and discuss the plausibility of this scenario.

*58. (**a**) Use the information in the caption to Figure 20-12 to calculate the diameter of the nebula SMC N76. Express your answer in parsecs. (**b**) How does your answer to part (a) compare to the diameters of the planetary nebulae depicted in Figure 20-7? Explain how this is consistent with the observation that gases ejected from a supergiant travel faster than gases ejected from an AGB star.

*59. The supergiant star depicted in Figure 20-12 is actually one member of a binary star system. The masses of the two stars are 18 $M_\odot$ and 34 $M_\odot$, and the orbital period is 19.56 days. (**a**) What is the average separation between the two stars? Give your answer in AU. (**b**) Compare your answer in part (a) to the sizes of the orbits of Mercury, Venus, and Earth around the Sun.

60. (**a**) What kinds of stars would you monitor if you wished to observe a core-collapse supernova explosion from its very beginning? (**b**) Examine Appendices 4 and 5, which list the nearest and brightest stars, respectively. Which, if any, of these stars are possible supernova candidates? Explain your answer.

*61. Consider a high-mass star just prior to a supernova explosion, with a core of diameter 20 km and density 4 $\times 10^{17}$ kg/m^3. (**a**) Calculate the mass of the core. Give your answer in kilograms and in solar masses. (**b**) Calculate the force of gravity on a 1-kg object at the surface of the core. How many times larger is this than the gravitational force on such an object at the surface of Earth, which is about 10 newtons?

(c) Calculate the escape speed from the surface of the star's core. Give your answer in meters per second and as a fraction of the speed of light. What does this tell you about how powerful a supernova explosion must be in order to blow material away from the star's core?

62. The shock wave that traveled through the progenitor star of SN 1987A took 2.8 hours to reach the star's surface (see Figure 20-15a). (a) Given the size of a blue supergiant star (see Section 20-7), estimate the speed with which the front-end of the shock wave traveled through the star's outer layers. (The core of the progenitor star was very small, so you may consider the shock wave to have started at the very center of the star.) Give your answer in meters per second.

63. The neutrinos from SN 1987A arrived 3 hours before the visible light. While they were en route to Earth, what was the distance between the neutrinos and the first photons from SN 1987A? Assume that neutrinos are massless and thus travel at the speed of light. Give your answers in kilometers and in AU.

*64. Compared to SN 1987A (see Figure 20-16), a supernova observed in 1993 called SN 1993J had a maximum apparent brightness only 9.1×10^{24} as great. SN 1993J was 3.6 Mpc away in the galaxy M81. Using the distances from Earth to each of these supernovae, determine the ratio of the maximum luminosity of SN 1993J to that of SN 1987A. Which of the two supernovae had the greater maximum luminosity?

*65. Suppose that the brightness of a star becoming a supernova increases by 20 magnitudes. Show that this corresponds to an increase of 10^8 in luminosity.

*66. Suppose that the red supergiant star Betelgeuse, which lies some 425 ly from Earth, becomes a Type II supernova. (a) At the height of the outburst, how bright would it appear in the sky? Give your answer as a fraction of the brightness of the Sun ($b_\odot$). (b) How would it compare with the brightness of Venus (about 10^{-9} $b_\odot$)?

*67. In July 1997, a supernova named SN 1997cw exploded in the galaxy NGC 105 in the constellation Cetus (the Whale). It reached an apparent magnitude of 116.5 at maximum brilliance, and its spectrum showed an absorption line of ionized silicon. Use this information to find the distance to NGC 105. (*Hint:* Inspect the light curves in Figure 20-20 to find the *absolute* magnitudes of typical supernovae at peak brightness.)

68. Figure 20-21 shows a portion of the Veil Nebula in Cygnus. Use the information given in the caption to find the average speed at which material has been moving away from the site of the supernova explosion over the past 15,000 years. Express your answer in km/s and as a fraction of the speed of light.

69. The images that open this chapter show two kinds of glowing gas clouds: a planetary nebula and a supernova remnant. (a) Explain what makes the planetary nebula glow and what makes the supernova remnant glow. (*Hint:* The explanations are different for the two kinds of gas clouds.) (b) Which of these two kinds of gas clouds continues to glow for a longer time? Why?

70. The planetary nebula and supernova remnant shown in the images that open this chapter are both about the same age. Both objects consist of glowing gases that have expanded away from a central star. Based on these images, in which of these objects have the gases expanded more rapidly? Explain your reasoning.

71. There are many more main-sequence stars of low mass (less than 8 $M_\odot$) than of high mass (8 $M_\odot$ or more). Use this fact to explain why white dwarf stars are far more common than neutron stars.

72. The distance to the Crab Nebula is about 2000 parsecs. In what year did the star actually explode? Explain your answer.

73. The Crab Nebula has an apparent size of about 5 arcmin, and this size is increasing at a rate of 0.23 arcsec per year. (a) Assume that the expansion rate has been constant over the entire history of the Crab Nebula. Based on this assumption, in what year would Earth observers have seen the supernova explosion that formed the nebula? (b) Does your answer to part (a) agree with the known year of the supernova, 1054 C.E? If not, can you point to assumptions you made in your computations that led to the discrepancies? Or do you think your calculations suggest additional physical effects are at work in the Crab Nebula, over and above a constant rate of expansion?

74. Emission lines in the spectrum of the Crab Nebula exhibit a Doppler shift, which indicates that gas in the part of the nebula closest to us is moving toward us at 1450 km/s. (a) Assume that the expanding gas has been moving at the same speed since the original supernova explosion, observed in 1054 C.E., and calculate what radius and what diameter (in light-years) we should observe the nebula to have today. (b) Compare your result in part (a) to the actual size of the nebula, given in the caption to Figure 20-26.

75. A neutron has a mass of about 1.7×10^{-27} kg and a radius of about 10^{-15} m. (a) Compare the density of matter in a neutron with the average density of a neutron star. (b) If the neutron star's density is more than that of a neutron, the neutrons within the star are overlapping; if it is less, the neutrons are not overlapping. Which of these seems to be the case for average neutrons within the star? Which do you think is the case at the center of the neutron star, where densities are higher than average?

76. In an X-ray burster, the surface of a neutron star 10 km in radius is heated to a temperature of 3×10^7 K. (a) Determine the wavelength of maximum emission of the heated surface (which you may treat as a blackbody). In what part of the electromagnetic spectrum does this lie? (See Figure 5-7.) (b) Find the luminosity of the heated neutron star. Give your answer in watts and in terms of the luminosity of the Sun,

given in Table 16-1. How does this compare with the peak luminosity of a nova? Of a Type Ia supernova?

Discussion Questions

77. Suppose that you discover a small, glowing disk of light while searching the sky with a telescope. How would you decide if this object is a planetary nebula? Could your object be something else? Explain.

78. Suppose the convective zone in AGB stars did *not* reach all the way down into their carbon-rich cores. How might this have affected the origin and evolution of life on Earth?

79. Imagine that our Sun was somehow replaced by a 1-$M_{\odot}$ white dwarf star, and that our Earth continued in an orbit of semimajor axis 1 AU around this star. Discuss what effects this would have on our planet. What would the white dwarf look like as seen from Earth? Could you look at it safely with the unaided eye? Would Earth's surface temperature remain the same as it is now?

80. The similar names *white dwarf, red dwarf,* and *brown dwarf* describe three very different kinds of objects. Suggest better names for these three kinds of objects, and describe how your names more accurately describe the objects' properties.

81. The major final product of silicon fusion is ^{56}Fe, an isotope of iron with 26 protons and 30 neutrons. This is also the most common isotope of iron found on Earth. Discuss what this tells you about the origin of the solar system.

82. SN 1987A did not agree with the theoretical picture outlined in Section 20-6. Does this mean that the theory was wrong? Discuss.

83. Imagine that we are somehow able to stand (and survive) at one of the magnetic poles of the Crab pulsar. Describe what you would see. How would the stars appear to move in the sky? What would you see if you looked straight up? What factors make this location a very unhealthy place to visit?

84. When neutrons are very close to one another, they repel one another through the strong nuclear force. If this repulsion were made even stronger, what effect might this have on the maximum mass of a neutron star? Explain your answer.

Web/eBook Questions

85. It has been claimed that the Dogon tribe in western Africa has known for thousands of years that Sirius is a binary star. Search the World Wide Web for information about these claims. What is the basis of these claims? Why are scientists skeptical, and how do they refute these claims?

86. Search the World Wide Web for information about SN 1994I, a supernova that occurred in the galaxy M51 (NGC 5194). Why was this supernova unusual? Was it bright enough to have been seen by amateur astronomers?

87. ANIMATION 20-2 **Convection Inside a Giant Star.** Access and view the animation "Convection Inside a Giant Star" in Chapter 20 of the *Universe* Web site or eBook. Describe the motion of material in the interior of the star. In what ways is this motion similar to convection within the present-day Sun (see Section 16-2)? In what ways is it different? Is a dredge-up taking place in this animation? How can you tell?

88. ANIMATION 20-3 ANIMATION 20-6 **Types of Supernovae.** Access and view the animations "In the Heart of a Core-Collapse (Type II, Ib, or Ic) Supernova" and "A ThermoNuclear (Type Ia) Supernova" in Chapter 20 of the *Universe* Web site or eBook. Describe how these two types of supernova are fundamentally different in their origin.

89. Search the World Wide Web for information about the latest observations of the stellar remnant at the center of SN 1987A. Has a neutron star been detected? Has a pulsar been detected? Has the supernova's debris thinned out enough to give a clear view of the neutron star?

ACTIVITIES

Observing Projects

Observing tips and tools

While planetary nebulae are rather bright objects, their brightness is spread over a relatively large angular size, which can make seeing them a challenge for the beginning observer. For example, the Helix Nebula (shown in the left-hand image on the page that opens this chapter) has the largest angular size of any planetary nebula but is also one of the most difficult to see. To improve your view, make your observations on a dark, moonless night from a location well shielded from city lights. Another useful trick, mentioned in Chapters 17 and 18, is to use "averted vision." Once you have the nebula centered in the telescope, you will get a brighter and clearer image if you look at the nebula out of the corner of your eye. The so-called Blinking Planetary in Cygnus affords an excellent demonstration of this effect; the nebula seems to disappear when you look straight at it, but it reappears as soon as you look toward the side of your field of view.

Another useful tip is to view the nebula through a green filter (a #58, or O III, filter available from telescope supply houses). Green light is emitted by excited, doubly ionized oxygen atoms, which are common in planetary nebulae but not in most other celestial objects. Using such a filter can make a planetary nebula stand out more distinctly against the sky. As a side benefit, it also helps to block out stray light from street lamps. The same tips also apply to observing supernova remnants.

90. Although they represent a fleeting stage at the end of a star's life, planetary nebulae are found all across the sky. Some of the brightest are listed in the accompanying table. Note that the distances to most of these nebulae are quite uncertain. Observe as many of these planetary nebulae as you can on a clear, moonless night using the largest telescope at your disposal. Note and compare the various shapes of the different

Planetary nebula	Distance (light-years)	Angular size	Constellation	Right ascension	Declination
Dumbbell (M27, NGC 6853)	490–3500	8.0 × 5.7	Vulpecula	19h 59.6m	+22° 43′
Ring (M57, NGC 6720)	1300–4100	1.4 × 1.0	Lyra	18h 53.6m	+33° 02′
Little Dumbbell (M76, NGC 650)	1700–15,000	2.7 × 1.8	Perseus	01h 42.4m	+51° 34′
Owl (M97, NGC 3587)	1300–12,000	3.4 × 3.3	Ursa Major	11h 14.8m	+55° 01′
Saturn (NGC 7009)	1600–3900	0.4 × 1.6	Aquarius	21h 04.2m	–11° 22′
Helix (NGC 7293)	450	41 × 41	Aquarius	22h 29.6m	–20° 48′
Eskimo (NGC 2392)	1400–10,000	0.5 × 0.5	Gemini	7h 29.2m	+20° 55′
Blinking Planetary (NGC 6826)	3300(?)	2.2 × 0.5	Cygnus	19h 44.8m	+50° 31′

nebulae. In how many cases can you see the central star? The central star in the Eskimo Nebula is supposed to be the "nose" of an Eskimo wearing a parka. Can you see this pattern?

91. Northern hemisphere observers with modest telescopes can see two supernova remnants, one in the winter sky and the other in the summer sky. Both are quite faint, however, so you should schedule your observations for a moonless night. The winter sky contains the Crab Nebula, which is discussed in detail in Chapter 23. The coordinates are R.A. = $5^h\ 34.5^m$ and Decl. = +22° 00′, which places the object near the star marking the eastern horn of Taurus (the Bull). Whereas the entire Crab Nebula easily fits in the field of view of an eyepiece, the Veil or Cirrus Nebula in the summer sky is so vast that you can see only a small fraction of it at a time. The easiest way to find the Veil Nebula is to aim the telescope at the star 52 Cygni (R.A. = $20^h\ 45.7^m$ and Decl. = 130° 43′), which lies on one of the brightest portions of the nebula. If you then move the telescope slightly north or south until 52 Cygni is just out of the field of view, you should see faint wisps of glowing gas.

92. Use a telescope to observe the remarkable triple star 40 Eridani, whose coordinates are R.A. = $4^h\ 15.3^m$ and Decl. = 27° 39′. The primary, a 4.4-magnitude yellowish star like the Sun, has a 9.6-magnitude white dwarf companion, the most easily seen white dwarf in the sky. On a clear, dark night with a moderately large telescope, you should also see that the white dwarf has an 11th-magnitude companion, which completes this most interesting trio.

93. Planetary nebulae represent the late stages of the evolution of stars whose masses are similar to that of the Sun and are found throughout our Galaxy. You can use *Starry Night*™ to explore the distribution of these objects in our sky and to view several of these spectacular nebulae. Set the view for your home location at some time in the evening with a field of view of about 100°. Open the **Options** pane and expand the **Deep Space** panel. Expand the **NGC-IC Database** list, click in its box to activate the display of the objects in this list, and click **Off** all entries in the list except **Planetary Nebula.** Use the hand tool to move around the sky. Note that these nebulae are mostly concentrated around the Milky Way in our sky. If you have access to a telescope, try to locate and observe several of these planetary nebulae, if possible on a clear, moonless night. Some of the more notable planetary nebulae include Little Dumbbell (M76), NGC 1535, Eskimo, Ghost of Jupiter, Owl (M97), Ring (M57), Blinking Planetary, Dumbbell (M27), Saturn Nebula, and NGC 7662.

If you do not have access to a telescope, use *Starry Night*™ to examine in detail two of these planetary nebulae, M57 (the Ring Nebula) and M27 (the Dumbbell Nebula), and compare their shapes and sizes. Select **Favourites > Explorations > Atlas** and use the **Find** pane to center upon and magnify these two nebulae in turn. You can compare a ground-based image of M57 with a high-resolution image taken by the Hubble Space Telescope. First, open the **Options** pane, expand the **Deep Space** layer and click on the **Messier Objects** to see an image taken with a ground-based telescope. Then replace this image by a space image by clicking in the Hubble Images box. Note that the Hubble image is displayed in a different alignment to that of the ground-based image. For each of these objects, note their distance from observer in the HUD, and then use the angular separation tool to measure the approximate angular radius of each of these nebulae. (**a**) How do you account for the difference in the shape of these two planetary nebulae? (**b**) What is the nature of the central star in each of these nebulae? (**c**) Calculate the physical size of these nebulae. (*Hint:* Translate angular size in arcseconds to radians and use the small-angle relationship: 1 radian = 206,265 arcseconds; 1 ly = 9.46×10^{12} km.) (**d**) Assuming that both of these nebulae have been expanding at the same rate (measured in km/s), which of the stars at the cores of these nebulae reached the end of its life cycle first (i.e., which of these nebulae is the oldest)? (**e**) Using an average rate of expansion of 20 km/s for the shell of gas that forms the Ring Nebula (M57), approximately how long ago, in years, did the star that formed this nebula initiate the expansion of its outer atmosphere?

94. The red supergiant Betelgeuse in the constellation Orion will explode as a supernova at some time in the future. Use the *Starry Night*™ program to investigate how the supernova might appear if the light from this explosion were to arrive at Earth tonight. Click the **Home** button in the toolbar to show the sky as seen from your location at the present time. Use the **Find** pane to locate Betelgeuse. If Betelgeuse is below the horizon, allow the program to reset the time to show this star. (**a**) At what time does Betelgeuse rise on today's date? At what time does it set? (**b**) What is the apparent magnitude (m_V) of Betelgeuse? (*Hint:* Use the HUD or the Info pane to find this information.) (**c**) If Betelgeuse became a supernova today, then at peak brightness it would be 11 magnitudes brighter than it is now. For comparison, $m_V = -4$ for Venus at its brightest and $m_V = -12.6$ for the full Moon. Would Betelgeuse be visible in the daytime? How would it appear at night? Do you think it would cast shadows? (**d**) Are Betelgeuse and the Moon both in the night sky tonight? (Use the **Find** pane to locate the Moon.) (**e**) If Betelgeuse were to become a supernova, how would the shadows cast by Betelgeuse differ from those cast by the Moon?

95. Use the *Starry Night*™ program to investigate the X-ray source and probable black hole, Cygnus X-1. This region of space is one of the brightest in the sky at X-ray wavelengths. Click the **Home** button in the toolbar and then use the **Find** pane to center the field of view on Cygnus X-1. If Cygnus X-1 is below the horizon, allow the program to reset the time to when it can best be seen. Click the checkbox to the left of the listing for Cygnus X-1 to apply a label to this object. Use the **Zoom** controls to set the field of view to 100 degrees. (**a**) Use the **Time** controls in the toolbar to determine when Cygnus X-1 rises and sets on today's date from your location. (**b**) **Zoom** in until you can see an object at the location indicated by the label. What apparent magnitude and radius does *Starry Night*™ give for this object (you can obtain this information from the HUD or by using the **Show Info** command from the contextual menu for this object)? Keeping in mind that the object that gives rise to this X-ray source is a black hole, to what must this apparent magnitude and radius refer? Explain.

96. Use the *Starry Night*™ program to examine in some detail the central regions of two galaxies that contain supermassive black holes at their centers. (**a**) The Milky Way is one such galaxy. Select **Favourites > Explorations > Milky Way Centre** to view our Galaxy from the equivalent of the center of a transparent Earth. The star HIP86919 is very closely aligned with the direction of the central core of the Galaxy and can be used as a guide when viewing this region. You can brighten the appearance of the Milky Way by opening **Options > Stars > Milky Way...** and moving the **Brightness** slider to the far right in the Milky Way Options panel. The Time and Date, August 30, 2009, at 6:30 A.M., have been chosen when the Moon is crossing the Milky Way plane, thereby providing a convenient angular scale, about ½° in diameter, for comparison with Milky Way features. (Click on **Options > Solar System > Planets-Moons...** to display the Planets-Moons Options window and ensure that the **Enlarge Moon Size at large FOVs** box under Other is **Off,** and click **OK.**) You will notice that, at visible wavelengths, this central region of the Milky Way appears to be dark. What explanation can you give for this dark band across the galactic plane? **Zoom** in to a field of view of about 5° around the galactic center to examine the region surrounding the black hole using the highly penetrating X-rays detected by the Chandra X-ray Space Telescope. If necessary, open the **Options** pane, expand the **Deep Space** layer, click on the **Chandra Images,** and move the slider to the right to display the image mosaic showing numerous hot and intense X-ray sources very close to the supermassive black hole. (Note that this image is slightly offset from the galactic center to avoid confusion.) (**b**) M87 is an active galaxy for which evidence is strong for the presence of a supermassive black hole at its center. Click **Home** to return to your sky and stop the advancement of time. Use the **Find** tab to center your view on **M87,** allowing the program to adjust the time to ensure that this object is in your sky. Open the **Options** pane, expand the **Deep Space** layer and click **Off** the **Chandra** and **Hubble Images,** leaving the **Messier Objects** displayed, and then **Zoom** in to a field of view about 1° wide to show this giant elliptical galaxy as seen from ground-based telescopes. (You can click on the **Info** tab and click on **Description** to read about this somewhat featureless but surprising elliptical galaxy.) In the **Options > Deep Space** panel, move the slider to the right for the **Chandra X-ray image** to see the structure of hot gases around M87. Finally, move the **Hubble Images** slider to the right to display the high-resolution Hubble Space Telescope image of the gas jet emanating from the black hole. Again, this image is displaced, to the upper right, from the galactic center position by the program. Right-click over this square image, click **Centre** and then **Zoom** in to a field of view of about 1 arcminute to see this spectacular jet as it interacts violently with the interstellar medium above the black hole. Comment on the suggestion that supermassive black holes were discovered only after relatively recent advances were made in telescope and detector technology.

97. Use the *Starry Night*™ program to observe the sky in July 1054, when the supernova that spawned the Crab Nebula (M1) would have been visible, probably even in daylight, from North America and may have been recorded as a pictograph at this time by inhabitants of Chaco Canyon, New Mexico. Open **Favourites > Explorations > Crab-Pictograph** to position yourself in Chaco Canyon, at latitude 36°N and Longitude 108°W at 5 A.M. on July 5, 1054 A.D., looking toward the east, just before sunrise. **Zoom** in to display a field of view of about 10°, centered on the Crab Nebula, when you can see the position of the Moon near to the nebula. You may find it helpful to turn daylight on or off (select **Show Daylight** or **Hide Daylight** in the **View** menu). (**a**) What is the phase of the Moon? (**b**) Investigate how the relative positions of the Moon and the Crab Nebula change when you set the date to July 4, 1054, or July 6, 1054. On which date do the relative positions of the Moon and the Crab Nebula give the best match to the pictograph shown in the textbook? You can now investigate the nebula more closely by zooming in to see a ground-based view of this expanding gas cloud from the violent supernova explosion. You can also compare this image with those from spacecraft at both visible and X-ray wavelengths. Open the **Options** pane, expand the **Deep Space** layer, and click the checkbox beside the **Hubble Images** option to turn on this feature. *Starry Night*™ can also superimpose an X-ray image of this active region from the Chandra Space Telescope onto the visible light image of this Messier object. Turn on the **Chandra Images** and compare the X-ray and visible light images by using the slide controls for the **Chandra Images** and **Messier Objects** options in the **Deep Space** layer of the **Options** pane. (Unfortunately, both the Hubble telescope and the Chandra X-ray images are misaligned with respect to the visible images. Nevertheless, you can see the very active regions surrounding the central spinning neutron star, the pulsar, at the center of the Crab Nebula.) (**c**) Open the **Info** pane for the Crab Nebula and obtain its distance from Earth in light-years under the **Position in Space** layer and its angular diameter from the **Other Data** layer. Using this information and the year in which the supernova was seen to explode, calculate the speed of expansion of the supernova remnant in km/s. (**d**) Taking the travel time of light into account, in which year did the star actually explode?

98. Use the *Starry Night*™ program to examine the Veil Nebula, a large supernova remnant. Open **Favourites > Explorations > Veil Nebula** to see a view of the nebula high in the sky of Calgary, Canada, at midnight on August 1, 2013. (**a**) What significant feature do you notice about this supernova remnant in this 5° field of view? (**b**) Use the angular separation tool to measure the approximate angular distance between the components of this nebula. What angular distance separates these components? What form of optical aid is best suited to observing this object? (**c**) It is believed that the supernova that produced this nebula would have been visible to people on Earth about 7000 years ago and that its distance from Earth is about 2500 ly. From these data, what is the approximate speed of expansion of this nebula in km/s? **Zoom** in on each component of the nebula to examine the fine details of this expanding remnant of a violent stellar explosion.

Collaborative Exercises

99. Imagine that a supernova originating from a close binary star system, both of whose stars have less than 4 solar masses, began (as seen from Earth) on the most recent birthday of the youngest person in your group. Using the light curves in Figure 20-20, what would its new luminosity be today and how bright would it appear in the sky (apparent magnitude) if it were located 10 parsecs (32.6 ly) away? How would your answers change if you were to discover that the supernova actually originated from an isolated star with a mass 15 times greater than our Sun?

100. Consider the graph showing a recording of a pulsar in Figure 20-27. Sketch and label similar graphs that your group estimates for: (1) a rapidly spinning, professional ice skater holding a flashlight; and (2) a siren on an emergency ambulance.

101. As stars go, pulsars are tiny, only about 20 km across. Name three specific things or places that have a size or a separation of about 20 km.

ANSWERS

ConceptChecks

ConceptCheck 20-1: The energy released from core helium fusion expands the star's core, and this slows down reactions for both the core helium fusion and the hydrogen-fusing shell. With a decreased rate of energy released in and around the star's core, the luminosity of the star decreases. These stars are called horizontal-branch stars.

ConceptCheck 20-2: Most of the carbon in your body came from stellar winds around carbon stars.

ConceptCheck 20-3: The visible colors of a planetary nebula do not come from the temperature of the nebula's gas. Rather, a planetary nebula's emission is powered by the ultraviolet light emitted by the central star.

ConceptCheck 20-4: For any atom with electrons, the electrons orbit very far from the nucleus, compared to the size of the nucleus itself. Thus, any atom on the periodic table of the elements consists mostly of empty space.

ConceptCheck 20-5: Yes. Electron degeneracy pressure can support white dwarf masses up to the Chandrasekhar limit of 1.4 $M_{\odot}$. There can be no white dwarfs beyond this limit. As shown in Figure 20-10, a 1.2- $M_{\odot}$ white dwarf is smaller than Earth.

ConceptCheck 20-6: No. The Sun's size can be read in the Cosmic Connections figure from the diagonal dashed lines. The Sun expands to about 100 $R_{\odot}$ as a red giant, but this phase ends when the Sun is about 12.23 billion years old (Stage 3 in the figure). At 12.365 billion years, the Sun is even larger (Stage 5). The Sun is also more luminous by this time, and remains so until nuclear reactions come to an end.

ConceptCheck 20-7: It is the loss of the star's mass going into the nebula that relieves pressure on the core. The pressure becomes too low for further nuclear reactions to occur (Stages 6 and 7 in the Cosmic Connections figure).

ConceptCheck 20-8: Yes. However, these reactions do not release energy; energy must be added for these reactions to take place.

ConceptCheck 20-9: No. In order for the iron core to collapse, it must overcome degenerate electron pressure, which requires over 1.4 $M_{\odot}$ (this is the Chandrasekhar limit).

ConceptCheck 20-10: For stars that had main-sequence masses over 8 $M_{\odot}$, iron cores larger than the Chandrasekhar limit can develop and collapse. During collapse, iron is converted into neutrons, which then exert a degenerate-neutron pressure that can halt further collapse.

ConceptCheck 20-11: Neutrinos deliver energy to these bubbles. These neutrinos were produced in the collapse when electrons combined with protons to produce neutrons and neutrinos.

ConceptCheck 20-12: Atoms heavier than iron such as uranium are produced during a supernova. Nuclear reactions producing uranium require an *input* of energy, and there is plenty of gravitational energy in the collapse that ultimately ends up producing atomic elements heavier than iron. Uranium does not release energy when produced.

ConceptCheck 20-13: The typical progenitor star for a core-collapse supernova is a red supergiant. The blue supergiant progenitor star that led to SN 1987A was still large in mass but smaller in size than a red supergiant. That means some of the supernova energy that would have normally gone into light was used just to push the smaller star's matter outward against its own gravity.

ConceptCheck 20-14: When a high-energy neutrino hits a proton in a big water tank, it can cause a positron to be emitted. The positron can travel faster than light travels within that same water, and this produces a shock wave of light that can be detected.

ConceptCheck 20-15: By comparison with the spectra in Figure 20-18, the lack of hydrogen, helium, or silicon lines matches the core-collapse of a Type Ic supernova.

ConceptCheck 20-16: By observing the expansion rate of the remnant (its speed), and how far it has expanded (its approximate radius), the time since the explosion can be estimated.

ConceptCheck 20-17: Just as a white dwarf is held up by degenerate electron pressure, a neutron star is held up by degenerate neutron pressure.

ConceptCheck 20-18: By shrinking during formation, neutron stars spin rapidly and have very large magnetic fields. Charged particles in these magnetic fields emit the light (often radio waves) observed at Earth. The light arrives in pulses the way a lighthouse beacon repeatedly sweeps by an observer.

ConceptCheck 20-19: In a supernova, the entire white dwarf undergoes nuclear reactions, whereas in a nova, only a surface layer undergoes nuclear fusion.

ConceptCheck 20-20: A type Ia supernova blows the object apart. For a nova or X-ray burster, only an outer layer of transferred mass undergoes nuclear reactions. When enough mass is again transferred from the binary partner, the detonation repeats itself.

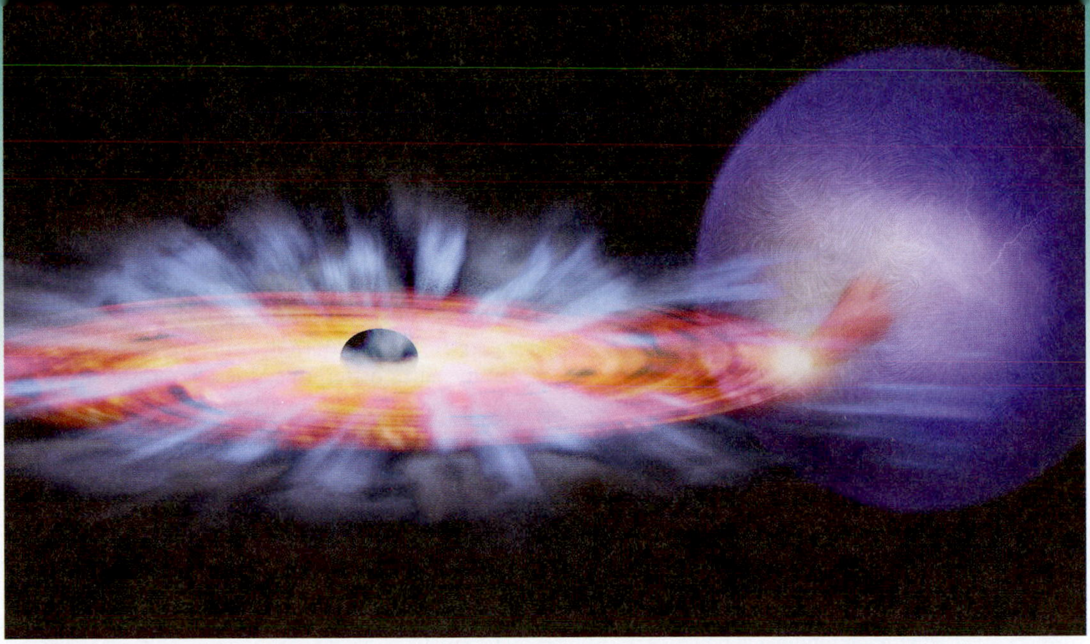

An artist's impression of a close binary star system that includes a black hole. The accretion disk around the black hole is made of material drawn from the blue companion star. (NASA/CXC/M. Weiss)

CHAPTER 21

Black Holes

LEARNING GOALS

By reading the sections of this chapter, you will learn

21-1 The main ideas of Einstein's special theory of relativity

21-2 How Einstein's general theory of relativity describes the nature of gravitation

21-3 The evidence for black holes in binary star systems

21-4 How the sudden formation of black holes can explain the mysterious gamma-ray bursts

21-5 How astronomers have detected supermassive black holes at the centers of galaxies

21-6 The simple structure of a nonrotating black hole

21-7 How just three numbers completely describe the properties of a black hole

21-8 What it might be like to approach a black hole

21-9 How black holes evaporate over time

Imagine a swirling disk of gas and dust, orbiting around an object that has more mass than the Sun but the object is so dark that it cannot be seen. Imagine the material in this disk being compressed and heated as it spirals into the unseen object, reaching temperatures so high that the material emits X-rays. And imagine that the unseen object has such powerful gravity that any material that falls into it simply disappears, never to be seen again.

Such hellish maelstroms, like the one shown in the illustration, really do exist. The disks are called *accretion disks,* and the unseen objects with immensely strong gravity are called *black holes.*

The matter that makes up a black hole has been so greatly compressed that it violently warps space and time. If you get too close to a black hole, the speed you would need to escape exceeds the speed of light. Because nothing can travel faster than light, nothing—not even light—can escape from a black hole.

Black holes, whose existence was predicted by Einstein's general theory of relativity (our best description of what gravity is and how it behaves), are both strange and simple. The structure of a black hole is completely specified by only three quantities—its mass, electric charge, and angular momentum. Perhaps strangest of all, there is compelling evidence that black holes really exist. In recent years, astronomers have found that certain binary star systems contain black holes. They have also found evidence that extremely massive stars form black holes at their centers, producing immense bursts of gamma rays in the process. Even more remarkable is the discovery that black holes of more than a million solar masses lie at the centers of many galaxies, including our own Milky Way.

21-1 The special theory of relativity changes our conceptions of space and time

The idea of a black hole is a truly strange one. To appreciate it, we must discard some "commonsense" notions about the nature of space and time.

The Special Theory of Relativity

According to the classical physics of Newton, space is perfectly uniform and fills the universe like a rigid framework. Similarly, time passes at a monotonous, unchanging rate. It is always possible to know exactly how fast you are moving through this rigid fabric of space and time.

Those ideas were upset forever in 1905 when Albert Einstein proposed his **special theory of relativity.** This theory describes how motion affects our measurements of distance and time. It is "special" in the sense of being specialized. In particular, it does not include the effects of gravity. The word "relativity" is used because one of the key ideas of the theory is that all measurements are made relative to an observer. Contrary to Newton, there is *no* absolute framework of space and time.

To understand black holes, we must first grasp the nature of space and time as described by Einstein's special theory of relativity

In particular, the distance between two points is not an absolute, nor is the time interval between two events. Instead, the values that you measure for these quantities depend on how you are moving, and these values are relative to you. Thus, one of Einstein's key discoveries is that someone moving in a different way would measure different lengths for objects and different durations of events. In our everyday experience we do not notice that time and distance measurements depend on the speed between an observer and the events being observed because, as we will see, these effects are most noticeable when that speed approaches the speed of light.

Remarkably, Einstein's theory is based on just two basic principles. The first is quite simple:

Your description of physical reality is the same, regardless of the constant velocity at which you move.

In other words, if you are moving in a straight line at a constant speed, you experience the same laws of physics as you would if you were moving at any other constant speed and in any other direction. As an example, suppose you were inside a closed railroad car moving due north in a straight line at 100 km/h. Any measurements you make inside the car—for example, how long your thumb is or how long it takes for a ball to drop to the floor—will give exactly the same results as if the car were moving in any other direction or at any other speed, or were not moving at all. This first principle is what we experience in everyday life when riding on a bus or flying on an airplane: Phenomena inside a flying airplane (such as dropping a book) appear the same as when the plane is on the ground even though we are moving, and this would still be true if the plane could fly near the speed of light.

CAUTION! You might think there is a contradiction between the two statements that (1) measurements of the same event (such as dropping a ball) inside the railroad car give the same results regardless of the train's speed, and that (2) motion between an observer and events changes the times and distances being observed, and that someone moving in a different way could measure different values. To understand these statements, think of the train car as a laboratory, and *one observer* is inside moving right along with the experiments being conducted. The first statement says that you can move the laboratory at any speed and it will not change what the observer inside measures. The second statement—a primary result of special relativity—refers to differences between *two observers.* If a second observer standing on the train tracks takes measurements of the experiments taking place in the train (perhaps looking through a window), the two observers are moving at different speeds with respect to the experiment, and they can measure different values for some of the phenomena. Again, this discrepancy between observers is most noticeable when the speed between them is near the speed of light.

Einstein's second principle is much more bizarre than the first:

Regardless of your speed or direction of motion, you always measure the speed of light to be the same.

To see what this implies, imagine that you are in a spaceship that is moving toward a flashlight. Even if you are moving at 99% of the speed of light, you will measure the photons from the flashlight to be moving at the same speed ($c = 3 \times 10^8$ m/s $= 3 \times 10^5$ km/s) as if your spaceship were motionless. This statement is in direct conflict with the Newtonian view that a stationary person and a moving person should measure different speeds (Figure 21-1).

Speed involves both distance and time. Since speed has a very different behavior in the special theory of relativity than in Newton's physics, it follows that both space and time behave differently as well. Indeed, in relativity, time proves to be so intimately intertwined with the three dimensions of space that we regard them as a single *four*-dimensional entity called **spacetime.**

Length Contraction

Einstein expressed his ideas about spacetime in a mathematical form and used this description to make a number of predictions about nature. All these predictions have been verified in innumerable experiments. One prediction is that the length you measure an object to have depends on how that object is moving; the faster it moves, the shorter its length along its direction of motion (Figure 21-2). This is called **length contraction.** In other words, if a railroad car moves past you at high speed, from your perspective on the ground you will actually measure it to be shorter than if it were at rest (Figure 21-2a). However, if you are on board the railroad car and moving with it, you will measure its length to be the same as measured on the ground when it was at rest. The word "relativity" emphasizes the importance of the relative speed between the observer and the object being measured.

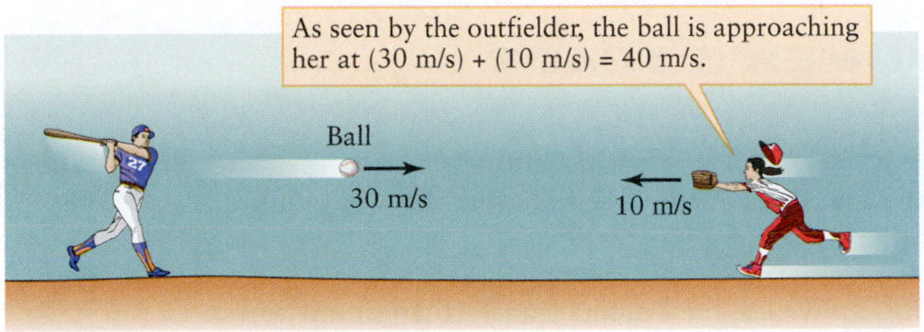

As seen by the outfielder, the ball is approaching her at (30 m/s) + (10 m/s) = 40 m/s.

(a)

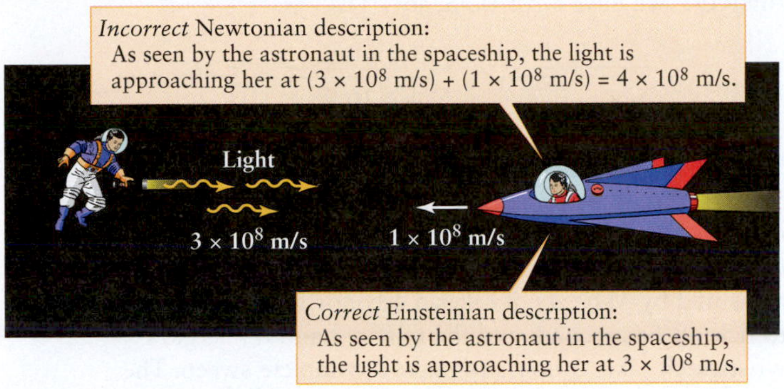

Incorrect Newtonian description:

Correct Einsteinian description:

(b)

FIGURE 21-1

The Speed of Light Is the Same to All Observers **(a)** The speed you measure for ordinary objects depends on how you are moving. Thus, the batter sees the ball moving at 30 m/s, but the outfielder sees it moving at 40 m/s relative to her. **(b)** Einstein showed that this commonsense principle does not apply to light. No matter how fast or in what direction the astronaut in the spaceship is moving, she and the astronaut holding the flashlight will always measure light to have the same speed.

If the idea of length contraction seems outrageous, it is because this effect is noticeable only at very high speeds, near the speed of light. But even the fastest spacecraft ever built by humans travels at a mere 1/25,000 of the speed of light. At this speed, a spacecraft 10 m long would be contracted in length by only 8 nm—a distance equal to the width of a single protein molecule! For moving cars, trains, and airplanes, length contraction is far too small to measure. As Box 21-1 describes, however, we can easily see the effects of length contraction for subatomic particles that do indeed travel at nearly the speed of light.

Time Dilation

A second result of relativity is no less strange: A clock moving past you runs more slowly than a clock that is at rest. Like length contraction, this **time dilation** is a very small effect unless the clock is moving at extremely high speeds (Figure 21-2b). Nonetheless, physicists have confirmed the existence of time dilation by using an extremely accurate atomic clock carried on a jet airliner. When the airliner landed, they found that the on-board clock had actually ticked off slightly less time than an identical, stationary clock on the ground. For the passengers on board the airliner during this experiment, however, time flowed at a normal rate: The on-board clock ticked off the seconds as usual, their hearts beat at a normal rate, and so on. You only measure a clock (or a beating heart) to be running slow if it is *moving* relative to you. Box 21-1 discusses time dilation in more detail.

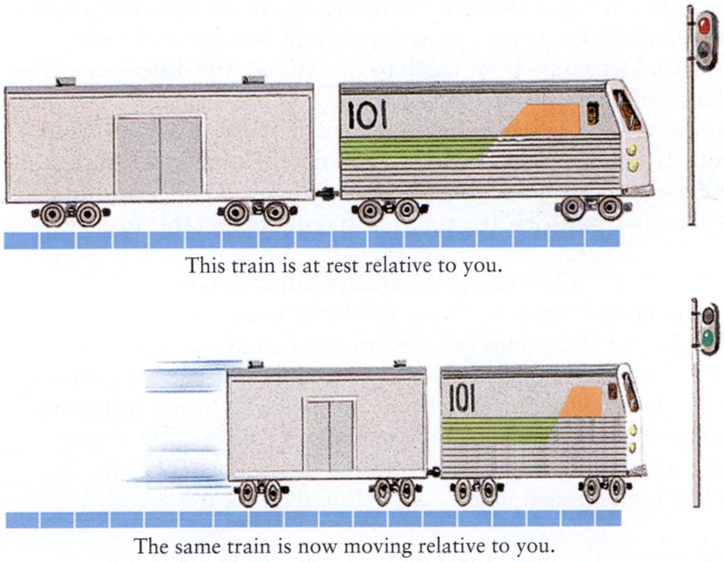

101
This train is at rest relative to you.
101
The same train is now moving relative to you.

(a) Length contraction

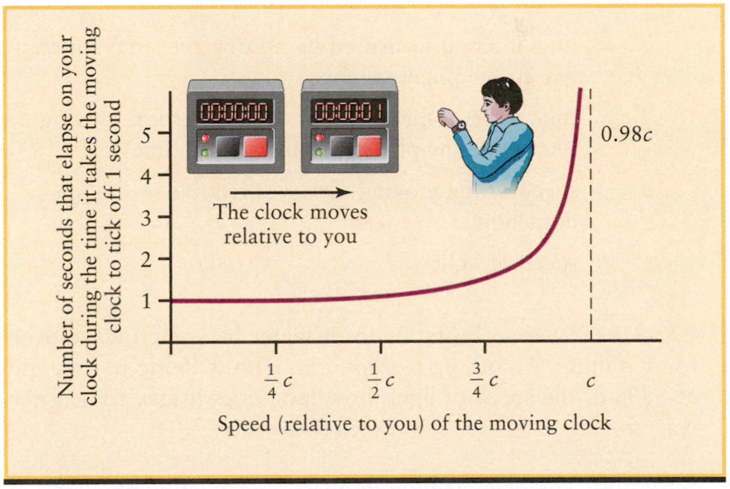

(b) Time dilation

FIGURE 21-2

Length Contraction and Time Dilation **(a)** The faster an object moves, the shorter it becomes along its direction of motion. Speed has no effect on the object's dimensions perpendicular to the direction of motion. **(b)** The faster a clock moves, the slower it runs. This graph shows how many seconds (as measured on your clock) it takes a clock that moves relative to you to tick off 1 second. The effect is pronounced only for speeds near the speed of light c.

BOX 21-1 TOOLS OF THE ASTRONOMER'S TRADE

Time Dilation and Length Contraction

The special theory of relativity describes how motion affects measurements of time and distance. By using the two basic principles of his theory—that no matter how fast you move, the laws of physics and the speed of light are the same—Einstein concluded that measurements of time and distance must depend on how the person making the measurements is moving.

To describe how measurements depend on motion, Einstein derived a series of equations named the **Lorentz transformations,** after the famous Dutch physicist Hendrik Antoon Lorentz. (Lorentz was a contemporary of Einstein who developed these equations independently but did not grasp their true meaning.) These equations tell us exactly how moving clocks slow down and how moving objects shrink.

To appreciate the Lorentz transformations, imagine two observers named Sergio and Majeeda. Sergio is at rest on Earth, while Majeeda is flying past in her spaceship at a speed v. Sergio and Majeeda both observe the same phenomenon on Earth—say, the beating of Sergio's heart or the ticking of Sergio's watch—which appears to occur over an interval of time. According to Sergio's clock (which is not moving relative to the phenomenon), the phenomenon lasts for T_0 seconds. This time period is called the **proper time** of the phenomenon. But according to Majeeda's clock (which *is* moving relative to the phenomenon), the same phenomenon lasts for a different length of time, T seconds. These two time intervals are related as follows:

Lorentz transformation for time

$$T = \frac{T_0}{\sqrt{1 - (v/c)^2}}$$

T = time interval measured by an observer moving relative to the phenomenon

T_0 = time interval measured by a observer *not* moving relative to the phenomenon (proper time)

v = speed of the moving observer relative to the phenomenon

c = speed of light

EXAMPLE: Sergio heats a cup of water in a microwave oven for 1 minute. According to Majeeda, who is flying past Sergio at 98% of the speed of light, how long does it take to heat the water?

Situation: The phenomenon in question is heating the water in the microwave oven. Sergio is not moving relative to this phenomenon (so he measures the *proper* time interval T_0). Majeeda is moving at speed $v = 0.98c$ relative to this phenomenon (so she measures a different time interval T). Our goal is to determine the value of T.

Tools: We use the Lorentz transformation for time to calculate T.

Answer: We have v/c 5 0.98, so

$$T = \frac{T_0}{\sqrt{1 - (v/c)^2}} = \frac{1 \text{ minute}}{\sqrt{1 - (0.98)^2}} = 5 \text{ minutes}$$

Review: A phenomenon that lasts for $T_0 = 1$ minute on Sergio's clock is stretched out to $T = 5T_0 = 5$ minutes as measured on Majeeda's clock moving at 98% of the speed of light. Other phenomena are affected in the same way: As measured by *Majeeda,* it takes 5 seconds for Sergio's wristwatch to tick off one second, and the minute hand on Sergio's wristwatch takes 5 hours to make a complete sweep. The converse is also true. As measured by *Sergio,* the minute hand on Majeeda's wristwatch will take 5 hours to make a complete sweep. This phenomenon, in which events moving relative to an observer happen at a slower pace, is called *time dilation.*

The Lorentz transformation for time is plotted in Figure 21-2b, which shows how 1 second measured on a stationary clock (say, Sergio's) is stretched out when measured using a clock carried by a moving observer (such as Majeeda). For speeds less than about half the speed of light, the mathematical factor $\sqrt{1 - (v/c)^2}$ is not too different from 1, and there is little difference between the recordings of the stationary and moving clocks. At the fastest speeds that humans have ever traveled (en route from Earth to the Moon), the difference between 1 and the factor $\sqrt{1 - (v/c)^2}$ is less than 10^{-9}. So, for any speed associated with human activities, stationary and slowly moving clocks tick at essentially the same rate. As the next example shows, however, time dilation is important for subatomic particles that travel at speeds comparable to c.

EXAMPLE: When unstable particles called *muons* (pronounced "mew-ons") are produced in experiments on Earth, they decay into other particles in an average time of 2.2×10^{-6} s. Muons are also produced by fast-moving protons from interstellar space when they collide with atoms in Earth's upper atmosphere. These muons typically move at 99.9% of the speed of light and are formed at an altitude of 10 km. How long do such muons last before they decay?

Situation: The phenomenon in question is the life of a muon, which lasts a time $T_0 = 2.2 \times 10^{-6}$ s as measured by an observer not moving with respect to the muon. Our goal is to find the time interval T measured by an observer on Earth, which is moving at $v = 0.999c$ relative to the muon (the same speed at which the muon is moving relative to Earth).

Tools: As in the previous example, we use the Lorentz transformation for time.

Answer: Using $v/c = 0.999$,

$$T = \frac{2.2 \times 10^{-6}\ \text{s}}{\sqrt{1-(0.999)^2}} = 4.9 \times 10^{-5}\ \text{s}$$

Review: At this speed, the muon's lifetime is slowed down by time dilation by a factor of more than 22. Note that as measured by an observer on Earth, the time that it takes a muon produced at an altitude of 10 km = 10,000 m to reach Earth's surface is

$$\frac{\text{distance}}{\text{speed}} = \frac{10{,}000\ \text{m}}{0.999 \times 3.00 \times 10^8\ \text{m/s}} = 3.3 \times 10^{-5}\ \text{s}$$

Were there no time dilation, such a muon would decay in just 2.2×10^{-6} s and would never reach Earth's surface. But in fact, these muons *are* detected by experiments on Earth's surface, because a muon moving at $0.999c$ lasts more than 3.3×10^{-5} s. The detection at Earth's surface of muons from the upper atmosphere is compelling evidence for the reality of time dilation.

In the same terminology as "proper time," we say that a ruler at rest measures **proper length** or **proper distance** (L_0). According to the Lorentz transformations, distances perpendicular to the direction of motion are unaffected. However, a ruler of proper length L_0 held parallel to the direction of motion shrinks to a length L, given by

Lorentz transformation for length

$$L = L_0\sqrt{1-(v/c)^2}$$

L = length of a moving object along the direction of motion

L_0 = length of the same object at rest (proper length)

v = speed of the moving object

c = speed of light

EXAMPLE: Again, imagine that Majeeda is traveling at 98% of the speed of light relative to Sergio. If Majeeda holds a 1-m ruler parallel to the direction of motion, how long is this ruler as measured by Sergio?

Situation: The ruler is at rest relative to Majeeda, so she measures its *proper* length $L_0 = 1$ m. Our goal is to determine its length L as measured by Sergio, who is moving at $v = 0.98c$ relative to Majeeda and her ruler.

Tools: We use the Lorentz transformation for length.

Answer: With $v/c = 0.98$, we find

$$L = L_0\sqrt{1-(v/c)^2} = (1\ \text{m})\sqrt{1-(0.98)^2}$$
$$= (1\ \text{m}) \times (0.20) = 0.20\ \text{m} = 20\ \text{cm}$$

Review: We saw in the first example that according to *Sergio*, Majeeda's clocks are ticking only one-fifth as fast as his. This example shows that he also measures Majeeda's 1-m ruler to be only one-fifth as long (20 cm) when held parallel to the direction of motion. Note that the converse is also true: If Sergio holds a 1-m ruler parallel to the direction of relative motion, *Majeeda* measures it to be only 20 cm long. This shrinkage of length is called *length contraction.*

EXAMPLE: Consider again the above example about muons created 10 km above Earth's surface. If a muon is traveling straight down, what is the distance to the surface as measured by an observer riding along with the muon?

Situation: Imagine a ruler that extends straight up from Earth's surface to where the muon is formed. This ruler is at rest relative to Earth, so its length of 10 km is the proper length L_0. Our goal is to find the length L of this ruler as measured by the observer riding with the muon.

Tools: As in the previous example, we use the Lorentz transformation for length.

Answer: With $v/c = 0.999$, we calculate

$$L = (1\ \text{km})\sqrt{1-(0.999)^2} = 0.45\ \text{km} = 450\ \text{m}$$

Review: The distance is contracted tremendously as measured by an observer riding with the muon. This result gives us another way to explain how such muons are able to reach Earth's surface. As measured by the muon, the time required to travel the contracted distance is

$$\frac{\text{distance}}{\text{speed}} = \frac{450\ \text{m}}{0.999 \times 3.00 \times 10^8\ \text{m/s}} = 1.5 \times 10^{-6}\ \text{s}$$

This time is less than the 2.2×10^{-6} s that an average muon takes to decay. Hence, muons can successfully reach Earth's surface.

As an example of the time dilation in Figure 21-2b, suppose a rider on the train has a heart that beats once per second. Any method used to measure the rider's heartbeat on the train gives this value—1 second between beats. Perhaps watching through a window, consider a nurse on the tracks observing the rider's heartbeat as the hypothetical train speeds by at 98% of the speed of light (labeled $0.98c$ in the figure). In Figure 21-2b, we see that from the tracks, the nurse would measure that the rider's heart only beat once every 5 seconds on the nurse's own clock. In other words, the stationary observer would see a slower heartbeat.

The special theory of relativity predicts that the astronaut with the flashlight in Figure 21-1b sees the astronaut flying in the spaceship as shortened along the direction of motion and as having slowly ticking clocks. Remarkably, it also says that the *flying* astronaut sees the astronaut with the flashlight (who is moving relative to her) as being shortened and as having slowly ticking clocks! These observations may seem contradictory, but they are not: Each spaceship is moving relative to the other, and in the special theory of relativity only relative motion matters. Each astronaut's measurements are correct relative to his or her frame of reference.

Now we see the significance of relative motion between observers. Relative motion changes what observers measure for certain quantities, such as length and time. **Objects observed in motion are shorter in their direction of motion, and clocks observed in motion tick more slowly.** However, relative motion does not change the speed of light (through empty space) and it does not change the laws used to describe nature.

CAUTION! You may have the idea that the special theory of relativity implies that there are *no* absolutes in nature and that everything is relative. But, in fact, the special theory is based on the principles that the laws of physics and the speed of light in a vacuum *are* absolutes. Only certain quantities, such as distance and time, depend on your state of motion. You may also have the idea that length contraction and time dilation are just optical illusions caused by high speeds. But these effects are *real*. A moving spaceship does not just look shorter or seem to be shorter—it really *is* shorter. Likewise, a moving clock does not merely look like it is ticking slowly, or seem to be ticking slowly—it really *does* tick more slowly. Einstein's theory is supported by every experiment designed to test it. Relativity is very strange, but it is also very real!

CONCEPTCHECK 21-1

Suppose you are on the hypothetical train in Figure 21-2a. If the train sped along at half the speed of light, would it appear to you that the train contracts in length? Would your heart beat more slowly?

CONCEPTCHECK 21-2

Suppose two spaceships travel directly toward each other at 90% the speed of light relative to an observer watching from a planet right in the middle of the spaceships. Each spaceship turns on its lights for everyone to see. At what speed does a traveler from one spaceship see light arriving from the other spaceship? At what speed does the observer on the planet see light arriving from the spaceship?

Answers appear at the end of the chapter.

Mass and Energy

Einstein's special theory of relativity predicts another famous relationship: Any object with mass (m) has energy (E) embodied in its mass. This idea is expressed by the well-known formula $E = mc^2$, where c is the speed of light. In thermonuclear reactions, part of the energy embodied in the mass of atomic nuclei is converted into other forms of energy. This released energy is what makes the Sun and the stars shine (see Section 16-1). The conversion of mass into energy results in a real loss of mass; as the Sun shines it loses about 100 times the mass of the largest Egyptian pyramid each second. The sunlight we see by day and the starlight that graces our nights are a brilliant verification of the special theory of relativity.

The special theory of relativity also explains why it is impossible for a spaceship to move at the speed of light. If it could, then a light beam traveling in the same direction as the spaceship would appear to the ship's crew to be stationary. But this would contradict the second of Einstein's principles, which says that all observers, including those on a spaceship, must see light traveling at the speed of light. Therefore, it cannot be possible for a spaceship to travel at the speed of light. **In fact, no object with mass can travel at the speed of light.** This conclusion has also been verified by experiments. Subatomic particles can be made to travel at 99.99999999% of the speed of light c, but they can never make it all the way to c.

21-2 The general theory of relativity predicts black holes

TUTORIAL 21-1

Einstein's special theory of relativity is a *special* case in the sense that it does not include the effects of gravity. Einstein's next goal was to develop an even more comprehensive theory that includes gravity and its effects on space and time. This is Einstein's **general theory of relativity,** which he published in 1915.

In the special theory of relativity, a relative velocity between observers leads to different observations of distance and time. In addition to all of the effects of the special theory, the general theory of relativity describes how observers in gravitational fields of different strengths also measure differences in distance and time.

The Equivalence Principle

According to Newton's theory of gravity, an apple falls to the floor because the force of gravity pulls the apple down. But Einstein pointed out that the apple would appear to behave in exactly the same way in space, far from the gravitational influence of any planet or star, if the floor were to accelerate upward (in other words, if the floor came up to meet the apple).

In Figure 21-3, two famous gentlemen are watching an apple fall toward the floor of their closed compartments. They have no way of telling who is at rest on Earth and who is in the hypothetical elevator moving upward through empty space at a constantly increasing speed. This is an example of Einstein's **equivalence principle,** which

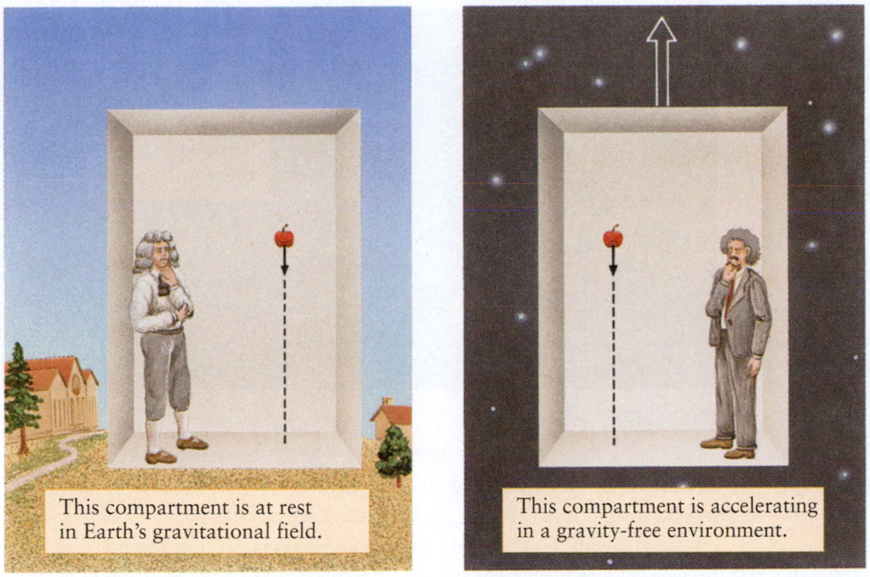

(a) The apple hits the floor of the compartment because Earth's gravity accelerates the apple downward.

(b) The apple hits the floor of the compartment because the compartment accelerates upward.

ANIMATION 21-1

FIGURE 21-3

The Equivalence Principle The equivalence principle asserts that you cannot distinguish between **(a)** being at rest in a gravitational field and **(b)** being accelerated upward in a gravity-free environment. This idea was an important step in Einstein's quest to develop the general theory of relativity.

states that in a small volume of space, the downward pull of gravity can be accurately and completely duplicated by an upward acceleration of the observer.

The equivalence principle is the key to the general theory of relativity. It allowed Einstein to focus entirely on motion, rather than force, in discussing gravity. A hallmark of gravity is that it causes the same acceleration no matter the mass of the object. For example, a baseball and a cannon ball have very different masses, but if you drop them side by side in a vacuum, they accelerate downward at exactly the same rate. To explain this observation, Einstein envisioned gravity as being caused by a *curvature of space.* In fact, his general theory of relativity describes gravity entirely in terms of the geometry of both space *and* time, that is, of spacetime. Far from a source of gravity, like a planet or a star, spacetime is "flat" and clocks tick at their normal rate. Closer to a source of gravity, however, space is curved and clocks slow down.

ANALOGY A useful analogy is to picture the spacetime near a massive object such as the Sun as being curved like the surface in Figure 21-4. Imagine a ball rolling along this surface. Far from the "well" that represents the Sun, the surface is fairly flat and the ball moves in a straight line. If the ball passes near the well, however, it curves in toward it. If the ball is moving at an appropriate speed, it might move in an orbit around the sides of the well. The curvature of the well has the same effect on a ball of any size, which explains why gravity produces the same acceleration on objects of different mass.

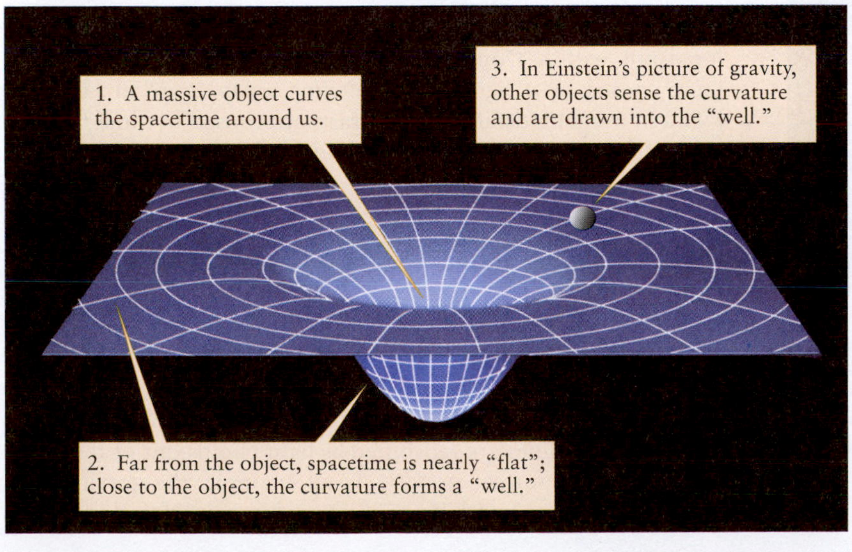

1. A massive object curves
the spacetime around us.
3. In Einstein's picture of gravity,
other objects sense the curvature
and are drawn into the "well."
2. Far from the object, spacetime is nearly "flat";
close to the object, the curvature forms a "well."

FIGURE 21-4

The Gravitational Curvature of Spacetime According to Einstein's general theory of relativity, spacetime becomes curved near a massive object. To help you visualize the curvature of four-dimensional spacetime, this figure shows the curvature of a two-dimensional space around a massive object. The curvature of spacetime has been measured directly around Earth, where Earth was the massive object in the center (location 1), and a satellite called *Gravity Probe B* was orbiting at location 3.

Testing the General Theory

Einstein's general theory of relativity and its picture of curved spacetime have been tested in a variety of different ways. In what follows, we discuss some key experimental tests of the theory.

Careful experiments have verified the key ideas of Einstein's theory of gravity

Experimental Test 1: The gravitational bending of light. Light rays naturally travel in straight lines. But if the space through which the rays travel is curved, as happens when light passes near the surface of a massive object like the Sun, the paths of the rays will likewise be curved (Figure 21-5). In other words, gravity should bend light rays. This prediction was first tested in 1919 during a total solar eclipse. During totality, when the Moon blocked out the Sun's disk, astronomers photographed the stars around the Sun. Careful measurements afterward revealed that the stars around the Sun were shifted from their usual positions by an amount consistent with Einstein's theory.

Experimental Test 2: The precession of Mercury's orbit. As the planet Mercury moves along its elliptical orbit, the orbit itself slowly changes orientation, or *precesses* (Figure 21-6). Most of Mercury's precession is caused by the gravitational pull of the other planets, as explained by Newtonian mechanics. But once the effects of all the other planets had been accounted for, there remained an unexplained excess rotation of Mercury's major axis of 43 arcsec per century. Although this discrepancy may seem very small, it frustrated astronomers for half a century. Some astronomers searched for a missing planet even closer to the Sun that might be tugging on Mercury; none has ever been found. Einstein showed that at Mercury's position close to the Sun, the general theory of relativity predicts a small correction to Newton's description of gravity. This correction is just enough to account for the excess precession.

Experimental Test 3: The gravitational slowing of time and the gravitational red shift (Figure 21-7). In the general theory of relativity, a massive object such as Earth warps time as well as space. Einstein predicted that clocks on the ground floor of a building should tick slightly more slowly than clocks on the top floor, which are farther from Earth (Figure 21-7a).

A light wave can be thought of as a clock; just as a clock makes a steady number of ticks per minute, an observer sees a steady number of complete cycles of a light wave passing by each second. If a light beam is aimed straight up from the ground floor of a building, an observer at the top floor will measure a "slow-ticking" light wave with a lower frequency, and thus a longer wavelength, than will an observer on the ground floor (Figure 21-7b). Because photons with longer wavelengths have lower energy, the increase in wavelength means that a photon reaching the top floor has less energy than when it left the ground floor. (You may want to review the discussion of frequency and wavelength in Section 5-2 as well as the description of photons in Section 5-5.) These effects, which have no counterpart in Newton's theory of gravity, are called the **gravitational redshift.**

CAUTION! Be careful not to confuse the gravitational redshift with a Doppler shift. In the Doppler effect, redshifts are caused by a light source moving away from an observer. Gravitational redshifts, by contrast, are caused by time flowing at different rates at different locations. No motion is involved.

The American physicists Robert Pound and Glen Rebka first measured the gravitational redshift in 1960 using gamma rays fired between the top and bottom of a shaft 20 m tall. Because Earth's gravity is relatively weak, the redshift that they measured was very small ($\Delta\lambda/\lambda = 2.5 \times 10^{-15}$) but was in complete agreement with Einstein's prediction.

Much larger shifts are seen in the spectra of white dwarfs, whose spectral lines are redshifted as light climbs out of the white dwarf's intense surface gravity. As an example, the gravitational redshift of the spectral lines of the white dwarf Sirius B (see

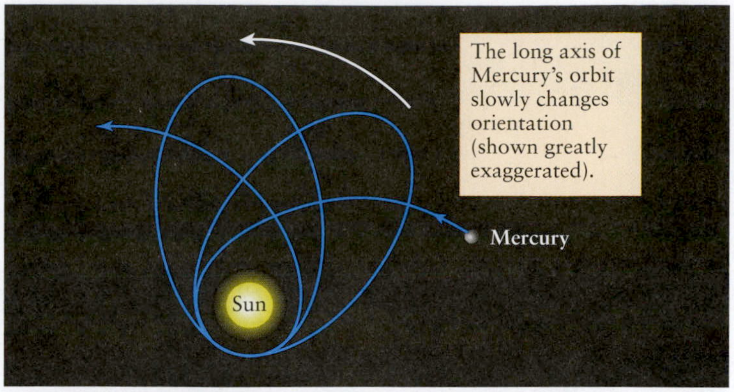

FIGURE 21-6

The Precession of Mercury's Orbit Mercury's orbit changes orientation at a rate of 574 arcsec (about one-sixth of a degree) per century. Newtonian mechanics predicts that the gravitational influences of the other planets should make the orientation change by only 531 arcsec per century. Einstein was able to explain the excess 43 arcsec per century using his general theory of relativity.

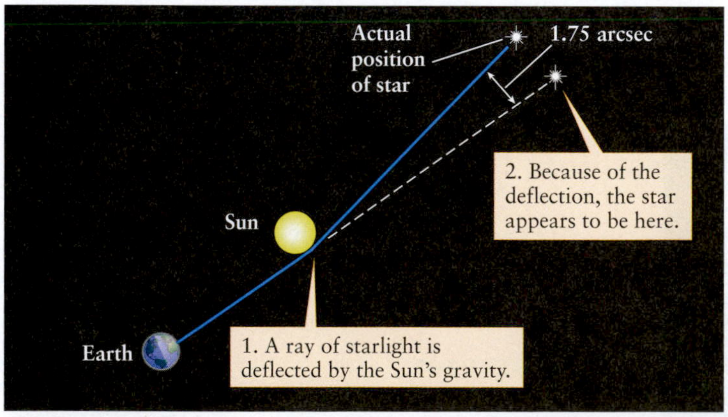

FIGURE 21-5

The Gravitational Deflection of Light Light rays are deflected by the curved spacetime around a massive object like the Sun. General relativity predicts a small deflection of 1.75 arcsec for a light ray grazing the Sun's surface, and this deflection could not be explained by Newton's theory of gravity. The deflection of starlight by the Sun was confirmed during a solar eclipse in 1919.

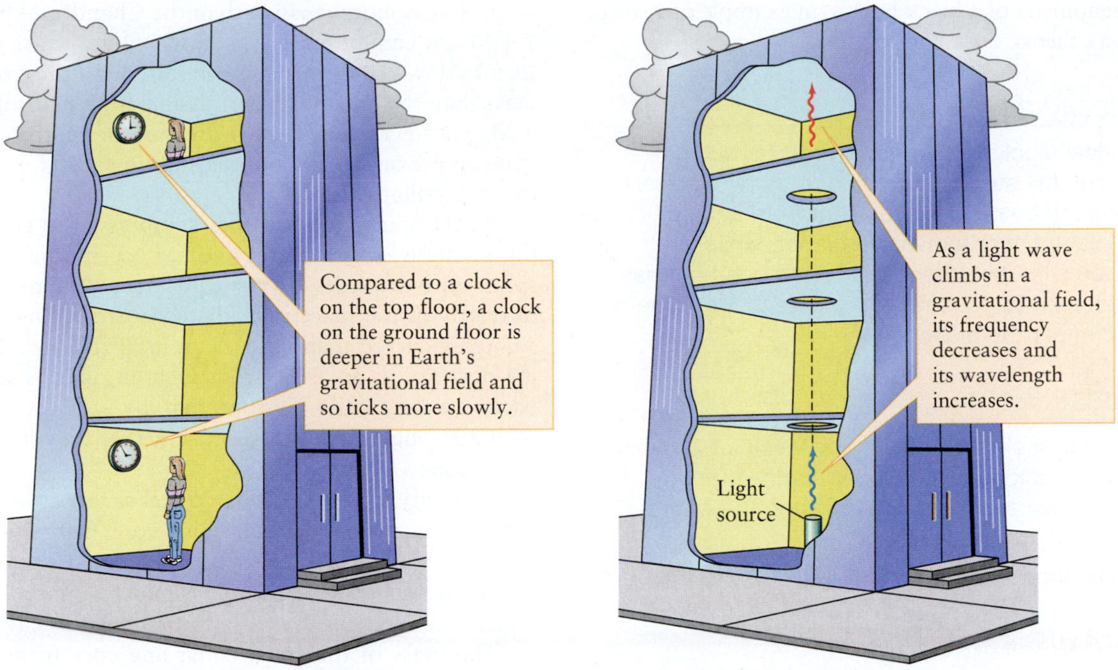

(a) The gravitational slowing of time

(b) The gravitational redshift

FIGURE 21-7

The Gravitational Slowing of Time and the Gravitational Redshift **(a)** Clocks at different heights in a gravitational field tick at different rates. Assume that identical clocks were first brought together and set to the same time. After being placed on different floors and left to tick for a while, the stronger gravity on the lower floor (closer to Earth) results in less time passing for that clock. In stronger gravity, time passes more slowly. **(b)** The oscillations of a light wave constitute a type of clock. As a light wave climbs from the ground floor toward the top floor, its oscillation frequency becomes lower and its wavelength becomes longer.

Figure 20-9) is $\Delta\lambda/\lambda = 3.0 \times 10^{-4}$, which also agrees with the general theory of relativity.

LOOKING DEEPER 21-1 *Experimental Test 4: Gravitational waves.* Electric charges oscillating up and down in a radio transmitter's antenna produce electromagnetic radiation. In a similar way, the general theory of relativity predicts that oscillating massive objects should produce **gravitational radiation,** or **gravitational waves.** (Newton's theory of gravity makes no such prediction.)

Gravitational radiation is exceedingly difficult to detect, because it is by nature much weaker than electromagnetic radiation. Although physicists have built a number of sensitive "antennas" for gravitational radiation, no confirmed detections have yet been made. But compelling *indirect* evidence for the existence of gravitational radiation has come from a binary system of two neutron stars. Russell Hulse and Joseph Taylor at the University of Massachusetts discovered that these stars are slowly spiraling toward each other and losing energy in the process. The rate at which they lose energy is just what would be expected if the two neutron stars are emitting gravitational radiation as predicted by Einstein. Hulse and Taylor shared the 1993 Nobel Prize in Physics for their discovery.

Recent Evidence for Einstein's Gravity Theory

The curvature of spacetime was measured around Earth in 2011. Analogous to Figure 21-4, a scientific satellite called *Gravity Probe B* carrying sensitive gyroscopes was sent to measure the curvature produced by Earth's gravity. Spacetime curvature would produce a subtle precession of a gyroscope's axis as it orbits Earth, which is an effect that is unique to Einstein's theory. Amazingly, scientists reported precise agreement between general relativity and *Gravity Probe B*'s detailed measurements.

Using the world's most accurate clocks in 2010, the gravitational slowing of time has been measured at two vertical positions *separated by only 1 foot* in a laboratory. Just as predicted, time slows down enough that the lower clock would read about 90 billionths of a second less over a 79-year period.

Another context in which the general theory of relativity shows up is when certain technology, by its very nature, is highly sensitive. For example, global positioning system (GPS) satellites must use extremely accurate clocks in order to triangulate an accurate position on Earth. The effects of gravity slow down time on Earth's surface, compared to the satellite's clocks, so general relativity *must* be taken into account for an accurate GPS result. Without general relativity included, GPS would accumulate errors of more than 10 km per day!

The general theory of relativity has never made an incorrect prediction. It now stands as our most accurate and complete description of gravity. Einstein demonstrated that Newtonian mechanics is accurate only when applied to low speeds and weak gravity. If extremely high speeds or powerful gravity are involved, only a calculation using relativity will give correct answers. As we will see next, the bizarre phenomena of a black hole is an example of strong gravity that Newton's theory does not describe.

CONCEPTCHECK 21-3

Suppose two identical clocks are synchronized far out in space and away from any strong sources of gravity. You learn what time they measure by radio waves sent from each clock. Then one clock is placed near a massive planet, while you stay floating in space with the other clock. After one hour has passed on the clock next to you, has more time or less time passed on the clock near the planet?

CONCEPTCHECK 21-4

Consider the same set of clocks in ConceptCheck 21-3. When radio waves from the clock near the planet reach you out in space, are they the same wavelength as when they were emitted near the planet?

Answers appear at the end of the chapter.

The Formation of Black Holes

Perhaps the most dramatic prediction of the general theory of relativity concerns what happens when a large amount of matter is concentrated in a small volume. We have seen that if a dying star is not too massive, it ends up as a white dwarf star. If the dying star is more massive than the Chandrasekhar limit of about 1.4 $M_\odot$, it cannot exist as a stable white dwarf star and, instead, shrinks down to form a neutron star. But if the dying star has more mass than the maximum permissible for a neutron star, about 2 to 3 $M_\odot$, not even the internal pressure of neutrons (the degenerate neutron pressure) can hold the star up against its own gravity, and the core collapses.

As the star's core becomes compressed to enormous densities, the strength of gravity at the surface of this rapidly shrinking sphere also increases dramatically. According to the general theory of relativity, the space immediately surrounding the core becomes so highly curved that it closes on itself (Figure 21-8). Photons flying outward at an angle from such an object's surface would arc back inward, while photons that fly straight outward undergo such a strong gravitational redshift that they lose all their energy and cease to exist.

Ordinary matter can never travel as fast as light. Hence, if light cannot escape from the collapsing core, neither can anything else. An object from which neither light nor matter can escape is called a **black hole.** In a sense, a hole is punched in the fabric of the universe, and the dying star disappears into this cavity (Figure 21-9).

The mass of the star's collapsing core forms the black hole; this mass gives the spacetime around the black hole its strong curvature. Thanks to this curvature, the black hole's gravitational influence can still be felt by other objects. However, most of the original mass of the dying star does not make it into the black

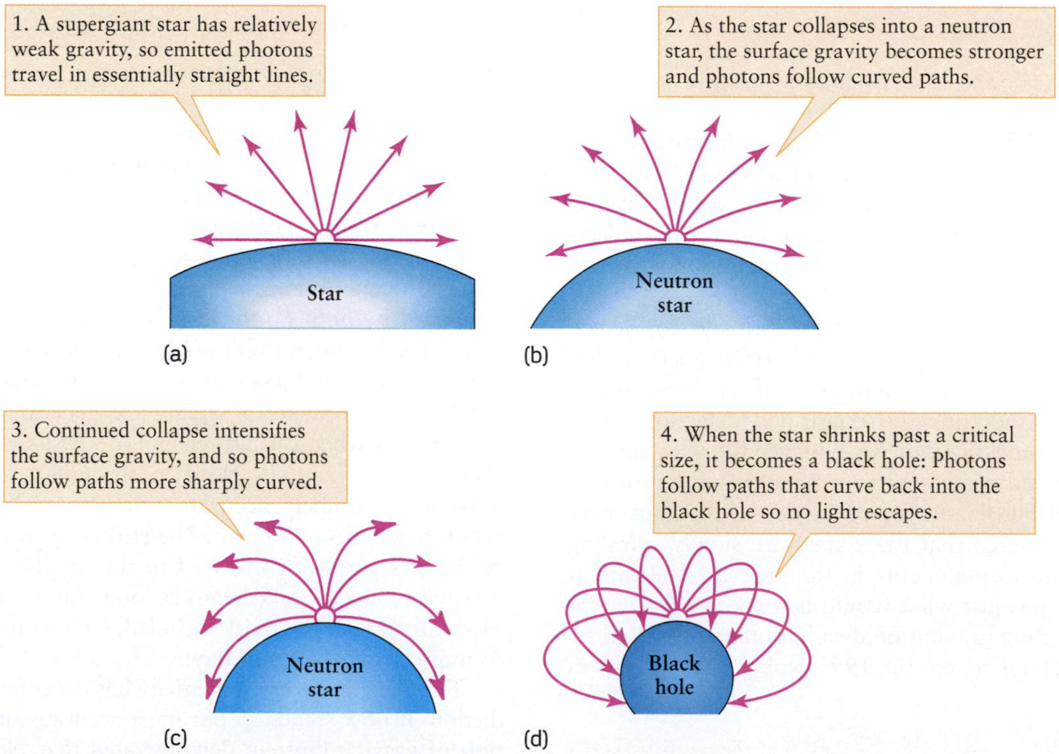

1. A supergiant star has relatively weak gravity, so emitted photons travel in essentially straight lines.
Star
(a)
2. As the star collapses into a neutron star, the surface gravity becomes stronger and photons follow curved paths.
Neutron star
(b)
3. Continued collapse intensifies the surface gravity, and so photons follow paths more sharply curved.
Neutron star
(c)
4. When the star shrinks past a critical size, it becomes a black hole: Photons follow paths that curve back into the black hole so no light escapes.
Black hole
(d)

FIGURE 21-8

The Formation of a Black Hole **(a)–(c)** These illustrations show four steps leading up to the formation of a black hole from a dying star. **(d)** When the star becomes a black hole, not even photons emitted directly upward from the surface can escape; they undergo an infinite gravitational redshift and disappear.

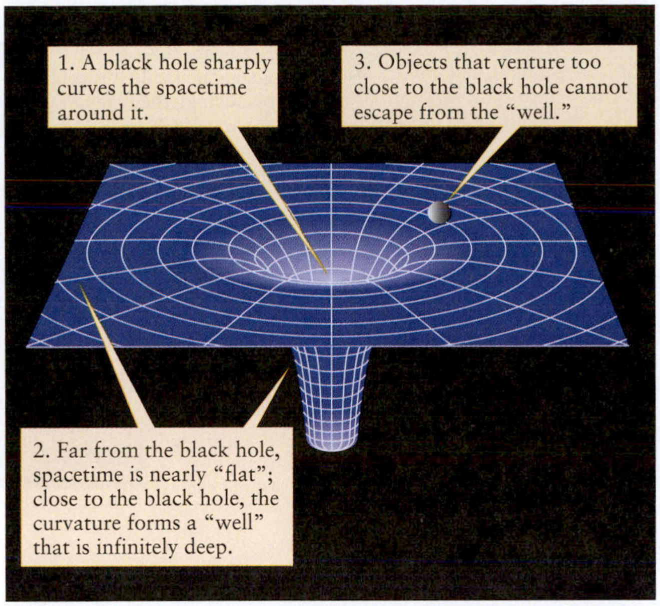

1. A black hole sharply curves the spacetime around it.
3. Objects that venture too close to the black hole cannot escape from the "well."
2. Far from the black hole, spacetime is nearly "flat"; close to the black hole, the curvature forms a "well" that is infinitely deep.

FIGURE 21-9

Curved Spacetime around a Black Hole This diagram suggests how spacetime is distorted by a black hole's mass. As in Figure 21-4, spacetime is represented as a two-dimensional surface. Unlike the situation in Figure 21-4, a black hole's gravitational "well" is infinitely deep.

hole. For example, a 5-$M_\odot$ black hole formed by core-collapse would probably come from a star of about 20 $M_\odot$ or more.

CAUTION! Some low-quality science-fiction movies and books suggest that black holes are evil things that go around gobbling up everything in the universe. Not so! The bizarre effects created by highly warped spacetime are limited to a region quite near the hole. For example, the effects of the general theory of relativity predominate only within 1000 km of a 10-$M_\odot$ black hole. Beyond 1000 km, gravity is weak enough that Newtonian physics can adequately describe everything. If our own Sun somehow turned into a black hole (an event that, happily, seems to be quite impossible), the orbits of the planets would hardly be affected at all.

CONCEPTCHECK 21-5

The exact number is not known, but approximately what is the lowest-mass black hole you might expect from a stellar core collapse?

Answer appears at the end of the chapter.

The Escape Speed from Black Holes is the Speed of Light

One way to understand black holes is to first consider the escape speed from a planet (see Section 4-7). The escape speed of a planet is the minimum ejection speed an object must have at the planet's surface in order to escape the planet and never return.

Very importantly, if you imagine compressing a planet's mass into a smaller radius, the planet's escape speed increases. It is even possible, if an object were compressed into a small enough radius, that its calculated escape speed would exceed the speed of light. Since nothing—not even light—can travel faster than the speed of light, nothing can escape such an unusual object. **An object with an escape speed exceeding the speed of light is a black hole.** Surrounding the black hole, where the escape speed from the hole just equals the speed of light, is the **event horizon.**

You can think of the event horizon as the "surface" of the black hole, although it is just a spherical boundary and not a physical surface. Once a massive dying star collapses to within its event horizon, it disappears permanently from the universe. The term "event horizon" is quite appropriate, because this surface is like a horizon beyond which we cannot see any events.

Perhaps the most remarkable aspect of black holes is that they really exist! As we will see, astronomers have located a number of black holes with masses between about 5 and 15 $M_\odot$. Black holes in this range (or even greater, if discovered) are called **stellar-mass black holes,** and they are assumed to form from the cores of supernovae. What is truly amazing is that astronomers have also discovered many enormous black holes containing millions or billions of solar masses; these were formed through a different and unknown mechanism. These discoveries are a resounding confirmation of the ideas of the general theory of relativity. In the next three sections we will see how astronomers hunt down black holes in space.

CONCEPTCHECK 21-6

Suppose that aliens could somehow add mass to our Moon without changing the Moon's radius. Does the Moon's escape speed increase or decrease? What happens if the Moon's escape speed exceeds the speed of light?

Answer appears at the end of the chapter.

21-3 Certain binary star systems probably contain black holes

TUTORIAL 21-2

Finding black holes is a difficult business. Because light cannot escape from inside the black hole, you cannot observe one directly in the same way that you can observe a star or a planet. The best you can hope for is to detect the effects of a black hole's powerful gravity.

One key to detecting black holes is to search for the radiation emitted by material as it falls into a hole

Close binary star systems offer a unique way to find **stellar-mass black holes.** Imagine that one of the stars in a binary system evolves into a black hole. The binary system is not destroyed when one star undergoes a supernova to produce a black hole, leaving the binary partner to aid in the detection of the black hole itself. If the black hole orbits close enough to the remaining ordinary star, tidal forces can draw matter from the star onto the black hole. Fortunately, we can detect this "stolen" matter as it accretes around the black hole.

Given the tremendous strength of gravity around stellar-mass black holes, their accreted matter would be compressed and heated up to very high temperatures before being sucked into the black hole. In fact, the temperatures are so high that the accreted matter would give off X-rays, so the search for these black holes is conducted with X-ray telescopes.

The Case of Cygnus X-1

The first sign of such emissions from a binary system with a black hole came shortly after the launch of the *Uhuru* X-ray–detecting satellite in 1971. Astronomers became intrigued with an X-ray source designated Cygnus X-1 (Figure 21-10). This source is quite unlike pulsating X-ray sources, which emit regular bursts of X-rays every few seconds (see Section 21-8). Instead, the X-ray emissions from Cygnus X-1 are highly variable and irregular; they flicker on timescales as short as one-hundredth of a second.

Is Cygnus X-1 big, like a regular star, or very compact, like a stellar remnant core? Amazingly, this can be answered very well without directly seeing the physical extent of Cygnus X-1. One of the fundamental concepts in physics is that nothing can travel faster than the speed of light (recall Section 21-1). Because of this limitation, different parts of an object cannot flicker in unison faster than the time required for light to travel across the object. Because light travels 3000 km in a hundredth of a second, Cygnus X-1 can be no more than 3000 km across, or about a quarter the size of Earth. This places a fundamental upper limit on the size of Cygnus X-1 and clearly rules it out as a star.

Cygnus X-1 occasionally emits bursts of radio emission, and in 1971 radio astronomers used these bursts to show that the source was nearly at the same location in the sky as the star HDE 226868. Spectroscopic observations revealed that HDE 226868 is a B0 supergiant with a surface temperature of about 31,000 K. Such stars do not emit significant amounts of X-rays, so HDE 226868 alone cannot be the X-ray source Cygnus X-1. Because binary stars are very common, astronomers began to suspect that the visible star and the X-ray source are in orbit about each other.

Further spectroscopic observations soon showed that the lines in the spectrum of HDE 226868 shift back and forth with a period of 5.6 days; this is the Doppler shift as illustrated in Figure 17-23. The clear implication is that HDE 226868 and Cygnus X-1 are the two members of a binary star system. From what we know about the masses of other supergiant stars, HDE 226868 is estimated to have a mass of roughly 20 $M_\odot$.

The rest of the puzzle involves estimating the mass of Cygnus X-1 based on the properties of the binary system. For example, the *magnitude* of the measured Doppler shifts—how much the spectral lines of the ordinary star shift back and forth—are influenced by the mass of Cygnus X-1. After decades of careful observation, the mass of Cygnus X-1 is estimated to be 14.8 $M_\odot$. Keep in mind that Cygnus X-1 is very compact—smaller than Earth. With a maximum possible white dwarf mass of 1.4 $M_\odot$, a maximum neutron star mass of 2–3 $M_\odot$, it is very likely that Cygnus X-1 is a 14.8-$M_\odot$ black hole.

If Cygnus X-1 does contain a black hole, the X-rays do not come from the black hole itself. Gas captured from HDE 226868 goes into orbit about the hole, forming an accretion disk about 4 million km in diameter (see Figure 21-10). As material in the disk gradually spirals in toward the hole, friction heats the gas to temperatures approaching 2×10^6 K. In the final 200 km above the hole, these extremely hot gases emit the X-rays that our satellites detect. Presumably the X-ray flickering is caused by small hot spots on the rapidly rotating inner edge of the accretion disk. In this way, the black hole's existence is revealed by hot gases just before they plunge to oblivion.

But is there really a black hole at the center of Cygnus X-1? With Cygnus X-1 so far away (about 6,100 ly), astronomers do not have the capability to detect the emission signatures unique to black holes expected near the event horizon. Furthermore, the size constraints mentioned above for Cygnus X-1 do not place its mass of 14.8 $M_\odot$ within its event horizon. Nonetheless, theoretical calculations provide two compelling results. First, there is no known type of object or state of matter that could contain 14.8 $M_\odot$ within the observed upper size limit of Cygnus X-1. Second, even if the observed 14.8 $M_\odot$ were due to some unknown object (or collection of objects), it would very quickly collapse into a black hole anyway. More than 40 years after its discovery, astronomers think the case for a black hole at the center of Cygnus X-1 is very strong.

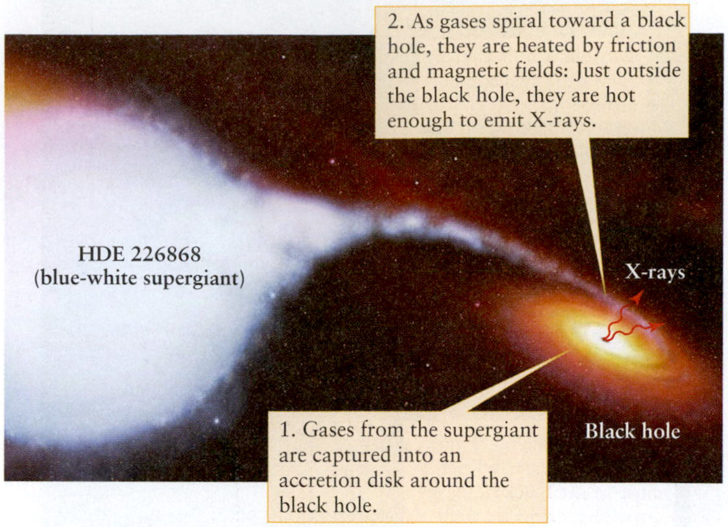

FIGURE 21-10

X-rays Generated by Accretion of Matter Near a Black Hole Stellar-remnant black holes, such as Cygnus X-1, V404 Cygni, and probably A0620–00, are detected in close binary star systems. This drawing (of the Cygnus X-1 system) shows how gas from the 30-$M_\odot$ companion star, HDE 226868, transfers to the black hole, which has about 14.8 $M_\odot$. This process creates an accretion disk. As the gas spirals inward, friction and compression heat it so much that the gas emits X-rays, which astronomers can detect. (NASA, ESA, Martin Kornmesser [ESA/Hubble])

CONCEPTCHECK 21-7

There are stars that are 14.8 $M_\odot$, so why is the mass found in Cygnus X-1 considered a black hole?

Answer appears at the end of the chapter.

Other Black Hole Candidates

Astronomers have found more than 20 other black hole candidates like Cygnus X-1. All are compact X-ray sources orbiting ordinary stars in a spectroscopic binary system. One of the best candidates is V404 Cygni in the constellation Cygnus. Doppler shift measurements reveal that as the visible star moves around its 6.47-day orbit, its radial velocity (see Section 17-10) varies by more than 400 km/s. These data indicate a black hole mass of 12 $M_{\odot}$ for the unseen companion of the visible star. It is probably impossible for a neutron star to be more massive than 3 $M_{\odot}$, so V404 Cygni must almost certainly be a black hole.

Another particularly convincing case is the flickering X-ray source A0620-00; at only 3000 ly away, it is the black hole candidate closest to Earth. (The A refers to the British satellite *Ariel 5*, which discovered the source; the numbers denote its position in the sky.) The visible companion to A0620-00 is an orange K5 main-sequence star called V616 Monocerotis, which orbits the X-ray source every 7.75 hours. Because this star is relatively faint, it is possible to observe the shifting spectral lines from *both* the visible companion and the X-ray source as they orbit each other. With this more complete information about the orbits, astronomers have determined the black hole mass of A0620-00 to be about 11 $M_{\odot}$.

Astronomers have seen jets of hot, glowing material extending several light-years from some black hole candidates. The ejected material emerges from the vicinity of the black hole at speeds approaching the speed of light. It is thought that these jets are formed by strong electric and magnetic fields in the material around a *rotating* black hole, much as occurs for rotating protostars (see Figure 18-16) and for rotating neutron stars (see Figure 20-29). Figure 21-11 is an artist's impression of the immediate vicinity of such a rotating black hole. (We will discuss the curious features of rotating black holes in Section 21-7.)

If there really are black holes in close binary systems, how did they get there? One possibility already mentioned is that one member of the binary explodes as a Type II supernova, leaving a burned-out core whose mass exceeds 3 $M_{\odot}$. This core undergoes gravitational collapse to form a black hole.

A white dwarf or a neutron star in a binary system could also become a black hole if it accretes enough matter from its companion star. This transformation can occur when the companion star becomes a red giant and dumps a significant part of its mass over its Roche lobe (see Figure 19-21b).

Finally, another possibility is two dead stars coalescing to form a black hole. For example, imagine two neutron stars orbiting each other, like the binary system discussed Section 21-2. Because the two stars emit gravitational radiation, they gradually spiral in toward each other and can eventually merge. If their total mass exceeds 2 to 3 $M_{\odot}$, the entire system may become a black hole.

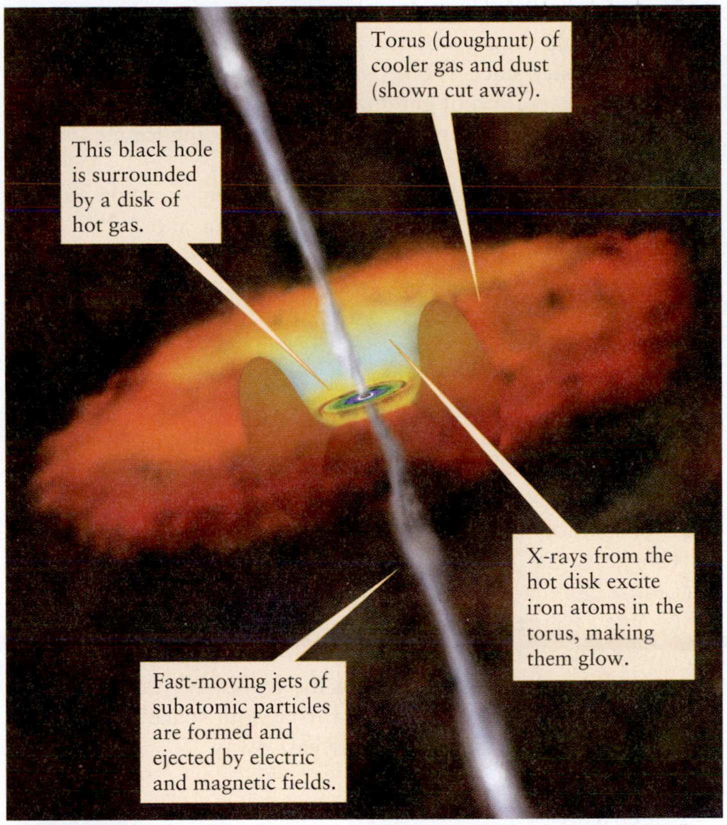

Torus (doughnut) of cooler gas and dust (shown cut away).
This black hole is surrounded by a disk of hot gas.
X-rays from the hot disk excite iron atoms in the torus, making them glow.
Fast-moving jets of subatomic particles are formed and ejected by electric and magnetic fields.

FIGURE 21-11
The Environment of an Accreting Black Hole If a black hole is rotating, it can generate strong electric and magnetic fields in its immediate vicinity. These fields draw material from the accretion disk around the black hole and accelerate it into oppositely directed jets along the black hole's rotation axis. This illustration also shows other features of the material surrounding such a black hole. (NASA/CXC/SAO)

21-4 The most intense radiation bursts in the universe may be caused by the formation of black holes

As we have seen, bursts of X-rays from binary systems such as Cygnus X-1 lead us to the remarkable conclusion that black holes are real, and that certain stars can evolve into black holes. Even more remarkable is that astronomers may actually be able to observe the moment that a black hole forms from a dying star. The key to this comes from **gamma-ray bursts,** mysterious objects that emit the most powerful bursts of high-energy radiation ever measured. To understand the connection between these objects and black holes, we begin by looking at how gamma-ray bursts were discovered and how astronomers struggled to determine their true nature.

Gamma-ray bursts are incredibly bright flashes of radiation from remarkably distant galaxies

A Gamma-Ray Mystery

Gamma-ray bursts were discovered in the late 1960s by the orbiting *Vela* satellites, whose detectors noticed flashes of gamma rays coming from random parts of the sky at random intervals. This discovery was an unexpected consequence of the Cold War. The *Vela* satellites were originally placed in orbit by the United States to look for high-energy photons coming from above-ground tests of nuclear weapons by the Soviet Union, tests that had been banned by treaty since 1963. (No such tests were ever detected.) With a new generation of gamma-ray telescopes currently in orbit (see Section 6-7), new gamma-ray bursts are being found at a rate of about one per day.

Gamma-ray bursts fall into two types. *Long-duration* gamma-ray bursts, which are more common, last from about 2 to about 1000 seconds before fading to invisibility. The less common *short-duration* gamma-ray bursts last from a few hundredths of a second to about 2 seconds, and tend to emit photons of shorter wavelength and hence higher energy. Unlike X-ray bursts (see Section 20-12), gamma-ray bursts of both types appear to emit only one burst in their entire history.

What are gamma-ray bursts, and how far away are they? These questions plagued astronomers for almost 30 years. One important clue is that gamma-ray bursts are seen with roughly equal probability in all parts of the sky, as Figure 21-12a shows. This suggests that they are not in the disk of our Galaxy, because then most gamma-ray bursts would be found in the plane of the Milky Way (Figure 21-12b). One idea to explain their uniformity on the sky was that gamma-ray bursts are relatively close to us and lie in a spherical halo surrounding the Milky Way. Alternatively, gamma-ray bursts could be strewn throughout space like galaxies, with some of them billions of light-years away.

Long-Duration Gamma-Ray Bursts and Supernovae

For many years there was no convincing way to decide between these competing models, which placed the bursts either quite close or very far away. This state of affairs changed dramatically in 1997, thanks to the Italian-Dutch BeppoSAX satellite. Unlike previous gamma-ray satellites, which could determine the position of a burst only to within a few degrees, BeppoSAX could pin down its position to within an arcminute. This made it far easier for astronomers using optical telescopes to search for a counterpart to the burst. Just 21 hours after BeppoSAX observed a long-duration gamma-ray burst on February 28, 1997, astronomers located a visible-light afterglow of the burst. Since then, astronomers using large visible-light and infrared telescopes have detected the afterglow of many gamma-ray bursts and found that they occur within galaxies (Figure 21-13). As of this writing, the most distant gamma-ray burst yet seen lies within a galaxy some 13 billion (1.3×10^{10}) ly away. (In Chapter 23 we will see how astronomers determine the distances to remote galaxies.) To be seen at such immense distances a gamma-ray burst must release a truly stupendous amount of energy in the form of radiation.

The relationship between gamma-ray bursts and galaxies strongly suggests that a burst is a cataclysmic event involving one of a galaxy's stars. Observations of a relatively bright (and, hence, relatively close) long-duration gamma-ray burst have confirmed this idea. On March 29, 2003, the orbiting gamma-ray telescope HETE-2 (short for High Energy Transient Explorer) detected this burst—denoted GRB 030329 for the date it was observed—in the constellation Leo. (GRB 030329 was so intense that its high-energy photons temporarily ionized part of Earth's atmosphere—the only event of its kind ever caused by an object beyond our own Galaxy.) Within 90 minutes a visible-light afterglow of GRB 030329 was found by astronomers in Australia and Japan. The afterglow was bright enough that astronomers in the United States and Chile were able to measure its detailed spectrum, which turned out to be that of a Type Ic supernova at a distance of some 2.65 billion (2.65×10^9) ly. We saw in Section 20-9 that a Type Ic supernova results when a massive star undergoes core collapse after having

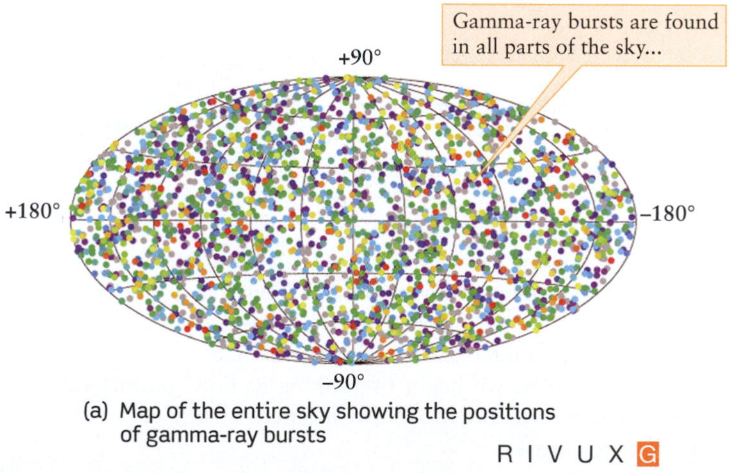

(a) Map of the entire sky showing the positions of gamma-ray bursts

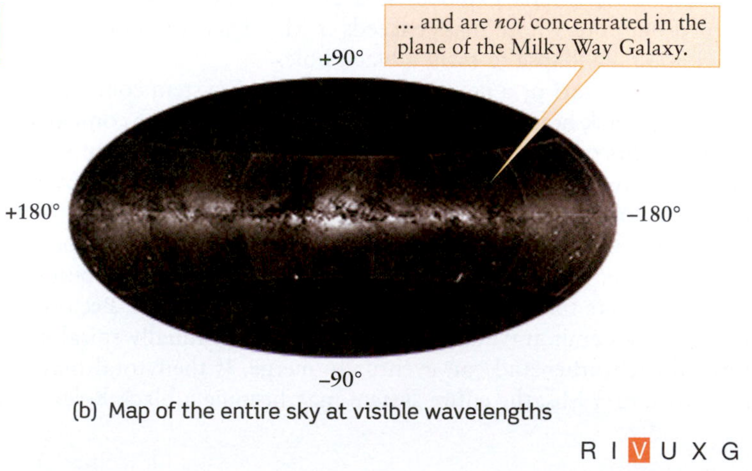

(b) Map of the entire sky at visible wavelengths

FIGURE 21-12

Gamma-Ray Bursts (a) This map shows the locations of 2704 gamma-ray bursts detected by the Compton Gamma Ray Observatory (see Section 6-7). The entire celestial sphere is mapped onto an oval. The colors in the order of the rainbow indicate the total amount of energy detected from each burst; bright bursts appear in red and weak bursts appear in violet. **(b)** This map shows the same sky as in part (a) but at visible wavelengths. Comparing part (b) with part (a) shows that unlike X-ray bursts, which originate in the disk of the Milky Way Galaxy, gamma-ray bursts are seen in all parts of the sky. (a: NASA)

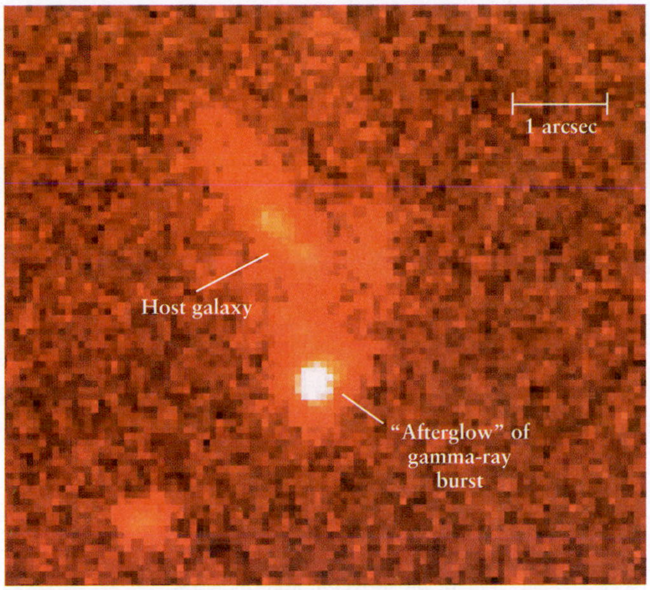

FIGURE 21-13 R I V U X G

The Host Galaxy of a Gamma-Ray Burst This false-color image was made on February 8, 1999, 16 days after a gamma-ray burst was observed at this location. The host galaxy of the gamma-ray burst has a very blue color (not shown in this image), indicating the presence of many recently formed stars. The gamma-ray burst may have been produced when one of the most massive of these stars became a supernova. (Andrew Fruchter, STScI; and NASA)

first shed the hydrogen and helium from its outer layers (see Figure 20-18c). Subsequent observations with the Hubble Space Telescope revealed an expanding shell of gas at the position of the gamma-ray burst. This shell is just what we would expect from a supernova explosion. Another long-duration gamma-ray burst with an associated Type Ic supernova was seen in 2006.

Beamed Radiation

A long-duration gamma-ray burst cannot simply be an ordinary Type Ic core-collapse supernova. Such supernovae release about 10^{46} joules of energy, of which only about 0.03% is released as electromagnetic radiation. (Most of the remaining energy goes into neutrinos, while some goes into accelerating the debris that expands away from the supernova.) This radiation streams outward from the supernova in all directions equally, like light from the Sun. If we assume that a gamma-ray burst also emits light equally in all directions, the inverse-square law that relates brightness, luminosity, and distance (see Section 17-2) tells us that the most energetic bursts would have to emit about 3×10^{47} joules of radiation in less than a minute. It would take 100,000 ordinary supernovae going off simultaneously to release this much radiation!

This dilemma can be resolved if we imagine a type of supernova that emits most of its radiation in narrow beams. If such a beam happened to be aimed toward Earth, we would detect a far more intense burst of radiation than we would from an ordinary supernova.

ANALOGY An ordinary flashlight is an example of beamed radiation. The lightbulb in a flashlight is very small and produces only a weak light if you remove it from the flashlight housing. But the flashlight's curved mirror behind the bulb channels the bulb's light into a narrow beam. Thus, a flashlight produces an intense beam of light with only a small input of energy from the batteries.

Collapsars and the Birth of a Black Hole

One theoretical model that would produce beamed radiation invokes a special type of supernova called a **collapsar** (also called a *hypernova*). These objects are thought to result from progenitor stars that are very massive (more than about 30 $M_\odot$), have lost their outer layers of hydrogen and helium, and are rotating rapidly. The core of such a star is too massive to be a white dwarf or a neutron star, so when thermonuclear reactions cease, the core will become a black hole. Figure 21-14 depicts what happens when thermonuclear reactions cease in the core of such a star. The black hole forms before the material outside the core has a chance to contract very much (Figure 21-14a). As a result, the black hole quickly forms an accretion disk from the surrounding stellar material as shown in Figure 21-14b. This process is aided because the progenitor star was rotating rapidly, which makes it easier for the infalling material to form a rotating disk around the black hole. Some of the infalling material does not fall into the black hole, but is ejected in powerful, back-to-back jets along the rotation axis of the accretion disk. (Much the same mechanism acts in close binary star systems in which one member is a black hole, as we described in Section 21-3.)

The jets are so energetic that they reach and break through the surface of the star within 5 to 10 seconds after being formed, as Figure 21-14c shows. (It helps that the star has already lost much of its outer layers, which makes it easier for the jets to break through what remains of the star.) As they travel outward, the jets produce powerful shock waves that blow the star to pieces.

If one of the jets from a collapsar happens to be aimed toward Earth, we see an intense burst of gamma rays. These are produced as the relativistic particles in the jet slow down and convert their kinetic energy (energy of motion) into radiation. The burst is short-lived because the jets have only a brief existence: It takes the black hole only about 20 seconds to accrete the entire inner core of the star, after which there is no longer an accretion disk to produce the jets. We do not see a burst if the jets are aimed away from Earth. Hence, for each gamma-ray burst that we observe, there may be hundreds or thousands of collapsars that go undetected.

The collapsar model cannot explain short-duration gamma-ray bursts, whose bursts of less than 2 seconds are far shorter than those of a collapsar's jets. A number of models for short-duration bursts have been proposed, such as the energy released by the merger of two neutron stars or a particularly intense magnetar burst.

The newest tool in the search for answers about gamma-ray bursts is a NASA satellite called *Swift,* which was launched in 2004, and has observed over 800 gamma-ray bursts. In addition to sensitive gamma-ray detectors to identify bursts, *Swift* carries

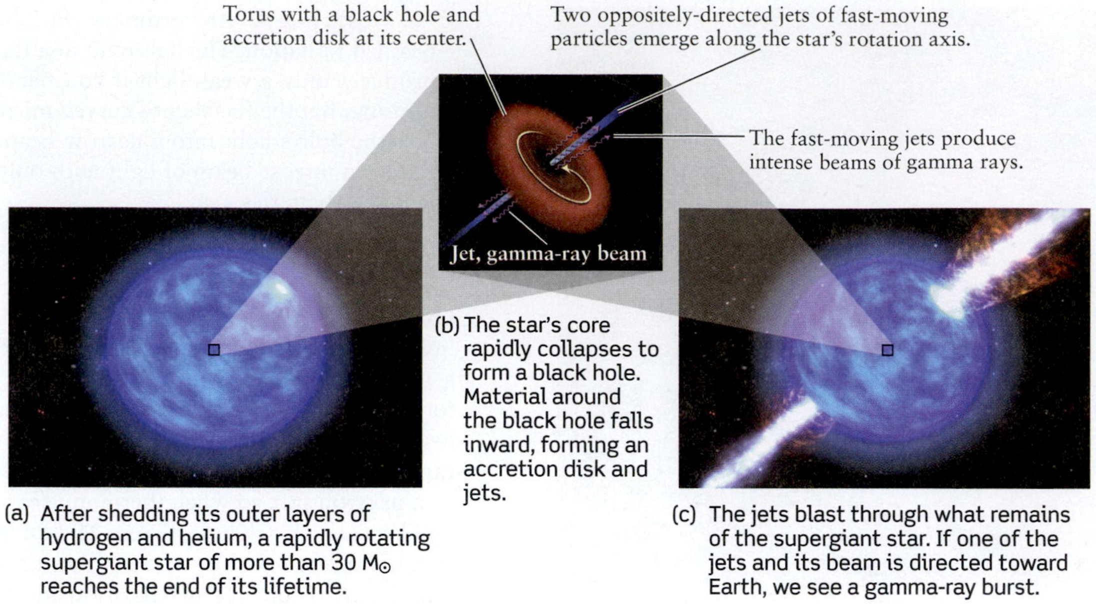

Torus with a black hole and accretion disk at its center.
Two oppositely-directed jets of fast-moving particles emerge along the star's rotation axis.
The fast-moving jets produce intense beams of gamma rays.
Jet, gamma-ray beam
(a) After shedding its outer layers of hydrogen and helium, a rapidly rotating supergiant star of more than 30 M⊙ reaches the end of its lifetime.
(b) The star's core rapidly collapses to form a black hole. Material around the black hole falls inward, forming an accretion disk and jets.
(c) The jets blast through what remains of the supergiant star. If one of the jets and its beam is directed toward Earth, we see a gamma-ray burst.

ANIMATION 21-2

FIGURE 21-14

The Collapsar Model of a Long-Duration Gamma-ray Burst These illustrations show the final few seconds in the life of a massive, rapidly rotating supergiant star that has lost its outer layers of hydrogen and helium. After the jets have subsided, what remains is a Type Ic supernova with a black hole at its center. (a, c: NASA/Sky Works Digital; b: NASA and A. Field, STScI)

telescopes that are sensitive to X-rays, ultraviolet light, and visible light. Thus, instead of having to wait for ground-based telescopes to make follow-up observations of gamma-ray bursts, *Swift* makes these observations itself as swiftly as possible (hence the spacecraft's name). Observations from *Swift* may help us better understand the nature of long-duration bursts, and provide important clues about the still enigmatic short-duration bursts.

CONCEPTCHECK 21-8

Why do supernovae models associated with gamma-ray bursts require jets?

Answer appears at the end of the chapter.

21-5 Supermassive black holes exist at the centers of most galaxies

Stellar evolution makes it possible to form black holes with masses several times that of the Sun. But calculations suggest other ways to form black holes that are either extremely large or extraordinarily small.

Supermassive Black Holes

The largest black holes have billions of times the mass of the Sun

As we will discuss in Chapter 25, galaxies formed in the early universe by the coalescence of gas clouds. During this formation process, some of a galaxy's gas might plunge straight toward the center of the galaxy, where it collects and compresses together under its own gravity. If this central clump has enough mass, it can form a black hole. Typical galaxies have masses of 10^{11} $M_\odot$, so even a tiny percentage of the galaxy's gas collected at its center would give rise to a **supermassive black hole** with a truly stupendous mass.

CAUTION! While stellar-mass–size black holes require incomprehensibly high-density material to form, supermassive black holes require much lower-density material to form. For example, if you *imagine filling our solar system with water* out to the orbital distance of Earth, the total mass of water in this sphere would be around 500 million $M_\odot$, and a supermassive black hole would form! In other words, a supermassive black hole can be formed with an average density equal to that of water. Because larger black holes require lower average densities to form compared to smaller black holes, supermassive black holes might form through *less extreme* processes than stellar-mass–size black holes.

Since the 1970s, astronomers have found evidence suggesting the existence of supermassive black holes in other galaxies. With the advent of a new generation of optical telescopes in the 1990s, it became possible to see the environments of these suspected black holes with unprecedented detail (Figure 21-15). Figure 21-15a shows radio emissions from two jets that extend some 15,000 pc (50,000 ly) from the center of the galaxy NGC 4261. These resemble an immensely magnified version of the jets seen emerging from some stellar-mass black holes (see Section 21-3). The highly magnified Hubble Space Telescope image in Figure 21-15b shows a disk some

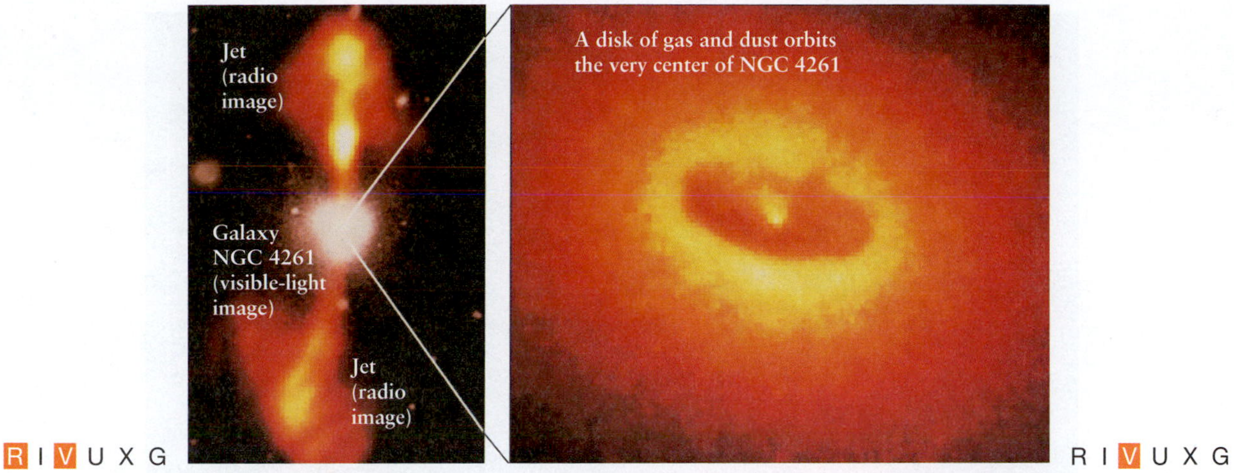

(a) Galaxy NGC 4261 (b) Evidence for a supermassive black hole in NGC 4261

ANIMATION 21-3

FIGURE 21-15

A Supermassive Black Hole The galaxy NGC 4261 lies some 30 million pc (100 million ly) from Earth in the constellation Virgo. **(a)** This composite view superimposes a visible-light image of the galaxy (white) with a radio image of the galaxy's immense jets (orange). **(b)** This Hubble Space Telescope image shows a disk of gas and dust about 250 pc (800 ly) across at the center of NGC 4261. Observations indicate that a supermassive black hole is at the center of the disk. (NASA, ESA)

250 pc (800 ly) in diameter at the very center of this galaxy. The jets are perpendicular to the plane of the disk, just as in Figure 21-11.

By measuring the Doppler shifts of light coming from the two sides of the disk, astronomers found that the disk material was orbiting around the bright object at the center of the disk at speeds of hundreds of kilometers per second. (By comparison, Earth orbits the Sun at 30 km/s.) Given the size of the material's orbit, and using Newton's form of Kepler's third law (see Section 4-7 and Box 4-4), they were able to calculate the mass of the central bright object. The answer is an amazing 1.2 *billion* (1.2×10^9) solar masses! What is more, the observations show that this object can be no larger than our solar system. The only known explanation is that the object at the center of NGC 4261 is a black hole.

Dozens of other black holes in the centers of galaxies have been identified by their gravitational effect on surrounding gas and dust. Surveys of galaxies have shown that supermassive black holes are not at all unusual; most large galaxies appear to have them at their centers. As we will see in the next chapter, a black hole with several million solar masses lies at the center of our own Milky Way Galaxy, some 8000 pc (26,000 ly) from Earth.

Other Black Holes

Theoretically, a black hole can have any mass, but the masses of black holes that exist reflect what can be produced through natural processes. The black hole masses we do observe seem to fall into two very different groups. Supermassive black holes have masses from 10^6 to 10^9 $M_\odot$, while stellar-mass black holes like those discussed in Section 21-3 have masses around 10 $M_\odot$. In 2000 a team of astronomers from Japan, the United States, and Great Britain reported finding a "missing link" between these two groups.

The new discovery was a fluctuating X-ray source located some 200 pc (600 ly) from the center of the galaxy M82 (Figure 21-16). Images do not reveal material orbiting around this source, so we cannot determine the source's mass using Newton's form of Kepler's third law (see Box 4-4). But by relating the source's X-ray luminosity to the rate at which matter would have to fall onto the source to produce that luminosity, it can be shown that the source must have a mass of at least 500 $M_\odot$ or so. The rapid fluctuations of the source show that it has a very small diameter, which suggests that it is a black hole (see Section 21-3). Hence, the source in M82 may be an **intermediate-mass black hole** (also called a **mid-mass black hole**) Additional intermediate-mass black hole candidates have been found in other galaxies. Such black holes might have formed by the coalescence of many normal stars or by the direct merger of stellar-mass black holes.

During the early 1970s, Stephen Hawking of Cambridge University proposed the existence of yet another type of black hole. He suggested that during the Big Bang some 13.7 billion years ago, local pockets of the universe could have been dense enough and under sufficient pressure to form small black holes. According to his theory, these **primordial black holes** could have had masses as large as Earth or as small as 5×10^{-8} kg, about 1/40 the mass of a single raindrop. To date, however, no evidence has been found for the existence of primordial black holes in nature.

CONCEPTCHECK 21-9

Do all black holes require extreme densities of matter to form?

CONCEPTCHECK 21-10

Can black holes only exist within certain mass ranges?

Answers appear at the end of the chapter.

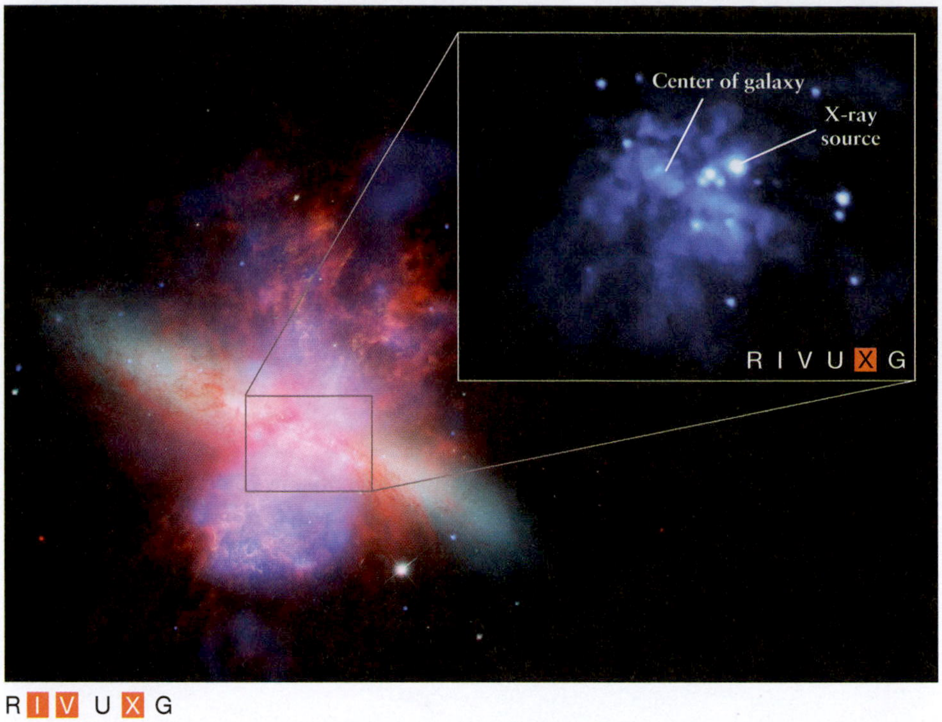

VIDEO 21-1

FIGURE 21-16

An Intermediate-mass Black Hole? The largest black holes have billions of times the mass of the Sun. M82 is an unusual galaxy in the constellation Ursa Major. The inset is an image of the central region of this galaxy from the Chandra X-ray Observatory. The bright, compact X-ray source varies in its light output over a period of months. The properties of this source suggest that it may be a black hole of roughly 500 solar masses or more. If correct, the intermediate-mass black hole is about 600 ly away from a supermassive black hole at the center of the galaxy that has about 30 million solar masses. (NASA/CXC/JHU/D. Strickland; optical: NASA/ESA/STScI/AURA/The Hubble Heritage Team; infrared: NASA/JPL-Caltech/U. of Arizona/C. Engelbracht; inset: NASA/SAO/CXC)

21-6 A nonrotating black hole has only a "center" and a "surface"

Understanding in detail how black holes form is a challenging task. But once a black hole forms, it has a remarkably simple structure.

Recall from Section 21-2, a black hole is an object with an escape speed greater than the speed of light. Since nothing can travel faster than the speed of light, nothing can escape a black hole. Even light can't escape, which is how black holes got their name. A black hole has no surface, but instead has a spherical boundary called the event horizon where the escape speed just equals the speed of light. Once the core of a massive dying star collapses to within its event horizon, it disappears permanently from the universe.

CAUTION! A black hole does *not* suck in everything from its surroundings. While anything that crosses the event horizon is lost forever, material outside the event horizon traveling at high speeds can pass by without getting trapped. Far away from a black hole, the black hole is like any other object. As mentioned earlier, if the Sun became a black hole, the planets are all too far away for any noticeable orbital effects.

If the dying star is not rotating before it collapses, the black hole will likewise not be rotating. We will consider these nonrotating black holes first, and then consider rotating black holes in the next section. The distance from the center of a nonrotating black hole to its event horizon is called the **Schwarzschild radius** (denoted R_{Sch}), after the German physicist Karl Schwarzschild who first determined its properties. Box 21-2 describes how to calculate the Schwarzschild radius, which depends only on the black hole's mass. The more massive the black hole, the larger its event horizon.

Inside a Black Hole

All of the mass of a nonrotating black hole is concentrated nearly to a point at the center called the singularity

Once an object (such as a stellar core) has contracted enough to form an event horizon, nothing can prevent the matter's further collapse. Equations describing the matter within the event horizon indicate that the object's entire mass is crushed nearly to a single point, known as the **singularity**, at the center of the black hole. We now can see that the structure of a nonrotating black hole is quite simple. As drawn in Figure 21-17, it has only two parts: a singularity at the center surrounded by a

BOX 21-2 TOOLS OF THE ASTRONOMER'S TRADE

The Schwarzschild Radius

The Schwarzschild radius (R_{Sch}) is the distance from the center of a nonrotating black hole to its event horizon. You can think of this radius as the "size" of the black hole. For an object of a given mass, the Schwarzschild radius is the size that mass would need to be concentrated within, in order to make the escape speed from that object equal to the speed of light. The Schwarzschild radius is related to the mass M of the black hole by

Schwarzschild radius

$$R_{Sch} = \frac{2GM}{c^2}$$

R_{Sch} = Schwarzschild radius of a black hole
G = universal constant of gravitation
M = mass of black hole
c = speed of light

When using this formula, be careful to express the mass in kilograms, not solar masses, because of the units in which G and c are commonly expressed ($G = 6.67 \times 10^{-11}$ N $\times$ m^2/kg^2 and $c = 3.00 \times 10^8$ m/s). If you use M in kilograms, the answer that you get for R_{Sch} will be in meters.

EXAMPLE: Find the Schwarzschild radius (in kilometers) of a black hole with 10 times the mass of the Sun.

Situation: We are given the mass of the black hole (10 $M_\odot$) and wish to determine its Schwarzschild radius R_{Sch}.

Tools: We use the above formula for R_{Sch}, being careful to first convert the mass from solar masses to kilograms.

Answer: One solar mass is 1.99×10^{30} kg, so in this case $M = 10 \times 1.99 \times 10^{30}$ kg $= 1.99 \times 10^{31}$ kg. The Schwarzschild radius of this black hole is then

$$R_{Sch} = \frac{2GM}{c^2} = \frac{2(6.67 \times 10^{-11})(1.99 \times 10^{31})}{(3.00 \times 10^8)^2}$$

$$= 3.0 \times 10^4 \text{ m}$$

One kilometer is 10^3 m, so the Schwarzschild radius of this 10-$M_\odot$ black hole is

$$R_{Sch} = (3.0 \times 10^4 \text{ m}) \times \frac{1 \text{ km}}{10^3 \text{ m}} = 30 \text{ km, or } 19 \text{ mi}$$

Review: The Schwarzschild radius is directly proportional to the mass of the black hole. Thus, a black hole with 10 times the mass (100 $M_\odot$) would have a Schwarzschild radius 10 times larger ($R_{Sch} = 10 \times 30$ km $= 300$ km); a black hole with half the mass (5 $M_\odot$) would have half as large a Schwarzschild radius ($R_{Sch} = 1/2 \times 30$ km $= 15$ km), and so on.

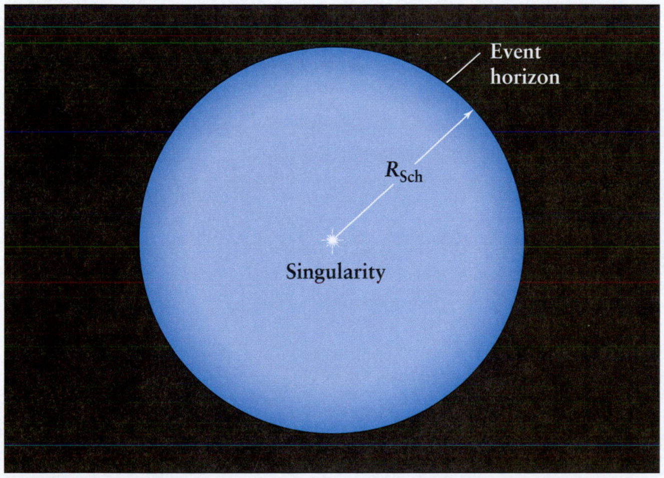

FIGURE 21-17
The Structure of a Nonrotating (Schwarzschild) Black Hole A nonrotating black hole has only two parts: a singularity, where all of the mass is located, and a surrounding event horizon. The distance from the singularity to the event horizon is called the Schwarzschild radius (R_{Sch}). The event horizon is a one-way surface: Things can fall in, but nothing can get out.

spherical event horizon. Note that while a black hole's effects on space and time away from the singularity are fairly well understood, the singularity itself—with its nearly infinite density—poses both mathematical and conceptual problems in physics.

As seen in Box 21-2, the size—called the Schwarzschild radius—is given for a massive object to form a black hole. For larger mass black holes, this size corresponds to lower average densities than for smaller mass black holes: As noted in Section 21-5, a supermassive black hole could be formed by a solar system filled with water. However, once a black hole forms, its mass rapidly collapses toward the center. So, while less extreme conditions are required to *form* supermassive black holes, the conditions inside all black holes are extreme.

To understand why the complete collapse of such a doomed star is inevitable, first think about your own life on Earth, far from any black holes. You have the freedom to move as you wish through the three dimensions of space: up and down, left and right, or forward and back. But you do not have the freedom to move at will through the dimension of time. Whether we like it or not, we are all carried inexorably through time from the cradle to the grave.

Inside a black hole, gravity distorts the structure of spacetime so severely that the directions of space and time become

COSMIC CONNECTIONS

Black Hole "Urban Legends"

Black holes are such intriguing objects that a folklore has developed about them. As often is the case with folklore, myth and reality can become confused. These illustrations depict two misconceptions or "urban legends" about black holes.

Urban Legend #1:
When a star becomes a black hole, its gravitational pull becomes stronger.

Reality: At great distances from a black hole, the gravitational force that it exerts is exactly the *same* as that exerted by an ordinary star of the same mass. The truly stupendous gravitational effects of a black hole appear only if you venture close to the black hole's event horizon.

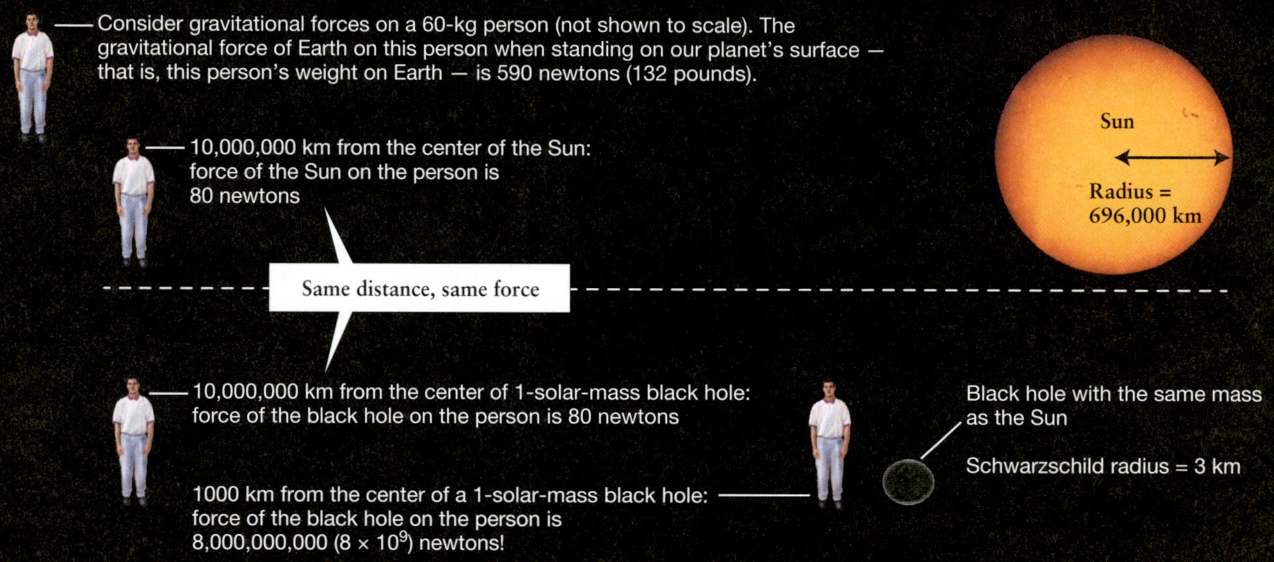
Consider gravitational forces on a 60-kg person (not shown to scale). The gravitational force of Earth on this person when standing on our planet's surface — that is, this person's weight on Earth — is 590 newtons (132 pounds).
10,000,000 km from the center of the Sun: force of the Sun on the person is 80 newtons
Sun
Radius = 696,000 km
Same distance, same force
10,000,000 km from the center of 1-solar-mass black hole: force of the black hole on the person is 80 newtons
Black hole with the same mass as the Sun
Schwarzschild radius = 3 km
1000 km from the center of a 1-solar-mass black hole: force of the black hole on the person is 8,000,000,000 (8 × 10^9) newtons!

Urban Legend #2:
The larger a black hole, the more powerful the gravity near its Schwarzschild radius.

Reality: If you increase the mass of a black hole, its Schwarzschild radius increases by the same factor. Newton's law of universal gravitation, $F = Gm_1m_2/r2$, tells us that the force that an object of mass m_1 exerts on a second object of mass m_2 is directly proportional to mass m_1 but *inversely* proportional to the *square* of the distance r between the two objects. Hence the force that a black hole exerts on an object a given distance outside its Schwarzschild radius actually *decreases* as the black hole's mass and Schwarzschild radius increase.

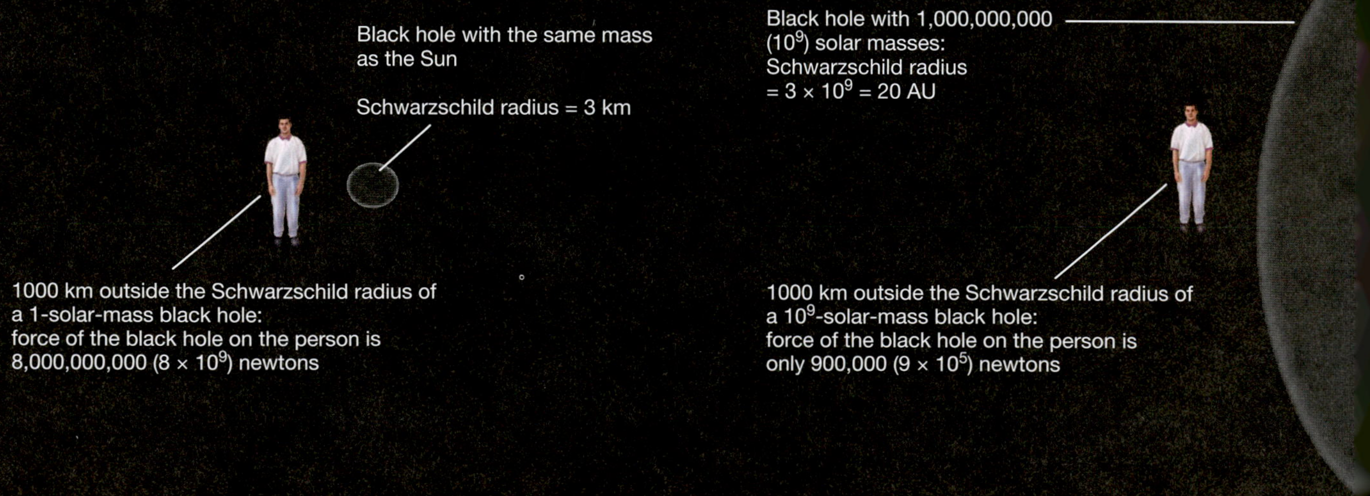
Black hole with the same mass as the Sun
Schwarzschild radius = 3 km
Black hole with 1,000,000,000 (10^9) solar masses: Schwarzschild radius = 3 × 10^9 = 20 AU
1000 km outside the Schwarzschild radius of a 1-solar-mass black hole: force of the black hole on the person is 8,000,000,000 (8 × 10^9) newtons
1000 km outside the Schwarzschild radius of a 10^9-solar-mass black hole: force of the black hole on the person is only 900,000 (9 × 10^5) newtons

interchanged. In a sense, inside a black hole you acquire a limited ability to affect the passage of time. This apparent gain does you no good, however, because you lose a corresponding amount of freedom to move through space. Whether you like it or not, you will be dragged inexorably from the event horizon toward the singularity. Just as no force in the universe can prevent the forward march of time from past to future outside a black hole, no force in the universe can prevent the inward march of matter from the event horizon to the singularity inside a black hole. Once an object dropped into a black hole crosses the event horizon, it is gone forever, like an object dropped into a bottomless pit.

The same is true for a light beam aimed at the black hole. Because all this light will be absorbed and none reflected back, a black hole is indeed black. By contrast, even the blackest object on Earth reflects *some* light when you shine a flashlight on it—a black hole would reflect none. (In Section 21-9 we will see that black holes are expected to emit some particles and light through a quantum physics process, but this radiation arises just *outside* the event horizon and would be negligible for black holes of stellar-mass size or larger.)

Although black holes are really very intriguing objects, there are some common misconceptions about their nature. The *Cosmic Connections: Black Hole "Urban Legends"* exposes two of these misconceptions.

CONCEPTCHECK 21-11

What is the connection between a black hole's Schwarzschild radius and its event horizon?

CONCEPTCHECK 21-12

Consult the *Cosmic Connections: Urban Legend* #1 figure. Consider being 10 million km from the Sun, and then the same distance from a black hole of 1 $M_\odot$. From which object do you feel a greater gravitational force?

CONCEPTCHECK 21-13

Consult the *Cosmic Connections: Urban Legend* #2 figure. If you had a personal rocket spacecraft that was low on gas, would it be safer to fly by the event horizon of a small black hole or a large black hole?

Answers appear at the end of the chapter.

21-7 Just three numbers completely describe the structure of a black hole

Because light and matter cannot escape the event horizon, we are prevented from ever knowing much about what has already fallen into a black hole. In this sense a black hole would appear to be an "information sink." Many properties of an object that falls into a black hole, such as the object's chemical composition, texture, color, shape, and size, would appear to vanish as soon as it crosses the event horizon.

Because this information has seemingly vanished, it cannot affect the basic structure or properties of the hole. For example, consider two hypothetical black holes, one made from the gravitational collapse of 10 $M_\odot$ of iron and the other made from the gravitational collapse of 10 $M_\odot$ of peanut butter. Obviously, quite different substances went into the creation of the two holes. However, once the event horizons of these two black holes have formed, both the iron and the peanut butter will have permanently disappeared from the universe. As seen from the outside, the two holes are identical, making it impossible for us to tell which ate the peanut butter and which ate the iron. A black hole is thus unaffected by the type of matter it consumes.

When an object falls into a black hole, all information about that object disappears except its mass, electric charge, and angular momentum

The Three Properties of a Black Hole

Because a black hole appears to be an information sink, it is reasonable to wonder whether we can determine anything at all about a black hole. What properties *do* characterize a black hole?

First, we can measure the *mass* of a black hole. One way to do this would be by placing a satellite into orbit around the hole. After measuring the size and period of the satellite's orbit, we could use Newton's form of Kepler's third law (recall Section 4-7 and Box 4-4) to determine the mass of the black hole. This mass is equal to the total mass of all the material that has gone into the hole.

Second, we can also measure the total *electric charge* possessed by a black hole. Like gravity, the electric force acts over long distances—it is a long-range interaction that is felt in the space around the hole. Appropriate equipment on a space probe passing near the hole could measure the strength of the electric force, and the electric charge could thus be determined.

In actuality, we would not expect a black hole to possess much, if any, electric charge. For example, if a hole did happen to start off with a sizable positive charge, it would vigorously attract vast numbers of negatively charged electrons from the interstellar medium, which would soon neutralize the hole's charge. For this reason, astronomers usually neglect electric charge when discussing real black holes.

Although a black hole might theoretically have a tiny electric charge, it can have no magnetic field of its own whatsoever. We discussed in Section 21-3 how magnetic fields are involved in producing jets from black holes. However, these fields are associated with the accretion disk around the black hole, not the black hole itself. When a black hole is created, however, the collapsing star from which it forms may possess an appreciable magnetic field. The star must therefore radiate this magnetic field away before it can settle down inside its event horizon. Theory predicts that the star does this by emitting electromagnetic and gravitational waves. As we saw in Section 21-2, gravitational waves are ripples in the overall geometry of space. Various experiments soon to be in operation may detect bursts of gravitational radiation emitted by massive stars as they collapse to form black holes.

Third, we can detect the effects of a black hole's rotation, that is, measure its *angular momentum*. An object's angular momentum is related to how fast it rotates and how the object's mass is distributed over its volume. As a dead star collapses into a black

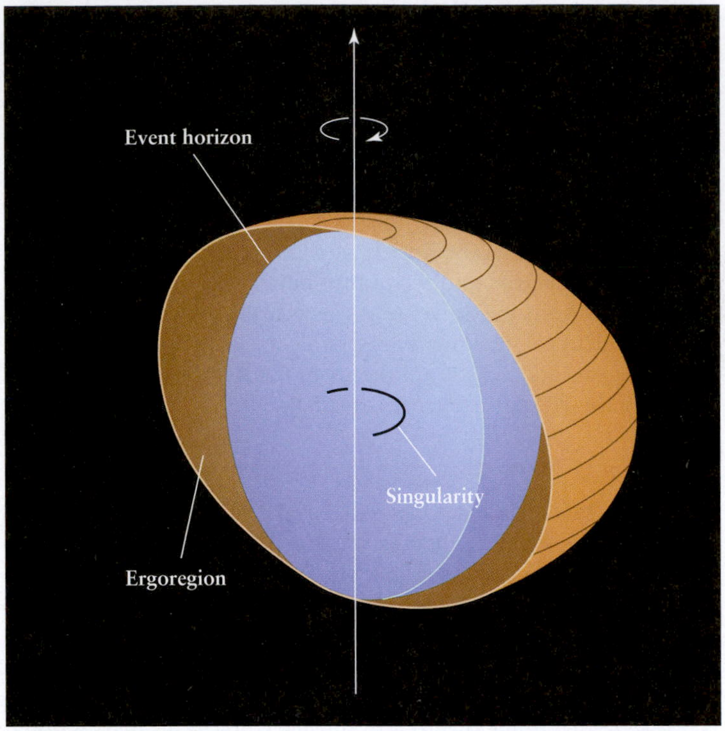

FIGURE 21-18
The Structure of a Rotating (Kerr) Black Hole The singularity of a rotating black hole is an infinitely thin ring centered on the geometrical center of the hole. (It appears as an arc in this cutaway diagram.) Outside the spherical event horizon is the doughnut-shaped ergoregion, where the dragging of spacetime around the hole is so severe that nothing can remain at rest. The tan-colored surface marks the outer boundary of the ergoregion.

hole, its rotation naturally speeds up as its mass moves toward the center, just as a figure skater rotates faster when she pulls her arms and legs in. This same effect explains the rotation of the solar nebula (see Section 8-4), as well as why neutron stars spin so fast (see Section 20-11). We expect a black hole that forms from a rotating star to be spinning very rapidly.

Rotating Black Holes

When the matter that collapses to form a black hole is rotating, that matter does not compress to a point. Instead, it collapses into a ring-shaped singularity located between the center of the hole and the event horizon (Figure 21-18). The structure of such rotating black holes was first worked out in 1963 by the New Zealand mathematician Roy Kerr. Most rotating black holes should be spinning thousands of times per second, even faster than the most rapid pulsars.

If a rotating black hole is surrounded by an accretion disk with a magnetic field, it may be possible for the magnetic field to steal some of the rotational energy and angular momentum from the black hole and transfer it to the disk. The magnetic field acts as a "brake" that retards the black hole's rotation and makes the disk's rotation speed up. (On Earth, magnetic braking is used to slow locomotives, amusement park rides, and hybrid cars to a smooth stop without generating excess heat.) Figure 21-19 shows an arching magnetic field that connects an accretion disk with a rotating black hole in just this way. This process may be taking place with the supermassive black hole at the center of the galaxy MCG–6-15-30. Observations with the XMM-Newton telescope (see Section 6-7) indicate that unusually intense X-rays are coming from the accretion disk around this black hole. The suspicion is that the energy to power this radiation may be extracted from the black hole's rotation. (This transfer of energy and angular momentum can also go the other way. The image that opens this chapter depicts a system in which magnetic fields are thought to transfer angular momentum *to* a black hole *from* its accretion disk. Robbing the accretion disk of its rotation makes it easier for its material to fall into the black hole.)

Even a rotating black hole without an accretion disk can transfer energy to other objects. This energy transfer is possible according to Einstein's general theory of relativity, which makes the startling prediction that a rotating body drags spacetime around it. (Very precise spacecraft measurements indicate that the rotating Earth drags spacetime in just this way.) Surrounding the event horizon of every rotating black hole is a region where this dragging of space and time is so severe that it is impossible to stay in the same place. No matter what you do, you get pulled around the hole, along with the rotating geometry of space and time. This region, where it is impossible to be at rest, is called the **ergoregion** (see Figure 21-18).

To measure a black hole's angular momentum, we could hypothetically place two satellites in orbit about the hole. Suppose that

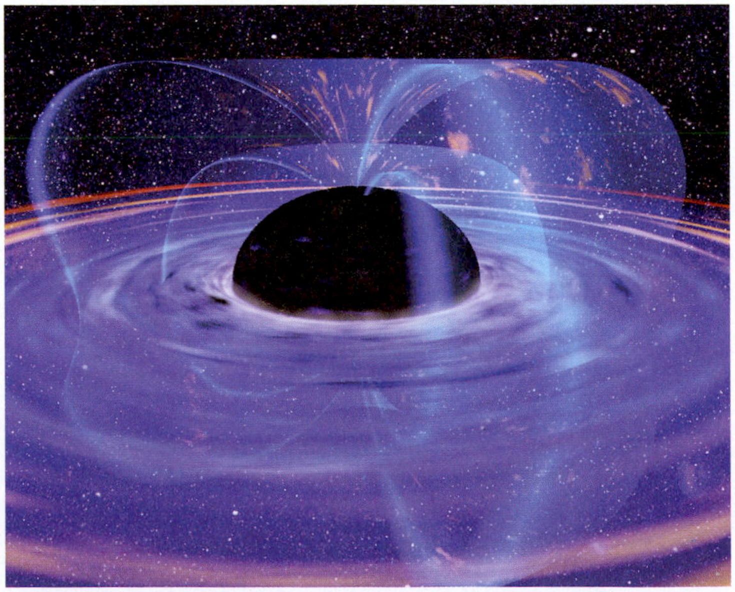

FIGURE 21-19
A Rotating Supermassive Black Hole This artist's impression shows the accretion disk around the supermassive black hole at the heart of the galaxy MCG–6-15-30. The arching magnetic field allows the accretion disk to extract energy and angular momentum from the black hole. (XMM-Newton/ESA/NASA)

one satellite circles the hole in the same direction the hole rotates and the other in the opposite direction. One satellite is thus carried along by the geometry of space and time, but the other is constantly fighting its way "upstream." The two satellites will thus have different orbital periods. By comparing these two periods, astronomers can deduce the total angular momentum of the hole.

Because the ergoregion is outside the event horizon, this bizarre region is accessible to us, and spacecraft could travel through it without disappearing into the black hole. According to detailed calculations, objects grazing the ergoregion could be catapulted back out into space at tremendous speeds. In other words, the ejected object could leave the ergoregion with more energy than it had initially, having extracted added energy from the hole's rotation. This process is called the *Penrose process*, after Roger Penrose, the British mathematician who proposed it.

Mass, charge, and angular momentum are the only three properties that a black hole possesses. This simplicity is the essence of the **no-hair theorem**, first formulated in the early 1970s: "Black holes have no hair." In other words, "hair" would be some property of a black hole other than mass, charge, or angular momentum, yet another property is not possible. Therefore, once an object has fallen into a black hole, except for the object's mass, charge, and angular momentum, any additional properties carried by the object disappear from the universe. However, the idea that black holes can completely remove information from the universe is contentious, as some scientists do not think information is ever truly lost. We will revisit black hole information in Section 21-9, where subtle quantum effects might help resolve the issue.

CONCEPTCHECK 21-14

Consider two black holes. They have the same mass and angular momentum, and the total electric charge for each of them is zero. According to the no-hair theorem, could we tell if one of these black holes was made from iron and the other from peanut butter?

Answer appears at the end of the chapter.

21-8 Falling into a black hole is an infinite voyage

TUTORIAL 21-3

Imagine that you are on board a spaceship at a safe distance from a 5-$M_\odot$ black hole. A distance of 1000 Schwarzschild radii, or 15,000 km, would suffice. You now release a space probe with a video camera and let it fall into the black hole. What will you see on the video as the probe falls?

Initially, at 1000 Schwarzschild radii from the black hole, the video camera would send back a rather normal view of space. But as the probe approaches the black hole, the bending of light by the black hole becomes more pronounced (Figure 21-20). Light rays passing close to the back hole are deflected so much that background stars are severely distorted. For the black hole in Figure 21-20, light from a bright object

The gravitational field of a black hole distorts the image of other stars

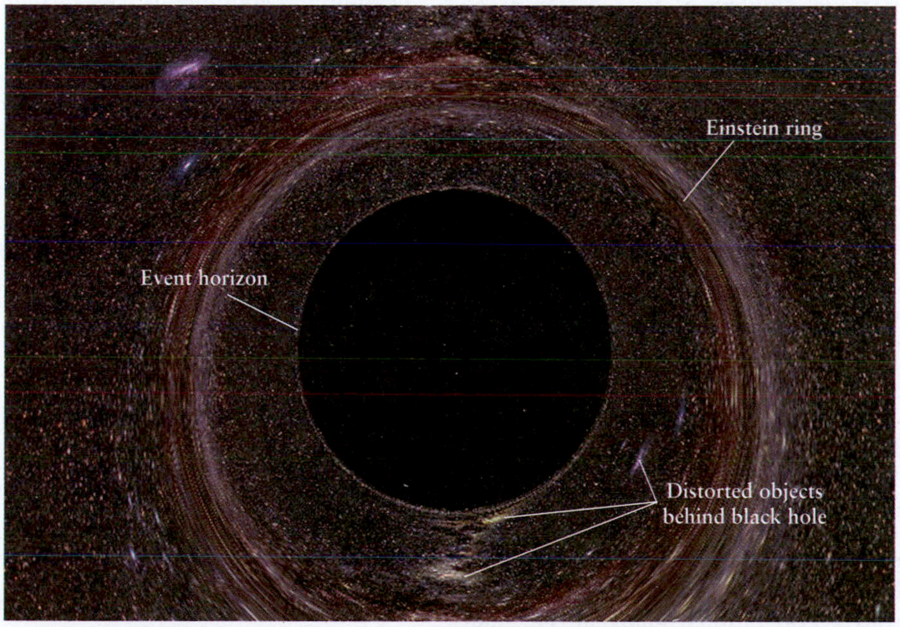

FIGURE 21-20
The View Approaching a Black Hole An image from a computer simulation of the appearance of the sky behind a black hole. For typical objects behind a black hole, lines of sight that are closer to the edge of the black hole result in greater light-bending and distortion of the object's image. In the rare case of a bright object right behind a black hole, a bright "Einstein ring" is formed. For illustration, this simulation places the extensive Milky Way behind the black hole, which leads to an unusually strong Einstein ring. (Andrew J. S. Hamilton/University of Colorado, Boulder)

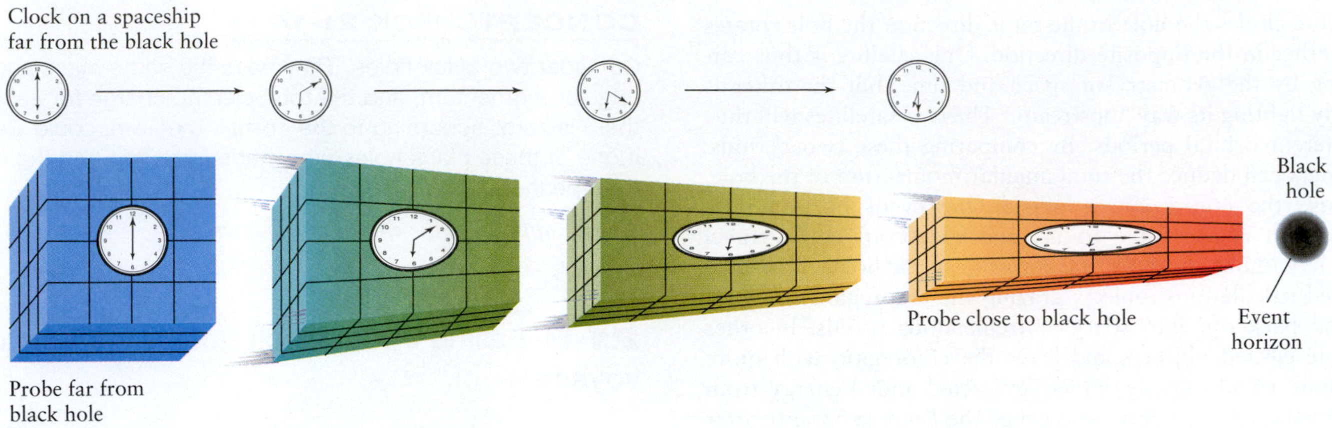

FIGURE 21-21
Effect of a Black Hole's Tidal Force on Infalling Matter A cube-shaped probe 1500 km from a 5-$M_\odot$ black hole. Near the Schwarzschild radius, the probe is pulled long and thin by the difference in the gravitational forces felt by its different sides. This tidal effect is a greatly magnified version of the Moon's gravitational force on Earth. The probe changes color as its photons undergo extreme gravitational redshift and time slows down on the probe, as seen from far away.

directly behind the black hole is warped into a feature called an "Einstein ring."

How will the probe itself look from your vantage point on the spaceship at a safe distance? To discuss changes in the probe, let's assume it emits blue light, and it has a clock that was initially synchronized with a clock on your spaceship (Figure 21-21). You might expect that as the probe falls, its speed should continue to increase. This expectation is true up to a point. But as the probe approaches the event horizon, where the black hole's gravity is extremely strong, the *gravitational slowing of time* becomes so pronounced that the probe will appear to slow down! From your point of view on the spaceship, the probe takes an infinite amount of time to reach the event horizon, where it will appear to remain suspended for all eternity. You can see the slower passage of time by comparing the clocks on the spaceship to the clocks on the probe in Figure 21-21.

To watch these effects, however, you will need special equipment. The reason for this is the gravitational redshift: As the probe falls, the light reaching you from the probe is shifted to longer and longer wavelengths. The probe's blue light will turn green, then yellow, then red, and eventually fade into infrared wavelengths that your unaided eye cannot detect. Therefore, while the object appears to never cross the event horizon to an outside observer, any light emitted by the object still disappears from view.

If you could view the falling probe with goggles that tracked light of ever-increasing wavelengths, you would see it come to an unpleasant end. Near the event horizon, the strength of the black hole's gravity increases dramatically as the probe moves just a short distance closer to the hole. In fact, the side of the probe nearest the black hole feels a much stronger gravitational pull than the side opposite the hole. These *tidal forces* are like those that the Moon exerts on Earth (see Section 4-8), but are tremendously stronger. Furthermore, the sides of the probe are pulled together, since the hole's gravity makes them fall in straight lines aimed at the center of the hole. The net effect is that the probe will be stretched out along the line pointing toward the hole, and squeezed together along the perpendicular directions. The stretching—sometimes called *spaghettification*—is so great that it can rip even the strongest materials apart.

Someone foolhardy enough to ride along with the probe would have a *very different view* of the fall than your view from the spaceship. From the point of view while riding the probe, there is no slowing of time. The probe and rider experience spaghettification and continue to pass through the event horizon into the singularity.

The size of the black hole matters: Tidal forces causing spaghettification are greater for smaller black holes. For stellar-mass black holes (like the 5-$M_\odot$ black hole considered here), tidal forces would squeeze and stretch a person to death even before they reach the event horizon. However, a person could fall right through the event horizon of a supermassive black hole, with tidal forces only pulling them apart closer to the center. Therefore, a person really could see what is inside a supermassive black hole, but at great cost for they could never share what they learned.

CONCEPTCHECK 21-15

If a person fell into a black hole feet first, would they feel a strong tidal force pulling on their feet before crossing the event horizon?

Answer appears at the end of the chapter.

21-9 Hawking radiation and black hole evaporation

With all the mass of a black hole hidden behind its event horizon and collapsed into a singularity, it may seem that there is no way of getting mass from the black hole back out into the universe. But in fact there is, as Stephen Hawking figured out in the 1970s. To understand how this is possible, we must look at the quantum mechanical behavior of matter at the microscopic scale of nuclei and electrons.

The **Heisenberg uncertainty principle** is a basic tenet of quantum physics. This principle states that you cannot determine precisely both the position and the speed of a subatomic particle. Over extremely short distances or times, a certain amount of "fuzziness" is built into the nature of reality.

The Heisenberg uncertainty principle leads to the concept of **virtual pairs:** *At every point in space, pairs of particles and antiparticles are constantly and spontaneously being created and destroyed.* An **antiparticle** is quite like an ordinary particle except that it has an opposite electric charge and can annihilate an ordinary particle so that both particles disappear. In the case of virtual pairs, the process of creation and annihilation occurs over such incredibly brief time intervals that these virtual particles and antiparticles cannot be observed directly. With some added energy, a virtual particle can become a real and observable particle, and a black hole helps this to happen.

With virtual particle pairs everywhere, now let's think about a black hole. Furthermore, think about the momentary creation of a virtual pair of an electron and a positron (the antiparticle of an electron) just *outside* the black hole's event horizon (Figure 21-22). It may happen that one of the virtual particles falls into the black hole. Its partner is then deprived of a counterpart with which to annihilate and ends up becoming a real particle. However, some energy is required for a virtual particle to become a real particle. The energy to accomplish this conversion comes from the black hole's mass according to $E = mc^2$. This decreases the mass of the black hole by a tiny amount, and the newly created real particle is free to radiate away from the black hole. In this way, real particles can quantum mechanically "leak" out from a microscopic region just outside the event horizon of a black hole, carrying some of the hole's mass with them. These particles would appear as a form of steady emission by a black hole and are called **Hawking radiation.**

A fundamental uncertainty in physical knowledge makes it possible for black holes to emit particles and radiation into space

Hawking radiation might also have implications for the information loss implied by the no-hair theorem. Some scientists suggest that the detailed information about objects entering a black hole is not lost, but is actually preserved in slight quantum variations of the Hawking radiation. While that might answer an important question within theoretical physics, for all practical purposes, the information is unobservable through any foreseeable techniques.

The Temperature of a Black Hole

Hawking radiation lead to another remarkable discovery that black holes have a temperature. Recall that the warmer an object, the more radiation it emits (Section 5-3). Detailed calculations show that Hawking radiation has the same properties as the radiation that is emitted from objects due to their temperature: black holes emit blackbody radiation. The result is that black holes actually have a temperature, and **the smaller a black hole's mass, the higher its temperature.**

As an example, a 1-trillion-ton (10^{15} kg) black hole would emit energy as if it were a blackbody with a temperature of nearly 10^9 K. That is hotter than the center of our Sun. However, the black holes that nature seems to produce are much larger and cooler. A 1-$M_\odot$ (2 = 10^{30} kg) black hole would have a miniscule temperature around 10^{-7} K. This temperature is barely above absolute zero. In other words, stellar mass black holes (and larger) have such low temperatures that they would emit very little Hawking radiation.

The story is different for very low-mass black holes, such as the proposed primordial black holes we discussed in Section 21-5. As particles escape from a very small black hole, the mass of the black hole decreases significantly, making its temperature go up. As its temperature rises, still more particles escape, further decreasing the hole's mass and forcing the temperature still higher. This runaway process of **black hole evaporation** finally causes the hole to vanish altogether! During its final seconds of evaporation, the hole gives up the last of its mass in a violent burst of energy equal to the detonation of a billion megaton hydrogen bombs!

A 10^{10}-kg primordial black hole (comparable to the mass of Mount Everest) would take about 15 billion years to evaporate.

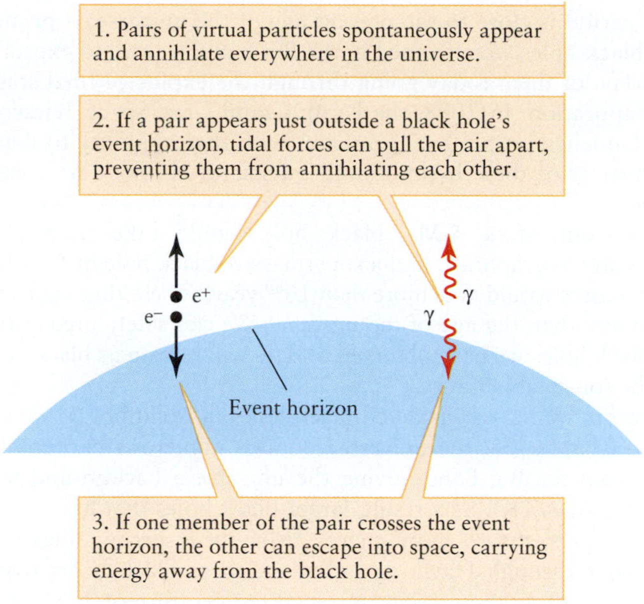

1. Pairs of virtual particles spontaneously appear and annihilate everywhere in the universe.
2. If a pair appears just outside a black hole's event horizon, tidal forces can pull the pair apart, preventing them from annihilating each other.
e+
e−
γ
γ
Event horizon
3. If one member of the pair crosses the event horizon, the other can escape into space, carrying energy away from the black hole.

FIGURE 21-22
Hawking Radiation and Evaporation of a Black Hole This illustration shows two pairs of virtual particles—an electron (e^-) and a positron (e^+), and a pair of virtual photons (γ)—appearing just outside the event horizon of a black hole. If one member of the pair becomes real and escapes, it carries a little energy away from the black hole, the black hole decreases in mass and the event horizon shrinks. The real particles radiating away are referred to as Hawking radiation, and the process is described as black hole evaporation.

This period is close to the present age of the universe. If primordial black holes were formed in the Big Bang, we would expect to see some of them today going through the explosive final stages of evaporation. In 2008, the Fermi Gamma-ray Space Telescope was launched on a 10-year mission with the capability to detect the nearest of these hypothesized bursts; so far, none have been found.

By contrast, a 5-$M_\odot$ black hole would take more than 10^{62} years to evaporate, and a supermassive black hole of 5 million solar masses would take more than 10^{80} years. These time spans are far longer than the age of the universe. We can safely predict that the black holes we have observed to date will remain as black holes for the foreseeable future.

So far, we have considered black holes in isolation. As we will see in Chapter 25, the universe contains numerous photons left over from the Big Bang (giving the universe a background temperature of 2.7 K). As a result, larger black holes that have cooler temperatures *absorb* more energy from their surroundings than they emit through Hawking radiation; these black holes would slowly grow! At the present universal temperature of 2.7 K, only black holes with masses less than that of the Moon (which would have a diameter about the width of a human hair) are hot enough to shrink by emitting more radiation than they absorb.

While Hawking radiation and black hole evaporation have not been observationally verified (yet), these predictions—based on established laws of physics—are being pursued. The Fermi Gamma-ray Space Telescope is already looking for evidence of evaporating black holes as noted above, and even particle accelerators have joined in the search. The Large Hadron Collider at CERN (which discovered the Higgs particle in 2012) has the energy to look for certain miniature black holes. They have not been detected, but if created by the collider, they would immediately evaporate by emitting observable Hawking radiation.

From supermassive black holes to stellar mass and primordial black holes, the physical parameters of black holes stretch across a wide range—from large, massive and cold to small, hot, and evaporating away with a burst. As new instruments are designed to detect these black holes, their exotic properties can be explored. As we will see in Chapter 24, supermassive black holes are already known to play a central role in powering objects called quasars, which are some of the brightest objects in the universe.

CONCEPTCHECK 21-16

Which object radiates more intensely, a supermassive black hole or a stellar mass black hole?

CONCEPTCHECK 21-17

Do black holes remain the same temperature forever?

CONCEPTCHECK 21-18

Suppose for a moment that a 1-kg black hole was detected. Is it likely that this miniature black hole was produced billions of years ago in the Big Bang or created more recently?

Answers appear at the end of the chapter.

KEY WORDS

Terms preceded by an asterisk (*) are discussed in the Boxes.

KEY IDEAS

The Special Theory of Relativity: This theory, published by Einstein in 1905, describes how measurements of lengths and durations of time depend on the motion of an observer and can appear different to different observers.

- The speed of light is the same to all observers, no matter how fast they are moving.
- An observer will note a slowing of clocks and a shortening of rulers that are moving with respect to the observer. This effect becomes significant only if the clock or ruler is moving at a substantial fraction of the speed of light.
- Space and time are not wholly independent of each other, but are aspects of a single entity called spacetime.

The General Theory of Relativity: Published by Einstein in 1915, this is a theory of gravity. Any massive object causes space to curve and time to slow down, and these effects manifest themselves as a gravitational force. These distortions of space and time are most noticeable in the vicinity of large masses or compact objects.

- The general theory of relativity is our most accurate description of gravitation. It predicts a number of phenomena, including the bending of light by gravity and the gravitational redshift, whose existence has been confirmed by observation and experiment.
- The general theory of relativity also predicts the existence of gravitational waves, which are ripples in the overall geometry of space and time produced by moving masses. Gravitational waves have been detected indirectly, and specialized antennas are under

construction to make direct measurement of the gravitational waves from cosmic cataclysms.

Black Holes: If a stellar corpse has a mass greater than about 2 to 3 $M_\odot$, gravitational compression will overwhelm any and all forms of internal pressure.

- A black hole has an escape speed greater than the speed of light.
- Nothing, not even light, can escape a black hole once it crosses into a black hole's event horizon.

Observing Black Holes: Black holes have been detected using indirect methods.

- Some binary star systems contain a black hole. In such a system, gases captured from the companion star by the black hole emit detectable X-rays. The emission comes from outside the event horizon.
- Many galaxies have supermassive black holes at their centers. These are detected by observing the motions of material around the black hole.

Gamma-ray Bursts: Short, intense bursts of gamma rays are observed at random times coming from random parts of the sky.

- By observing the afterglow of long-duration gamma-ray bursts, astronomers find that these objects have very large redshifts and appear to be located within distant galaxies. The bursts are correlated with supernovae and may be due to an exotic type of supernova called a collapsar.
- The origin of short-duration gamma-ray bursts is unknown.

Properties of Black Holes: The entire mass of a black hole gets quickly concentrated in a nearly infinitely dense singularity.

- A black hole has only three physical properties: mass, electric charge, and angular momentum.
- A rotating black hole (one with angular momentum) has an ergoregion around the outside of the event horizon. In the ergoregion, space and time themselves are dragged along with the rotation of the black hole.
- Black holes emit Hawking radiation from a quantum physics process that occurs just outside of their event horizons.
- Black holes have temperatures and their Hawking radiation is emitted like the blackbody radiation that is emitted by any object with a temperature.
- Black holes can evaporate, but they do so at extremely slow rates unless the black holes are very small.

QUESTIONS

Review Questions

1. You drop a ball inside a car traveling at a steady 50 km/h in a straight line on a smooth road. Does it fall in the same way as it does inside a stationary car? How does this question relate to Einstein's special theory of relativity?
2. In Einstein's special theory of relativity, two different observers moving at different speeds will measure the same value of the speed of light. Will these same observers measure the same value of, say, the speed of an airplane? Explain your answer.
3. A friend summarizes the special theory of relativity by saying "Everything is relative." Explain why this statement is inaccurate.
4. Serena flies past Michael in her spaceship at nearly the speed of light. According to Michael, Serena's clock runs slow. According to Serena, does Michael's clock run slow, fast, or at the normal rate? Explain your answer.
5. Ole flies past Lena in a spherical spaceship at nearly the speed of light. According to Lena, how does the distance from the front to the back of Ole's spaceship (that is, measured along the direction of motion) compare to the distance from the top to the bottom (that is, measured perpendicular to the direction of motion)? Explain your answer.
6. Why does the speed of light represent an ultimate speed limit?
7. TUTORIAL 21-1 Why is Einstein's general theory of relativity a better description of gravity than Newton's universal law of gravitation? Under what circumstances is Newton's description of gravity adequate?
8. Describe two different predictions of the general theory of relativity and how these predictions were tested experimentally. Do the results of the experiments agree with the theory?
9. How does a gravitational redshift differ from a Doppler shift?
10. In what circumstances are degenerate electron pressure and degenerate neutron pressure incapable of preventing the complete gravitational collapse of a dead star?
11. Should we worry about Earth being pulled into a black hole? Why or why not?
12. How does rapid flickering of an X-ray source provide evidence that the source is small (see discussion of Cygnus X-1)?
13. TUTORIAL 21-2 All the stellar-mass black hole candidates mentioned in the text are members of very short-period binary systems. Explain how this makes it possible to detect the presence of the black hole.
14. Astronomers cannot actually see the black hole candidates in close binary systems. How, then, do they know that these candidates are not white dwarfs or neutron stars?
15. Describe two ways in which a member of a binary star system could become a black hole.
16. What is a gamma-ray burst? What is the evidence that gamma-ray bursts are not located in the disk of our Galaxy or in a halo surrounding our Galaxy?
17. Summarize the evidence that gamma-ray bursts result from a process involving a star in a distant galaxy.
18. What is a collapsar? How does the collapsar model account for the existence of long-duration gamma-ray bursts?

19. How do astronomers locate supermassive black holes in galaxies?
20. What is an intermediate-mass black hole? How are such objects thought to form?
21. When we say that the Moon has a radius of 1738 km, we mean that this is the smallest radius that encloses all of the Moon's material. In this sense, is it correct to think of the Schwarzschild radius as the radius of a black hole? Why or why not?
22. TUTORIAL 21-3 A twenty-third-century instructor at Starfleet Academy tells her students, "If someday your starship falls into a black hole, it'll be your own fault." Explain why it would require careful piloting to direct a spacecraft into a black hole.
23. In what way is a black hole blacker than black ink or a black piece of paper?
24. If the Sun suddenly became a black hole, how would Earth's orbit be affected? Explain your answer.
25. According to the general theory of relativity, why can't some sort of yet-undiscovered degenerate pressure prevent the matter inside a black hole from collapsing all the way down to a singularity?
26. Is it possible to tell which chemical elements went into a black hole? Why or why not?
27. Why is it unlikely that a black hole has a large electric charge?
28. What kind of black hole is surrounded by an ergoregion? What happens inside the ergoregion?
29. What is the no-hair theorem?
30. As seen by a distant observer, how long does it take an object dropped from a great distance to fall through the event horizon of a black hole? Explain your answer.
31. Summarize what leads to Hawking radiation.
32. If even light cannot escape from a black hole, how is it possible for black holes to evaporate?
33. Why do smaller black holes evaporate more quickly than larger black holes?

Advanced Questions

Questions preceded by an asterisk (*) involve topics discussed in the Boxes.

Problem-solving tips and tools

Remember that the time to travel a certain distance is equal to the distance divided by the speed, and the density of an object is its mass divided by its volume. The volume of a sphere of radius r is $4\pi r^3/3$. Section 4-7 describes Newton's law of universal gravitation. Box 4-4 shows how to use Newton's formulation of Kepler's third law, which explicitly includes masses; when using this formula, note that the period P must be expressed in seconds, the semimajor axis a in meters, and the masses in kilograms. For another version of this formula, in which period is in years, semimajor axis in AU, and masses in solar masses, see Section 17-9.

*34. A spaceship flies from Earth to a distant star at a constant speed. Upon arrival, a clock on board the spaceship shows a total elapsed time of 8 years for the trip. An identical clock on Earth shows that the total elapsed time for the trip was 10 years. What was the speed of the spaceship relative to Earth?

*35. An unstable particle called a positive pion (pronounced "pie-on") decays in an average time of 2.6×10^{-8} s. On average, how long will a positive pion last if it is traveling at 95% of the speed of light?

*36. How fast should a meter stick be moving in order to appear to be only 60 cm long?

*37. An astronaut flies from Earth to a distant star at 80% of the speed of light. As measured by the astronaut, the one-way trip takes 15 years. (**a**) How long does the trip take as measured by an observer on Earth? (**b**) What is the distance from Earth to the star (in light-years) as measured by an Earth observer? As measured by the astronaut?

38. In the binary system of two neutron stars discovered by Hulse and Taylor (Section 21-2), one of the neutron stars is a pulsar. The distance between the two stars varies between 1.1 and 4.8 times the radius of the Sun. The time interval between pulses from the pulsar is not constant: It is greatest when the two stars are closest to each other and least when the two stars are farthest apart. Explain why this is consistent with the gravitational slowing of time (Figure 21-7a).

39. Find the total mass of the neutron star binary system discovered by Hulse and Taylor (Section 21-2), for which the orbital period is 7.75 hours and the average distance between the neutron stars is 2.8 solar radii. Is your result reasonable for a pair of neutron stars? Explain your answer.

40. Estimate how long it will be until the two neutron stars that make up the binary system discovered by Hulse and Taylor collide with each other. Assume that the distance between the two stars will continue to decrease at its present rate of 3 mm every 7.75 hours, and use the data given in Question 39. (You can assume that the two stars are very small, so they will collide when the distance between them is equal to zero.)

41. The orbital period of the binary system containing A0620-00 is 0.32 day, and Doppler shift measurements reveal that the radial velocity of the X-ray source peaks at 457 km/s (about 1 million miles per hour). (**a**) Assuming that the orbit of the X-ray source is a circle, find the radius of its orbit in kilometers. (This is actually an estimate of the semimajor axis of the orbit.) (**b**) By using Newton's form of Kepler's third law, prove that the mass of the X-ray source must be at least 3.1 times the mass of the Sun. (*Hint:* Assume that the mass of the K5V visible star—about 0.5 $M_{\odot}$ from the mass-luminosity relationship—is negligible compared to that of the invisible companion.)

42. Contrast gamma-ray bursts with X-ray bursts (discussed in Section 20-12). From our models of what causes these energetic phenomena, explain why X-ray bursts emit repeated pulses but gamma-ray bursts apparently emit just once.

43. Long-duration gamma-ray bursts are only observed in galaxies where there is ongoing star formation. Explain how this is consistent with the collapsar model of how these bursts occur.

44. The spectrum of a Type Ic supernova lacks absorption lines of hydrogen and helium. This means that when a black hole formed at the center of the progenitor star, the resulting jets were more easily able to escape into space. Explain why this is so.

45. Find the orbital period of a star moving in a circular orbit of radius 500 AU around the supermassive black hole in the galaxy NGC 4261 (Section 21-5).

*46. Find the Schwarzschild radius for an object having a mass equal to that of the planet Saturn.

*47. What is the Schwarzschild radius of a black hole whose mass is that of (a) Earth, (b) the Sun, (c) the supermassive black hole in NGC 4261 (Section 21-5)? In each case, also calculate what the density would be if the matter were spread uniformly throughout the volume of the event horizon.

*48. What is the mass in kilograms of a black hole whose Schwarzschild radius is 11 km?

*49. To what density must the matter of a dead 8-$M_\odot$ star be compressed in order for the star to disappear inside its event horizon? How does this compare with the density at the center of a neutron star, about 3×10^{18} kg/m^3?

*50. Prove that the density of matter needed to produce a black hole is inversely proportional to the square of the mass of the hole. If you wanted to make a black hole from matter compressed to the density of water (1000 kg/m^3), how much mass would you need?

Discussion Questions

51. The speed of light is the same for all observers, regardless of their motion. Discuss why this requires us to abandon the Newtonian view of space and time.

52. Describe the kinds of observations you might make in order to locate and identify black holes.

53. Speculate on the effects you might encounter on a trip to the center of a black hole (assuming that you could survive the journey).

Web/eBook Questions

54. Search the World Wide Web for information about a stellar-mass black hole candidate named V4641 Sgr. In what ways does it resemble other black hole candidates such as Cygnus X-1 and V404 Cygni? In what ways is it different and more dramatic? How do astronomers explain why V4641 Sgr is different?

55. Search the World Wide Web for information about supernova SN 2006aj, which was associated with gamma-ray burst GRB 060218. In what ways were this supernova and gamma-ray burst unusual? Are the observations of these objects consistent with the collapsar model?

56. Search the World Wide Web for information about the intermediate-mass black hole candidate in M82. Is this still thought to be an intermediate-mass black hole? What new evidence has been used to either support or oppose the idea that this object is an intermediate-mass black hole?

57. ANIMATION 21-1 **The Equivalence Principle.** Access the animation "The Equivalence Principle" in Chapter 21 of the *Universe* Web site or eBook. View the animation and answer the following questions. (a) Describe what is happening as viewed from the frame of reference of the elevator. What causes the apple to fall to the floor of each elevator? (b) Describe what is happening as viewed from the frame of reference of the stars. What causes the apple to fall to the floor of each elevator? (c) Think of another experiment you could perform with the apple. Describe what would happen during this experiment as seen by Newton (in the left-hand box) and by Einstein (in the right-hand box).

ACTIVITIES

Observing Projects

58. You cannot see a black hole with a telescope. Nevertheless, you might want to observe the visible companion of Cygnus X-1. The epoch 2000 coordinates of this ninth-magnitude star are R.A. = 19^h 58.4^m and Decl. = 135° 12′, which is quite near the bright star η (eta) Cygni. Compare what you see with the photograph in Figure 21-10.

59. Use the *Starry Night*™ program to investigate the X-ray source and probable black hole, Cygnus X-1. This region of space is one of the brightest in the sky at X-ray wavelengths. Click the **Home** button in the toolbar and then use the **Find** pane to center the field of view on Cygnus X-1. If Cygnus X-1 is below the horizon, allow the program to reset the time to when it can best be seen. Click the checkbox to the left of the listing for Cygnus X-1 to apply a label to this object. Use the **Zoom** controls to set the field of view to 100 degrees. (a) Use the **Time** controls in the toolbar to determine when Cygnus X-1 rises and sets on today's date from your location. (b) **Zoom** in until you can see an object at the location indicated by the label. What apparent magnitude and radius does *Starry Night*™ give for this object (you can obtain this information from the **HUD** or by using the **Show Info** command from the contextual menu for this object)? Keeping in mind that the object that gives rise to this X-ray source is a black hole, to what must this apparent magnitude and radius refer? Explain.

60. Use the *Starry Night*™ program to examine in some detail the central regions of two galaxies that contain supermassive black holes at their centers. (a) The Milky Way is one such galaxy. Select **Favourites > Explorations > Milky Way Centre** to view our Galaxy from the equivalent of the center of a transparent Earth. The star HIP86919 is very closely aligned with the direction of the central core of the galaxy and can be used as a guide when viewing this region. You can brighten the appearance of the Milky Way by opening **Options > Stars > Milky Way...**

and moving the **Brightness** slider to the far right in the Milky Way Options panel. The **Time** and **Date,** August 30, 2009, at 6:30 A.M., have been chosen when the Moon is crossing the Milky Way plane, thereby providing a convenient angular scale, about ½° in diameter, for comparison with Milky Way features. (Click on **Options > Solar System > Planets-Moons...** to display the Planets-Moons Options window and make sure that the **Enlarge Moon Size at large FOVs** box under Other is **Off,** and click **OK.**) You will notice that, at visible wavelengths, this central region of the Milky Way appears to be dark. What explanation can you give for this dark band across the galactic plane? **Zoom** in to a field of view of about 5° around the galactic center to examine the region surrounding the black hole using the highly penetrating X-rays detected by the Chandra X-ray Space Telescope. If necessary, open the **Options** pane, expand the **Deep Space** layer, click On the **Chandra Images** and move the slider to the right to display the image mosaic showing numerous hot and intense X-ray sources very close to the supermassive black hole. (Note that this image is slightly offset from the galactic center to avoid confusion.) (**b**) M87 is an active galaxy for which evidence is strong for the presence of a supermassive black hole at its center. Click **Home** to return to your sky and stop the advancement of time. Use the **Find** tab to center your view on **M87,** allowing the program to adjust the time to ensure that this object is in your sky. Open the **Options** pane, expand the **Deep Space** layer and click Off the **Chandra** and **Hubble Images,** leaving the **Messier Objects** displayed, and then **Zoom** in to a field of view about 1° wide to show this giant elliptical galaxy as seen from ground-based telescopes. (You can click on the **Info** tab and click on **Description** to read about this somewhat featureless but surprising elliptical galaxy.) In the **Options > Deep Space** panel, move the slider to the right for the **Chandra X-ray image** to see the structure of hot gases around M87. Finally, move the **Hubble Images** slider to the right to display the high-resolution Hubble Space Telescope image of the gas jet emanating from the black hole. Again, this image is displaced to the upper-right from the galactic center position by the program. Right-click over this square image, click **Centre** and then zoom in to a field of view of about 1 arcminute to see this spectacular jet as it interacts violently with the interstellar medium above the black hole. Comment on the suggestion that supermassive black holes were discovered only after relatively recent advances were made in telescope and detector technology.

Collaborative Exercises

61. Using Einstein's special theory of relativity, estimate (1) the length of your pencil or pen at constant speed at the speeds of a bicycle rider, a car traveling on the highway, and a commercial jet liner at cruising altitude; and (2) the speed of a light beam emitted by a spaceship traveling at 200,000 km/s toward another spaceship traveling at the same speed.

ANSWERS

ConceptChecks

ConceptCheck 21-1: For anyone riding on the speeding train, the train is not moving relative to the riders. They observe no length contraction or time dilation, so their hearts do not beat more slowly.

ConceptCheck 21-2: All observers, whether moving or not, see light travel at the same speed through the vacuum of space: They all see light travel at $c = 3310^8$ m/s.

ConceptCheck 21-3: Less time has passed on the clock near the planet—the planet's gravity slows down time; the stronger the gravity, the more time slows down.

ConceptCheck 21-4: No. As the radio waves "climb" out of the gravitational field of the planet, they experience a gravitational redshift—the frequency decreases and the wavelength increases.

ConceptCheck 21-5: Neutron stars can support up to 2 to 3 $M_\odot$ without collapsing further. Masses larger than this can collapse into black holes, so these are the lowest black hole masses expected from stellar core-collapse.

ConceptCheck 21-6: If an object is compressed, its escape speed increases, and the escape speed also increases if more mass is added to the object while maintaining its size. When an object's escape speed exceeds the speed of light, no light can escape, and the object becomes a black hole.

ConceptCheck 21-7: Stars do not emit strongly in X-rays, as Cygnus X-1 does. More importantly, fast variations in the X-ray brightness tell us that the mass of Cygnus X-1 is smaller than Earth—far too small for any type of star or other object.

ConceptCheck 21-8: The collapsar supernova model for gamma-ray bursts involves the special case in which jets are formed during the collapse, and one jet of gamma rays points toward Earth. Without the formation and enhanced intensity of jets, a regular supernova (of Type I or II) would be too weak to account for the strength of the observed gamma rays.

ConceptCheck 21-9: No. The larger the mass of a black hole, the lower the average density required to form the black hole. If it could be carried out, filling our solar system with water would produce a very large black hole.

ConceptCheck 21-10: No. Theoretically speaking, the laws of gravity allow a black hole to exist with any mass. However, *natural processes* in the cosmos might only *produce* black holes in certain mass ranges.

ConceptCheck 21-11: A black hole's event horizon is a spherical boundary with a radius given by the Schwarzschild radius. Objects getting closer to the Schwarzschild radius have fallen irreversibly within the event horizon.

ConceptCheck 21-12: At such large distances, the gravitational force is the same from each object. Stronger forces and other effects only show up much closer to a black hole.

ConceptCheck 21-13: It would be safer to fly by the event horizon of the larger black hole, where the gravitational force is actually weaker.

ConceptCheck 21-14: No. Once matter enters a black hole, the only properties of the black hole are its mass, charge, and angular momentum.

ConceptCheck 21-15: Maybe, as it depends on the mass of the black hole. For a supermassive black hole with weak tidal forces, they would not notice their feet being tugged with a greater force than any other part of their body. However, for a stellar-mass black hole, their feet could be pulled relative to their legs by deadly forces.

ConceptCheck 21-16: The smaller the black hole, the larger its temperature. Therefore, the stellar-mass black hole is hotter and radiates more intensely.

ConceptCheck 21-17: No. As black holes lose mass through Hawking radiation, they heat up to even greater temperatures. This causes them to lose even more mass in a process of evaporation, which is more noticeable for smaller black holes.

ConceptCheck 21-18: Small black holes evaporate more quickly and a black hole of this extraordinarily small size would have evaporated long ago if it were produced during the Big Bang. Therefore, the black hole would have been created more recently.

CHAPTER

22

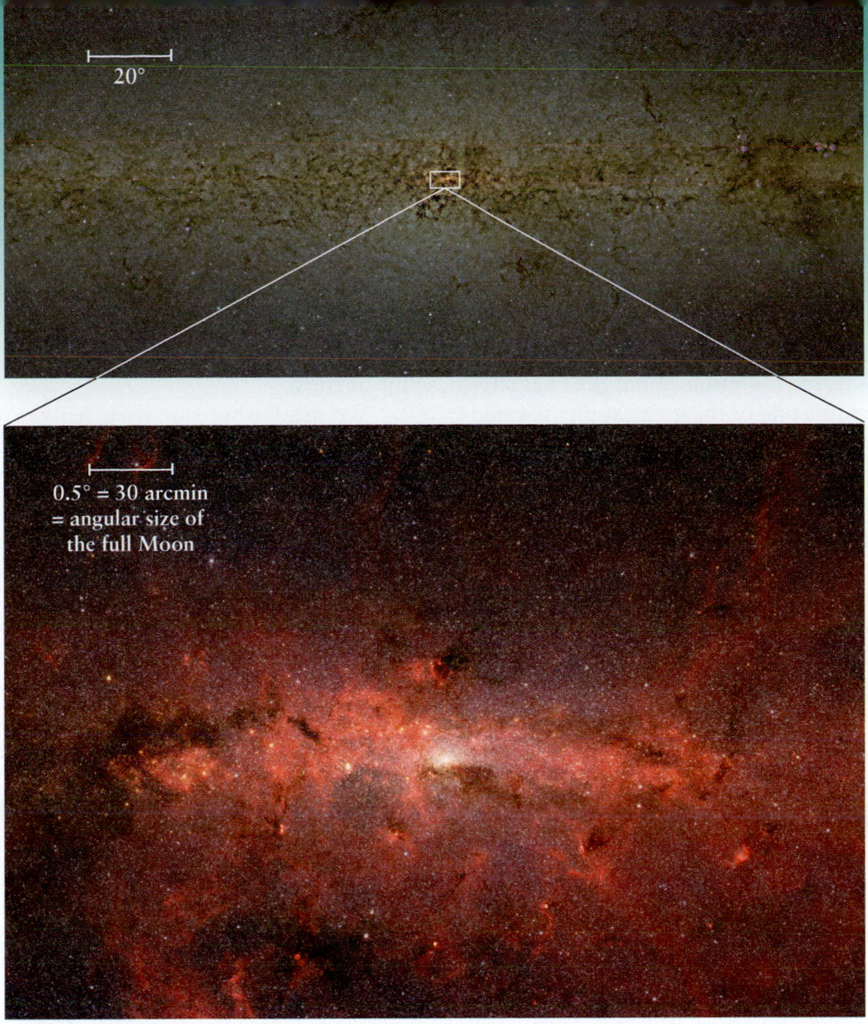

R I V U X G

Two views of the Milky Way: a wide-angle infrared image (upper) and a close-up infrared image (lower). *(upper:* Ohainaut/ESO/Handout/dpa/Corbis; *lower:* NASA/JPL-Caltech/S. Stolovy [SSC/Caltech])

Our Galaxy

LEARNING GOALS

By reading the sections of this chapter, you will learn

22-1 How astronomers discovered the solar system's location within the Milky Way Galaxy

22-2 The shape and size of our Galaxy

22-3 How the Milky Way's spiral structure was discovered

22-4 The evidence for the existence of dark matter in our Galaxy

22-5 What causes the Milky Way's spiral arms to form and persist

22-6 How astronomers discovered a supermassive black hole at the galactic center

On a clear, moonless night, away from the glare of city lights, you can often see a hazy, luminous band stretching across the sky. This band, called the Milky Way, extends all the way around the celestial sphere. The upper of the two accompanying photographs is centered on the brightest part of the Milky Way, in the constellation Sagittarius.

Galileo, the first person to view the Milky Way with a telescope, discovered that it is composed of countless dim stars. Today, we realize that the Milky Way is actually a disk tens of thousands of parsecs across containing hundreds of billions of stars. (It is hard to picture hundreds of billions of stars, but it is about the number of grains in a few trashcans full of beach sand.) Between the stars, there are also large quantities of gas and dust. One of these stars is our Sun, and this vast assemblage of matter—our home galaxy—is collectively called the Milky Way Galaxy.

Just as Galileo's telescope revealed aspects of the Milky Way that the naked eye could not, modern astronomers use telescopes at nonvisible wavelengths to peer through our Galaxy's obscuring dust and observe what visible-light telescopes never could. For example, the lower of the two accompanying photographs is an infrared image that shows hundreds of thousands of stars near the center of the Galaxy. As we will see, radio, infrared, and X-ray observations reveal that at the very center of the Galaxy lies a black hole with a mass of 4.1 million Suns.

Modern astronomers have also discovered that most of the Milky Way's mass is not in its stars, gas, or dust, but in a halo of *dark matter* that emits no measurable radiation. What the character of this dark matter could be remains one of the greatest unanswered questions in astronomy and physics.

The Milky Way is just one of myriad *galaxies,* or systems of stars and interstellar matter, that are spread across the observable universe. By studying our home galaxy, the Milky Way Galaxy, we begin to explore the universe on a grand scale. Instead of focusing on individual stars, we look at the overall arrangement and history of a huge stellar community of which the Sun is a member. In this way, we gain insights into galaxies in general and prepare ourselves to ask fundamental questions about the cosmos.

22-1 The Sun is located in the disk of our Galaxy, about 8000 parsecs from the galactic center

TUTORIAL 22-1 Eighteenth-century astronomers were the first to suspect that because the Milky Way completely encircles us, all the stars in the sky are part of an enormous disk of stars—the **Milky Way Galaxy.** As we learned in Section 1-4, a **galaxy** is an immense collection of stars—as well as gas and dust referred to as **interstellar matter.**

Our Galaxy is both large and mostly empty. To put our Galaxy into perspective, consider this scale model suggested by astronomer Mark Whittle: Suppose that our Galaxy was shrunk down to the size of the continental United States. Then, the average distance between the shrunken stars would be about the size of a football field. But what about the stars themselves? The average size of the shrunken stars would be about the size of human cells! Clearly, the stars of the Milky Way are widely separated, and there are a lot of them.

We are within our own Milky Way Galaxy, which makes it hard to view. Our solar system is located inside the pancakelike disk of our Galaxy, which is why the Milky Way appears as a band around the sky (Figure 22-1).

Locating the Sun Within the Galaxy: Early Attempts

But where within this disk is our own Sun? Until the twentieth century, the prevailing opinion was that the Sun and planets lie at the Galaxy's center. One of the first to come to this conclusion was the eighteenth-century English astronomer William Herschel, who discovered the planet Uranus and was a pioneering cataloger of binary star systems (see Section 17-9). Herschel's approach to determining the Sun's position within the Galaxy was to count the number of stars in each of 683 regions of the sky. He reasoned that he should see the greatest number of stars toward the Galaxy's center and a lesser number toward the Galaxy's edge.

Herschel found approximately the same density of stars all along the Milky Way. Therefore, he concluded that we are at the center of our Galaxy (Figure 22-2). In the early 1900s, the Dutch astronomer Jacobus Kapteyn came to essentially the same conclusion by analyzing the brightness and proper motions of a large number of stars. According to Kapteyn, the Milky Way is about 17 kpc (17 kiloparsecs = 17,000 parsecs or 55,000 light-years) in diameter, with the Sun near its center.

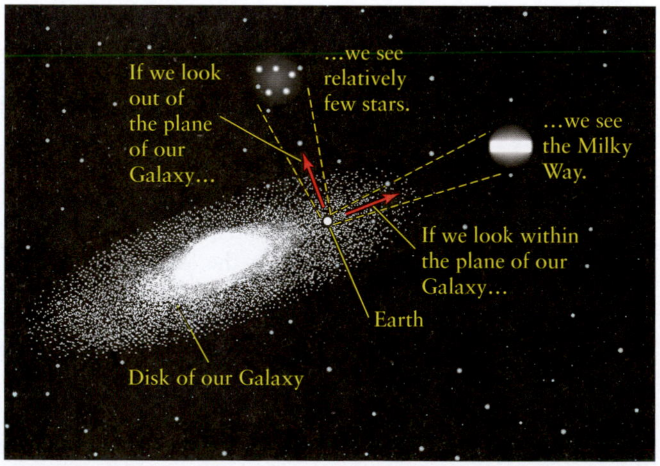

(a)

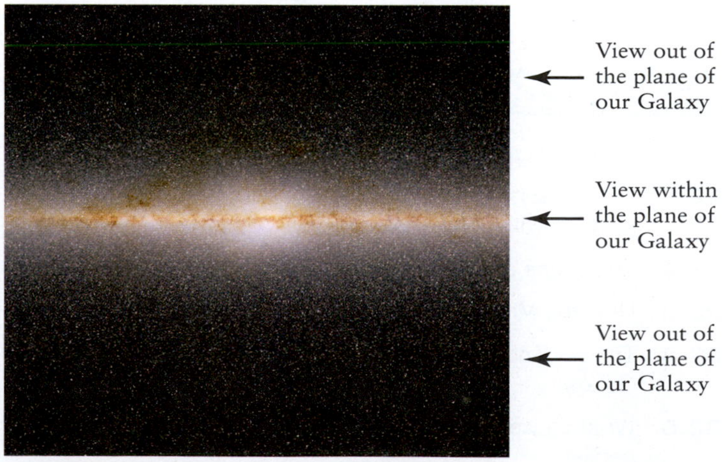

(b) R I V U X G

FIGURE 22-1

Our View of the Milky Way (a) The Milky Way Galaxy is a disk-shaped collection of stars. When we look out at the night sky in the plane of the disk, the stars appear as a band of light that stretches all the way around the sky. When we look perpendicular to the plane of the Galaxy, we see only those relatively few stars that lie between us and the "top" or "bottom" of the disk. (b) This wide-angle photograph shows a 180° view of the Milky Way centered on the constellation Sagittarius (compare with the photograph that opens this chapter). The dark streaks across the Milky Way are due to interstellar dust in the plane of our Galaxy. (b: Stocktrek Images/Getty Images)

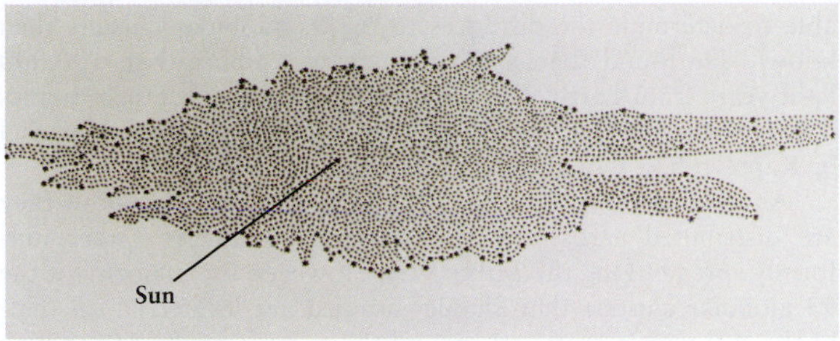

FIGURE 22-2
Herschel's Map of Our Galaxy In a paper published in 1785, the English astronomer William Herschel presented this map of the Milky Way Galaxy. He determined the Galaxy's shape by counting the numbers of stars in various parts of the sky. Herschel's conclusions were flawed because interstellar dust blocked his view of distant stars, leading him to the erroneous idea that the Sun is at the center of the Galaxy. (Dr. Jeremy Burgess/Science Source)

The Problem: Interstellar Extinction

Both Herschel and Kapteyn were wrong about the Sun being at the center of our Galaxy. The reason for their mistake was discerned in 1930 by Robert J. Trumpler of Lick Observatory. While studying star clusters, Trumpler discovered that the more remote clusters appear unusually dim—more so than would be expected from their distances alone. As a result, Trumpler concluded that interstellar space must not be a perfect vacuum: It must contain dust that absorbs or scatters light from distant stars.

Like the stars themselves, interstellar dust is concentrated in the plane of the Galaxy (see Section 18-2). As a result, it obscures our view within the plane and makes distant objects appear dim, an effect called **interstellar extinction.** Great patches of interstellar dust are clearly visible in wide-angle photographs such as the ones that open this chapter. Thanks to interstellar extinction, Herschel and Kapteyn were actually seeing only the nearest stars in the Galaxy. Hence, they had no idea of either the enormous size of the Galaxy or of the vast number of stars concentrated around the galactic center.

ANALOGY Herschel and Kapteyn faced much the same dilemma as a lost motorist on a foggy night. Unable to see more than a city block in any direction, the motorist would have a hard time deciding what part of town he was in. If the fog layer were relatively shallow, however, our motorist would be able to see the lights from tall buildings that extend above the fog, and in that way he could determine his location (Figure 22-3a).

The same principle applies to our Galaxy. While interstellar dust in the plane of our Galaxy hides the sky covered by the Milky Way, we have an almost unobscured view out of the plane (that is, to either side of the Milky Way). To find our location in the Galaxy, we need to locate bright objects that are part of the Galaxy but lie outside its plane in unobscured regions of the sky.

The Breakthrough: Globular Clusters and Variable Stars

Fortunately, bright objects of the sort we need do in fact exist. They are the **globular clusters,** a class of star clusters associated with the Galaxy but which lie outside its plane (Figure 22-3b). As we saw in Section 19-4, a typical globular cluster is a spherical distribution of roughly 10^6 stars packed in a volume only a few hundred light-years across (see Figure 19-12).

Observations of pulsating variable stars revealed the immense size of the Milky Way

However, to use globular clusters to determine our location in the Galaxy, we must first determine the distances from Earth to these clusters. (Think

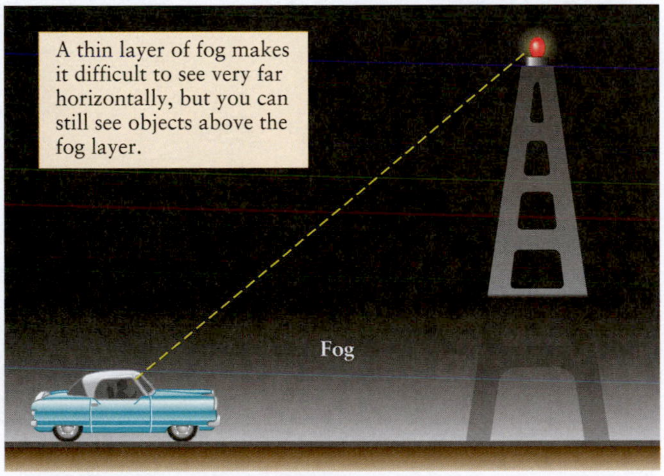

A thin layer of fog makes it difficult to see very far horizontally, but you can still see objects above the fog layer.
Fog

(a) Determining your position in the fog

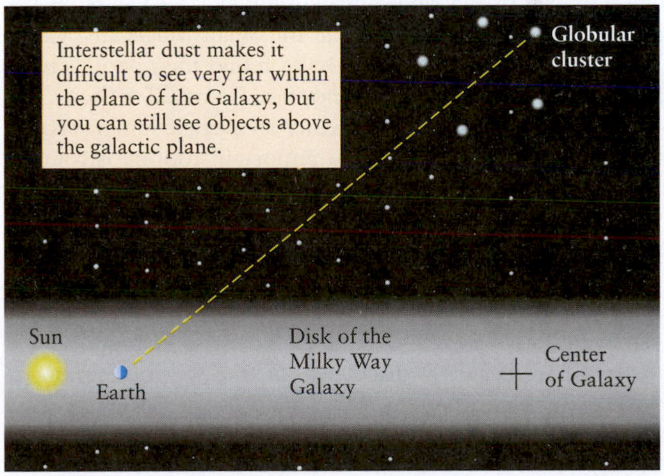

(b) Determining your position in the Galaxy

FIGURE 22-3
Finding the Center of the Galaxy **(a)** A motorist lost on a foggy night can determine his location by looking for tall buildings that extend above the fog. **(b)** In the same way, astronomers determine our location in the Galaxy by observing globular clusters that are part of the Galaxy but lie outside the obscuring material in the galactic disk. The globular clusters form a spherical halo centered on the center of the Galaxy.

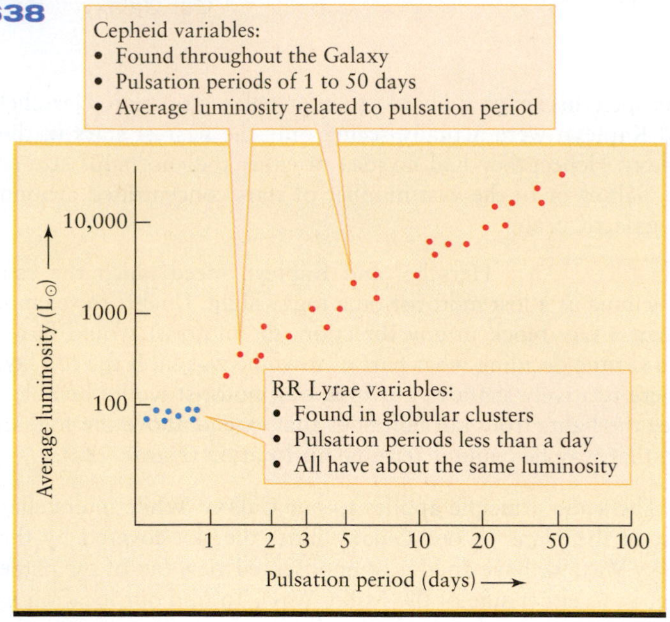

FIGURE 22-4

Period and Luminosity for Cepheid and RR Lyrae Variables This graph shows the relationship between period and average luminosity for Cepheid variables and RR Lyrae variables. Cepheids come in a broad range of luminosities: The more luminous the Cepheid, the longer its pulsation period. By contrast, RR Lyrae variables are horizontal branch stars that all have roughly the same average luminosity of about 100 $L_\odot$.

again of our lost motorist—glimpsing the lights of a skyscraper through the fog may be useful to the motorist, but only if he can tell how far away the skyscraper is.) Pulsating variable stars in globular clusters provide the distances, giving astronomers the key to the dimensions of our Galaxy.

In 1912, the American astronomer Henrietta Leavitt reported her important discovery of the period-luminosity relation for Cepheid variables. As we saw in Section 19-6, Cepheid variables are pulsating stars that vary periodically in brightness (see Figure 19-18). Leavitt studied numerous Cepheids in the Small Magellanic Cloud (a small galaxy near the Milky Way) and found their periods to be directly related to their average luminosities. Figure 22-4 shows that the longer a Cepheid's period, the greater its average luminosity.

The period-luminosity law is an important tool in astronomy because it can be used to determine distances. For example, suppose you find a Cepheid variable in the sky. By measuring its period and using a graph like Figure 22-4, you can determine the star's average luminosity. Knowing the star's average luminosity, you can find out how far away the star must be in order to give the observed brightness. (Box 17-2 explains how this is done.)

Shortly after Leavitt's discovery of the period-luminosity law, Harlow Shapley, a young astronomer at the Mount Wilson Observatory in California, began studying a family of pulsating stars closely related to Cepheid variables called **RR Lyrae variables.** The light curve of an RR Lyrae variable is similar to that of a Cepheid, but RR Lyrae variables have shorter pulsation periods and lower peak luminosities (see Figure 22-4).

The tremendous importance of RR Lyrae variables is that they are commonly found in globular clusters (Figure 22-5). By using the period-luminosity relationship for these stars, Shapley was able to determine the distances to the 93 globular clusters then known. He found that some of them were more than 100,000 light-years from Earth. The large values of these distances immediately suggested that the Galaxy was much larger than Herschel or Kapteyn had thought.

Another striking property of globular clusters is how they are distributed across the sky. Ordinary stars are rather uniformly spread along the Milky Way. However, the majority of the 93 globular clusters that Shapley studied are located in one-half of the sky, widely scattered around the portion of the Milky Way that is in the constellation Sagittarius.

From the directions to the globular clusters and their distances from us, Shapley mapped out the three-dimensional distribution of these clusters in space. In 1920 he concluded that the globular clusters form a huge spherical distribution centered not on Earth but rather about a point in the Milky Way several kiloparsecs away in the direction of Sagittarius (see Figure 22-3b). This point, reasoned Shapley, must coincide with the center of our Galaxy, because of gravitational forces between the disk of the Galaxy and the "halo" of globular clusters. Therefore, by locating the center of the distribution of globular clusters, Shapley was in effect measuring the location of the galactic center.

Modern-Day Measurements

Since Shapley's pioneering observations, many astronomers have measured the distance from the Sun to the **galactic nucleus,** the center of our Galaxy. Shapley's estimate of this distance was too large by about a factor of 2, because he did not take into account the effects of interstellar extinction (which were not well understood at the time). Today, the generally accepted distance to the

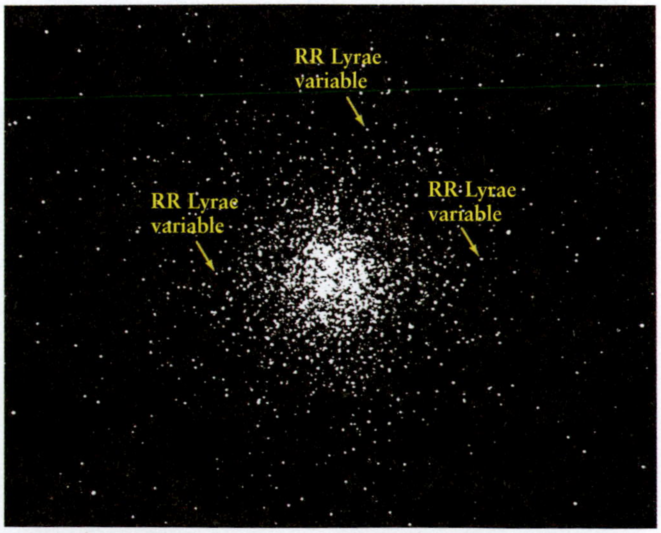

VIDEO 22.1 **FIGURE 22-5** R I V U X G

RR Lyrae Variables in a Globular Cluster The arrows point to three RR Lyrae variables in the globular cluster M55, located in the constellation Sagittarius. From the average apparent brightness (as seen in this photograph) and average luminosity (known to be roughly 100 $L_\odot$) of these variable stars, astronomers have deduced that the distance to M55 is 6500 pc (20,000 ly). (Harvard-Smithsonian Center for Astrophysics) galactic nucleus is about 8 kpc (26,000 ly); the actual distance could be greater or less than that value by about 1 kpc (3300 ly).

Just as Copernicus and Galileo showed that Earth was not at the center of the solar system, Shapley and his successors showed that the solar system lies nowhere near the center of the Galaxy. We see that Earth indeed occupies no special position in the universe.

CONCEPTCHECK 22-1

If astronomers observed that globular clusters were evenly spread across all parts of the sky, what would astronomers assume about the location of our Sun and Earth in our Galaxy?

Answer appears at the end of the chapter.

22-2 Observations at nonvisible wavelengths reveal the shape of the Galaxy

At visible wavelengths, light suffers so much interstellar extinction that the center of the galaxy is totally obscured from view. But the amount of interstellar extinction is roughly inversely proportional to wavelength of light. In other words, the longer the wavelength of light, the farther that light can travel through interstellar dust without being scattered or absorbed. As a result, we can see farther into the plane of the Milky Way at infrared wavelengths than at visible wavelengths, and radio waves can travel all the way through the Galaxy! For this reason, telescopes sensitive to these nonvisible wavelengths are important tools for studying the structure of our Galaxy.

Exploring the Milky Way in the Infrared

Our Galaxy's dust and stars—including the Sun—lie mostly in a relatively thin disk

Infrared light is particularly useful for tracing the location of interstellar dust in the Galaxy. Starlight warms the dust grains to temperatures in the range of 10 to 90 K; thus, in accordance with Wien's law (see Section 5-4), the dust emits radiation predominately at wavelengths from about 30 to 300 μm. These are called **far-infrared** wavelengths, because they lie in the part of the infrared spectrum most different in wavelength from visible light (see Figure 5-7). At these wavelengths, interstellar dust radiates more strongly than stars, so a far-infrared view of the sky is principally a view of where the dust is. In 1983 the Infrared Astronomical Satellite (IRAS) scanned the sky with a 60-cm reflecting telescope at far-infrared wavelengths, giving the panoramic view of the Milky Way's dust shown in Figure 22-6a.

In 1990 an instrument on the Cosmic Background Explorer (COBE) satellite scanned the sky at **near-infrared** wavelengths, that is, relatively short wavelengths closer to the visible spectrum. Figure 22-6b shows the resulting near-infrared view of the plane of the Milky Way. At near-infrared wavelengths, interstellar dust

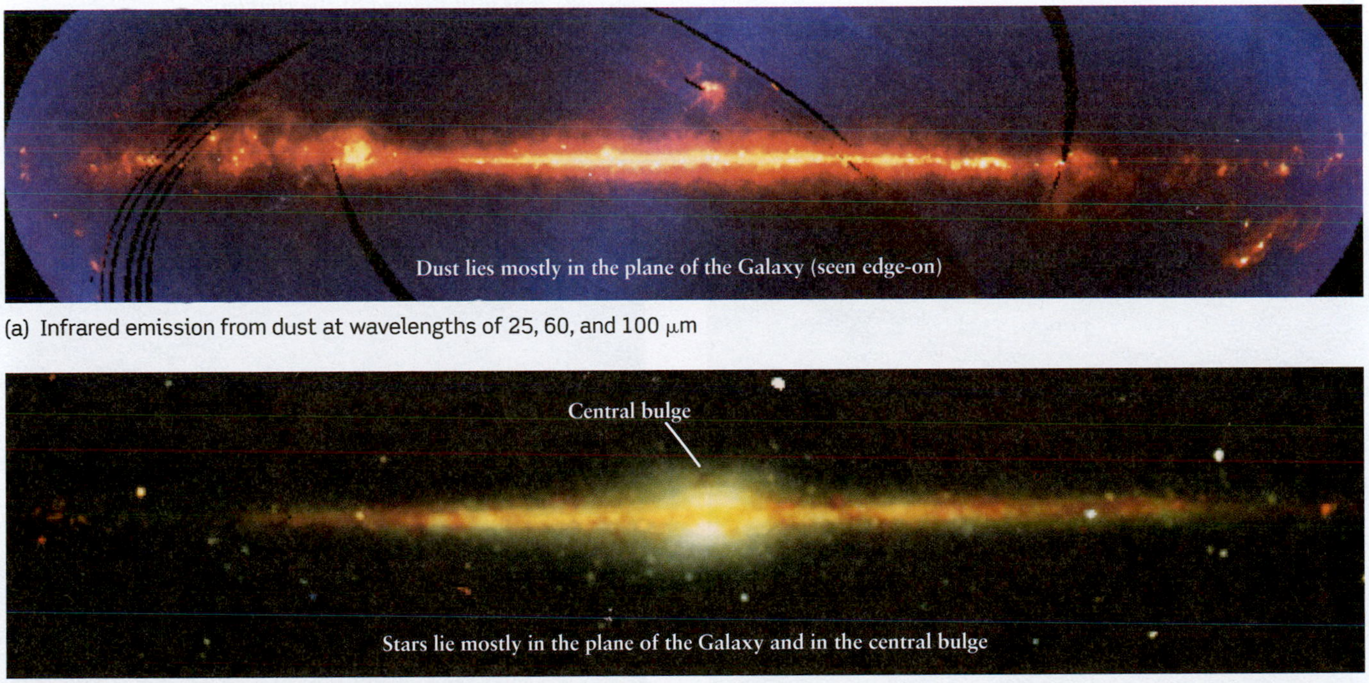
Dust lies mostly in the plane of the Galaxy (seen edge-on)
(a) Infrared emission from dust at wavelengths of 25, 60, and 100 μm
Central bulge
Stars lie mostly in the plane of the Galaxy and in the central bulge
(b) Infrared emission from dust at wavelengths of 1.2, 2.2, and 3.4 μm

FIGURE 22-6 R I V U X G

The Infrared Milky Way **(a)** This view was constructed from observations made at far-infrared wavelengths by the IRAS spacecraft. Interstellar dust, which is mostly confined to the plane of the Galaxy, is the principal source of radiation in this wavelength range. **(b)** Observing at near-infrared wavelengths, as in this composite of COBE data, allows us to see much farther through interstellar dust than we can at visible wavelengths. Light in this wavelength range comes mostly from stars in the plane of the Galaxy and in the bulge at the Galaxy's center. (NASA)

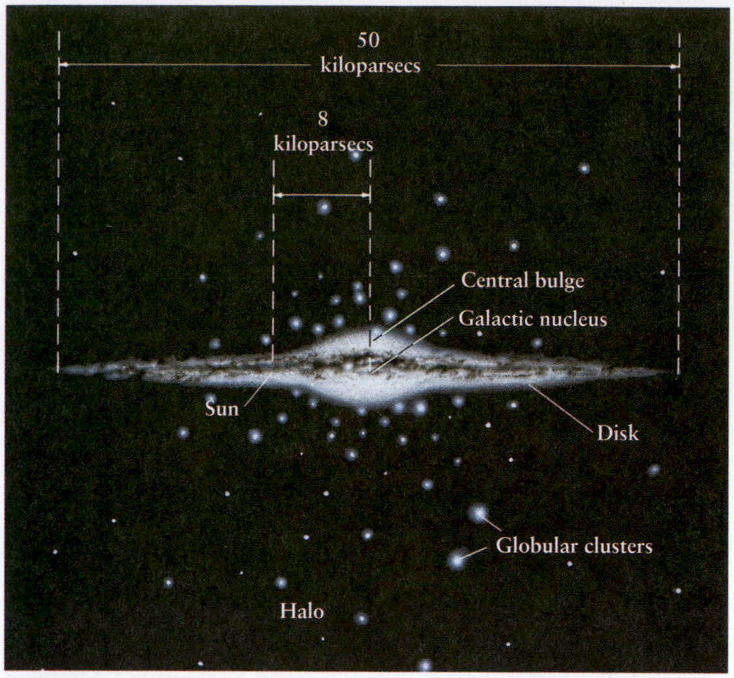

FIGURE 22-7
Our Galaxy (Schematic Edge-on View) There are three major components of our Galaxy: a disk, a central bulge, and a halo. The disk contains gas and dust along with metal-rich (Population I) stars. The halo is composed almost exclusively of old, metal-poor (Population II) stars. The central bulge is a mixture of Population I and Population II stars.

does not emit very much light. Hence, the light sources in Figure 22-6b are stars, which do emit strongly in the near-infrared (especially the cool stars, such as red giants and supergiants). Because interstellar dust causes little interstellar extinction in the near-infrared, many of the stars whose light is recorded in Figure 22-6b lie deep within the Milky Way.

Observations such as those shown in Figure 22-6, along with the known distance to the center of the Galaxy, have helped astronomers establish the dimensions of the Galaxy. The **disk** of our Galaxy is about 50 kpc (160,000 ly) in diameter and about 0.6 kpc (2000 ly) thick, as shown in Figure 22-7. The center of the Galaxy is surrounded by a distribution of stars, called the **central bulge,** which is about 2 kpc (6500 ly) in diameter. This central bulge is clearly visible in Figure 22-6b. The spherical distribution of globular clusters traces the **halo** of the Galaxy.

This structure is not unique to our Milky Way Galaxy. Figure 22-8 shows another galaxy whose dust and stars lie in a disk and that has a central bulge of stars, just like the Milky Way. In the same way that our Sun is a rather ordinary member of the stellar community that makes up the Milky Way, the Milky Way turns out to be a rather common variety of galaxy.

CONCEPTCHECK 22-2

If an astronomer wanted to study the nature of cool stars hiding inside dust clouds, which wavelength would be the best choice, near-infrared (with shorter wavelengths that are nearer to the visible spectrum) or far-infrared (with longer wavelengths farther from the visible spectrum)?

CONCEPTCHECK 22-3

Would an imaginary space traveler have a better view of our Sun and its planets by standing at the galactic center or in the halo?

Answers appear at the end of the chapter.

The Milky Way's Distinct Stellar Populations

It is estimated that our Galaxy contains about 200 billion (2×10^{11}) stars. Remarkably, different kinds of stars are found in the various

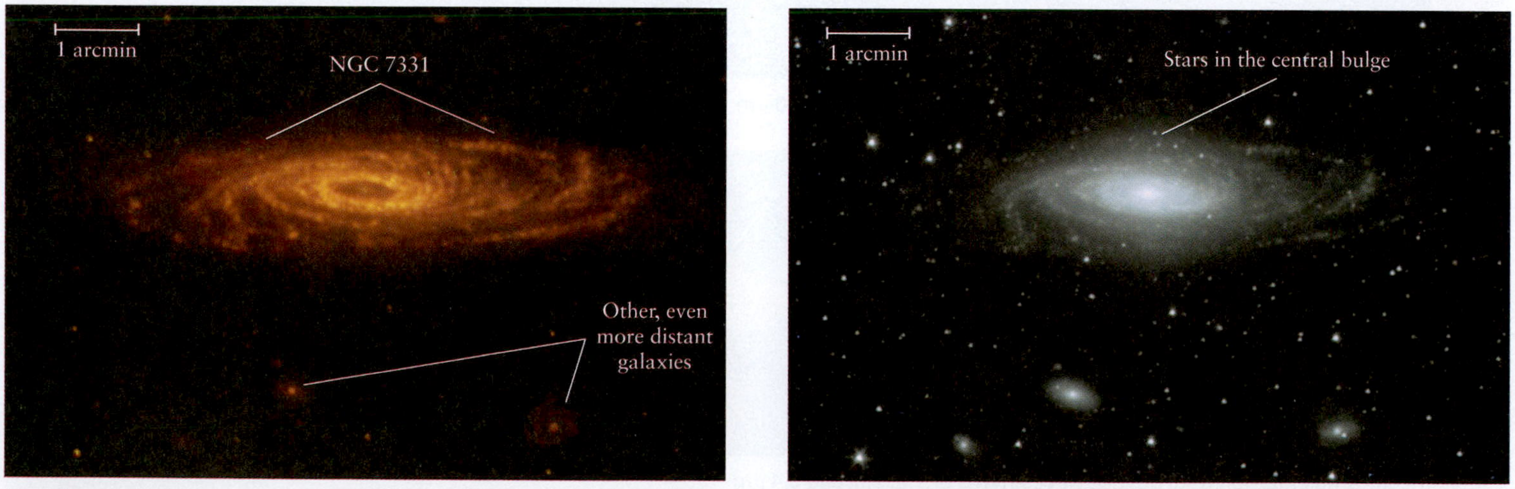

(a) Infrared emission from dust in NGC 7331 at 5.8 and 8.0 μm

(b) Infrared emission from stars in NGC 7331 at 3.6 and 4.5 μm

FIGURE 22-8 R I V U X G
NGC 7331: A Near-Twin of the Milky Way If we could view our Galaxy from a great distance, it would probably look like this galaxy in the constellation Pegasus. As in Figure 22-6, the far-infrared image **(a)** reveals the presence of dust in the galaxy's plane, while the near-infrared image **(b)** shows the distribution of stars. These images of NGC 7331, which is about 15 million pc (50 million ly) from Earth, were made with the Spitzer Space Telescope (see Section 6-7, especially Figure 6-26). (NASA; JPL-Caltech; M. Regan [STScI]; and the SINGS Team)

FIGURE 22-9 R I V U X G

Star Orbits in the Milky Way The different populations of stars in our Galaxy travel along different sorts of orbits. The galaxy in this visible-light image is the Milky Way's near-twin NGC 7331, the same galaxy shown at infrared wavelengths in Figure 22-8. (Russell Croman/Science Source)

components of the Galaxy. The globular clusters in the halo are composed of old, metal-poor, Population II stars (see Section 19-5). Although globular clusters are conspicuous, they contain only about 1% of the total number of stars in the halo; most halo stars are single Population II stars in isolation. These ancient stars orbit the Galaxy along paths tilted at random angles to the disk of the Milky Way, as do the globular clusters. By contrast, stars in the disk travel along orbits that remain in the disk (Figure 22-9).

Unlike the halo, the stars in the disk are mostly young, metal-rich, Population I stars like the Sun. The disk of a galaxy like the Milky Way appears bluish because its light is dominated by radiation from hot O and B main-sequence stars. Such stars have very short main-sequence lifetimes (see Section 19-1, especially Table 19-1), so they must be quite young by astronomical standards. Hence, their presence shows that there must be active star formation in the galactic disk. By contrast, no O or B stars are present in the halo, which implies that star formation ceased there long ago.

The central bulge contains *both* Population I stars and metal-poor Population II stars. Since Population II stars are thought to have formed early in the history of the universe, this suggests that some of the stars in the bulge are quite ancient while others were created more recently. The central bulge looks yellowish or reddish because it contains many red giants and red supergiants (see Figure 1-7), but does *not* contain luminous, short-lived, blue O or B stars. Hence, there cannot be ongoing star formation in the central bulge. The same is true for other galaxies whose structure is similar to that of the Milky Way (Figure 22-10).

Why are there such different populations of stars in the halo, disk, and central bulge? Why has star formation stopped in some regions of the Galaxy but continues in other regions? The answers to these questions are related to the way that stars, as well as the gas and dust from which stars form, move within the Galaxy.

Only Population I (metal-rich) stars are found in the disk. The presence of hot, blue, young O and B stars indicates active star formation.
Both Population I (metal-rich) and Population II (metal-poor) stars are found in the central bulge — but there are *no* young blue stars. Hence star formation is not active there.

FIGURE 22-10 R I V U X G

Stellar Populations: Disk Versus Central Bulge The disk and central bulge of the Milky Way contain rather different populations of stars. The same is true for the galaxy NGC 1309, which has a similar structure to the Milky Way Galaxy and happens to be oriented face-on to us. NGC 1309 is about 30 million pc (100 million ly) from us in the constellation Eridanus. (NASA, ESA, The Hubble Heritage Team [STScI/AURA] and A. Riess [STScI])

CONCEPTCHECK 22-4

If you were looking to find the most recently formed stars in the Galaxy, where would you most likely look?

Answer appears at the end of the chapter.

22-3 Observations of cold hydrogen clouds and star-forming regions reveal that our Galaxy has spiral arms

The galaxies shown in Figure 22-8 and Figure 22-10 both have **spiral arms,** spiral-shaped concentrations of gas and dust that extend outward from the center in a shape reminiscent of a pinwheel. Assuming our Galaxy is similar to other galaxies, observations of spiral arms would lead us to suspect that our own Milky Way Galaxy has this feature. However, because interstellar dust obscures our visible-light view in the pancakelike plane of our Galaxy, a detailed understanding of the structure of our galactic disk had to wait until the development of radio astronomy. Thanks to their long wavelengths, radio waves can penetrate the interstellar medium even more easily than infrared light and can travel without being scattered or absorbed. As we shall see in this section, both radio and optical observations reveal that our Galaxy does indeed have spiral arms.

Mapping Hydrogen in the Milky Way

LOOKING DEEPER 22-1

Our Galaxy's dust and stars—including the Sun—lie mostly in a relatively thin disk

Hydrogen is by far the most abundant element in the universe (see Figure 8-4 in Section 8-2). Hence, by looking for concentrations of hydrogen gas, we should be able to detect important clues about the distribution of matter in our Galaxy. Unfortunately, ordinary visible-light telescopes are of little use in this quest, because hydrogen atoms can only emit visible light if they are first excited to high energy levels (see Section 5-8, and especially Figure 5-24). This excitation is quite unlikely to occur in the cold depths of interstellar space. Furthermore, even if there are some hydrogen atoms that glow strongly at visible wavelengths, interstellar extinction due to dust (see Section 22-1) would make it impossible to see this glow from distant parts of the Galaxy.

What makes it possible to map out the distribution of hydrogen in our Galaxy is that even cold hydrogen clouds emit *radio* waves. As we saw in Section 22-2, radio waves can easily penetrate the interstellar medium, so we can detect the radio emission from such cold clouds no matter where in the Galaxy they lie. The hydrogen in these clouds is neutral—that is, not ionized—and is called **H I.** (This distinguishes it from ionized hydrogen, which is designated H II.) To understand how H I clouds can emit radio waves, we must probe a bit more deeply into the structure of protons and electrons, the particles of which hydrogen atoms are made.

In addition to having mass and charge, particles such as protons and electrons possess a tiny amount of angular momentum (that is, rotational motion) commonly called **spin.** Very roughly, you can visualize a proton or electron as a tiny, electrically charged sphere that spins on its axis. Because electric charges in motion generate magnetic fields, a proton or electron behaves like a tiny magnet with a north pole and a south pole (Figure 22-11).

If you have ever played with magnets, you know that two magnets attract when the north pole of one magnet is next to the south pole of the other and repel when two like poles (both north or both south) are next to each other (Figure 22-11a). This behavior can also be described in terms of magnetic energy: The energy of the two magnets is least when opposite poles are together and highest

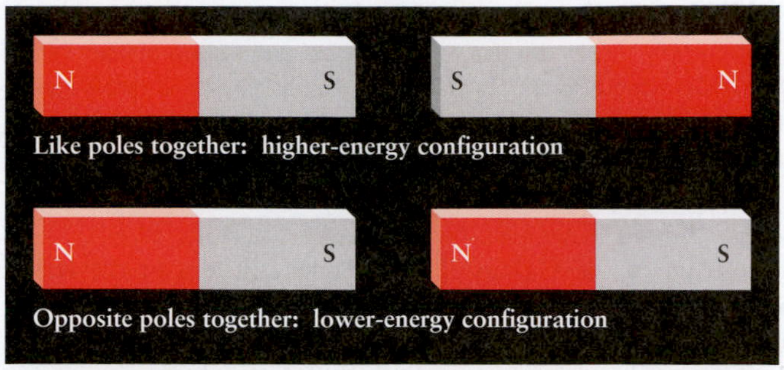

(a) The magnetic energy of two bar magnets depends on their relative orientation

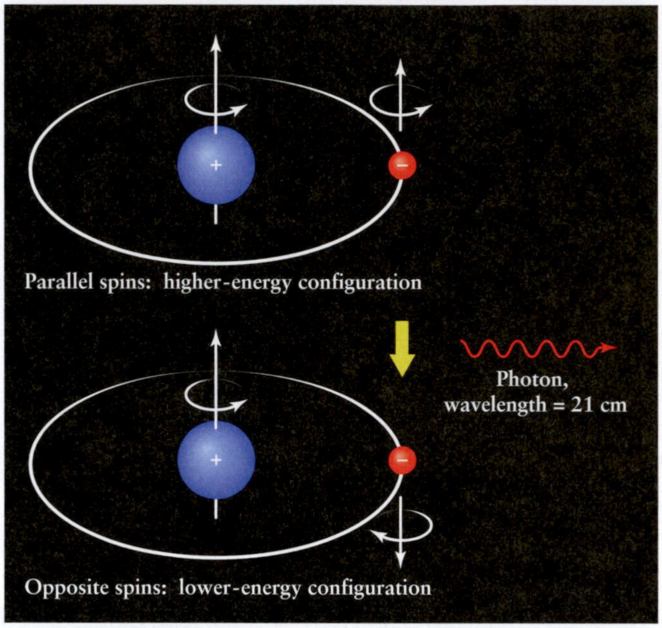

(b) The magnetic energy of a proton and electron depends on the relative spin orientation

FIGURE 22-11

Magnetic Interactions in the Hydrogen Atom **(a)** The energy of a pair of magnets is high when their north poles or their south poles are near each other, and low when they have opposite poles near each other. **(b)** Thanks to their spin, electrons and protons are both tiny magnets. When the electron flips from the higher-energy configuration (with its spin in the same direction as the proton's spin) to the lower-energy configuration (with its spin opposite to the proton's spin), the atom loses a tiny amount of energy and emits a radio photon with a wavelength of 21 cm.

when like poles are together. Hence, as shown in Figure 22-11b, the energy of a hydrogen atom is slightly different depending on whether the spins of the proton and electron are in the same direction or opposite directions. (According to the laws of quantum mechanics, these are the only two possibilities; the spins cannot be at random angles.)

If the spin of the electron changes its orientation from the higher-energy configuration to the lower-energy one—called a **spin-flip transition**—a photon is emitted. The energy difference between the two spin configurations is very small, only about 10^{-6} as great as those between different electron orbits (see Figure 5-24). Therefore, the photon emitted in a spin-flip transition between these configurations has only a small energy, and thus its wavelength is a relatively long 21 cm—a radio wavelength. These spin-flip transitions occur spontaneously in our Milky Way's diffuse hydrogen gas, which allows the possibility of mapping hydrogen through this method.

The spin-flip transition in neutral hydrogen was first predicted in 1944 by the Dutch astronomer Hendrik van de Hulst. His calculations suggested that it should be possible to detect the **21-cm radio emission** from interstellar hydrogen, although a very sensitive radio telescope would be required. In 1951, Harold Ewen and Edward Purcell at Harvard University first succeeded in detecting this faint emission from hydrogen between the stars.

Figure 22-12 shows the results of a more recent 21-cm survey of the entire sky. Neutral hydrogen gas (H I) in the plane of the Milky Way stands out prominently as a bright band across the middle of this image.

LOOKING DEEPER 22-2

The distribution of gas in the Milky Way is not uniform but is actually quite frothy. In fact, our Sun lies near the edge of an irregularly shaped region within which the interstellar medium is very thin but at very high temperatures (about 10^6 K, but so thin it would feel cold). This region, called the **Local Bubble**, is several hundred parsecs across. The Local Bubble may have been carved out by a supernova that exploded nearby some 300,000 years ago.

Remarkably, spin-flip transitions are used not only to map our Galaxy but also to map the internal structure of the human body. Box 22-1 discusses this application, called magnetic resonance imaging.

CONCEPTCHECK 22-5

If a spin-flip transition released substantially more energy, such that a photon was emitted in the far-infrared range, would the Milky Way stretching across the night sky appear brighter or dimmer?

CALCULATIONCHECK 22-1

If a 21-cm photon is observed when a single hydrogen atom undergoes a spin-flip transition, what wavelength is observed when 10 hydrogen atoms undergo spin-flip transitions?

Answers appear at the end of the chapter.

Detecting Our Galaxy's Spiral Arms

The detection of 21-cm radio emission was a major breakthrough that permitted astronomers to reveal the presence of spiral arms in the galactic disk. Figure 22-13 shows how spiral arms were detected. Suppose that you aim a radio telescope along a particular line of sight across the Galaxy. Your radio receiver, located at *S* (the position of the solar system), picks up 21-cm emission from H I clouds at points 1, 2, 3, and 4. However, the radio waves from these various clouds are Doppler shifted by slightly different amounts, because the clouds are moving at different speeds as they travel with the rotating Galaxy.

It is important to remember that the Doppler shift reveals only motion along the line of sight (review Figure 5-26). In Figure 22-13, cloud 2 has the highest speed along our line of sight, because it is moving directly toward us. Consequently, the radio waves from cloud 2 exhibit a larger Doppler shift than those from the other three clouds along our line of sight. Because clouds 1 and 3 are at the same distance from the galactic center, they have the same orbital speed. The fraction of their velocity parallel to our line of sight is also the same, so their radio waves exhibit the

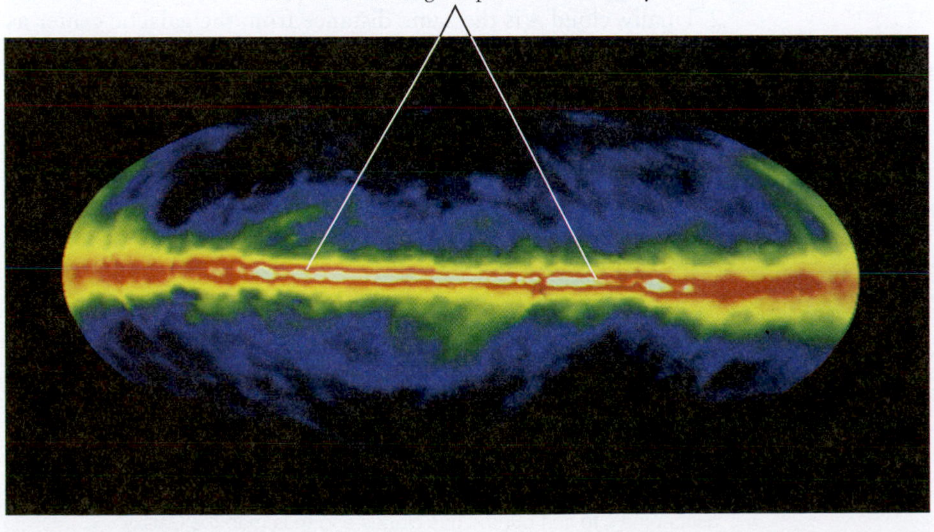

21-cm emission shows that hydrogen gas is concentrated along the plane of the Galaxy

FIGURE 22-12 R I V U X G
The Sky at 21 Centimeters This image was made by mapping the sky with radio telescopes tuned to the 21-cm wavelength emitted by neutral interstellar hydrogen (H I). The entire sky has been mapped onto an oval, and the plane of the Galaxy extends horizontally across the image as in Figure 22-6. Black and blue represent the weakest emission, and red and white the strongest. (Courtesy of C. Jones and W. Forman, Harvard-Smithsonian Center for Astrophysics)

BOX 22-1 ASTRONOMY DOWN TO EARTH

Spin-Flip Transitions in Medicine

Thanks to their spin, protons and electrons act like microscopic bar magnets. In a hydrogen atom, the interaction between the magnetism of the electron and that of the proton gives rise to the 21-cm radio emission. But these particles can also interact with outside magnetic fields, such as that produced by a large electromagnet. This physical principle is behind **magnetic resonance imaging,** an important diagnostic tool of modern medicine.

Much of the human body is made of water. Every water molecule has two hydrogen atoms, each of which has a nucleus made of a single proton. If a person's body is placed in a strong magnetic field, the spins of the protons in the hydrogen atoms of their body can either be in the same direction as the field ("aligned") or in the direction opposite to the field ("opposed"). The aligned orientation has lower energy, and therefore the majority of protons end up with their spins in this orientation. But if a radio wave of just the right wavelength is now sent through the person's body, an aligned proton can absorb a radio photon and flip its spin into the higher-energy, opposed orientation. How much of the radio wave is absorbed depends on the number of protons in the body, which in turns depends on how much water (and, thus, how much water-containing tissue) is in the body.

In magnetic resonance imaging, a magnetic field is used whose strength varies from place to place. The difference in energy between the opposed and aligned orientations of a proton depends on the strength of the magnetic field, so radio waves will only be absorbed at places where this energy difference is equal to the energy of a radio photon. (This equality is called *resonance,* which is how magnetic resonance imaging gets its name.) By varying the magnetic field strength over the body and the wavelength of the radio waves, and by measuring how much of the radio wave is absorbed by different parts of the body, it is possible to map out the body's tissues. The accompanying false-color image shows such a map of a patient's head.

Unlike X-ray images, which show only the densest parts of the body, such as bones and teeth, magnetic resonance imaging can be used to view less dense (but water-containing) soft tissue. Just as the 21-cm radio emission has given astronomers a clear view of what were hidden regions of our Galaxy, magnetic resonance imaging allows modern medicine to see otherwise invisible parts of the human body.

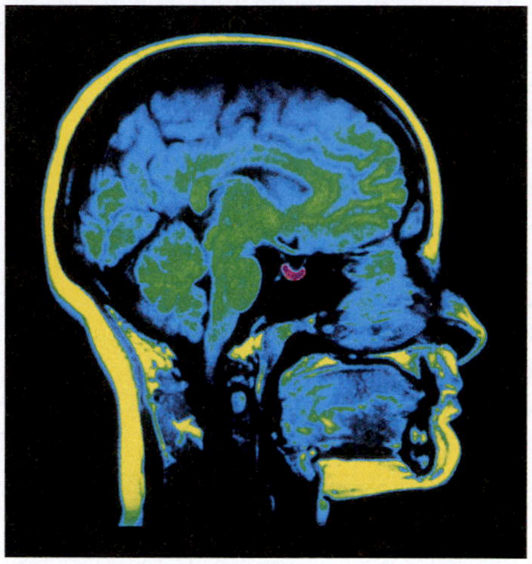

R I V U X G

(Scott Camazine/Science Source)

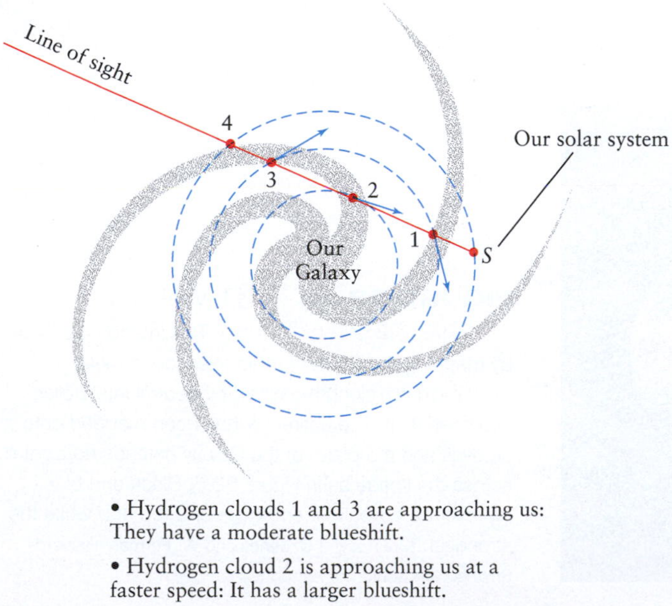
Line of sight
4
3
2
1
S
Our solar system
Our Galaxy
• Hydrogen clouds 1 and 3 are approaching us: They have a moderate blueshift.
• Hydrogen cloud 2 is approaching us at a faster speed: It has a larger blueshift.

same Doppler shift, which is less than the Doppler shift of cloud 2. Finally, cloud 4 is the same distance from the galactic center as the Sun. This cloud is thus orbiting the Galaxy at the same speed as the Sun, resulting in no net motion along the line of sight. Radio waves from cloud 4, as well as from hydrogen gas near the Sun, are not Doppler shifted at all.

These various Doppler shifts cause radio waves from gases in different parts of the Galaxy to arrive at our radio telescopes with wavelengths slightly different from 21 cm. It is therefore possible

FIGURE 22-13
A Technique for Mapping Our Galaxy If we look within the plane of our Galaxy from our position at *S*, hydrogen clouds at different locations (shown as 1, 2, 3, and 4) along our line of sight are moving at slightly different speeds relative to us. As a result, radio waves from these various gas clouds are subjected to slightly different Doppler shifts. This permits radio astronomers to sort out the gas clouds and thus map the Galaxy.

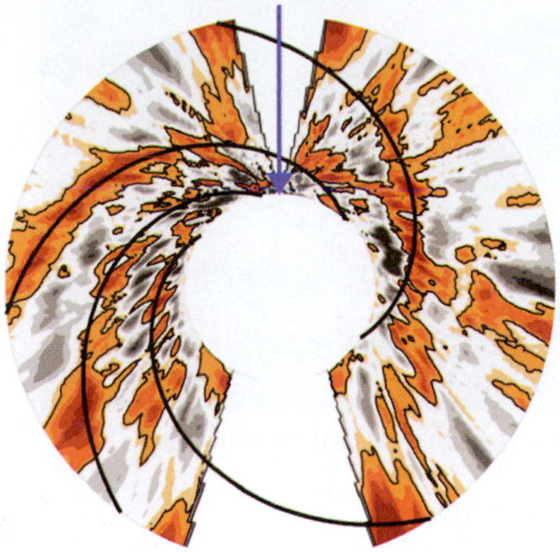

FIGURE 22-14 R I V U X G

A Map of Neutral Hydrogen in Our Galaxy This map, constructed from radio-telescope surveys of 21-cm radiation, shows the distribution of hydrogen gas in a reconstructed (or hypothetical) face-on view of our Galaxy. The map suggests a spiral structure. Details in the blank, wedge-shaped region at the bottom of the map are unknown. Gas in this part of the Galaxy is moving perpendicular to our line of sight and thus does not exhibit a detectable Doppler shift. (Image courtesy of Leo Blitz, Ph.D.)

to sort out the various gas clouds and thus produce a map of the Galaxy like that shown in Figure 22-14.

Figure 22-14 shows that neutral hydrogen gas is not spread uniformly around the disk of the Galaxy but is concentrated into numerous arched lanes. Similar features are seen in other galaxies beyond the Milky Way. Unlike the view from within our own Galaxy, we can view other galaxies face-on to easily see their distribution of stars, gas, and dust. As an example, the galaxy in Figure 22-15a has prominent spiral arms outlined by hot, luminous, blue main-sequence stars and the red emission nebulae (H II regions) found near many such stars. Stars of this sort are very short-lived, so these features indicate that spiral arms are sites of active, ongoing star formation. The 21-cm radio image of this same galaxy, shown in Figure 22-15b, shows that spiral arms are also regions where neutral hydrogen gas is concentrated, similar to the structures in our own Galaxy visible in Figure 22-14. This similarity is a strong indication that our Galaxy also has spiral arms.

CAUTION! Photographs such as Figure 22-15a can lead to the impression that there are very few stars between the spiral arms of a galaxy. Nothing could be further from the truth! In fact, stars are distributed rather uniformly throughout the disk of a galaxy like the one in Figure 22-15a; the density of stars in the spiral arms is only about 5% higher than in the rest of the disk. The spiral arms stand out nonetheless because they are where hot, blue O and B stars are found. One such star is about 10^4 times more luminous than an average star in the disk, so the light from O and B stars completely dominates the visible appearance of a spiral galaxy.

CONCEPTCHECK 22-6

When looking at neutral hydrogen gas that is moving away from you, how is its wavelength changed as compared to neutral hydrogen gas that is not moving relative to you?

Answer appears at the end of the chapter.

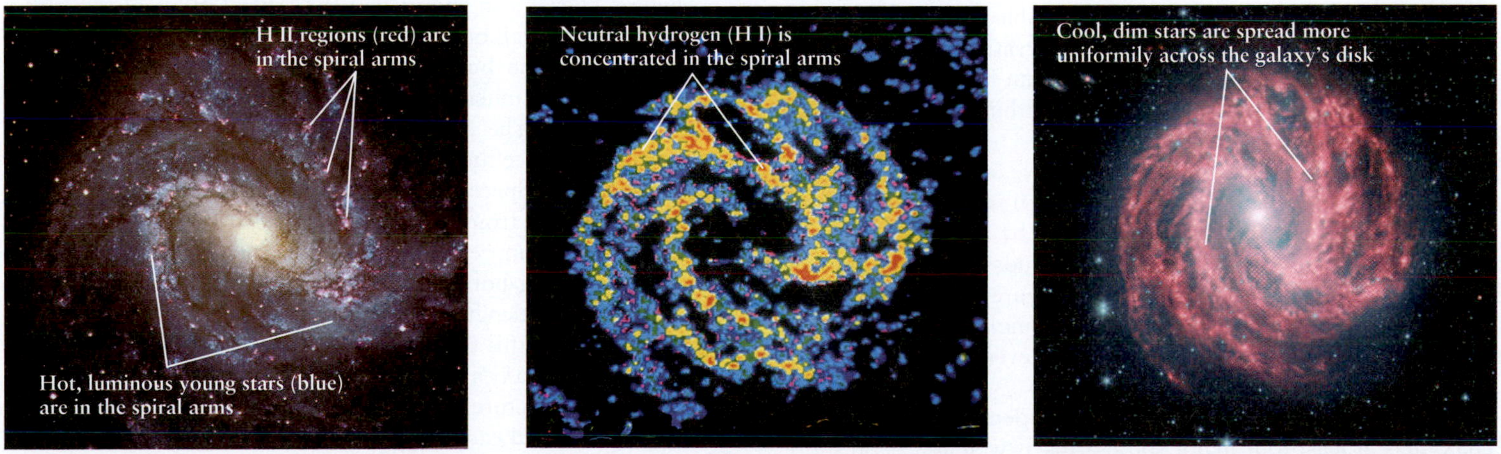

H II regions (red) are in the spiral arms
Hot, luminous young stars (blue) are in the spiral arms
Neutral hydrogen (H I) is concentrated in the spiral arms
Cool, dim stars are spread more uniformly across the galaxy's disk

(a) Visible-light view of M83 R I V U X G (b) 21-cm radio view of M83 R I V U X G (c) Near-infrared view of M83 R I V U X G

FIGURE 22-15

A Spiral Galaxy The galaxy M83 lies in the southern constellation Hydra about 5 million pc (15 million ly) from Earth. **(a)** This visible-light image clearly shows the spiral arms. The presence of young stars and H II regions indicates that star formation takes place in spiral arms. **(b)** This radio view at a wavelength of 21 cm shows the emission from neutral interstellar hydrogen gas (H I). Note that essentially the same pattern of spiral arms is traced out in this image as in the visible-light photograph. **(c)** M83 has a different appearance in this near-infrared view. The starlight has been removed in this image to reveal the infrared emission of dust throughout the galaxy. (a: ©Australian Astronomical Observatory/David Malin Images; b: VLA, NRAO; c: NASA/JPL-Caltech)

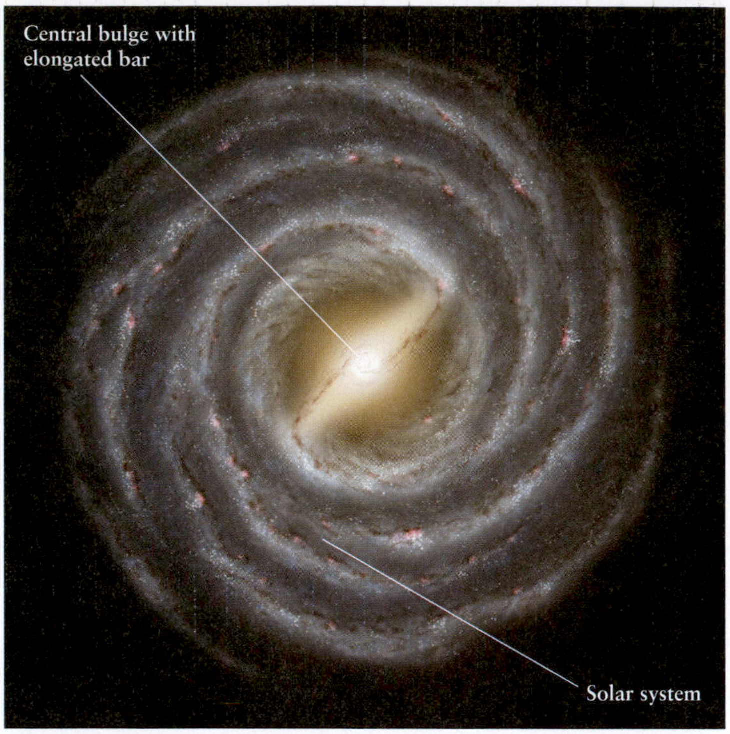

(a) The structure of the Milky Way's disk

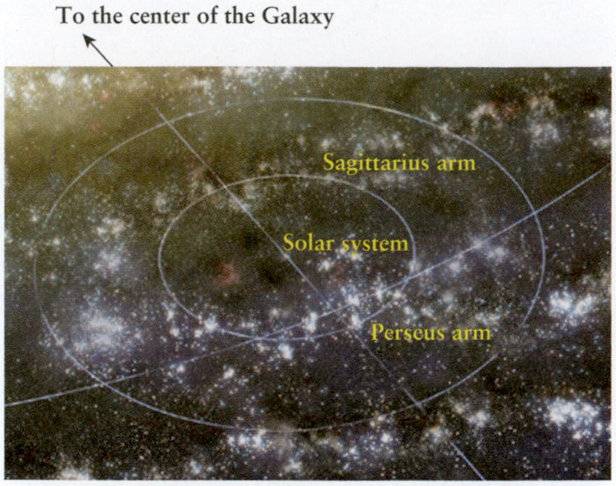

(b) Closeup of the Sun's galactic neighborhood

FIGURE 22-16

Our Galaxy Seen Face-on: Artist's Impressions **(a)** The Galaxy's diameter is about 50,000 pc (160,000 ly), and our solar system is about 8000 pc (26,000 ly) from the galactic center. The elongated central bulge is about 8300 pc (27,000 ly) long and is oriented at approximately 45° to a line running from the solar system to the galactic center. **(b)** Our solar system is located between the Sagittarius and Perseus arms, two of the major spiral arms in the Milky Way. (a: NASA/JPL-Caltech/R. Hurt, SSC; b: NG Maps/National Geographic Creative)

Mapping the Spiral Arms and the Central Bulge

Figure 22-15a suggests that we can confirm the presence of spiral structure in our own Galaxy by mapping the locations of star-forming regions. Such regions are marked by OB associations, H II regions, and molecular clouds (see Section 18-7). Unfortunately, the first two of these are best observed using visible light, and interstellar extinction limits the range of visual observations in the plane of the Galaxy to less than 3 kpc (10,000 ly) from Earth. But there are enough OB associations and H II regions within this range to plot the spiral arms in the vicinity of the Sun.

Molecular clouds are easier to observe at great distances, because molecules of carbon monoxide (CO) in these clouds emit radio waves that are relatively unaffected by interstellar extinction. Hence, the positions of molecular clouds have been plotted even in remote regions of the Galaxy, as Figure 18-21 shows. (We saw in Section 18-7 that CO molecules in molecular clouds emit more strongly than the hydrogen atoms do, even though hydrogen is the principal constituent of these clouds.)

Taken together, all these observations demonstrate that our Galaxy has at least four major spiral arms as well as several short arm segments (Figure 22-16). The Sun is located just outside a relatively short arm segment called the Orion arm, which includes the Orion Nebula and neighboring sites of vigorous star formation in that constellation.

Two major spiral arms border either side of the Sun's position. The Sagittarius arm is on the side toward the galactic center. You see this arm on June and July nights when you look at the portion of the Milky Way stretching across Scorpius and Sagittarius, near the center of the upper photograph that opens this chapter. In December and January, when our nighttime view is directed away from the galactic center, we see the Perseus arm. The other major spiral arms cannot be seen at visible wavelengths due to the obscuring effects of dust.

Figure 22-16a also shows that the central bulge of the Milky Way is not spherical, but is elongated like a bar. The Milky Way's bulge is unlike the bulge of the galaxy NGC 7331 shown in Figure 22-8, but similar to the bulge of the galaxy M83 shown in Figure 22-15. The elongated shape of the central bulge had been suspected since the 1980s; this shape was confirmed in 2005 using the Spitzer Space Telescope, which was used to survey the infrared emissions from some 3 million stars in the central bulge. Thus, the artist's impression shown in Figure 22-16a is based on observations using both radio wavelengths (for the spiral arms) and infrared wavelengths (for the central bulge). We will see in Section 22-5 that this elongated shape may play a crucial role in sustaining the Galaxy's spiral structure.

To get a full picture of our Galaxy, you need to know not only where our stars and gas are located, but you must also understand their motion. We now look at the rotational motion of our Galaxy and how it reveals one of the greatest mysteries in all of science—dark matter.

CONCEPTCHECK 22-7

If you traveled from Earth in a direction *away* from the center of the Galaxy, what is the first major arm you would reach?

Answer appears at the end of the chapter.

22-4 The rotation of our Galaxy reveals the presence of dark matter

The Doppler shifts observed in the disk of our Galaxy tell us that the disk rotates. This rotation shows that the stars, gas, and dust in our Galaxy are all orbiting the galactic center. Indeed, if this were not the case, mutual gravitational attraction would cause the entire Galaxy to collapse into the galactic center. In the same way, the Moon is kept from crashing into Earth and the planets from crashing into the Moon because of their orbital motion (see Section 4-7).

Measuring the rotation of our Galaxy accurately is a difficult business. But such challenging measurements have been made, as we shall see, and the results led to a remarkable conclusion: Most of the mass of the Galaxy is in the form of *dark matter,* a mysterious sort of material that emits no light at all.

Measuring How the Milky Way Rotates

Radio observations of 21-cm radiation from hydrogen gas provide important clues about our Galaxy's rotation. Doppler shift measurements of this radiation indicate that stars and gas all orbit in the same direction around the galactic center, just as the planets all orbit in the same direction around the Sun.

Measurements also show that the orbital speed of stars and gas around the galactic center is fairly uniform throughout much of the Galaxy's disk (Figure 22-17). As a result, stars orbiting between the Sun and the galactic center complete a trip around the galactic center more quickly than the Sun, because the stars have a shorter distance to travel. Conversely, stars outside the Sun's orbit take longer to go once around the galactic center because they have farther to travel. As seen by Earth-based astronomers moving along with the Sun, stars inside the Sun's orbit overtake and pass us, while we overtake and pass stars outside the Sun's orbit (Figure 22-17a).

CAUTION! Note that when we say that objects in different parts of the Galaxy orbit at the same speed, we do *not* mean that the Galaxy rotates like a solid disk. All parts of a rotating solid disk—a CD or DVD, for example—take the same time to complete one rotation. Because the outer part of the disk has to travel around a larger circle than the inner part, the speed (distance per time) is greater in the outer part (Figure 22-17b). By contrast, the orbital speed of material in our Galaxy is roughly the *same* at all distances from the galactic center.

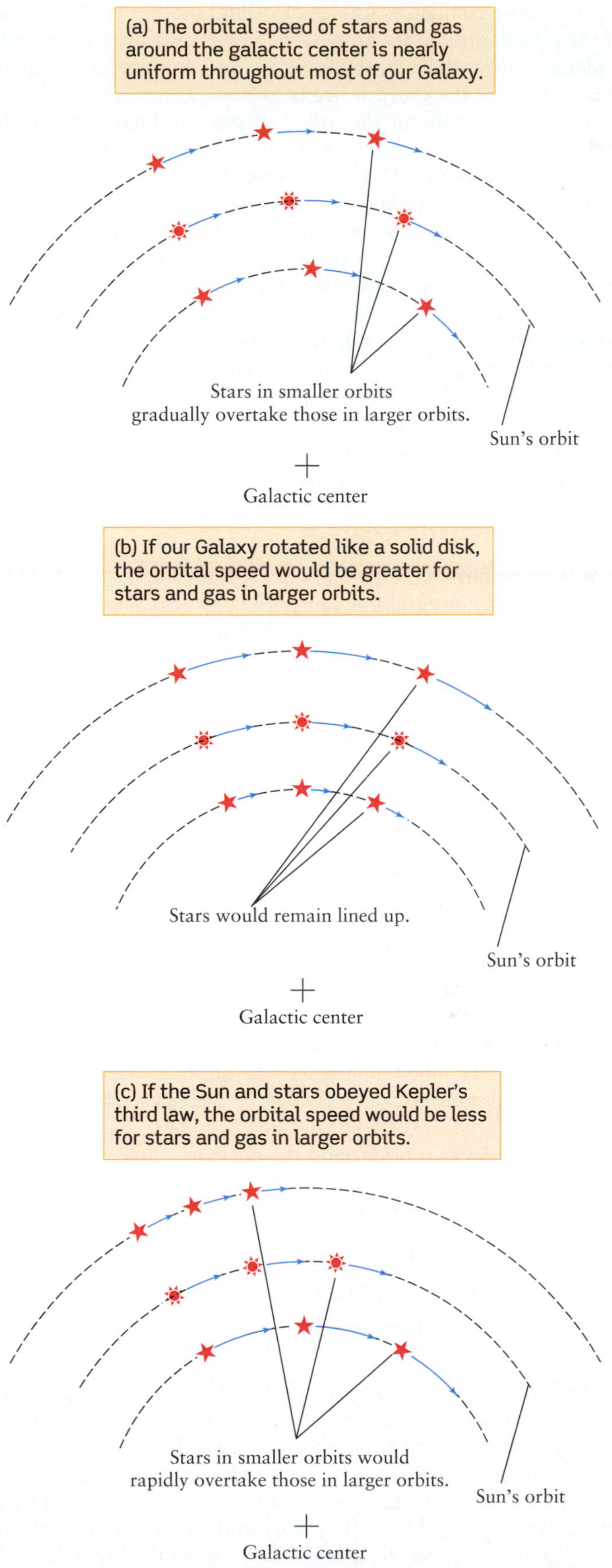
(a) The orbital speed of stars and gas around the galactic center is nearly uniform throughout most of our Galaxy.
Stars in smaller orbits gradually overtake those in larger orbits.
Sun's orbit
Galactic center
(b) If our Galaxy rotated like a solid disk, the orbital speed would be greater for stars and gas in larger orbits.
Stars would remain lined up.
Sun's orbit
Galactic center
(c) If the Sun and stars obeyed Kepler's third law, the orbital speed would be less for stars and gas in larger orbits.
Stars in smaller orbits would rapidly overtake those in larger orbits.
Sun's orbit
Galactic center

FIGURE 22-17

The Rotation of Our Galaxy Each schematic diagram follows three stars (the Sun and two others) orbiting the center of the Galaxy at different distances from the galactic center. **(a)** This case of stars with similar, nearly uniform, orbital speeds is what we have in our Galaxy. Although they start off lined up in this illustration, the stars become increasingly separated as they move along their orbits. With a uniform speed for all stars, stars inside the Sun's orbit overtake and move ahead of the Sun, while stars far from the galactic center lag behind the Sun. **(b)** The stars would remain lined up if the Galaxy rotated like a solid disk. This orientation is not what is observed. **(c)** If stars orbited the galactic center in the same way that planets orbit the Sun, stars inside the Sun's orbit would overtake us faster than they are observed to do.

The most familiar examples of orbital motion are the motions of the planets around the Sun. As we saw in Section 4-7, the farther a planet is from the Sun, the less gravitational force it experiences and the slower the speed it needs to have to remain in orbit. The same would be true for the orbits of stars and gas in the Galaxy *if* they were held in orbit by a single, massive object at the galactic center (see Figure 22-17c). Hence, the 21-cm H I wavelength observations of our Galaxy, which show that the speed of orbiting objects does *not* decrease with increasing distance from the galactic center, demonstrate that there is no such single, massive object holding objects in their galactic orbits.

Instead, what keeps a star in its orbit around the center of the Galaxy is the combined gravitational force exerted on it by all of the mass (including stars, gas, and dust) that lies *within* the star's orbit. (It turns out that the gravitational force from matter *outside* a star's orbit balances out and has little or no net effect on the star's motion around the galactic center.) This gives us a tool for determining the Galaxy's mass and how that mass is distributed.

CONCEPTCHECK 22-8

Which would have a greater impact on changing the length of time it takes the Sun to orbit the galactic center—adding more stars to the galactic center or adding more stars to the galactic halo?

Answer appears at the end of the chapter.

The Sun's Orbital Motion and the Mass of the Galaxy

An important example is the orbital motion of the Sun (and the solar system) around the center of the Galaxy. If we know the semimajor axis and period of the Sun's orbit, we can use Newton's form of Kepler's third law (described in Section 4-7) to determine the mass of that portion of the Galaxy that lies within the orbit. We saw in Section 22-2 that the Sun is about 8000 pc (26,000 ly) from the galactic center. The orbit is in fact nearly circular, so we can regard 8000 pc as the radius r of the orbit. (This radius is also the semimajor axis of the Sun's orbit.) In one complete trip around the Galaxy, the Sun travels a distance equal to the circumference of its orbit, which is $2\pi r$. The time required for one orbit, or orbital period P, is equal to the distance traveled divided by the Sun's orbital speed v:

Period of the Sun's orbit around the galactic center

$$P = \frac{2\pi r}{v}$$

P = orbital period of the Sun

r = distance from the Sun to the galactic center

v = orbital speed of the Sun

Unfortunately, we cannot tell the Sun's orbital speed from 21-cm observations, because these reveal only how fast things are moving relative to the Sun. Instead, we need to measure how the Sun is moving relative to a background that is not rotating along with the rest of the Galaxy. Such a background is provided by distant galaxies beyond the Milky Way and by the globular clusters. (Since globular clusters lie outside the plane of the Milky Way, they do not take part in the rotation of our Galaxy's disk.) By measuring the Doppler shifts of these objects and averaging their velocities, astronomers deduce that the Sun is moving along its orbit around the galactic center at about 220 km/s—about 790,000 kilometers per hour or 490,000 miles per hour!

Using this information, we find that the Sun's orbital period is

$$P = \frac{2\pi \times 8000\text{ pc}}{220\text{ km/s}} \times \frac{3.09 \times 10^{13}\text{ km}}{1\text{ pc}} = 7.1 \times 10^{15}\text{ s}$$
$$= 2.2 \times 10^{8}\text{ years}$$

Traveling at 790,000 kilometers per hour, it takes the Sun about 220 million years to complete one trip around the Galaxy. (In the 65 million years since the demise of the dinosaurs, our solar system has traveled less than a third of the way around its orbit.) The Galaxy is a very large place!

Box 22-2 shows how to combine the radius and period of the Sun's orbit to calculate the total mass of all the matter that lies inside the Sun's orbit. Such calculations give an answer of $9.0 \times 10^{10}\ M_\odot$ (90 billion solar masses). As Figure 22-7 shows, the Galaxy extends well beyond the Sun's orbit, so the mass of the entire Galaxy must be larger than this mass.

CALCULATIONCHECK 22-2

Since it formed 4.5 billion years ago, how many orbits around the Galaxy has the Sun made?

Answer appears at the end of the chapter.

Rotation Curves and the Mystery of Dark Matter

In recent years, astronomers have been astonished to discover how much matter may lie outside the Sun's orbit. The clues come from 21-cm radiation emitted by hydrogen in spiral arms that extend to the outer reaches of the Galaxy. Because we know the true speed of the Sun, we can convert the Doppler shifts of this radiation into actual speeds for the spiral arms. This calculation gives us a **rotation curve,** a graph of the speeds of galactic rotation measured outward from the galactic center (Figure 22-18). We would expect that for gas clouds beyond the confines of most of the Galaxy's mass, the orbital speed should decrease with increasing distance from the Galaxy's center, just as the orbital speeds of the planets decrease with increasing distance from the Sun (see Figure 22-17c). But as Figure 22-18 shows, the Galaxy's rotation curve is quite flat, indicating roughly uniform orbital speeds well beyond the visible edge of the galactic disk.

Unlike the Galaxy's stars and dust, its dark matter forms a roughly spherical halo

To explain these nearly uniform orbital speeds in the outer parts of the Galaxy, astronomers conclude that a large amount of matter must lie outside the Sun's orbit. When this matter is included, the total mass of our Galaxy could exceed $10^{12}\ M_\odot$ or more, of which about 10% is in the form of stars. This percentage implies that our Galaxy contains roughly 200 billion stars.

These observations lead to a profound mystery. Stars, gas, and dust account for only about 10% of the Galaxy's total mass. What, then, makes up the remaining 90% of the matter in our Galaxy? Whatever it's made of, it's dark. It does not show up on photographs nor indeed in images made in any part of the electromagnetic spectrum. This unseen material, which is by far the predominant constituent of our Galaxy, is called **dark matter.** We sense its presence only through its gravitational influence on the orbits of stars and gas clouds.

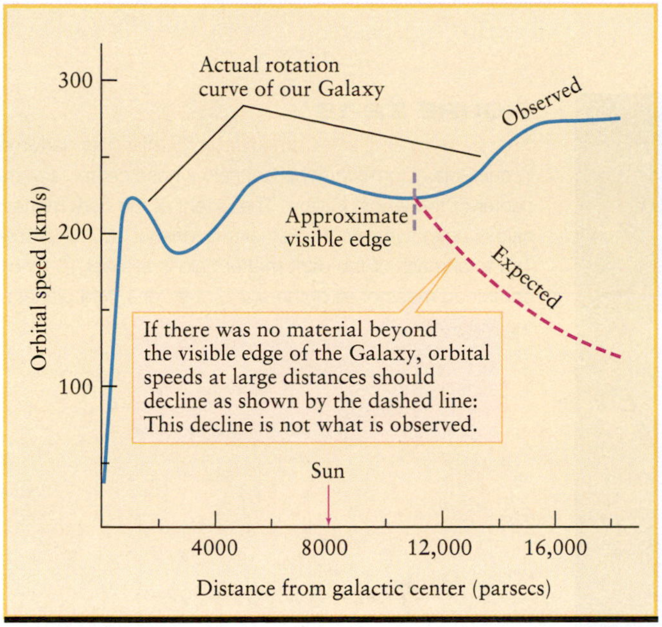
Actual rotation curve of our Galaxy
Observed
Approximate visible edge
Expected
If there was no material beyond the visible edge of the Galaxy, orbital speeds at large distances should decline as shown by the dashed line: This decline is not what is observed.
Sun
300
200
100
Orbital speed (km/s)
4000
8000
12,000
16,000
Distance from galactic center (parsecs)

FIGURE 22-18

The Galaxy's Rotation Curve The blue curve shows the orbital speeds of stars and gas in the disk of the Galaxy out to a distance of 18,000 parsecs from the galactic center. (Very few stars are found beyond this distance.) The dashed red curve indicates how this orbital speed should decline beyond the confines of most of the Galaxy's visible mass. Because there is no such decline, there must be an abundance of invisible dark matter that extends to great distances from the galactic center.

CAUTION! Be careful not to confuse dark *matter* with dark *nebulae*. A dark nebula like the one in Figure 18-4 emits no visible light, but does radiate at longer wavelengths. By contrast, no electromagnetic radiation of any kind has yet been discovered coming from dark matter.

Observations of star groupings outside the Milky Way suggest that our Galaxy's dark matter forms a spherical halo, with the galactic nucleus at its center, like the halo stars and globular clusters shown in Figure 22-7. However, the dark matter halo is much larger; it may extend to a distance of 100–200 kpc from the center of our Galaxy, some 2 to 4 times the extent of the visible halo (Figure 22-19). Analysis of the rotation curve in Figure 22-18 shows that the density of the dark matter halo decreases with increasing distance from the center of the Galaxy.

CONCEPTCHECK 22-9

How would the motion of stars at the edge of our Galaxy be different if there were no dark matter present?

Answer appears at the end of the chapter.

Dark Matter Speculations

What is the nature of this mysterious dark matter? One proposal is that the dark matter halo is composed, at least in part, of dim objects with masses less than 1 $M_\odot$. These objects, which could

BOX 22-2 TOOLS OF THE ASTRONOMER'S TRADE

Estimating the Mass Inside the Sun's Orbit

The force that keeps the Sun in orbit around the center of the Galaxy is the gravitational pull of all the matter *interior* to the Sun's orbit—this force is described by Newton's law of gravity. We can estimate the total mass of all of this matter using Newton's form of Kepler's third law (see Section 4-7 and Box 4-4):

$$P^2 = \frac{4\pi^2 a^3}{G(M + M_\odot)}$$

In this equation P is the orbital period of the Sun, a is the semimajor axis of the Sun's orbit around the galactic center, G is the universal constant of gravitation, M is the amount of mass inside the Sun's orbit, and $M_\odot$ is the mass of the Sun.

Because the Sun is only one of more than 10^{11} stars in the Galaxy, the Sun's mass is minuscule compared to M. Hence, we can safely replace the sum $M + M_\odot$ in the above equation with just M. If we now assume that the Sun's orbit is a circle, the semimajor axis a of the orbit is just the radius of this circle, which we call r. The period P of the orbit is equal to $2\pi r/v$, where v is the Sun's orbital speed. You can then show that

$$M = \frac{rv^2}{G}$$

(We leave the derivation of this equation as an exercise at the end of this chapter.)

Now we can insert known values to obtain the mass inside the Sun's orbit. Being careful to express distance in meters and speed in meters per second, we have v = 220 km/s = 2.2×10^5 m/s, $G = 6.67 \times 10^{-11}$ newton • m²/kg², and

$$r = 26{,}000 \text{ light-years} \times \frac{9.46 \times 10^{12}\text{ km}}{1 \text{ light-year}} \times \frac{10^3\text{ km}}{1\text{ km}}$$
$$= 2.5 \times 10^{20}\text{ m}$$

Hence, we find that

$$M = \frac{2.5 \times 10^{20} \times (2.2 \times 10^5)^2}{6.67 \times 10^{-11}} = 1.8 \times 10^{41}\text{ kg}$$

or, in terms of the mass of the Sun,

$$M = 1.8 \times 10^{41}\text{ kg} \times \frac{1 M_\odot}{1.99 \times 10^{30}\text{ kg}}$$
$$= 9.0 \times 10^{10} M_\odot$$

This estimate involves only mass that is interior to the Sun's orbit. Matter outside the Sun's orbit has no net gravitational effect on the Sun's motion and thus does not enter into Kepler's third law. (This is strictly true only if the matter outside our orbit is distributed over a sphere rather than a disk. In fact, the dark matter that dominates our Galaxy seems to have a spherical distribution.)

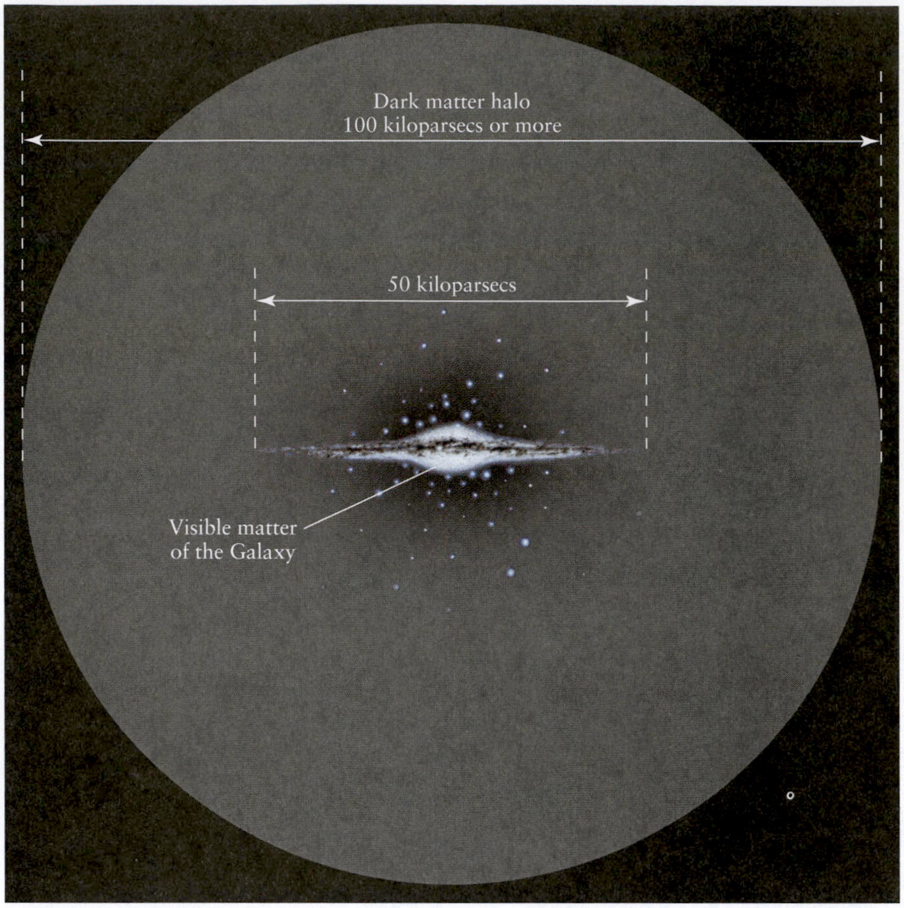

FIGURE 22-19
The Galaxy and Its Dark Matter Halo The dark matter in our Galaxy forms a spherical halo whose center is at the center of the visible Galaxy. The extent of the dark matter halo is unknown, but its diameter is at least 100 kiloparsecs. The total mass of the dark matter halo is at least 10 times the combined mass of all of the stars, dust, gas, and planets in the Milky Way.

include brown dwarfs, white dwarfs, or black holes, are called **massive compact halo objects,** or **MACHOs.** While some brown dwarfs and white dwarfs emit measurable light, it is possible that many more are simply too dim to detect, making them candidates for dark matter. Astronomers have searched for MACHOs by monitoring the light from distant stars. If a MACHO passes between us and the star, its gravity will bend the light coming from the star. (In Section 23-2 we describe how gravity can bend starlight.) As Figure 22-20 shows, the MACHO's gravity acts like a lens that focuses the light from the star. This effect, called **microlensing,** makes the star appear to brighten substantially for a few days.

Astronomers have indeed detected MACHOs in this way, but not nearly enough to solve the dark matter mystery. MACHOs with very low mass (10^{-6} to 0.1 $M_{\odot}$ each) do not appear to be a significant part of the dark matter halo. MACHOs of roughly 0.5 $M_{\odot}$ are more prevalent, but account for less than 10% of the dark matter halo.

The remainder of the dark matter is thought to be more exotic than the regular atoms (consisting of electrons, protons, and neutrons) that make up known astrophysical objects. One idea proposed for dark matter is the *neutrino*. As we saw in Section 16-4, one type of neutrino can transform into another, and these transformations can only take place if neutrinos have mass. While we do not yet know what masses the neutrinos have, we have come to understand that they are unlikely to account for halo dark matter. As we will see in Chapter 26 (Section 26-6), fast-moving neutrinos cannot "clump" enough to form a halo around our Galaxy. The remainder of the dark matter is thought to be much more exotic.

Another possibility that has been proposed is a new class of subatomic particles called **weakly interacting massive particles,** or **WIMPs.** These particles are suggested by certain theories that have not yet been confirmed experimentally. In these theories, WIMPS do not emit or absorb light (neither visible light nor any other electromagnetic radiation), but interact with regular matter through the same "weak" force that neutrinos do.

A billion or so WIMPs might be passing through you each second (compared to a trillion or so neutrinos). While dark matter makes up 90% of the matter in our Galaxy, dark matter particles, such as WIMPs, would be spread out much more than regular matter. In fact, the amount of dark matter passing through Earth at any given time would only be about 1 kg!

Physicists are attempting to detect these curious particles, which would have masses 10 to 10,000 times greater than a proton or neutron, by using a large crystal cooled almost to absolute zero. If a WIMP should enter this crystal and collide with one of its atoms, the collision will deposit a tiny but measurable amount of heat in the crystal.

Another approach for detecting WIMPs searches for the light (and electrons) that are produced after the WIMPs collide with xenon atoms. One such experiment that began collecting data in 2013 is called LUX and is at the bottom of the Homestake Mine in South Dakota (Figure 22-21). The earth above the experiment

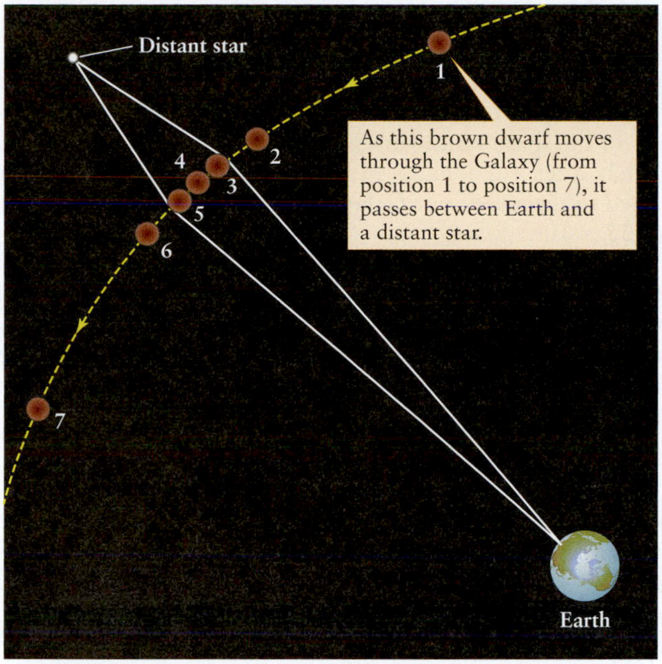

(a)

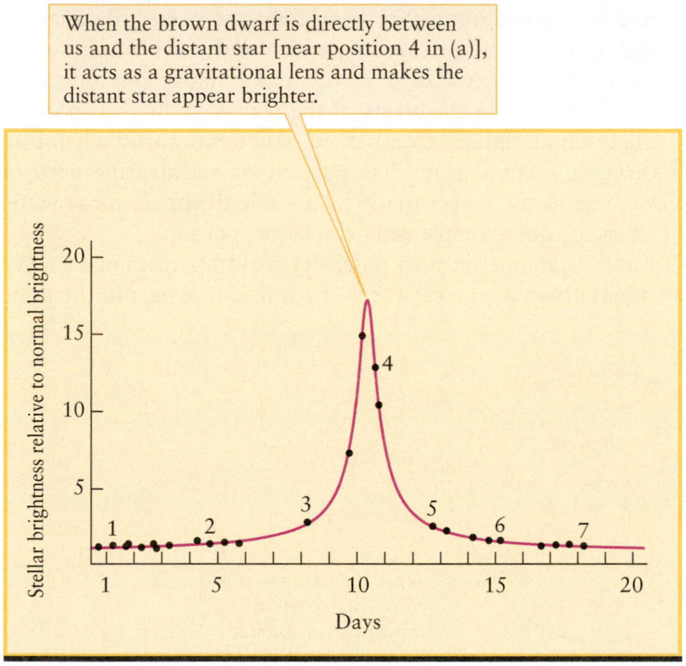

(b)

FIGURE 22-20

Microlensing by Dark Matter in the Galactic Halo **(a)** If a dense object such as a brown dwarf or black hole passes between Earth and a distant star, the gravitational curvature of space around the dense object deflects the starlight and focuses it in our direction. This effect is called microlensing. **(b)** This light curve shows the gravitational microlensing of light from a star in the Galaxy's central bulge. Astronomers do not know the nature of the object that passed between Earth and this star to cause the microlensing. (Courtesy of the MACHO and GMAN Collaborations)

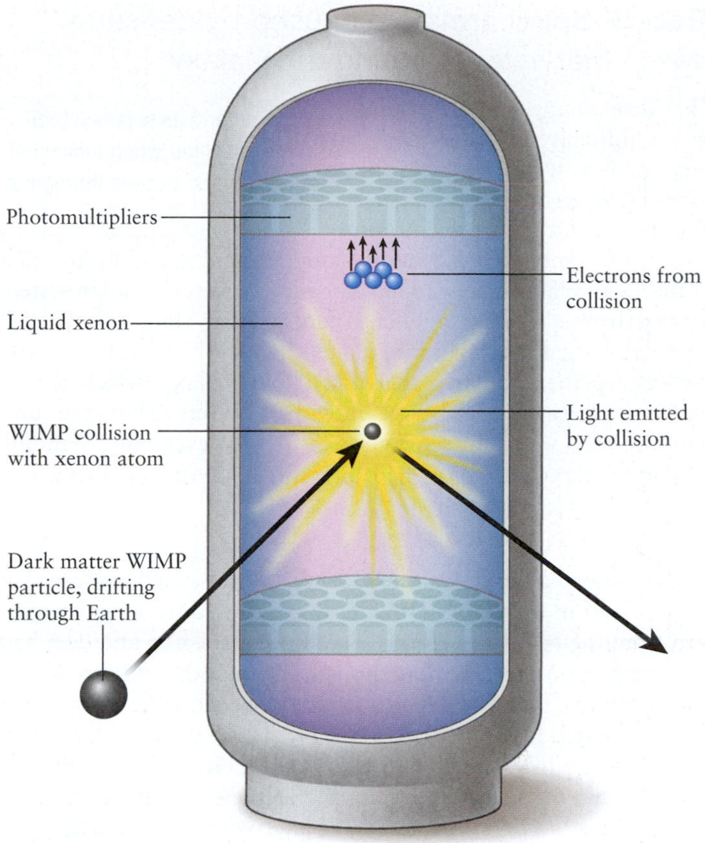

FIGURE 22-21

Detecting Dark Matter WIMPs This detector sits near the bottom of an old gold mine 4850 feet underground. If a WIMP collides with a xenon particle, the so-called weak interaction ends up producing light that can be detected by photomultipliers. Electrons are also produced in collisions to varying degrees, which helps to determine if the collisions were actually due to WIMPs, and not some other particle. While Earth shields out cosmic rays that would otherwise overwhelm the many fewer WIMP detections, the tank shown here is also surrounded by 72,000 gallons of water that helps to shield neutrons (also a source of noise) that are emitted by natural radioactivity from the surrounding rock.

shields out unwanted particles coming in from space, while the WIMPs should easily pass through toward the xenon-filled detector. With WIMPs as the main hypothesis for dark matter, even larger experiments are being built for their discovery.

As yet, the true nature of dark matter remains a mystery. Furthermore, this mystery is not confined to our own Galaxy. In Chapter 23 we will find that other galaxies have the same sort of rotation curve as in Figure 22-18, indicating that they also contain vast amounts of dark matter. Indeed, dark matter appears to make up most of the mass in our universe. Hence, dark matter is one of the most important unsolved problems in physics and astronomy.

CONCEPTCHECK 22-10

If MACHOs have been observed but WIMPs have not, why do you suppose scientists think that most of dark matter might come from WIMPs?

Answer appears at the end of the chapter.

22-5 Spiral arms are caused by density waves that sweep around the Galaxy

Our Sun and its solar system were spawned when a cloud of gas and dust passed through a spiral arm

The disk shape of our Galaxy is not difficult to understand. In Section 8-4 we described what happens when a large number of objects are put into orbit around a common center: Over time the objects tend naturally to orbit in the same plane. This is what happened when our solar system formed from the solar nebula. There a giant cloud of material eventually organized itself into planets, all of which orbit in nearly the same plane. Similarly, the disk of our Galaxy, which is also made up of a large number of individual objects orbiting a common center, is very flat (see Figure 22-7). However, understanding why our Galaxy has spiral arms presents more of a challenge.

The Winding Dilemma

One early explanation for the Galaxy's spiral structure was that the material in the Galaxy somehow condensed into a spiral pattern from the very start. In this view, once stars, gas, and dust had become concentrated within the spiral arms, the pattern would remain fixed. This fixed pattern would be possible only if the Galaxy rotated like a solid disk (see Figure 22-17b); the fixed pattern would be like the spokes on a rotating bicycle wheel. But the reality is that the Galaxy is not a solid disk. As we have seen, stars, gas, and dust all orbit the galactic center with approximately the same speed, as shown in Figure 22-17a. Let us see why this makes it impossible for a rigid spiral pattern to persist.

Imagine four stars, *A, B, C,* and *D,* that originally lie on a line extending outward from the galactic center (Figure 22-22a). In a given amount of time, each of the stars travels the same distance around its orbit. But because the innermost star has a smaller orbit than the others, it takes less time to complete one orbit. As a result, a line connecting the four stars is soon bent into a spiral (Figure 22-22b). Moreover, the spiral becomes tighter and tighter with the passage of time (Figures 22-22c and 22-22d). This "winding up" of the spiral arms causes the spiral structure to disappear completely after a few hundred million years—a very brief time compared to the age of our Galaxy, thought to be about 13.5 billion (1.35×10^{10}) years.

Figure 22-22 suggests that the Milky Way's spiral arms ought to have disappeared by now. The fact that they have not is called the **winding dilemma.** It shows that the spiral arms cannot simply be assemblages of stars and interstellar matter that travel around the Galaxy together, like a troop of soldiers marching in formation around a flagpole. What, then, can the spiral arms be?

The Density-Wave Model

In the 1940s, the Swedish astronomer Bertil Lindblad proposed that the spiral arms of a galaxy are actually a pattern that moves through the Galaxy like ripples on water. This idea was elaborated on greatly in the 1960s by the American astronomers Chia Chiao Lin and Frank Shu. In this picture, spiral arms are a kind of wave, like the waves that move across the surface of a pond when you toss a stone into the water. Water molecules pile up at a crest of the wave but spread out again when the crest passes. By analogy, Lindblad, Lin, and Shu pictured a pattern of **density waves** sweeping around the Galaxy. These waves make matter pile up in the spiral arms, which are the crests of the waves. Individual parts of the Galaxy's material are compressed only temporarily when they pass through a spiral arm. The pattern of spiral arms persists, however, just as the waves made by a stone dropped in the water can persist for quite awhile after the stone has sunk.

To understand better how a density wave operates in a galaxy, think again about a water wave in a pond. If one part of the pond

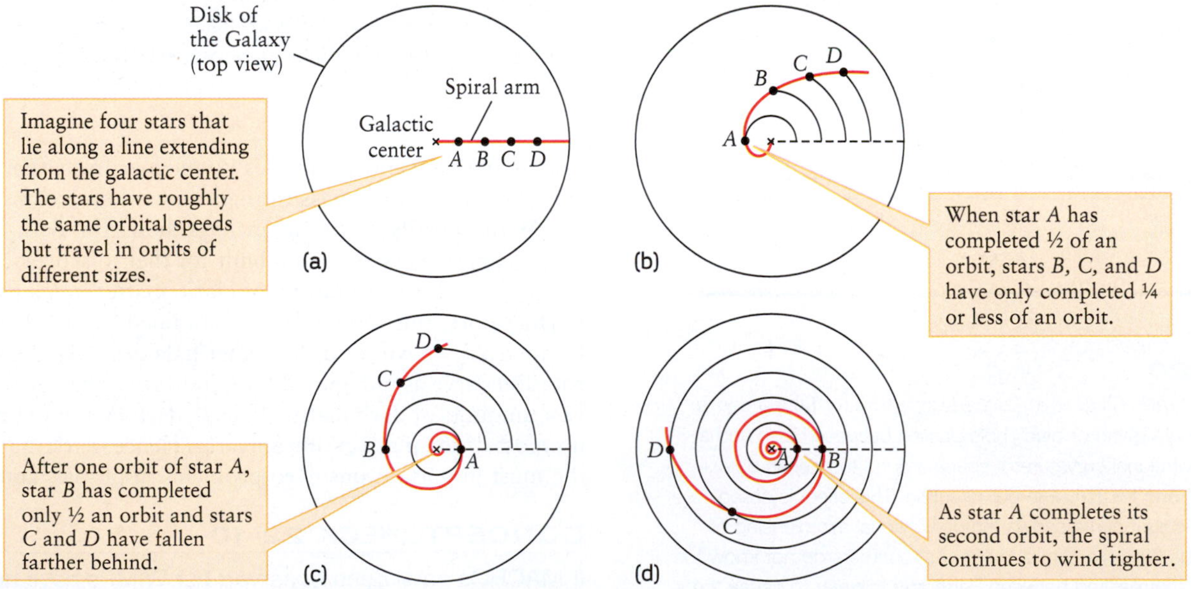

Disk of the Galaxy (top view)
Spiral arm
Galactic center
A B C D
Imagine four stars that lie along a line extending from the galactic center. The stars have roughly the same orbital speeds but travel in orbits of different sizes.
(a)
B C D
A
(b)
When star A has completed ½ of an orbit, stars B, C, and D have only completed ¼ or less of an orbit.
D
C
B
A
After one orbit of star A, star B has completed only ½ an orbit and stars C and D have fallen farther behind.
(c)
D
A
B
C
(d)
As star A completes its second orbit, the spiral continues to wind tighter.

FIGURE 22-22
The Winding Dilemma This series of drawings shows that spiral arms in galaxies like the Milky Way cannot simply be assemblages of stars. If they were, the spiral arms would "wind up" and disappear in just a few hundred million years.

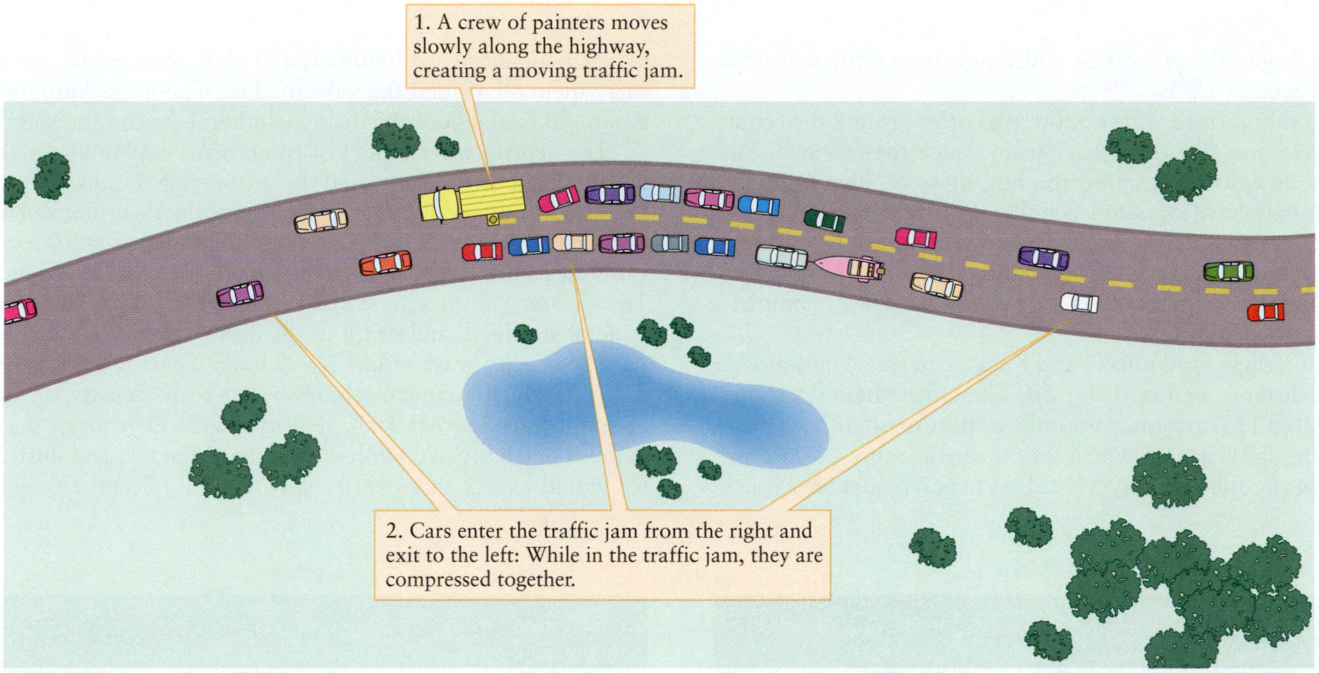
1. A crew of painters moves slowly along the highway, creating a moving traffic jam.
2. Cars enter the traffic jam from the right and exit to the left: While in the traffic jam, they are compressed together.

FIGURE 22-23

A Density Wave on the Highway A density wave in a spiral galaxy is analogous to a crew of painters moving slowly along the highway, creating a moving traffic jam. Like such a traffic jam, a density wave in a spiral galaxy is a slow-moving region where stars, gas, and dust are more densely packed than in the rest of the galaxy. As the material of the galaxy passes through the density wave, it is compressed. This triggers star formation, as Figure 22-24 shows.

is disturbed by dropping a stone into it, the molecules in that part will be displaced a bit. They will nudge the molecules next to them, causing those molecules to be displaced and to nudge the molecules beyond them. In this way the wave disturbance spreads throughout the pond.

In a galaxy, stars play the role of water molecules. Although stars and interstellar clouds of gas and dust are separated by vast distances, they can nonetheless exert forces on each other because they are affected by each other's gravity. If a region of above-average density should form, its gravitational attraction will draw nearby material into it. The displacement of this material will change the gravitational force that it exerts on other parts of the galaxy, causing additional displacements. In this way a spiral-shaped density wave can travel around the disk of a galaxy.

ANALOGY A key feature of density waves is that they move more slowly around a galaxy than do stars or interstellar matter. To visualize this movement, imagine workers painting a line down a busy freeway (Figure 22-23). The cars normally cruise along the freeway at high speed, but the crew of painters is moving much more slowly. When the cars come up on the painters, they must slow down temporarily to avoid hitting anyone. As seen from the air, cars are jammed together around the painters. An individual car spends only a few moments in the traffic jam before resuming its usual speed, but the traffic jam itself lasts all day, inching its way along the road as the painters advance.

A similar crowding takes place when the matter between the stars, called the interstellar medium, enters a spiral arm. This crowding plays a key role in the formation of stars and the recycling of the interstellar medium. As interstellar gas and dust moves through a spiral arm, it is compressed into new nebulae (Figure 22-24). This

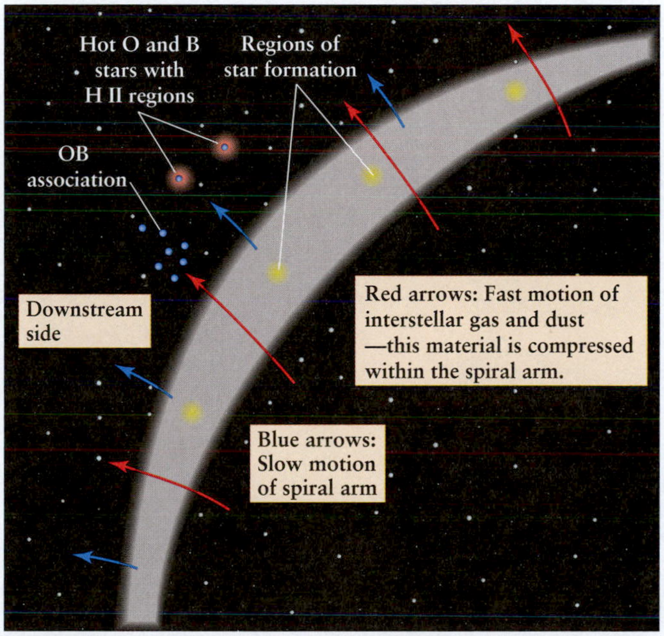
Hot O and B stars with H II regions
Regions of star formation
OB association
Downstream side
Red arrows: Fast motion of interstellar gas and dust —this material is compressed within the spiral arm.
Blue arrows: Slow motion of spiral arm

FIGURE 22-24

Star Formation in the Density-Wave Model A spiral arm is a region where the density of material is higher than in the surrounding parts of a galaxy. Interstellar matter moves around the galactic center rapidly (shown by the red arrows) and is compressed as it passes through the slow-moving spiral arms (whose motion is shown by the blue arrows). This compression triggers star formation in the interstellar matter, so that new stars appear on the "downstream" side of the densest part of the spiral arms.

compression begins the process by which new stars form, which we described in Section 18-3.

These freshly formed stars continue to orbit around the center of their galaxy, just like the matter from which they formed. The most luminous among these are the hot, massive, blue O and B stars, which may have emission nebulae (H II regions) associated with them. These stars have main-sequence lifetimes of only 3 to 15 million years (see Table 19-1), which is very short compared to the 220 million years required for the Sun to make a complete orbit around the Galaxy.

As a result, these luminous O and B stars can travel only a relatively short distance before dying off. Therefore, these stars, and their associated H II regions, are only seen in or slightly "downstream" of the spiral arm in which they formed. Figure 22-25 illustrates this for the spiral galaxy M51. Less massive stars have much longer main-sequence lifetimes, and thus their orbits are able to carry them all around the galactic disk. These less-luminous stars are found throughout the disk, including between the spiral arms.

The density-wave model of spiral arms explains why the disk of our Galaxy is dominated by metal-rich Population I stars. Because the material left over from the death of ancient stars is enriched in heavy elements, new generations of stars formed in spiral arms are likely to be more metal-rich than their ancestors. The *Cosmic Connections: Stars in the Milky Way* illustrates this cycle of star birth and death in the disk of our Galaxy.

The density-wave model is still under development. One problem is finding a driving mechanism that keeps density waves going in spiral galaxies. After all, density waves expend an enormous amount of energy to compress the interstellar gas and dust. Hence, we would expect that density waves should eventually die away,

The densest part of this spiral arm (indicated by the presence of dust, shown in red) has not yet moved past Star A...
...but some of the recently formed bright blue stars in this spiral arm have already moved past Star A.
Star A
Rotation of M51
Star B
...but some of the recently formed bright blue stars in this spiral arm have already moved past Star B.
Rotation of M51
The densest part of this spiral arm (indicated by the presence of dust shown in red) has not yet moved past Star B...

(a) An infrared view of M51 shows the locations of dust

(b) A visible-light view of M51 shows the locations of young stars

FIGURE 22-25 R **I** V U X G / R I **V** U X G

Star Formation in the Whirlpool Galaxy The spiral galaxy M51 (called the Whirlpool) is a real-life example of the density-wave model illustrated in Figure 22-24. **(a)** This infrared image shows where dust has piled up as the material within M51 passes through its spiral arms. Radio images of M51 show that hydrogen gas also piles up in the same locations, thus beginning the formation of new stars. **(b)** By the time stars complete their formation process, their motion around the galaxy has swept them "downstream" of the positions of greatest dust density, just as depicted in Figure 22-24. (a: NASA, JPL-Caltech, and R. Kennicutt [Univ. of Arizona]; b: DSS)

Stars in the Milky Way

Different populations of stars are found in different neighborhoods of our home galaxy. (The galaxy shown here is another spiral galaxy similar to our own.) The variations from one galactic region to another are due to the presence or absence of ongoing star formation.

1. All the stars, gas clouds, and dust particles within the disk orbit the center of the Galaxy. They are held in orbit by the combined gravitational effects of the other parts of the Galaxy.
Orbital motion
2. Dark dust lanes indicate where spiral density waves cause material to pile up as it orbits the galactic center. Gas is compressed as it passes through these regions, triggering the formation of new stars.
3. Hot, blue, massive O and B stars begin their formation in the dark dust lanes, then appear fully formed some distance "downstream" of the dust lanes. These stars have short main-sequence lifetimes, so they die out before they travel too far around the disk from their birthplace.
4. Most of the stars produced in the dust lanes are relatively faint, with 1 solar mass or less. These long-lived stars complete many round trips around the Galaxy during a main-sequence lifetime. (Our Sun has made more than 20 round trips in the 4.56 billion years since it formed.)
5. As a star in the disk evolves, it produces metals. Much of a star's metal-enriched material is returned to the interstellar medium when it dies. When this material is next compressed by a density wave, it is incorporated into a new generation of stars. Thus the disk contains many metal-rich stars of Population I. Our Sun is one example.
6. The density waves do not penetrate to the center of the Galaxy, so new stars are not presently forming there. The blue O and B stars have long since died out, so the bulge has a yellowish color. It contains both old Population I stars and even older Population II stars (metal-poor stars from the first generation of stars to form in the Galaxy).
R I V U X G
7. Also orbiting the Galaxy are the globular clusters. These star clusters have no mechanism to trigger star formation, so they contain only very old, metal-poor Population II stars. Studies of these clusters suggest that the first stars formed in them about 13.6 billion years ago—a mere 100 million years after the Big Bang.
8. Not visible to any telescope is the Galaxy's dark matter, which emits no electromagnetic radiation of any kind. The spherical halo of dark matter that envelops our Galaxy has at least 10 times the combined mass of all of the Galaxy's stars, planets, gas, and dust. Its fundamental nature remains a mystery.

just as do ripples on a pond. The American astronomers Debra and Bruce Elmegreen have suggested that gravity can supply that needed energy. As mentioned in Section 22-3, the central bulge of our Galaxy is elongated into a bar shape, much like the central bulge of the galaxy M83 (see Figure 22-15a and Figure 22-16a). The asymmetric gravitational field of such a bar pulls on the stars and interstellar matter of a galaxy to generate density waves. Another factor that may help to generate and sustain spiral arms is the gravitational interactions *between* galaxies. We will discuss this in Chapter 23.

CONCEPTCHECK 22-11

If our Galaxy did not have density waves, how would it be different?

Answer appears at the end of the chapter.

The Self-Propagating Star-Formation Model

Spiral density waves may not be the whole story behind spiral arms in our Galaxy and other galaxies. The reason is that spiral density waves should produce very well defined spiral arms. We do indeed see many so-called **grand-design spiral galaxies** (Figure 22-26a), with thin, graceful, and well-defined spiral arms. But in some galaxies, called **flocculent spiral galaxies** (Figure 22-26b), the spiral arms are broad, fuzzy, chaotic, and poorly defined. ("Flocculent" means "resembling wool.")

To explain such flocculent spirals, M. W. Mueller and W. David Arnett in 1976 proposed a theory of **self-propagating star formation.** Imagine that star formation begins in a dense interstellar cloud within the disk of a galaxy that does not yet have spiral arms. As soon as hot, massive stars form, their radiation and stellar winds compress nearby matter, triggering the formation of additional stars in that gas. When massive stars become supernovae, they produce shock waves that further compress the surrounding interstellar medium, thus encouraging still more star formation (see Figure 18-25).

Although all parts of this broad, star-forming region have approximately the same orbital speed about the galaxy's center, the inner regions have a shorter distance to travel to complete one orbit than the outer regions. As a result, the inner edges of the star-forming region move ahead of the outer edges as the galaxy rotates. The bright O and B stars and their nearby glowing nebulae soon become stretched out, forming what appears as a segment of a spiral arm. These spiral arm segments come and go essentially at random across a galaxy. Bits and pieces of spiral arms appear where star formation has recently begun but fade and disappear at other locations where all the massive stars have died off. Self-propagating star formation therefore tends to produce flocculent spiral galaxies that have a chaotic appearance with poorly defined spiral arms, like the galaxy in Figure 22-26b.

The two theories presented here are very different in character. In the density-wave model, star formation is caused by the spiral arms; in the self-propagating star formation model, by contrast, the spiral arms are caused by star formation. A complete and correct description of spiral arms in our Galaxy remains a topic of active research.

CONCEPTCHECK 22-12

If astronomers learned that the earliest galaxies in the universe were flocculent spiral galaxies, what would this suggest about which came first—stars or spiral arms?

Answer appears at the end of the chapter.

(a) Grand-design spiral galaxy

(b) Flocculent spiral galaxy

FIGURE 22-26 R I V U X G

Variety in Spiral Arms The differences from one spiral galaxy to another suggest that more than one process can create spiral arms. **(a)** NGC 628 is a grand-design spiral galaxy with thin, well-defined spiral arms. **(b)** NGC 4414 is a flocculent spiral galaxy with fuzzy, broken, and poorly defined spiral arms. (a: Gemini Observatory—GMOS Team; b: Olivier Vallejo [Observatoire de Bordeaux], HST, ESA, NASA)

IV Galaxies and Cosmology

This Hubble Ultra Deep Field image shows the distant background galaxies of our universe with more sensitivity than any other image. To accomplish this, the Hubble Space Telescope was pointed at a patch of space with fewer stars in our own Galaxy or obscuring nearby galaxies. While there are some stars in this image, every object that doesn't look perfectly round is a galaxy. Because it takes so long for the light to reach us from these distant galaxies, looking far away is also looking back in time. While our universe is 13.7 billion years old, some of these galaxies are only about half a billion years old. If a 75-year-old person looked back that far in time, they would be seeing themselves before their third birthday. (NASA)

R I **V** U X G

The Whirlpool Galaxy has well-defined spiral arms and a small companion galaxy on its right. These galaxies are about 20 million light-years away and are bright and large enough to be viewed with binoculars. The small red patches are typically a few hundred light-years across: Ultraviolet light from hotter stars kicks electrons off from hydrogen atoms, and as the electrons recombine with the hydrogen, the atoms emit red light. Computer modeling carried out to account for the speed and distance of the companion relative to the larger galaxy suggests that the companion may have passed right through the spiral plane of the Whirlpool. Rather than disturb the spiral arms, gravitational interactions with the companion may have given the arms their distinctive structure. (NASA)

R I **V** U X G

22-6 Infrared, radio, X-ray, and gamma-ray observations are used to probe the galactic center

TUTORIAL 22-2

At the very center of the Milky Way Galaxy is a maelstrom of activity centered on a supermassive black hole

The innermost region of our Galaxy is an active, crowded place. If you lived on a planet near the galactic center, you could see a million stars as bright as Sirius, the brightest single star in our own night sky. The total intensity of starlight from all those nearby stars would be equivalent to 200 of our full moons. In effect, night would never really fall on a planet near the center of the Milky Way. At the center of this empire of light, however, lies the darkest of all objects in the universe—a black hole millions of times more massive than the Sun.

Sagittarius A*: Heart of Darkness

Because of the severe interstellar absorption at visual wavelengths, most of our information about the galactic center comes from infrared and radio observations. Figure 22-27 shows three infrared views of the center of our Galaxy. Figure 22-27a is a wide-angle view covering a 50° segment of the Milky Way from Sagittarius through Scorpius. (The top photograph that opens this chapter shows this same region at visible wavelengths, viewed at a different angle.) The prominent reddish band through the center of this false-color infrared image is a layer of dust in the plane of the Galaxy. Figure 22-27b is an IRAS view of the galactic center. It is surrounded by numerous streamers of dust (shown in blue). The strongest infrared emission (shown in white) comes from sources that also emit radio waves. One of these sources, **Sagittarius A*** (say "A star"), lies at the very center of the Galaxy. (Its position, pinpointed with simultaneous observations by radio telescopes scattered around the world, seems to be very near the gravitational center of the Galaxy.) The high-resolution infrared view in Figure 22-27c, made using adaptive optics (see Section 6-3), shows hundreds of stars crowded within 1 ly (0.3 pc) of Sagittarius A*. Compare this region to our region of the Galaxy, where the average distance between stars is more than a light-year.

Sagittarius A* itself does not appear in infrared images. Nonetheless, astronomers have used infrared observations to make truly startling discoveries about this object. Since the 1990s, two research groups—one headed by Reinhard Genzel of the Max Planck Institute for Extraterrestrial Physics in Garching, Germany, and another led by Andrea Ghez at the University of California, Los Angeles—have been using infrared detectors to monitor the motions of stars in the immediate vicinity of Sagittarius A*. They have found a number of stars orbiting Sagittarius A* at speeds in excess of 1500 km/s (Figure 22-28). (By comparison, Earth orbits the Sun at a lackadaisical 30 km/s.) In 2000 the UCLA group observed one such star, called S0-16, as its elliptical orbit brought it within a mere 45 AU from Sagittarius A* (1½ times the distance from the Sun to Neptune). At its closest approach, S0-16 was traveling at a breathtaking speed of 12,000 km/s, or 4% of the speed of light!

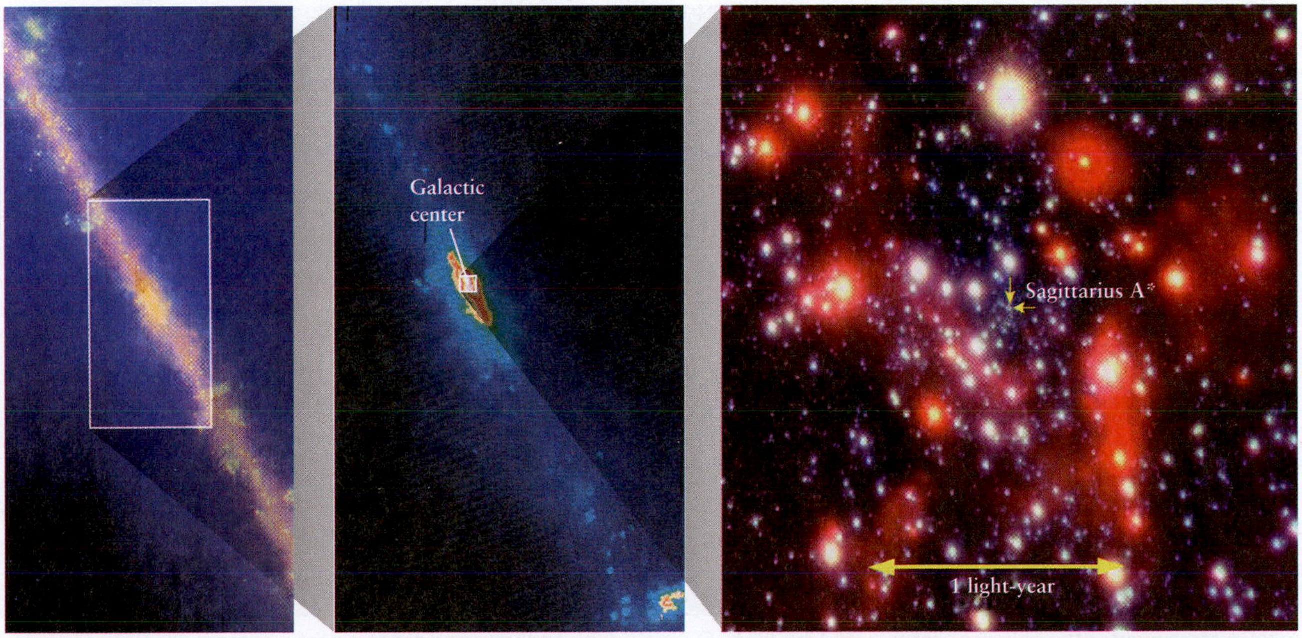

(a) A wide-angle (50°) infrared view

(b) A close-up view shows a more luminous region at the galactic center

(c) An extreme close-up view centered on Sagittarius A*, a radio source at the very center of the Milky Way Galaxy, shows hundreds of stars within 1 ly (0.3 pc)

FIGURE 22-27 R I V U X G

The Galactic Center **(a)** In this false-color infrared image, the reddish band is dust in the plane of the Galaxy and the fainter bluish blobs are interstellar clouds heated by young O and B stars. **(b)** This close-up infrared view covers the area outlined by the white rectangle in (a). **(c)** Adaptive optics reveals stars densely packed around the galactic center. (a, b: NASA; c: European Southern Observatory)

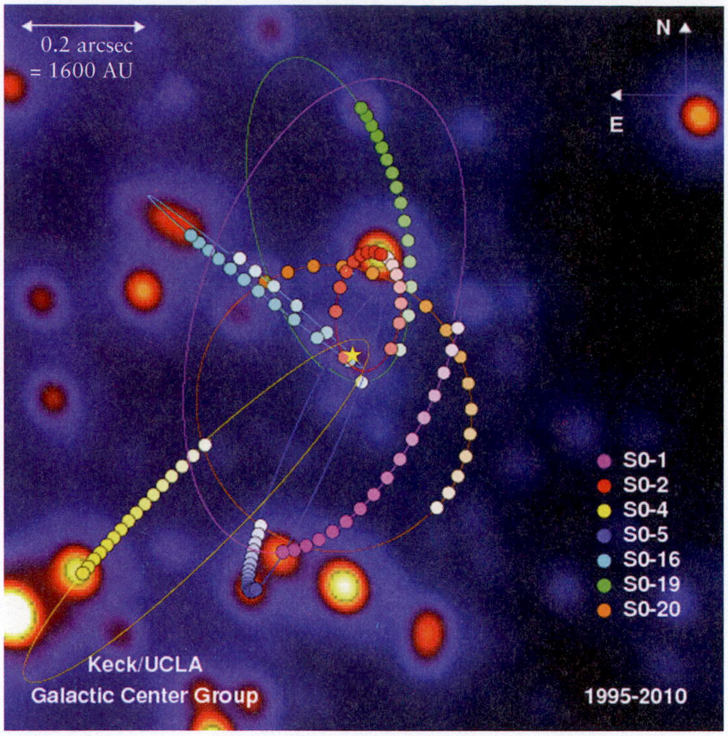

FIGURE 22-28 R I V U X G

Stars Orbiting Sagittarius A* The colored dots superimposed on this infrared image show the motion of seven stars in the vicinity of the unseen massive object (denoted by the yellow five-pointed star) at the position of the radio source Sagittarius A*. The orbits were measured over an 15-year period. Analysis of the orbits indicates that the stars are held in orbit by a black hole of 4.1 million solar masses. The blue dots for S0-16 show this star reached 4% the speed of light at its closest approach to the black hole. (Keck/UCLA Galactic Center Group)

In order to keep stars like S0-16 in such small, rapid orbits, Sagittarius A* must exert a powerful gravitational force and hence must be very massive. By applying Newton's form of Kepler's third law to the motions of these stars around Sagittarius A*, the UCLA group calculates the mass of Sagittarius A* to be a remarkable 4.1 *million* solar masses (4.1×10^6 $M_\odot$). Furthermore, the small separation between S0-16 and Sagittarius A* at closest approach shows that Sagittarius A* can be no more than 45 AU in radius. An object this massive and this compact can only be one thing—a supermassive black hole (see Section 22-4).

CONCEPTCHECK 22-13

How would stars around Sagittarius A* be moving if a supermassive black hole were not found at the galactic center?

Answer appears at the end of the chapter.

X-rays from Around a Supermassive Black Hole

Evidence in favor of this picture comes from the Chandra X-ray Observatory, which has observed X-ray flares coming from Sagittarius A*. The flares brighten dramatically over the

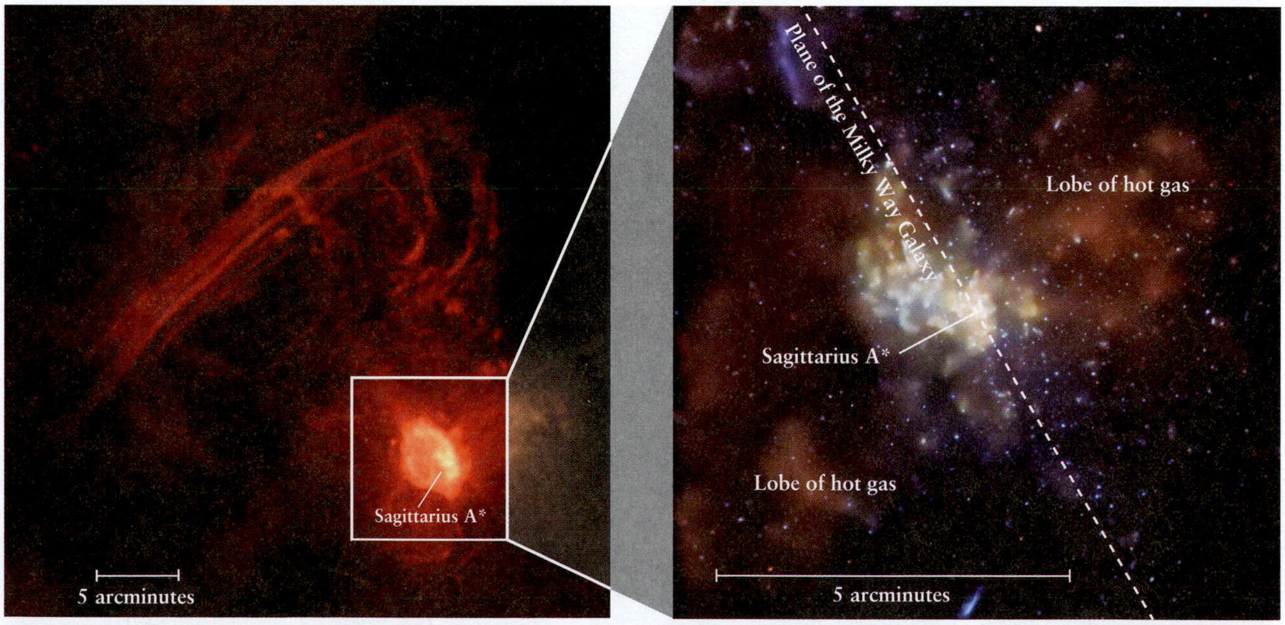

(a) A radio view of the galactic center R I V U X G

(b) An X-ray view of the galactic center R I V U X G

FIGURE 22-29

The Energetic Center of the Galaxy **(a)** The area shown in this radio image has the same angular size as the full moon. Sagittarius A*, at the very center of the Galaxy, is one of the brightest radio sources in the sky. Magnetic fields shape nearby interstellar gas into immense, graceful arches. **(b)** This composite of images at X-ray wavelengths from 0.16 to 0.62 nm shows lobes of gas on either side of Sagittarius A*. The character of the X-ray emission shows that the gas temperature is as high as 2×10^7 K. (VLA, F. Sadeh et al./NRAO; b: F. K. Baganoff et al./CXC/MIT/NASA) space of just 10 minutes, which shows that the size of the flare's source can be no larger than the distance that light travels in 10 minutes. (We used a similar argument in Section 21-3 to show that the flickering X-ray source Cygnus X-1 must be very small.) In 10 minutes light travels a distance of 1.8×10^8 km or 1.2 AU, and only a black hole could pack a mass of $4.1 \times 10^6\ M_\odot$ into a volume that size or smaller. The X-ray flares were presumably emitted by blobs of material that were compressed and heated as they fell into the black hole (see Section 21-3).

The X-ray flares from Sagittarius A* are relatively feeble, which suggests that the supermassive black hole is swallowing only relatively small amounts of material. But the region around Sagittarius A* is nonetheless an active and dynamic place. Figure 22-29a is a wide-angle radio image of the galactic center covering an area more than 60 pc (200 ly) across. Huge filaments of gas stretch for 20 pc (65 ly) northward of the galactic center (to the right and upward in Figure 22-29a), then abruptly arch southward (down and to the left in the figure). The orderly arrangement of these filaments is reminiscent of prominences on the Sun (see Section 18-10, especially Figure 18-28). This suggests that, as on the Sun, there is ionized gas at the galactic center that is being controlled by a powerful magnetic field. Indeed, much of the radio emission from the galactic center is synchrotron radiation: As we saw in Section 22-4, such radiation is produced by high-energy electrons spiraling in a magnetic field.

The false-color X-ray image in Figure 22-29b shows the immediate vicinity of Sagittarius A*. The black hole is flanked by lobes of hot, ionized, X-ray–emitting gas that extend for dozens of light-years. These lobes are thought to be the relics of immense explosions that may have taken place over the past several thousand years. Perhaps these past explosions cleared away much of the material around Sagittarius A*, leaving only small amounts to fall into the black hole. This could explain why the X-ray flares from Sagittarius A* are so weak.

There is nonetheless evidence that Sagittarius A* has been more active in the recent past. Between 2002 and 2005, Michael Muno and his colleagues at the California Institute of Technology observed that certain nebulae within about 50 ly of Sagittarius A* suddenly became very bright at X-ray wavelengths. The character of the X-ray light was such that it could not have originated from the nebulae themselves. Muno and colleagues concluded that X-rays emitted from Sagittarius A* about 50 years earlier had struck and excited the nebulae, causing the nebulae to glow intensely at X-ray wavelengths. To produce such an intense X-ray glow, an object the size of the planet Mercury must have fallen into the supermassive black hole.

CONCEPTCHECK 22-14

Why would the sudden brightening of a nebula near the galactic center suggest the presence of a supermassive black hole?

Answer appears at the end of the chapter.

Gamma Rays Offer Clues to Possible Black Hole Activity

In 2010, astronomers with the Fermi Gamma-ray Space Telescope (FGST) reported seeing enormous gamma-ray "bubbles" above and below the plane of our Galaxy (Figure 22-30). The gamma rays are produced when high-speed electrons collide with infrared or visible photons and bump the photon energies up to gamma-ray levels. What remains unclear is what created the high-speed electrons. One idea involves a hot "wind" of electrons and other particles blown away from an accretion disk of matter (see Figure 21-11) around the black hole at our Galaxy's center.

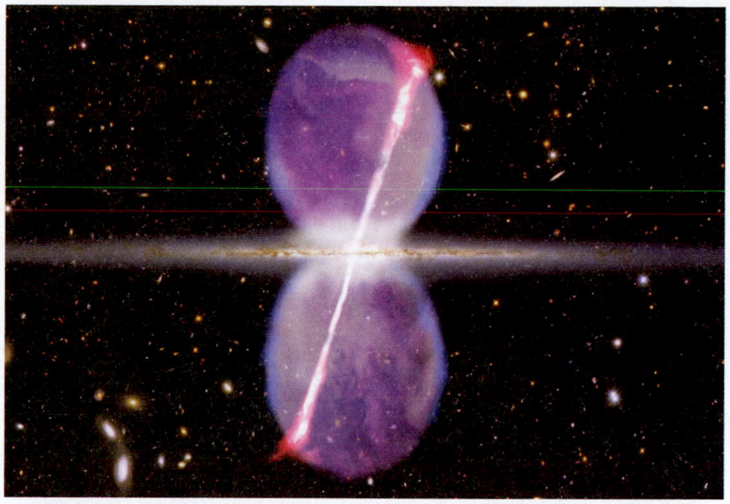

FIGURE 22-30
Galactic Gamma-Ray Bubbles This is an artist's illustration of giant gamma-ray "bubbles" seen above and below our galactic plane, possibly due to the black hole at the galactic center. Each of the bubbles extends about 25,000 light-years away from the plane. The bubbles are estimated to be a few million years old. The gamma ray data also contain hints of linear jetlike features, which are sometimes found around black holes, but the existence of the jets is far from clear. If the jets are real, producing them might have required the black hole to consume a giant hydrogen cloud of about 10,000 solar masses, with only a small portion escaping into the jets. (David A. Aguilar [CfA])

It is too early to say if the gamma-ray bubbles are caused by a galactic black hole. However, if these interpretations hold up, they certainly point to interesting behavior by the black hole at the center of our Galaxy. As intriguing as it is, the supermassive black hole at the center of our Galaxy is not unique. Observations show that such titanic black holes are a feature of most large galaxies. In Chapter 24 we will see how black holes of this kind power *quasars*, the most luminous sustained light sources in the cosmos.

KEY WORDS

Word preceded by an asterisk (*) is discussed in Box 22-1.

central bulge (of a galaxy), p. 640
dark matter, p. 648
density wave, p. 652
disk (of a galaxy), p. 640
far-infrared, p. 639
flocculent spiral galaxy, p. 656
galactic nucleus, p. 638
galaxy, p. 636
globular cluster, p. 637
grand-design spiral galaxy, p. 656
H I, p. 6420
halo (of a galaxy), p. 640
interstellar extinction, p. 637
interstellar matter, p. 636
Local Bubble, p. 643
*magnetic resonance imaging (MRI), p. 644

massive compact halo object (MACHO), p. 650
microlensing, p. 650
Milky Way Galaxy, p. 636
near-infrared, p. 639
rotation curve, p. 648
RR Lyrae variable, p. 638
Sagittarius A*, p. 657
self-propagating star formation, p. 656
spin (of a particle), p. 642
spin-flip transition, p. 643
spiral arm, p. 642
21-cm radio emission, p. 643
weakly interacting massive particle (WIMP), p. 650
winding dilemma, p. 652

KEY IDEAS

The Shape and Size of the Galaxy: Our Galaxy has a disk about 50 kpc (160,000 ly) in diameter and about 600 pc (2000 ly) thick, with a high concentration of interstellar dust and gas in the disk.

• The galactic center is surrounded by a large distribution of stars called the central bulge. This bulge is not perfectly symmetrical, but may have a bar shape.

• The disk of the Galaxy is surrounded by a spherical distribution of globular clusters and old stars, called the galactic halo.

• There are about 200 billion (2×10^{11}) stars in the Galaxy's disk, central bulge, and halo.

The Sun's Location in the Galaxy: Our Sun lies within the galactic disk, some 8000 pc (26,000 ly) from the center of the Galaxy.

• Interstellar dust obscures our view at visible wavelengths along lines of sight that lie in the plane of the galactic disk. As a result, the Sun's location in the Galaxy was unknown for many years. This dilemma was resolved by observing parts of the Galaxy outside the disk.

• The Sun orbits the center of the Galaxy at a speed of about 790,000 km/h. It takes about 220 million years to complete one orbit.

The Rotation of the Galaxy and Dark Matter: From studies of the rotation of the Galaxy, astronomers estimate that the total mass of the Galaxy is about $10^{12}\ M_\odot$. Only about 10% of this mass is in the form of visible stars, gas, and dust. The remaining 90% is in some nonvisible form, called dark matter, that extends beyond the edge of the luminous material in the Galaxy.

• Our Galaxy's dark matter contains only a small amount of MACHOs (dim, star-sized objects) and is hypothesized to consist mostly of WIMPs (relatively massive subatomic particles).

The Galaxy's Spiral Structure: OB associations, H II regions, and molecular clouds in the galactic disk outline huge spiral arms.

• Spiral arms can be traced from the positions of clouds of atomic hydrogen. These can be detected throughout the galactic disk by the 21-cm radio waves emitted by the spin-flip transition in hydrogen. These emissions easily penetrate the intervening interstellar dust.

Theories of Spiral Structure: There are two leading theories of spiral structure in galaxies.

• According to the density-wave theory, spiral arms are created by density waves that sweep around the Galaxy. The gravitational field of this spiral pattern compresses the interstellar clouds through which it passes, thereby triggering the formation of the OB associations and H II regions that illuminate the spiral arms.

• According to the theory of self-propagating star formation, spiral arms are caused by the birth of stars over an extended region in a galaxy. Differential rotation of the galaxy stretches the star-forming region into an elongated arch of stars and nebulae.

The Galactic Nucleus: The innermost part of the Galaxy, or galactic nucleus, has been studied through its radio, infrared, X-ray, and gamma-ray emissions (which are able to pass through interstellar dust).

• A strong radio source called Sagittarius A* is located at the galactic center. This marks the position of a supermassive black hole with a mass of about $4.1 \times 10^6\ M_\odot$.

QUESTIONS

Review Questions

1. Why do the stars of the Galaxy appear to form a bright band that extends around the sky?
2. TUTORIAL 22-1 How did interstellar extinction mislead astronomers into believing that we are at the center of our Galaxy?
3. How did observations of globular clusters help astronomers determine our location in the Galaxy?
4. What are RR Lyrae stars? Why are they useful for determining the distance from our solar system to the center of the Galaxy?
5. Why are infrared telescopes useful for exploring the structure of the Galaxy? Why is it important to make observations at both near-infrared and far-infrared wavelengths?
6. The galactic halo is dominated by Population II stars, whereas the galactic disk contains predominantly Population I stars. In which of these parts of the Galaxy has star formation taken place recently? Explain your answer.
7. O or B main-sequence stars are found in the galactic disk but not in globular clusters. Why is this so?
8. What must happen within a hydrogen atom for it to emit a photon of wavelength 21 cm?
9. Most interstellar hydrogen atoms emit only radio waves at a wavelength of 21 cm, but some hydrogen clouds emit profuse amounts of visible light (see, for example, Figure 18-1b and Figure 18-2). What causes this difference?
10. How do astronomers determine the distances to H I (neutral hydrogen) clouds?
11. The radio map in Figure 22-14 has a large gap on the side of the Galaxy opposite to ours. Why is this?

12. In a spiral galaxy, are stars in general concentrated in the spiral arms? Why are spiral arms so prominent in visible-light images of spiral galaxies?

13. Many old black-and-white photographs of spiral galaxies were taken using film that was most sensitive to blue light. Explain why the spiral arms were particularly prominent in such photographs.

14. What kinds of objects (other than H I clouds) do astronomers observe to map out the Galaxy's spiral structure? What is special about these objects? Which of these can be observed at great distances?

15. Why don't astronomers detect 21-cm radiation from the hydrogen in giant molecular clouds?

16. In what way are the orbits of stars in the galactic disk different from the orbits of planets in our solar system? What does this difference imply about the way that matter is distributed in the Galaxy?

17. How do astronomers determine how fast the Sun moves in its orbit around the Galaxy? How does this speed tell us about the amount of mass inside the Sun's orbit? Does this speed tell us about the amount of mass outside the Sun's orbit?

18. How do astronomers conclude that vast quantities of dark matter surround our Galaxy? How is this dark matter distributed in space?

19. Another student tells you that the Milky Way Galaxy is made up "mostly of stars." Is this statement accurate? Why or why not?

20. What is the difference between dark matter and dark nebulae?

21. What proposals have been made to explain the nature of dark matter? What experiments or observations have been made to investigate these proposals? What are the results of this research?

22. What is the winding dilemma? What does it tell us about the nature of spiral arms?

23. Do density waves form a stationary pattern in a galaxy? If not, do they move more rapidly, less rapidly, or at the same speed as stars in the disk?

24. In our Galaxy, why are stars of spectral classes O and B found only in or near the spiral arms? Is the same true for stars of other spectral classes? Explain why or why not.

25. Compare the kinds of spiral arms produced by density waves with those produced by self-propagating star formation. By examining Figure 22-16, cite evidence that both processes

26. TUTORIAL 22-2 What is the evidence that there is a supermassive black hole at the center of our Galaxy? How is it possible to determine the mass of this black hole?

27. What is the evidence that material has been falling into the supermassive black hole at the galactic center?

Advanced Questions

Questions preceded by an asterisk (*) involve topics discussed in the Boxes.

Problem-solving tips and tools

Several of the following questions make extensive use of Newton's form of Kepler's third law, and you might find it helpful to review Box 4-4. Another useful version of Kepler's third law is given in Section 17-9. Box 19-2 discusses the relationship between luminosity, apparent brightness, and distance. We discussed the relationship between the energy and wavelength of a photon in Section 5-5. According to the Pythagorean theorem, an isosceles right triangle has a hypotenuse that is longer than its sides by a factor of $\sqrt{2} = 1.414$. The formula for the Schwarzschild radius of a black hole is given in Box 21-2. You will find the small-angle formula in Box 1-1 useful. It is also helpful to remember that an object 1 AU across viewed at a distance of 1 parsec has an angular size of 1 arcsecond. Remember, too, that the volume of a cylinder is equal to its height multiplied by the area of its base, the area of a circle of radius r is πr^2, and the volume of a sphere of radius r is $4\pi r^3/3$. You can find other geometrical formulae in Appendix 8.

28. Discuss how the Milky Way would appear to us if the Sun were relocated to (**a**) the edge of the Galaxy; (**b**) the galactic halo; (**c**) the galactic bulge.

29. Explain why globular clusters spend most of their time in the galactic halo, even though their eccentric orbits take them close to the galactic center.

30. The disk of the Galaxy is about 50 kpc in diameter and 600 pc thick. (**a**) Find the volume of the disk in cubic parsecs. (**b**) Find the volume (in cubic parsecs) of a sphere 300 pc in radius centered on the Sun. (**c**) If supernovae occur randomly throughout the volume of the Galaxy, what is the probability that a given supernova will occur within 300 pc of the Sun? If there are about three supernovae each century in our Galaxy, how often, on average, should we expect to see one within 300 pc of the Sun?

*31. An RR Lyrae star whose peak luminosity is 100 $L_\odot$ is in a globular cluster. At its peak luminosity, this star appears from Earth to be only 1.47×10^{-18} as bright as the Sun. Determine the distance to this globular cluster (**a**) in AU and (**b**) in parsecs.

32. A typical hydrogen atom in interstellar space undergoes a spin-flip transition only once every 10^7 years. How, then, is it at all possible to detect the 21-cm radio emission from interstellar hydrogen?

33. Calculate the energy of the photon emitted when a hydrogen atom undergoes a spin-flip transition. How many such photons would it take to equal the energy of a single H_α photon of wavelength 656.3 nm?

34. Suppose you were to use a radio telescope to measure the Doppler shift of 21-cm radiation in the plane of the Galaxy. (**a**) If you observe 21-cm radiation from clouds of atomic hydrogen at an angle of 45° from the galactic center, you will see the highest Doppler shift from a cloud that is as far from the galactic center as it is from the Sun. Explain this statement using a diagram. (**b**) Find the distance from the Sun to the particular cloud mentioned in (a).
35. Sketch the rotation curve you would obtain if the Galaxy were rotating like a rigid body.
36. The mass of our Galaxy inside the Sun's orbit is calculated from the radius of the Sun's orbit and its orbital speed. By how much would this estimate be in error if the calculated distance to the galactic center were off by 10%? By how much would this estimate be in error if the calculated orbital velocity were off by 10%? Explain your reasoning.
37. A gas cloud located in the spiral arm of a distant galaxy is observed to have an orbital velocity of 400 km/s. If the cloud is 20,000 pc from the center of the galaxy and is moving in a circular orbit, find (**a**) the orbital period of the cloud and (**b**) the mass of the galaxy contained within the cloud's orbit.
38. According to the Galaxy's rotation curve in Figure 22-18, a star 16 kpc from the galactic center has an orbital speed of about 270 km/s. Calculate the mass within that star's orbit.
39. Speculate on the reasons for the rapid rise in the Galaxy's rotation curve (see Figure 22-18) at distances close to the galactic center.

*40. Show that the form of Kepler's third law stated in Box 22-2, $P^2 = 4\pi^2 a^3/G\,(M + M_\odot)$, is equivalent to $M = rv^2/G$, provided the orbit is a circle. (*Hint:* The mass of the Sun ($M_\odot$) is much less than the mass of the Galaxy inside the Sun's orbit (M).)

41. The accompanying image shows the spiral galaxy M74, located about 55 million light-years from Earth in the constellation Pisces (the Fish). It is actually a superposition of two false-color images. The red portion is an optical image taken at visible wavelengths, while the blue portion is an ultraviolet image made by NASA's Ultraviolet Imaging Telescope, which was carried into orbit by the space shuttle *Columbia* during the *Astro-1* mission in 1990. Compare the visible and ultraviolet images and, from what you know about stellar evolution and spiral structure, explain the differences you see.

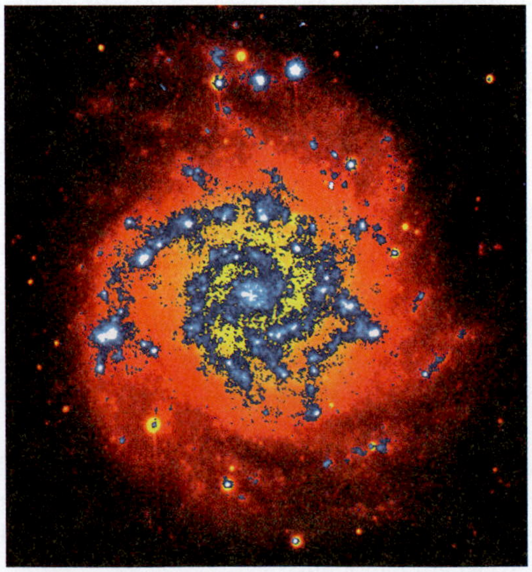

R I V U X G
(NASA, UIT)

42. The accompanying figure shows infrared images of two spiral galaxies. Explain which of these is a grand-design spiral galaxy and which is a flocculent spiral galaxy. Explain your reasoning.

(a)

(b)

R I V U X G
(NASA; JPL-Caltech; R. Kennicutt [University of Arizona]; and the SINGS Team)

*43. (**a**) Calculate the Schwarzschild radius of a supermassive black hole of mass $4.1 \times 10^6\ M_\odot$, the estimated mass of the black hole at the galactic center. Give your answer in both kilometers and astronomical units. (**b**) What is the angular diameter of such a black hole as seen at a distance of 8 kpc, the distance from Earth to the galactic center? Give your answer in arcseconds. Observing an object with such a small angular size will be a challenge indeed! (**c**) What is the angular diameter of such a black hole as seen from a distance of 45 AU, the closest that the star S0-16 comes to Sagittarius A*? Again, give your answer in arcseconds. Would it be discernible to the naked eye at that distance? (A normal human eye can see details as small as about 60 arcseconds.)

*44. (**a**) The scale bar in Figure 22-28 shows that at the distance of Sagittarius A*, a length of 1600 AU has an angular size of 0.2 arcsecond. Use this information to calculate the distance to Sagittarius A*. (**b**) The star S0-16 moves around Sagittarius A* in an elliptical orbit with semimajor axis 1680 AU. Use this and the information given in the caption to Figure 22-28 to find the orbital period of S0-16. (**c**) Given the period and the semimajor axis of the star's orbit, is it possible to calculate the mass of S0-16 itself? If it is, explain how this could be done; if not, explain why not.

45. The stars S0-2 and S0-19 orbit Sagittarius A with orbital periods of 14.5 and 37.3 years, respectively. (**a**) Assuming that the supermassive black hole in Sagittarius A* has a mass of $4.1 \times 10^6\ M_\odot$, determine the semimajor axes of the orbits of these two stars. Give your answers in AU. (**b**) Calculate the angular size of each orbit's semimajor axis as seen from Earth. (See Section 22-1 for the distance from Earth to the center of the Galaxy.) Explain why extremely high-resolution infrared images are required to observe the motions of these stars.

46. Consider a star that orbits around Sagittarius A* in a circular orbit of radius 530 AU. (**a**) If the star's orbital speed is 2500 km/s, what is its orbital period? Give your answer in years. (**b**) Determine the sum of the masses of Sagittarius A* and the star. Give your answer in solar masses. (Your answer is an estimate of the mass of Sagittarius A*, because the mass of a single star is negligibly small by comparison.

Discussion Questions

47. From what you know about stellar evolution, the interstellar medium, and the density-wave theory, explain the appearance and structure of the spiral arms of grand-design spiral galaxies.

48. What observations would you make to determine the nature of the dark matter in our Galaxy's halo?

49. Describe how the appearance of the night sky might change if dark matter were visible to our eyes.

50. Discuss how a supermassive black hole could have formed at the center of our Galaxy.

Web/eBook Questions

51. Some scientists have suggested that the rotation curve of the Galaxy can be explained by modifying the laws of physics rather than by positing the existence of dark matter. Search the World Wide Web for information about these proposed modifications, called MOND (for *MO*dified *N*ewtonian *D*ynamics). Under what circumstances do MOND and conventional physics predict different behavior? What evidence is there that MOND might be correct?

52. VIDEO 22.2 **Fast-Moving Stars at the Galactic Center.** Access and view the video "Stars Orbiting Sagittarius A*" in Chapter 22 of the *Universe* Web site or eBook. Explain how you can tell which of the stars in the video are actually close to Sagittarius A* and which just happen to lie along our line of sight to the center of the Galaxy.

ACTIVITIES

Observing Projects

53. Use the *Starry Night*™ program to explore the alignment of the Milky Way with respect to our Earth-based coordinate system and its appearance at different electromagnetic wavelengths. Select **Favourites > Explorations > Milky Way** to display a wide-field view of our spiral galaxy from the center of a transparent Earth at 6:30 A.M. on September 1, 2014. The view is similar to that seen by observers on Earth and shows an edge-on view of this galaxy in a direction toward the galactic center, represented by the marked position of a foreground star, HIP86948. (**a**) You can use this view to estimate the alignment of the galactic plane with our reference plane in the sky, the **Celestial Equator,** which is the projection of the Earth's equator on to the sky. Display the plane of the galaxy by clicking on **View > Galactic Guides > Equator** and then show the Celestial Equator by clicking on **View > Celestial Guides > Equator.** Estimate the angle between these two planes. (**b**) This visible-light image of the Milky Way shows significant structure, with many dark regions. These dark regions do not show the absence of material in these directions but the presence of dust and gas clouds obscuring the distant stars. Visible light is scattered and absorbed by these dense clouds. You can examine this galactic plane at other electromagnetic wavelengths. Open the **Options** menu and choose **Stars > Milky Way…** to display **Milky Way Options.** Use the brightness slider to show maximum brightness. Click the **Wavelength** box to expand the list of possible wavelengths and display the image of the galactic plane at each wavelength. At which wavelengths does the **Galactic Center** show up most prominently? Why do you think this region shows up brighter at these wavelengths?

54. There are two distinct populations of stars in our galaxy, the **Population I** young, hot stars, associated with dust and gas clouds in which active star formation continues, and the **Population II** older stars with no associated dust and gas. You can explore the distribution of these populations of stars by examining the distribution of clusters of stars of each population. For example, **Open Clusters** are composed of Population I stars while **Globular Clusters** consist of older

Population II stars. You can use the *Starry Night*™ program to explore the distribution of clusters in and around our galaxy. Select **Favourites > Explorations > Milky Way** to view the galaxy edge-on from the center of a transparent Earth. Before displaying the clusters, ensure that each type of cluster is assigned a different color by opening **Options > Stars > Globular Clusters...** and, in the **Globular Clusters Options** panel, change the **Outline Colour** to red and close the panel. Repeat this process to change the color of all clusters by opening **Options > Stars > Star Clusters...** and changing the **Outline Color** in the **Star Clusters Options** panel to blue. Open the **View** menu and click on **Stars > Globular Clusters** to display the distribution of these old Population II star clusters around the galactic center. (Early measurements of distances to many of these clusters, using variable stars with known intrinsic brightness as beacons, were used to determine the position of the center of the galaxy.) You can now display clusters of all types, both the **Open Clusters** of young stars (represented by dotted circles) and the older **Globular Clusters** (represented by solid circles), by selecting **View > Stars > Star Clusters** from the menu. With the above color selections for clusters, the globular clusters will change to purple and will be distinguishable from the remaining open clusters depicted in blue. (**a**) What do you notice about the distribution of the globular clusters compared to the open clusters? Use the hand tool to move along the Milky Way plane to examine the distribution of each type of cluster, and hence each population of stars, across the sky. (**b**) Note in particular the two extra concentrations of clusters off the galactic plane. Why do you think that these extra concentrations are in these positions in our sky? (*Hint:* Move the cursor over these regions to identify them.) (**c**) You can examine several of these star clusters in detail. Use the **Find** facility to look at and zoom in on the globular clusters M13, M80, M5, M2, NGC4590, and NGC7089 and the open clusters, the nearby Pleiades, NGC2264 and M7. Comment on the noted differences in the nature and structure of these different types of clusters.

55. Use the *Starry Night*™ program to measure the dimensions of the Milky Way Galaxy. Select **Favourites > Explorations > Milky Way Galaxy** to display a face-on view of a simulation of the Milky Way Galaxy from a distance of 0.128 Mly from the Earth. The position of the Sun is labeled, directly below the center of the Galaxy in the view on the screen. Note the spiral arm structure of the galaxy. (Recent research has revealed that our Galaxy appears to have a bar structure surrounding the galactic center.) (**a**) Click on the cursor selection tool to the left of the toolbar and activate the angular separation tool to measure the angular separation between the Sun and the center of the Galaxy as seen from this vantage point, and find the corresponding distance in light years (ly). (Both values are shown in the display beside the line drawn by the angular separation tool.) Convert the angular separation to a decimal number in degrees ($1° = 60' = 3600''$); then find the scale of the image of the Galaxy in ly/degree. (**b**) Change the cursor to the location scroller and use it to view the Galaxy edge-on and oriented vertically on the screen, with the Sun still below the center. To do this, place the location scroller tool at the center of the right-hand edge of the screen, hold down the mouse button (on a two-button mouse, hold down the left mouse button), and move the location scroller directly to the left, toward the center of the Galaxy. Then change to the angular separation tool and use it to measure the angular separation of the Sun from the center of the Galaxy. This value should be approximately the same as you found in part a. (**c**) Use the angular separation tool to find the total angular diameter of the Milky Way Galaxy as seen from this viewpoint, measured from one end to the other of this edge-on view (zoom out if necessary). Round off the measurement to the nearest degree and then use the scale that you calculated in part (a) to find the approximate diameter of the Galaxy in ly. (**d**) In a similar fashion, use this viewpoint to measure the distance in degrees and ly for the following quantities: the diameter (in the plane of the Galaxy) of the central bulge, the thickness (perpendicular to the plane of the Galaxy) of the central bulge, and the thickness (perpendicular to the plane of the Galaxy) of the disk of the Galaxy at the location of the Sun. (**e**) In the edge-on view of this simulated galaxy, note the effect of the dust and gas clouds that obscure the light from the more distant spiral arms. Compare this simulated view with that of the edge-on view of a distant galaxy by clicking on **Home** to return to your present sky and using the **Find** facility to display NGC4565. (*Note:* The image of this galaxy is slightly offset from the position found by *Starry Night*™.) How similar is this galaxy to the Milky Way Galaxy, as represented by the *Starry Night*™ simulation?

Collaborative Exercises

56. Student book bags often contain a wide collection of odd-shaped objects. Each person in your group should rummage through their own book bags and find one object that is most similar to the Milky Way Galaxy in shape. List the items from each group member's belongings and describe what about the items is similar to the shape of our Galaxy and what about the items is not similar; then indicate which of the items is the closest match.

57. One strategy for identifying a central location is called *triangulation.* In triangulation, a central position can be pinpointed by knowing the distance from each of three different places. First, on a piece of paper, create a rough map showing where each person in your group lives. Second, create a circle around each person's home that has a radius equal to the distance that each home is from your classroom. Label the place where the circles intersect as your classroom. Why can you not identify the position of the classroom with only two people's circles?

58. Figure 22-13 shows how emission spectra from hydrogen clouds would be shifted due to their motion around the Galaxy. Create a similar sketch showing an oval automobile

racetrack with four cars moving on the track and a stationary observer outside the track at one end. Position and label the four moving cars all sounding their horns: (1) one that would have its horn sound shifted to longer wavelengths; (2) one that would have its horn sound shifted to shorter wavelengths; and (3) two cars moving in opposite directions so that their horn sounds would have no Doppler shift at all.

ANSWERS

ConceptChecks

ConceptCheck 22-1: Looking at other galaxies, astronomers have observed that globular clusters tend to form a spherical distribution centered on the center of galaxies. If our own globular clusters appeared in every direction equally, then astronomers would assume that we were near the center of the globular clusters and, subsequently, near the center of our Galaxy.

ConceptCheck 22-2: Far-infrared wavelengths best reveal the presence of warmed dust grains, and weak emission from stars is not easily detected with dust around. Near-infrared wavelengths, which can pass through dust, are emitted more strongly from cool stars than far-infrared, making near-infrared a better choice.

ConceptCheck 22-3: As the view between the galactic center and the Sun is obscured by intervening dust, the better vantage point would be from outside the disk in the halo, where the path of light is largely unobstructed by dust.

ConceptCheck 22-4: There is virtually no star formation occurring in the surrounding halo, so the youngest stars would be found actively forming in the rich dusty regions along the Galaxy's disk.

ConceptCheck 22-5: Because human eyes are only sensitive in the visible range, we would observe virtually no difference in the appearance of the Milky Way.

ConceptCheck 22-6: Neutral hydrogen gas emits 21-cm photons, and this is the wavelength observed when there is no relative motion between the gas and the observer. However, if the gas is moving away from the observer, its emitted light will be Doppler shifted and detected at longer wavelengths.

ConceptCheck 22-7: The Perseus arm (see Figure 22-16).

ConceptCheck 22-8: The most important factor for how fast a star moves around the galaxy is how much mass is closer to the galactic center than the star.

ConceptCheck 22-9: In the absence of dark matter, the stars most distant from the galactic center should be moving much slower than those stars orbiting closer to the center.

ConceptCheck 22-10: Astronomers detected only a small amount of matter in the form of MACHOs, yet they would have detected more if it were there. Therefore, the source of dark matter still remains to be discovered. Until WIMPs are detected, or ruled out, they remain a good candidate.

ConceptCheck 22-11: Without density waves causing areas of the galaxy to bunch up here and there, the galaxy would have no discernible arms.

ConceptCheck 22-12: The arms of flocculent spiral galaxies are likely the result of star formation; this suggests that stars came before spiral arms in the early universe.

ConceptCheck 22-13: The gravitational effects of a supermassive black hole cause surrounding stars to move at very high speeds; if there were no supermassive black hole, then nearby stars would be moving much more slowly.

ConceptCheck 22-14: The gases in the nebula would need to be energized by an external energy source, and it is proposed that a sudden intense release of X-rays from heated material moving into a supermassive black hole could have externally energized the nebula.

CalculationChecks

CalculationCheck 22-1: No matter how many hydrogen atoms are undergoing spin-flip transitions, only 21-cm photons are released from each atom. Fortunately, greater numbers of hydrogen atoms results in more photons and a stronger signal received at radio telescopes.

CalculationCheck 22-2: Remembering that it takes 2.2×10^8 years for the Sun to make one orbit around the Galaxy, 4.5×10^9 years $\times$ (1 orbit / 2.2×10^8 years) = 20 orbits.

CHAPTER

23

The two galaxies NGC 1531 and NGC 1532 are so close together that they exert strong gravitational forces on each other. Both galaxies are about 17 million pc (55 million ly) from us in the constellation Eridanus. (Gemini Observatory/Travis Rector, University of Alaska, Anchorage)

R I V U X G

Galaxies

LEARNING GOALS

By reading the sections of this chapter, you will learn

23-1 How astronomers first observed other galaxies

23-2 How astronomers determined the distances to other galaxies

23-3 The basic types of galaxies

23-4 What techniques astronomers use to determine distances to remote galaxies

23-5 How the velocities of remote galaxies tell us that the universe is expanding

23-6 How galaxies are grouped into clusters and larger structures

23-7 What happens when galaxies collide

23-8 What observations indicate the presence of dark matter in other galaxies and clusters

23-9 How galaxies formed and evolved

A century ago, most astronomers thought that the entire universe was only a few thousand light-years across and that nothing lay beyond our Milky Way Galaxy. One of the most important discoveries of the twentieth century was that this conception was utterly wrong. We now understand that the Milky Way is just one of billions of galaxies strewn across billions of light-years. The accompanying image shows two of them, denoted by rather mundane catalog numbers (NGC 1531 and NGC 1532) that give no hint to these galaxies' magnificence.

Some galaxies are spirals like NGC 1532 or the Milky Way, with arching spiral arms that are active sites of star formation. (The bright pink bands in NGC 1532 are H II regions, clouds of excited hydrogen that are set aglow by ultraviolet radiation from freshly formed massive stars.) Others, like NGC 1531, are featureless, ellipse-shaped agglomerations of stars, virtually devoid of interstellar gas and dust. Some galaxies are only one one-hundredth the size and one ten-thousandth the mass of the Milky Way. Others are giants, with 5 times the size and 50 times the mass of the Milky Way. Only about 10% of a typical galaxy's mass emits radiation of any kind; the remainder is made up of the mysterious dark matter.

Just as most stars are found within galaxies, most galaxies are located in groups and clusters. These clusters of galaxies stretch across the universe, forming huge, lacy patterns. Remarkably, remote clusters of galaxies are receding from us; the greater their distance, the more rapidly they are moving away. This relationship between distance and recessional velocity, called the Hubble law, reveals that our immense universe is expanding. In Chapters 25 and 26 we will learn what this implies about the distant past and remote future of the universe.

23-1 When galaxies were first discovered, it was not clear that they lie far beyond the Milky Way

As early as 1755, the German philosopher Immanuel Kant suggested that vast collections of stars lie outside the confines of the Milky Way. Less than a century later, an Irish astronomer observed the structure of some of the "island universes" that Kant proposed.

> As late as the 1920s it was unclear whether spiral nebulae were very remote "island universes" or simply nearby parts of our Galaxy

William Parsons, the third Earl of Rosse, was a wealthy amateur astronomer who used his fortune to build immense telescopes. His largest telescope, completed in February 1845, had an objective mirror 1.8 meters (6 feet) in diameter (Figure 23-1). The mirror was mounted at one end of a 60-foot tube controlled by cables, straps, pulleys, and cranes (Figure 23-1a). For many years, this triumph of nineteenth-century engineering was the largest telescope in the world.

With this new telescope, Rosse examined many of the nebulae that had been discovered and cataloged by William Herschel. He observed that some of these nebulae have a distinct spiral structure. One of the best examples is M51, also called NGC 5194. (The "M" designations of galaxies and nebulae come from a catalog compiled by the French astronomer Charles Messier between 1758 and 1782. The "NGC" designations come from the much more extensive "New General Catalogue" of galaxies and nebulae published in 1888 by J. L. E. Dreyer, a Danish astronomer who lived and worked in Ireland.)

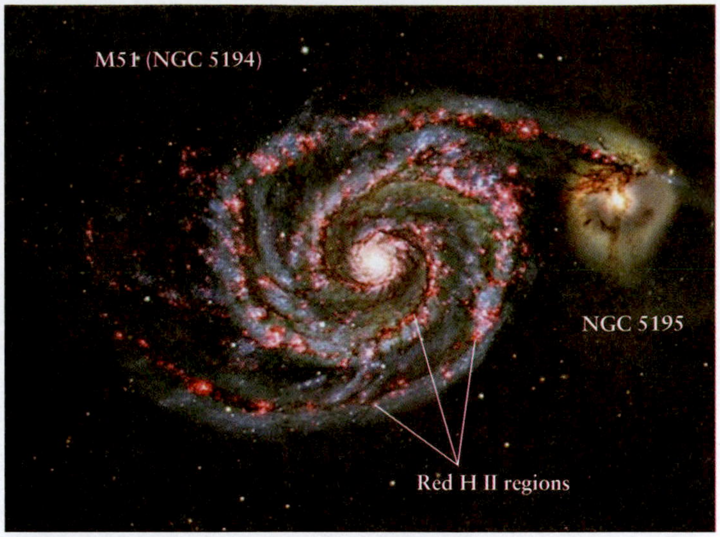

FIGURE 23-2 R I **V** U X G
A Modern View of the Spiral Galaxy M51 This galaxy, also called NGC 5194, has spiral arms that are outlined by glowing H II regions. These regions reveal the sites of star formation (see Section 18-2). One spiral arm extends toward the companion galaxy NGC 5195. (CFHT)

Lacking photographic equipment, Lord Rosse had to make drawings of what he saw. Figure 23-1b shows a drawing he made of M51, which compares favorably with modern photographs, as shown in Figure 23-2. Views such as this inspired Lord Rosse to echo Kant's proposal that such nebulae are actually "island universes."

Many astronomers of the nineteenth century disagreed with this notion of island universes. A considerable number of nebulae are in fact scattered throughout the Milky Way. (Figures 18-2

(a) Rosse's "Leviathan of Parsonstown"

(b) M51 as viewed through the "Leviathan"

FIGURE 23-1 R I **V** U X G
A Pioneering View of Another Galaxy **(a)** Built in 1845, this structure housed the largest telescope of its day. The telescope itself (the black cylinder pointing at a 45° angle above the horizontal) was restored to its original state during 1996–1998. **(b)** Using his telescope, Lord Rosse made this sketch of spiral structure in M51. This galaxy, whose angular size is 8 × 11 arcminutes (about a third the angular size of the full moon), is today called the Whirlpool Galaxy because of its distinctive appearance. (a: Richard T. Nowitz/Corbis)

and 18-4 show some examples.) The safest assumption was that "spiral nebulae," even though they are very different in shape from other sorts of nebulae, could also be components of our Galaxy.

The astronomical community became increasingly divided over the nature of the spiral nebulae. In April 1920, two opposing ideas were presented before the National Academy of Sciences in Washington, D.C. On one side was Harlow Shapley from the Mount Wilson Observatory, renowned for his recent determination of the size of the Milky Way Galaxy (see Section 22-1). Shapley thought the spiral nebulae were relatively small, nearby objects scattered around our Galaxy like the globular clusters he had studied. Opposing Shapley was Heber D. Curtis of the University of California's Lick Observatory. Curtis championed the island universe theory, arguing that each of these spiral nebulae is a rotating system of stars much like our own Galaxy.

The Shapley-Curtis "debate" generated much heat but little light. Nothing was decided, because no one could present conclusive evidence to demonstrate exactly how far away the spiral nebulae are. Astronomy desperately needed a definitive determination of the distance to a spiral nebula. Such a measurement became the first great achievement of a young man who studied astronomy at the Yerkes Observatory, near Chicago. His name was Edwin Hubble.

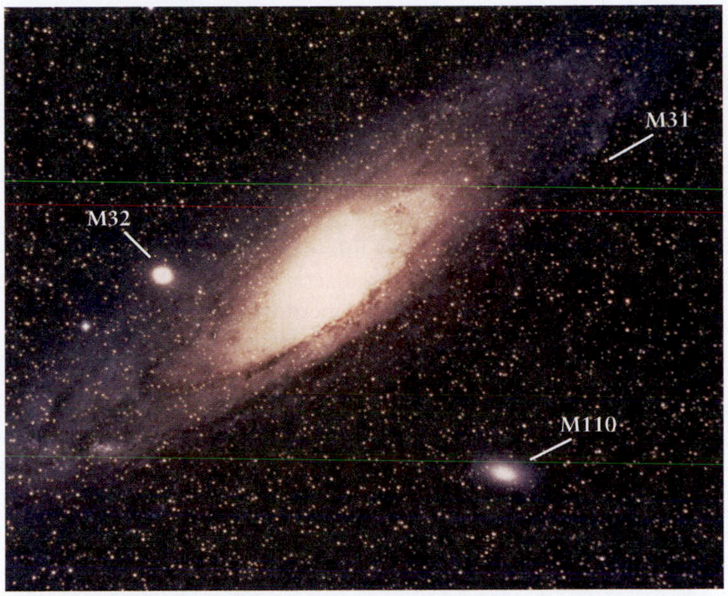

FIGURE 23-3 R I V U X G
The Andromeda Galaxy Also known as M31, the Andromeda Galaxy can be seen with a small telescope, or even the naked eye on a clear night. Edwin Hubble was the first to demonstrate that M31 is actually a galaxy that lies far beyond the Milky Way. M31 is the largest galactic neighbor near our Milky Way. A collision between our Milky Way and the Andromeda Galaxy is expected in about 3.75 billion years. M32 and M110 are two small satellite galaxies that orbit M31. (Science Source)

CONCEPTCHECK 23-1

If you were observing one of the "spiral nebulae" and determined that it was closer than some of the stars of our Galaxy, would you be providing evidence in support of Shapley's argument or Curtis's argument?

Answer appears at the end of the chapter.

23-2 Hubble proved that the spiral nebulae are far beyond the Milky Way

After completing his studies, Edwin Hubble joined the staff of the Mount Wilson Observatory in Pasadena, California. On October 6, 1923, he took an historic photograph of the Andromeda "Nebula," one of the spiral nebulae around which controversy raged. (Figure 23-3 is a modern photograph of this object.)

Hubble carefully examined his photographic plate and discovered a *new* bright spot—what he at first thought to be a nova. Referring to previous plates of that region, he soon realized that the object was actually a Cepheid variable star. Further scrutiny of additional plates over the next several months revealed several more Cepheids. Figure 23-4 shows modern observations of a Cepheid in another "spiral nebula."

> Just as RR Lyrae variable stars demonstrated the extent of the Milky Way, Cepheid variables revealed the immense distances to other galaxies

As we saw in Section 19-6, Cepheid variables help astronomers determine distances. An astronomer begins by carefully measuring the variations in apparent brightness of a Cepheid variable, then recording the results in the form of a plot of brightness versus time, or light curve (see Figure 19-18a). This graph gives the variable star's period and average brightness. Given the star's period,

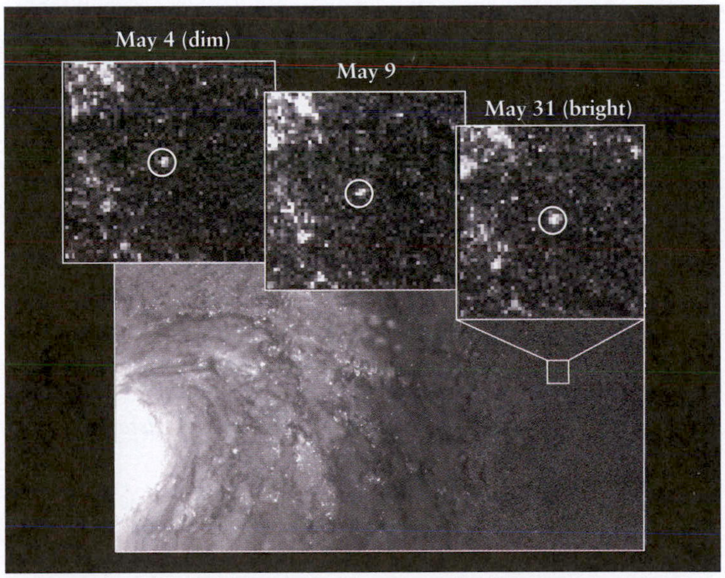

VIDEO 23-1 **FIGURE 23-4** R I V U X G
Measuring Galaxy Distances with Cepheid Variables By observing Cepheid variable stars in M100, the galaxy shown here, astronomers have found that it is about 17 Mpc (56 million ly) from Earth. The insets show one of the Cepheids in M100 at different stages in its brightness cycle, which lasts several weeks. (Wendy L. Freedman, Carnegie Institution of Washington, and NASA)

BOX 23-1 TOOLS OF THE ASTRONOMER'S TRADE

Cepheids and Supernovae as Indicators of Distance

Because their periods are directly linked to their luminosities, Cepheid variables are one of the most reliable tools astronomers have for determining the distances to galaxies. To this day, astronomers use this link—much as Hubble did back in the 1920s—to measure intergalactic distances. More recently, they have begun to use Type Ia supernovae, which are far more luminous and thus can be seen much farther away, to determine the distances to very remote galaxies.

EXAMPLE: In 1992 a team of astronomers observed Cepheid variables in a galaxy called IC 4182 to deduce this galaxy's distance from Earth. One such Cepheid has a period of 42.0 days and an average apparent magnitude (m) of +22.0. (See Box 17-3 for an explanation of the apparent magnitude scale.) By comparison, the dimmest star you can see with the naked eye has $m = +6$; this Cepheid in IC 4182 appears less than one one-millionth as bright.

According to the period-luminosity relation shown in Figure 19-20, such a Cepheid with a period of 42.0 days has an average luminosity of 33,000 $L_\odot$. An equivalent statement is that this Cepheid has an average absolute magnitude (M) of −6.5. (This compares to $M = +4.8$ for the Sun.) Use this information to determine the distance to IC 4182.

Situation: We are given the apparent magnitude $m = +22.0$ and the absolute magnitude $M = -6.5$ of the Cepheid variable star in IC 4182. Our goal is to calculate the distance to this star, and hence the distance to the galaxy of which it is part.

Tools: We use the relationship between apparent magnitude, absolute magnitude, and distance given in Box 17-3.

Answer: In Box 17-3, we saw that the apparent magnitude of a star is related to its absolute magnitude and distance in parsecs (d) by

$$m - M = 5 \log d - 5$$

This equation can be rewritten as

$$d = 10^{(m - M + 5)/5} \text{ parsecs}$$

We have $m - M = (+22.0) - (-6.5) = 22.0 + 6.5 = 28.5$. (Recall from Box 17-3 that $m - M$ is called the *distance modulus.*) Hence, our equation becomes

$$d = 10^{(28.5 + 5)/5} \text{ parsecs} = 10^{6.7} \text{ parsecs} = 5 \times 10^6 \text{ parsecs}$$

(A calculator is needed to calculate the quantity $10^{6.7}$.)

Review: Our result tells us that the galaxy is 5 million parcsecs, or 5 Mpc, from Earth (1 Mpc = 10^6 pc). This distance can also be expressed as 16 million light-years.

EXAMPLE: Astronomers are interested in IC 4182 because a Type Ia supernova was observed there in 1937. All Type Ia supernovae are exploding white dwarfs that reach nearly the same maximum brightness at the peak of their outburst (see Section 20-9). Once astronomers know the peak absolute magnitude of Type Ia supernovae, they can use these supernovae as distance indicators. Because the distance to IC 4182 is known from its Cepheids, the 1937 observations of the supernova in that galaxy help us calibrate Type Ia supernovae as distance indicators.

At maximum brightness, the 1937 supernova reached an apparent magnitude of $m = +8.6$. What was its absolute magnitude at maximum brightness?

Situation: We are given the supernova's apparent magnitude m, and we know its distance from the previous example. Our goal is to calculate its absolute magnitude M.

Tools: We again use the relationship $m - M = 5 \log d - 5$.

Answer: We could plug in the value of d found in the previous example. But it is simpler to note that the distance modulus $m - M$ has the same value no matter whether it refers to a Cepheid, a supernova, or any other object, just so it is at the same distance d. From the Cepheid example we have $m - M = 28.5$ for IC 4182, so

$$M = m - (m - M) = 8.6 - (28.5) = -19.9$$

This absolute magnitude corresponds to a remarkable peak luminosity of 10^{10} $L_\odot$.

Review: Whenever astronomers find a Type Ia supernova in a remote galaxy, they can combine this absolute magnitude with the observed maximum apparent magnitude to get the galaxy's distance modulus, from which the galaxy's distance can be easily calculated (just as we did above for the Cepheids in IC 4182). This technique has been used to determine the distances to galaxies hundreds of millions of parsecs away (see Section 25-4).

the astronomer then uses the period-luminosity relation shown in Figure 19-20 to find the Cepheid's average luminosity. Knowing both the apparent brightness and luminosity of the Cepheid, the astronomer can then use the inverse-square law to calculate the distance to the star (see Box 17-2). Box 23-1 presents an example of this calculation. This procedure is very similar to that used by Harlow Shapley to measure the distances to the Milky Way's globular clusters using RR Lyrae variable stars (see Section 22-1).

Cepheid variables are intrinsically quite luminous, with average luminosities that can exceed $10^4\ L_\odot$. Hubble realized that for these luminous stars to appear as dim as they were on his photographs of the Andromeda "Nebula," they must be extremely far away. Straightforward calculations using modern data reveal that M31 is some 750 kiloparsecs (2.5 million light-years) from Earth. Based on its angular size, M31 has a diameter of 70 kiloparsecs—larger than the diameter of our own Milky Way Galaxy!

These results prove that the Andromeda "Nebula" is actually an enormous stellar system, far beyond the confines of the Milky Way. Today, this system is properly called the Andromeda Galaxy. (Under good observing conditions, you can actually see this galaxy's central bulge with the naked eye. If you could see the entire Andromeda Galaxy, it would cover an area of the sky roughly 5 times as large as the full moon.) Galaxies are so far away that their distances from us are usually given in millions of parsecs, or *megaparsecs* (Mpc): 1 Mpc = 10^6 pc. For example, the distance to the galaxies in the image that opens this chapter is 17 Mpc.

Hubble's results, which were presented at a meeting of the American Astronomical Society on December 30, 1924, settled the Shapley-Curtis "debate" once and for all. The universe was recognized to be far larger and populated with far bigger objects than anyone had seriously imagined. Hubble had discovered the realm of the galaxies.

CAUTION! In everyday language, many people use the words "galaxy" and "universe" interchangeably. It is true that before Hubble's discoveries our Milky Way Galaxy was thought to constitute essentially the entire universe. But we now know that the universe contains literally billions of galaxies. A single galaxy, vast though it may be, is just a tiny part of the entire observable universe.

CONCEPTCHECK 23-2

If a nearby galaxy were discovered, why would astronomers immediately look for Cepheids?

Answer appears at the end of the chapter.

23-3 Galaxies are classified according to their appearance

Millions of galaxies are visible across every unobscured part of the sky. Although all galaxies are made up of large numbers of stars, they come in a variety of shapes and sizes.

Hubble classified galaxies into four broad categories based on their appearance. These categories form the basis for the **Hubble classification,** a scheme that is still used today. The four classes of galaxies are the spirals, classified S; barred spirals, or SB; ellipticals, E; and irregulars, Irr. Table 23-1 summarizes some key properties of each class. These various types of galaxies differ not only in their shapes but also in the kinds of processes taking place within them.

Spiral Galaxies: Stellar Birthplaces

M51 (Figure 23-2), M31 (Figure 23-3), and M100 (Figure 23-4) are examples of **spiral galaxies.** Figure 23-5 shows that spiral galaxies are characterized by arched lanes of stars, just as is our own Milky Way Galaxy (see Section 22-3). The spiral arms contain young, hot, blue stars and their associated H II regions, indicating ongoing star formation.

Thermonuclear reactions within stars create *metals,* that is, elements heavier than hydrogen or helium (see Section 17-5). These metals are dispersed into space as the stars evolve and die. So, if new stars are being formed from the interstellar matter in spiral galaxies, they will incorporate these metals and be Population I stars (see Section 19-5 and Section 22-2). Indeed, the visible-light spectrum of the disk of a spiral galaxy has strong metal absorption lines. Such a spectrum is a composite of the spectra of many stars and shows that the stars in the disk are principally of Population I. By contrast, there is relatively little star formation in the central bulges of spiral galaxies, and these regions are dominated by old Population II stars that have a low metal content. The lack of star formation also explains why the central bulges of spiral galaxies have a yellowish or reddish color; as a population of stars ages, the massive, luminous blue stars die off first, leaving only the longer-lived, low-mass red stars.

TABLE 23-1 Some Properties of Galaxies

	Spiral (S) and barred spiral (SB) galaxies	Elliptical galaxies (E)	Irregular galaxies (Irr)
Mass ($M_\odot$)	10^9 to 4×10^{11}	10^5 to 10^{13}	10^8 to 3×10^{10}
Luminosity ($L_\odot$)	10^8 to 2×10^{10}	3×10^5 to 10^{11}	10^7 to 10^9
Diameter (kpc)	5 to 250	1 to 200	1 to 10
Stellar populations	Spiral arms: young Population I Nucleus and throughout disk: Population II and old Population I	Population II and old Population I	mostly Population 1
Percentage of observed galaxies	77%	20%*	3%

This percentage does not include dwarf elliptical galaxies that are as yet too dim and distant to detect. Hence, the actual percentage of galaxies that are ellipticals may be higher than shown here.

(a) Sa (NGC 1357) (b) Sb (M81) (c) Sc (NGC 4321)

FIGURE 23-5 R I V U X G

Spiral Galaxies Edwin Hubble classified spiral galaxies according to the texture of their spiral arms and the relative size of their central bulges. Sa galaxies have smooth, broad spiral arms and the largest central bulges, while Sc galaxies have narrow, well-defined arms and the smallest central bulges. (a: Adam Block/Steve Mandel/Jim Rada and Students/NOAO/AURA/NSF; b: Robert Gendler/Science Source; c: FORS Team, 8.2-meter VLT, ESO)

Hubble further classified spiral galaxies according to the texture of the spiral arms and the relative size of the central bulge. Spirals with smooth, broad spiral arms and a fat central bulge are called Sa galaxies, for spiral type *a* (Figure 23-5a); those galaxies with moderately well-defined spiral arms and a moderate-sized central bulge, like M31 and M51, are Sb galaxies (Figure 23-5b); and galaxies with narrow, well-defined spiral arms and a tiny central bulge are Sc galaxies (Figure 23-5c).

The differences between Sa, Sb, and Sc galaxies may be related to the relative amounts of gas and dust they contain. Observations with infrared telescopes (which detect the emission from interstellar dust) and radio telescopes (which detect radiation from interstellar gases such as hydrogen and carbon monoxide) show that about 4% of the mass of a Sa galaxy is in the form of gas and dust. This percentage is 8% for Sb galaxies and 25% for Sc galaxies.

Interstellar gas and dust is the material from which new stars are formed, so an Sc galaxy has a greater proportion of its mass involved in star formation than an Sb or Sa galaxy. Hence, a Sc galaxy has a large disk (where star formation occurs) and a small central bulge (where there is little or no star formation). By comparison, a Sa galaxy, which has relatively little gas and dust, and thus less material from which to form stars, has a large central bulge and only a small star-forming disk.

The central bulge contains more than just stars: It is also thought to contain a supermassive black hole at its center. The mass of the black hole is about 1/1000 the mass of the central bulge, which corresponds to black hole masses of millions to hundreds of millions of solar masses. (The mass of the black hole at the center of our Milky Way is just over 4 million solar masses.) Chapter 24 discusses how these black holes are the "central engines" of quasars and active galaxies.

(a) SBa (NGC 1291) (b) SBb (M83) (c) SBc (NGC 1365)

FIGURE 23-6 R I V U X G

Barred Spiral Galaxies As with spiral galaxies, Hubble classified barred spirals according to the texture of their spiral arms (which correlates to the sizes of their central bulges). SBa galaxies have the smoothest spiral arms and the largest central bulges, while SBc galaxies have narrow, well-defined arms and the smallest central bulges. (a: NASA/JPL-Caltech/CTIO; b, c: FORS Team, 8.2-meter VLT, ESO)

CONCEPTCHECK 23-3

When making a sketch of a spiral galaxy with considerable star formation, which areas would you label with *more* star formation? Which areas would have *less?*

Answer appears at the end of the chapter.

Barred Spiral Galaxies: Spirals with an Extra Twist

In **barred spiral galaxies,** such as those shown in Figure 23-6, the spiral arms originate at the ends of a bar-shaped region running through the galaxy's nucleus rather than from the nucleus itself. As with ordinary spirals, Hubble subdivided barred spirals according to the relative size of their central bulge and the character of their spiral arms. A SBa galaxy has a large central bulge and thin, tightly wound spiral arms (Figure 23-6a). Likewise, a SBb galaxy is a barred spiral with a moderate central bulge and moderately wound spiral arms (Figure 23-6b), while a SBc galaxy has lumpy, loosely wound spiral arms and a tiny central bulge (Figure 23-6c). As for ordinary spiral galaxies, the difference between SBa, SBb, and SBc galaxies may be related to the amount of gas and dust in the galaxy.

Bars appear to form naturally in many spiral galaxies. This conclusion comes from computer simulations of galaxies, which set hundreds of thousands of simulated "stars" into orbit around a common center. As the "stars" orbit and exert gravitational forces on one another, a bar structure forms in most cases. Indeed, barred spiral galaxies outnumber ordinary spirals by about two to one. (As we saw in Section 22-2, there is evidence that the Milky Way Galaxy is a barred spiral.)

Why don't all spiral galaxies have bars? According to calculations by Jeremiah Ostriker and P. J. E. Peebles of Princeton University, a bar will not develop if a galaxy is surrounded by a sufficiently massive halo of nonluminous *dark matter.* (In Section 22-4 we saw evidence that our Milky Way Galaxy is surrounded by such a dark matter halo.) The difference between barred spirals and ordinary spirals may thus lie in the amount of dark matter the galaxy possesses. In Section 23-8 we will see compelling evidence for the existence of dark matter in spiral galaxies.

CONCEPTCHECK 23-4

How are galaxies with loosely wrapped arms categorized using Hubble's scheme?

Answer appears at the end of the chapter.

Elliptical Galaxies: From Giants to Dwarfs

Elliptical galaxies, so named because of their distinctly elliptical shapes, have no spiral arms. Hubble subdivided these galaxies according to how round or flattened they look. The roundest elliptical galaxies are called E0 galaxies and the flattest, E7 galaxies. Elliptical galaxies with intermediate amounts of flattening are given designations between these extremes (Figure 23-7).

CAUTION! Unlike the designations for spirals and barred spirals, the classifications E0 through E7 may not reflect the true shape of elliptical galaxies. An E1 or E2 galaxy might actually be a very flattened disk of stars that we just happen to view face-on, and a cigar-shaped E7 galaxy might look spherical if seen end-on. The Hubble scheme classifies galaxies entirely by how they *appear* to us on Earth.

Elliptical galaxies look far less dramatic than their spiral and barred spiral cousins. The reason, as shown by radio and infrared observations, is that ellipticals are virtually devoid of interstellar gas and dust. Consequently, there is little material from which stars could have recently formed, and indeed there is no evidence of young stars in most elliptical galaxies. For the most part, star formation in elliptical galaxies ended long ago. Hence, these galaxies are composed of old, red, Population II stars with only small amounts of metals.

Elliptical galaxies come in a wide range of sizes and masses. Both the largest and the smallest galaxies in the known universe are elliptical. Figure 23-8 shows two **giant elliptical galaxies** that are about 20 times larger than an average galaxy. These giant ellipticals are located near the middle of a large cluster of galaxies in the constellation Virgo. (We will discuss this and other clusters of galaxies in Section 23-6.)

(a) E0 (M105) (b) E3 (NGC 4406) (c) E6 (NGC 3377)

FIGURE 23-7 R I V U X G

Elliptical Galaxies Hubble classified elliptical galaxies according to how round or flattened they look. A galaxy that appears round is labeled E0, and the flattest-appearing elliptical galaxies are designated E7. (a: Karl Gebhardt (University of Michigan), Tod Lauer (NOAO), and NASA; b: Jean-Charles Cuillandre, Hawaiian Starlight, CFHT; c: Karl Gebhardt (University of Michigan), Tod Lauer (NOAO), and NASA)

FIGURE 23-8 R I V U X G

Giant Elliptical Galaxies The Virgo cluster is a rich, sprawling collection of more than 2000 galaxies about 17 Mpc (56 million ly) from Earth. Only the center of this huge cluster appears in this photograph. The two largest members of this cluster are the giant elliptical galaxies M84 and M86. These galaxies have angular sizes of 5 to 7 arcmin. (Jean-Charles Cuillandre, Hawaiian Starlight, CFHT)

FIGURE 23-9 R I V U X G

A Dwarf Elliptical Galaxy This diffuse cloud of stars is a nearby E4 dwarf elliptical called Leo I. It actually orbits the Milky Way at a distance of about 180 kpc (600,000 ly). Leo I is about 1 kpc (3000 ly) in diameter but contains so few stars that you can see through the galaxy's center. (NASA and The Hubble Heritage Team [STScI/AURA])

Giant ellipticals are rather rare, but **dwarf elliptical galaxies** are quite common. Dwarf ellipticals are only a fraction the size of their normal counterparts and contain so few stars—only a few million, compared to more than 100 billion (10^{11}) stars in our Milky Way Galaxy—that these galaxies are completely transparent. You can actually see straight through the center of a dwarf galaxy and out the other side, as Figure 23-9 shows.

The visible light from a galaxy is emitted by its stars, so the visible-light spectrum of a galaxy has absorption lines. But because a galaxy's stars are in motion, with some approaching us and others moving away, the Doppler effect smears out and broadens the absorption lines. The average motions of stars in a galaxy can be deduced from the details of this broadening.

For elliptical galaxies, studies of this kind show that star motions are quite random. In a very round (E0) elliptical galaxy, this randomness is **isotropic,** meaning "equal in all directions." Because the stars are whizzing around equally in all directions, the galaxy is genuinely spherical. In a flattened (E7) elliptical galaxy, the randomness of the stellar motions is **anisotropic,** which means that the range of star speeds is different in different directions.

Hubble also identified galaxies that are midway in appearance between ellipticals and the two kinds of spirals. These are denoted as S0 and SB0 galaxies, also called **lenticular galaxies.** Although they look somewhat elliptical, lenticular ("lens-shaped") galaxies have both a central bulge and a disk like spiral galaxies, but no discernible spiral arms. They are therefore sometimes referred to as "armless spirals." Figure 23-10 shows an example of an SB0 lenticular galaxy.

Edwin Hubble summarized his classification scheme for spiral, barred spiral, and elliptical galaxies in a diagram, now called the **tuning fork diagram** for its shape (Figure 23-11).

CAUTION! When Hubble first drew his tuning fork diagram, he had the idea that it represented an evolutionary sequence. He thought that galaxies evolved over time from the left to the right of the diagram, beginning as ellipticals and eventually becoming either spiral or barred spiral galaxies. We now understand that this evolution of galaxies is not the case at all! For one thing, elliptical galaxies have little or no overall rotation, while spiral and barred spiral galaxies have a substantial amount of overall rotation. There is no way that an elliptical galaxy could suddenly start rotating, which means that it could not evolve into a spiral galaxy.

A more modern interpretation of the Hubble tuning fork diagram is that it is an arrangement of galaxies according to their overall rotation. A rapidly rotating collection of matter in space tends to form a disk, while a slowly rotating collection does not. Thus, the elliptical galaxies at the far left of the tuning fork diagram have little internal rotation and hence no disk. Sa and SBa galaxies have enough overall rotation to form a disk, though their central bulges are still dominant. The galaxies with the greatest amount of overall rotation are Sc and SBc galaxies, in which the central bulges are small and most of the gas, dust, and stars are in the disk.

Irregular Galaxies: Deformed and Dynamic

Galaxies that do not fit into the scheme of spirals, barred spirals, and ellipticals are usually referred to as **irregular galaxies.** They are generally rich in interstellar gas and dust, and have both young and old stars. For lack of any better scheme, the irregular galaxies are sometimes placed between the ends of the tines of the Hubble tuning fork diagram, as in Figure 23-11.

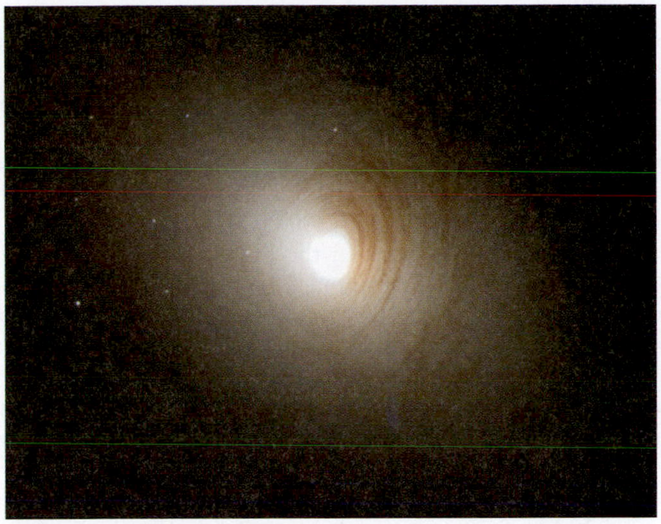

FIGURE 23-10 R I **V** U X G

A Lenticular Galaxy NGC 2787 is classified as a lenticular galaxy because it has a disk but no discernible spiral arms. Its nucleus displays a faint bar (not apparent in this image), so NGC 2787 is denoted as an SB0 galaxy. It lies about 7.4 Mpc (24 million ly) from Earth in the constellation Ursa Major. (NASA and The Hubble Heritage Team, STScI/AURA)

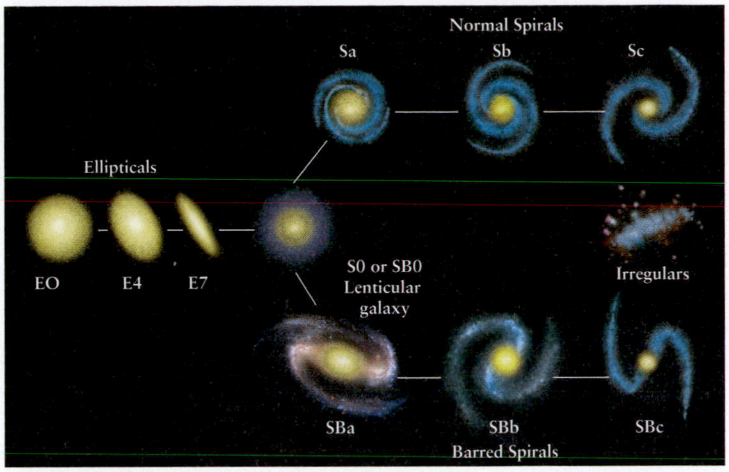

FIGURE 23-11

Hubble's Tuning Fork Diagram Edwin Hubble's classification of regular galaxies is shown in his tuning fork diagram. An elliptical galaxy is classified by how flattened it appears. A spiral or barred spiral galaxy is classified by the texture of its spiral arms and the size of its central bulge. A lenticular galaxy, is intermediate between ellipticals and spirals. Irregular galaxies do not fit into this simple classification scheme. The ordering of galaxies does not represent an evolutionary sequence.

Hubble defined two types of irregulars. Irr I galaxies have only hints of organized structure, and have many OB associations and H II regions. The best-known examples of Irr I galaxies are the Large Magellanic Cloud (Figure 23-12) and the Small Magellanic Cloud. Both are nearby companions of our Milky Way and can be seen with the naked eye from southern latitudes. Both these galaxies contain substantial amounts of interstellar gas. Tidal forces exerted on these irregular galaxies by the Milky Way help to compress the gas, which is why both of the Magellanic clouds are sites of active star formation.

The other type of irregulars, called Irr II galaxies, have asymmetrical, distorted shapes that seem to have been caused by collisions with other galaxies or by violent activity in their nuclei. M82, shown in Figure 21-16 and in the image that opens Chapter 6, is an example of an Irr II galaxy.

FIGURE 23-12 R I **V** U X G

The Large Magellanic Cloud (LMC) At a distance of only 55 kpc (179,000 ly), this Irr I galaxy is the third closest known companion of our Milky Way Galaxy. About 19 kpc (62,000 ly) across, the LMC spans 22° across the sky, or about 50 times the size of the full moon. One sign that star formation is ongoing in the LMC is the Tarantula Nebula, whose diameter of 250 pc (800 ly) and mass of 5 × 106 M make it the largest known H II region. (NOAO/AURA/NSF)

CONCEPTCHECK 23-5

Which type of galaxy has almost no ongoing active star formation?

Answer appears at the end of the chapter.

23-4 Astronomers use various techniques to determine the distances to remote galaxies

The various methods of distance determination are interrelated because one is used to calibrate another

A key question that astronomers ask about galaxies is "How far away are they?" Knowing the distances to galaxies is essential for learning the structure and history of the universe. Unfortunately, many of the techniques that are used to measure distances within our Milky Way Galaxy cannot be used for the far greater distances to other galaxies. The extremely accurate parallax method that we described in Section 17-1 can be used only for stars within about 500 pc. Beyond that distance, parallax angles become too small to measure. Spectroscopic parallax, in which the distance to a star or star cluster is found with the help of the H-R diagram (see Section 17-8), is accurate only out to roughly 10 kpc from Earth; more distant stars or clusters are too dim to give reliable results.

Standard Candles: Variable Stars and Type Ia Supernovae

To determine the distance to a remote galaxy, astronomers look instead for a **standard candle**—an object, such as a star, that lies within that galaxy and for which we know the luminosity (or, equivalently, the absolute magnitude, described in Section 17-3). By measuring how bright the standard candle appears, astronomers can calculate its distance—and hence the distance to the galaxy of which it is part—using the inverse-square law.

The challenge is to find standard candles that are luminous enough to be seen across the tremendous distances to galaxies. To be useful, standard candles should have four properties:

1. They should be luminous so that we can see them out to great distances.

2. We should be fairly certain about their luminosities so that we can be equally certain of any distance calculated from a standard candle's apparent brightness and luminosity.

3. They should be easily identifiable—for example, by the shape of the light curve of a variable star.

4. They should be relatively common so that astronomers can use them to determine the distances to many different galaxies.

For nearby galaxies, Cepheid variable stars make reliable standard candles. These variables can be seen out to about 30 Mpc (100 million ly) using the Hubble Space Telescope, and their luminosity can be determined from their period through the period-luminosity relation depicted in Figure 19-20. Box 23-1 gives an example of using Cepheid variables to determine distances. RR Lyrae stars, which are Population II variable stars often found in globular clusters, can be used as standard candles in a similar way. (We saw in Section 22-1 how RR Lyrae variables helped determine the size of our Galaxy.) Because they are less luminous than Cepheids, RR Lyrae variables can be seen only out to 100 kpc (300,000 ly).

Beyond about 30 Mpc even the brightest Cepheid variables, which have luminosities of about $2 \times 10^4\ L_\odot$, fade from view. Astronomers have tried to use even more luminous stars such as blue supergiants to serve as standard candles. However, this idea is based on the assumption that there is a fixed upper limit on the luminosities of stars, which may not be the case. Hence, these standard candles are not very "standard," and distances measured in this way are somewhat uncertain.

One class of standard candles that astronomers have used beyond 30 Mpc is Type Ia supernovae. As we described in Section 20-9, these supernovae occur when a white dwarf in a close binary system accretes enough matter from its companion to blow itself apart in a thermonuclear conflagration. A Type Ia supernova can reach a maximum luminosity of about $3 \times 10^9\ L_\odot$ (Figure 23-13). If a Type Ia supernova is seen in a distant galaxy and its maximum apparent brightness measured, the inverse-square law can be used to find the galaxy's distance (see Box 23-1).

One complication is that not all Type Ia supernovae are equally luminous. Fortunately, there is a simple relationship between the peak luminosity of a Type Ia supernova and the rate at which the luminosity decreases after the peak: The more slowly the brightness decreases, the more luminous the supernova. Using this relationship, astronomers have measured distances to supernovae more than 1000 Mpc (3 billion ly) from Earth.

R I V U X G

(a) M100 in March 2002

R I V U X G

(b) M100 in February 2006, showing Supernova 2006X

FIGURE 23-13

A Supernova in a Spiral Galaxy These images from the Very Large Telescope show the spiral galaxy M100 **(a)** before and **(b)** after a Type Ia supernova exploded within the galaxy in 2006. (The two images were made with different color filters, which gives them different appearances.) Such luminous supernovae, which can be seen at extreme distances, are important standard candles used to determine the distances to faraway galaxies. The distance to M100 is also known from observations of Cepheid variables (see Figure 23-4), so this particular supernova can help calibrate Type Ia supernovae as distance indicators. (European Southern Observatory)

Unfortunately, this technique can be used only for galaxies in which we happen to observe a Type Ia supernova. But telescopic surveys now identify many dozens of these supernovae every year, so the number of galaxies whose distances can be measured in this way is continually increasing.

CONCEPTCHECK 23-6

Imagine that a thin cloud of intergalactic dust reduced the observed brightness of a Type Ia supernova found in a distant galaxy. Would astronomers who did not know the dust was there mistakenly assume that the galaxy is farther or closer than it actually is?

Answer appears at the end of the chapter.

Distance Determination without Standard Candles

Other methods for determining the distances to galaxies do not make use of standard candles. One was discovered in the 1970s by the astronomers Brent Tully and Richard Fisher. They found that the width of the hydrogen 21-cm emission line of a spiral galaxy (see Section 22-3) is related to the galaxy's luminosity. This correlation is the **Tully-Fisher relation**—the broader the line, the more luminous the galaxy.

Such a relationship exists because radiation from the approaching side of a rotating galaxy is blueshifted while that from the galaxy's receding side is redshifted. Thus, the 21-cm line is Doppler broadened by an amount directly related to how fast a galaxy is rotating. Rotation speed is related to the galaxy's mass by Newton's form of Kepler's third law. The more massive a galaxy, the more stars it contains and thus the more luminous it is. Consequently, the width of a galaxy's 21-cm line is directly related to its luminosity.

Because line widths can be measured quite accurately, astronomers can use the Tully-Fisher relation to determine the luminosity of a distant spiral galaxy. By combining this information with measurements of apparent brightness, they can calculate the distance to the galaxy. This technique can be used to measure distances of 100 Mpc or more.

Elliptical galaxies do not rotate, so the Tully-Fisher relation cannot be used to determine their distances. But in 1987 the American astronomers Marc Davis and George Djorgovski pointed out a correlation between the size of an elliptical galaxy, the average motions of its stars, and how the galaxy's brightness appears distributed over its surface.

In geometry, three points define a plane, so the relationship among size, motion, and brightness is called the **fundamental plane.** By measuring the last two quantities, an astronomer can use the fundamental plane relationship to determine a galaxy's actual size. And by comparing this to the galaxy's apparent size, the astronomer can calculate the distance to the galaxy using the small-angle formula (see Box 1-1). Ellipticals can be substantially larger and more luminous than spirals, so the fundamental plane can be used at somewhat greater distances than can the Tully-Fisher relation.

CONCEPTCHECK 23-7

Is a galaxy that appears to be quite bright, yet has a narrow width for its emitted hydrogen light, relatively close or very distant from our own Galaxy?

Answer appears at the end of the chapter.

The Distance Ladder

Figure 23-14 shows the ranges of applicability of several important means of determining astronomical distances. Because these ranges overlap, one technique can be used to calibrate another. As an example, astronomers have studied Cepheids in nearby galaxies that have also been host to Type Ia supernovae. The Cepheids provide the distances to these nearby galaxies, making it possible to determine the peak luminosity of each supernova using its maximum apparent brightness and the inverse-square law. Once the peak luminosity is known, it can be used to determine the

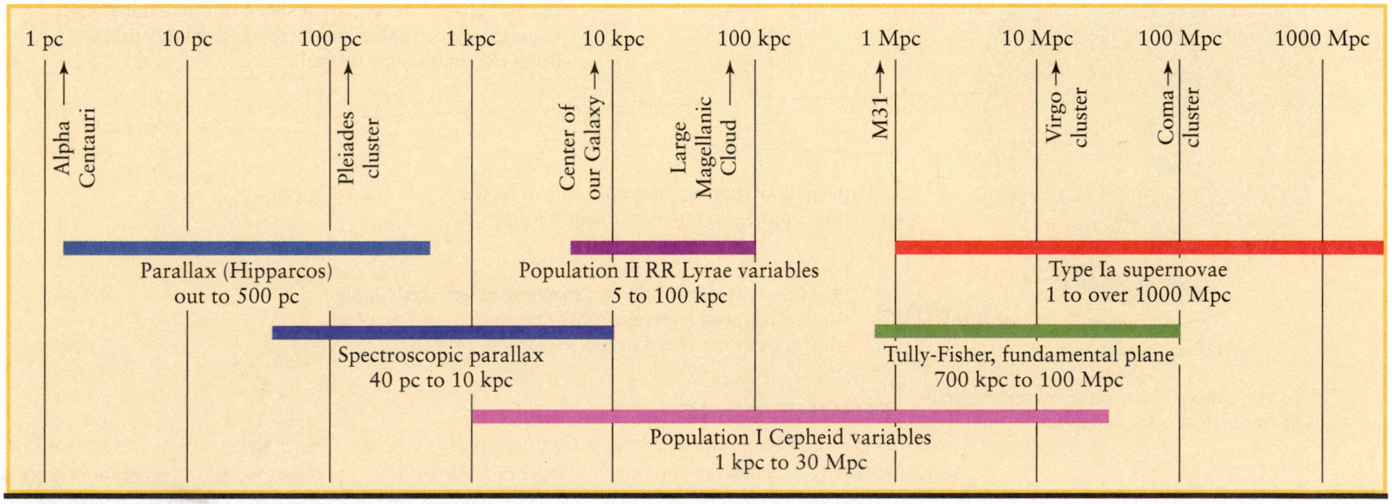

FIGURE 23-14

The Distance Ladder Astronomers employ a variety of techniques for determining the distances to objects beyond the solar system. Because their ranges of applicability overlap, one technique can be used to calibrate another. The arrows indicate distances to several important objects. Note that each division on the scale indicates a tenfold increase in distance, such as from 1 to 10 Mpc.

distance to Type Ia supernovae in more distant galaxies. Because one measuring technique leads us to the next one like rungs on a ladder, the techniques shown in Figure 23-14 (along with others) are referred to collectively as the **distance ladder.**

ANALOGY If you give a slight shake to the bottom of a tall ladder, the top can wobble back and forth alarmingly. A change in distance-measuring techniques used for nearby objects can also have substantial effects on the distances to remote galaxies. For instance, if astronomers discovered that the distances to nearby Cepheids were in error, distance measurements using any technique that is calibrated by Cepheids would be affected as well. (As an example, the distance to the galaxy M100 shown in Figure 23-4 is determined using Cepheids. A Type Ia supernova has been seen in M100, as Figure 23-13 shows, and its luminosity is determined using the Cepheid-derived distance to M100. Any change in the calculated distance to M100 would change the calculated luminosity of the Type Ia supernova, and so it would have an effect on all distances derived from observations of how bright these supernovae appear in other galaxies.) For this reason, astronomers go to great lengths to check the accuracy and reliability of their standard candles.

One distance-measuring technique that has broken free of the distance ladder uses observations of molecular clouds called **masers.** ("Maser" is an acronym for "microwave amplification by stimulated emission of radiation.") Just as an electric current stimulates a laser to emit an intense beam of visible light, nearby luminous stars can stimulate water molecules in a maser to emit intensely at microwave wavelengths. This radiation is so intense that masers can be detected millions of parsecs away.

During the 1990s, Jim Herrnstein and his collaborators used the Very Long Baseline Array (see Section 6-6) to observe a number of masers orbiting in a disk around the center of the spiral galaxy M106. They determined the orbital speed of the masers from the Doppler shift of masers near the edges of the disk, where they are moving most directly toward or away from Earth. They also measured the apparent change in position of masers moving across the face of M106. By relating this apparent speed to the true speed determined from the Doppler shift, they calculated that the masers and the galaxy of which they are part are 7.2 Mpc (23 million ly) from Earth (**Figure 23-15**).

The maser technique is still in its infancy. But because this technique is independent of all other distance-measuring methods, it is likely to play an important role in calibrating the rungs of the distance ladder.

CONCEPTCHECK 23-8

Which of the rungs on the distance ladder depends on an accurate measurement of parallax?

Answer appears at the end of the chapter.

23-5 The Hubble law relates the redshifts of remote galaxies to their distances from Earth

Whenever an astronomer finds an object in the sky that can be seen or photographed, the natural inclination is to attach a

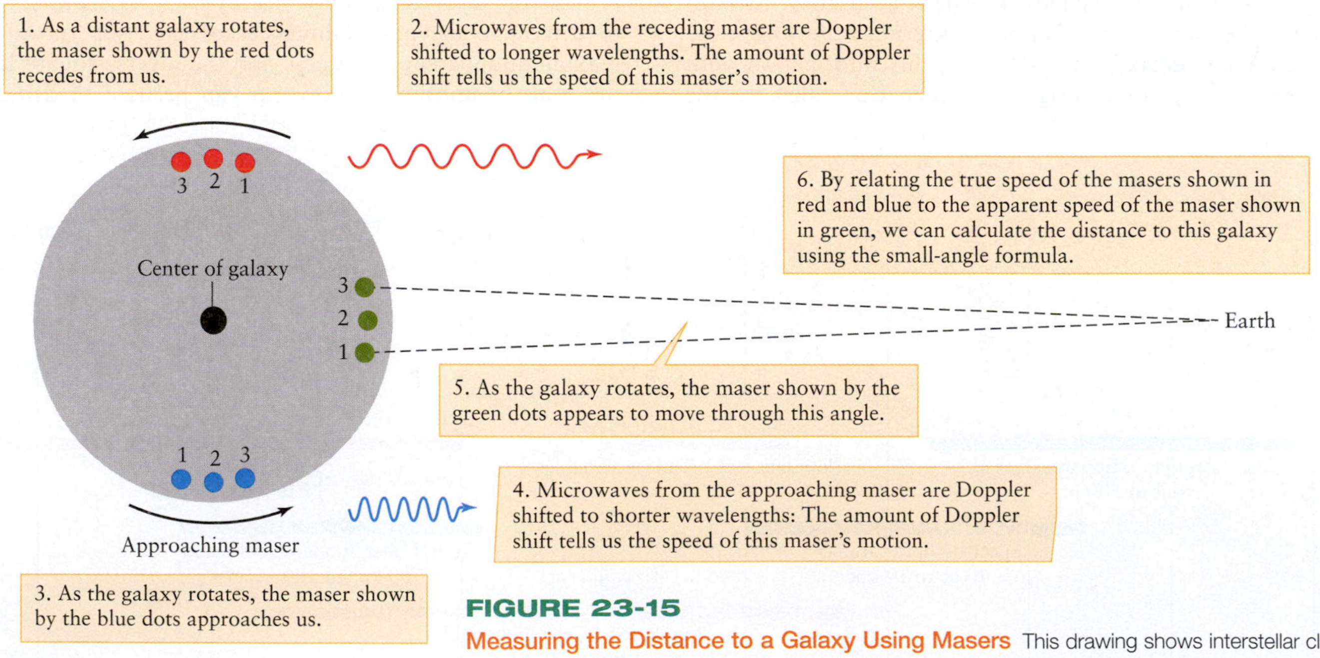

1. As a distant galaxy rotates, the maser shown by the red dots recedes from us.
2. Microwaves from the receding maser are Doppler shifted to longer wavelengths. The amount of Doppler shift tells us the speed of this maser's motion.
3 2 1
6. By relating the true speed of the masers shown in red and blue to the apparent speed of the maser shown in green, we can calculate the distance to this galaxy using the small-angle formula.
Center of galaxy
3
2
1
Earth
5. As the galaxy rotates, the maser shown by the green dots appears to move through this angle.
1 2 3
4. Microwaves from the approaching maser are Doppler shifted to shorter wavelengths: The amount of Doppler shift tells us the speed of this maser's motion.
Approaching maser
3. As the galaxy rotates, the maser shown by the blue dots approaches us.

FIGURE 23-15

Measuring the Distance to a Galaxy Using Masers This drawing shows interstellar clouds called masers (the colored dots) moving from position 1 to 2 to 3 as they orbit the center of a galaxy. The redshift and blueshift of microwaves from the masers shown in red and blue tell us their orbital speed. By relating this to the angle through which the masers shown in green appear to move in a certain amount of time, we can calculate the distance to the galaxy.

spectrograph to a telescope and record the spectrum. As long ago as 1914, Vesto M. Slipher, working at the Lowell Observatory in Arizona, began taking spectra of "spiral nebulae"—a name used before they were known to be galaxies. He was surprised to discover that of the 15 spiral nebulae he studied, the spectral lines of 11 were shifted toward the red end of the spectrum, indicating that they were moving away from Earth.

This marked dominance of redshifts over blueshifts was presented by Curtis in the 1920 Shapley-Curtis "debate" as evidence that these spiral nebulae could not be ordinary nebulae in our Milky Way Galaxy. It was only later that astronomers realized that the redshifts of spiral nebulae—that is, galaxies—reveal a basic law of our expanding universe.

Redshift, Distance, and the Hubble Law

During the 1920s, Edwin Hubble and Milton Humason photographed the spectra of many galaxies with the 100-inch (2.5-meter) telescope on Mount Wilson in California. By observing the apparent brightnesses and pulsation periods of Cepheid variables in these galaxies, they were also able to measure the distance to each galaxy (see Section 23-2). Hubble and Humason found that most galaxies show a redshift in their spectrum. They also found a direct correlation between the distance to a galaxy and its redshift:

The greater the redshift of a distant galaxy, the farther away it is

The more distant a galaxy, the greater its redshift and the more rapidly it is receding from us.

In other words, nearby galaxies are moving away from us slowly, and more distant galaxies are rushing away from us much more rapidly. Figure 23-16 shows this relationship for five representative elliptical galaxies. This universal recessional movement is referred to as the **Hubble flow.**

Hubble estimated the distances to a number of galaxies and the redshifts of those galaxies. The **redshift,** denoted by the symbol z, is found by taking the wavelength (λ) observed for a given spectral line, subtracting from it the ordinary, unshifted wavelength of that line (λ_0) to get the wavelength difference ($\Delta\lambda$), and then dividing that difference by λ_0:

Redshift of a receding object

$$z = \frac{\lambda - \lambda_0}{\lambda_0} = \frac{\Delta\lambda}{\lambda_0}$$

z = redshift of an object

λ_0 = ordinary, unshifted wavelength of a spectral line

λ = wavelength of that spectral line that is actually observed from the object

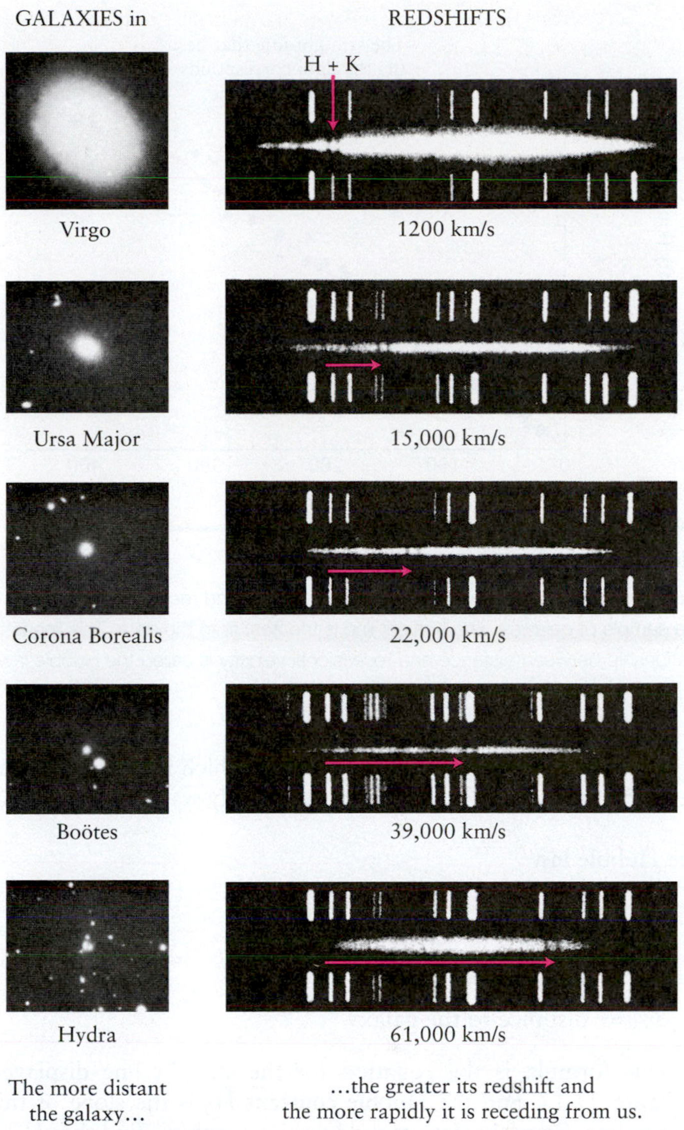

FIGURE 23-16 R I V U X G

Relating the Distances and Redshifts of Galaxies These five galaxies are arranged, from top to bottom, in order of increasing distance from us. All are shown at the same magnification. Each galaxy's spectrum is a bright band with dark absorption lines; the bright lines above and below it are a comparison spectrum of a light source at the observatory on Earth. The horizontal red arrows show how much the H and K lines of singly ionized calcium are redshifted in each galaxy's spectrum. Below each spectrum is the recessional velocity calculated from the redshift. The more distant a galaxy is, the greater its redshift. (Carnegie Observatories)

From the redshifts, Hubble used the Doppler formula to calculate the speed at which these galaxies are receding from us. Box 23-2 describes this calculation. Plotting the data on a graph of distance versus speed, Hubble found that the points lie near a straight line. Figure 23-17 is a modern version of Hubble's graph based on recent data.

This relationship between the distances to galaxies and their redshifts was one of the most important astronomical discoveries of the twentieth century. As we will see in Chapter 26, this relationship tells us that we are living in an expanding universe.

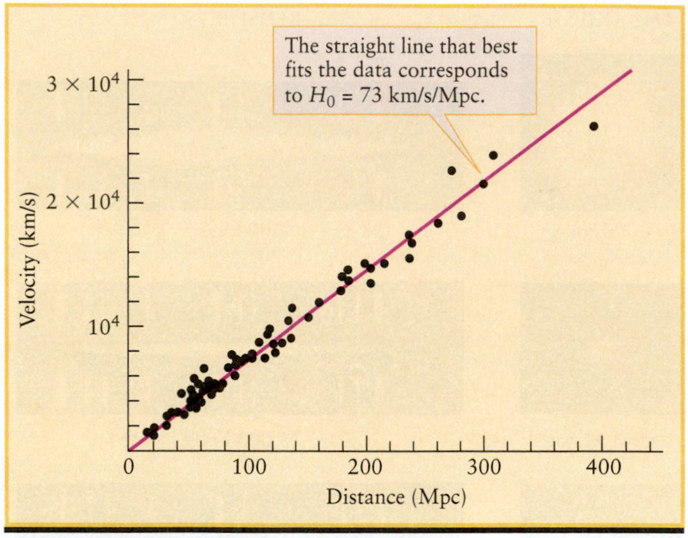

FIGURE 23-17

The Hubble Law This graph plots the distances and recessional velocities of a sample of galaxies. The straight line is the best fit to the data. This linear relationship between distance and recessional velocity is called the Hubble law.

In 1929, Hubble published this discovery, which is now known as the **Hubble law.** The Hubble law is most easily stated as a formula:

The Hubble law

$$v = H_0 d$$

v = recessional velocity of a galaxy

H_0 = Hubble constant

d = distance to the galaxy

This formula is the equation for the straight line displayed in Figure 23-17, and the **Hubble constant** H_0 is the slope of this straight line. From the data plotted on this graph we find that H_0 = 73 km/s/Mpc (say "73 kilometers per second per megaparsec"). In other words, for each million parsecs to a galaxy, the galaxy's speed away from us increases by 73 km/s. For example, a galaxy located 100 million parsecs from Earth should be rushing away from us with a speed of 7300 km/s. (In other books you may see the units of the Hubble constant written with exponents: 73 km s^{-1} Mpc^{-1}.)

CAUTION! A common misconception about the Hubble law is that *all* galaxies are moving away from the Milky Way. The reality is that galaxies have their own neighbor-induced motions relative to one another, due to their mutual gravitational attraction of neighboring galaxies. Astronomers call this neighbor-induced motion *intrinsic velocity.* For nearby galaxies, the speed of the Hubble flow is small compared to these intrinsic velocities. Hence, some of the nearest galaxies, including the Andromeda Galaxy (shown in Figure 23-3), are actually approaching us and have blueshifts rather than redshifts. But for distant galaxies, the Hubble speed $v = H_0 d$ is much greater than any intrinsic motion that the galaxies might have. Even if the intrinsic velocity of such a distant galaxy is toward the Milky Way, the fast-moving Hubble flow sweeps that galaxy away from us.

CONCEPTCHECK 23-9

In Figure 23-17, which follows Hubble's law, consider galaxies at 100 Mpc and 200 Mpc. Which galaxy has the larger redshift?

Answer appears at the end of the chapter.

Pinning Down the Hubble Constant

A precise value of the Hubble constant has been a topic of heated debate among astronomers for several decades. The problem is that while redshifts are relatively easy to measure in a reliable way, distances to galaxies (especially remote galaxies) are not, as we saw in Section 23-4. Hence, astronomers who use different methods of determining galactic distances have obtained different values of H_0. To see why different values are measured, it is helpful to rewrite the Hubble law as

$$H_0 = \frac{v}{d}$$

This equation shows that if galaxies of a given recessional velocity (v) are far away (so d is large), the Hubble constant H_0 must be relatively small. But if these galaxies are relatively close (so d is small), then H_0 must have a larger value.

In the past, astronomers who used Type Ia supernovae for determining galactic distances found galaxies to be farther away than their colleagues who employed the Tully-Fisher relation. Therefore, the supernova adherents found values of H_0 in the range from about 40 to 65 km/s/Mpc, while the values from the Tully-Fisher relation ranged from about 80 to 100 km/s/Mpc.

In the past few years, the Hubble Space Telescope has been used to observe Cepheid variables with unprecedented precision and in galaxies as far away as 30 Mpc (100 million ly). These observations and others suggest a value of the Hubble constant of about 73 km/s/Mpc, with an uncertainty of no more than 10%. At the same time, reanalysis of the supernova and Tully-Fisher results have brought the values of H_0 from these techniques closer to the Hubble Space Telescope Cepheid value. We will adopt the value H_0 = 73 km/s/Mpc in this book.

Determining the value of H_0 has been an important task of astronomers for a very simple reason: The Hubble constant is one of the most important numbers in all astronomy. It expresses the rate at which the universe is expanding and, as we will see in Chapter 25, even helps give the age of the universe. Furthermore, the Hubble law can be used to determine the distances to extremely remote galaxies. If the redshift of a galaxy is known, the Hubble law can be used to determine its distance from Earth. Thus, the value of the Hubble constant helps determine the distances of the most remote objects in the universe that astronomers can observe.

Because the value of H_0 remains somewhat uncertain, astronomers often express the distance to a remote galaxy simply in terms of its redshift z (which can be measured very accurately). Given the redshift, the distance to this galaxy can be calculated from the Hubble law, but the distance obtained in this way will depend on the particular value of H_0 adopted. Rather than going through these calculations, an astronomer might simply say that a certain galaxy is "at $z = 0.128$." From the Hubble law relating redshift and distance, this redshift makes it clear that the galaxy in question is more distant than one at $z = 0.120$ but not as distant as one

BOX 23-2 TOOLS OF THE ASTRONOMER'S TRADE

The Hubble Law, Redshifts, and Recessional Velocities

Suppose that you aim a telescope at a distant galaxy. The galaxy will be moving away from you in the Hubble flow. You take a spectrum of the galaxy and find that the spectral lines are shifted toward the red end of the spectrum. For example, a particular spectral line whose normal wavelength is λ_0 appears in the galaxy's spectrum at a longer wavelength λ. The spectral line has thus been shifted by an amount $\Delta\lambda = \lambda - \lambda_0$. The redshift of the galaxy, z, is given by

$$z = \frac{\lambda - \lambda_0}{\lambda_0} = \frac{\Delta\lambda}{\lambda_0}$$

The redshift means that the galaxy is receding from us. According to the Hubble law, the recessional velocity v of a galaxy is related to its distance d from Earth by

$$v = H_0 d$$

where H_0 is the Hubble constant. We can rewrite this equation

$$d = \frac{v}{H_0}$$

Given the value of H_0, we can find the distance d to the galaxy if we know how to determine the recessional velocity v from the redshift z.

If the redshift is not too great (so that the redshift z is much less than 1), we can use the following Doppler shift equation to find the recessional speed:

$$z = \frac{v}{c} \text{ (valid for low speeds only)}$$

where c is the speed of light. For example, a 5% shift in wavelength ($z = 0.05$) corresponds to a recessional velocity of 5% of the speed of light ($v = 0.05c$).

For larger redshifts (around $z = 1$ or greater), a more complicated equation is needed to determine the recessional velocity. A small redshift can be explained by the Doppler shift as we have used here. However, as we will see in Chapter 25, larger redshifts arise from a large *expansion of space* and cannot be described as a simple Doppler shift. In fact, for galaxies at redshifts greater than about $z > 1.5$, recessional speeds exceed the speed of light. Due to the unique attributes of expanding space, faster-than-light recession is not a violation of Einstein's laws.

EXAMPLE: When measured in a laboratory on Earth, the so-called K line of singly ionized calcium has a wavelength $\lambda_0 = 393.3$ nm. But when you observe the spectrum of the giant elliptical galaxy NGC 4889, you find this spectral line at $\lambda = 401.8$ nm. Using $H_0 = 73$ km/s/Mpc, find the distance to this galaxy.

Situation: We are given the values of λ and λ_0 for a line in this galaxy's spectrum. Our goal is to determine the galaxy's distance d.

Tools: We use the relationship $z = (\lambda - \lambda_0)/\lambda_0$ to determine the redshift. We then use the appropriate formula to determine the galaxy's recessional velocity v, and finally use the Hubble law to determine the distance to the galaxy.

Answer: The redshift of this galaxy is

$$z = \frac{401.8 \text{ nm} - 393.3 \text{ nm}}{393.3 \text{ nm}} = 0.0216$$

This value is substantially less than 0.1, so we can safely use the low-redshift relationship between recessional speed and redshift: $v = zc$. So NGC 4889 is moving away from us with a speed

$$v = zc = (0.0216)(3 \times 10^5 \text{ km/s}) = 6480 \text{ km/s}$$

Using $H_0 = 73$ km/s/Mpc in the Hubble law, the distance to this galaxy is

$$d = \frac{v}{H_0} = \frac{6480 \text{ km/s}}{73 \text{ km/s/Mpc}} = 89 \text{ Mpc}$$

Review: This galaxy is receding from us at 0.0216 (2.16%) of the speed of light, and it is 89 megaparsecs (290 million light-years) away. Thus the light we see from NGC 4889 today left the galaxy 290 million years ago, even before the first dinosaurs appeared on Earth.

at $z = 0.130$. When astronomers use redshift to describe distance, they are making use of the following general rule:

The greater the redshift of a distant galaxy, the greater its distance.

CONCEPTCHECK 23-10

Although nearly all distant galaxies have measureable redshifts, the relatively nearby Andromeda Galaxy exhibits an overall blueshift. What does this mean about the Andromeda Galaxy's movement?

CALCULATIONCHECK 23-1

What is the redshift z-value for a galaxy that has a galaxy spectral line shifted to 725.6 nm when a stationary object would emit the line at 656.3 nm?

CALCULATIONCHECK 23-2

What is the distance to a galaxy that is observed to have a recessional velocity of 10,000 km/s?

Answers appear at the end of the chapter.

23-6 Galaxies are grouped into clusters and superclusters

Galaxies are not scattered randomly throughout the universe but are found in **clusters.** Figure 23-18 shows one such cluster. Like stars within a star cluster, the members of a cluster of galaxies are in continual motion around one another. They appear to be at rest only because they are so distant from us.

FIGURE 23-18 R I V U X G

The Hercules Cluster This irregular cluster of galaxies is about 200 Mpc (650 million ly) from Earth. The Hercules cluster contains many spiral galaxies, often associated in pairs and small groups. (NOAO)

Clusters of Galaxies: Poor and Rich, Regular and Irregular

A cluster is said to be either **poor** or **rich,** depending on how many galaxies it contains. Poor clusters, which far outnumber rich ones, are often called **groups.** For example, the Milky Way Galaxy, the Andromeda Galaxy (M31), and the Large and Small Magellanic clouds belong to a poor cluster familiarly known as the **Local Group.** The Local Group contains more than 40 galaxies, most of which are dwarf ellipticals (see Figure 23-9). Figure 23-19 is a map of a portion of the Local Group.

In recent years, astronomers have discovered several previously unknown dwarf elliptical galaxies in the Local Group. Two recently discovered dwarf elliptical galaxies are Leo T and the Canis Major Dwarf. Tidal forces exerted by the Milky Way on the Canis Major Dwarf are causing this dwarf galaxy to gradually disintegrate and leave a trail of debris behind it (Figure 23-20). This galaxy is about 13 kpc (42,000 ly) from the center of the Milky Way Galaxy and a mere 8 kpc (25,000 ly) from Earth (about the same as the distance from Earth to the center of the Milky Way).

We may never know the total number of galaxies in the Local Group, because dust in the plane of the Milky Way obscures our view over a considerable region of the sky. Nevertheless, we can be certain that no additional large spiral galaxies are hidden by the Milky Way. Radio astronomers would have detected 21-cm radiation from them, even though their visible light is completely absorbed by interstellar dust.

The nearest fairly rich cluster is the Virgo cluster, a collection of more than 2000 galaxies covering a 10° × 12° area of the sky. Figure 23-8 shows a portion of this cluster. One member of this cluster not shown in Figure 23-8 is the spiral galaxy M100;

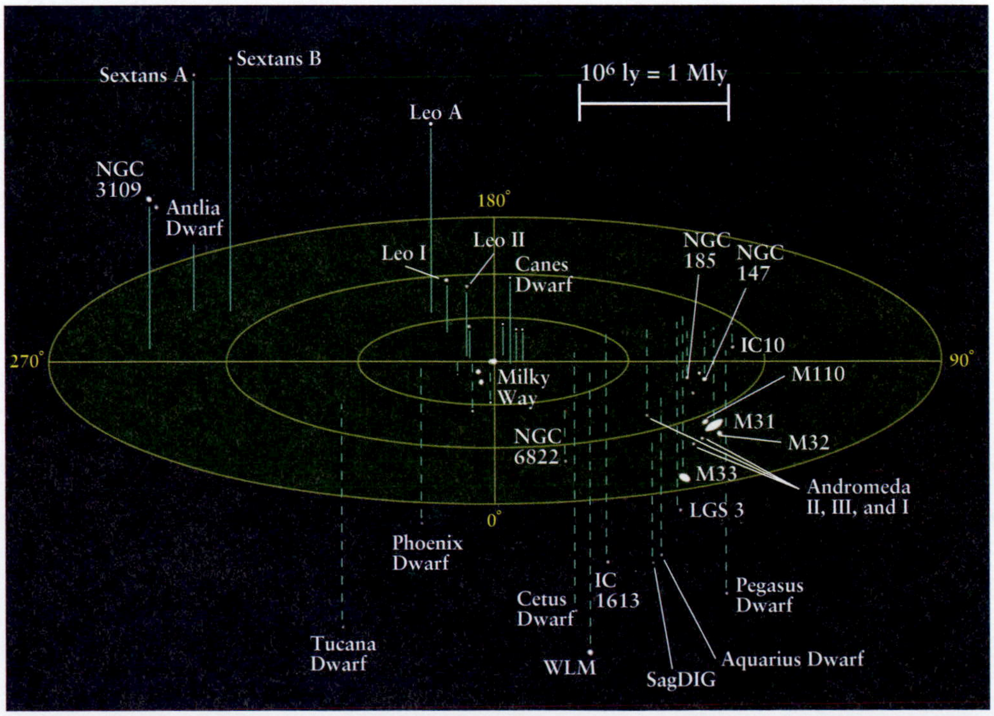

FIGURE 23-19

The Local Group This illustration shows the relative positions of the galaxies that comprise the Local Group, a poor, irregular cluster of which our Galaxy is part. (The blue rings represent the plane of the Milky Way's disk; 0° is the direction from Earth toward the Milky Way's center. Solid and dashed lines point to galaxies above and below the plane, respectively.) The largest and most massive galaxy in the Local Group is M31, the Andromeda Galaxy; in second place is the Milky Way, followed by the spiral galaxy M33. Both the Milky Way and M31 are surrounded by a number of small satellite galaxies. (Adapted from © Richard Powell, www.atlasoftheuniverse.com)

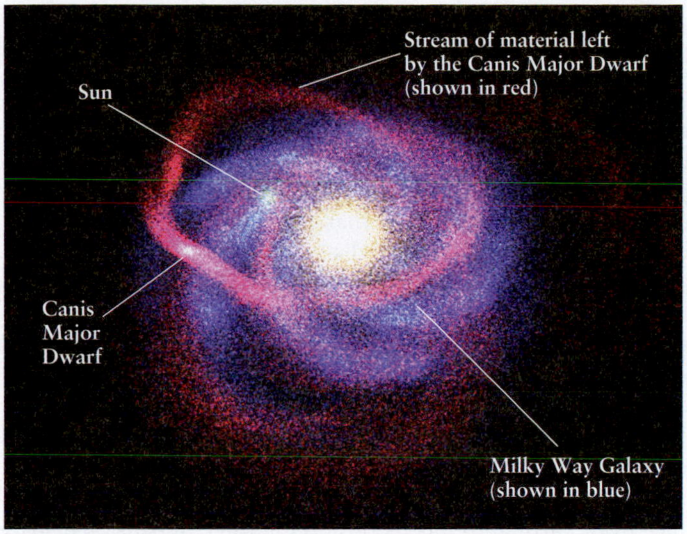

ANIMATION 23-1 **FIGURE 23-20** R I V U X G

The Canis Major Dwarf Discovered in 2003, this dwarf elliptical galaxy is actually slightly closer to Earth than is the center of the Milky Way Galaxy. This illustration shows the stream of material left behind by the Canis Major Dwarf as it orbits the Milky Way. This material is torn away by the Milky Way's tidal forces (see Section 4-8). (R. Ibata et al., Observatoire de Strasbourg/Université Louis Pasteur; 2MASS; and NASA)

FIGURE 23-21 R I V U X G

The Coma Cluster This rich, regular cluster is about 90 Mpc (300 million light-years) from Earth. Almost all of the spots of light in this image are individual galaxies of the cluster. Two giant elliptical galaxies, NGC 4889 and NGC 4874, dominate the center of the cluster. The bright star at the upper right is within our own Milky Way Galaxy, a million times closer than any of the galaxies shown here. (O. Lopez-Cruz and I. K. Shelton, University of Toronto/NOAO/AURA/NSF)

measurements of Cepheid variables in M100 (see Figure 23-4) give a distance of about 17 Mpc (56 million ly). The Tully-Fisher relation and observations of Type Ia supernovae (see Figure 23-13) give similar distances to this cluster. The overall diameter of the cluster is about 3 Mpc (9 million ly).

The center of the Virgo cluster is dominated by three giant elliptical galaxies. You can see two of these, M84 and M86, in Figure 23-8. The diameter of each of these enormous galaxies is comparable to the 750-kpc distance between the Milky Way and the Andromeda Galaxy. In other words, one giant elliptical is approximately the same size as the entire Local Group!

Astronomers also categorize clusters of galaxies as regular or irregular, depending on the overall shape of the cluster. The Virgo cluster, for example, is called an **irregular cluster,** because its galaxies are scattered throughout a sprawling region of the sky. Our own Local Group is also an irregular cluster. In contrast, a **regular cluster** has a distinctly spherical appearance, with a marked concentration of galaxies at its center.

The nearest example of a rich, regular cluster is the Coma cluster, located about 90 Mpc (300 million light-years) from us toward the constellation Coma Berenices (Berenice's Hair) (Figure 23-21). Despite its great distance, telescopic images of this cluster show more than 1000 galaxies. The Coma cluster almost certainly contains many thousands of dwarf ellipticals, so the total membership of the cluster may be as many as 10,000 galaxies. The core of the Coma cluster is dominated by two giant ellipticals surrounded by many normal-sized galaxies.

The overall shape of a cluster is related to the dominant types of galaxies it contains. Rich, regular clusters contain mostly elliptical and lenticular galaxies. For example, about 80% of the brightest galaxies in the Coma cluster (see Figure 23-21) are ellipticals; only a few spiral galaxies are scattered around the cluster's outer regions. Irregular clusters, such as the Virgo cluster and the Hercules cluster shown in Figure 23-18, have a more even mixture of galaxy types.

CONCEPTCHECK 23-11

How would a photograph of a rich cluster's appearance compare to that of a poor cluster?

Answer appears at the end of the chapter.

Superclusters: Clusters of Clusters of Galaxies

Clusters of galaxies are themselves grouped together in huge associations called **superclusters.** A typical supercluster contains dozens of individual clusters spread over a region of space up to 45 Mpc (150 million ly) across. Figure 23-22 shows the distribution of clusters in our part of the universe. The nearer ones out to the Virgo cluster, including our own Local Group, are members of the *Local Supercluster.* The other clusters shown in Figure 23-22 belong to other superclusters. The most massive cluster in the local universe is called the *Great Attractor;* its gravity is so great that the Milky Way, as well as the rest of the Local Supercluster, is moving toward it at speeds of several hundred kilometers per second.

Observations indicate that unlike clusters, superclusters are not bound together by gravity. That is, most clusters in each supercluster are drifting away from most of the other clusters in that same supercluster. Furthermore, the superclusters are all moving away from each other due to the Hubble flow.

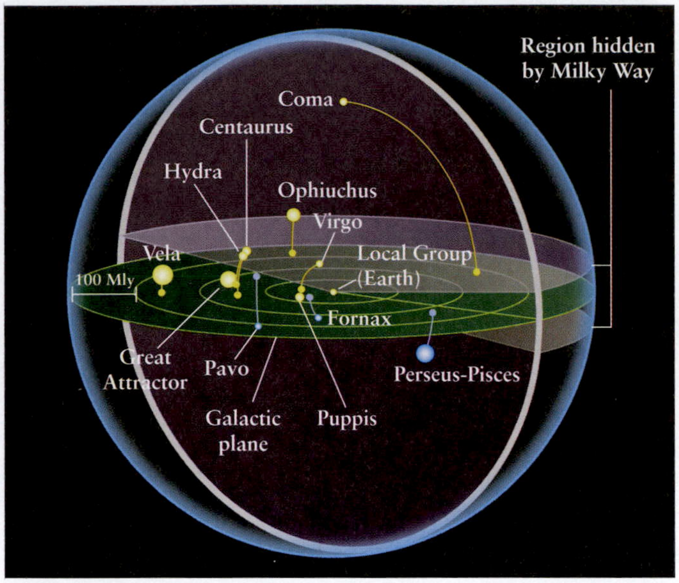

FIGURE 23-22
Nearby Clusters of Galaxies This illustration shows a sphere of space 800 million ly (250 Mpc) in diameter centered on Earth in the Local Group. Each spherical dot represents a cluster of galaxies. To better see the three-dimensionality of this figure, colored arcs are drawn from each cluster to the green plane, which is an extension of the plane of the Milky Way outward into the universe. Note that clusters of galaxies are unevenly distributed here, as they are elsewhere in the universe.

Cosmic Voids and Sheets: The Distribution of Superclusters

Superclusters of galaxies are not spread uniformly across the universe, but are found in vast sheets separated by truly immense voids

Since the 1980s, astronomers have been working to understand how superclusters are distributed in space. Some of this structure is revealed by maps such as that shown in Figure 23-23, which displays the positions on the sky of 1.6 million galaxies. Such maps reveal that superclusters are not randomly distributed, but seem to lie along filaments. But to comprehend more fully the distribution of superclusters, it is necessary to map their positions in three dimensions. Their positions are mapped by measuring both the position of a galaxy on the sky as well as the galaxy's redshift. Using the Hubble law (see Section 23-5), astronomers can use each galaxy's redshift to estimate its distance from Earth and thus its position in three-dimensional space.

The first three-dimensional maps of this kind were made in the 1970s and included measurements of a few hundred galaxies. Collecting the data for such maps required many months or years of telescopic observations. Technology for astronomy has advanced tremendously since then, and it is now possible to measure the redshifts of 400 galaxies in a single hour!

Two extensive galaxy maps are from the Sloan Digital Sky Survey and from the Two Degree Field Galactic Redshift Survey (2dF). (The name refers to the 2° field of view of the telescope used for the observations, which is unusually wide for a research

FIGURE 23-23 R I V U X G
Structure in the Nearby Universe This composite infrared image from the 2MASS (Two-Micron All-Sky Survey) project shows the light from 1.6 million galaxies. In this image, the entire sky is projected onto an oval; the blue band running vertically across the center of the image is light from the plane of the Milky Way. Note that galaxies form a lacy, filamentary structure. Note also the large, dark voids that contain few galaxies. (Two Micron All Sky Survey [2MASS])

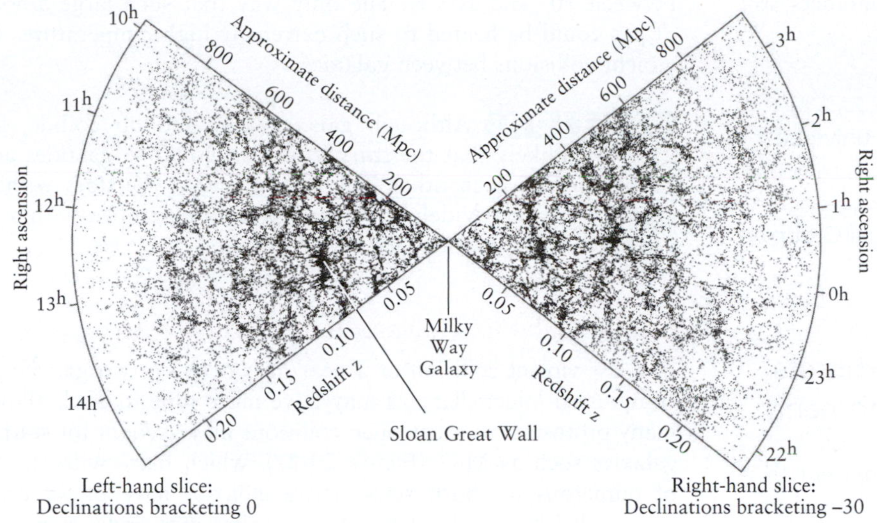

(a) The 2dF galaxy survey

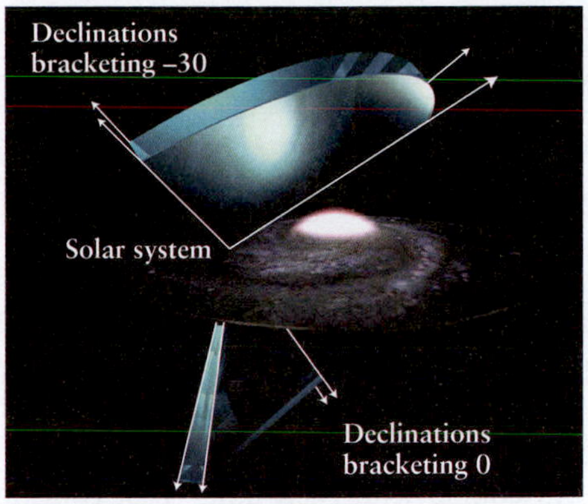

(b) Fields of view in the 2dF survey

ANIMATION 23-2

FIGURE 23-24

The Large-Scale Distribution of Galaxies (a) This map shows the distribution of 62,559 galaxies in two wedges extending out to redshift $z = 0.25$. (For an explanation of right ascension and declination, see Box 2-1.) Note the prominent voids surrounded by thin regions full of galaxies. (b) The two wedges shown in (a) lie roughly perpendicular to the plane of the Milky Way. These were chosen to avoid the obscuring dust that lies in our Galaxy's plane. (Courtesy of the 2dF Galaxy Redshift Survey Team/Australian Astronomical Observatory)

telescope.) Figure 23-24a shows a map made from 2dF measurements of more than 60,000 galaxies. This particular map encompasses two wedge-shaped slices of the universe, one on either side of the plane of the Milky Way (Figure 23-24b). Earth (in the Milky Way) is at the apex of the wedge-shaped map; each dot represents a galaxy. The measurements used to create this map included galaxies with redshifts as large as $z = 0.25$, corresponding to a recessional velocity of 66,000 km/s. Using a Hubble constant of 73 km/s/Mpc, you can determine that the map in Figure 23-24a extends out to a distance of nearly 1000 Mpc, or 3 *billion* light-years from Earth.

Maps such as that shown in Figure 23-24a reveal enormous **voids** where exceptionally few galaxies are found. (These voids were first discovered in 1978 in a pioneering study by Stephen Gregory and Laird Thompson at the Kitt Peak National Observatory.) These voids are roughly spherical and measure 30 to 120 Mpc (100 million to 400 million ly) in diameter. They are not entirely empty, however. There is evidence for hydrogen clouds in some voids, while others may be subdivided by strings of dim galaxies.

Figure 23-24a shows that most galaxies are concentrated in sheets on the surfaces between voids. These sheets can be more than 100 Mpc long and several megaparsecs thick. This pattern is similar to the structure of a sponge or soap bubbles, with sheets of soap film (analogous to galaxies) surrounding air bubbles (analogous to voids) (Figure 23-25). These titanic sheets of galaxies are the largest structures known in the universe: On scales much larger than 100 Mpc, the distribution of galaxies in the universe appears to be roughly uniform. As we will see in Chapter 26, this pattern of sheets and voids contains important clues about how clusters of galaxies formed in the early universe.

There is an unusually large and dense sheet in Figure 23-24a called the Sloan Great Wall; this is the largest known structure in the universe! This structure is about 1.4 billion light-years in length (which is almost 2% of the visible universe). The Sloan

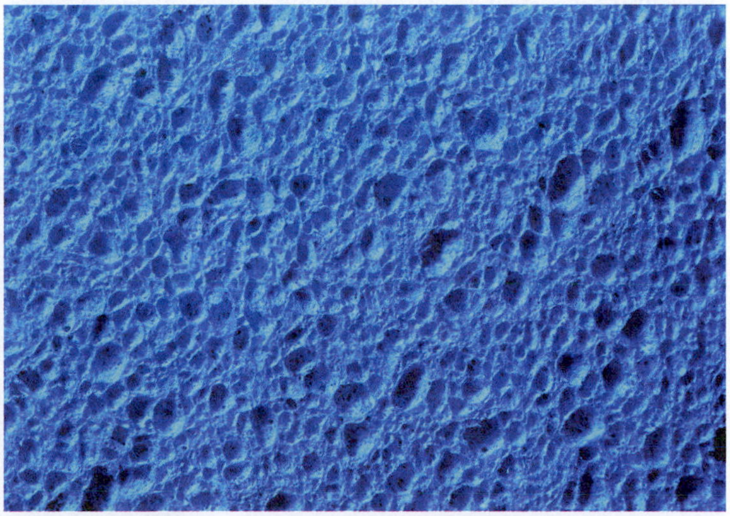

FIGURE 23-25

Foamy Structure of the Universe A sponge that recreates the distribution of bright clusters of galaxies throughout the universe. The empty spaces in the foam are analogous to the voids found throughout the universe. The spongy regions are analogous to the locations of most of the superclusters of galaxies. (Image Source/Super Stock)

Great Wall contains several superclusters, and astronomers still have not figured out how this large structure formed.

CONCEPTCHECK 23-12

Are superclusters evenly distributed throughout the universe?

CONCEPTCHECK 23-13

What is the most common type of galaxy in our Local Group?

Answers appear at the end of the chapter.

23-7 Colliding galaxies produce starbursts, spiral arms, and other spectacular phenomena

Occasionally, two galaxies within a cluster or from adjacent clusters can collide with each other. Past collisions have hurled vast numbers of stars into intergalactic space. In some cases, we can even observe a collision in progress, a cosmic catastrophe that gives birth to new stars. And astronomers can predict collisions that will not take place for billions of years, such as the collision that is fated to occur between the Andromeda Galaxy and our own Milky Way Galaxy.

High-Speed Galaxy Collisions: Shredding Gas and Dust

When two galaxies collide at high speed, the huge clouds of interstellar gas and dust in the galaxies slam into each other—like two cars locking bumpers in a collision. In this way, two colliding galaxies can leave behind their interstellar gas and dust as the stars in each galaxy pass through the impact site.

The best evidence that such collisions take place is that many rich clusters of galaxies are strong sources of X-rays (Figure 23-26). This emission reveals the presence of substantial amounts of hot **intracluster gas** (that is, gas within the cluster) at temperatures between 10^7 and 108 K. The only way that such large amounts of gas could be heated to such extremely high temperatures is in violent collisions between galaxies.

CAUTION! Although galaxies can and do collide, it is highly unlikely that the *stars* from two colliding galaxies actually run into each other. The reason is that the stars within a galaxy are very widely separated from one another, with a tremendous amount of space between them.

Gentle Galactic Collisions and Starbursts

In a less violent collision or a near miss between two galaxies, the compressed interstellar gas may have more time to cool, allowing many protostars to form. Such collisions may account for **starburst galaxies** such as M82 (Figure 23-27), which blaze with the light of numerous newborn stars. These galaxies have bright centers surrounded by clouds of warm interstellar dust, indicating recent, vigorous star birth. Their warm dust is so abundant that starburst galaxies are among the most luminous objects in the universe at infrared wavelengths. (The right-hand image at the opening of Chapter 6 shows the infrared emission from M82's warm dust.)

The starburst galaxy M82 shown in Figure 23-27 also shows the effects of strong winds from young, luminous stars. It also contains a number of luminous globular clusters. Unlike the globular clusters in our Galaxy, whose stars are about 12.5 billion years old, those in M82 are no more than 600 million years old. These young globular clusters are another sign of recent star formation.

M82 is one member of a nearby cluster of galaxies that includes the beautiful spiral galaxy M81 and a fainter elliptical companion called NGC 3077 (Figure 23-28a). Radio surveys of that region of the sky reveal enormous streams of hydrogen gas connecting the three galaxies (Figure 23-28b). The loops and twists in these streamers suggest that the three galaxies have had several close encounters over the ages. A similar stream of hydrogen gas connects our Galaxy with its second nearest neighbor, the

(a) An X-ray image of Abell 2029 shows emission from hot gas.

R I V U **X** G

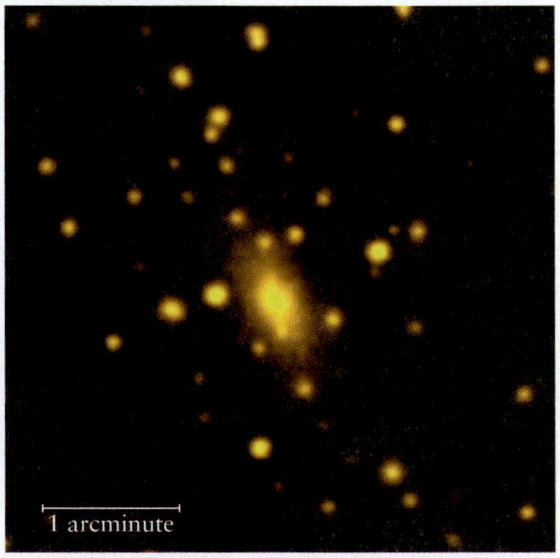

(b) A visible-light image of Abell 2029 shows the cluster's galaxies.

R I **V** U X G

FIGURE 23-26

X-ray Emission from a Cluster of Galaxies **(a)** An X-ray image of this cluster of galaxies shows emission from hot gas between the galaxies. The gas was heated by collisions between galaxies within the cluster. **(b)** The galaxies themselves are too dim at X-ray wavelengths to be seen in (a), but they are apparent at visible wavelengths. This cluster, one of many cataloged by the UCLA astronomer George O. Abell, is about 300 Mpc (1 billion ly) from Earth in the constellation Serpens. (a: NASA, CXC, and University of California, Irvine/A. Lewis et al. b: Palomar Observatory DSS)

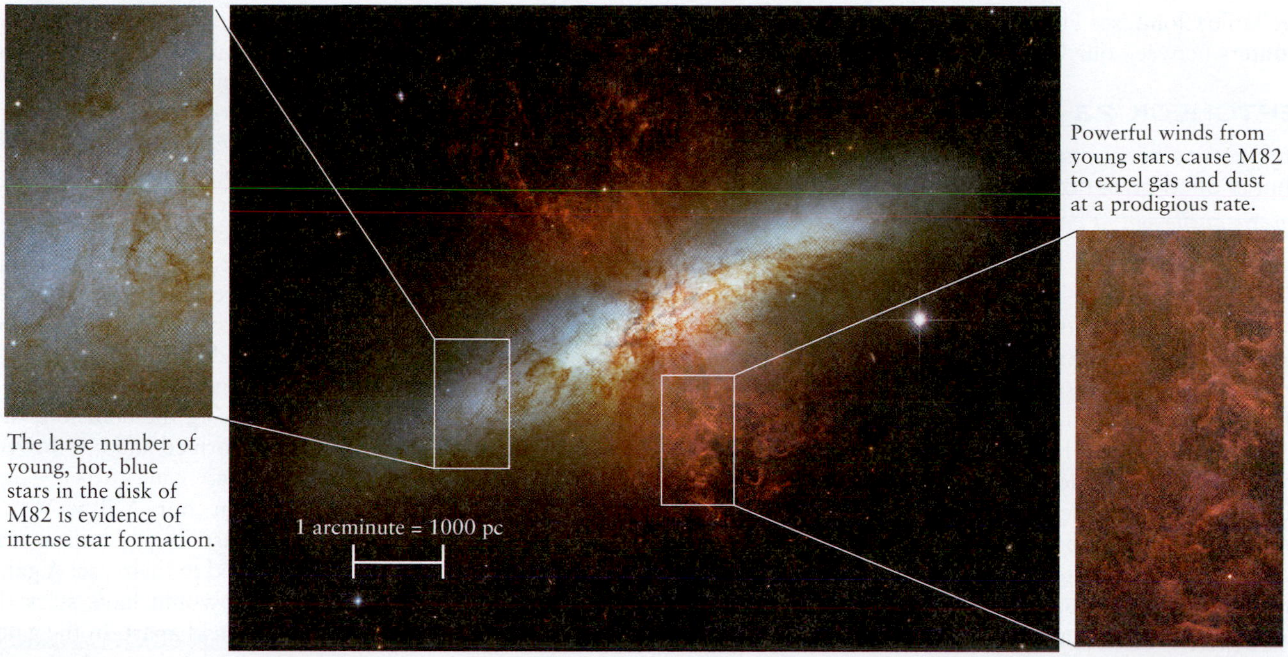

The large number of young, hot, blue stars in the disk of M82 is evidence of intense star formation.
1 arcminute = 1000 pc
Powerful winds from young stars cause M82 to expel gas and dust at a prodigious rate.

VIDEO 23-3

FIGURE 23-27 R I V U X G

A Starburst Galaxy Prolific star formation is occurring at the center of the irregular galaxy M82, which lies about 3.6 Mpc (12 million ly) from Earth in the constellation Ursa Major. M82 also contains an unusual X-ray source, shown in Figure 21-16. (NASA; ESA; and the Hubble Heritage Team, STScI/AURA)

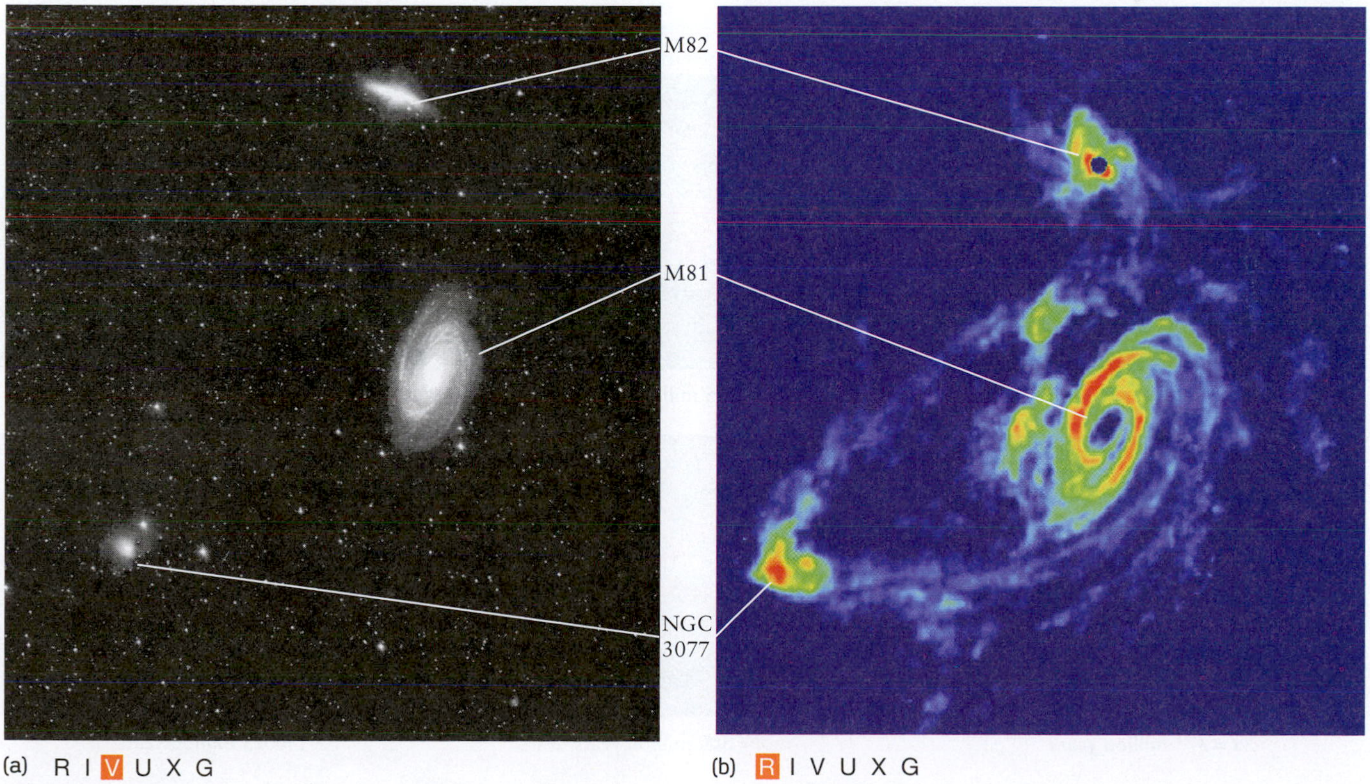

FIGURE 23-28

The M81 Group **(a)** The starburst galaxy M82 is part of a cluster of about a dozen galaxies. This wide-angle visible-light photograph shows the three brightest galaxies of the cluster. The area shown is about 1° across. **(b)** This false-color radio image of the same region, created from data taken by the Very Large Array, shows streamers of hydrogen gas that connect the three bright galaxies as well as several dim ones. (a: Palomar Sky Survey; b: Image courtesy of NRAO/AUI)

Large Magellanic Cloud (see Figure 23-12), suggesting a history of close encounters between our Galaxy and the LMC.

CONCEPTCHECK 23-14

What energy source accounts for intergalactic gas between galaxies that is often quite hot?

Answer appears at the end of the chapter.

Tidal Forces and Galaxy Mergers

> Galaxies need not actually collide to exert strong forces on each other

Tidal forces between colliding galaxies can deform the galaxies from their original shapes, just as the tidal forces of the Moon on Earth deform the oceans and help give rise to the tides (see Section 4-8, especially Figure 4-27). The galactic deformation is so great that thousands of stars can be hurled into intergalactic space along huge, arching streams. (This same effect has stripped material away from the Canis Major Dwarf Galaxy as it orbits the Milky Way, as shown in Figure 23-20.) Supercomputer simulations of such collisions show that while some of the stars are flung far and wide, other stars slow down and the galaxies may merge.

Figure 23-29 shows one such simulation of stars (but no gas or dust). As the two galaxies pass through each other, they are severely distorted by gravitational interactions and throw out a pair of extended tails. The interaction also prevents the galaxies in the simulation from continuing on their original paths. Instead, they fall back together for a second encounter (at 625 million years). The simulated galaxies merge soon thereafter, leaving a single object. *Cosmic Connections: When Galaxies Collide* explores a real-life example of two galaxies that are colliding in just this way.

Our own Milky Way Galaxy is expected to undergo a galactic collision like that shown in Figure 23-28. The Milky Way and the Andromeda Galaxy, shown in Figure 23-3, are actually approaching each other today fast enough to travel the Earth-Moon distance in one hour and should collide in 3.75 billion years or so (Figure 23-30). (Recall that our Sun will not become a red giant for more than 5 billion years.) During the collision, the sky will light up with a plethora of newly formed stars, followed in rapid succession by a string of supernovae.

The complete merger between our Milky Way and the Andromeda Galaxy is expected to last about 2 billion years. Recall that stars are very far apart compared to their size: A galaxy scaled down in size to the United States would have stars the size of human cells, each about a football field apart. In the galactic collision, stars are not expected to hit each other and our solar system should remain intact.

When two galaxies merge, the result is a bigger galaxy. If this new galaxy is located in a rich cluster, it may capture and devour additional galaxies, growing to enormous dimensions

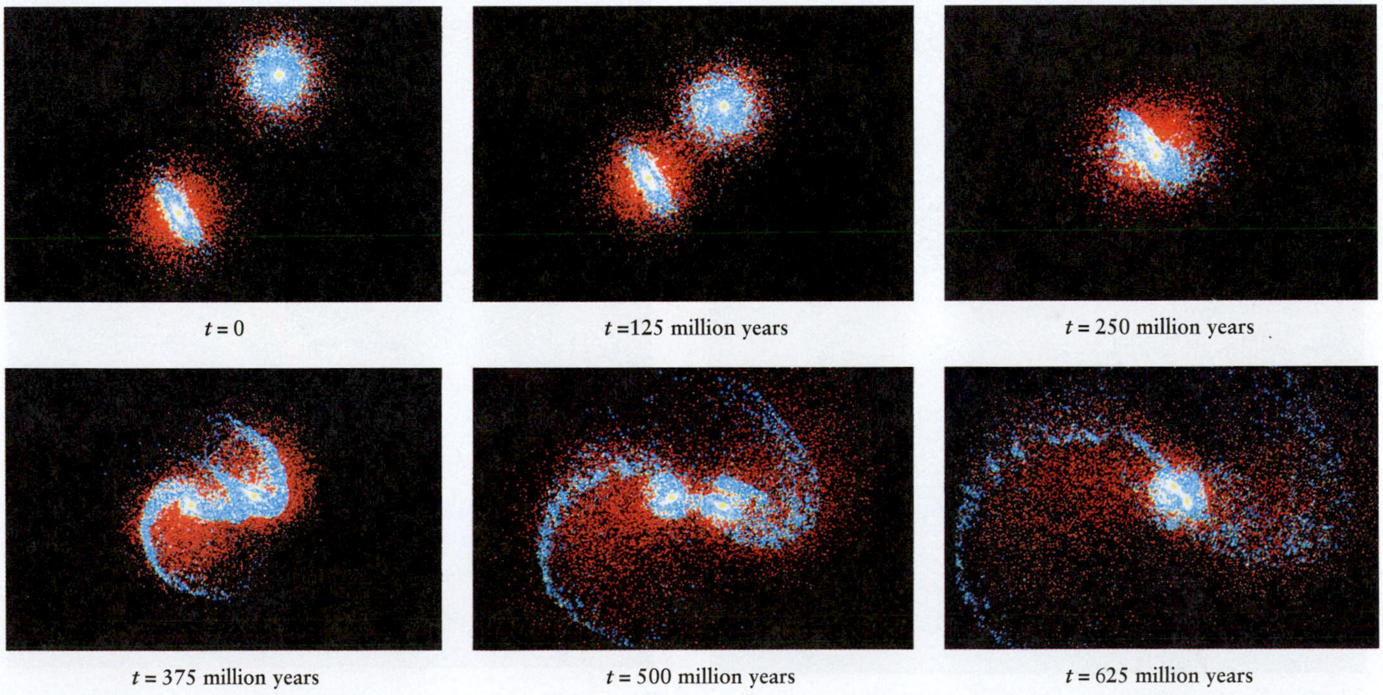

FIGURE 23-29
A Simulated Collision Between Two Galaxies These frames from a supercomputer simulation show the collision and merger of two galaxies accompanied by an ejection of stars into intergalactic space. Stars in the disk of each galaxy are colored blue, while stars in their central bulges are yellow-white. Red indicates dark matter that surrounds each galaxy. The frames progress at 125-million-year intervals. Compare the bottom frames with the image of the Antennae galaxies in the *Cosmic Connections: When Galaxies Collide*. (Joshua Edward Barnes, Institute for Astronomy, University of Hawaii)

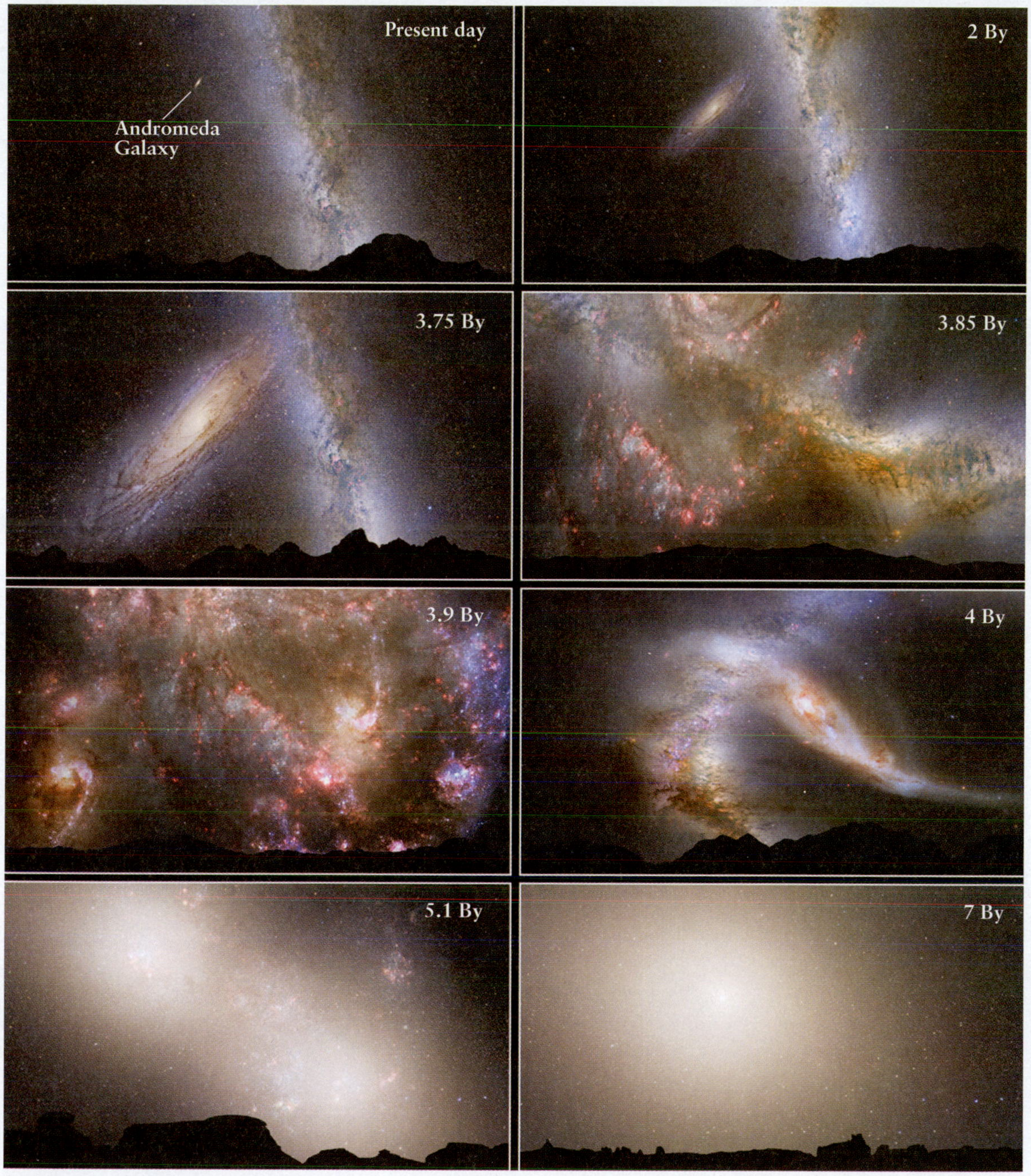

FIGURE 23-30
Collision Between Andromeda and the Milky Way Computer simulations of the merger show what the night sky might look like from Earth. Two billion years from now, Andromeda will look much bigger, but the real collision will not begin until about 3.75 billion years from now. In 3.85 billion years from now, the sky will be engulfed in new star formation, which will have largely ended by 4 billion years, leaving the galaxies to completely merge. After 7 billion years, our solar system is expected to end up much farther from the galactic core than it is today. (NASA; ESA; Z. Levay and R. van der Marel, STScI; T. Hallas, and A. Mellinger)

by **galactic cannibalism.** Cannibalism differs from mergers in that the galaxy that does the devouring is bigger than its "meal," whereas merging galaxies are about the same size.

Many astronomers suspect that galactic cannibalism is the reason that giant ellipticals are so huge. As we have seen, giant galaxies typically occupy the centers of rich clusters. In many cases, smaller galaxies are located around these giants (see Figure 23-8 and Figure 23-21). As they pass through the extended halo of a giant elliptical, these smaller galaxies slow down and are eventually devoured by the larger galaxy.

COSMIC CONNECTIONS

When Galaxies Collide

VIDEO 23.4

Although galaxies can collide at very high speeds by Earth standards, they are so vast that a collision can last hundreds of millions of years. Understanding what happens during a galactic collision requires ideas about tidal forces (Chapter 4), star formation (Chapter 18), and stellar evolution (Chapter 19).

1. One example of a galactic collision is the pair of galaxies called the Antennae, which lie 19 Mpc (16 million ly) from Earth in the constellation Corvus (the Crow). They probably began to interact several hundred million years ago.

Tidal forces between the galaxies pulled out these long "tidal tails" 200 to 300 million years ago.

R I V U X G

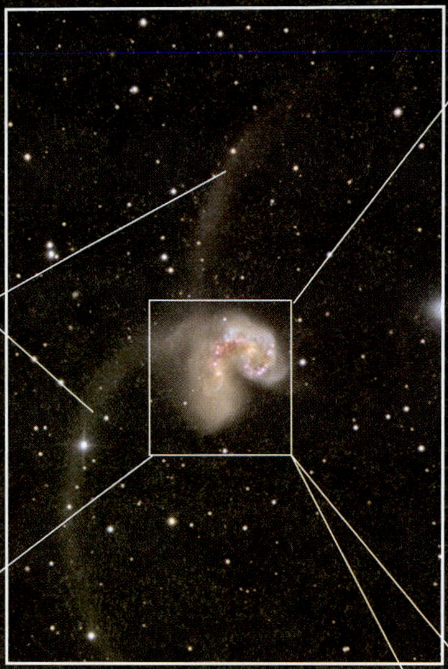

2. As the gas and dust clouds of the two galaxies collide with each other, they are greatly compressed. This compression causes stars to form in tremendous numbers.

R I V U X G

R I V U X G

3. This composite infrared and visible-light image of the Antennae allows us to see inside the two galaxies and reveals clouds of dust warmed by the light of hot young stars.

4. The globular clusters that orbit our Milky Way Galaxy contain only old stars; all of the short-lived blue stars have long since died. But some of the globular clusters that orbit the Antennae galaxies *do* have hot blue stars. Hence these clusters must be young. These, too, are a result of the compression of gas and dust that takes place in a collision between galaxies.

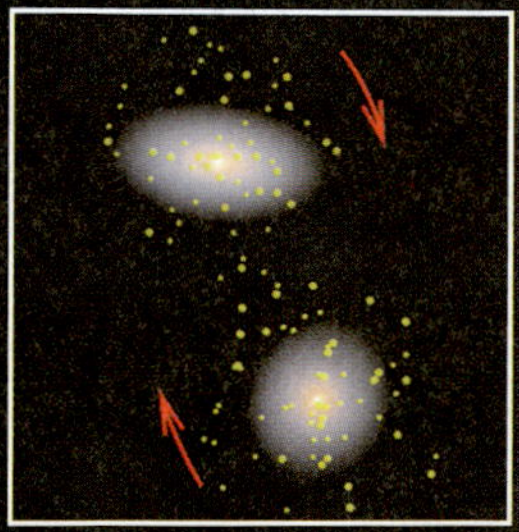

a. Two galaxies, each with old globular clusters (yellow), begin to interact.

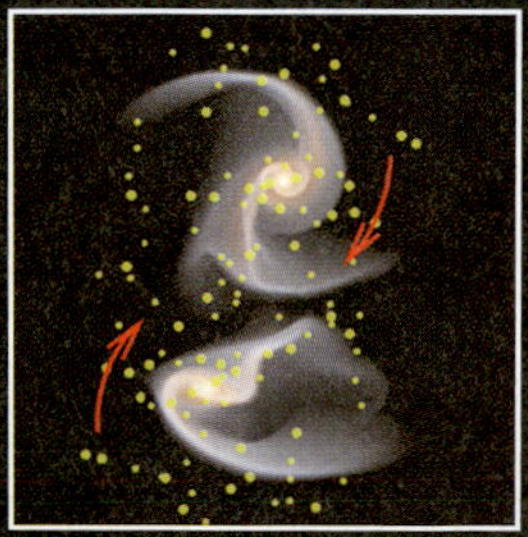

b. The two galaxies swoop past each other before finally settling down as a single merged galaxy. Gas and dust is compressed in both galaxies in the process, creating new star clusters.

(John Dubinski, University of Toronto)

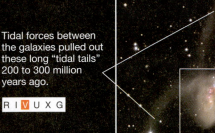

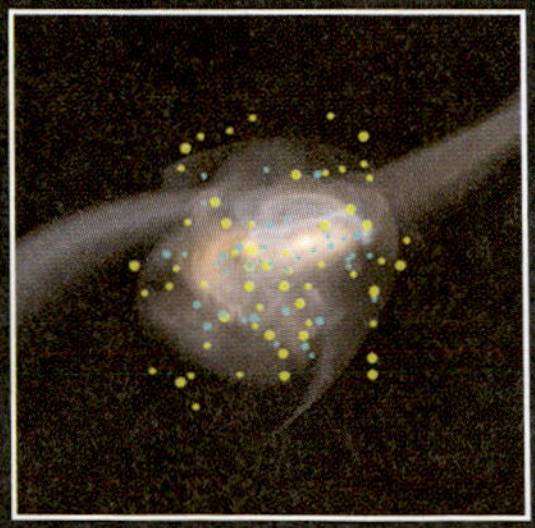

c. The combined galaxy has the original globular clusters (shown in yellow) as well as new ones (shown in blue).

Galaxy Interactions and Spiral Arms

ANIMATION 23-5 Close encounters between galaxies provide a third way of forming spiral arms (in addition to density waves and self-propagating star formation, discussed in Section 23-5). Computer simulations clearly demonstrate that spiral arms can be created during a collision, either by drawing out long streamers of stars or by compressing clouds of interstellar gas. For example, the spiral arms of M51 (examine Figure 23-2) may have been produced by a close encounter with a second galaxy. The disruptive galaxy, NGC 5195, is now located at the end of one of the spiral arms created by the collision. The two galaxies shown in the image that opens this chapter are thought to be interacting in the same way.

The very fact of our existence may be intimately related to interactions between galaxies. Some astronomers argue that the spiral arms of our Milky Way Galaxy were produced by a close encounter with the Large Magellanic Cloud. As we saw in Section 22-5, spiral arms compress the interstellar medium in the Milky Way's disk to form Population I stars like our own Sun, which have enough heavy elements to produce Earthlike planets. Thus, the chain of events that led to the formation of our Sun, our solar system, and life on our planet may have been initiated by a long-ago interaction between two galaxies.

CONCEPTCHECK 23-15

When the Milky Way Galaxy and the Andromeda Galaxy finish colliding with each other, what will be left over?

Answer appears at the end of the chapter.

23-8 Most of the matter in the universe is mysterious dark matter

Galaxies in a cluster typically move at large speeds and must be held in the cluster by gravity. In other words, there must be enough matter in the cluster to prevent the galaxies from drifting away. Nevertheless, careful examination of a rich cluster, like the Coma cluster shown in Figure 23-21, reveals that the mass of the visually luminous matter (principally the stars in the galaxies) is not at all sufficient to bind the cluster gravitationally. The observed line-of-sight speeds of the galaxies, measured by Doppler shifts, are so large that the cluster should have broken apart long ago. Considerably more mass than is visible is needed to keep the galaxies bound in orbit about the center of the cluster.

We encountered a similar situation in studying our own Milky Way Galaxy in Section 22-4: The total mass of our Galaxy is more than the amount of visible mass. As for our Galaxy, we conclude that clusters of galaxies must contain significant amounts of nonluminous *dark matter.* If this dark matter were not there, the galaxies would have long ago dispersed in random directions and the cluster would no longer exist today. Analyses demonstrate that the total mass needed to bind a typical rich cluster is about 10 times greater than the mass of material that shows up on visible-light images.

The Dark-Matter Problem and Rotation Curves

As for our Galaxy, the problem is to determine what form the invisible mass takes. A resolution to this **dark-matter problem,** which dates from the 1930s, was provided by the discovery in the late 1970s of hot, X-ray–emitting gas within clusters of galaxies (see Figure 23-26a). By measuring the amount of X-ray emission, astronomers find that the total mass of intracluster gas (mostly ionized hydrogen) in a typical rich cluster can be greater than the combined mass of all the stars in all the cluster's galaxies. The mass of intracluster gas is sufficient to account for only about 10% of the invisible mass, however. The remainder is dark matter of unknown composition.

Vast assemblages of dark matter reveal their presence by bending passing rays of light

Although we do not know what dark matter is made of, it is possible to investigate how dark matter is distributed in galaxies and clusters of galaxies. It appears that dark matter lies within and immediately surrounding galaxies, not in the vast spaces between galaxies. The evidence for this dark matter distribution comes principally from observations of the rotation curves of galaxies and of the gravitational bending of light by clusters of galaxies.

As we saw in Section 22-4, a rotation curve is a graph that shows how fast stars in a galaxy are moving at different distances from that galaxy's center. For example, Figure 23-18 is the rotation curve for our Galaxy. As Figure 23-31 illustrates, many other spiral galaxies have similar rotation curves that remain remarkably flat out to surprisingly great distances from each galaxy's center. In other words, the orbital speed of the stars remains roughly constant out to the visible edges of these galaxies.

This observation tells us that we still have not detected the *true* edges of these galaxies (and many similar ones). Near the true edge of a galaxy we should see a decline in orbital speed, in accordance with Kepler's third law (see Figure 22-18). Because this decline has not been observed, astronomers conclude that there must be a considerable amount of dark material that extends well beyond the visible portion of the disk.

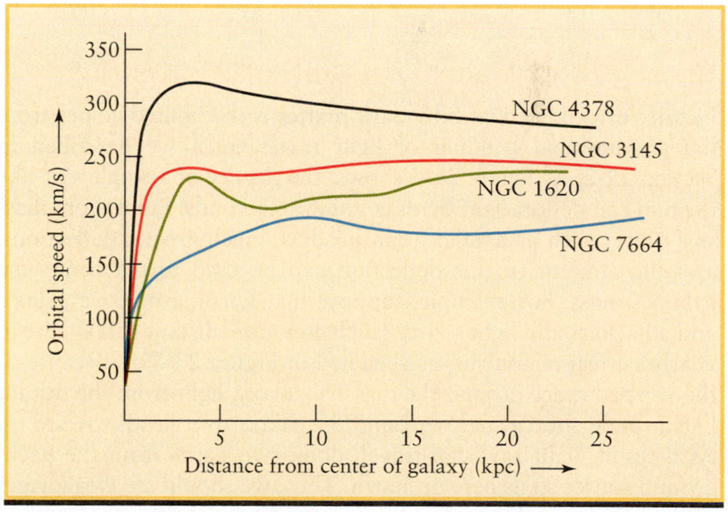

ANIMATION 23-6 **FIGURE 23-31**

The Rotation Curves of Four Spiral Galaxies This graph shows how the orbital speed of material in the disks of four spiral galaxies varies with the distance from the center of each galaxy. If most of each galaxy's mass were concentrated near its center, these curves would fall off at large distances. But these and many other galaxies have flat rotation curves that do not fall off. This indicates the presence of extended halos of dark matter. (Adapted from V. Rubin and K. Ford)

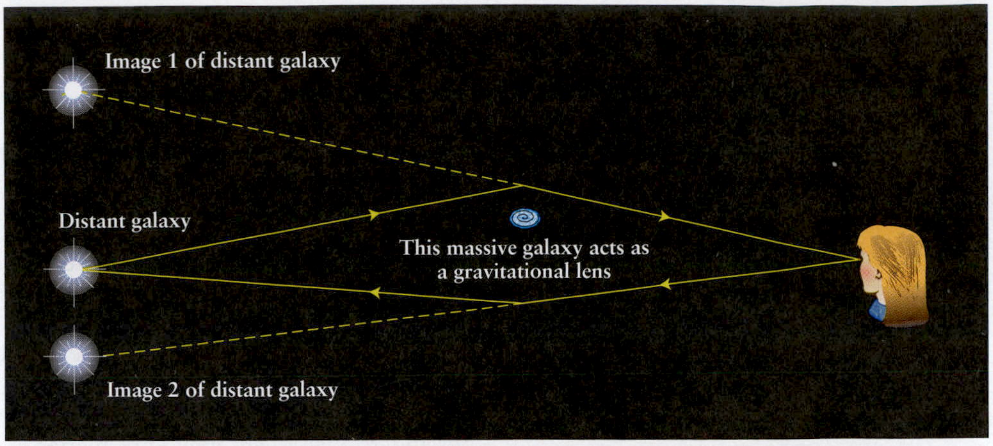
Image 1 of distant galaxy
Distant galaxy
This massive galaxy acts as a gravitational lens
Image 2 of distant galaxy

(a) How gravitational lensing happens

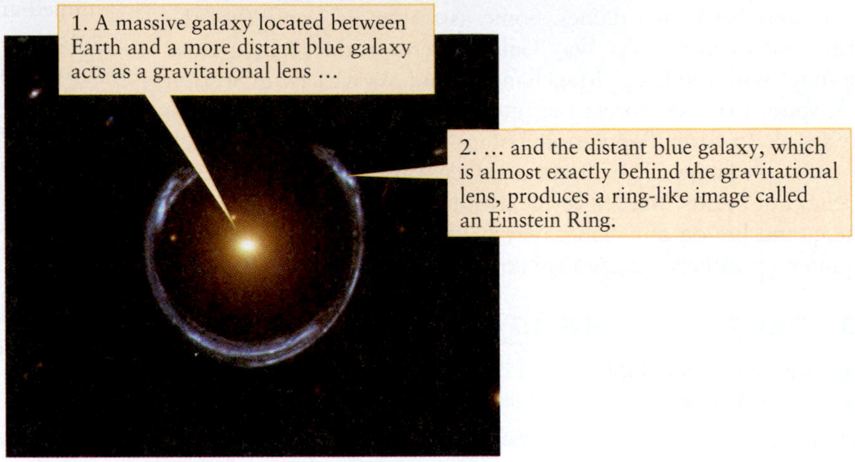
1. A massive galaxy located between Earth and a more distant blue galaxy acts as a gravitational lens ...
2. ... and the distant blue galaxy, which is almost exactly behind the gravitational lens, produces a ring-like image called an Einstein Ring.

(b) An Einstein ring

VIDEO 23-5 **FIGURE 23-32** R I V U X G

Gravitational Lensing **(a)** A massive object such as a galaxy can deflect light rays like a lens so that an observer sees more than one image of a more distant galaxy. **(b)** An Einstein Ring is produced when a distant object is lined up almost exactly behind the lens. (b: ESA/Hubble & NASA)

"Seeing" Dark Matter with Gravitational Lensing

Further evidence about how dark matter is distributed comes from the gravitational bending of light rays, which we described in Section 21-2. As Figure 21-5 shows, the gravity of a single star like the Sun can deflect light by only a few arcseconds. But a more massive object such as a galaxy can produce much greater deflections, and the amount of this deflection can be used to determine the galaxy's mass. For example, suppose that Earth, a massive galaxy, and a background light source (such as a more distant galaxy) are in nearly perfect alignment, as sketched in Figure 23-32a. Because of the warped space around the massive galaxy, light from the distant background source curves around the galaxy as it heads toward us. As a result, light rays can travel along two paths from the background source to us here on Earth. Thus, we should see two images of the background source.

A powerful source of gravity that distorts background images is called a **gravitational lens.** For gravitational lensing to work, the alignment must be very good between Earth, the massive galaxy acting as a lens, and a remote background light source. As the alignment worsens, the second image of the background galaxy is too faint to be noticeable, leaving a regular view of the background galaxy. There is even a case that produces four images of the same distant object. Beginning in 1979, astronomers have discovered a great number of examples of gravitational lensing.

When the alignment between Earth, a massive gravitational lens, and a distant object becomes nearly perfect, a ring-like image is formed that is called an Einstein Ring (Figure 23-32b). As discussed next, beyond the interesting images, gravitational lensing can also be used as a tool to study dark matter.

The observed geometry of a gravitational lens and the resulting distorted image allows astronomers to deduce the mass of the gravitational lensing galaxy. The results agree with studies that use rotation curves: ***Most of the galactic matter is dark.*** This agreement leads to an important point. Early on, some astronomers proposed that there might not be any dark matter and that the unexpected rotation curves were actually telling us that Newton's law of gravity needed to be changed (see Box 23-2). However, because gravitational lensing measures the same amount of missing matter without relying on Newton's law of gravity, all of these observations (and more described next) are best explained in terms of dark matter.

Figure 23-33 shows a situation in which an entire cluster of galaxies acts as a gravitational lens. The image shows an

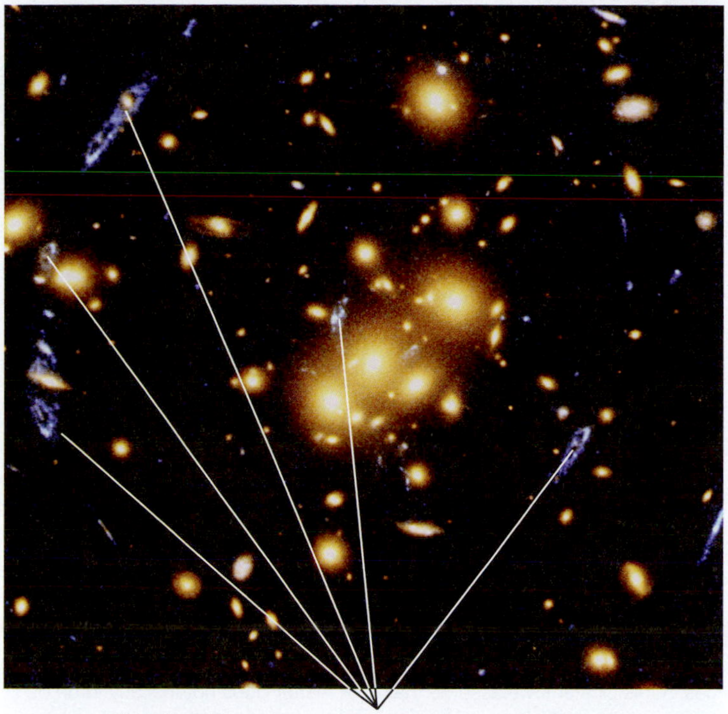
All of these blue arcs are images of the same distant galaxy.

FIGURE 23-33 R I V U X G

Gravitational Lensing by a Cluster of Galaxies The blue arcs in this image of the rich cluster CL 0024+1654 are distorted multiple images of a single more distant galaxy. These images are the result of gravitational lensing by the matter in CL 0024+1654. The cluster is about 1600 Mpc (5 billion ly) from Earth; the blue galaxy is about twice as distant. The blue color of the remote galaxy suggests that it is very young and is actively forming stars. (Science Source)

ordinary-looking rich cluster of yellowish elliptical and spiral galaxies, but with a number of curious blue arcs. Reconstruction of the light paths through the cluster shows that all these blue arcs are actually distorted images of a single galaxy that lies billions of light-years beyond the cluster.

By measuring the distortion of the images of such background galaxies, J. Anthony Tyson of Bell Laboratories and his colleagues have determined that dark matter, which constitutes about 90% of the cluster's mass, is distributed much like the visible matter in the cluster. In other words, the overall arrangement of visible galaxies seems to trace the location of dark matter.

CONCEPTCHECK 23-16

If dark matter cannot be seen, what are the two primary lines of evidence that it exists?

Answer appears at the end of the chapter.

The Nature of Dark Matter

Many proposals have been made to explain the composition of dark matter. One reasonable suggestion was that clusters might contain a large number of faint, red, low-mass (0.2 $M_\odot$ or lower) stars. These faint stars could be located in extended halos surrounding individual galaxies or scattered throughout the spaces between the galaxies of a cluster. They would have escaped detection because their luminosity and hence apparent brightness would be very low. (The mass-luminosity relation for main-sequence stars, which we discussed in Section 17-9, tells us that low-mass stars are intrinsically very faint.) Searches for these stars around other galaxies as well as around the Milky Way have been carried out using the Hubble Space Telescope. None has yet been detected, so it is thought that faint stars are unlikely to constitute the majority of the dark matter in the universe.

As we described in Section 22-4, other dark matter candidates include massive neutrinos, subatomic particles called WIMPs (weakly interacting massive particles), and MACHOs (massive compact halo objects, such as small black holes or brown dwarfs). Although WIMPs have never been detected, they are the favored hypothesis. Interacting very weakly with regular matter, about a billion WIMPs might pass through you each second, and about 1 kg of WIMPs might be passing through Earth at any moment. To date, however, the true nature of dark matter remains unknown.

An important clue about the nature of dark matter was discovered in 2006 by examining a rich cluster of galaxies called the Bullet Cluster. Remarkably, the visible matter and dark matter in this cluster do *not* have the same distribution (Figure 23-34a). The best explanation for how this could have come about is that the Bullet Cluster is the result of a collision between two galaxy clusters, one larger than the other (Figure 23-34b). (In Section 23-7 we asked you to visualize collisions between entire galaxies; now you must imagine a vastly more immense collision between entire clusters of galaxies!) During the collision, the gas from one cluster slams into the gas from the other cluster and slows down due to *fluid resistance*. (You feel the force of fluid resistance pushing against you whenever you try to move through a liquid or gas—for example, when you swim in a lake or put your hand outside the window of a fast-moving car.) Fluid resistance is a consequence of the electric forces between adjacent atoms and molecules in a fluid. But if dark matter is made up of some curious material that responds only to *gravitational* forces, it is unaffected by fluid resistance. As a result, during the collision sketched in Figure 23-34b the gas is slowed by fluid resistance but the dark matter is not. The agreement of the simulation in Figure 23-34b with the observations in Figure 23-34a strongly reinforces the idea that dark matter, though mysterious, is quite real.

The study of dark matter forces us to reconsider our impression of the universe based on luminous matter alone. As we will see in Chapter 25, we now know that there is about 5 times as much dark matter in the observable universe as there is visible matter. Because dark matter is so dominant, "ordinary" visible matter—including this book, the air that you breathe, all the planets and stars, and your own body—is in fact relatively rare. Thus, the vast majority of mass in the universe is of completely unknown composition.

Exotic Dark Matter

The mystery of dark matter is even greater for another reason. If dark matter consisted of unseen but otherwise regular objects made of atoms from the periodic table of the elements (objects such as

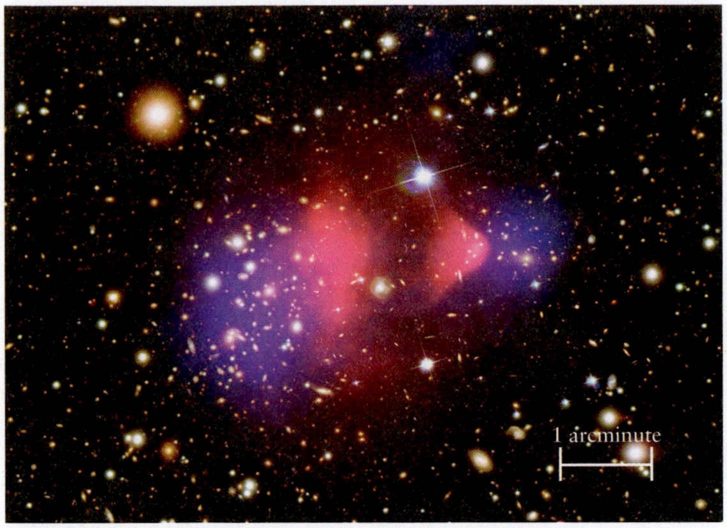

(a) Composite image of galaxy cluster 1E0657-56 showing visible galaxies, X-ray-emitting gas (red) and dark matter (blue)

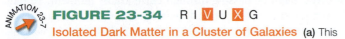

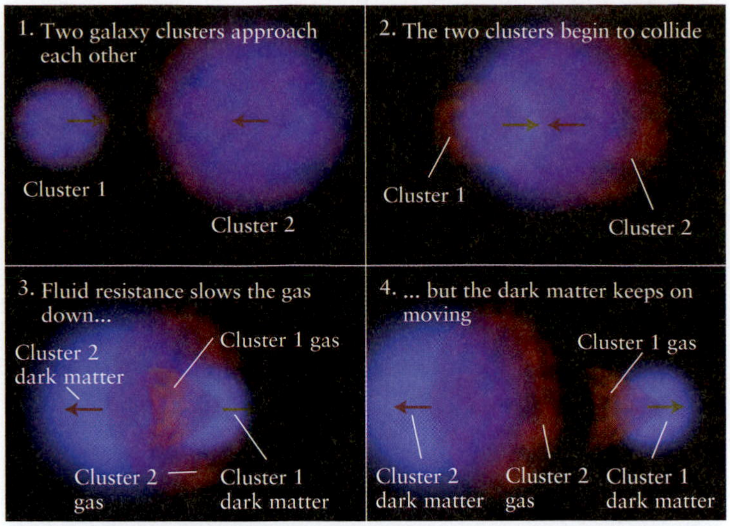

(b) A model of how the gas and dark matter in 1E0657-56 could have become separated

ANIMATION 23-7

FIGURE 23-34 R I V U X G

Isolated Dark Matter in a Cluster of Galaxies **(a)** This visible-light image of the Bullet Cluster shows more than a thousand galaxies. The superimposed image in red shows the distribution of the cluster's hot, X-ray emitting gas, and the blue image shows the distribution of dark matter as determined by gravitational lensing of numerous background galaxies (see Figure 23-30). **(b)** We can understand the separation of dark matter and gas in this cluster if we assume that dark matter does not feel any force of fluid resistance. This is what we would expect if dark matter responds to gravitational forces only. (a: NASA/STScI; Magellan/U. Arizona/D. Clew et al.; b: NASA/CXC/M. Weiss.)

brown dwarfs), dark matter would still be a familiar form of matter. However, there is very strong evidence that dark matter is not made up of this "ordinary" matter consisting of protons and neutrons.

Over the eons, collisions and mergers have dramatically altered the population of galaxies

As we will see in Chapter 25 (on the cosmic microwave background radiation), the amount of matter in the form of protons and neutrons has been measured. Because the amount of dark matter is much greater than the amount of atomic matter, most of the dark matter is in a truly unknown form. The WIMPs, if ever discovered, would be an example of a completely new form of matter. We use the term **exotic dark matter** for these hypothesized new forms of matter that are not made of protons and neutrons.

CONCEPTCHECK 23-17

If some dark matter consists of neutron stars, would this be "ordinary" matter or exotic dark matter?

Answer appears at the end of the chapter.

23-9 Galaxies formed from the merger of smaller objects

How do galaxies form and how do they evolve? Astronomers can gain important clues about galactic evolution simply by looking deep into space. The more distant a galaxy is, the longer its light takes to reach us. As we examine galaxies that are at increasing distances from Earth, we are actually looking further and further back in time. By looking into the past, we can see galaxies in their earliest stages.

Building Galaxies from the "Bottom Up"

The Hubble Space Telescope images in Figure 23-35 provide a glimpse of galaxy formation in the early universe. Figure 23-35 shows a number of galaxylike objects some 13 billion light-years (3988 Mpc) away and are thus seen as they were 13 billion years ago. These objects are between one-tenth and one-half the size of our Milky Way Galaxy and have unusual, irregular shapes. Furthermore, computer simulations of these objects show that they will likely collide to form a large elliptical galaxy.

Images of the young universe such as those in Figure 23-35 lead astronomers to conclude that galaxies formed "from the bottom up"—that is, by the merger of smaller objects to form full-size galaxies. (These same images provide evidence against an older idea that galaxies formed "from the top down"—that is, fragmenting or breaking apart from immense, cluster-sized clouds of material.) We will see evidence in Chapter 26 that the matter in the universe formed "clumps" even earlier than this. These clumps evolved into objects like those shown in Figure 23-35, which in turn merged to form the population of galaxies that we see today.

CONCEPTCHECK 23-18

Did today's galaxies form from combining smaller galaxies or from the separation of larger galaxies?

Answer appears at the end of the chapter.

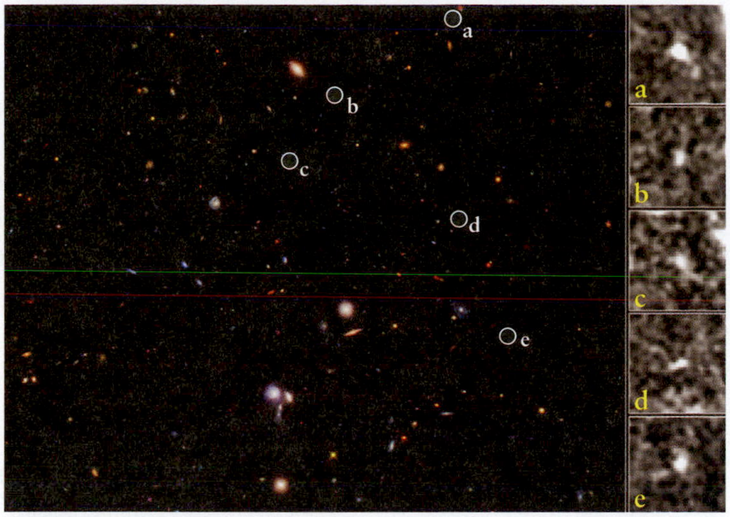

FIGURE 23-35 R I V U X G

The Building Blocks of Galaxies In this Hubble Space Telescope image, the objects outlined by circles are about 13 billion ly from Earth and only a fraction of the size of our Milky Way. They are the building blocks that merge to form larger galaxies and clusters. While barely visible, these are the brightest objects at this great distance and early time, and numerous objects even smaller and dimmer are expected. (NASA, ESA, M. Trenti (University of Colorado, Boulder, and Institute of Astronomy, University of Cambridge, UK), L. Bradley (STScI), and the BoRG team)

Forming Spirals, Lenticulars, and Ellipticals

Once a number of subgalactic units combine, they make an object called a *protogalaxy.* The rate at which stars form within a protogalaxy may determine whether this protogalaxy becomes a spiral or an elliptical. If stars form relatively slowly, the gas surrounding them has enough time to settle by collisions to form a flattened disk, much as happened on a much smaller scale in the solar nebula (see Section 8-4). Star formation continues because the disk contains an ample amount of hydrogen from which to make new stars. The result is a spiral or lenticular galaxy (Figure 23-36a). But if stars initially form in the protogalaxy at a rapid rate, virtually all of the available gas is used up to make stars before a disk can form. In this case what results is an elliptical galaxy (Figure 23-36b).

Figure 23-36c compares the stellar birthrate in the two types of galaxies. This graph helps us understand some of the differences between spiral and elliptical galaxies that we described in Section 23-3. Protogalaxies are thought to have been composed almost exclusively of hydrogen and helium gas, so the first stars were Population II stars with hardly any metals (that is, heavy elements). As stars die and form planetary nebulae or supernovae, they eject gases enriched in metals into the interstellar medium. In a spiral galaxy there is ongoing star formation in the disk, so these metals are incorporated into new generations of stars, making relatively metal-rich Population I stars like the Sun. By contrast, an elliptical galaxy has a single flurry of star formation when it is young, after which star formation ceases. Elliptical galaxies therefore contain only metal-poor Population II stars.

Figure 23-36c shows that both elliptical and spiral galaxies form stars most rapidly when they are young. This idea is borne out by the observation that very distant galaxies tend to be blue, which means that galaxies were bluer in the distant past than they are today. (Note the very blue colors of the distant, gravitationally lensed galaxies shown in Figure 23-32 and Figure 23-33.) Spectroscopic studies of such galaxies demonstrate that most owe their blue color to vigorous star formation, often occurring in intense, episodic bursts. The hot, luminous, and short-lived O and

(a) Formation of a spiral galaxy

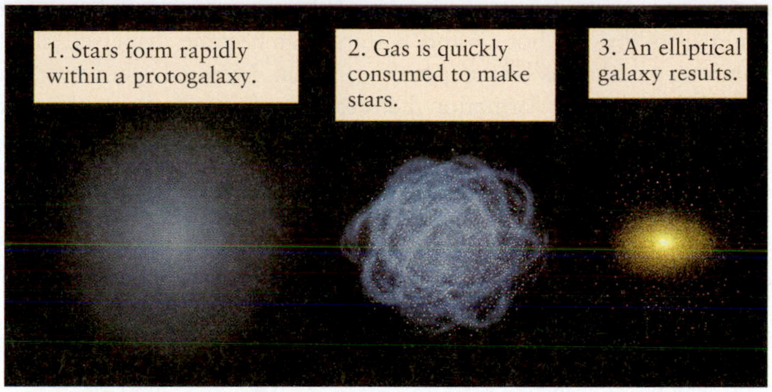

(b) Formation of an elliptical galaxy

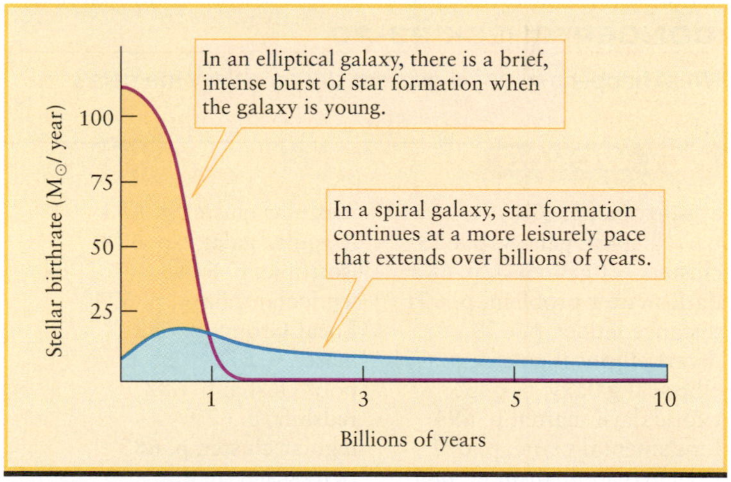

(c) The stellar birthrate in galaxies

FIGURE 23-36

The Formation of Spiral and Elliptical Galaxies **(a)** If the initial star formation rate in a protogalaxy is low, it can evolve into a spiral galaxy with a disk. **(b)** If the initial star formation rate is rapid, no gas is left to form a disk. The result is an elliptical galaxy. **(c)** This graph shows how the rate of star birth (in solar masses per year) varies with age in spiral and elliptical galaxies.

B stars produced in these bursts of star formation give blue galaxies their characteristic color.

CONCEPTCHECK 23-19

Why do elliptical galaxies contain only Population II stars?

Answer appears at the end of the chapter.

An Evolving Universe of Galaxies

In addition to changes in galaxy colors, the character of the galactic population has also changed over the past several billion years. In nearby rich clusters, only about 5% of the galaxies are spirals. But observations of rich clusters at a redshift of $z = 0.4$—which corresponds to looking about 4 billion years into the past—show that about 30% of their galaxies were spirals.

Why were spiral galaxies more common in rich clusters in the distant past? Galactic collisions and mergers are probably responsible. During a collision, interstellar gas in the colliding galaxies is vigorously compressed, triggering a burst of star formation (see Cosmic Connections: When Galaxies Collide in Section 23-7). A succession of collisions produces a series of star-forming episodes that create numerous bright, hot O and B stars that become dispersed along arching spiral arms by the galaxy's rotation. Eventually, however, the gas is used up; star formation then ceases and the spiral arms become less visible. Furthermore, tidal forces tend to disrupt colliding galaxies, strewing their stars across intergalactic space until the galaxies are completely disrupted (see Figure 23-28).

A full description of galaxy formation and evolution must include the effects of dark matter. As we have seen, only about 10% of the mass of a galaxy—its stars, gas, and dust—emits electromagnetic radiation of any kind. As yet, we have no idea what the remaining 90% looks like or what it is made of. The dilemma of dark matter is one of the most challenging problems facing astronomers today.

CONCEPTCHECK 23-20

What happens to spiral galaxies during galaxy mergers?

KEY WORDS

anisotropic, p. 674
barred spiral galaxy, p. 673
clusters (of galaxies), p. 682
dark-matter problem, p. 691
distance ladder, p. 678
dwarf elliptical galaxy, p. 674
elliptical galaxy, p. 673
exotic dark matter, p. 694
fundamental plane, p. 677
galactic cannibalism, p. 689
giant elliptical galaxy, p. 673
gravitational lens, p. 692
groups (of galaxies), p. 682
Hubble classification, p. 671
Hubble constant, p. 680
Hubble flow, p. 679
Hubble law, p. 680
intracluster gas, p. 686
irregular cluster, p. 683
irregular galaxy, p. 674
isotropic, p. 674
lenticular galaxy, p. 674
Local Group, p. 682
maser, p. 678
poor cluster, p. 682
redshift, p. 679
regular cluster, p. 683
rich cluster, p. 682
spiral galaxy, p. 671
standard candle, p. 676
starburst galaxy, p. 686
supercluster, p. 683
Tully-Fisher relation, p. 677
tuning fork diagram, p. 674
void, p. 685

KEY IDEAS

The Hubble Classification: Galaxies can be grouped into four major categories: spirals, barred spirals, ellipticals, and irregulars.

• The disks of spiral and barred spiral galaxies are sites of active star formation.

• Elliptical galaxies are nearly devoid of interstellar gas and dust, so star formation is severely limited.

• Lenticular galaxies are intermediate between spiral and elliptical galaxies.

• Irregular galaxies have ill-defined, asymmetrical shapes. They are often found associated with other galaxies.

Distance to Galaxies: Standard candles, such as Cepheid variables and the most luminous supergiants, globular clusters, H II regions, and supernovae in a galaxy, are used in estimating intergalactic distances.

• The Tully-Fisher relation, which correlates the width of the 21-cm line of hydrogen in a spiral galaxy with its luminosity, can also be used for determining distance. A method that can be used for elliptical galaxies is the fundamental plane, which relates the galaxy's size to its surface brightness distribution and to the motions of its stars.

The Hubble Law: There is a simple linear relationship between the distance from Earth to a remote galaxy and the redshift of that galaxy (which is a measure of the speed with which it is receding from us). This relationship is the Hubble law, $v = H_0 d$.

• The value of the Hubble constant, H_0, is not known with certainty but is close to 73 km/s/Mpc.

Clusters and Superclusters: Galaxies are grouped into clusters rather than being scattered randomly throughout the universe.

• A rich cluster contains hundreds or even thousands of galaxies; a poor cluster, often called a group, may contain only a few dozen.

• A regular cluster has a nearly spherical shape with a central concentration of galaxies; in an irregular cluster, galaxies are distributed asymmetrically.

• Our Galaxy is a member of a poor, irregular cluster called the Local Group.

• Rich, regular clusters contain mostly elliptical and lenticular galaxies; irregular clusters contain spiral, barred spiral, and irregular galaxies along with ellipticals.

• Giant elliptical galaxies are often found near the centers of rich clusters.

Galactic Collisions and Mergers: When two galaxies collide, their stars pass each other, but their interstellar media collide violently, either stripping the gas and dust from the galaxies or triggering prolific star formation.

• The gravitational effects during a galactic collision can throw stars out of their galaxies into intergalactic space.

• Galactic mergers may occur; a large galaxy in a rich cluster may tend to grow steadily through galactic cannibalism, perhaps producing in the process a giant elliptical galaxy.

The Dark-Matter Problem: The luminous mass of a cluster of galaxies is not large enough to account for the observed motions of the galaxies; a large amount of unobserved mass must also be present. This situation is called the dark-matter problem.

• Hot intergalactic gases in rich clusters account for a small part of the unobserved mass. These gases are detected by their X-ray emission. The remaining unobserved mass is probably in the form of dark-matter halos that surround the galaxies in these clusters. Particles called WIMPs are the favored hypothesis for dark matter.

• Gravitational lensing of remote galaxies by a foreground cluster enables astronomers to glean information about the distribution of dark matter in the foreground cluster.

Formation and Evolution of Galaxies: Observations indicate that galaxies arose from mergers of smaller collections of stars.

• Whether a protogalaxy evolves into a spiral galaxy or an elliptical galaxy depends on its initial rate of star formation.

QUESTIONS

Review Questions

1. Why did many nineteenth-century astronomers think that the "spiral nebulae" are part of the Milky Way?
2. What was the Shapley-Curtis "debate" all about? Was a winner declared at the end of the "debate"? Whose ideas turned out to be correct?
3. How did Edwin Hubble prove that the Andromeda "Nebula" is not a nebula within our Milky Way Galaxy?
4. Are any galaxies besides our own visible with the naked eye from Earth? If so, which one(s)?
5. An educational publication for children included the following statement: "The Sun is in fact the only star in our galaxy. All of the other stars in the sky are located in other galaxies." How would you correct this statement?
6. TUTORIAL 23-1 What is the Hubble classification scheme? Which category includes the largest galaxies? Which includes the smallest? Which category of galaxy is the most common?
7. Which is more likely to have a blue color, a spiral galaxy or an elliptical galaxy? Explain why.
8. Which types of galaxies are most likely to have new stars forming? Describe the observational evidence that supports your answer.
9. Explain why the apparent shape of an elliptical galaxy may be quite different from its real shape.
10. Why do astronomers suspect that the Hubble tuning fork diagram does not depict the evolutionary sequence of galaxies?
11. Why are Cepheid variable stars useful for finding the distances to galaxies? Are there any limitations on their use for this purpose?
12. Why are Type Ia supernovae useful for finding the distances to very remote galaxies? Can they be used to find the distance to any galaxy you might choose? Explain your answers.
13. What is the Tully-Fisher relation? How is it used for measuring distances? Can it be used for galaxies of all kinds? Why or why not?
14. What are masers? How can they be used to measure the distance to a galaxy?
15. What is the Hubble law? How can it be used to determine distances?
16. How did the discovery of the Hubble Law reinforce the idea that the spiral "nebulae" could not be part of the Milky Way?
17. Why do you suppose it has been so difficult to determine the value of H_0?
18. Some galaxies in the Local Group exhibit blueshifted spectral lines. Why aren't these blueshifts violations of the Hubble law?
19. What are the differences between regular and irregular clusters?
20. What is the difference between a cluster and a supercluster? Are both clusters and superclusters held together by their gravity?
21. What measurements do astronomers make to construct three-dimensional maps of the positions of galaxies in space?
22. What is the Sloan Great Wall, and why is it interesting to astronomers?
23. Describe what voids are and what they tell us about the large-scale structure of the universe.
24. Why is the intracluster gas in galaxy clusters at such high temperatures?
25. What are starburst galaxies? How can they be produced by collisions between galaxies?
26. Why do giant elliptical galaxies dominate rich clusters but not poor clusters?
27. What evidence is there for the existence of dark matter in clusters of galaxies?
28. What is gravitational lensing? Why don't we notice the gravitational lensing of light by ordinary objects on Earth?
29. How do observations of the Bullet Cluster help constrain the nature of dark matter?
30. What is the favored hypothesis for what dark matter is made of? What properties do these particles have?
31. Would the existence of WIMPs be a form of exotic dark matter or "ordinary" matter? What is exotic dark matter?
32. What observations suggest that present-day galaxies formed from smaller assemblages of matter?
33. On what grounds do astronomers think that in the past, spiral galaxies were more numerous in rich clusters than they are today? What could account for this excess of spiral galaxies in the past?

Advanced Questions

Questions preceded by an asterisk (*) involve topics discussed in the Boxes.

Problem-solving tips and tools

Box 1-1 explains the small-angle formula, and Box 17-3 discusses the relationship among apparent magnitude, absolute magnitude, and distance. As Box 22-2 explains, a useful form of Kepler's third law is $M = rv^2/G$, where M is the mass within an orbit of radius r, v is the orbital speed, and G is the gravitational constant. Another form of Kepler's third law, particularly useful for two stars or two galaxies orbiting each other, is given in Section 17-9. The volume of a sphere of radius r is $4\pi r^3/3$. The mass of a hydrogen atom (^1H) is given in Appendix 7.

34. Hubble made his observations of Cepheids in the Andromeda Galaxy (M31) using the 100-inch (2.5-meter) telescope on Mount Wilson. Completed in 1917, this was the largest telescope in the world when Hubble carried out his observations in 1923. Why was it helpful to use such a large telescope?

35. The following image shows the Small Magellanic Cloud (SMC), an irregular galaxy that orbits the Milky Way. The SMC is 63 kpc (200,000 ly) from Earth and 8 kpc (26,000 ly) across, and can be seen with the naked eye from southern latitudes. What features of this image indicate that there has been recent star formation in the SMC? Explain your answer.

R I **V** U X G

(Australian Astronomical Observatory/David Malin Images)

***36.** Astronomers often state the distance to a remote galaxy in terms of its distance modulus, which is the difference between the apparent magnitude m and the absolute magnitude M (see Box 17-3). (**a**) By measuring the brightness of supernova 1994I in the galaxy M51 (see Figure 23-2), the distance modulus for this galaxy was determined to be $m - M = 29.2$. Find the distance to M51 in megaparsecs (Mpc). (**b**) A separate distance determination, which involved measuring the brightnesses of planetary nebulae in M51, found $m - M = 29.6$. What is the distance to M51 that you calculate from this information? (**c**) What is the difference between your answers to parts (a) and (b)? Compare this difference with the 750-kpc distance from Earth to M31, the Andromeda Galaxy. The difference between your answers illustrates the uncertainties involved in determining the distances to galaxies!

***37.** Suppose you discover a Type Ia supernova in a distant galaxy. At maximum brilliance, the supernova reaches an apparent magnitude of +10. How far away is the galaxy? (*Hint:* See Box 23-1.)

38. The masers that orbit the center of the spiral galaxy M106 travel at an orbital speed of about 1000 km/s. Astronomers observed these masers at intervals of 4 months. (**a**) What distance does a single maser move during a 4-month period? Give your answer in kilometers and in AU. (**b**) During this period, a maser moving across the line of sight (like the maser shown in green in Figure 23-15) appeared to move through an angle of only 10^{-5} arcsec. Calculate the distance to the galaxy.

39. The average radial velocity of galaxies in the Hercules cluster pictured in Figure 23-18 is 10,800 km/s. (**a**) Using $H_0 =$ 73 km/s/Mpc, find the distance to this cluster. Give your answer in megaparsecs and in light-years. (**b**) How would your answer to (a) differ if the Hubble constant had a smaller value? A larger value? Explain your answers.

40. A certain galaxy is observed to be receding from the Sun at a rate of 7500 km/s. The distance to this galaxy is measured independently and found to be 1.4×10^8 pc. Using these data, what is the value of the Hubble constant?

***41.** In the spectrum of the galaxy NGC 4839, the K line of singly ionized calcium has a wavelength 403.2 nm. (**a**) What is the redshift of this galaxy? (*Hint:* See Box 23-2.) (**b**) Determine the distance to this galaxy using the Hubble law with $H_0 =$ 73 km/s/Mpc.

***42.** The galaxy RD1 has a redshift of $z = 5.34$. (**a**) Determine its recessional velocity v in km/s and as a fraction of the speed of light. (**b**) What recessional velocity would you have calculated if you had erroneously used the low-speed formula relating z and v? Would using this formula have been a small or large error? (**c**) According to the Hubble law, what is the distance from Earth to RD1? Use $H_0 =$ 73 km/s/Mpc for the Hubble constant, and give your answer in both megaparsecs and light-years.

43. It is estimated that the Coma cluster (see Figure 23-21) contains about 10^{13} $M_\odot$ of intracluster gas. (**a**) Assuming that this gas is made of hydrogen atoms, calculate the total number of intracluster gas atoms in the Coma cluster. (**b**) The Coma cluster is roughly spherical in shape, with a radius of about 3 Mpc. Calculate the number of intracluster gas atoms per cubic centimeter in the Coma cluster. Assume that the gas fills the cluster uniformly. (**c**) Compare the intracluster gas in the Coma cluster with the gas in our atmosphere (3×10^{19} molecules per cubic centimeter, temperature 300 K); a typical gas cloud within our own Galaxy (a few hundred molecules per cubic centimeter, temperature 50 K or less); and the corona of the Sun (10^5 atoms per cubic centimeter, temperature 10^6 K).

*44. Two galaxies separated by 600 kpc are orbiting each other with a period of 40 billion years. What is the total mass of the two galaxies?

*45. Figure 23-31 shows the rotation curve of the Sa galaxy NGC 4378. Using data from that graph, calculate the orbital period of stars 20 kpc from the galaxy's center. How much mass lies within 20 kpc from the center of NGC 4378?

46. How might you determine what part of a galaxy's redshift is caused by the galaxy's orbital motion about the center of mass of its cluster?

47. The accompanying images show the unusual elliptical galaxy NGC 5128 in visible and infrared wavelengths. Explain how the properties of this galaxy seen in the infrared image can be explained if NGC 5128 is the result of a merger of an elliptical galaxy and a spiral galaxy.

NGC 5128 is an elliptical galaxy...
...with a dark dust lane

R I V U X G

Yellow and green: Dust warmed by starlight
Pink: Star-forming regions

R I V U X G

(Visible image: Eric Peng, Herzberg Institute of Astrophysics and NOAO/AURA/NSF; Infrared image: Jocelyn Keene, NASA/JPL and Caltech)

48. Explain why the dark matter in galaxy clusters could not be neutral hydrogen.

49. According to Figure 23-36c, elliptical galaxies continue to form stars for about a billion years after they form. Give an argument why we might expect to find some Population I stars in an elliptical galaxy. (*Hint:* Table 19-1 gives the main-sequence lifetimes for stars of different masses.)

Discussion Questions

50. Earth is composed principally of heavy elements, such as silicon, nickel, and iron. Would you be likely to find such planets orbiting stars in the disk of a spiral galaxy? In the nucleus of a spiral galaxy? In an elliptical galaxy? In an irregular galaxy? Explain your answers.

51. Discuss what observations you might make to determine whether or not the various Hubble types of galaxies represent some sort of evolutionary sequence.

52. Discuss the advantages and disadvantages of using the various standard candle distance indicators to obtain extragalactic distances.

53. How would you distinguish star images from unresolved images of remote galaxies on a CCD?

54. Describe what sorts of observations you might make to search for as-yet-undiscovered galaxies in our Local Group. How is it possible that such galaxies might still remain to be discovered? In what part of the sky would these galaxies be located? What sorts of observations might reveal these galaxies?

Web/eBook Questions

55. When galaxies pass close to one another, as should happen frequently in a rich cluster, tidal forces between the galaxies can strip away their outlying stars. The result should be a loosely dispersed sea of "intergalactic stars" populating the space between galaxies in a cluster. Search the World Wide Web for information about intergalactic stars. Have they been observed? If so, where are they found? What would our nighttime sky look like if our Sun were an intergalactic star?

56. The Hubble Space Telescope (HST) has made extensive observations of very distant galaxies. Visit the HST Web site to learn about these investigations. How far back in time has HST been able to look? What sorts of early galaxies are observed? What is the current thinking about how galaxies formed and evolved?

57. ANIMATION 23-6 **Radiation from a Rotating Galaxy.** Access and view the animation "Radiation from a Rotating Galaxy" in Chapter 23 of the *Universe* Web site or eBook. Describe how the animation would have to be changed for a spiral galaxy of the same size but (**a**) greater mass and (**b**) smaller mass.

ACTIVITIES

Observing Projects

58. Using a telescope with an aperture of at least 30 cm (12 in.), observe as many of the spiral galaxies listed in the following table as you can. If you have a copy of the *Starry Night*™ program, use it to help determine when these galaxies can best be viewed. Many of these galaxies are members of the Virgo cluster, which can best be seen from March through June.

Because all galaxies are quite faint, be sure to schedule your observations for a moonless night. The best view is obtained when a galaxy is near the meridian. While at the eyepiece, make a sketch of what you see. Can you distinguish any spiral structure? After completing your observations, compare your sketches with photographs found in an online catalog of Messier objects (the "M" in the galaxy designations stands for Messier).

Spiral galaxy	Right ascension	Declination	Hubble type
M31 (NGC 224)	$0^h\ 42.7^m$	+41° 16′	Sb
M58 (NGC 4579)	12 37.7	+11 49	Sb
M61 (NGC 4303)	12 21.9	+4 28	Sc
M63 (NGC 5055)	13 15.8	+42 02	Sb
M64 (NGC 4826)	12 56.7	+21 41	Sb
M74 (NGC 628)	1 36.7	+15 47	Sc
M83 (NGC 5236)	13 37.0	−29 52	Sc
M88 (NGC 4501)	12 32.0	+14 25	Sb
M90 (NGC 4569)	12 36.8	+13 10	Sb
M91 (NGC 4548)	12 35.4	+14 30	SBb
M94 (NGC 4736)	12 50.9	+41 07	Sb
M98 (NGC 4192)	12 13.8	+14 54	Sb
M99 (NGC 4254)	12 18.8	+14 25	Sc
M100 (NGC 4321)	12 22.9	+15 49	Sc
M101 (NGC 5457)	14 03.2	+54 21	Sc
M104 (NGC 4594)	12 40.0	−11 37	Sa
M108 (NGC 3556)	11 11.5	+55 40	Sc

Note: The right ascensions and declinations are given for epoch 2000.

59. Using a telescope with an aperture of at least 30 cm (12 in.), observe as many of the following elliptical galaxies as you can. Six of these galaxies are in the Virgo cluster, which is conveniently located in the evening sky from March through June. If you have a copy of the *Starry Night*™ program, use it to help determine when these galaxies can best be viewed. As in the previous exercise, be sure to schedule your observations for a moonless night, when the galaxies you wish to observe will be near the meridian. Do these elliptical galaxies differ in appearance from spiral galaxies?

Elliptical galaxy	Right ascension	Declination	Hubble type
M49 (NGC 4472)	$12^h\ 29.8^m$	+8° 009	E4
M59 (NGC 4621)	12 42.0	+11 39	E3
M60 (NGC 4649)	12 43.7	+11 33	E1
M84 (NGC 4374)	12 25.1	+12 53	E1
M86 (NGC 4406)	12 26.2	+12 57	E3
M89 (NGC 4552)	12 35.7	+12 33	E0
M110 (NGC 205)	00 40.4	+41 41	E6

Note: The right ascensions and declinations are given for epoch 2000

60. Using a telescope with an aperture of at least 30 cm (12 in.), observe as many of the following interacting galaxies as you can. If you have a copy of the *Starry Night*™ program, use it to help determine when these galaxies can best be viewed. As in the previous exercises, be sure to schedule your observations for a moonless night, when the galaxies you wish to observe will be near the meridian. While at the eyepiece, make a sketch of each galaxy. Can you distinguish hints of interplay among the galaxies? After completing your observations, compare your sketches with photographs found in an online catalog of Messier objects (the "M" in the galaxy designations stands for Messier).

Interacting galaxies	Right ascension	Declination
M51 (NGC 5194)	$13^h\ 29.9^m$	+47° 12′
NGC 5195	13 30.0	+47 16
M65 (NGC 3623)	11 18.9	+13 05
M66 (NGC 3627)	11 20.2	+12 59
NGC 3628	11 20.3	+13 36
M81 (NGC 3031)	9 55.6	+69 04
M82 (NGC 3034)	9 55.8	+69 41
M95 (NGC 3351)	10 44.0	+11 42
M96 (NGC 3368)	10 46.8	+11 49
M105 (NGC 3379)	10 47.8	+12 35

Note: The right ascensions and declinations are given for epoch 2000.

61. Use *Starry Night*™ to visit a variety of galaxies and determine whether they are spiral, barred spiral, elliptical, or irregular. Click on **Home** to see the sky from your home location. Click on the **Options** tab, expand the **Deep Space** layer and click **Off** all images except **Messier Objects** and **Bright NGC Objects.** Type Ctrl-H (Cmd-H on a Mac) or select **View > Hide Horizon** from the menu to remove the horizon. Also select **View > Hide Daylight** to remove daylight from the view. Use the **Find** pane to visit each of the galaxies listed below. For each object, type its name in the search box of the **Find** pane and press the **Enter** key. (*Hint:* To go to the galaxy without slewing, press the spacebar.) (**a**) Use the **Zoom** buttons to examine each galaxy in detail and then classify it as a spiral (S), barred spiral (SB), elliptical (E), or irregular (Irr), and the subclassification of each galaxy (e.g., Sa, E5): M33, M58, M74, M81, M83, M94, M109, Large Magellanic Cloud, Small Magellanic Cloud, NGC1232, M84, M86, M59. (**b**) Use the **Find** pane and locate M51. What is the classification of this galaxy? (**c**) Examine this galaxy carefully and comment on anything unusual about its structure and/or its neighboring objects.

62. Clusters of galaxies contain different numbers and distributions of galaxies and harbor significant amounts of the mysterious dark matter. In this exercise you can use *Starry Night*™ to compare a few of these groupings and see the gravitational effect of dark matter. You can start by looking at one of the largest galaxy clusters, the Virgo cluster. Select **Favourites > Explorations > Virgo Cluster-Milky Way** from the menu. You are looking at this group of galaxies from a very large distance out in space, at about 66 Mly from the Sun. Our own Milky Way Galaxy is labeled at the bottom left of the view, across a void in space from this cluster. Use the location scroller to move around the Virgo cluster and consider its overall shape and its relationship to neighboring galaxies. (**a**) What is the general shape of the Virgo cluster? **Zoom** in toward this cluster until individual galaxies are shown and use the location scroller to help you to identify each classification of galaxy (elliptical, spiral and irregular) within the group. Select **File > Revert** to return to the original view and identify several other groups of galaxies. Select one or two clusters of galaxies in turn, move the cursor over a galaxy within the selected group, and right-click the mouse to open the object contextual menu and select the **Highlight** option to identify this group. You can select the **Centre** option to move the selected cluster to the center of the view and examine the cluster's extent across space. Again, use the location scroller and **Zoom** in to examine this cluster from various viewpoints. (**b**) Describe the distribution of the galaxies within the cluster, compared to the distribution in the Virgo Cluster. For example, what are their shapes and relative sizes compared to Virgo and to each other? See if you can recognize the walls of galaxies surrounding large voids in space that link these concentrated regions of galaxies. (**c**) Click on the **Home** button to return to your sky. Click on the **Find** tab and ensure that the search box is empty. Click on the magnifying glass icon in the search box to open a dropdown list and click on **Hubble Images.** In the list of Hubble images, click on **Gravitational Lens** to center on this image of a cluster of galaxies known as CL0024+1654. **Zoom** in to a field of view of about 1 arcminute. This Hubble Space Telescope image shows a rich cluster of ordinary-looking yellowish galaxies surrounded by blue arcs of light. These are multiple images of a very distant galaxy, as seen through the gravitational "lens" of dark matter pervading the cluster of galaxies. The blue color of these images suggests that this distant galaxy is composed largely of young, blue stars. The distribution of this mysterious substance within galactic clusters can be inferred from these types of images of distant galaxies.

Collaborative Exercises

63. In the early twentieth century, there was considerable debate about the nature of spiral nebulae and their distance from us, but the debate was resolved by improvements in technology. As a group, list three issues that we, as a culture, did not understand in the past but understand today, and explain why we now have that understanding.

64. Even though there are billions of galaxies, there are not billions of different kinds. In fact, galaxies are classified according to their appearance. As a group, dig into your book bags and put all of the writing implements (pens, pencils, highlighters, and so on) you have in a central pile. Remember which ones are yours! Determine a classification scheme that sorts the writing implements into at least three to six piles. Write down the scheme and the number of items in each pile. Ask the group next to you to use your scheme and sort your materials. Correct any ambiguities before submitting your classification scheme.

65. Imagine your company, Astronomical Artistry, has been contracted by the local marching band to create a football halftime show about spiral galaxies. How exactly would you design the positions of the band members on the field to represent the different spiral galaxies of classes Sa, Sb, and Sc? Create two columns on your paper by drawing a line from top to bottom, drawing sketches in the left-hand column and writing a description of each sketch in the right-hand column. Also include what the band's opening formation and final formation should be.

ANSWERS

ConceptChecks

ConceptCheck 23-1: If spiral nebulae are closer than some of the stars of our Galaxy, then the evidence presented supports Shapley's argument that spiral nebulae are within our Galaxy and are not themselves large galaxies very far away.

ConceptCheck 23-2: Cepheid variable stars have well-known luminosities, and can even be identified in other galaxies. Comparing a Cepheid's luminosity to apparent brightness reveals a precise distance to the galaxy.

ConceptCheck 23-3: Sc galaxies have the most active star formation, so a sketch would have a smaller central bulge and a relatively large star-forming disk with considerable gas and dust.

ConceptCheck 23-4: In both spiral and barred spirals, the designation *a* is used for tightly wound arms and *c* for loosely wrapped arms.

ConceptCheck 23-5: Elliptical galaxies have almost no gas or dust available for the formation of stars.

ConceptCheck 23-6: If the Type Ia supernova appeared to be dimmer because of intervening dust, then astronomers would mistakenly believe that the supernova was farther away and, subsequently, that the galaxy was farther away than it really is.

ConceptCheck 23-7: A galaxy has a narrow width to its emitted hydrogen light if it rotates slowly. This is because the width arises from the Doppler shift of emitted light from approaching and receding sides of the rotating galaxy. Smaller galaxies rotate more slowly than larger galaxies, which means that galaxies with narrower emission line widths are also smaller. If, as in this question, the galaxy appears bright, than it must be close to our own Galaxy.

ConceptCheck 23-8: Because the base of the distance ladder depends on parallax, all rungs of the ladder are completely dependent on an accurate understanding of brightness as determined by parallax.

ConceptCheck 23-9: From Figure 23-17, the farther away the galaxy, the larger its recessional velocity in the Hubble flow. Furthermore, a larger recessional velocity produces a larger redshift, so the farther galaxy has the largest redshift.

ConceptCheck 23-10: The spectral shift toward the shorter blue wavelengths means that the Andromeda Galaxy is moving toward our Milky Way Galaxy.

ConceptCheck 23-11: Rich clusters are observed to contain significantly more galaxies within them than poor clusters.

ConceptCheck 23-12: No, although on the largest scales, galaxies appear in every direction. Observations of superclusters show they are clumped into uneven groups and into long filaments.

ConceptCheck 23-13: The Local Group, dominated by the Milky Way Galaxy and the Andromeda Galaxy, contains about 40 dwarf ellipticals.

ConceptCheck 23-14: Collisions between gas and dust in colliding galaxies will warm gas, causing it to glow in X-rays.

ConceptCheck 23-15: Simulations suggest that a single elliptical galaxy will form with many newly formed stars.

ConceptCheck 23-16: The best evidence is that the rotation curves in galaxies cannot be accounted for by the observed mass and the ability of galaxies and galaxy clusters to gravitationally lens more distant sources of light.

ConceptCheck 23-17: All matter consisting of neutrons and protons is called "ordinary" matter, so a neutron star is still made of ordinary matter, even though a neutron star is not an ordinary object like a planet or star.

ConceptCheck 23-18: Earlier galaxies appear to have been much smaller, suggesting that today's large galaxies formed by collisions and combinations of smaller galaxies.

ConceptCheck 23-19: Elliptical galaxies create most of their stars during the initial formation of the galaxy. Metal-rich, Population I stars can only form from the remains of Population II stars, which can only occur in later generations of star formation, which do not occur in elliptical galaxies.

ConceptCheck 23-20: The arms become distorted during collisions, resulting in astronomers observing far fewer spiral galaxies today than in the past.

CalculationChecks

CalculationCheck 23-1: Because z = the change in wavelength $(\lambda - \lambda_0)$ divided by the original λ_0, $z = (725.6\text{ nm} - 656.3\text{ nm}) \div 656.3\text{ nm} = 0.11$.

CalculationCheck 23-2: Hubble's Law, $v = H_0 d$, can be rearranged as $d = v \div H_0$. Using $H_0 = 73$ km/s/Mpc, $d = v \div H_0 = 10{,}000\text{ km/s} \div 73\text{ km/s/Mpc} = 134$ Mpc (million parsecs).

CHAPTER

24

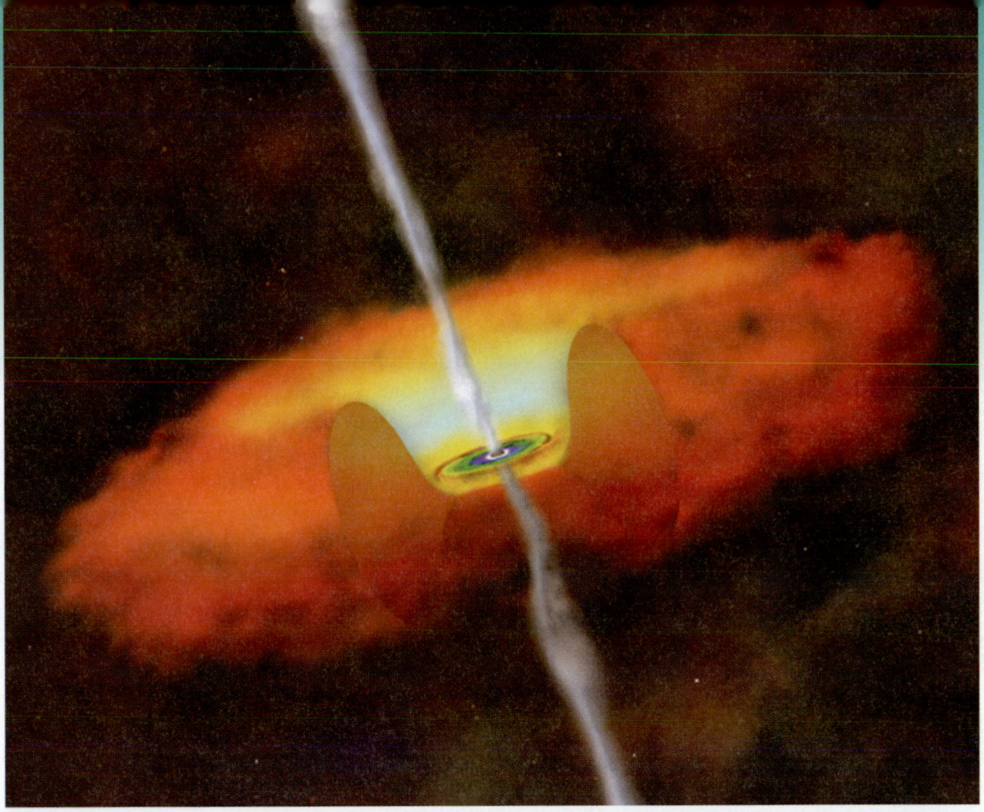

This artist's conception of a quasar shows a thick doughnut-shaped torus surrounding a small bright inner region of hot swirling gas. The inner region is about the size of our solar system and is called an accretion disk. At the very center is a supermassive black hole that steadily consumes, or accretes, the surrounding gas. As illustrated here, some quasars have strong jets beaming away perpendicularly to the disk of swirling matter. While the outer torus can sometimes block much of the light emitted by the interior, when the interior light is visible, it can be a hundred thousand times as bright as our Milky Way Galaxy! (NASA/CXC/M.Weiss)

Quasars and Active Galaxies

LEARNING GOALS

By reading the sections of this chapter, you will learn

24-1 The distinctive features of quasars

24-2 The role of supermassive black holes in powering active galactic nuclei (AGN)

24-3 How quasars can form accretion disks and jets

24-4 That active galactic nuclei can look very different depending on their orientation

24-5 The fate of active galactic nuclei and their potential to occasionally flare up

An ordinary star emits radiation primarily at ultraviolet, visible, and infrared wavelengths, in varying proportions that are governed by the star's surface temperature. Composed of many stars, ordinary galaxies, too, emit most strongly in these wavelength regions. But the object illustrated here, called a quasar, is outrageously different: It emits strongly over an immense range of wavelengths from radio to X-ray.

The name quasar refers to one of the most fantastic types of objects in all of astronomy. At a quasar's core is a supermassive black hole that can be a *billion* times more massive than our Sun! Hot gas around this black hole can give off a tremendous amount of visible light—enough to shine with a luminosity *ten thousand* times more than a typical galaxy.

As in this illustration, some quasars even have strong jets. Unlike young stars that form smaller and slower-speed jets, matter in quasar jets travels near the speed of light and can shoot well beyond the galaxy that hosts the quasar. Galaxies that hold a quasar—always at the galaxy's center—are called *active galaxies.* Untangling the mysterious properties of quasars has been in process for more than half a century and is still a major topic of research today. Physicist George Gamow captured the intriguing nature of quasars in his 1964 revised lullaby:

Twinkle, twinkle quasi-star
Biggest puzzle from afar
How unlike the other ones
Brighter than a billion suns.
Twinkle, twinkle quasi-star
How I wonder what you are.

24-1 Quasars are the ultraluminous centers of distant galaxies

Examining the spectra of quasars revealed that they are immensely distant

The discovery of quasars begins in the Illinois backyard of Grote Reber, a radio engineer who built the first true radio telescope in 1936. One of the objects he discovered is called Cygnus A; a modern radio and optical image of this quasar is shown in Figure 24-1. The object's thin jets shoot matter well beyond its host galaxy; to appreciate the scale, consider that about four Milky Way galaxies could fit between the ends of the two lobes in Cygnus A.

In a visible-wavelength image, a distant quasar often appears as a point source of light, like a star, because the host galaxy containing the quasar is so far away that the galaxy's light is not easily observed (Figure 24-2). Along with intense radio emission, these properties led to the name "**quasar,** " which means *quasi-stellar radio source.*

While the visible light from a quasar might appear pointlike, the way a star does, its spectrum tells a very different story. Stars, and the galaxies they make up, show absorption lines in their spectra. Quasars, on the other hand, show strong *emission* lines.

Clues from Emission Lines

Although quasars were clearly oddballs, many astronomers thought they were just strange stars in our own Galaxy. A breakthrough occurred in 1963, when Maarten Schmidt at Caltech took another look at the emission lines of the quasar 3C 273 (Figure 24-3). Emission lines are caused by excited atoms, which emit radiation at specific wavelengths unique to each atom (see Figure 6-21). However, the lines of quasar 3C 273 did not match the known emission line wavelengths of hydrogen (or any other known atoms or molecules), so astronomers could not identify what quasars were made of.

Schmidt realized that four of 3C 273's brightest emission lines are positioned relative to one another in precisely the same way as four common emission lines of hydrogen (these are the Balmer lines; see Figure 5-24). However, these emission lines from 3C 273 were all shifted to *much longer wavelengths* than had ever been seen in the Balmer lines before. Were these lines really from hydrogen, but Doppler-shifted from an object receding away at great speed? Schmidt determined that 3C 273 has a redshift of $z = 0.158$, corresponding to a recessional velocity equal to 16% of the speed of light (44,000 km/s). No star could be moving this fast and remain within our Galaxy for very long. Hence, Schmidt concluded that 3C 273 could not be a nearby star, but must lie outside the Milky Way.

According to the Hubble law, the recessional velocity of 3C 273 implies that its present distance from us is 2 billion light-years; that is over 800 times farther away from us than the Andromeda Galaxy! To be detected at such distances, 3C 273

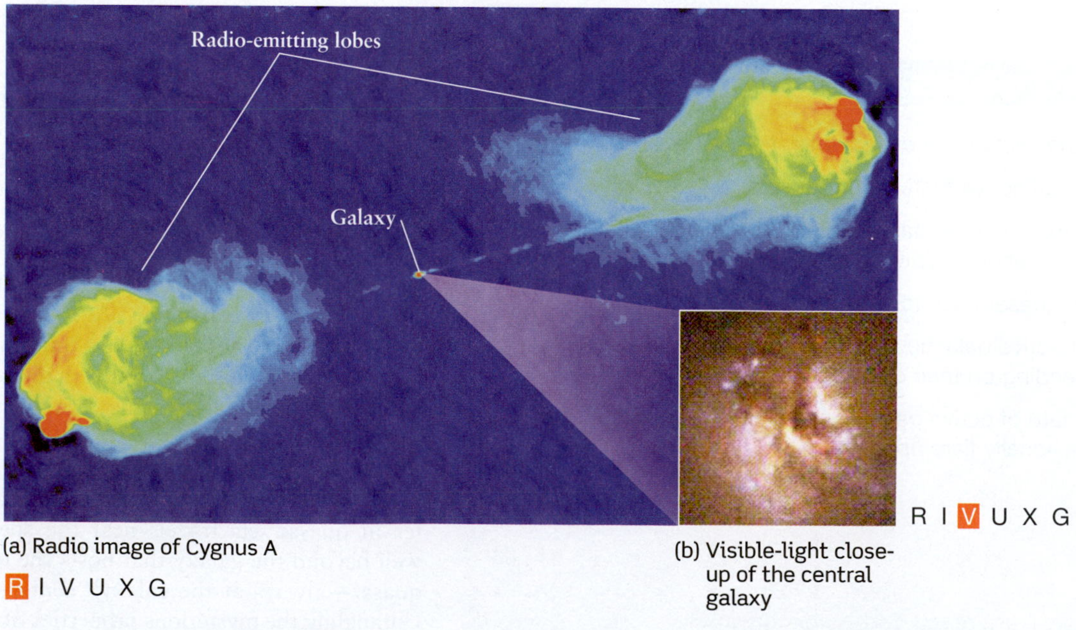

FIGURE 24-1
Cygnus A **(a)** This false-color radio image from the Very Large Array shows that most of the emission from Cygnus A comes from luminous radio lobes located on either side of a peculiar galaxy. (Red indicates the strongest radio emission, while blue indicates the faintest). Each lobe extends about 230,000 ly) from the galaxy. **(b)** The galaxy at the heart of Cygnus A has a substantial redshift, so it must be extremely far from Earth (it is about 740 million ly). Only the very brightest central region of the galaxy is visible here. To be so distant and yet be one of the brightest radio sources in the sky, Cygnus A must have an enormous energy output. (a: NRAO/AUI; b: Hubble Space Telescope, cropped from a mosaic of three images by Bill Keel)

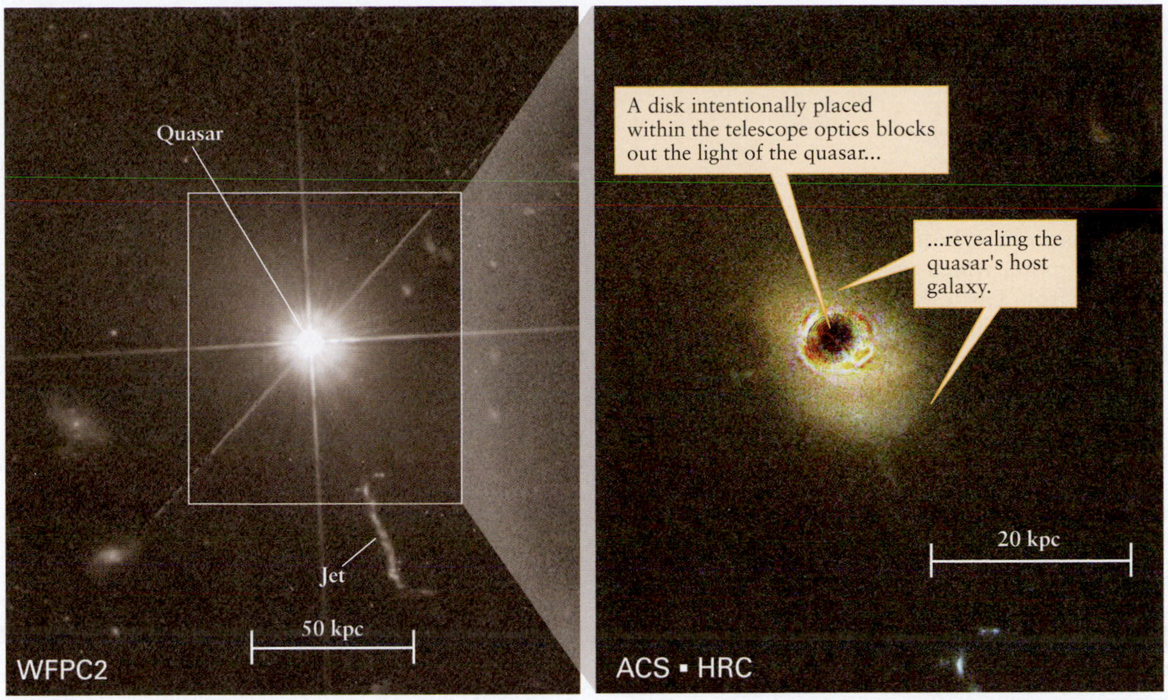

FIGURE 24-2 R I **V** U X G

A Quasar and Its Host Galaxy **(a)** Appearing pointlike, a quasar can look like a star in images. In this HST image, the bright glare of a quasar named 3C 273 hides its fainter host galaxy. Quasar 3C 273 also has a jet seen in visible light, although quasar jets are usually seen at radio wavelengths. **(b)** Using a special technique, the quasar is blocked out to reveal the host galaxy. (NASA, A. Martel [JHU], the ACS Science Team, J. Bahcall [IAS] and ESA)

must be an extraordinarily powerful source of both visible light and radio emission.

The most luminous quasars emit 100,000 times more radiation than the entire Milky Way Galaxy

A quasar's luminosity can be calculated from its apparent brightness and distance using the inverse-square law (see Section 17-2). For example, 3C 273 has a luminosity of about 10^{40} watts, which is equivalent to 2.5×10^{13} (25 trillion) Suns. (The Sun's luminosity is $L_{\odot} = 3.90 \times 10^{26}$ watts.) Generally, quasar luminosities range from about 10^{38} watts up to nearly 10^{42} watts. For comparison, a typical large galaxy, like our own Milky Way, shines with a luminosity of 10^{37} watts, which equals 2.5×10^{10} (25 billion) Suns. Thus, a bright quasar can be many thousands of times more luminous than our entire Milky Way Galaxy!

The rapid variability of active galactic nuclei tells us that they must be very small

Quasars are members of a larger class of objects called **active galactic nuclei,** or **AGN** (the singular form, active galactic nucleus, is also called an AGN). AGN include a range of similar objects that, technically speaking, are not all luminous enough to be called quasars. Nonetheless, all AGN are powered by hot gas accreting around a supermassive black hole, and astronomers often use the term quasar and AGN interchangeably. If the host galaxy that contains an AGN can be easily observed, the object is called an **active galaxy.** If the active galaxy emits strongly at radio wavelengths, as in the case of Cygnus A in Figure 24-1, the AGN is called a **radio galaxy.**

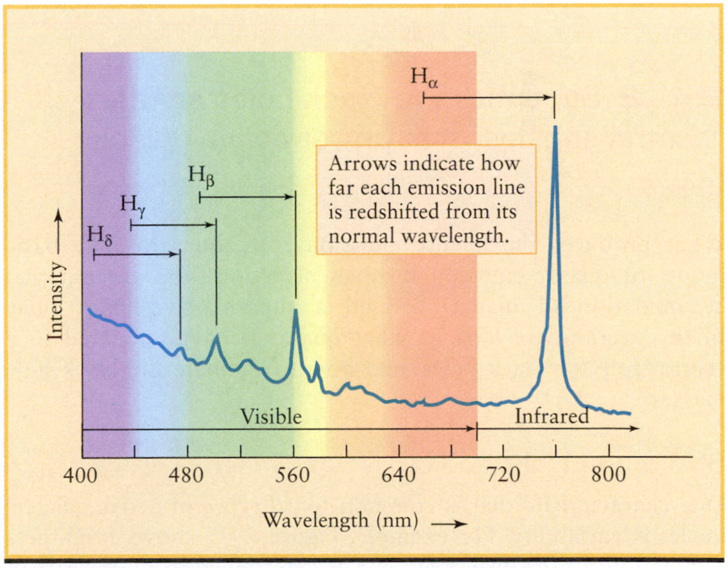

FIGURE 24-3

The Spectrum of a Quasar The visible-light and infrared spectrum of quasar 3C 273 is dominated by four bright emission lines of hydrogen (see Section 5-8). The redshift is $z = 0.158$, so the wavelength of each line is 15.8% greater than for a sample of hydrogen on Earth. For example, the wavelength of H_β is shifted from 486 nm (a blue-green wavelength) to $1.158 \times (486 \text{ nm}) = 563$ nm (a yellow wavelength).

CONCEPTCHECK 24-1

What are two main differences between the spectra of a star and a quasar?

Answer appears at the end of the chapter.

Quasars: High Redshifts, Extreme Distances

More than 200,000 quasars have been discovered. Most quasars have redshifts of 0.3 or more, which implies that they are more than 3 billion light-years from Earth. Because there are no quasars with small redshifts, it follows that *there are no nearby quasars.* The nearest one is some 800 million ly from Earth. Furthermore, light takes time to travel across space, so when we observe a very remote object with a large redshift, we are seeing an image propagating from the remote past. Hence, the absence of nearby quasars means that *the era of quasars ended long ago*. Indeed, the number of quasars began to decline precipitously roughly 10 billion years ago. Quasars were a common feature of the universe in the distant past, but there are none in the present-day universe (Figure 24-4). Quasars not only occurred early in our universe, but whenever they occurred, their brief lifetimes were only about 1% as long as the lives of their host galaxies.

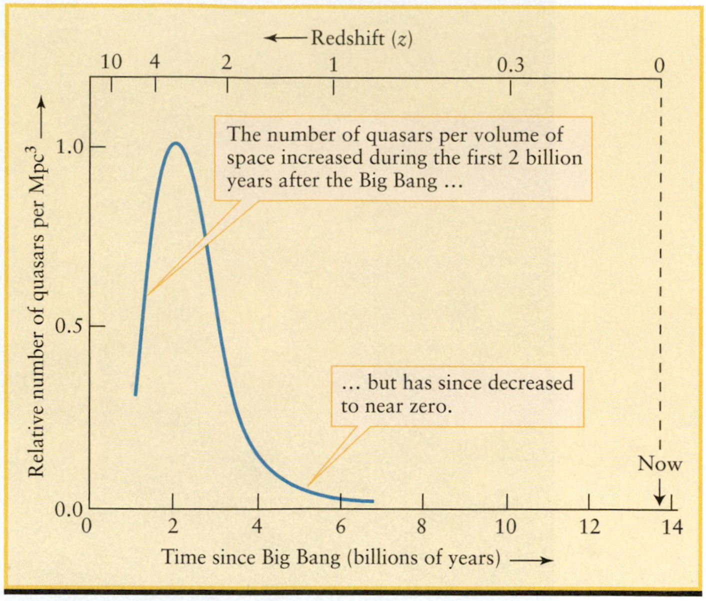

FIGURE 24-4

Quasars Are Extinct The greater the redshift of a quasar, the farther it is from Earth and the farther back in time we are seeing it. By observing the number of quasars found at different redshifts, astronomers can calculate how the density of quasars in the universe has changed over the history of the universe. The peak of quasar activity occurred more than 10 billion years ago, and there is no significant quasar activity today. The history of quasars is reminiscent of the history of the dinosaurs, which once populated the entire Earth but today are extinct. (Peter Shaver, European Southern Observatory)

CONCEPTCHECK 24-2

From Figure 24-4, about how old was the universe when quasars were most abundant? What is the approximate redshift of objects from this time?

CONCEPTCHECK 24-3

Why can we say that any quasar observed from Earth today must have been very luminous when its light was emitted?

Answers appear at the end of the chapter.

24-2 Supermassive black holes are the "central engines" that power active galactic nuclei

What produces the intense emission from quasars? The basic model of quasar emission involves **accretion**—the gravitational accumulation of matter—around a supermassive black hole. In this section, we look at what quasar behavior pointed to a central role for black holes, and how to estimate the black hole masses.

Size of the Light Source

One characteristic that is common to *all* types of active galactic nuclei is variability. For example, Figure 24-5 shows brightness fluctuations of the quasar 3C 273 as determined from 29 years of observations. The brightness of 3C 273 increased by 60% from the beginning to the end of 1982, then declined to the starting value in just five months. Other AGN undergo even greater fluctuations in brightness (by a factor of 25 or more) that occur even more rapidly (X-ray observations reveal that some AGN vary in brightness over time intervals as short as 3 hours).

The crucial aspect of these fluctuations in brightness is that they allow astronomers to place fundamental limits on the maximum size of a light source. This strict limit arises because *an object cannot vary in brightness faster than light can travel across that object.* For example, an object that is 1 light-year in diameter cannot vary significantly in brightness over a period of less than 1 year.

To understand this limitation, imagine an object that measures 1 light-year across, as in Figure 24-6. Suppose the entire object suddenly brightens, emitting a brief flash of light. Photons from that part of the object nearest Earth arrive at our telescopes first. Photons from the middle of the object arrive at Earth 6 months later. Finally, light from the far side of the object arrives a year after the first photons. Although the object emitted a sudden flash of light, we observe only a gradual increase in brightness that lasts a full year. In other words, the flash is stretched out over an interval equal to the difference in the light travel time between the nearest and farthest observable regions of the object.

The rapid flickering exhibited by active galactic nuclei means that they emit their energy from a small volume—in some cases less than half a light-day across (which is about the diameter of Neptune's orbit). Thus, from quasar variability, astronomers discovered that *a region about the size of our solar system can emit more energy per second than a thousand galaxies!* More than anything else, this size-constraint points to supermassive black holes, as no other known object can power the release of so much energy from such a small volume.

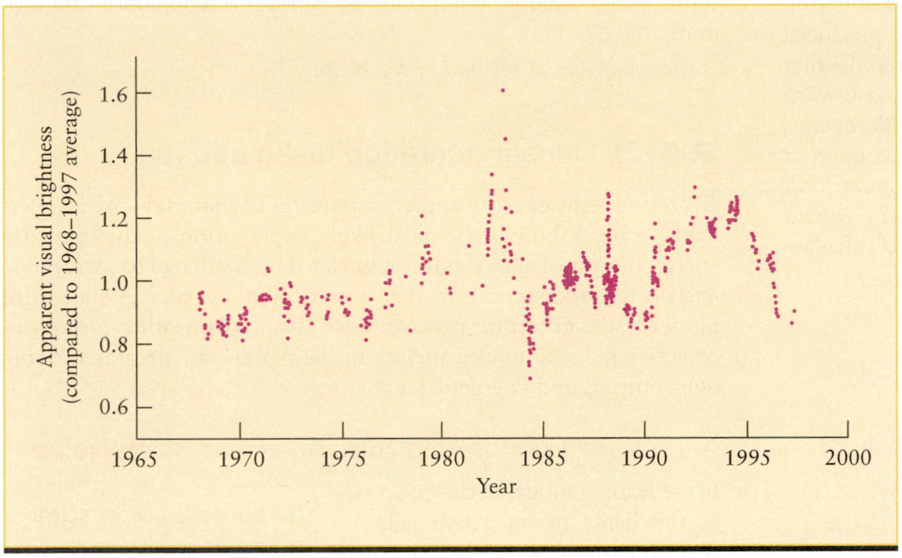

FIGURE 24-5

Brightness Variations of an AGN This graph shows variations over a 29-year period in the apparent brightness of the quasar 3C 273 (see Figure 24-2a, b). Note the large outburst in 1982–1983 and the somewhat smaller ones in 1988 and 1992. (Adapted from M. Türler, S. Paltani, and T. J.-L. Courvoisier)

CAUTION! Quasars are intensely luminous, but their central black holes do not emit this light. Instead, the light is emitted by hot gas swirling around the black hole. The black holes are still considered the "central engines" of the quasar because it is the black hole's gravity that pulls in the surrounding gas and heats it up.

While the light-emitting region in a quasar might be about the size of our solar system, the black hole is smaller still. Just how big and massive are the black holes in quasars? Surprisingly, the size and mass of the black hole can be estimated from great distances where nothing is seen but the light from its hot accreting gas.

The Eddington Limit and Black Hole Sizes

There is a relationship between the size of a black hole and the luminosity emitted by the hot gas falling into it. Even if there is plenty of gas around to act as "fuel," there is a natural limit to the luminosity that can be radiated by accretion onto a compact object like a black hole. This limit is called the **Eddington limit,** after the British astrophysicist Sir Arthur Eddington. The Eddington limit applies to any object held together by its own gravity and applies to stars as well as quasars.

The accretion of fuel by a black hole is thought to be a smooth and steady process, but let's imagine what might happen if a quasar

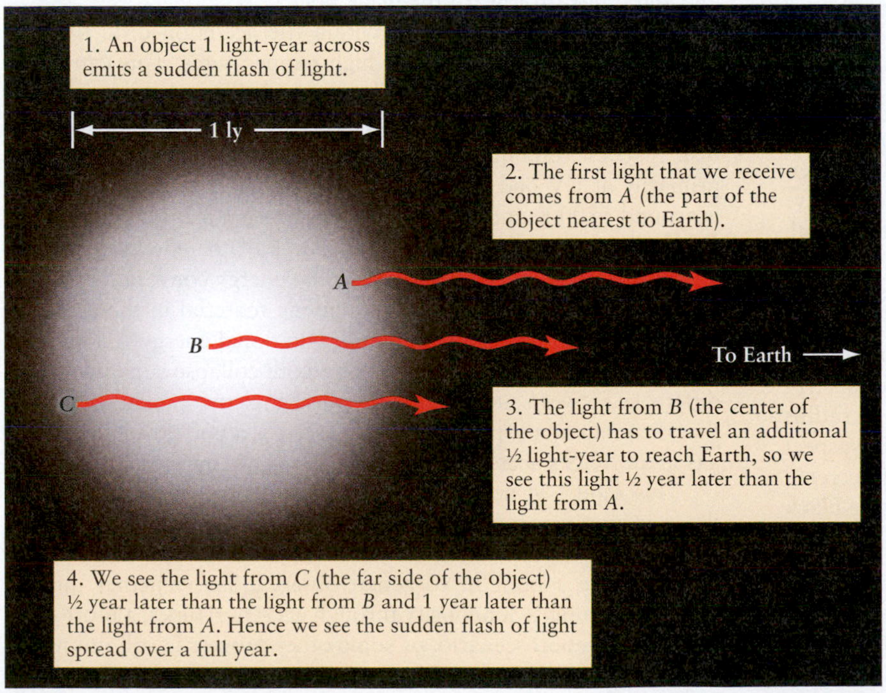

1. An object 1 light-year across emits a sudden flash of light.
1 ly
2. The first light that we receive comes from *A* (the part of the object nearest to Earth).
A
B
C
To Earth
3. The light from *B* (the center of the object) has to travel an additional ½ light-year to reach Earth, so we see this light ½ year later than the light from *A*.
4. We see the light from *C* (the far side of the object) ½ year later than the light from *B* and 1 year later than the light from *A*. Hence we see the sudden flash of light spread over a full year.

FIGURE 24-6

A Limit on the Speed of Variations in Brightness The rapidity with which the brightness of an object can vary significantly is limited by the time it takes light to travel across the object. If an object 1 light-year in size emits a sudden flash of light, the flash will be observed from Earth to last a full year. If the object is 2 light-years in size, brightness variations will last at least 2 years as seen from Earth, and so on.

were to suddenly brighten. If the luminosity exceeds the Eddington limit, there is so much *radiation pressure*—the pressure produced by photons streaming outward from the infalling material—that the surrounding gas is pushed outward rather than falling inward onto the black hole. Without a source of gas to provide energy, the luminosity naturally decreases to below the Eddington limit, at which point gas can again fall inward. Thus, *the Eddington limit represents the balance between radiation pushing gas-fuel outward and gravity pulling the gas inward.* This limit allows us to calculate the minimum mass of the black hole in a quasar.

Numerically, the Eddington limit is

The Eddington limit

$$L_{\text{Edd}} = 30{,}000 \left(\frac{M}{M_\odot}\right) L_\odot$$

L_{Edd} = maximum luminosity that can be radiated by accretion around a black hole

M = mass of the black hole

$M_\odot$ = mass of the Sun

$L_\odot$ = luminosity of the Sun

The tremendous luminosity of a quasar must be less than or equal to its Eddington limit (or it would blow away its accreting gas), so this limit must be very high indeed. Hence, the mass of the black hole must also be quite large. For example, consider the quasar 3C 273, which has a luminosity of about $3 \times 10^{13}\ L_\odot$. To calculate the minimum mass of a black hole that could continue to attract gas to power the quasar, assume that the quasar's luminosity equals the Eddington limit. Inserting $L_{\text{Edd}} = 3 \times 10^{13}\ L_\odot$ into the above equation, we find that $M = 10^9\ M_\odot$. Therefore, if a black hole is responsible for the energy output of 3C 273, its mass must be greater than a billion Suns!

Astronomers have indeed found evidence for such **supermassive black holes** at the centers of many nearby normal galaxies (see Section 22-5), and these are thought to be the remnants of AGN. As we saw in Section 22-6, at the center of our own Milky Way Galaxy lies what is almost certainly a black hole of about 4.1×10^6 solar masses—supermassive in comparison to a star, but less than 1% the mass of the behemoth black hole at the center of 3C 273.

> Most or all large galaxies have supermassive black holes at their centers

Unlike stellar-mass black holes, which require a supernova to produce them, supermassive black holes can be produced without extreme densities. Recall (Section 21-5) that matter with an average density no greater than water could form a 500 million $M_\odot$ black hole if this matter coalesced into a sphere with a radius about the Earth–Sun distance (1 AU). Some astronomers suspect that galaxy collisions produced massive black holes that accreted their way to even larger sizes, but the origin of supermassive black holes, and the AGN they powered, remains a mystery.

CONCEPTCHECK 24-4

Suppose a quasar's luminosity is initially right at the Eddington limit. What would happen if the quasar were somehow able to temporarily become even brighter without a significant change in its mass?

Answer appears at the end of the chapter.

24-3 Quasar accretion disks and jets

Accretion—the capture and consumption of material—by a supermassive black hole is the most likely explanation of the immense energy output of active galactic nuclei. The challenge to astrophysicists is to understand how that accretion takes place. A successful model of the accretion process must also explain other properties of active galactic nuclei, including their unusual spectra, variable light output, and energetic jets.

Accretion Disks around Supermassive Black Holes

In the leading model of this process, at the heart of an active galaxy is a supermassive black hole surrounded by an **accretion disk,** a solar system–sized disk of matter captured by the hole's gravity and spiraling into it. We saw in Section 18-5 that accretion disks are found around stars in the process of formation. The accretion disks that astrophysicists envision around supermassive black holes are similar but far larger and far more dynamic. The *Cosmic Connections: Accretion Disks* figure depicts the physics of such accretion disks.

> The power source for active galaxies is gravitational energy released by material falling toward a central black hole

Imagine a billion-solar-mass black hole sitting at the center of a galaxy, surrounded by a rotating accretion disk. According to Kepler's third law, the inner regions of this accretion disk would orbit the hole more rapidly than would the outer parts. Thus, the rapidly spinning inner regions would constantly rub against the slower moving gases in the outer regions. This friction, aided by magnetic forces within the disk, would cause the gases to lose energy and spiral inward toward the black hole.

As the gases move inward within the accretion disk, they are compressed and heated to very high temperatures. This causes the accretion disk to glow, thus producing the brilliant luminosity of an active galactic nucleus. The temperature of material in the accretion disk reaches 100,000 K or more, which emits thermal radiation primarily at UV and visible wavelengths. (By comparison, the Sun is around 5800 K and the hottest blue stars are around 50,000 K.)

In this model, the fundamental source of the energy output by an active galactic nucleus is *gravitational* energy converted into, and released as, thermal radiation by infalling material in the accretion disk. We saw in Section 20-6 that gravitational energy is also what powers the immense light output of a core-collapse supernova. The difference between such a supernova and an active galactic nucleus is that the infall of a supernova's outer layers is a one-time event, while gas in an AGN's accretion disk spirals inward continuously.

This model of accretion works well for smaller disks such as those around neutron stars or stellar mass black holes. However, as much as it is the favored model, detailed accretion disk calculations for AGN are a poor fit to actual observations. Whether quasars derive their intense luminosity through accretion disks, a more complicated accretion scenario, or some other process is still unknown. Until the emission mechanism for quasars is understood, it is likely to remain on the frontier of modern research.

COSMIC CONNECTIONS

Accretion Disks

Quasars and active galaxies are thought to be powered by gravitational energy released when gas falls toward a central supermassive black hole. Due to conservation of angular momentum, the gas does not fall straight into the black hole. Instead, it must negotiate its way through an accretion disk.

How an accretion disk forms around a black hole

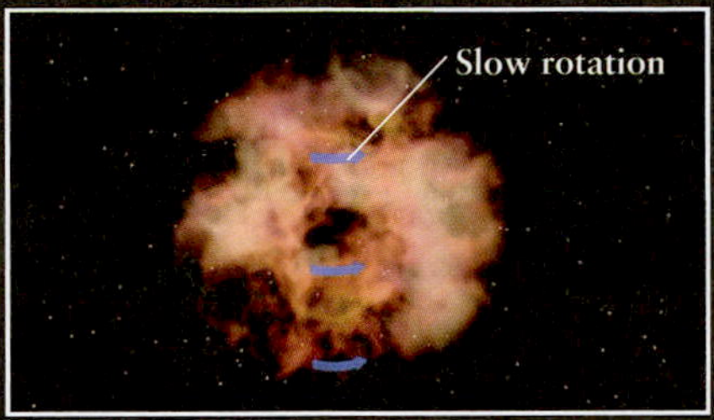
Slow rotation

1. A slowly rotating cloud of gas coalesces around the black hole.

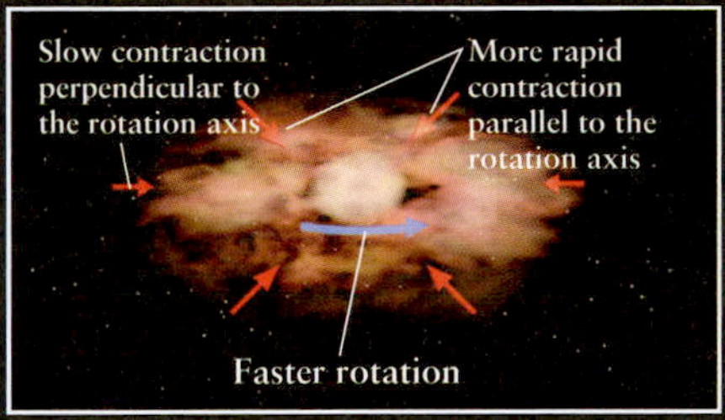
Slow contraction perpendicular to the rotation axis
More rapid contraction parallel to the rotation axis
Faster rotation

2. Due to conservation of angular momentum, the gas moves inward at different rates in different directions.

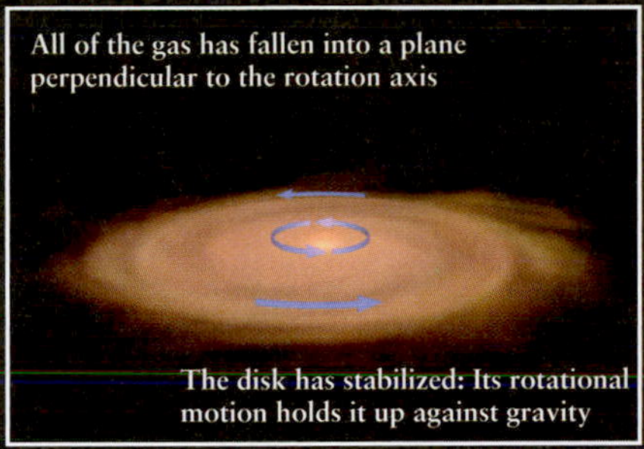
All of the gas has fallen into a plane perpendicular to the rotation axis
The disk has stabilized: Its rotational motion holds it up against gravity

3. The final configuration of the gas is a flattened disk.

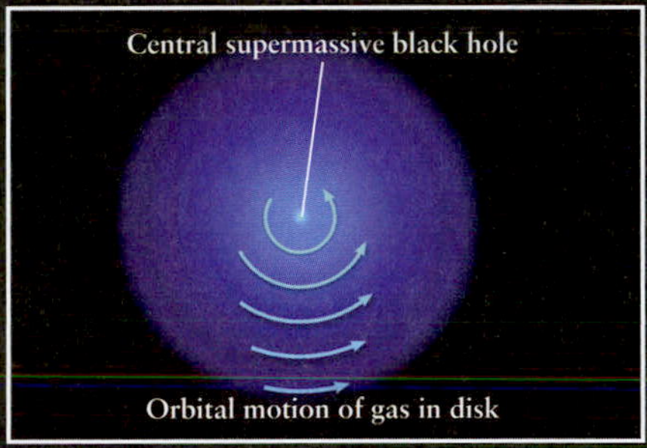
Central supermassive black hole
Orbital motion of gas in disk

4. Like planets orbiting around the Sun, the closer a blob of gas is to the center of the disk, the faster its orbital speed.

Why gas in an accretion disk spirals in and radiates

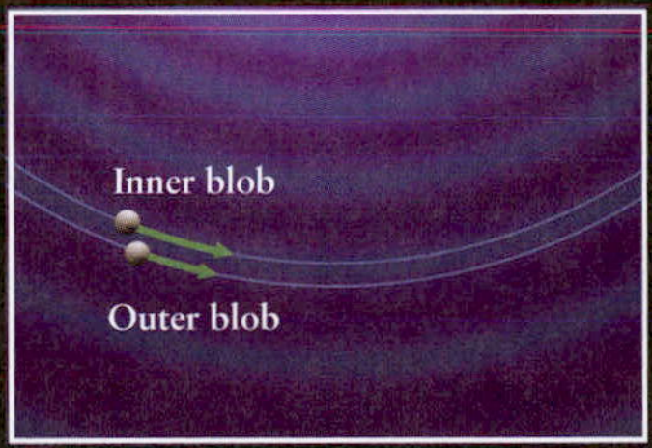
Inner blob
Outer blob

5. Consider two adjacent blobs of gas in the accretion disk. The inner blob moves a bit faster than the outer blob (see 4).

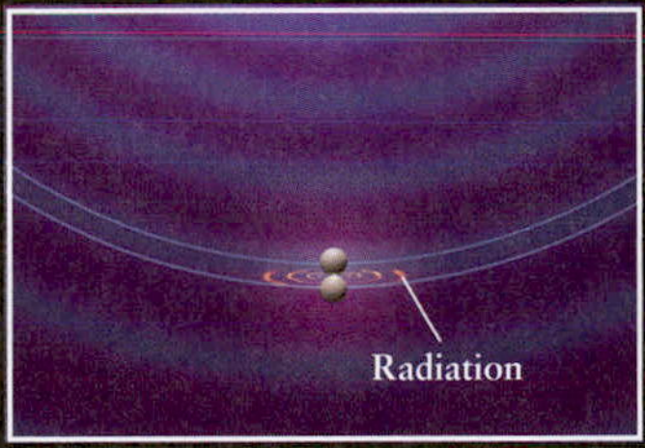
Radiation

6. When the two blobs rub past each other, energy is transferred from the inner blob to the outer blob. The heated gas emits radiation.

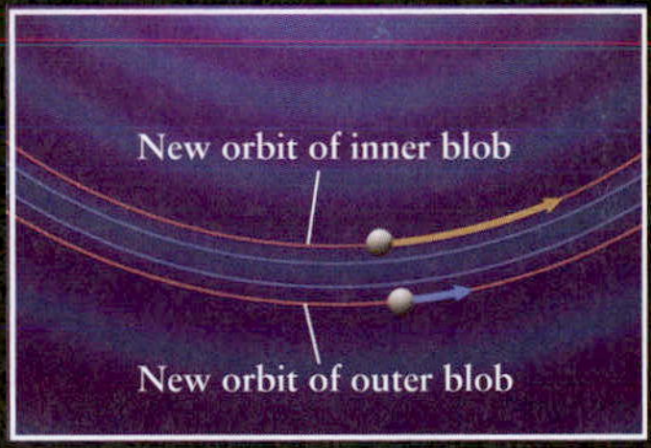
New orbit of inner blob
New orbit of outer blob

7. Having lost energy, the inner blob moves inward toward the center of the disk and speeds up. The outer blob moves outward and slows down.

8. Conservation of angular momentum slows the inward motion of the blobs of gas.

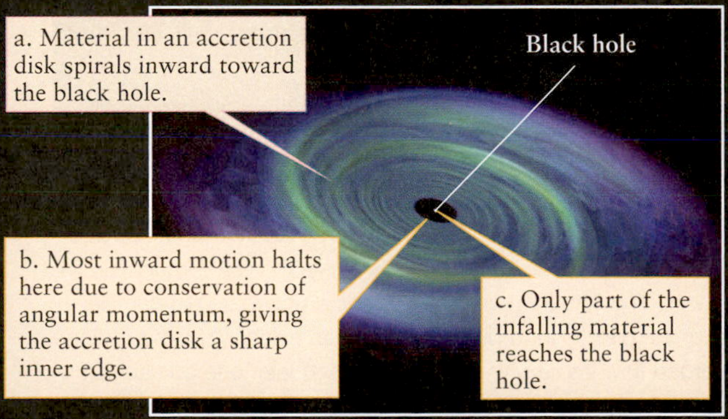
a. Material in an accretion disk spirals inward toward the black hole.
Black hole
b. Most inward motion halts here due to conservation of angular momentum, giving the accretion disk a sharp inner edge.
c. Only part of the infalling material reaches the black hole.

Jets of Charged Particles

Because of the constant inward crowding of hot gas in the plane of the accretion disk, pressures climb rapidly near the center. These pressures expel matter at extremely high speeds. This ejected material escapes at right angles to the plane of the accretion disk and the material forms into two jets (Figure 24-7). However, the active galaxy M87 only shows a single jet. Similar to a flashlight beam, we can clearly see the jet pointed toward us, but the jet pointing away is not visible (Figure 24-8) For AGN, the matter in the jets travels at nearly the speed of light, and this leads to beamed emission that is only visible when pointed somewhat toward the observer.

Magnetic forces play a crucial role in steering these fast-moving particles. These forces arise because the hot gases in the accretion disk are ionized, forming a plasma, and the motions of this plasma generate a magnetic field (see Section 16-9). As the plasma in the disk rotates around the black hole, it pulls the magnetic field along with it. But because the disk rotates faster in its inner regions than at its outer rim, the magnetic field becomes severely twisted. This twisted field forms two helix shapes, one on either side of the plane of the disk.

Charged particles flowing outward from the accretion disk tend to follow these magnetic field lines. The result is that the outflowing beams of particles are focused into two jets oriented perpendicular to the plane of the accretion disk. The figure that opens this chapter is an artist's conception of the accretion disk and jets. The highly collimated jets shoot well beyond the host galaxy, and very diffuse gas in the space between galaxies spreads the jets into radio lobes, as in Figure 24-1. Jets are primarily observed at radio wavelengths but some jets can also be seen at visible wavelengths and even X-rays (Figure 24-9). It is not known if the charged particles in jets are composed of electrons and protons, or electrons and their antimatter partners, the positrons.

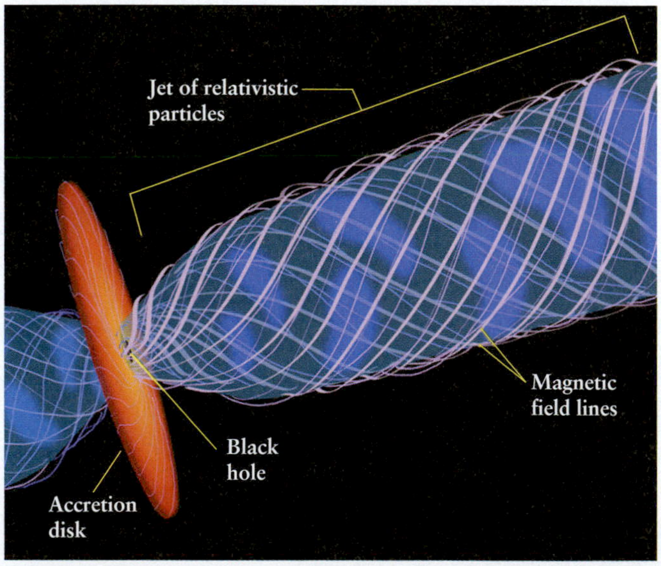

FIGURE 24-7

Jets from a Supermassive Black Hole The rotation of the accretion disk surrounding a supermassive black hole twists the disk's magnetic field lines into a helix. The field then channels the flow of subatomic particles pouring out of the disk. Over a distance less than a light-year, this channeling changes a broad flow into a pair of tightly focused jets, one on each side of the disk. Figure 18-16 shows a similar process that takes place on a much smaller scale in protostars. (NASA and Ann Feild, Space Telescope Science Institute)

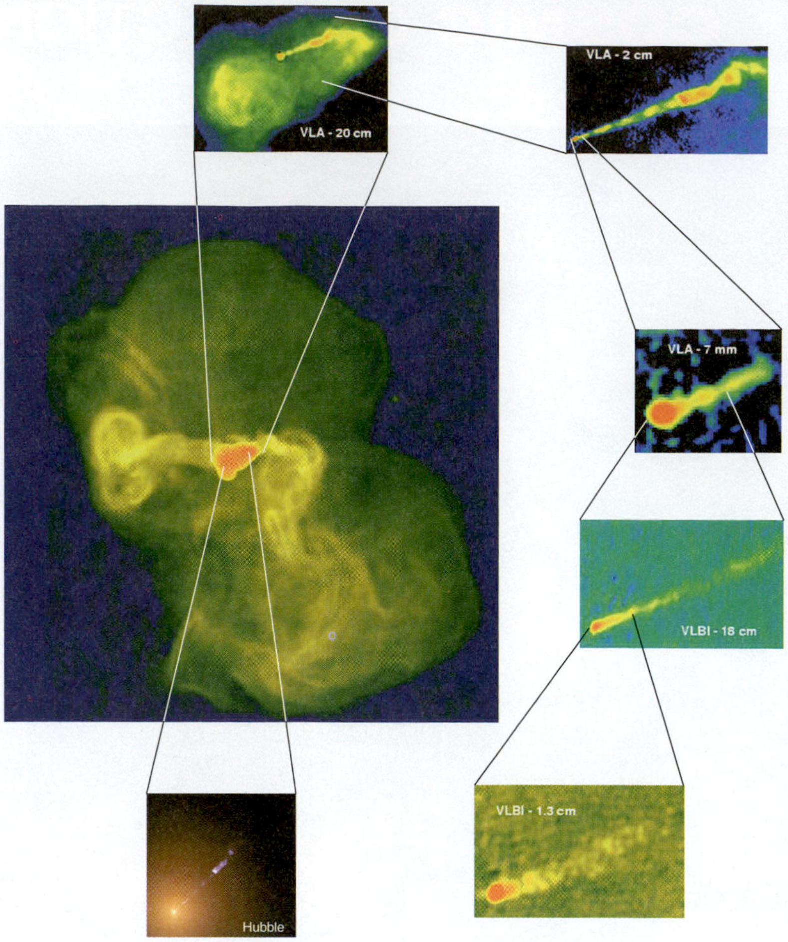

FIGURE 24-8 R I V U X G

Giant Elliptical Galaxy M87 M87 is located near the center of the sprawling, rich Virgo cluster, which is about 50 million ly from Earth. Embedded in this radio image of gas is the galaxy M87 from which the gas has been ejected (bottom inset). The bottom inset is a visible light image from the Hubble Space Telescope, and the rest are at radio wavelengths. As the images zoom into smaller scales, they reveal a variety of details about the structure of the jets of gas. M87's extraordinarily bright nucleus and the gas jets result from a 3-billion-$M_\odot$ black hole, whose gravity causes huge amounts of gas and an enormous number of stars to crowd around it. (NASA and the Hubble Heritage Team [STScI/AURA]; Frazer Owen [NRAO], John Biretta [STScI] and colleagues)

CONCEPTCHECK 24-5

In the accretion disk model, what happens when an inner blob of gas passes an outer blob of gas during accretion (see Cosmic Connections: Accretion Disks)? How is light produced in an accretion disk?

Answer appears at the end of the chapter.

The Strange Case of Superluminal Motion

Nothing travels faster than the speed of light, but if something did, it would be called **superluminal motion.** Initially quite a mystery,

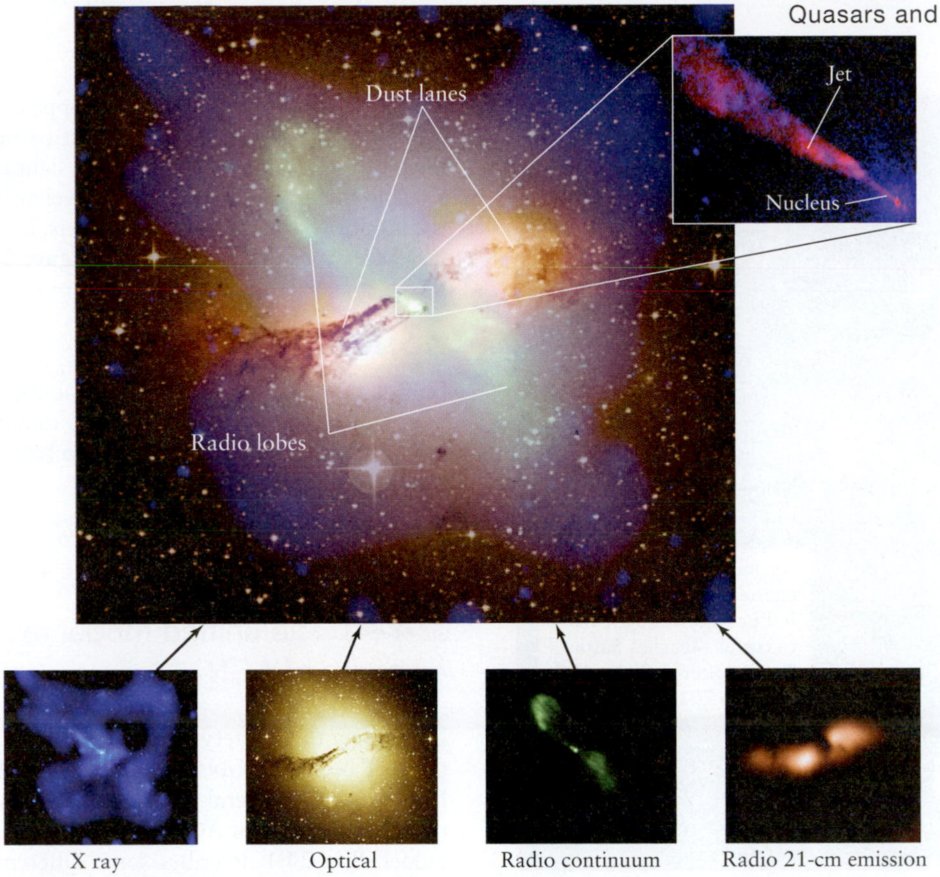

FIGURE 24-9 R I V U X G
Peculiar Galaxy NGC 5128 (Centaurus A) This extraordinary radio galaxy is located in the constellation Centaurus, 11 million ly from Earth. At visible wavelengths a dust lane crosses the face of the galaxy. Superimposed on this visible image is a false-color radio image (green) showing that vast quantities of radio emission pour from matter ejected from the galaxy perpendicular to the dust lane, along with radio emission (rose-colored) along the dust lane, and X-ray emission (blue) detected by NASA's Chandra X-ray Observatory. The X-rays may be from material ejected by the black hole or from the collision of Centaurus A with a smaller galaxy. Upper right inset: This X-ray image from the Einstein Observatory shows that NGC 5128 has a bright X-ray nucleus. An X-ray jet protrudes from the nucleus along a direction perpendicular to the galaxy's dust lane. (X-ray: NASA/CXC/M. Karovska et al.; radio 21-cm: NRAO/VLA/J. Van Gorkom/Schminovich et al.; radio continuum: NRAO/VLA/J. Conden et al.; optical: Digitized Sky Survey U.K. Schmidt Image/STScI; inset: X-ray: NASA/CXC/Bristol U./M. Hardcastle; radio: RAO/VLA/Bristol U./M. Hardcastle)

some quasars seemed to show superluminal motion when looking very closely at "blobs" of radio-emitting particles near the very base of their radio jets. For example, Figure 24-10 shows four high-resolution images of the quasar 3C 273 spanning three years. During this interval, a blob moved away from the quasar at a rate of almost 0.001 arcsec per year. Taking into account the distance

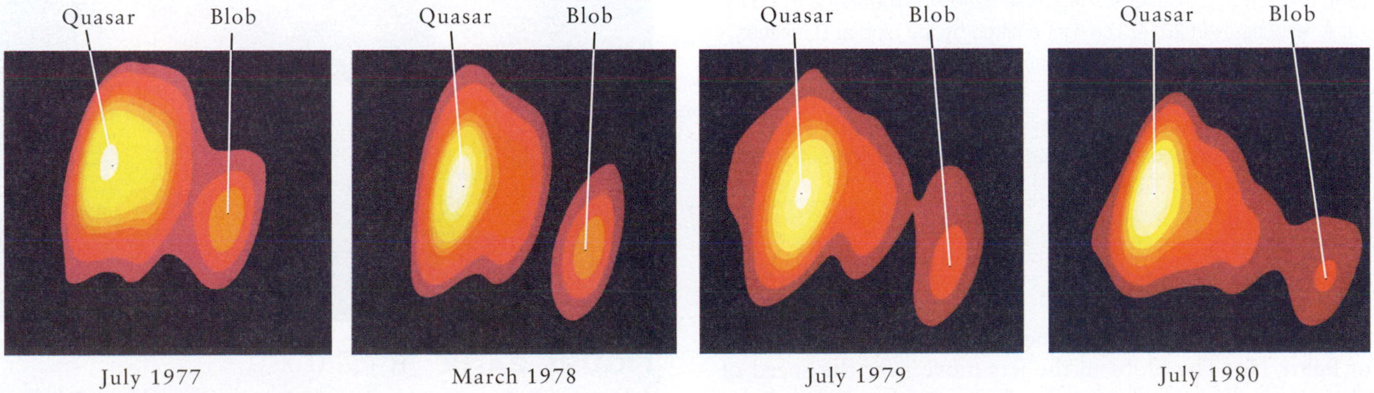

VIDEO 24-1 **FIGURE 24-10** R I V U X G
Superluminal Motion in 3C 273 These four images are false-color, high-resolution radio maps of the quasar 3C 273 (shown in the image that opens this chapter). They show a blob that seems to move away from the quasar at 10 times the speed of light. In fact, a beam of relativistic particles from 3C 273 is aimed almost directly at Earth, giving the illusion of faster-than-light motion. Each image is about 7 milliarcseconds (0.007 arcsec) across. (NRAO)

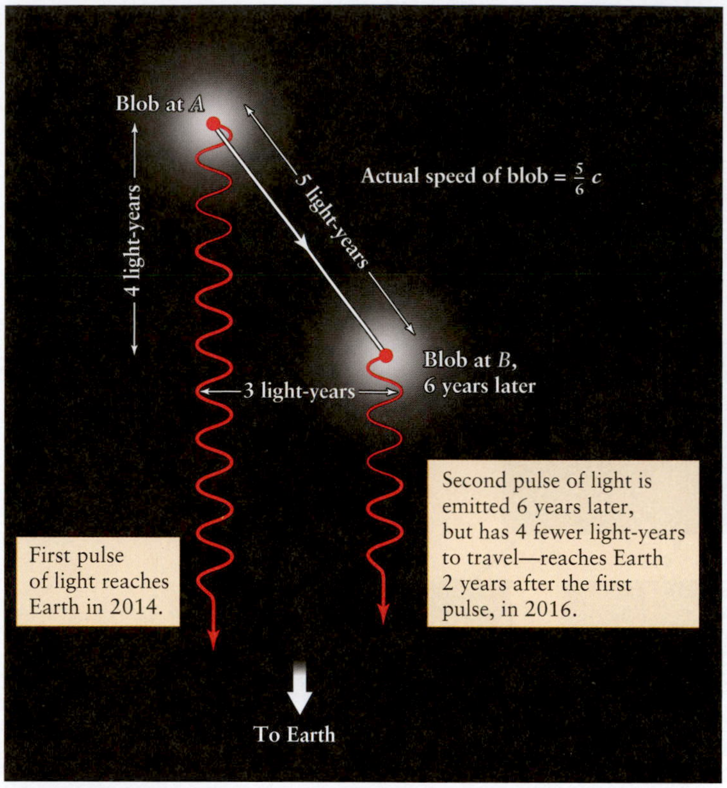

(a) View from above

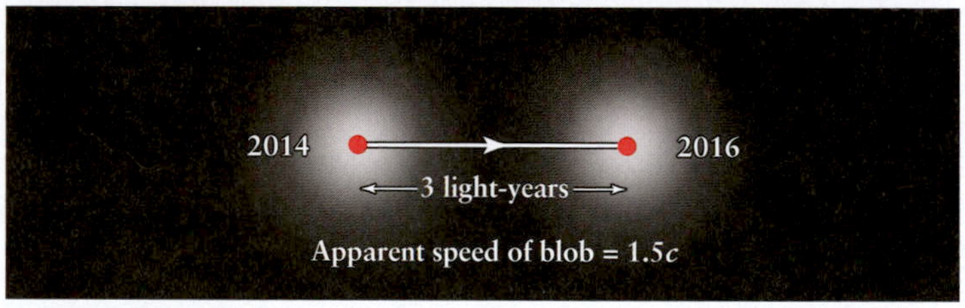

(b) View from Earth

FIGURE 24-11

An Explanation of Superluminal Motion **(a)** If a blob of material ejected from a quasar moves at five-sixths of the speed of light, it covers the 5 ly from point A to point B in 6 years. In the case shown here, it moves 4 ly toward Earth and 3 ly in a sideways, transverse direction. The light emitted by the blob at A reaches us in 2014. The light emitted by the blob at B reaches us in 2016. The light left the blob at B 6 years later than the light from A but had 4 fewer light-years to travel to reach us. **(b)** From Earth we can see only the blob's sideways, transverse motion across the sky, as in Figure 24-10. It appears that the blob has traveled 3 ly in just 2 years, so its apparent speed is 3/2 of the speed of light, or 1.5c.

to the source, this rate of angular separation corresponds to a speed 10 times that of light!

Was Albert Einstein wrong that nothing travels faster than the speed of light? No. The blobs in the jets move near the speed of light—these are called relativistic jets—but never faster than light. Astronomers soon realized that apparent superluminal motion can be explained by movement slower than light—once we take into account the *angle* at which we view the relativistic jet.

In order to see material that appears to move faster than the speed of light, two conditions must be met: The beam of material must be aimed close to your line of sight (within about 5 degrees or so), and material in the jet must travel near the speed of light. Given the number of quasars with radio jets, it is not surprising that a few just happen to point toward us. Figure 24-11 shows how jets and their geometry can produce apparent superluminal motion.

CONCEPTCHECK 24-6

In Figure 24-11, a bright blob appears to move faster than the speed of light by traversing a 3-ly distance across the sky in only 2 light-years. How far did the blob actually travel, and how long did it take?

Answer appears at the end of the chapter.

24-4 The unified model explains much of the diversity of AGN

Untangling the mystery of active galactic nuclei was extraordinarily difficult because AGN can appear very different from each other. In fact, it took several decades before researchers determined that the diverse group of AGN could all be "unified" into one physical model. Box 24-1 describes some different properties of AGN and their naming conventions. For example, one AGN, such as 3C 273, might have a visible spectrum with unusually strong and broad

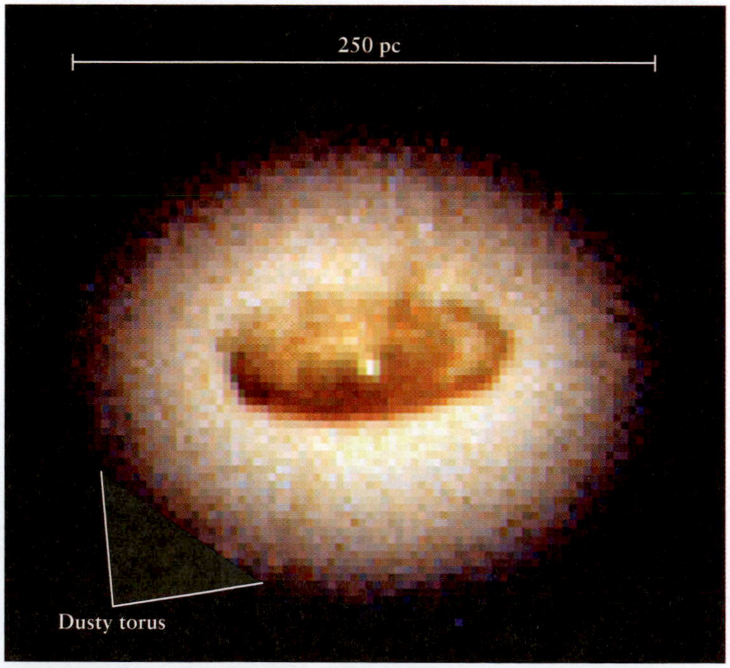

FIGURE 24-12 R I V U X G

A Dusty Torus Around a Supermassive Black Hole This immense doughnut of dust and gas orbits the black hole at the center of the active galaxy NGC 4261. Radio jets emerge perpendicular to the plane of this torus (see Figure 21-16). (L. Ferrarese/Johns Hopkins University, NASA)

BOX 24-1 TOOLS OF THE ASTRONOMER'S TRADE

The Diversity of AGN

Active galactic nuclei, or AGN, come in two unique varieties: those with radio jets and lobes (called *radio-loud*), and those without (called *radio-quiet*). Radio galaxies, like those in Figures 24-1 and 24-8, are radio-loud AGN. About 10% of quasars are radio-loud, and, when observable, their host galaxies are elliptical. The remaining quasars are radio-quiet and, when observable, their hosts are spiral galaxies. (It is not understood why radio-loud and radio-quiet AGN are each found in ellipticals and spirals, respectively—this remains a mystery.)

The light emitted from jets is focused into a strong beam because the jets' emitting plasma travels near the speed of light. In cases where the jets are pointed almost directly toward Earth, the jets' beamed emission swamps the usual AGN emission, and these objects have a very different spectrum (often without any detected emission lines). These objects are called **blazars,** and while originally thought to be distinct from quasars due to their different appearance, they are now understood as special cases of jet alignment with our line of sight from Earth.

Quite similar to quasars, **Seyfert galaxies** also show bright, compact nuclei with emission lines. Eventually, it was realized that both quasars and Seyfert galaxies contain the same type of object, and both are examples of AGN. Historically, bright and distant AGN, where the host galaxies were not initially observed, are called quasars; nearby cases where the host galaxies can be easily seen are called Seyfert galaxies.

Seyfert galaxies and quasars also come in two further varieties: those with strong broad hydrogen emission lines (called Type 1) and those with weak narrow hydrogen emission lines (called Type 2). Furthermore, the continuum emission in the Type 1 AGN can vary significantly in brightness, whereas the continuum in the Type 2 AGN does not vary.

Explaining Type 1 and Type 2 AGN in terms of the *same central engine* was a major advance in understanding quasars, and the explanation is called the unified model. Ultimately, appearing as Type 1 or Type 2 depends on the line of sight, as shown in Figure 24-13. The table below summarizes some properties of AGN; the modern distinction of Type 1 or Type 2 is in parenthesis after the object's name.

Properties of Active Galactic Nuclei (AGN)

Object (Type)	Found in which type of galaxy	Strength of radio emission	Type of emission lines in spectrum	Intrinsic Luminosity (watts)	Intrinsic Luminosity (Milky Way Galaxy = 1)
Blazar (1)	Elliptical	Strong	None	10^{38} to 10^{42}	10 to 10^5
Radio-loud quasar (1)	Elliptical	Strong	Broad	10^{38} to 10^{42}	10 to 10^5
Radio galaxy (2)	Elliptical	Strong	Narrow	10^{36} to 10^{38}	0.1 to 10
Radio-quiet quasar (1)	Spiral or elliptical	Weak	Broad	10^{38} to 10^{42}	10 to 10^5
Type 1 Seyfert	Spiral	Weak	Broad	10^{36} to 10^{38}	0.1 to 10
Type 2 Seyfert	Spiral	Weak	Narrow	10^{36} to 10^{38}	0.1 to 10

hydrogen emission lines (Figure 24-3) and show strong variability (see Figure 24-5). Another AGN might have very weak and narrow hydrogen emission lines with little variability. Given the difficulty of explaining how one type of "central engine" powering their luminosity could explain such diverse behavior, these observations suggested—incorrectly—that these two AGN contain different emission mechanisms. In fact, the **unified model** explains how different AGN, at their core, are fundamentally the same.

The astronomer Robert Antonucci (now at University of California, Santa Barbara) put together the main picture of the unified model and uncovered its key ingredient: a thick dusty doughnut-shaped torus around the central engine's accretion disk. The presence of such a torus seems to be a natural result of the overall accretion process. One such dusty torus is shown in Figure 24-12. Rather than emitting visible light, the dusty torus can *block* light from passing through it from the central engine. In contrast to the torus, the accretion disk within the torus is much too small to be seen—its emission appears pointlike—so no structural features of an accretion disk have ever been imaged.

If there were no dusty torus around an accreting supermassive black hole, an observer could view such a black hole from any angle and see the intense radiation from the accretion disk. But,

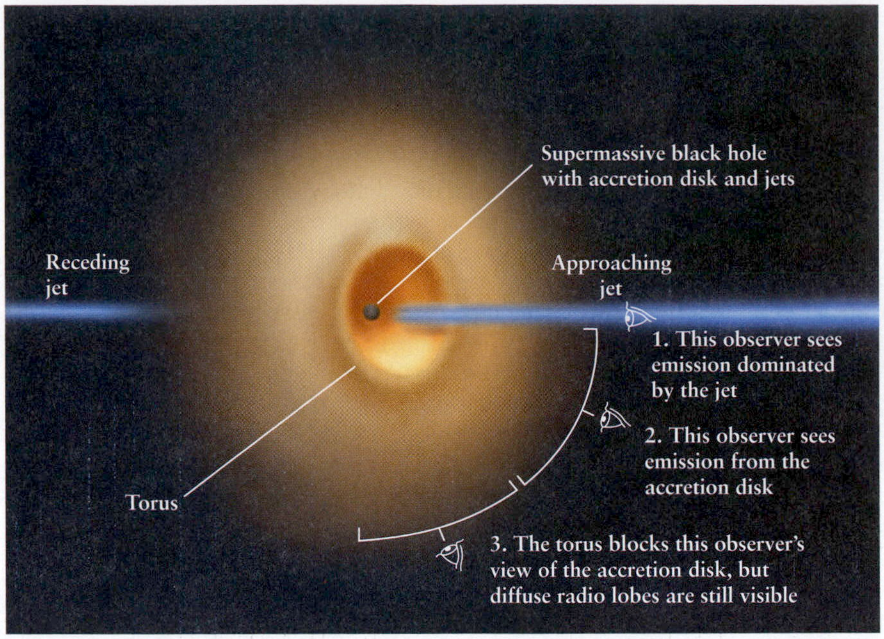

Supermassive black hole with accretion disk and jets
Receding jet
Approaching jet
1. This observer sees emission dominated by the jet
2. This observer sees emission from the accretion disk
Torus
3. The torus blocks this observer's view of the accretion disk, but diffuse radio lobes are still visible

FIGURE 24-13

A Unified Model of Active Galaxies The AGN model consists of a supermassive black hole surrounded by an intensely luminous accretion disk. Farther out is a dusty doughnut-shaped torus. A pair of jets is also shown on this object, although not every AGN has jets. While this basic structure is at the heart of the randomly oriented AGN in space, each AGN can appear different depending on the viewing angle.

as a result of the torus, from certain angles the torus blocks the view of the accretion disk. This idea offers a single explanation for several types of seemingly different active galaxies.

Figure 24-13 illustrates the unified model for a luminous active galactic nucleus with radio jets. If an imaginary observer looks straight down the axis of the jet (#1), the observed radiation is dominated by the jet and has a unique spectrum with emission lines often undetected. (When viewed from this angle, an AGN is a blazar, as described in Box 24-1.)

At an intermediate angle (#2), the observer gets a clear view of the luminous accretion disk and the turbulent region around the black hole. Because gases move at many different velocities in this turbulent region, the observer sees hydrogen spectral lines (and other lines) that have been broadened by the Doppler effect. The observer also sees intense thermal radiation from the hot accretion disk. This line of sight, and the resulting features of the emission, account for the properties of quasar 3C 273 and the majority of quasars. (The Type 1 objects referred to in Box 24-1 are viewed from these intermediate angles.)

If the observer looks nearly edge-on at the torus (#3), the accretion disk is completely hidden. Some of the light reaching the observer comes from hot gas flowing out of the accretion disk, and this light has an emission-line spectrum. But given the edge-on orientation, this gas is not moving rapidly toward or away from the observer, so there is little Doppler shift and the hydrogen emission lines are narrow. (From this edge-on view, the AGN would be a Type 2 object as described in Box 24-1.) No radio jets are visible at this angle, but diffuse radio lobes are visible.

Unlike our imaginary observer, we cannot move the vast distances through space that would be needed to view a given active galaxy from different angles. Instead, we may see a given active galaxy differently, depending on how the accretion disk and torus are oriented to our line of sight. In other words, "What you see depends on the line of sight you get!"

CAUTION! Note that we are really making use of *two* different but complementary models here. The unified model says that different types of active galactic nuclei are really different views of the same type of "central engine," while the accretion-disk model explains how the "central engine" works.

CONCEPTCHECK 24-7

Is the large object in Figure 24-12 an accretion disk? Does it emit intense visible light?

CONCEPTCHECK 24-8

From which viewing angle in Figure 24-13 would an AGN appear bright in visible light and show broad emission lines?

Answers appear at the end of the chapter.

24-5 The evolution of AGN

The accretion-disk idea can explain what happens when active galaxies collide or merge with other galaxies. Such collisions and mergers can transfer gas and dust from one galaxy to another, providing more "fuel" for a supermassive black hole. For example, the image in Figure 24-14 shows a close-up view of a supermassive black hole about to take in more "fuel" from a merger. Thus, it is not surprising that many of the most luminous active galaxies are also those that have recently undergone collisions, such as the radio galaxy Centaurus A shown in Figure 24-15.

The accretion disk idea also helps to explain why there are no nearby quasars (see Figure 24-4). Over time, much of the available gas and dust surrounding a quasar passes into the accretion disk and then enters the black hole. Eventually, the central engine has less and less infalling matter to act as fuel, and the quasar becomes

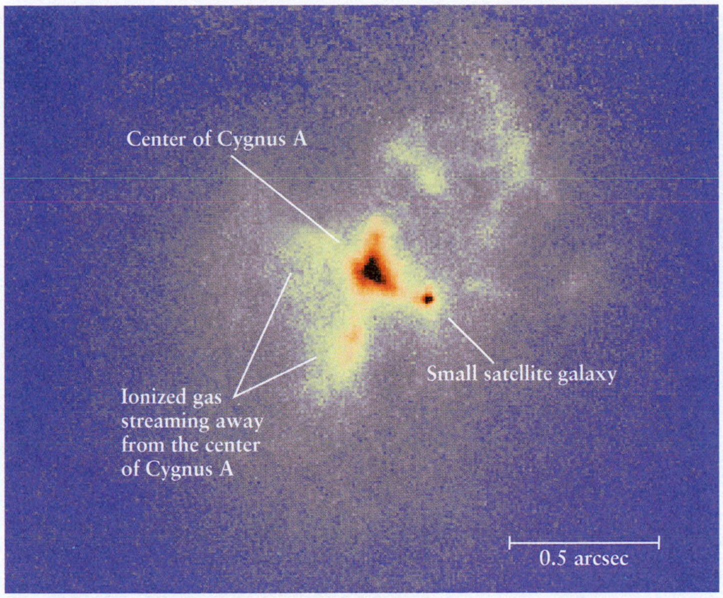

FIGURE 24-14 R I V U X G

Providing Fresh "Fuel" for a Supermassive Black Hole A small satellite galaxy has fallen into the central regions of the radio galaxy Cygnus A (see Figure 24-1). Its material may eventually accrete onto the supermassive black hole at Cygnus A's very center. This false-color extreme close-up was made using adaptive optics, described in Section 6-3. (Courtesy G. Canalizo, C. Max, D. Whysong, R. Antonucci, and S. E. Dahm)

inactive. Since quasars started shining in the early universe (about 12 billion years ago), and they consumed their fuel rather quickly (in a few hundred million years), the only quasars visible today in our expanding universe are very far away.

Dead Quasars and Flares

When a quasar runs out of accreting fuel altogether and becomes inactive, some astronomers call the leftover supermassive black hole a **dead quasar.** The vast majority of galaxies, including our own Milky Way, appear to have supermassive black holes at their centers that are not presently active. Therefore, dead quasars seem to be in almost all luminous galaxies, so most galaxies once led much more dramatic lives as quasars when they were young.

Is there a way to test the idea that most galaxies harbor a supermassive black hole? Yes. Fortunately, there is a chance that a dead quasar can light up once again, even if only briefly. If a star wanders too close to the supermassive black hole, tidal forces (which act to stretch the star) can tear the star apart and pull in some of its matter. This newly acquired stellar gas acts as temporary accretion fuel for the otherwise dormant black hole. Models of this process suggest that it might occur about once every 10,000 years in a typical galaxy. As the star's matter is accreted and is consumed by the black hole over a period of a year or so, it can shine brightly and is called a **dead quasar flare.**

Very few events provide convincing evidence for a flare. After all, there are other objects that can give a burst of light. One case for a flare came in March 2011, when a burst of X-rays (named Swift J1644+57) was observed at the center of a galaxy. This flare lasted only two days, but was observed from beginning to end. The event was much brighter and shorter than expected for a typical flare, and astronomers think the emission came from temporarily formed jets, as opposed to an accretion disk (Figure 24-16).

Much is still unknown about quasars. How do they form? Does their visible light really come from an accretion disk? Why do some AGN have strong radio emission, while others do not? What happens if a star wanders too close to a dead quasar—does its captured gas form an accretion disk or jets? All of

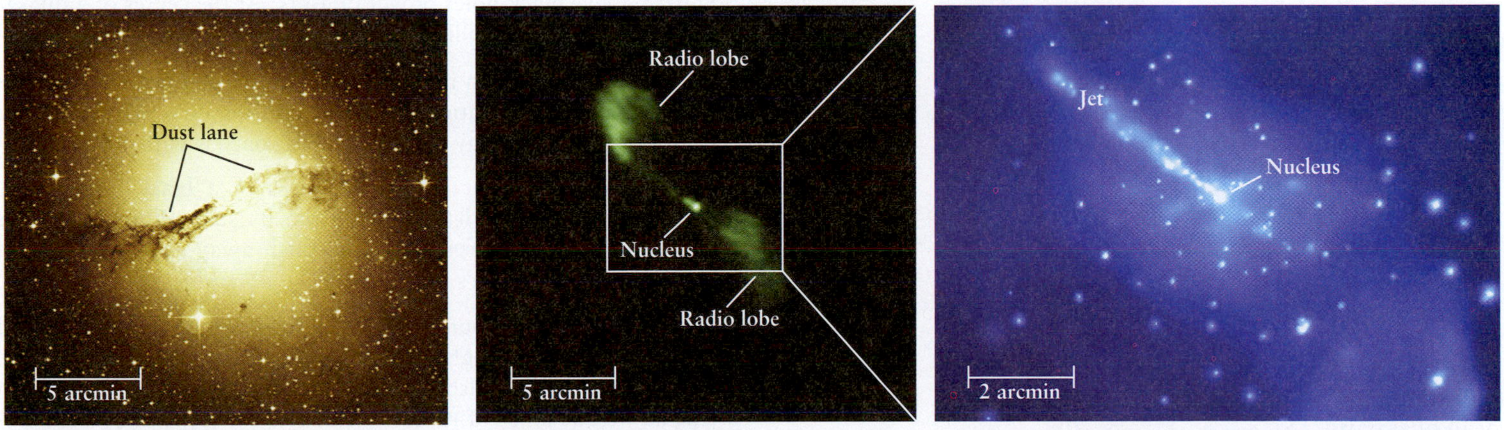

(a) Centaurus A: light from stars R I V U X G

(b) Centaurus A: radio lobes R I V U X G

(c) An X-ray-emitting jet emanates from the nucleus R I V U X G

FIGURE 24-15

The Radio Galaxy Centaurus A **(a)** This elliptical galaxy, called NGC 5128, lies about 4 Mpc (13 million ly) from Earth at the location of the radio source Centaurus A. The dust lane is evidence of a collision with another galaxy. **(b)** At radio wavelengths we see synchrotron radiation from two radio lobes centered on the galaxy's nucleus. **(c)** A luminous jet extends from the nucleus directly toward one of the lobes. (a: Digitized Sky Survey/UK Schmidt Image/STScI; b: J. Condon et al./VLA/NRAO; c: R. Kraft et al./SAO/NASA)

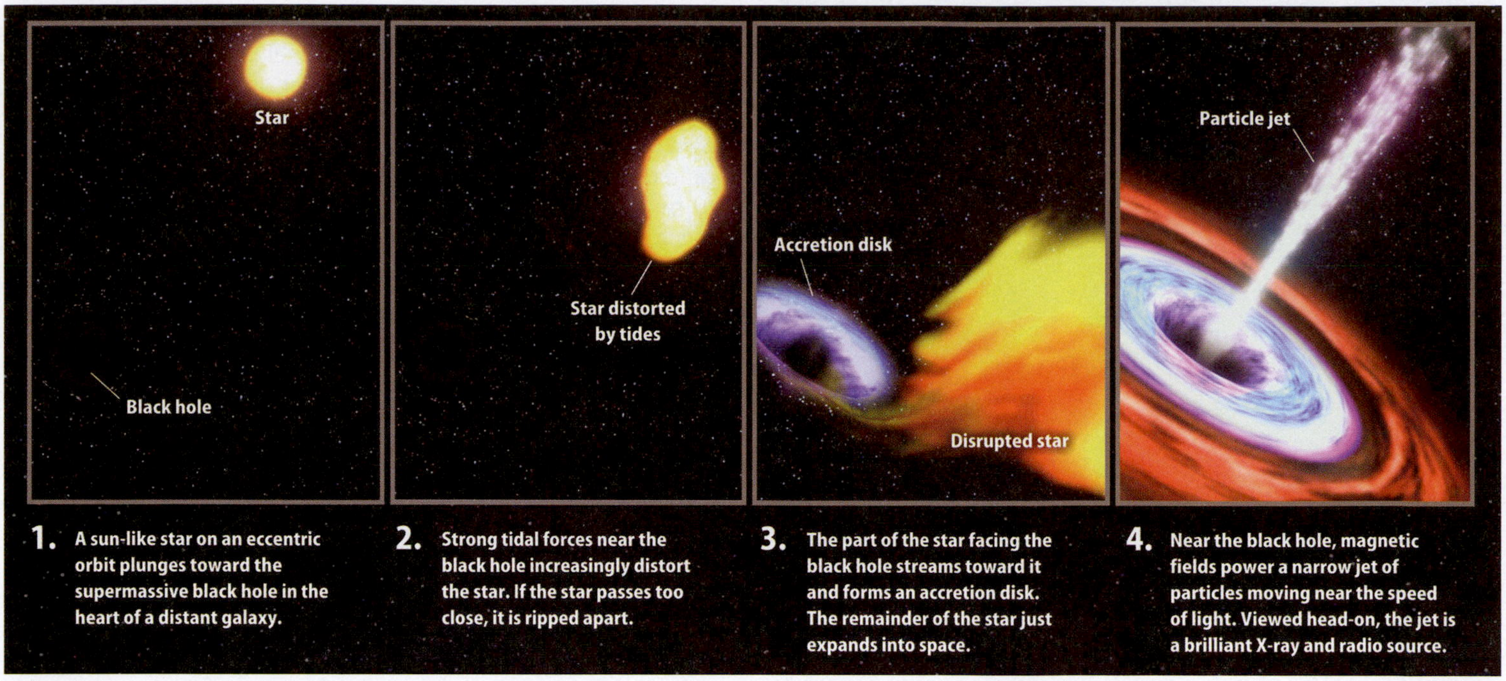

Star
Black hole
Star distorted by tides
Accretion disk
Disrupted star
Particle jet
1. A sun-like star on an eccentric orbit plunges toward the supermassive black hole in the heart of a distant galaxy.
2. Strong tidal forces near the black hole increasingly distort the star. If the star passes too close, it is ripped apart.
3. The part of the star facing the black hole streams toward it and forms an accretion disk. The remainder of the star just expands into space.
4. Near the black hole, magnetic fields power a narrow jet of particles moving near the speed of light. Viewed head-on, the jet is a brilliant X-ray and radio source.

FIGURE 24-16

Dead Quasar Flare A star wandering too close to a supermassive black hole can provide temporary fuel for the dead quasar. Modeling indicates that this event in 2012, called Swift J1644+57, occurred around a black hole about twice the mass of the 4 million solar mass black hole at the center of our Milky Way Galaxy. The X-ray emission from this flare lost most of its initial intensity over a period of several months. (NASA/Goddard Space Flight Center/Swift)

these questions are part of ongoing research in the field of active galaxies.

CONCEPTCHECK 24-9

Could our own Milky Way Galaxy have once hosted a quasar? If so, are we likely to witness a dead quasar flare from the center of our Milky Way Galaxy?

Answer appears at the end of the chapter.

KEY WORDS

Words preceded by an asterisk (*) are discussed in Box 24-1.

accretion, p. 706
accretion disk, p. 708
active galactic nucleus (AGN), p. 705
active galaxy, p. 705
*blazar, p. 713
dead quasar, p. 715
dead quasar flare, p. 715
Eddington limit, p. 707
quasar, p. 704
radio galaxy, p. 705
*Seyfert galaxy, p. 713
superluminal motion, p. 710
supermassive black hole, p. 708
unified model, p. 713

KEY IDEAS

Quasars or active galactic nuclei (AGN): A quasar looks like a star but has a huge redshift. According to the Hubble law, these redshifts show that quasars are typically billions of light-years away from Earth.

• To be seen at such large distances, quasars must be very luminous, typically about 1000 times brighter than an ordinary galaxy.

• The peak of quasar activity took place when the universe was just over 2 billion years old; the era of quasar activity has long since passed.

Supermassive black holes power quasars: At the center of a quasar is a black hole with millions or even billions of solar masses.

• Rapid fluctuations in the brightness of quasars indicate that the region that emits radiation is quite small; the only known way to produce so much energy in such a small space is with a black hole.

• The Eddington limit relates the mass of a black hole to the maximum luminosity it can emit from its surroundings; if more luminous, radiation pressure would push away accreting gas and shut off the black hole's fuel source.

Accretion and jets around black holes: Hot gas accreting around the black hole is thought to give a quasar its intense luminosity. About 10% of AGN also have strong radio emission coming from jets, as well as lobes of radio emission as the jets collide with surrounding gas on scales much larger than the host galaxy.

• Models for a thin accretion disk describe heat and emission produced as portions of gas on different orbits rub past each other, although no specific model of accretion has been verified.

• As gases spiral in toward the supermassive black hole, some of the gas may be redirected to form two jets of high-speed particles that are aligned perpendicularly to the accretion disk. When a jet is directed toward us, we can observe apparent superluminal motion.

Unified model of AGN: A thick dusty doughnut-shaped torus surrounds a quasar that, when viewed edge-on, blocks a direct view of the luminous accreting gas.

• A quasar is only visible if our viewing angle allows us to see the region in the center of the torus.

• When viewing the torus somewhat edge-on, the most luminous part of a quasar is blocked. Some emission is still observed, and the object appears as an active galaxy with a different emission spectrum.

• While the viewing angle changes what we observe, the central engine—an accreting black hole—is fundamentally the same. This insight unifies a diverse group of AGN.

QUESTIONS

Review Questions

Questions preceded by an asterisk (*) involve topics discussed in Box 24-1.

1. When quasi-stellar radio sources were first discovered and named, why were they called "quasi-stellar"?
2. How were quasars first discovered? How was it known that they are very distant objects?
3. Suppose you saw an object in the sky that you suspected might be a quasar. What sort of observations might you perform to test your suspicion?
4. How does the spectrum of a quasar differ from that of an ordinary galaxy? How do spectral lines help astronomers determine the distances to quasars?
5. Since quasars lie at the centers of galaxies, why don't we see strong absorption lines from a galaxy's stars when we look at the spectrum of a quasar (like that shown in Figures 24-3)?
6. It was suggested in the 1960s that quasars might be compact objects ejected at high speeds from the centers of nearby ordinary galaxies. Explain why the absence of blueshifted quasars disproves this hypothesis. (We see only quasars with redshifted lines.)
7. How do astronomers know that quasars are located in galaxies? Are the host galaxies easy to observe for distant quasars?
8. What is a radio galaxy?
9. What is the main difference between a Type 1 and a Type 2 AGN?
10. Some quasars appear to be ejecting material at speeds faster than light. Is the material really moving that fast? If so, how is this possible? If not, why does the material appear to be traveling so fast?
11. What do the brightness fluctuations of a particular active galaxy tell us about the size of the energy-emitting region within that galaxy?
12. How could a supermassive black hole, from which nothing—not even light—can escape, be responsible for the extraordinary luminosity of a quasar?
13. What is the Eddington limit? Explain how it can be used to set a limit on the mass of a supermassive black hole, and explain why this limit represents a minimum mass for the black hole.
14. How does matter falling inward toward a central black hole find itself being ejected outward in a high-speed jet?
15. In the unified model, why do some AGN have broad spectral lines while others do not?

*16. How does the unified model of active galaxies explain that quasars and blazars are the same kind of object viewed from different angles?

17. What is it about fueling of quasars that there are no quasars relatively near to our Galaxy?

Advanced Questions

Questions preceded by an asterisk (*) involve topics discussed in the Boxes in Chapters 4, 21, 23.

> **Problem-solving tips and tools**
>
> Box 21-1 describes time dilation in the special theory of relativity. You may find it useful to recall from Section 1-7 that 1 light-year = 63,240 AU. A speed of 1 km/s is the same as 0.211 AU/yr. You can find Newton's form of Kepler's third law in Box 4-4 and the formula for the Schwarzschild radius in Box 21-2.

18. When we observe a quasar with redshift $z = 1.0$, about how far into the past are we looking (see Figure 24-5)? If we could see that object as it really is right now (that is, if the light from the object could somehow reach us instantaneously), would it still look like a quasar? Explain why or why not.
19. In the image that opens this chapter, the close-up view of the Cygnus A jet shows that different wavelengths are preferentially emitted at different locations along the jet's length. Explain why, using the following principle: As an individual particle moves in a magnetic field, the greater its speed, the shorter the wavelength of the radiation that it emits.
20. Suppose the distance from point A to point B in Figure 24-11a is 26 light-years and the blob moves at 13/15 of the speed of light. As the blob moves from A to B, it moves 24 light-years toward Earth and 10 light-years in a sideways, transverse direction. **(a)** How long does it take the blob to travel from A to B? **(b)** If the light from the blob at A reaches Earth in 2020, in what year does the light from B reach Earth? **(c)** As seen from Earth, at what speed does the blob appear to move across the sky?

*21. Suppose a blazar at $z = 1.00$ goes through a fluctuation in brightness that lasts one week (168 hours) as seen from Earth. **(a)** At what speed does the blazar seem to be moving away from us? **(b)** Using the idea of time dilation, determine how long this fluctuation lasted as measured by an astronomer within the blazar's host galaxy. **(c)** What is the maximum size (in AU) of the region from which this blazar emits energy?

22. **(a)** Calculate the maximum luminosity that could be generated by accretion onto a black hole of 3.7×10^6 solar

masses. (This is the size of the black hole found at the center of the Milky Way.) Compare this to the total luminosity of the Milky Way, about 2.5×10^{10} $L_\odot$. (**b**) Speculate on what we might see if the center of our Galaxy became an active galactic nucleus with the luminosity you calculated in (a).

*23. Observations of a certain galaxy show that stars at a distance of 16 pc from the center of the galaxy orbit the center at a speed of 200 km/s. Use Newton's form of Kepler's third law to determine the mass of the central black hole.

*24. Calculate the Schwarzschild radius of a 10^9 solar mass black hole. How does your answer compare with the size of our solar system (given by the diameter of Pluto's orbit)?

*25. Figure 24-15 shows the double radio source Centaurus A. Is it possible that somewhere in the universe there is an alien astronomer who observes this same object as a blazar? Explain your answer with a drawing showing the relative positions of Earth, the alien astronomer, and Centaurus A.

*26. TUTORIAL 24-1 What is a blazar? What is unique about its spectrum? How is it related to other active galaxies?

*27. What is a Type 1 Seyfert galaxy? Does the active nucleus of a Type 1 Seyfert galaxy differ from that of a quasar?

Discussion Questions

28. The accompanying image from the Very Large Array (VLA) shows the radio galaxy 3C 75 in the constellation Cetus. This galaxy has several radio-emitting jets. High-resolution optical photographs reveal that the galaxy has two nuclei, which are the two red spots near the center of the VLA image. Propose a scenario that might explain the appearance of 3C 75.

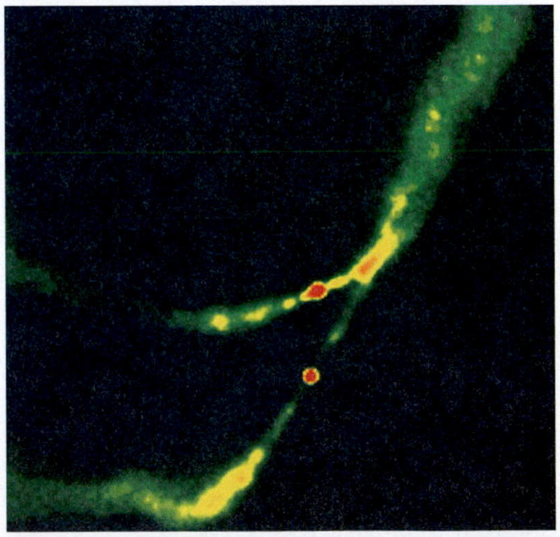

R I V U X G

(NRAO)

29. Some quasars show several sets of absorption lines whose redshifts are less than the redshifts of their emission lines. For example, the quasar PKS 0237-23 has five sets of absorption lines with redshifts in the range from 1.364 to 2.202, whereas the quasar's emission lines have a redshift of 2.223. Propose an explanation for these sets of absorption lines.

30. The Milky Way Galaxy is in the process of absorbing the satellite galaxy called the Canis Major Dwarf (see Section 23-6). Discuss whether this process could cause the Milky Way to someday become an active galaxy.

31. Figure 24-14 shows ionized gas streaming away from the "central engine" of the radio galaxy Cygnus A. Instead of spreading outward equally in all directions, the gas appears to be funneled into two oppositely directed cones. Discuss how this could be caused by a dusty torus surrounding a supermassive black hole at the center of Cygnus A.

Web/eBook Questions

32. Search the World Wide Web for information about "microquasars." These are objects that are found *within* the Milky Way Galaxy. How are they detected? What are the similarities and differences between these objects and true quasars? Are they long-lasting or short-lived?

33. The Lockman hole is a region in the constellation Ursa Minor where the Milky Way's interstellar hydrogen is the thinnest. By observing in this part of the sky, astronomers get the clearest possible view of distant galaxies and quasars. Search the World Wide Web for information about observations of the Lockman hole. What has been learned through X-ray observations made by spacecraft such as ROSAT and XMM-Newton?

ACTIVITIES

Observing Projects

34. Use a telescope with an aperture of at least 20 cm (8 in.) to observe the Seyfert Galaxy NGC 1068 (also known as M77). Located in the constellation Cetus (the Whale), this galaxy is most easily seen from September through January. The epoch 2000 coordinates are R.A. = $2^h\ 2.7^m$ and Decl. = $-0°\ 01'$. Sketch what you see. Is the galaxy's nucleus diffuse or starlike? How does this compare with other galaxies you have observed?

35. Use a telescope with an aperture of at least 20 cm (8 in.) to observe the two companions of the Andromeda Galaxy, M32 and M110. Both are small elliptical galaxies, but only M32 is suspected of harboring a supermassive black hole. They are located on opposite sides of the Andromeda Galaxy and are most easily seen from September through January. The epoch 2000 coordinates are

Galaxy	Right ascension	Declination
M32 (NGC 221)	$0^h\ 42.7^m$	$+40°\ 52'$
M110 (NGC 205)	0 40.4	+41 41

Make a sketch of each galaxy. Can you see any obvious difference in the appearance of these galaxies? How does this difference correlate with what you know about these galaxies?

36. If you have access to a telescope with an aperture of at least 30 cm (12 in.), you should definitely observe the Sombrero Galaxy, M104, which is suspected to have a supermassive black

hole at its center. It is located in Virgo and can most easily be seen from March through July. The epoch 2000 coordinates are R.A. = 12^h 40.0^m and Decl. −11° 37′. Make a sketch of the galaxy. Can you see the dust lane? How does the nucleus of M104 compare with the centers of other galaxies you have observed?

37. If you have access to a telescope with an aperture of at least 30 cm (12 in.), observe M87 and compare it with the other two giant elliptical galaxies, M84 and M86, that dominate the central regions of the Virgo cluster. If you have access to the *Starry Night*™ program, use it to help you plan your observations. The epoch 2000 coordinates are:

Galaxy	Right ascension	Declination
M84 (NGC 4374)	12^h 25.1^m	+12° 53′
M86 (NGC 4406)	12 26.2	+12 57
M87 (NGC 4486)	12 30.8	+12 24

38. If you have access to a telescope with an aperture of at least 40 cm (16 in.), you might try to observe the brightest-appearing quasar, 3C 273, which has an apparent magnitude of nearly +13. It is located in Virgo at coordinates R.A. = 12^h 29^m 07^s and Decl. +2° 3′ 8″.

39. Energetic jets of material moving at relativistic speeds are observed to be blasting outward in opposite directions from the centers of many active galaxies. These are thought to be projected at right angles to the accretion disks of supermassive black holes at the centers of these galaxies. Use the *Starry Night*™ program to examine an example of one of these energetic jets in the vicinity of the galaxy M87. Select **Favourites > Explorations > Atlas** to view the sky from the center of a transparent Earth. Use the **Find** facility to locate M87 and **Zoom** in to a field of view of about 25 arc minutes. (**a**) What type of galaxy is M87? (**b**) Place the cursor over the galaxy, right-click to open the dropdown menu, click on **Show Info**, and obtain the distance to M87 from the **Position in Space** layer in the **Info** pane. What is the distance of M87 from the Earth? (**c**) Click on the **Options** tab, expand the **Deep Space** layer, and ensure that the **Hubble Images** are switched **On.** A high-resolution Hubble Space Telescope image of M87 is shown in the view, offset from the galaxy position to the upper right to avoid confusion, close to the position of the star HIP61051. Position the cursor over this image, right-click to open the dropdown list, and **Centre** on this image. **Zoom** in to a field of view of about 1 arcminute to view one of the jets emanating from the galaxy. Use the angular separation tool to measure the angular extent of this jet. What is this angle? (**d**) What is the projected length of this jet into space? (*Hint:* Use the small-angle equation, the distance to the galaxy, and the measured angle, remembering that 1 radian = 2.06×10^5 arcseconds). (**e**) Click on **Favourites > Explorations > Milky Way Galaxy** and use the location scroller to move the simulated image of the Milky Way Galaxy to imagine how such a jet would appear to us from Earth if our Galaxy were to produce a jet like that from M87. The Sun is about 26,000 ly from the galactic center. Calculate how far across our sky this jet would extend if it were perpendicular to the galactic plane, again using the small-angle approximation.

40. Use the *Starry Night*™ program to examine a very distant pair of quasars. Most quasars are starlike in appearance but their spectra show very large redshifts. If these redshifts are cosmological (i.e., caused by the expansion of the universe), then quasars obey Hubble's law, and their large redshifts show that they must be very far away. Since they appear bright in our sky, they must be very luminous, with outputs equivalent to that of billions and even trillions of Suns. They are part of a group of objects known as active galactic nuclei. The source of energy output of these very luminous objects is probably a supermassive black hole at the object's center. Select **Favourites > Explorations > Quasar Pair** and **Zoom** in to a field of view about 30 arcseconds wide. Open the **Info** pane for this pair of quasars and note the **Distance from observer** from the **Position in Space** layer. (**a**) What is the distance from Earth of these quasars? (**b**) Using a Hubble constant of 73 km/s/Mpc, calculate the expected recessional velocity of this quasar pair. (*Note:* 1 pc = 3.26 ly) (**c**) What is the ratio of the recessional velocity of these quasars to the speed of light?

Collaborative Exercises

41. Make a labeled sketch clearly showing how the spectrum of 3C 273, shown in Figure 24-3, would be different if the object were moving toward us at the same velocity. Compare your sketch to that of another group and resolve any inconsistencies.

42. Two dramatic images of M87 are shown in Figure 24-8. Invent an imaginary scenario that would be analogous to a local dance club where two pictures would show different aspects of the event and write a description of each. Be sure to include a description of how this is analogous to the images in Figure 24-8.

ANSWERS

ConceptChecks

ConceptCheck 24-1: Here are three differences: A quasar has emission lines, whereas a star has absorption lines. A quasar also emits more strongly than a star in the radio and X-ray wavelengths. The Doppler shift of spectral lines (whether emission or absorption) is much larger for quasars than for the stars in our Galaxy, indicating that quasars recede at much higher speeds than stars in our Galaxy.

ConceptCheck 24-2: The universe was around 2 billion years old when quasar activity peaked, and that time corresponds to a redshift z of around 3.

ConceptCheck 24-3: The farther away an object is observed, the lower its apparent brightness. The redshifts in Figure 24-5 all correspond to very distant objects that can only be observed because they were very luminous at the early times and remote distances when they shined bright.

ConceptCheck 24-4: Radiation pressure would push outward on potentially accreting gas-fuel more than gravity could pull it in. This would slow the inward flow of gas (slowing accretion), and since accretion of gas fuels the quasar's luminosity, the quasar would soon become less luminous. This reduction in luminosity would actually allow accretion to then rise to its original Eddington limit.

ConceptCheck 24-5: Orbits closer to the central mass have higher speeds so that an inner blob rubs against an outer blob in the disk, causing the inner blob to lose energy and spiral inward. Also, as the blobs rub past each other they produce heat. Thermal radiation from this heat produces much of a quasar's visible light.

ConceptCheck 24-6: The blob travels 5 light-years (even though we see only the 3 light-years of its sideways, transverse motion). The blob takes 6 years to travel this distance (and it travels slower than the speed of light).

ConceptCheck 24-7: The answer is "no" to both questions. The object is a dim dusty torus that can block light, depending on the viewing angle. The intense visible light from a quasar comes from the much-smaller accretion disk, which appears pointlike in images.

ConceptCheck 24-8: Viewing angle #2. At these intermediate angles, visible light can be seen directly from the accretion disk without getting blocked by the torus. This line of sight also receives Doppler-shifted light from fast-moving clouds in a turbulent region near the black hole; these clouds produce the broad emission lines.

ConceptCheck 24-9: Yes, the supermassive black hole at our Galaxy's center is probably a dead quasar. However, since a supermassive black hole is expected to capture a star and produce a dead quasar flare only about once every 10,000 years, we are unlikely to witness this event.

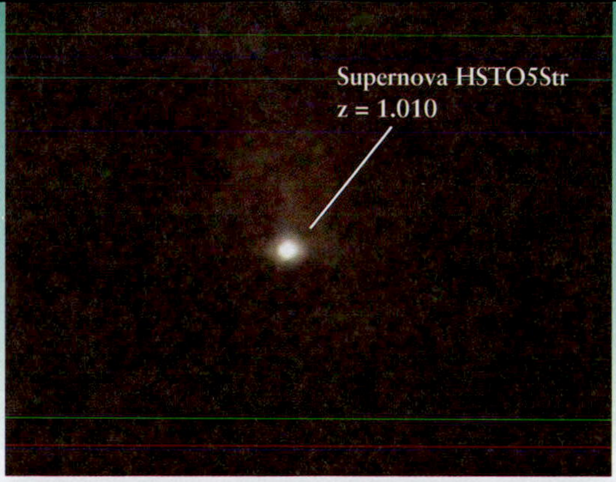

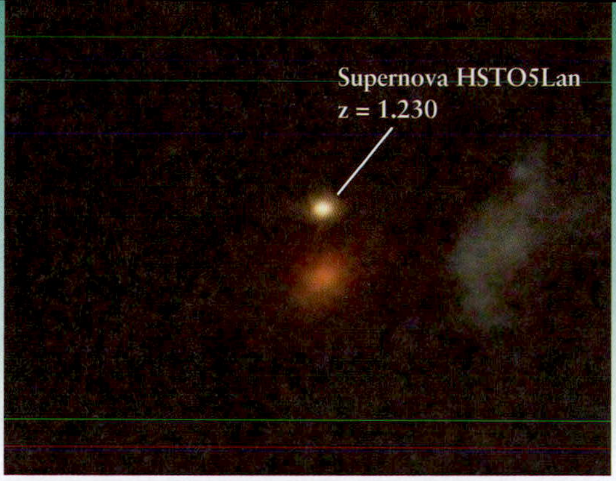

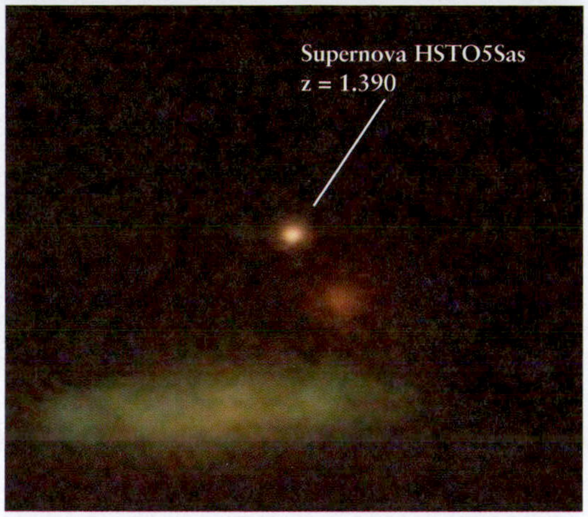

Very distant supernovae—which we see as they were billions of years ago—help us understand the evolution of the universe. (NASA; ESA; and A. Riess, STScI)

R **I** **V** U X G

CHAPTER 25

Cosmology: The Origin and Evolution of the Universe

LEARNING GOALS

By reading the sections of this chapter, you will learn

25-1 Why the darkness of the night sky presents a mystery

25-2 What it means to say that the universe is expanding

25-3 How to estimate the age of the universe from its expansion rate

25-4 How astronomers detect the afterglow of the Big Bang

25-5 What the universe was like during its first 380,000 years

25-6 How the curvature of the universe reveals its matter and energy content

25-7 What distant supernovae tell us about the expansion history of the universe

25-8 How cosmic sound waves reveal details of our universe

So far in this book we have cataloged the contents of the universe. Our scope has ranged from subatomic objects to superclusters of galaxies hundreds of millions of light-years across. In between, we have studied planets, moons, and stars.

But now we turn our focus beyond the objects we find in the universe to the nature of the universe itself—the subject of the science called *cosmology.* How large is the universe? What is its structure? How long has it existed, and how has it changed over time?

In this chapter we will see that the universe is expanding. This expansion began with an event at the beginning of time called the *Big Bang.* We will see direct evidence of the Big Bang in the form of microwave radiation from space. This radiation is the faint afterglow of a primordial fireball that filled all space shortly after the beginning of the universe.

Will the universe continue to expand forever, or will it eventually collapse back on itself? We will find that to predict the future of the universe, we must first understand what happened in the remote past. To this end, astronomers study luminous supernovae like the example shown in the above images. These can be seen across billions of light-years and so can tell us about conditions in the universe billions of years ago. We will see how recent results from such supernovae, as well as from studies of the Big Bang's afterglow, have revolutionized our understanding of cosmology and given us new insights into our place in the cosmos.

25-1 The darkness of the night sky tells us about the nature of the universe

Cosmology is the science concerned with the structure and evolution of the universe as a whole. One of the most profound and basic questions in cosmology may at first seem foolish: Why is the sky dark at night? This question, which haunted Johannes Kepler as long ago as 1610, was brought to public attention in the early 1800s by the German amateur astronomer Heinrich Olbers.

Olbers' Paradox and Newton's Static Universe

Olbers and his contemporaries pictured a universe of stars scattered more or less randomly throughout infinite space. Isaac Newton himself thought that no other model made sense. The gravitational forces between any *finite* number of stars, he argued, would in time cause them all to fall together, and the universe would soon be a compact blob.

Obviously, this has not happened. Newton concluded that we must be living amid a static, infinite expanse of stars. In this model, the universe is infinitely old, and it will exist forever without major changes in its structure. Olbers noticed, however, that a static, infinite universe presents a major puzzle.

If space goes on forever, with stars scattered throughout it, then any line of sight must eventually hit a star. In this case, no matter where you look in the night sky, you should ultimately see a star. The entire sky should be as bright as an average star, so, even at night, the sky should blaze like the surface of the Sun. **Olbers's paradox** is that the night sky is actually *dark* (Figure 25-1).

FIGURE 25-1 R I V U X G
The Dark Night Sky If the universe were infinitely old and filled uniformly with stars that were fixed in place, the night sky would be ablaze with light. In fact the night sky is dark, punctuated only by the light from isolated stars and galaxies, Hence, this simple picture of an infinite, static universe cannot be correct. (NASA; ESA; and the Hubble Heritage Team, STScI/AURA)

Olbers's paradox suggests that something is wrong with Newton's infinite, static universe. According to the classical, Newtonian picture of reality, space is like a gigantic flat sheet of inflexible, rectangular graph paper. (Space is actually three-dimensional, but it is easier to visualize just two of its three dimensions. In a similar way, an ordinary map represents the three-dimensional surface of Earth, with its hills and valleys, as a flat, two-dimensional surface.)

This rigid, flat, Newtonian space extends unchanged, on and on, totally independent of stars or galaxies or anything else. The same is true of time in Newton's view of the universe; a Newtonian clock ticks steadily and monotonously forever, never slowing down or speeding up. Furthermore, Newtonian space and time are unrelated, in that a clock runs at the same rate no matter where in the universe it is located.

Einstein's Revolution and His "Greatest Blunder"

Albert Einstein overturned this view of space and time. His special theory of relativity (recall Section 21-1 and Box 21-1) shows that measurements with clocks and rulers depend on the motion of the observer. What is more, Einstein's general theory of relativity (Section 21-2) tells us that gravity curves the fabric of space. As a result, just as one massive object can bend the space around it, the matter that occupies the universe influences the overall shape of space throughout the universe.

If we represent the universe as a sheet of graph paper, the sheet is not perfectly flat but has a dip wherever there is a concentration of mass, such as a person, a planet, or a star (see Figure 21-4). Also, because of gravitational effects, clocks run at different rates depending on whether they are close to or far from a massive object, as Figure 21-7a shows.

Einstein narrowly missed predicting that our universe is not static

What does the general theory of relativity, with its many differences from the Newtonian picture, have to say about the structure of the universe as a whole? Einstein attacked this problem shortly after formulating his general theory in 1915. At that time, the prevailing view was that the universe was static, just as Newton had thought.

Einstein was therefore dismayed to find that his calculations could not produce a truly static universe. According to general relativity, the universe must be either expanding or contracting. In a desperate move to force his theory to predict a static universe, he added to the equations of general relativity a term called the **cosmological constant** (denoted by Λ, the capital Greek letter lambda). The cosmological constant was intended to represent a pressure that tends to make the universe expand as a whole. Einstein's idea was that this pressure would just exactly balance gravitational attraction, so that the universe would be static and not collapse.

ANALOGY Einstein's cosmological constant is analogous to the pressure of gas inside a balloon once the balloon has already been inflated. This pressure exactly balances the inward force exerted by the stretched rubber of the balloon itself, so the balloon maintains the same size.

Unlike other aspects of Einstein's theories, the cosmological constant did not have a firm basis in physics. He just added it to make the general theory of relativity agree with the prejudice that the universe is static.

Because Einstein doubted his original equations, he missed an incredible opportunity: He could have postulated that we live in an expanding universe. Einstein has been quoted as saying in his later years that the cosmological constant was "the greatest blunder of my life." (In fact, the cosmological constant plays an important role in modern cosmology, although a very different one from what Einstein proposed. We will explore this in Section 25-7.)

Instead, the first hint that we live in an expanding universe came more than a decade later from the observations of Edwin Hubble. As we will see in Section 25-3, Hubble's discovery provides the resolution of Olbers's paradox.

CONCEPTCHECK 25-1

In an infinite, static universe uniformly filled with stars, would space be bright or dark when viewed from the Moon (looking away from the Sun)?

Answer appears at the end of the chapter.

25-2 The universe is expanding

TUTORIAL 25-1

Hubble is usually credited with discovering that our universe is expanding. He found a simple linear relationship between the distances to remote galaxies and the redshifts of the spectral lines of those galaxies (review Section 23-5, especially Figures 23-16 and 23-17). This relationship, now called the *Hubble law,* states that the greater the distance to a galaxy, the greater is the galaxy's redshift. Thus, remote galaxies are moving away from us with speeds proportional to their distances. Now let us see how Hubble's law reveals an expanding universe.

The Hubble Law and the Expanding Universe

The Hubble law can be stated as a very simple equation that relates the recessional velocity v of a galaxy to its distance d from Earth:

$$v = H_0 d$$

where H_0 is the Hubble constant. Because remote galaxies are getting farther and farther apart as time goes on, astronomers say that the universe is expanding. We say these galaxies are in the Hubble flow.

ANALOGY What does it actually mean to say that the universe is expanding? According to general relativity, space itself is not rigid. The amount of space between widely separated locations increases over time. A good analogy is that of baking a chocolate chip cake, as in Figure 25-2. As the cake expands during baking, the chocolate chips do not move through the cake but the amount of space between the chocolate chips gets larger and larger. In the same way, as the universe expands, galaxies in the Hubble flow do not move through space, but the amount of space between widely separated galaxies increases. The expansion of the universe *is* the expansion of space.

CAUTION! It is important to realize that the expansion of the universe occurs primarily in the vast spaces that separate clusters of galaxies. Just as the individual chocolate chips in Figure 25-2 do not expand as the cake expands during baking, galaxies themselves do not expand. Einstein and others have established that an object that is held together by its own gravity, such as a galaxy or a cluster of galaxies, creates a region of nonexpanding space. Likewise, molecular forces, which are much stronger than gravity, prevent expansion. For example, Earth and your body, which are held together by molecular forces, are not getting any bigger. Only the distance between widely separated galaxies increases with time. This expansion actually creates space. The *Cosmic Connections: "Urban Legends" about the Expanding Universe* has more to say about several misconceptions concerning the expanding universe.

The Hubble law is a direct proportionality—that is, a galaxy twice as far away is receding from us twice as fast. This property is just what we would expect in an expanding universe. To see why this property results from uniform expansion, imagine a grid of parallel lines (as on a piece of graph paper) crisscrossing the universe. Figure 25-3a shows a series of such gridlines 100 Mpc apart, along with five galaxies labeled A, B, C, D, and E that happen to lie where gridlines cross. As the universe expands in all directions, the gridlines and the attached galaxies spread apart. (This is just what would happen if the universe were a two-dimensional rubber sheet that was being pulled equally on all sides. Alternatively, you can imagine that Figure 25-3a depicts a very small portion of the chocolate chip cake in Figure 25-2, with galaxies taking the place of chocolate chips.)

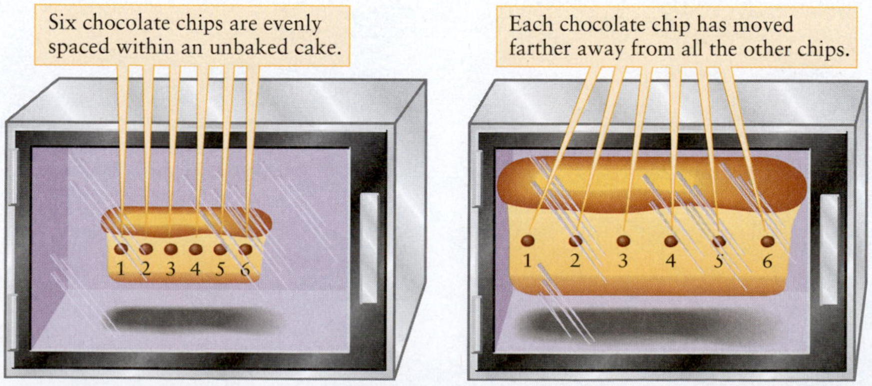

Six chocolate chips are evenly spaced within an unbaked cake.
1 2 3 4 5 6
Each chocolate chip has moved farther away from all the other chips.
1 2 3 4 5 6

FIGURE 25-2
The Expanding Chocolate Chip Cake Analogy The expanding universe can be compared to what happens inside a chocolate chip cake as the cake expands during baking. All of the chocolate chips in the cake recede from one another as the cake expands, just as all the galaxies recede from one another as the universe expands.

“Urban Legends” about the Expanding Universe

Incorrect Urban Legend #1:
The expansion of the universe means that as time goes by, galaxies move away from each other through empty space. In this picture, space is simply a background upon which the galaxies act out their parts .

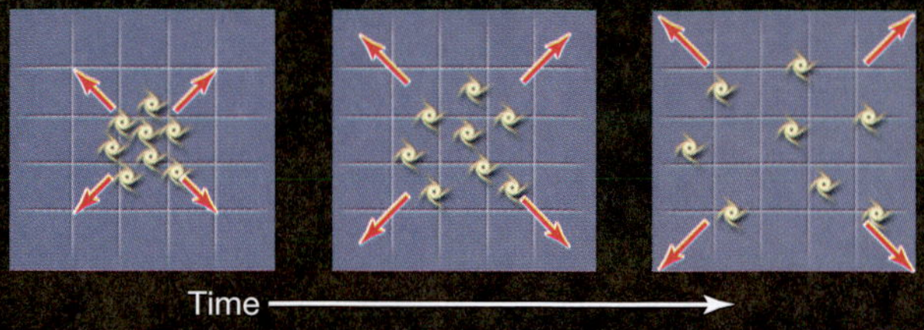

Time

Reality:
The expansion of the universe means that as time goes by, space itself expands (shown here by the expanding grid). As it expands, it carries the galaxies along with it.

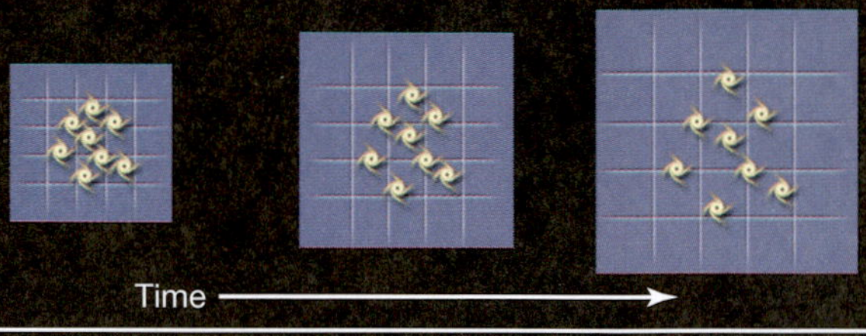

Time

Incorrect Urban Legend #2:
The redshift of light from distant galaxies is a Doppler shift produced by galaxies moving away from us through an unchanging space. Once the light is emitted with its Doppler shift, the wavelength remains the same during its journey to us.

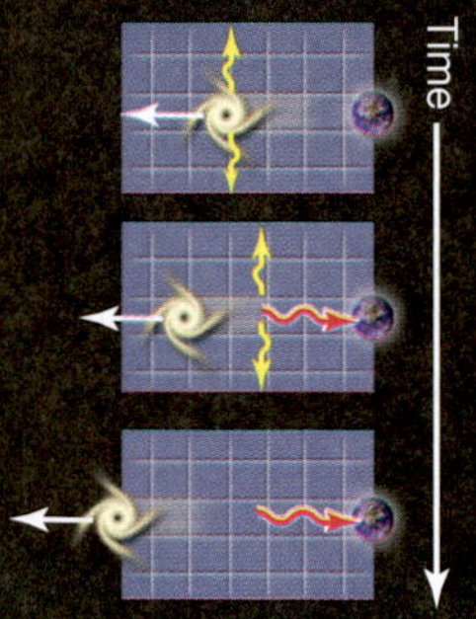

Time

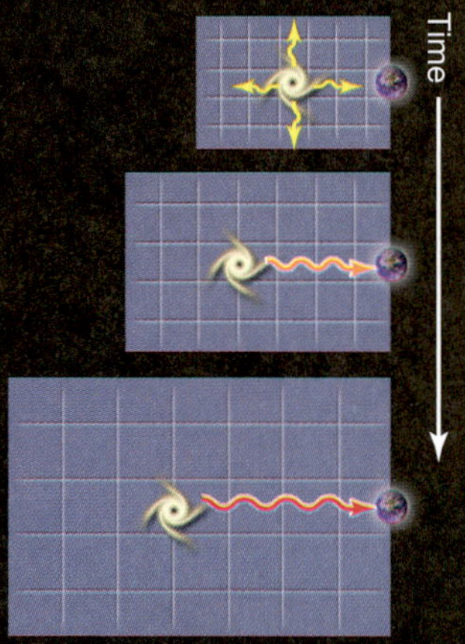

Time

Reality:
As light travels through intergalactic space, its wavelength gradually expands as the
space through which it is travelling expands. This is called a cosmological redshift.

Incorrect Urban Legend #3:
As the universe expands, so do objects within the universe. Hence galaxies within a cluster are now more spread out than they were billions of years ago.

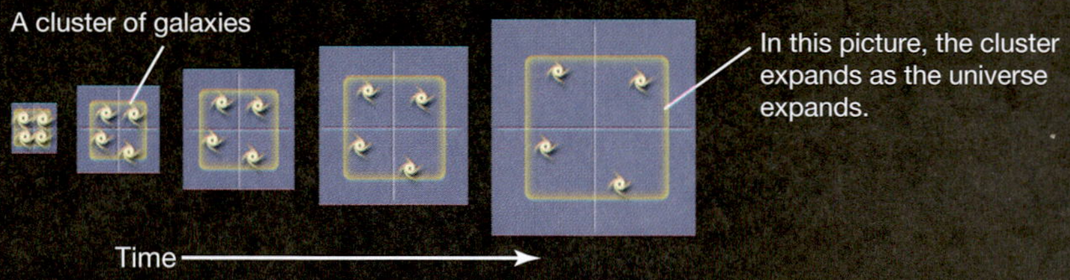

A cluster of galaxies
Time
In this picture, the cluster expands as the universe expands.

Reality:
At first the expansion of the universe tends to pull the galaxies of a cluster away from each other. But the force of gravitational attraction binds the members of the clusters together, so the cluster stabilizes at a certain size.

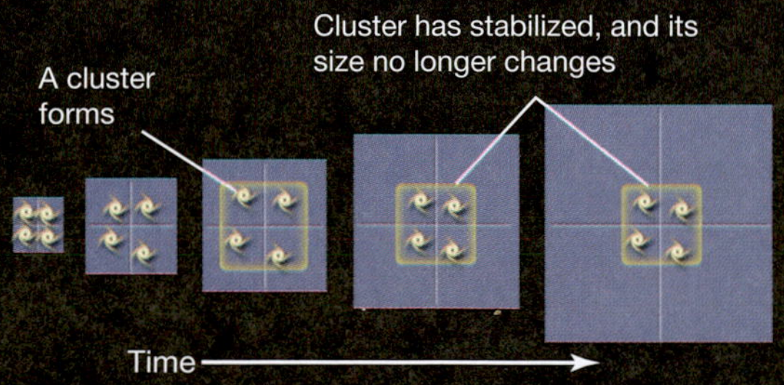

Cluster has stabilized, and its size no longer changes
A cluster forms
Time

(Illustrations by Alfred T. Kamajian, from C. H. Lineweaver and T. M. Davis, “Misconceptions about the Big Bang,” *Scientific American*, March 2005)

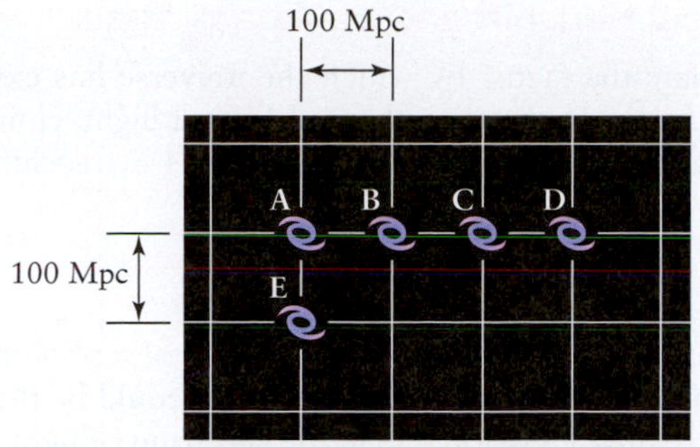

(a) Five galaxies spaced 100 Mpc apart

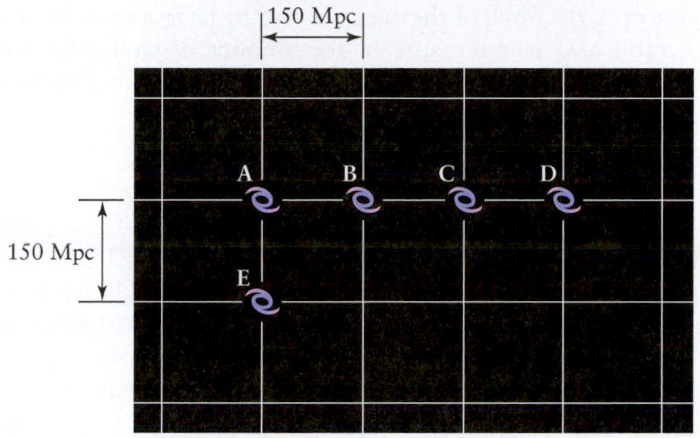

(b) The expansion of the universe spreads the galaxies apart

	Original distance (Mpc)	Later distance (Mpc)	Change in distance (Mpc)
A–B	100	150	50
A–C	200	300	100
A–D	300	450	150
A–E	100	150	50

ANIMATION 25-1

FIGURE 25-3

The Expanding Universe and the Hubble Law **(a)** Imagine five galaxies labeled A, B, C, D, and E. At the time shown here, adjacent galaxies are 100 Mpc apart. **(b)** As the universe expands, by some later time the spacing between adjacent galaxies has increased to 150 Mpc. The table shows that the greater the original distance between galaxies, the greater the amount that distance has increased. This agrees with the Hubble law.

Figure 25-3b shows the universe at a later time, when the gridlines are 50% farther apart (150 Mpc) and all the distances between galaxies are 50% greater than in Figure 25-3a. Imagine that A represents our Galaxy, the Milky Way. The table accompanying Figure 25-3 shows how far each of the other galaxies has moved away from us during the expansion: Galaxies A and B and galaxies A and E were originally 100 Mpc apart and have moved away from each other by an additional 50 Mpc; A and C, which were originally 200 Mpc apart, have increased their separation by an additional 100 Mpc; and the distance between A and D, originally 300 Mpc, has increased by an additional 150 Mpc. In other words, the increase in distance between any pair of galaxies is in direct proportion to the original distance; if the original distance is twice as great, the increase in distance is also twice as great.

To see what these distances tell us about the recessional velocities of galaxies, remember that velocity is equal to the distance moved divided by the elapsed time. (For example, if you traveled in a straight line for 360 kilometers in 4 hours, your velocity was (360 km)/(4 h) = 90 kilometers per hour.) Because the distance that each galaxy moves away from A during the expansion is directly proportional to its original distance from A, it follows that the velocity v at which each galaxy moves away from A is also directly proportional to the original distance d. This result is just the Hubble law, $v = H_0 d$.

CAUTION! It may seem that if the universe is expanding, and if we see all the distant galaxies rushing away from us, then we must be in a special position at the very center of the universe. In fact, the expansion of the universe looks the same from the vantage point of *any* galaxy. For example, as seen from galaxy D in Figure 25-3, the initial distances to galaxies A, B, and C are 300 Mpc, 200 Mpc, and 100 Mpc, respectively. Between parts (a) and (b) of the figure, these distances increase by 150 Mpc, 100 Mpc, and 50 Mpc, respectively. So, as seen from D as well, the recessional velocity increases in direct proportion with the distance, and in the same proportion as seen from A. In other words, no matter which galaxy you call home, you will see all the other galaxies receding from you in accordance with the same Hubble law (and the same Hubble constant) that we observe from Earth.

Figure 25-2 also shows that the expansion of the universe looks the same from one galaxy as from any other. An insect sitting on any one of the chocolate chips would see all the other chips moving away. If the cake were infinitely long, it would not actually have a center; as seen from any chocolate chip within such a cake, the cake would extend off to infinity to the left and to the right, and the expansion of the cake would appear to be centered on that chip. Likewise, because *every* point in the universe appears to be at the center of the expansion, it follows that our universe has no center at all.

CAUTION! "If the universe is expanding, what is it expanding into?" This commonly asked question arises only if we take our chocolate chip cake analogy too literally. In Figure 25-2, the cake (representing the universe) expands in three-dimensional space into the surrounding air. But the actual universe includes *all* space; there is no extra or additional space for the universe to expand into. Asking "What lies beyond the universe?" is like asking "Where on Earth is north of the North Pole?"—there just isn't a place on Earth that this point could be.

The ongoing expansion of space explains why the light from remote galaxies is redshifted. Imagine light or a photon coming toward us from a distant galaxy. As the photon travels through space, the space is expanding, so the photon's wavelength

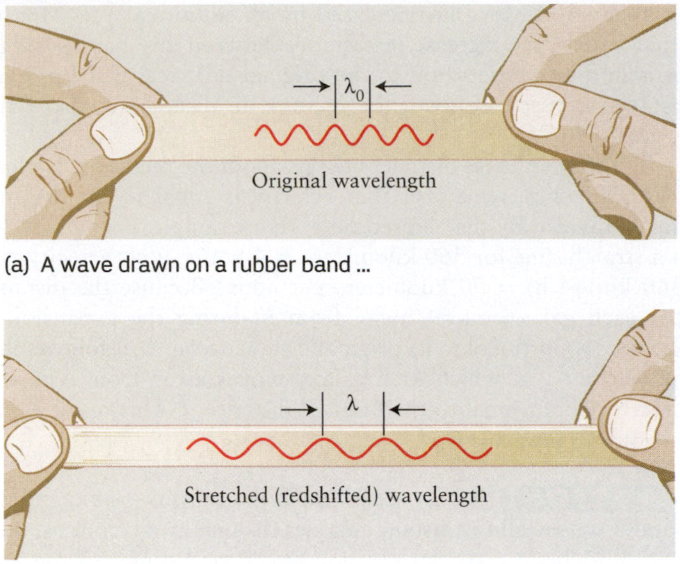

λ_0
Original wavelength
(a) A wave drawn on a rubber band ...
λ
Stretched (redshifted) wavelength
(b) ... increases in wavelength as the rubber band is stretched.

FIGURE 25-4

Cosmological Redshift A wave drawn on a rubber band stretches along with the rubber band. In an analogous way, a light wave traveling through an expanding universe "stretches"; that is, its wavelength increases.

becomes stretched. When the photon reaches our eyes, we see an increased wavelength: The photon has been redshifted. The longer the photon's journey, the more its wavelength will have been stretched along the way. Thus, photons from distant galaxies have larger redshifts than those of photons from nearby galaxies, as expressed by the Hubble law.

> The redshifts of distant galaxies are not Doppler shifts; they are caused by the expansion of space itself

A redshift caused by the expansion of the universe is properly called a **cosmological redshift.** It is *not* the same as a Doppler shift. Doppler shifts are caused by an object's *motion through space,* whereas a cosmological redshift is caused by the *expansion of space* (Figure 25-4).

In Hubble's law, the original interpretation of redshifts was that they were due to Doppler shifts by receding galaxies, and this works because for small redshifts, the Doppler shift can account for the recession of galaxies in expanding space. However, the full properties of expanding space, described by Einstein's general theory of relativity, must be taken into account for larger redshifts. In fact, at redshifts (z) greater than about $z > 1.5$, galaxies are observed with recessional velocities greater than the speed of light. While this would violate the laws of physics in a static universe, recessional speeds can exceed the speed of light in an expanding universe.

CONCEPTCHECK 25-2

Although nearly all distant galaxies have measureable redshifts, the relatively nearby Andromeda Galaxy exhibits an overall blue shift. What does this mean about the Andromeda Galaxy's movement?

Answer appears at the end of the chapter.

Cosmological Redshift and Lookback Time

We can calculate the factor by which the universe has expanded since some ancient time from the redshift of light emitted by objects at that time. As we saw in Section 24-5, redshift (z) is defined as

$$z = \frac{\lambda - \lambda_0}{\lambda_0}$$

In this equation λ_0 is the unshifted wavelength of a photon and λ is the wavelength we observe. For example, λ_0 could be the wavelength of a particular emission line in the spectrum of light leaving a remote galaxy. As the galaxy's light travels through space, its wavelength is stretched by the expansion of the universe. Thus, at our telescopes we observe the spectral line to have a wavelength λ. The ratio λ/λ_0 is a measure of the amount of stretching. By rearranging terms in the preceding equation, we can solve for this ratio to obtain

$$\frac{\lambda}{\lambda_0} = 1 + z$$

For example, consider a galaxy with a redshift $z = 3$. Since the time that light left that galaxy, the universe has expanded by a factor of $1 + z = 1 + 3 = 4$. In other words, when the light left that galaxy, representative distances between widely separated galaxies were only one-quarter as large as they are today. A representative volume of space, which is proportional to the cube of its dimensions, was only $(1/4)^3 = 1/64$ as large as it is today. Thus, the density of matter in such a volume was 64 times greater than it is today.

If you know the redshift z of a distant object such as a remote galaxy, you can calculate that object's recessional velocity v (see Box 23-2 for this calculation). Then, using the Hubble law, you can determine the distance d to that object if you know the value of the Hubble constant H_0. This distance also tells you the **lookback time** of that object, that is, how far into the past you are looking when you see that object. For example, if the lookback time for a distant galaxy is a billion years, that means the light from the galaxy took a billion years to reach us, so we are seeing it as it was a billion years ago. The images that open this chapter show supernovae in distant galaxies with lookback times from 7.7 to 9.0 billion years.

Distances and lookback times determined in this way are somewhat uncertain, because there is some uncertainty in the value of the Hubble constant. Furthermore, as we will see in Section 25-8, the universe has not always expanded at the same rate. This variable rate of expansion means that the value of the Hubble constant H_0 was different in the distant past. (In other words, the Hubble "constant" is not actually constant in time, but at a given time it is still constant throughout space.) Furthermore, the correct distance d to use in the Hubble law is not the distance at which we *see* the object, but rather the distance between us and the object *now.* This latter distance is larger because during the time that it takes light to reach us from a distant object, that object has moved farther away due to the expansion of the universe.

To avoid dealing with these uncertainties and complications, astronomers commonly refer to times in the distant past in terms of redshift rather than years. For example, instead of asking, "How common were quasars 5 billion years ago?," an astronomer might ask, "How common were quasars at $z = 1.0$?" In this question, "at $z = 1.0$" is a shorthand way of saying "at the lookback time that corresponds to objects at $z = 1.0$." We will use this terminology later in this chapter. Remember that the greater the redshift, the greater the lookback time, and hence the further back into time we are peering.

The Cosmological Principle

In cosmology, there is only one universe that we can observe. Unlike other sciences, we cannot carry out controlled experiments or even make comparisons between two similar systems, when the system is the universe itself. But then a question arises: Are we observing the universe from a special location, resulting in a unique, nonrepresentative appearance? Or would observers anywhere see a universe with the same properties on large scales? The answer to this question determines how we interpret Hubble's law, because receding galaxies could be interpreted to mean that we are at the center of the universe. The idea that we are at the center of the universe was rejected, however, because it violates a cosmological extension of Copernicus's argument that we do not occupy a special location in space. An expanding universe, on the other hand, has no center; at all locations in space, observers would see galaxies recede according to Hubble's law.

When Einstein began applying his general theory of relativity to cosmology, he made a daring assumption: Over very large distances the universe is **homogeneous,** meaning that every region is the same as every other region, and **isotropic,** meaning that the universe looks the same in every direction. In other words, if you could stand back and look at a very large region of space, any one part of the universe would look basically the same as any other part, with the same kinds of galaxies distributed through space in the same way. The assumption that the universe is homogeneous and isotropic constitutes the **cosmological principle.** It gives precise meaning to the idea that we do not occupy a special location in space.

Models of the universe based on the cosmological principle have proven remarkably successful in describing the structure and evolution of the universe and in interpreting observational data. More recently, direct observations reveal a homogeneous and isotropic universe on the largest scales—a validation of the cosmological principle. A homogenous and isotropic universe is revealed in maps made of large-scale structure, as shown in Figures 23-23 and 23-24. As we will see in Section 25-5, radiation left over from the Big Bang produces a similar result for the early universe.

CONCEPTCHECK 25-3

If we notice in the night sky that there are more stars along the Milky Way than in other regions of the sky, is this consistent or inconsistent with the cosmological principle?

CALCULATIONCHECK 25-1

What is the redshift z-value for a galaxy that has a galaxy spectral line shifted to 725.6 nm when a stationary object would emit the line at from 656.3 nm?

CALCULATIONCHECK 25-2

What is the distance to a galaxy that is observed to have a recessional velocity of 10,000 km/s?

Answers appear at the end of the chapter.

25-3 The expanding universe emerged from a cataclysmic event called the Big Bang

The Hubble flow shows that the universe has been expanding for billions of years. Therefore, we can conclude that in the past the matter in the universe must have been closer together and therefore denser than it is today. Indeed, as we look farther into the past, we clearly see the density of the universe increasing. There is even strong evidence of increasing density back to the first moments of time. Therefore, some sort of tremendous event caused high-density matter to begin the expansion that continues to the present day. This event, called the **Big Bang,** marks the creation of the universe.

CAUTION! It is not correct to think of the Big Bang as an explosion. When a bomb explodes, pieces of debris fly off *into space* from a central location. If you could trace all the pieces back to their origin, you could find out exactly where the bomb had been. This process is not possible with the universe, however, because the universe itself always has and always will consist of all space, with no center. There is no single or central location in space where the Big Bang occurred, because the Big Bang is the expansion of all space (see the discussion of the expanding chocolate chip cake in Section 25-2).

Estimating the Age of the Universe

How long ago did the Big Bang take place? To estimate an answer, imagine two galaxies that today are separated by a distance d and receding from each other with a velocity v. A movie of these galaxies would show them flying apart. If you run the movie backward, you would see the two galaxies approaching each other as time runs in reverse. We can calculate the time T_0 it will take for the reverse-run galaxies to collide by using the equation

$$T_0 = \frac{d}{v}$$

This expression says that the time to travel a distance d at velocity v is equal to the ratio d/v. (As an example, to travel a distance of 360 km at a velocity of 90 km/h takes (360 km)/(90 km/h) = 4 hours.) If we use the Hubble law, $v = H_0 d$, to replace the velocity v in this equation, we get

$$T_0 = \frac{d}{H_0 d} = \frac{1}{H_0}$$

Note that the distance of separation, d, has canceled out and does not appear in the final expression. Distance not appearing in the expression means that T_0 is the same for *all* galaxies. This period is the time in the past when all the matter in galaxies was crushed together, the time back to the Big Bang. In other words, the reciprocal (or inverse) of the Hubble constant H_0 gives us an estimate of the age of the universe called the **Hubble time;** this is one reason why H_0 is such an important quantity in cosmology.

Observations suggest that $H_0 = 73$ km/s/Mpc to within a few percent, and this is the value we choose as our standard (see Section 23-5). Using this value, our estimate for the age of the universe is

$$T_0 = \frac{1}{73 \text{ km/s/Mpc}}$$

To convert this into units of time, we simply need to remember that 1 Mpc equals 3.09×10^{19} km and 1 year equals 3.156×10^{7} s. Using the technique we discussed in Box 1-3 for converting units, we get

$$T_0 = \frac{1}{73}\ \frac{\text{Mpc s}}{\text{km}} \times \frac{3.09 \times 10^{19}\ \text{km}}{1\ \text{Mpc}} \times \frac{1\ \text{year}}{3.156 \times 10^{7}\text{s}}$$

$$= 1.34 \times 10^{10}\ \text{years} = 13.4\ \text{billion years}$$

By comparison, the age of our solar system is only 4.56 billion years, or about one-third the age of the universe. Thus, the formation of our home planet is a relatively recent event in the history of the cosmos.

The value of H_0 has an uncertainty of about 5%, so our simple estimate of the age of the universe is likewise uncertain by at least 5%. Furthermore, the formula $T_0 = 1/H_0$ is at best an approximation, because in deriving it we assumed that the universe expands at a constant rate. In Section 25-8 we will discuss how the expansion rate of the universe has changed over its history. When these factors are taken into consideration, we find that the age of the universe is 13.7 billion years, with an uncertainty of about 0.2 billion years. This time period is remarkably close to our simple estimate.

Whatever the true age of the universe, it must be at least as old as the oldest stars. The oldest stars that we can observe readily lie in the Milky Way's globular clusters (see Section 19-4 and Section 22-1). The most recent observations, combined with calculations based on the theory of stellar evolution, indicate that these stars are about 13.4 billion years old (time period with an uncertainty of about 6%). This time period is less than the calculated age of the universe: As required, the oldest stars in our universe are younger than the universe itself.

What was our universe like 15 billion years ago? The passage of time (as we understand time) in the only universe we know of did not exist prior to the Big Bang. Nor did space (as we understand space) exist in the universe until expansion from the Big Bang. Time began to pass and space began to exist at the Big Bang. Since the universe, as we understand it, is only 13.7 billion years old, there does not appear to be physical meaning to the question of what things were like in the universe 15 billion years ago.

CONCEPTCHECK 25-4

If the Hubble constant were smaller than it actually is, would the universe be younger or older than 13.7 billion years?

Answer appears at the end of the chapter.

Our Observable Universe and the Dark Night Sky

The Big Bang helps resolve Olbers's paradox, which we discussed in Section 25-1. We know that the universe had a definite beginning, and thus its age is finite (as opposed to infinite). If the universe is 13.7 billion years old, then the most distant objects that we can see are those whose light has traveled 13.7 billion years to reach us. (Due to the expansion of the universe, these objects are now more than 13.7 billion light-years away.) As a result, we can only see objects that lie within an immense sphere centered on Earth (Figure 25-5). This is true even if the universe is infinite, with galaxies scattered throughout its limitless expanse.

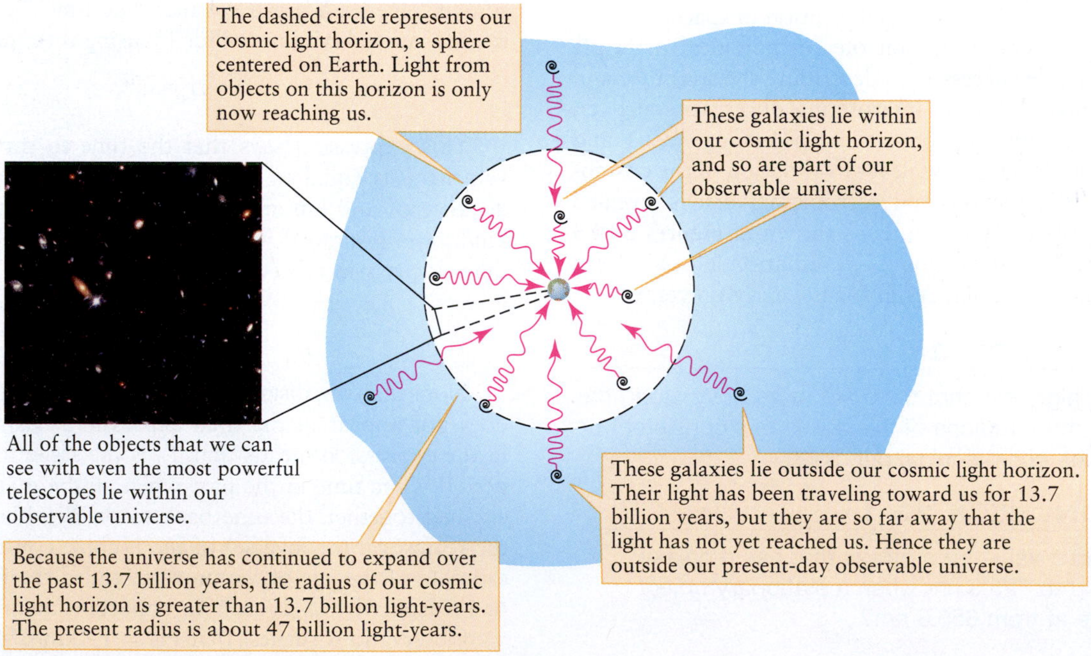

The dashed circle represents our cosmic light horizon, a sphere centered on Earth. Light from objects on this horizon is only now reaching us.
These galaxies lie within our cosmic light horizon, and so are part of our observable universe.
All of the objects that we can see with even the most powerful telescopes lie within our observable universe.
These galaxies lie outside our cosmic light horizon. Their light has been traveling toward us for 13.7 billion years, but they are so far away that the light has not yet reached us. Hence they are outside our present-day observable universe.
Because the universe has continued to expand over the past 13.7 billion years, the radius of our cosmic light horizon is greater than 13.7 billion light-years. The present radius is about 47 billion light-years.

FIGURE 25-5 R I V U X G

Our Observable Universe The part of the universe that we can observe lies within our cosmic light horizon. The galaxies that we can just barely make out with our most powerful telescopes lie inside our cosmic light horizon; we see them as they were less than a billion years after the Big Bang. We cannot see objects beyond our cosmic light horizon, because in the 13.7 billion years since the Big Bang their light has not had enough time to reach us. (Inset: Robert Williams and the Hubble Deep Field Team, STScI; NASA)

The surface of the sphere depicted in Figure 25-5 is called our **cosmic light horizon.** Our entire **observable universe** is located inside this sphere. We cannot see anything beyond our cosmic light horizon, because the time required for light to reach us from these incredibly remote distances is greater than the present age of the universe. As time goes by, light from more distant parts of the universe reaches us for the first time, and the size of the cosmic light horizon—the size of our observable universe—increases. The finite size of the observable universe, with a finite number of stars and galaxies, also helps to resolve Olber's paradox: Galaxies are distributed sparsely enough in our observable universe that there are no stars along most of our lines of sight. This sparse distribution of visible stars is one reason why the night sky is dark.

Besides the finite age of the universe, a second effect also contributes significantly to the darkness of the night sky—the redshift. According to the Hubble law, the greater the distance to a galaxy, the greater the redshift. When a photon is redshifted, its wavelength becomes longer, and its energy—which is inversely proportional to its wavelength (see Section 5-5)—decreases. Consequently, even though there are many galaxies far from Earth, they have large redshifts and their light does not carry much energy. A galaxy nearly at the cosmic light horizon has a nearly infinite redshift, meaning that the light we receive from that galaxy carries practically no energy at all. This decrease in photon energy because of the expansion of the universe decreases the brilliance of remote galaxies, helping to make the night sky dark.

Measuring the recessional velocities of galaxies allows us to estimate the age of the universe

The concept of a Big Bang origin for the universe is a straightforward, logical consequence of an expanding universe. In Section 25-4, we will see direct evidence of the primordial fireball associated with the Big Bang and other confirmations of the Big Bang back to the first few moments. But how far back in time can our laws describe the universe? If you can just imagine far enough back into the past, you can arrive at a time 13.7 billion years ago, when the density throughout the universe was nearly infinite. As a result, throughout the universe space and time were jumbled up with nearly infinite curvature. A full description of this earliest instant requires (as do aspects of black holes) a valid mathematical theory of quantum gravity, which is a work in progress. But when in the past did the known laws of physics begin to apply?

A very short time after the Big Bang, space and time began to behave in the way we think of them today. This short time interval, called the **Planck time** (t_P), is given by the following expression:

The Planck time

$$t_P = \sqrt{\frac{Gh}{c^5}} = 1.35 \times 10^{-43}\ \text{s}$$

t_P = Planck time
G = universal constant of gravitation
h = Planck's constant
c = speed of light

We do not yet understand how space, time, and matter behaved in that brief but important interval from the beginning of the Big Bang to the Planck time, about 10^{-43} seconds later. (Indeed, the laws of physics suggest that it might be impossible ever to know what happened during this extremely short time interval.) Hence, the Planck time represents a limit to our knowledge of conditions at the very beginning of the universe.

CONCEPTCHECK 25-5

Consider two hypothetical stars of equal luminosity that are initially at the same remote distance from Earth, but each is in a different type of universe. Imagine one star in a static universe so that the star is not receding away. Imagine that the other star is receding away due to cosmic expansion, just as galaxies do in our actual universe. When this imaginary experiment begins and both stars are observed from Earth, which star, if any, appears dimmer?

CONCEPTCHECK 25-6

Why does our observable universe get larger over time?

Answer appears at the end of the chapter.

25-4 The microwave radiation that fills all space is evidence of a hot Big Bang

TUTORIAL 25-2

One of the major advances in twentieth-century astronomy was the discovery of the origin of the heavy elements. We know today that essentially all the heavy elements are created by thermonuclear reactions at the centers of stars and in supernovae (see Chapter 20). The starting point of all these reactions is the fusion of hydrogen into helium, which we described in Section 16-1. But as astronomers began to understand the details of thermonuclear synthesis in the 1960s, they were faced with a dilemma: There is far more helium in the universe than could have been created by hydrogen fusion in stars.

For example, the Sun consists of about 74% hydrogen and 25% helium by mass, and the universe as a whole is also about 25% helium. Some helium was produced by the thermonuclear fusion of hydrogen within stars, but calculations show that the amount of helium produced in this way is not nearly enough to account for the large fraction of helium observed in the universe. Because it was thought that the universe originally contained only hydrogen—the simplest of all the chemical elements—the presence of so much helium posed a major dilemma.

A Hot Big Bang and the Cosmic Microwave Background

Shortly after World War II, Ralph Alpher and Robert Hermann proposed that the universe immediately following the Big Bang must have been so incredibly hot that thermonuclear reactions occurred everywhere throughout space. Calculations for the amount of helium (and some other elements) produced by thermonuclear reactions in the first few minutes after the Big Bang are in extraordinary agreement with observations. This astounding prediction of the Big Bang was confirmed by detailed observations, which means that strong evidence for the Big Bang goes all the way back to the first few minutes of the universe.

Fortunately, predictions coming from the Big Bang theory did not stop there. Princeton University physicists Robert Dicke and P. J. E. Peebles calculated that in order for the entire universe to undergo the required thermonuclear reactions, in addition to charged particles, the early universe must have been very hot and filled with many high-energy, short-wavelength photons. With these photons interacting strongly with the "soup" of charged particles, the intensity distribution of photons formed blackbody radiation (see Figure 5-12).

The universe has expanded so much since those ancient times that all those short-wavelength photons have had their wavelengths stretched by a tremendous factor. As a result, they have become low-energy, long-wavelength photons. The temperature of this cosmic radiation field is now only a few degrees above absolute zero. By Wien's law, radiation at such a low temperature should have its peak intensity at microwave wavelengths of approximately 1 millimeter. Hence, this radiation field, which fills all of space, is called the **cosmic microwave background (CMB),** or **cosmic background radiation.** In the early 1960s, Dicke and his colleagues began designing an antenna to detect this microwave radiation.

Meanwhile, just a few miles from Princeton University, Arno Penzias and Robert Wilson of Bell Telephone Laboratories in New Jersey were working on a new microwave horn antenna designed to relay telephone calls to Earth-orbiting communications satellites (Figure 25-6). Penzias and Wilson were deeply puzzled when, no matter where in the sky they pointed their antenna, they detected faint background noise. Thanks to a colleague, they learned about the work of Dicke and Peebles and realized that they had discovered the cooled-down cosmic background radiation left over from the hot Big Bang. Penzias and Wilson shared the 1978 Nobel Prize in Physics for their discovery.

FIGURE 25-6 R I V U X G
The Bell Labs Horn Antenna Using this microwave horn antenna, originally built for communications purposes, Arno Penzias and Robert Wilson detected a signal that seemed to come from all parts of the sky. After carefully removing all potential sources of electronic "noise" (including bird droppings inside the antenna) that could create a false signal, Penzias and Wilson realized that they were actually detecting radiation from space. This radiation is the afterglow of the Big Bang. (Bell Labs)

A TV using an antenna for signal reception (as opposed to cable or a satellite dish) can actually detect cosmic background radiation. This radiation is responsible for about 1% of the random noise or "hash" that appears on the screen when you tune a television to a station that is off the air. Using far more sophisticated detectors than TV sets, scientists have made many measurements of the intensity of the background radiation at a variety of wavelengths. Unfortunately, Earth's atmosphere is almost totally opaque to wavelengths between about 10 mm and 1 cm (see Figure 6-25), which is just the wavelength range in which the background radiation is most intense. As a result, scientists have had to place detectors either on high-altitude balloons (which can fly above the majority of the obscuring atmosphere) or, even better, on board orbiting spacecraft.

The afterglow of the Big Bang was first discovered by a happy coincidence—and can be detected with an ordinary television

A Detailed Look at the Cosmic Microwave Background

The first high-precision measurements of the cosmic microwave background came from the *Cosmic Background Explorer* (COBE, pronounced "coe-bee") satellite, which was placed in Earth orbit in 1989 (Figure 25-7a). Data from COBE's spectrometer, shown in Figure 25-7b, demonstrate that this ancient radiation has the spectrum of a blackbody with a temperature of 2.725 K. In recognition of this discovery, as well as others that we will discuss in Section 25-5, COBE team leaders John Mather of NASA and George Smoot of the University of California, Berkeley, were awarded the 2006 Nobel Prize in Physics.

An important feature of the microwave background is that its intensity is almost perfectly isotropic, that is, the same in all directions. In other words, we detect nearly the same background intensity from all parts of the sky. This is a striking confirmation of Einstein's assumption that the universe is isotropic (see Section 25-2). However, extremely accurate measurements, first made from high-flying airplanes and later from high-altitude balloons and from COBE, reveal a very slight variation in temperature across the sky. The microwave background appears slightly warmer than average toward the constellation of Leo and slightly cooler than average in the opposite direction toward Aquarius. Between the warm spot in Leo and the cool spot in Aquarius, the background temperature declines smoothly across the sky. Figure 25-8 is a map of the microwave sky showing this variation.

This apparent variation in temperature is caused by Earth's motion through the cosmos. If we were at rest with respect to the microwave background, the radiation would be even more nearly isotropic. Because we are moving through this radiation field, however, we see a Doppler shift. Specifically, we see shorter-than-average wavelengths in the direction toward which we are moving, as drawn in Figure 25-9. A decrease in wavelength corresponds to an increase in photon energy and thus an increase in the temperature of the radiation. The slight temperature excess observed, about 0.00337 K, corresponds to a speed of 371 km/s. Conversely, we see longer-than-average wavelengths in that part

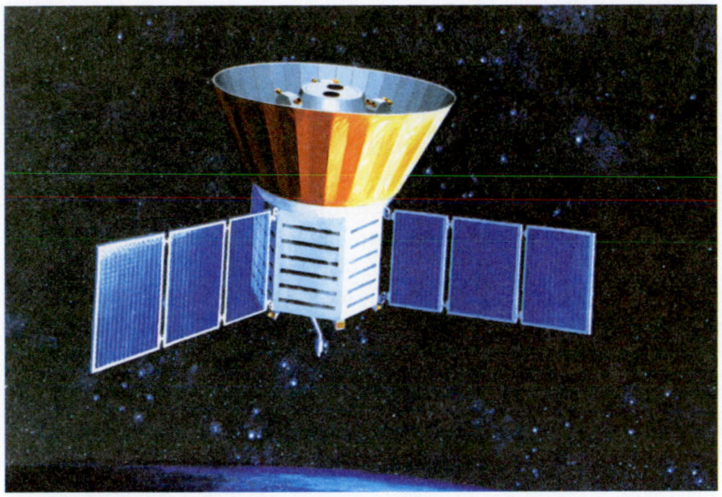

(a) The COBE spacecraft

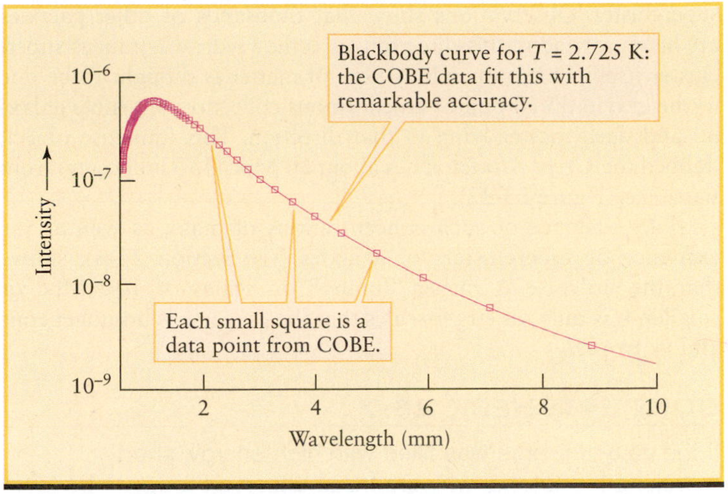

(b) The spectrum of the cosmic microwave background

FIGURE 25-7

COBE and the Spectrum of the Cosmic Microwave Background **(a)** The *Cosmic Background Explorer* (COBE), launched in 1989, measured the spectrum and angular distribution of the cosmic microwave background over a wavelength range from 1 mm to 1 cm. **(b)** A blackbody curve gives an excellent match to the COBE data. (a: Courtesy of J. Mather/NASA; b: Courtesy of E. Cheng/NASA COBE Science Team)

of the sky from which we are receding. An increase in wavelength corresponds to a decline in photon energy and, hence, a decline in radiation temperature.

Our solar system is thus traveling away from Aquarius and toward Leo at a speed of 371 km/s. Taking into account the known velocity of the Sun around the center of our Galaxy, we find that the entire Local Group of galaxies, including our Milky Way Galaxy, is moving at about 620 km/s toward the Hydra-Centaurus

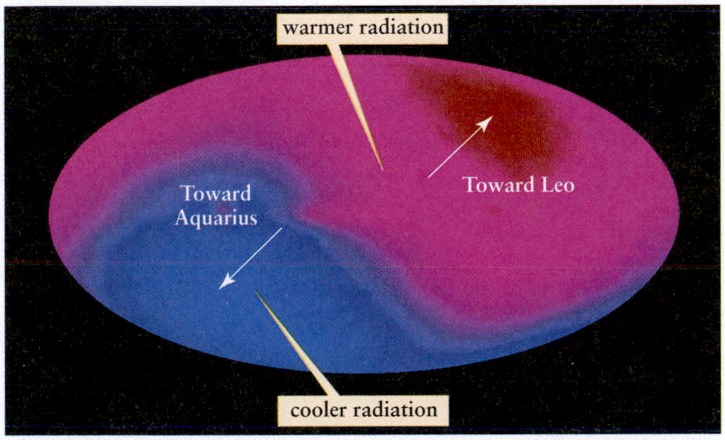

FIGURE 25-8 R I V U X G

The Microwave Sky In this map of the entire sky made from COBE data, the plane of the Milky Way runs horizontally across the map, with the galactic center in the middle. Color indicates temperature—red is warm and blue is cool. The small temperature variation across the sky—only 0.0033 K above or below the average radiation temperature of 2.725 K—is caused by Earth's motion through the microwave background. (NASA)

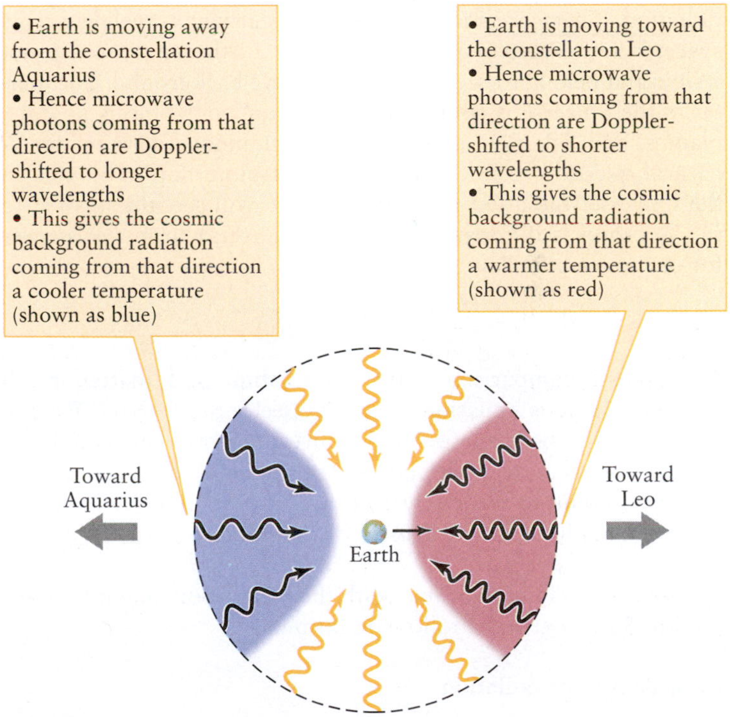

FIGURE 25-9

Our Motion Through the Microwave Background Because of the Doppler effect, we detect shorter wavelengths in the microwave background and a higher temperature of radiation in that part of the sky toward which we are moving. This part of the sky is the area shown in red in Figure 25-8. In the opposite part of the sky, shown in blue in Figure 25-8, the microwave radiation has longer wavelengths and a cooler temperature.

supercluster. Observations show that thousands of other galaxies are being carried in this direction, as is the Hydra-Centaurus supercluster itself. This tremendous flow of matter is thought to be due to the gravitational pull of an enormous collection of visible galaxies and dark matter lying in that direction. This immense object, dubbed the *Great Attractor,* lies about 50 Mpc (150 million ly) from Earth (see Figure 23-22).

The existence of such concentrations of mass, as well as the existence of superclusters of galaxies (see Section 21-6), shows that the universe is rather "lumpy" on scales of 100 Mpc or smaller. It is only on larger scales that the universe is homogeneous and isotropic.

CONCEPTCHECK 25-7

If the early universe was filled with high-energy, short-wavelength photons, why are these observed today to be low-energy, long-wavelength microwave photons?

Answer appears at the end of the chapter.

25-5 The universe was a hot, opaque plasma during its first 380,000 years

Energy in the universe falls into one of two categories—radiation or matter. (We will encounter another type of energy in Section 25-7.) Photons are massless particles of light and are a form of radiation. There are many photons of starlight traveling across space, but the vast majority of photons in the universe are not visible and belong to the cosmic microwave background. The matter in the universe is contained in such luminous objects as stars, planets, and galaxies, as well as in nonluminous dark matter. A natural question to ask is which plays a more important role in the universe: radiation or matter? As we will see, the answer to this question is different for the early universe from the answer for our universe today.

Radiation and Matter in the Universe

To make a comparison between radiation and matter, recall Einstein's famous equation $E = mc^2$ (see Section 16-1). We can think of the photon energy in the universe (E) as being equal to a quantity of mass (m) multiplied by the square of the speed of light (c). The amount of this equivalent mass in a volume V, divided by that volume, is the **mass density of radiation** (ρ_{rad}; say "rho sub rad").

We can combine $E = mc^2$ with the Stefan-Boltzmann law (see Section 5-4) to give the following formula:

Mass density of radiation

$$\rho_{rad} = \frac{4\sigma T^4}{c^3}$$

ρ_{rad} = mass density of radiation

T = temperature of radiation

σ = Stefan-Boltzmann constant = 5.67×10^{-8} W m^{-2} K^{-4}

c = speed of light = 3.00×10^8 m/s

For the present-day temperature of the cosmic background radiation, $T = 2.725$ K, this equation yields

$$\rho_{rad} = 4.6 \times 10^{-31} \text{ kg/m}^3$$

The **average density of matter** (ρ_m; say "rho sub em") in the universe is harder to determine. To find this density, we look at a large volume (V) of space, determine the total mass (M) of all the stars, galaxies, and dark matter in that volume, and divide the volume into the mass: $\rho_m = M/V$. (We emphasize that this quantity is the *average* density of matter. It would be the actual density if all the matter in the universe were spread out uniformly rather than being clumped into galaxies and clusters of galaxies.) Determining how much mass is in a large volume of space is a challenging task. A major part of the challenge involves dark matter, which emits no electromagnetic radiation and can be detected only by its gravitational influence (see Section 22-4 and Section 23-8).

One method that appears to deal successfully with this challenge is to observe clusters of galaxies, within which most of the luminous mass in the universe is concentrated and the contribution from dark matter can also be estimated. Rich clusters are surrounded by halos of hot, X-ray–emitting gas, typically at temperatures of 10^7 to 10^8 K (see Figure 23-26). Such a halo should be in hydrostatic equilibrium, so that it neither expands nor contracts but remains the same size (see Section 16-2, especially Figure 16-2). The outward gas pressure associated with the halo's high temperature would then be balanced by the inward gravitational pull due to the total mass of the cluster. Thus, by measuring the temperature of the halo—which can be determined from the properties of the halo's X-ray emission—astronomers can infer the cluster's mass, including the contribution from dark matter!

From galaxy clusters and other measurements, the present-day average density of matter in the universe is estimated to be

$$\rho_m = 2.4 \times 10^{-27} \text{ kg/m}^3$$

with an uncertainty of about 15%. The mass of a single hydrogen atom is 1.67×10^{-27} kg. Hence, if the mass of the universe were spread uniformly over space, there would be the equivalent of 1½ hydrogen atoms per cubic meter of space. By contrast, there are 5×10^{25} atoms in a cubic meter of the air you breathe! The very small value of ρ_m shows that our universe has a very low average density.

Furthermore, by counting galaxies and other measurements, astronomers determine that the average density of *luminous* matter (that is, the stars and gas within clusters of galaxies) is about 4.2×10^{-28} kg/m^3. This density is only about 17% of the average density of matter of all forms. Thus, nonluminous dark matter is actually the predominant form of matter in our universe. The "ordinary" matter (protons and neutrons) of which the stars, the planets, and ourselves are made is only about one-fifth of the total matter!

The fact that there is more dark matter than regular matter does not make it any easier to detect. By interacting with light, regular matter can cool and condense into stars and planets. However, dark matter is more evenly and sparsely distributed. The actual amount of dark matter within Earth's volume at any given time is only about 1 kg. It is thought that dark matter particles stream right through the Earth with only rare interactions. Due to its elusive nature, yet dominance over regular matter, the detection of dark matter particles is considered one of the "Holy Grails" of physics and astronomy and would surely result in a Nobel Prize.

When Radiation Held Sway Over Matter

Although the average density of matter in the universe is tiny by Earth standards, it is thousands of times larger than ρ_{rad}, the mass density of radiation. However, this ratio was not always the case. Matter prevails over radiation today only because the energy now carried by microwave photons is so small. Nevertheless, the number of photons in the microwave background is astounding. From the physics of blackbody radiation it can be calculated that there are today 410 million (4.1×10^8) photons in every cubic meter of space. In other words, the photons in space outnumber atoms by roughly a billion (10^9) to one. In terms of total number of particles, the universe thus consists almost entirely of microwave photons. This radiation field no longer has much effect on the universe however, because its photons have been redshifted to long wavelengths and low energies after 13.7 billion years of being stretched by the expansion of the universe.

In contrast, think back toward the Big Bang. The universe becomes increasingly compressed, and thus the density of matter increases as we go back in time. The photons in the background radiation also become more crowded together as we go back in time. But, in addition, the photons become less redshifted and thus have shorter wavelengths and higher energy than they do today. Because of this added energy, the mass density of radiation (ρ_{rad}) increases more quickly as we go back in time than does the average density of matter (ρ_m). In fact, as Figure 25-10 shows, there was a time in the ancient past when ρ_{rad} equaled ρ_m. Before this time, ρ_{rad} was greater than ρ_m, and radiation thus held sway over matter. Astronomers call this state a **radiation-dominated universe.** After ρ_m became greater than ρ_{rad}, so that matter prevailed over radiation, our universe became a **matter-dominated universe.**

This transition from a radiation-dominated universe to a matter-dominated universe occurred about 24,000 years after the Big Bang, at a time that corresponds to a redshift of about $z = 5200$. Since that time the wavelengths of photons have been stretched by a factor of $1 + z$, or about 5200. Today these microwave photons typically have wavelengths of about 1 mm. But when the universe was about 24,000 years old, they had wavelengths of about 190 nm in the ultraviolet part of the spectrum.

The temperature of the background radiation has declined over the eons thanks to the expansion of the universe

To calculate the temperature of the cosmic background radiation at the time of this transition from a radiation-dominated universe to a matter-dominated one, we use Wien's law (see Section 5-4). This law says that the wavelength of maximum emission (ρ_{max}) of a blackbody is inversely proportional to its temperature (T): A *decrease* of ρ_{max} by a factor of 2 corresponds to an *increase* of T by a factor of 2.

The present-day peak wavelength of the cosmic background radiation corresponds to a blackbody temperature of 2.725 K. Hence, a peak wavelength 5200 times smaller corresponds to a temperature 5200 times greater: $T = 5200 \times 2.725\ \text{K} = 14{,}000\ \text{K}$. In other words, the radiation temperature at redshift z was greater than the present-day radiation temperature by a factor of $1 + z$.

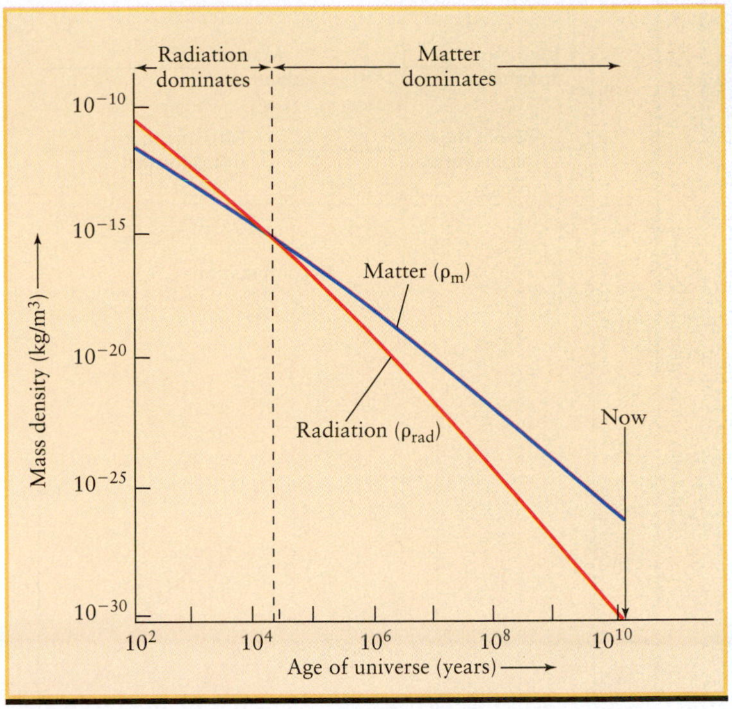

FIGURE 25-10

The Evolution of Density For approximately 24,000 years after the Big Bang, the mass density of radiation (ρ_{rad}, shown in red) exceeded the matter density (ρ_m, shown in blue), and the universe was radiation-dominated. Later, however, continued expansion of the universe caused ρ_{rad} to become less than ρ_m, at which point the universe became matter-dominated. (Graph courtesy of Clem Pryke, University of Minnesota)

Therefore, the temperature of the radiation background was once much greater and has been declining over the ages, as Figure 25-11 shows.

CONCEPTCHECK 25-8

At the time Earth formed, was the universe dominated by energy or by matter?

Answer appears at the end of the chapter.

When the First Atoms Formed

The nature of the universe changed again in a fundamental way about 380,000 years after the Big Bang, when z was roughly 1100 and the temperature of the radiation background was about $1100 \times 2.725\ \text{K} = 3000\ \text{K}$. (At a temperature of 3000 K, the entire universe was filled with visible orange light.) To see the significance of this moment in cosmic history, recall that hydrogen is by far the most abundant element in the universe—hydrogen atoms outnumber helium atoms by about 12 to 1. A hydrogen atom consists of a single proton orbited by a single electron, and it takes relatively little energy to knock the electron completely out of its orbit around the proton. In fact, ultraviolet radiation warmer than about 3000 K easily ionizes hydrogen. Thus, neutral hydrogen atoms could not survive in the universe that existed before $z = 1100$. That is, in the first 380,000 years after the Big Bang, the background photons

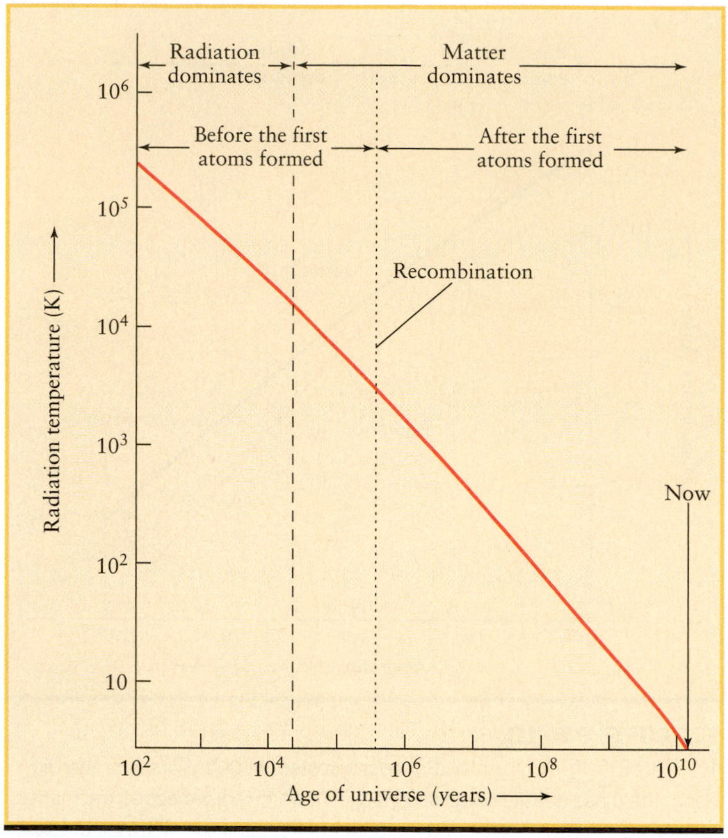

FIGURE 25-11

The Evolution of Radiation Temperature As the universe expanded, the photons in the radiation background became increasingly redshifted and the temperature of the radiation fell. Approximately 380,000 years after the Big Bang, when the temperature fell below 3000 K, hydrogen atoms formed and the radiation field "decoupled" from the matter in the universe. After that point, the temperature of matter in the universe was not the same as the temperature of radiation. The time when the first atoms formed is called the era of recombination (see Figure 25-12). (Graph courtesy of Clem Pryke, University of Minnesota)

had energies great enough to prevent electrons and protons from binding to form hydrogen atoms (Figure 25-12a). Only since $z = 1100$ (that is, since $t = 380{,}000$ years) have the energies of these photons been low enough to permit hydrogen atoms to exist (Figure 25-12b).

The epoch when atoms first formed at $t = 380{,}000$ years is called the **era of recombination.** This name refers to electrons "recombining" to form atoms. (The name is a bit misleading, because the electrons and protons had never before combined into atoms.)

Prior to $t = 380{,}000$ years, the universe was completely filled with a shimmering expanse of high-energy photons colliding vigorously with protons and electrons. This state of matter, called a **plasma,** is opaque, which means that light cannot pass through it without being strongly scattered and absorbed. The surface and interior of the Sun are also a hot, glowing, opaque plasma (see Section 16-9). P. J. E. Peebles coined the term **primordial fireball** to describe the universe during its first 380,000 years of existence.

After $t = 380{,}000$ years, the photons no longer had enough energy to keep the protons and electrons apart. As soon as the temperature of the radiation field fell below about 3000 K, protons and electrons began combining to form hydrogen atoms. These atoms do not absorb low-energy photons, so space became transparent! All the photons that heretofore had been vigorously colliding with charged particles could now stream unimpeded across space. Today, these same photons constitute the microwave background.

The significance of the era of recombination becomes clear by considering how it affects an astronomer's ability to form images of the early universe. After recombination, the photons from a large clump of matter can stream to us while having very little interaction with the matter they pass through along their journey. These undisturbed photons can carry images of different clumps of matter in the early universe. On the other hand, photons emitted prior to recombination are scattered so strongly that any image from that early time is completely blurred out before its photons reach us. Therefore, *the era of recombination marks the*

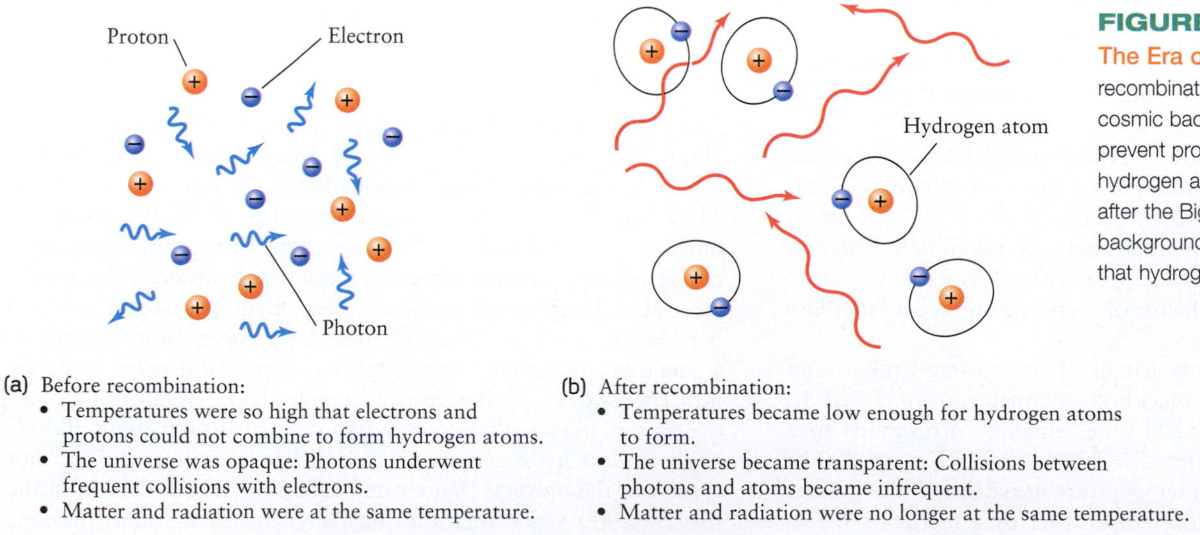

Proton
Electron
Photon
Hydrogen atom
(a) Before recombination:
• Temperatures were so high that electrons and protons could not combine to form hydrogen atoms.
• The universe was opaque: Photons underwent frequent collisions with electrons.
• Matter and radiation were at the same temperature.
(b) After recombination:
• Temperatures became low enough for hydrogen atoms to form.
• The universe became transparent: Collisions between photons and atoms became infrequent.
• Matter and radiation were no longer at the same temperature.

FIGURE 25-12

The Era of Recombination **(a)** Before recombination, the energy of photons in the cosmic background was high enough to prevent protons and electrons from forming hydrogen atoms. **(b)** Some 380,000 years after the Big Bang, the energy of the cosmic background radiation became low enough that hydrogen atoms could survive.

earliest time and farthest distance that astronomers can collect images of the universe.

Before recombination, matter and the radiation field had the same temperature, because photons, electrons, and protons were all in continuous interaction with one another. After recombination, photons and atoms hardly interacted at all, and thus the temperature of matter in the universe was no longer the same as the temperature of the background radiation. Thus, $T = 2.725$ K is the temperature of the present-day background radiation field, *not* the temperature of the matter in the universe. Note that while the temperature of the background radiation is very uniform, the temperature of matter in the universe is anything but: It ranges from hundreds of millions of kelvins in the interiors of giant stars to a few tens of kelvins in the interstellar medium.

ANALOGY A good analogy is the behavior of a glass of cold water. If you hold the glass in your hand, the water will get warmer and your hand will get colder until both the water and your hand are at the same temperature. But if you set the glass down and do not touch it, so that the glass and your hand do not interact, their temperatures are decoupled: The water will stay cold and your hand will stay warm for much longer.

Because the universe was opaque prior to $t = 380{,}000$ years, we cannot see any further into the past than the era of recombination. In particular, we cannot see back to the era when the universe was radiation-dominated. The microwave background, whose photons have suffered a redshift of $z = 1100$, contains the most ancient photons we will ever be able to observe.

CONCEPTCHECK 25-9

If the universe had cooled more slowly, would the first atoms have appeared more quickly or more slowly?

Answer appears at the end of the chapter.

Nonuniformities in the Early Universe and the Origin of Galaxies

The COBE data show that the universe is full of blackbody radiation with a temperature of 2.725 K. (This is the cosmic microwave background radiation, or CMB, discussed in Section 25-4.) However, more sensitive observations show that there are slight temperature variations spread across the sky (Figure 25-13). What do these temperature variations tell us?

The distribution of matter in the early universe was not perfectly uniform, and these temperature variations reveal the nonuniformities—the clumps and voids—in the mass distribution. It is important to understand that the temperature fluctuations do not arise from the temperature of the matter. Instead, clumps in the distribution of matter produce a gravitational redshift (see Figure 21-7b) on the CMB photons as they leave a denser region. Redshifted CMB photons not only have longer wavelengths and lower energy, but also correspond to lower CMB temperatures. Thus, in Figure 25-13, the cooler bluer spots correspond to denser regions.

Astronomers place great importance on studying temperature and density variations in the cosmic background radiation. The reason is that clumps of matter seen at early times are the "seeds" of structures that continued to "grow" into today's superclusters of galaxies through gravitational contraction. Within these immense concentrations formed the galaxies, stars, and planets. Thus, by studying these nonuniformities, we are really studying the origins of our present-day structure.

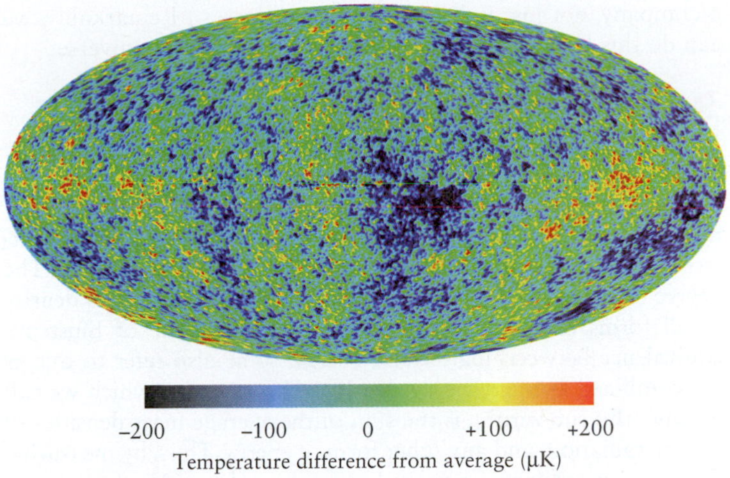

ANIMATION 25-2 ANIMATION 25-3 **FIGURE 25-13** R I V U X G **Temperature Variations in the Cosmic Microwave Background** This map from WMAP data shows small variations in the temperature of the cosmic background radiation across the entire sky. (The variations due to Earth's motions through space, shown in Figure 25-8, have been factored out.) Lower-temperature regions (shown in blue) show where the early universe was slightly denser than average; warmer regions (shown in red) correspond to regions where the density of matter was less dense than average. Note that the temperature variations in this figure are no more than 200 μK, or 2×10^{-4} K; these are tiny fluctuations around the blackbody radiation measured by COBE in Figure 25-7. (NASA/WMAP Science Team)

Temperature variations in the cosmic background radiation do more than show us the origins of large-scale structure in the universe. As we will see in the next two sections, these temperature variations actually reveal the shape of the universe as a whole, and many of its most fundamental properties.

CONCEPTCHECK 25-10

Consider a blue patch in Figure 25-13. If astronomers looked at the same region with a visible-light telescope, what are they likely to see?

Answer appears at the end of the chapter.

25-6 The shape of the universe indicates its matter and energy content

We have seen that by following the mass densities of radiation (ρ_{rad}) and of matter (ρ_m), we can learn about the evolution of the universe. But it is equally important to know the combined mass density of *all* forms of matter and energy. (In an analogous way, an accountant needs to know the overall financial status of a company, not just individual profits or losses.) Remarkably, we can do this by investigating the overall shape of the universe.

The Curvature of the Universe

Einstein's general theory of relativity explains that gravity curves the fabric of space. Furthermore, the equivalence between matter and energy, expressed by Einstein's equation $E = mc^2$, tells that either matter or energy produces gravity. Thus, the matter and energy scattered across space should give the universe an overall curvature. The degree of curvature depends on the **combined average mass density** of all forms of matter and energy. (Again, because of Einstein's equivalence between matter and energy, some also refer to this as the combined average energy density.) This quantity, which we call ρ_0 (say "rho sub zero"), is the sum of the average mass densities of matter, radiation, and any other form of energy. Thus, by measuring the curvature of space, we should be able to determine the value of ρ_0 and, hence, learn about the content of the universe as a whole.

To see what astronomers mean by the curvature of the universe, imagine shining two powerful laser beams out into space so that they are perfectly parallel as they leave Earth. Furthermore, suppose that nothing gets in the way of these two beams, so we can follow them for billions of light-years as they travel across the universe and across the space whose curvature we wish to detect. Figure 25-14 illustrates the only three possibilities:

1. We might find that our two beams of light remain perfectly parallel, even after traversing billions of light-years. In this case, space would not be curved: The universe would have **zero curvature,** and space would be **flat.**

2. Alternatively, we might find that our two beams of light gradually converge. In such a case, space would not be flat. Recall that lines of longitude on Earth's surface are parallel at the equator but intersect at the poles. Thus, in this case the three-dimensional geometry of the universe would be analogous to the two-dimensional geometry of a spherical surface. We would then say that space is **spherical** and that the universe has **positive curvature.** Such a universe is also called **closed,** because if you travel in a straight line in any direction in such a universe, you will eventually return to your starting point.

3. Finally, we might find that the two initially parallel beams of light would gradually diverge, becoming farther and farther apart as they moved across the universe. In this case, the universe would still have to be curved, but in the opposite sense from the spherical model. We would then say that the universe has **negative curvature.** In the same way that a sphere is a positively curved two-dimensional surface, a saddle is a good example of a negatively curved two-dimensional surface. Parallel lines drawn on a sphere always converge, but parallel lines drawn on a saddle always diverge. Mathematicians say that saddle-shaped surfaces are hyperbolic. Thus, in a negatively curved universe, we would describe space as **hyperbolic.** Such a universe is also called **open** because if you were to travel in a straight line in any direction, you would never return to your starting point.

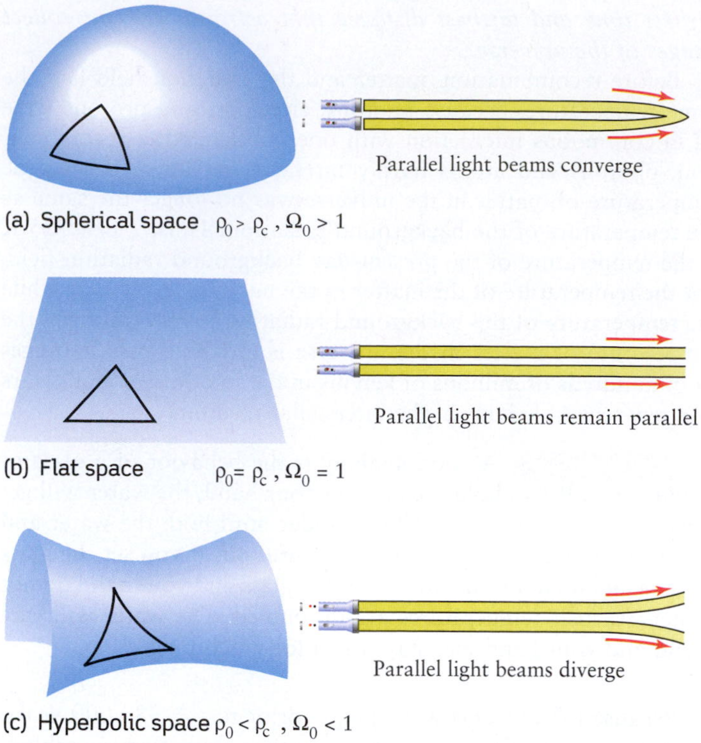

FIGURE 25-14

The Geometry of the Universe The curvature of the universe is either (a) positive, (b) zero, or (c) negative. The curvature depends on whether the combined mass density is greater than, equal to, or less than the critical density, or, equivalently, on whether the density parameter Ω_0 is greater than, equal to, or less than 1. In theory, the curvature could be determined by seeing whether two laser beams initially parallel to each other would converge, remain parallel, or spread apart.

Figure 25-14 summarizes the three cases of positive curvature, zero curvature, and negative curvature. Real space is three-dimensional, but we have drawn the three cases as analogous, more easily visualized two-dimensional surfaces. Therefore, as you examine the drawings in Figure 25-14, remember that the real universe has one more dimension. For example, if the universe is in fact hyperbolic, then the geometry of space must be the (difficult to visualize) three-dimensional analog of the two-dimensional surface of a saddle.

Note that in accordance with the cosmological principle, none of these models of the universe has an "edge" or a "center." This is clearly the case for both the flat and hyperbolic universes, because they are infinite and extend forever in all directions. A spherical universe is finite, but it also lacks a center and an edge. You could walk forever around the surface of a sphere (like the surface of Earth) without finding a center or an edge.

CONCEPTCHECK 25-11

If two lasers are aligned to be perfectly parallel out in a region of space far from any galaxies, could the geometry of space ever cause their light beams to cross?

Answer appears at the end of the chapter.

Density Determines Curvature

The curvature of the universe is determined by the actual value of the combined mass density ρ_0 and a reference density called the **critical density** ρ_c (say "rho sub cee"). If ρ_0 is greater than the critical density ρ_c, the universe has positive curvature and is closed. If ρ_0 is less than ρ_c, the universe has negative curvature and is open. In the special case that ρ_0 is exactly equal to ρ_c, the universe is flat. Clearly, the critical density plays a crucial role in determining the geometry of the universe. It is given by the expression

Critical density of the universe

$$\rho_c = \frac{3H_0^2}{8\pi G}$$

ρ_c = critical density of the universe

H_0 = Hubble constant

G = universal constant of gravitation

Using a Hubble constant $H_0 = 73$ km/s/Mpc, we get

$$\rho_c = 1.0 \times 10^{-26}\ \text{kg/m}^3$$

This critical density amounts to only about 5 hydrogen atoms per cubic meter! Even with the vast empty spaces between the stars, this density is about a million times less than the density of matter from stars and gas in the disk in our Milky Way Galaxy.

Many astronomers prefer to characterize the combined average mass density of the universe in terms of the **density parameter** Ω_0 (say "omega sub zero"). This parameter is just the ratio of the combined average mass density to the critical density:

Density parameter

$$\Omega_0 = \frac{\rho_0}{\rho_c}$$

Ω_0 = density parameter

ρ_0 = combined average mass density

ρ_c = critical density

An open universe (negative curvature) has a density parameter Ω_0 between 0 and 1, and a closed universe (positive curvature) has Ω_0 greater than 1. In a flat universe, Ω_0 is equal to 1. Thus, we can use the value of Ω_0 as a measure of the curvature of the universe (Table 25-1).

CALCULATIONCHECK 25-3

What is the combined average mass density (in kg/m^3) in a flat universe?

Answer appears at the end of the chapter.

Measuring the Cosmic Curvature

How can we determine the curvature of space across the universe? In theory, if you drew an enormous triangle whose sides were each a billion light-years long (see Figure 25-14), you could determine the curvature of space by measuring the three angles of the triangle. If their sum equaled 180°, space would be flat. If the sum was greater than 180°, space would be spherical. And if the sum of the three angles was less than 180°, space would be hyperbolic. Unfortunately, this direct method for measuring the curvature of space is not practical.

A way to determine the curvature of the universe that is both practical and precise is to see if light rays bend toward or away from each other, as shown in Figure 25-14. The greater the distance a pair of light rays has traveled, and hence, the longer the time the light has been in flight, the more pronounced any such bending should be. Therefore, astronomers test for the presence of such bending by examining the oldest radiation in the universe—the cosmic microwave background.

If the cosmic microwave background were truly isotropic, so that equal amounts of radiation reached us from all directions in the sky, it would be impossible to tell whether individual light rays have been bent. However, as we saw in Section 25-5, there are localized "hot spots" in the cosmic microwave background due to density variations in the early universe. The apparent size of these hot spots depends on the curvature of the universe (Figure 25-15). If the universe is closed, the bending of light rays from a hot spot will make the spot appear larger (Figure 25-15a); if the universe is open, the light rays will bend the other way and the hot spots will appear smaller (Figure 25-15c). Only in a flat universe will the light rays travel along straight lines, so that the hot spots appear with their true size (Figure 25-15b).

By calculating what conditions were like in the primordial fireball, astrophysicists find that in a flat universe, the dominant "hot spots" in the cosmic background radiation should have an angular size of about 1°. (In Section 25-9 we will learn how this is deduced.) This is just what observations of the CMB have confirmed (see Figure 25-13). Hence, the curvature of the universe must be very close to zero, and the universe must be either flat or very nearly so.

TABLE 25-1 The Geometry and Average Density of the Universe

Geometry of space	Curvature of space	Type of universe	Combined average mass density (ρ_0)	Density parameter (Ω_0)
Spherical	positive	closed	$\rho_0 > \rho_c$	$\Omega_0 > 1$
Flat	zero	flat	$\rho_0 = \rho_c$	$\Omega_0 = 1$
Hyperbolic	negative	open	$\rho_0 < \rho_c$	$\Omega_0 < 1$

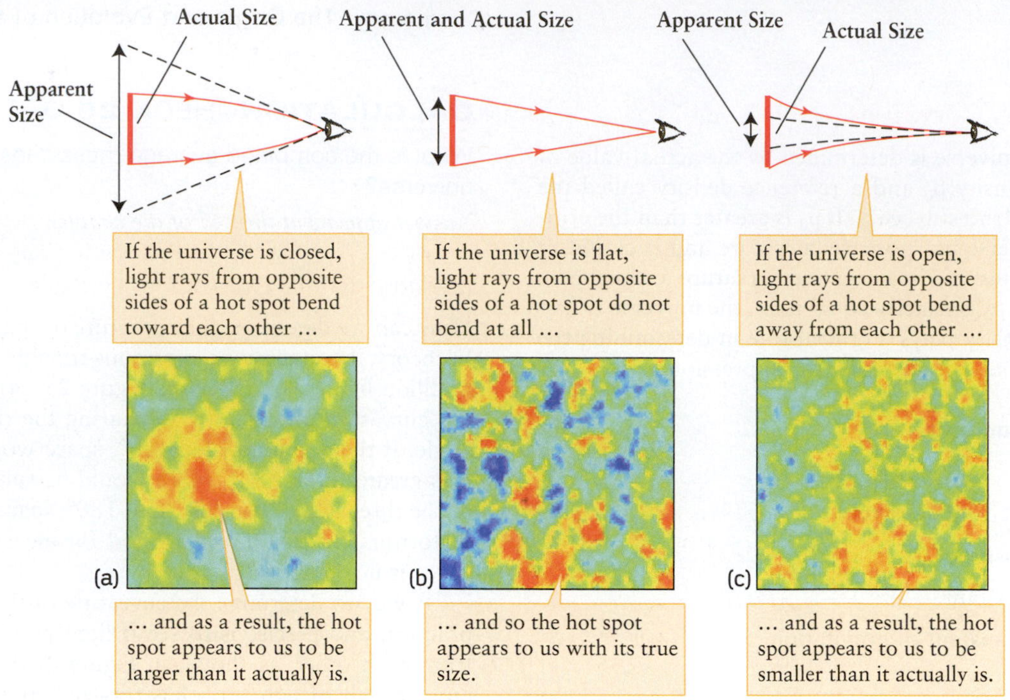

FIGURE 25-15 R I V U X G

The Cosmic Microwave Background and the Curvature of Space Temperature variations in the early universe appear as "hot spots" in the cosmic microwave background. The apparent size of these spots depends on the curvature of space. (The BOOMERANG Group, University of California, Santa Barbara)

As Table 25-1 shows, once we know the curvature of the universe, we can determine the density parameter Ω_0 and hence the combined average mass density ρ_0. By analyzing the data shown in Figure 25-13, astrophysicists find that $\Omega_0 = 1.0$ with an uncertainty of about 2%. In other words, ρ_0 is within 2% of the critical density ρ_c.

The flatness of the universe poses a major dilemma: Even if you include dark matter, there is not enough matter to make the universe flat. We saw in Section 25-5 that the average mass density of matter in the universe, ρ_m, is 2.4×10^{-27} kg/m^3. This density is only 0.24 of the critical density ρ_c. We can express this ratio in terms of the **matter density parameter** Ω_m (say "omega sub em"), equal to the ratio of ρ_m to the critical density:

$$\Omega_m = \frac{\rho_m}{\rho_c} = 0.24$$

If matter and radiation were all there is in the universe, the combined average mass density ρ_0 would be equal to ρ_m (plus a tiny contribution from radiation, which we can neglect because the average mass density of radiation is only about 0.02% that of matter). Then the density parameter Ω_0 would be equal to Ω_m—that is, equal to 0.24—and the universe would be open. But the temperature variations in the cosmic microwave background clearly show that the universe is either flat or very nearly so. These variations also show that the density parameter Ω_0, which includes the effects of *all* kinds of matter and energy, is equal to 1.0. In other words, radiation and matter, including dark matter, together account for only 24% of the total density of the universe! The dilemma is to account for the rest of the density.

CONCEPTCHECK 25-12

In a closed universe, a 1-meter stick is moved from Earth out to a specific location in the distant cosmos. Assuming it could be observed from Earth, would the stick appear larger, smaller, or the same in size than in a flat universe at that same distance?

Answer appears at the end of the chapter.

Dark Energy

> The geometry of space reveals that the universe is filled with dark energy

The source of the missing energy density must be some form of energy that we cannot detect through the various gravitational effects that astronomers use to detect dark matter. The missing energy density also does not appear to emit detectable radiation of any kind, so we cannot directly detect it with light. With these properties in mind, we refer to this mysterious energy as **dark energy.**

Just as we express the average density of matter and radiation by the matter density parameter Ω_m, we can express the average density of dark energy in terms of the **dark energy density parameter** Ω_Λ (say "omega sub lambda"). This parameter is equal to the average mass density of dark energy, ρ_Λ, divided by the critical density ρ_c:

$$\Omega_\Lambda = \frac{\rho_\Lambda}{\rho_c}$$

We can determine the value of Ω_Λ by noting that the combined average mass density ρ_0 must be the sum of the average mass densities of matter, radiation, and dark energy. As we have seen, the contribution of radiation is so small that we can ignore it, so we have

$$\rho_0 = \rho_m + \rho_\Lambda$$

If we divide this through by the critical density ρ_c, we obtain

$$\Omega_0 = \Omega_m + \Omega_\Lambda$$

That is, the density parameter Ω_0 is the sum of the matter density parameter Ω_m and the dark energy density parameter Ω_Λ. Solving for Ω_Λ, we find

$$\Omega_\Lambda = \Omega_0 - \Omega_m$$

Since Ω_0 is close to 1.0 and Ω_m is 0.24, we conclude that Ω_Λ must be $1.0 - 0.24 = 0.76$. Thus, whatever dark energy is, it accounts for 76% (about three-quarters) of the energy content of the universe!

To put the dark energy density into context, consider the kinetic energy of a jumping flea. The flea's kinetic energy is very small (about the same amount of electrical energy used in turning on a single LED light for one ten-millionth of a second). Since an energy density is the amount of energy contained within a certain volume, to put the dark energy density into context, we also need to indicate a corresponding volume. Thus, *the dark energy density is equivalent to a single jumping flea within the volume of a big football stadium* (dome and all). Since most of the universe is devoid of matter, this energy density, filling all of space, ends up being the dominant form of energy in our universe. (The CMB contains many photons, but the expansion of space has significantly lowered their energy and their density.)

The concept of dark energy is actually due to Einstein. When he proposed the existence of a cosmological constant, he was suggesting that the universe is filled with a form of energy that by itself tends to make the universe expand (see Section 25-1). Unlike gravity, which tends to make objects attract, the energy associated with a cosmological constant would provide a form of "antigravity." Hence, it would not be detected in the same way as matter. (The subscript Λ in the symbol for the dark energy density parameter pays homage to the symbol that Einstein chose for the cosmological constant.)

If dark energy is in fact due to a cosmological constant, the value of this constant must be far larger than Einstein suggested. This change in the constant is needed if we are to explain why Ω_Λ has a large value of 0.76. If Einstein felt he erred by introducing the idea of a cosmological constant, his error was giving it too small a value!

These ideas concerning dark energy are extraordinary, and extraordinary claims require extraordinary evidence to confirm them. As we will see in the next section, another way to measure dark energy is to examine how the rate of expansion of the universe has evolved over the eons.

CONCEPTCHECK 25-13

If we do not know what makes up dark energy, how can we estimate its energy density?

Answer appears at the end of the chapter.

25-7 Observations of distant supernovae reveal that we live in an accelerating universe

We have seen that the universe is expanding. But does the rate of expansion stay the same? Because there is matter in the universe, and because gravity tends to pull the bits of matter in the universe toward one another, we would expect that the expansion should slow down with time. (In the same way, a cannonball shot upward from the surface of Earth will slow down as it ascends because of Earth's gravitational pull.) If there is a cosmological constant, however, its associated dark energy will exert an outward pressure that tends to accelerate the expansion. Which of these effects is more important?

Modeling the Expansion History of the Universe

To determine whether the expansion of the universe is slowing down or speeding up, astronomers study the relationship between redshift and distance for extremely remote galaxies. We see these galaxies as they were billions of years ago. If the rate of expansion was the same in the distant past as it is now, the same Hubble law should apply to distant galaxies as to nearby ones. But if the rate of expansion has either increased or decreased, we will find important deviations from the Hubble law.

To see how astronomers approach this problem, first imagine two different parallel universes. Both Universe #1 and Universe #2 are expanding at constant rates, so for both universes there is a direct proportion between recessional velocity v and distance d as expressed by the Hubble law $v = H_0 d$. Hence, a graph of distance versus recessional velocity for either universe is a straight line, as Figure 25-16a shows. The only difference is that Universe #1 is expanding at a slower rate than Universe #2. Hence, a galaxy at a certain distance from Earth in Universe #1 will have a slower recessional velocity than a galaxy at the same distance from Earth in Universe #2. As a result, the graph of distance versus recessional velocity for slowly expanding Universe #1 (shown in blue) has a steeper slope than the graph for rapidly expanding Universe #2 (shown in green). Keep this observation in mind: A slower expansion means a steeper slope on a graph of distance versus recessional velocity.

Now consider *our* universe and allow for the possibility that the expansion rate may change over time. If we observe very remote galaxies, we are seeing them as they were in the remote past. If the expansion of the universe in the remote past was slower or faster than it is now, the slope of the graph of distance versus recessional velocity will be different for those remote galaxies. If the expansion was slower, then the slope will be steeper for distant galaxies (shown in blue in Figure 25-16b); if the expansion was faster, the slope will be shallower for distant galaxies (shown in green in Figure 25-16b). In either case, there will be a deviation from the straight-line Hubble law (shown in red in Figure 25-16b).

Measuring Ancient Expansion with Type Ia Supernovae

Which of the possibilities shown in Figure 25-16b represents the actual history of the expansion of our universe? In Section 23-5 we looked at the observed relationship between distance and

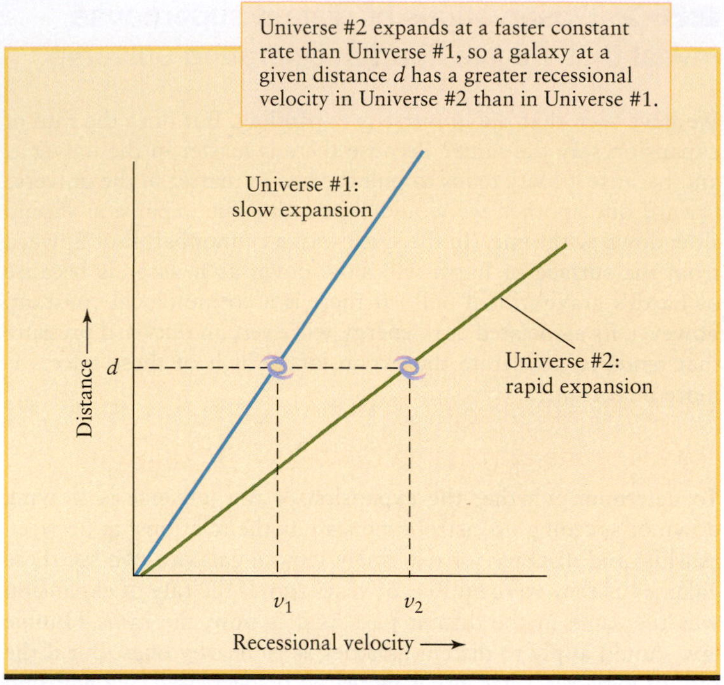

(a) Two universes with different expansion rates

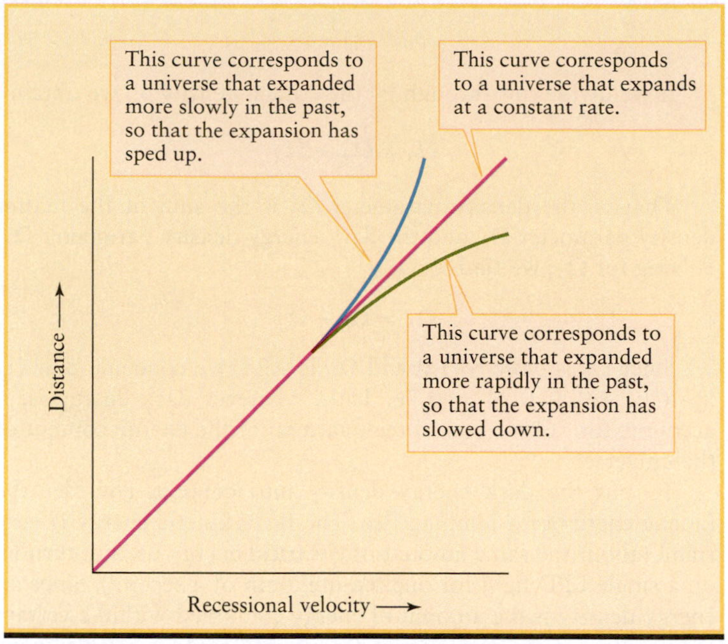

(b) Possible expansion histories of the universe

FIGURE 25-16

Varying Rates of Cosmic Expansion **(a)** Imagine two universes, #1 and #2. Each expands at its own constant rate. For a galaxy at a given distance, the recessional velocity will be greater in the more rapidly expanding universe. Hence, the graph of distance *d* versus recessional velocity *v* will have a shallower slope for the rapidly expanding universe and a steeper slope for the slowly expanding one. **(b)** If the rate of expansion of our universe was more rapid in the distant past, corresponding to remote distances, the graph of *d* versus *v* will have a shallower slope for large distances (green curve). If the expansion rate was slower in the distant past, the graph will have a steeper slope for large distances (blue curve).

recessional velocity for galaxies. Figure 23-16 is a plot of some representative data. The data points appear to lie along a straight line, suggesting that the rate of cosmological expansion has not changed. (Figure 23-16 is actually a graph of recessional velocity versus distance, not the other way around. But a straight line on one kind of graph will be a straight line on the other, because in either case there is a direct proportion between the two quantities being graphed.) However, the graph in Figure 23-16 was based on measurements of galaxies no farther than 400 Mpc (1.3 billion ly) from Earth, which means we are looking only 1.3 billion years into the past. The straightness of the line in Figure 23-16 means only that the expansion of the universe has been relatively constant over the past 1.3 billion years—only 10% of the age of the universe and a relatively brief interval on the cosmic scale.

Now suppose that you were to measure the redshifts and distances of galaxies *several* billion light-years from Earth. The light from these galaxies has taken billions of years to arrive at your telescope, so your measurements will reveal how fast the universe was expanding billions of years ago. To determine the expansion, we need a technique that will allow us to find the distances to these very remote galaxies. We saw in Section 23-4 that one way to determine distances is to identify Type Ia supernovae in such galaxies. These supernovae are among the most luminous objects in the universe, and hence can be detected even at extremely large distances (see Figure 23-14). The maximum brightness of a supernova tells astronomers its distance through the inverse-square law for light, and the redshift of the supernova's spectrum tells them its recessional velocity. As an example, the image that opens this chapter shows Type Ia supernovae with redshifts $z = 1.010$, 1.230, and 1.390, corresponding to recessional velocities of 60%, 67%, and 70% of the speed of light. We see these supernovae as they were 7.7 to 9.0 billion years ago, when the universe was less than half of its present age.

In 1998, two large research groups—the Supernova Cosmology Project, led by Saul Perlmutter of Lawrence Berkeley National Laboratory, and the High-Z Supernova Search Team, led by Brian Schmidt of the Mount Stromlo and Siding Springs Observatories in Australia—reported their results from a survey of Type Ia supernovae in galaxies at redshifts of 0.2 or greater, corresponding to distances beyond 750 Mpc (2.4 billion ly). Figure 25-17 shows some of their data, along with more recent observations, on a graph of apparent magnitude versus redshift. Recall that a greater apparent magnitude corresponds to a dimmer supernova (see Section 17-3), which means that the supernova is more distant. A greater redshift implies a greater recessional velocity. Hence, this graph is basically the same as those in Figure 25-16 (distance versus recessional velocity).

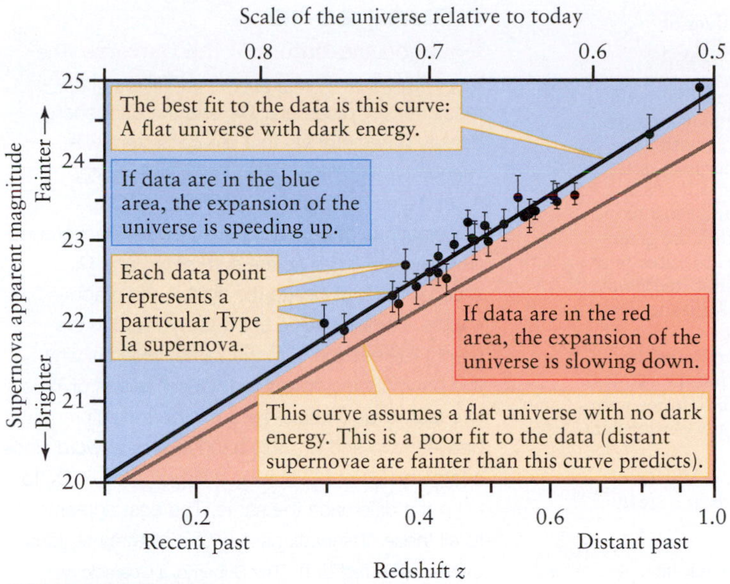

FIGURE 25-17

The Hubble Diagram for Distant Supernovae This graph shows apparent magnitude versus redshift for supernovae in distant galaxies. The greater the apparent magnitude, the dimmer the supernova and the greater the distance to it and its host galaxy. If the expansion of the universe is speeding up, the data will lie in the blue area; if it is slowing down, the data will lie in the red area. The data show that the expansion is in fact speeding up. (The Supernova Cosmology Project/S. Perlmutter)

To interpret these results, we need guidance from **relativistic cosmology.** This field provides a theoretical description of the universe and its expansion, based on Einstein's general theory of relativity, and was developed in the 1920s by Alexander Friedmann in Russia, Georges Lemaître in France, and Willem de Sitter in the Netherlands. Given values of the mass density parameter Ω_m and the dark energy density parameter Ω_Λ, cosmologists can use the equations of relativistic cosmology to predict how the expansion rate of the universe should change over time. Such predictions can be expressed as curves on a graph of distance versus redshift such as Figure 25-17.

The lower, gray curve in Figure 25-17 shows what would be expected in a flat universe with $\Omega_0 = 1.00$ but with no dark energy, so that $\Omega_\Lambda = 0$ and $\Omega_m = \Omega_0$ (that is, the density parameter is due to matter and radiation alone). In this model, and in fact in any model whose curve lies in the red area in Figure 25-17, the absence of dark energy means that gravitational attraction between galaxies would cause the expansion of the universe to slow down with time. Hence, the expansion rate would have been greater in the past (compare with the green curve in Figure 25-16b).

In fact, the data points in Figure 25-17 are almost all in the *blue* region of the graph, and agree very well with the curve shown in black. This curve also assumes a flat universe, but with an amount of dark energy consistent with the results from the cosmic microwave background ($\Omega_m = 0.24$, $\Omega_\Lambda = 0.76$, $\Omega_0 = \Omega_m + \Omega_\Lambda = 1.00$). In this model, and indeed in any model whose curve lies in the blue region of Figure 25-17, dark energy has made the expansion of the universe speed up over time. Hence, the expansion of the universe was slower in the distant past, which means that we live in an *accelerating* universe.

Just like the blue curve in Figure 25-16b, the data in Figure 25-17 show that supernovae of a certain brightness (and hence a given distance) have smaller redshifts (and hence smaller recessional velocities) than would be the case if the expansion rate had always been the same. These data provide compelling evidence of the existence of dark energy.

Roughly speaking, the data in Figure 25-17 indicate the relative importance of dark energy (which tends to make the expansion speed up) and gravitational attraction between galaxies (which tends to make the attraction slow down). Thus, these data tell us about the *difference* between the values of the dark energy density parameter Ω_Λ and the matter density parameter Ω_m. By contrast, measurements of the cosmic microwave background (Section 25-6) give information about Ω_0, equal to the *sum* of Ω_Λ and Ω_m. Observations of galaxy clusters (Section 25-5) set limits on the value of Ω_m by itself (which includes visible and dark matter). By combining these three very different kinds of observations as shown in Figure 25-18, we can set more stringent limits on both Ω_Λ and Ω_m.

Taken together and combined with other observations, all these data suggest the following values.

$$\Omega_m = 0.241 \pm 0.034$$
$$\Omega_\Lambda = 0.759 \pm 0.034$$
$$\Omega_0 = \Omega_m + \Omega_\Lambda = 1.02 \pm 0.02$$

In each case, the number after the ± sign is the uncertainty in the value.

> Dark energy became the dominant form of energy in the universe about the same time that our solar system formed

This collection of numbers points to a radically different model of the universe from what was suspected just a few years ago. In the 1980s there was no compelling evidence for an accelerating expansion of the universe, so it was widely assumed that $\Omega_\Lambda = 0$. Evidence from distant galaxies suggested a flat universe, so it was presumed that $\Omega_m = 1$. Figure 25-18 shows that modern data rule out this model.

The model we are left with is one in which the universe is suffused with a curious dark energy due to a cosmological constant. Unlike matter or radiation, whose average densities decrease as the universe expands and thins out, the average density of this dark energy remains constant throughout the history of the universe (Figure 25-19). The dark energy was relatively unimportant over most of the early history of the universe. Today, however, the density of dark energy is greater than that of matter (Ω_Λ is greater than Ω_m). In other words, we live in a **dark-energy–dominated universe.**

CONCEPTCHECK 25-14

What assumption must one make about Type Ia supernovae in order to use them to measure distances to galaxies?

Answer appears at the end of the chapter.

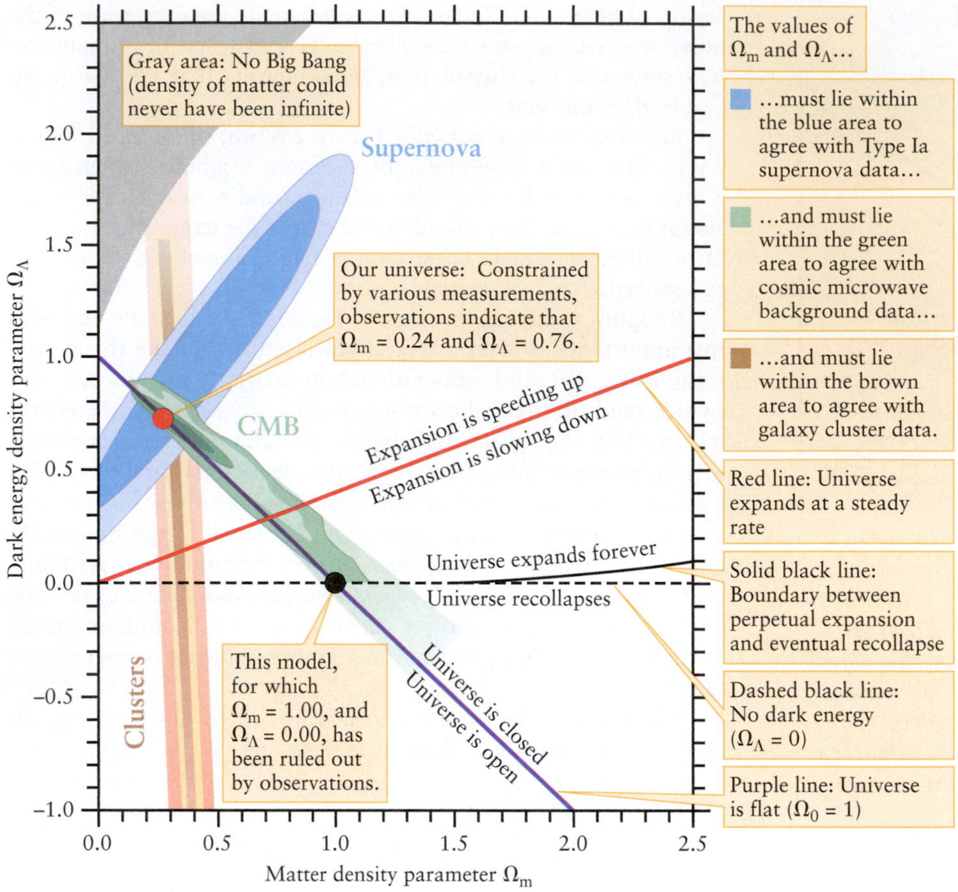

FIGURE 25-18

Limits on the Nature of the Universe The three regions on this graph show values of the mass density parameter Ω_m and the dark energy density parameter Ω_Λ that are consistent with various types of observations. Galaxy cluster measurements (in brown) set limits on Ω_m. Observations of the cosmic microwave background (in green) set limits on the sum of Ω_m and Ω_Λ: A larger value of Ω_m (to the right in the graph) implies a smaller value of Ω_Λ (downward in the graph) to keep the sum the same, which is why this band slopes downward. Observations of Type Ia supernovae (in blue) set limits on the difference between Ω_m and Ω_Λ; this band slopes upward since a larger value of Ω_m implies a larger value of Ω_Λ to keep the difference the same. The best agreement to all these observations is where all three regions overlap (the red dot). (The Supernova Cosmology Project/S. Perlmutter)

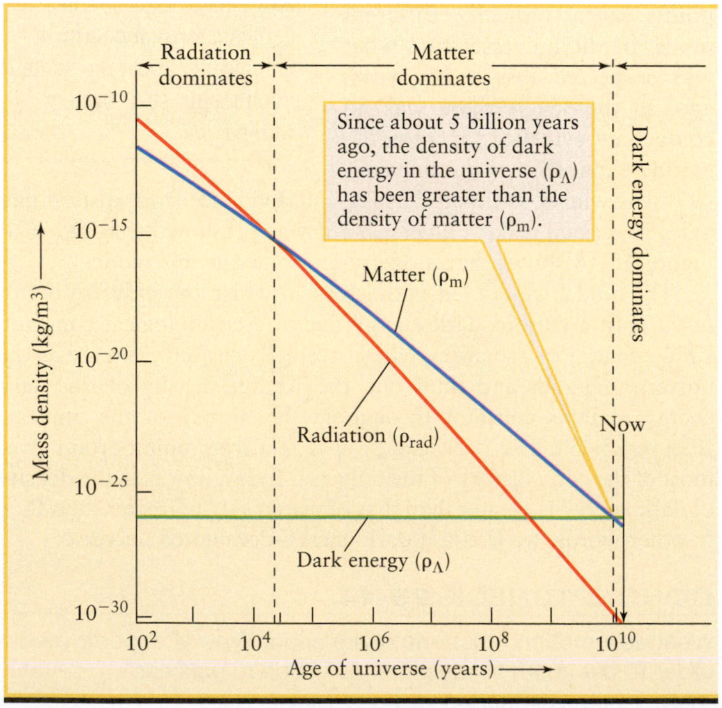

Cosmic Expansion: From Slowing Down to Speeding Up

ANIMATION 25-4 ANIMATION 25-5 As Figure 25-19 shows, the dominance of dark energy is a relatively recent development in the history of the universe. Prior to about 5 billion years ago, the density of matter should have been greater than that of dark energy. Hence, we would expect that up until about 5 billion years ago, the expansion of the universe should have been slowing down rather than speeding up. Recently, astronomers have found evidence of this picture by using the Hubble Space Telescope to observe extremely distant Type Ia supernovae with redshifts z greater than 1. (The image that opens this chapter shows three of these supernovae.) When astronomers compare the distance to these supernovae

FIGURE 25-19

The Evolution of Density, Revisited The average mass density of matter, ρ_m, and the average mass density of radiation, ρ_{rad}, both decrease as the universe expands and becomes more tenuous. But if the dark energy is due to a cosmological constant, its average mass density ρ_Λ remains constant. In this model, our universe became dominated by dark energy about 5 billion years ago.

(determined from their brightness) to their redshifts, they find that the redshifts are *greater* than would be the case if the expansion of the universe had always been at the same rate or had always been speeding up (see Figure 25-16). This is just what would be expected if the expansion was slowing down in the very early universe. After about 5 billion years ago, the effects of dark energy became dominant and the expansion began to speed up.

ANALOGY If you see a red light up ahead while driving, you would probably apply the brakes to make the car slow down. But if the light then turns green before your car comes to a stop, and the road ahead is clear, you would step on the gas to make the car speed up again. The expansion of the universe has had a similar history. The mutual gravitational attraction of all the matter in the universe means that "the brakes were on" for about the first 9 billion years after the Big Bang, so that the expansion slowed down. But for about the past 5 billion years, dark energy has "had its foot on the gas," and the expansion has been speeding up.

So far, observations of distant supernovae support dark energy that is described by a cosmological constant. If dark energy truly is a cosmological constant, the density of dark energy will continue to remain constant, as shown in Figure 25-19. Due to the effects of this dark energy, the universe will keep on expanding forever, and the rate of expansion will continue to accelerate. Eventually, some 30 billion years from now, the universe will have expanded so much that only a thousand or so of the nearest galaxies will still be visible. The billions of other galaxies that we can observe today will have moved so far away from us that their light will have faded to invisibility. Furthermore, they will be moving away from us so rapidly that what light we do receive from them will have been redshifted out of the visible range.

There may be other explanations for dark energy besides a cosmological constant, however. Several physicists have proposed a type of dark energy whose density decreases slowly as the universe expands. Depending on how the density of dark energy evolves over time, the universe could continue to expand or could eventually recollapse on itself. Future observations, including space-based measurements of both the cosmic background radiation and of Type Ia supernovae, should help resolve the nature of the mysterious dark energy.

What Is Dark Energy?

While the evidence for an accelerating universe is strong enough that its discoverers were awarded the Nobel Prize in Physics in 2011, the nature of the dark energy implied by this acceleration is far from clear. The physical interpretation of the cosmological constant is that there is an energy associated with space itself. With a cosmological constant, the more space there is, the more energy there is. In other words, as the universe expands, the additional space leads to more energy. Just as the energy content in the universe determines its shape (Section 25-6), the energy content can also cause the universe to accelerate its expansion, and this is what dark energy is doing. But, how does energy arise from seemingly empty space?

The answer might come from **virtual particle pairs**—imperceptible particles made of matter and antimatter. As mentioned in Section 21-9, quantum physics reveals that seemingly empty space is not so empty after all. Everywhere, at all times, virtual particle pairs pop up from nothing and then disappear again on very short timescales. While the virtual particles themselves are not directly observable, the phenomenon is real and secondary effects of the virtual particles have been observed. In fact, virtual particle effects between microscopic nanotechnology devices are strong enough that this phenomenon can produce practical problems—forces that cause some devices to clump together. In Chapter 26, we will also discuss virtual particles for their possible role in the early universe.

Because virtual particles appear in the otherwise empty "vacuum" of space, they are called vacuum fluctuations. Overall, the effect of virtual particles, or vacuum fluctuations, is to produce an energy associated with space itself. This seems to be just what is needed to produce a cosmological constant. However, there is a big problem with this explanation for dark energy. When the effects of virtual particles are calculated, they predict a cosmological constant that is 100 orders of magnitude larger than what is actually observed; that number has a one with a *hundred zeroes* after it! With such a large numerical discrepancy between theory and observation, our current understanding of virtual particles and vacuum fluctuations cannot explain dark energy. Perhaps a modification of the vacuum fluctuation model can explain dark energy, but each theoretical modification also introduces new problems.

There are other theories offered to explain dark energy. Some propose that Einstein's general theory of relativity needs modifying on very large scales. Others propose that we are in a special part of the universe that only gives the appearance of accelerated expansion, so that there is actually no dark energy at all. However, each theory has its own difficulties and no explanation has gained wide acceptance. While the observations march forward, they are difficult, and dark energy is likely to remain one of the universe's greatest mysteries for some time to come.

25-8 Primordial sound waves help reveal the character of the universe

We have seen how studying the "hot spots" in the cosmic background radiation reveals that we live in a flat universe. In fact, temperature variations reveal more: They give us a window on conditions in the early universe, and actually help us pin down the values of other important quantities such as the Hubble constant and the density of matter in the universe. The key to extracting this additional information from the cosmic background radiation is recognizing that the hot and cold spots in a map such as Figure 25-13 actually result from sound waves.

Sound Waves in the Early Universe

Sound waves can travel in gases and fluids of all kinds. Sound waves in air are used in human speech and hearing, while whales communicate using high-frequency underwater clicks and whistles. If you could take a snapshot of a sound wave, you would see that at any moment there are some regions, called **compressions,** where the gas or fluid is squeezed together, and other regions, called **rarefactions,** where the gas or fluid is thinned out or rarefied (Figure 25-20).

Immense sound waves in the early universe left their imprint as variations in the cosmic microwave background

There would also have been sound waves in the early universe before recombination. During the first 380,000 years after the Big Bang, the universe was filled with a fluidlike medium composed primarily of photons, electrons, and protons, with a density more than 10^9 times greater than that of our present-day universe. Just as water molecules in a glass of water collide with each other, photons and particles collided frequently with each other in this primordial fluid, triggering random sound waves with compressions and rarefactions.

Before discussing the sound waves in more depth, it is worth revisiting the relationship discussed in Section 25-5 between temperature and density variations. Because there was more mass in a compression than in a rarefaction, photons emerging from a compression experienced a greater gravitational redshift than did photons emerging from a rarefaction (see Section 21-2). As a result, the light from a compression is shifted to slightly longer wavelengths. We saw in Sections 5-3 and 5-4 that a blackbody spectrum dominated by longer wavelengths corresponds to a lower blackbody temperature (see Figure 5-11). Hence, we see compressions as the blue cold spots in Figure 25-13, and we see rarefactions as the red hot spots. The overall pattern of cold and hot spots is thus a record of the density variations due to sound waves that were present just as the universe became transparent, some 380,000 years after the Big Bang. From the denser compressions arose our present-day population of galaxies.

The nature of a sound wave depends on the material through which it passes. For example, sound waves travel faster in helium than they do in air (because helium is less dense) and faster still in water (which, while denser than air, is much more resistant to compression). So, by studying the primordial sound waves recorded in Figure 25-13, we can learn about the properties of the fluid that made up the early universe. These properties include the average densities of matter and dark energy in the fluid, as well as the value of the Hubble constant (which helps determine how rapidly the fluid was expanding and thinning out as the universe expanded). We can also determine the age of the universe at the time that the cosmic background radiation was emitted, since this determines the maximum size to which a hot spot (rarefaction) or cold spot (compression) could have grown since the Big Bang.

Figure 25-21 shows an important way in which astronomers systematize their data about hot and cold spots in the cosmic background radiation. This graph shows the number of observed hot or cold spots of different angular sizes, with larger spots on the left and smaller spots on the right. The presence of peaks in the graph shows that spots of certain sizes are more common than others. The largest peak tells us that the predominant angular size is about 1°, which corresponds to a region of compression or rarefaction that was about a million (10^6) light-years across at the time of recombination at $z = 1100$. (By contrast, the compressions and rarefactions in the sound waves most used in speech are a few meters across.) Since then the universe has expanded by a factor of about 1100, so that same region is now about a billion (10^9) light-years across.

The peaks in Figure 25-21 tell us the size scales that matter formed into clumps, and these over-dense regions were the seeds of structure formation. As predicted, we see the imprint of these early sound waves in today's distribution of mass. Figure 25-22 illustrates this evolution from variations in the CMB to the distribution of galaxies.

Different cosmological models predict different shapes for the curve shown in Figure 25-21. Astronomers determine the best model by seeing which one gives a curve that best fits the data points. For example, the peak of the curve at an angular size of 1° is just what would be expected for a flat universe with $\Omega_0 = 1$. It should be emphasized that Figure 25-21 represents an enormous success for Big Bang cosmology in making detailed predictions, matched by observations, for features covering the entire sky—some only fractions of a degree wide and others covering tens of degrees. By fitting detailed models to the CMB data, the temperature fluctuations reveal the Hubble constant, the combined (total) energy density parameter, the ordinary matter density, and the dark energy density. Table 25-2 summarizes the results of a flat-universe model that yields the particular curve shown in Figure 25-21.

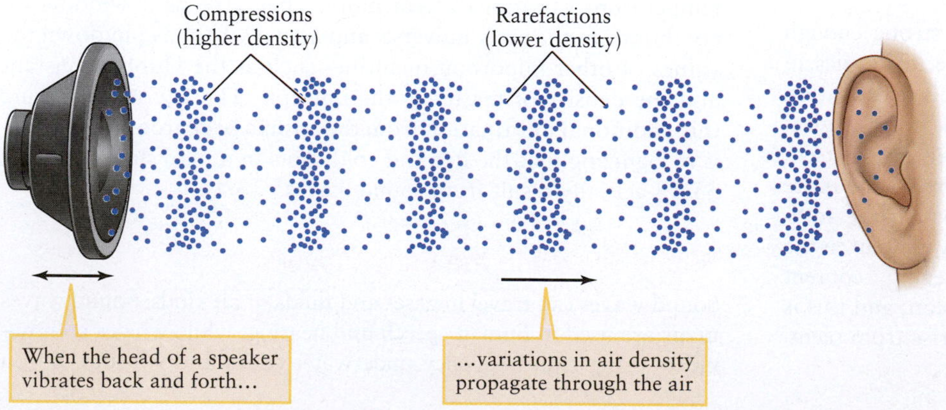

Compressions (higher density)
Rarefactions (lower density)
When the head of a speaker vibrates back and forth...
...variations in air density propagate through the air

FIGURE 25-20
Snapshot of a Sound Wave When the head of a speaker oscillates back and forth, higher-density compressions are created that propagate through the air. The sound from clapping your hands is produced in a similar way, but for a shorter length of time. When a sound wave enters a human ear, the air next to the eardrum is alternately compressed and rarefied, which makes the eardrum flex back and forth. This flexing is translated into an electrical signal that is sent to the brain. Sound waves produced shortly after the Big Bang (inaudible to the human ear) also produced variations in the density of matter throughout the universe.

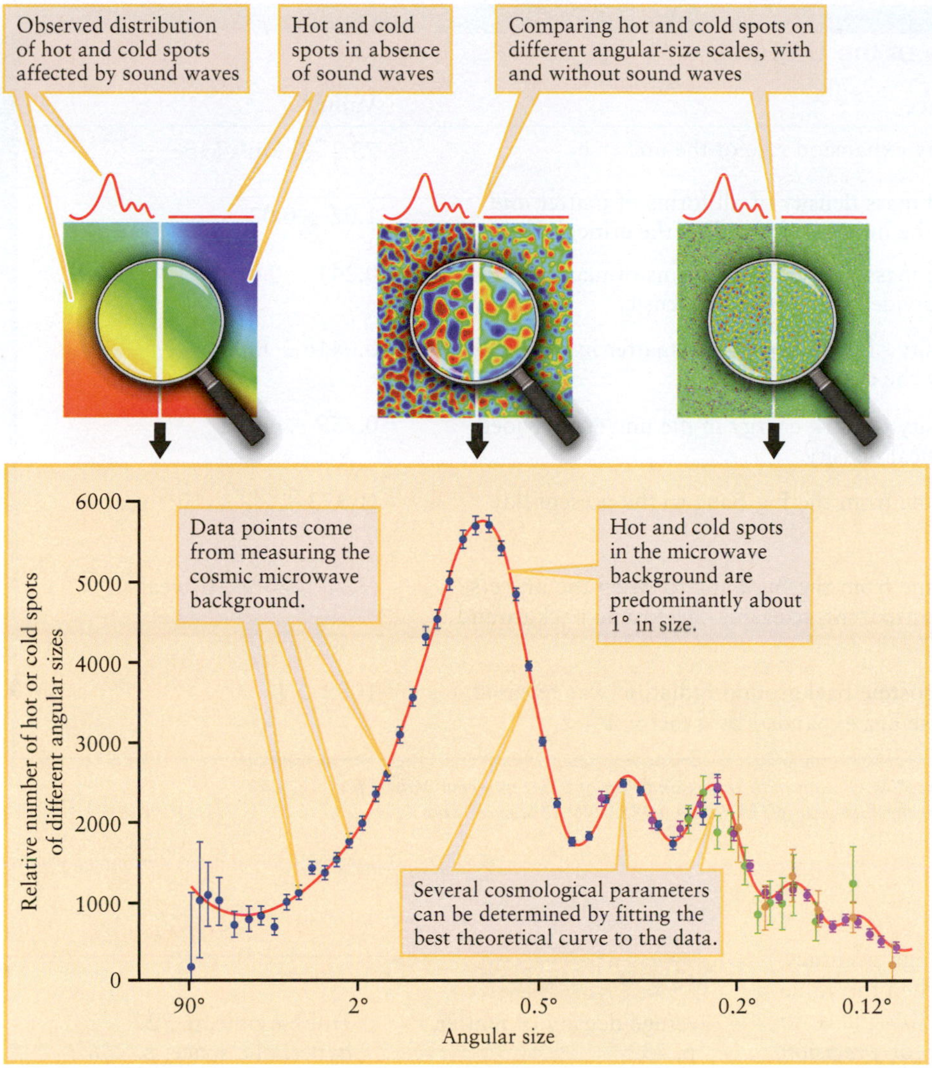
Observed distribution of hot and cold spots affected by sound waves
Hot and cold spots in absence of sound waves
Comparing hot and cold spots on different angular-size scales, with and without sound waves
Data points come from measuring the cosmic microwave background.
Hot and cold spots in the microwave background are predominantly about 1° in size.
Several cosmological parameters can be determined by fitting the best theoretical curve to the data.
Relative number of hot or cold spots of different angular sizes
6000
5000
4000
3000
2000
1000
0
90°
2°
0.5°
0.2°
0.12°
Angular size

FIGURE 25-21
Sound Waves in the Early Universe Top panel: At three different angular sizes (90°, 1°, and 0.25°) a comparison is made between the observed temperature variations in the cosmic microwave background resulting from sound waves and the temperature variations in a hypothetical universe with no sound waves. Sound waves affect the distribution of temperature fluctuations, but do not entirely create them. Bottom panel: Observations of the cosmic background radiation show that hot and cold spots of certain angular sizes are more common than others. Matching a model that describes these observations (the red curve) helps to determine the values of important cosmological parameters. Most of the data shown here are from WMAP; the data for the smallest angles (at the right of the graph) come from the CBI detector in the Chilean Andes and the ACBAR and BOOMERANG detectors at the south pole. (top panel: NASA/WMAP Science Team; graph: Adapted from "The Hubble Constant," Wendy L. Freedman and Barry F. Madore, Carnegie Observatories)

Our understanding of the universe as a whole has increased tremendously over the past several years. We have found compelling evidence that dark energy exists and that it is the dominant form of energy in the universe. Studies of supernovae, galaxy clusters, and the cosmic background radiation have provided us with so much high-quality data that we can now express the key

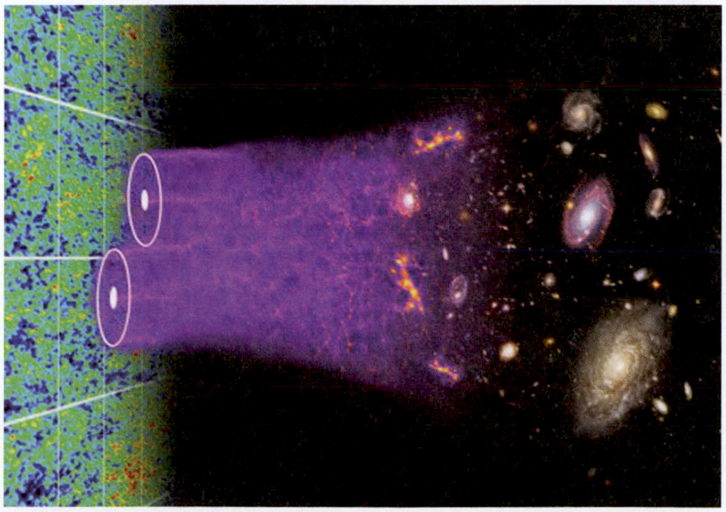

FIGURE 25-22
Structure Evolution from Sound Waves This illustration shows the evolution of early density variations seen in the CMB to the present-day distribution of galaxies. The size scales of the resulting variations are much larger than galaxies or clusters, and the variations show up as filaments and voids traced out by the distribution of superclusters of galaxies. The Baryon Oscillation Spectroscopic Survey (BOSS) has observed these variations at their predicted size scale of around 150 Mpc. (Illustration courtesy of Chris Blake and Sam Moorfield)

TABLE 25-2 Some Key Properties of the Universe

Quantity	Significance	Value*
Hubble constant, H_0	Present-day expansion rate of the universe	$73.2^{+3.1}_{-3.2}$ km/s/Mpc
Density parameter, Ω_0	Combined mass density of all forms of matter *and* energy in the universe divided by the critical density	1.02 ± 0.02
Matter density parameter, Ω_m	Combined mass density of all forms of matter in the universe, divided by the critical density	0.241 ± 0.034
Density parameter for ordinary matter, Ω_b	Mass density of ordinary atomic matter in the universe divided by the critical density	0.0416 ± 0.001
Dark energy density parameter, Ω_Λ	Mass density of dark energy in the universe divided by the critical density	0.759 ± 0.034
Age of the universe, T_0	Elapsed time from the Big Bang to the present day	$(1.373^{+0.016}_{-0.015}) \times 10^{10}$ years
Age of the universe at the time of recombination	Elapsed time from the Big Bang to when the universe became transparent, releasing the cosmic background radiation	$(3.79^{+0.08}_{-0.07}) \times 10^5$ years
Redshift z at the time of recombination	Since the cosmic background radiation was released, the universe has expanded by a factor $1 + z$	1089 ± 1

**Values for H_0, Ω_m, Ω_b, and T_0 are based on the three-year WMAP data with the assumption of a flat universe. Values for the time and redshift of recombination and for Ω_0 are from the first-year WMAP data.* (NASA/WMAP Science Team)

parameters of the universe (Table 25-2) with very high accuracy. When we look back to the situation in the 1980s, when the value of the Hubble constant was uncertain by at least 50%, it is no exaggeration to say that we have entered an age of *precision* cosmology.

Yet many questions remain unanswered. What is the nature of dark matter? What actually is dark energy? Together, the unknown nature of dark matter and dark energy indicate that we do not understand about 96% of the energy content in our universe. Can either of these mysterious entities be detected and studied in the laboratory? These and other questions will continue to occupy cosmologists for many years to come.

CONCEPTCHECK 25-15

Consider Figure 25-21. The colorful pictures along the top compare how temperature fluctuations appear in the presence of sound waves with how they would appear in the absence of sound waves. On what size scale are the temperature fluctuations most affected by the presence of sound waves in the early universe?

CALCULATIONCHECK 25-4

Consult Table 25-2. What percent of the energy content of the universe comes from ordinary matter (such as the matter making planets and stars)? What percent comes from dark energy?

Answers appear at the end of the chapter.

KEY WORDS

average density of matter, p. 732
Big Bang, p. 727
closed universe, p. 736
combined average mass density, p. 736
compression, p. 744
cosmic background radiation, p. 730
cosmic light horizon, p. 729
cosmic microwave background (CMB), p. 730
cosmological constant, p. 722
cosmological principle, p. 727
cosmological redshift, p. 726
cosmology, p. 722
critical density, p. 737
dark energy, p. 738
dark energy density parameter, p. 738
dark-energy–dominated universe, p. 741
density parameter, p. 737
era of recombination, p. 734
flat space, p. 736
homogeneous, p. 727
Hubble time, p. 727
hyperbolic space, p. 736
isotropic, p. 727
lookback time, p. 726
mass density of radiation, p. 732
matter density parameter, p. 738
matter-dominated universe, p. 733
negative curvature, p. 736
observable universe, p. 729
Olbers's paradox, p. 722
open universe, p. 736
Planck time, p. 729
plasma, p. 734
positive curvature, p. 736
primordial fireball, p. 734
radiation-dominated universe, p. 733
rarefaction, p. 744
relativistic cosmology, p. 741
spherical space, p. 736
virtual particle pairs, p. 743
zero curvature, p. 736

KEY IDEAS

The Expansion of the Universe: The Hubble law describes the continuing expansion of space. On large scales, galaxies spread out in this cosmic flow, also called the Hubble flow.

- In the Hubble flow, galaxies are not actually moving through space, but the space between them is increasing as space expands.
- The redshifts that we see from distant galaxies are caused by this cosmic expansion and are called cosmological redshifts. These redshifts develop as photons travel through expanding space, getting stretched along their journey.
- The redshift of a distant galaxy is a measure of the relative size and age of the universe at the time the galaxy emitted its light.
- There is no center or edge of the known universe.

The Cosmological Principle: Cosmological theories are based on the idea that on large scales, the universe looks roughly the same at all locations and in every direction.

The Big Bang: The universe began with nearly infinite density and began its expansion in the event called the Big Bang, which can be described as the beginning of time and space.

- The observable universe extends about 14 billion light-years in every direction from Earth. We cannot see objects beyond this distance because light from these objects has not had enough time to reach us.
- During the first 10^{-43} second after the Big Bang, the universe was too dense to be described by the known laws of physics.

Cosmic Background Radiation and the Evolution of the Universe: The cosmic microwave background radiation, corresponding to radiation from a blackbody at a temperature of nearly 3 K, is the greatly redshifted remnant of the hot universe as it existed about 380,000 years after the Big Bang.

- The background radiation was hotter and more intense in the past. During the first 380,000 years of the universe, radiation and matter formed an opaque plasma called the primordial fireball.
- Light from this early era gets scattered before it can carry images to us, but then there is an abrupt change. When the temperature of the radiation fell below 3000 K, protons and electrons could combine to form hydrogen atoms and the universe became transparent. We can look all the way back to this transition, which is what we see in images of the cosmic microwave background radiation.
- The universal abundance of helium is much more than stars can produce. Most helium was produced by thermonuclear reactions occurring throughout the universe during its first few minutes. It was the high temperatures required for these thermonuclear reactions that led to the prediction of the cosmic microwave background radiation, which was much hotter in the early universe.

The Geometry of the Universe: The curvature of the universe as a whole depends on how the combined average mass density ρ_0 compares to a critical density ρ_c. Due to the equivalence of mass and energy, ρ_0 also includes contributions from dark energy.

- If ρ_0 is greater than ρ_c, the density parameter Ω_0 has a value greater than 1, the universe is closed, and space is spherical (with positive curvature).
- If ρ_0 is less than ρ_c, the density parameter Ω_0 has a value less than 1, the universe is open, and space is hyperbolic (with negative curvature).
- If ρ_0 is equal to ρ_c, the density parameter Ω_0 is equal to 1 and space is flat (with zero curvature).

Cosmological Parameters and Dark Energy: Observations of temperature variations in the cosmic microwave background indicate that the universe is flat or nearly so, with a combined average mass density equal to the critical density. Observations of galaxy clusters suggest that the average density of matter in the universe is about 0.24 of the critical density. The remaining contribution to the average density is called dark energy. Therefore, about 0.76, or 76%, of the total energy content of the universe consists of some unknown entity.

- Measurements of Type Ia supernovae in distant galaxies show that the expansion of the universe is speeding up. This may be due to the presence of dark energy in the form of a cosmological constant, which provides a pressure that pushes the universe outward.

Cosmological Parameters and Primordial Sound Waves: Temperature variations in the cosmic background radiation are a record of sound waves in the early universe. Studying the character of these sound waves helps to determine that the universe is flat and other fundamental properties of the universe.

QUESTIONS

Review Questions

1. Why did Isaac Newton conclude that the universe was static? Was he correct?
2. What is Olbers's paradox? How can it be resolved?
3. What is a cosmological constant? Why did Einstein introduce it into cosmology?
4. TUTORIAL 25-1 What does it mean when astronomers say that we live in an expanding universe? What is actually expanding?
5. Describe how the expansion of the universe explains Hubble's law.
6. Would it be correct to say that due to the expansion of the universe, Earth is larger today than it was 4.56 billion years ago? Why or why not?
7. Using a diagram, explain why the expansion of the universe as seen from a distant galaxy would look the same as seen from our Galaxy.
8. How does modern cosmology preclude the possibility of either a center or an edge to the known universe?
9. Explain the difference between a Doppler shift and a cosmological redshift.
10. Explain how redshift can be used as a measure of lookback time. In what ways is it superior to time measured in years?
11. By what factor has the universe expanded since $z = 1$? Explain your reasoning.

12. What does it mean to say that the universe is homogeneous? That it is isotropic?
13. What is the cosmological principle? How is it justified?
14. How was the Big Bang different from an ordinary explosion? Where in the universe did it occur?
15. Some people refer to the Hubble constant as "the Hubble variable." In what sense is this justified?
16. What is meant by "the observable universe"?
17. (**a**) Explain why the radius of the observable universe is continually increasing. (**b**) Although the universe is 13.7 billion years old, the observable universe includes objects that are more than 13.7 billion light-years away from Earth. Explain why.
18. Imagine an astronomer living in a galaxy a billion light-years away. Is the observable universe for that astronomer the same as for an astronomer on Earth? Why or why not?
19. TUTORIAL 25-2 How did the abundance of helium in the universe suggest the existence of the cosmic background radiation?
20. Can you see the cosmic background radiation with the naked eye? With a visible-light telescope? Explain why or why not.
21. If the universe continues to expand forever, what will eventually become of the cosmic background radiation?
22. How can astronomers measure the average mass density of the universe?
23. What does it mean to say that the universe was once radiation-dominated? What happened when the universe changed from being radiation-dominated to being matter-dominated? When did this happen?
24. What was the era of recombination? What significant events occurred in the universe during this era? Was the universe matter-dominated or radiation-dominated during this era?
25. (**a**) Was there ever an era when the universe was radiation-dominated and matter and radiation were at the same temperature? If so, approximately when was this, and were there atoms during that era? If not, explain why not. (**b**) Was there ever an era when the universe was radiation-dominated and matter and radiation were *not* at the same temperature? If so, approximately when was this, and were there atoms during that era? If not, explain why not.
26. Describe two different ways in which the cosmic microwave background is not isotropic.
27. What is meant by the critical density of the universe? Why is this quantity important to cosmologists?
28. Describe how astronomers use the cosmic background radiation to determine the geometry of the universe.
29. Explain why it is important to measure how the expansion rate of the universe has changed over time. How is this rate measured?
30. What is dark energy? Describe two ways that we can infer its presence.
31. What does it mean to say that the universe is dark energy–dominated? What happened when the universe changed from being matter-dominated to being dark-energy–dominated?
32. How can we detect the presence of sound waves in the early universe? What do these sound waves tell us?

Advanced Questions

Problem-solving tips and tools

We discussed Wien's law in Section 5-4. You may find it useful to recall that 1 parsec equals 3.26 light-years, that 1 Mpc equals 3.09×10^{19} km, and that a year contains 3.16×10^7 seconds.

33. (**a**) For what value of the redshift z were representative distances between galaxies only 20% as large as they are now? (**b**) Compared to representative distances between galaxies in the present-day universe, how large were such distances at $z = 8$? Compared to the density of matter in the present-day universe, what was the density of matter at $z = 8$? (**c**) If dark energy is in the form of a cosmological constant, how does its present-day density compare to the density of dark energy at $z = 2$? At $z = 5$? Explain your answers.
34. The host galaxy of the supernova HST04Sas (see the image that opens this chapter) has a redshift $z = 1.390$. The light from this galaxy includes the Lyman-alpha (L_α) spectral line of hydrogen, with an unshifted wavelength of 121.6 nm. Calculate the wavelength at which we detect the Lyman-alpha photons from this galaxy. In what part of the electromagnetic spectrum does this wavelength lie?
35. Estimate the age of the universe for a Hubble constant of (**a**) 50 km/s/Mpc, (**b**) 75 km/s/Mpc, and (**c**) 100 km/s/Mpc. On the basis of your answers, explain how the ages of globular clusters could be used to place a limit on the maximum value of the Hubble constant.
36. Some people claim that the universe came into being about 6000 years ago. Find the value of the Hubble constant for such a universe. Is this a reasonable value for H_0? Explain your answer.
37. The quasar HS 194617658 has a redshift $z = 3.02$. At the time when the light we see from HS 194617658 left the quasar, how many times more dense was the matter in the universe than it is today?
38. Use Wien's law (Section 5-4) to calculate the wavelength at which the cosmic microwave background ($T = 2.725$ K) is most intense.
39. If the mass density of radiation in the universe were 625 times larger than it is now, what would the background temperature be?

40. Suppose that the present-day temperature of the cosmic background radiation were somehow increased by a factor of 100, from 2.725 K to 272.5 K. (a) Calculate ρ_{rad} in this situation. (b) If the average density of matter (ρ_m) remained unchanged, would it be more accurate to describe our universe as matter-dominated or radiation-dominated? Explain your answer.

41. Calculate the mass density of radiation (ρ_{rad}) in each of the following situations, and explain whether each situation is matter-dominated or radiation-dominated: (a) the photosphere of the Sun ($T = 5800$ K, $\rho_m = 3 \times 10^{-4}$ kg/m^3); (b) the center of the Sun ($T = 1.55 \times 10^7$ K, $\rho_m = 1.6 \times 10^5$ kg/m^3); (c) the solar corona ($T = 2 \times 10^6$ K, $\rho_m = 5 \times 10^{-13}$ kg/m^3).

42. If a photon from the cosmic microwave background had wavelength λ_0 when it was emitted at redshift z, its wavelength today is $\lambda = \lambda_0/(1 + z)$. (a) Let T be the symbol for the temperature of the cosmic microwave background today. Explain why the radiation temperature was $T_0 = T(1 + z)$ at redshift z. (b) What was the radiation temperature at $z = 1$? (c) At what redshift was the radiation temperature equal to 293 K (a typical room temperature)?

43. What would be the critical density of matter in the universe (ρ_c) if the value of the Hubble constant were (a) 50 km/s/Mpc? (b) 100 km/s/Mpc?

44. Consider the quasar HS 194617658 (see Advanced Question 37), which has $z = 3.02$. (a) Suppose that in the present-day universe, two clusters of galaxies are 500 Mpc apart. At the time that the light was emitted from HS 194617658 to produce an image on Earth tonight, how far apart were those two clusters? (b) What was the average density of matter (ρ_m) at that time? Assume that in today's universe, $\rho_m = 2.4 \times 10^{-27}$ kg/m^3. (c) What were the temperatures of the cosmic background radiation and the mass density of radiation (ρ_{rad}) at that time? (d) At this time in the remote past, was the universe matter-dominated, radiation-dominated, or dark-energy–dominated? Explain your answer.

45. Whether the expansion of the universe is speeding up or slowing down, it can be expressed in terms of a quantity called the *deceleration parameter,* denoted by q_0. The expansion is slowing down if q_0 is positive and speeding up if q_0 is negative; if $q_0 = 0$, the expansion proceeds at a constant rate. If we assume that the dark energy is due to a cosmological constant, the deceleration parameter can be calculated using the formula

$$q_0 = \frac{1}{2}\Omega_0 - \frac{3}{2}\Omega_\Lambda$$

(Recall that the density parameter Ω_0 is equal to $\Omega_m + \Omega_\Lambda$.) (a) Show that if there is no cosmological constant, the expansion of the universe must slow down. (b) Using the values of Ω_m and Ω_Λ given in Table 25-2, find the value of the deceleration parameter for our present-day universe. Based on this, is the expansion of the universe speeding up or slowing down? (c) Imagine a universe that has the same value of Ω_m as our universe but in which the expansion of the universe is neither speeding up nor slowing down. What would be the value of Ω_Λ in such a universe? Which would be dominant in such a universe, matter or dark energy? Explain your answer.

46. In general, the deceleration parameter (see Advanced Question 45) is not constant but varies with time. For a flat universe, the deceleration parameter at a redshift z is given by the formula

$$q_z = \frac{1}{2} - \frac{3}{2}\left[\frac{\Omega_\Lambda}{\Omega_\Lambda + (1 - \Omega_\Lambda)(1 + z)^3}\right]$$

where Ω_Λ is the dark energy density parameter. Using the value of Ω_Λ given in Table 25-2, find the value of q_z for (a) $z = 0.5$ and (b) $z = 1.0$. (c) Explain how your results show that the expansion of the universe was actually decelerating at $z = 1.0$, but changed from deceleration to acceleration between $z = 1.0$ and $z = 0.5$.

47. The dark energy density parameter Ω_Λ is related to the value of the cosmological constant Λ by the formula

$$\Omega_\Lambda = \frac{\Lambda c^2}{3H_0^2}$$

where $c = 3.00 \times 10^8$ m/s is the speed of light. Determine the value of Λ if Ω_Λ and H_0 have the values given in Table 25-2. (*Hint:* You will need to convert units to eliminate kilometers and megaparsecs.)

Discussion Questions

48. Suppose we were living in a radiation-dominated universe. Discuss how such a universe would be different from the universe we now observe.

49. How can astronomers be certain that the cosmic microwave background fills the entire cosmos, not just the vicinity of Earth?

50. Do you think there can be "other universes," regions of space and time that are not connected to our universe? Should astronomers be concerned with such possibilities? Why or why not?

Web/eBook Questions

51. Before the discovery of the cosmic microwave background, it seemed possible that we might be living in a "steady-state universe" with overall properties that do not change with time. The steady-state model, like the Big Bang model, assumes an expanding universe, but it does not assume a "creation event" such as the Big Bang. Instead, matter is assumed to be created continuously everywhere in space to ensure that the average density of the universe remains constant. Search the World Wide Web for information about the steady-state

theory. Explain why the existence of the cosmic microwave background was a fatal blow to the steady-state theory.

ACTIVITIES

Observing Projects

52. In an attempt to explore the far reaches of the universe, the Hubble Space Telescope (HST) took long-exposure images of very dark regions of space that appear to contain no bright stars or galaxies. These images, known as the Hubble Deep Field and Hubble Ultra Deep Field images, reveal very rich fields of faint and very distant galaxies. The light now arriving at Earth from some of these galaxies has traveled for more than 13 billion years and was collected by the HST at a rate of a few photons per minute! This light was emitted very early in the life of the universe, only a few hundred million years after the Big Bang. You can examine and measure these two images. (a) In *Starry Night*™, open the **Options** pane and ensure that the **Hubble Images** option is checked in the **Deep Space** layer. Open the **Find** pane, ensure that the search edit box is empty, and click on the icon in this box to display a list of image sources. Click on **Hubble Images** and double-click on **Hubble Deep Field** to center the view on this dark region of space. Note its position with respect to the Big Dipper. (*Note:* If you cannot identify this region of the northern sky, click on **View > Constellations > Asterisms** and **View > Constellations > Labels.** Remove these indicators after you have identified the region.) **Zoom in** to a field of view about 3° wide and note that the region still appears to be devoid of objects. **Zoom in** again until the **Hubble Deep Field** (HDF) fills the view. One-quarter of the full HDF, with dimensions of 1.15′ × 1.15′, is displayed in *Starry Night*™. The bright object with spikes radiating from it is a star in our own galaxy (the spikes are caused by diffraction by the supports for Hubble's secondary mirror), but all of the other objects that appear on this long-exposure image are galaxies containing millions of stars. Examine this image carefully and attempt to identify some examples of each kind of galaxy—spiral, barred spiral, elliptical and irregular—in this field. Choose 5 or 6 of the largest galaxies in this field, record their shapes and galaxy types, and use the angular separation tool to measure carefully and record their angular dimensions. (**b**) Click the **Zoom** panel in the toolbar and select **90°** from the dropdown menu. Return to the **Find** pane and the list of **Hubble Images** and double-click on **Hubble Ultra Deep Field** (HUDF) to center the view of this "dark" region of the sky. **Zoom in** on this region and note that, even at a field of view as small as 2°, no objects can be seen in the position of this long-exposure image. **Zoom in** further until the full HUDF, with dimensions of 3.3′ × 3.3′, fills the field of view to see this rich field of faint and very distant galaxies. Examine this image carefully and attempt to identify each kind of galaxy—spiral, barred spiral, elliptical and irregular—in this field. Again, select 5 or 6 of the largest galaxies in this field, record their shapes and galaxy types and use the angular separation tool to measure their dimensions. (**c**) Consider the mix of different kinds of galaxies and assess whether the proportions of different kinds are the same in these two images. Compare the angular sizes of the largest galaxies in these two images.

53. Use *Starry Night*™ to compare the distances of objects in the **Tully Database** with the radius of the limit of our observable universe, the Cosmic Light Horizon. As you will find, the most distant galaxies in this database are a long way away from Earth and yet these distances are only a small fraction of the distances from which we can see light in our universe. Select **Favourites > Explorations > Tully Database** to display this collection of galaxies in their correct 3-dimensional positions in space around the Milky Way Galaxy, as seen from a distance of about 0.5 Mly from the Sun. Note the Zone of Avoidance surrounding the Milky Way, where material within the arms of the galaxy prevents us from seeing objects in these regions of our sky. Now use the **Increase current elevation** button (to the left of Home button) in the toolbar to increase the distance from the Milky Way to about 900 Mly, when the outer extent of the Tully Database becomes visible. **Open File > Preferences,** select **Cursor Tracking (HUD)** in the dropdown menu and ensure that **Distance from observer, Name,** and **Object type** are selected. Select the angular separation tool from the cursor tool selection at the upper-left of the screen. (**a**) Use this tool to measure the distances from the Milky Way to a selection of outer galaxies. (**b**) How do these distances compare with the Cosmic Light Horizon? (**c**) What fraction of the radius of the observable universe is covered by the Tully Database?

Collaborative Exercises

54. As a group, create a four- to six-panel cartoon strip showing a discussion between two individuals describing why the sky is dark at night.

55. Imagine your firm, Creative Cosmologists Coalition, has been hired to create a three-panel, folded brochure describing the principal observations that astronomers use to infer the existence of a Big Bang. Create this brochure on an 8½ × 11 piece of paper. Be sure each member of your group supervises the development of a different portion of the brochure and that the small print acknowledges who in your group was primarily responsible for which portion.

56. The three potential geometries of the universe are shown in Figure 25-14. To demonstrate this, ask one member of your group to hold a piece of paper in one of the positions while another member draws two parallel lines that never change in one geometry, eventually cross in another geometry, and eventually diverge in another.

ANSWERS

ConceptChecks

ConceptCheck 25-1: Bright. Even though more distant stars would individually appear dimmer, that loss of intensity is made up for by there being a greater number of distant stars in any given patch of the sky. Taken together, these two effects would tend to make the universe appear bright, instead of dark as it actually does. This is Olber's paradox.

ConceptCheck 25-2: The spectral shift toward the shorter blue wavelengths means that the Andromeda Galaxy is moving toward our Milky Way Galaxy.

ConceptCheck 25-3: The cosmological principle applies over very large regions of outer space; it does not apply to the distribution of stars in objects as small as a galaxy.

ConceptCheck 25-4: Older. A smaller Hubble constant corresponds to slower expansion. That means it would have taken longer for galaxies to reach their present separation distances.

ConceptCheck 25-5: The receding star appears dimmer. As its light travels through expanding space, the wavelength of its light is stretched. This lowers the light's energy and the star appears dimmer.

ConceptCheck 25-6: The observable universe becomes larger as more light from never-before-seen distant galaxies has sufficient time to reach us.

ConceptCheck 25-7: The expansion of the universe over the past 13.7 billion years has stretched and redshifted these photons to such a great degree that they are now low-energy, long-wavelength microwave photons.

ConceptCheck 25-8: Earth formed about 4.5 billion years ago, when the universe was about 9 billion years old, and the universe has been dominated by matter for most of its history.

ConceptCheck 25-9: The universe had to be sufficiently cool for atoms to form, so if the early universe was hotter for longer, the first atoms would have appeared more slowly.

ConceptCheck 25-10: Cooler blue regions correspond to denser concentrations of mass in the early universe. These are often the seeds of structures that form later in that same region. A visible-light telescope would probably find a supercluster of galaxies in that same region.

ConceptCheck 25-11: Yes, as in Figure 25-14a. In a universe with the geometry of spherical space, parallel light beams will eventually cross. In this case, the universe is considered a *closed* space and has positive curvature.

ConceptCheck 25-12: It would appear larger in size. As illustrated in Figure 25-15a, since the light rays are bent inward as they travel from the object toward an observer on Earth, the light rays appear to be coming from ends of the stick that are farther apart than as would be observed in a flat universe (Figure 25-15b). This makes the stick appear larger in the closed universe.

ConceptCheck 25-13: The total energy density of the universe determines whether the universe is open, closed, or flat. The cosmic microwave background reveals that the universe is nearly flat with Ω_0 close to 1.0. Since matter can only account for about 0.24 of this energy density (and radiation is negligible), the remaining dark energy density is around 0.76.

ConceptCheck 25-14: Astronomers must make the assumption that all Type Ia supernovae have identical luminosities, regardless of the galaxy in which they are found.

ConceptCheck 25-15: The comparison in the middle, corresponding to size scales of around 1 degree, shows the greatest difference between the actual fluctuations (which include sound waves) and the variations that would be observed without sound waves. This is consistent with the peak in the lower part of the figure, showing that the effects of sound waves are strongest on scales of 1 degree.

CalculationChecks

CalculationCheck 25-1: Because z = the change in wavelength $(\lambda - \lambda_0)$ divided by the original λ_0, $z = (725.6\text{nm} - 656.3\text{nm}) \div 656.3 \text{ nm} = 0.11$.

CalculationCheck 25-2: Hubble's Law, $v = H_0 d$ can be rearranged as $d = v \div H_0$. Using $H_0 = 73$ km/s/Mpc, $d = v \div H_0 = 10{,}000$ km/s $\div$ 73 km/s/Mpc = 134 Mpc (million parsecs).

CalculationCheck 25-3: In a flat universe, Ω_0 is equal to 1. That means ρ_0 = (the combined average mass density) equals ρ_c = (the critical density). Therefore, $\rho_0 = 1.0 \times 10^{-26}$ kg/m^3.

CalculationCheck 25-4: The combined total density parameter for both mass and energy has a value of nearly $\Omega_0 = 1$. The density parameter for ordinary matter is $\Omega_b = 0.04$, so ordinary matter contributes only 4% to the total energy content of the universe. For dark energy, $\Omega_\Lambda = 0.76$, so dark energy makes up about 76% of the universe's energy content.

INSIGHTS

COSMOLOGY

Dark Forces at Work

Ten years ago two teams discovered that the universe will expand forever at an ever faster rate, thanks to an unseen energy. The leader of one of the groups, Saul Perlmutter, expects that new observations will soon illuminate the universe's dark side.

BY DAVID APPELL

(from David Appell, "Dark Forces at Work," *Scientific American,* May 2008, 100–102)

One of the chief astrophysicists behind the discovery of the acceleration of the expansion of the universe, among the most startling revelations in the history of cosmology, delights in the confusion about the observation. In fact, he wonders if the acceleration will end up being the most important feature in the ultimate explanation. "It might be something unexpected that looks like acceleration," says Saul Perlmutter, leader of the Supernova Cosmology Project (SCP), which first announced the astonishing fact in 1998. Ever the experimentalist, the 48-year-old Perlmutter is waiting, and planning, for more observations: "Until we go for a long run of more data, this just isn't a mature field."

Perlmutter philosophizes about the strangeness of the cosmos from his office at Lawrence Berkeley National Laboratory, high in the western hills of the San Francisco Bay Area. The room is the scientist's amalgam of too many computer screens, too many piles of papers and an equation-filled whiteboard that would have done Einstein proud. The spectacular view of the Golden Gate Bridge in the distance cannot help but promote lofty thinking.

It has been a decade since the science community learned of the shocking discovery made by Perlmutter's group and, independently, by the High-Z Supernova Search Team led by Brian Schmidt of the Australian National University (with analyses pioneered by Adam Riess of the Space Telescope Science Institute). The cosmos, the researchers found, is not just expanding; for unknown reasons, it is speeding up in its expansion.

The discovery took years of innovation and problem solving. The key was supernovae—specifically, those called type Ia. Such events are surprisingly invariable—the explosions have an intrinsic brightness that predictably fades over time, enabling astronomers to use them as "standard candles" and thus determine their distances from Earth. Perlmutter worked with Carl Pennypacker of the University of California, Berkeley, in the 1980s to robotically search for supernovae at relatively nearby distances. The field was then so young that their main competition came from Robert Evans, an amateur astronomer in Australia who identified supernovae with a backyard telescope.

In the beginning, the difficulty for Perlmutter's group lay in obtaining telescope time, always precious in the astronomical community. How would the researchers convince allocators to give them the chance to look for something—a supernova explosion—that had not yet taken place? So they worked out methods to predict and automatically detect supernovae in a given patch of the sky. But their goal of determining the universe's dynamics—then thought to be a decelerating expansion dominated by matter—still required additional observation to plot supernovae's brightness peaks and declines, which take place over a few weeks. Perlmutter twisted arms and begged colleagues for an hour or two on short notice, calling frantically around the world at all times of the day. Everyone knew him, he says, as half-annoying. "I was always worried about something that had to happen in the next 24 hours or sometimes the next two hours. It was a terrible way to lead an ordinary life," he recalls.

But persistence paid off. Observations of distant type Ia supernovae found them to be dimmer than expected. After eliminating the possibility of intergalactic dust and after years of painstaking data gathering and analysis at telescopes around the world (and in orbit), Perlmutter's team came to the conclusion that, incredibly, the universe is not only expanding, as Edwin Hubble discovered in 1929, but that its expansion rate is increasing. Some unknown force with negative pressure seems to be pushing the universe apart.

Subsequent balloon-borne observations of the cosmic microwave background made two years later showed that the universe is spatially flat—it was stretched out by an exponential expansion, called inflation, right after the big bang. The equations behind these experiments complemented those of the supernova teams taken a few years earlier, and together the results enabled scientists to calculate separately the density of dark energy in the universe and the density of matter.

But on the other hand, the discovery opened a mystery the size of, well, the universe. The simplest explanation is that dark energy is Einstein's famed "cosmological constant," an energy that permeates space but does not interact with any type of matter. Today astronomers have homed in on the details of this scenario; if true, then the universe consists of 72 percent antigravity dark energy, 23 percent dark matter (unseen and uncharacterized, but susceptible to gravity), and 5 percent normal matter (protons, neutrons, electrons). We would be just a small part of totality, surrounded by perplexity.

"It could well be that there's some big piece of reality that we don't fully understand," says astronomer Christopher Stubbs of Harvard University, who in a paper likened the new universe to "living in a bad episode of Star Trek." Physicist Steven Weinberg of the University of Texas at Austin calls it simply "a bone in the throat of theoretical physics."

Magic has not yet been proposed to explain the accelerating universe, but almost everything else has. In the past few years, physicists have widened their search beyond vacuum energy to include possible modifications to general relativity, spinless energy fields that vary with time and space, massive gravitons, brane worlds and extra dimensions. "All of them are so exciting, and any is going to rewrite the textbooks," says Eric Linder, a cosmologist at Lawrence Berkeley and U.C. Berkeley. The hypothetical repulsive dark energy field may well not survive in the final explanation.

"It's true the theorists right now are stuck," Perlmutter says. "But from an experimentalist's point of view, this is great: we have a mystery, and we have ways to get at it"—namely, in the form of new telescopes and satellites to look even farther across the universe (and, hence, farther back in time).

Ground-based projects are already gathering more data, looking for hundreds of type Ia events (instead of Perlmutter's and Schmidt's five dozen) to determine the relation between the pressure and density of the universe, akin to the ideal gas law. A galaxy like our Milky Way exhibits about one type Ia supernova every few hundred years, and its brightness fades in weeks, making the search for them quite a challenge. By observing the cosmic background radiation, the soon-to-launch Planck satellite will contribute more details about the universe's expansion.

Dark energy aficionados look especially to the Joint Dark Energy Mission, now in the planning stages in the U.S. for a possible launch in 2014. The probe will host a device that could find thousands of supernovae a year and provide far smaller error bars than anything done so far. One candidate is the SuperNova Acceleration Probe (SNAP), for which Perlmutter is the lead scientist and Linder the head theorist. It would host a telescope about two meters wide and have a gigapixel camera.

The discovery of cosmic acceleration will assuredly win a Nobel Prize, and over the years there has been some dispute over which team deserves priority. Perlmutter's SCP team announced the discovery first, but Schmidt's High-Z team beat the SCP group in publishing the finding. Both Perlmutter and Schmidt shared one fourth of the 2007 Gruber Cosmology Prize, with the remaining fraction going to their two teams collectively.

Gregarious and talkative, Perlmutter attributes his success to being able to convey his excitement and convince other researchers to join his team. An amateur violinist who also teaches an undergraduate physics and music course, he draws an orchestral analogy. "As a violinist, I always love the moments when a group of people are creatively tuned in together."

CHAPTER 26

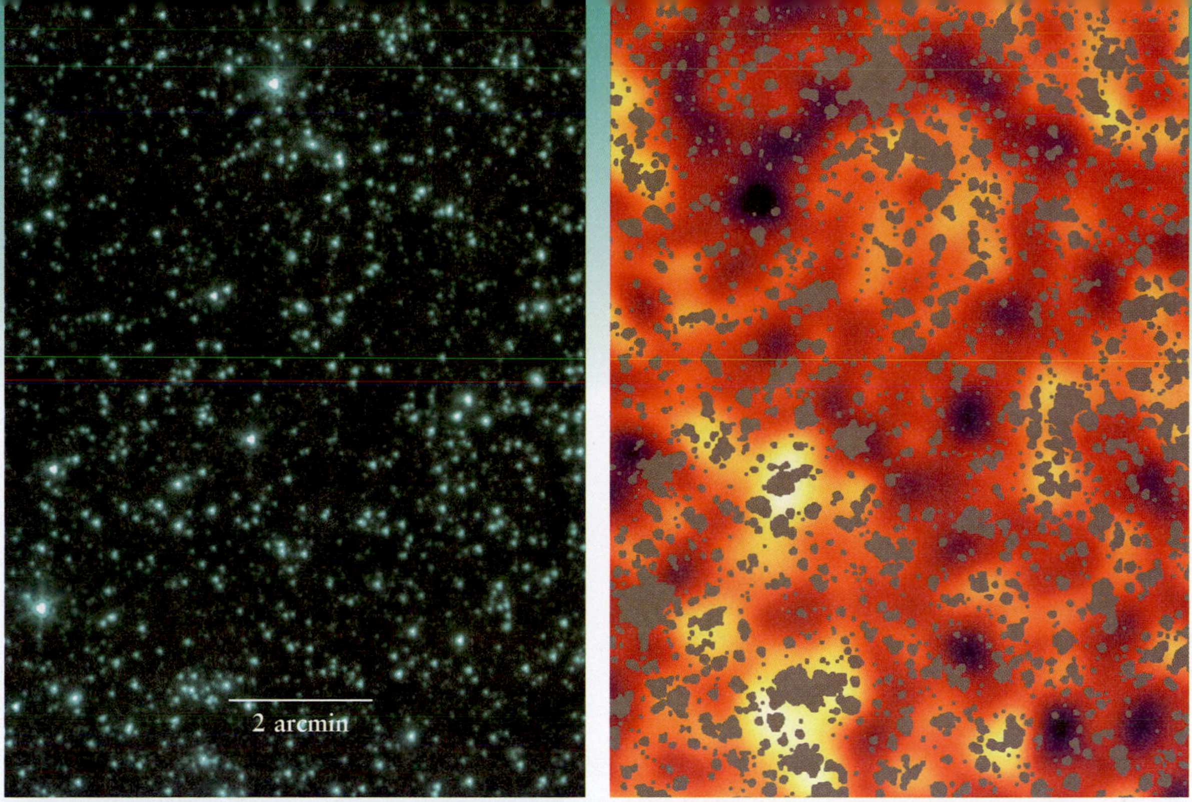

Nearby stars

Infrared light from very distant, primordial stars

These Spitzer Space Telescope images show (left) light from nearby stars and (right) light from a remote population of ancient stars in the same patch of sky with the nearby stars in our Galaxy removed. (NASA; JPL-Caltech; and A. Kashlinsky, Goddard Space Flight Center)

R I V U X G

Exploring the Early Universe

LEARNING GOALS

By reading the sections of this chapter, you will learn

26-1 How the fundamental forces of nature and the properties of empty space changed during the first second after the Big Bang

26-2 How the fundamental forces of nature and the properties of empty space changed during the first second after the Big Bang

26-3 How the physics of subatomic particles affected the evolution of the early universe

26-4 As the early universe expanded and cooled, most of the matter and antimatter annihilated each other

26-5 Which chemical elements in today's universe are remnants of the primordial fireball

26-6 How the first stars and galaxies formed in the early universe

26-7 What steps scientists are taking in the quest toward an all-encompassing "theory of everything"

The two images here show a patch of sky near the Big Dipper in the constellation Ursa Major. The left-hand image is dominated by relatively nearby stars in our Galaxy. But when these stars are removed digitally, what remains in the right-hand, false-color image is an intriguing pattern of highly redshifted light from objects much farther away. This radiation is thought to be some of the oldest starlight in the universe: It was emitted by members of the very first generation of stars, born when our universe was less than a billion years old.

Only recently have new telescopes begun to reveal the story of the first stars and first galaxies. But no telescope can ever hope to directly observe events during the first 380,000 years after the Big Bang, when the universe was so opaque that it blocked the free passage of light. We can nonetheless reconstruct some of the events of that hidden epoch, because many aspects of today's universe are relics of the earliest events in the cosmos.

In this chapter we will see evidence that during the first minuscule fraction of a second after the Big Bang, the universe inflated in size by a stupendously large factor of about 10^{50}. During the next 15 minutes, after inflation came to an end, the universe was so dense and hot that particles were constantly colliding at high speeds. As we will see, the events of those 15 violent minutes set the stage for all that came afterward—from the formation of atoms 380,000 years later, to the appearance of the first stars and galaxies some 400 million years after the Big Bang, down to the diverse present-day universe of which we are part.

26-1 The newborn universe underwent a brief period of vigorous expansion

With the discovery of the cosmic microwave background, astronomers had direct evidence that the universe began with a hot Big Bang (see Section 25-4). Remarkably, the microwave background is incredibly uniform, or *isotropic*, across the sky. If we subtract the effects of our own motion through the microwave background (see Figure 25-8), we find that the temperature of the microwave background is the same in all parts of the sky with deviations typically less than 0.01%. These small deviations, or nonuniformities, in the microwave background are also remarkable, in part because their angular sizes help indicate that our universe has a flat geometry (see Section 25-6, especially Figure 25-16). As striking as these observations are, they pose substantial challenges to the standard theory of the expansion of the universe.

The Isotropy Problem: Why Is the Microwave Background So Uniform?

To appreciate why the uniformity of the microwave background poses a problem, think about two opposite parts of the sky, labeled *A* and *B* in Figure 26-1. Both of these points lie on a sphere centered on Earth; our **observable universe** lies within the sphere, and the surface is called our **cosmic light horizon** (see Section 25-3, especially Figure 25-5). We can see light coming from any object on or inside our cosmic light horizon, but we cannot yet see objects beyond this horizon. Even traveling at the speed of light—the ultimate speed limit—no information of any kind from objects beyond our cosmic light horizon has had time to reach us over the entire 13.7-billion-year history of the universe. Thus, the cosmic light horizon defines the limits of our observable universe. (As time passes, our cosmic light horizon expands, so eventually we will receive light from objects that are beyond the present-day horizon.)

Points *A* and *B* in Figure 26-1 lie near the distant edge of the observable universe, just inside our cosmic light horizon, so when we look at these points we are seeing about as far back into the past as possible—that is, the light we see from these points is the cosmic background radiation. In order for the radiation that reaches us from *A* and *B* to be nearly the same, the material of the early universe at *A* must have had the same temperature as at *B*. But for two objects to be at the same temperature, they should have been in contact and able to exchange heat with each other. (A hot cup of coffee and a cold spoonful of cream reach the same temperature only after you pour the cream into the coffee.)

The problem is that the widely separated regions at *A* and *B* have absolutely no connection with each other. As Figure 26-1 shows, point *A* lies outside the cosmic light horizon for point *B*, and point *B* lies outside the cosmic light horizon for point *A*. (Put another way, point *A* lies outside the observable universe of point *B* and vice versa.) Hence, no information has had time to travel between points *A* and *B* over the entire history of the universe. How is it possible, then, that these unrelated parts of the universe have almost exactly the same temperature? This dilemma is called the **isotropy problem** or the **horizon problem.** (*Isotropy* means uniformity *in all directions.*)

Even when we consider that the contents of the universe were closer together at earlier times in the universe, the isotropy problem does not go away. While it takes less time for closer regions to come into contact with each other, there is also less time for this to occur because the universe is not as old. When both of these effects are taken into account, there does not seem to be a way for the most distant regions to have reached the same temperature.

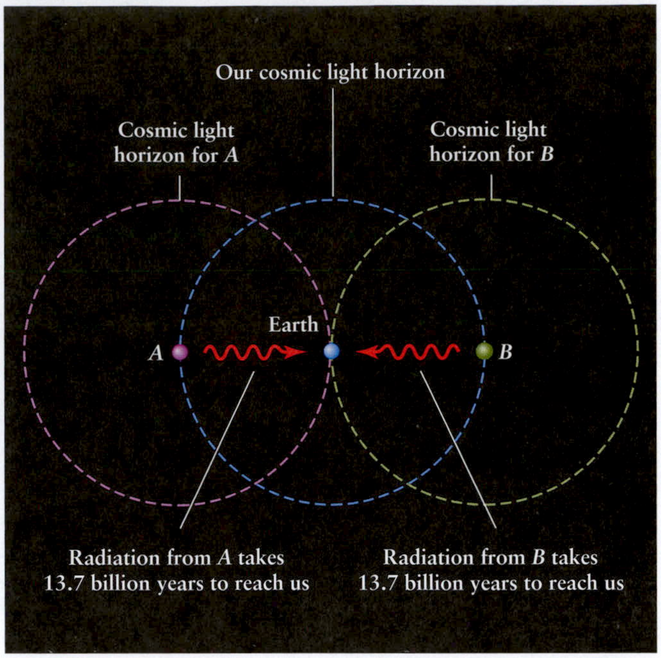

FIGURE 26-1
The Isotropy Problem Regions A and B, both of which lie on our cosmic light horizon, are so far apart that they seem never to have been in communication over the lifetime of the universe. Yet the cosmic background radiation from *A* and *B* and from all other parts of the sky shows that these disconnected regions have almost exactly the same temperature. The dilemma of why this should be is called the isotropy problem.

The Flatness Problem: Why Is $\Omega_0 = 1$?

The flatness of our universe presents us with a second enigma. Recall that the geometry of our universe depends on the density parameter Ω_0, which is the ratio of the combined average mass density in the universe (ρ_0) to the critical density (ρ_c) (this includes dark energy; see Section 25-6). Observations of temperature variations in the cosmic microwave background indicate that Ω_0 is very close to 1, which corresponds to a flat universe.

For the density parameter Ω_0 to be close to 1 today, it must have been *extremely* close to 1 during the Big Bang. In other words, the density of the early universe was almost exactly equal to the critical density. (The density was much higher than it is today, but the value of the critical density was also much higher. In a flat universe, the average mass density and the critical density decrease together as the universe expands, so that they remain equal at all times.)

The equations for an expanding universe show that any deviation from exact equality would have mushroomed within a fraction of a second. Had the average mass density been slightly less than ρ_c, the universe would have expanded so rapidly that matter could never have clumped together to form galaxies. Without galaxies, there would be no stars or planets, and humans would never have evolved. If, on the other hand, the density had been slightly greater than ρ_c, the universe would soon have become tightly packed with matter. Had this been the case, the gravitational attraction of this matter would long ago have collapsed the entire cosmos in a reversed Big Bang or "Big Crunch," and again humans would never have evolved.

In other words, immediately after the Big Bang the fate of the universe hung in the balance. The tiniest deviation from the precise equality $\rho_0 = \rho_c$ would have rapidly propelled the universe away from the special case of $\Omega_0 = 1$. Had there been such a deviation, we would not be here to contemplate the nature of the universe.

Happily, our universe is one in which galaxies, stars, planets, and humans do exist, a Big Crunch has not taken place, and Ω_0 is very close to 1. These conditions in the universe today tell us that the density of the universe immediately after the Big Bang must have been equal to the critical density to an incredibly high order of precision. In order for space to be as flat as it is today, the value of ρ_0 right after the Big Bang must have been equal to ρ_c to more than 50 decimal places! Unless there was a physical mechanism to bring these two parameters into such close agreement, this extraordinary match is an unappealing characteristic for a scientific theory because it seems to depend on a lucky coincidence.

What could have happened during the first few moments of the universe to ensure that $\rho_0 = \rho_c$ to such an astounding degree of accuracy? Because $\rho_0 = \rho_c$ means that space is flat, this enigma is called the **flatness problem.**

Solving the Problems: The Inflationary Model

In the early 1980s, a remarkable solution was proposed to both the isotropy problem and the flatness problem. Several physicists suggested that the universe might have experienced a brief period of **inflation,** or extremely rapid expansion, shortly after the Planck time. (As we saw in Section 25-3, before the Planck time the normal laws of physics do not properly describe the behavior of space, time, and matter.) During this **inflationary epoch,** the universe expanded outward in all directions by a factor of about 10^{50}. This epoch of dramatic expansion may have lasted only about 10^{-32} of a second (Figure 26-2).

The inflationary model explains not only why the cosmic microwave background is so uniform, but also how stars, galaxies, and humans can exist

Inflation accounts for the isotropy of the microwave background. During the inflationary epoch, much of the material that was originally near our location was moved out to tremendous distances. Over the past 13.7 billion years, our cosmic light horizon has expanded so that we can see radiation from these distant regions. Hence, when we examine microwaves from opposite parts of the sky, we are seeing radiation from parts of the universe that were originally in intimate contact with one another. This contact allowed for the exchange of heat and is why all parts of the sky have almost exactly the same temperature.

ANALOGY An inflationary epoch can also account for the flatness of the universe. To see why, think about a small portion of Earth's spherical surface, such as a small lake. For all practical purposes, it is impossible to detect Earth's curvature over such a small area, and a small lake looks flat. Similarly, the observable universe is such a tiny fraction of the entire inflated universe that any overall curvature in it is virtually undetectable (Figure 26-3). Like a small lake, the segment of space we can observe looks flat.

Not only does tremendous inflation produce a nearly flat universe, but it does so for just about *any* original degree of curvature prior to inflation. Let us consider why this is very appealing for a scientific theory. Recall that the flatness of our present universe could be explained (in an unsatisfactory way) if, by extraordinary coincidence alone, the value of ρ_0 right after the Big Bang was equal to ρ_c for the first 50 decimal places. Also recall that curvature is determined by ρ_0 (Section 25-6 and Table 25-1), where $\rho_0 = \rho_c$ is flat. Since Figure 26-3 shows that inflation flattens out just about

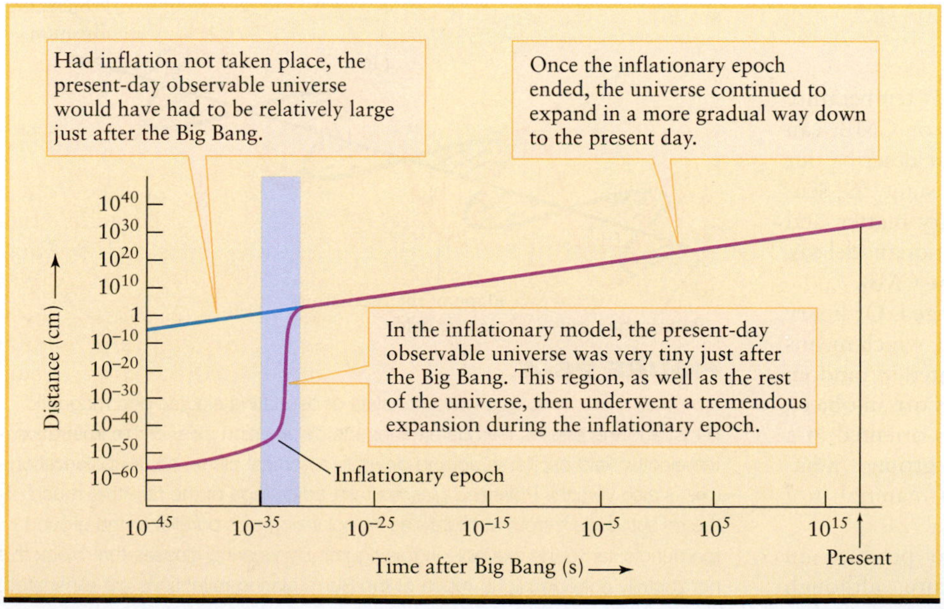
Had inflation not taken place, the present-day observable universe would have had to be relatively large just after the Big Bang.
Once the inflationary epoch ended, the universe continued to expand in a more gradual way down to the present day.
In the inflationary model, the present-day observable universe was very tiny just after the Big Bang. This region, as well as the rest of the universe, then underwent a tremendous expansion during the inflationary epoch.
Inflationary epoch
Distance (cm)
10^40
10^30
10^20
10^10
1
10^-10
10^-20
10^-30
10^-40
10^-50
10^-60
10^-45
10^-35
10^-25
10^-15
10^-5
10^5
10^15
Present
Time after Big Bang (s)

FIGURE 26-2
The Observable Universe With and Without Inflation According to the inflationary model (shown in purple), the universe expanded by a factor of about 10^{50} shortly after the Big Bang. This growth in the size of the present-day observable universe—that portion of the universe that lies within our present cosmic light horizon—occurred during a very brief interval, as indicated by the vertical shaded area on the graph. The blue line shows the projected size of the present-day observable universe soon after the Big Bang if inflation had not taken place. (Adapted from A. Guth)

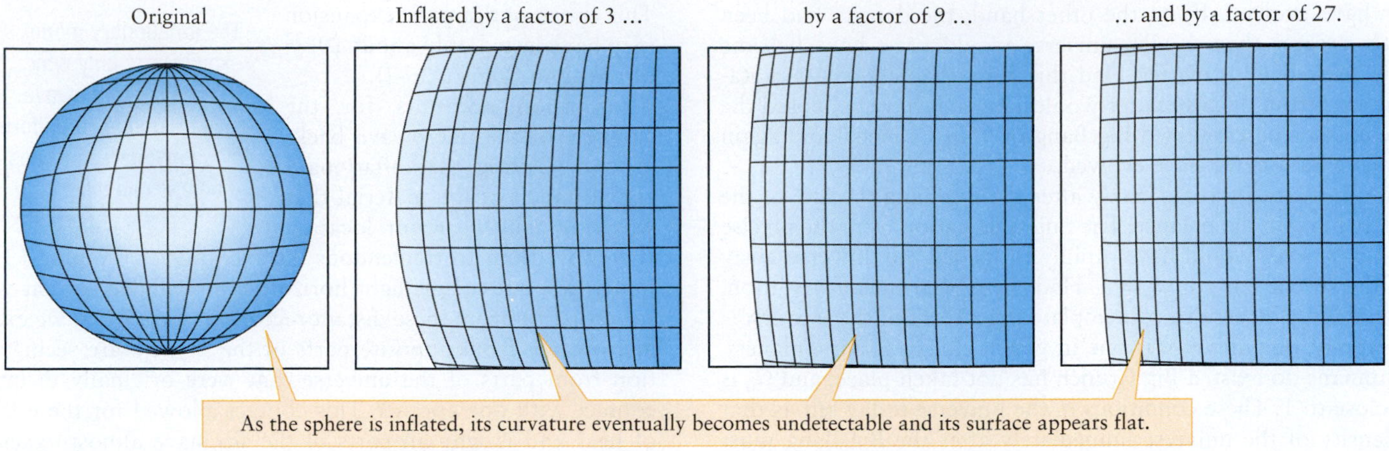
Original
Inflated by a factor of 3 ...
by a factor of 9 ...
... and by a factor of 27.
As the sphere is inflated, its curvature eventually becomes undetectable and its surface appears flat.

FIGURE 26-3
Inflation Solves the Flatness Problem This sequence of drawings shows how inflation can produce a locally flat geometry. In each successive frame, the sphere is inflated by a factor of 3 (and the number of grid lines on the sphere is increased by the same factor). Note how the curvature of the surface quickly becomes undetectable on the scale of the illustration. Inflation models expand the universe by about a factor 10^{50}, so nearly any original degree of curvature (indicated by the radius of the sphere on the left), will end up nearly flat after inflation. (Adapted from A. Guth and P. Steinhardt)

any original degree of curvature, inflation also *drives any initial value* of ρ_0 toward ρ_c, and removes the need for an unlikely coincidence alone to account for a flat universe.

CAUTION! It is important to note that the concept of inflation does not violate Einstein's law that nothing can travel faster than the speed of light. Remember that the expansion of the universe is the expansion of space itself, not the motion of objects through space. During the inflationary epoch, the distances between particles increased faster than the speed of light, but this was entirely the result of a sudden vigorous *expansion of space.* Particles did not move through space faster than the speed of light; space itself inflated to increase the distance between the particles. Even today, galaxies at redshifts greater than about $z = 1.5$ are receding faster than the speed of light.

Inflation and Polarization of the Cosmic Microwave Background

As we saw in Section 25-8, the detailed properties of temperature fluctuations in the cosmic microwave background, or CMB, can reveal a number of the fundamental parameters that describe our universe. This impressive list includes the Hubble constant, the density parameters for the total energy and for ordinary matter, and also for dark energy. In the near future, even the inflation model can be tested by measuring the *polarization* of light in the CMB.

As illustrated in Figure 26-4, light can be polarized. Ordinary light from a lightbulb or from the Sun is *unpolarized,* which means that the electric fields of many light waves are oriented in random directions. But when light collides with and bounces off an object, it tends to become *polarized,* with its electric fields oriented in a specific direction. In a similar way, the cosmic background radiation acquires a polarization when it scatters from material in a dense cold spot (Figure 26-5).

The effects of inflation are also expected to produce an additional and uniquely polarized CMB signature, although measurem≈ from WMAP (Figure 26-5) were unable to detect these weak signals. However, there are several experiments that hope to detect CMB polarization produced by inflation (these include the *Planck* satellite, BICEP, and POLARBEAR). These are crucial experiments, as inflation is one of the most profound predictions about our early universe, and CMB polarization appears to be our best hope of testing the inflation model.

CONCEPTCHECK 26-1

In Figure 26-1, points *A* and *B* appear too far apart to reach the same temperature by exchanging energy with each other. Nonetheless, distant locations like this in the universe do show nearly the same 2.7 K temperatures. How does the inflation model solve this problem?

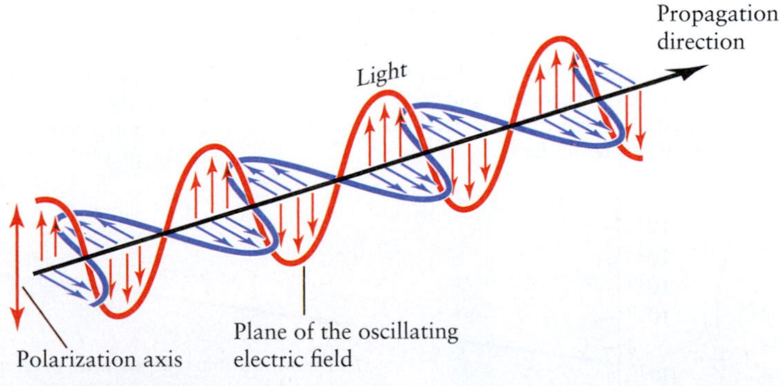
Propagation direction
Light
Plane of the oscillating electric field
Polarization axis

FIGURE 26-4
Polarized Light A light wave consists of oscillating electric and magnetic fields, and the electric field determines the polarization axis. In this illustration, the electric field oscillates up and down in a vertical plane, so the polarization axis is also vertical. Polarized glasses take advantage of the fact that much of the reflective glare from sunlight has its polarizing axis parallel to the ground (perpendicular to the example in this figure). By wearing glasses that block this horizontally polarized light, much of the Sun's glaring reflections are eliminated.

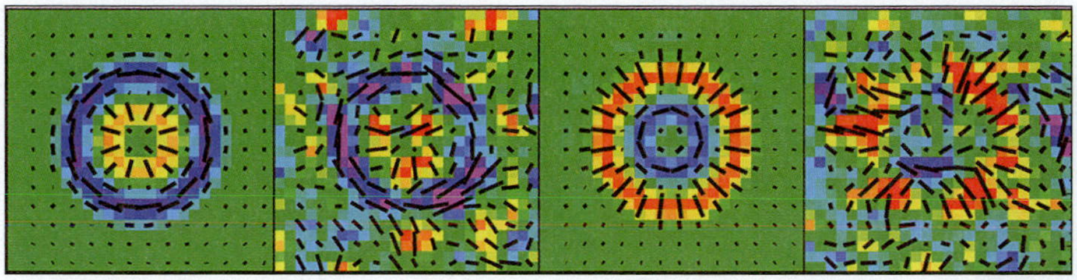

FIGURE 26-5 R I V U X G

Polarization of the Cosmic Microwave Background The polarization axes of CMB light are indicated by the orientation of the small black lines. In the two spots on the left, as CMB radiation scatters off of denser regions in the early universe, a unique ring-structured polarization pattern is created. These are the regions that show up as cooler blue spots in images of the CMB temperature fluctuations (such as Figure 25-13). In the two spots on the right, a unique pattern is also created when the CMB scatters off of warmer regions of matter. While these polarization measurements agree with predictions, they are not sensitive enough to test the inflation model, and more sensitive experiments have already begun. (NASA/WMAP Science Team)

CONCEPTCHECK 26-2

Suppose the curvature of the universe before inflation was represented by the surface of a sphere that was 1 billion times smaller than the sphere on the left of Figure 26-3. With this high degree of initial curvature, could inflation have produced a universe that appears nearly flat today?

Answers appear at the end of the chapter.

26-2 Inflation extends the principles that govern the fundamental forces of nature

If the universe went through an episode of extreme inflation, what could have triggered it? Our understanding is that inflation was one of a sequence of remarkable events during the first 10^{-12} seconds after the Big Bang. In each of these events there was a fundamental transformation of the basic physical properties of the universe. To understand what happened during that brief moment of time, when the universe was a hot, dense sea of fast-moving particles and energetic photons colliding with each other, we must first understand how particles interact at very high energies.

The Fundamental Forces of Nature and the Standard Model

Just *four* fundamental forces—*gravitation, electromagnetism,* and the *strong* and *weak forces*—explain the interactions of everything in the universe. Of these forces, gravitation is the most familiar (Figure 26-6a). It is a long-range force that dominates the universe over astronomical distances. Electromagnetism, which accounts for the electric and magnetic forces, creates a long-range force, but it is intrinsically much stronger than the gravitational force. For example, the electric force between an electron and a proton is about 10^{39} times stronger than the gravitational force between those two particles. Because the electromagnetic force is stronger by a factor of 10^{39}, electromagnetism, not the gravitation, holds electrons in orbit about the nuclei in atoms.

We do not generally observe longer-range effects of the electromagnetic force, because there is usually a negative electric charge for every positive charge and a south magnetic pole for every north magnetic pole. Thus, over great volumes of space the effects of electromagnetism effectively cancel out. No similar canceling occurs with gravity because there is no equivalent "negative mass." Because the effects of gravity do not cancel out, the force that holds Earth in orbit around the Sun is gravitational, not electromagnetic.

The **strong force** holds protons and neutrons together to form the nuclei of atoms (Figure 26-6b). It is said to be a *short-range force:* Its influence extends only over distances less than about 10^{-15} m, about the diameter of a proton. Without the strong force, nuclei would disintegrate because of the electric repulsion of the positively charged protons. In fact, the strong force overpowers the electric forces inside nuclei. While there are technical differences, the strong force is also referred to as the **nuclear force.**

The **weak force,** which also is a short-range force, is at work in certain kinds of radioactive decay (Figure 26-6c). An example is the transformation of a neutron (n) into a proton (p), in which an electron (e^-) is released along with a nearly massless particle called an antineutrino ($\overline{\nu}$):

$$n \rightarrow p + e^- + \overline{\nu}$$

Numerous experiments in nuclear physics demonstrate that protons and neutrons are themselves composed of even more basic particles called **quarks,** the most common varieties being "up" (u) quarks and "down" (d) quarks. A proton is composed of two up quarks and one down quark, a neutron of two down quarks and one up quark.

In the 1970s the concept of quarks led to a breakthrough in our understanding of the strong and weak forces. The strong force holds quarks together, while the weak force is at work whenever a quark changes from one variety to another. For example, when a neutron decays into a proton, one of the neutron's down quarks changes into an up quark, as shown in Figure 26-6c. Thus, the weak force is responsible for transformations such as

$$d \rightarrow u + e^- + \overline{\nu}$$

In the 1940s, physicists Richard P. Feynman and Julian S. Schwinger (working independently in the United States) and Sin-Itiro Tomonaga (in Japan) succeeded in developing a basic description of what we mean by force. Focusing their attention on the electromagnetic force, they tried to describe exactly what happens when two charged particles interact. According to their theory, now called **quantum electrodynamics,** charged particles interact

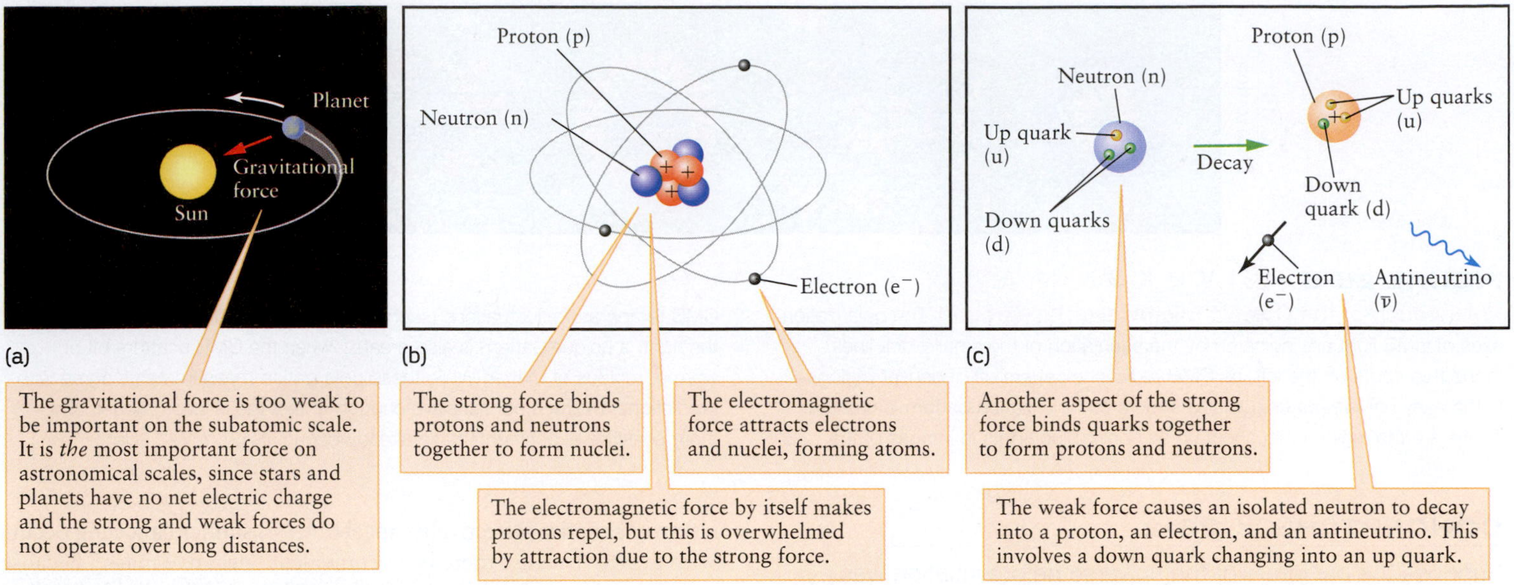
Planet
Gravitational force
Sun
(a)
Proton (p)
Neutron (n)
Electron (e⁻)
(b)
Proton (p)
Neutron (n)
Up quark (u)
Down quarks (d)
Decay
Up quarks (u)
Down quark (d)
Electron (e⁻)
Antineutrino (ν̄)
(c)
The gravitational force is too weak to be important on the subatomic scale. It is *the* most important force on astronomical scales, since stars and planets have no net electric charge and the strong and weak forces do not operate over long distances.
The strong force binds protons and neutrons together to form nuclei.
The electromagnetic force attracts electrons and nuclei, forming atoms.
The electromagnetic force by itself makes protons repel, but this is overwhelmed by attraction due to the strong force.
Another aspect of the strong force binds quarks together to form protons and neutrons.
The weak force causes an isolated neutron to decay into a proton, an electron, and an antineutrino. This involves a down quark changing into an up quark.

Force	Relative strength	Particles exchanged	Particles on which the force can act	Range	Example
Strong	1	gluons	quarks	10^{-15} m	holding protons, neutrons, and nuclei together
Electromagnetic	1/137	photons	charged particles	infinite	holding atoms together
Weak	10^{-4}	intermediate vector bosons	quarks, electrons, neutrinos	10^{-16} m	radioactive decay
Gravitational	6×10^{-39}	gravitons	everything	infinite	holding the solar system together

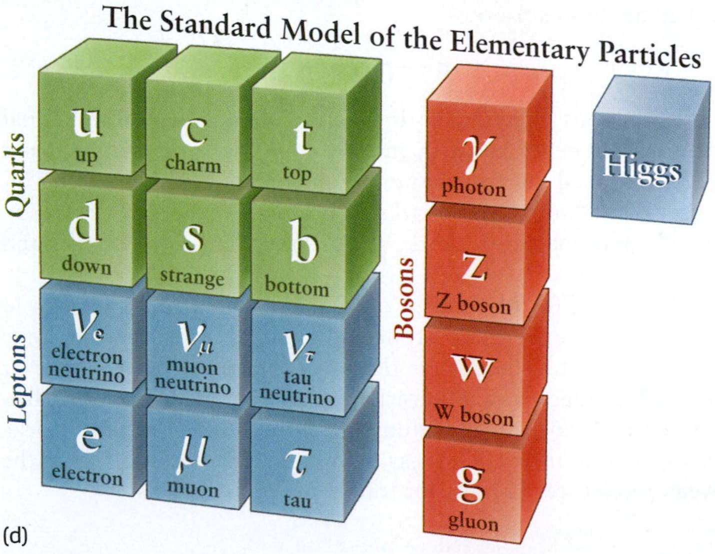
The Standard Model of the Elementary Particles
Quarks
u up
c charm
t top
d down
s strange
b bottom
Leptons
νe electron neutrino
νμ muon neutrino
ντ tau neutrino
e electron
μ muon
τ tau
Bosons
γ photon
Z Z boson
W W boson
g gluon
Higgs
(d)

FIGURE 26-6

The Four Forces and the Standard Model **(a)** Gravitation is dominant on the scales of planets, star systems, and galaxies, while **(b), (c)** the strong, electromagnetic, and weak forces hold sway on the scale of atoms and nuclei. **(d)** The Standard Model describes the particles involved in the electromagnetic, weak, and strong forces. Bosons transmit the different forces; the photon transmits the electromagnetic force between charged particles (holding electrons in atoms); the gluon transmits the force between quarks (holding together atomic nuclei); and the intermediate vector bosons (Z,W) transmit the weak force (resulting in certain radioactive decays). These are all the known fundamental particles found in nature.

by exchanging *virtual* photons that cannot be directly observed. (These virtual photons will be explored further in Section 26-3.)

Quantum electrodynamics has proved the most successful theory in modern physics. It describes with remarkable accuracy many details of electromagnetic forces and interactions between charged particles. Inspired by these successes, physicists developed similar theories for the other three forces. A key feature of a microscopic description of force is that two particles exert a force on each other by exchanging a third and different particle (Figure 26-7). For example, the weak force occurs when particles exchange particles called **intermediate vector bosons,** and quarks stick together by exchanging elementary particles called **gluons.** These theories

Experimental and theoretical research into the basic forces of nature helps us understand the evolution of the cosmos

FIGURE 26-7

Forces Due to Particle Exchange Imagine each skateboarder as a particle that can move freely. By tossing a basketball, the right skateboarder is pushed backward, and by catching the ball, the left skateboarder is pushed away as well. This example of particle exchange illustrates a repulsive force, and in quantum physics, particle exchange can produce both repulsive and attractive forces.

have been verified by experiments, and our overall understanding of the known particles is called the **Standard Model** (Figure 26-6d). The table in Figure 26-6 summarizes these features of the four fundamental forces.

CONCEPTCHECK 26-3

A neutron consists of one up quark (u) and two down quarks (d) as in Figure 26-6c. Are there any other particles in the Standard Model that you would expect to be involved in holding a neutron together?

Answer appears at the end of the chapter.

Experimentally Verified Unification Theories and the Higgs Particle

Physicists made important progress in understanding the weak force during the 1970s. Steven Weinberg, Sheldon Glashow, and Abdus Salam proposed a theory with three types of intermediate vector bosons, which are exchanged in various manifestations of the weak force. These three particles were actually discovered in experiments in the 1980s, providing strong support for the theory.

A startling prediction of the Weinberg-Glashow-Salam theory is that the weak force and the electromagnetic force should be identical to each other for particles with energies greater than 100 GeV. (One GeV equals 10^9, or 1 billion, electron volts; see Section 5-5.) In other words, if particles are slammed together with a total energy greater than 100 GeV, then electromagnetic interactions become indistinguishable from weak interactions. We say that above 100 GeV the electromagnetic force and the weak force are "unified" into a single **electroweak force.** This is our first example of the "unification" of forces, and the only one to be experimentally verified.

This unification occurs because the three types of intermediate vector bosons behave just like photons above 100 GeV. At such high energies, these three particles actually lose their mass, and the weak force becomes a long-range force with the same intrinsic strength as electromagnetism. Physicists describe this similarity by saying that "symmetry is restored" above 100 GeV.

In the world around us, however, the typical energies with which particles interact are much lower, on the order of 1 eV or less. Below 100 GeV, intermediate vector bosons behave like massive particles, but photons have no mass. Because intermediate vector bosons and photons are not similar at low energies, we say that "symmetry is broken" below 100 GeV, which is why the electromagnetic and the weak forces behave so differently in the lower-energy world around us. In the language of physics, the electroweak force experiences a **spontaneous symmetry breaking;** above 100 GeV the electroweak force represents a form of symmetry, and below 100 GeV, the symmetry is broken and two distinct forces emerge: the electromagnetic and weak forces.

ANALOGY As abstract as spontaneous symmetry breaking sounds, it is a familiar phenomenon when describing water. At room temperature, water is liquid. If you were to take a vantage point from within liquid water, it would look the same in all directions. In this sense, liquid water is uniform and symmetric. At lower temperatures, however, water freezes into ice crystals and no longer appears the same in every direction (Figure 26-8). While commonly referred to as a "phase transition," this change can also be described as the spontaneous symmetry breaking of water when it freezes.

FIGURE 26-8 R I V U X G

Spontaneous Symmetry Breaking of Water When liquid water freezes it transitions from a highly uniform and symmetric state to ice crystals that break this uniformity. Since lower temperatures correspond to lower average energy for the water's particles, we say that water experiences a spontaneous symmetry breaking at lower energy. By analogy, the high temperature symmetric water is like the electroweak force, and the ice crystals are like the electromagnetic and weak forces. (Frank Siteman/Aflo Relax/Corbis)

(a)

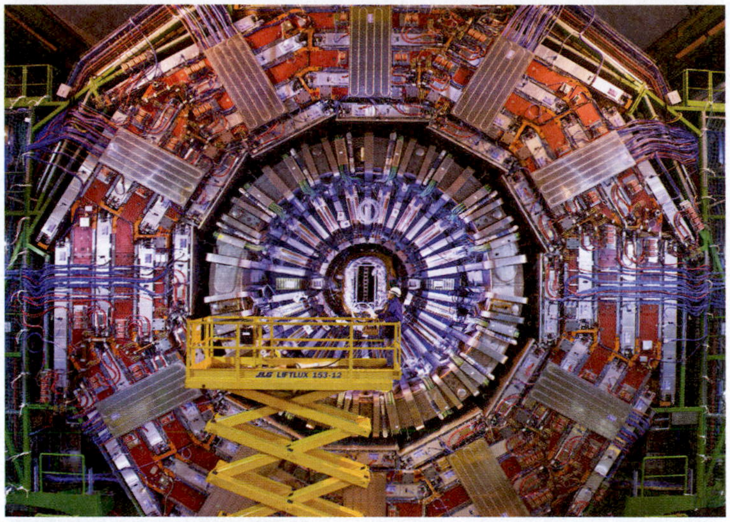

(b)

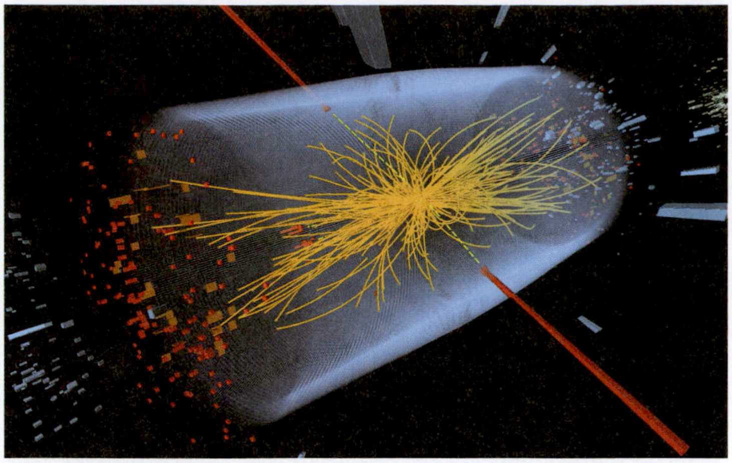

(c)

FIGURE 26-9 R I **V** U X G

Discovery of the Higgs Particle **(a)** The CERN particle collider is 27 km in diameter and straddles the border between Switzerland and France. Protons are accelerated to speeds very close to the speed of light in a tube (shown in red) that is around 100 m below ground. **(b)** When these protons collide, their energy produces numerous additional particles that are tracked and detected in enormous particle detectors as shown here (under construction). This is a direct result of converting energy to mass via $E = mc^2$. **(c)** By analyzing the particle tracks (in yellow), the mass, charge, and various other properties of the particles can be determined. This is how new particles are discovered, including the Higgs particle. (a, b: © 2001 CERN; c: HANDOUT/AFP/Getty Images/Newscom)

As strange as these ideas sound, they have been tested to great precision in particle accelerators that smash particles together at extremely high energies. The world's largest particle accelerator is CERN in Europe (Figure 26-9), which made a monumental discovery in 2012. For more than 40 years, there was one major prediction of the Standard Model that had not been observed—the Higgs particle (see Figure 26-6d). This key particle was predicted as part of the electroweak symmetry breaking that gives the Standard Model some of the properties we observe. In fact, one of these properties is mass: The Higgs particle "gives mass" to the other particles (except for the massless photons and gluons) by interacting with them. Since you and I, the planets, and the stars are all made of these particles, we owe our heft to the Higgs particle.

ANALOGY How does the Higgs particle give other particles mass? First of all, in the realm of quantum physics, every particle is associated with an extended entity called a "field." Thus, the Higgs particle is associated with a Higgs field, and this field is present even at low energy when the Higgs particle itself is absent. To acquire the property of mass, particles in the Standard Model interact with this Higgs field, which produces a force on them. This force is somewhat similar to what a crumb experiences when it's surrounded by syrup; analogously, the Higgs field is like a syrup that pervades all of space. The interactions of particles with this "syrup" make it harder to set particles into motion, giving them a property of mass. (Interestingly, it is not known why different particles interact more strongly to have different masses; this is an unsolved problem in physics.)

Long before the Higgs particle was observed, the Standard Model's overall success at unifying the electromagnetic force with the weak force led to attempts at unifying the remaining two forces: the strong force and gravity.

CONCEPTCHECK 26-4

Are the photons, the W and the Z bosons of the Standard Model in Figure 26-6d, "symmetric," meaning that they all have the same physical properties?

CONCEPTCHECK 26-5

Symmetry breaking, which refers to the fact that below 100 GeV, the electroweak force appears as two distinct forces (electromagnetic and weak) is said to be "spontaneous." What is spontaneous about symmetry breaking?

Answers appear at the end of the chapter.

Proposed Unification Theories of all the Forces

In the 1970s, several physicists proposed **grand unified theories** (or **GUTs**), which predict that the strong force becomes unified with the weak and electromagnetic forces (but not gravity) at energies above 10^{14} GeV. In other words, if particles were to collide at energies greater than 10^{14} GeV, the strong, weak, and electromagnetic interactions would all be long-range forces and would be indistinguishable from each other.

Many physicists suspect that all four forces, including gravitation, may be unified at energies greater than 10^{19} GeV (Figure 26-10). That is, if particles were to collide at these colossal energies, there would be no difference between the gravitational, electromagnetic, and nuclear forces. However, no one has yet succeeded in working out the details of such a **supergrand unified theory,** which is sometimes called a **theory of everything** (or **TOE**). The favored TOE to describe all of the forces is called string theory, which we will discuss further in Section 26-7.

A supergrand unified theory or TOE that includes gravitation would also be a theory of *quantum gravity* (which would describe the effects of strong gravity on microscopic scales). We have already seen with black holes and the Big Bang that developing the answers to some questions requires a theory of quantum gravity (such as what happens to the information about material after it falls into a black hole). Similar to the particle exchange mechanism between the other forces, in a quantum gravity theory the gravitational force is predicted to arise when particles exchange **gravitons,** although gravitons themselves have not been observed.

The Fundamental Forces in the Early Universe

Figure 26-10 shows how the various forces are thought to have changed during the first fraction of a second after the Big Bang. Before the Planck time (from $t = 0$ second to $t = 10^{-43}$ second), particles collided with energies greater than 10^{19} GeV, and all four forces were unified. Because we do not yet have a TOE that properly describes the behavior of gravity, we remain ignorant of what was going on during the first 10^{-43} second of the universe's existence. We know, however, that by the end of the Planck time, the expansion and cooling of the universe had caused the energy of particles to fall to 10^{19} GeV. At energies below this level, gravity is not unified with the other three forces.

In the language of physics, at $t = 10^{-43}$ second there was a spontaneous symmetry breaking in which gravity was "frozen out" of the otherwise unified hot soup that filled all space. In such a "soup," the typical energy of a particle (E) is related to temperature (T) by $E = kT$, where k is the Boltzmann constant (about 10^{-4} eV/K, or 10^{-13} GeV/K). Thus, the temperature of the universe was 10^{32} K when gravity emerged as a separate force.

As the universe expanded, its temperature decreased and the energy of particles decreased as well. (We discussed this property of gases in Box 19-1.) At $t = 10^{-35}$ second, the energy of particles in the universe had fallen to 10^{14} GeV, equivalent to a temperature of 10^{27} K. Below these energies and temperatures, the strong force is no longer unified with the electromagnetic and weak forces. Thus, at $t = 10^{-35}$ second, there was a second spontaneous symmetry breaking, at which time the strong force "froze out."

The inflationary epoch is thought to have begun at this point. Physicists hypothesize that before the strong force decoupled from the electroweak force, the universe was in an unstable state called a **false vacuum.** In this state, physicists hypothesize that the energy associated with a quantity called the *inflaton field* (or inflation field) had a nonzero value (Figure 26-11a). (Just as the space around a magnet is permeated by a magnetic field, like that shown in Figure 7-13a, the entire universe is thought to be permeated by

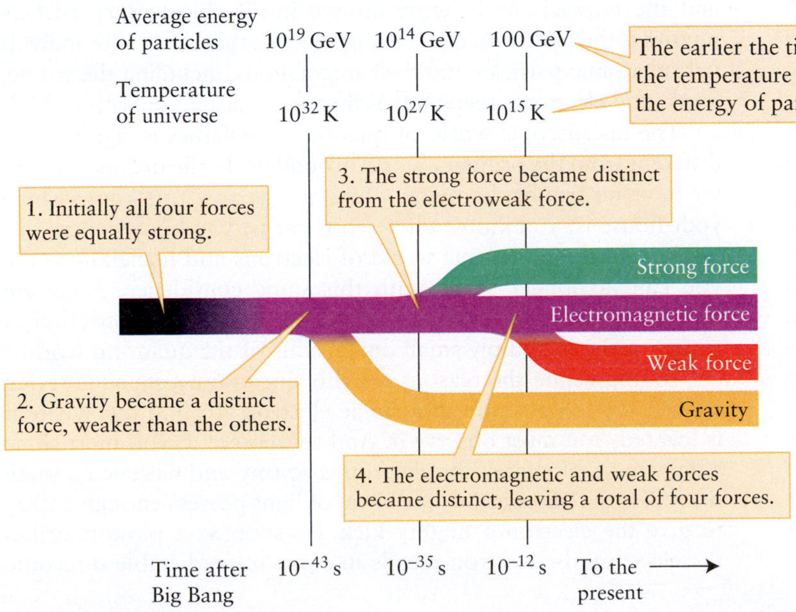

Average energy of particles
10^{19} GeV
10^{14} GeV
100 GeV
Temperature of universe
10^{32} K
10^{27} K
10^{15} K
The earlier the time, the higher the temperature and the greater the energy of particles.
1. Initially all four forces were equally strong.
3. The strong force became distinct from the electroweak force.
Strong force
Electromagnetic force
Weak force
Gravity
2. Gravity became a distinct force, weaker than the others.
4. The electromagnetic and weak forces became distinct, leaving a total of four forces.
Time after Big Bang
10^{-43} s
10^{-35} s
10^{-12} s
To the present

FIGURE 26-10
Unification of the Four Forces The strength of the four fundamental forces depends on the energy of the particles that interact. As shown in this schematic diagram, the higher the energy, the more the forces resemble each other. Also included here are the temperature of the universe and the time after the Big Bang when the strengths of the forces are thought to have been equal. At energies between 10^{14} and 10^{19} GeV, the forces might be described by a grand unified theory (GUT), and at energies above 10^{19} GeV, the forces might all appear similar in a supergrand unified theory.

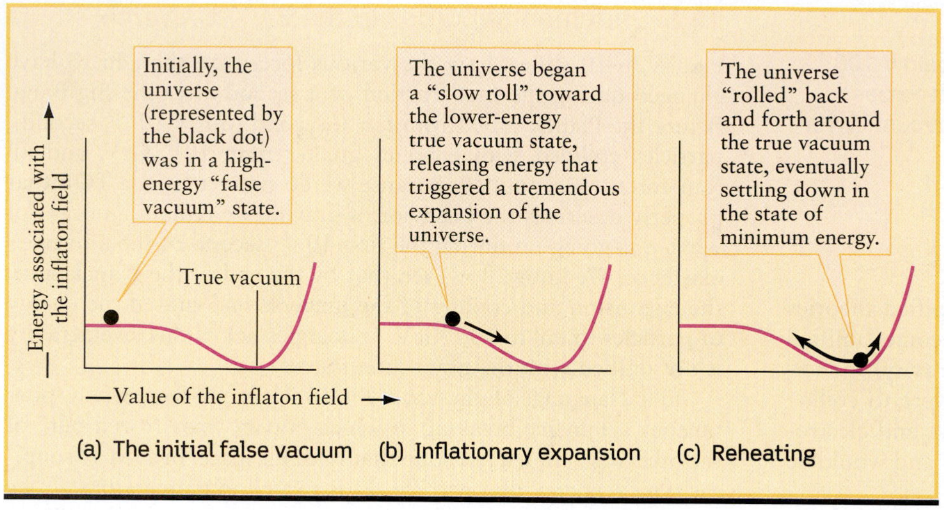

Initially, the universe (represented by the black dot) was in a high-energy "false vacuum" state.
True vacuum
The universe began a "slow roll" toward the lower-energy true vacuum state, releasing energy that triggered a tremendous expansion of the universe.
The universe "rolled" back and forth around the true vacuum state, eventually settling down in the state of minimum energy.
Energy associated with the inflaton field
Value of the inflaton field
(a) The initial false vacuum
(b) Inflationary expansion
(c) Reheating

FIGURE 26-11

Inflation: Transitioning from the False Vacuum to the True Vacuum **(a)** The energy of the vacuum is thought to be determined by the value of a quantity called the inflaton field. As shown by the red curve in this diagram, this energy is at a minimum for a certain value of this field. The state of lowest energy is called the true vacuum. **(b)** Inflation is associated with the universe making a transition from its initial false-vacuum state to a true-vacuum state. This expansion took place over a period of about 10^{-32} seconds, during which the universe cooled to a temperature of about 3 K. **(c)** As the universe "settled in" to the true vacuum state, energy was released that reheated the universe to a temperature of 10^{27} K.

the inflaton field.) This state was unstable in the same sense as a ball perched atop a cone with a pointed top: The ball will stay there if left undisturbed, but will roll downhill if even slightly disturbed. In an analogous way, it is thought that at the time that the strong and electroweak forces decoupled, the universe "rolled downhill" to the *true vacuum,* a state of lower energy. The universe's transition to the true vacuum released energy that caused it to expand tremendously in a brief interval of time (Figure 26-11b). By the time the inflationary epoch had ended, about 10^{-32} second after the Big Bang, the universe had increased in scale by a factor of roughly 10^{50}.

The rapid expansion of the universe also gave rise to a rapid cooling. At the end of the inflationary epoch, the temperature of the universe may have dropped to about 3 K, about the same as the temperature that the cosmic background radiation has today. But as the universe finally settled into the vacuum-energy state, an additional amount of energy was released that went into *reheating* the universe to a temperature of 10^{27} K—about the same as it had before inflation began (Figure 26-11c). Thus, inflation caused the universe to expand tremendously while having no net effect on its temperature.

After the end of the inflationary epoch at $t = 10^{-32}$ second, the universe continued to expand and cool at a more sedate rate. At $t = 10^{-12}$ second, the temperature of the universe had dropped to 10^{15} K, the energy of the particles had fallen to 100 GeV, and there was a final spontaneous symmetry breaking as described by the Standard Model. That last symmetry breaking separated the electromagnetic force from the weak force, and from that moment on, all four forces have interacted with particles essentially as they do today.

The electroweak symmetry breaking around 100 GeV has been verified, including the predicted Higgs particle. While the two symmetry breaking events predicted to occur at higher energies (and earlier times) have not been observed, they follow a similar mechanism for what has been observed at lower energies with the electroweak force. This lower-energy success certainly does not prove that these higher energy theories are correct, but it provides motivation for formulating and testing them.

26-3 During inflation, all the mass and energy in the universe burst forth from the vacuum of space

If the ideas of inflation are correct, it was a brief but stupendous expansion of the universe soon after the beginning of time. Physicists now realize that inflation helps explain where all the matter and radiation in the universe came from. To see how a violent expansion of space could create particles, we must first understand what quantum mechanics tells us about space.

Quantum Mechanics and the Heisenberg Uncertainty Principle

Quantum mechanics is the branch of physics that explains the behavior of nature on the atomic scale and smaller. For example, quantum mechanics tells us how to calculate the structure of atoms and the interactions between atomic nuclei. **Elementary particle physics** is the branch of quantum mechanics that deals with individual subatomic particles and their interactions, including the strong, weak, and electromagnetic forces that we discussed in Section 26-2.

The microscopic world of quantum mechanics is significantly different from the ordinary world around us. In the ordinary world we have no trouble knowing where things are. You know where your house is; you know where your car is; you know where this book is. In the subatomic world of electrons and nuclei, however, you can no longer speak with this same confidence. A certain amount of fuzziness, or uncertainty, enters into the description of reality at the incredibly small dimensions of the quantum world.

To appreciate the reasons for this uncertainty, imagine trying to measure the position of a single electron. To find out where it is located, you must observe it. And to observe it, you must shine a light on it. However, the electron is so tiny and has such a small mass that the photons in your beam of light possess enough energy to give the electron a mighty kick. As soon as a photon strikes the electron, the electron recoils in some unpredictable direction.

Consequently, no matter how carefully you try to measure the precise location of an electron, you necessarily introduce some uncertainty into the speed of that electron.

These ideas are at the heart of the **Heisenberg uncertainty principle,** first formulated in 1927 by the German physicist Werner Heisenberg, one of the founders of quantum mechanics. This principle states that there is a reciprocal uncertainty between position and momentum (momentum is equal to the mass of a particle multiplied by its velocity). The more precisely you try to measure the position of a particle, the more unsure you become of how the particle is moving. Conversely, the more accurately you determine the momentum of a particle, the less sure you are of its location. These restrictions are not a result of errors in making measurements; they are fundamental limitations inherent to all natural phenomena.

There is an analogous uncertainty involving energy and time. You cannot know the energy of a system with infinite precision at every moment in time. Over short time intervals, there can be great uncertainty about the amounts of energy in a subatomic system. Specifically, let ΔE be the smallest possible uncertainty in energy measured over a short interval of time Δt. (Astronomers and physicists often use the capital Greek letter delta, Δ, as a prefix to denote a small quantity or a small change in a quantity.)

Heisenberg uncertainty principle for energy and time

$$\Delta E \times \Delta t = \frac{h}{2\pi}$$

ΔE = uncertainty in energy

Δt = time interval over which energy is measured

h = Planck's constant = 6.625×10^{-34} J s

This equation says that the shorter the time interval Δt, the greater the energy uncertainty ΔE must be in order to ensure that the product of ΔE and Δt is equal to $h/2\pi$.

We might look upon the Heisenberg uncertainty principle as merely an unfortunate limitation on our ability to know everything with infinite precision. But, in fact, this principle provides startling insights into the nature of the universe.

Spontaneously Created Matter and Antimatter

We have seen that one of the important conclusions of Einstein's special theory of relativity is the equivalence of mass and energy: $E = mc^2$ (see Section 16-1). There is nothing uncertain about the speed of light (c), which is an absolute constant. Therefore, any uncertainty in the energy of a physical system can be attributed to an uncertainty Δm in the mass. Thus,

$$\Delta E = \Delta m \times c^2$$

Combining this expression with the previous equation, we obtain

Heisenberg uncertainty principle for mass and time

$$\Delta m \times \Delta t = \frac{h}{2\pi c^2}$$

Δm = uncertainty in mass

Δt = time interval over which mass is measured

h = Planck's constant = 6.625×10^{-34} J s

c = speed of light = 3.00×10^8 m/s

This result is astonishing. It means that over a very brief interval Δt of time, we cannot be sure how much matter there is in a particular location, even in "empty space." During this brief moment, matter can spontaneously appear and then disappear. The greater the amount of matter (Δm) that appears spontaneously, the shorter the time interval (Δt) during which it can exist before disappearing into nothingness. This bizarre state of affairs is a natural consequence of quantum mechanics.

> Quantum mechanics reveals that "empty" space is not empty, but is seething with particles and antiparticles that appear and then annihilate

No particle can appear spontaneously by itself, however. For each particle created, so is a second, almost identical **antiparticle;** particles are made of matter, and antiparticles are made of antimatter. In other words, equal amounts of matter and **antimatter** come into existence and then disappear. (These are the virtual particle pairs discussed in Section 25-7, and in more depth below.)

Despite its exotic name, there is actually nothing terribly mysterious about antimatter. A particle and an antiparticle are identical in almost every respect; their main distinction is that they carry opposite electric charges. For example, an ordinary electron (e^-) carries a negative charge; the corresponding antiparticle has the same mass but a positive charge, which is why it is called a **positron** (e^+, and also called an antielectron). Protons (with positive charge) also have an antimatter partner called an **antiproton** (with negative charge). Particles that have no electric charge can also have corresponding antiparticles. An example is the neutrino (ν); we met its antiparticle, the antineutrino ($\overline{\nu}$), in Section 26-2. The antineutrino is also electrically neutral, but differs from the neutrino in having opposite values of other, more subtle physical properties.

A spontaneously created particle-antiparticle pair lasts for only an incredibly brief time. For example, consider an electron and a positron, each with a mass of 9.11×10^{-31} kg. If we rewrite the Heisenberg uncertainty principle for mass and time to solve for Δt and then substitute the combined mass of $2 \times 9.11 \times 10^{-31}$ kg, we find that a spontaneously created electron-positron pair can last for a time

$$\Delta t = \frac{1}{\Delta m}\frac{h}{2\pi c^2} = \frac{1}{2 \times 9.11 \times 10^{-31}} \times \frac{6.625 \times 10^{-34}}{2\pi(3.00 \times 10^8)^2}$$
$$= 6.43 \times 10^{-22}\ \text{s}$$

In other words, an electron and a positron can spontaneously appear and then disappear without violating any laws of physics—but they can remain in existence for no longer than 6.43×10^{-22} second. In fact, the laws of quantum physics *require* that electrons and positrons constantly pop in and out of existence. The same behavior applies to all massive particles, but the more massive the

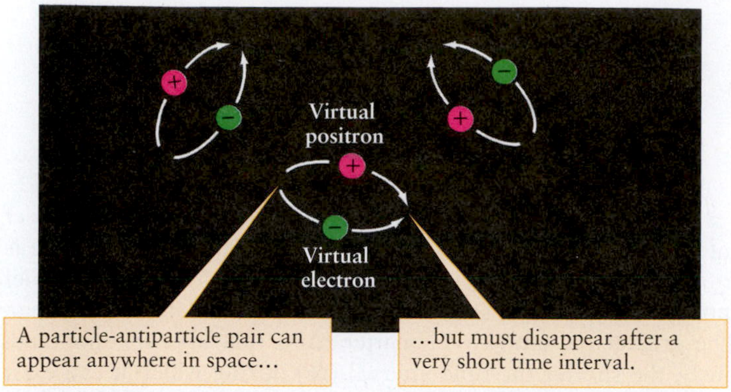

FIGURE 26-12

Virtual Pairs Pairs of particles and antiparticles can appear and then disappear anywhere in space provided that each pair exists only for a very short time interval, as dictated by the uncertainty principle. In this sketch, electrons are shown in green and positrons are shown in red.

particle, the shorter the time interval it can exist. Next, we look at this process in more detail.

Virtual Pairs

Spontaneous creation can and does happen absolutely anywhere and at any time, not just under the unusual conditions of the early universe. (It is happening right now in the space between this book and your eyes.) Quantum mechanics tells us that if a process is not strictly forbidden, then it must occur. Pairs of every conceivable particle and antiparticle are constantly being created and destroyed at every location across the universe. However, we have no way of observing these pairs directly without violating the uncertainty principle. For this reason, they are called **virtual pairs.** They do not "really" exist in the same sense as ordinary particles; they "virtually" exist. The particles that are exchanged in the four fundamental forces (see Section 26-2) are also virtual particles.

Although virtual pairs of particles and antiparticles cannot be observed directly, their effects have nonetheless been detected. Imagine, for example, an electron in orbit about the nucleus of an atom, such as a hydrogen atom. Ideally, the electron should follow its orbit in a smooth, unhampered fashion. However, because of the constant brief appearance and disappearance of pairs of particles and antiparticles, minuscule electric fields exist for extremely short intervals of time. These tiny, fleeting electric fields cause the electron to jiggle slightly in its orbit. This jiggling produces slight changes in the energies of different electron orbits in the hydrogen atom, which manifest themselves as a minuscule shift in the wavelengths of the hydrogen spectral lines. (We discussed the connection between the energies of electron orbits in hydrogen and the hydrogen spectrum in Section 5-8.)

This shift was first detected in 1947 and today is known as the **Lamb shift.** The Lamb shift and more recent experiments provide powerful evidence that every point in space, all across the universe, is seething with virtual pairs of particles and antiparticles. In this sense, "empty space" is actually not empty at all. Figure 26-12 sketches the constant appearance and disappearance of virtual particles and antiparticles.

Pair production is routinely observed in high-energy particle accelerators. Indeed, it is one of the ways in which physicists manufacture exotic species of particles and antiparticles. The only requirement is that nature's balance sheet be satisfied. To create a particle and an antiparticle having a total mass M, the incoming gamma-ray photons must possess an amount of energy E that is greater than or equal to Mc^2. If the photons carry too little energy (less than Mc^2), pair production will not occur. Likewise, the more energetic the photons, the more massive the particles and antiparticles that can be manufactured.

Not all pairs of particles and antiparticles are virtual. In a phenomenon called **pair production,** pairs of highly energetic gamma rays (photons) can convert their energy into pairs of particles and antiparticles (Figure 26-13a). Quite simply, the gamma-ray photons disappear upon colliding into each other and convert their energy into a particle and an antiparticle. As real particles, they are directly observable and do not have to annihilate. While Figure 26-13a shows this process of pair production, Figure 26-13b shows the inverse process of **annihilation,** in which a particle and antiparticle collide with each other and are converted into high-energy gamma rays.

Around 1980, physicists began applying these ideas to their thinking about the creation of the universe. During the inflationary epoch, space was expanding with explosive vigor. As we have seen, however, all space is seething with virtual pairs of particles and antiparticles. Normally, a particle and an antiparticle have no trouble getting back together in a time interval (Δt) short enough to be in compliance with the uncertainty principle. During inflation, however, the universe expanded so fast that particles were rapidly separated from their corresponding antiparticles. Deprived of the opportunity to recombine and annihilate, these virtual particles became *real* particles in the real world. In this way, the universe was flooded with particles and antiparticles created by the violent expansion of space (Figure 26-14). For a "report card" on cosmology theories, including inflation, see the article at the end of this chapter.

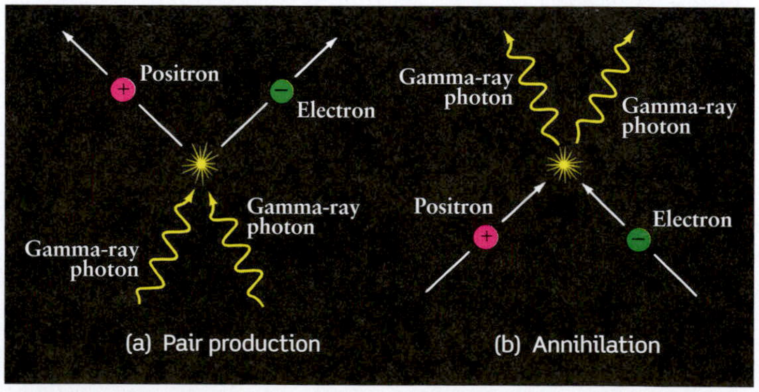

FIGURE 26-13

Pair Production and Annihilation **(a)** Pairs of virtual particles can be converted into real particles by high-energy gamma-ray photons. In this illustration, an electron (shown in green) and a positron (in red) are produced. This process can take place only if the combined energy of the two photons is no less than Mc^2, where M is the total mass of the electron and positron. **(b)** Conversely, a particle and an antiparticle can annihilate each other and be transformed into energy in the form of gamma rays.

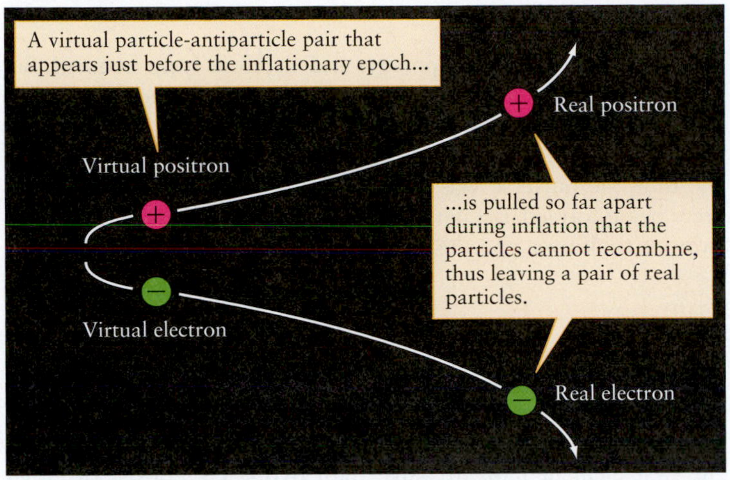

FIGURE 26-14
Inflation: From Virtual to Real Particles The universe expanded by such a tremendous factor during the inflationary epoch that the members of virtual particle-antiparticle pairs could no longer find each other. As a result, these virtual particles and antiparticles became real particles and antiparticles.

CONCEPTCHECK 26-6

Protons are about 2000 times more massive than electrons. If a virtual proton and antiproton spontaneously appear out of otherwise empty space, would this virtual pair last longer or shorter than a pair of virtual electrons before annihilating?

CONCEPTCHECK 26-7

Does the appearance of electrically charged virtual pairs violate the physical law that the total charge of an isolated system remains constant? (This law is also referred to as *conservation of charge.*)

Answers appear at the end of the chapter.

26-4 As the early universe expanded and cooled, most of the matter and antimatter annihilated each other

As soon as the flood of matter and antimatter appeared in the universe, collisions between particles and antiparticles began to produce numerous high-energy gamma rays. As these gamma rays collided, they promptly turned back into the particles and antiparticles from which they came. As a result, the rate of pair production soon equaled the rate of annihilation. For example, for every electron and positron that annihilated each other to create gamma rays (Figure 26-13b), two gamma rays collided elsewhere to produce an electron and a positron (Figure 26-13a). In other words, annihilation and pair production reactions proceeded with equal vigor, and as many particles and antiparticles were being created as were being destroyed.

As the universe continued to expand, all the gamma-ray photons became increasingly redshifted. As a result, the temperature of the radiation field fell. Due to their frequent interaction, radiation and particles of all kinds were in **thermal equilibrium:** All particle species, including photons, were at the same temperature. Hence, as the radiation temperature decreased, the temperature of particles of different types decreased as well.

From Quark Confinement to Particle-Antiparticle Annihilation

The first change in the population of particles and antiparticles occurred at $t = 10^{-6}$ second, when the temperature was 10^{13} K and particles were colliding with energies of roughly 1 GeV. Prior to this moment, particles collided so violently that individual protons and neutrons could not exist, being constantly fragmented into quarks. After this time, appropriately called the period of **quark confinement,** quarks were finally able to stick together and became confined within individual protons and neutrons.

LOOKING DEEPER 26-1

The cosmic background radiation we see today was spawned from a vast sea of particles and antiparticles in the early universe

As the universe continued to expand, temperatures eventually became so low that the gamma rays no longer had enough energy to create particular kinds of particles and antiparticles. We say that the temperature dropped below the particular particle's **threshold temperature.** Collisions between these types of particles and antiparticles continued to add photons to the cosmic-radiation background, but collisions between photons could no longer replenish the supply of particles and antiparticles.

At the same time that quark confinement became possible so that protons and neutrons appeared, the universe also became cooler than the 10^{13}-K threshold temperatures of both protons and neutrons. No new protons or neutrons were formed by pair production, but the annihilation of protons by antiprotons and of neutrons by antineutrons continued vigorously everywhere throughout space. This wholesale annihilation dramatically lowered the matter content (particles and antiparticles) of the universe, while simultaneously increasing the radiation (photon) content.

A little later, when the universe was about 1 second old, its temperature fell below 6×10^9 K, the threshold temperature for electrons and positrons. A similar annihilation of pairs of electrons and positrons further decreased the matter content of the universe while raising its radiation content. This radiation field, which fills all space, is the *primordial fireball* discussed in Section 25-5. This fireball, which dominated the universe for the next 380,000 years, therefore derived much of its energy from the annihilation of particles and antiparticles during the first second after the Big Bang.

Now we have a dilemma. If there had been perfect symmetry between particles and antiparticles, then for every proton there should have been an antiproton. For every electron, there should likewise have been a positron. Consequently, by the time the universe was 1 second old, every particle would have been annihilated by an antiparticle, leaving no matter at all in the universe.

Obviously, a total annihilation of all matter and antimatter never happened. The planets, stars, and galaxies we see in the sky are made of matter, not antimatter. If there were still substantial amounts of antimatter in the universe, it would eventually collide with ordinary matter. We would then see copious amounts of gamma rays being emitted from the entire sky. While we do observe gamma-ray photons from various locations in the universe, they are neither numerous enough nor of the right energy to indicate the presence of much antimatter. Thus, there must have been an excess of matter over antimatter immediately after the Big Bang so that the particles outnumbered the antiparticles.

Quite remarkably, we can estimate the extent of this initial asymmetry between matter and antimatter. In other words, while we essentially see a universe consisting of matter today, we can still infer how much antimatter there was before the antimatter portion of the universe annihilated with most of the matter portion. The key is light: Each annihilation of two particles produces exactly two photons (Figure 26-13b), and after being stretched to longer wavelengths, these are the photons we observe in the cosmic microwave background. As noted in Section 25-5, there are roughly 10^9 (one billion) photons today in the microwave background for each proton and neutron in the universe. Thus, for every 10^9 antiprotons, there must have been 10^9 plus one ordinary protons, leaving one surviving proton after annihilation. Similarly, for every 10^9 positrons, there must have been 10^9 plus one ordinary electrons.

While we can infer that the initial matter-antimatter asymmetry was about one additional particle made of regular matter per billion matter-antimatter pairs, we do not understand what process led to this initial asymmetry. There are several ideas for explaining the asymmetry but no hard evidence or agreement on a likely solution, and this initial asymmetry remains one of the great unsolved problems in physics and astronomy.

CONCEPTCHECK 26-8

Why does the following hypothesis seem unlikely: There is a perfect symmetry between matter and antimatter, but the antimatter is "somewhere else" and avoids annihilating with matter?

Answer appears at the end of the chapter.

26-5 A background of neutrinos and most of the helium in the universe are relics of the primordial fireball

The early universe must have been populated with vast numbers of neutrinos (ν) and their antiparticles, the antineutrinos ($\overline{\nu}$). These particles have a very small mass, so their threshold temperature is quite low. These particles take part in the nuclear reactions that transform neutrons into protons and vice versa. For example, a neutron can decay into a proton by emitting an electron and an antineutrino:

$$n \rightarrow p + e^- + \overline{\nu}$$

This radioactive decay happens quickly (its half-life is about 10.5 minutes), which is why we do not find free neutrons floating around in the universe today. In the first 2 seconds after the Big Bang, however, neutrons were also created by collisions between protons and electrons:

$$p + e^- \rightarrow n + \nu$$

This reaction kept the number of neutrons approximately equal to the number of protons. This balance was maintained only as long as collisions between protons and electrons were frequent. But the number of electrons decreased precipitously as the temperature of the universe fell below 6×10^9 K and electrons and positrons annihilated each other without being replenished (Section 26-4). By the time the universe was about 2 seconds old, no new neutrons were being formed, the natural tendency for neutrons to decay into protons took over, and the number of neutrons began to decline.

Spawning the First Nuclei

> The first atomic nuclei formed within a quarter-hour after the Big Bang

Before many of the neutrons could decay into protons, they began to combine with protons to form nuclei. Nuclei of helium, the first element more massive than hydrogen, consist of either two protons and two neutrons (^{4}He) or two protons and a single neutron (^{3}He). It is exceedingly improbable that two protons and one or two neutrons should all simultaneously collide with one another to form a helium nucleus. Instead, helium nuclei are built in a series of steps. The first step is to have a single proton and a single neutron combine to form deuterium (^{2}H), sometimes called "heavy hydrogen." A photon (γ) is emitted in this process, so we write this reaction as

$$p + n \rightarrow {}^2H + \gamma$$

Forming deuterium, however, does not immediately lead to the formation of helium. The problem is that deuterium nuclei are easily destroyed, because a proton and a neutron do not stick together very well. Indeed, in the early universe, high-energy gamma rays easily broke deuterium nuclei back down into independent protons and neutrons. As a result, the synthesis of helium could not get beyond the first step. This block to the creation of helium is called the **deuterium bottleneck.**

When the universe was about 3 minutes old, the background radiation had cooled enough that its photons no longer had enough energy to break up the deuterium. By this time, most of the neutrons had decayed into protons, and protons outnumbered neutrons by about 6 to 1. Because deuterium nuclei could now survive, the remaining neutrons combined with protons and rapidly produced helium. (The Cosmic Connections: The Proton-Proton Chain figure for Chapter 16 depicts a similar sequence of reactions that take place in the core of the present-day Sun.)

The result was what we find in the universe today—about 1 helium atom for every 10 hydrogen atoms. In addition to helium, nuclei of lithium (Li, which has 3 protons) and beryllium (Be, which has 4) were also produced in small numbers. The process of building up nuclei such as deuterium and helium from protons and neutrons is called **nucleosynthesis** (Figure 26-15).

Because nuclei have positive electric charges, bringing them together to form more massive nuclei requires that they overcome their mutual electric repulsion. They are unable to do so if they are moving too slowly, which will be the case if the temperature is too low. As a result, by about 15 minutes after the Big Bang the universe was no longer hot enough for nucleosynthesis to take place. Only the four lightest elements (hydrogen, helium, lithium, and beryllium) were present in appreciable numbers. The heavier elements would be formed only much later, once stars had formed and nuclear reactions within those stars could manufacture carbon, nitrogen, oxygen, and all the other elements.

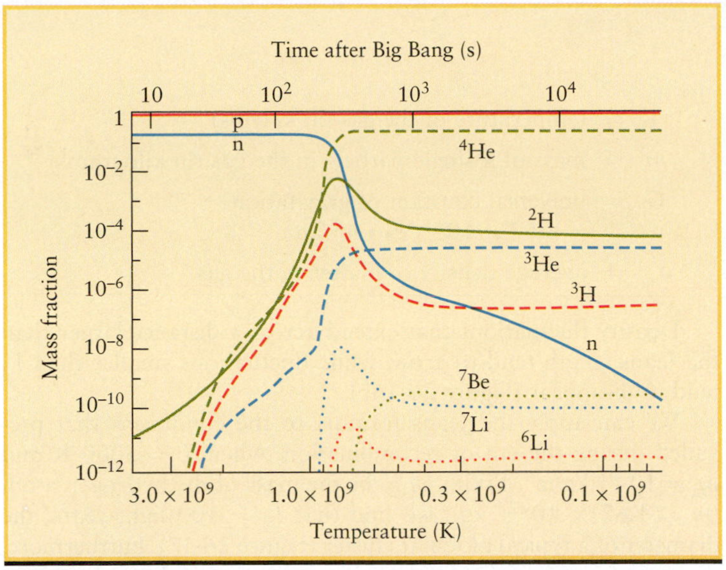

FIGURE 26-15

Nucleosynthesis in the Early Universe This graph shows how nuclei were produced between 10 seconds and 10 hours after the Big Bang. The vertical axis shows the fraction of the total mass that was in each type of particle or nucleus (p = proton, n = neutron, ^2H = deuterium, ^3He and ^4He = helium, ^6Li and ^7Li = lithium, ^7Be = beryllium). Very few nuclei were formed before the universe was 10 seconds old, due to the phenomenon of the deuterium bottleneck, which occurred at times earlier than those shown here. By about 10^3 seconds (roughly 15 minutes) after the Big Bang, the temperature had dropped below 4×10^8 K, no further nucleosynthesis was possible, and the relative amounts of different nuclei stabilized. The number of free neutrons declined rapidly as these particles decayed into protons, electrons, and antineutrinos. Agreement between these predicted mass fractions with what is observed today provides strong evidence for nuclear reactions in the early universe. (Adapted from R. V. Wagoner)

Nucleosynthesis was one of the great early successes of the Big Bang theory. At lower temperatures (to the right in Figure 26-15), the amounts of different nuclei leveled off and remains unchanged. That allows us to compare the amount of each predicted atomic element to the amount observed today, and the agreement is excellent. Furthermore, these nucleosynthesis calculations also predicted the density of ordinary matter in the universe, and this prediction has been verified by the CMB fluctuations as in Figure 25-22 and Table 25-2. Taken together, the success of nucleosynthesis provides very strong evidence for a hot Big Bang, and that during its first few minutes the entire universe was one big nuclear reactor.

CAUTION! Keep in mind that only *nuclei* formed in the first 15 minutes of the history of the universe. It would be another 380,000 years before temperatures became low enough for these nuclei to combine with electrons to form neutral atoms.

The Neutrino-Antineutrino Background

While nuclei were being formed in the early universe, what happened to all those primordial neutrinos and antineutrinos that had interacted so vigorously with the protons and neutrons before the universe was 2 seconds old? The answer is that by $t = 2$ seconds, matter was sufficiently spread out so that the universe became transparent to neutrinos and antineutrinos. From that time on, neutrinos and antineutrinos could travel across the universe unimpeded. Their interaction with matter is so weak that about a trillion neutrinos pass right through our bodies every second. Even Earth itself is virtually transparent to neutrinos from the Sun (Section 16-4).

The neutrinos and antineutrinos that were liberated at $t = 2$ seconds should now fill the universe much as the cosmic microwave background does. Indeed, these ancient neutrinos and antineutrinos may be about as populous today as the photons in the microwave background (of which there are 4.1×10^8 per cubic meter). The *neutrino-antineutrino background* should be slightly cooler than the photon background, which received extra energy from electron-positron annihilations. Physicists estimate that the current temperature of the neutrino-antineutrino background is about 2 K, as opposed to 2.725 K for the microwave background. Unfortunately, because neutrinos and antineutrinos are so difficult to detect, we do not yet have direct evidence of the neutrino-antineutrino background.

CONCEPTCHECK 26-9

How does the changing temperature of the universe confine the production of the elements helium, lithium, and beryllium to a period ranging from about 3 minutes to 15 minutes after the Big Bang?

Answer appears at the end of the chapter.

26-6 Galaxies and the first stars formed from density fluctuations in the early universe

The distribution of matter in the universe today is quite lumpy. Stars are grouped together in galaxies, galaxies into clusters, and clusters into superclusters that stretch across 50 Mpc (150 million ly) or more (see Section 23-6). Furthermore, galaxies seem to be concentrated along enormous sheets, which in turn surround voids measuring 30 to 120 Mpc (100 million to 400 million ly) across. Figures 23-23 and 23-24 show these features, which characterize the large-scale structure of the universe. How did this large-scale structure arise from the chaos of the primordial fireball? When did stars first appear in the universe? And when and how did galaxies first form?

Density Fluctuations and the Jeans Length

At first glance the origin of large-scale structure seems puzzling, because the early universe must have been exceedingly smooth. To see why, think back to the era of recombination that occurred 380,000 years after the Big Bang (see Section 25-5). Before recombination, high-energy photons were constantly and vigorously colliding with charged particles throughout all space. After recombination, the universe became transparent, and these photons stopped interacting with the matter in the universe. Astronomers say that matter "decoupled" from radiation during the era of recombination. Because the cosmic microwave background is extremely isotropic, we can conclude that the matter with which these photons once collided so frequently must also have been spread smoothly across space.

The distribution of matter during the early universe could not have been *perfectly* uniform, however. If it had been, it would still have to be absolutely uniform today; there would now be only a few atoms per cubic meter of space, with no stars and no galaxies. Consequently, there must have been slight lumpiness, or **density**

fluctuations, in the distribution of matter in the early universe. These fluctuations are thought to have originated in the very early universe, even before the inflationary epoch. Infinitesimally small quantum fluctuations in density, which are required by the Heisenberg uncertainty principle (see Section 26-3), were stretched during inflation to appreciable size. Through the action of gravity, these fluctuations initiated the clumping that eventually grew to become the galaxies and clusters of galaxies that we see today throughout the universe. As we saw in Section 25-5 and Section 25-8, the pattern of density fluctuations became imprinted on the cosmic background radiation during the era of recombination. Figure 25-13 shows a map of these fluctuations obtained from the WMAP microwave background spacecraft.

Our understanding of how gravity can amplify density fluctuations dates back to 1902, when the British physicist James Jeans solved a problem first proposed by Isaac Newton. Suppose that you have a gas with only very tiny fluctuations in density, as shown in Figure 26-16a. These regions of higher density will then gravitationally attract nearby material and thus gain mass. As the regions become more massive, however, the pressure of the gas inside these regions will also increase, which can make these regions expand and disperse. Jeans analyzed these opposing effects and asked the question: Under what conditions does gravity overwhelm gas pressure so that a permanent object can form?

Jeans proved that an object will grow due to a density fluctuation provided that the fluctuation extends over a distance that exceeds the so-called **Jeans length** (L_J):

Jeans length for density fluctuations

$$L_J = \sqrt{\frac{\pi k T}{m G \rho_m}}$$

L_J = Jeans length

k = Boltzmann constant = 1.38×10^{-23} J/K

T = temperature of the gas (in kelvins)

m = mass of a single particle in the gas (in kilograms)

G = universal constant of gravitation = 6.67×10^{-11} N • m^2/kg^2

ρ_m = average density of matter in the gas

Density fluctuations that extend across a distance larger than the Jeans length tend to grow, while fluctuations smaller than L_J tend to disappear (Figure 26-16b).

We can apply the Jeans formula to the conditions that prevailed during the era of recombination, when $T = 3000$ K and $\rho_m = 10^{-18}$ kg/m^3. Taking m to be the mass of the hydrogen atom ($m = 1.67 \times 10^{-27}$ kg), we find that $L_J = 100$ light-years, the diameter of a typical globular cluster (Figure 26-17). Furthermore, the mass contained in a cube whose sides are 1 Jeans length in size (equal to the product of the density, ρ_m, and the volume of the cube, $L_J{}^3$) is about 5×10^5 $M_\odot$, equal to the mass of a typical globular cluster. These calculations suggest that globular clusters were among the first objects to form after recombination.

CONCEPTCHECK 26-10

Consider a density fluctuation around the era of recombination when the universe had a temperature of $T = 3000$ K. If the density fluctuation is about 10 light-years across, will the matter within that fluctuation begin clumping together? What causes or prevents clumping?

Answers appear at the end of the chapter.

Population III Stars: The "Zeroth" Generation

We saw in Section 19-4 that globular clusters contain the most ancient stars we can find in the present-day universe. These are Population II stars with a low percentage of metals (elements heaver than hydrogen and helium), and are of an earlier stellar

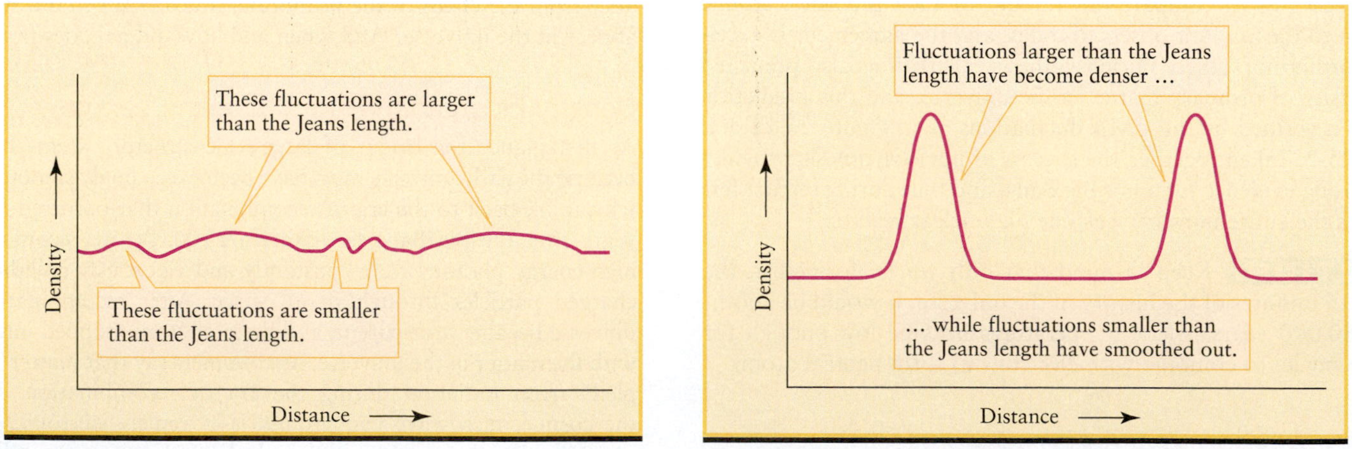

(a) At an early time (b) At a later time

FIGURE 26-16

The Growth of Density Fluctuations **(a)** This conceptual illustration shows small density fluctuations in the distribution of matter shortly after the era of recombination. **(b)** If the size of a fluctuation is greater than the Jeans length (L_J), it becomes gravitationally unstable and can grow in amplitude.

FIGURE 26-17 R I V U X G

A Globular Cluster A typical globular cluster contains 10^5 to 10^6 stars, each with an average mass of about 1 $M_\odot$, so the total mass of a typical cluster is 10^5–10^6 $M_\odot$. Cluster diameters range from about 6 to 120 pc (20 to 400 ly). Because these masses and diameters are comparable to the Jeans length (L_J) during the era of recombination, astronomers suspect that globular clusters were among the first objects to form in the universe. (Hubble Heritage Team/AURA/STScI/NASA)

generation than metal-rich Population I stars like the Sun (see Section 19-5). However, the Population II stars in globular clusters *cannot* be the very first stars to have formed after the Big Bang. Those first stars could have contained only hydrogen, helium, and tiny amounts of lithium and beryllium; as Figure 26-15 shows, these were the only elements whose nuclei formed in the early universe. Hence, these original stars would have contained an even smaller percentage of metals than the Population II stars found in globular clusters. Such "zeroth-generation" stars are called **Population III stars.**

ANIMATION 26-1

Like stars in the present-day universe, Population III stars would have formed from clouds of gas. These ancient gas clouds were composed almost exclusively of hydrogen and helium atoms, and such clouds have higher internal pressures than do metal-rich clouds of the same temperature. A star can form only when the mutual gravitation of the various parts of a cloud (which tends to make the cloud collapse) overcomes the internal pressure of the cloud (which tends to prevent collapse). Hence, Population III stars could form only if their mass (and hence their mutual gravitation) was rather large. Calculations suggest that these stars had masses from 30 to 1000 $M_\odot$, compared to the range of 0.4 to roughly 120 $M_\odot$ for modern stars. Even the smallest Population III star would rank among the largest stars observed today.

The "zeroth generation" of stars were much more massive and luminous than stars today

Although no Population III stars have yet been observed directly, we have at least indirect evidence that they existed. Infrared images from the Spitzer Space Telescope like the one that opens this chapter reveal a distant infrared background of starlight that is what we would expect to see from these zeroth-generation stars.

Another bit of evidence follows from the tremendous energy output of these stars: A 1000-$M_\odot$ Population III star would have been millions of times more luminous than the Sun and have a surface temperature in excess of 10^5 K, causing it to emit a flood of short-wavelength, high-energy photons. The photons from even a small number of such stars would have ionized most of the atoms in the universe, leaving electrons and nuclei of hydrogen and helium. This process is called **reionization,** a name that reminds us the universe had previously been ionized prior to recombination at $t = 380{,}000$ years. We saw in Section 25-8 that the photons of the cosmic microwave background are scattered by free electrons, and the effects of this scattering can be detected in maps of the background radiation. Data from the WMAP spacecraft suggest that reionization took place around 400 million years after the Big Bang, which in turn suggests that Population III stars formed around that time.

Since Population III stars were all very massive, their lifetimes were short and none could have survived to the present day. But during their short lifetimes, thermonuclear reactions within these stars produced elements heavier than beryllium for the first time in the history of the universe. What's more, calculations suggest that when these stars exploded—and due to their great mass, all of them did—they did not leave a white dwarf, neutron star, or black hole behind. Instead, all of their mass was ejected into space to be incorporated into the next generation of stars. The presence of heavy elements in the ejected material meant that when this material subsequently formed into clouds, the internal pressure of these clouds was lowered substantially and it became possible for low-mass stars to form. This laid the foundation for today's universe, in which dim, low-mass stars are common and luminous, massive stars are the exception (see Figure 17-5).

The era from recombination at $t = 380{,}000$ years to the first stars at $t = 400$ million years is called the **dark ages.** The only photons present at that time were those that make up today's cosmic background radiation. The dark ages ended when the universe was filled for the first time with the light from stars (Figure 26-18).

Forming Large-Scale Structure

Once clumps the size of globular clusters had formed in the universe, how did they form into galaxies, clusters of galaxies, and larger structures? One issue that complicates this matter is the presence of dark energy, which acts to accelerate the expansion of the universe (see Section 25-6). This accelerated expansion pulls clumps of material away from each other and makes it more difficult for them to coalesce into larger structures. Another complication is that about 85% of the mass in the universe is in the form of dark matter, whose nature is not known (see Section 23-8). Researchers

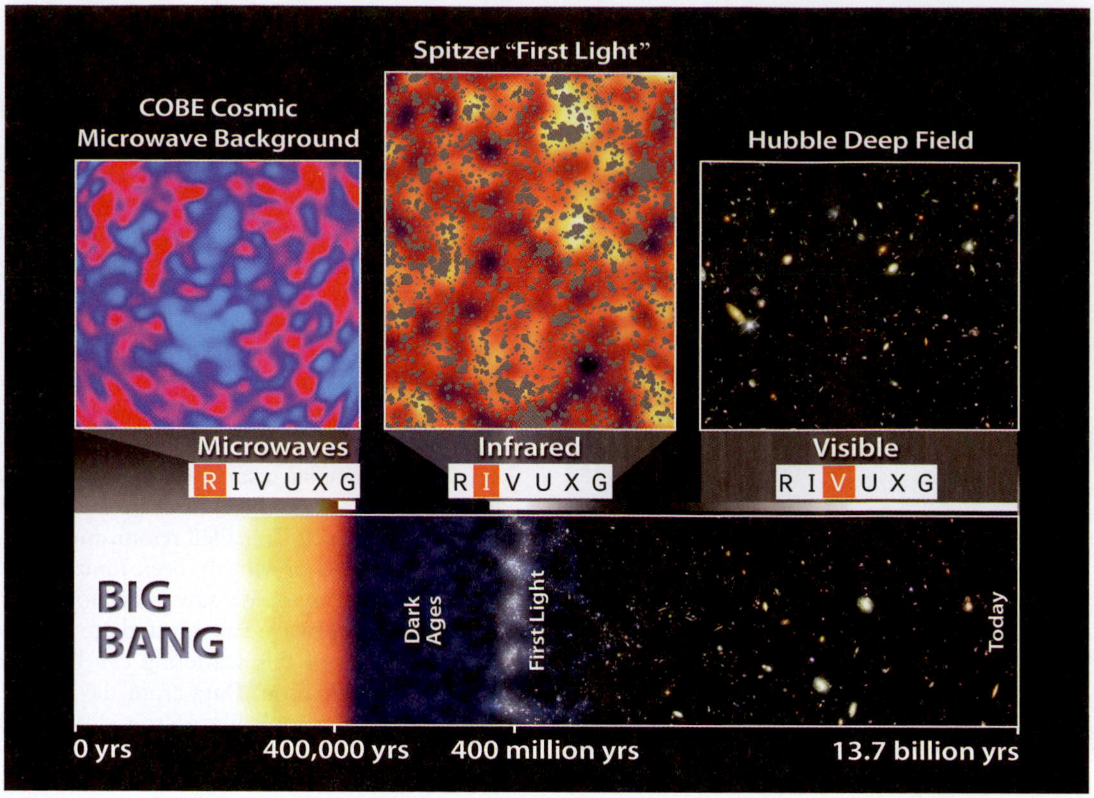

FIGURE 26-18

A Timeline of Light in the Universe The oldest light that we can see today is the cosmic background radiation, which comes from a time 380,000 years after the Big Bang when the universe first became transparent. This light has a redshift of about $z = 1100$ and appears in the microwave spectrum. Some 400 million years later at a redshift of about $z = 11$ the first stars appeared; their light is now redshifted to infrared wavelengths. Galaxies formed more recently and can be seen at visible wavelengths. (NASA; JPL-Caltech; and A. Kashlinsky, Goddard Space Flight Center)

have hypothesized different types of dark matter in the hope of explaining the large-scale structure that we see. Neutrinos are an example of **hot dark matter,** so named because it consists of lightweight particles traveling at high speeds. **Cold dark matter,** on the other hand, consists of massive particles traveling at slow speeds. Examples include WIMPs (which we discussed in Section 22-4) as well as other exotic, speculative particles.

Scientists use supercomputer simulations to see how different types of dark matter would influence the development of large-scale structure. Figure 26-19 shows the results of such a simulation for a flat universe with dark energy and *cold* dark matter. The simulation follows the motions of 2 million particles of cold dark matter in a box that expands as the universe expands. The box at the lower right of the figure, representing the present time (redshift $z = 0$) is 43 Mpc (160 million ly) a side. At earlier times, the box represents a volume whose side is smaller by a factor $1/(1 + z)$. For example, each side of the box for $z = 0.99$ is actually 1/1.99 as long as the box for $z = 0$.

The simulation begins 120 million years after the Big Bang with an almost perfectly uniform distribution of particles, mimicking the tiny density fluctuations that must have been present just after inflation. A supercomputer then calculates how these particles move, based on Newton's laws in an expanding universe. As time goes on, the fluctuations grow into small, bright clumps whose sizes and masses are similar to those of galaxies. A large filament also forms, spanning the entire box from left to right. The simulation shows that no additional structures formed after the universe was about 6 billion years old, corresponding to redshift $z = 1$. The explanation is that after this time, the accelerating expansion of the universe becomes more important than gravitational attraction. The final frame of the simulation strongly resembles actual maps of galaxies in our present-day universe (see Figure 23-24).

Simulations similar to those in Figure 26-19 have also been carried out using *hot* dark matter in the form of neutrinos. A massless neutrino would always travel at the speed of light, just as a photon does. However, experiments show that neutrinos do have a small mass. (This nonzero mass is what allows one type of neutrino to transform into another. We saw in Section 16-4 that such transformations provided the explanation to the long-standing solar neutrino problem.) Hence, neutrinos travel slower than light, and slow down as the universe expands and cools. Slow-moving neutrinos would accumulate over time within density fluctuations,

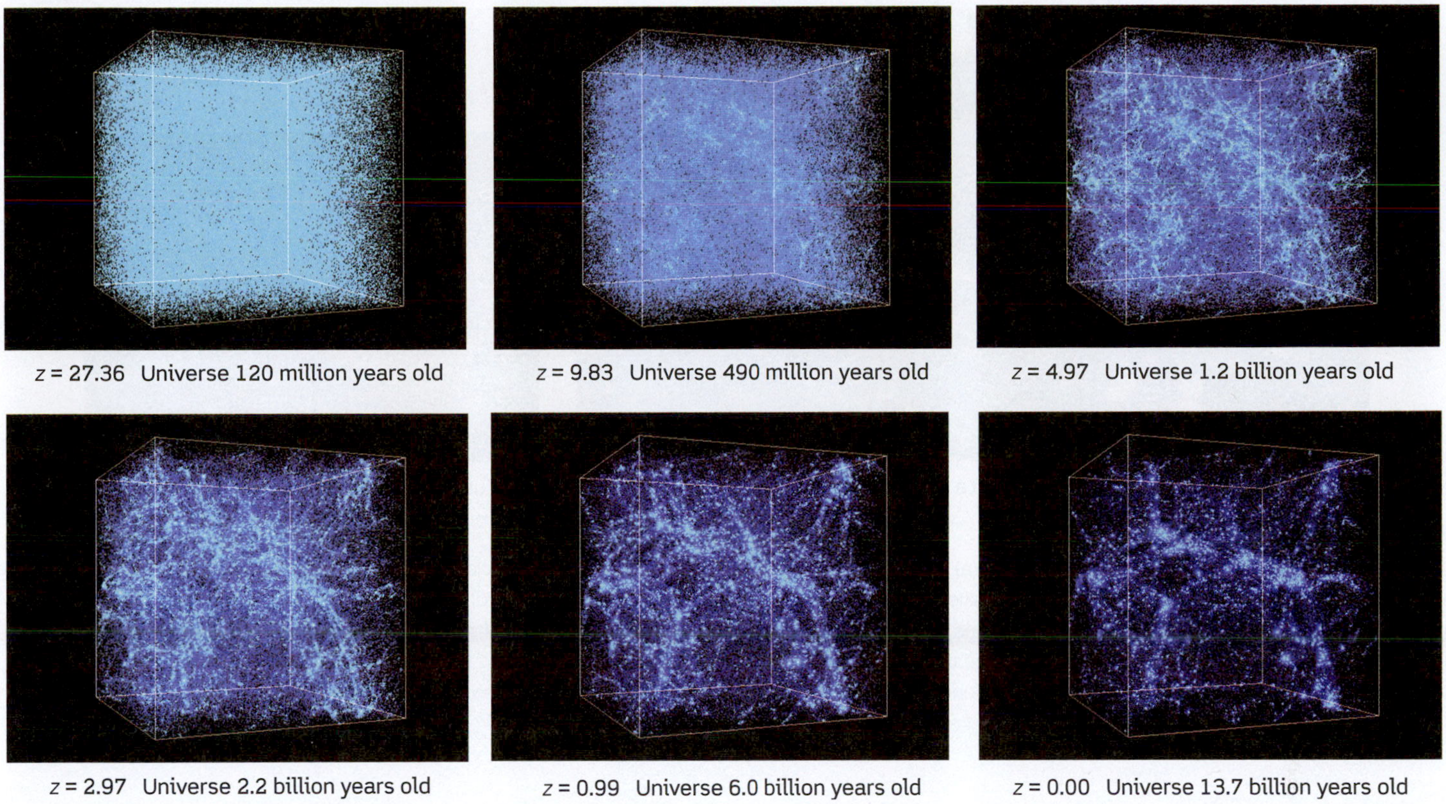

ANIMATION 26-2 ANIMATION 26-3

FIGURE 26-19

A Cold Dark Matter Simulation with Dark Energy These six views show the evolution of dark matter particles in a large, box-shaped volume of space. The box actually expands with time to follow the expansion of the universe and holds the same number of particles. In this figure, the boxes have been rescaled so they all appear at the same size. Small fluctuations in density are put into the simulation at the beginning (at upper left); these evolve over time to form structures that resemble those actually observed in our present-day universe ($z = 0.00$, shown at the lower right). (Applications by Andrey Kravtsov, University of Chicago, and Anatoly Klypin, New Mexico State University; visualizations by Andrey Kravtsov)

and the gravitational pull of these neutrinos on surrounding matter could eventually lead to the formation of clusters of galaxies.

A primary difference between simulations based on cold and hot dark matter is the way in which galaxies form. In calculations based on cold dark matter, the formation of galaxies takes place from the "bottom up." In these simulations, the densest gas undergoes collapse early in the history of the universe and stars begin to form. The regions of star formation stream along the filaments (Figure 26-20). When they meet at the intersections between

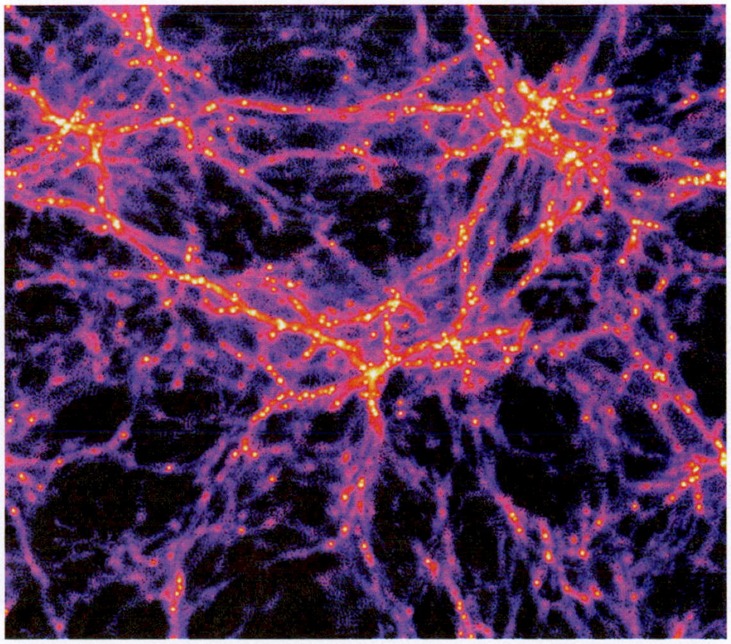

FIGURE 26-20

"Bottom-Up" Galaxy Formation: Simulation This image is taken from a cold dark matter simulation like that shown in Figure 26-15. A portion of the universe is shown at a time 2.2 billion years after the Big Bang, corresponding to redshift $z = 3.04$. The colors indicate the density of gas: Yellow is highest, red is medium, and blue is the lowest density. Over time, the gas tends to pile up at points where filaments intersect, forming galaxies and clusters of galaxies. (T. Theuns, MPA Garching/ESO)

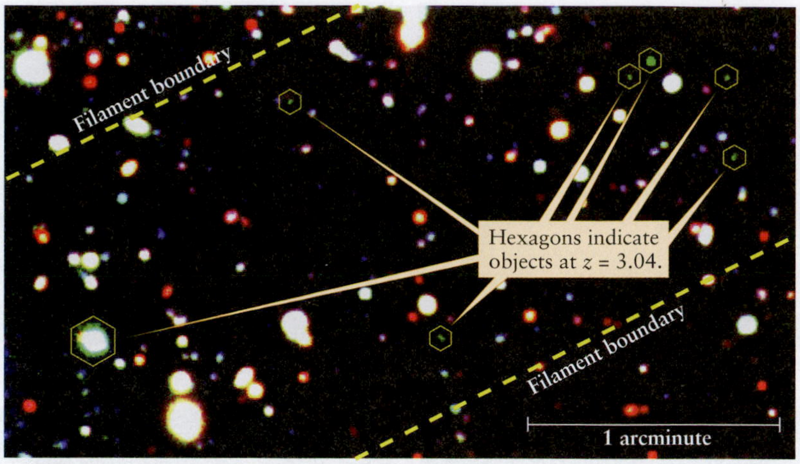

(a) High-redshift objects that lie within a filament

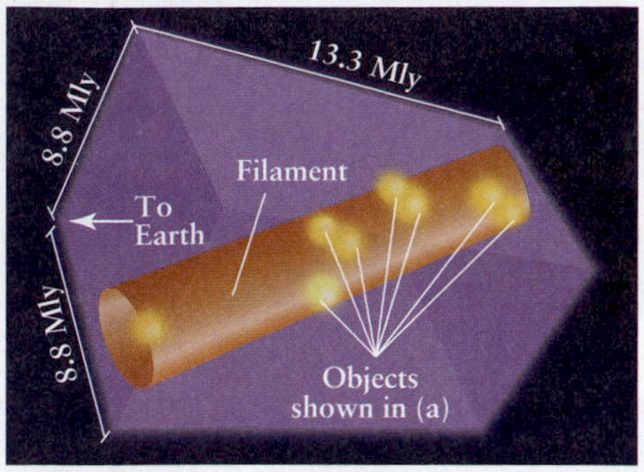

(b) Illustration of the filament

FIGURE 26-21 R I V U X G

"Bottom-Up" Galaxy Formation: Observation **(a)** The hexagons in this image from the Very Large Telescope show the positions of a number of sub-galaxy-sized objects at a redshift $z = 3.04$, the same as in the simulation shown in Figure 26-16. Excited hydrogen atoms in these objects emit ultraviolet photons, which are redshifted to visible wavelengths. This redshifted emission gives these objects a characteristic green color. (The object at lower left actually lies in front of a much brighter quasar.) **(b)** The objects in (a) all lie within an immense filament. The purple box shows the volume of space studied in this observing program. The dimensions are given in millions of light-years (Mly). (European Southern Observatory)

filaments, they merge and group together into galaxies, then clusters of galaxies, then superclusters. But in calculations based on hot dark matter, galaxies form from the "top down." Huge supercluster-sized sheets of matter form first and then fragment into galaxies. Observations of remote galaxies show that galaxies actually formed from the "bottom up" scenario. One piece of evidence for this is the image in Figure 26-21a, which shows a handful of "galaxy building blocks" at $z = 3.04$ (when the universe was 2.2 billion years old). These "building blocks," which have not yet coalesced into galaxies, lie within a long filament (see Figure 26-21b) that resembles those shown in the simulation of Figure 26-20. Figure 26-22 shows a collection of "building blocks" at a later stage in the process of merging into a galaxy. These observations strongly suggest that the dominant form of dark matter is cold, not hot.

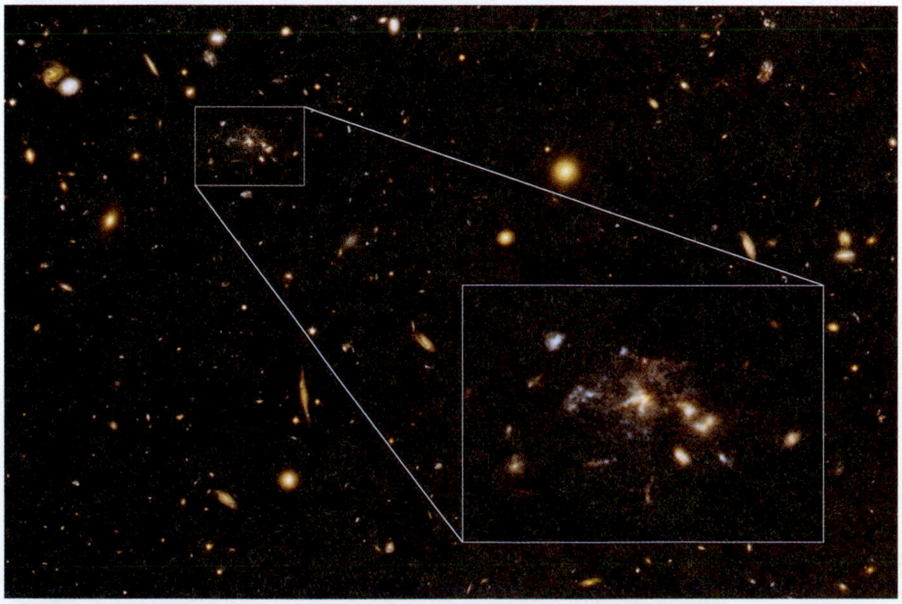

FIGURE 26-22 R I V U X G

A Galaxy Under Construction This Hubble Space Telescope image shows dozens of small galaxies in the process of merging into a single large galaxy. We see this galaxy at a redshift $z = 2.2$, corresponding to a time 3.1 billion years after the Big Bang. (NASA; ESA; G. Miley and R. Overzier, Leiden Observatory; and the ACS Science Team)

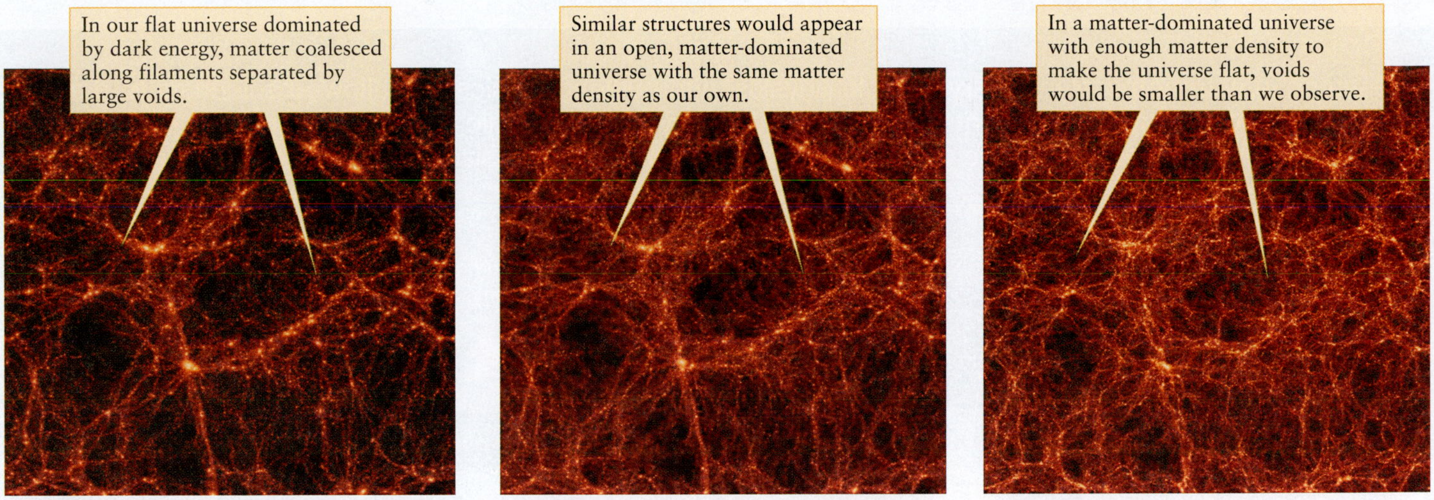

In our flat universe dominated by dark energy, matter coalesced along filaments separated by large voids.
Similar structures would appear in an open, matter-dominated universe with the same matter density as our own.
In a matter-dominated universe with enough matter density to make the universe flat, voids would be smaller than we observe.

(a) A flat universe with dark energy: $\Omega_m = 0.3, \Omega_\Lambda = 0.7$

(b) A open universe without dark energy: $\Omega_m = 0.3, \Omega_\Lambda = 0$

(c) A flat universe without dark energy: $\Omega_m = 1.0, \Omega_\Lambda = 0$

FIGURE 26-23

Using Simulations to Constrain the Matter Density of the Universe Cold dark matter simulations like those in Figures 26-15 and 26-16 help astronomers determine the value of the matter density parameter Ω_m. These three simulations show a portion of the universe at $z = 0$. **(a)** A simulation with $\Omega_m = 0.3$ and $\Omega_\Lambda = 0.7$, close to the values for our universe, gives a good match to the observed distribution of filaments and voids. **(b)** Nearly as good a match is obtained if we keep $\Omega_m = 0.3$, but eliminate dark energy so that $\Omega_\Lambda = 0$. **(c)** If we use a larger value of Ω_m, the distribution of matter in the simulation is a poor match to our universe. (Simulation by the Virgo Supercomputing Consortium using computers based at the Computer Centre of the Max Planck Society in Garching and at the Edinburgh Parallel Computer Centre)

What Large-Scale Structure Reveals

How might the universe have evolved if it had contained different amounts of cold dark matter and dark energy? Figure 26-23 shows some simulations designed to explore these possibilities. If the density of matter in the universe is kept constant for various simulations, the simulations predict approximately the same structure for different values of the dark energy density parameter Ω_Λ defined in Section 25-6 (see Figure 26-23a and Figure 26-23b). But if too large a matter density is used in the simulation, the voids between galaxies are smaller than what we actually observe in our universe (Figure 26-23c). Hence, observations of galaxy clustering coupled with supercomputer simulations of galaxy formation help determine the matter density of our universe. (We made use of this idea in Section 25-7. The brown band in Figure 25-18 shows the constraints on cosmological parameters from these observations and simulations of galaxies.)

The best match to the observed distribution of galaxy clusters and to the cosmic background radiation data (see Figure 25-22) is a model like that shown in Figure 26-19, with dark energy and cold dark matter in the proportions listed in Table 25-2.

Cosmic Connections: The History of the Universe summarizes the past history of our universe down to the present day. As we discussed in Section 25-7, the *future* of our universe is less certain and depends on the detailed character of dark energy. More detailed data about galaxy clusters and the cosmic background radiation will be needed to pin down the future evolution of our universe.

CONCEPTCHECK 26-11

How does structure formation in the early universe provide clues to the nature of dark matter?

Answer appears at the end of the chapter.

26-7 String theory attempts to unify the fundamental forces and predicts that the universe may have 11 dimensions

While we have a growing understanding of the early universe, there remains a veil obscuring the first 10^{-43} seconds after the Big Bang. These very *first* moments in the history of the universe, whose duration was the Planck time, set the stage for what would come after. To understand this brief interval, we need to construct a quantum-mechanical theory that unifies gravity with the other fundamental forces of nature and that reconciles quantum mechanics with gravity—a theory of quantum gravity. While unification remains an unfinished task, remarkable progress has been made in recent years. One major development is that physicists have abandoned the idea that there are only three dimensions of space.

COSMIC CONNECTIONS

The History of the Universe

Research in astronomy, elementary particle physics, and nuclear physics has allowed scientists to piece together the grand sweep of events over the first billion years of cosmic history. The greatest drama occurred in the first few minutes, and these graphs have been drawn to emphasize those earliest moments.

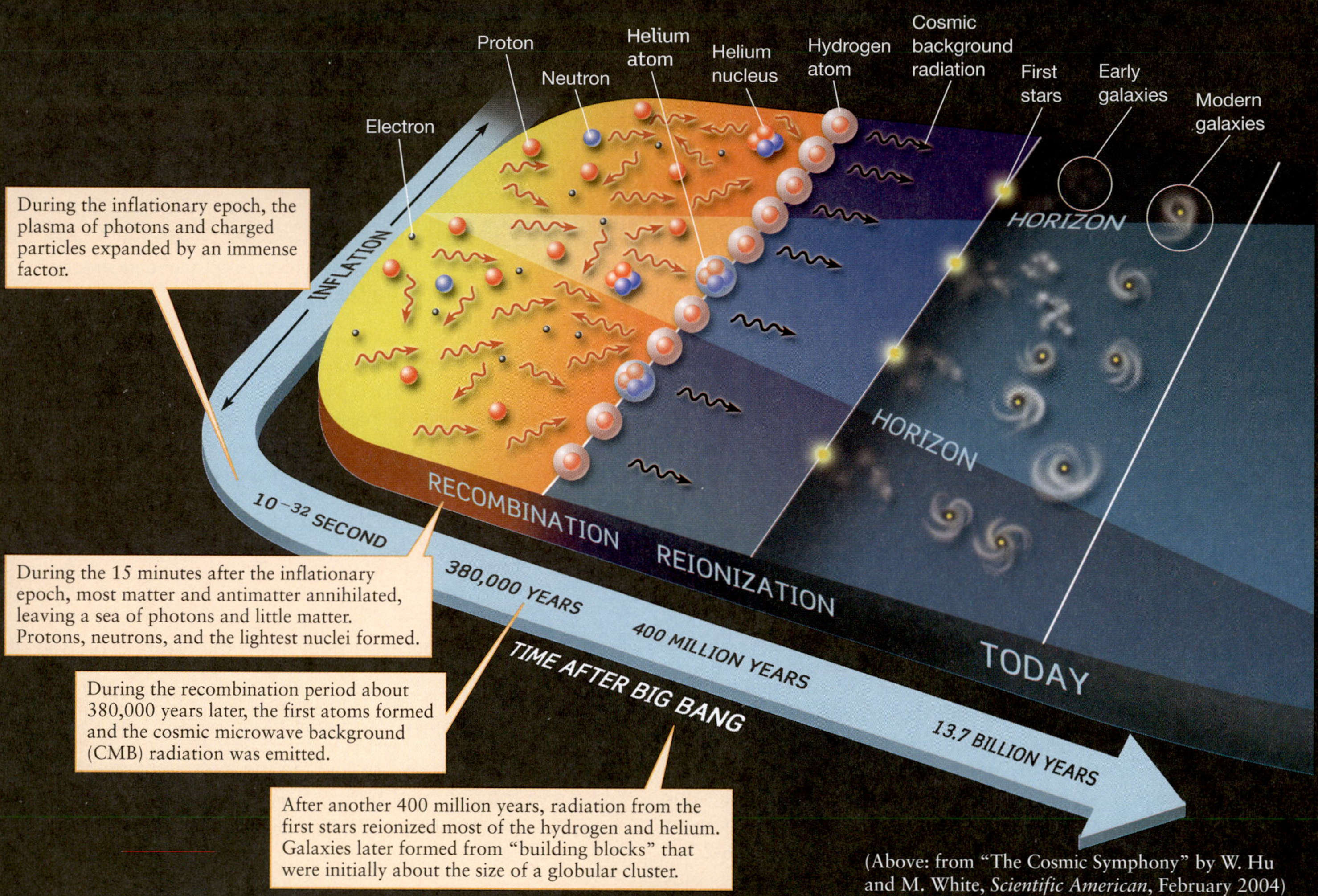

Proton
Helium atom
Helium nucleus
Hydrogen atom
Cosmic background radiation
Neutron
First stars
Early galaxies
Modern galaxies
Electron
HORIZON
HORIZON
INFLATION
During the inflationary epoch, the plasma of photons and charged particles expanded by an immense factor.
RECOMBINATION
REIONIZATION
TODAY
10^{-32} SECOND
380,000 YEARS
400 MILLION YEARS
13.7 BILLION YEARS
TIME AFTER BIG BANG
During the 15 minutes after the inflationary epoch, most matter and antimatter annihilated, leaving a sea of photons and little matter. Protons, neutrons, and the lightest nuclei formed.
During the recombination period about 380,000 years later, the first atoms formed and the cosmic microwave background (CMB) radiation was emitted.
After another 400 million years, radiation from the first stars reionized most of the hydrogen and helium. Galaxies later formed from "building blocks" that were initially about the size of a globular cluster.

(Above: from "The Cosmic Symphony" by W. Hu and M. White, *Scientific American*, February 2004)

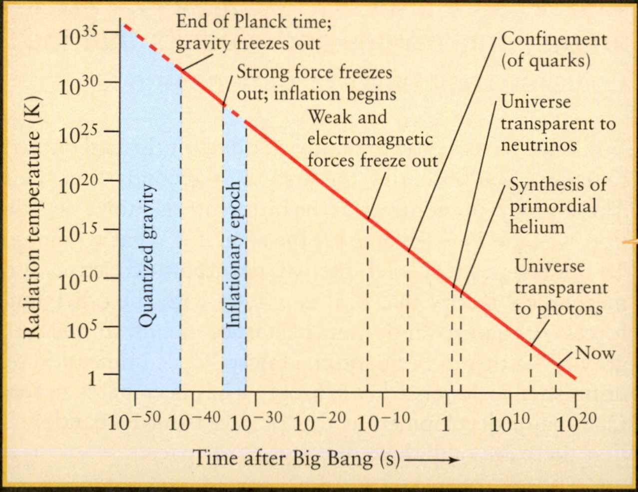

10^{35}
10^{30}
10^{25}
10^{20}
10^{15}
10^{10}
10^{5}
1
Radiation temperature (K)
End of Planck time; gravity freezes out
Strong force freezes out; inflation begins
Weak and electromagnetic forces freeze out
Confinement (of quarks)
Universe transparent to neutrinos
Synthesis of primordial helium
Universe transparent to photons
Now
Quantized gravity
Inflationary epoch
10^{-50} 10^{-40} 10^{-30} 10^{-20} 10^{-10} 1 10^{10} 10^{20}
Time after Big Bang (s)

This graph shows the temperature history of the universe.
- As the universe cooled, the four forces "froze out" of their unified state as a result of spontaneous symmetry breaking.
- Neutrons and protons froze out of the hot "quark soup" during the quark confinement stage, which occured 10^{-6} second after the Big Bang.

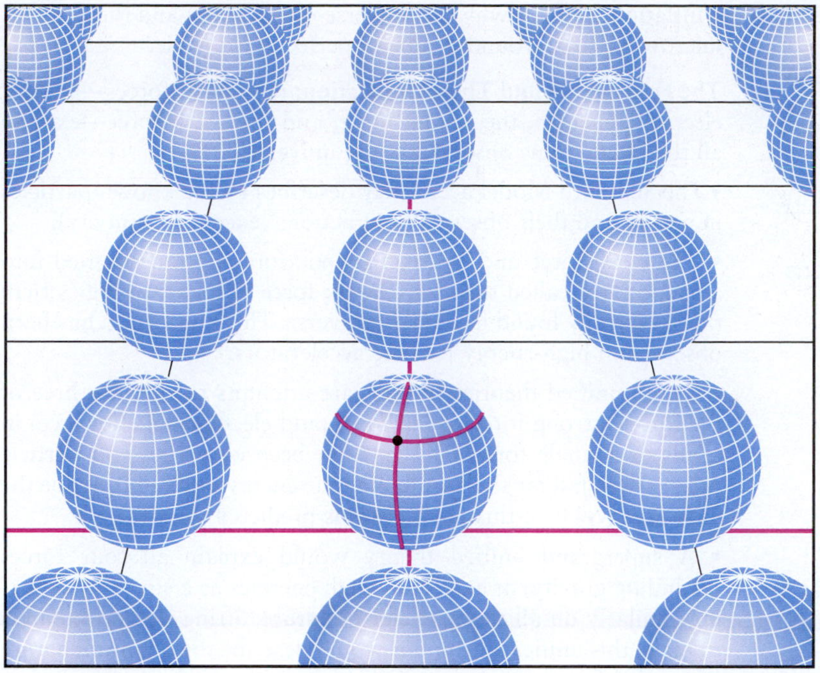

FIGURE 26-24

Hidden Dimensions of Space Hidden dimensions of space might exist provided they are curled up so tightly that we cannot observe them. This drawing shows how an ordinary two-dimensional plane might contain two additional dimensions. At every point on the plane, there is a very tiny sphere so small that it cannot be seen. To pinpoint a particular location, you need to give not only a position on the plane but also a position on the sphere. Thus, if at every point, space really has additional curled up dimensions represented by these spheres, then space has more than the three regular dimensions of space we are familiar with. Like a hamster running endlessly on its wheel, moving in a curled up dimension is not the same as traveling through regular space. (Adapted from D. Freedman and P. Van Nieuwenhuizen)

Beyond Four Dimensions

In his special and general theories of relativity, Einstein combined time with the three known dimensions of ordinary space, resulting in a four-dimensional combination called *spacetime* (see Section 21-1). In 1919, the Polish physicist Theodor Kaluza proposed the existence of a *fifth* dimension. Kaluza hoped to describe both gravity and electromagnetism in terms of the curvature of five-dimensional spacetime, just as Einstein had explained gravity by itself in terms of the curvature of four-dimensional spacetime (see Section 21-2).

In Kaluza's theory, a particle always follows the straightest possible path in the four space dimensions he proposed. But in the three dimensions of ordinary space, the path appears curved. Hence, it appears to us that the particle has been deflected—by force—and this is Kaluza's mechanism to describe gravitational and electromagnetic forces. Kaluza's hypothetical fifth dimension exists at every point in ordinary space but is curled up so tightly, like a very tiny sphere, that it is not directly observable (Figure 26-24).

In 1926, the Swedish physicist Oskar Klein attempted to make Kaluza's five-dimensional theory compatible with quantum mechanics. While he was not successful, Klein discovered that particles of different masses could be identified with different vibrations occurring in the tiny sphere of Kaluza's fifth dimension.

String Theory and Speculative Models of the Universe

When Kaluza and Klein developed their theories, gravity and electromagnetism were the only known forces of nature. Today we know of four fundamental forces, which suggests that modern unification theories might require even more than five dimensions. Edward Witten at Princeton University has argued that a geometric theory for describing all four forces would work best with 11 dimensions, 10 of space and 1 of time. These models that attempt to unify all the forces of nature, including a quantum mechanical description of gravity, are called **string theory.** (There are also related names, such as superstring theory and **M-theory.**)

Subatomic particles may actually be multidimensional membranes

We see only the regular three dimensions of space, and one dimension of time. As with Kaluza's theory, particles traveling straight in 11-dimensional space would appear to take curved paths in ordinary three-dimensional space. From our perspective, these curved paths would be the result of the four known forces acting on the particles.

Theoretical physicists have shown that if there are indeed 11 dimensions, there must also exist very massive particles that have not yet been discovered. The more massive the particle, the more energy is required in particle accelerators to create and observe it, and particle accelerators might not ever be able to reach these very high energies. Some of these speculative particles may be the dark matter that pervades the universe. Even more bizarre, the new theories no longer regard fundamental particles, such as electrons and quarks, as tiny points of mass. Instead, these particles may actually be multidimensional strings or membranes, wrapped so tightly around the extra dimensions of space that they appear to us as points. Just as a guitar can vibrate in different ways to make different sounds, vibrations of these fundamental strings correspond to different particles and the forces between them.

One criticism of string theory is that its predictions are too far beyond the energies that can foreseeably be tested. However, when the extremely large energies of the Big Bang are considered, proponents of string theory hope that the early universe itself will have acted as a powerful cosmic particle accelerator that left behind telling evidence. In this manner, astronomy might be used to gather evidence for strings whose vibrations lead to all the known forces in nature.

CONCEPTCHECK 26-12

If, as suggested by string theory, there really are 11 dimensions, why don't we notice them? How many dimensions appear to be hidden?

Answer appears at the end of the chapter.

We shall not cease from exploration
And the end of all our exploring
Will be to arrive where we started
And know the place for the first time.

T. S. Eliot, *Four Quartets*

KEY WORDS

KEY IDEAS

Cosmic Inflation: A brief period of rapid expansion, called inflation, is thought to have occurred immediately after the Big Bang. During a tiny fraction of a second, the universe expanded to a size many times larger than it would have reached through its normal expansion rate.

• Inflation explains why the universe is nearly flat and the 2.725-K microwave background is almost perfectly isotropic.

The Four Forces and Their Unification: Four basic forces—gravity, electromagnetism, the strong force, and the weak force—explain all the interactions observed in the universe.

• The Standard Model accurately describes all the known particles in nature and their observed interactions (except for gravity).

• The weak force and electromagnetic force become unified into a single force called the electroweak force at higher energies than those typically found in today's universe. This unification has been observed in high-energy particle accelerators.

• Grand unified theories (GUTs) are attempts to explain three of the forces (strong force, weak force, and electromagnetic force) in terms of a single force. This has not been observed, and particle accelerators fall far short of having the energy to directly probe the high energy where this unification is predicted to occur.

• A supergrand unified theory would explain all four forces (including gravity) at extremely high energies as a single force acting similarly on all the particles in nature. String theory attempts to make this unification, and it would describe the quantum nature of gravity. Supergrand unification is hypothesized to occur before the Planck time ($t = 10^{-43}$ seconds after the Big Bang).

Spontaneous Symmetry Breaking: As the universe expands and cools, the unified forces break into separate forces. Starting around the Planck time, gravity became a distinct force through a spontaneous symmetry breaking. During a second spontaneous symmetry breaking, the strong nuclear force became a distinct force. A final spontaneous symmetry breaking separated the electromagnetic force from the weak nuclear force; from that moment on, the universe behaved as it does today.

Particles and Antiparticles: Heisenberg's uncertainty principle states that the amount of uncertainty in the mass of a subatomic particle increases as it is observed for shorter and shorter time periods.

• Because of the uncertainty principle, particle-antiparticle pairs can spontaneously form and disappear within a fraction of a second. These pairs, whose presence can be detected only indirectly, are called virtual pairs.

• The collision of two high-energy photons can produce a real particle-antiparticle pair. In this process, called pair production, the photons disappear, and their energy is transformed into the masses of the particle-antiparticle pair. In the process of annihilation, a colliding particle-antiparticle pair disappears and two high-energy photons appear.

The Origin of Matter: Just after the inflationary epoch, the universe was filled with particles and antiparticles formed from numerous high-energy photons. The particles also annihilated to produce a state of thermal equilibrium between the particles and the photons.

• As the universe expanded, its temperature decreased. When the temperature fell below the threshold temperature required to produce each kind of particle, annihilation of that kind of particle began to dominate over production.

• Matter is much more prevalent than antimatter in the present-day universe, because particles and antiparticles were not created or maintained in exactly equal numbers. What caused this initial asymmetry is not known.

Nucleosynthesis: Helium could not have been produced until the cosmological redshift eliminated most of the high-energy photons. These photons created a deuterium bottleneck by breaking protons apart from neutrons before they could combine further to form helium.

Density Fluctuations and the Origin of Stars and Galaxies: The large-scale structure of the universe arose from primordial density fluctuations.

• The first stars were much more massive and luminous than stars in the present-day universe. The material that they ejected into space seeded the cosmos for all later generations of stars.

• Galaxies are generally located on the surfaces of roughly spherical voids. Models based on dark energy and cold dark matter give good agreement with details of this large-scale structure.

The Frontier of Knowledge: The search for a theory that unifies gravity with the other fundamental forces suggests that the universe might actually have 11 dimensions (10 of space and 1 of time), 7 of which are folded on themselves so that we cannot see them. The fundamental objects of our universe may be very small strings, rather than pointlike particles.

QUESTIONS

Review Questions

1. TUTORIAL 25-1 What is the horizon problem? What is the flatness problem? How can these problems be resolved by the idea of inflation?
2. The inflationary epoch lasted a mere 10^{-32} second. Why, then, is it worthy of so much attention by scientists?
3. In what ways is inflation similar to the present-day expansion of the universe? In what ways is it different?
4. Explain why the inflationary model does not violate the principle that the speed of light in a vacuum represents the ultimate speed limit.
5. Describe an example of each of the four basic types of interactions in the physical universe. Do you think it possible that a fifth force might be discovered someday? Explain your answer.
6. If gravity is intrinsically so weak compared to the strong force, why do we say that gravity rather than the strong force keeps the planets in orbit around the Sun? (*Hint:* Forces have a strength and a range.)
7. Explain how changes in the energy of the vacuum can account for the rapid expansion during the inflationary epoch.
8. What is the Heisenberg uncertainty principle? How does it lead to the idea that all space is filled with virtual particle-antiparticle pairs?
9. What is the difference between an electron and a positron?
10. What happens when an electron and a positron annihilate?
11. Is it possible for photons to create a particle pair of matter and antimatter? If so, what kind of photons?
12. Is it possible for a single hydrogen atom, with a positively charged proton and a negatively charged electron, to be created as a virtual pair? Why or why not?
13. Which can exist for a longer time, a virtual electron-positron pair or a virtual proton-antiproton pair? Explain your reasoning.
14. Explain why antimatter was present in copious amounts in the early universe but is very rare today.
15. What is meant by the threshold temperature of a particle?
16. Explain the connection between the fact that humans exist and the imbalance between matter and antimatter in the early universe.
17. Explain the connection between particles and antiparticles in the early universe and the cosmic microwave background that we observe today.
18. What is the deuterium bottleneck? Why was it important during the formation of nuclei in the early universe?
19. How does nucleosynthesis provide strong evidence for a hot big bang?
20. Why were only the four lightest chemical elements produced in the early universe?
21. The first stars in the universe are thought to have appeared some 400 million (4×10^8) years after the Big Bang. Once these stars formed, thermonuclear fusion reactions began in their interiors. Explain why these were the first fusion reactions to occur since the universe was 15 minutes old.
22. Why is it reasonable to suppose that all space is filled with a neutrino background analogous to the cosmic microwave background?
23. What is the Jeans length? Why is it significant for the formation of structure in the universe?
24. Why do astronomers suspect that globular clusters were among the first objects to form in the history of the universe? Why not something larger and more massive?
25. What are Population III stars? How do they differ from stars found in the present-day universe? Why are they so difficult to detect directly?
26. Describe the large-scale structure of the universe as revealed by the distribution of clusters and superclusters of galaxies.
27. What is the difference between hot and cold dark matter? How do astronomers decide which was more important in the formation of large-scale structures such as clusters of galaxies?
28. How did the presence of dark energy help to "turn off" the process of structure formation in the universe?
29. Describe the observational evidence, if any, for (**a**) the Big Bang, (**b**) the inflationary epoch, (**c**) the confinement of quarks, and (**d**) the era of recombination.

Advanced Questions

Problem-solving tips and tools

We described Wien's law for blackbody radiation in Section 5-4. If light with wavelength λ_0 is emitted by an object at redshift z, the wavelength that we measure is $\lambda = \lambda_0\,(1 + z)$. As explained in Section 26-6, the critical density ρ_c is equal to $3H_0^2/8\pi G$; that is, ρ_c is proportional to the square of the Hubble constant H_0. If $H_0 = 73$ km/s/Mpc, the critical density is equal to about 1.0×10^{-26} kg/m^3. It is also useful to know that the mass of a proton is 1.67×10^{-27} kg, the mass of an electron is 9.11×10^{-31} kg, the mass of the Sun is 1.99×10^{30} kg, that 1 m^3 = 10^6 cm^3, that 1 light-year = 9.46×10^{15} m, and that 1 GeV = 10^3 MeV = 10^9 eV.

30. An electron has a lifetime of 1.0×10^{-8} seconds in a given energy state before it makes a transition to a lower state. What is the uncertainty in the energy of the photon emitted in this process?

31. How many times stronger than the weak force is the electromagnetic force? How many times stronger than the electromagnetic force is the strong force? Use this information to suggest one reason why the electromagnetic and weak forces can become unified at a lower energy than do the electroweak and strong forces.

32. How long can a proton-antiproton pair exist without violating the principle of the conservation of mass?

33. The mass of the intermediate vector boson W^+ (and of its antiparticle, the W^-) is 85.6 times the mass of the proton. The weak nuclear force involves the exchange of the W^+ and the W^-. (**a**) Find the rest energy of the W. Give your answer in GeV. (**b**) Find the threshold temperature for the W^+ and W^-. (**c**) From Figure 26-10, how long after the Big Bang did W^+ and W^- particles begin to disappear from the universe? Explain your answer.

34. Using the physical conditions present in the universe during the era of recombination ($T = 3000$ K and $\rho_m = 10^{-18}$ kg/m^3), show by calculation that the Jeans length for the universe at that time was about 100 ly and that the total mass contained in a sphere with this diameter was about $4 \times 10^5\ M_\odot$.

35. (**a**) If a Population III star had a surface temperature of 10^5 K, what was its wavelength of maximum emission? In what part of the electromagnetic spectrum does this wavelength lie? (**b**) To ionize a hydrogen atom requires a photon of wavelength 91.2 nm or shorter. Explain how Population III stars caused reionization. (**c**) If reionization occurred at $z = 11$, what do we measure the wavelength of maximum emission of a Population III star to be? In what part of the electromagnetic spectrum does this wavelength lie? (**d**) The image that opens this chapter was made using infrared wavelengths. Suggest why these wavelengths were chosen.

36. (**a**) If the Hubble constant is 73 km/s/Mpc, the critical density ρ_c is 1.0×10^{-26} kg/m^3. The average density of dark matter is known to be about 0.20 times the critical density. Suppose that massive neutrinos constitute this dark matter, and the average density of neutrinos throughout space is 100 neutrinos per cubic centimeter. (In fact, the density of neutrinos is far less than this.) Under these assumptions, what must be the mass of the neutrino? Give your answers in kilograms and as a fraction of the mass of the electron. (**b**) Why do astronomers think that massive neutrinos are *not* the dominant type of dark matter in the universe?

37. A typical dark nebula (see Figure 18-4) has a temperature of 30 K and a density of about 10^{-12} kg/m^3. (**a**) Calculate the Jeans length for such a dark nebula, assuming that the nebula is composed mostly of hydrogen. Express your answer in meters and in light-years. (**b**) A typical dark nebula is several light-years across. Is it likely that density fluctuations within such nebulae will grow with time? (**c**) Explain how your answer to (b) relates to the idea that protostars form within dark nebulae (see Section 18-3).

38. At the time labeled $z = 4.97$ in Figure 26-19, how large was the length of each side of the box used in the simulation compared to its size in the present day ($z = 0$)? How much greater was the density at $z = 4.97$ than the present-day density?

Discussion Questions

39. If you hold an iron rod next to a strong magnet, the rod will become magnetized; one end will be a north pole and the other a south pole. But if you heat the iron rod to 1043 K (770°C = 1418°F) or higher, it will lose its magnetization and there will be no preferred magnetic direction in the rod. This demagnetization is an example of *restoring* a spontaneously broken symmetry. Explain why.

40. Some GUTs predict that the proton is unstable, although with a half-life far longer than the present age of the universe. What would it be like to live at a time when protons were decaying in large numbers?

Web/eBook Questions

41. Search the World Wide Web for information about the top quark. What kind of particle is it? How does it compare with the up and down quarks found in protons and neutrons? Why did physicists work so hard to try to find it?

42. Search the World Wide Web for information about primordial deuterium (that is, deuterium that was formed in the very early universe). Why are astronomers interested in knowing how abundant primordial deuterium is in the universe? What techniques do they use to detect it?

43. Search the World Wide Web for information about the South Pole Telescope. What is the purpose of this telescope? Why is it sited at the South Pole? How will it help us understand the early universe?

ACTIVITIES

Observing Projects

44. Use the *Starry Night*™ program to observe globular clusters, which contain some of the oldest stars in our universe. Display the **Milky Way** as seen from the center of a transparent Earth by selecting **Favourites > Explorations > Milky Way center.** Click on the **Options** tab and click on **Stars > Globular Clusters** and use the hand tool to move the view to explore the distribution of these old groups of stars around the galactic center. This distribution is that of a halo around the Galaxy, a situation matched by most galaxies. The age of these clusters is determined by examining the Hertzsprung-Russell diagrams of their stars and by determining the relative amounts of heavy elements compared to that of the lighter elements, hydrogen and helium. While these clusters contain the oldest stars still shining in our sky, there would have been older stars whose heavy-element content would have been less than that of globular cluster stars, the so-called Population III stars that have contributed to background radiation in space. Open the **Find** pane and locate and examine the following globular clusters, noting particularly their symmetry, another sign of their great age. (**a**) In each case, find the approximate angular diameter of the cluster: (i) M3; (ii) M12; (iii) M13. (**b**) In view of the expected conditions that would have been found in the universe at the time of recombination, 380,000 years after the Big Bang, how would these clusters have appeared to us in the unlikely event that we were present at this time?

Collaborative Exercises

45. The four fundamental forces of nature are the strong force, the weak force, the gravitational force, and the electromagnetic force. List four things at your school that rely on one of these fundamental forces, and explain how each thing is dependent on one of the fundamental forces.
46. Consider the following hypothetical scenario adapted from a daytime cable television talk show. Chris states that Pat borrowed Chris's telescope without permission. Tyler purchased balloons and a new telescope eyepiece without telling Chris. Sean borrowed star maps from the library, with the library's permission, but without telling Pat. Eventually, when the four met on Sunday evening, Chris was crying and speechless. Can you create a "grand unified theory" that explains this entire situation?
47. The *Cosmic Connections: The History of the Universe* figure shows the history of the universe in the form of a graph of the temperature versus the time after the Big Bang. Create a similar history of your class, starting with estimated outside temperature on the vertical axis and number of days since the beginning of the academic term on the horizontal axis. Include dates for major exams and assignments up through today. In different color ink, show your predictions for temperatures, days, and events from today until the end of the course.

ANSWERS

ConceptChecks

ConceptCheck 26-1: Early expansion is different in the inflation model compared to the gradual Hubble expansion. As shown in Figure 26-2, the inflationary model predicts an even smaller universe at early times than that described by gradual expansion. In this more compact condition, much more material would be in contact and reach the same temperature. Then, during inflation, this matter would get spread out so much that today this matter is at opposite ends of our observable universe (such as points *A* and *B*).

ConceptCheck 26-2: Yes. Inflation expands the universe by about a factor of 10^{50}. With the curved space of the universe represented on the surface of a sphere, after the radius of the sphere increases during inflation by a factor of 10^{50}, the surface (and universe) would appear flat. Due to tremendous inflation, just about any initial curvature ends up nearly flat after inflation.

ConceptCheck 26-3: Yes. The up and down quarks are held together by the strong force. Since the strong force is transmitted by gluons, these gluons are also involved in holding a neutron together.

ConceptCheck 26-4: No. Photons are massless, whereas W and Z particles have mass. The particles of the Standard Model appear *after* spontaneous symmetry breaking has changed the electroweak force into two distinct forces (electromagnetic and weak), which are transmitted by distinct particles (photons and intermediate vector bosons).

ConceptCheck 26-5: As the universe expands, it cools down, and the average energy of colliding particles decreases. Therefore, as the universe expands, there is a spontaneous symmetry breaking when particles naturally begin to collide with energies below 100 GeV.

ConceptCheck 26-6: Shorter. The greater the mass of the virtual particles, the shorter they last. A virtual pair of protons lasts about 1/2000th as long as a virtual pair of electrons.

ConceptCheck 26-7: No. Electrically charged virtual pairs come with one positively charged particle and one negative particle, so the pair's total charge is zero.

ConceptCheck 26-8: First, because the amount of antimatter would be equal to the amount of matter in our universe, this hypothesis would imply a lot of "hidden" antimatter. Second, near the boundary or boundaries of the hidden cache of antimatter, matter and antimatter would annihilate to produce gamma rays. And third, the gamma rays produced by these annihilations would be very easy to recognize because they all occur at the same energy determined by $E = mc^2$. Taken together, these effects would make it very hard to hide so much antimatter.

ConceptCheck 26-9: The temperature of the universe steadily decreases with time as the universe expands. When the universe was less than 3 minutes old, hot radiation filling the universe was energetic enough to break apart protons and neutrons before they could begin to build atomic nuclei. However, the temperature also determines the average speed at which atomic nuclei collide into each other. After 15 minutes, the temperature was not high enough for colliding nuclei, with their charged protons, to overcome their electric repulsion and build larger nuclei.

ConceptCheck 26-10: A clump of matter will not form. When the universe reaches $T = 3000$ K, the Jeans length is $L_J = 100$ light-years. Fluctuations shorter in length will not contract into clumps because there is not enough gravitational attraction to overwhelm the increased pressure that clumping would produce.

ConceptCheck 26-11: Cold dark matter particles (which are heavier and move at lower speeds) lead to "bottom up" formation, which is what we observe. Hot dark matter, such as neutrinos, leads to "top down" formation, which is not supported by observations.

ConceptCheck 26-12: We only notice four dimensions—three for space and one for time. That would leave seven hidden dimensions of space. If these dimensions are curled up tightly, as in Figure 26-24, we would not notice them until experiments could probe the tiny microscopic sizes scales in which they are curled.

BRAVE NEW COSMOS

Confused by all those theories? Good.

Making Sense of Modern Cosmology

BY P. JAMES E. PEEBLES

(From P. James E. Peebles, "Making Sense of Modern Cosmology," *Scientific American,* January 2001, 54–55)

This is an exciting time for cosmologists: findings are pouring in, ideas are bubbling up, and research to test those ideas is simmering away. But it is also a confusing time. All the ideas under discussion cannot possibly be right; they are not even consistent with one another. How is one to judge the progress? Here is how I go about it.

For all the talk of overturned theories, cosmologists have firmly established the foundations of our field. Over the past 70 years we have gathered abundant evidence that our universe is expanding and cooling. First, the light from distant galaxies is shifted toward the red, as it should be if space is expanding and galaxies are pulled away from one another. Second, a sea of thermal radiation (called thermal cosmic background radiation in this chapter) fills space, as it should if space used to be denser and hotter. Third, the universe contains large amounts of deuterium and helium, as it should if temperatures were once much higher. Fourth, galaxies billions of years ago look distinctly younger, as they should if they are closer to the time when no galaxies existed. Finally, the curvature of spacetime seems to be related to the material content of the universe, as it should be if the universe is expanding according to the predictions of Einstein's gravity theory, the general theory of relativity.

That the universe is expanding and cooling is the essence of the big bang theory. You will notice I have said nothing about an "explosion"—the big bang theory describes how our universe is evolving, not how it began.

I compare the process of establishing such compelling results, in cosmology or any other science, to the assembly of a framework. We seek to reinforce each piece of evidence by adding cross bracing from diverse measurements. Our framework for the expansion of the universe is braced tightly enough to be solid. The big bang theory is no longer seriously questioned; it fits together too well. Even the most radical alternative—the latest incarnation of the steady state theory—does not dispute that the universe is expanding and cooling. You still hear differences of opinion in cosmology, to be sure, but they concern additions to the solid part.

For example, we do not know what the universe was doing before it was expanding. A leading theory, inflation, is an attractive addition to the framework, but it lacks cross bracing. That is precisely what cosmologists are now seeking. If measurements in progress agree with the unique signatures of inflation, then we will count them as a persuasive argument for this theory. But until that time, I would not settle any bets on whether inflation really happened. I am not criticizing the theory; I simply mean that this is brave, pioneering work still to be tested.

More solid is the evidence that most of the mass of the universe consists of dark matter clumped around the outer parts of galaxies. We also have a reasonable case for Einstein's infamous cosmological constant or something similar; it would be the agent of the acceleration that the universe now seems to be undergoing. A decade ago cosmologists generally welcomed dark matter as an elegant way to account for the motions of stars and gas within galaxies. Most researchers, however, had a real distaste for the cosmological constant. Now the majority accepts it, or its allied concept, quintessence. Particle physicists have come to welcome the challenge that the cosmological constant poses for quantum theory. This shift in opinion is not a reflection of some inherent weakness; rather it shows the subject in a healthy state of chaos around a slowly growing fixed framework. We are students of nature, and we adjust our concepts as the lessons continue.

The lessons, in this case, include the signs that cosmic expansion is accelerating: the brightness of supernovae near and far; the ages of the oldest stars; the bending of light around distant masses; and the fluctuations of the temperature of the thermal radiation across the sky [see "Special Report: Revolution in Cosmology," *Scientific American,* January 1999]. The evidence is impressive, but I am still uneasy about details of the case for the cosmological constant, including possible contradictions with the evolution of galaxies and their spatial distribution. The theory of the accelerating universe is a work in progress. I admire the architecture, but I would not want to move in just yet.

How might one judge reports in the media on the progress of cosmology? I feel uneasy about articles based on an interview with just one person. Research is a complex and messy business. Even the most experienced scientist finds it hard to keep everything in perspective. How do I know that this individual has managed it well? An entire community of scientists can head off in the wrong direction, too, but it happens less often. That is why I feel better when I can see that the journalist has consulted a cross section of the community and has found agreement that a certain result is worth considering. The result becomes more interesting when others

Report Card for Major Theories

Concept	Grade	Comments
The universe evolved from a hotter, denser state	A–	Compelling evidence drawn from many corners of astronomy and physics
The universe expands as the general theory of relativity predicts		Passes the tests so far, but few of the tests have been tight
Dark matter made of exotic particles dominates galaxies	B+	Many lines of indirect evidence, but the particles have yet to be found and alternative theories have yet to be ruled out
Most of the mass of the universe is smoothly distributed; It acts like Einstein's cosmological constant, causing the expansion to accelerate	B–	Encouraging fit from recent measurements, but more must be done to improve the evidence and resolve the theoretical conundrums
The universe grew out of inflation	Inc	Elegant, but lacks direct evidence and requires huge extrapolation of the laws of physics

reproduce it. It starts to become convincing when independent lines of evidence point to the same conclusion. To my mind, the best media reports on science describe not only the latest discoveries and ideas but also the essential, if sometimes tedious, process of testing and installing the cross bracing.

Over time, inflation, quintessence and other concepts now under debate either will be solidly integrated into the central framework or will be abandoned and replaced by something better. In a sense, we are working ourselves out of a job. But the universe is a complicated place, to put it mildly, and it is silly to think we will run out of productive lines of research anytime soon. Confusion is a sign that we are doing something right: it is the fertile commotion of a construction site.

P. JAMES E. PEEBLES is one of the world's most distinguished cosmologists, a key player in the early analysis of the cosmic microwave background radiation and the bulk composition of the universe. He has received some of the highest awards in astronomy, including the 1982 Heineman Prize, the 1993 Henry Norris Russell Lectureship of the American Astronomical Society, and the 1995 Bruce Medal of the Astronomical Society of the Pacific. Peebles is currently an emeritus professor at Princeton University

Further Information

"The Evolution of the Universe." P. James E. Peebles, David N. Schramm, Edwin L. Turner, and Richard G. Kron in *Scientific American* 271, no. 4 (October 1994): 52–57.

The Inflationary Universe: The Quest for a New Theory of Cosmic Origins. Alan H. Guth. Perseus Press, 1997.

Before The Beginning: Our Universe and Others. Martin Rees. Perseus Press, 1998.

The Accelerating Universe: Infinite Expansion, the Cosmological Constant, and the Beauty of the Cosmos. Mario Livio and Allan Sandage. John Wiley & Sons, 2000.

"Concluding Remarks on New Cosmological Data and the Values of the Fundamental Parameters." P. James E. Peebles in *IAU Symposium 201: New Cosmological Data and the Values of the Fundamental Parameters,* edited by A. N. Lasenby, A. W. Jones, and A. Wilkinson; August 2000.

An "alien" life-form on Earth: pink, eyeless worms in an underwater mound of solid methane. (Courtesy of Okeanos Explorer Program/NOAA)

R I V U X G

CHAPTER 27

The Search for Extraterrestrial Life

LEARNING GOALS

By reading the sections of this chapter, you will learn

27-1 How comets and meteorites could have aided the origin of life on Earth

27-2 On which other worlds of our solar system life might have evolved

27-3 About the controversy over proposed fossilized life from Mars

27-4 How astronomers estimate the number of other civilizations in the Galaxy

27-5 Why radio telescopes may be a useful tool for contacting alien civilizations

27-6 How astronomers hope to detect Earthlike planets orbiting other stars

One of the most compelling questions in science is also one of the simplest: Are we alone? That is, does life exist beyond Earth? As yet, we have no definitive answer to this question. None of our spacecraft has found life elsewhere in the solar system, and radio telescopes have yet to detect signals of intelligent origin coming from space. Claims of aliens visiting our planet and abducting humans make compelling science fiction, but none of these stories has ever been verified.

Yet there are reasons to suspect that life might indeed exist beyond Earth. One is that biologists find living organisms in some of the most "unearthly" environments on our planet. An example (shown on this page) is at the bottom of the Gulf of Mexico, where the crushing pressure and low temperature cause methane—normally a gas—to form solid, yellowish mounds. Amazingly, these mounds teem with colonies of pink, eyeless, alien-looking worms the size of your thumb. If life can flourish here, might it not also flourish in the seemingly hostile conditions found on other worlds?

In this chapter we will look for places in our solar system where life may once have originated, and where it may exist today. We will see how scientists estimate the chances of finding life beyond our solar system, and how they search for signals from other intelligent species. And we will learn how a new generation of telescopes may make it possible to detect the presence of even single-celled organisms on worlds many light-years away.

27-1 The chemical building blocks of life are found in space

Suppose you were the first visitor to a new and alien planet. If you saw a three-headed lizard running by, you would be sure it was an alien life-form. But how can you distinguish an alien microbe—or even just a fossil-like remnant of a microbe—from a dust grain? What might alien life look like? Questions such as these are central to **astrobiology,** the study of life in the universe. Most astrobiologists suspect that if we find living organisms on other worlds, they will be "life as we know it"—that is, their biochemistry will be based on the unique properties of the carbon atom, as is the case for all life on Earth.

Organic Molecules in the Universe

Why carbon? The reason is that carbon has the most versatile chemistry of any element. Carbon atoms can form chemical bonds to create especially long and complex molecules (**Figure 27-1**). These carbon-based compounds, called **organic molecules,** include all the molecules of which living organisms are made. (Silicon has some chemical similarities to carbon, and it can also form complex molecules. But as **Figure 27-2** shows, complex silicon molecules do not have the right properties to make up complex systems such as living organisms.)

Organic molecules can be linked together to form elaborate structures, such as chains, lattices, and fibers. Some of these structures are capable of complex, self-regulating chemical reactions. Furthermore, the primary constituents of organic molecules—carbon, hydrogen, nitrogen, oxygen, sulfur, and phosphorus—are among the most abundant elements in the universe. The versatility and abundance of carbon suggest that extraterrestrial life is also likely to be based on organic chemistry.

If life is based on organic molecules, then these molecules must initially be present on a planet in order for life to arise from

$$Z-X-X-X-X-Y$$

(a) Linear molecule

(b) Glucose

FIGURE 27-1

Complex Molecules and Carbon **(a)** Atoms that can bond to only two other atoms, like the atoms denoted X shown here, can form a chain of atoms called a linear molecule. The chain stops where we introduce an atom, such as those labeled Y and Z, that can bond to only one other atom. **(b)** A carbon atom (denoted C) can bond with up to four other atoms. Hence, carbon atoms can form more complex, nonlinear molecules like glucose. All organic molecules that are found in living organisms have backbones of carbon atoms.

nonliving matter. We now understand that many carbon-based molecules originate from nonbiological processes in interstellar space. One such molecule is carbon monoxide (CO), which is made when a carbon atom and an oxygen atom collide and bond together. Carbon monoxide is found in abundance within giant interstellar clouds that lie along the spiral arms of our Milky Way Galaxy (see Figure 1-7) as well as in other galaxies (see Figure 1-9). Carbon atoms have also combined with other elements to produce an impressive array of interstellar organic molecules, including ethyl alcohol (CH_3CH_2OH), formaldehyde (H_2CO), methylcyanoacetylene (CH_3C_3N), and acetaldehyde (CH_3CHO). Radio astronomers

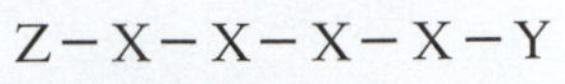

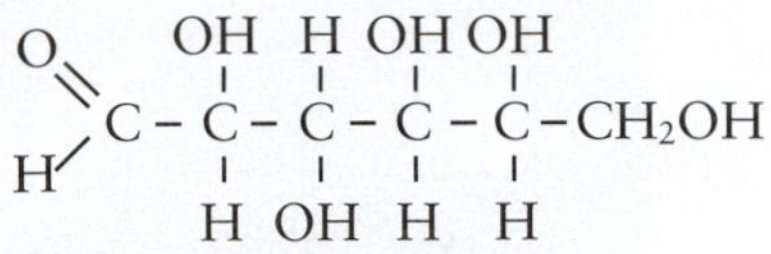

(a)

(b)

FIGURE 27-2 R I V U X G

Why Silicon Is Unsuitable for Making Living Organisms Like carbon, silicon atoms can bond with up to four other atoms. However, the resulting compounds are either too soft or too hard, or react too much or too little, to be suitable for use in living organisms. **(a)** Silicone has a backbone of silicon and oxygen atoms. The molecules form a gel or liquid rather than a solid, and react too slowly to undergo the rapid chemical changes required of molecules in organisms. **(b)** Molecules can also be made with a silicon-carbon-oxygen backbone, but the results (like this quartz crystal) are too rigid for use in organisms. (a: Richard Megna/Fundamental Photographs; b: Ispace/Shutterstock)

FIGURE 27-3 R I V U X G

A Carbonaceous Chondrite Carbonaceous chondrites are primitive meteorites that date back to the very beginning of the solar system. This sample is a piece of the Allende meteorite, a large carbonaceous chondrite that fell in Mexico in 1969. Chemical analyses of newly fallen specimens disclose that they are rich in organic molecules, many of which are the chemical building blocks of life. (Detlev van Ravenswaay/Science Source)

have detected these molecules by looking for the telltale microwave emission lines of carbon-based chemicals in interstellar clouds.

The planets of our solar system formed out of interstellar material (see Section 8-5), and some of the organic molecules in that material must have ended up on the planets' surfaces. Evidence for this comes from meteorites called **carbonaceous chondrites,** like the one shown in Figure 27-3. These meteorites are ancient, date from the formation of the solar system, and are often found to contain a variety of carbon-based molecules. The Murchison meteorite, which fell on Australia in 1969, contains more than 70 amino acids, and these organic molecules are some of the building blocks of life.

The spectra of comets (see Section 7-5)—which are also among the oldest objects in the solar system—show that they, too, contain an assortment of organic compounds. In 2006, NASA's *Stardust* mission returned samples from the atmosphere (or coma) of Comet Wild 2. The samples were very limited, but one type of amino acid was found. In 2014, a probe from the European Space Agency is scheduled to land on a comet's surface for direct samples of its nucleus. Astrobiologists are hoping the lander finds a variety of amino acids, and the results will help them estimate how much of Earth's organic building blocks came from comets.

Comets and meteoroids were much more numerous in the early solar system than they are today, and they were correspondingly more likely to collide with a planet. These collisions would have seeded the planets with organic compounds from the very beginning of our solar system's history. Organic compounds are also found in interplanetary dust particles (see Section 15-5), which continually rain down on the planets. Once meteorites, comets, and interplanetary dust particles bring simple organic chemicals to a planet's surface, additional chemical reactions can produce an even wider range of the complex organic compounds needed for life. Similar processes are thought to take place in other planetary systems, which are thought to form in basically the same way as did our own (see Figure 8-13 and the image that opens Chapter 8).

CONCEPTCHECK 27-1

Why would life-forms throughout the cosmos likely be based on carbon?

CONCEPTCHECK 27-2

How can organic molecules end up on the surfaces of planets?

Answers appear at the end of the chapter.

The Miller-Urey Experiment

Comets and meteorites would not have been the only sources of organic material on the young planets of our solar system. In 1952, the American chemists Stanley Miller and Harold Urey demonstrated that under conditions that are thought to have prevailed on the young Earth, simple chemicals can combine to form the chemical building blocks of life. In a closed container, they prepared a sample of "atmosphere": a mixture of hydrogen (H_2), ammonia (NH_3), methane (CH_4), and water vapor (H_2O), the most common molecules in the solar system. Miller and Urey then exposed this mixture of gases to an electric arc (to simulate atmospheric lightning) for a week. At the end of this period, the inside of the container had become coated with a reddish-brown substance rich in amino acids and other compounds essential to life.

Organic molecules do not have to come from living organisms—they can also be synthesized in nature from simple chemicals

Since Miller and Urey's original experiment, most scientists have come to the conclusion that Earth's primordial atmosphere was composed of carbon dioxide (CO_2), nitrogen (N_2), and water vapor outgassed from volcanoes, along with some hydrogen. Modern versions of the Miller-Urey experiment (Figure 27-4) using these common gases have also succeeded in synthesizing a wide variety of organic compounds. The combination of comets and meteorites falling from space and chemical synthesis in

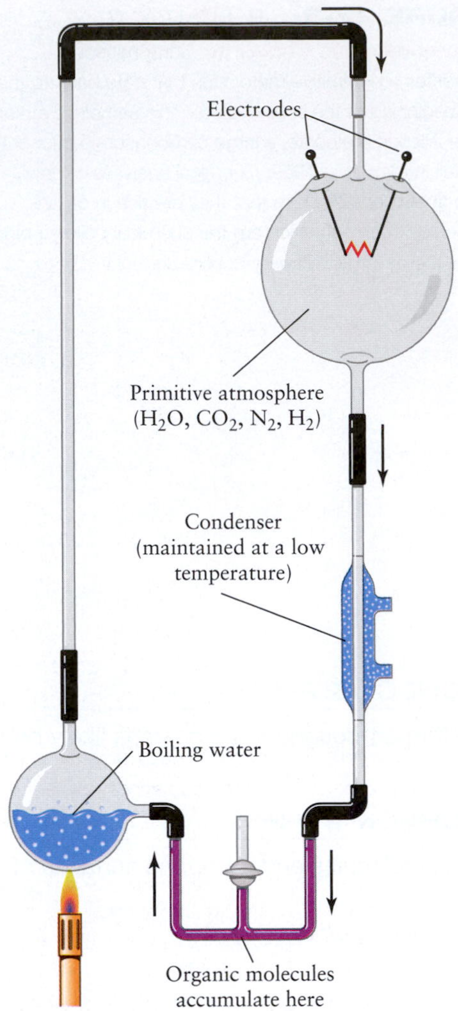

FIGURE 27-4

An Updated Miller-Urey Experiment Modern versions of this classic experiment prove that numerous organic compounds important to life can be synthesized from gases that were present in Earth's primordial atmosphere. This experiment supports the hypothesis that life on Earth arose as a result of ordinary chemical reactions.

the atmosphere could have made the chemical building blocks of life available in substantial quantities on the young Earth.

CAUTION! It is important to emphasize that scientists have *not* created life in a test tube. While organic molecules may have been available on the ancient Earth, biologists have yet to figure out how these molecules gathered themselves into cells and developed systems for self-replication. Nevertheless, because so many chemical components of life are so easily synthesized under conditions that simulate the primordial Earth, it is reasonable to suppose that life could have originated as the result of chemical processes. Furthermore, because the molecules that combine to form these compounds are rather common, it seems equally reasonable that life could have originated in the same way on other planets.

Organic building blocks are commonplace throughout the universe, but their abundance does not guarantee that life is equally commonplace. If a planet's environment is hostile, life may never get started or may quickly be extinguished. But we now have evidence that nearly Earth-sized planets orbit other stars (see Section 8-7) and that additional planetary systems are forming around young stars (see Section 8-4, especially Figure 8-8). It seems increasingly likely that Earthlike planets will be found orbiting other stars, and that conditions on some of these worlds may be suitable for life as we know it. The moons of some planets, in our solar system or associated with other stars, might also have environments suitable for life.

CONCEPTCHECK 27-3

What did Miller and Urey create when they passed electricity through their sample of "atmosphere" containing a mixture of hydrogen (H_2), ammonia (NH_3), methane (CH_4), and water vapor (H_2O)?

Answer appears at the end of the chapter.

27-2 Water and the potential for life

If life evolved on Earth from nonliving organic molecules, might the same process have taken place elsewhere in our solar system? Scientists are carefully scrutinizing the planets and their moons in an attempt to answer this question.

The Importance of Liquid Water

One major constraint is that liquid water is essential for the survival of life as we know it. However, the water need not be pleasant by human standards—terrestrial organisms have been found in water that is boiling hot, fiercely acidic, or ice cold (see the image that opens this chapter)—but it must be liquid. Organisms living in such extreme conditions are called **extremophiles,** such as the thermophiles that thrive at high temperatures (Figure 27-5). Other extremophiles include microorganisms that are resistant to high levels of nuclear radiation, and those that live *within* the tiny porous spaces inside of rocks. All known life-forms require at least some liquid water, and even the extremophile microbes living within ice contain a form of antifreeze to keep their internal water liquid.

In order for water on a planet's surface to remain liquid, the temperature cannot be too hot or too cold. Furthermore, there must be a relatively thick atmosphere to provide enough pressure to keep liquid water from evaporating. Of all the worlds in our present-day solar system, only Earth has the right conditions for water to remain liquid on its surface.

The Oceans of Europa and Enceladus

There is now compelling evidence that Europa, one of the large satellites of Jupiter (see Table 7-2 and Figure 7-4), has an ocean of water *beneath* its icy surface. As it orbits Jupiter, Europa is caught in a tug-of-war between gravitational forces from Jupiter and Jupiter's other large satellites. These forces flex the interior of Europa, and this flexing generates enough heat to keep subsurface water from freezing. Images indicate that chunks of ice have

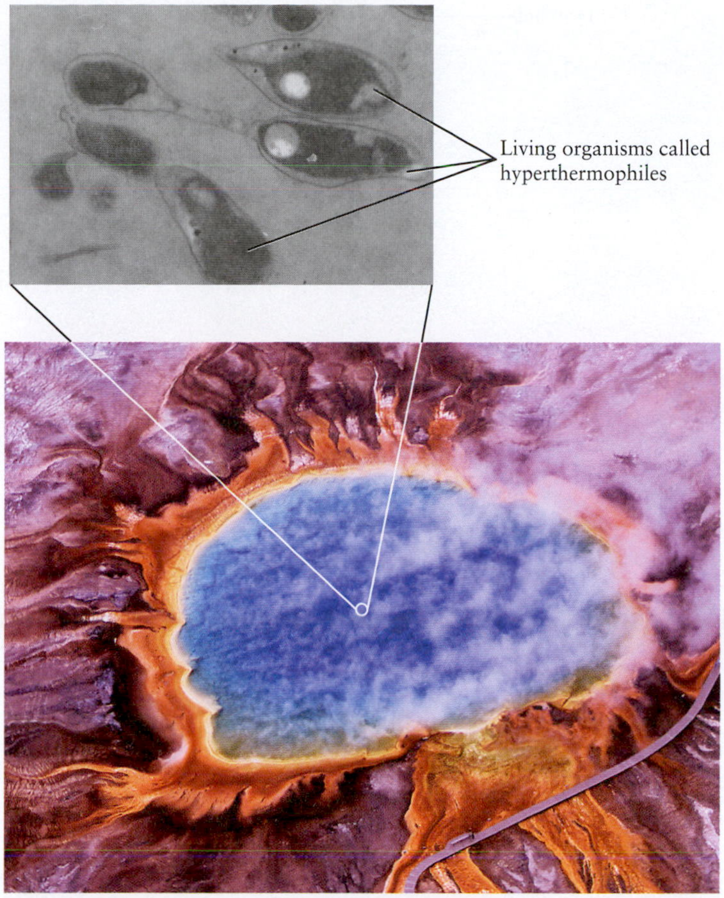

(a) Grand prismatic spring, Yellowstone National Park.

(b)

FIGURE 27-5 R I V U X G

Extremophiles Can Take the Heat **(a)** These microscopic thermophiles (heat-loving organisms) live in water that is between 80°C and 100°C (85°F–140°F). **(b)** Tube worms (light-green tubes) with hemoglobin-rich red plumes. They reside around black smokers—vents in the ocean bottom that are in the same temperature range as the hot springs shown in (a). These vents are more than 3 km (2 mi) under water. (a: Jim Peaco; July 2001 Yellowstone National Park image by NPS Photo; inset: NASA; b: Verena Tunnicliffe, University of Victoria, School of Earth & Ocean Sciences, Department of Biology)

floated around on the surface, and magnetic field measurements reveal that Europa has an underground ocean (see Section 13-6).

No one knows whether life exists in Europa's ocean. But interest in this exotic little world is great, and scientists have proposed several missions to explore Europa in more detail. Unfortunately, the ice layer is probably too thick (several kilometers) for a robotic probe to penetrate the ice and sample the underlying ocean anytime soon. However, just as microorganisms on Earth can survive within ice, perhaps life on Europa has spread beyond its ocean toward the surface.

A mission to search Europa's surface ice is still very challenging, as high levels of radiation on the atmosphere-free moon would likely destroy organisms within a few meters of the surface. A thermal probe designed to simply melt its way down a few meters also faces challenges because of the fine dust (regolith) that is expected to exist near the surface but does not melt. To search for life about 10 m down, scientists are testing a number of ideas, including a thermal probe with a drill tip to get through any patches of dust.

Saturn's moon Enceladus has revealed a large subsurface ocean through salty water spewing out in plumes from ice volcanoes (Figure 27-6 and Figure 13-29). The salts, which are composed of a variety of dissolved minerals, form when the water makes extensive contact with the rocky interior. This evidence points strongly toward a large interior ocean. Enceladus's ocean of liquid water with dissolved minerals, along with some organic material detected in its plumes, suggests that its ocean could make a suitable environment for life.

While Enceladus's ocean could be kilometers beneath the surface and difficult to explore, its ice volcanoes offer a location where microorganisms seeping up from the ocean might be ejected from the surface. In fact, after initially favoring Europa's deep ocean as the primary target in the search for extraterrestrial life,

FIGURE 27-6 R I V U X G

Subsurface Ocean on Enceladus Plumes containing water vapor and dissolved minerals indicate a subsurface ocean on Saturn's moon Enceladus. Organic compounds, in concentrations 20 times higher than expected, were also detected in the plumes. As on Earth, liquid water with dissolved minerals and organic compounds could be a suitable setting for life. (NASA/JPL/Space Science Institute) many scientists now favor Enceladus. Without the challenges of drilling or melting through ice, as on Europa, a probe can pass right through the plumes of Enceladus's ice volcanoes and test samples for life.

Even if there is no life in the Enceladean ocean, the ice volcanoes themselves might harbor life. The ice volcanoes contain liquid water, organic compounds, and are about 200 K warmer than the surrounding terrain. Within these ice volcanoes, water, food, and energy could make a hospitable environment for life.

CONCEPTCHECK 27-4

Why do dissolved salts (or minerals) ejected from Enceladus suggest a large ocean beneath its surface?

Answer appears at the end of the chapter.

Life in Similar Environments on Earth

Here on Earth, there are environments somewhat similar to those expected in the subsurface oceans of Europa and Enceladus. By investigating these environments, we learn if life might be possible in such extremes. One such place is Lake Vida in Antarctica. Its liquid water is covered by 20 m of frozen ice. At that depth virtually no sunlight reaches the water. The water is also 6 times saltier than seawater, which is too salty for most organisms. However, the high concentration of salt also prevents the water from freezing, where it drops to temperatures of –13.5°C. Incredibly, in 2012, researchers found that bacteria are thriving in these subfreezing waters.

The most extreme liquid water environment known on Earth is Lake Vostok in Antarctica (Figure 27-7). This enormous subsurface lake is completely buried by a whopping 2.2 miles of glacial ice. With this lake so cut off from its surroundings, its water has probably been isolated for more than 15 million years! In 2012, Russian scientists very carefully drilled more than 2 miles into Lake Vostok, piercing its icy depths to sample the waters beneath. When water was finally reached, it rushed up and gushed out of the borehole. In 2013, pristine samples of the lake water were collected and are being investigated for signs of life. Thus, while astronomers look to the skies, they also keep one eye on Lake Vostok.

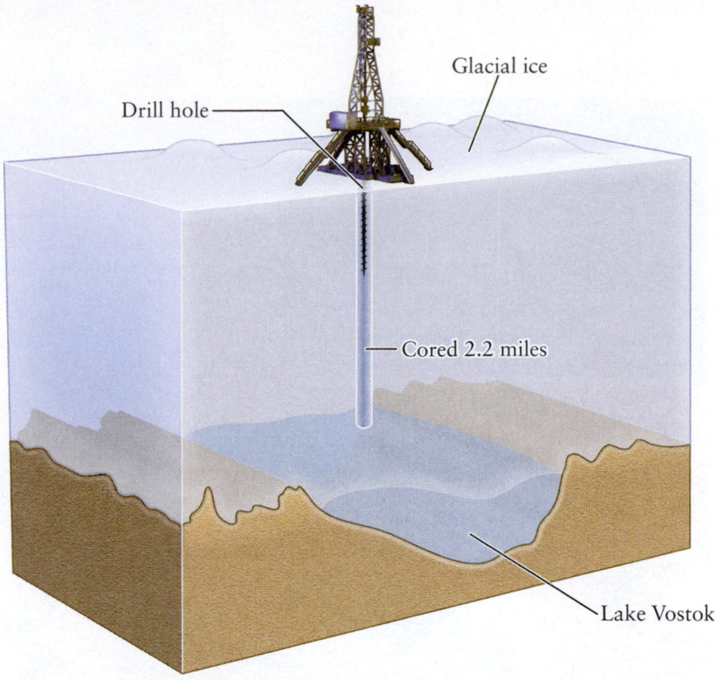

FIGURE 27-7

Exploring Lake Vostok Russian scientists search for life in an environment that might be similar to the oceans of Europa or Enceladus. The coldest temperature recorded on Earth was above Lake Vostok, at –89°C (–128°F). This is similar to typical Martian surface temperatures. However, the lake's liquid water below is much warmer at only about –3°C. This is encouraging because thick ice can act as insulation to keep in a planet or moon's internal heat. (Adapted from Nicolle Rager-Fuller/NSF)

Searching for Life on Mars

> Some areas on Mars were covered with liquid water for extended periods

Another possibility for the existence of life in our solar system is Mars. The present-day Martian atmosphere is so thin that water can exist only as ice or as a vapor. However, images made from Martian orbit show dried-up rivers, and the *Curiosity* rover has confirmed that one riverbed must have held water for thousands to millions of years (see Section 11-8 and the *Scientific American* article "Reading the Red Planet" at the end of Chapter 11). These features are evidence that the Martian atmosphere was once thicker and that liquid water once coursed over the planet's surface. Could life have evolved on Mars during its "wet" period? If so, could life—in the form of extremophile microorganisms—have survived as the Martian atmosphere thinned and the surface water either froze or evaporated? While *Curiosity* is designed to find suitable habitats where life may have existed, the rover is not equipped to test for the presence of life itself. Fortunately, some tests for Martian life have already been carried out, and future tests for life will likely involve similar experiments.

In 1976, the *Viking Lander 1* and *Viking Lander 2* landed on different parts of Mars. Each spacecraft carried a scoop at the end of a mechanical arm to retrieve surface samples (Figure 27-8). These samples were deposited into a compact on-board biological laboratory that carried out three different tests for Martian microorganisms.

1. The *gas-exchange experiment* was designed to detect gases released due to chemical processes involving nutrients. For example, to extract energy from food, many organisms produce carbon dioxide (which is why we exhale this gas in our breath). The experiment involved a surface sample that was placed in a sealed container along with a controlled amount of gas and nutrient-rich water. The gases in the container were then monitored to check for any microbial *exhalation*.

2. The *labeled-release experiment* was designed to detect metabolic processes. A sample was moistened with nutrients containing low levels of radioactive carbon atoms. If any organisms in the sample consumed the nutrients, their waste products should include gases containing the easily detectable radioactive carbon.

3. The *pyrolytic-release experiment* was designed to detect photosynthesis, the biological process by which terrestrial plants use solar energy to help synthesize organic compounds from carbon dioxide. A surface sample was placed in a container along with slightly radioactive carbon dioxide and exposed

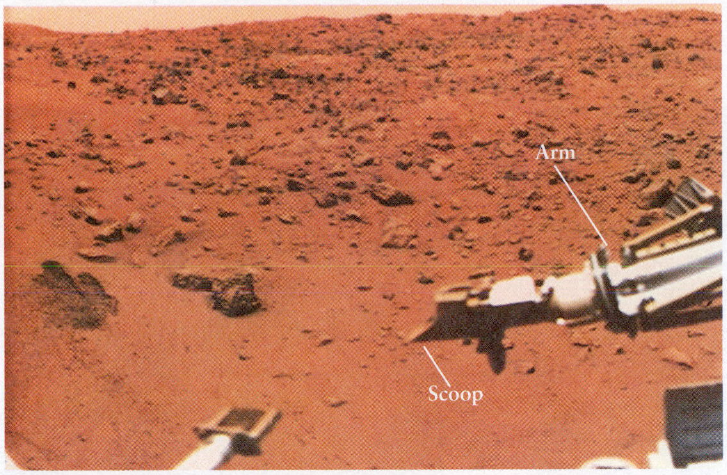

FIGURE 27-8 R I V U X G

Digging in the Martian Surface This view from the *Viking Lander 1* spacecraft shows the mechanical arm with its small scoop against the backdrop of the Martian terrain. The scoop was able to dig about 30 cm (12 in.) beneath the surface. (NASA)

to artificial sunlight. If plantlike photosynthesis occurred, microorganisms in the sample would take in some of the radioactive carbon from the gas.

The first data returned from these experiments caused great excitement, for in almost every case, rapid and extensive changes were detected inside the sealed containers. Further analysis of the data, however, led to the conclusion that these changes were due solely to nonbiological chemical processes. It appears that the Martian surface is rich in unstable chemicals that react with water to release oxygen gas. Because the present-day surface of Mars is bone-dry, these chemicals had nothing to react with until they were placed inside the moist interior of the *Viking Lander* laboratory.

At best, the results from the *Viking Lander* biological experiments were inconclusive. Perhaps life never existed on Mars at all. Or perhaps it did originate there, but failed to survive the thinning of the Martian atmosphere, the unstable chemistry of the planet's surface, and exposure to ultraviolet radiation from the Sun. (Unlike Earth, Mars has no ozone layer to block ultraviolet rays.) Another possibility is that Martian microorganisms have survived only in certain locations that the *Viking Landers* did not sample, such as isolated spots on the surface or beneath the ground. And yet another option is that there is life on Mars, but the experimental apparatus on board the *Viking Lander* spacecraft was not sophisticated enough to detect it.

The Power of Measuring Carbon Isotopes

An entirely different set of biological experiments were designed for the British spacecraft *Beagle 2,* which landed on Mars in December 2003. To test for life on Mars, *Beagle 2* was to sample an ancient lakebed region potentially containing the carbon-rich molecules of organisms–*dead or alive*. Then, signs of past or present life might be found by measuring how many of the organic molecules contain the isotope ^{12}C, which appears preferentially in biological molecules, and how many contain ^{13}C (which does not). (See Box 5-5 for a description of isotopes.) Thus, not only can ^{12}C isotopes indicate the existence of life, but this test works even when the organisms have died long ago.

Sadly, scientists on Earth were unable to establish contact with *Beagle 2* after its descent through the Martian atmosphere. However, the *Curiosity* rover presently exploring Mars can also measure the abundance of carbon isotopes and will search for areas of excess ^{12}C. One experiment hopes to examine the carbon in methane gas (CH_4), previously discovered in the Martian atmosphere. Microorganisms on Earth can gain energy by converting carbon dioxide to methane, and if Martian microorganisms do the same, it could explain the unexpected presence of Martian methane. Once it appears, methane rapidly decomposes in the Martian atmosphere, so we know that Mars presently has a source of methane. What we do not know is if this source of methane is geochemical or biological, and *Curiosity* can potentially answer this question by finding an excess of ^{12}C. As of this writing, *Curiosity* has yet to detect Martian methane.

CONCEPTCHECK 27-5

By measuring the abundance of the carbon isotope ^{12}C, how can *Curiosity* gather clues that life once existed on Mars?

Answer appears at the end of the chapter.

The Martian Civilization That Never Was

In 1976, while the *Viking Landers* were carrying out their biological experiments on the Martian surface, the companion *Viking Orbiter* spacecraft photographed some surface features that at first glance seemed to have been crafted by *intelligent* life on Mars. The *Viking Orbiter 1* image in Figure 27-9a shows what appears to be a humanlike face, perhaps the product of an advanced and artistic civilization. However, when the more advanced *Mars Global Surveyor* spacecraft viewed the surface in 1998 using a superior camera (Figure 27-9b), it found no evidence for facial features.

Scientists are universally convinced that the "face" and other apparent patterns in the *Viking Orbiter* images were created by shadows on windblown hills. Microscopic life may once have existed on Mars, and may yet exist today, but there is no evidence that the red planet has ever been the home of intelligent beings.

27-3 Meteorites from Mars have been scrutinized for life-forms

While spacecraft can carry biological experiments to other worlds such as Mars, many astrobiologists look forward to the day when a spacecraft will return Martian samples to laboratories on Earth. Until that day arrives, we have the next best thing: More than a hundred meteorites that appear to have formed on Mars have been found at a variety of locations on Earth.

The feature that identifies meteorites as having come from Mars is the chemical composition of trace amounts of gas trapped within them. This composition is very different from that of Earth's atmosphere, but is a nearly perfect match to the composition of the Martian atmosphere found by the *Viking Landers.*

How could a rock have traveled from Mars to Earth? When an asteroid collides with a planet's surface and forms an impact crater, most of the material thrown upward by the impact falls back onto the planet's surface. But some extraordinarily powerful

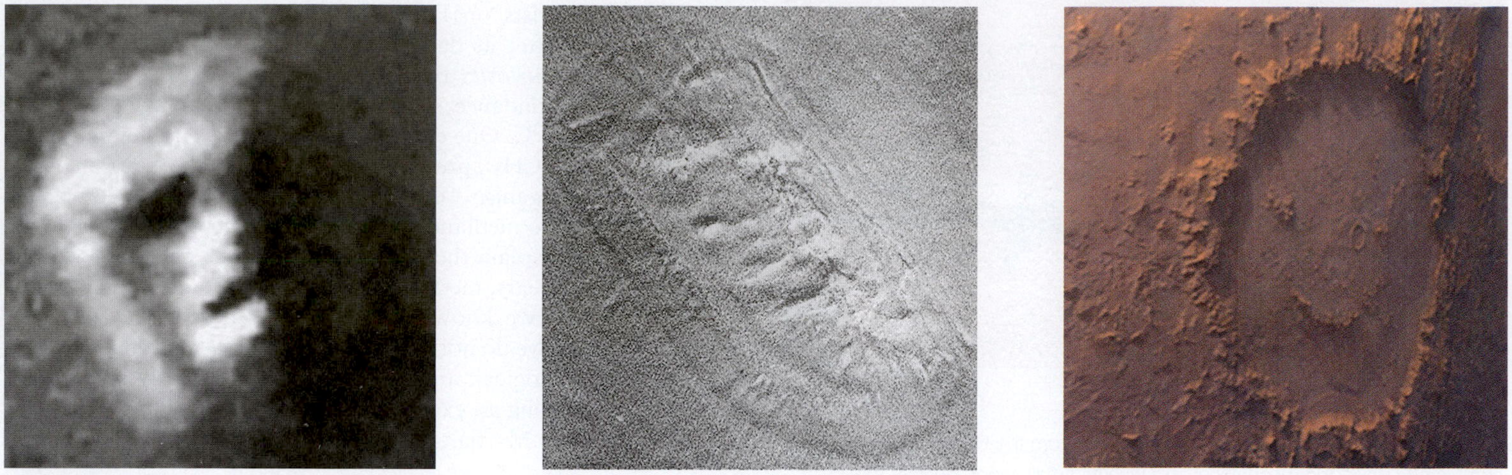

FIGURE 27-9 R I V U X G

A "Face" on Mars? **(a)** This 1976 image from *Viking Orbiter 1* shows a Martian surface feature that resembles a human face. Some suggested that this feature might have been made by intelligent beings. **(b)** This 1998 *Mars Global Surveyor* (MGS) image, made under different lighting conditions with a far superior camera, reveals the "face" to be just an eroded hill. **(c)** This MGS image shows features of natural origin within a 215-km (134-mi) wide crater on Mars. Can you see this "face"? (a: NSSDC/NASA and Dr. Michael H. Carr; b, c: Malin Space Science Systems/NASA)

impacts have produced large craters on Mars—roughly 100 km in diameter or larger. These tremendous impacts eject some rocks with such speed that they escape the planet's gravitational attraction and fly off into space.

There are numerous large craters on Mars, so a good number of Martian rocks have probably been blasted into space over the planet's history. These ejected rocks then go into elliptical orbits around the Sun. Many such rocks will have orbits that put them on a collision course with Earth, and these are the ones that scientists find as meteorites from Mars. In fact, estimates are that several tons of Martian rocks land on Earth *each year* (equaling around a cubic meter of rock).

Using the radioactive age-dating technique (see Section 8-3), scientists find that most meteorites from Mars are between 200 million and 1.3 billion years old, much younger than the 4.56-billion-year age of the solar system. But one meteorite from Mars, denoted by the serial number ALH 84001 and found in Antarctica in 1984, was found to be 4.5 billion years old (Figure 27-10a). Thus, ALH 84001 is a truly ancient piece of Mars. Further analysis of its radioactivity suggests that ALH 84001 was ejected from Mars by an impact about 16 million years ago and landed in Antarctica a mere 13,000 years ago.

ALH 84001 was on Mars during the era when liquid water existed on the planet's surface. Scientists have therefore investigated this rock carefully for clues about Martian water and possible life. One such clue is the presence of rounded grains of minerals called *carbonates*. Analysis of these carbonates indicates that they formed in liquid water (at a comfortable temperature of about 64°F).

In 1996, David McKay and Everett Gibson of the NASA Johnson Space Center, along with several collaborators, began reporting the results from studies of the carbonate grains in ALH 84001. They provided several pieces of evidence that life may have once existed within the rock's cracks while it was still on Mars:

1. There are large numbers of elongated, tubelike structures around the carbonate grains that could be fossilized microorganisms (Figure 27-10b).
2. Microscopic cracks contain organic molecules—consistent with what is expected from the decay of microorganisms.
3. The carbonate grains contain small magnetic crystals called magnetite, which are similar to magnetite produced by bacteria on Earth.

> Claims that scientists have found Martian microorganisms are intriguing but very controversial

Are McKay and Gibson's conclusions correct? Their claims of ancient life on Mars are extraordinary, and they require extraordinary proof. Further studies found that the organic compounds detected are not a very good match for decayed organisms. Also, it was later found that the tubelike structures and magnetite crystals found in ALH 84001, while consistent with life, could have been formed through geological processes on Mars. Thus, while some of the evidence is consistent with Martian life, it is far from compelling.

CONCEPTCHECK 27-6

If bacteria on Earth are known to create magnetite crystals like those from Mars, why isn't this proof of life on Mars?

Answer appears at the end of the chapter.

Possible Microbial Tunnels

The most recent indications of possible Martian microbial remnants involve microscopic tunnels found in meteorites from Mars. In 2006, Martin Fisk from Oregon State University led a team that discovered small tunnels in a Martian meteorite named Nakhla. These tunnels are similar to structures found in rocks on Earth that are

(a) (b)

FIGURE 27-10 R I V U X G

A Meteorite from Mars **(a)** This 1.9-kg meteorite, known as ALH 84001, formed on Mars some 4.5 billion years ago. About 16 million years ago a massive impact blasted it into space, where it drifted in orbit around the Sun until landing in Antarctica 13,000 years ago. The small cube at lower right is 1 cm (0.4 in.) across. **(b)** This electron microscope image, magnified some 100,000 times, shows tubular structures about 100 nanometers (10^{27} m) in length found within the Martian meteorite ALH 84001. One controversial interpretation is that these are the fossils of microorganisms that lived on Mars billions of years ago. (a: NASA Johnson Space Center; b: *Science,* NASA)

thought to be produced by rock-eating microbes. The Nakhla meteorite is not the only one with tunnels; they are also found in ALH 84001, and a 2013 study at the Johnson Space Center led by Lauren White finds them in the Martian meteorite Y000593 (Figure 27-11).

Could the tunnels have been formed after the meteorites hit Earth? That is possible for ALH 84001, and Y000593, which sat on Earth for thousands of years before being collected. But the Nakhla meteorite was observed falling on June 28, 1911, and its debris was collected too quickly for microbes on Earth to have created its tunnels. Are the microtunnels produced by Martian microbes? As in the case of the tubelike structures and magnetite crystals from ALH 84001, features that *appear* biological in origin might simply occur naturally through Martian geology. Thus, the Martian tunnels lead us back to Earth, where we must determine if these tunnels can instead be created through geological processes. For now, the origin of these tunnels is uncertain. Though they only provide weak evidence at this early stage of analysis, these microscopic tunnels open up a whole new line of inquiry in the search for Martian life.

CONCEPTCHECK 27-7

Why are scientists convinced that these meteorites actually came from Mars and not from our Moon?

Answer appears at the end of the chapter.

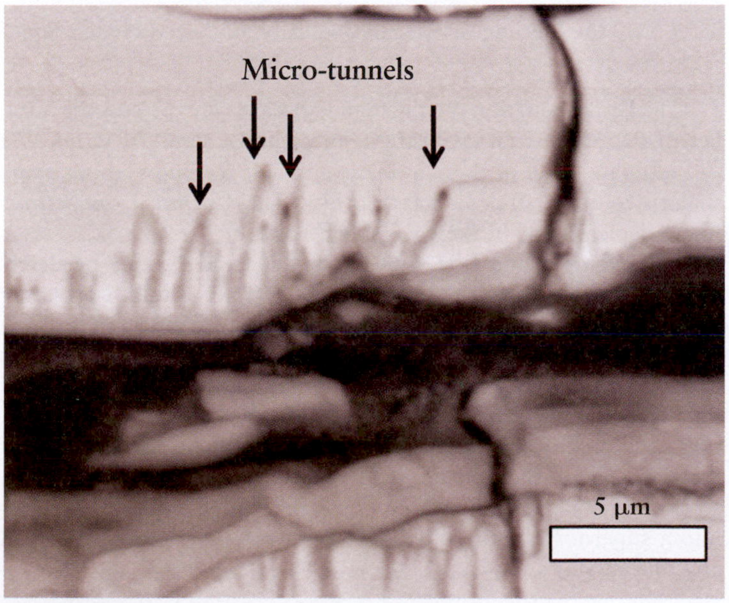

FIGURE 27-11 R I V U X G

Microscopic Tunnels in Martian Meteorite This figure shows microtunnels in the Martian meteorite Y000593. Although not common, very similar microscopic tunnels are found in rocks on Earth that contain DNA and are thought to be produced by rock-eating microbes. Even though Y000593 sat on Earth for thousands of years before being collected, the tunnels themselves were likely created on Mars: the tunnels in Y000593 are filled with Martian clay as determined by an analysis of the clay's composition. The microtunnels emanate from larger fractures in the rock and are about 1–5 micrometers long. (Lauren White)

27-4 The Drake equation helps scientists estimate how many civilizations may inhabit our Galaxy

TUTORIAL 27-1

Are we alone, or does the Galaxy teem with intelligent life? Or is the truth somewhere in between?

We have seen that only a few locations in our solar system may have been suitable for the origin of life. But what about planets and moons around other stars? The existence of life on Earth seems to suggest that extraterrestrial life, possibly including intelligent species, might evolve on terrestrial planets around other stars, given sufficient time and hospitable conditions. How can we learn whether such worlds exist, given the tremendous distances that separate us from them? This is the great challenge facing the **search for extraterrestrial intelligence,** or **SETI.**

Close Encounters Versus Remote Communication

A common belief is that alien civilizations do exist and that their spacecraft have visited Earth. Indeed, surveys show that between one-third and one-half of all Americans believe in unidentified flying objects (UFOs) of alien origin. A somewhat smaller percentage believes that aliens have landed on Earth. But, in fact, there is absolutely *no* scientifically verifiable evidence of alien visitations. As an example, many UFO proponents believe that the U.S. government is hiding evidence of an alien spacecraft that crashed near Roswell, New Mexico, in 1947. However, the bits of "spacecraft wreckage" found near Roswell turned out to be nothing more than remnants of an unmanned research balloon. Atomic isotopes are one way to tell if a material comes from Earth or not, because we know that the abundances of many isotopes vary in our solar system and beyond, yet no alien spacecraft debris with a non-Earth isotope signature has been found.

While there is no evidence of alien visits to Earth, alien civilizations could exist around some of the billions of stars in our Galaxy. To find real evidence for intelligent civilizations on other worlds, we must look to the stars.

With our present technology, sending even a small unmanned spacecraft to another star requires a flight time of tens of thousands of years. Speculative design studies have been made for unmanned probes that could reach other stars within a century or less, but these probes are prohibitively expensive. Instead, many astronomers hope to discover extraterrestrial civilizations by detecting radio transmissions from them. Radio waves are a logical choice for interstellar communication because they can travel immense distances without being significantly degraded by the interstellar medium, the thin gas and dust found between the stars (see Section 8-1).

Over the past several decades, astronomers have proposed various ways to search for alien radio transmissions, and several searches have been undertaken. In 1960, Frank Drake first used a radio telescope at the National Radio Astronomy Observatory in West Virginia to listen to two Sunlike stars, Tau Ceti and Epsilon Eridani, without success. Since then, many SETI searches have taken place using radio telescopes around the world. Occasionally, a search has detected an unusual or powerful signal. But none has ever repeated, as a signal of intelligent origin might be expected to do. To date, we have no confirmed evidence of radio transmissions from another world.

CONCEPTCHECK 27-8

Why might the use of radio waves for exploration for life in the Galaxy be more fruitful than using unmanned interstellar spaceships?

Answer appears at the end of the chapter.

Is There Anybody Out There?

Should we be discouraged by this failure to make contact? What are the chances that a radio astronomer might someday detect radio signals from an extraterrestrial civilization? The first person to tackle this issue was Frank Drake, who proposed that the number of technologically advanced civilizations in the Galaxy could be estimated by a simple equation. This is now called the **Drake equation:**

Drake equation

$$N = R_* f_p n_e f_l f_i f_c L$$

N = number of technologically advanced civilizations in the Galaxy whose messages we might be able to detect

R_* = the rate at which solar-type stars form in the Galaxy

f_p = the fraction of stars that have planets

n_e = the number of planets per solar system that are Earthlike (that is, suitable for life)

f_l = the fraction of those Earthlike planets on which life actually arises

f_i = the fraction of those life-forms that evolve into intelligent species

f_c = the fraction of those species that develop adequate technology and then choose to send messages out into space

L = the lifetime of a technologically advanced civilization

The Drake equation is enlightening because it expresses the number of extraterrestrial civilizations in a simple series of terms. We can estimate some of these terms from what we know about stars, stellar evolution, and planetary orbits. The *Cosmic Connections* figure depicts some of these considerations.

For example, the first two factors, R_* and f_p, can be determined by observation. In estimating R_*, we should probably exclude stars with masses greater than about 1.5 times that of the Sun. These more massive stars use up the hydrogen in their cores in 3 billion (3×10^9) years or less. On Earth, by contrast, human intelligence developed only within the last million years or so, some 4.56 billion years after the formation of the solar system. If that is typical of the time needed to evolve higher life-forms, then a star of 1.5 solar masses or more probably fades away or explodes into a supernova before intelligent creatures like us can evolve on any of that star's planets.

COSMIC CONNECTIONS

Habitable Zones for Life

Intelligent civilizations in our Milky Way Galaxy can evolve only in a certain region called the galactic habitable zone. In that zone, a suitable planet must lie within the planetary habitable zone of its parent star. (After C. H. Lineweaver, Y. Fenner, and B. K. Gibson)

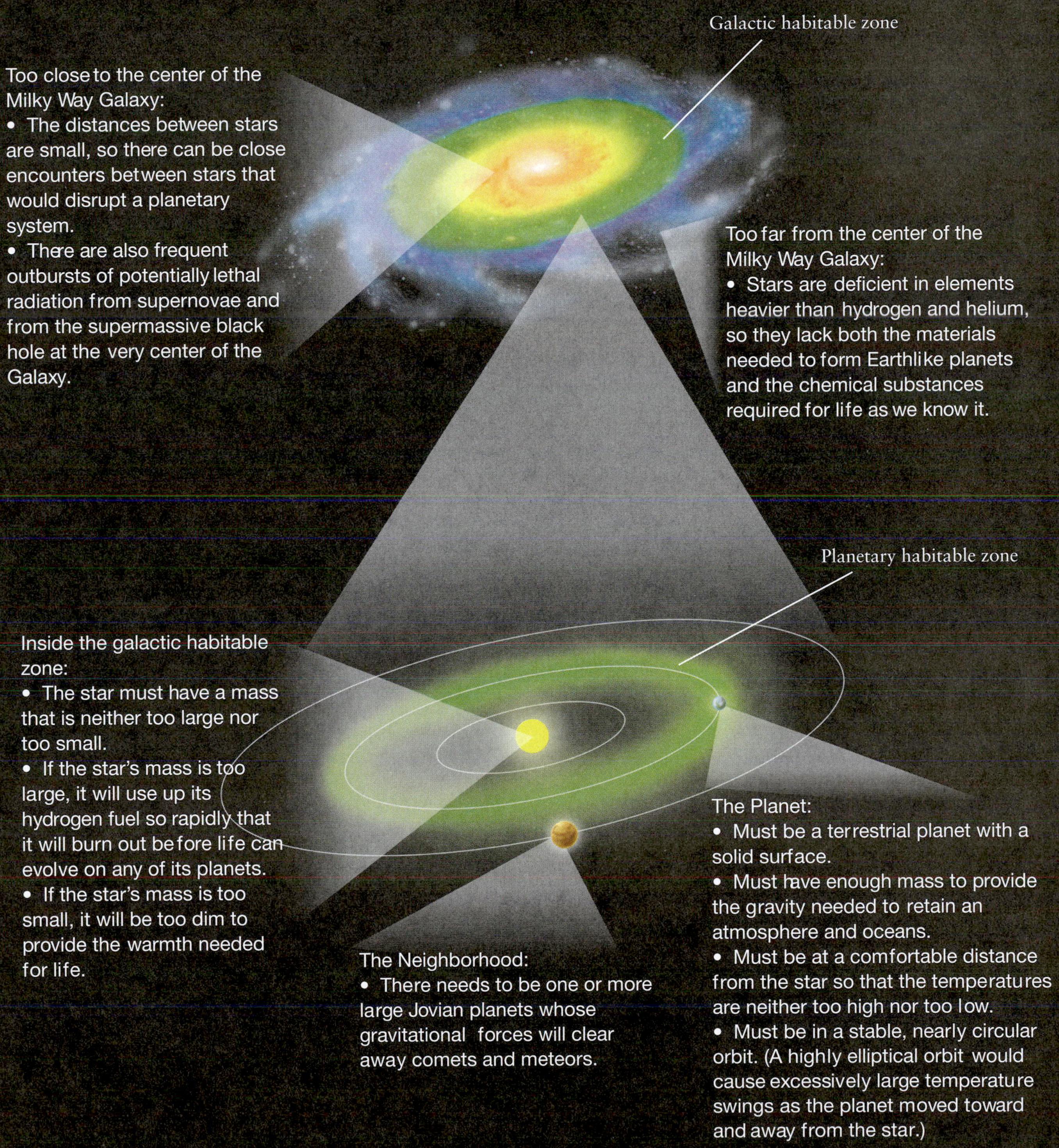

Galactic habitable zone
Too close to the center of the Milky Way Galaxy:
• The distances between stars are small, so there can be close encounters between stars that would disrupt a planetary system.
• There are also frequent outbursts of potentially lethal radiation from supernovae and from the supermassive black hole at the very center of the Galaxy.
Too far from the center of the Milky Way Galaxy:
• Stars are deficient in elements heavier than hydrogen and helium, so they lack both the materials needed to form Earthlike planets and the chemical substances required for life as we know it.
Planetary habitable zone
Inside the galactic habitable zone:
• The star must have a mass that is neither too large nor too small.
• If the star's mass is too large, it will use up its hydrogen fuel so rapidly that it will burn out before life can evolve on any of its planets.
• If the star's mass is too small, it will be too dim to provide the warmth needed for life.
The Neighborhood:
• There needs to be one or more large Jovian planets whose gravitational forces will clear away comets and meteors.
The Planet:
• Must be a terrestrial planet with a solid surface.
• Must have enough mass to provide the gravity needed to retain an atmosphere and oceans.
• Must be at a comfortable distance from the star so that the temperatures are neither too high nor too low.
• Must be in a stable, nearly circular orbit. (A highly elliptical orbit would cause excessively large temperature swings as the planet moved toward and away from the star.)

Although stars less massive than the Sun have much longer lifetimes, they also seem unsuited for life because they are so dim. Only planets very near a low-mass star would be sufficiently warm for life as we know it, and a planet that close is subject to strong tidal forces from its star. We saw in Section 4-8 how Earth's tidal forces keep the Moon locked in synchronous rotation, with one face continually facing Earth. In the same way, a planet that orbits too close to its star would have one hemisphere that always faced the star, while the other hemisphere would be in perpetual, frigid darkness.

These considerations leave us with stars not too different from the Sun. (Like Goldilocks sampling the three bears' porridge, we must have a star that is not too hot and not too cold but just right.) Based on statistical studies of star formation in the Milky Way, some astronomers estimate that roughly one of these Sunlike stars forms in the Galaxy each year in the galactic **habitable zone** (see the *Cosmic Connections* figure). This result sets R_* at 1 per year.

As we saw in Sections 8-4 and 8-5, the planets in our solar system formed as a natural consequence of the birth of the Sun. We have also seen evidence suggesting that planetary formation may be commonplace around single stars (see Figure 8-8). Many astronomers suspect that most Sunlike stars probably have planets, and so they give f_p a value of 1.

Unfortunately, the rest of the terms in the Drake equation are very uncertain. Let's play with some hypothetical values. The chances that a planetary system has an Earthlike world suitable for life are not known. (But observations of planets around other stars made by the *Kepler* spacecraft will help determine this number in the coming years.) Were we to consider our own solar system as representative, we could put n_e at 1. Let's be more conservative, however, and suppose that 1 in 10 solar-type stars is orbited by a habitable planet, making $n_e = 0.1$.

What about the fraction of Earthlike planets that develop life? Given the existence of life on Earth, we might assume that, given appropriate conditions, the development of life is quite likely, and we could set $f_l = 1$. However, it is also possible that life on Earthlike planets is rare. With only one Earthlike planet (our own Earth) to observe, there is no way to deduce the probability of Earthlike planets in general to form life. Furthermore, biochemists do not understand the detailed steps of life's origins and therefore cannot predict the probability for life to form. This is a topic of intense interest to astrobiologists, but we will be optimistic and assume here that $f_l = 1$.

For the sake of argument, we might also assume that evolution might naturally lead to the development of intelligence (a conjecture that is hotly debated) and also make $f_i = 1$. It is anyone's guess as to whether these intelligent extraterrestrial beings would attempt communication with other civilizations in the Galaxy, but were we to assume that they would, f_c would also be put at 1.

The last variable, L, involving the longevity of an advanced civilization, might be the most uncertain of all. Looking at our own example, we see a planet whose atmosphere and oceans are increasingly polluted by creatures that possess nuclear weapons. If we are typical, perhaps L is as short as 100 years. Putting all these numbers together, we arrive at

$$N = 1/\text{year} \times 1 \times 0.1 \times 1 \times 1 \times 1 \times 100 \text{ years} = 10$$

In other words, out of the hundreds of billions of stars in the Galaxy, we would estimate that there are only 10 technologically advanced civilizations from which we might receive communications.

A wide range of values has been proposed for the terms in the Drake equation, and these various guesses produce vastly different estimates of N. Some scientists argue that there is exactly one advanced civilization in the Galaxy and that we are it. Others speculate that there may be hundreds or thousands of planets inhabited by intelligent creatures. If we wish to know whether our Galaxy is devoid of other intelligence, teeming with civilizations, or something in between, we must keep searching the skies.

CONCEPTCHECK 27-9

What makes the longevity of an advanced civilization, L, so difficult to estimate?

Answer appears at the end of the chapter.

27-5 Radio searches for alien civilizations are underway

Even if only a few alien civilizations are scattered across the Galaxy, we have the technology to detect radio transmissions from them. On the one hand, the signals would be too weak to detect if they were similar in strength to the radio waves we use for communication and radio stations here on Earth. Waves like this are like ripples on a pond that diminish as they spread outward. On the other hand, if an advanced civilization wanted to make its presence known, it *might* sweep the skies with a focused beam of radio waves, which is much easier to detect. Even if we could not decode such a signal, it should be easy to tell that it results from an advanced technology.

But if other civilizations are trying to communicate with us using radio waves, what frequency are they using? This is an important question, because if we fail to tune our radio telescopes to the right frequency, we might never know whether the aliens are out there.

A reasonable choice would be a frequency that is fairly free of interference from extraneous sources. SETI pioneer Bernard Oliver was the first to draw attention to a range of relatively noise-free frequencies in the neighborhood of the microwave emission lines of hydrogen (H) and hydroxide (OH) (Figure 27-12). This region of the spectrum can be called microwave or radio emission and is called the **water hole,** because H and OH together make H_2O, or water.

In 1989, NASA began work on the High Resolution Microwave Survey (HRMS), an ambitious project to scan the entire sky at frequencies spanning the water hole from 10^3 to 10^4 MHz. HRMS would have observed more than 800 nearby solar-type stars, but, unfortunately, its funding was cut just one year after becoming operational.

Even though NASA no longer funds searches for extraterrestrial intelligence, several teams of scientists remain actively involved in SETI programs. Funding for these projects has come from nongovernmental organizations such as the Planetary Society and from private individuals. In 1995 the SETI Institute in California began Project Phoenix, the direct successor to HRMS. In 2004, after careful observation of the nearest 800 solar-type stars, scientists for Project Phoenix announced that they had not found any promising signals.

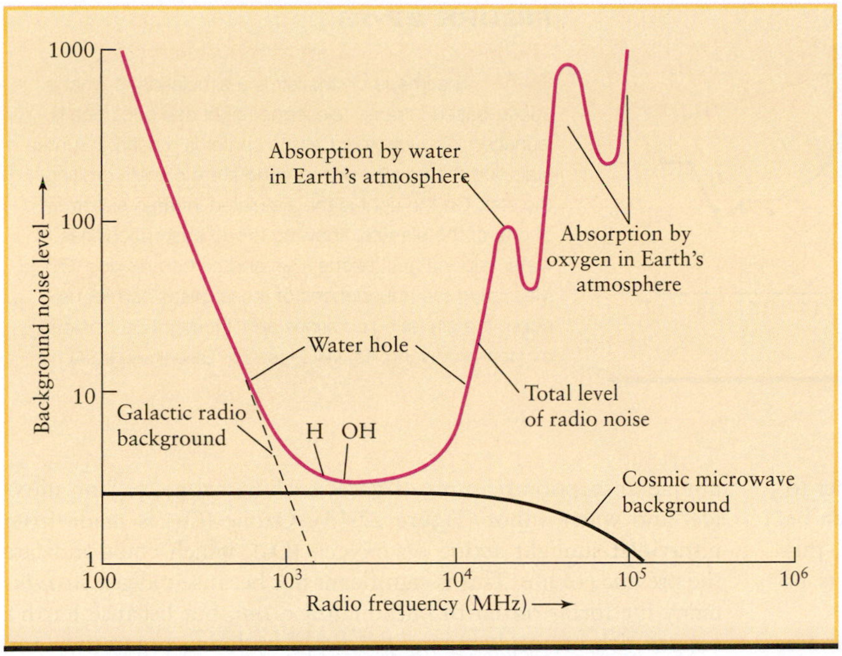

FIGURE 27-12
The Water Hole This graph shows the background noise level from the sky at various radio and microwave frequencies. The so-called water hole is a range of radio frequencies from about 10^3 to 10^4 megahertz (MHz) in which there is little noise and little absorption by Earth's atmosphere. Some scientists suggest that this noise-free region would be well suited for interstellar communication. Within the water hole itself, the principal source of noise is the afterglow of the Big Bang, called the cosmic microwave background. To put this graph in perspective, a frequency of 100 MHz corresponds to "100" on a FM radio, and 10^3 MHz is a frequency used for various types of radar. (Adapted from C. Sagan and F. Drake)

A major challenge facing SETI is the tremendous amount of computer time needed to analyze the mountains of data returned by radio searches. To this end, scientists at the University of California, Berkeley, have recruited more than 5 million personal computer users to participate in a project called SETI@home. Each user receives actual data from a detector called SERENDIP IV (*S*earch for *E*xtraterrestrial *R*adio *E*missions from *N*earby, *D*eveloped, *I*ntelligent *P*opulations) and a data analysis program that also acts as a screensaver. When the computer's screensaver is on, the program runs, the data are analyzed, and the results are reported via the Internet to the researchers at Berkeley. The program then downloads new data to be analyzed. Since 1999, SETI@home users have provided the equivalent of more than 3 million years of computing time!

> Millions of personal computers have helped scan for alien radio transmissions

In 2007, the SETI Institute began operation of a large radio telescope dedicated to the search for intelligent signals. This telescope, called the Allen Telescope Array (ATA), is actually hundreds of relatively small and inexpensive radio dishes working together. Designed for rapid and sensitive searches, the ATA hopes to search the nearest 100,000 solar-type stars. If astronomers knew which stars had Earthlike planets, SETI's telescopes could focus their attention on these objects, and the effort to find these stars is underway.

CONCEPTCHECK 27-10

Why should the range of frequencies in the water hole of Figure 27-12 be an attractive range of frequencies for an alien civilization to send out messages?

Answer appears at the end of the chapter.

27-6 Telescopes have begun searching for Earthlike planets

Although no longer involved in SETI, NASA is carrying out a major effort to search for Earthlike planets suitable for the evolution of an advanced civilization. Such a search poses a major challenge. One problem is that Earth-sized planets are too dim to be seen in visible light against the glare of their parent star. Using a different approach, astronomers have discovered many Jupiter-sized planets by detecting the "wobble" that these planets produce in their parent star (see Section 8-7). But a planet the size of Earth exerts only a weak gravitational force on its parent star. In 2009, the first Earth-sized planet (at 1.9 Earth masses) was discovered through a tiny wobble. However, this planet, named Gliese 581 e, orbits too close to its parent star to allow for liquid water. Furthermore, the wobble is so tiny that few Earth-sized planets are expected to be found using this technique.

Searching for Planetary Transits

An alternative technique was being used by an orbiting telescope called *Kepler,* which began detecting planets in 2009, and discovered hundreds of Earth-sized planets. Recall from Section 8-7, that if a star is orbited by a planet whose orbital plane is oriented edge-on to our line of sight, once per orbit the planet will pass in front of the star in an event called a *transit* (Figure 8-19). By blocking some of the star's light, this transit causes a temporary dimming of the light we see from that star.

Once a transiting planet is detected, astronomers determine the transiting planet's size (inferred from how much dimming takes place), as well as the size of its orbit. Recall from Johannes Kepler's third law that the size of the orbit can be calculated using the orbital period of the planet, which is the same as the time interval

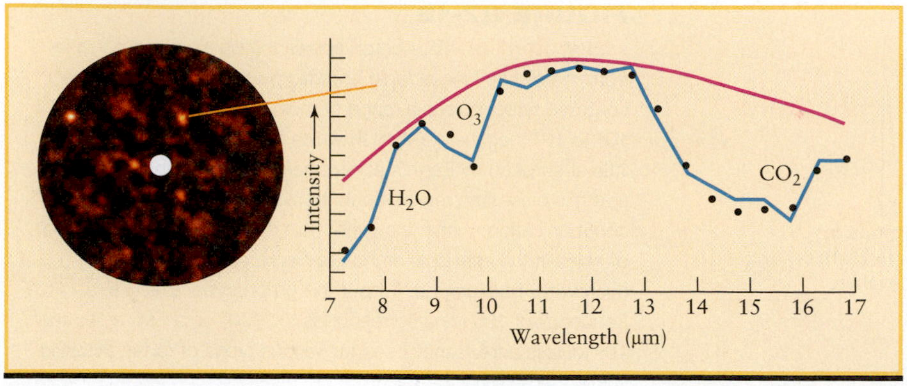

FIGURE 27-13
The Infrared Spectrum of a Simulated Planet The image on the left is a simulation of what a space-based infrared telescope might see once one is launched. The white dot at the center is a nearby Sunlike star, and the smaller dots around it are planets orbiting the star. On the right is the simulated infrared spectrum of one of the planets, showing broad absorption lines of water vapor (H_2O), ozone (O_3), and carbon dioxide (CO_2). Measuring the abundances of these gases can tell us about the planet's conditions and if it might be harboring oxygen-producing life. (Jet Propulsion Laboratory/NASA)

between successive transits. Given the distance from the planet to its parent star and the star's luminosity, astronomers will even be able to estimate the planet's average temperature. However, this estimated temperature would not include the warming effects of an atmosphere without additional observations.

To determine a planet's atmosphere, detailed spectra are required and these observations are only just beginning. It is important to keep in mind that just because a planet is Earth-sized and at a certain distance from its sun, it does not mean it is necessarily Earthlike. Finding a truly Earthlike planet with liquid water is difficult but could be discovered with the current generation of telescopes, or the next generation designed specifically for this goal.

CONCEPTCHECK 27-11

What exactly was the *Kepler* telescope watching for as it searched for planets around stars?

CONCEPTCHECK 27-12

Which planets were the *Kepler* telescope not able to find, even if quite nearby?

Answers appear at the end of the chapter.

Searches Using Images and Spectra

The results from *Kepler* may help SETI and other astronomers select stars to study in more detail. But what comes next? NASA and the European Space Agency have studied how the next generation of telescopes might continue investigating Earthlike planets. These space agencies have proposed the Terrestrial Planet Finder (a system of telescopes) and *Darwin* (a spacecraft mission), but there is no funding for these missions as of this writing.

Studies suggest that a future telescope might search for Earthlike planets by detecting their infrared radiation. The rationale is that stars like the Sun emit much less infrared radiation than visible light, while planets are relatively strong emitters of infrared. Hence, observing in the infrared makes it less difficult (although still technically challenging) to detect planets orbiting a star.

This next generation of telescopes might also be able to measure a "biosignature" of life in a planet's atmosphere. By analyzing the infrared spectra of potentially Earthlike planets, the telescopes could reveal the atmospheric composition from the unique absorption features of gases such as ozone, carbon dioxide, and water vapor (Figure 27-13). Ozone (O_3) is made from ultraviolet sunlight acting on oxygen (O_2), which could indicate the presence of life. This is significant not because oxygen *sustains* many life-forms on Earth (although it does), but because Earth's oxygen-rich atmosphere was *produced* by living organisms, and does not seem attainable through nonbiological means. Therefore, oxygen is a key atmospheric biosignature that will be the target of future studies.

Sometime during the twenty-first century, we will probably answer the question "Are there worlds like Earth orbiting other stars?" If the answer is yes, then searches for life around other stars will gain even more impetus.

Future telescope technology may make it possible to resolve details on planets orbiting other stars

The potential rewards from such searches are great. Detecting a message from an alien civilization could dramatically change the course of our own civilization, through the sharing of scientific information with another species or an awakening of social or humanistic enlightenment. In only a few years our technology, industry, and social structure might advance the equivalent of centuries into the future. Such changes would touch every person on Earth. Mindful of these profound implications, scientists push ahead with the search for extraterrestrial intelligence.

CONCEPTCHECK 27-13

How does ozone act as a biosignature for life?

CONCEPTCHECK 27-14

Why not look for oxygen gas (O_2) directly? Consult Figure 27-13.

Answers appear at the end of a chapter.

KEY WORDS

astrobiology, p. 786
carbonaceous chondrite, p. 787
Drake equation, p. 794
extremophile, p. 788
habitable zone, p. 796
organic molecules, p. 786
search for extraterrestrial intelligence (SETI), p. 794
water hole, p. 796

KEY IDEAS

Organic Molecules in the Universe: All life on Earth, and presumably on other worlds, depends on organic (carbon-based) molecules. These molecules occur naturally throughout interstellar space.

• The organic molecules needed for life to originate were probably brought to the young Earth by comets, meteorites, and interplanetary dust particles. Another likely source for organic molecules is chemical reactions in Earth's primitive atmosphere. Similar processes may occur on other worlds.

Life in the Solar System: Besides Earth, other worlds in our solar system—the planet Mars, Jupiter's satellite Europa, and Saturn's satellite Enceladus—may have had the right conditions for the origin of life.

• Europa appears to have extensive liquid water beneath its icy surface. Future missions may search for the presence of life there. Enceladus also appears to contain a large subsurface ocean, and some of this material is ejected through ice volcanoes for easier sampling.

• Mars once had liquid water on its surface, though it has none today. Life may have originated on Mars during the liquid water era.

• The *Viking Lander* spacecraft searched for microorganisms on the Martian surface, but found no conclusive sign of their presence. The Mars Science Laboratory is designed to assess the past and present suitability of Mars for microbial life.

• An ancient Martian rock that came to Earth as a meteorite shows features that have been suggested as remnants of Martian life. Most of these features can be explained by geologic processes.

Radio Searches for Extraterrestrial Intelligence: Astronomers have carried out a number of searches for radio signals from other stars. No signs of intelligent life have yet been detected, but searches are continuing and use increasingly sophisticated techniques.

• The Drake equation is a tool for estimating the number of intelligent, communicative civilizations in our Galaxy.

Telescope Searches for Earthlike Planets: A future generation of orbiting telescopes may be able to detect numerous terrestrial planets around nearby stars. If such planets are found, their infrared spectra may reveal the presence or absence of life.

QUESTIONS

Review Questions

1. Why are extreme life-forms on Earth, such as those shown in the photograph that opens this chapter, of interest to astrobiologists?
2. What is meant by "life as we know it"? Why do astrobiologists suspect that extraterrestrial life is likely to be of this form?
3. How have astronomers discovered organic molecules in interstellar space? Does this discovery mean that life of some sort exists in the space between the stars?
4. Mercury, Venus, and the Moon are all considered unlikely places to find life. Suggest why this should be.
5. Why do astronomers think Enceladus could support life? What would be the easiest way to search for this life?
6. Why are scientists interested in searching for life in Lake Vostok, Antarctica?
7. Many science-fiction stories and movies—including *The War of the Worlds, Invaders from Mars, Mars Attacks!,* and *Martians, Go Home*—involve invasions of Earth by intelligent beings from Mars. Why Mars rather than any of the other planets?
8. Summarize the three tests performed by the *Viking Landers* to search for Martian microorganisms.
9. What arguments can you give against the idea that the "face" on Mars (Figure 27-9) is of intelligent origin? What arguments can you give in favor of this idea?
10. Suppose the *Curiosity* rover measured an enhanced abundance of the carbon isotope ^{12}C in an ancient riverbed. What might this tell us?
11. Suppose someone brought you a rock that he claimed was a Martian meteorite. What scientific tests would you recommend be done to test this claim?
12. The Martian meteorite ALH 84001 contains magnetite, which can be produced by microorganisms on Earth. Why isn't this good evidence for Martian life?
13. Why are astronomers interested in the microtunnels within several Martian meteorites? Why don't these prove the existence of Martian life?
14. Why are most searches for extraterrestrial intelligence made using radio telescopes? Why are most of these carried out at frequencies between 10^3 MHz and 10^4 MHz?
15. Explain why telescopes that would look for the infrared spectra of other planets need to be placed in space (*Hint:* See spectrum in Figure 27-13).

Advanced Questions

Problem-solving tips and tools

The small-angle formula, discussed in Box 1-1, will be useful. Section 5-2 gives the relationship between wavelength and frequency, while Section 5-9 and Box 5-6 discuss the Doppler effect. Section 6-3 gives the relationship between the angular resolution of a telescope, the telescope diameter, and the wavelength used. You will find useful data about the planets in Appendix 1.

16. In 1802, when it seemed likely to many scholars that there was life on Mars, the German mathematician Karl Friedrich Gauss proposed that we signal the Martian inhabitants by drawing huge geometric patterns in the snows of Siberia. His plan was never carried out. (a) Suppose patterns had been drawn that were 1000 km across. What minimum diameter would the objective of a Martian telescope need to have to

be able to resolve these patterns? Assume that the observations are made at a wavelength of 550 nm, and assume that Earth and Mars are at their minimum separation. (**b**) Ideally, the patterns used would be ones that could not be mistaken for natural formations. They should also indicate that they were created by an advanced civilization. What sort of patterns would you have chosen?

17. TUTORIAL 27-1 Assume that all the terms in the Drake equation have the values given in the text, except for *N* and *L*. (**a**) If there are 1000 civilizations in the Galaxy today, what must be the average lifetime of a technological civilization? (**b**) What if there are a million such civilizations?

18. (**a**) Of the visually brightest stars in the sky listed in Appendix 5, which might be candidates for having Earthlike planets on which intelligent civilizations have evolved? Explain your selection criteria. (**b**) Repeat part (a) for the nearest stars, listed in Appendix 4.

19. It has been suggested that extraterrestrial civilizations would choose to communicate at a wavelength of 21 cm. Hydrogen atoms in interstellar space naturally emit at this wavelength, so astronomers studying the distribution of hydrogen around the Galaxy would already have their radio telescopes tuned to receive extraterrestrial signals. (**a**) Calculate the frequency of this radiation in megahertz. Is this inside or outside the water hole? (**b**) Discuss the merits of this suggestion.

20. Imagine that a civilization in another planetary system is sending a radio signal toward Earth. As our planet moves in its orbit around the Sun, the wavelength of the signal we receive will change due to the Doppler effect. This gives SETI scientists a way to distinguish stray signals of terrestrial origin (which will not show this kind of wavelength change) from interstellar signals. (**a**) Use the data in Appendix 1 to calculate the speed of Earth in its orbit. For simplicity, assume the orbit is circular. (**b**) If the alien civilization is transmitting at a frequency of 3000 MHz, what wavelength (in meters) would we receive if Earth were moving neither toward nor away from their planet? (**c**) The maximum Doppler shift occurs if Earth's orbital motion takes it directly toward or directly away from the alien planet. How large is that maximum wavelength shift? Express your answer both in meters and as a percentage of the unshifted wavelength you found in (b). (**d**) Discuss why it is important that SETI radio receivers be able to measure frequency and wavelength to very high precision.

21. Astronomers have proposed using interferometry to make an extremely high-resolution telescope. This proposal involves placing a number of infrared telescopes in space, separating them by thousands of kilometers, and combining the light from the individual telescopes. One design of this kind has an effective diameter of 6000 km and uses infrared radiation with a wavelength of 10 mm. If it is used to observe an Earthlike planet orbiting the star Epsilon Eridani, 3.22 parsecs (10.5 light-years) from Earth, what is the size of the smallest detail that this system will be able to resolve on the face of that planet? Give your answer in kilometers.

Discussion Questions

22. Suppose someone told you that the *Viking Landers* failed to detect life on Mars simply because the tests were designed to detect terrestrial life-forms, not Martian life-forms. How would you respond?

23. Science-fiction television shows and movies often depict aliens as looking very much like humans. Discuss the likelihood that intelligent creatures from another world would have (**a**) a biochemistry similar to our own, (**b**) two legs and two arms, and (**c**) about the same dimensions as a human.

24. The late, great science-fiction editor John W. Campbell exhorted his authors to write stories about organisms that think as well as humans but not *like* humans. Discuss the possibility that an intelligent being from another world might be so alien in its thought processes that we could not communicate with it.

25. If a planet always kept the same face toward its star, just as the Moon always keeps the same face toward Earth, most of the planet's surface would be uninhabitable. Discuss why.

26. How do you think our society would respond to the discovery of intelligent messages coming from a civilization on a planet orbiting another star? Explain your reasoning.

27. What do you think will set the limit on the lifetime of our technological civilization? This factor is *L* in the Drake equation. Explain your reasoning.

28. The first of all Earth spacecraft to venture into interstellar space were *Pioneer 10* and *Pioneer 11*, which were launched in 1972 and 1973, respectively. Their missions took them past Jupiter and Saturn and eventually beyond the solar system. Both spacecraft carry a metal plaque with artwork (reproduced below) that shows where the spacecraft is from and what sort of creatures designed it. If an alien civilization were someday to find one of these spacecraft, which of the features on the plaque do you think would be easily understandable to them? Explain your reasoning.

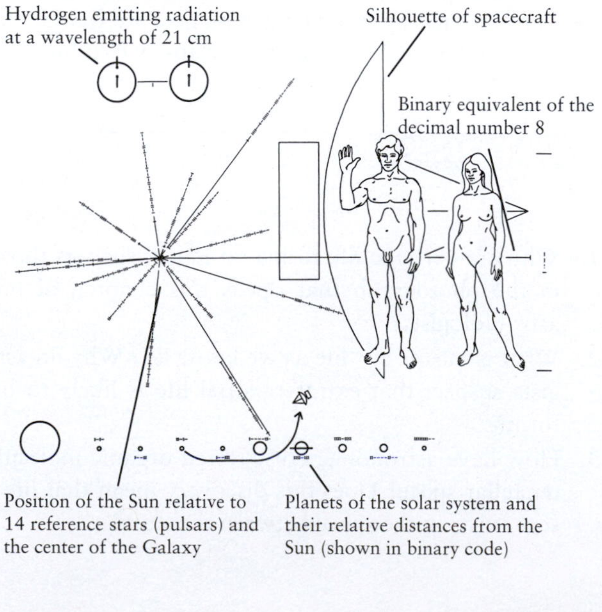

Web/eBook Questions

29. Any living creatures in the subsurface ocean of Europa would have to survive without sunlight. Instead, they might obtain energy from Europa's inner heat. Search the World Wide Web for information about "black smokers," which are associated with high-temperature vents at the bottom of Earth's oceans. What kind of life is found around black smokers? How do these life-forms differ from the more familiar organisms found in the upper levels of the ocean?
30. Search the World Wide Web for information about the *Mars Express* orbiter and the *Spirit* and *Opportunity* rovers. What discoveries have these missions made about water on Mars? Have they found any evidence that liquid water has existed on Mars in the recent past? Describe the evidence, if any.
31. Like other popular media, the World Wide Web is full of claims of the existence of "extraterrestrial intelligence," namely, UFO sightings and alien abductions. (**a**) Choose a Web site of this kind and analyze its content using the idea of Occam's razor, the principle that if there is more than one viable explanation for a phenomenon, one should choose the simplest explanation that fits all the observed facts. (**b**) Read what a skeptical Web site has to say about UFO sightings. A good example is the Web site of the Committee for Skeptical Inquiry, or CSI. After considering what you have read on both sides of the UFO debate, discuss your opinions about whether intelligent aliens really have landed on Earth.

ACTIVITIES

Observing Projects

32. Use the *Starry Night*™ program to view Earth from space. Click the **Home** button in the toolbar to move your viewing location to your home town or city. Click the **Increase current elevation** button beneath the **Viewing Location** panel in the toolbar to raise your position to about 11,000 km above your home location. Use the hand tool to move Earth into the center of the view. Locate the position of your home and zoom in on it, using the **Zoom** tool on the right of the toolbar to set the field of view to about $16° \times 11°$. Use the location scroller to move over various regions of Earth's surface. (**a**) Can you see any evidence of the presence of life or of man-made objects? (**b**) **Right-click** on Earth (**Ctrl-click** on a Mac) and select **Google Maps** from the contextual menu to examine regions of Earth in great detail on these images from an orbiting satellite. (You will need to have access to the Internet in order to use Google Maps.) You can attempt to locate your own home on these maps. As an exercise, find the country of Panama. **Zoom** in progressively upon the Panama Canal and search for the ships traversing this incredible waterway and the massive locks that lift these ships into the upper waterways. What does this suggest about the importance of sending spacecraft to planets to explore their surfaces at close range?
33. Use *Starry Night*™ to examine the planet Mars. Select **Favourites > Explorations > Mars. Zoom** in or out on Mars and use the location scroller to examine the planet's surface. Based on what you observe, where on the Martian surface would you choose to land a spacecraft to search for the presence of life? Explain the reasons for your choice.
34. Use the *Starry Night*™ program to examine Jupiter's moon Europa. Select **Favourites > Explorations > Europa** to view this enigmatic moon of our largest planet. Use the location scroller cursor and zoom controls to examine the surface of this moon. Are there any surface features that suggest that this moon might harbor life? Is there any other evidence, perhaps not visible in this simulation, that suggests the possibility that life exists there?
35. Use *Starry Night*™ to investigate the likelihood of the existence of life in the universe beyond that found upon Earth. Select **Favourites > Explorations > Atlas** and then open the **Options** pane. Expand the **Stars** layer and the **Stars** heading and click the checkbox to turn on the **Mark stars with extrasolar planets** option. Then use the hand tool or cursor keys to look around the sky. Each marked star in this view has at least one planet orbiting around it. Click and hold the **Increase current elevation** button in the toolbar until the viewpoint is about 1000 ly from Earth. Note that the stars with extrasolar planets are contained in a small knot of stars in the region of our Sun, our solar neighborhood. This is because present techniques place a limit upon the distance to which we can find extrasolar planets in space. There is no logical reason why the same proportion of stars in the rest of the Milky Way, or indeed the rest of the universe, should not have companion planets. **Increase current elevation** again to about 90,000 ly from Earth to see a view of the Milky Way galaxy. Use the location scroller to view the Milky Way face-on and compare the clump of stars representing the solar neighborhood to the size of the Milky Way Galaxy. Click the **Increase current elevation** button again until the **Viewing Location** panel indicates a distance to Earth of about 0.300 Mly (300,000 light-years). In this view, each point of light represents a separate galaxy containing hundreds of billions of stars. Now, gradually increase current elevation to about 1500 Mly (1.5 billion light-years from Earth) to see the entire database of 28,000 galaxies included in the *Starry Night*™ data bank. The borders of this cube of galaxies reaching out to about 500 Mly from Earth is defined by the limits of the methods for determining distance to galaxies in deep space. In practice, this volume of space represents less than 5% of the size of the observable universe and galaxies abound beyond the limits of this view. To view images of some of these very distant galaxies, open the view named **Hubble Deep Field** from the **Explorations** folder in the **Favourites** pane. This view from the center of Earth is centered upon a seemingly empty patch of sky. Zoom in to see this image of more than 1500 very distant galaxies that were found by the Hubble Space Telescope in a small region of our sky. Then open the view named **Hubble Ultra Deep Field** and zoom in on an image of over 10,000 galaxies that

extends to the limits of our observable universe. (**a**) Do these views influence your thoughts on the likelihood that life, including intelligent life, exists elsewhere in the universe other than on Earth? Explain your reasoning using the Drake equation as a guide. (**b**) Why is it unlikely that we will be able to communicate with life forms that might exist on planets in these very distant regions of space?

Collaborative Exercise

36. Imagine that astronomers have discovered intelligent life in a nearby star system. Your group is submitting a proposal for who on Earth should speak for the planet and what 50-word message should be conveyed. Prepare a maximum one-page proposal that states (**a**) who should speak for Earth and why; (**b**) what this person should say in 50 words; and (**c**) why this message is the most important compared to other things that could be said. Only serious responses receive full credit.

ANSWERS

ConceptChecks

ConceptCheck 27-1: Carbon atoms can combine with other elements to form an impressively wide array of different molecules, and this complexity is needed for even the smallest living organisms. Furthermore, carbon is produced in stars and is widely available.

ConceptCheck 27-2: Organic molecules are commonly found in meteorites and comets, which frequently crashed onto planet surfaces during the early formation of the solar system. Also, interplanetary dust grains containing organic compounds steadily fall to a planet's surface.

ConceptCheck 27-3: They created amino acids and other compounds that living organisms need—they did not create living entities.

ConceptCheck 27-4: The salts dissolve in water when the water is in contact with rock. Rather than a small pocket of water within ice, the salts suggest a much larger ocean is in contact with a rocky mantle.

ConceptCheck 27-5: Living organisms that depend on photosynthesis concentrate ^{12}C compared to their surroundings. This increased abundance can remain in the environment long after the organisms are dead.

ConceptCheck 27-6: Geological processes (involving no biology) can also produce this type of magnetite, so there is no compelling reason to assume that the magnetite was produced by Martian microbes. (*Note:* The claims that this magnetite was of biological origin came *before* it was figured out that this magnetite can be produced through geology.)

ConceptCheck 27-7: The Martian meteorites have trace amounts of gas trapped within them that are nearly identical to the unique Martian atmosphere.

ConceptCheck 27-8: At the speed of light, radio waves travel at more than 100,000 km every second, whereas our fastest spacecraft can only travel at tens of thousands of kilometers every hour. Therefore, the vast distances between the stars make it seemingly impossible for spacecraft to carry out exploration of the stars.

ConceptCheck 27-9: L is particularly difficult to estimate because at about the same time the technology to communicate beyond one's own planet is developed, the technology of weapons of mass destruction can develop, which could end the civilization just as it starts (alternatively, it can learn to live without war and exist for a very long time).

ConceptCheck 27-10: At other frequencies, there is more background noise, and more energy would need to be used to reliably transmit the same message across the Galaxy. Any advanced civilization could figure this out and would see the benefits of sending messages in this narrow range.

ConceptCheck 27-11: The *Kepler* telescope was looking for an ever-so-slight dimming of a star as an orbiting planet passes in front of the star, partially blocking the star from our view. This is called a transit.

ConceptCheck 27-12: If the orbital plane of a planet around its star is tilted up or down from our perspective, then the planet will not block out any of the star's light during its orbit. Since *Kepler* looked for the dimming of stars as they were blocked by planets, *Kepler* would not have detected a planet with this kind of orbit.

ConceptCheck 27-13: Ozone comes from oxygen gas, O_2, which is produced by living organisms on Earth. Scientists do not know of any plausible nonbiological way to create a dense oxygen atmosphere.

ConceptCheck 27-14: In the infrared where a planet's spectrum can be more easily measured, an absorption line from oxygen gas is not present, but absorption from ozone (O_3) is.

A Biologist's View of Astrobiology by Kevin W. Plaxco

In the early 1960s, flush with the excitement of Sputnik and the first manned space missions, the Nobel Prize-winning biologist Joshua Lederberg coined the word "exobiology" to describe the scientific study of extraterrestrial life. But after a brief flurry of popularity in the 1960s and 1970s (due partly to the pioneering research and public outreach work of Carl Sagan), exobiology fell out of fashion among scientists; it was, after all, the only field of scientific study without any actual subject material to research.

In the last decade, however, the field of astrobiology has filled the void left by exobiology's demise by being simultaneously more encompassing and more practical. Astrobiology removes exobiology's distinction between life on Earth and life "out there" and focuses on broader, but more tractable, questions regarding the relationship between life (any life, anywhere) and the universe. In a nutshell, astrobiology uses our significant (if incomplete) knowledge of life on Earth to address three broad questions about life in the universe:

- Which of the physical attributes of our universe were necessary in order for it to support the origins and further evolution of life?
- How did life arise and evolve on Earth, and how might it have arisen and evolved elsewhere?
- Where else might life have arisen in our universe, and how do we best search for it?

Much of the worth and appeal of astrobiology lie in the fact that these questions, which are among the most fundamental posed by science, address the most profound issues regarding our origins and our nature.

Perhaps not surprisingly, some aspects of astrobiology are far better understood than others. We understand, for example, the broad details of how the almost metal-free hydrogen and helium left over after the Big Bang was enriched in the metallic elements by fusion reactions occurring in the more massive stars and how these elements—absolutely critical for biology because only they support the complex chemistry required for life—were then returned to the cosmos in supernova explosions. Similarly, we understand in detail how rocky planets—which we see as potentially habitable—form out of the dust and gas of prestellar nebulae. In contrast, however, much of the *biology* in astrobiology remains speculative at best. There is, for example, nothing even remotely resembling consensus regarding how life first arose on Earth much less how—or if—it might have arisen elsewhere in the cosmos.

Perhaps equally unsurprisingly, some branches of the astrobiology community are more "optimistic" than others. Astronomers generally view astrobiology in terms of the processes by which potentially habitable planets form. As these processes are reasonably well understood, astronomers tend to be rather upbeat regarding the events surrounding life's origins. The radio astronomer Frank Drake, who pioneered the search for extraterrestrial intelligence (SETI) in the 1960s, is a relatively extreme example of this hopefulness: When trying to estimate the number of intelligent, "communicating" civilizations in the Galaxy, he assumed that rocky, water-soaked planets almost invariably give rise to life. Biochemists, in contrast, tend to view the key issues in astrobiology as relating to the first steps in the origins of life—that is, the detailed processes by which inanimate chemicals first combine to form a living, evolving, self-replicating system. To date, about a half dozen theories have been put forth to explain how this might have occurred on Earth, but all are fraught with enormous scientific difficulties. Specifically, every single theory of the origins of life postulated to date either utterly lacks experimental evidence to back it up or requires events that are so improbable they make even the "astronomical" number of planets in the universe look small by comparison. When faced with what is, in effect, a complete unknown, biochemists tend to be much more cautious about estimating the probability of life arising and the chances of our being alone in the cosmos. To quote Francis Crick, codiscoverer of the structure of DNA, "We cannot decide whether the origin of life on Earth was an extremely unlikely event or almost a certainty, or any possibility in between these two extremes."

Irrespective of our degree of personal optimism or pessimism, though, no one would be more thrilled than us biologists by the detection of life that had arisen independently of life on Earth. Such a find would revolutionize our field by providing the first, tangible evidence of whether there are viable routes to the evolution of life other than the DNA/proteins/cells path that it took on Earth. But even barring such a momentous discovery as the detection of extraterrestrial life, astrobiology still provides a significant service to biology: By placing life into a broader context, astrobiology provides the perspective that we need in order to truly understand who we are and where we come from.

Prior to becoming a professor of biochemistry at the University of California at Santa Barbara, **Kevin Plaxco** earned his BS in chemistry from the University of California at Riverside and his Ph.D. in molecular biology from the California Institute of Technology. While his research focuses predominantly on the study of proteins, the "macromolecular machines" that perform the lion's share of the action in our cells, he has also performed NASA-funded research into methods for the detection of extraterrestrial life. In addition, he teaches a course in astrobiology and has published a popular science book on the subject.

Appendices

Appendix 1 The Planets: Orbital Data

Planet	Semimajor axis (10[6] km)	Semimajor axis (AU)	Sidereal period (years)	Sidereal period (days)	Synodic period (days)	Average orbital speed (km/s)	Orbital eccentricity	Inclination of orbit to ecliptic (°)
Mercury	57.9	0.387	0.241	87.969	115.88	47.9	0.206	7.00
Venus	108.2	0.723	0.615	224.70	583.92	35.0	0.007	3.39
Earth	149.6	1.000	1.000	365.256	—	29.79	0.017	0.00
Mars	227.9	1.524	1.88	686.98	779.94	24.1	0.093	1.85
Jupiter	778.3	5.203	11.86		398.9	13.1	0.048	1.30
Saturn	1429	9.554	29.46		378.1	9.64	0.053	2.48
Uranus	2871	19.194	84.10		369.7	6.83	0.043	0.77
Neptune	4498	30.066	164.86		367.5	5.5	0.010	1.77

Appendix 2 The Planets: Physical Data

Planet	Equatorial diameter (km)	Equatorial diameter (Earth = 1)	Mass (kg)	Mass (Earth = 1)	Average density (kg/m^3)	Rotation period* (solar days)	Inclination of equator to orbit (°)	Surface gravity Earth = 1	Albedo	Escape speed (km/s)
Mercury	4880	0.383	3.302 × 1023	0.0553	5430	58.646	0.5	0.38	0.12	4.3
Venus	12,104	0.949	4.868 × 1024	0.8149	5243	243.01[R]	177.4	0.91	0.59	10.4
Earth	12,756	1.000	5.974 × 1024	1.000	5515	0.997	23.45	1.000	0.39	11.2
Mars	6794	0.533	6.418 × 1023	0.107	3934	1.026	25.19	0.38	0.15	5.0
Jupiter	142,984	11.209	1.899 × 1027	317.8	1326	0.414	3.12	2.36	0.44	59.5
Saturn	120,536	9.449	5.685 × 1026	95.16	687	0.444	26.73	1.1	0.47	35.5
Uranus	51,118	4.007	8.682 × 1025	14.53	1318	0.718[R]	97.86	0.92	0.56	21.3
Neptune	49,528	3.883	1.024 × 1026	17.15	1638	0.671	29.56	1.1	0.51	23.5

**For Jupiter, Saturn, Uranus, and Neptune, the internal rotation period is given. A superscript* R *means that the rotation is retrograde (opposite the planet's orbital motion).*

Appendix 3 Major Satellites of the Planets by Mass

Planet	Satellite	Date of discovery	Average distance from center of planet (km)	Orbital (sidereal) period (days)*	Orbital eccentricity	Size of satellite (km)**	Mass (kg)
EARTH	Moon	—	384,400	27.322	0.0549	3476	7.349×10^{22}
MARS	Phobos	1877	9378	0.319	0.01	28 × 23 × 20	1.1×10^{22}
	Deimos	1877	23,460	1.263	0.00	16 × 12 × 10	1.1×10^{22}
JUPITER	Ganymede	1610	1,070,000	7.155	0.0011	5268	1.482×10^{23}
	Callisto	1610	1,883,000	16.689	0.0074	4800	1.077×10^{23}
	Io	1610	421,600	1.769	0.0041	3642	8.932×10^{22}
	Europa	1610	670,900	3.551	0.0094	3120	4.791×10^{22}
	Himalia	1904	11,461,000	250.56	0.1623	184	6.7×10^{18}
	Amalthea	1892	181,400	0.498	0.0031	270 × 200 × 155	2.1×10^{18}
	Elara	1905	11,741,000	259.64	0.2174	78	8.7×10^{17}
	Thebe	1979	221,900	0.675	0.0177	98	1.5×10^{18}
	Pasiphaë	1908	23,624,000	743.63^{R}	0.4090	58	3.0×10^{17}
	Carme	1938	23,404,000	734.17^{R}	0.2533	46	1.3×10^{17}
SATURN	Titan	1655	1,221,870	15.95	0.0288	5150	1.34×10^{23}
	Rhea	1672	527,070	4.518	0.001	1528	2.3×10^{21}
	Iapetus	1671	3,560,840	79.33	0.0283	1472	2.0×10^{21}
	Dione	1684	377,420	2.737	0.0022	1123	1.1×10^{21}
	Tethys	1684	294,670	1.888	0.0001	1066	6.2×10^{20}
	Enceladus	1789	238,040	1.37	0.0047	504	1.1×10^{20}
	Mimas	1789	185,540	0.942	0.0196	397	3.8×10^{19}
	Phoebe	1898	12,947,780	550.31^{R}	0.1635	230 × 220 × 210	8.3×10^{18}
	Hyperion	1848	1,500,880	21.28	0.0274	360 × 280 × 225	5.7×10^{18}
	Janus	1980	151,460	0.695	0.0068	193 × 173 × 137	1.9×10^{18}
URANUS	Titania	1787	436,300	8.706	0.0011	1578	3.53×10^{21}
	Oberon	1787	583,500	13.46	0.0014	1522	3.01×10^{21}
	Ariel	1851	190,900	2.52	0.0012	1158	1.35×10^{21}
	Umbriel	1851	266,000	4.144	0.0039	1169	1.2×10^{21}
	Miranda	1948	129,900	1.413	0.0013	471	6.59×10^{19}
	Puck	1985	86,000	0.762	0.0001	162	2.9×10^{18}
	Sycorax	1997	12,179,000	1288.3^{R}	0.5224	150	5.4×10^{18}
	Portia	1986	66,100	0.513	0.0001	156 × 126	1.7×10^{18}
	Juliet	1986	64,400	0.493	0.0007	150 × 74	5.6×10^{17}
	Belinda	1986	75,300	0.624	0.0001	128 × 64	3.6×10^{17}

Appendix 3 Major Satellites of the Planets by Mass

Planet	Satellite	Date of discovery	Average distance from center of planet (km)	Orbital (sidereal) period (days)*	Orbital eccentricity	Size of satellite (km)**	Mass (kg)
NEPTUNE	Triton	1846	354,800	5.877^R	0	2706	2.15×10^{22}
	Proteus	1989	117,647	1.122	0.0005	440 × 416 × 404	4.4 3 10^{19}
	Nereid	1949	5,513,400	360.14	0.7512	340	3.1×10^{19}
	Larissa	1989	73,548	0.555	0.0014	216 × 204 × 164	4.2×10^{18}
	Galatea	1989	61,953	0.429	0	204 × 184 × 144	2.1×10^{18}
	Despina	1989	52,526	0.335	0.0002	180 × 150 × 130	2.1×10^{18}
	Thalassa	1989	50,075	0.311	0.0002	108 × 100 × 52	3.5×10^{17}
	Naiad	1989	48,227	0.294	0.0004	96 × 60 × 52	1.9×10^{17}
	Halimede	2002	15,728,000	1879.71^R	0.5711	62	1.8×10^{17}
	Neso	2002	22,422,000	2914.07	0.2931	44	6.3×10^{16}

This table was compiled from data provided by the Jet Propulsion Laboratory.
**A superscript R means that the satellite orbits in a retrograde direction (opposite to the planet's rotation).*

***The size of a spherical satellite is equal to its diameter.*

Appendix 4 The Nearest Stars

Name	Parallax (arcsec)	Distance (parsecs)	Distance (light-years)	Spectral Type	Proper motion (arcsec/yr)	Apparent visual magnitude	Absolute visual magnitude	Mass (Sun = 1)
Proxima Centauri	0.772	1.30	4.22	M5.5 V	3.853	+11.09	+15.53	0.107
Alpha Centauri A	0.747	1.34	4.36	G2 V	3.710	−0.01	+4.36	1.144
Alpha Centauri B	0.747	1.34	4.36	K0 V	3.724	+1.34	+5.71	0.916
Barnard's Star	0.547	1.83	5.96	M4.0 V	10.358	+9.53	+13.22	0.166
Wolf 359	0.419	2.39	7.78	M6.0 V	4.696	+13.44	+16.55	0.092
Lalande 21185	0.393	2.54	8.29	M2.0 V	4.802	+7.47	+10.44	0.464
Sirius A	0.380	2.63	8.58	A1 V	1.339	−1.43	+1.47	1.991
Sirius B	0.380	2.63	8.58	white dwarf	1.339	+8.44	+11.34	0.500
UV Ceti	0.374	2.68	8.73	M5.5 V	3.368	+12.54	+15.40	0.109
BL Ceti	0.374	2.68	8.73	M6.0 V	3.368	+12.99	+15.85	0.102
Ross 154	0.337	2.97	9.68	M3.5 V	0.666	+10.43	+13.07	0.171
Ross 248	0.316	3.16	10.32	M5.5 V	1.617	+12.29	+14.79	0.121
Epsilon Eridani	0.310	3.23	10.52	K2 V	0.977	+3.73	+6.19	0.850
Lacailee 9352	0.304	3.29	10.74	M1.5 V	6.896	+7.34	+9.75	0.529
Ross 128	0.299	3.35	10.92	M4.0 V	1.361	+11.13	+13.51	0.156
EZ Aquarii A	0.290	3.45	11.27	M5.0 V	3.254	+13.33	+15.64	0.105
EZ Aquarii B	0.290	3.45	11.27	—	3.254	+13.27	+15.58	0.106
EZ Aquarii C	0.290	3.45	11.27	—	3.254	+14.03	+16.34	0.095
Procyon A	0.286	3.50	11.40	F5 IV-V	1.259	+0.38	+2.66	1.569
Procyon B	0.286	3.50	11.40	white dwarf	1.259	+10.70	+12.98	0.500
61 Cygni A	0.286	3.50	11.40	K5.0 V	5.281	+5.21	+7.49	0.703
61 Cygni B	0.286	3.50	11.40	K7.0 V	5.172	+6.03	+8.31	0.630
GJ725 A	0.283	3.53	11.53	M3.0 V	2.238	+8.90	+11.16	0.351
GJ725 B	0.283	3.53	11.53	M3.5 V	2.313	+9.69	+11.95	0.259
GX Andromedae	0.281	3.56	11.62	M1.5 V	2.918	+8.08	+10.32	0.486
GQ Andromedae	0.281	3.56	11.62	M3.5 V	2.918	+11.06	+13.30	0.163
Epsilon Indi A	0.276	3.63	11.82	K5 V	4.704	+4.69	+6.89	0.766
Epsilon Indi B	0.276	3.63	11.82	T1.0	4.823			0.044
Epsilon Indi C	0.276	3.63	11.82	T6.0	4.823			0.028
DX Cancri	0.276	3.63	11.83	M6.5 V	1.290	+14.78	+16.98	0.087
Tau Ceti	0.274	3.64	11.89	G8 V	1.922	+3.49	+5.68	0.921
RECONS 1	0.272	3.68	11.99	M5.5 V	0.814	+13.03	+15.21	0.113
YZ Ceti	0.269	3.72	12.13	M4.5 V	1.372	+12.02	+14.17	0.136
Luyten's Star	0.264	3.79	12.37	M3.5 V	3.738	+9.86	+11.97	0.257
Kapteyn's Star	0.255	3.92	12.78	M1.5 V	8.670	+8.84	+10.87	0.393
AX Microscopium	0.253	3.95	12.87	M0.0 V	3.455	+6.67	+8.69	0.600

This table, compiled from data reported by the Research Consortium on Nearby Stars, lists all known stars within 4.00 parsecs (13.05 light-years).

**Stars that are components of multiple star systems are labeled A, B, and C.*

Appendix 5 The Visually Brightest Stars

Name	Designation	Distance (parsecs)	Distance (light-years)	Spectral Type	Radial velocity (km/s)*	Proper motion (arcsec/yr)	Apparent visual magnitude	Apparent visual brightness (Sirius = 1)**	Mass (Sun = 1)
Sirius A	α CMa A	2.63	8.58	A1 V	−7.6	1.34	−1.43	1.000	+1.46
Canopus	α Car	95.9	313	F0 II	+20.5	0.03	−0.72	0.520	−5.63
Arcturus	α Boo	11.3	36.7	K1.5 III	−5.2	2.28	−0.04	0.278	−0.30
Alpha Centauri A	α Cen A	1.34	4.36	G2 V	−25	3.71	−0.01	0.270	+4.36
Vega	α Lyr	7.76	25.3	A0 V	−13.9	0.35	+0.03	0.261	+0.58
Capella	α Aur	12.9	42.2	G5 III	+30.2	0.43	+0.08	0.249	−0.48
Rigel	α Ori A	237	773	B8 Ia	+20.7	0.002	+0.12	0.240	−6.75
Procyon	α CMi A	3.50	11.4	F5 IV-V	−3.2	1.26	+0.34	0.196	+2.62
Achernar	α Eri	44.1	144	B3 V	+16	0.10	+0.50	0.169	−2.72
Betelgeuse	α Ori	131	427	M1 Iab	+21	0.03	+0.58	0.157	−5.01
Hadar	β Cen	161	525	B1 III	+5.9	0.04	+0.60	0.154	−5.43
Altair	α Aql	51.4	168	A7 V	−26.1	0.66	+0.77	0.132	−2.79
Aldebaran	α Tau A	20.0	65.1	K5 III	+54.3	0.20	+0.85	0.122	−0.65
Spica	α Vir	80.4	262	B1 III-IV	+1	0.05	+1.04	0.103	−3.49
Antares	α Sco A	185	604	M1.5 Iab	−3.4	0.03	+1.09	0.098	−5.25
Pollux	β Gem	10.3	33.7	K0 IIIb	+3.3	0.63	+1.15	0.093	+1.08
Fomalhaut	α PsA	7.69	25.1	A3 V	+6.5	0.37	+1.16	0.092	+1.73
Deneb	α Cyg	990	3230	A2 Ia	−4.5	0.002	+1.25	0.085	−8.73
Mimosa	β Cru	108	353	B0.5 IV	+15.6	0.05	+1.297	0.081	−3.87
Regulus	α Leo A	23.8	77.5	B7 V	+5.9	0.25	+1.35	0.077	−0.53

Data in this table were compiled from SIMBAD database operated at the Centre de Données Astronomiques de Strasbourg, France.

**A positive radial velocity means that the star is receding; a negative radial velocity means that the star is approaching.*

***This is a ratio of the star's apparent brightness to that of Sirius, the brightest star in the night sky.*

Note: Acrux, or α Cru (the brightest star in Crux, the Southern Cross) appears to the naked eye as a star of apparent magnitude +0.87, the same as Aldebaran. However, it does not appear in this table because Acrux is actually a binary star system. The blue-white component stars of this binary system have apparent magnitudes of +1.4 and +1.9, and so they are dimmer than any of the stars listed here.

Appendix 6 Some Important Astronomical Quantities

Astronomical Unit	1 AU = 1.4960×10^{11} m = 1.4960×10^{8} km
Light-year	1 pc = 3.2616 ly = 63,240 AU
Parsec	1 pc = 3.2616 ly = 3.0857×10^{16} m = 206,265 AU
Year	1 y = 365.2564 days = 3.156×10^{7} s
Solar Mass	$1\ M_{\odot} = 1.989 \times 10^{30}$ kg
Solar Radius	$1\ R_{\odot} = 6.9599 \times 10^{8}$ m
Solar Luminosity	$1\ L_{\odot} = 3.90 \times 10^{26}$ W

Appendix 7 Some Important Physical Constants

Speed of light	$c = 2.9979 \times 10^{8}$ m/s
Gravitational constant	$G = 6.673 \times 10^{-11}$ N m^2/kg^2
Planck's constant	$h = 6.6261 \times 10^{-34}$ J s $= 4.1357 \times 10^{-15}$ eV s
Boltzmann constant	$k = 1.3807 \times 10^{-23}$ J/K $= 8.6173 \times 10^{-5}$ eV/K
Stefan-Boltzmann constant	$\sigma = 5.6704 \times 10^{-8}$ W m^{-2} K^{-4}
Mass of electron	$m_e = 9.1094 \times 10^{-31}$ kg
Mass of proton	$m_p = 1.6726 \times 10^{-27}$ kg
Mass of neutron	$m_n = 1.6749 \times 10^{-27}$ kg
Mass of hydrogen atom	$m_H = 1.6735 \times 10^{-27}$ kg
Rydberg constant	$R = 1.0973 \times 10^{-7}$ kg m^{-1}
Electron volt	1 eV = 1.6022×10^{-19} J

Appendix 8 Some Useful Mathematics

Area of a rectangle of sides *a* and *b*	$A = ab$
Volume of a rectangular solid of sides *a*, *b*, and *c*	$V = abc$
Hypotenuse of a right triangle whose other sides are *a* and *b*	$c = \sqrt{a^2+b^2}$
Circumference of a circle of radius *r*	$C = 2\pi r$
Area of a circle of radius *r*	$A = \pi r^2$
Surface areas of a sphere of radius *r*	$A = 4\pi r^2$
Volume of a sphere of radius *r*	$V = 4\pi r^3/3$
Value of π	$\pi = 3.1415926536$

Glossary

You can find more information about each term in the indicated chapter or chapters. See the Key Words section in each chapter for the specific page on which the meaning of a given term is described.

A ring One of three prominent rings encircling Saturn. (Chapter 12)

absolute magnitude The apparent magnitude that a star would have if it were at a distance of 10 parsecs. (Chapter 17)

absolute zero A temperature of –273°C (or 0 K), at which all molecular motion stops; the lowest possible temperature. (Chapter 5)

absorption line spectrum Dark lines superimposed on a continuous spectrum. (Chapter 5)

acceleration The rate at which an object's velocity changes due to a change in speed, a change in direction, or both. (Chapter 4)

accretion The gradual accumulation of matter in one location, typically due to the action of gravity. (Chapter 8, Chapter 18, Chapter 24)

accretion disk A disk of gas orbiting a star or black hole. (Chapter 24)

active galactic nucleus (AGN) The center of an active galaxy. (Chapter 24)

active galaxy A galaxy that is emitting exceptionally large amounts of energy; a Seyfert galaxy, radio galaxy, or quasar. (Chapter 24)

active optics A technique for improving a telescopic image by altering the telescope's optics to compensate for variations in air temperature or flexing of the telescope mount. (Chapter 6)

adaptive optics A technique for improving a telescopic image by altering the telescope's optics in a way that compensates for distortion caused by Earth's atmosphere. (Chapter 6)

aerosol Tiny droplets of liquid dispersed in a gas. (Chapter 13)

AGB star See *asymptotic giant branch star.*

AGN See *active galactic nucleus.*

air drag A force that opposes the motion of an object through air. A portion of the object's energy is transferred to the air molecules. (Chapter 4)

albedo The fraction of sunlight that a planet, asteroid, or satellite reflects. (Chapter 9)

alpha particle The nucleus of a helium atom, consisting of two protons and two neutrons. (Chapter 19)

amino acids The chemical building blocks of proteins. (Chapter 15)

angle The opening between two lines that meet at a point. (Chapter 1)

angular diameter The angle subtended by the diameter of an object. (Chapter 1)

angular distance The angle between two points in the sky. (Chapter 1)

angular momentum See *conservation of angular momentum.*

angular resolution The angular size of the smallest feature that can be distinguished with a telescope. (Chapter 6)

angular size See *angular diameter.*

anisotropic Possessing unequal properties in different directions; not isotropic. (Chapter 23)

annihilation The process by which the masses of a particle and antiparticle are converted into energy. (Chapter 26)

annular eclipse An eclipse of the Sun in which the Moon is too distant to cover the Sun completely so that a ring of sunlight is seen around the Moon at mid-eclipse. (Chapter 3)

anorthosite Rock commonly found in ancient, cratered highlands on the Moon. (Chapter 10)

Antarctic Circle A circle of latitude 23½° north of Earth's south pole. (Chapter 2)

antimatter Matter consisting of antiparticles such as antiprotons, antineutrons, and positrons. (Chapter 26)

antiparticle A particle with the same mass as an ordinary particle but with other properties, such as electric charge, reversed. (Chapter 21, Chapter 26)

antipode The point on a planet exactly opposite a specific location. (Chapter 11)

antiproton A particle with the same mass as a proton but with a negative electric charge. (Chapter 26)

aphelion The point in its orbit where a planet is farthest from the Sun. (Chapter 4)

apogee The point in its orbit where a satellite or the Moon is farthest from Earth. (Chapter 3)

apparent brightness The flux of a star's light arriving at Earth. (Chapter 17)

apparent magnitude A measure of the brightness of light from a star or other object as measured from Earth. (Chapter 17)

apparent solar day The interval between two successive transits of the Sun's center across the local meridian. (Chapter 2)

apparent solar time Time reckoned by the position of the Sun in the sky. (Chapter 2)

arcminute One-sixtieth (1/60) of a degree, designated by the symbol ′. (Chapter 1)

arcsecond One-sixtieth (1/60) of an arcminute or 1/3600 of a degree, designated by the symbol ″. (Chapter 1)

Arctic Circle A circle of latitude 23½° south of Earth's north pole. (Chapter 2)

asteroid One of tens of thousands of small, rocky, planetlike objects in orbit about the Sun. Also called minor planet. (Chapter 7, Chapter 15)

asteroid belt A region between the orbits of Mars and Jupiter that encompasses the orbits of many asteroids. (Chapter 7, Chapter 15)

asthenosphere A warm, plastic layer of the mantle beneath the lithosphere of Earth. (Chapter 9)

astrobiology The study of life in the universe. (Chapter 27)

astrometric method A technique for detecting extrasolar planets by looking for stars that "wobble" periodically. (Chapter 8)

astronomical unit (AU) The semimajor axis of Earth's orbit; the average distance between Earth and the Sun. (Chapter 1)

asymptotic giant branch A region on the Hertzsprung-Russell diagram occupied by stars in their second and final red-giant phase. (Chapter 20)

asymptotic giant branch star (AGB star) A red giant star that has ascended the giant branch for the second and final time. (Chapter 20)

atmosphere (atm) A unit of atmospheric pressure. (Chapter 9)

atmospheric pressure The force per unit area exerted by a planet's atmosphere. (Chapter 9)

atom The smallest particle of an element that has the properties characterizing that element. (Chapter 5)

atomic number The number of protons in the nucleus of an atom of a particular element. (Chapter 5, Chapter 8)

AU See *astronomical unit.*

aurora (plural **aurorae**) Light radiated by atoms and ions in Earth's upper atmosphere, mostly in the polar regions. (Chapter 9)

aurora australis Aurorae seen from southern latitudes; the southern lights. (Chapter 9)

aurora borealis Aurorae seen from northern latitudes; the northern lights. (Chapter 9)

autumnal equinox The intersection of the ecliptic and the celestial equator where the Sun crosses the equator from north to south. Also used to refer to the date on which the Sun passes through this intersection. (Chapter 2)

average density The mass of an object divided by its volume. (Chapter 7)

average density of matter In cosmology, the average density of all the material substance in the universe. (Chapter 25)

B ring One of three prominent rings encircling Saturn. (Chapter 12)

Balmer line An emission or absorption line in the spectrum of hydrogen caused by an electron transition between the second and higher energy levels. (Chapter 5)

Balmer series The entire pattern of Balmer lines. (Chapter 5)

Barnard object One of a class of dark nebulae discovered by E. E. Barnard. (Chapter 18)

barred spiral galaxy A spiral galaxy in which the spiral arms begin from the ends of a "bar" running through the nucleus rather than from the nucleus itself. (Chapter 23)

baseline In interferometry, the distance between two telescopes whose signals are combined to give a higher-resolution image. (Chapter 6)

belt A dark band in Jupiter's atmosphere. (Chapter 12)

Big Bang An explosion of all space that took place roughly 13.7 billion years ago and that marks the beginning of the universe. (Chapter 1, Chapter 25)

binary star (also **binary star system** or **binary**) Two stars orbiting each other. (Chapter 17)

biosphere The layer of soil, water, and air surrounding Earth in which living organisms thrive. (Chapter 9)

bipolar outflow Oppositely directed jets of gas expelled from a young star. (Chapter 18)

black hole An object whose gravity is so strong that the escape speed exceeds the speed of light. (Chapter 1, Chapter 25)

black hole evaporation The process by which black holes emit particles. (Chapter 21)

blackbody A hypothetical perfect radiator that absorbs and re-emits all radiation falling upon it. (Chapter 5)

blackbody curve The intensity of radiation emitted by a blackbody plotted as a function of wavelength or frequency. (Chapter 5)

blackbody radiation The radiation emitted by a perfect blackbody. (Chapter 5)

blazar A type of active galaxy whose nucleus has a featureless spectrum without prominent spectral lines. (Chapter 24)

blueshift A decrease in the wavelength of photons emitted by an approaching source of light. (Chapter 5)

Bohr orbits In the model of the atom described by Niels Bohr, the only orbits in which electrons are allowed to move about the nucleus. (Chapter 5)

Bok globule A small, roundish, dark nebula. (Chapter 18)

bow shock A region where charged particles in the solar wind abruptly slow down as they encounter a planet's magnetosphere. (Chapter 12)

bright terrain (on Ganymede) Young, reflective, relatively crater-free terrain on the surface of Ganymede. (Chapter 13)

brightness See *apparent brightness.*

brown dwarf A starlike object that is not massive enough to sustain hydrogen fusion in its core. (Chapter 17)

brown ovals Elongated, brownish features usually seen in Jupiter's northern hemisphere. (Chapter 12)

C ring One of three prominent rings encircling Saturn. (Chapter 12)

capture theory The hypothesis that the Moon was gravitationally captured by Earth. (Chapter 10)

carbon fusion The thermonuclear fusion of carbon to produce heavier nuclei. (Chapter 20)

carbon star A peculiar red giant star whose spectrum shows strong absorption lines of carbon and carbon compounds. (Chapter 20)

carbonaceous chondrite A type of meteorite that has a high abundance of carbon and volatile compounds. (Chapter 15, Chapter 27)

Cassegrain focus An optical arrangement in a reflecting telescope in which light rays are reflected by a secondary mirror to a focus behind the primary mirror. (Chapter 6)

Cassini division An apparent gap between Saturn's A and B rings. (Chapter 12)

CCD See *charge-coupled device.*

celestial equator A great circle on the celestial sphere 90° from the celestial poles. (Chapter 2)

celestial sphere An imaginary sphere of very large radius centered on an observer; the apparent sphere of the sky. (Chapter 2)

center of mass The point between a star and a planet, or between two stars, around which both objects orbit. (Chapter 8, Chapter 10, Chapter 17)

central bulge (of a galaxy) A spherical distribution of stars around the nucleus of a spiral galaxy. (Chapter 22)

Cepheid variable A type of yellow, supergiant, pulsating star. (Chapter 19)

Cerenkov radiation The radiation emitted by a particle traveling through a substance at a speed greater than the speed of light in that substance. (Chapter 20)

Chandrasekhar limit The maximum mass of a white dwarf. (Chapter 20)

charge-coupled device (CCD) A type of solid-state device designed to detect photons. (Chapter 6)

chemical composition A description of which chemical substances make up a given object. (Chapter 7)

chemical differentiation The process by which the heavier elements in a planet sink toward its center while lighter elements rise toward its surface. (Chapter 8)

chemical element See *element.*

chondrule A glassy, roughly spherical blob found within meteorites. (Chapter 8)

chromatic aberration An optical defect whereby different colors of light passing through a lens are focused at different locations. (Chapter 6)

chromosphere A layer in the atmosphere of the Sun between the photosphere and the corona. (Chapter 16)

circumpolar A term describing a star that neither rises nor sets but appears to rotate around one of the celestial poles. (Chapter 2)

circumstellar accretion disk An accretion disk that surrounds a protostar. (Chapter 18)

close binary A binary star system in which the stars are separated by a distance roughly comparable to their diameters. (Chapter 19)

closed universe A universe with positive curvature, so that its geometry is analogous to that of the surface of a sphere. (Chapter 25)

cluster (of galaxies) A collection of galaxies containing a few to several thousand member galaxies. (Chapter 23)

cluster (of stars) A group of stars that formed together and that have remained together because of their mutual gravitational attraction. (Chapter 18)

CNO cycle A series of nuclear reactions in which carbon is used as a catalyst to transform hydrogen into helium. (Chapter 16)

cocoon nebula The nebulosity surrounding a protostar. (Chapter 18)

co-creation theory The hypothesis that Earth and the Moon formed at the same time from the same material. (Chapter 10)

cold dark matter Slowly moving, weakly interacting particles presumed to constitute the bulk of matter in the universe. (Chapter 26)

collapsar A proposed type of supernova in which a black hole forms at the center of a dying star before the outer layers of the star have time to collapse. (Chapter 21)

collisional ejection theory The hypothesis that the Moon formed from material ejected from Earth by the impact of a large asteroid. (Chapter 10)

color ratio The ratio of the apparent brightness of a star measured in one spectral region to its brightness measured in a different region. (Chapter 17)

color-magnitude diagram A plot of the apparent magnitudes (that is, apparent brightnesses) of stars in a cluster versus their color indices (a measure of their surface temperatures). (Chapter 19)

coma (of a comet) The diffuse gaseous component of the head of a comet. (Chapter 15)

coma (optical) The distortion of off-axis images formed by a parabolic mirror. (Chapter 6)

combined average mass density The average density in the observable universe of all forms of matter and energy, measured in units of mass per volume. (Chapter 25)

comet A small body of ice and dust in orbit about the Sun. While passing near the Sun, a comet's vaporized ices give rise to a coma and tail. (Chapter 7, Chapter 15)

compound A substance consisting of two or more chemical elements in a definite proportion. (Chapter 5)

compression A region in a sound wave where the medium carrying the wave is compressed. (Chapter 25)

condensation temperature The temperature at which a particular substance in a low-pressure gas condenses into a solid. (Chapter 8)

conduction The transfer of heat by directly passing energy from atom to atom. (Chapter 16)

conic section The curve of intersection between a circular cone and a plane; this curve can be a circle, ellipse, parabola, or hyperbola. (Chapter 4)

conjunction The geometric arrangement of a planet in the same part of the sky as the Sun, so that the planet is at an elongation of 0°. (Chapter 4)

conservation of angular momentum A law of physics stating that in an isolated system, the total amount of angular momentum—a measure of the amount of rotation—remains constant. (Chapter 8)

conservation of energy A law of physics stating that energy can change forms and be transferred from one object to another but can't be created or destroyed. (Chapter 4)

constellation A configuration of stars in the same region of the sky. (Chapter 2)

contact binary A binary star system in which both members fill their Roche lobes. (Chapter 19)

continuous spectrum A spectrum of light over a range of wavelengths without any spectral lines. (Chapter 5)

convection The transfer of energy by moving currents of fluid or gas containing that energy. (Chapter 9, Chapter 16)

convection cell A circulating loop of gas or liquid that transports heat from a warm region to a cool region. (Chapter 9)

convection current The pattern of motion in a gas or liquid in which convection is taking place. (Chapter 9)

convective zone The region in a star where convection is the dominant means of energy transport. (Chapter 16)

core (of Earth) The iron-rich inner region of Earth's interior. (Chapter 9)

core accretion model The hypothesis that each of the Jovian planets formed by accretion of gas onto a rocky core. (Chapter 8)

core helium fusion The thermonuclear fusion of helium at the center of a star. (Chapter 19, Chapter 20)

core hydrogen fusion The thermonuclear fusion of hydrogen at the center of a star. (Chapter 19)

core-collapse supernova A supernova that occurs at the end of a massive star's lifetime when the star's core collapses to high density and the star's outer layers are expelled into space. (See Type Ib supernova, Type Ic supernova, and Type II supernova) (Chapter 20)

corona (of the Sun) The Sun's outer atmosphere, which has a high temperature and a low density. (Chapter 16)

coronal hole A region in the Sun's corona that is deficient in hot gases. (Chapter 16)

coronal mass ejection An event in which billions of tons of gas from the Sun's corona is suddenly blasted into space at high speed. (Chapter 16)

cosmic background radiation See *cosmic microwave background (CMB).*

cosmic light horizon An imaginary sphere, centered on Earth, whose radius equals the distance light has traveled since the Big Bang. (Chapter 25, Chapter 26)

cosmic microwave background (CMB) An isotropic radiation field with a blackbody temperature of about 2.725 K that permeates the entire universe. (Chapter 25)

cosmic rays Fast-moving subatomic particles that enter our solar system from interstellar space. (Chapter 11)

cosmological constant (L) In the equations of general relativity, a quantity signifying a pressure throughout all space that helps the universe expand; a type of dark energy. (Chapter 25)

cosmological principle The assumption that the universe is homogeneous and isotropic on the largest scale. (Chapter 25)

cosmological redshift A redshift that is caused by the expansion of the universe. (Chapter 25)

cosmology The study of the structure and evolution of the universe. (Chapter 25)

coudé focus An optical arrangement with a reflecting telescope. A series of mirrors is used to direct light to a remote focus away from the moving parts of the telescope. (Chapter 6)

crater See *impact crater.*

critical density The value of the combined average mass density throughout the universe for which space would be flat. (Chapter 25)

crust (of a planet) The surface layer of a terrestrial planet. (Chapter 9)

crustal dichotomy (on Mars) The contrast between the young northern lowlands and older southern highlands on Mars. (Chapter 11)

cryovolcano An ice-volcano that erupts or oozes water instead of lava. (Chapter 13)

crystal A material in which atoms are arranged in orderly rows. (Chapter 9)

D ring One of several faint rings encircling Saturn. (Chapter 12)

dark ages The era between the formation of the first atoms (when the universe was about 380,000 years old) and the formation of the first stars (when the universe was about 400 million years old). (Chapter 26)

dark energy A form of energy that appears to pervade the universe and causes the expansion of the universe to accelerate but has no discernible gravitational effect. (Chapter 25)

dark energy density parameter (V_L) The ratio of the average mass density of dark energy in the universe to the critical density. (Chapter 25)

dark matter Nonluminous matter that is the dominant form of matter in galaxies and throughout the universe. (Chapter 22)

dark nebula A cloud of interstellar gas and dust that obscures the light of more distant stars. (Chapter 18)

dark terrain (on Ganymede) Older, heavily cratered, dark-colored terrain on the surface of Ganymede. (Chapter 13)

dark-energy–dominated universe A universe in which the mass density of dark energy exceeds both the average density of matter and the mass density of radiation. (Chapter 25)

dark-matter problem The enigma that most of the matter in the universe seems not to emit radiation of any kind and is detectable by its gravity alone. (Chapter 23)

dead quasar A supermassive black hole that has already consumed the gas and dust that previously provided fuel for quasar activity. Any supermassive black hole in the center of a galaxy is presumed to have gone through a brief period when it exhibited quasar activity. (Chapter 24)

dead quasar flare The flare that results if a star wanders too close to a supermassive black hole in the center of a galaxy (which is presumably a dead quasar); gravitational tidal forces can tear some matter away from the star and provide a small source of fuel to briefly "light up" the dead quasar with an approximately year-long flare of light. (Chapter 24)

declination Angular distance of a celestial object north or south of the celestial equator. (Chapter 2)

deferent A stationary circle in the Ptolemaic system along which another circle (an epicycle) moves, carrying a planet, the Sun, or the Moon. (Chapter 4)

degeneracy The phenomenon, due to quantum mechanical effects, whereby the pressure exerted by a gas does not depend on its temperature. (Chapter 19)

degenerate-electron (-neutron) pressure The pressure exerted by degenerate electrons (neutrons). (Chapter 19, Chapter 20)

degree A basic unit of angular measure, designated by the symbol °. (Chapter 1)

degree Celsius A basic unit of temperature, designated by the symbol °C and used on a scale where water freezes at 0° and boils at 100°. (Chapter 5)

degree Fahrenheit A basic unit of temperature, designated by the symbol °F and used on a scale where water freezes at 32° and boils at 212°. (Chapter 5)

density See *average density.*

density fluctuation A variation of density from one place to another. (Chapter 26)

density parameter (V_0) The ratio of the average density of matter and energy in the universe to the critical density. (Chapter 25)

density wave In a spiral galaxy, a localized region in which matter piles up as it orbits the center of the galaxy; this is a proposed explanation of spiral structure. (Chapter 22)

detached binary A binary star system in which neither star fills its Roche lobe. (Chapter 19)

deuterium bottleneck The situation during the first 3 minutes after the Big Bang when deuterium (an isotope of hydrogen whose nucleus contains one proton and one neutron) inhibited the formation of heavier elements. (Chapter 26)

differential rotation The rotation of a nonrigid object in which parts adjacent to each other at a given time do not always stay close together. (Chapter 12, Chapter 16)

differentiated asteroid An asteroid in which chemical differentiation has taken place so that denser material is toward the asteroid's center. (Chapter 15)

differentiation See *chemical differentiation.*

diffraction The spreading out of light passing through an aperture or opening in an opaque object. (Chapter 6)

diffraction grating An optical device, consisting of thousands of closely spaced lines etched in glass or metal, that disperses light into a spectrum. (Chapter 6)

direct motion The apparent eastward movement of a planet seen against the background stars. (Chapter 4)

disk (of a galaxy) The disk-shaped distribution of Population I stars that dominates the appearance of a spiral galaxy. (Chapter 22)

distance ladder The sequence of techniques used to determine the distances to very remote galaxies. (Chapter 23)

distance modulus The difference between the apparent and absolute magnitudes of an object. (Chapter 17)

diurnal motion Any apparent motion in the sky that repeats on a daily basis, such as the rising and setting of stars. (Chapter 2)

Doppler effect The apparent change in wavelength of radiation due to relative motion between the source and the observer along the line of sight. (Chapter 5)

double star A pair of stars located at nearly the same position in the night sky. Some, but not all, double stars are binary stars. (Chapter 17)

Drake equation An equation used to estimate the number of intelligent civilizations in the Galaxy with whom we might communicate. (Chapter 27)

dredge-up The transporting of matter from deep within a star to its surface by convection. (Chapter 20)

dust devil Whirlwind found in dry or desert areas on both Earth and Mars. (Chapter 11)

dust grain A microscopic bit of solid matter found in interplanetary or interstellar space. (Chapter 18)

dust tail The tail of a comet that is composed primarily of dust grains. (Chapter 15)

dwarf elliptical galaxy A low-mass elliptical galaxy that contains only a few million stars. (Chapter 23)

dwarf planet A solar system body that is large enough to be spherical in shape and have a circular orbit around the Sun, but not large enough to clear its own path of other bodies. The term is used for Ceres, Pluto, and Eris. (Chapter 14)

dynamo The mechanism whereby electric currents within an astronomical body generate a magnetic field. Also referred to as a magnetic dynamo. (Chapter 7, Chapter 8)

E ring A very broad, faint ring encircling Saturn. (Chapter 12)

earthquake A sudden vibratory motion of Earth's surface. (Chapter 9)

eccentricity A number between 0 and 1 that describes the shape of an ellipse. (Chapter 4)

eclipse path The track of the tip of the Moon's shadow along Earth's surface during a total or annular solar eclipse. (Chapter 3)

eclipse The cutting off of part or all the light from one celestial object by another. (Chapter 3)

eclipse year The interval between successive passages of the Sun through the same node of the Moon's orbit. (Chapter 3)

eclipsing binary A binary star system in which, as seen from Earth, stars periodically pass in front of each other. (Chapter 17)

ecliptic The apparent annual path of the Sun on the celestial sphere. (Chapter 2)

ecliptic plane The plane of Earth's orbit around the Sun. (Chapter 2)

Eddington limit A constraint on the rate at which a black hole can accrete matter. (Chapter 24)

electromagnetic radiation Radiation consisting of oscillating electric and magnetic fields. Examples include gamma rays, X rays, visible light, ultraviolet and infrared radiation, radio waves, and microwaves. (Chapter 5)

electromagnetic spectrum The entire array of electromagnetic radiation. (Chapter 5)

electromagnetism Electric and magnetic phenomena, including electromagnetic radiation. (Chapter 5)

electron A subatomic particle with a negative charge and a small mass, usually found in orbits about the nuclei of atoms. (Chapter 5)

electron volt (eV) The energy acquired by an electron accelerated through an electric potential of one volt. (Chapter 5)

electroweak force A unification of the electromagnetic and weak forces that occurs at high energies. (Chapter 26)

element A chemical that cannot be broken down into more basic chemicals. (Chapter 5)

elementary particle physics The branch of quantum mechanics that deals with individual subatomic particles and their interactions. (Chapter 26)

ellipse A conic section obtained by cutting completely through a circular cone with a plane. (Chapter 4)

elliptical galaxy A galaxy with an elliptical shape and no conspicuous interstellar material. (Chapter 23)

elongation The angular distance between a planet and the Sun as viewed from Earth. (Chapter 4)

emission line spectrum A spectrum that contains bright emission lines. (Chapter 5)

emission nebula A glowing gaseous nebula whose spectrum has bright emission lines. (Chapter 18)

Encke gap A narrow gap in Saturn's A ring. (Chapter 12)

energy flux The rate of energy flow, usually measured in joules per square meter per second. (Chapter 5)

energy level In an atom, a particular amount of energy possessed by an atom above the atom's least energetic state. (Chapter 5)

energy-level diagram A diagram showing the arrangement of an atom's energy levels. (Chapter 5)

epicenter The location on Earth's surface directly over the focus of an earthquake. (Chapter 9)

epicycle A moving circle in the Ptolemaic system about which a planet revolves. (Chapter 4)

epoch The date used to define the coordinate system for objects on the sky. (Chapter 2)

equal areas, law of See *Kepler's second law.*

equinox One of the intersections of the ecliptic and the celestial equator. Also used to refer to the date on which the Sun passes through such an intersection. (Chapter 2) See also *autumnal equinox* and *vernal equinox.*

equivalence principle In the general theory of relativity, the principle that in a small volume of space, the downward pull of gravity can be accurately and completely duplicated by an upward acceleration of the observer. (Chapter 21)

era of recombination The moment, approximately 380,000 years after the Big Bang, when the universe had cooled sufficiently to permit the formation of hydrogen atoms. (Chapter 25)

ergoregion The region of space immediately outside the event horizon of a rotating black hole where it is impossible to remain at rest. (Chapter 21)

escape speed The speed needed by an object (such as a spaceship) to leave a second object (such as a planet or star) permanently and to escape into interplanetary space. (Chapter 7)

eV See *electron volt.*

event horizon The location around a black hole where the escape speed equals the speed of light; the surface of a black hole. (Chapter 21)

evolutionary track The path on an H-R diagram followed by a star as it evolves. (Chapter 18)

excited state A state of an atom, ion, or molecule with a higher energy than the ground state. (Chapter 5)

exoplanet Planets orbiting a star other than our Sun. (Chapter 8)

exotic dark matter Dark matter that is not made up of regular matter. Regular matter is made of protons, neutrons, and electrons, including all atoms and molecules. There is evidence that the vast majority of dark matter is not made of regular matter, but little else is known about what it might be. (Chapter 23)

exponent A number placed above and after another number to denote the power to which the latter is to be raised, as n in 10^n. (Chapter 1)

extinction (interstellar) See *interstellar extinction.*

extrasolar planet A planet orbiting a star other than the Sun. (Chapter 8)

extremophile Microorganisms that live in extreme conditions such as hot or cold temperatures, high acidity, or high radioactivity. (Chapter 27)

eyepiece lens A magnifying lens used to view the image produced at the focus of a telescope. (Chapter 6)

F ring A thin, faint ring encircling Saturn just beyond the A ring. (Chapter 12)

false color In astronomical images, color used to denote different values of intensity, temperature, or other quantities. (Chapter 6)

false vacuum Empty space with an abnormally high energy density that is thought to have filled the universe immediately after the Big Bang. (Chapter 26)

far side (of the Moon) he side of the Moon that faces perpetually away from Earth. (Chapter 10)

far-infrared The part of the infrared spectrum most different in wavelength from visible light. (Chapter 22)

filament A portion of the Sun's chromosphere that arches to high altitudes. (Chapter 16)

first quarter moon The phase of the Moon that occurs when the Moon is 90° east of the Sun. (Chapter 3)

fission theory The hypothesis that the Moon was pulled out of a rapidly rotating proto-Earth. (Chapter 10)

flake tectonics A model of a planetary interior, particularly Venus, in which a thin crust remains stationary but wrinkles and flakes in response to interior convection currents. (Chapter 11)

flare See *solar flare.*

flat space Space that is not curved; space with zero curvature. (Chapter 25)

flatness problem The dilemma posed by the fact that the combined average mass density of the universe is very nearly equal to the critical density. (Chapter 26)

flocculent spiral galaxy A spiral galaxy with fuzzy, poorly defined spiral arms. (Chapter 22)

fluorescence A process in which high-energy ultraviolet photons are absorbed and the absorbed energy is radiated as lower-energy photons of visible light. (Chapter 18)

focal length The distance from a lens or mirror to the point where converging light rays meet. (Chapter 6)

focal plane The plane in which a lens or mirror forms an image of a distant object. (Chapter 6)

focal point The point at which a lens or mirror forms an image of a distant point of light. (Chapter 6)

focus (of a lens or mirror) The point to which light rays converge after passing through a lens or being reflected from a mirror. (Chapter 6)

focus (of an ellipse) (plural **foci**) One of two points inside an ellipse such that the combined distance from the two foci to any point on the ellipse is a constant. (Chapter 4)

force A push or pull that acts on an object. (Chapter 4)

frequency The number of crests or troughs of a wave that cross a given point per unit time. Also, the number of vibrations per unit time. (Chapter 5)

full moon A phase of the Moon during which its full daylight hemisphere can be seen from Earth. (Chapter 3)

fundamental plane A relationship among the size of an elliptical galaxy, the average motions of its stars, and how the galaxy's brightness appears distributed over its surface. (Chapter 23)

fusion crust The coating on a stony meteorite caused by the heating of the meteorite as it descended through Earth's atmosphere. (Chapter 15)

G ring A thin, faint ring encircling Saturn. (Chapter 12)

galactic cannibalism A collision between two galaxies of unequal mass and size in which the smaller galaxy seems to be absorbed by the larger galaxy. (Chapter 23)

galactic cluster See *open cluster.*

galactic nucleus The center of a galaxy. (Chapter 22)

galaxy A large assemblage of stars, nebulae, and interstellar gas and dust. (Chapter 1, Chapter 22)

Galilean satellites The four large moons of Jupiter. (Chapter 13)

gamma rays The most energetic form of electromagnetic radiation. (Chapter 5)

gamma-ray bursters Objects found in all parts of the sky that emit a one-time intense burst of high-energy radiation. (Chapter 21)

general theory of relativity A description of gravity formulated by Albert Einstein. It states that gravity affects the geometry of space and the flow of time. (Chapter 21)

geocentric model An Earth-centered theory of the universe. (Chapter 4)

giant A star whose diameter is typically 10 to 100 times that of the Sun and whose luminosity is roughly that of 100 Suns. (Chapter 17)

giant elliptical galaxy A large, massive, elliptical galaxy containing many billions of stars. (Chapter 23)

giant molecular cloud A large cloud of interstellar gas and dust in which temperatures are low enough and densities high enough for atoms to form into molecules. (Chapter 18)

global warming The upward trend of Earth's average temperature caused by increased amounts of greenhouse gases in the atmosphere. (Chapter 9)

globular cluster A large spherical cluster of stars, typically found in the outlying regions of a galaxy. (Chapter 19, Chapter 22)

gluon A particle that is exchanged between quarks. (Chapter 26)

Grand Tack model A model in which the Jovian planets migrate toward the Sun in the first few million years of the solar system and then move back outward. This model can help explain Mars's small mass and icy objects in the asteroid belt. (Chapter 8)

grand unified theory (GUT) A theory that describes and explains the four physical forces. (Chapter 26)

grand-design spiral galaxy A galaxy with well-defined spiral arms. (Chapter 22)

granulation The rice grain-like structure found in the solar photosphere. (Chapter 16)

granule A convective cell in the solar photosphere. (Chapter 16)

grating See *diffraction grating.*

gravitational force See *gravity.*

gravitational lens A massive object that deflects light rays from a remote source, forming an image much as an ordinary lens does. (Chapter 23)

gravitational potential energy Energy associated with Earth's pull on an object toward Earth's center. This energy increases with an object's distance from Earth. (Chapter 4)

gravitational radiation (gravitational waves) Oscillations of space produced by changes in the distribution of matter. (Chapter 21)

gravitational redshift The increase in the wavelength of a photon as it climbs upward in a gravitational field. (Chapter 21)

graviton The particle that is responsible for the gravitational force. (Chapter 26)

gravity The force with which all matter attracts all other matter. (Chapter 4)

Great Red Spot A prominent high-pressure system in Jupiter's southern hemisphere. (Chapter 12)

greatest eastern elongation The configuration of an inferior planet at its greatest angular distance east of the Sun. (Chapter 4, Chapter 11)

greatest western elongation The configuration of an inferior planet at its greatest angular distance west of the Sun. (Chapter 4, Chapter 11)

greenhouse effect The trapping of infrared radiation near a planet's surface by the planet's atmosphere. (Chapter 9)

greenhouse gas A substance whose presence in a planet's atmosphere enhances the greenhouse effect. (Chapter 9)

ground state The state of an atom, ion, or molecule with the least possible energy. (Chapter 5)

group (of galaxies) A poor cluster of galaxies. (Chapter 23)

GUT See *grand unified theory.*

H I Neutral, unionized hydrogen. (Chapter 22)

H II region A region of ionized hydrogen in interstellar space. (Chapter 18)

habitable zone Regions of a galaxy or around a star in which conditions may be suitable for life to have developed. (Chapter 8, Chapter 27)

half-life The time required for one-half of a quantity of a radioactive substance to decay. (Chapter 8)

halo (of a galaxy) A spherical distribution of globular clusters and Population II stars that surround a spiral galaxy. (Chapter 22)

Hawking radiation Light and particles made of matter and antimatter that emanate away from a region just outside a black hole's event horizon. The radiation occurs when one member of a virtual particle pair is captured by the black hole; the other member of the pair becomes a real particle and carries away some of the black hole's energy (and therefore, some of its mass). This loss of energy leads to black hole evaporation. (Chapter 21)

Heisenberg uncertainty principle A principle of quantum mechanics that places limits on the precision of simultaneous measurements. (Chapter 21, Chapter 26)

heliocentric model A Sun-centered theory of the universe. (Chapter 4)

helioseismology The study of the vibrations of the Sun as a whole. (Chapter 16)

helium flash The nearly explosive beginning of helium fusion in the dense core of a red giant star. (Chapter 19)

helium fusion The thermonuclear fusion of helium to form carbon and oxygen. (Chapter 19)

helium shell flash A brief thermal runaway that occurs in the helium-fusing shell of a red supergiant. (Chapter 20)

Herbig-Haro object A small, luminous nebula associated with the end point of a jet emanating from a young star. (Chapter 18)

Hertzsprung-Russell (H-R) diagram A plot of the luminosity (or absolute magnitude) of stars against their surface temperature (or spectral type). (Chapter 17)

highlands (on Mars) See *southern highlands.*

highlands (on the Moon) See *lunar highlands.*

Hirayama family A group of asteroids that have nearly identical orbits about the Sun. (Chapter 15)

homogeneous Having the same property in one region as in every other region. (Chapter 25)

horizon problem See *isotropy problem.*

horizontal branch A group of stars on the color-magnitude diagram of a typical globular cluster that lies near the main sequence and has roughly constant luminosity. (Chapter 20)

horizontal-branch star A low-mass, post–helium-flash star on the horizontal branch. (Chapter 19)

hot dark matter Dark matter consisting of low-mass particles moving at high speeds. (Chapter 26)

hot Jupiter A Jupiter-sized planet orbiting a star other than our Sun with an orbital distance similar to Mercury's, and thus a very hot planet. (Chapter 8)

hot-spot volcanism Volcanic activity that occurs over a hot region buried deep within a planet. (Chapter 11)

H-R diagram See *Hertzsprung-Russell diagram.*

Hubble classification A method of classifying galaxies as spirals, barred spirals, ellipticals, or irregulars according to their appearance. (Chapter 23)

Hubble constant (H_0) In the Hubble law, the constant of proportionality between the recessional velocities of remote galaxies and their distances. (Chapter 23)

Hubble flow The recessional motions of remote galaxies caused by the expansion of the universe. (Chapter 23)

Hubble law The empirical relationship stating that the redshifts of remote galaxies are directly proportional to their distances from Earth. (Chapter 23)

Hubble time A fairly accurate estimate of the age of the universe. It is equal to the inverse of the Hubble constant. This age estimate makes a simplifying assumption that the expansion of the universe has occurred at precisely the same rate since the Big Bang, but the expansion rate was slightly different in the past. (Chapter 25)

hydrocarbon Any one of a variety of chemical compounds composed of hydrogen and carbon. (Chapter 13)

hydrogen envelope A huge, tenuous sphere of gas surrounding the head of a comet. (Chapter 15)

hydrogen fusion The thermonuclear conversion of hydrogen into helium. (Chapter 16)

hydrostatic equilibrium A balance between the weight of a layer in a star and the pressure that supports it. (Chapter 16)

hyperbola A conic section formed by cutting a circular cone with a plane at an angle steeper than the sides of the cone. (Chapter 4)

hyperbolic space Space with negative curvature. (Chapter 25)

hypothesis An idea or collection of ideas that seems to explain a specified phenomenon; a conjecture. (Chapter 1)

ice rafts (Europa) Segments of Europa's icy crust that have been moved by tectonic disturbances. (Chapter 13)

ices Solid materials with low condensation temperatures, including ices of water, methane, and ammonia. (Chapter 7)

ideal gas A gas in which the pressure is directly proportional to both the density and the temperature of the gas; an idealization of a real gas. (Chapter 19)

igneous rock A rock that formed from the solidification of molten lava or magma. (Chapter 9)

imaging The process of recording the image made by a telescope of a distant object. (Chapter 6)

impact breccia A type of rock formed from other rocks that were broken apart, mixed, and fused together by a series of meteoritic impacts. (Chapter 10)

impact crater A circular depression on a planet or satellite caused by the impact of a meteoroid. (Chapter 7, Chapter 10)

induced electric current Electric current resulting from a magnetic field sweeping over an electrically conductive material. (Chapter 13)

induced magnetic field A magnetic field resulting from an induced electric current. (Chapter 13)

inertia, law of See *Newton's first law of motion.*

inferior conjunction The configuration when an inferior planet is between the Sun and Earth. (Chapter 4)

inferior planet A planet that is closer to the Sun than Earth is. (Chapter 4)

inflation A sudden expansion of space. (Chapter 26)

inflationary epoch A brief period shortly after the Big Bang during which the scale of the universe increased very rapidly. (Chapter 26)

infrared radiation Electromagnetic radiation of wavelength longer than visible light but shorter than radio waves. (Chapter 5)

inner core (of Earth) The solid innermost portion of Earth's iron-rich core. (Chapter 9)

inner Lagrangian point The point between the two stars constituting a binary system where their Roche lobes touch; the point across which mass transfer can occur. (Chapter 19)

instability strip A region of the H-R diagram occupied by pulsating stars. (Chapter 19)

interferometry A technique of combining the observations of two or more telescopes to produce images better than one telescope alone could make. (Chapter 6)

intermediate vector boson The particle that is responsible for the weak force. (Chapter 26)

intermediate-mass black hole See *mid-mass black hole.*

interstellar extinction The dimming of starlight as it passes through the interstellar medium. (Chapter 18, Chapter 22)

interstellar matter Diffuse gas and dust spread out in the galaxy between the stars. Interstellar matter is mostly hydrogen. (Chapter 22)

interstellar medium Gas and dust in interstellar space. (Chapter 8, Chapter 18)

interstellar reddening The reddening of starlight passing through the interstellar medium as a result of blue light being scattered more than red. (Chapter 18)

intracluster gas Gas found between the galaxies that make up a cluster of galaxies. (Chapter 23)

inverse-square law The statement that the apparent brightness of a light source varies inversely with the square of the distance from the source. (Chapter 17)

Io torus A doughnut-shaped ring of gas circling Jupiter at the distance of Io's orbit. (Chapter 13)

ion tail The relatively straight tail of a comet produced by the solar wind acting on ions. (Chapter 15)

ionization The process by which a neutral atom becomes an electrically charged ion through the loss or gain of electrons. (Chapter 5)

iron meteorite A meteorite composed primarily of iron. (Chapter 15)

irregular cluster (of galaxies) A sprawling collection of galaxies whose overall distribution in space does not exhibit any noticeable spherical symmetry. (Chapter 23)

irregular galaxy An asymmetrical galaxy having neither spiral arms nor an elliptical shape. (Chapter 23)

isotope Any of several forms for the same chemical element whose nuclei all have the same number of protons but different numbers of neutrons. (Chapter 5)

isotropic Having the same property in all directions. (Chapter 23, Chapter 25)

isotropy problem The dilemma posed by the fact that the cosmic microwave background is isotropic; also called the horizon problem. (Chapter 26)

Jeans length The smallest scale over which a density fluctuation in a medium will contract to form a gravitationally bound object. (Chapter 26)

joule (J) A unit of energy. (Chapter 5)

Jovian planet Low-density planet composed primarily of hydrogen and helium; the Jovian planets include Jupiter, Saturn, Uranus, and Neptune. (Chapter 7)

Jupiter-family comet A comet with an orbital period of less than 20 years. (Chapter 15)

kelvin (K) A unit of temperature on the Kelvin temperature scale, equivalent to a degree Celsius. (Chapter 5)

Kelvin-Helmholtz contraction The contraction of a gaseous body, such as a star or nebula, during which gravitational energy is transformed into thermal energy. (Chapter 8)

Kepler's first law The statement that each planet moves around the Sun in an elliptical orbit with the Sun at one focus of the ellipse. (Chapter 4)

Kepler's second law The statement that a planet sweeps out equal areas in equal times as it orbits the Sun; also called the law of equal areas. (Chapter 4)

Kepler's third law A relationship between the period of an orbiting object and the semimajor axis of its elliptical orbit. (Chapter 4)

kiloparsec (kpc) One thousand parsecs; about 3260 light-years. (Chapter 1)

kinetic energy The energy possessed by an object because of its motion. (Chapter 4, Chapter 7)

Kirchhoff's laws Three statements about circumstances that produce absorption lines, emission lines, and continuous spectra. (Chapter 5)

Kirkwood gaps Gaps in the spacing of asteroid orbits, discovered by Daniel Kirkwood. (Chapter 15)

Kuiper belt A region that extends from around the orbit of Pluto to about 500 AU from the Sun where many icy objects orbit the Sun. (Chapter 7, Chapter 14, Chapter 15)

Lamb shift A tiny shift in the spectral lines of hydrogen caused by the presence of virtual particles. (Chapter 26)

Late Heavy Bombardment A brief, intense period of large impacts on all bodies of the solar system about 600 million years after formation of the solar system. (Chapter 8)

lava Molten rock flowing on the surface of a planet. (Chapter 9)

law of equal areas See *Kepler's second law.*

law of inertia See *Newton's first law of motion.*

law of universal gravitation A formula deduced by Isaac Newton that expresses the strength of the force of gravity that two masses exert on each other. (Chapter 4)

laws of physics A set of physical principles with which we can understand natural phenomena and the nature of the universe. (Chapter 1)

length contraction In the special theory of relativity, the shrinking of an object's length along its direction of motion. (Chapter 21)

lens A piece of transparent material (usually glass) that can bend light and bring it to a focus. (Chapter 6)

lenticular galaxy A galaxy with a central bulge and a disk but no spiral structure; an S0 galaxy. (Chapter 23)

libration An apparent rocking of the Moon whereby an Earth-based observer can, over time, see slightly more than one-half the Moon's surface. (Chapter 10)

light curve A graph that displays how the brightness of a star or other astronomical object varies over time. (Chapter 17)

light pollution Light from cities and towns that degrades telescope images. (Chapter 6)

light scattering The process by which light bounces off particles in its path. (Chapter 5, Chapter 12)

light-gathering power A measure of the amount of radiation brought to a focus by a telescope. (Chapter 6)

light-year (ly) The distance light travels in a vacuum in one year. (Chapter 1)

limb darkening The phenomenon whereby the Sun looks darker near its apparent edge, or limb, than near the center of its disk. (Chapter 16)

line of nodes The line where the plane of Earth's orbit intersects the plane of the Moon's orbit. (Chapter 3)

liquid metallic hydrogen Hydrogen compressed to such a density that it behaves like a liquid metal. (Chapter 7, Chapter 12)

lithosphere The solid, upper layer of Earth; essentially Earth's crust. (Chapter 9)

Local Bubble A large cavity in the interstellar medium within which the Sun and nearby stars are located. (Chapter 22)

Local Group The cluster of galaxies of which our Galaxy is a member. (Chapter 23)

local meridian See *meridian.*

long-period comet A comet that takes hundreds of thousands of years or more to complete one orbit of the Sun. (Chapter 15)

long-period variable A variable star with a period longer than about 100 days. (Chapter 19)

lookback time How far into the past we are looking when we see a particular object. (Chapter 25)

Lorentz transformations Equations that relate the measurements of different observers who are moving relative to each other at high speeds. (Chapter 21)

lower meridian The half of the meridian that lies below the horizon. (Chapter 2)

lowlands (on Mars) See *northern lowlands.*

luminosity The rate at which electromagnetic radiation is emitted from a star or other object. (Chapter 5, Chapter 16, Chapter 17)

luminosity class A classification of a star of a given spectral type according to its luminosity. (Chapter 17)

luminosity function The numbers of stars of differing brightness per cubic parsec. (Chapter 17)

lunar eclipse An eclipse of the Moon by Earth; a passage of the Moon through Earth's shadow. (Chapter 3)

lunar highlands Ancient, high-elevation, heavily cratered terrain on the Moon. (Chapter 10)

lunar month See *synodic month.*

lunar phase The appearance of the illuminated area of the Moon as seen from Earth. (Chapter 3)

Lyman series A series of spectral lines of hydrogen produced by electron transitions to and from the lowest energy state of the hydrogen atom. (Chapter 5)

MACHO See *massive compact halo object.*

magma Molten rock beneath a planet's surface. (Chapter 9)

magnetic axis A line connecting the north and south magnetic poles of a planet or star possessing a magnetic field. (Chapter 14)

magnetic dynamo The mechanism whereby electric currents within an astronomical body generate a magnetic field. Sometimes simply referred to as a dynamo. (Chapter 9)

magnetic reconnection An event where two oppositely directed magnetic fields approach and cancel, thus releasing energy. (Chapter 16)

magnetic resonance imaging (MRI) A technique for viewing the interior of the human body that makes use of the way protons respond to magnetic fields. (Chapter 22)

magnetic-dynamo model A theory that explains the solar cycle as a result of the Sun's differential rotation acting on the Sun's magnetic field. (Chapter 16)

magnetogram An image of the Sun that shows regions of different magnetic polarity. (Chapter 16)

magnetometer A device for measuring magnetic fields. (Chapter 7)

magnetopause That region of a planet's magnetosphere where the magnetic field counterbalances the pressure from the solar wind. (Chapter 9)

magnetosphere The region around a planet occupied by its magnetic field. (Chapter 9)

magnification The factor by which the apparent angular size of an object is increased when viewed through a telescope. (Chapter 6)

magnifying power See *magnification.*

magnitude scale A system for denoting the brightnesses of astronomical objects. (Chapter 17)

main sequence A grouping of stars on the Hertzsprung-Russell diagram extending diagonally across the graph from hot, luminous stars to cool, dim stars. (Chapter 17)

main-sequence lifetime The total time that a star spends fusing hydrogen in its core, and hence the total time that it will spend as a main-sequence star. (Chapter 19)

main-sequence star A star whose luminosity and surface temperature place it on the main sequence on an H-R diagram; a star that derives its energy from core hydrogen fusion. (Chapter 17)

major axis (of an ellipse) The longest diameter of an ellipse. (Chapter 4)

mantle (of a planet) That portion of a terrestrial planet located between its crust and core. (Chapter 9)

mare (plural **maria**) Latin for "sea"; a large, relatively crater-free plain on the Moon. (Chapter 10)

mare basalt A type of lunar rock commonly found in the mare basins. (Chapter 10)

maser An interstellar cloud in which water molecules emit intensely at microwave wavelengths. (Chapter 23)

mass A measure of the total amount of material in an object. (Chapter 4)

mass density of radiation The energy possessed by a radiation field per unit volume divided by the square of the speed of light. (Chapter 25)

mass loss A process by which a star gently loses matter. (Chapter 19)

mass transfer The flow of gases from one star in a binary system to the other. (Chapter 19)

massive compact halo object (MACHO) A dim star or low-mass black hole that may constitute part of the unseen dark matter. (Chapter 22)

mass-luminosity relation A relationship between the masses and luminosities of main-sequence stars. (Chapter 17)

mass-radius relation A relationship between the masses and radii of white dwarf stars. (Chapter 20)

matter density parameter (V_m) The ratio of the average density of matter in the universe to the critical density. (Chapter 25)

matter-dominated universe A universe in which the average density of matter exceeds both the mass density of radiation and the mass density of dark energy. (Chapter 25)

mean solar day The interval between successive meridian passages of the mean Sun; the average length of a solar day. (Chapter 2)

mean sun A fictitious object that moves eastward at a constant speed along the celestial equator, completing one circuit of the sky with respect to the vernal equinox in one tropical year. (Chapter 2)

medium (plural **media**) A material through which light travels. (Chapter 6)

medium, interstellar See *interstellar medium.*

megaparsec (Mpc) One million parsecs. (Chapter 1)

melting point The temperature at which a substance changes from solid to liquid. (Chapter 9)

meridian (or **local meridian**) The great circle on the celestial sphere that passes through an observer's zenith and the north and south celestial poles. (Chapter 2)

meridian transit The crossing of the meridian by any astronomical object. (Chapter 2)

mesosphere A layer in Earth's atmosphere above the stratosphere. (Chapter 9)

metal In reference to stars and galaxies, any element other than hydrogen and helium; in reference to planets, a material such as iron, silver, and aluminum that is a good conductor of electricity and of heat. (Chapter 17)

metal-poor star A star that, compared to the Sun, is deficient in elements heavier than helium; also called a Population II star. (Chapter 19)

metal-rich star A star whose abundance of heavy elements is roughly comparable to that of the Sun; also called a Population I star. (Chapter 19)

metamorphic rock A rock whose properties and appearance have been transformed by the action of pressure and heat beneath Earth's surface. (Chapter 9)

meteor The luminous phenomenon seen when a meteoroid enters Earth's atmosphere; a "shooting star." (Chapter 15)

meteor shower Many meteors that seem to radiate from a common point in the sky. (Chapter 15)

meteorite A fragment of a meteoroid that has survived passage through Earth's atmosphere. (Chapter 1, Chapter 8, Chapter 15)

meteoritic swarm A collection of meteoroids moving together along an orbit about the Sun. (Chapter 15)

meteoroid A small rock in interplanetary space. (Chapter 7, Chapter 15)

microlensing A phenomenon in which a compact object such as a MACHO acts as a gravitational lens, focusing the light from a distant star. (Chapter 8, Chapter 22)

microwaves Short-wavelength radio waves. (Chapter 5)

mid-mass black hole A black hole with a mass of hundreds of Suns. (Chapter 21)

migration Changes in a planet's average orbital distance from the Sun. (Chapter 8)

Milky Way Galaxy Our Galaxy; the band of faint stars seen from Earth in the plane of our Galaxy's disk. (Chapter 22)

mineral A naturally occurring solid composed of a single element or chemical combination of elements, often in the form of crystals. (Chapter 9)

minor planet See *asteroid.*

minute of arc See *arcminute.*

model (1) A hypothesis that has withstood experimental or observational tests. (2) The results of a theoretical calculation that gives the values of temperature, pressure, density, and so forth throughout the interior of an object such as a planet or star. (Chapter 1)

molecule A combination of two or more atoms. (Chapter 5)

moonquake Sudden, vibratory motion of the Moon's surface. (Chapter 10)

MRI See *magnetic resonance imaging.*

M-theory A theory that attempts to describe fundamental particles as 11-dimensional membranes. (Chapter 26)

nanometer (nm) One billionth of a meter: 1 nm = 10^{-9} meter = 10^{-6} millimeter = 10^{-3} μm. (Chapter 5)

neap tide An ocean tide that occurs when the Moon is near first-quarter or third-quarter phase. (Chapter 4)

near-Earth object (NEO) An asteroid whose orbit lies wholly or partly within the orbit of Mars. (Chapter 15)

near-infrared The part of the infrared spectrum closest in wavelength to visible light. (Chapter 22)

nebula (plural **nebulae**) A cloud of interstellar gas and dust. (Chapter 1, Chapter 18)

nebular hypothesis The idea that the Sun and the rest of the solar system formed from a cloud of interstellar material. (Chapter 8)

nebulosity See *nebula.*

negative curvature The curvature of a surface or space in which parallel lines diverge and the sum of the angles of a triangle is less than 180°. (Chapter 25)

negative hydrogen ion A hydrogen atom that has acquired a second electron. (Chapter 16)

NEO See *near-Earth object.*

neon fusion The thermonuclear fusion of neon to produce heavier nuclei. (Chapter 20)

neutrino A subatomic particle with no electric charge and very little mass, yet one that is important in many nuclear reactions. (Chapter 16)

neutrino oscillation The spontaneous transformation of one type of neutrino into another type. (Chapter 16)

neutron A subatomic particle with no electric charge and with a mass nearly equal to that of the proton. (Chapter 5)

neutron capture The buildup of neutrons inside nuclei. (Chapter 20)

neutron star A very compact, dense star composed almost entirely of neutrons. (Chapter 20)

new moon The phase of the Moon when the dark hemisphere of the Moon faces Earth. (Chapter 3)

Newton's first law of motion The statement that a body remains at rest, or moves in a straight line at a constant speed, unless acted upon by a net outside force; the law of inertia. (Chapter 4)

Newton's form of Kepler's third law A relationship between the period of two objects orbiting each other, the semimajor axis of their orbit, and the masses of the objects. (Chapter 4)

Newton's second law of motion A relationship between the acceleration of an object, the object's mass, and the net outside force acting on the mass. (Chapter 4)

Newton's third law of motion The statement that whenever one body exerts a force on a second body, the second body exerts an equal and opposite force on the first body. (Chapter 4)

Newtonian mechanics The branch of physics based on Newton's laws of motion. (Chapter 1, Chapter 4)

Newtonian reflector A reflecting telescope that uses a small mirror to deflect the image to one side of the telescope tube. (Chapter 6)

Nice model A model in which the Jovian planets began in a more compact configuration in the early solar system and then migrated outward. This model can explain aspects of the Kuiper belt, the Oort cloud, and the Late Heavy Bombardment. (Chapter 8)

noble gas An element whose atoms do not combine into molecules. (Chapter 12)

node See *line of nodes.*

no-hair theorem A statement of the simplicity of black holes. (Chapter 21)

north celestial pole The point directly above Earth's north pole where Earth's axis of rotation, if extended, would intersect the celestial sphere. (Chapter 2)

northern lights See *aurora borealis.*

northern lowlands (on Mars) Relatively young and crater-free terrain in the Martian northern hemisphere. (Chapter 11)

nova (plural **novae**) A star that experiences a sudden outburst of radiant energy, temporarily increasing its luminosity roughly a thousandfold. (Chapter 20)

nuclear density The density of matter in an atomic nucleus; about $4 \cdot 10^{17}$ kg/m^3. (Chapter 20)

nuclear force The force that holds protons and neutrons together in an atom. It arises from the strong force. It is strong enough to hold protons closely together despite their electrical repulsion. (Chapter 26)

nucleosynthesis The process of building up nuclei such as deuterium and helium from protons and neutrons. (Chapter 26)

nucleus (of an atom) The massive part of an atom, composed of protons and neutrons, about which electrons revolve. (Chapter 5)

nucleus (of a comet) A collection of ices and dust that constitute the solid part of a comet. (Chapter 15)

nucleus (of a galaxy) See *galactic nucleus.*

OB association A grouping of hot, young, massive stars, predominantly of spectral types O and B. (Chapter 18)

OBAFGKM The temperature sequence (from hot to cold) of spectral classes. (Chapter 17)

objective lens The principal lens of a refracting telescope. (Chapter 6)

objective mirror The principal mirror of a reflecting telescope. (Chapter 6)

oblate Flattened at the poles. (Chapter 12)

oblateness A measure of how much a flattened sphere (or spheroid) differs from a perfect sphere. (Chapter 12)

observable universe That portion of the universe inside our cosmic light horizon. (Chapter 25, Chapter 26)

Occam's razor The notion that a straightforward explanation of a phenomenon is more likely to be correct than a convoluted one. (Chapter 4)

occultation The eclipsing of an astronomical object by the Moon or a planet. (Chapter 13, Chapter 14)

oceanic rift A crack in the ocean floor that exudes lava. (Chapter 9)

Olbers's paradox The dilemma associated with the fact that the night sky is dark. (Chapter 25)

1-to-1 spin-orbit coupling See *synchronous rotation.*

Oort cloud A presumed accumulation of comets and cometary material surrounding the Sun at distances of roughly 50,000 AU. (Chapter 7, Chapter 15)

opaque When light cannot travel through an object without being scattered many times or absorbed. Unlike glass or our atmosphere, an opaque material is not transparent. (Chapter 5)

open cluster A loose association of young stars in the disk of our Galaxy; a galactic cluster. (Chapter 18, Chapter 19)

open universe A universe with negative curvature, so that its geometry is analogous to a saddle-shaped hyperbolic surface. (Chapter 25)

opposition The configuration of a planet when it is at an elongation of 180° and thus appears opposite the Sun in the sky. (Chapter 4)

optical double star Two stars that lie along nearly the same line of sight but are actually at very different distances from us. (Chapter 17)

optical telescope A telescope designed to detect visible light. (Chapter 6)

optical window The range of visible wavelengths to which Earth's atmosphere is transparent. (Chapter 6)

orbital energy The net energy of an orbiting object due to its kinetic and gravitational potential energy. The farther a planet orbits from the Sun, the greater its orbital energy. (Chapter 4)

orbital resonance The resonance that occurs whenever the orbital periods of two objects are related by a ratio of small integers—for example, when one moon has twice the orbital period of another. The rhythmic gravitational interactions in an orbital resonance can have either a stabilizing or a destabilizing effect. (Chapter 12, Chapter 13, Chapter 15)

organic molecules Molecules containing carbon, some of which are the molecules of which living organisms are made. (Chapter 27)

outer core (of Earth) The outer, molten portion of Earth's iron-rich core. (Chapter 9)

outgassing The release of gases into a planet's atmosphere by volcanic activity. (Chapter 9)

overcontact binary A close binary system in which the two stars share a common atmosphere. (Chapter 19)

oxygen fusion The thermonuclear fusion of oxygen to produce heavier nuclei. (Chapter 20)

ozone A type of oxygen whose molecules contain three oxygen atoms. (Chapter 9)

ozone hole A region of Earth's atmosphere over Antarctica where the concentration of ozone is abnormally low. (Chapter 9)

ozone layer A layer in Earth's upper atmosphere where the concentration of ozone is high enough to prevent much ultraviolet light from reaching the surface. (Chapter 9)

P wave One of three kinds of seismic waves produced by an earthquake; a primary wave. (Chapter 9)

pair production The creation of a particle and its antiparticle from energy. (Chapter 26)

parabola A conic section formed by cutting a circular cone at an angle parallel to one of the sides of the cone. (Chapter 4)

parallax The apparent displacement of an object due to the motion of the observer. (Chapter 4, Chapter 17)

parsec (pc) A unit of distance; 3.26 light-years. (Chapter 1, Chapter 17)

partial lunar eclipse A lunar eclipse in which the Moon does not appear completely covered. (Chapter 3)

partial solar eclipse A solar eclipse in which the Sun does not appear completely covered. (Chapter 3)

Paschen series A series of spectral lines of hydrogen produced by electron transitions between the third and higher energy levels. (Chapter 5)

Pauli exclusion principle A principle of quantum mechanics stating that no two electrons can have the same position and momentum. (Chapter 19)

penumbra (of a shadow) (plural **penumbrae**) The portion of a shadow in which only part of the light source is covered by an opaque body. (Chapter 3)

penumbral eclipse A lunar eclipse in which the Moon passes only through Earth's penumbra. (Chapter 3)

perigee The point in its orbit where a satellite or the Moon is nearest Earth. (Chapter 3)

perihelion The point in its orbit where a planet or comet is nearest the Sun. (Chapter 4)

period (of a planet) The interval of time between successive geometric arrangements of a planet and an astronomical object, such as the Sun. (Chapter 4)

periodic table A listing of the chemical elements according to their properties, invented by Dmitri Mendeleev. (Chapter 5)

period-luminosity relation A relationship between the period and average density of a pulsating star. (Chapter 18)

photodisintegration The breakup of nuclei by high-energy gamma rays. (Chapter 20)

photometry The measurement of light intensities. (Chapter 6, Chapter 17)

photon A discrete unit of electromagnetic energy. (Chapter 5)

photosphere The region in the solar atmosphere from which most of the visible light escapes into space. (Chapter 16)

photosynthesis A biochemical process in which solar energy is converted into chemical energy, carbon dioxide and water are absorbed, and oxygen is released. (Chapter 9)

physics, laws of See *laws of physics.*

pixel A picture element. (Chapter 6)

plage A bright region in the solar atmosphere as observed in the monochromatic light of a spectral line. (Chapter 16)

Planck time A fundamental interval of time defined by basic physical constants. (Chapter 25)

Planck's law The relationship between the energy of a photon and its wavelength or frequency; $E = hc/\lambda = h\nu$. (Chapter 5)

planetary nebula A luminous shell of gas ejected from an old, low-mass star. (Chapter 20)

planetesimal One of many small bodies of primordial dust and ice that combined to form the planets. (Chapter 8)

plasma A hot ionized gas. (Chapter 13, Chapter 16, Chapter 25)

plastic The attribute of being nearly solid yet able to flow. (Chapter 9)

plate A large section of Earth's lithosphere that moves as a single unit. (Chapter 9)

plate tectonics The motions of large segments (plates) of Earth's surface over the underlying mantle. (Chapter 7, Chapter 9)

plutino One of about 100 objects in the Kuiper belt that orbit the Sun with nearly the same semimajor axis as Pluto. (Chapter 14)

poor cluster (of galaxies) A cluster of galaxies with very few members; a group of galaxies. (Chapter 23)

Population I star A star whose spectrum exhibits spectral lines of many elements heavier than helium; a metal-rich star. (Chapter 19)

Population II star A star whose spectrum exhibits comparatively few spectral lines of elements heavier than helium; a metal-poor star. (Chapter 19)

Population III star One of the first stars to form after the Big Bang, composed of only hydrogen, helium, and tiny amounts of lithium and beryllium; a "zeroth-generation" star. (Chapter 26)

positional astronomy The study of the apparent positions of the planets and stars and how those positions change. (Chapter 2)

positive curvature The curvature of a surface or space in which parallel lines converge and the sum of the angles of a triangle is greater than 180°. (Chapter 25)

positron An electron with a positive rather than negative electric charge; the antiparticle of the electron. (Chapter 16, Chapter 26)

power of ten The exponent n in 10^n. (Chapter 1)

powers-of-ten notation A shorthand method of writing numbers, involving 10 followed by an exponent. (Chapter 1)

precession (of Earth) A slow, conical motion of Earth's axis of rotation caused by the gravitational pull of the Moon and Sun on Earth's equatorial bulge. (Chapter 2)

precession of the equinoxes The slow westward motion of the equinoxes along the ecliptic due to precession of Earth. (Chapter 2)

primary mirror See *objective mirror.*

prime focus The point in a telescope where the objective focuses light. (Chapter 6)

primitive asteroid See *undifferentiated asteroid.*

primordial black hole A type of black hole that may have formed in the very early universe. (Chapter 21)

primordial fireball The extremely hot gas that filled the universe immediately following the Big Bang. (Chapter 25)

principle of equivalence See *equivalence principle.*

progenitor star A star that later explodes into a supernova. (Chapter 20)

prograde orbit An orbit of a satellite around a planet that is in the same direction as the rotation of the planet. (Chapter 13)

prograde rotation A situation in which an object (such as a planet) rotates in the same direction that it orbits around another object (such as the Sun). (Chapter 11)

projection The projection of a point on Earth is made by an imaginary line from the point, extending perpendicular to the surface of Earth, until the line intersects the imaginary celestial sphere. (Chapter 2)

prominence Flamelike protrusions seen near the limb of the Sun and extending into the solar corona. (Chapter 16)

proper distance See *proper length.*

proper length A length measured by a ruler at rest with respect to an observer. (Chapter 21)

proper motion The angular rate of change in the location of a star on the celestial sphere, usually expressed in arcseconds per year. (Chapter 17)

proper time A time interval measured with a clock at rest with respect to an observer. (Chapter 21)

proplyd See *protoplanetary disk.*

proton A heavy, positively charged subatomic particle that is one of two principal constituents of atomic nuclei. (Chapter 5)

proton-proton chain A sequence of thermonuclear reactions by which hydrogen nuclei are built up into helium nuclei. (Chapter 16)

protoplanet A Moon-sized object formed by the coalescence of planetesimals. (Chapter 8)

protoplanetary disk (proplyd) A disk of material encircling a protostar or a newborn star. (Chapter 8, Chapter 18)

protostar A star in its earliest stages of formation. (Chapter 18)

protosun The part of the solar nebula that eventually developed into the Sun. (Chapter 8)

Ptolemaic system The definitive version of the geocentric cosmogony of ancient Greece. (Chapter 4)

pulsar A pulsating radio source thought to be associated with a rapidly rotating neutron star. (Chapter 1, Chapter 20)

pulsating variable star A star that pulsates in size and luminosity. (Chapter 19)

quantum electrodynamics A theory that describes details of how charged particles interact by exchanging photons. (Chapter 26)

quantum mechanics The branch of physics dealing with the structure and behavior of atoms and their constituents as well as their interaction with light. (Chapter 5, Chapter 26)

quark confinement The permanent bonding of quarks to form particles like protons and neutrons. (Chapter 26)

quark One of several particles thought to be the internal constituents of certain heavy subatomic particles such as protons and neutrons. (Chapter 26)

quasar A very luminous object with a very large redshift and a starlike appearance. (Chapter 1, Chapter 24)

radial velocity That portion of an object's velocity parallel to the line of sight. (Chapter 5, Chapter 17)

radial velocity curve A plot showing the variation of radial velocity with time for a binary star or variable star. (Chapter 17)

radial velocity method A technique used to detect extrasolar planets by observing Doppler shifts in the spectrum of the planet's star. (Chapter 8)

radiation darkening The darkening of methane ice by electron impacts. (Chapter 14)

radiation pressure Pressure exerted on an object by radiation falling on the object. (Chapter 15)

radiation-dominated universe A universe in which the mass density of radiation exceeds both the average density of matter and the mass density of dark energy. (Chapter 25)

radiative diffusion The random migration of photons from a star's center toward its surface. (Chapter 16)

radiative zone A region within a star where radiative diffusion is the dominant mode of energy transport. (Chapter 16)

radio galaxy A galaxy that emits an unusually large amount of radio waves. (Chapter 24)

radio telescope A telescope designed to detect radio waves. (Chapter 6)

radio waves The longest-wavelength electromagnetic radiation. (Chapter 5)

radio window The range of radio wavelengths to which Earth's atmosphere is transparent. (Chapter 6)

radioactive dating A technique for determining the age of a rock sample by measuring the radioactive elements and their decay products in the sample. (Chapter 8)

radioactive decay The process whereby certain atomic nuclei spontaneously transform into other nuclei. (Chapter 8)

rarefaction A region in a sound wave where the medium carrying the wave is spread out or rarefied. (Chapter 25)

recombination The process in which an electron combines with a positively charged ion. (Chapter 18)

red dwarf A main-sequence star with a mass between about 0.08 $M_{\odot}$ and 0.4 $M_{\odot}$ that has a fully convective interior and that never goes through a red giant stage. (Chapter 19)

red giant A large, cool star of high luminosity. (Chapter 17, Chapter 19)

reddening (interstellar) See *interstellar reddening.*

red-giant branch The region of an H-R diagram occupied by stars in their first red-giant phase. (Chapter 20)

redshift The shifting to longer wavelengths of the light from remote galaxies and quasars; the Doppler shift of light from a receding source. (Chapter 5, Chapter 23)

reflecting telescope A telescope in which the principal optical component is a concave mirror. (Chapter 6)

reflection The return of light rays by a surface. (Chapter 6)

reflection nebula A comparatively dense cloud of dust in interstellar space that is illuminated by a star. (Chapter 18)

reflector A reflecting telescope. (Chapter 6)

refracting telescope A telescope in which the principal optical component is a lens. (Chapter 6)

refraction The bending of light rays when they pass from one transparent medium to another. (Chapter 6)

refractor A refracting telescope. (Chapter 6)

regolith The layer of rock fragments covering the surface of the Moon. (Chapter 10)

regular cluster (of galaxies) A spherical cluster of galaxies. (Chapter 23)

reionization The process whereby high-energy photons emitted by the most massive of the first stars (Population III stars) ionized most of the atoms in the universe. (Chapter 26)

relativistic cosmology A cosmology based on the general theory of relativity. (Chapter 25)

residual polar cap An ice-covered polar region on Mars that does not completely evaporate during the Martian summer. (Chapter 11)

respiration A biological process that produces energy by consuming oxygen and releasing carbon dioxide. (Chapter 9)

retrograde motion The apparent westward motion of a planet with respect to background stars. (Chapter 4)

retrograde orbit An orbit of a satellite around a planet that is in the direction opposite to which the planet rotates. (Chapter 13)

retrograde rotation A situation in which an object (such as a planet) rotates in the direction opposite to which it orbits around another object (such as the Sun). (Chapter 11)

rich cluster (of galaxies) A cluster of galaxies containing many members. (Chapter 23)

ridge push A force on the oceanic plate that occurs when molten rock emerging from an oceanic rift is pulled by gravity down the ridge's slopes. This segment of ocean seafloor on the slope pushes on the larger seafloor plate and is one of the main forces producing plate tectonics. (Chapter 9)

right ascension A coordinate for measuring the east-west positions of objects on the celestial sphere. (Chapter 2)

ring particles Small particles that constitute a planetary ring. (Chapter 12)

ringlet One of many narrow bands of particles of which Saturn's ring system is composed. (Chapter 12)

Roche limit The smallest distance from a planet or other object at which a second object can be held together by purely gravitational forces. (Chapter 12)

Roche lobe A teardrop-shaped volume surrounding a star in a binary inside which gases are gravitationally bound to that star. (Chapter 19)

rock A mineral or combination of minerals. (Chapter 9)

rotation curve A plot of the orbital speeds of stars and nebulae in a galaxy versus distance from the center of the galaxy. (Chapter 22)

RR Lyrae variable A type of pulsating star with a period of less than one day. (Chapter 19, Chapter 22)

runaway greenhouse effect A greenhouse effect in which the temperature continues to increase. (Chapter 11)

runaway icehouse effect A situation in which a decrease in atmospheric temperature causes a further decrease in temperature. (Chapter 11)

S wave One of three kinds of seismic waves produced by an earthquake; a secondary wave. (Chapter 9)

Sagittarius A* A powerful radio source at the center of our Galaxy. (Chapter 22)

saros A particular cycle of similar eclipses that recur about every 18 years. (Chapter 3)

scarp A line of cliffs formed by the faulting or fracturing of a planet's surface. (Chapter 11)

scattering of light See *light scattering.*

Schwarzschild radius (R_{Sch}) The distance from the singularity to the event horizon in a nonrotating black hole. (Chapter 21)

scientific method The basic procedure used by scientists to investigate phenomena. (Chapter 1)

seafloor spreading The separation of plates under the ocean due to lava emerging in an oceanic rift. (Chapter 9)

search for extraterrestrial intelligence (SETI) The scientific search for evidence of intelligent life on other planets. (Chapter 27)

second of arc See *arcsecond.*

sedimentary rock A rock that is formed from material deposited on land by rain or winds, or on the ocean floor. (Chapter 9)

seeing disk The angular diameter of a star's image. (Chapter 6)

seismic wave A vibration traveling through a terrestrial planet, usually associated with earthquake-like phenomena. (Chapter 9)

seismograph A device used to record and measure seismic waves, such as those produced by earthquakes. (Chapter 9)

selection effect The bias that occurs when an observational technique is more sensitive to one type of system than another. For example, the radial velocity method for detecting exoplanets can more easily detect large planets orbiting closely to their stars. This method is unlikely to detect many planets with Earth's characteristics, but it does not mean they are not there. (Chapter 8)

self-propagating star formation The process by which the formation of stars in one location in a galaxy stimulates the formation of stars in a neighboring location. (Chapter 22)

semidetached binary A binary star system in which one star fills its Roche lobe. (Chapter 19)

semimajor axis One-half of the major axis of an ellipse. For an orbiting object, the semimajor axis is also equal to the average orbital distance. (Chapter 4)

SETI See *search for extraterrestrial intelligence.*

Seyfert galaxy A spiral galaxy with a bright nucleus whose spectrum exhibits emission lines. (Chapter 24)

shell helium fusion The thermonuclear fusion of helium in a shell surrounding a star's core. (Chapter 20)

shell hydrogen fusion The thermonuclear fusion of hydrogen in a shell surrounding a star's core. (Chapter 19)

shepherd satellite A satellite whose gravity restricts the motions of particles in a planetary ring, preventing them from dispersing. (Chapter 12)

shield volcano A volcano with long, gently sloping sides. (Chapter 11)

shock wave An abrupt, localized region of compressed gas caused by an object traveling through the gas at a speed greater than the speed of sound. (Chapter 9)

short-period comet A comet that originated in the Kuiper belt and has a period of less than 200 years. (Chapter 15)

SI units The International System of Units, based on the meter (m), the second (s), and the kilogram (kg). (Chapter 1)

sidereal clock A clock that measures sidereal time. (Chapter 2)

sidereal day The interval between successive meridian passages of the vernal equinox. (Chapter 2)

sidereal month The period of the Moon's revolution about Earth with respect to the stars. (Chapter 3)

sidereal period The orbital period of one object about another as measured with respect to the stars. (Chapter 4)

sidereal time Time reckoned by the location of the vernal equinox. (Chapter 2)

sidereal year The orbital period of Earth about the Sun with respect to the stars. (Chapter 2)

silicon fusion The thermonuclear fusion of silicon to produce heavier elements, especially iron. (Chapter 20)

singularity A place of infinite space-time curvature; the center of a black hole. (Chapter 21)

slab pull A force on seafloor plates that occurs when a portion (or slab) of the seafloor plate sinks down into the mantle at a subduction zone. Gravity pulls on this higher-density slab more than on the surrounding mantle material. The pull on this slab results in a pull on the entire seafloor plate, which is the main force driving plate tectonics. (Chapter 9)

small-angle formula A relationship between the angular and linear sizes of a distant object. (Chapter 1)

snow line The distance from the Sun at which water vapor solidifies into ice or frost. Formation of this solid material beyond the snow line helped to build the large Jovian planets. (Chapter 8)

solar constant The average amount of energy received from the Sun per square meter per second, measured just above Earth's atmosphere. (Chapter 5)

solar corona Hot, faintly glowing gases seen around the Sun during a total solar eclipse; the uppermost regions of the solar atmosphere. (Chapter 3)

solar cycle See *22-year solar cycle.*

solar eclipse An eclipse of the Sun by the Moon; a passage of Earth through the Moon's shadow. (Chapter 3)

solar flare A sudden, temporary outburst of light from an extended region of the solar surface. (Chapter 16)

solar nebula The cloud of gas and dust from which the Sun and solar system formed. (Chapter 8)

solar neutrino A neutrino emitted from the core of the Sun. (Chapter 16)

solar neutrino problem The discrepancy between the predicted and observed numbers of solar neutrinos. (Chapter 16)

solar system The Sun, planets and their satellites, asteroids, comets, and related objects that orbit the Sun. (Chapter 1)

solar wind An outward flow of particles (mostly electrons and protons) from the Sun. (Chapter 9, Chapter 16)

south celestial pole The point directly above Earth's south pole where Earth's axis of rotation, if extended, would intersect the celestial sphere. (Chapter 2)

southern highlands (on Mars) Older, cratered terrain in the Martian southern hemisphere. (Chapter 11)

southern lights See *aurora australis.*

space velocity The speed and direction in which a star moves through space. (Chapter 17)

space weather Variations in the solar wind and magnetic field, which can affect satellites and astronauts. (Chapter 16)

spacetime A four-dimensional combination of time and the three dimensions of space. (Chapter 21)

special theory of relativity A description of mechanics and electromagnetic theory formulated by Albert Einstein, which explains that measurements of distance, time, and mass are affected by the observer's motion. (Chapter 21)

spectral analysis The identification of chemical substances from the patterns of lines in their spectra. (Chapter 5)

spectral class A classification of stars according to the appearance of their spectra. (Chapter 17)

spectral line In a spectrum, an absorption or emission feature that is at a particular wavelength. (Chapter 5)

spectral type A subdivision of a spectral class. (Chapter 17)

spectrograph An instrument for photographing a spectrum. (Chapter 6)

spectroscopic binary A binary star system whose binary nature is deduced from the periodic Doppler shifting of lines in its spectrum. (Chapter 17)

spectroscopic parallax The distance to a star derived by comparing its apparent brightness to a luminosity inferred from the star's spectrum. (Chapter 17)

spectroscopy The study of spectra and spectral lines. (Chapter 5, Chapter 6, Chapter 7)

spectrum (plural spectra) **The result of dispersing a beam of electromagnetic radiation so that components with different wavelengths are separated in space. (Chapter 5)**

spectrum binary A binary star whose binary nature is deduced from the presence of two sets of incongruous spectral lines. (Chapter 17)

speed Distance traveled divided by the time elapsed to cover that distance. (Chapter 4)

spherical aberration The distortion of an image formed by a telescope due to differing focal lengths of the optical system. (Chapter 6)

spherical space Space with positive curvature. (Chapter 25)

spicule A narrow jet of rising gas in the solar chromosphere. (Chapter 16)

spin (of a particle) The intrinsic angular momentum possessed by certain particles. (Chapter 22)

spin-flip transition A transition in the ground state of the hydrogen atom, which occurs when the orientation of the electron's spin changes. (Chapter 22)

spin-orbit coupling See *1-to-1 spin-orbit coupling* and *3-to-2 spin-orbit coupling.*

spiral arms Lanes of interstellar gas, dust, and young stars that wind outward in a plane from the central regions of a galaxy. (Chapter 22)

spiral galaxy A flattened, rotating galaxy with pinwheel-like spiral arms winding outward from the galaxy's nucleus. (Chapter 23)

spontaneous symmetry breaking A process by which certain symmetries in the mathematics of particle physics are suddenly altered to produce new particles and forces. (Chapter 26)

spring tide An ocean tide that occurs at new moon and full moon phases. (Chapter 4)

stable Lagrange points Locations along Jupiter's orbit where the combined gravitational effects of the Sun and Jupiter cause asteroids to collect. (Chapter 15)

standard candle An astronomical object of known intrinsic brightness that can be used to determine the distances to other galaxies. (Chapter 23)

Standard Model The detailed theory, verified by experiments, that explains all of the known forces between all of the known particles (not including the force of gravity). Phenomena such as dark matter and dark energy may or may not be related to the Standard Model. (Chapter 26)

star cluster See *cluster (of stars).*

starburst galaxy A galaxy that is experiencing an exceptionally high rate of star formation. (Chapter 23)

stationary absorption line An absorption line in the spectrum of a binary star that does not show the same Doppler shift as other lines, indicating that it originates in the interstellar medium. (Chapter 18)

Stefan-Boltzmann law A relationship between the temperature of a blackbody and the rate at which it radiates energy. (Chapter 5)

stellar association A loose grouping of young stars. (Chapter 18)

stellar evolution The changes in size, luminosity, temperature, and so forth that occur as a star ages. (Chapter 18)

stellar parallax The apparent displacement of a star due to Earth's motion around the Sun. (Chapter 17)

stellar-mass black hole A black hole with a mass comparable to that of a star. (Chapter 21)

stony iron meteorite A meteorite composed of both stone and iron. (Chapter 15)

stony meteorite A meteorite composed of stone. (Chapter 15)

stratosphere A layer in Earth's atmosphere directly above the troposphere. (Chapter 9)

string theory The currently favored theory for combining all the known forces in nature. (Chapter 26)

strong force The force that binds protons and neutrons together in nuclei. (Chapter 26)

subduction zone A location where colliding tectonic plates cause Earth's crust to be pulled down into the mantle. (Chapter 9)

subtend To extend over an angle. (Chapter 1)

summer solstice The point on the ecliptic where the Sun is farthest north of the celestial equator. Also used to refer to the date on which the Sun passes through this point. (Chapter 2)

sunspot A temporary cool region in the solar photosphere. (Chapter 16)

sunspot cycle The semiregular 11-year period with which the number of sunspots fluctuates. (Chapter 16)

sunspot maximum/minimum That time during the sunspot cycle when the number of sunspots is highest/lowest. (Chapter 16)

supercluster A collection of clusters of galaxies. (Chapter 23)

supergiant A very large, extremely luminous star of luminosity class I. (Chapter 17, Chapter 20)

supergrand unified theory A complete description of all forces and particles, as well as the structure of space and time; a "theory of everything" (TOE). (Chapter 26)

supergranule A large convective feature in the solar atmosphere, usually outlined by spicules. (Chapter 16)

superior conjunction The configuration of a planet being behind the Sun as viewed from Earth. (Chapter 4)

superior planet A planet that is more distant from the Sun than is Earth. (Chapter 4)

superluminal motion Motion that appears to involve speeds greater than the speed of light. (Chapter 24)

supermassive black hole A black hole with a mass of a million or more Suns. (Chapter 21, Chapter 24)

supernova (plural **supernovae**) A stellar outburst during which a star suddenly increases its brightness roughly a millionfold. (Chapter 1, Chapter 15, Chapter 20)

supernova remnant The gases ejected by a supernova. (Chapter 18, Chapter 20)

supersonic Faster than the speed of sound. (Chapter 18)

surface wave A type of seismic wave that travels only over Earth's surface. (Chapter 9)

synchronous rotation The rotation of a body with a period equal to its orbital period; also called 1-to-1 spin-orbit coupling. (Chapter 3, Chapter 10)

synodic month The period of revolution of the Moon with respect to the Sun; the length of one cycle of lunar phases. Also called the lunar month. (Chapter 3)

synodic period The interval between successive occurrences of the same configuration of a planet. (Chapter 4)

T Tauri stars Young variable stars associated with interstellar matter that show erratic changes in luminosity. (Chapter 18)

tail (of a comet) Gas and dust particles from a comet's nucleus that have been swept away from the comet's head by the radiation pressure of sunlight and the solar wind. (Chapter 15)

tangential velocity That portion of an object's velocity perpendicular to the line of sight. (Chapter 17)

temperature See *degree Celsius, degree Fahrenheit,* and *kelvin.*

terminator The line dividing day and night on the surface of the Moon or a planet; the line of sunset or sunrise. (Chapter 10)

terrae Cratered lunar highlands. (Chapter 10)

terrestrial planet High-density worlds with solid surfaces, including Mercury, Venus, Earth, and Mars. (Chapter 7)

theory A hypothesis that has withstood experimental or observational tests. (Chapter 1)

theory of everything (TOE) See *supergrand unified theory.*

thermal energy An energy that arises from the kinetic energy of the individual atoms and molecules in a substance. The hotter an object, the greater its thermal energy. (Chapter 5)

thermal equilibrium A balance between the input and outflow of heat in a system. (Chapter 16, Chapter 26)

thermal pulse A brief burst in energy output from the helium-fusing shell of an aging low-mass star. (Chapter 20)

thermonuclear Describes nuclear reactions that fuse atoms together as a result of very high temperatures. (Chapter 20)

thermonuclear fusion The combining of nuclei under conditions of high temperature in a process that releases substantial energy. (Chapter 16)

thermonuclear supernova A supernova that occurs when a white dwarf in a close binary system accretes so much matter from its companion star that the white dwarf explodes in a blast of thermonuclear fusion; a Type Ia supernova. (Chapter 20)

thermosphere A region in Earth's atmosphere between the mesosphere and the exosphere. (Chapter 9)

third quarter moon The phase of the Moon that occurs when the Moon is 90° west of the Sun. (Chapter 3)

3-to-2 spin-orbit coupling The rotation of Mercury, which makes three complete rotations on its axis for every two complete orbits around the Sun. (Chapter 11)

threshold temperature The temperature above which photons spontaneously produce particles and antiparticles of a particular type. (Chapter 26)

tidal force A gravitational force whose strength and/or direction varies over a body and thus tends to deform the body. (Chapter 4, Chapter 12)

tidal heating The heating of the interior of a satellite by continually varying tidal stresses. (Chapter 13)

time dilation The slowing of time due to relativistic motion. (Chapter 21)

time zone A region on Earth where, by agreement, all clocks have the same time. (Chapter 2)

TOE See *supergrand unified theory.*

total lunar eclipse A lunar eclipse during which the Moon is completely immersed in Earth's umbra. (Chapter 3)

total solar eclipse A solar eclipse during which the Sun is completely hidden by the Moon. (Chapter 3)

totality (of lunar eclipse) The period during a total lunar eclipse when the Moon is entirely within Earth's umbra. (Chapter 3)

totality (of solar eclipse) The period during a total solar eclipse when the disk of the Sun is completely hidden. (Chapter 3)

transient lunar phenomena A flash of light that can occur when meteorites as small as 10 cm hit the lunar surface. Enough heat is generated in the impact to produce light visible from Earth by the naked eye. (Chapter 10)

transit An event in which an astronomical body moves in front of another. See also *meridian transit.* (Chapter 8)

transit method A method for detecting extrasolar planets that come between us and their parent star, dimming the star's light. (Chapter 8)

Trans-Neptunian object Any small body of rock and ice that orbits the Sun within the solar system, but beyond the orbit of Neptune. (Chapter 7, Chapter 14)

triple alpha process A sequence of two thermonuclear reactions in which three helium nuclei combine to form one carbon nucleus. (Chapter 19)

Trojan asteroid One of several asteroids that share Jupiter's orbit about the Sun. (Chapter 15)

Tropic of Cancer A circle of latitude 23½° north of Earth's equator. (Chapter 2)

Tropic of Capricorn A circle of latitude 23½° south of Earth's equator. (Chapter 2)

tropical year The period of revolution of Earth about the Sun with respect to the vernal equinox. (Chapter 2)

troposphere The lowest level in Earth's atmosphere. (Chapter 9)

Tully-Fisher relation A correlation between the width of the 21-cm line of a spiral galaxy and the total luminosity of that galaxy. (Chapter 23)

tuning fork diagram A diagram that summarizes Edwin Hubble's classification scheme for spiral, barred spiral, and elliptical galaxies. (Chapter 23)

turnoff point The point on an H-R diagram where the stars in a cluster are leaving the main sequence. (Chapter 19)

21-cm radio emission Radio radiation emitted by neutral hydrogen atoms in interstellar space. (Chapter 22)

22-year solar cycle The semiregular 22-year interval between successive appearances of sunspots at the same latitude and with the same magnetic polarity. (Chapter 16)

Type I supernova A supernova whose spectrum lacks hydrogen lines. Type I supernovae are further classified as Type Ia, Ib, or Ic. (Chapter 20)

Type Ia supernova A supernova whose spectrum lacks hydrogen lines but has a strong absorption line of ionized silicon. (Chapter 20)

Type Ib supernova A supernova whose spectrum lacks hydrogen and silicon lines but has a strong helium absorption line. (Chapter 20)

Type Ic supernova A supernova whose spectrum is almost devoid of emission or absorption lines. (Chapter 20)

Type II supernova A supernova with hydrogen emission lines in its spectrum, caused by the explosion of a massive star. (Chapter 20)

UBV photometry A system for determining the surface temperature of a star by measuring the star's brightness in the ultraviolet (U), blue (B), and visible (V) spectral regions. (Chapter 17)

ultraviolet radiation Electromagnetic radiation of wavelengths shorter than those of visible light but longer than those of X rays. (Chapter 5)

umbra (of a shadow) (plural **umbrae**) The central, completely dark portion of a shadow. (Chapter 3)

undifferentiated asteroid An asteroid within which chemical differentiation did not occur. (Chapter 15)

unified model Describes a physical system at the core of active galactic nuclei (AGN) that appears very different depending on the line of sight with which it viewed from Earth. The basic model involves a bright accretion disk that may or may not be obscured by a thick dusty torus. (Chapter 24)

universal constant of gravitation (*G*) The constant of proportionality in Newton's law of gravitation. (Chapter 4)

upper meridian The half of the meridian that lies above the horizon. (Chapter 2)

Van Allen belts Two doughnut-shaped regions around Earth where many charged particles (protons and electrons) are trapped by Earth's magnetic field. (Chapter 9)

velocity The speed and direction of an object's motion. (Chapter 4)

vernal equinox The point on the ecliptic where the Sun crosses the celestial equator from south to north. Also used to refer to the date on which the Sun passes through this intersection. (Chapter 2)

very-long-baseline interferometry (VLBI) A method of connecting widely separated radio telescopes to make very high-resolution observations. (Chapter 6)

virtual pair (also **virtual particle pair**) A particle and antiparticle that exist for such a brief interval that they cannot be observed. (Chapter 21, Chapter 25, Chapter 26)

virtual particle pair See *virtual pair.*

visible light Electromagnetic radiation detectable by the human eye. (Chapter 5)

visual binary A binary star in which the two components can be resolved through a telescope. (Chapter 17)

VLBI See *very-long-baseline interferometry.*

void A large volume of space, typically 30 to 120 Mpc (100 to 400 million light-years) in diameter, that contains very few galaxies. (Chapter 23)

volatile element An element with low melting and boiling points. (Chapter 10, Chapter 11)

waning crescent moon The phase of the Moon that occurs between third quarter and new moon. (Chapter 3)

waning gibbous moon The phase of the Moon that occurs between full moon and third quarter. (Chapter 3)

water hole A range of frequencies in the microwave spectrum suitable for interstellar radio communication. (Chapter 27)

watt A unit of power equal to one joule of energy per second. (Chapter 5)

wavelength of maximum emission The wavelength at which a heated object emits the greatest intensity of radiation. (Chapter 5)

wavelength The distance between two successive wave crests. (Chapter 5)

waxing crescent moon The phase of the Moon that occurs between new moon and first quarter. (Chapter 3)

waxing gibbous moon The phase of the Moon that occurs between first quarter and full moon. (Chapter 3)

weak force The short-range force that is responsible for transforming certain particles into other particles, such as the decay of a neutron into a proton. (Chapter 26)

weakly interacting massive particle (WIMP) A hypothetical massive particle that may make up part of the unseen dark matter. (Chapter 22)

weight The force with which gravity acts on a body. (Chapter 4)

white dwarf A low-mass star that has exhausted all its thermonuclear fuel and contracted to a size roughly equal to the size of Earth. (Chapter 17, Chapter 20)

white ovals Round, whitish features usually seen in Jupiter's southern hemisphere. (Chapter 12)

Widmanstätten patterns Crystalline structure seen in certain types of meteorites. (Chapter 15)

Wien's law A relationship between the temperature of a blackbody and the wavelength at which it emits the greatest intensity of radiation. (Chapter 5)

WIMP See *weakly interacting massive particle.*

winding dilemma The problem that the spiral arms of a galaxy like the Milky Way do not disappear. (Chapter 22)

winter solstice The point on the ecliptic where the Sun reaches its greatest distance south of the celestial equator. Also used to refer to the date on which the Sun passes through this point. (Chapter 2)

X-ray burster A nonperiodic X-ray source that emits powerful bursts of X rays. (Chapter 20)

X-rays Electromagnetic radiation whose wavelength is between that of ultraviolet light and gamma rays. (Chapter 5)

ZAMS See *zero-age main sequence.*

Zeeman effect A splitting or broadening of spectral lines due to a magnetic field. (Chapter 16)

zenith The point on the celestial sphere directly overhead an observer. (Chapter 2)

zero curvature The curvature of a surface or space in which parallel lines remain parallel and the sum of the angles of a triangle is exactly 180°. (Chapter 25)

zero-age main sequence The main sequence of young stars that have just begun to burn hydrogen at their cores. (Chapter 19)

zero-age main-sequence star A newly formed star that has just arrived on the main sequence. (Chapter 19)

zodiac A band of 12 constellations around the sky centered on the ecliptic. (Chapter 2)

zonal winds The pattern of alternating eastward and westward winds found in the atmospheres of Jupiter and Saturn. (Chapter 12)

zone A light-colored band in Jupiter's atmosphere. (Chapter 12)

Answers to Selected Questions

CAUTION! Only mathematical answers are given, not answers that require interpretation or discussion. Your instructor will expect you to show the steps required to reach each mathematical answer.

Chapter 1

24. 8.5×10^3 km **25.** 2.8×10^7 Suns **26.** About 3×10^{36} times larger **27.** 8.94×10^{56} hydrogen atoms **28.** (a) 1.581×10^{-5} ly (b) 4.848×10^{-6} pc **29.** 4.99×10^2 s **30.** 4.3×10^9 km **31.** (a) 1.59×10^{14} km (b) 16.8 years **32.** 4.32×10^{17} s **34.** (a) 1.5 m (b) 89 m (c) 5.4×10^3 m **35.** 6.9 m **36.** 3.4×10^3 km **37.** 3.7 km **38.** 0.320 arcmin

Chapter 2

28. Around 8:02 P.M. **33.** (a) About 9 hours **46.** October 25, 1917 **49.** 50° **50.** 3:50 A.M. local time **52.** (a) 6:00 P.M. (b) September 21

Chapter 3

31. (a) 0.91 hour (b) 11° **32.** 49 arcsec

Chapter 4

18. Semimajor axis = 0.25 AU, period = 0.125 year **19.** Average = 25 AU, farthest = 50 AU **23.** 6 newtons, 4 m/s² **26.** 1/9 as strong **38.** 87.97 days **43.** (a) 4 AU (b) 8 years **44.** (a) 16.0 AU (b) 0.5 AU **45.** (a) 1.26 AU (b) 258 days **46.** Earth exerts a force of 1.98×10^{20} newtons on the Moon. **47.** Forces are approximately the same; Earth's acceleration is 100 times greater. **48.** ¼ as much as on Earth **49.** 62 newtons, 0.13 **53.** 0.5 year **54.** (a) 24 hours (b) 43,200 km **55.** 119 minutes **57.** (a) 5 years (b) 2.92 AU **58.** (a) 3.43×10^{-5} newton (b) 3.21×10^{-5} newton (c) 2.2×10^{-6} newton

Chapter 5

2. 7.5 times **3.** 500 s **7.** 0.340 m **8.** 2.61×10^{14} Hz **17.** 135 nm **18.** 3400 K **27.** 9400 nm **28.** 9980°F **29.** About 10 μm **31.** 2.9 nm **32.** 3.9×10^{26} W **33.** 190 times more **34.** (a) 4890 nm = 4.89 μm (b) 540 times more **35.** (a) 5.75×10^8 W/m (b) 10,000 K **36.** 2.43×10^{-3} nm **37.** (a) 1005 nm **41.** Possible transitions: energy = 1 eV, λ = 1240 nm; energy = 2 eV, λ = 620 nm; energy = 3 eV, λ = 414 nm **42.** 8.6×10^4 km/s **43.** Coming toward us at 13.0 km/s

Chapter 6

31. $1/25 = 0.04$ **33.** (a) 222× (b) 100× (c) 36× (d) 0.75 arcsec **36.** 300 km (Hubble Space Telescope, Jupiter's moons); 110 km (human eye, our Moon) **37.** (a) 34 ly (b) 37 km **40.** (a) 5.39×10^{-4} m (c) 0.56 m

Chapter 7

23. Mass = 6.4×10^{23} kg, average density = 3900 kg/m³ **25.** (a) 1.0×10^{13} kg (b) 1.2 m/s **26.** (a) 3.3×10^{21} J (b) Equivalent to 3.9×10^7 Hiroshima-type weapons **27.** (a) 2.02 km/s (b) 19.4 km/s **28.** 1.2 km/s **29.** (a) 618 km/s **31.** 1.76 years **33.** (a) 1000 years (b) 17 days **35.** In 100 years, probability is 8.3×10^{-8} (one chance in 12 million); in 10^6 years, probability is 8.3×10^{-4} (one chance in 1200)

Chapter 8

41. 0.40 kg after 1.3 billion years; 0.20 kg after 2.6 billion years; 0.10 kg after 3.9 billion years **31.** 2.6 billion years **44.** (a) About 180 AU **45.** 2.9×10^7 AU = 140 pc = 460 ly **46.** (a) About 600 AU (b) About 5×10^{40} cubic meters (c) About 10^{55} atoms (d) About 3×10^{14} atoms per cubic meter

Chapter 16

29. (a) 1.8×10^{-9} J (b) 9.0×10^{16} J (c) 5.4×10^{41} J **30.** (a) 4.6×10^{-36} s (b) 2.3×10^{10} s (c) 1.4×10^{15} s = 4.4×10^7 years **31.** 0.048 (4.8%) of the Sun's mass will be converted from hydrogen to helium; chemical composition of the Sun (by mass) will be 69% hydrogen, 30% helium **32.** (a) 8.8×10^{29} kg of hydrogen consumed, 6.2×10^{26} kg lost **34.** (a) 1.64×10^{-13} J (b) 2.43×10^{-3} nm **35.** 1.4×10^{13} kg/s **39.** 98,600 kg/m³ **40.** 1.9×10^{-7} nm **42.** (a) 1700 nm **44.** For the photosphere, 500 nm; for the chromosphere, 58 nm; for the corona, 1.9 nm **47.** For the umbra, 670 nm; for the penumbra, 580 nm **48.** (a) (Flux from patch of penumbra)/(flux from patch of photosphere) = 0.55 (b) (Flux from patch of penumbra)/(flux from patch of umbra) = 1.8

Chapter 17

34. (a) 9.7 pc (b) 0.10 arcsec **35.** 6.54 pc **36.** (a) 161 km/s (b) 294 km/s **37.** Distance = 105 pc; tangential velocity would have to be about 5200 km/s **38.** 110 km/s **39.** (a) +59.4 km/s (c) 486.23 nm **45.** 37.0 AU **46.** 6.1 $L_\odot$ **47.** 0.38 pc = 7.9×10^4 AU **48.** (a) +13.8 (b) (Luminosity of HIP 72509)/(luminosity of Sun) = 2.4×10^{-4} **49.** +17 **50.** 6300 pc **51.** (a) Brightest star has $M = -2.37$; dimmest star has $M = +4.32$ (b) $M = +0.79$ **54.** (b) $m_B - m_V = -0.23$ (Bellatrix), +0.68 (Sun), +1.86 (Betelgeuse) **55.** 99 $R_\odot$ **59.** The radius of star X is 17 times larger than the radius of star Y. **60.** Radius increases by a factor of 2, luminosity increases by a factor of 64 **62.** T = 10,000 K, $R = 3.8\ R_\odot$ **63.** $L = 35\ L_\odot$, so distance = 14 pc **64.** (b) $0.26\ R_\odot = 1.8 \times 10^5$ km **65.** 9700 years **66.** (a) 5.0 pc (b) 22.5 AU (c) 1.5 $M_\odot$ **67.** (a) 40 $M_\odot$ **68.** About 2500 pc, about 125 times greater volume

Chapter 18

30. 0.34% **32.** 3.4×10^4 atoms per cm³ **35.** 100 $R_\odot = 7.4 \times 10^8$ km = 4.9 AU **37.** 250 AU = 3.7×10^{10} km **39.** 1.3×10^{31} m³ **41.** 2.2×10^3 km/s, or 7.5×10^{-3} (0.75%) of the speed of light

Chapter 19

29. 657,000 km **30.** (a) 618 km/s (b) 61.8 km/s **31.** (a) 12.0 km/s (b) 9.3 km/s **32.** 2.3×10^{29} kg, or 0.15 (15%) of the original mass of hydrogen **36.** (a) 4.9×10^7 years (b) 3.8×10^{11} years **38.** About 1900 K (1600°C, or 2900°F) **39.** 3.4×10^7 years **44.** About 650 pc **45.** About 370 pc **49.** (a) 1.65×10^6 km

Chapter 20

49. 0.11 $R_\odot$ **52.** We see the nebula about 5900 to 8200 years after the central star shed its outer layers **54.** (a) 97 nm **56.** (a) 1.8×10^9 kg/m^3 (b) 6.5×10^3 km/s **57.** About 4×10^6 kg **58.** (a) 1.7×10^{30} kg $= 0.84\ M_\odot$ (b) 1.1×10^{12} newtons (c) 1.5×10^8 m/s, or 0.5 (50%) of the speed of light **61.** (a) 6.4×10^5 ms/ **63.** 3.2×10^9 km = 22 AU **64.** (Maximum luminosity of SN 1993J)/(maximum luminosity of SN 1987A) = 4.5 **66.** (a) $7 \times 10^{-7}\ b_\odot$ (b) It would be about 700 times brighter than Venus **67.** 1.3×10^8 pc = 130 Mpc

Chapter 21

34. 0.6 of the speed of light **35.** 8.3×10^{-8} s **36.** 0.8 of the speed of light **37.** (a) 25 years (b) 20 ly as measured by an Earth observer, 12 ly as measured by the astronaut **39.** 2.8 $M_\odot$ **40.** 5.7×10^8 years **41.** (a) 2.01×10^6 km **45.** 0.32 year **46.** 0.84 m **47.** (a) R_{Sch} = 8.9 mm, density = 2.0×10^{30} kg/m^3 (b) R_{Sch} = 3.0 km, density = 1.8×10^{19} kg/m^3 (c) $R_{Sch} = 3.5 \times 10^9$ km = 24 AU, density = 13 kg/m^3 **48.** 7.4×10^{30} kg **49.** 2.9×10^{17} kg/m^3 **50.** 2.7×10^{38} kg = $1.4 \times 10^8\ M_\odot$

Chapter 22

30. (a) 1.2×10^{12} cubic parsecs (b) 1.1×10^8 cubic parsecs (c) Probability = 9.6×10^{-5}; we can expect to see a supernova within 300 pc once every 350,000 years **31.** (a) 8.3×10^9 AU (b) 4.0×10^4 pc **33.** 9.5×10^{-25} J; it takes 3.2×10^5 such photons to equal the energy of one H_α photon **34.** (b) 5700 pc **36.** A 10% error in radius results in a 10% error in mass; a 10% error in velocity results in a 20% error in mass. **37.** (a) 3.1×10^8 years (b) $7.4 \times 10^{11}\ M_\odot$ **38.** $2.7 \times 10^{11}\ M_\odot$ **46.** (a) 6.3 years (b) $3.7 \times 10^6\ M_\odot$

Chapter 23

36. (a) 6.9 Mpc (b) 8.3 Mpc (c) 1.4 Mpc **37.** 9.5 Mpc **38.** (a) 1.1×10^{10} km = 70 AU (b) 7.0 Mpc **39.** (a) 152 Mpc = 4.96×10^8 ly **40.** 54 km/s/Mpc **41.** (a) $z = 0.0252$ (b) 106 Mpc **42.** (a) 2.85×10^5 km/s, or 0.951 (95.1%) of the speed of light (b) 1.60×10^6 km/s (c) 4020 Mpc = 1.31×10^{10} ly **43.** (a) 1.2×10^{70} atoms (b) 3.6×10^{-6} atom per cm^3 **44.** $1.2 \times 10^{12}\ M_\odot$ **45.** Period = 4.4×10^8 years; mass = $3.6 \times 10^{11}\ M_\odot$

Chapter 24

20. (a) 30 years (b) 2014 (c) 5/3 of the speed of light **21.** (a) 1.80×10^5 km/s, or 0.600 (60.0%) of the speed of light (b) 134 hours (c) 970 AU **22.** (a) $1.2 \times 10^{11}\ L_\odot$ **23.** $1.5 \times 10^8\ M_\odot$ **24.** 2.9×10^9 km = 20 AU

Chapter 25

33. $z = 4$ (b) Distances were 1/9 as great and the density of matter was 729 times greater. **35.** (a) 20 billion years (b) 13 billion years (c) 9.7 billion years **36.** 1.6×10^8 km/s/Mpc **37.** 65.0 times denser **38.** 1.06×10^{-3} m = 1.06 mm **39.** 13.6 K **40.** (a) 4.6×10^{-23} kg/m^3 **41.** (a) 9.5×10^{-18} kg/m^3 (b) 4.8×10^{-18} kg/m^3 (c) 1.3×10^{-7} kg/m^3 **43.** (a) 4.7×10^{-27} kg/m^3 (b) 1.9×10^{-26} kg/m^3 **45.** (b) $q_0 = -0.595$ (c) $\Omega_\Lambda = 0.135$ **46.** (a) −0.17 (b) +0.12 **47.** 1.3×10^{-53} m^{-2}

Chapter 26

30. 1.1×10^{-26} J **32.** 3.5×10^{-25} s **33.** (a) 80.5 GeV (b) 9.34×10^{14} K **37.** (a) 1.1×10^{14} m = 0.011 ly **38.** Length was 0.168 of its present-day value; density was 213 times the present-day value.

Chapter 27

16. (a) 3.7 cm **17.** (a) 10,000 years (b) 10 million years **19.** (a) 1430 MHz **20.** (a) 29.8 km/s (b) 0.10 m (c) 9.9×10^{-6} m, or 9.9×10^{-3} % of the unshifted wavelength **21.** 200 km

Index

Page numbers in *italics* indicate figures and captions.

Star Charts

The following set of star charts, one for each month of the year, are from *Griffith Observer* magazine. They are useful in the northern hemisphere only. For a set of star charts suitable for use in the southern hemisphere, see the *Universe* Web site (www.whfreeman.com/universe).

To use these charts, first select the chart that best corresponds to the date and time of your observations. Hold the chart vertically as shown in the illustration below and turn it so that the direction you are facing is shown at the bottom.

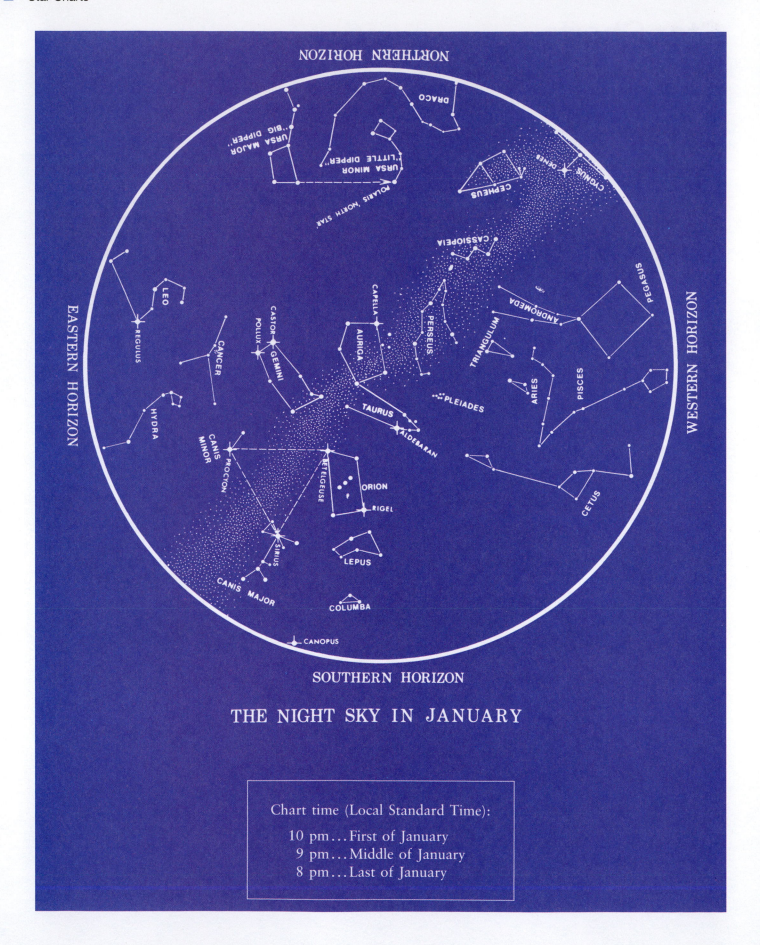

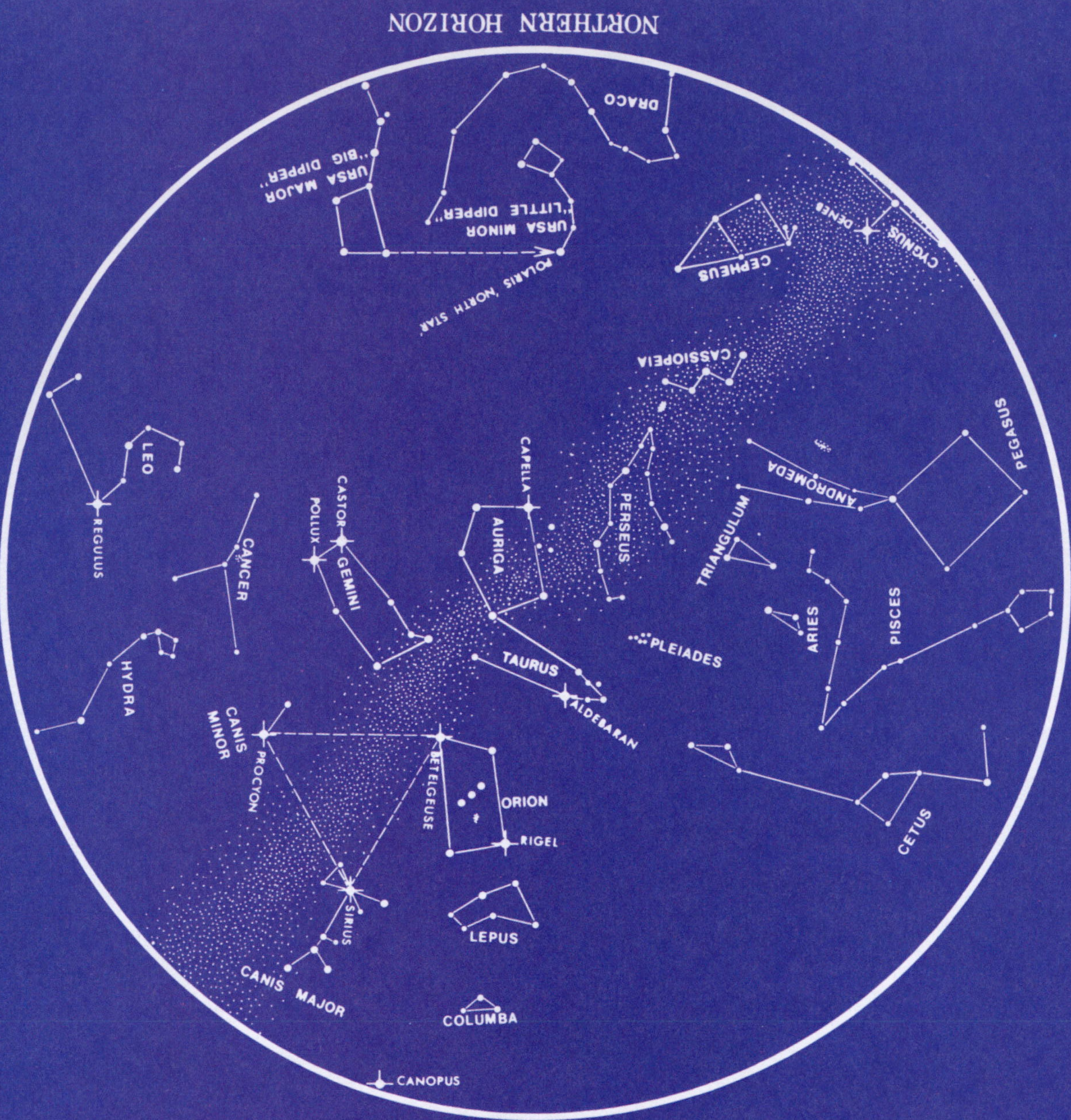

THE NIGHT SKY IN JANUARY

Chart time (Local Standard Time):
10 pm . . . First of January
9 pm . . . Middle of January
8 pm . . . Last of January

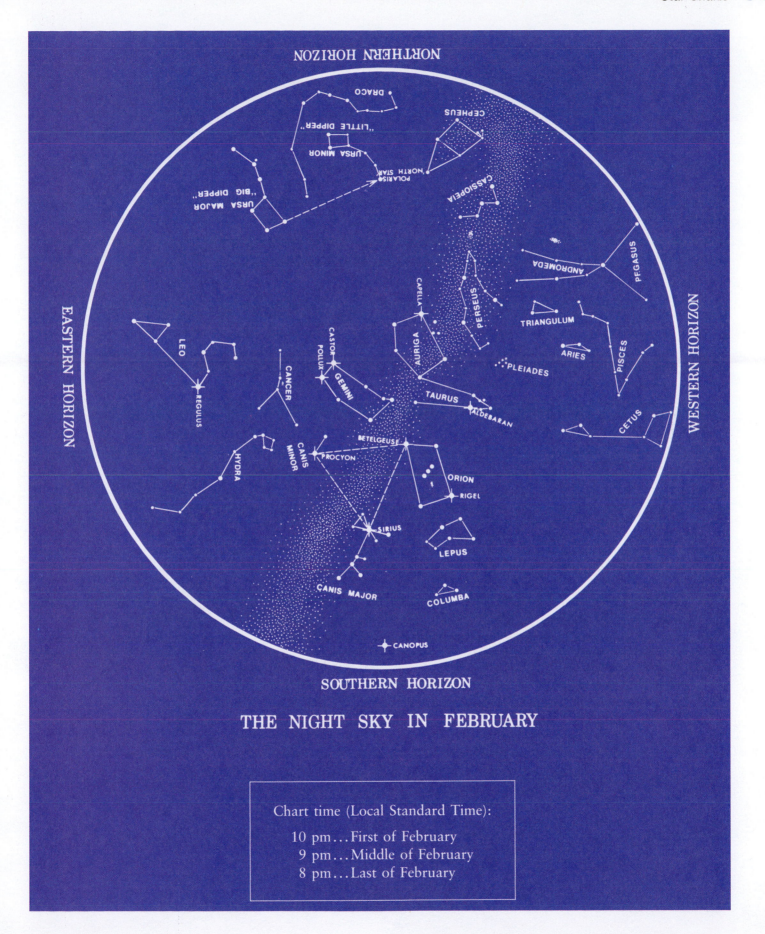

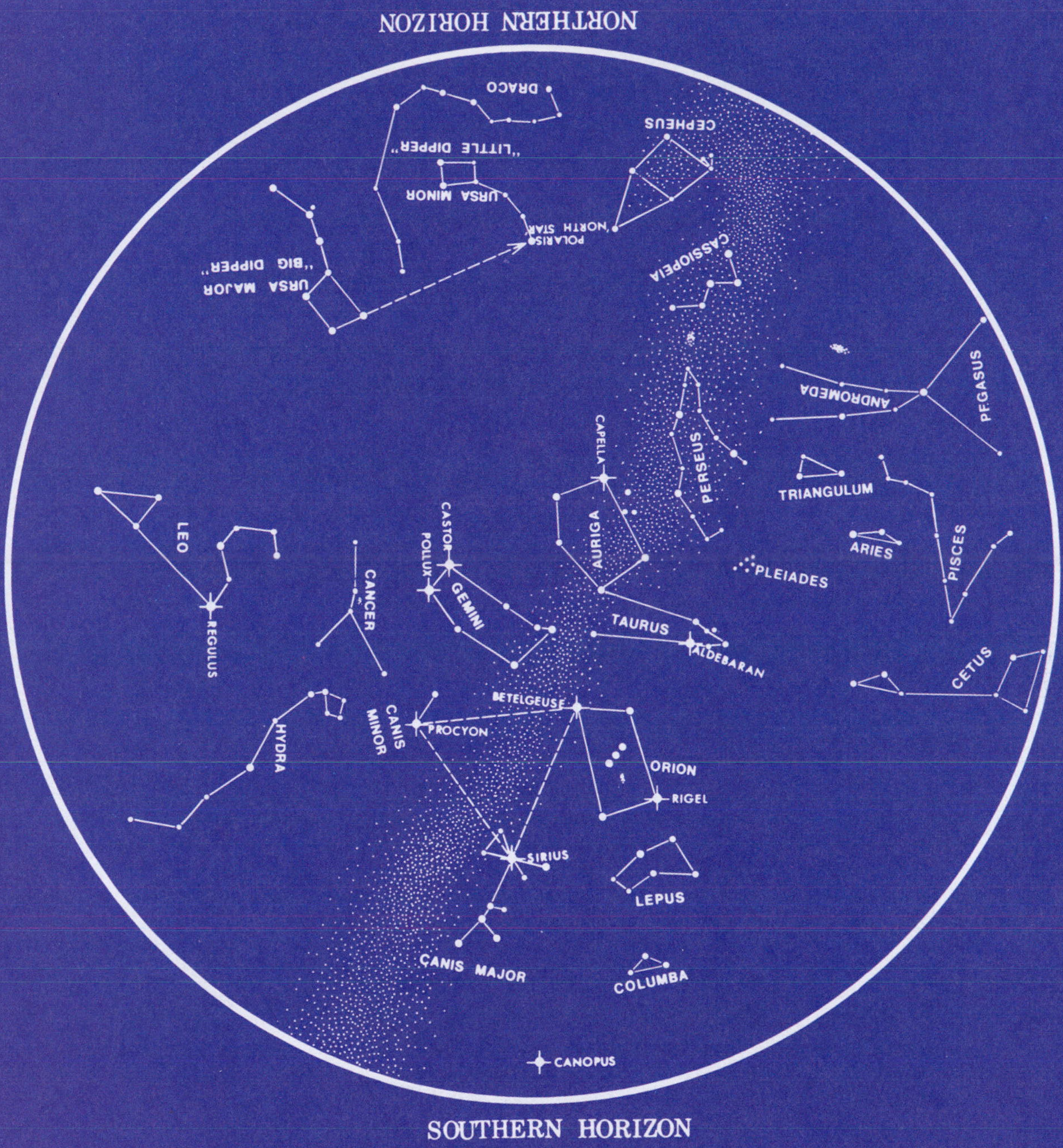

THE NIGHT SKY IN FEBRUARY

Chart time (Local Standard Time):

10 pm . . . First of February
9 pm . . . Middle of February
8 pm . . . Last of February

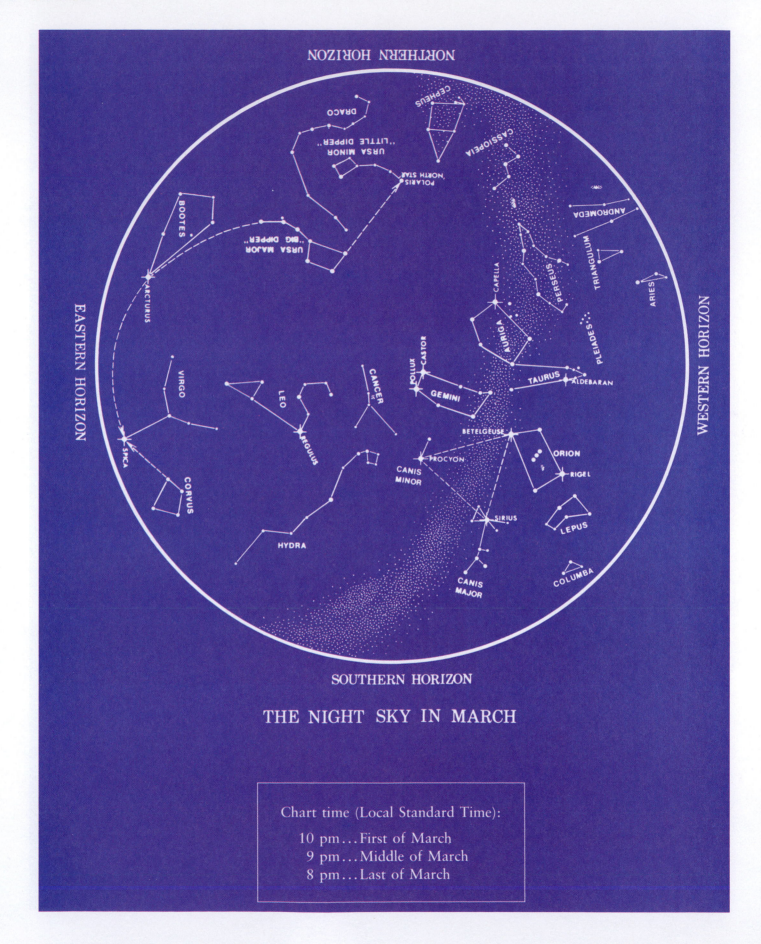

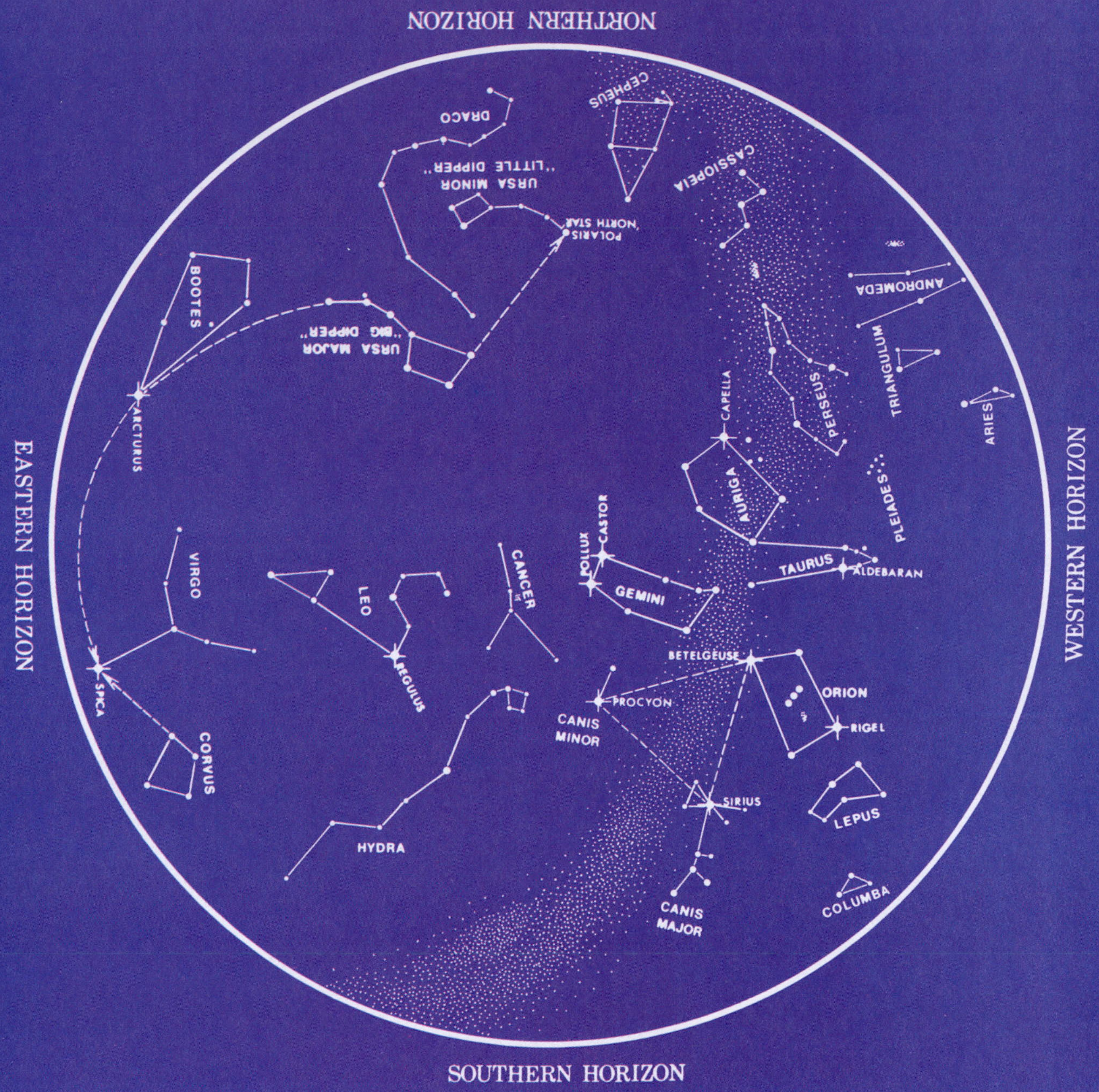

THE NIGHT SKY IN MARCH

Chart time (Local Standard Time):

10 pm . . . First of March
9 pm . . . Middle of March
8 pm . . . Last of March

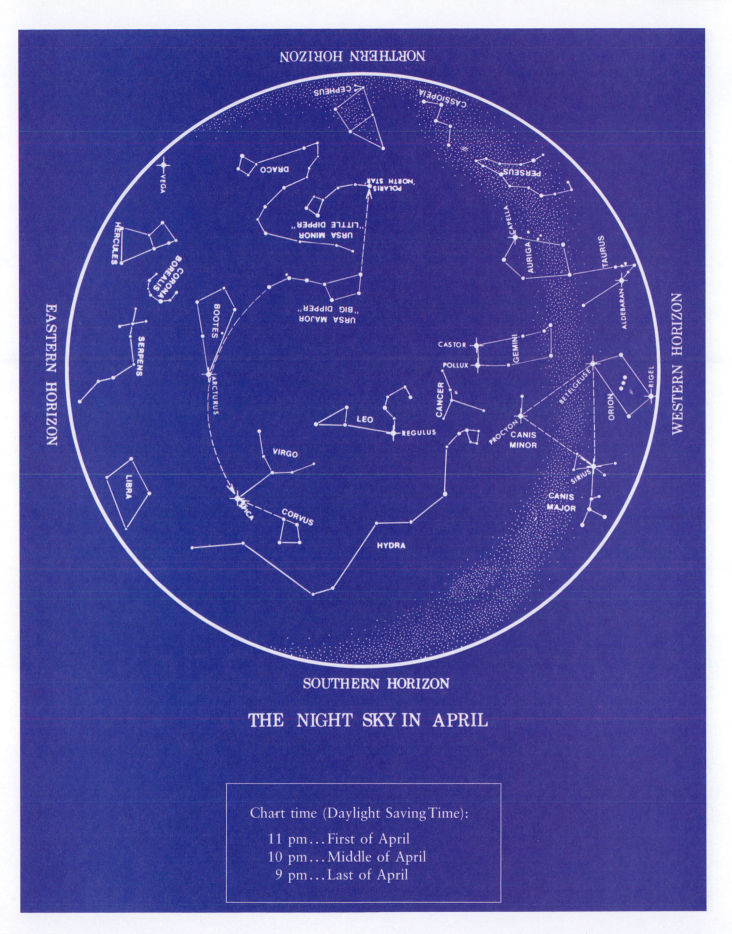

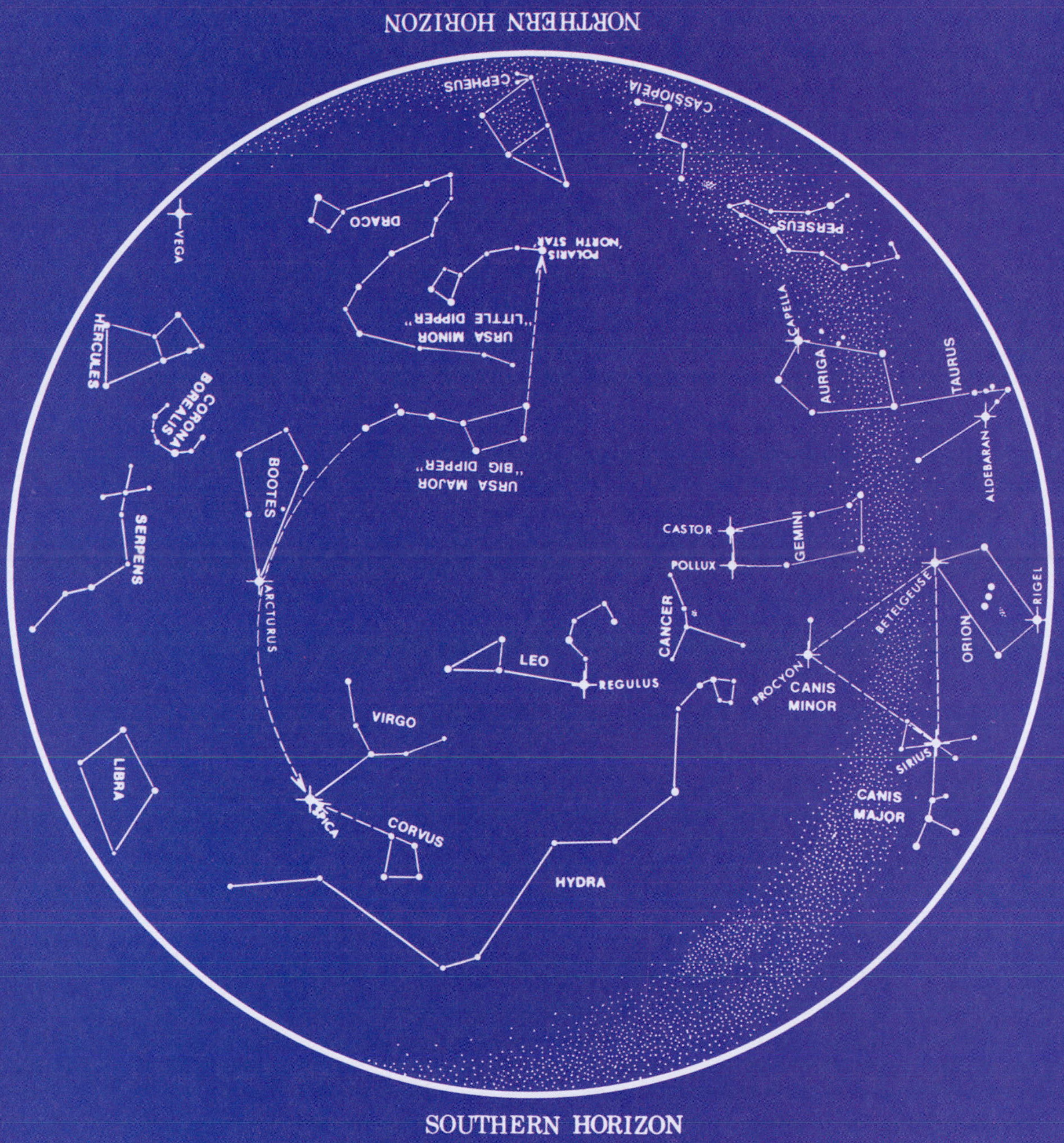

THE NIGHT SKY IN APRIL

Chart time (Daylight Saving Time):

11 pm . . . First of April
10 pm . . . Middle of April
9 pm . . . Last of April

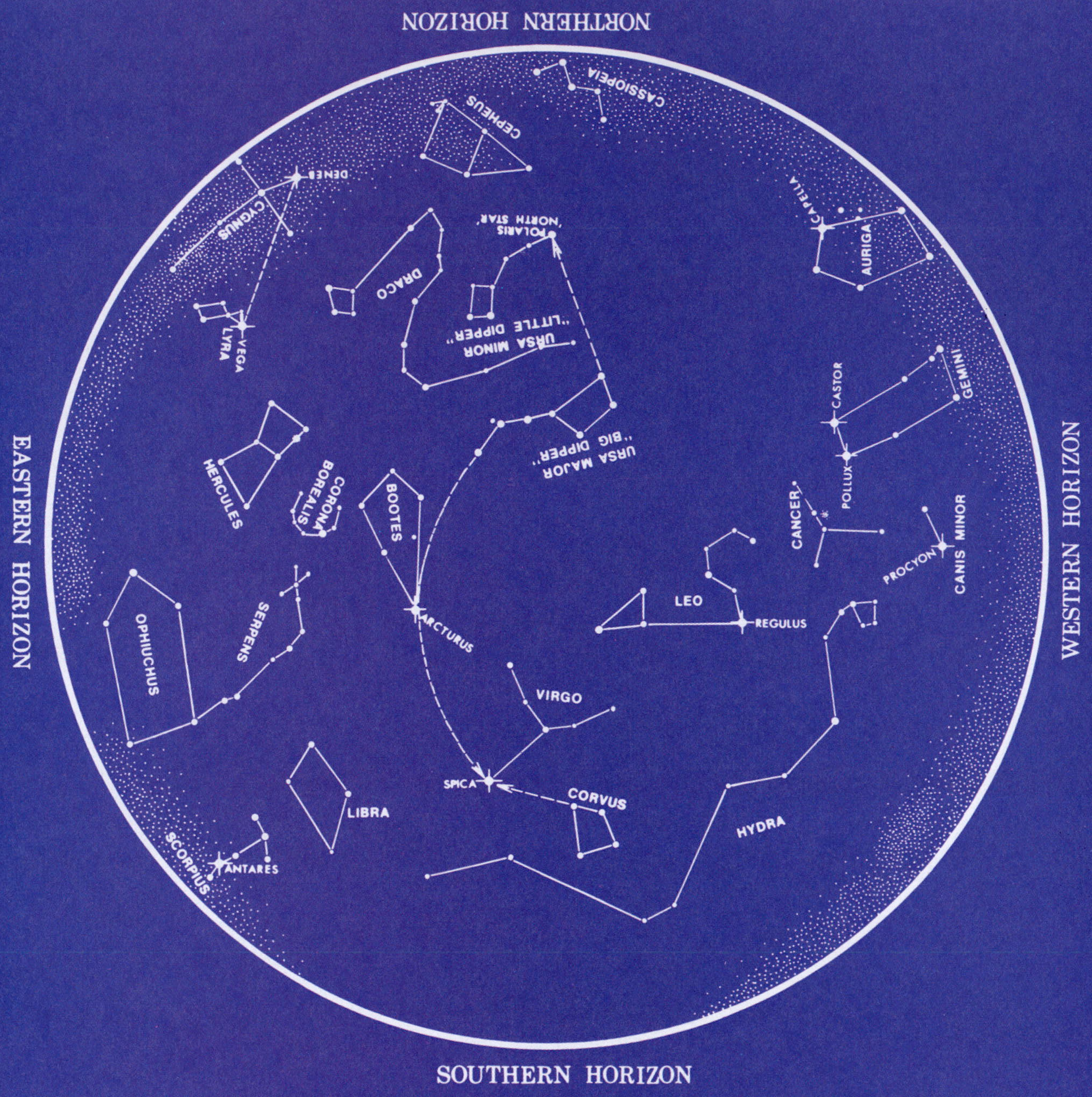

THE NIGHT SKY IN MAY

Chart time (Daylight Saving Time):

11 pm . . . First of May
10 pm . . . Middle of May
9 pm . . . Last of May

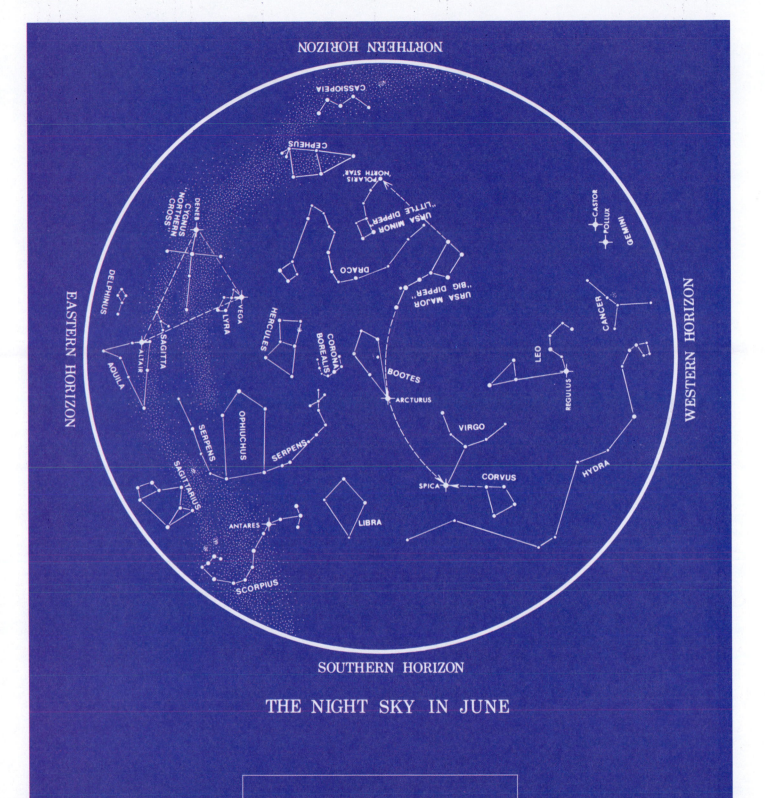

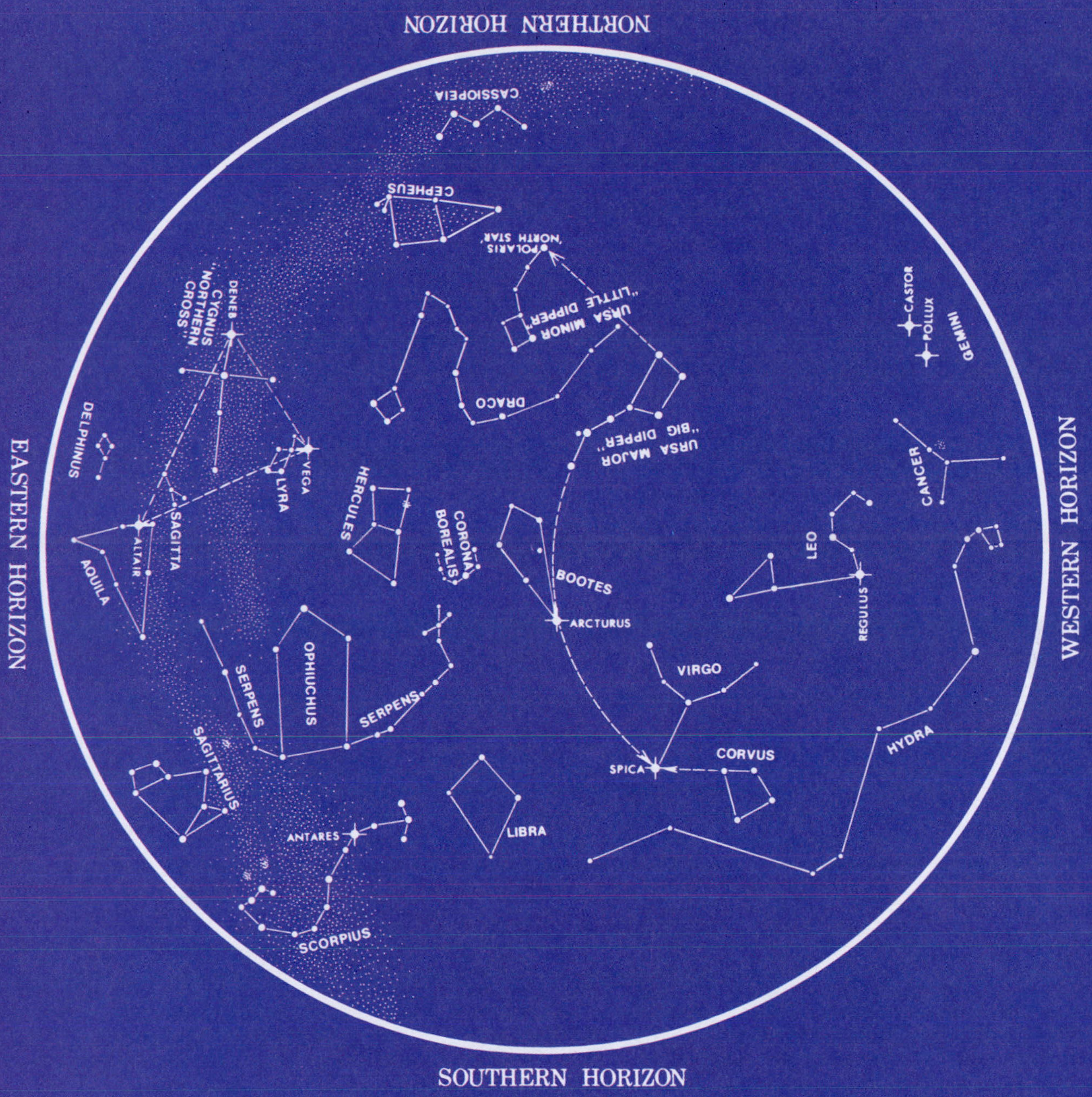

THE NIGHT SKY IN JUNE

Chart time (Daylight Saving Time):

11 pm . . . First of June
10 pm . . . Middle of June
9 pm . . . Last of June

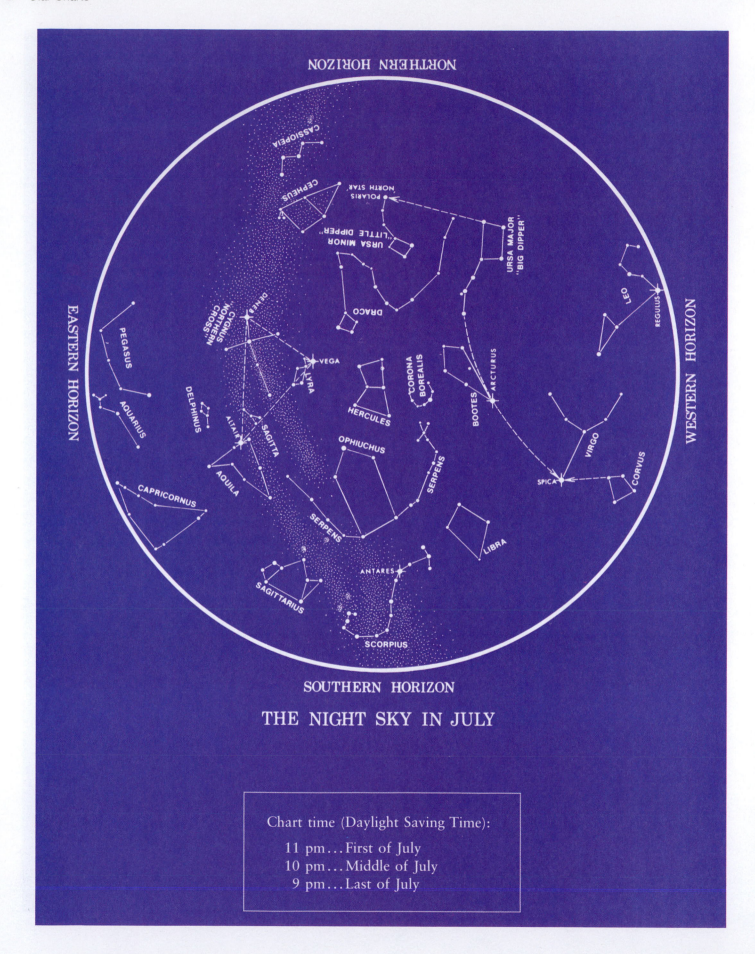

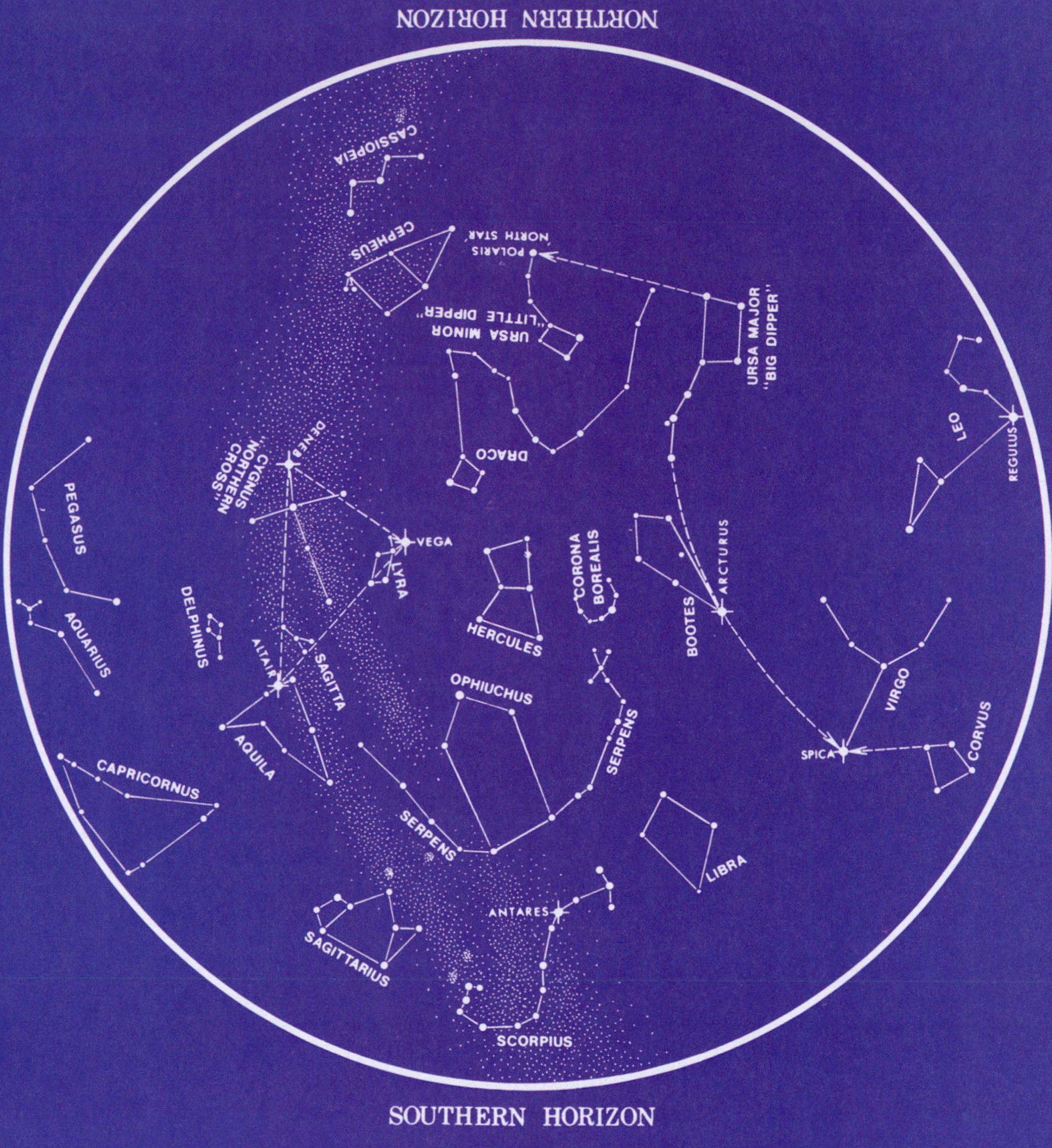

THE NIGHT SKY IN JULY

Chart time (Daylight Saving Time):

11 pm . . . First of July
10 pm . . . Middle of July
9 pm . . . Last of July

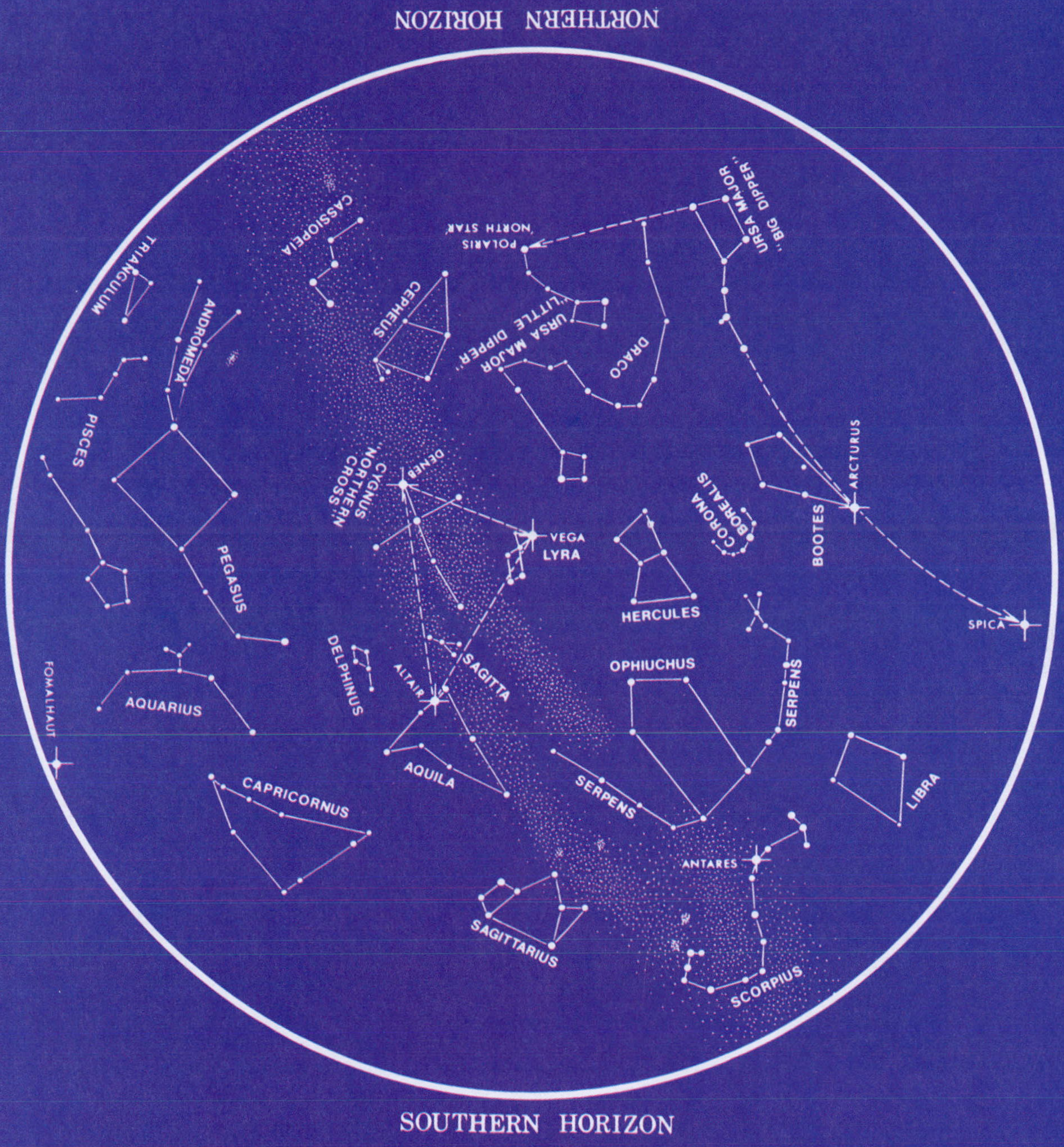

THE NIGHT SKY IN AUGUST

Chart time (Daylight Saving Time):

11 pm... First of August
10 pm... Middle of August
9 pm... Last of August

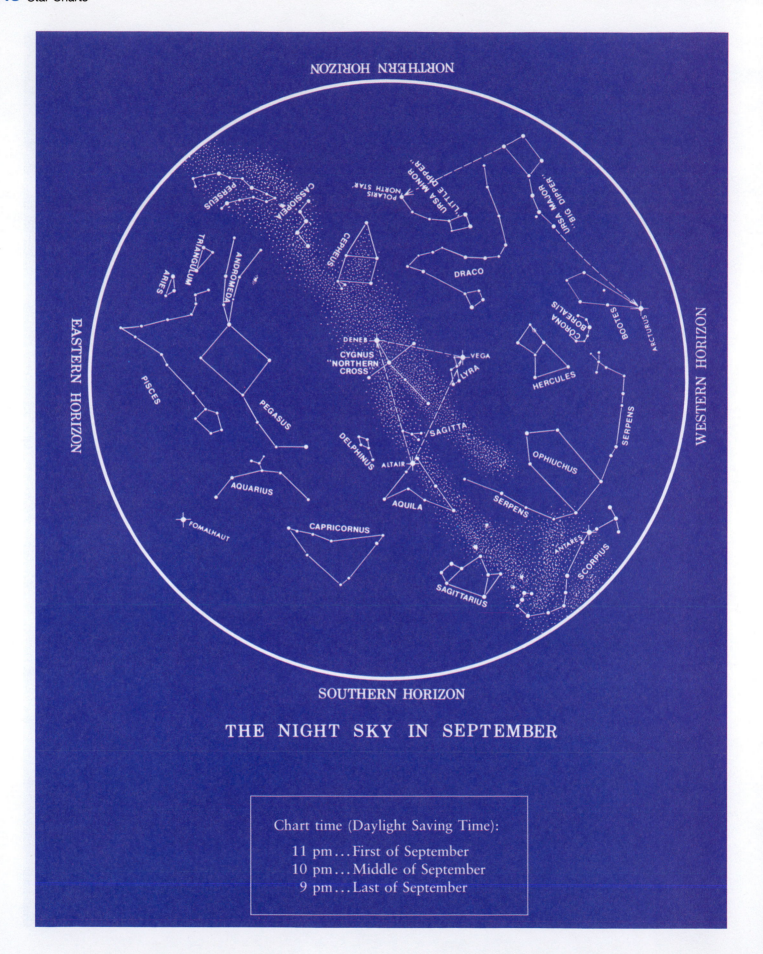

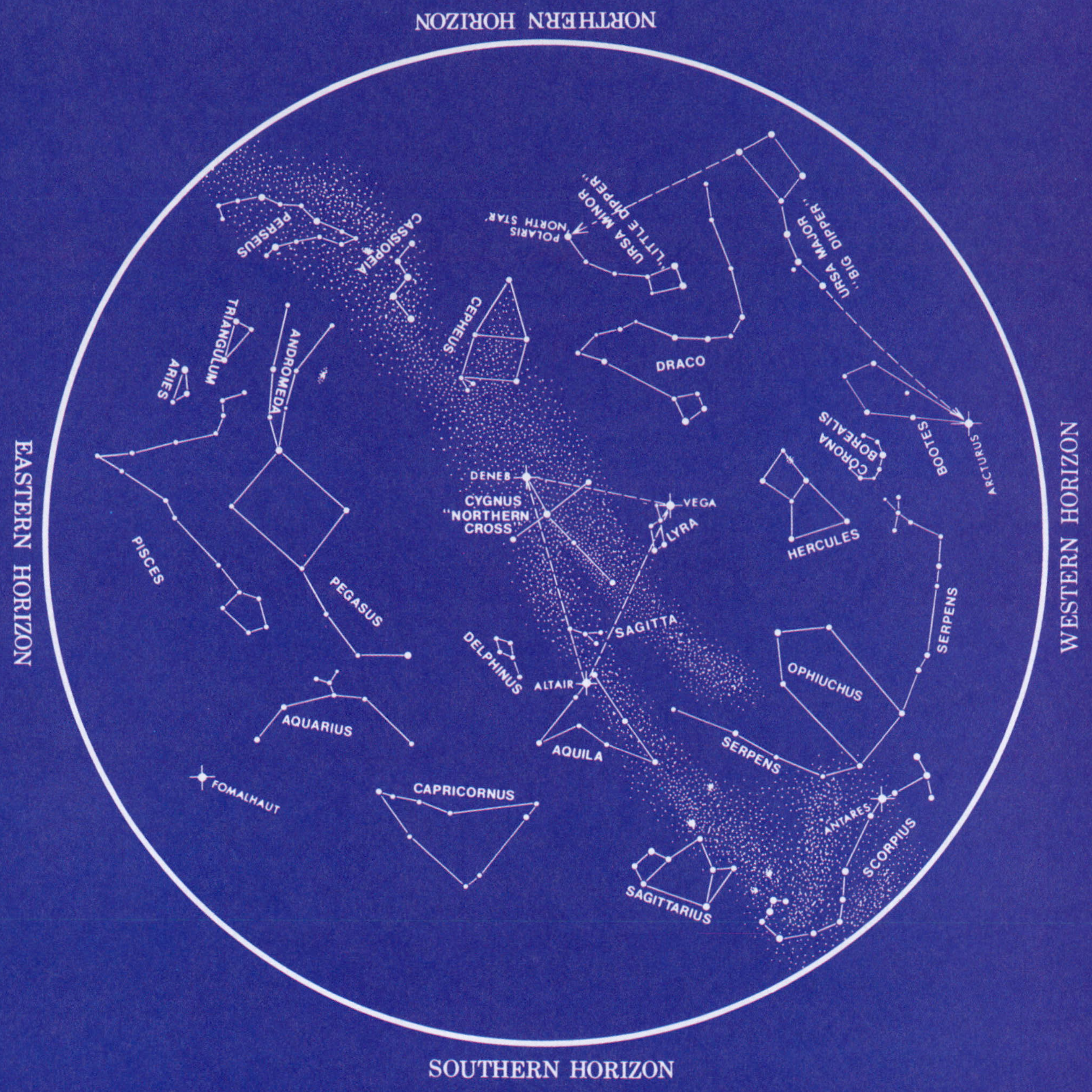

THE NIGHT SKY IN SEPTEMBER

Chart time (Daylight Saving Time):

11 pm . . . First of September
10 pm . . . Middle of September
9 pm . . . Last of September

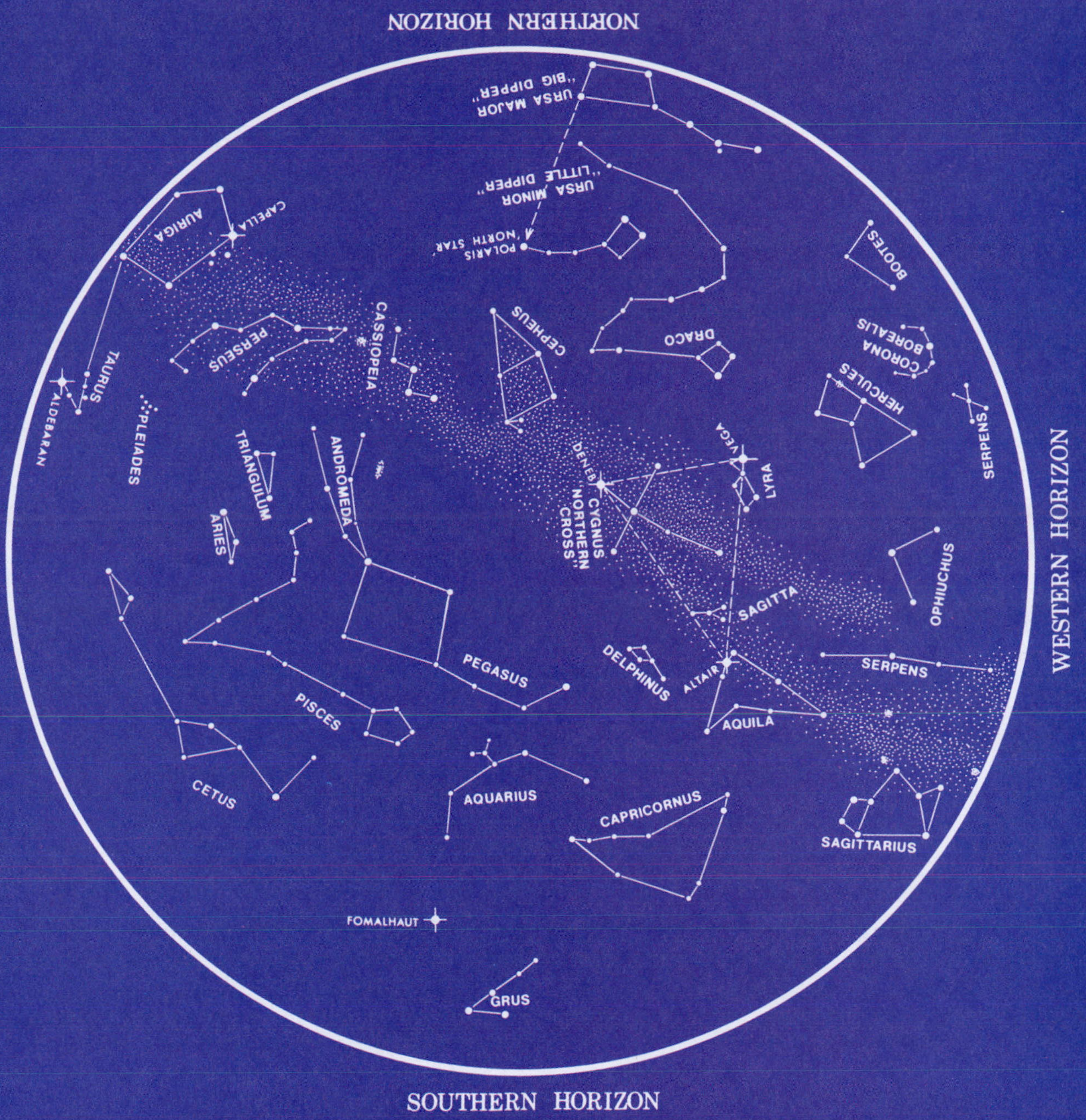

THE NIGHT SKY IN OCTOBER

Chart time (Daylight Saving Time):

11 pm... First of October
10 pm... Middle of October
9 pm... Last of October

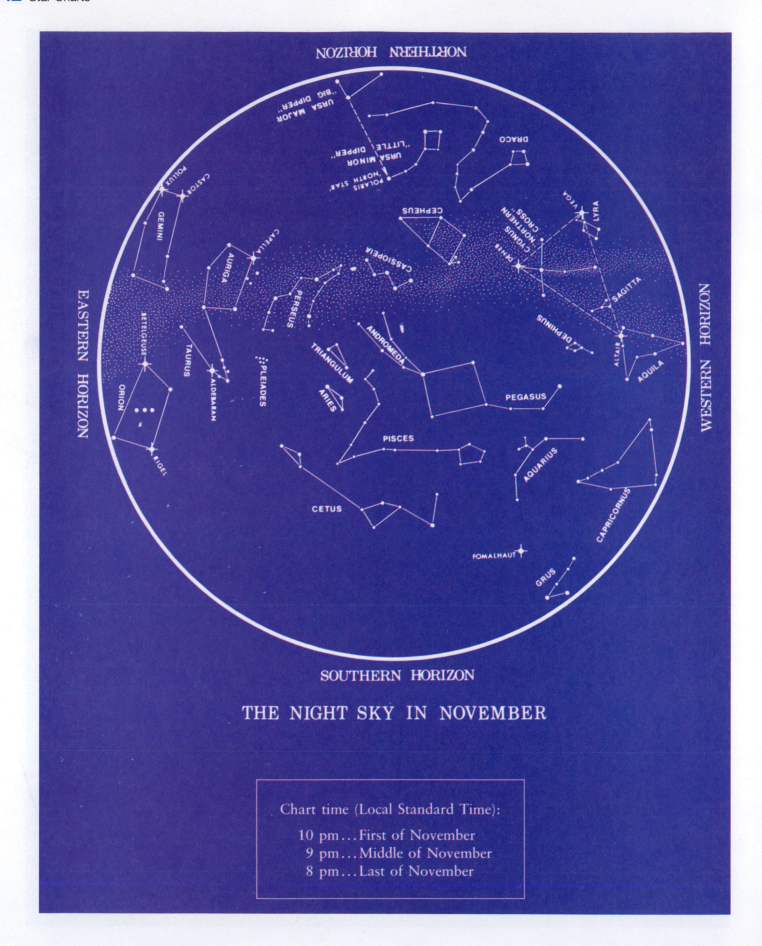

THE NIGHT SKY IN NOVEMBER

Chart time (Local Standard Time):

10 pm . . . First of November
9 pm . . . Middle of November
8 pm . . . Last of November

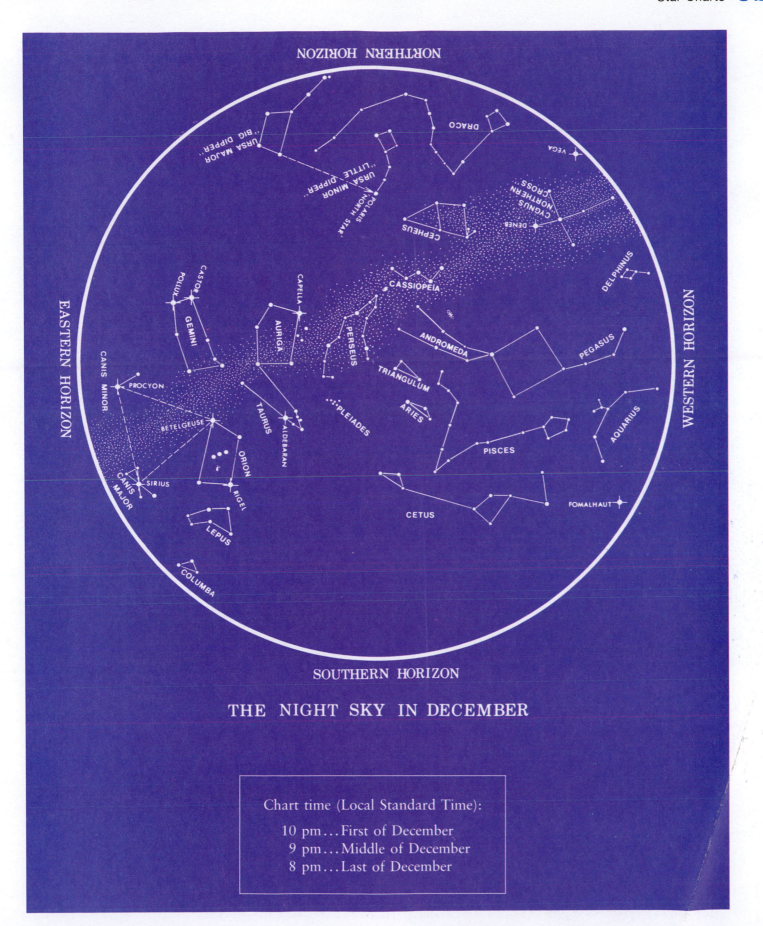

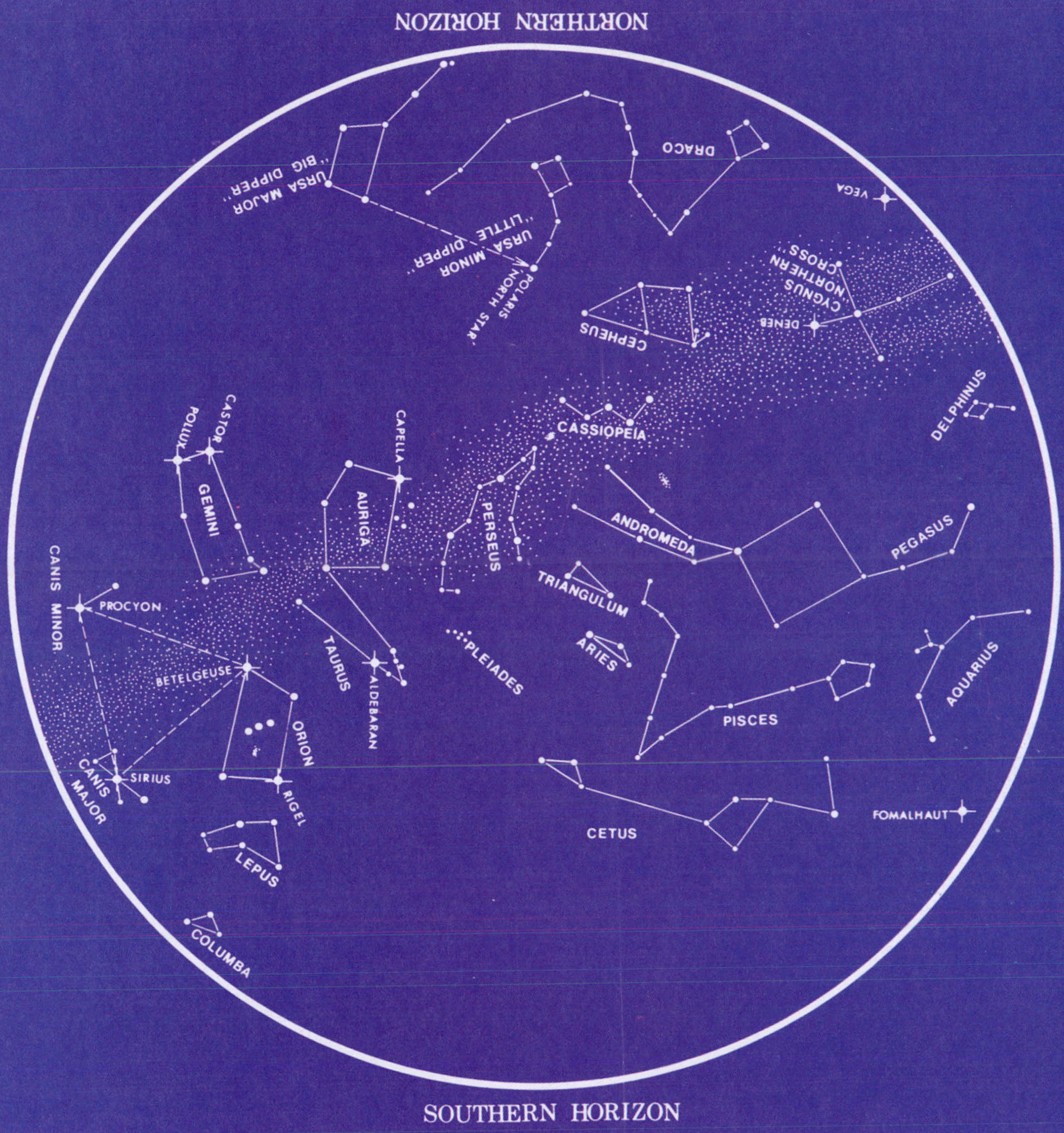

THE NIGHT SKY IN DECEMBER

Chart time (Local Standard Time):

10 pm . . . First of December
9 pm . . . Middle of December
8 pm . . . Last of December